T5-BAP-796

$LT(f)$, $LT(I)$	the leading term of the polynomial f, the ideal of leading terms
$M_n(R)$, $M_{n \times m}(R)$	the $n \times n$, and the $n \times m$ matrices over R
$M_{\mathcal{B}}^{\mathcal{E}}(\varphi)$	the matrix of the linear transformation φ with respect to bases $\mathcal{B}$ (domain) and $\mathcal{E}$ (range)
$\mathrm{tr}\,(A)$	the trace of the matrix A
$\mathrm{Hom}_R(A, B)$	the R-module homomorphisms from A to B
$\mathrm{End}(M)$	the endomorphism ring of the module M
$\mathrm{Tor}(M)$	the torsion submodule of M
$\mathrm{Ann}(M)$	the annihilator of the module M
$M \otimes_R N$	the tensor product of modules M and N over R
$\mathcal{T}^k(M)$, $\mathcal{T}(M)$	the kth tensor power, and the tensor algebra of M
$\mathcal{S}^k(M)$, $\mathcal{S}(M)$	the kth symmetric power, and the symmetric algebra of M
$\bigwedge^k(M)$, $\bigwedge(M)$	the kth exterior power, and the exterior algebra of M
$m_T(x)$, $c_T(x)$	the minimal, and characteristic polynomial of T
$\mathrm{ch}(F)$	the characteristic of the field F
K/F	the field K is an extension of the field F
$[K : F]$	the degree of the field extension K/F
$F(\alpha)$, $F(\alpha, \beta)$, etc.	the field generated over F by α or α, β, etc.
$m_{\alpha, F}(x)$	the minimal polynomal of α over the field F
$\mathrm{Aut}(K)$	the group of automorphisms of a field K
$\mathrm{Aut}(K/F)$	the group of automorphisms of a field K fixing the field F
$\mathrm{Gal}(K/F)$	the Galois group of the extension K/F
$\mathbb{A}^n$	affine n-space
$k[\mathbb{A}^n]$, $k[V]$	the coordinate ring of $\mathbb{A}^n$, and of the affine algebraic set V
$\mathcal{Z}(I)$, $\mathcal{Z}(f)$	the locus or zero set of I, the locus of an element f
$\mathcal{I}(A)$	the ideal of functions that vanish on A
$\mathrm{rad}\,I$	the radical of the ideal I
$\mathrm{Ass}_R(M)$	the associated primes for the module M
$\mathrm{Supp}(M)$	the support of the module M
$D^{-1}R$	the ring of fractions (localization) of R with respect to D
R_P, R_f	the localization of R at the prime ideal P, and at the element f
$\mathcal{O}_{v,V}$, $\mathbb{T}_{v,V}$	the local ring, and the tangent space of the variety V at the point v
$\mathrm{m}_{v,V}$	the unique maximal ideal of $\mathcal{O}_{v,V}$
$\mathrm{Spec}\,R$, $\mathrm{mSpec}\,R$	the prime spectrum, and the maximal spectrum of R
$\mathcal{O}_X$	the structure sheaf of $X = \mathrm{Spec}\,R$
$\mathcal{O}(U)$	the ring of sections on an open set U in $\mathrm{Spec}\,R$
$\mathcal{O}_P$	the stalk of the structure sheaf at P
$\mathrm{Jac}\,R$	the Jacobson radical of the ring R
$\mathrm{Ext}_R^n(A, B)$	the nth cohomology group derived from Hom_R
$\mathrm{Tor}_n^R(A, B)$	the nth cohomology group derived from the tensor product over R
A^G	the fixed points of G acting on the G-module A
$H^n(G, A)$	the nth cohomology group of G with coefficients in A
Res, Cor	the restriction, and corestriction maps on cohomology
$\mathrm{Stab}(1 \trianglelefteq A \trianglelefteq G)$	the stability group of the series $1 \trianglelefteq A \trianglelefteq G$
$\|\theta\|$	the norm of the character θ
$\mathrm{Ind}_H^G(\psi)$	the character of the representation ψ induced from H to G

ABSTRACT ALGEBRA
Third Edition

David S. Dummit
University of Vermont

Richard M. Foote
University of Vermont

John Wiley & Sons, Inc.

ASSOCIATE PUBLISHER	Laurie Rosatone
ASSISTANT EDITOR	Jennifer Battista
FREELANCE DEVELOPMENTAL EDITOR	Anne Scanlan-Rohrer
SENIOR MARKETING MANAGER	Julie Z. Lindstrom
SENIOR PRODUCTION EDITOR	Ken Santor
COVER DESIGNER	Michael Jung

This book was typeset using the Y&Y TeX System with DVIWindo. The text was set in Times Roman using *MathTime* from Y&Y, Inc. Titles were set in OceanSans. This book was printed by Malloy Inc. and the cover was printed by Phoenix Color Corporation.

This book is printed on acid-free paper.

Copyright © 2004 John Wiley and Sons, Inc. All rights reserved.

No part of this publication may be reproduced, stored in a retrieval system or transmitted in any form or by any means, electronic, mechanical, photocopying, recording, scanning, or otherwise, except as permitted under Sections 107 or 108 of the 1976 United States Copyright Act, without either the prior written permission of the Publisher, or authorization through payment of the appropriate per-copy fee to the Copyright Clearance Center, Inc., 222 Rosewood Drive, Danvers, MA 01923, (508) 750-8400, fax (508) 750-4470. Requests to the Publisher for permission should be addressed to the Permissions Department, John Wiley & Sons, Inc., 111 River Street, Hoboken, NJ 07030, (201)748-6011, fax (201)748-6008, E-mail: PERMREQ@WILEY.COM.

To order books or for customer service please call 1-800-CALL WILEY (225-5945).

ISBN 978-0-471-43334-7

WIE 978-0-471-45234-8

Printed in the United States of America.

10 9

Dedicated to our families
especially
Janice, Evan, and Krysta
and
Zsuzsanna, Peter, Karoline, and Alexandra

Contents

Preface to the Third Edition

The principal change from the second edition is the addition of Gröbner bases to this edition. The basic theory is introduced in a new Section 9.6. Applications to solving systems of polynomial equations (elimination theory) appear at the end of this section, rounding it out as a self-contained foundation in the topic. Additional applications and examples are then woven into the treatment of affine algebraic sets and k-algebra homomorphisms in Chapter 15. Although the theory in the latter chapter remains independent of Gröbner bases, the new applications, examples and computational techniques significantly enhance the development, and we recommend that Section 9.6 be read either as a segue to or in parallel with Chapter 15. A wealth of exercises involving Gröbner bases, both computational and theoretical in nature, have been added in Section 9.6 and Chapter 15. Preliminary exercises on Gröbner bases can (and should, as an aid to understanding the algorithms) be done by hand, but more extensive computations, and in particular most of the use of Gröbner bases in the exercises in Chapter 15, will likely require computer assisted computation.

Other changes include a streamlining of the classification of simple groups of order 168 (Section 6.2), with the addition of a uniqueness proof via the projective plane of order 2. Some other proofs or portions of the text have been revised slightly. A number of new exercises have been added throughout the book, primarily at the ends of sections in order to preserve as much as possible the numbering schemes of earlier editions. In particular, exercises have been added on free modules over noncommutative rings (10.3), on Krull dimension (15.3), and on flat modules (10.5 and 17.1).

As with previous editions, the text contains substantially more than can normally be covered in a one year course. A basic introductory (one year) course should probably include Part I up through Section 5.3, Part II through Section 9.5, Sections 10.1, 10.2, 10.3, 11.1, 11.2 and Part IV. Chapter 12 should also be covered, either before or after Part IV. Additional topics from Chapters 5, 6, 9, 10 and 11 may be interspersed in such a course, or covered at the end as time permits.

Sections 10.4 and 10.5 are at a slightly higher level of difficulty than the initial sections of Chapter 10, and can be deferred on a first reading for those following the text sequentially. The latter section on properties of exact sequences, although quite long, maintains coherence through a parallel treatment of three basic functors in respective subsections.

Beyond the core material, the third edition provides significant flexibility for students and instructors wishing to pursue a number of important areas of modern algebra, either in the form of independent study or courses. For example, well integrated one-semester courses for students with some prior algebra background might include the following: Section 9.6 and Chapters 15 and 16; or Chapters 10 and 17; or Chapters 5, 6 and Part VI. Each of these would also provide a solid background for a follow-up course delving more deeply into one of many possible areas: algebraic number theory, algebraic topology, algebraic geometry, representation theory, Lie groups, etc.

The choice of new material and the style for developing and integrating it into the text are in consonance with a basic theme in the book: the power and beauty that accrues from a rich interplay between different areas of mathematics. The emphasis throughout has been to motivate the introduction and development of important algebraic concepts using as many examples as possible. We have not attempted to be encyclopedic, but have tried to touch on many of the central themes in elementary algebra in a manner suggesting the very natural development of these ideas.

A number of important ideas and results appear in the exercises. This is not because they are not significant, rather because they did not fit easily into the flow of the text but were too important to leave out entirely. Sequences of exercises on one topic are prefaced with some remarks and are structured so that they may be read without actually doing the exercises. In some instances, new material is introduced first in the exercises—often a few sections before it appears in the text—so that students may obtain an easier introduction to it by doing these exercises (e.g., Lagrange's Theorem appears in the exercises in Section 1.7 and in the text in Section 3.2). All the exercises are within the scope of the text and hints are given [in brackets] where we felt they were needed. Exercises we felt might be less straightforward are usually phrased so as to provide the answer to the exercise; as well many exercises have been broken down into a sequence of more routine exercises in order to make them more accessible.

We have also purposely minimized the functorial language in the text in order to keep the presentation as elementary as possible. We have refrained from providing specific references for additional reading when there are many fine choices readily available. Also, while we have endeavored to include as many fundamental topics as possible, we apologize if for reasons of space or personal taste we have neglected any of the reader's particular favorites.

We are deeply grateful to and would like here to thank the many students and colleagues around the world who, over more than 15 years, have offered valuable comments, insights and encouragement—their continuing support and interest have motivated our writing of this third edition.

David Dummit
Richard Foote
June, 2003

Preliminaries

Some results and notation that are used throughout the text are collected in this chapter for convenience. Students may wish to review this chapter quickly at first and then read each section more carefully again as the concepts appear in the course of the text.

0.1 BASICS

The basics of set theory: sets, $\cap, \cup, \in$, etc. should be familiar to the reader. Our notation for subsets of a given set A will be

$$B = \{a \in A \mid \ldots \text{(conditions on a)} \ldots\}.$$

The *order* or *cardinality* of a set A will be denoted by $|A|$. If A is a finite set the order of A is simply the number of elements of A.

It is important to understand how to test whether a particular $x \in A$ lies in a subset B of A (cf. Exercises 1-4). The *Cartesian product* of two sets A and B is the collection $A \times B = \{(a, b) \mid a \in A, b \in B\}$, of ordered pairs of elements from A and B.

We shall use the following notation for some common sets of numbers:

(1) $\mathbb{Z} = \{0, \pm 1, \pm 2, \pm 3, \ldots\}$ denotes the *integers* (the $\mathbb{Z}$ is for the German word for numbers: "Zahlen").

(2) $\mathbb{Q} = \{a/b \mid a, b \in \mathbb{Z}, b \neq 0\}$ denotes the *rational numbers* (or *rationals*).

(3) $\mathbb{R} = \{$ all decimal expansions $\pm d_1 d_2 \ldots d_n.a_1 a_2 a_3 \ldots\}$ denotes the *real numbers* (or *reals*).

(4) $\mathbb{C} = \{a + bi \mid a, b \in \mathbb{R}, i^2 = -1\}$ denotes the *complex numbers*.

(5) $\mathbb{Z}^+, \mathbb{Q}^+$ and $\mathbb{R}^+$ will denote the positive (nonzero) elements in $\mathbb{Z}, \mathbb{Q}$ and $\mathbb{R}$, respectively.

We shall use the notation $f : A \to B$ or $A \overset{f}{\to} B$ to denote a function f from A to B and the value of f at a is denoted $f(a)$ (i.e., we shall apply all our functions on the left). We use the words *function* and *map* interchangeably. The set A is called the *domain* of f and B is called the *codomain* of f. The notation $f : a \mapsto b$ or $a \mapsto b$ if f is understood indicates that $f(a) = b$, i.e., the function is being specified on *elements*.

If the function f is not specified on elements it is important in general to check that f is *well defined*, i.e., is unambiguously determined. For example, if the set A is the union of two subsets A_1 and A_2 then one can try to specify a function from A

to the set $\{0, 1\}$ by declaring that f is to map everything in A_1 to 0 and is to map everything in A_2 to 1. This unambiguously defines f unless A_1 and A_2 have elements in common (in which case it is not clear whether these elements should map to 0 or to 1). Checking that this f is well defined therefore amounts to checking that A_1 and A_2 have no intersection.

The set

$$f(A) = \{b \in B \mid b = f(a), \text{ for some } a \in A\}$$

is a subset of B, called the *range* or *image* of f (or the *image of A under f*). For each subset C of B the set

$$f^{-1}(C) = \{a \in A \mid f(a) \in C\}$$

consisting of the elements of A mapping into C under f is called the *preimage* or *inverse image* of C under f. For each $b \in B$, the preimage of $\{b\}$ under f is called the *fiber* of f over b. Note that f^{-1} is not in general a function and that the fibers of f generally contain many elements since there may be many elements of A mapping to the element b.

If $f : A \to B$ and $g : B \to C$, then the composite map $g \circ f : A \to C$ is defined by

$$(g \circ f)(a) = g(f(a)).$$

Let $f : A \to B$.
(1) f is *injective* or is an *injection* if whenever $a_1 \neq a_2$, then $f(a_1) \neq f(a_2)$.
(2) f is *surjective* or is a *surjection* if for all $b \in B$ there is some $a \in A$ such that $f(a) = b$, i.e., the image of f is *all* of B. Note that since a function always maps onto its range (by definition) it is necessary to specify the codomain B in order for the question of surjectivity to be meaningful.
(3) f is *bijective* or is a *bijection* if it is both injective and surjective. If such a bijection f exists from A to B, we say A and B are in *bijective correspondence*.
(4) f has a *left inverse* if there is a function $g : B \to A$ such that $g \circ f : A \to A$ is the identity map on A, i.e., $(g \circ f)(a) = a$, for all $a \in A$.
(5) f has a *right inverse* if there is a function $h : B \to A$ such that $f \circ h : B \to B$ is the identity map on B.

Proposition 1. Let $f : A \to B$.
(1) The map f is injective if and only if f has a left inverse.
(2) The map f is surjective if and only if f has a right inverse.
(3) The map f is a bijection if and only if there exists $g : B \to A$ such that $f \circ g$ is the identity map on B and $g \circ f$ is the identity map on A.
(4) If A and B are finite sets with the same number of elements (i.e., $|A| = |B|$), then $f : A \to B$ is bijective if and only if f is injective if and only if f is surjective.

Proof: Exercise.

In the situation of part (3) of the proposition above the map g is necessarily unique and we shall say g is the *2-sided inverse* (or simply the *inverse*) of f.

A *permutation* of a set A is simply a bijection from A to itself.

If $A \subseteq B$ and $f : B \to C$, we denote the *restriction* of f to A by $f|_A$. When the domain we are considering is understood we shall occasionally denote $f|_A$ again simply as f even though these are formally different functions (their domains are different).

If $A \subseteq B$ and $g : A \to C$ and there is a function $f : B \to C$ such that $f|_A = g$, we shall say f is an *extension* of g to B (such a map f need not exist nor be unique).

Let A be a nonempty set.

(1) A *binary relation* on a set A is a subset R of $A \times A$ and we write $a \sim b$ if $(a, b) \in R$.

(2) The relation $\sim$ on A is said to be:

    (a) *reflexive* if $a \sim a$, for all $a \in A$,

    (b) *symmetric* if $a \sim b$ implies $b \sim a$ for all $a, b \in A$,

    (c) *transitive* if $a \sim b$ and $b \sim c$ implies $a \sim c$ for all $a, b, c \in A$.

    A relation is an *equivalence relation* if it is reflexive, symmetric and transitive.

(3) If $\sim$ defines an equivalence relation on A, then the *equivalence class* of $a \in A$ is defined to be $\{x \in A \mid x \sim a\}$. Elements of the equivalence class of a are said to be *equivalent* to a. If C is an equivalence class, any element of C is called a *representative* of the class C.

(4) A *partition* of A is any collection $\{A_i \mid i \in I\}$ of nonempty subsets of A (I some indexing set) such that

    (a) $A = \cup_{i \in I} A_i$, and

    (b) $A_i \cap A_j = \emptyset$, for all $i, j \in I$ with $i \neq j$

        i.e., A is the disjoint union of the sets in the partition.

The notions of an equivalence relation on A and a partition of A are the same:

Proposition 2. Let A be a nonempty set.

    (1) If $\sim$ defines an equivalence relation on A then the set of equivalence classes of $\sim$ form a partition of A.

    (2) If $\{A_i \mid i \in I\}$ is a partition of A then there is an equivalence relation on A whose equivalence classes are precisely the sets $A_i, i \in I$.

Proof: Omitted.

Finally, we shall assume the reader is familiar with proofs by induction.

EXERCISES

In Exercises 1 to 4 let $\mathcal{A}$ be the set of 2×2 matrices with real number entries. Recall that matrix multiplication is defined by

$$\begin{pmatrix} a & b \\ c & d \end{pmatrix} \begin{pmatrix} p & q \\ r & s \end{pmatrix} = \begin{pmatrix} ap + br & aq + bs \\ cp + dr & cq + ds \end{pmatrix} \quad .$$

Let

$$M = \begin{pmatrix} 1 & 1 \\ 0 & 1 \end{pmatrix}$$

and let
$$\mathcal{B} = \{X \in \mathcal{A} \mid MX = XM\}.$$

1. Determine which of the following elements of $\mathcal{A}$ lie in $\mathcal{B}$:

$$\begin{pmatrix} 1 & 1 \\ 0 & 1 \end{pmatrix}, \quad \begin{pmatrix} 1 & 1 \\ 1 & 1 \end{pmatrix}, \quad \begin{pmatrix} 0 & 0 \\ 0 & 0 \end{pmatrix}, \quad \begin{pmatrix} 1 & 1 \\ 1 & 0 \end{pmatrix}, \quad \begin{pmatrix} 1 & 0 \\ 0 & 1 \end{pmatrix}, \quad \begin{pmatrix} 0 & 1 \\ 1 & 0 \end{pmatrix}.$$

2. Prove that if $P, Q \in \mathcal{B}$, then $P + Q \in \mathcal{B}$ (where $+$ denotes the usual sum of two matrices).

3. Prove that if $P, Q \in \mathcal{B}$, then $P \cdot Q \in \mathcal{B}$ (where $\cdot$ denotes the usual product of two matrices).

4. Find conditions on p, q, r, s which determine precisely when $\begin{pmatrix} p & q \\ r & s \end{pmatrix} \in \mathcal{B}$.

5. Determine whether the following functions f are well defined:
    (a) $f : \mathbb{Q} \to \mathbb{Z}$ defined by $f(a/b) = a$.
    (b) $f : \mathbb{Q} \to \mathbb{Q}$ defined by $f(a/b) = a^2/b^2$.

6. Determine whether the function $f : \mathbb{R}^+ \to \mathbb{Z}$ defined by mapping a real number r to the first digit to the right of the decimal point in a decimal expansion of r is well defined.

7. Let $f : A \to B$ be a surjective map of sets. Prove that the relation

$$a \sim b \text{ if and only if } f(a) = f(b)$$

is an equivalence relation whose equivalence classes are the fibers of f.

0.2 PROPERTIES OF THE INTEGERS

The following properties of the integers $\mathbb{Z}$ (many familiar from elementary arithmetic) will be proved in a more general context in the ring theory of Chapter 8, but it will be necessary to use them in Part I (of course, none of the ring theory proofs of these properties will rely on the group theory).

(1) (Well Ordering of $\mathbb{Z}$) If A is any nonempty subset of $\mathbb{Z}^+$, there is some element $m \in A$ such that $m \leq a$, for all $a \in A$ (m is called a *minimal element* of A).

(2) If $a, b \in \mathbb{Z}$ with $a \neq 0$, we say a *divides* b if there is an element $c \in \mathbb{Z}$ such that $b = ac$. In this case we write $a \mid b$; if a does not divide b we write $a \nmid b$.

(3) If $a, b \in \mathbb{Z} - \{0\}$, there is a unique positive integer d, called the *greatest common divisor of a and b* (or g.c.d. of a and b), satisfying:
    (a) $d \mid a$ and $d \mid b$ (so d is a common divisor of a and b), and
    (b) if $e \mid a$ and $e \mid b$, then $e \mid d$ (so d is the greatest such divisor).
The g.c.d. of a and b will be denoted by (a, b). If $(a, b) = 1$, we say that a and b are *relatively prime*.

(4) If $a, b \in \mathbb{Z} - \{0\}$, there is a unique positive integer l, called the *least common multiple of a and b* (or l.c.m. of a and b), satisfying:
    (a) $a \mid l$ and $b \mid l$ (so l is a common multiple of a and b), and
    (b) if $a \mid m$ and $b \mid m$, then $l \mid m$ (so l is the least such multiple).
The connection between the greatest common divisor d and the least common multiple l of two integers a and b is given by $dl = ab$.

(5) The *Division Algorithm*: if $a, b \in \mathbb{Z}$ and $b \neq 0$, then there exist unique $q, r \in \mathbb{Z}$ such that

$$a = qb + r \quad \text{and} \quad 0 \leq r < |b|,$$

where q is the *quotient* and r the *remainder*. This is the usual "long division" familiar from elementary arithmetic.

(6) The *Euclidean Algorithm* is an important procedure which produces a greatest common divisor of two integers a and b by iterating the Division Algorithm: if $a, b \in \mathbb{Z} - \{0\}$, then we obtain a sequence of quotients and remainders

$$a = q_0 b + r_0 \tag{0}$$
$$b = q_1 r_0 + r_1 \tag{1}$$
$$r_0 = q_2 r_1 + r_2 \tag{2}$$
$$r_1 = q_3 r_2 + r_3 \tag{3}$$
$$\vdots$$
$$r_{n-2} = q_n r_{n-1} + r_n \tag{n}$$
$$r_{n-1} = q_{n+1} r_n \tag{$n+1$}$$

where r_n is the last nonzero remainder. Such an r_n exists since $|b| > |r_0| > |r_1| > \cdots > |r_n|$ is a decreasing sequence of strictly positive integers if the remainders are nonzero and such a sequence cannot continue indefinitely. Then r_n is the g.c.d. (a, b) of a and b.

Example

Suppose $a = 57970$ and $b = 10353$. Then applying the Euclidean Algorithm we obtain:

$$57970 = (5)10353 + 6205$$
$$10353 = (1)6205 + 4148$$
$$6205 = (1)4148 + 2057$$
$$4148 = (2)2057 + 34$$
$$2057 = (60)34 + 17$$
$$34 = (2)17$$

which shows that $(57970, 10353) = 17$.

(7) One consequence of the Euclidean Algorithm which we shall use regularly is the following: if $a, b \in \mathbb{Z} - \{0\}$, then there exist $x, y \in \mathbb{Z}$ such that

$$(a, b) = ax + by$$

that is, *the g.c.d. of a and b is a $\mathbb{Z}$-linear combination of a and b*. This follows by recursively writing the element r_n in the Euclidean Algorithm in terms of the previous remainders (namely, use equation (n) above to solve for $r_n = r_{n-2} - q_n r_{n-1}$ in terms of the remainders r_{n-1} and r_{n-2}, then use equation $(n-1)$ to write r_n in terms of the remainders r_{n-2} and r_{n-3}, etc., eventually writing r_n in terms of a and b).

Example

Suppose $a = 57970$ and $b = 10353$, whose greatest common divisor we computed above to be 17. From the fifth equation (the next to last equation) in the Euclidean Algorithm applied to these two integers we solve for their greatest common divisor: $17 = 2057 - (60)34$. The fourth equation then shows that $34 = 4148 - (2)2057$, so substituting this expression for the previous remainder 34 gives the equation $17 = 2057 - (60)[4148 - (2)2057]$, i.e., $17 = (121)2057 - (60)4148$. Solving the third equation for 2057 and substituting gives $17 = (121)[6205 - (1)4148] - (60)4148 = (121)6205 - (181)4148$. Using the second equation to solve for 4148 and then the first equation to solve for 6205 we finally obtain

$$17 = (302)57970 - (1691)10353$$

as can easily be checked directly. Hence the equation $ax + by = (a, b)$ for the greatest common divisor of a and b in this example has the solution $x = 302$ and $y = -1691$. Note that it is relatively unlikely that this relation would have been found simply by guessing.

The integers x and y in (7) above are not unique. In the example with $a = 57970$ and $b = 10353$ we determined one solution to be $x = 302$ and $y = -1691$, for instance, and it is relatively simple to check that $x = -307$ and $y = 1719$ also satisfy $57970x + 10353y = 17$. The general solution for x and y is known (cf. the exercises below and in Chapter 8).

(8) An element p of $\mathbb{Z}^+$ is called a *prime* if $p > 1$ and the only positive divisors of p are 1 and p (initially, the word prime will refer only to positive integers). An integer $n > 1$ which is not prime is called *composite*. For example, 2,3,5,7,11,13,17,19,... are primes and 4,6,8,9,10,12,14,15,16,18,... are composite.

An important property of primes (which in fact can be used to *define* the primes (cf. Exercise 3)) is the following: if p is a prime and $p \mid ab$, for some $a, b \in \mathbb{Z}$, then either $p \mid a$ or $p \mid b$.

(9) The *Fundamental Theorem of Arithmetic* says: if $n \in \mathbb{Z}$, $n > 1$, then n can be factored uniquely into the product of primes, i.e., there are distinct primes $p_1, p_2, \ldots, p_s$ and positive integers $\alpha_1, \alpha_2, \ldots, \alpha_s$ such that

$$n = p_1^{\alpha_1} p_2^{\alpha_2} \ldots p_s^{\alpha_s}.$$

This factorization is unique in the sense that if $q_1, q_2, \ldots, q_t$ are any distinct primes and $\beta_1, \beta_2, \ldots, \beta_t$ positive integers such that

$$n = q_1^{\beta_1} q_2^{\beta_2} \ldots q_t^{\beta_t},$$

then $s = t$ and if we arrange the two sets of primes in increasing order, then $q_i = p_i$ and $\alpha_i = \beta_i$, $1 \le i \le s$. For example, $n = 1852423848 = 2^3 3^2 11^2 19^3 31$ and this decomposition into the product of primes is unique.

Suppose the positive integers a and b are expressed as products of prime powers:

$$a = p_1^{\alpha_1} p_2^{\alpha_2} \ldots p_s^{\alpha_s}, \quad b = p_1^{\beta_1} p_2^{\beta_2} \ldots p_s^{\beta_s}$$

where $p_1, p_2, \ldots, p_s$ are distinct and the exponents are ≥ 0 (we allow the exponents to be 0 here so that the products are taken over the same set of primes — the exponent will be 0 if that prime is not actually a divisor). Then the greatest common divisor of a and b is

$$(a, b) = p_1^{\min(\alpha_1, \beta_1)} p_2^{\min(\alpha_2, \beta_2)} \ldots p_s^{\min(\alpha_s, \beta_s)}$$

(and the least common multiple is obtained by instead taking the maximum of the α_i and β_i instead of the minimum).

Example

In the example above, $a = 57970$ and $b = 10353$ can be factored as $a = 2 \cdot 5 \cdot 11 \cdot 17 \cdot 31$ and $b = 3 \cdot 7 \cdot 17 \cdot 29$, from which we can immediately conclude that their greatest common divisor is 17. Note, however, that for large integers it is extremely difficult to determine their prime factorizations (several common codes in current use are based on this difficulty, in fact), so that this is not an effective method to determine greatest common divisors in general. The Euclidean Algorithm will produce greatest common divisors quite rapidly without the need for the prime factorization of a and b.

(10) The *Euler φ–function* is defined as follows: for $n \in \mathbb{Z}^+$ let $\varphi(n)$ be the number of positive integers $a \leq n$ with a relatively prime to n, i.e., $(a, n) = 1$. For example, $\varphi(12) = 4$ since 1, 5, 7 and 11 are the only positive integers less than or equal to 12 which have no factors in common with 12. Similarly, $\varphi(1) = 1$, $\varphi(2) = 1$, $\varphi(3) = 2$, $\varphi(4) = 2$, $\varphi(5) = 4$, $\varphi(6) = 2$, etc. For primes p, $\varphi(p) = p - 1$, and, more generally, for all $a \geq 1$ we have the formula

$$\varphi(p^a) = p^a - p^{a-1} = p^{a-1}(p - 1).$$

The function φ is *multiplicative* in the sense that

$$\varphi(ab) = \varphi(a)\varphi(b) \qquad \text{if } (a, b) = 1$$

(note that it is important here that a and b be relatively prime). Together with the formula above this gives a general formula for the values of φ : if $n = p_1^{\alpha_1} p_2^{\alpha_2} \ldots p_s^{\alpha_s}$, then

$$\varphi(n) = \varphi(p_1^{\alpha_1})\varphi(p_2^{\alpha_2}) \ldots \varphi(p_s^{\alpha_s})$$
$$= p_1^{\alpha_1 - 1}(p_1 - 1)p_2^{\alpha_2 - 1}(p_2 - 1) \ldots p_s^{\alpha_s - 1}(p_s - 1).$$

For example, $\varphi(12) = \varphi(2^2)\varphi(3) = 2^1(2 - 1)3^0(3 - 1) = 4$. The reader should note that we shall use the letter φ for many different functions throughout the text so when we want this letter to denote Euler's function we shall be careful to indicate this explicitly.

EXERCISES

1. For each of the following pairs of integers a and b, determine their greatest common divisor, their least common multiple, and write their greatest common divisor in the form $ax + by$ for some integers x and y.
   (a) $a = 20, b = 13$.
   (b) $a = 69, b = 372$.
   (c) $a = 792, b = 275$.
   (d) $a = 11391, b = 5673$.
   (e) $a = 1761, b = 1567$.
   (f) $a = 507885, b = 60808$.

2. Prove that if the integer k divides the integers a and b then k divides $as + bt$ for every pair of integers s and t.

3. Prove that if n is composite then there are integers a and b such that n divides ab but n does not divide either a or b.

4. Let a, b and N be fixed integers with a and b nonzero and let $d = (a, b)$ be the greatest common divisor of a and b. Suppose x_0 and y_0 are particular solutions to $ax + by = N$ (i.e., $ax_0 + by_0 = N$). Prove for any integer t that the integers

$$x = x_0 + \frac{b}{d}t \quad \text{and} \quad y = y_0 - \frac{a}{d}t$$

are also solutions to $ax + by = N$ (this is in fact the general solution).

5. Determine the value $\varphi(n)$ for each integer $n \leq 30$ where φ denotes the Euler φ-function.

6. Prove the Well Ordering Property of $\mathbb{Z}$ by induction and prove the minimal element is unique.

7. If p is a prime prove that there do not exist nonzero integers a and b such that $a^2 = pb^2$ (i.e., $\sqrt{p}$ is not a rational number).

8. Let p be a prime, $n \in \mathbb{Z}^+$. Find a formula for the largest power of p which divides $n! = n(n-1)(n-2)\ldots 2 \cdot 1$ (it involves the greatest integer function).

9. Write a computer program to determine the greatest common divisor (a, b) of two integers a and b and to express (a, b) in the form $ax + by$ for some integers x and y.

10. Prove for any given positive integer N there exist only finitely many integers n with $\varphi(n) = N$ where φ denotes Euler's φ-function. Conclude in particular that $\varphi(n)$ tends to infinity as n tends to infinity.

11. Prove that if d divides n then $\varphi(d)$ divides $\varphi(n)$ where φ denotes Euler's φ-function.

0.3 $\mathbb{Z}/n\,\mathbb{Z}$: THE INTEGERS MODULO n

Let n be a fixed positive integer. Define a relation on $\mathbb{Z}$ by

$$a \sim b \text{ if and only if } n \mid (b - a).$$

Clearly $a \sim a$, and $a \sim b$ implies $b \sim a$ for any integers a and b, so this relation is trivially reflexive and symmetric. If $a \sim b$ and $b \sim c$ then n divides $a - b$ and n divides $b - c$ so n also divides the sum of these two integers, i.e., n divides $(a - b) + (b - c) = a - c$, so $a \sim c$ and the relation is transitive. Hence this is an equivalence relation. Write $a \equiv b \pmod{n}$ (read: a is *congruent* to b mod n) if $a \sim b$. For any $k \in \mathbb{Z}$ we shall denote the equivalence class of a by $\bar{a}$ — this is called the *congruence class* or *residue class* of a mod n and consists of the integers which differ from a by an integral multiple of n, i.e.,

$$\bar{a} = \{a + kn \mid k \in \mathbb{Z}\}$$
$$= \{a, a \pm n, a \pm 2n, a \pm 3n, \ldots\}.$$

There are precisely n distinct equivalence classes mod n, namely

$$\bar{0}, \bar{1}, \bar{2}, \ldots, \overline{n-1}$$

determined by the possible remainders after division by n and these residue classes partition the integers $\mathbb{Z}$. The set of equivalence classes under this equivalence relation will be denoted by $\mathbb{Z}/n\mathbb{Z}$ and called the *integers modulo n* (or the *integers mod n*). The motivation for this notation will become clearer when we discuss quotient groups and quotient rings. Note that for different *n*'s the equivalence relation and equivalence classes are different so we shall always be careful to fix *n* first before using the bar notation. The process of finding the equivalence class mod *n* of some integer *a* is often referred to as *reducing a mod n*. This terminology also frequently refers to finding the smallest nonnegative integer congruent to *a* mod *n* (the *least residue* of *a* mod *n*).

We can define an addition and a multiplication for the elements of $\mathbb{Z}/n\mathbb{Z}$, defining *modular arithmetic* as follows: for $\bar{a}, \bar{b} \in \mathbb{Z}/n\mathbb{Z}$, define their sum and product by

$$\bar{a} + \bar{b} = \overline{a + b} \quad \text{and} \quad \bar{a} \cdot \bar{b} = \overline{ab}.$$

What this means is the following: given any two elements $\bar{a}$ and $\bar{b}$ in $\mathbb{Z}/n\mathbb{Z}$, to compute their sum (respectively, their product) take *any representative* integer *a* in the *class* $\bar{a}$ and *any representative* integer *b* in the *class* $\bar{b}$ and add (respectively, multiply) the integers *a* and *b* as usual in $\mathbb{Z}$ and then take the equivalence class containing the result. The following Theorem 3 asserts that this is well defined, i.e., does not depend on the choice of representatives taken for the elements $\bar{a}$ and $\bar{b}$ of $\mathbb{Z}/n\mathbb{Z}$.

Example

Suppose $n = 12$ and consider $\mathbb{Z}/12\mathbb{Z}$, which consists of the twelve residue classes

$$\bar{0}, \bar{1}, \bar{2}, \ldots, \overline{11}$$

determined by the twelve possible remainders of an integer after division by 12. The elements in the residue class $\bar{5}$, for example, are the integers which leave a remainder of 5 when divided by 12 (the integers *congruent to* 5 mod 12). Any integer congruent to 5 mod 12 (such as 5, 17, 29, ... or $-7, -19, \ldots$) will serve as a representative for the residue class $\bar{5}$. Note that $\mathbb{Z}/12\mathbb{Z}$ consists of the twelve *elements* above (and each of these elements of $\mathbb{Z}/12\mathbb{Z}$ consists of an infinite number of usual integers).

Suppose now that $\bar{a} = \bar{5}$ and $\bar{b} = \bar{8}$. The most obvious representative for $\bar{a}$ is the integer 5 and similarly 8 is the most obvious representative for $\bar{b}$. Using *these* representatives for the residue classes we obtain $\bar{5} + \bar{8} = \overline{13} = \bar{1}$ since 13 and 1 lie in the same class modulo $n = 12$. Had we instead taken the representative 17, say, for $\bar{a}$ (note that 5 and 17 do lie in the same residue class modulo 12) and the representative -28, say, for $\bar{b}$, we would obtain $\bar{5} + \bar{8} = \overline{(17 - 28)} = \overline{-11} = \bar{1}$ and as we mentioned the result does not depend on the choice of representatives chosen. The product of these two classes is $\bar{a} \cdot \bar{b} = \bar{5} \cdot \bar{8} = \overline{40} = \bar{4}$, also independent of the representatives chosen.

Theorem 3. The operations of addition and multiplication on $\mathbb{Z}/n\mathbb{Z}$ defined above are both well defined, that is, they do not depend on the choices of representatives for the classes involved. More precisely, if $a_1, a_2 \in \mathbb{Z}$ and $b_1, b_2 \in \mathbb{Z}$ with $\overline{a_1} = \overline{b_1}$ and $\overline{a_2} = \overline{b_2}$, then $\overline{a_1 + a_2} = \overline{b_1 + b_2}$ and $\overline{a_1 a_2} = \overline{b_1 b_2}$, i.e., if

$$a_1 \equiv b_1 \pmod{n} \quad \text{and} \quad a_2 \equiv b_2 \pmod{n}$$

then

$$a_1 + a_2 \equiv b_1 + b_2 \pmod{n} \quad \text{and} \quad a_1 a_2 \equiv b_1 b_2 \pmod{n}.$$

Proof: Suppose $a_1 \equiv b_1 \pmod{n}$, i.e., $a_1 - b_1$ is divisible by n. Then $a_1 = b_1 + sn$ for some integer s. Similarly, $a_2 \equiv b_2 \pmod{n}$ means $a_2 = b_2 + tn$ for some integer t. Then $a_1 + a_2 = (b_1 + b_2) + (s + t)n$ so that $a_1 + a_2 \equiv b_1 + b_2 \pmod{n}$, which shows that the sum of the residue classes is independent of the representatives chosen. Similarly, $a_1 a_2 = (b_1 + sn)(b_2 + tn) = b_1 b_2 + (b_1 t + b_2 s + stn)n$ shows that $a_1 a_2 \equiv b_1 b_2 \pmod{n}$ and so the product of the residue classes is also independent of the representatives chosen, completing the proof.

We shall see later that the process of adding equivalence classes by adding their representatives is a special case of a more general construction (the construction of a *quotient*). This notion of adding equivalence classes is already a familiar one in the context of adding rational numbers: each rational number a/b is really a class of expressions: $a/b = 2a/2b = -3a/-3b$ etc. and we often change representatives (for instance, take common denominators) in order to add two fractions (for example $1/2 + 1/3$ is computed by taking instead the equivalent representatives $3/6$ for $1/2$ and $2/6$ for $1/3$ to obtain $1/2 + 1/3 = 3/6 + 2/6 = 5/6$). The notion of modular arithmetic is also familiar: to find the hour of day after adding or subtracting some number of hours we reduce mod 12 and find the least residue.

It is important to be able to think of the equivalence classes of some equivalence relation as *elements* which can be manipulated (as we do, for example, with fractions) rather than as sets. Consistent with this attitude, we shall frequently denote the elements of $\mathbb{Z}/n\mathbb{Z}$ simply by $\{0, 1, \ldots, n-1\}$ where addition and multiplication are *reduced mod n*. It is important to remember, however, that the elements of $\mathbb{Z}/n\mathbb{Z}$ are *not* integers, but rather collections of usual integers, and the arithmetic is quite different. For example, $5 + 8$ is not 1 in the integers $\mathbb{Z}$ as it was in the example of $\mathbb{Z}/12\mathbb{Z}$ above.

The fact that one can define arithmetic in $\mathbb{Z}/n\mathbb{Z}$ has many important applications in elementary number theory. As one simple example we compute the last two digits in the number 2^{1000}. First observe that the last two digits give the remainder of 2^{1000} after we divide by 100 so we are interested in the residue class mod 100 containing 2^{1000}. We compute $2^{10} = 1024 \equiv 24 \pmod{100}$, so then $2^{20} = (2^{10})^2 \equiv 24^2 = 576 \equiv 76 \pmod{100}$. Then $2^{40} = (2^{20})^2 \equiv 76^2 = 5776 \equiv 76 \pmod{100}$. Similarly $2^{80} \equiv 2^{160} \equiv 2^{320} \equiv 2^{640} \equiv 76 \pmod{100}$. Finally, $2^{1000} = 2^{640} 2^{320} 2^{40} \equiv 76 \cdot 76 \cdot 76 \equiv 76 \pmod{100}$ so the final two digits are 76.

An important subset of $\mathbb{Z}/n\mathbb{Z}$ consists of the collection of residue classes which have a multiplicative inverse in $\mathbb{Z}/n\mathbb{Z}$:

$$(\mathbb{Z}/n\mathbb{Z})^{\times} = \{\bar{a} \in \mathbb{Z}/n\mathbb{Z} \mid \text{there exists } \bar{c} \in \mathbb{Z}/n\mathbb{Z} \text{ with } \bar{a} \cdot \bar{c} = \bar{1}\}.$$

Some of the following exercises outline a proof that $(\mathbb{Z}/n\mathbb{Z})^{\times}$ is also the collection of residue classes whose representatives are relatively prime to n, which proves the following proposition.

Proposition 4. $(\mathbb{Z}/n\mathbb{Z})^{\times} = \{\bar{a} \in \mathbb{Z}/n\mathbb{Z} \mid (a, n) = 1\}$.

It is easy to see that if *any* representative of $\bar{a}$ is relatively prime to n then *all* representatives are relatively prime to n so that the set on the right in the proposition is well defined.

Example

For $n = 9$ we obtain $(\mathbb{Z}/9\mathbb{Z})^{\times} = \{\overline{1}, \overline{2}, \overline{4}, \overline{5}, \overline{7}, \overline{8}\}$ from the proposition. The multiplicative inverses of these elements are $\{\overline{1}, \overline{5}, \overline{7}, \overline{2}, \overline{4}, \overline{8}\}$, respectively.

If a is an integer relatively prime to n then the Euclidean Algorithm produces integers x and y satisfying $ax + ny = 1$, hence $ax \equiv 1 \pmod{n}$, so that $\bar{x}$ is the multiplicative inverse of $\bar{a}$ in $\mathbb{Z}/n\mathbb{Z}$. This gives an efficient method for computing multiplicative inverses in $\mathbb{Z}/n\mathbb{Z}$.

Example

Suppose $n = 60$ and $a = 17$. Applying the Euclidean Algorithm we obtain

$$60 = (3)17 + 9$$
$$17 = (1)9 + 8$$
$$9 = (1)8 + 1$$

so that a and n are relatively prime, and $(-7)17 + (2)60 = 1$. Hence $\overline{-7} = \overline{53}$ is the multiplicative inverse of $\overline{17}$ in $\mathbb{Z}/60\mathbb{Z}$.

EXERCISES

1. Write down explicitly all the elements in the residue classes of $\mathbb{Z}/18\mathbb{Z}$.

2. Prove that the distinct equivalence classes in $\mathbb{Z}/n\mathbb{Z}$ are precisely $\overline{0}, \overline{1}, \overline{2}, \ldots, \overline{n-1}$ (use the Division Algorithm).

3. Prove that if $a = a_n 10^n + a_{n-1} 10^{n-1} + \cdots + a_1 10 + a_0$ is any positive integer then $a \equiv a_n + a_{n-1} + \cdots + a_1 + a_0 \pmod{9}$ (note that this is the usual arithmetic rule that the remainder after division by 9 is the same as the sum of the decimal digits mod 9 – in particular an integer is divisible by 9 if and only if the sum of its digits is divisible by 9) [note that $10 \equiv 1 \pmod{9}$].

4. Compute the remainder when 37^{100} is divided by 29.

5. Compute the last two digits of 9^{1500}.

6. Prove that the squares of the elements in $\mathbb{Z}/4\mathbb{Z}$ are just $\overline{0}$ and $\overline{1}$.

7. Prove for any integers a and b that $a^2 + b^2$ never leaves a remainder of 3 when divided by 4 (use the previous exercise).

8. Prove that the equation $a^2 + b^2 = 3c^2$ has no solutions in nonzero integers a, b and c. [Consider the equation mod 4 as in the previous two exercises and show that a, b and c would all have to be divisible by 2. Then each of a^2, b^2 and c^2 has a factor of 4 and by dividing through by 4 show that there would be a smaller set of solutions to the original equation. Iterate to reach a contradiction.]

9. Prove that the square of any odd integer always leaves a remainder of 1 when divided by 8.

10. Prove that the number of elements of $(\mathbb{Z}/n\mathbb{Z})^{\times}$ is $\varphi(n)$ where φ denotes the Euler φ-function.

11. Prove that if $\bar{a}, \bar{b} \in (\mathbb{Z}/n\mathbb{Z})^{\times}$, then $\bar{a} \cdot \bar{b} \in (\mathbb{Z}/n\mathbb{Z})^{\times}$.

12. Let $n \in \mathbb{Z}$, $n > 1$, and let $a \in \mathbb{Z}$ with $1 \leq a \leq n$. Prove if a and n are not relatively prime, there exists an integer b with $1 \leq b < n$ such that $ab \equiv 0 \pmod{n}$ and deduce that there cannot be an integer c such that $ac \equiv 1 \pmod{n}$.

13. Let $n \in \mathbb{Z}$, $n > 1$, and let $a \in \mathbb{Z}$ with $1 \leq a \leq n$. Prove that if a and n are relatively prime then there is an integer c such that $ac \equiv 1 \pmod{n}$ [use the fact that the g.c.d. of two integers is a $\mathbb{Z}$-linear combination of the integers].

14. Conclude from the previous two exercises that $(\mathbb{Z}/n\mathbb{Z})^{\times}$ is the set of elements $\bar{a}$ of $\mathbb{Z}/n\mathbb{Z}$ with $(a, n) = 1$ and hence prove Proposition 4. Verify this directly in the case $n = 12$.

15. For each of the following pairs of integers a and n, show that a is relatively prime to n and determine the multiplicative inverse of $\bar{a}$ in $\mathbb{Z}/n\mathbb{Z}$.
    (a) $a = 13, n = 20$.
    (b) $a = 69, n = 89$.
    (c) $a = 1891, n = 3797$.
    (d) $a = 6003722857$, $n = 77695236973$. [The Euclidean Algorithm requires only 3 steps for these integers.]

16. Write a computer program to add and multiply mod n, for any n given as input. The output of these operations should be the least residues of the sums and products of two integers. Also include the feature that if $(a, n) = 1$, an integer c between 1 and $n - 1$ such that $\bar{a} \cdot \bar{c} = \bar{1}$ may be printed on request. (Your program should not, of course, simply quote "mod" functions already built into many systems).

Part I

GROUP THEORY

The modern treatment of abstract algebra begins with the disarmingly simple abstract definition of a *group*. This simple definition quickly leads to difficult questions involving the structure of such objects. There are many specific examples of groups and the power of the abstract point of view becomes apparent when results for *all* of these examples are obtained by proving a *single* result for the abstract group.

The notion of a group did not simply spring into existence, however, but is rather the culmination of a long period of mathematical investigation, the first formal definition of an abstract group in the form in which we use it appearing in 1882.[1] The definition of an abstract group has its origins in extremely old problems in algebraic equations, number theory, and geometry, and arose because very similar techniques were found to be applicable in a variety of situations. As Otto Hölder (1859–1937) observed, one of the essential characteristics of mathematics is that after applying a certain algorithm or method of proof one then considers the scope and limits of the method. As a result, properties possessed by a number of interesting objects are frequently abstracted and the question raised: can one determine *all* the objects possessing these properties? Attempting to answer such a question also frequently adds considerable understanding of the original objects under consideration. It is in this fashion that the definition of an abstract group evolved into what is, for us, the starting point of abstract algebra.

We illustrate with a few of the disparate situations in which the ideas later formalized into the notion of an abstract group were used.

(1) In number theory the very object of study, the set of integers, is an example of a group. Consider for example what we refer to as "Euler's Theorem" (cf. Exercise 22 of Section 3.2), one extremely simple example of which is that a^{40} has last two digits 01 if a is any integer not divisible by 2 nor by 5. This was proved in 1761 by Leonhard Euler (1707–1783) using "group-theoretic" ideas of Joseph Louis Lagrange (1736–1813), long before the first formal definition of a group. From our perspective, one now proves "Lagrange's Theorem" (cf. Theorem 8 of Section 3.2), applying these techniques abstracted to an arbitrary group, and then *recovers* Euler's Theorem (and many others) as a *special case*.

---

[1] For most of the historical comments below, see the excellent book *A History of Algebra*, by B. L. van der Waerden, Springer-Verlag, 1980 and the references there, particularly *The Genesis of the Abstract Group Concept: A Contribution to the History of the Origin of Abstract Group Theory* (translated from the German by Abe Shenitzer), by H. Wussing, MIT Press, 1984. See also *Number Theory, An Approach Through History from Hammurapai to Legendre*, by A. Weil, Birkhäuser, 1984.

(2) Investigations into the question of rational solutions to algebraic equations of the form $y^2 = x^3 - 2x$ (there are infinitely many, for example $(0, 0)$, $(-1, 1)$, $(2, 2)$, $(9/4, -21/8)$, $(-1/169, 239/2197)$) showed that connecting any two solutions by a straight line and computing the intersection of this line with the curve $y^2 = x^3 - 2x$ produces another solution. Such "Diophantine equations," among others, were considered by Pierre de Fermat (1601–1655) (this one was solved by him in 1644), by Euler, by Lagrange around 1777, and others. In 1730 Euler raised the question of determining the indefinite integral $\int dx/\sqrt{1 - x^4}$ of the "lemniscatic differential" $dx/\sqrt{1 - x^4}$, used in determining the arc length along an ellipse (the question had also been considered by Gottfried Wilhelm Leibniz (1646–1716) and Johannes Bernoulli (1667–1748)). In 1752 Euler proved a "multiplication formula" for such elliptic integrals (using ideas of G.C. di Fagnano (1682–1766), received by Euler in 1751), which shows how two elliptic integrals give rise to a third, bringing into existence the theory of elliptic functions in analysis. In 1834 Carl Gustav Jacob Jacobi (1804–1851) observed that the work of Euler on solving certain Diophantine equations amounted to writing the multiplication formula for certain elliptic integrals. Today the curve above is referred to as an "elliptic curve" and these questions are viewed as two different aspects of the same thing — the fact that this geometric operation on points can be used to give the set of points on an elliptic curve the structure of a group. The study of the "arithmetic" of these groups is an active area of current research.[2]

(3) By 1824 it was known that there are formulas giving the roots of quadratic, cubic and quartic equations (extending the familiar quadratic formula for the roots of $ax^2 + bx + c = 0$). In 1824, however, Niels Henrik Abel (1802–1829) proved that such a formula for the roots of a quintic is impossible (cf. Corollary 40 of Section 14.7). The proof is based on the idea of examining what happens when the roots are permuted amongst themselves (for example, interchanging two of the roots). The collection of such permutations has the structure of a group (called, naturally enough, a "permutation group"). This idea culminated in the beautiful work of Evariste Galois (1811–1832) in 1830–32, working with explicit groups of "substitutions." Today this work is referred to as Galois Theory (and is the subject of the fourth part of this text). Similar explicit groups were being used in geometry as collections of geometric transformations (translations, reflections, etc.) by Arthur Cayley (1821–1895) around 1850, Camille Jordan (1838–1922) around 1867, Felix Klein (1849–1925) around 1870, etc., and the application of groups to geometry is still extremely active in current research into the structure of 3-space, 4-space, etc. The same group arising in the study of the solvability of the quintic arises in the study of the rigid motions of an icosahedron in geometry and in the study of elliptic functions in analysis.

The precursors of today's abstract group can be traced back many years, even before the groups of "substitutions" of Galois. The formal definition of an abstract group which is our starting point appeared in 1882 in the work of Walter Dyck (1856–1934), an assistant to Felix Klein, and also in the work of Heinrich Weber (1842–1913)

---

[2]See *The Arithmetic of Elliptic Curves* by J. Silverman, Springer-Verlag, 1986.

in the same year.

It is frequently the case in mathematics research to find specific application of an idea before having that idea extracted and presented as an item of interest in its own right (for example, Galois used the notion of a "quotient group" implicitly in his investigations in 1830 and the definition of an abstract quotient group is due to Hölder in 1889). It is important to realize, with or without the historical context, that the reason the abstract definitions are made is because it is useful to isolate specific characteristics and consider what structure is imposed on an object having these characteristics. The notion of the structure of an algebraic object (which is made more precise by the concept of an isomorphism — which considers when two apparently different objects are in some sense the same) is a major theme which will recur throughout the text.

CHAPTER 1

Introduction to Groups

1.1 BASIC AXIOMS AND EXAMPLES

In this section the basic algebraic structure to be studied in Part I is introduced and some examples are given.

Definition.

(1) A *binary operation* $\star$ on a set G is a function $\star : G \times G \to G$. For any $a, b \in G$ we shall write $a \star b$ for $\star(a, b)$.

(2) A binary operation $\star$ on a set G is *associative* if for all $a, b, c \in G$ we have $a \star (b \star c) = (a \star b) \star c$.

(3) If $\star$ is a binary operation on a set G we say elements a and b of G *commute* if $a \star b = b \star a$. We say $\star$ (or G) is *commutative* if for all $a, b \in G, a \star b = b \star a$.

Examples

(1) $+$ (usual addition) is a commutative binary operation on $\mathbb{Z}$ (or on $\mathbb{Q}, \mathbb{R}$, or $\mathbb{C}$ respectively).

(2) $\times$ (usual multiplication) is a commutative binary operation on $\mathbb{Z}$ (or on $\mathbb{Q}, \mathbb{R}$, or $\mathbb{C}$ respectively).

(3) $-$ (usual subtraction) is a noncommutative binary operation on $\mathbb{Z}$, where $-(a, b) = a - b$. The map $a \mapsto -a$ is not a binary operation (not binary).

(4) $-$ is not a binary operation on $\mathbb{Z}^+$ (nor $\mathbb{Q}^+, \mathbb{R}^+$) because for $a, b \in \mathbb{Z}^+$ with $a < b$, $a - b \notin \mathbb{Z}^+$, that is, $-$ does not map $\mathbb{Z}^+ \times \mathbb{Z}^+$ into $\mathbb{Z}^+$.

(5) Taking the vector cross-product of two vectors in 3-space $\mathbb{R}^3$ is a binary operation which is not associative and not commutative.

Suppose that $\star$ is a binary operation on a set G and H is a subset of G. If the restriction of $\star$ to H is a binary operation on H, i.e., for all $a, b \in H, a \star b \in H$, then H is said to be *closed* under $\star$. Observe that if $\star$ is an associative (respectively, commutative) binary operation on G and $\star$ restricted to some subset H of G is a binary operation on H, then $\star$ is automatically associative (respectively, commutative) on H as well.

Definition.

(1) A *group* is an ordered pair $(G, \star)$ where G is a set and $\star$ is a binary operation on G satisfying the following axioms:

(i) $(a \star b) \star c = a \star (b \star c)$, for all $a, b, c \in G$, i.e., $\star$ is *associative*,

(ii) there exists an element e in G, called an *identity* of G, such that for all $a \in G$ we have $a \star e = e \star a = a$,

(iii) for each $a \in G$ there is an element a^{-1} of G, called an *inverse* of a, such that $a \star a^{-1} = a^{-1} \star a = e$.

(2) The group $(G, \star)$ is called *abelian* (or *commutative*) if $a \star b = b \star a$ for all $a, b \in G$.

We shall immediately become less formal and say G is a group under $\star$ if $(G, \star)$ is a group (or just G is a group when the operation $\star$ is clear from the context). Also, we say G is a *finite group* if in addition G is a finite set. Note that axiom (ii) ensures that a group is always nonempty.

Examples

(1) $\mathbb{Z}$, $\mathbb{Q}$, $\mathbb{R}$ and $\mathbb{C}$ are groups under $+$ with $e = 0$ and $a^{-1} = -a$, for all a.

(2) $\mathbb{Q} - \{0\}$, $\mathbb{R} - \{0\}$, $\mathbb{C} - \{0\}$, $\mathbb{Q}^+$, $\mathbb{R}^+$ are groups under $\times$ with $e = 1$ and $a^{-1} = \dfrac{1}{a}$, for all a. Note however that $\mathbb{Z} - \{0\}$ is *not* a group under $\times$ because although $\times$ is an associative binary operation on $\mathbb{Z} - \{0\}$, the element 2 (for instance) does not have an inverse in $\mathbb{Z} - \{0\}$.

We have glossed over the fact that the associative law holds in these familiar examples. For $\mathbb{Z}$ under $+$ this is a consequence of the axiom of associativity for addition of natural numbers. The associative law for $\mathbb{Q}$ under $+$ follows from the associative law for $\mathbb{Z}$ — a proof of this will be outlined later when we rigorously construct $\mathbb{Q}$ from $\mathbb{Z}$ (cf. Section 7.5). The associative laws for $\mathbb{R}$ and, in turn, $\mathbb{C}$ under $+$ are proved in elementary analysis courses when $\mathbb{R}$ is constructed by completing $\mathbb{Q}$ — ultimately, associativity is again a consequence of associativity for $\mathbb{Z}$. The associative axiom for multiplication may be established via a similar development, starting first with $\mathbb{Z}$. Since $\mathbb{R}$ and $\mathbb{C}$ will be used largely for illustrative purposes and we shall not construct $\mathbb{R}$ from $\mathbb{Q}$ (although we shall construct $\mathbb{C}$ from $\mathbb{R}$) we shall take the associative laws (under $+$ and $\times$) for $\mathbb{R}$ and $\mathbb{C}$ as given.

Examples (continued)

(3) The axioms for a vector space V include those axioms which specify that $(V, +)$ is an abelian group (the operation $+$ is called vector addition). Thus any vector space such as $\mathbb{R}^n$ is, in particular, an additive group.

(4) For $n \in \mathbb{Z}^+$, $\mathbb{Z}/n\mathbb{Z}$ is an abelian group under the operation $+$ of addition of residue classes as described in Chapter 0. We shall prove in Chapter 3 (in a more general context) that this binary operation $+$ is well defined and associative; for now we take this for granted. The identity in this group is the element $\bar{0}$ and for each $\bar{a} \in \mathbb{Z}/n\mathbb{Z}$, the inverse of $\bar{a}$ is $\overline{-a}$. Henceforth, when we talk about the group $\mathbb{Z}/n\mathbb{Z}$ it will be understood that the group operation is addition of classes mod n.

(5) For $n \in \mathbb{Z}^+$, the set $(\mathbb{Z}/n\mathbb{Z})^\times$ of equivalence classes $\bar{a}$ which have multiplicative inverses mod n is an abelian group under *multiplication* of residue classes as described in Chapter 0. Again, we shall take for granted (for the moment) that this operation is well defined and associative. The identity of this group is the element $\bar{1}$ and, by definition of $(\mathbb{Z}/n\mathbb{Z})^\times$, each element has a multiplicative inverse. Henceforth, when we talk about the group $(\mathbb{Z}/n\mathbb{Z})^\times$ it will be understood that the group operation is multiplication of classes mod n.

(6) If $(A, \star)$ and $(B, \diamond)$ are groups, we can form a new group $A \times B$, called their *direct product*, whose elements are those in the Cartesian product

$$A \times B = \{(a, b) \mid a \in A, \ b \in B\}$$

and whose operation is defined componentwise:

$$(a_1, b_1)(a_2, b_2) = (a_1 \star a_2, b_1 \diamond b_2).$$

For example, if we take $A = B = \mathbb{R}$ (both operations addition), $\mathbb{R} \times \mathbb{R}$ is the familiar Euclidean plane. The proof that the direct product of two groups is again a group is left as a straightforward exercise (later) — the proof that each group axiom holds in $A \times B$ is a consequence of that axiom holding in both A and B together with the fact that the operation in $A \times B$ is defined componentwise.

There should be no confusion between the groups $\mathbb{Z}/n\mathbb{Z}$ (under addition) and $(\mathbb{Z}/n\mathbb{Z})^\times$ (under multiplication), even though the latter is a subset of the former — the superscript $\times$ will always indicate that the operation is multiplication.

Before continuing with more elaborate examples we prove two basic results which in particular enable us to talk about *the* identity and *the* inverse of an element.

Proposition 1. If G is a group under the operation $\star$, then
  (1) the identity of G is unique
  (2) for each $a \in G$, a^{-1} is uniquely determined
  (3) $(a^{-1})^{-1} = a$ for all $a \in G$
  (4) $(a \star b)^{-1} = (b^{-1}) \star (a^{-1})$
  (5) for any $a_1, a_2, \ldots, a_n \in G$ the value of $a_1 \star a_2 \star \cdots \star a_n$ is independent of how the expression is bracketed (this is called the *generalized associative law*).

Proof: (1) If f and g are both identities, then by axiom (ii) of the definition of a group $f \star g = f$ (take $a = f$ and $e = g$). By the same axiom $f \star g = g$ (take $a = g$ and $e = f$). Thus $f = g$, and the identity is unique.

(2) Assume b and c are both inverses of a and let e be the identity of G. By axiom (iii), $a \star b = e$ and $c \star a = e$. Thus

$$
\begin{aligned}
c &= c \star e && \text{(definition of e - axiom (ii))}\\
  &= c \star (a \star b) && \text{(since $e = a \star b$)}\\
  &= (c \star a) \star b && \text{(associative law)}\\
  &= e \star b && \text{(since $e = c \star a$)}\\
  &= b && \text{(axiom (ii)).}
\end{aligned}
$$

(3) To show $(a^{-1})^{-1} = a$ is exactly the problem of showing a is the inverse of a^{-1} (since by part (2) a has a unique inverse). Reading the definition of a^{-1}, with the roles of a and a^{-1} mentally interchanged shows that a satisfies the defining property for the inverse of a^{-1}, hence a is the inverse of a^{-1}.

(4) Let $c = (a \star b)^{-1}$ so by definition of c, $(a \star b) \star c = e$. By the associative law

$$a \star (b \star c) = e.$$

Multiply both sides on the left by a^{-1} to get

$$a^{-1} \star (a \star (b \star c)) = a^{-1} \star e.$$

The associative law on the left hand side and the definition of e on the right give

$$(a^{-1} \star a) \star (b \star c) = a^{-1}$$

so

$$e \star (b \star c) = a^{-1}$$

hence

$$b \star c = a^{-1}.$$

Now multiply both sides on the left by b^{-1} and simplify similarly:

$$b^{-1} \star (b \star c) = b^{-1} \star a^{-1}$$
$$(b^{-1} \star b) \star c = b^{-1} \star a^{-1}$$
$$e \star c = b^{-1} \star a^{-1}$$
$$c = b^{-1} \star a^{-1},$$

as claimed.

(5) This is left as a good exercise using induction on n. First show the result is true for $n = 1, 2$, and 3. Next assume for any $k < n$ that any bracketing of a product of k elements, $b_1 \star b_2 \star \cdots \star b_k$ can be reduced (without altering the value of the product) to an expression of the form

$$b_1 \star (b_2 \star (b_3 \star (\cdots \star b_k)) \ldots).$$

Now argue that any bracketing of the product $a_1 \star a_2 \star \cdots \star a_n$ must break into 2 subproducts, say $(a_1 \star a_2 \star \cdots \star a_k) \star (a_{k+1} \star a_{k+2} \star \cdots \star a_n)$, where each sub-product is bracketed in some fashion. Apply the induction assumption to each of these two sub-products and finally reduce the result to the form $a_1 \star (a_2 \star (a_3 \star (\cdots \star a_n)) \ldots)$ to complete the induction.

Note that throughout the proof of Proposition 1 we were careful not to change the *order* of any products (unless permitted by axioms (ii) and (iii)) since G may be non-abelian.

Notation:

(1) For an abstract group G it is tiresome to keep writing the operation $\star$ throughout our calculations. Henceforth (except when necessary) our abstract groups G, H, *etc.* will always be written with the operation as $\cdot$ and $a \cdot b$ will always be written as ab. In view of the generalized associative law, products of three or more group elements will not be bracketed (although the operation is still a binary operation). Finally, for an abstract group G (operation $\cdot$) we denote the identity of G by 1.

(2) For any group G (operation $\cdot$ implied) and $x \in G$ and $n \in \mathbb{Z}^+$ since the product $xx \cdots x$ (n terms) does not depend on how it is bracketed, we shall denote it by x^n. Denote $x^{-1}x^{-1} \cdots x^{-1}$ (n terms) by x^{-n}. Let $x^0 = 1$, the identity of G.

This new notation is pleasantly concise. Of course, when we are dealing with specific groups, we shall use the natural (given) operation. For example, when the operation is $+$, the identity will be denoted by 0 and for any element a, the inverse a^{-1} will be written $-a$ and $a + a + \cdots + a$ ($n > 0$ terms) will be written na; $-a - a \cdots - a$ (n terms) will be written $-na$ and $0a = 0$.

Proposition 2. Let G be a group and let $a, b \in G$. The equations $ax = b$ and $ya = b$ have unique solutions for $x, y \in G$. In particular, the left and right cancellation laws hold in G, i.e.,
  (1) if $au = av$, then $u = v$, and
  (2) if $ub = vb$, then $u = v$.

Proof: We can solve $ax = b$ by multiplying both sides on the left by a^{-1} and simplifying to get $x = a^{-1}b$. The uniqueness of x follows because a^{-1} is unique. Similarly, if $ya = b$, $y = ba^{-1}$. If $au = av$, multiply both sides on the left by a^{-1} and simplify to get $u = v$. Similarly, the right cancellation law holds.

One consequence of Proposition 2 is that if a is any element of G and for some $b \in G$, $ab = e$ or $ba = e$, then $b = a^{-1}$, i.e., we do not have to show both equations hold. Also, if for some $b \in G$, $ab = a$ (or $ba = a$), then b must be the identity of G, i.e., we do not have to check $bx = xb = x$ for all $x \in G$.

Definition. For G a group and $x \in G$ define the *order* of x to be the smallest positive integer n such that $x^n = 1$, and denote this integer by $|x|$. In this case x is said to be of order n. If no positive power of x is the identity, the order of x is defined to be infinity and x is said to be of infinite order.

The symbol for the order of x should not be confused with the absolute value symbol (when $G \subseteq \mathbb{R}$ we shall be careful to distinguish the two). It may seem injudicious to choose the same symbol for order of an element as the one used to denote the cardinality (or order) of a set, however, we shall see that the order of an element in a group is the same as the cardinality of the set of all its (distinct) powers so the two uses of the word "order" are naturally related.

Examples
  (1) An element of a group has order 1 if and only if it is the identity.
  (2) In the additive groups $\mathbb{Z}, \mathbb{Q}, \mathbb{R}$ or $\mathbb{C}$ every nonzero (i.e., nonidentity) element has infinite order.
  (3) In the multiplicative groups $\mathbb{R} - \{0\}$ or $\mathbb{Q} - \{0\}$ the element -1 has order 2 and all other nonidentity elements have infinite order.
  (4) In the additive group $\mathbb{Z}/9\mathbb{Z}$ the element $\bar{6}$ has order 3, since $\bar{6} \neq \bar{0}, \bar{6} + \bar{6} = \overline{12} = \bar{3} \neq \bar{0}$, but $\bar{6} + \bar{6} + \bar{6} = \overline{18} = \bar{0}$, the identity in this group. Recall that in an *additive* group the powers of an element are the integer multiples of the element. Similarly, the order of the element $\bar{5}$ is 9, since 45 is the smallest positive multiple of 5 that is divisible by 9.

(5) In the multiplicative group $(\mathbb{Z}/7\mathbb{Z})^{\times}$, the powers of the element $\bar{2}$ are $\bar{2}, \bar{4}, \bar{8} = \bar{1}$, the identity in this group, so $\bar{2}$ has order 3. Similarly, the element $\bar{3}$ has order 6, since 3^6 is the smallest positive power of 3 that is congruent to 1 modulo 7.

Definition. Let $G = \{g_1, g_2, \ldots, g_n\}$ be a finite group with $g_1 = 1$. The *multiplication table* or *group table* of G is the $n \times n$ matrix whose i, j entry is the group element $g_i g_j$.

For a finite group the multiplication table contains, in some sense, all the information about the group. Computationally, however, it is an unwieldly object (being of size the square of the group order) and visually it is not a very useful object for determining properties of the group. One might think of a group table as the analogue of having a table of all the distances between pairs of cities in the country. Such a table is useful and, in essence, captures all the distance relationships, yet a map (better yet, a map with all the distances labelled on it) is a much easier tool to work with. Part of our initial development of the theory of groups (finite groups in particular) is directed towards a more conceptual way of visualizing the internal structure of groups.

EXERCISES

Let G be a group.

1. Determine which of the following binary operations are associative:
   (a) the operation $\star$ on $\mathbb{Z}$ defined by $a \star b = a - b$
   (b) the operation $\star$ on $\mathbb{R}$ defined by $a \star b = a + b + ab$
   (c) the operation $\star$ on $\mathbb{Q}$ defined by $a \star b = \dfrac{a + b}{5}$
   (d) the operation $\star$ on $\mathbb{Z} \times \mathbb{Z}$ defined by $(a, b) \star (c, d) = (ad + bc, bd)$
   (e) the operation $\star$ on $\mathbb{Q} - \{0\}$ defined by $a \star b = \dfrac{a}{b}$.

2. Decide which of the binary operations in the preceding exercise are commutative.

3. Prove that addition of residue classes in $\mathbb{Z}/n\mathbb{Z}$ is associative (you may assume it is well defined).

4. Prove that multiplication of residue classes in $\mathbb{Z}/n\mathbb{Z}$ is associative (you may assume it is well defined).

5. Prove for all $n > 1$ that $\mathbb{Z}/n\mathbb{Z}$ is not a group under multiplication of residue classes.

6. Determine which of the following sets are groups under addition:
   (a) the set of rational numbers (including $0 = 0/1$) in lowest terms whose denominators are odd
   (b) the set of rational numbers in lowest terms whose denominators are even together with 0.
   (c) the set of rational numbers of absolute value < 1
   (d) the set of rational numbers of absolute value ≥ 1 together with 0
   (e) the set of rational numbers with denominators equal to 1 or 2
   (f) the set of rational numbers with denominators equal to 1, 2 or 3.

7. Let $G = \{x \in \mathbb{R} \mid 0 \leq x < 1\}$ and for $x, y \in G$ let $x \star y$ be the fractional part of $x + y$ (i.e., $x \star y = x + y - [x + y]$ where $[a]$ is the greatest integer less than or equal to a). Prove that $\star$ is a well defined binary operation on G and that G is an abelian group under $\star$ (called the *real numbers mod 1*).

8. Let $G = \{z \in \mathbb{C} \mid z^n = 1 \text{ for some } n \in \mathbb{Z}^+\}$.
  (a) Prove that G is a group under multiplication (called the group of *roots of unity* in $\mathbb{C}$).
  (b) Prove that G is not a group under addition.

9. Let $G = \{a + b\sqrt{2} \in \mathbb{R} \mid a, b \in \mathbb{Q}\}$.
  (a) Prove that G is a group under addition.
  (b) Prove that the nonzero elements of G are a group under multiplication. ["Rationalize the denominators" to find multiplicative inverses.]

10. Prove that a finite group is abelian if and only if its group table is a symmetric matrix.

11. Find the orders of each element of the additive group $\mathbb{Z}/12\mathbb{Z}$.

12. Find the orders of the following elements of the multiplicative group $(\mathbb{Z}/12\mathbb{Z})^\times$: $\overline{1}, \overline{-1}, \overline{5}, \overline{7}, \overline{-7}, \overline{13}$.

13. Find the orders of the following elements of the additive group $\mathbb{Z}/36\mathbb{Z}$: $\overline{1}, \overline{2}, \overline{6}, \overline{9}, \overline{10}, \overline{12}, \overline{-1}, \overline{-10}, \overline{-18}$.

14. Find the orders of the following elements of the multiplicative group $(\mathbb{Z}/36\mathbb{Z})^\times$: $\overline{1}, \overline{-1}, \overline{5}, \overline{13}, \overline{-13}, \overline{17}$.

15. Prove that $(a_1 a_2 \ldots a_n)^{-1} = a_n^{-1} a_{n-1}^{-1} \ldots a_1^{-1}$ for all $a_1, a_2, \ldots, a_n \in G$.

16. Let x be an element of G. Prove that $x^2 = 1$ if and only if $|x|$ is either 1 or 2.

17. Let x be an element of G. Prove that if $|x| = n$ for some positive integer n then $x^{-1} = x^{n-1}$.

18. Let x and y be elements of G. Prove that $xy = yx$ if and only if $y^{-1}xy = x$ if and only if $x^{-1}y^{-1}xy = 1$.

19. Let $x \in G$ and let $a, b \in \mathbb{Z}^+$.
  (a) Prove that $x^{a+b} = x^a x^b$ and $(x^a)^b = x^{ab}$.
  (b) Prove that $(x^a)^{-1} = x^{-a}$.
  (c) Establish part (a) for arbitrary integers a and b (positive, negative or zero).

20. For x an element in G show that x and x^{-1} have the same order.

21. Let G be a finite group and let x be an element of G of order n. Prove that if n is odd, then $x = (x^2)^k$ for some integer $k \geq 1$.

22. If x and g are elements of the group G, prove that $|x| = |g^{-1}xg|$. Deduce that $|ab| = |ba|$ for all $a, b \in G$.

23. Suppose $x \in G$ and $|x| = n < \infty$. If $n = st$ for some positive integers s and t, prove that $|x^s| = t$.

24. If a and b are *commuting* elements of G, prove that $(ab)^n = a^n b^n$ for all $n \in \mathbb{Z}$. [Do this by induction for positive n first.]

25. Prove that if $x^2 = 1$ for all $x \in G$ then G is abelian.

26. Assume H is a nonempty subset of $(G, \star)$ which is closed under the binary operation on G and is closed under inverses, i.e., for all h and $k \in H$, hk and $h^{-1} \in H$. Prove that H is a group under the operation $\star$ restricted to H (such a subset H is called a *subgroup* of G).

27. Prove that if x is an element of the group G then $\{x^n \mid n \in \mathbb{Z}\}$ is a subgroup (cf. the preceding exercise) of G (called the *cyclic subgroup* of G generated by x).

28. Let $(A, \star)$ and $(B, \diamond)$ be groups and let $A \times B$ be their direct product (as defined in Example 6). Verify all the group axioms for $A \times B$:
  (a) prove that the associative law holds: for all $(a_i, b_i) \in A \times B, i = 1, 2, 3$
  $$(a_1, b_1)[(a_2, b_2)(a_3, b_3)] = [(a_1, b_1)(a_2, b_2)](a_3, b_3),$$

(b) prove that $(1, 1)$ is the identity of $A \times B$, and

(c) prove that the inverse of (a, b) is (a^{-1}, b^{-1}).

29. Prove that $A \times B$ is an abelian group if and only if both A and B are abelian.

30. Prove that the elements $(a, 1)$ and $(1, b)$ of $A \times B$ commute and deduce that the order of (a, b) is the least common multiple of $|a|$ and $|b|$.

31. Prove that any finite group G of even order contains an element of order 2. [Let $t(G)$ be the set $\{g \in G \mid g \neq g^{-1}\}$. Show that $t(G)$ has an even number of elements and every nonidentity element of $G - t(G)$ has order 2.]

32. If x is an element of finite order n in G, prove that the elements $1, x, x^2, \ldots, x^{n-1}$ are all distinct. Deduce that $|x| \leq |G|$.

33. Let x be an element of finite order n in G.
   (a) Prove that if n is odd then $x^i \neq x^{-i}$ for all $i = 1, 2, \ldots, n - 1$.
   (b) Prove that if $n = 2k$ and $1 \leq i < n$ then $x^i = x^{-i}$ if and only if $i = k$.

34. If x is an element of infinite order in G, prove that the elements x^n, $n \in \mathbb{Z}$ are all distinct.

35. If x is an element of finite order n in G, use the Division Algorithm to show that *any* integral power of x equals one of the elements in the set $\{1, x, x^2, \ldots, x^{n-1}\}$ (so these are all the distinct elements of the cyclic subgroup (cf. Exercise 27 above) of G generated by x).

36. Assume $G = \{1, a, b, c\}$ is a group of order 4 with identity 1. Assume also that G has no elements of order 4 (so by Exercise 32, every element has order ≤ 3). Use the cancellation laws to show that there is a unique group table for G. Deduce that G is abelian.

1.2 DIHEDRAL GROUPS

An important family of examples of groups is the class of groups whose elements are symmetries of geometric objects. The simplest subclass is when the geometric objects are regular planar figures.

For each $n \in \mathbb{Z}^+$, $n \geq 3$ let D_{2n} be the set of symmetries of a regular n-gon, where a symmetry is any rigid motion of the n-gon which can be effected by taking a copy of the n-gon, moving this copy in any fashion in 3-space and then placing the copy back on the original n-gon so it exactly covers it. More precisely, we can describe the symmetries by first choosing a labelling of the n vertices, for example as shown in the following figure.

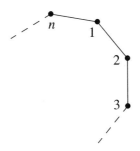

Then each symmetry s can be described uniquely by the corresponding permutation σ of $\{1, 2, 3, \ldots, n\}$ where if the symmetry s puts vertex i in the place where vertex j was originally, then σ is the permutation sending i to j. For instance, if s is a rotation of $2\pi/n$ radians clockwise about the center of the n-gon, then σ is the permutation sending i to $i + 1$, $1 \leq i \leq n - 1$, and $\sigma(n) = 1$. Now make D_{2n} into a group by defining st for $s, t \in D_{2n}$ to be the symmetry obtained by first applying t then s to the n-gon (note that we are viewing symmetries as functions on the n-gon, so st is just function composition — read as usual from right to left). If s, t effect the permutations σ, τ, respectively on the vertices, then st effects $\sigma \circ \tau$. The binary operation on D_{2n} is associative since composition of functions is associative. The identity of D_{2n} is the identity symmetry (which leaves all vertices fixed), denoted by 1, and the inverse of $s \in D_{2n}$ is the symmetry which reverses all rigid motions of s (so if s effects permutation σ on the vertices, s^{-1} effects σ^{-1}). In the next paragraph we show

$$|D_{2n}| = 2n$$

and so D_{2n} is called the *dihedral group of order* $2n$. In some texts this group is written D_n; however, D_{2n} (where the subscript gives the order of the group rather than the number of vertices) is more common in the group theory literature.

To find the order $|D_{2n}|$ observe that given any vertex i, there is a symmetry which sends vertex 1 into position i. Since vertex 2 is adjacent to vertex 1, vertex 2 must end up in position $i + 1$ or $i - 1$ (where $n + 1$ is 1 and $1 - 1$ is n, i.e., the integers labelling the vertices are read mod n). Moreover, by following the first symmetry by a reflection about the line through vertex i and the center of the n-gon one sees that vertex 2 can be sent to either position $i + 1$ or $i - 1$ by some symmetry. Thus there are $n \cdot 2$ positions the ordered pair of vertices 1, 2 may be sent to upon applying symmetries. Since symmetries are rigid motions one sees that once the position of the ordered pair of vertices 1, 2 has been specified, the action of the symmetry on all remaining vertices is completely determined. Thus there are exactly $2n$ symmetries of a regular n-gon. We can, moreover, explicitly exhibit $2n$ symmetries. These symmetries are the n rotations about the center through $2\pi i/n$ radian, $0 \leq i \leq n - 1$, and the n reflections through the n lines of symmetry (if n is odd, each symmetry line passes through a vertex and the mid-point of the opposite side; if n is even, there are $n/2$ lines of symmetry which pass through 2 opposite vertices and $n/2$ which perpendicularly bisect two opposite sides). For example, if $n = 4$ and we draw a square at the origin in an x, y plane, the lines of symmetry are

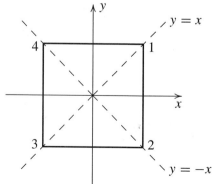

the lines $x = 0$ (y-axis), $y = 0$ (x-axis), $y = x$ and $y = -x$ (note that "reflection" through the origin is not a reflection but a rotation of π radians).

Since dihedral groups will be used extensively as an example throughout the text we fix some notation and mention some calculations which will simplify future computations and assist in viewing D_{2n} as an abstract group (rather than having to return to the geometric setting at every instance). Fix a regular n-gon centered at the origin in an x, y plane and label the vertices consecutively from 1 to n in a clockwise manner. Let r be the rotation clockwise about the origin through $2\pi/n$ radian. Let s be the reflection about the line of symmetry through vertex 1 and the origin (we use the same letters for each n, but the context will always make n clear). We leave the details of the following calculations as an exercise (for the most part we shall be working with D_6 and D_8, so the reader may wish to try these exercises for $n = 3$ and $n = 4$ first):

(1) $1, r, r^2, \ldots, r^{n-1}$ are all distinct and $r^n = 1$, so $|r| = n$.

(2) $|s| = 2$.

(3) $s \neq r^i$ for any i.

(4) $sr^i \neq sr^j$, for all $0 \leq i, j \leq n - 1$ with $i \neq j$, so

$$D_{2n} = \{1, r, r^2, \ldots, r^{n-1}, s, sr, sr^2, \ldots, sr^{n-1}\}$$

i.e., each element can be written *uniquely* in the form $s^k r^i$ for some $k = 0$ or 1 and $0 \leq i \leq n - 1$.

(5) $rs = sr^{-1}$. [First work out what permutation s effects on $\{1, 2, \ldots, n\}$ and then work out separately what each side in this equation does to vertices 1 and 2.] This shows in particular that r and s do not commute so that D_{2n} is non-abelian.

(6) $r^i s = s r^{-i}$, for all $0 \leq i \leq n$. [Proceed by induction on i and use the fact that $r^{i+1} s = r(r^i s)$ together with the preceding calculation.] This indicates how to commute s with powers of r.

Having done these calculations, we now observe that the complete multiplication table of D_{2n} can be written in terms r and s alone, that is, all the elements of D_{2n} have a (unique) representation in the form $s^k r^i$, $k = 0$ or 1 and $0 \leq i \leq n - 1$, and any product of two elements in this form can be reduced to another in the same form using only "relations" (1), (2) and (6) (reducing all exponents mod n). For example, if $n = 12$,

$$(sr^9)(sr^6) = s(r^9 s)r^6 = s(sr^{-9})r^6 = s^2 r^{-9+6} = r^{-3} = r^9.$$

Generators and Relations

The use of the generators r and s for the dihedral group provides a simple and succinct way of computing in D_{2n}. We can similarly introduce the notions of generators and relations for arbitrary groups. It is useful to have these concepts early (before their formal justification) since they provide simple ways of describing and computing in many groups. Generators will be discussed in greater detail in Section 2.4, and both concepts will be treated rigorously in Section 6.3 when we introduce the notion of free groups.

A subset S of elements of a group G with the property that every element of G can be written as a (finite) product of elements of S and their inverses is called a set of *generators* of G. We shall indicate this notationally by writing $G = \langle S \rangle$ and say G *is generated by S or S generates G.* For example, the integer 1 is a generator for the additive group $\mathbb{Z}$ of integers since every integer is a sum of a finite number of $+1$'s and -1's, so $\mathbb{Z} = \langle 1 \rangle$. By property (4) of D_{2n} the set $S = \{r, s\}$ is a set of generators of D_{2n}, so $D_{2n} = \langle r, s \rangle$. We shall see later that in a finite group G the set S generates G if every element of G is a finite product of elements of S (i.e., it is not necessary to include the inverses of the elements of S as well).

Any equations in a general group G that the generators satisfy are called *relations* in G. Thus in D_{2n} we have relations: $r^n = 1$, $s^2 = 1$ and $rs = sr^{-1}$. Moreover, in D_{2n} these three relations have the additional property that *any* other relation between elements of the group may be derived from these three (this is not immediately obvious; it follows from the fact that we can determine exactly when two group elements are equal by using only these three relations).

In general, if some group G is generated by a subset S and there is some collection of relations, say $R_1, R_2, \ldots, R_m$ (here each R_i is an equation in the elements from $S \cup \{1\}$) such that any relation among the elements of S can be deduced from these, we shall call these generators and relations a *presentation* of G and write

$$G = \langle S \mid R_1, R_2, \ldots, R_m \rangle.$$

One presentation for the dihedral group D_{2n} (using the generators and relations above) is then

$$D_{2n} = \langle r, s \mid r^n = s^2 = 1, \ rs = sr^{-1} \rangle. \tag{1.1}$$

We shall see that using this presentation to describe D_{2n} (rather than always reverting to the original geometric description) will greatly simplify working with these groups.

Presentations give an easy way of describing many groups, but there are a number of subtleties that need to be considered. One of these is that in an arbitrary presentation it may be difficult (or even impossible) to tell when two elements of the group (expressed in terms of the given generators) are equal. As a result it may not be evident what the order of the presented group is, or even whether the group is finite or infinite! For example, one can show that $\langle x_1, y_1 \mid x_1^2 = y_1^2 = (x_1 y_1)^2 = 1 \rangle$ is a presentation of a group of order 4 (cf. the exercises), whereas $\langle x_2, y_2 \mid x_2^3 = y_2^3 = (x_2 y_2)^3 = 1 \rangle$ is a presentation of an infinite group (cf. Exercise 14 in Section 6.3).

Another subtlety is that even in quite simple presentations, some "collapsing" may occur because the relations are intertwined in some unobvious way, i.e., there may be "hidden," or implicit, relations that are not explicitly given in the presentation but rather are consequences of the specified ones. This collapsing makes it difficult in general to determine even a lower bound for the size of the group being presented. For example, suppose one mimicked the presentation of D_{2n} in an attempt to create another group by defining:

$$X_{2n} = \langle x, y \mid x^n = y^2 = 1, \ xy = yx^2 \rangle. \tag{1.2}$$

The "commutation" relation $xy = yx^2$ determines how to commute y and x (i.e., how to "move" y from the right of x to the left), so that just as in the group D_{2n} every element in this group can be written in the form $y^k x^i$ with all the powers of y on the left and all the powers of x on the right. Also, by the first two relations any powers of x and y can be reduced so that i lies between 0 and $n - 1$ and k is 0 or 1. One might therefore suppose that X_{2n} is again a group of order $2n$. This is not the case because in this group there is a "hidden" relation obtained from the relation $x = xy^2$ (since $y^2 = 1$) by applying the commutation relation and the associative law repeatedly to move the y's to the left:

$$x = xy^2 = (xy)y = (yx^2)y = (yx)(xy) = (yx)(yx^2)$$
$$= y(xy)x^2 = y(yx^2)x^2 = y^2x^4 = x^4.$$

Since $x^4 = x$ it follows by the cancellation laws that $x^3 = 1$ in X_{2n}, and from the discussion above it follows that X_{2n} has order at most 6 for any n. Even more collapsing may occur, depending on the value of n (see the exercises).

As another example, consider the presentation

$$Y = \langle u, v \mid u^4 = v^3 = 1, \ uv = v^2u^2 \rangle. \tag{1.3}$$

In this case it is tempting to guess that Y is a group of order 12, but again there are additional implicit relations. In fact this group Y degenerates to the trivial group of order 1, i.e., u and v satisfy the additional relations $u = 1$ and $v = 1$ (a proof is outlined in the exercises).

This kind of collapsing does not occur for the presentation of D_{2n} because we showed by independent (geometric) means that there *is* a group of order $2n$ with generators r and s and satisfying the relations in (1). As a result, a group with only these relations must have order at *least* $2n$. On the other hand, it is easy to see (using the same sort of argument for X_{2n} above and the commutation relation $rs = sr^{-1}$) that any group defined by the generators and relations in (1) has order at *most* $2n$. It follows that the group with presentation (1) has order exactly $2n$ and also that this group is indeed the group of symmetries of the regular n–gon.

The additional information we have for the presentation (1) is the existence of a group of known order satisfying this information. In contrast, we have no independent knowledge about any groups satisfying the relations in either (2) or (3). Without such independent "lower bound" information we might not even be able to determine whether a given presentation just describes the trivial group, as in (3).

While in general it is necessary to be extremely careful in prescribing groups by presentations, the use of presentations for known groups is a powerful conceptual and computational tool. Additional results about presentations, including more elaborate examples, appear in Section 6.3.

EXERCISES

In these exercises, D_{2n} has the usual presentation $D_{2n} = \langle r, s \mid r^n = s^2 = 1, \ rs = sr^{-1} \rangle$.

1. Compute the order of each of the elements in the following groups:
   (a) D_6 (b) D_8 (c) D_{10}.

2. Use the generators and relations above to show that if x is any element of D_{2n} which is not a power of r, then $rx = xr^{-1}$.

3. Use the generators and relations above to show that every element of D_{2n} which is not a

power of r has order 2. Deduce that D_{2n} is generated by the two elements s and sr, both of which have order 2.

4. If $n = 2k$ is even and $n \geq 4$, show that $z = r^k$ is an element of order 2 which commutes with all elements of D_{2n}. Show also that z is the only nonidentity element of D_{2n} which commutes with all elements of D_{2n}. [cf. Exercise 33 of Section 1.]

5. If n is odd and $n \geq 3$, show that the identity is the only element of D_{2n} which commutes with all elements of D_{2n}. [cf. Exercise 33 of Section 1.]

6. Let x and y be elements of order 2 in any group G. Prove that if $t = xy$ then $tx = xt^{-1}$ (so that if $n = |xy| < \infty$ then x, t satisfy the same relations in G as s, r do in D_{2n}).

7. Show that $\langle a, b \mid a^2 = b^2 = (ab)^n = 1 \rangle$ gives a presentation for D_{2n} in terms of the two generators $a = s$ and $b = sr$ of order 2 computed in Exercise 3 above. [Show that the relations for r and s follow from the relations for a and b and, conversely, the relations for a and b follow from those for r and s.]

8. Find the order of the cyclic subgroup of D_{2n} generated by r (cf. Exercise 27 of Section 1).

In each of Exercises 9 to 13 you can find the order of the group of rigid motions in $\mathbb{R}^3$ (also called the group of rotations) of the given Platonic solid by following the proof for the order of D_{2n}: find the number of positions to which an adjacent pair of vertices can be sent. Alternatively, you can find the number of places to which a given face may be sent and, once a face is fixed, the number of positions to which a vertex on that face may be sent.

9. Let G be the group of rigid motions in $\mathbb{R}^3$ of a tetrahedron. Show that $|G| = 12$.

10. Let G be the group of rigid motions in $\mathbb{R}^3$ of a cube. Show that $|G| = 24$.

11. Let G be the group of rigid motions in $\mathbb{R}^3$ of an octahedron. Show that $|G| = 24$.

12. Let G be the group of rigid motions in $\mathbb{R}^3$ of a dodecahedron. Show that $|G| = 60$.

13. Let G be the group of rigid motions in $\mathbb{R}^3$ of an icosahedron. Show that $|G| = 60$.

14. Find a set of generators for $\mathbb{Z}$.

15. Find a set of generators and relations for $\mathbb{Z}/n\mathbb{Z}$.

16. Show that the group $\langle x_1, y_1 \mid x_1^2 = y_1^2 = (x_1 y_1)^2 = 1 \rangle$ is the dihedral group D_4 (where x_1 may be replaced by the letter r and y_1 by s). [Show that the last relation is the same as: $x_1 y_1 = y_1 x_1^{-1}$.]

17. Let X_{2n} be the group whose presentation is displayed in (1.2).
    (a) Show that if $n = 3k$, then X_{2n} has order 6, and it has the same generators and relations as D_6 when x is replaced by r and y by s.
    (b) Show that if $(3, n) = 1$, then x satisfies the additional relation: $x = 1$. In this case deduce that X_{2n} has order 2. [Use the facts that $x^n = 1$ and $x^3 = 1$.]

18. Let Y be the group whose presentation is displayed in (1.3).
    (a) Show that $v^2 = v^{-1}$. [Use the relation: $v^3 = 1$.]
    (b) Show that v commutes with u^3. [Show that $v^2 u^3 v = u^3$ by writing the left hand side as $(v^2 u^2)(uv)$ and using the relations to reduce this to the right hand side. Then use part (a).]
    (c) Show that v commutes with u. [Show that $u^9 = u$ and then use part (b).]
    (d) Show that $uv = 1$. [Use part (c) and the last relation.]
    (e) Show that $u = 1$, deduce that $v = 1$, and conclude that $Y = 1$. [Use part (d) and the equation $u^4 v^3 = 1$.]

1.3 SYMMETRIC GROUPS

Let Ω be any nonempty set and let S_Ω be the set of all bijections from Ω to itself (i.e., the set of all permutations of Ω). The set S_Ω is a group under function composition: $\circ$. Note that $\circ$ is a binary operation on S_Ω since if $\sigma : \Omega \to \Omega$ and $\tau : \Omega \to \Omega$ are both bijections, then $\sigma \circ \tau$ is also a bijection from Ω to Ω. Since function composition is associative in general, $\circ$ is associative. The identity of S_Ω is the permutation 1 defined by $1(a) = a$, for all $a \in \Omega$. For every permutation σ there is a (2-sided) inverse function, $\sigma^{-1} : \Omega \to \Omega$ satisfying $\sigma \circ \sigma^{-1} = \sigma^{-1} \circ \sigma = 1$. Thus, all the group axioms hold for $(S_\Omega, \circ)$. This group is called the *symmetric group on the set Ω*. It is important to recognize that the elements of S_Ω are the *permutations* of Ω, not the elements of Ω itself.

In the special case when $\Omega = \{1, 2, 3, \dots, n\}$, the symmetric group on Ω is denoted S_n, the *symmetric group of degree n*.[1] The group S_n will play an important role throughout the text both as a group of considerable interest in its own right and as a means of illustrating and motivating the general theory.

First we show that the order of S_n is $n!$. The permutations of $\{1, 2, 3, \dots, n\}$ are precisely the injective functions of this set to itself because it is finite (Proposition 0.1) and we can count the number of injective functions. An injective function σ can send the number 1 to any of the n elements of $\{1, 2, 3, \dots, n\}$; $\sigma(2)$ can then be any one of the elements of this set except $\sigma(1)$ (so there are $n - 1$ choices for $\sigma(2)$); $\sigma(3)$ can be any element except $\sigma(1)$ or $\sigma(2)$ (so there are $n - 2$ choices for $\sigma(3)$), and so on. Thus there are precisely $n \cdot (n - 1) \cdot (n - 2) \dots 2 \cdot 1 = n!$ possible injective functions from $\{1, 2, 3, \dots, n\}$ to itself. Hence there are precisely $n!$ permutations of $\{1, 2, 3, \dots, n\}$ so there are precisely $n!$ elements in S_n.

We now describe an efficient notation for writing elements σ of S_n which we shall use throughout the text and which is called the *cycle decomposition*.

A *cycle* is a string of integers which represents the element of S_n which cyclically permutes these integers (and fixes all other integers). The cycle $(a_1\, a_2 \dots a_m)$ is the permutation which sends a_i to a_{i+1}, $1 \le i \le m - 1$ and sends a_m to a_1. For example $(2\,1\,3)$ is the permutation which maps 2 to 1, 1 to 3 and 3 to 2. In general, for each $\sigma \in S_n$ the numbers from 1 to n will be rearranged and grouped into k cycles of the form

$$(a_1\, a_2 \dots a_{m_1})(a_{m_1+1}\, a_{m_1+2} \dots a_{m_2}) \dots (a_{m_{k-1}+1}\, a_{m_{k-1}+2} \dots a_{m_k})$$

from which the action of σ on any number from 1 to n can easily be read, as follows. For any $x \in \{1, 2, 3, \dots, n\}$ first locate x in the above expression. If x is not followed immediately by a right parenthesis (i.e., x is not at the right end of one of the k cycles), then $\sigma(x)$ is the integer appearing immediately to the right of x. If x is followed by a right parenthesis, then $\sigma(x)$ is the number which is at the start of the cycle ending with x (i.e., if $x = a_{m_i}$, for some i, then $\sigma(x) = a_{m_{i-1}+1}$ (where m_0 is taken to be 0)). We can represent this description of σ by

---

[1]We shall see in Section 6 that the structure of S_Ω depends only on the cardinality of Ω, not on the particular elements of Ω itself, so if Ω is any finite set with n elements, then S_Ω "looks like" S_n.

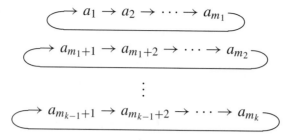

$$a_1 \rightarrow a_2 \rightarrow \cdots \rightarrow a_{m_1}$$

$$a_{m_1+1} \rightarrow a_{m_1+2} \rightarrow \cdots \rightarrow a_{m_2}$$

$$\vdots$$

$$a_{m_{k-1}+1} \rightarrow a_{m_{k-1}+2} \rightarrow \cdots \rightarrow a_{m_k}$$

The product of all the cycles is called the *cycle decomposition* of σ.

We now give an algorithm for computing the cycle decomposition of an element σ of S_n and work through the algorithm with a specific permutation. We defer the proof of this algorithm and full analysis of the uniqueness aspects of the cycle decomposition until Chapter 4.

Let $n = 13$ and let $\sigma \in S_{13}$ be defined by

$$\sigma(1) = 12, \quad \sigma(2) = 13, \quad \sigma(3) = 3, \quad \sigma(4) = 1, \quad \sigma(5) = 11,$$
$$\sigma(6) = 9, \quad \sigma(7) = 5, \quad \sigma(8) = 10, \quad \sigma(9) = 6, \quad \sigma(10) = 4,$$
$$\sigma(11) = 7, \quad \sigma(12) = 8, \quad \sigma(13) = 2.$$

Cycle Decomposition Algorithm

Method	Example
To start a new cycle pick the smallest element of $\{1, 2, \ldots, n\}$ which has not yet appeared in a previous cycle — call it a (if you are just starting, $a = 1$); begin the new cycle: $(a$	$(1$
Read off $\sigma(a)$ from the given description of σ — call it b. If $b = a$, close the cycle with a right parenthesis (without writing b down); this completes a cycle — return to step 1. If $b \neq a$, write b next to a in this cycle: $(a\,b$	$\sigma(1) = 12 = b$, $12 \neq 1$ so write: $(1\,12$
Read off $\sigma(b)$ from the given description of σ — call it c. If $c = a$, close the cycle with a right parenthesis to complete the cycle — return to step 1. If $c \neq a$, write c next to b in this cycle: $(a\,b\,c$ Repeat this step using the number c as the new value for b until the cycle closes.	$\sigma(12) = 8$, $8 \neq 1$ so continue the cycle as: $(1\,12\,8$

Naturally this process stops when all the numbers from $\{1, 2, \ldots, n\}$ have appeared in some cycle. For the particular σ in the example this gives

$$\sigma = (1\,12\,8\,10\,4)(2\,13)(3)(5\,11\,7)(6\,9).$$

The *length* of a cycle is the number of integers which appear in it. A cycle of length t is called a *t-cycle*. Two cycles are called *disjoint* if they have no numbers in common.

Thus the element σ above is the product of 5 (pairwise) disjoint cycles: a 5-cycle, a 2-cycle, a 1-cycle, a 3-cycle, and another 2-cycle.

Henceforth we adopt the convention that 1-cycles will not be written. Thus if some integer, i, does not appear in the cycle decomposition of a permutation τ it is understood that $\tau(i) = i$, i.e., that τ fixes i. The identity permutation of S_n has cycle decomposition $(1)(2)\dots(n)$ and will be written simply as 1. Hence the final step of the algorithm is:

Cycle Decomposition Algorithm (cont.)

Final Step: Remove all cycles of length 1	

The cycle decomposition for the particular σ in the example is therefore

$$\sigma = (1\ 12\ 8\ 10\ 4)(2\ 13)(5\ 11\ 7)(6\ 9)$$

This convention has the advantage that the cycle decomposition of an element τ of S_n is also the cycle decomposition of the permutation in S_m for $m \geq n$ which acts as τ on $\{1, 2, 3, \dots, n\}$ and fixes each element of $\{n+1, n+2, \dots, m\}$. Thus, for example, $(1\ 2)$ is the permutation which interchanges 1 and 2 and fixes all larger integers whether viewed in S_2, S_3 or S_4, etc.

As another example, the 6 elements of S_3 have the following cycle decompositions:

The group S_3

Values of σ_i	Cycle Decomposition of σ_i
$\sigma_1(1) = 1, \sigma_1(2) = 2, \sigma_1(3) = 3$	1
$\sigma_2(1) = 1, \sigma_2(2) = 3, \sigma_2(3) = 2$	$(2\ 3)$
$\sigma_3(1) = 3, \sigma_3(2) = 2, \sigma_3(3) = 1$	$(1\ 3)$
$\sigma_4(1) = 2, \sigma_4(2) = 1, \sigma_4(3) = 3$	$(1\ 2)$
$\sigma_5(1) = 2, \sigma_5(2) = 3, \sigma_5(3) = 1$	$(1\ 2\ 3)$
$\sigma_6(1) = 3, \sigma_6(2) = 1, \sigma_6(3) = 2$	$(1\ 3\ 2)$

For any $\sigma \in S_n$, the cycle decomposition of σ^{-1} is obtained by writing the numbers in each cycle of the cycle decomposition of σ in reverse order. For example, if $\sigma = (1\ 12\ 8\ 10\ 4)(2\ 13)(5\ 11\ 7)(6\ 9)$ is the element of S_{13} described before then

$$\sigma^{-1} = (4\ 10\ 8\ 12\ 1)(13\ 2)(7\ 11\ 5)(9\ 6).$$

Computing products in S_n is straightforward, keeping in mind that when computing $\sigma \circ \tau$ in S_n one reads the permutations from *right to left*. One simply "follows" the elements under the successive permutations. For example, in the product $(1\ 2\ 3) \circ (1\ 2)(3\ 4)$ the number 1 is sent to 2 by the first permutation, then 2 is sent to 3 by the second permutation, hence the composite maps 1 to 3. To compute the cycle decomposition of the product we need next to see what happens to 3. It is sent first to 4, then 4 is fixed, so 3 is mapped to 4 by the composite map. Similarly, 4 is first mapped to 3 then 3 is mapped to 1, completing this cycle in the product: (1 3 4). Finally, 2 is sent to 1, then 1 is sent to 2 so 2 is fixed by this product and so $(1\ 2\ 3) \circ (1\ 2)(3\ 4) = (1\ 3\ 4)$ is the cycle decomposition of the product.

As additional examples,

$$(12) \circ (13) = (1\ 3\ 2) \qquad \text{and} \qquad (1\ 3) \circ (1\ 2) = (1\ 2\ 3).$$

In particular this shows that

S_n is a non-abelian group for all $n \geq 3$.

Each cycle $(a_1\ a_2 \ldots a_m)$ in a cycle decomposition can be viewed as the permutation which cyclically permutes $a_1, a_2, \ldots, a_m$ and fixes all other integers. Since disjoint cycles permute numbers which lie in disjoint sets it follows that

disjoint cycles commute.

Thus rearranging the cycles in any product of disjoint cycles (in particular, in a cycle decomposition) does not change the permutation.

Also, since a given cycle, $(a_1\ a_2 \ldots a_m)$, permutes $\{a_1, a_2, \ldots, a_m\}$ cyclically, the numbers in the cycle itself can be cyclically permuted without altering the permutation, i.e.,

$$(a_1\ a_2 \ldots a_m) = (a_2\ a_3 \ldots a_m\ a_1) = (a_3\ a_4 \ldots a_m\ a_1\ a_2) = \ldots$$
$$= (a_m\ a_1\ a_2 \ldots a_{m-1}).$$

Thus, for instance, $(1\ 2) = (2\ 1)$ and $(1\ 2\ 3\ 4) = (3\ 4\ 1\ 2)$. By convention, the smallest number appearing in the cycle is usually written first.

One must exercise some care working with cycles since a permutation may be written in many ways as an arbitrary product of cycles. For instance, in S_3, $(1\ 2\ 3) = (1\ 2)(2\ 3) = (1\ 3)(1\ 3\ 2)(1\ 3)$ *etc.* But, (as we shall prove) the cycle decomposition of each permutation is the *unique* way of expressing a permutation as a product of disjoint cycles (up to rearranging its cycles and cyclically permuting the numbers within each cycle). Reducing an arbitrary product of cycles to a product of disjoint cycles allows us to determine at a glance whether or not two permutations are the same. Another advantage to this notation is that it is an exercise (outlined below) to prove that *the order of a permutation is the l.c.m. of the lengths of the cycles in its cycle decomposition.*

EXERCISES

1. Let σ be the permutation

$$1 \mapsto 3 \qquad 2 \mapsto 4 \qquad 3 \mapsto 5 \qquad 4 \mapsto 2 \qquad 5 \mapsto 1$$

and let τ be the permutation

$$1 \mapsto 5 \qquad 2 \mapsto 3 \qquad 3 \mapsto 2 \qquad 4 \mapsto 4 \qquad 5 \mapsto 1.$$

Find the cycle decompositions of each of the following permutations: $\sigma, \tau, \sigma^2, \sigma\tau, \tau\sigma$, and $\tau^2\sigma$.

2. Let σ be the permutation

$1 \mapsto 13$	$2 \mapsto 2$	$3 \mapsto 15$	$4 \mapsto 14$	$5 \mapsto 10$
$6 \mapsto 6$	$7 \mapsto 12$	$8 \mapsto 3$	$9 \mapsto 4$	$10 \mapsto 1$
$11 \mapsto 7$	$12 \mapsto 9$	$13 \mapsto 5$	$14 \mapsto 11$	$15 \mapsto 8$

and let τ be the permutation

$1 \mapsto 14$	$2 \mapsto 9$	$3 \mapsto 10$	$4 \mapsto 2$	$5 \mapsto 12$
$6 \mapsto 6$	$7 \mapsto 5$	$8 \mapsto 11$	$9 \mapsto 15$	$10 \mapsto 3$
$11 \mapsto 8$	$12 \mapsto 7$	$13 \mapsto 4$	$14 \mapsto 1$	$15 \mapsto 13.$

Find the cycle decompositions of the following permutations: $\sigma, \tau, \sigma^2, \sigma\tau, \tau\sigma,$ and $\tau^2\sigma$.

3. For each of the permutations whose cycle decompositions were computed in the preceding two exercises compute its order.

4. Compute the order of each of the elements in the following groups: **(a)** S_3 **(b)** S_4.

5. Find the order of $(1\ 12\ 8\ 10\ 4)(2\ 13)(5\ 11\ 7)(6\ 9)$.

6. Write out the cycle decomposition of each element of order 4 in S_4.

7. Write out the cycle decomposition of each element of order 2 in S_4.

8. Prove that if $\Omega = \{1, 2, 3, \ldots\}$ then S_Ω is an infinite group (do not say $\infty! = \infty$).

9. **(a)** Let σ be the 12-cycle $(1\ 2\ 3\ 4\ 5\ 6\ 7\ 8\ 9\ 10\ 11\ 12)$. For which positive integers i is σ^i also a 12-cycle?
   (b) Let τ be the 8-cycle $(1\ 2\ 3\ 4\ 5\ 6\ 7\ 8)$. For which positive integers i is τ^i also an 8-cycle?
   (c) Let ω be the 14-cycle $(1\ 2\ 3\ 4\ 5\ 6\ 7\ 8\ 9\ 10\ 11\ 12\ 13\ 14)$. For which positive integers i is ω^i also a 14-cycle?

10. Prove that if σ is the m-cycle $(a_1\ a_2\ \ldots\ a_m)$, then for all $i \in \{1, 2, \ldots, m\}$, $\sigma^i(a_k) = a_{k+i}$, where $k + i$ is replaced by its least positive residue mod m. Deduce that $|\sigma| = m$.

11. Let σ be the m-cycle $(1\ 2\ \ldots\ m)$. Show that σ^i is also an m-cycle if and only if i is relatively prime to m.

12. **(a)** If $\tau = (1\ 2)(3\ 4)(5\ 6)(7\ 8)(9\ 10)$ determine whether there is a n-cycle σ $(n \geq 10)$ with $\tau = \sigma^k$ for some integer k.
   (b) If $\tau = (1\ 2)(3\ 4\ 5)$ determine whether there is an n-cycle σ $(n \geq 5)$ with $\tau = \sigma^k$ for some integer k.

13. Show that an element has order 2 in S_n if and only if its cycle decomposition is a product of commuting 2-cycles.

14. Let p be a prime. Show that an element has order p in S_n if and only if its cycle decomposition is a product of commuting p-cycles. Show by an explicit example that this need not be the case if p is not prime.

15. Prove that the order of an element in S_n equals the least common multiple of the lengths of the cycles in its cycle decomposition. [Use Exercise 10 and Exercise 24 of Section 1.]

16. Show that if $n \geq m$ then the number of m-cycles in S_n is given by

$$\frac{n(n-1)(n-2)\ldots(n-m+1)}{m}.$$

[Count the number of ways of forming an m-cycle and divide by the number of representations of a particular m-cycle.]

17. Show that if $n \geq 4$ then the number of permutations in S_n which are the product of two disjoint 2-cycles is $n(n-1)(n-2)(n-3)/8$.

18. Find all numbers n such that S_5 contains an element of order n. [Use Exercise 15.]

19. Find all numbers n such that S_7 contains an element of order n. [Use Exercise 15.]

20. Find a set of generators and relations for S_3.

1.4 MATRIX GROUPS

In this section we introduce the notion of matrix groups where the coefficients come from fields. This example of a family of groups will be used for illustrative purposes in Part I and will be studied in more detail in the chapters on vector spaces.

A *field* is the "smallest" mathematical structure in which we can perform all the arithmetic operations $+$, $-$, $\times$, and $\div$ (division by nonzero elements), so in particular every nonzero element must have a multiplicative inverse. We shall study fields more thoroughly later and in this part of the text the only fields F we shall encounter will be $\mathbb{Q}$, $\mathbb{R}$ and $\mathbb{Z}/p\mathbb{Z}$, where p is a prime. The example $\mathbb{Z}/p\mathbb{Z}$ is a finite field, which, to emphasize that it is a field, we shall denote by $\mathbb{F}_p$. For the sake of completeness we include here the precise definition of a field.

Definition.

(1) A *field* is a set F together with two commutative binary operations $+$ and $\cdot$ on F such that $(F, +)$ is an abelian group (call its identity 0) and $(F - \{0\}, \cdot)$ is also an abelian group, and the following *distributive* law holds:
$$a \cdot (b + c) = (a \cdot b) + (a \cdot c), \qquad \text{for all } a, b, c \in F.$$

(2) For any field F let $F^\times = F - \{0\}$.

All the vector space theory, the theory of matrices and linear transformations and the theory of determinants when the scalars come from $\mathbb{R}$ is true, *mutatis mutandis*, when the scalars come from an arbitrary field F. When we use this theory in Part I we shall state explicitly what facts on fields we are assuming.

For each $n \in \mathbb{Z}^+$ let $GL_n(F)$ be the set of all $n \times n$ matrices whose entries come from F and whose determinant is nonzero, i.e.,

$$GL_n(F) = \{A \mid A \text{ is an } n \times n \text{ matrix with entries from } F \text{ and } \det(A) \neq 0 \},$$

where the determinant of any matrix A with entries from F can be computed by the same formulas used when $F = \mathbb{R}$. For arbitrary $n \times n$ matrices A and B let AB be the product of these matrices as computed by the same rules as when $F = \mathbb{R}$. This product is associative. Also, since $\det(AB) = \det(A) \cdot \det(B)$, it follows that if $\det(A) \neq 0$ and $\det(B) \neq 0$, then $\det(AB) \neq 0$, so $GL_n(F)$ is closed under matrix multiplication. Furthermore, $\det(A) \neq 0$ if and only if A has a matrix inverse (and this inverse can be computed by the same adjoint formula used when $F = \mathbb{R}$), so each $A \in GL_n(F)$ has an inverse, A^{-1}, in $GL_n(F)$:

$$AA^{-1} = A^{-1}A = I,$$

where I is the $n \times n$ identity matrix. Thus $GL_n(F)$ is a group under matrix multiplication, called the *general linear group of degree n*.

The following results will be proved in Part III but are recorded now for convenience:

(1) if F is a field and $|F| < \infty$, then $|F| = p^m$ for some prime p and integer m

(2) if $|F| = q < \infty$, then $|GL_n(F)| = (q^n - 1)(q^n - q)(q^n - q^2) \dots (q^n - q^{n-1})$.

EXERCISES

Let F be a field and let $n \in \mathbb{Z}^+$.

1. Prove that $|GL_2(\mathbb{F}_2)| = 6$.

2. Write out all the elements of $GL_2(\mathbb{F}_2)$ and compute the order of each element.

3. Show that $GL_2(\mathbb{F}_2)$ is non-abelian.

4. Show that if n is not prime then $\mathbb{Z}/n\mathbb{Z}$ is not a field.

5. Show that $GL_n(F)$ is a finite group if and only if F has a finite number of elements.

6. If $|F| = q$ is finite prove that $|GL_n(F)| < q^{n^2}$.

7. Let p be a prime. Prove that the order of $GL_2(\mathbb{F}_p)$ is $p^4 - p^3 - p^2 + p$ (do not just quote the order formula in this section). [Subtract the number of 2×2 matrices which are *not* invertible from the total number of 2×2 matrices over $\mathbb{F}_p$. You may use the fact that a 2×2 matrix is not invertible if and only if one row is a multiple of the other.]

8. Show that $GL_n(F)$ is non-abelian for any $n \geq 2$ and any F.

9. Prove that the binary operation of matrix multiplication of 2×2 matrices with real number entries is associative.

10. Let $G = \{ \begin{pmatrix} a & b \\ 0 & c \end{pmatrix} \mid a, b, c \in \mathbb{R}, \ a \neq 0, \ c \neq 0 \}$.

    (a) Compute the product of $\begin{pmatrix} a_1 & b_1 \\ 0 & c_1 \end{pmatrix}$ and $\begin{pmatrix} a_2 & b_2 \\ 0 & c_2 \end{pmatrix}$ to show that G is closed under matrix multiplication.

    (b) Find the matrix inverse of $\begin{pmatrix} a & b \\ 0 & c \end{pmatrix}$ and deduce that G is closed under inverses.

    (c) Deduce that G is a subgroup of $GL_2(\mathbb{R})$ (cf. Exercise 26, Section 1).

    (d) Prove that the set of elements of G whose two diagonal entries are equal (i.e., $a = c$) is also a subgroup of $GL_2(\mathbb{R})$.

The next exercise introduces the *Heisenberg group* over the field F and develops some of its basic properties. When $F = \mathbb{R}$ this group plays an important role in quantum mechanics and signal theory by giving a group theoretic interpretation (due to H. Weyl) of Heisenberg's Uncertainty Principle. Note also that the Heisenberg group may be defined more generally — for example, with entries in $\mathbb{Z}$.

11. Let $H(F) = \{ \begin{pmatrix} 1 & a & b \\ 0 & 1 & c \\ 0 & 0 & 1 \end{pmatrix} \mid a, b, c \in F \}$ — called the *Heisenberg group* over F. Let

$X = \begin{pmatrix} 1 & a & b \\ 0 & 1 & c \\ 0 & 0 & 1 \end{pmatrix}$ and $Y = \begin{pmatrix} 1 & d & e \\ 0 & 1 & f \\ 0 & 0 & 1 \end{pmatrix}$ be elements of $H(F)$.

    (a) Compute the matrix product XY and deduce that $H(F)$ is closed under matrix multiplication. Exhibit explicit matrices such that $XY \neq YX$ (so that $H(F)$ is always non-abelian).

(b) Find an explicit formula for the matrix inverse X^{-1} and deduce that $H(F)$ is closed under inverses.

(c) Prove the associative law for $H(F)$ and deduce that $H(F)$ is a group of order $|F|^3$. (Do not assume that matrix multiplication is associative.)

(d) Find the order of each element of the finite group $H(\mathbb{Z}/2\mathbb{Z})$.

(e) Prove that every nonidentity element of the group $H(\mathbb{R})$ has infinite order.

1.5 THE QUATERNION GROUP

The *quaternion group*, Q_8, is defined by

$$Q_8 = \{1, -1, i, -i, j, -j, k, -k\}$$

with product $\cdot$ computed as follows:

$$1 \cdot a = a \cdot 1 = a, \qquad \text{for all } a \in Q_8$$

$$(-1) \cdot (-1) = 1, \qquad (-1) \cdot a = a \cdot (-1) = -a, \qquad \text{for all } a \in Q_8$$

$$i \cdot i = j \cdot j = k \cdot k = -1$$

$$i \cdot j = k, \qquad j \cdot i = -k$$

$$j \cdot k = i, \qquad k \cdot j = -i$$

$$k \cdot i = j, \qquad i \cdot k = -j.$$

As usual, we shall henceforth write ab for $a \cdot b$. It is tedious to check the associative law (we shall prove this later by less computational means), but the other axioms are easily checked. Note that Q_8 is a non-abelian group of order 8.

EXERCISES

1. Compute the order of each of the elements in Q_8.

2. Write out the group tables for S_3, D_8 and Q_8.

3. Find a set of generators and relations for Q_8.

1.6 HOMOMORPHISMS AND ISOMORPHISMS

In this section we make precise the notion of when two groups "look the same," that is, have exactly the same group-theoretic structure. This is the notion of an *isomorphism* between two groups. We first define the notion of a *homomorphism* about which we shall have a great deal more to say later.

Definition. Let $(G, \star)$ and $(H, \diamond)$ be groups. A map $\varphi : G \to H$ such that

$$\varphi(x \star y) = \varphi(x) \diamond \varphi(y), \qquad \text{for all } x, y \in G$$

is called a *homomorphism*.

When the group operations for G and H are not explicitly written, the homomorphism condition becomes simply

$$\varphi(xy) = \varphi(x)\varphi(y)$$

but it is important to keep in mind that the product xy on the left is computed in G and the product $\varphi(x)\varphi(y)$ on the right is computed in H. Intuitively, a map φ is a homomorphism if it respects the group structures of its domain and codomain.

Definition. The map $\varphi : G \to H$ is called an *isomorphism* and G and H are said to be *isomorphic* or of the same *isomorphism type*, written $G \cong H$, if
    (1) φ is a homomorphism (i.e., $\varphi(xy) = \varphi(x)\varphi(y)$), and
    (2) φ is a bijection.

In other words, the groups G and H are isomorphic if there is a bijection between them which preserves the group operations. Intuitively, G and H are the same group except that the elements and the operations may be written differently in G and H. Thus any property which G has which depends only on the group structure of G (i.e., can be derived from the group axioms — for example, commutativity of the group) also holds in H. Note that this formally justifies writing all our group operations as $\cdot$ since changing the symbol of the operation does not change the isomorphism type.

Examples

    (1) For any group G, $G \cong G$. The identity map provides an obvious isomorphism but not, in general, the *only* isomorphism from G to itself. More generally, let $\mathcal{G}$ be any nonempty collection of groups. It is easy to check that the relation $\cong$ is an equivalence relation on $\mathcal{G}$ and the equivalence classes are called *isomorphism classes*. This accounts for the somewhat symmetric wording of the definition of "isomorphism."

    (2) The exponential map $\exp : \mathbb{R} \to \mathbb{R}^+$ defined by $\exp(x) = e^x$, where e is the base of the natural logarithm, is an isomorphism from $(\mathbb{R}, +)$ to $(\mathbb{R}^+, \times)$. Exp is a bijection since it has an inverse function (namely $\log_e$) and exp preserves the group operations since $e^{x+y} = e^x e^y$. In this example both the elements and the operations are different yet the two groups are isomorphic, that is, as groups they have identical structures.

    (3) In this example we show that the isomorphism type of a symmetric group depends only on the cardinality of the underlying set being permuted.

        Let Δ and Ω be nonempty sets. The symmetric groups S_Δ and S_Ω are isomorphic if $|\Delta| = |\Omega|$. We can see this intuitively as follows: given that $|\Delta| = |\Omega|$, there is a bijection θ from Δ onto Ω. Think of the elements of Δ and Ω as being glued together via θ, i.e., each $x \in \Delta$ is glued to $\theta(x) \in \Omega$. To obtain a map $\varphi : S_\Delta \to S_\Omega$ let $\sigma \in S_\Delta$ be a permutation of Δ and let $\varphi(\sigma)$ be the permutation of Ω which moves the elements of Ω in the same way σ moves the corresponding glued elements of Δ; that is, if $\sigma(x) = y$, for some $x, y \in \Delta$, then $\varphi(\sigma)(\theta(x)) = \theta(y)$ in Ω. Since the set bijection θ has an inverse, one can easily check that the map between symmetric groups also has an inverse. The precise technical definition of the map φ and the straightforward, albeit tedious, checking of the properties which ensure φ is an isomorphism are relegated to the following exercises.

        Conversely, if $S_\Delta \cong S_\Omega$, then $|\Delta| = |\Omega|$; we prove this only when the underlying sets are finite (when both Δ and Ω are infinite sets the proof is harder and will be given as an exercise in Chapter 4). Since any isomorphism between two groups G and H is, a priori, a bijection between them, a necessary condition for isomorphism is $|S_\Delta| = |S_\Omega|$. When Δ is a finite set of order n, then $|S_\Delta| = n!$. We actually only proved this for S_n, however the same reasoning applies for S_Δ. Similarly, if Ω is a finite set of order m, then $|S_\Omega| = m!$. Thus if S_Δ and S_Ω are isomorphic then $n! = m!$, so $m = n$, i.e., $|\Delta| = |\Omega|$.

Many more examples of isomorphisms will appear throughout the text. When we study different structures (rings, fields, vector spaces, etc.) we shall formulate corresponding notions of isomorphisms between respective structures. One of the central problems in mathematics is to determine what properties of a structure specify its isomorphism type (i.e., to prove that if G is an object with some structure (such as a group) and G has property $\mathcal{P}$, then any other similarly structured object (group) X with property $\mathcal{P}$ is isomorphic to G). Theorems of this type are referred to as *classification theorems*. For example, we shall prove that

any non-abelian group of order 6 is isomorphic to S_3

(so here G is the group S_3 and $\mathcal{P}$ is the property "non-abelian and of order 6"). From this classification theorem we obtain $D_6 \cong S_3$ and $GL_2(\mathbb{F}_2) \cong S_3$ without having to find explicit maps between these groups. Note that it is not true that any group of order 6 is isomorphic to S_3. In fact we shall prove that up to isomorphism there are precisely two groups of order 6: S_3 and $\mathbb{Z}/6\mathbb{Z}$ (i.e., any group of order 6 is isomorphic to one of these two groups and S_3 is not isomorphic to $\mathbb{Z}/6\mathbb{Z}$). Note that the conclusion is less specific (there are two possible types); however, the hypotheses are easier to check (namely, check to see if the order is 6). Results of the latter type are also referred to as classifications. Generally speaking it is subtle and difficult, even in specific instances, to determine whether or not two groups (or other mathematical objects) are isomorphic — constructing an explicit map between them which preserves the group operations or proving no such map exists is, except in tiny cases, computationally unfeasible as indicated already in trying to prove the above classification of groups of order 6 without further theory.

It is occasionally easy to see that two given groups are *not* isomorphic. For example, the exercises below assert that if $\varphi : G \to H$ is an isomorphism, then, in particular,

    (a) $|G| = |H|$

    (b) G is abelian if and only if H is abelian

    (c) for all $x \in G$, $|x| = |\varphi(x)|$.

Thus S_3 and $\mathbb{Z}/6\mathbb{Z}$ are not isomorphic (as indicated above) since one is abelian and the other is not. Also, $(\mathbb{R} - \{0\}, \times)$ and $(\mathbb{R}, +)$ cannot be isomorphic because in $(\mathbb{R} - \{0\}, \times)$ the element -1 has order 2 whereas $(\mathbb{R}, +)$ has no element of order 2, contrary to (c).

Finally, we record one very useful fact that we shall prove later (when we discuss free groups) dealing with the question of homomorphisms and isomorphisms between two groups given by generators and relations:

Let G be a finite group of order n for which we have a presentation and let $S = \{s_1, \dots, s_m\}$ be the generators. Let H be another group and $\{r_1, \dots, r_m\}$ be elements of H. Suppose that any relation satisfied in G by the s_i is also satisfied in H

when each s_i is replaced by r_i. Then there is a (unique) homomorphism $\varphi : G \to H$ which maps s_i to r_i. If we have a presentation for G, then we need only check the relations specified by this presentation (since, by definition of a presentation, every relation can be deduced from the relations given in the presentation). If H is generated by the elements $\{r_1, \ldots, r_m\}$, then φ is surjective (any product of the r_i's is the image of the corresponding product of the s_i's). If, in addition, H has the same (finite) order as G, then any surjective map is necessarily injective, i.e., φ is an isomorphism: $G \cong H$. Intuitively, we can map the generators of G to any elements of H and obtain a homomorphism provided that the relations in G are still satisfied.

Readers may already be familiar with the corresponding statement for vector spaces. Suppose V is a finite dimensional vector space of dimension n with basis S and W is another vector space. Then we can specify a linear transformation from V to W by mapping the elements of S to arbitrary vectors in W (here there are no relations to satisfy). If W is also of dimension n and the chosen vectors in W span W (and so are a basis for W) then this linear transformation is invertible (a vector space isomorphism).

Examples

(1) Recall that $D_{2n} = \langle r, s \mid r^n = s^2 = 1, sr = r^{-1}s \rangle$. Suppose H is a group containing elements a and b with $a^n = 1, b^2 = 1$ and $ba = a^{-1}b$. Then there is a homomorphism from D_{2n} to H mapping r to a and s to b. For instance, let k be an integer dividing n with $k \geq 3$ and let $D_{2k} = \langle r_1, s_1 \mid r_1^k = s_1^2 = 1, s_1 r_1 = r_1^{-1} s_1 \rangle$. Define

$$\varphi : D_{2n} \to D_{2k} \quad \text{by} \quad \varphi(r) = r_1 \text{ and } \varphi(s) = s_1.$$

If we write $n = km$, then since $r_1^k = 1$, also $r_1^n = (r_1^k)^m = 1$. Thus the three relations satisfied by r, s in D_{2n} are satisfied by r_1, s_1 in D_{2k}. Thus φ extends (uniquely) to a homomorphism from D_{2n} to D_{2k}. Since $\{r_1, s_1\}$ generates D_{2k}, φ is surjective. This homomorphism is not an isomorphism if $k < n$.

(2) Following up on the preceding example, let $G = D_6$ be as presented above. Check that in $H = S_3$ the elements $a = (1\,2\,3)$ and $b = (1\,2)$ satisfy the relations: $a^3 = 1$, $b^2 = 1$ and $ba = a^{-1}b$. Thus there is a homomorphism from D_6 to S_3 which sends $r \mapsto a$ and $s \mapsto b$. One may further check that S_3 is generated by a and b, so this homomorphism is surjective. Since D_6 and S_3 both have order 6, this homomorphism is an isomorphism: $D_6 \cong S_3$.

Note that the element a in the examples above need not have *order n* (i.e., n need not be the *smallest* power of a giving the identity in H) and similarly b need not have order 2 (for example b could well be the identity if $a = a^{-1}$). This allows us to more easily construct homomorphisms and is in keeping with the idea that the generators and relations for a group G constitute a complete set of data for the group structure of G.

EXERCISES

Let G and H be groups.

1. Let $\varphi : G \to H$ be a homomorphism.
   (a) Prove that $\varphi(x^n) = \varphi(x)^n$ for all $n \in \mathbb{Z}^+$.
   (b) Do part (a) for $n = -1$ and deduce that $\varphi(x^n) = \varphi(x)^n$ for all $n \in \mathbb{Z}$.

2. If $\varphi : G \to H$ is an isomorphism, prove that $|\varphi(x)| = |x|$ for all $x \in G$. Deduce that any two isomorphic groups have the same number of elements of order n for each $n \in \mathbb{Z}^+$. Is the result true if φ is only assumed to be a homomorphism?

3. If $\varphi : G \to H$ is an isomorphism, prove that G is abelian if and only if H is abelian. If $\varphi : G \to H$ is a homomorphism, what additional conditions on φ (if any) are sufficient to ensure that if G is abelian, then so is H?

4. Prove that the multiplicative groups $\mathbb{R} - \{0\}$ and $\mathbb{C} - \{0\}$ are not isomorphic.

5. Prove that the additive groups $\mathbb{R}$ and $\mathbb{Q}$ are not isomorphic.

6. Prove that the additive groups $\mathbb{Z}$ and $\mathbb{Q}$ are not isomorphic.

7. Prove that D_8 and Q_8 are not isomorphic.

8. Prove that if $n \neq m$, S_n and S_m are not isomorphic.

9. Prove that D_{24} and S_4 are not isomorphic.

10. Fill in the details of the proof that the symmetric groups S_Δ and S_Ω are isomorphic if $|\Delta| = |\Omega|$ as follows: let $\theta : \Delta \to \Omega$ be a bijection. Define

$$\varphi : S_\Delta \to S_\Omega \qquad \text{by} \qquad \varphi(\sigma) = \theta \circ \sigma \circ \theta^{-1} \quad \text{for all } \sigma \in S_\Delta$$

and prove the following:

(a) φ is well defined, that is, if σ is a permutation of Δ then $\theta \circ \sigma \circ \theta^{-1}$ is a permutation of Ω.

(b) φ is a bijection from S_Δ onto S_Ω. [Find a 2-sided inverse for φ.]

(c) φ is a homomorphism, that is, $\varphi(\sigma \circ \tau) = \varphi(\sigma) \circ \varphi(\tau)$.

Note the similarity to the *change of basis* or *similarity* transformations for matrices (we shall see the connections between these later in the text).

11. Let A and B be groups. Prove that $A \times B \cong B \times A$.

12. Let A, B, and C be groups and let $G = A \times B$ and $H = B \times C$. Prove that $G \times C \cong A \times H$.

13. Let G and H be groups and let $\varphi : G \to H$ be a homomorphism. Prove that the image of φ, $\varphi(G)$, is a subgroup of H (cf. Exercise 26 of Section 1). Prove that if φ is injective then $G \cong \varphi(G)$.

14. Let G and H be groups and let $\varphi : G \to H$ be a homomorphism. Define the *kernel* of φ to be $\{g \in G \mid \varphi(g) = 1_H\}$ (so the kernel is the set of elements in G which map to the identity of H, i.e., is the fiber over the identity of H). Prove that the kernel of φ is a subgroup (cf. Exercise 26 of Section 1) of G. Prove that φ is injective if and only if the kernel of φ is the identity subgroup of G.

15. Define a map $\pi : \mathbb{R}^2 \to \mathbb{R}$ by $\pi((x, y)) = x$. Prove that π is a homomorphism and find the kernel of π (cf. Exercise 14).

16. Let A and B be groups and let G be their direct product, $A \times B$. Prove that the maps $\pi_1 : G \to A$ and $\pi_2 : G \to B$ defined by $\pi_1((a, b)) = a$ and $\pi_2((a, b)) = b$ are homomorphisms and find their kernels (cf. Exercise 14).

17. Let G be any group. Prove that the map from G to itself defined by $g \mapsto g^{-1}$ is a homomorphism if and only if G is abelian.

18. Let G be any group. Prove that the map from G to itself defined by $g \mapsto g^2$ is a homomorphism if and only if G is abelian.

19. Let $G = \{z \in \mathbb{C} \mid z^n = 1 \text{ for some } n \in \mathbb{Z}^+\}$. Prove that for any fixed integer $k > 1$ the map from G to itself defined by $z \mapsto z^k$ is a surjective homomorphism but is not an isomorphism.

20. Let G be a group and let $\mathrm{Aut}(G)$ be the set of all isomorphisms from G onto G. Prove that $\mathrm{Aut}(G)$ is a group under function composition (called the *automorphism group* of G and the elements of $\mathrm{Aut}(G)$ are called *automorphisms* of G).

21. Prove that for each fixed nonzero $k \in \mathbb{Q}$ the map from $\mathbb{Q}$ to itself defined by $q \mapsto kq$ is an automorphism of $\mathbb{Q}$ (cf. Exercise 20).

22. Let A be an abelian group and fix some $k \in \mathbb{Z}$. Prove that the map $a \mapsto a^k$ is a homomorphism from A to itself. If $k = -1$ prove that this homomorphism is an isomorphism (i.e., is an automorphism of A).

23. Let G be a finite group which possesses an automorphism σ (cf. Exercise 20) such that $\sigma(g) = g$ if and only if $g = 1$. If σ^2 is the identity map from G to G, prove that G is abelian (such an automorphism σ is called *fixed point free* of order 2). [Show that every element of G can be written in the form $x^{-1}\sigma(x)$ and apply σ to such an expression.]

24. Let G be a finite group and let x and y be distinct elements of order 2 in G that generate G. Prove that $G \cong D_{2n}$, where $n = |xy|$. [See Exercise 6 in Section 2.]

25. Let $n \in \mathbb{Z}^+$, let r and s be the usual generators of D_{2n} and let $\theta = 2\pi/n$.

    (a) Prove that the matrix $\begin{pmatrix} \cos\theta & -\sin\theta \\ \sin\theta & \cos\theta \end{pmatrix}$ is the matrix of the linear transformation which rotates the x, y plane about the origin in a counterclockwise direction by θ radians.

    (b) Prove that the map $\varphi : D_{2n} \to GL_2(\mathbb{R})$ defined on generators by

$$\varphi(r) = \begin{pmatrix} \cos\theta & -\sin\theta \\ \sin\theta & \cos\theta \end{pmatrix} \qquad \text{and} \qquad \varphi(s) = \begin{pmatrix} 0 & 1 \\ 1 & 0 \end{pmatrix}$$

    extends to a homomorphism of D_{2n} into $GL_2(\mathbb{R})$.

    (c) Prove that the homomorphism φ in part (b) is injective.

26. Let i and j be the generators of Q_8 described in Section 5. Prove that the map φ from Q_8 to $GL_2(\mathbb{C})$ defined on generators by

$$\varphi(i) = \begin{pmatrix} \sqrt{-1} & 0 \\ 0 & -\sqrt{-1} \end{pmatrix} \qquad \text{and} \qquad \varphi(j) = \begin{pmatrix} 0 & -1 \\ 1 & 0 \end{pmatrix}$$

extends to a homomorphism. Prove that φ is injective.

1.7 GROUP ACTIONS

In this section we introduce the precise definition of a group acting on a set and present some examples. Group actions will be a powerful tool which we shall use both for proving theorems for abstract groups and for unravelling the structure of specific examples. Moreover, the concept of an "action" is a theme which will recur throughout the text as a method for studying an algebraic object by seeing how it can act on other structures.

Definition. A *group action* of a group G on a set A is a map from $G \times A$ to A (written as $g \cdot a$, for all $g \in G$ and $a \in A$) satisfying the following properties:
    (1) $g_1 \cdot (g_2 \cdot a) = (g_1 g_2) \cdot a$,   for all $g_1, g_2 \in G$, $a \in A$, and
    (2) $1 \cdot a = a$,   for all $a \in A$.

We shall immediately become less formal and say G is a group acting on a set A. The expression $g \cdot a$ will usually be written simply as ga when there is no danger of confusing this map with, say, the group operation (remember, $\cdot$ is not a binary operation and ga is always a member of A). Note that on the left hand side of the equation in property (1) $g_2 \cdot a$ is an element of A so it makes sense to act on this by g_1. On the right hand side of this equation the product $(g_1 g_2)$ is taken in G and the resulting group element acts on the set element a.

Before giving some examples of group actions we make some observations. Let the group G act on the set A. For each fixed $g \in G$ we get a map σ_g defined by

$$\sigma_g : A \to A$$
$$\sigma_g(a) = g \cdot a.$$

We prove two important facts:

(i) for each fixed $g \in G$, σ_g is a *permutation* of A, and
(ii) the map from G to S_A defined by $g \mapsto \sigma_g$ is a homomorphism.

To see that σ_g is a permutation of A we show that as a set map from A to A it has a 2-sided inverse, namely $\sigma_{g^{-1}}$ (it is then a permutation by Proposition 1 of Section 0.1). For all $a \in A$

$$
\begin{aligned}
(\sigma_{g^{-1}} \circ \sigma_g)(a) = \sigma_{g^{-1}}(\sigma_g(a)) \quad &\text{(by definition of function composition)} \\
= g^{-1} \cdot (g \cdot a) \quad &\text{(by definition of } \sigma_{g^{-1}} \text{ and } \sigma_g) \\
= (g^{-1}g) \cdot a \quad &\text{(by property (1) of an action)} \\
= 1 \cdot a = a \quad &\text{(by property (2) of an action).}
\end{aligned}
$$

This proves $\sigma_{g^{-1}} \circ \sigma_g$ is the identity map from A to A. Since g was arbitrary, we may interchange the roles of g and g^{-1} to obtain $\sigma_g \circ \sigma_{g^{-1}}$ is also the identity map on A. Thus σ_g has a 2-sided inverse, hence is a permutation of A.

To check assertion (ii) above let $\varphi : G \to S_A$ be defined by $\varphi(g) = \sigma_g$. Note that part (i) shows that σ_g is indeed an element of S_A. To see that φ is a homomorphism we must prove $\varphi(g_1 g_2) = \varphi(g_1) \circ \varphi(g_2)$ (recall that S_A is a group under function composition). The permutations $\varphi(g_1 g_2)$ and $\varphi(g_1) \circ \varphi(g_2)$ are equal if and only if their values agree on every element $a \in A$. For all $a \in A$

$$
\begin{aligned}
\varphi(g_1 g_2)(a) = \sigma_{g_1 g_2}(a) \quad &\text{(by definition of } \varphi) \\
= (g_1 g_2) \cdot a \quad &\text{(by definition of } \sigma_{g_1 g_2}) \\
= g_1 \cdot (g_2 \cdot a) \quad &\text{(by property (1) of an action)} \\
= \sigma_{g_1}(\sigma_{g_2}(a)) \quad &\text{(by definition of } \sigma_{g_1} \text{ and } \sigma_{g_2}) \\
= (\varphi(g_1) \circ \varphi(g_2))(a) \quad &\text{(by definition of } \varphi).
\end{aligned}
$$

This proves assertion (ii) above.

Intuitively, a group action of G on a set A just means that every element g in G acts as a permutation on A in a manner consistent with the group operations in G; assertions (i) and (ii) above make this precise. The homomorphism from G to S_A given above is called the *permutation representation* associated to the given action. It is easy to see that this process is reversible in the sense that if $\varphi : G \rightarrow S_A$ is any homomorphism from a group G to the symmetric group on a set A, then the map from $G \times A$ to A defined by

$$g \cdot a = \varphi(g)(a) \qquad \text{for all } g \in G, \text{ and all } a \in A$$

satisfies the properties of a group action of G on A. Thus actions of a group G on a set A and the homomorphisms from G into the symmetric group S_A are in bijective correspondence (i.e., are essentially the same notion, phrased in different terminology).

We should also note that the definition of an action might have been more precisely named a *left* action since the group elements appear on the left of the set elements. We could similarly define the notion of a *right* action.

Examples

Let G be a group and A a nonempty set. In each of the following examples the check of properties (1) and (2) of an action are left as exercises.

(1) Let $ga = a$, for all $g \in G$, $a \in A$. Properties (1) and (2) of a group action follow immediately. This action is called the *trivial action* and G is said to *act trivially* on A. Note that *distinct* elements of G induce the *same* permutation on A (in this case the identity permutation). The associated permutation representation $G \rightarrow S_A$ is the trivial homomorphism which maps every element of G to the identity.

If G acts on a set B and distinct elements of G induce *distinct* permutations of B, the action is said to be *faithful*. A faithful action is therefore one in which the associated permutation representation is injective.

The *kernel* of the action of G on B is defined to be $\{g \in G \mid gb = b \text{ for all } b \in B\}$, namely the elements of G which fix *all* the elements of B. For the trivial action, the kernel of the action is all of G and this action is not faithful when $|G| > 1$.

(2) The axioms for a vector space V over a field F include the two axioms that the multiplicative group $F^\times$ act on the set V. Thus vector spaces are familiar examples of actions of multiplicative groups of fields where there is even more structure (in particular, V must be an abelian group) which can be exploited. In the special case when $V = \mathbb{R}^n$ and $F = \mathbb{R}$ the action is specified by

$$\alpha(r_1, r_2, \ldots, r_n) = (\alpha r_1, \alpha r_2, \ldots, \alpha r_n)$$

for all $\alpha \in \mathbb{R}$, $(r_1, r_2, \ldots, r_n) \in \mathbb{R}^n$, where αr_i is just multiplication of two real numbers.

(3) For any nonempty set A the symmetric group S_A acts on A by $\sigma \cdot a = \sigma(a)$, for all $\sigma \in S_A$, $a \in A$. The associated permutation representation is the identity map from S_A to itself.

(4) If we fix a labelling of the vertices of a regular n-gon, each element α of D_{2n} gives rise to a permutation σ_α of $\{1, 2, \ldots, n\}$ by the way the symmetry α permutes the corresponding vertices. The map of $D_{2n} \times \{1, 2, \ldots, n\}$ onto $\{1, 2, \ldots, n\}$ defined by $(\alpha, i) \rightarrow \sigma_\alpha(i)$ defines a group action of D_{2n} on $\{1, 2, \ldots, n\}$. In keeping with our notation for group actions we can now dispense with the formal and cumbersome notation $\sigma_\alpha(i)$ and write αi in its place. Note that this action is faithful: distinct symmetries of a regular n-gon induce distinct permutations of the vertices.

When $n = 3$ the action of D_6 on the three (labelled) vertices of a triangle gives an injective homomorphism from D_6 to S_3. Since these groups have the same order, this map must also be surjective, i.e., is an isomorphism: $D_6 \cong S_3$. This is another proof of the same fact we established via generators and relations in the preceding section. Geometrically it says that any permutation of the vertices of a triangle is a symmetry. The analogous statement is not true for any n-gon with $n \geq 4$ (just by order considerations we cannot have D_{2n} isomorphic to S_n for any $n \geq 4$).

(5) Let G be any group and let $A = G$. Define a map from $G \times A$ to A by $g \cdot a = ga$, for each $g \in G$ and $a \in A$, where ga on the right hand side is the product of g and a in the group G. This gives a group action of G on itself, where each (fixed) $g \in G$ permutes the elements of G by *left multiplication*:

$$g : a \mapsto ga \qquad \text{for all } a \in G$$

(or, if G is written additively, we get $a \mapsto g + a$ and call this *left translation*). This action is called the *left regular action* of G on itself. By the cancellation laws, this action is faithful (check this).

Other examples of actions are given in the exercises.

EXERCISES

1. Let F be a field. Show that the multiplicative group of nonzero elements of F (denoted by $F^{\times}$) acts on the set F by $g \cdot a = ga$, where $g \in F^{\times}$, $a \in F$ and ga is the usual product in F of the two field elements (state clearly which axioms in the definition of a field are used).

2. Show that the additive group $\mathbb{Z}$ acts on itself by $z \cdot a = z + a$ for all $z, a \in \mathbb{Z}$.

3. Show that the additive group $\mathbb{R}$ acts on the x, y plane $\mathbb{R} \times \mathbb{R}$ by $r \cdot (x, y) = (x + ry, y)$.

4. Let G be a group acting on a set A and fix some $a \in A$. Show that the following sets are subgroups of G (cf. Exercise 26 of Section 1):
 (a) the kernel of the action,
 (b) $\{g \in G \mid ga = a\}$ — this subgroup is called the *stabilizer* of a in G.

5. Prove that the kernel of an action of the group G on the set A is the same as the kernel of the corresponding permutation representation $G \to S_A$ (cf. Exercise 14 in Section 6).

6. Prove that a group G acts faithfully on a set A if and only if the kernel of the action is the set consisting only of the identity.

7. Prove that in Example 2 in this section the action is faithful.

8. Let A be a nonempty set and let k be a positive integer with $k \leq |A|$. The symmetric group S_A acts on the set B consisting of all subsets of A of cardinality k by $\sigma \cdot \{a_1, \ldots, a_k\} = \{\sigma(a_1), \ldots, \sigma(a_k)\}$.
 (a) Prove that this is a group action.
 (b) Describe explicitly how the elements (1 2) and (1 2 3) act on the six 2-element subsets of $\{1, 2, 3, 4\}$.

9. Do both parts of the preceding exercise with "ordered k-tuples" in place of "k-element subsets," where the action on k-tuples is defined as above but with set braces replaced by parentheses (note that, for example, the 2-tuples (1,2) and (2,1) are different even though the sets $\{1, 2\}$ and $\{2, 1\}$ are the same, so the sets being acted upon are different).

10. With reference to the preceding two exercises determine:
 (a) for which values of k the action of S_n on k-element subsets is faithful, and
 (b) for which values of k the action of S_n on ordered k-tuples is faithful.

44

11. Write out the cycle decomposition of the eight permutations in S_4 corresponding to the elements of D_8 given by the action of D_8 on the vertices of a square (where the vertices of the square are labelled as in Section 2).

12. Assume n is an even positive integer and show that D_{2n} acts on the set consisting of pairs of opposite vertices of a regular n-gon. Find the kernel of this action (label vertices as usual).

13. Find the kernel of the left regular action.

14. Let G be a group and let $A = G$. Show that if G is non-abelian then the maps defined by $g \cdot a = ag$ for all $g, a \in G$ do *not* satisfy the axioms of a (left) group action of G on itself.

15. Let G be any group and let $A = G$. Show that the maps defined by $g \cdot a = ag^{-1}$ for all $g, a \in G$ do satisfy the axioms of a (left) group action of G on itself.

16. Let G be any group and let $A = G$. Show that the maps defined by $g \cdot a = gag^{-1}$ for all $g, a \in G$ do satisfy the axioms of a (left) group action (this action of G on itself is called *conjugation*).

17. Let G be a group and let G act on itself by left conjugation, so each $g \in G$ maps G to G by
$$x \mapsto gxg^{-1}.$$
For fixed $g \in G$, prove that conjugation by g is an isomorphism from G onto itself (i.e., is an automorphism of G — cf. Exercise 20, Section 6). Deduce that x and gxg^{-1} have the same order for all x in G and that for any subset A of G, $|A| = |gAg^{-1}|$ (here $gAg^{-1} = \{gag^{-1} \mid a \in A\}$).

18. Let H be a group acting on a set A. Prove that the relation $\sim$ on A defined by
$$a \sim b \qquad \text{if and only if} \qquad a = hb \quad \text{for some } h \in H$$
is an equivalence relation. (For each $x \in A$ the equivalence class of x under $\sim$ is called the *orbit* of x under the action of H. The orbits under the action of H partition the set A.)

19. Let H be a subgroup (cf. Exercise 26 of Section 1) of the finite group G and let H act on G (here $A = G$) by left multiplication. Let $x \in G$ and let $\mathcal{O}$ be the orbit of x under the action of H. Prove that the map
$$H \to \mathcal{O} \qquad \text{defined by} \qquad h \mapsto hx$$
is a bijection (hence all orbits have cardinality $|H|$). From this and the preceding exercise deduce *Lagrange's Theorem*:
if G is a finite group and H is a subgroup of G then $|H|$ divides $|G|$.

20. Show that the group of rigid motions of a tetrahedron is isomorphic to a subgroup (cf. Exercise 26 of Section 1) of S_4.

21. Show that the group of rigid motions of a cube is isomorphic to S_4. [This group acts on the set of four pairs of opposite vertices.]

22. Show that the group of rigid motions of an octahedron is isomorphic to S_4. [This group acts on the set of four pairs of opposite faces.] Deduce that the groups of rigid motions of a cube and an octahedron are isomorphic. (These groups are isomorphic because these solids are "dual" — see *Introduction to Geometry* by H. Coxeter, Wiley, 1961. We shall see later that the groups of rigid motions of the dodecahedron and icosahedron are isomorphic as well — these solids are also dual.)

23. Explain why the action of the group of rigid motions of a cube on the set of three pairs of opposite faces is not faithful. Find the kernel of this action.

CHAPTER 2

Subgroups

2.1 DEFINITION AND EXAMPLES

One basic method for unravelling the structure of any mathematical object which is defined by a set of axioms is to study *subsets* of that object which also *satisfy the same axioms*. We begin this program by discussing subgroups of a group. A second basic method for unravelling structure is to study quotients of an object; the notion of a quotient group, which is a way (roughly speaking) of collapsing one group onto a smaller group, will be dealt with in the next chapter. Both of these themes will recur throughout the text as we study subgroups and quotient groups of a group, subrings and quotient rings of a ring, subspaces and quotient spaces of a vector space, etc.

Definition.   Let G be a group. The subset H of G is a *subgroup* of G if H is nonempty and H is closed under products and inverses (i.e., $x, y \in H$ implies $x^{-1} \in H$ and $xy \in H$). If H is a subgroup of G we shall write $H \leq G$.

Subgroups of G are just subsets of G which are themselves groups with respect to the operation defined in G, i.e., the binary operation on G restricts to give a binary operation on H which is associative, has an identity in H, and has inverses in H for all the elements of H.

When we say that H is a subgroup of G we shall always mean that the operation for the group H is the operation on G restricted to H (in general it is possible that the subset H has the structure of a group with respect to some operation other than the operation on G restricted to H, cf. Example 5(a) following). As we have been doing for functions restricted to a subset, we shall denote the operation for G and the operation for the subgroup H by the same symbol. If $H \leq G$ and $H \neq G$ we shall write $H < G$ to emphasize that the containment is proper.

If H is a subgroup of G then, since the operation for H is the operation for G restricted to H, any equation in the subgroup H may also be viewed as an equation in the group G. Thus the cancellation laws for G imply that the identity for H is the same as the identity of G (in particular, every subgroup must contain 1, the identity of G) and the inverse of an element x in H is the same as the inverse of x when considered as an element of G (so the notation x^{-1} is unambiguous).

46

Examples

(1) $\mathbb{Z} \leq \mathbb{Q}$ and $\mathbb{Q} \leq \mathbb{R}$ with the operation of addition.

(2) Any group G has two subgroups: $H = G$ and $H = \{1\}$; the latter is called the *trivial subgroup* and will henceforth be denoted by 1.

(3) If $G = D_{2n}$ is the dihedral group of order $2n$, let H be $\{1, r, r^2, \ldots, r^{n-1}\}$, the set of all rotations in G. Since the product of two rotations is again a rotation and the inverse of a rotation is also a rotation it follows that H is a subgroup of D_{2n} of order n.

(4) The set of even integers is a subgroup of the group of all integers under addition.

(5) Some examples of subsets which are *not* subgroups:

    (a) $\mathbb{Q} - \{0\}$ under multiplication is not a subgroup of $\mathbb{R}$ under addition even though both are groups and $\mathbb{Q} - \{0\}$ is a subset of $\mathbb{R}$; the operation of multiplication on $\mathbb{Q} - \{0\}$ is not the restriction of the operation of addition on $\mathbb{R}$.

    (b) $\mathbb{Z}^+$ (under addition) is not a subgroup of $\mathbb{Z}$ (under addition) because although $\mathbb{Z}^+$ is closed under $+$, it does not contain the identity, 0, of $\mathbb{Z}$ and although each $x \in \mathbb{Z}^+$ has an additive inverse, $-x$, in $\mathbb{Z}$, $-x \notin \mathbb{Z}^+$, i.e., $\mathbb{Z}^+$ is not closed under the operation of taking inverses (in particular, $\mathbb{Z}^+$ is not a group under addition). For analogous reasons, $(\mathbb{Z} - \{0\}, \times)$ is not a subgroup of $(\mathbb{Q} - \{0\}, \times)$.

    (c) D_6 is not a subgroup of D_8 since the former is not even a subset of the latter.

(6) The relation "is a subgroup of" is transitive: if H is a subgroup of a group G and K is a subgroup of H, then K is also a subgroup of G.

As we saw in Chapter 1, even for easy examples checking that all the group axioms (especially the associative law) hold for any given binary operation can be tedious at best. Once we know that we have a group, however, checking that a subset of it is (or is not) a subgroup is a much easier task, since all we need to check is closure under multiplication and under taking inverses. The next proposition shows that these can be amalgamated into a single test and also shows that for *finite* groups it suffices to check for closure under multiplication.

Proposition 1. *(The Subgroup Criterion)* A subset H of a group G is a subgroup if and only if

    (1) $H \neq \varnothing$, and

    (2) for all $x, y \in H$, $xy^{-1} \in H$.

Furthermore, if H is finite, then it suffices to check that H is nonempty and closed under multiplication.

Proof: If H is a subgroup of G, then certainly (1) and (2) hold because H contains the identity of G and the inverse of each of its elements and because H is closed under multiplication.

It remains to show conversely that if H satisfies both (1) and (2), then $H \leq G$. Let x be any element in H (such x exists by property (1)). Let $y = x$ and apply property (2) to deduce that $1 = xx^{-1} \in H$, so H contains the identity of G. Then, again by (2), since H contains 1 and x, H contains the element $1x^{-1}$, i.e., $x^{-1} \in H$ and H is closed under taking inverses. Finally, if x and y are any two elements of H, then H contains x and y^{-1} by what we have just proved, so by (2), H also contains $x(y^{-1})^{-1} = xy$. Hence H is also closed under multiplication, which proves H is a subgroup of G.

Suppose now that H is finite and closed under multiplication and let x be any element in H. Then there are only finitely many distinct elements among $x, x^2, x^3, \ldots$ and so $x^a = x^b$ for some integers a, b with $b > a$. If $n = b - a$, then $x^n = 1$ so in particular every element $x \in H$ is of finite order. Then $x^{n-1} = x^{-1}$ is an element of H, so H is automatically also closed under inverses.

EXERCISES

Let G be a group.

1. In each of (a) – (e) prove that the specified subset is a subgroup of the given group:
   (a) the set of complex numbers of the form $a + ai$, $a \in \mathbb{R}$ (under addition)
   (b) the set of complex numbers of absolute value 1, i.e., the unit circle in the complex plane (under multiplication)
   (c) for fixed $n \in \mathbb{Z}^+$ the set of rational numbers whose denominators divide n (under addition)
   (d) for fixed $n \in \mathbb{Z}^+$ the set of rational numbers whose denominators are relatively prime to n (under addition)
   (e) the set of nonzero real numbers whose square is a rational number (under multiplication).

2. In each of (a) – (e) prove that the specified subset is *not* a subgroup of the given group:
   (a) the set of 2-cycles in S_n for $n \geq 3$
   (b) the set of reflections in D_{2n} for $n \geq 3$
   (c) for n a composite integer > 1 and G a group containing an element of order n, the set $\{x \in G \mid |x| = n\} \cup \{1\}$
   (d) the set of (positive and negative) odd integers in $\mathbb{Z}$ together with 0
   (e) the set of real numbers whose square is a rational number (under addition).

3. Show that the following subsets of the dihedral group D_8 are actually subgroups:
   (a) $\{1, r^2, s, sr^2\}$, (b) $\{1, r^2, sr, sr^3\}$.

4. Give an explicit example of a group G and an infinite subset H of G that is closed under the group operation but is not a subgroup of G.

5. Prove that G cannot have a subgroup H with $|H| = n - 1$, where $n = |G| > 2$.

6. Let G be an abelian group. Prove that $\{g \in G \mid |g| < \infty\}$ is a subgroup of G (called the *torsion subgroup* of G). Give an explicit example where this set is not a subgroup when G is non-abelian.

7. Fix some $n \in \mathbb{Z}$ with $n > 1$. Find the torsion subgroup (cf. the previous exercise) of $\mathbb{Z} \times (\mathbb{Z}/n\mathbb{Z})$. Show that the set of elements of infinite order together with the identity is *not* a subgroup of this direct product.

8. Let H and K be subgroups of G. Prove that $H \cup K$ is a subgroup if and only if either $H \subseteq K$ or $K \subseteq H$.

9. Let $G = GL_n(F)$, where F is any field. Define
   $$SL_n(F) = \{A \in GL_n(F) \mid \det(A) = 1\}$$
   (called the *special linear group*). Prove that $SL_n(F) \leq GL_n(F)$.

10. (a) Prove that if H and K are subgroups of G then so is their intersection $H \cap K$.
    (b) Prove that the intersection of an arbitrary nonempty collection of subgroups of G is again a subgroup of G (do not assume the collection is countable).

11. Let A and B be groups. Prove that the following sets are subgroups of the direct product $A \times B$:

(a) $\{(a, 1) \mid a \in A\}$

(b) $\{(1, b) \mid b \in B\}$

(c) $\{(a, a) \mid a \in A\}$, where here we assume $B = A$ (called the *diagonal subgroup*).

12. Let A be an abelian group and fix some $n \in \mathbb{Z}$. Prove that the following sets are subgroups of A:

(a) $\{a^n \mid a \in A\}$

(b) $\{a \in A \mid a^n = 1\}$.

13. Let H be a subgroup of the additive group of rational numbers with the property that $1/x \in H$ for every nonzero element x of H. Prove that $H = 0$ or $\mathbb{Q}$.

14. Show that $\{x \in D_{2n} \mid x^2 = 1\}$ is not a subgroup of D_{2n} (here $n \geq 3$).

15. Let $H_1 \leq H_2 \leq \cdots$ be an ascending chain of subgroups of G. Prove that $\cup_{i=1}^{\infty} H_i$ is a subgroup of G.

16. Let $n \in \mathbb{Z}^+$ and let F be a field. Prove that the set $\{(a_{ij}) \in GL_n(F) \mid a_{ij} = 0 \text{ for all } i > j\}$ is a subgroup of $GL_n(F)$ (called the group of *upper triangular* matrices).

17. Let $n \in \mathbb{Z}^+$ and let F be a field. Prove that the set $\{(a_{ij}) \in GL_n(F) \mid a_{ij} = 0 \text{ for all } i > j,$ and $a_{ii} = 1 \text{ for all } i\}$ is a subgroup of $GL_n(F)$.

2.2 CENTRALIZERS AND NORMALIZERS, STABILIZERS AND KERNELS

We now introduce some important families of subgroups of an arbitrary group G which in particular provide many examples of subgroups. Let A be any nonempty subset of G.

Definition. Define $C_G(A) = \{g \in G \mid gag^{-1} = a \text{ for all } a \in A\}$. This subset of G is called the *centralizer* of A in G. Since $gag^{-1} = a$ if and only if $ga = ag$, $C_G(A)$ is the set of elements of G which commute with every element of A.

We show $C_G(A)$ is a subgroup of G. First of all, $C_G(A) \neq \emptyset$ because $1 \in C_G(A)$: the definition of the identity specifies that $1a = a1$, for all $a \in G$ (in particular, for all $a \in A$) so 1 satisfies the defining condition for membership in $C_G(A)$. Secondly, assume $x, y \in C_G(A)$, that is, for all $a \in A$, $xax^{-1} = a$ and $yay^{-1} = a$ (note that this does *not* mean $xy = yx$). Observe first that since $yay^{-1} = a$, multiplying both sides of this first on the left by y^{-1}, then on the right by y and then simplifying gives $a = y^{-1}ay$, i.e., $y^{-1} \in C_G(A)$ so that $C_G(A)$ is closed under taking inverses. Now

$$(xy)a(xy)^{-1} = (xy)a(y^{-1}x^{-1}) \quad \text{(by Proposition 1.1(4) applied to } (xy)^{-1} \text{)}$$
$$= x(yay^{-1})x^{-1} \quad \text{(by the associative law)}$$
$$= xax^{-1} \quad \text{(since } y \in C_G(A) \text{)}$$
$$= a \quad \text{(since } x \in C_G(A) \text{)}$$

so $xy \in C_G(A)$ and $C_G(A)$ is closed under products, hence $C_G(A) \leq G$.

In the special case when $A = \{a\}$ we shall write simply $C_G(a)$ instead of $C_G(\{a\})$. In this case $a^n \in C_G(a)$ for all $n \in \mathbb{Z}$.

For example, in an abelian group G, $C_G(A) = G$, for all subsets A. One can check by inspection that $C_{Q_8}(i) = \{\pm 1, \pm i\}$. Some other examples are specified in the exercises.

We shall shortly discuss how to minimize the calculation of commutativities between single group elements which appears to be inherent in the computation of centralizers (and other subgroups of a similar nature).

Definition. Define $Z(G) = \{g \in G \mid gx = xg \text{ for all } x \in G\}$, the set of elements commuting with all the elements of G. This subset of G is called the *center* of G.

Note that $Z(G) = C_G(G)$, so the argument above proves $Z(G) \le G$ as a special case. As an exercise, the reader may wish to prove $Z(G)$ is a subgroup directly.

Definition. Define $gAg^{-1} = \{gag^{-1} \mid a \in A\}$. Define the *normalizer* of A in G to be the set $N_G(A) = \{g \in G \mid gAg^{-1} = A\}$.

Notice that if $g \in C_G(A)$, then $gag^{-1} = a \in A$ for all $a \in A$ so $C_G(A) \le N_G(A)$. The proof that $N_G(A)$ is a subgroup of G follows the same steps which demonstrated that $C_G(A) \le G$ with appropriate modifications.

Examples

(1) If G is abelian then all the elements of G commute, so $Z(G) = G$. Similarly, $C_G(A) = N_G(A) = G$ for *any* subset A of G since $gag^{-1} = gg^{-1}a = a$ for every $g \in G$ and every $a \in A$.

(2) Let $G = D_8$ be the dihedral group of order 8 with the usual generators r and s and let $A = \{1, r, r^2, r^3\}$ be the subgroup of rotations in D_8. We show that $C_{D_8}(A) = A$. Since all powers of r commute with each other, $A \le C_{D_8}(A)$. Since $sr = r^{-1}s \ne rs$ the element s does not commute with all members of A, i.e., $s \notin C_{D_8}(A)$. Finally, the elements of D_8 that are not in A are all of the form sr^i for some $i \in \{0, 1, 2, 3\}$. If the element sr^i were in $C_{D_8}(A)$ then since $C_{D_8}(A)$ is a *subgroup* which contains r we would also have the element $s = (sr^i)(r^{-i})$ in $C_{D_8}(A)$, a contradiction. This shows $C_{D_8}(A) = A$.

(3) As in the preceding example let $G = D_8$ and let $A = \{1, r, r^2, r^3\}$. We show that $N_{D_8}(A) = D_8$. Since, in general, the centralizer of a subset is contained in its normalizer, $A \le N_{D_8}(A)$. Next compute that

$$s As^{-1} = \{s1s^{-1}, srs^{-1}, sr^2s^{-1}, sr^3s^{-1}\} = \{1, r^3, r^2, r\} = A,$$

so that $s \in N_{D_8}(A)$. (Note that the *set* sAs^{-1} equals the *set* A even though the elements in these two sets appear in different orders — this is because s is in the normalizer of A but not in the centralizer of A.) Now both r and s belong to the *subgroup* $N_{D_8}(A)$ and hence $s^i r^j \in N_{D_8}(A)$ for all integers i and j, that is, every element of D_8 is in $N_{D_8}(A)$ (recall that r and s *generate* D_8). Since $D_8 \le N_{D_8}(A)$ we have $N_{D_8}(A) = D_8$ (the reverse containment being obvious from the definition of a normalizer).

(4) We show that the center of D_8 is the subgroup $\{1, r^2\}$. First observe that the center of any group G is contained in $C_G(A)$ for any subset A of G. Thus by Example 2 above $Z(D_8) \le C_{D_8}(A) = A$, where $A = \{1, r, r^2, r^3\}$. The calculation in Example 2 shows that r and similarly r^3 are not in $Z(D_8)$, so $Z(D_8) \le \{1, r^2\}$. To show the

reverse inclusion note that r commutes with r^2 and calculate that s also commutes with r^2. Since r and s generate D_8, every element of D_8 commutes with r^2 (and 1), hence $\{1, r^2\} \leq Z(D_8)$ and so equality holds.

(5) Let $G = S_3$ and let A be the subgroup $\{1, (1\,2)\}$. We explain why $C_{S_3}(A) = N_{S_3}(A) = A$. One can compute directly that $C_{S_3}(A) = A$, using the ideas in Example 2 above to minimize the calculations. Alternatively, since an element commutes with its powers, $A \leq C_{S_3}(A)$. By Lagrange's Theorem (Exercise 19 in Section 1.7) the order of the subgroup $C_{S_3}(A)$ of S_3 divides $|S_3| = 6$. Also by Lagrange's Theorem applied to the subgroup A of the group $C_{S_3}(A)$ we have that $2 \mid |C_{S_3}(A)|$. The only possibilities are: $|C_{S_3}(A)| = 2$ or 6. If the latter occurs, $C_{S_3}(A) = S_3$, i.e., $A \leq Z(S_3)$; this is a contradiction because $(1\,2)$ does not commute with $(1\,2\,3)$. Thus $|C_{S_3}(A)| = 2$ and so $A = C_{S_3}(A)$.

Next note that $N_{S_3}(A) = A$ because $\sigma \in N_{S_3}(A)$ if and only if

$$\{\sigma 1 \sigma^{-1}, \ \sigma(1\,2)\sigma^{-1}\} = \{1, (1\,2)\}.$$

Since $\sigma 1 \sigma^{-1} = 1$, this equality of sets occurs if and only if $\sigma(1\,2)\sigma^{-1} = (1\,2)$ as well, i.e., if and only if $\sigma \in C_{S_3}(A)$.

The center of S_3 is the identity because $Z(S_3) \leq C_{S_3}(A) = A$ and $(1\,2) \notin Z(S_3)$.

Stabilizers and Kernels of Group Actions

The fact that the normalizer of A in G, the centralizer of A in G, and the center of G are all subgroups can be deduced as special cases of results on group actions, indicating that the structure of G is reflected by the sets on which it acts, as follows: if G is a group acting on a set S and s is some fixed element of S, the *stabilizer* of s in G is the set

$$G_s = \{g \in G \mid g \cdot s = s\}$$

(see Exercise 4 in Section 1.7). We show briefly that $G_s \leq G$: first $1 \in G_s$ by axiom (2) of an action. Also, if $y \in G_s$,

$$
\begin{aligned}
s = 1 \cdot s &= (y^{-1}y) \cdot s \\
&= y^{-1} \cdot (y \cdot s) &&\text{(by axiom (1) of an action)} \\
&= y^{-1} \cdot s &&\text{(since } y \in G_s\text{)}
\end{aligned}
$$

so $y^{-1} \in G_s$ as well. Finally, if $x, y \in G_s$, then

$$
\begin{aligned}
(xy) \cdot s &= x \cdot (y \cdot s) &&\text{(by axiom (1) of an action)} \\
&= x \cdot s &&\text{(since } y \in G_s\text{)} \\
&= s &&\text{(since } x \in G_s\text{).}
\end{aligned}
$$

This proves G_s is a subgroup[1] of G. A similar (but easier) argument proves that the *kernel* of an action is a subgroup, where the kernel of the action of G on S is defined as

$$\{g \in G \mid g \cdot s = s, \text{ for all } s \in S\}$$

(see Exercise 4(b) in Section 1.7).

---

[1] Notice how the steps to prove G_s is a subgroup are the same as those to prove $C_G(A) \leq G$ with axiom (1) of an action taking the place of the associative law.

Examples

(1) The group $G = D_8$ acts on the set A of four vertices of a square (cf. Example 4 in Section 1.7). The stabilizer of any vertex a is the subgroup $\{1, t\}$ of D_8, where t is the reflection about the line of symmetry passing through vertex a and the center of the square. The kernel of this action is the identity subgroup since only the identity symmetry fixes every vertex.

(2) The group $G = D_8$ also acts on the set A whose elements are the two unordered pairs of opposite vertices (in the labelling of Figure 2 in Section 1.2, $A = \{\{1, 3\}, \{2, 4\}\}$). The kernel of the action of D_8 on this set A is the subgroup $\{1, s, r^2, sr^2\}$ and for either element $a \in A$ the stabilizer of a in D_8 equals the kernel of the action.

Finally, we observe that the fact that centralizers, normalizers and centers are subgroups is a special case of the fact that stabilizers and kernels of actions are subgroups (this will be discussed further in Chapter 4). Let $S = \mathcal{P}(G)$, the collection of all subsets of G, and let G act on S by *conjugation*, that is, for each $g \in G$ and each $B \subseteq G$ let

$$g : B \to gBg^{-1} \quad \text{where} \quad gBg^{-1} = \{gbg^{-1} \mid b \in B\}$$

(see Exercise 16 in Section 1.7). Under this action, it is easy to check that $N_G(A)$ is precisely the stabilizer of A in G (i.e., $N_G(A) = G_s$ where $s = A \in \mathcal{P}(G)$), so $N_G(A)$ is a subgroup of G.

Next let the group $N_G(A)$ act on the set $S = A$ by conjugation, i.e., for all $g \in N_G(A)$ and $a \in A$

$$g : a \mapsto gag^{-1}.$$

Note that this does map A to A by the definition of $N_G(A)$ and so gives an action on A. Here it is easy to check that $C_G(A)$ is precisely the kernel of this action, hence $C_G(A) \leq N_G(A)$; by transitivity of the relation "$\leq$," $C_G(A) \leq G$. Finally, $Z(G)$ is the kernel of G acting on $S = G$ by conjugation, so $Z(G) \leq G$.

EXERCISES

1. Prove that $C_G(A) = \{g \in G \mid g^{-1}ag = a \text{ for all } a \in A\}$.

2. Prove that $C_G(Z(G)) = G$ and deduce that $N_G(Z(G)) = G$.

3. Prove that if A and B are subsets of G with $A \subseteq B$ then $C_G(B)$ is a subgroup of $C_G(A)$.

4. For each of S_3, D_8, and Q_8 compute the centralizers of each element and find the center of each group. Does Lagrange's Theorem (Exercise 19 in Section 1.7) simplify your work?

5. In each of parts (a) to (c) show that for the specified group G and subgroup A of G, $C_G(A) = A$ and $N_G(A) = G$.
   (a) $G = S_3$ and $A = \{1, (1\,2\,3), (1\,3\,2)\}$.
   (b) $G = D_8$ and $A = \{1, s, r^2, sr^2\}$.
   (c) $G = D_{10}$ and $A = \{1, r, r^2, r^3, r^4\}$.

6. Let H be a subgroup of the group G.
   (a) Show that $H \leq N_G(H)$. Give an example to show that this is not necessarily true if H is not a subgroup.
   (b) Show that $H \leq C_G(H)$ if and only if H is abelian.

7. Let $n \in \mathbb{Z}$ with $n \geq 3$. Prove the following:
   (a) $Z(D_{2n}) = 1$ if n is odd

(b) $Z(D_{2n}) = \{1, r^k\}$ if $n = 2k$.

8. Let $G = S_n$, fix an $i \in \{1, 2, \ldots, n\}$ and let $G_i = \{\sigma \in G \mid \sigma(i) = i\}$ (the stabilizer of i in G). Use group actions to prove that G_i is a subgroup of G. Find $|G_i|$.

9. For any subgroup H of G and any nonempty subset A of G define $N_H(A)$ to be the set $\{h \in H \mid hAh^{-1} = A\}$. Show that $N_H(A) = N_G(A) \cap H$ and deduce that $N_H(A)$ is a subgroup of H (note that A need not be a subset of H).

10. Let H be a subgroup of order 2 in G. Show that $N_G(H) = C_G(H)$. Deduce that if $N_G(H) = G$ then $H \le Z(G)$.

11. Prove that $Z(G) \le N_G(A)$ for any subset A of G.

12. Let R be the set of all polynomials with integer coefficients in the independent variables x_1, x_2, x_3, x_4 i.e., the members of R are finite sums of elements of the form $ax_1^{r_1} x_2^{r_2} x_3^{r_3} x_4^{r_4}$, where a is any integer and $r_1, \ldots, r_4$ are nonnegative integers. For example,

$$12x_1^5 x_2^7 x_4 - 18x_2^3 x_3 + 11x_1^6 x_2 x_3^3 x_4^{23} \qquad (*)$$

is a typical element of R. Each $\sigma \in S_4$ gives a permutation of $\{x_1, \ldots, x_4\}$ by defining $\sigma \cdot x_i = x_{\sigma(i)}$. This may be extended to a map from R to R by defining

$$\sigma \cdot p(x_1, x_2, x_3, x_4) = p(x_{\sigma(1)}, x_{\sigma(2)}, x_{\sigma(3)}, x_{\sigma(4)})$$

for all $p(x_1, x_2, x_3, x_4) \in R$ (i.e., σ simply permutes the indices of the variables). For example, if $\sigma = (1\ 2)(3\ 4)$ and $p(x_1, \ldots, x_4)$ is the polynomial in $(*)$ above, then

$$\sigma \cdot p(x_1, x_2, x_3, x_4) = 12x_2^5 x_1^7 x_3 - 18x_1^3 x_4 + 11x_2^6 x_1 x_4^3 x_3^{23}$$
$$= 12x_1^7 x_2^5 x_3 - 18x_1^3 x_4 + 11x_1 x_2^6 x_3^{23} x_4^3.$$

(a) Let $p = p(x_1, \ldots, x_4)$ be the polynomial in $(*)$ above, let $\sigma = (1\ 2\ 3\ 4)$ and let $\tau = (1\ 2\ 3)$. Compute $\sigma \cdot p$, $\tau \cdot (\sigma \cdot p)$, $(\tau \circ \sigma) \cdot p$, and $(\sigma \circ \tau) \cdot p$.

(b) Prove that these definitions give a (left) group action of S_4 on R.

(c) Exhibit all permutations in S_4 that stabilize x_4 and prove that they form a subgroup isomorphic to S_3.

(d) Exhibit all permutations in S_4 that stabilize the element $x_1 + x_2$ and prove that they form an abelian subgroup of order 4.

(e) Exhibit all permutations in S_4 that stabilize the element $x_1 x_2 + x_3 x_4$ and prove that they form a subgroup isomorphic to the dihedral group of order 8.

(f) Show that the permutations in S_4 that stabilize the element $(x_1 + x_2)(x_3 + x_4)$ are exactly the same as those found in part (e). (The two polynomials appearing in parts (e) and (f) and the subgroup that stabilizes them will play an important role in the study of roots of quartic equations in Section 14.6.)

13. Let n be a positive integer and let R be the set of all polynomials with integer coefficients in the independent variables $x_1, x_2, \ldots, x_n$, i.e., the members of R are finite sums of elements of the form $ax_1^{r_1} x_2^{r_2} \cdots x_n^{r_n}$, where a is any integer and $r_1, \ldots, r_n$ are nonnegative integers. For each $\sigma \in S_n$ define a map

$$\sigma : R \to R \qquad \text{by} \qquad \sigma \cdot p(x_1, x_2, \ldots, x_n) = p(x_{\sigma(1)}, x_{\sigma(2)}, \ldots, x_{\sigma(n)}).$$

Prove that this defines a (left) group action of S_n on R.

14. Let $H(F)$ be the Heisenberg group over the field F introduced in Exercise 11 of Section 1.4. Determine which matrices lie in the center of $H(F)$ and prove that $Z(H(F))$ is isomorphic to the additive group F.

2.3 CYCLIC GROUPS AND CYCLIC SUBGROUPS

Let G be any group and let x be any element of G. One way of forming a subgroup H of G is by letting H be the set of all integer (positive, negative and zero) powers of x (this guarantees closure under inverses and products at least as far as x is concerned). In this section we study groups which are generated by one element.

Definition. A group H is *cyclic* if H can be generated by a single element, i.e., there is some element $x \in H$ such that $H = \{x^n \mid n \in \mathbb{Z}\}$ (where as usual the operation is multiplication).

In additive notation H is cyclic if $H = \{nx \mid n \in \mathbb{Z}\}$. In both cases we shall write $H = \langle x \rangle$ and say H is *generated* by x (and x is a *generator* of H). A cyclic group may have more than one generator. For example, if $H = \langle x \rangle$, then also $H = \langle x^{-1} \rangle$ because $(x^{-1})^n = x^{-n}$ and as n runs over all integers so does $-n$ so that

$$\{x^n \mid n \in \mathbb{Z}\} = \{(x^{-1})^n \mid n \in \mathbb{Z}\}.$$

We shall shortly show how to determine all generators for a given cyclic group H. One should note that the elements of $\langle x \rangle$ are powers of x (or multiples of x, in groups written additively) and not integers. It is not necessarily true that all powers of x are distinct. Also, by the laws for exponents (Exercise 19 in Section 1.1) cyclic groups are abelian.

Examples

(1) Let $G = D_{2n} = \langle r, s \mid r^n = s^2 = 1, rs = sr^{-1} \rangle$, $n \geq 3$ and let H be the subgroup of all rotations of the n-gon. Thus $H = \langle r \rangle$ and the distinct elements of H are $1, r, r^2, \ldots, r^{n-1}$ (these are all the distinct powers of r). In particular, $|H| = n$ and the generator, r, of H has order n. The powers of r "cycle" (forward and backward) with period n, that is,

$$r^n = 1, \ r^{n+1} = r, \ r^{n+2} = r^2, \ldots$$
$$r^{-1} = r^{n-1}, \ r^{-2} = r^{n-2}, \ldots \quad \text{etc.}$$

In general, to write any power of r, say r^t, in the form r^k, for some k between 0 and $n - 1$ use the Division Algorithm to write

$$t = nq + k, \qquad \text{where } 0 \leq k < n,$$

so that

$$r^t = r^{nq+k} = (r^n)^q r^k = 1^q r^k = r^k.$$

For example, in D_8, $r^4 = 1$ so $r^{105} = r^{4(26)+1} = r$ and $r^{-42} = r^{4(-11)+2} = r^2$. Observe that D_{2n} itself is not a cyclic group since it is non-abelian.

(2) Let $H = \mathbb{Z}$ with operation $+$. Thus $H = \langle 1 \rangle$ (here 1 is the integer 1 and the identity of H is 0) and each element in H can be written uniquely in the form $n \cdot 1$, for some $n \in \mathbb{Z}$. In contrast to the preceding example, multiples of the generator are all distinct and we need to take both positive, negative and zero multiples of the generator to obtain all elements of H. In this example $|H|$ and the order of the generator 1 are both ∞. Note also that $H = \langle -1 \rangle$ since each integer x can be written (uniquely) as $(-x)(-1)$.

Before discussing cyclic groups further we prove that the various properties of finite and infinite cyclic groups we observed in the preceding two examples are generic. This proposition also validates the claim (in Chapter 1) that the use of the terminology for "order" of an element and the use of the symbol | | are consistent with the notion of order of a set.

Proposition 2. If $H = \langle x \rangle$, then $|H| = |x|$ (where if one side of this equality is infinite, so is the other). More specifically
  (1) if $|H| = n < \infty$, then $x^n = 1$ and $1, x, x^2, \ldots, x^{n-1}$ are all the distinct elements of H, and
  (2) if $|H| = \infty$, then $x^n \neq 1$ for all $n \neq 0$ and $x^a \neq x^b$ for all $a \neq b$ in $\mathbb{Z}$.

Proof: Let $|x| = n$ and first consider the case when $n < \infty$. The elements $1, x, x^2, \ldots, x^{n-1}$ are distinct because if $x^a = x^b$, with, say, $0 \leq a < b < n$, then $x^{b-a} = x^0 = 1$, contrary to n being the smallest positive power of x giving the identity. Thus H has at least n elements and it remains to show that these are all of them. As we did in Example 1, if x^t is any power of x, use the Division Algorithm to write $t = nq + k$, where $0 \leq k < n$, so

$$x^t = x^{nq+k} = (x^n)^q x^k = 1^q x^k = x^k \in \{1, x, x^2, \ldots, x^{n-1}\},$$

as desired.

Next suppose $|x| = \infty$ so no positive power of x is the identity. If $x^a = x^b$, for some a and b with, say, $a < b$, then $x^{b-a} = 1$, a contradiction. Distinct powers of x are distinct elements of H so $|H| = \infty$. This completes the proof of the proposition.

Note that the proof of the proposition gives the method for reducing arbitrary powers of a generator in a finite cyclic group to the "least residue" powers. It is not a coincidence that the calculations of distinct powers of a generator of a cyclic group of order n are carried out via arithmetic in $\mathbb{Z}/n\mathbb{Z}$. Theorem 4 following proves that these two groups are isomorphic.

First we need an easy proposition.

Proposition 3. Let G be an arbitrary group, $x \in G$ and let $m, n \in \mathbb{Z}$. If $x^n = 1$ and $x^m = 1$, then $x^d = 1$, where $d = (m, n)$. In particular, if $x^m = 1$ for some $m \in \mathbb{Z}$, then $|x|$ divides m.

Proof: By the Euclidean Algorithm (see Section 0.2 (6)) there exist integers r and s such that $d = mr + ns$, where d is the g.c.d. of m and n. Thus

$$x^d = x^{mr+ns} = (x^m)^r (x^n)^s = 1^r 1^s = 1.$$

This proves the first assertion.

If $x^m = 1$, let $n = |x|$. If $m = 0$, certainly $n \mid m$, so we may assume $m \neq 0$. Since some nonzero power of x is the identity, $n < \infty$. Let $d = (m, n)$ so by the preceding result $x^d = 1$. Since $0 < d \leq n$ and n is the smallest positive power of x which gives the identity, we must have $d = n$, that is, $n \mid m$, as asserted.

Theorem 4. Any two cyclic groups of the same order are isomorphic. More specifically,

(1) if $n \in \mathbb{Z}^+$ and $\langle x \rangle$ and $\langle y \rangle$ are both cyclic groups of order n, then the map

$$\varphi : \langle x \rangle \to \langle y \rangle$$
$$x^k \mapsto y^k$$

is well defined and is an isomorphism

(2) if $\langle x \rangle$ is an infinite cyclic group, the map

$$\varphi : \mathbb{Z} \to \langle x \rangle$$
$$k \mapsto x^k$$

is well defined and is an isomorphism.

Proof: Suppose $\langle x \rangle$ and $\langle y \rangle$ are both cyclic groups of order n. Let $\varphi : \langle x \rangle \to \langle y \rangle$ be defined by $\varphi(x^k) = y^k$; we must first prove φ is well defined, that is,

$$\text{if } x^r = x^s, \text{ then } \varphi(x^r) = \varphi(x^s).$$

Since $x^{r-s} = 1$, Proposition 3 implies $n \mid r - s$. Write $r = tn + s$ so

$$\varphi(x^r) = \varphi(x^{tn+s})$$
$$= y^{tn+s}$$
$$= (y^n)^t y^s$$
$$= y^s = \varphi(x^s).$$

This proves φ is well defined. It is immediate from the laws of exponents that $\varphi(x^a x^b) = \varphi(x^a)\varphi(x^b)$ (check this), that is, φ is a homomorphism. Since the element y^k of $\langle y \rangle$ is the image of x^k under φ, this map is surjective. Since both groups have the same finite order, any surjection from one to the other is a bijection, so φ is an isomorphism (alternatively, φ has an obvious two-sided inverse).

If $\langle x \rangle$ is an infinite cyclic group, let $\varphi : \mathbb{Z} \to \langle x \rangle$ be defined by $\varphi(k) = x^k$. Note that this map is already well defined since there is no ambiguity in the representation of elements in the domain. Since (by Proposition 2) $x^a \neq x^b$, for all distinct $a, b \in \mathbb{Z}$, φ is injective. By definition of a cyclic group, φ is surjective. As above, the laws of exponents ensure φ is a homomorphism, hence φ is an isomorphism, completing the proof.

We chose to use the rotation group $\langle r \rangle$ as our prototypical example of a finite cyclic group of order n (instead of the isomorphic group $\mathbb{Z}/n\mathbb{Z}$) since we shall usually write our cyclic groups multiplicatively:

Notation: For each $n \in \mathbb{Z}^+$, let Z_n be the cyclic group of order n (written multiplicatively).

Up to isomorphism, Z_n is the unique cyclic group of order n and $Z_n \cong \mathbb{Z}/n\mathbb{Z}$. On occasion when we find additive notation advantageous we shall use the latter group as our representative of the isomorphism class of cyclic groups of order n. We shall occasionally say "let $\langle x \rangle$ be the infinite cyclic group" (written multiplicatively), however we shall always use $\mathbb{Z}$ (additively) to represent the infinite cyclic group.

As noted earlier, a given cyclic group may have more than one generator. The next two propositions determine precisely which powers of x generate the group $\langle x \rangle$.

Proposition 5. Let G be a group, let $x \in G$ and let $a \in \mathbb{Z} - \{0\}$.
  (1) If $|x| = \infty$, then $|x^a| = \infty$.
  (2) If $|x| = n < \infty$, then $|x^a| = \dfrac{n}{(n, a)}$.
  (3) In particular, if $|x| = n < \infty$ and a is a positive integer dividing n, then $|x^a| = \dfrac{n}{a}$.

Proof: (1) By way of contradiction assume $|x| = \infty$ but $|x^a| = m < \infty$. By definition of order

$$1 = (x^a)^m = x^{am}.$$

Also,

$$x^{-am} = (x^{am})^{-1} = 1^{-1} = 1.$$

Now one of am or $-am$ is positive (since neither a nor m is 0) so some positive power of x is the identity. This contradicts the hypothesis $|x| = \infty$, so the assumption $|x^a| < \infty$ must be false, that is, (1) holds.

(2) Under the notation of (2) let

$$y = x^a, \quad (n, a) = d \quad \text{and write} \quad n = db, \ a = dc,$$

for suitable $b, c \in \mathbb{Z}$ with $b > 0$. Since d is the greatest common divisor of n and a, the integers b and c are relatively prime:

$$(b, c) = 1.$$

To establish (2) we must show $|y| = b$. First note that

$$y^b = x^{ab} = x^{dcb} = (x^{db})^c = (x^n)^c = 1^c = 1$$

so, by Proposition 3 applied to $\langle y \rangle$, we see that $|y|$ divides b. Let $k = |y|$. Then

$$x^{ak} = y^k = 1$$

so by Proposition 3 applied to $\langle x \rangle$, $n \mid ak$, i.e., $db \mid dck$. Thus $b \mid ck$. Since b and c have no factors in common, b must divide k. Since b and k are positive integers which divide each other, $b = k$, which proves (2).

(3) This is a special case of (2) recorded for future reference.

Proposition 6. Let $H = \langle x \rangle$.
  (1) Assume $|x| = \infty$. Then $H = \langle x^a \rangle$ if and only if $a = \pm 1$.
  (2) Assume $|x| = n < \infty$. Then $H = \langle x^a \rangle$ if and only if $(a, n) = 1$. In particular, the number of generators of H is $\varphi(n)$ (where φ is Euler's φ-function).

Proof: We leave (1) as an exercise. In (2) if $|x| = n < \infty$, Proposition 2 says x^a generates a subgroup of H of order $|x^a|$. This subgroup equals all of H if and only if $|x^a| = |x|$. By Proposition 5,

$$|x^a| = |x| \quad \text{if and only if} \quad \frac{n}{(a,n)} = n, \quad \text{i.e. if and only if } (a,n) = 1.$$

Since $\varphi(n)$ is, by definition, the number of $a \in \{1, 2, \ldots, n\}$ such that $(a, n) = 1$, this is the number of generators of H.

Example

Proposition 6 tells precisely which residue classes mod n generate $\mathbb{Z}/n\mathbb{Z}$: namely, $\bar{a}$ generates $\mathbb{Z}/n\mathbb{Z}$ if and only if $(a, n) = 1$. For instance, $\bar{1}, \bar{5}, \bar{7}$ and $\overline{11}$ are the generators of $\mathbb{Z}/12\mathbb{Z}$ and $\varphi(12) = 4$.

The final theorem in this section gives the complete subgroup structure of a cyclic group.

Theorem 7. Let $H = \langle x \rangle$ be a cyclic group.
  (1) Every subgroup of H is cyclic. More precisely, if $K \le H$, then either $K = \{1\}$ or $K = \langle x^d \rangle$, where d is the smallest positive integer such that $x^d \in K$.
  (2) If $|H| = \infty$, then for any distinct nonnegative integers a and b, $\langle x^a \rangle \neq \langle x^b \rangle$. Furthermore, for every integer m, $\langle x^m \rangle = \langle x^{|m|} \rangle$, where $|m|$ denotes the absolute value of m, so that the nontrivial subgroups of H correspond bijectively with the integers $1, 2, 3, \ldots$.
  (3) If $|H| = n < \infty$, then for each positive integer a dividing n there is a unique subgroup of H of order a. This subgroup is the cyclic group $\langle x^d \rangle$, where $d = \dfrac{n}{a}$. Furthermore, for every integer m, $\langle x^m \rangle = \langle x^{(n,m)} \rangle$, so that the subgroups of H correspond bijectively with the positive divisors of n.

Proof: (1) Let $K \le H$. If $K = \{1\}$, the proposition is true for this subgroup, so we assume $K \neq \{1\}$. Thus there exists some $a \neq 0$ such that $x^a \in K$. If $a < 0$ then since K is a group also $x^{-a} = (x^a)^{-1} \in K$. Hence K always contains some positive power of x. Let

$$\mathcal{P} = \{b \mid b \in \mathbb{Z}^+ \text{ and } x^b \in K\}.$$

By the above, $\mathcal{P}$ is a nonempty set of positive integers. By the Well Ordering Principle (Section 0.2) $\mathcal{P}$ has a minimum element — call it d. Since K is a subgroup and $x^d \in K$, $\langle x^d \rangle \le K$. Since K is a subgroup of H, any element of K is of the form x^a for some integer a. By the Division Algorithm write

$$a = qd + r \qquad 0 \le r < d.$$

Then $x^r = x^{(a-qd)} = x^a (x^d)^{-q}$ is an element of K since both x^a and x^d are elements of K. By the minimality of d it follows that $r = 0$, i.e., $a = qd$ and so $x^a = (x^d)^q \in \langle x^d \rangle$. This gives the reverse containment $K \le \langle x^d \rangle$ which proves (1).

We leave the proof of (2) as an exercise (the reasoning is similar to and easier than the proof of (3) which follows).

(3) Assume $|H| = n < \infty$ and $a \mid n$. Let $d = \dfrac{n}{a}$ and apply Proposition 5(3) to obtain that $\langle x^d \rangle$ is a subgroup of order a, showing the existence of a subgroup of order a. To show uniqueness, suppose K is any subgroup of H of order a. By part (1) we have

$$K = \langle x^b \rangle$$

where b is the smallest positive integer such that $x^b \in K$. By Proposition 5

$$\frac{n}{d} = a = |K| = |x^b| = \frac{n}{(n, b)},$$

so $d = (n, b)$. In particular, $d \mid b$. Since b is a multiple of d, $x^b \in \langle x^d \rangle$, hence

$$K = \langle x^b \rangle \le \langle x^d \rangle.$$

Since $|\langle x^d \rangle| = a = |K|$, we have $K = \langle x^d \rangle$.

The final assertion of (3) follows from the observation that $\langle x^m \rangle$ is a subgroup of $\langle x^{(n,m)} \rangle$ (check this) and, it follows from Proposition 5(2) and Proposition 2 that they have the same order. Since (n, m) is certainly a divisor of n, this shows that every subgroup of H arises from a divisor of n, completing the proof.

Examples

(1) We can use Proposition 6 and Theorem 7 to list all the subgroups of $\mathbb{Z}/n\mathbb{Z}$ for any given n. For example, the subgroups of $\mathbb{Z}/12\mathbb{Z}$ are

(a) $\mathbb{Z}/12\mathbb{Z} = \langle \bar{1} \rangle = \langle \bar{5} \rangle = \langle \bar{7} \rangle = \langle \overline{11} \rangle$ (order 12)
(b) $\langle \bar{2} \rangle = \langle \overline{10} \rangle$ (order 6)
(c) $\langle \bar{3} \rangle = \langle \bar{9} \rangle$ (order 4)
(d) $\langle \bar{4} \rangle = \langle \bar{8} \rangle$ (order 3)
(e) $\langle \bar{6} \rangle$ (order 2)
(f) $\langle \bar{0} \rangle$ (order 1).

The inclusions between them are given by

$$\langle \bar{a} \rangle \le \langle \bar{b} \rangle \quad \text{if and only if } (b, 12) \mid (a, 12), \quad 1 \le a, b \le 12.$$

(2) We can also combine the results of this section with those of the preceding one. For example, we can obtain subgroups of a group G by forming $C_G(\langle x \rangle)$ and $N_G(\langle x \rangle)$, for each $x \in G$. One can check that an element g in G commutes with x if and only if g commutes with all powers of x, hence

$$C_G(\langle x \rangle) = C_G(x).$$

As noted in Exercise 6, Section 2, $\langle x \rangle \le N_G(\langle x \rangle)$ but equality need not hold. For instance, if $G = Q_8$ and $x = i$,

$$C_G(\langle i \rangle) = \{\pm 1, \pm i\} = \langle i \rangle \quad \text{and} \quad N_G(\langle i \rangle) = Q_8.$$

Note that we already observed the first of the above two equalities and the second is most easily computed using the result of Exercise 24 following.

1. Find all subgroups of $Z_{45} = \langle x \rangle$, giving a generator for each. Describe the containments between these subgroups.

2. If x is an element of the finite group G and $|x| = |G|$, prove that $G = \langle x \rangle$. Give an explicit example to show that this result need not be true if G is an infinite group.

3. Find all generators for $\mathbb{Z}/48\mathbb{Z}$.

4. Find all generators for $\mathbb{Z}/202\mathbb{Z}$.

5. Find the number of generators for $\mathbb{Z}/49000\mathbb{Z}$.

6. In $\mathbb{Z}/48\mathbb{Z}$ write out all elements of $\langle \bar{a} \rangle$ for every $\bar{a}$. Find all inclusions between subgroups in $\mathbb{Z}/48\mathbb{Z}$.

7. Let $Z_{48} = \langle x \rangle$ and use the isomorphism $\mathbb{Z}/48\mathbb{Z} \cong Z_{48}$ given by $\bar{1} \mapsto x$ to list all subgroups of Z_{48} as computed in the preceding exercise.

8. Let $Z_{48} = \langle x \rangle$. For which integers a does the map φ_a defined by $\varphi_a : \bar{1} \mapsto x^a$ extend to an *isomorphism* from $\mathbb{Z}/48\mathbb{Z}$ onto Z_{48}.

9. Let $Z_{36} = \langle x \rangle$. For which integers a does the map ψ_a defined by $\psi_a : \bar{1} \mapsto x^a$ extend to a *well defined homomorphism* from $\mathbb{Z}/48\mathbb{Z}$ into Z_{36}. Can ψ_a ever be a surjective homomorphism?

10. What is the order of $\overline{30}$ in $\mathbb{Z}/54\mathbb{Z}$? Write out all of the elements and their orders in $\langle \overline{30} \rangle$.

11. Find all cyclic subgroups of D_8. Find a proper subgroup of D_8 which is not cyclic.

12. Prove that the following groups are *not* cyclic:
    (a) $Z_2 \times Z_2$
    (b) $Z_2 \times \mathbb{Z}$
    (c) $\mathbb{Z} \times \mathbb{Z}$.

13. Prove that the following pairs of groups are *not* isomorphic:
    (a) $\mathbb{Z} \times Z_2$ and $\mathbb{Z}$
    (b) $\mathbb{Q} \times Z_2$ and $\mathbb{Q}$.

14. Let $\sigma = (1\ 2\ 3\ 4\ 5\ 6\ 7\ 8\ 9\ 10\ 11\ 12)$. For each of the following integers a compute σ^a: $a = 13, 65, 626, 1195, -6, -81, -570$ and -1211.

15. Prove that $\mathbb{Q} \times \mathbb{Q}$ is not cyclic.

16. Assume $|x| = n$ and $|y| = m$. Suppose that x and y *commute*: $xy = yx$. Prove that $|xy|$ divides the least common multiple of m and n. Need this be true if x and y do *not* commute? Give an example of commuting elements x, y such that the order of xy is not equal to the least common multiple of $|x|$ and $|y|$.

17. Find a presentation for Z_n with one generator.

18. Show that if H is any group and h is an element of H with $h^n = 1$, then there is a unique homomorphism from $Z_n = \langle x \rangle$ to H such that $x \mapsto h$.

19. Show that if H is any group and h is an element of H, then there is a unique homomorphism from $\mathbb{Z}$ to H such that $1 \mapsto h$.

20. Let p be a prime and let n be a positive integer. Show that if x is an element of the group G such that $x^{p^n} = 1$ then $|x| = p^m$ for some $m \le n$.

21. Let p be an odd prime and let n be a positive integer. Use the Binomial Theorem to show that $(1 + p)^{p^{n-1}} \equiv 1 \pmod{p^n}$ but $(1 + p)^{p^{n-2}} \not\equiv 1 \pmod{p^n}$. Deduce that $1 + p$ is an element of order p^{n-1} in the multiplicative group $(\mathbb{Z}/p^n\mathbb{Z})^\times$.

22. Let n be an integer ≥ 3. Use the Binomial Theorem to show that $(1 + 2^2)^{2^{n-2}} \equiv 1 \pmod{2^n}$ but $(1 + 2^2)^{2^{n-3}} \not\equiv 1 \pmod{2^n}$. Deduce that 5 is an element of order 2^{n-2} in the multiplicative group $(\mathbb{Z}/2^n\mathbb{Z})^\times$.

23. Show that $(\mathbb{Z}/2^n\mathbb{Z})^\times$ is not cyclic for any $n \geq 3$. [Find two distinct subgroups of order 2.]

24. Let G be a finite group and let $x \in G$.
   (a) Prove that if $g \in N_G(\langle x \rangle)$ then $gxg^{-1} = x^a$ for some $a \in \mathbb{Z}$.
   (b) Prove conversely that if $gxg^{-1} = x^a$ for some $a \in \mathbb{Z}$ then $g \in N_G(\langle x \rangle)$. [Show first that $gx^kg^{-1} = (gxg^{-1})^k = x^{ak}$ for any integer k, so that $g \langle x \rangle g^{-1} \leq \langle x \rangle$. If x has order n, show the elements gx^ig^{-1}, $i = 0, 1, \ldots, n - 1$ are distinct, so that $|g \langle x \rangle g^{-1}| = |\langle x \rangle| = n$ and conclude that $g \langle x \rangle g^{-1} = \langle x \rangle$.]

   Note that this cuts down some of the work in computing normalizers of cyclic subgroups since one does not have to check $ghg^{-1} \in \langle x \rangle$ for every $h \in \langle x \rangle$.

25. Let G be a cyclic group of order n and let k be an integer relatively prime to n. Prove that the map $x \mapsto x^k$ is surjective. Use Lagrange's Theorem (Exercise 19, Section 1.7) to prove the same is true for any finite group of order n. (For such k each element has a kth root in G. It follows from Cauchy's Theorem in Section 3.2 that if k is not relatively prime to the order of G then the map $x \mapsto x^k$ is not surjective.)

26. Let Z_n be a cyclic group of order n and for each integer a let

$$\sigma_a : Z_n \to Z_n \quad \text{by} \quad \sigma_a(x) = x^a \text{ for all } x \in Z_n.$$

   (a) Prove that σ_a is an automorphism of Z_n if and only if a and n are relatively prime (automorphisms were introduced in Exercise 20, Section 1.6).
   (b) Prove that $\sigma_a = \sigma_b$ if and only if $a \equiv b \pmod{n}$.
   (c) Prove that *every* automorphism of Z_n is equal to σ_a for some integer a.
   (d) Prove that $\sigma_a \circ \sigma_b = \sigma_{ab}$. Deduce that the map $\overline{a} \mapsto \sigma_a$ is an isomorphism of $(\mathbb{Z}/n\mathbb{Z})^\times$ onto the automorphism group of Z_n (so $\text{Aut}(Z_n)$ is an abelian group of order $\varphi(n)$).

2.4 SUBGROUPS GENERATED BY SUBSETS OF A GROUP

The method of forming cyclic subgroups of a given group is a special case of the general technique where one forms the subgroup generated by an arbitrary subset of a group. In the case of cyclic subgroups one takes a singleton subset $\{x\}$ of the group G and forms all integral powers of x, which amounts to closing the set $\{x\}$ under the group operation and the process of taking inverses. The resulting subgroup is the smallest subgroup of G which contains the set $\{x\}$ (smallest in the sense that if H is any subgroup which contains $\{x\}$, then H contains $\langle x \rangle$). Another way of saying this is that $\langle x \rangle$ is the unique minimal element of the set of subgroups of G containing x (ordered under inclusion). In this section we investigate analogues of this when $\{x\}$ is replaced by an arbitrary subset of G.

Throughout mathematics the following theme recurs: given an object G (such as a group, field, vector space, etc.) and a subset A of G, is there a unique minimal subobject of G (subgroup, subfield, subspace, etc.) which contains A and, if so, how are the elements of this subobject computed? Students may already have encountered this question in the study of vector spaces. When G is a vector space (with, say, real number scalars) and $A = \{v_1, v_2, \ldots, v_n\}$, then there is a unique smallest subspace of

G which contains A, namely the (linear) span of $v_1, v_2, \ldots, v_n$ and each vector in this span can be written as $k_1 v_1 + k_2 v_2 + \cdots + k_n v_n$, for some $k_1, \ldots, k_n \in \mathbb{R}$. When A is a single nonzero vector, v, the span of $\{v\}$ is simply the 1-dimensional subspace or line containing v and every element of this subspace is of the form kv for some $k \in \mathbb{R}$. This is the analogue in the theory of vector spaces of cyclic subgroups of a group. Note that the 1-dimensional subspaces contain kv, where $k \in \mathbb{R}$, not just kv, where $k \in \mathbb{Z}$; the reason being that a subspace must be closed under *all* the vector space operations (e.g., scalar multiplication) not just the group operation of vector addition.

Let G be any group and let A be any subset of G. We now make precise the notion of the subgroup of G generated by A. We prove that because the intersection of any set of subgroups of G is also a subgroup of G, the subgroup generated by A is the unique smallest subgroup of G containing A; it is " smallest" in the sense of being the minimal element of the set of all subgroups containing A. We show that the elements of this subgroup are obtained by closing the given subset under the group operation (and taking inverses). In succeeding parts of the text when we develop the theory of other algebraic objects we shall refer to this section as the paradigm in proving that a given subset is contained in a unique smallest subobject and that the elements of this subobject are obtained by closing the subset under the operations which define the object. Since in the latter chapters the details will be omitted, students should acquire a solid understanding of the process at this point.

In order to proceed we need only the following.

Proposition 8. If $\mathcal{A}$ is any nonempty collection of subgroups of G, then the intersection of all members of $\mathcal{A}$ is also a subgroup of G.

Proof: This is an easy application of the subgroup criterion (see also Exercise 10, Section 1). Let

$$K = \bigcap_{H \in \mathcal{A}} H.$$

Since each $H \in \mathcal{A}$ is a subgroup, $1 \in H$, so $1 \in K$, that is, $K \neq \emptyset$. If $a, b \in K$, then $a, b \in H$, for all $H \in \mathcal{A}$. Since each H is a group, $ab^{-1} \in H$, for all H, hence $ab^{-1} \in K$. Proposition 1 gives that $K \leq G$.

Definition. If A is any subset of the group G define

$$\langle A \rangle = \bigcap_{\substack{A \subseteq H \\ H \leq G}} H.$$

This is called the *subgroup of G generated by A*.

Thus $\langle A \rangle$ is the intersection of all subgroups of G containing A. It is a subgroup of G by Proposition 8 applied to the set $\mathcal{A} = \{H \leq G \mid A \subseteq H\}$ ($\mathcal{A}$ is nonempty since $G \in \mathcal{A}$). Since A lies in each $H \in \mathcal{A}$, A is a subset of their intersection, $\langle A \rangle$. Note that $\langle A \rangle$ is the unique minimal element of $\mathcal{A}$ as follows: $\langle A \rangle$ is a subgroup of G containing A, so $\langle A \rangle \in \mathcal{A}$; and any element of $\mathcal{A}$ contains the intersection of all elements in $\mathcal{A}$, i.e., contains $\langle A \rangle$.

When A is the finite set $\{a_1, a_2, \ldots, a_n\}$ we write $\langle a_1, a_2, \ldots, a_n \rangle$ for the group generated by $a_1, a_2, \ldots, a_n$ instead of $\langle \{a_1, a_2, \ldots, a_n\} \rangle$. If A and B are two subsets of G we shall write $\langle A, B \rangle$ in place of $\langle A \cup B \rangle$.

This "top down" approach to defining $\langle A \rangle$ proves existence and uniqueness of the smallest subgroup of G containing A but is not too enlightening as to how to construct the elements in it. As the word "generates" suggests we now define the set which is the closure of A under the group operation (and the process of taking inverses) and prove this set equals $\langle A \rangle$. Let

$$\overline{A} = \{a_1^{\epsilon_1} a_2^{\epsilon_2} \ldots a_n^{\epsilon_n} \mid n \in \mathbb{Z},\ n \geq 0 \text{ and } a_i \in A, \epsilon_i = \pm 1 \text{ for each } i\}$$

where $\overline{A} = \{1\}$ if $A = \emptyset$, so that $\overline{A}$ is the set of all finite products (called *words*) of elements of A and inverses of elements of A. Note that the a_i's need not be distinct, so a^2 is written aa in the notation defining $\overline{A}$. Note also that A is not assumed to be a finite (or even countable) set.

Proposition 9. $\overline{A} = \langle A \rangle$.

Proof: We first prove $\overline{A}$ is a subgroup. Note that $\overline{A} \neq \emptyset$ (even if $A = \emptyset$). If $a, b \in \overline{A}$ with $a = a_1^{\epsilon_1} a_2^{\epsilon_2} \ldots a_n^{\epsilon_n}$ and $b = b_1^{\delta_1} b_2^{\delta_2} \ldots b_m^{\delta_m}$, then

$$ab^{-1} = a_1^{\epsilon_1} a_2^{\epsilon_2} \ldots a_n^{\epsilon_n} \cdot b_m^{-\delta_m} b_{m-1}^{-\delta_{m-1}} \ldots b_1^{-\delta_1}$$

(where we used Exercise 15 of Section 1.1 to compute b^{-1}). Thus ab^{-1} is a product of elements of A raised to powers ± 1, hence $ab^{-1} \in \overline{A}$. Proposition 1 implies $\overline{A}$ is a subgroup of G.

Since each $a \in A$ may be written a^1, it follows that $A \subseteq \overline{A}$, hence $\langle A \rangle \subseteq \overline{A}$. But $\langle A \rangle$ is a group containing A and, since it is closed under the group operation and the process of taking inverses, $\langle A \rangle$ contains each element of the form $a_1^{\epsilon_1} a_2^{\epsilon_2} \ldots a_n^{\epsilon_n}$, that is, $\overline{A} \subseteq \langle A \rangle$. This completes the proof of the proposition.

We now use $\langle A \rangle$ in place of $\overline{A}$ and may take the definition of $\overline{A}$ as an equivalent definition of $\langle A \rangle$. As noted above, in this equivalent definition of $\langle A \rangle$, products of the form $a \cdot a, a \cdot a \cdot a, a \cdot a^{-1}$, etc. could have been simplified to $a^2, a^3, 1$, etc. respectively, so another way of writing $\langle A \rangle$ is

$$\langle A \rangle = \{a_1^{\alpha_1} a_2^{\alpha_2} \ldots a_n^{\alpha_n} \mid \text{for each } i, \quad a_i \in A, \alpha_i \in \mathbb{Z}, a_i \neq a_{i+1} \text{ and } n \in \mathbb{Z}^+\}.$$

In fact, when $A = \{x\}$ this was our definition of $\langle A \rangle$.

If G is *abelian*, we could commute the a_i's and so collect all powers of a given generator together. For instance, if A were the finite subset $\{a_1, a_2, \ldots, a_k\}$ of the abelian group G, one easily checks that

$$\langle A \rangle = \{a_1^{\alpha_1} a_2^{\alpha_2} \ldots a_k^{\alpha_k} \mid \alpha_i \in \mathbb{Z} \text{ for each } i\}.$$

If in this situation we further assume that each a_i has finite order d_i, for all i, then since there are exactly d_i distinct powers of a_i, the total number of distinct products of the form $a_1^{\alpha_1} a_2^{\alpha_2} \ldots a_k^{\alpha_k}$ is at most $d_1 d_2 \ldots d_k$, that is,

$$|\langle A \rangle| \leq d_1 d_2 \ldots d_k.$$

It may happen that $a^\alpha b^\beta = a^\gamma b^\delta$ even though $a^\alpha \neq a^\gamma$ and $b^\beta \neq b^\delta$. We shall explore exactly when this happens when we study direct products in Chapter 5.

When G is *non-abelian* the situation is much more complicated. For example, let $G = D_8$ and let r and s be the usual generators of D_8 (note that the notation $D_8 = \langle r, s \rangle$ is consistent with the notation introduced in Section 1.2). Let $a = s$, let $b = rs$ and let $A = \{a, b\}$. Since both s and r ($= rs \cdot s$) belong to $\langle a, b \rangle$, $G = \langle a, b \rangle$, i.e., G is also generated by a and b. Both a and b have order 2, however D_8 has order 8. This means that it is *not* possible to write every element of D_8 in the form $a^\alpha b^\beta$, $\alpha, \beta \in \mathbb{Z}$. More specifically, the product aba cannot be simplified to a product of the form $a^\alpha b^\beta$. In fact, if $G = D_{2n}$ for any $n > 2$, and r, s, a, b are defined in the same way as above, it is still true that

$$|a| = |b| = 2, \quad D_{2n} = \langle a, b \rangle \quad \text{and} \quad |D_{2n}| = 2n.$$

This means that for large n, long products of the form $abab \ldots ab$ cannot be further simplified. In particular, this illustrates that, unlike the abelian (or, better yet, cyclic) group case, the order of a (finite) group cannot even be bounded once we know the orders of the elements in some generating set.

Another example of this phenomenon is S_n:

$$S_n = \langle (1\,2), (1\,2\,3 \ldots n) \rangle.$$

Thus S_n is generated by an element of order 2 together with one of order n, yet $|S_n| = n!$ (we shall prove these statements later after developing some more techniques).

One final example emphasizes the fact that if G is non-abelian, subgroups of G generated by more than one element of G may be quite complicated. Let

$$G = GL_2(\mathbb{R}), \quad a = \begin{pmatrix} 0 & 1 \\ 1 & 0 \end{pmatrix}, \quad b = \begin{pmatrix} 0 & 2 \\ 1/2 & 0 \end{pmatrix}$$

so $a^2 = b^2 = 1$ but $ab = \begin{pmatrix} 1/2 & 0 \\ 0 & 2 \end{pmatrix}$. It is easy to see that ab has infinite order, so $\langle a, b \rangle$ is an *infinite* subgroup of $GL_2(\mathbb{R})$ which is generated by two elements of order 2.

These examples illustrate that when $|A| \geq 2$ it is difficult, in general, to compute even the order of the subgroup generated by A, let alone any other structural properties. It is therefore impractical to gather much information about subgroups of a non-abelian group created by taking random subsets A and trying to write out the elements of (or other information about) $\langle A \rangle$. For certain "well chosen" subsets A, even of a non-abelian group G, we shall be able to make both theoretical and computational use of the subgroup generated by A. One example of this might be when we want to find a subgroup of G which contains $\langle x \rangle$ properly; we might search for some element y which commutes with x (i.e., $y \in C_G(x)$) and form $\langle x, y \rangle$. It is easy to check that the latter group is abelian, so its order is bounded by $|x||y|$. Alternatively, we might instead take y in $N_G(\langle x \rangle)$ — in this case the same order bound holds and the structure of $\langle x, y \rangle$ is again not too complicated (as we shall see in the next chapter).

The complications which arise for non-abelian groups are generally not quite as serious when we study other basic algebraic systems because of the additional algebraic structure imposed.

EXERCISES

1. Prove that if H is a subgroup of G then $\langle H \rangle = H$.

2. Prove that if A is a subset of B then $\langle A \rangle \le \langle B \rangle$. Give an example where $A \subseteq B$ with $A \ne B$ but $\langle A \rangle = \langle B \rangle$.

3. Prove that if H is an abelian subgroup of a group G then $\langle H, Z(G) \rangle$ is abelian. Give an explicit example of an abelian subgroup H of a group G such that $\langle H, C_G(H) \rangle$ is not abelian.

4. Prove that if H is a subgroup of G then H is generated by the set $H - \{1\}$.

5. Prove that the subgroup generated by any two distinct elements of order 2 in S_3 is all of S_3.

6. Prove that the subgroup of S_4 generated by (1 2) and (1 2)(3 4) is a noncyclic group of order 4.

7. Prove that the subgroup of S_4 generated by (1 2) and (1 3)(2 4) is isomorphic to the dihedral group of order 8.

8. Prove that $S_4 = \langle (1\ 2\ 3\ 4), (1\ 2\ 4\ 3) \rangle$.

9. Prove that $SL_2(\mathbb{F}_3)$ is the subgroup of $GL_2(\mathbb{F}_3)$ generated by $\begin{pmatrix} 1 & 1 \\ 0 & 1 \end{pmatrix}$ and $\begin{pmatrix} 1 & 0 \\ 1 & 1 \end{pmatrix}$. [Recall from Exercise 9 of Section 1 that $SL_2(\mathbb{F}_3)$ is the subgroup of matrices of determinant 1. You may assume this subgroup has order 24 — this will be an exercise in Section 3.2.]

10. Prove that the subgroup of $SL_2(\mathbb{F}_3)$ generated by $\begin{pmatrix} 0 & -1 \\ 1 & 0 \end{pmatrix}$ and $\begin{pmatrix} 1 & 1 \\ 1 & -1 \end{pmatrix}$ is isomorphic to the quaternion group of order 8. [Use a presentation for Q_8.]

11. Show that $SL_2(\mathbb{F}_3)$ and S_4 are two nonisomorphic groups of order 24.

12. Prove that the subgroup of upper triangular matrices in $GL_3(\mathbb{F}_2)$ is isomorphic to the dihedral group of order 8 (cf. Exercise 16, Section 1). [First find the order of this subgroup.]

13. Prove that the multiplicative group of positive rational numbers is generated by the set $\{\frac{1}{p} \mid p \text{ is a prime }\}$.

14. A group H is called *finitely generated* if there is a finite set A such that $H = \langle A \rangle$.
    (a) Prove that every finite group is finitely generated.
    (b) Prove that $\mathbb{Z}$ is finitely generated.
    (c) Prove that every finitely generated subgroup of the additive group $\mathbb{Q}$ is cyclic. [If H is a finitely generated subgroup of $\mathbb{Q}$, show that $H \le \langle \frac{1}{k} \rangle$, where k is the product of all the denominators which appear in a set of generators for H.]
    (d) Prove that $\mathbb{Q}$ is not finitely generated.

15. Exhibit a proper subgroup of $\mathbb{Q}$ which is not cyclic.

16. A subgroup M of a group G is called a *maximal subgroup* if $M \ne G$ and the only subgroups of G which contain M are M and G.
    (a) Prove that if H is a proper subgroup of the finite group G then there is a maximal subgroup of G containing H.
    (b) Show that the subgroup of all rotations in a dihedral group is a maximal subgroup.
    (c) Show that if $G = \langle x \rangle$ is a cyclic group of order $n \ge 1$ then a subgroup H is maximal if and only $H = \langle x^p \rangle$ for some prime p dividing n.

17. This is an exercise involving Zorn's Lemma (see Appendix I) to prove that every nontrivial finitely generated group possesses maximal subgroups. Let G be a finitely generated

group, say $G = \langle g_1, g_2, \ldots, g_n \rangle$, and let $\mathcal{S}$ be the set of all proper subgroups of G. Then $\mathcal{S}$ is partially ordered by inclusion. Let $\mathcal{C}$ be a chain in $\mathcal{S}$.

(a) Prove that the union, H, of all the subgroups in $\mathcal{C}$ is a subgroup of G.

(b) Prove that H is a *proper* subgroup. [If not, each g_i must lie in H and so must lie in some element of the chain $\mathcal{C}$. Use the definition of a chain to arrive at a contradiction.]

(c) Use Zorn's Lemma to show that $\mathcal{S}$ has a maximal element (which is, by definition, a maximal subgroup).

18. Let p be a prime and let $Z = \{z \in \mathbb{C} \mid z^{p^n} = 1 \text{ for some } n \in \mathbb{Z}^+\}$ (so Z is the multiplicative group of all p-power roots of unity in $\mathbb{C}$). For each $k \in \mathbb{Z}^+$ let $H_k = \{z \in Z \mid z^{p^k} = 1\}$ (the group of p^kth roots of unity). Prove the following:

(a) $H_k \le H_m$ if and only if $k \le m$

(b) H_k is cyclic for all k (assume that for any $n \in \mathbb{Z}^+$, $\{e^{2\pi i t/n} \mid t = 0, 1, \ldots, n-1\}$ is the set of all n^{th} roots of 1 in $\mathbb{C}$)

(c) every proper subgroup of Z equals H_k for some $k \in \mathbb{Z}^+$ (in particular, every proper subgroup of Z is finite and cyclic)

(d) Z is not finitely generated.

19. A nontrivial abelian group A (written multiplicatively) is called *divisible* if for each element $a \in A$ and each nonzero integer k there is an element $x \in A$ such that $x^k = a$, i.e., each element has a kth root in A (in additive notation, each element is the kth multiple of some element of A).

(a) Prove that the additive group of rational numbers, $\mathbb{Q}$, is divisible.

(b) Prove that no finite abelian group is divisible.

20. Prove that if A and B are nontrivial abelian groups, then $A \times B$ is divisible if and only if both A and B are divisible groups.

2.5 THE LATTICE OF SUBGROUPS OF A GROUP

In this section we describe a graph associated with a group which depicts the relationships among its subgroups. This graph, called the lattice[2] of subgroups of the group, is a good way of "visualizing" a group — it certainly illuminates the structure of a group better than the group table. We shall be using lattice diagrams, or parts of them, to describe both specific groups and certain properties of general groups throughout the chapters on group theory. Moreover, the lattice of subgroups of a group will play an important role in Galois Theory.

The lattice of subgroups of a given finite group G is constructed as follows: plot all subgroups of G starting at the bottom with 1, ending at the top with G and, roughly speaking, with subgroups of larger order positioned higher on the page than those of smaller order. Draw paths upwards between subgroups using the rule that there will be a line upward from A to B if $A \le B$ and there are no subgroups properly between A and B. Thus if $A \le B$ there is a path (possibly many paths) upward from A to B passing through a chain of intermediate subgroups (and a path downward from B to A if $B \ge A$). The initial positioning of the subgroups on the page, which is, a priori, somewhat arbitrary, can often (with practice) be chosen to produce a simple picture. Notice that for any pair of subgroups H and K of G the unique smallest subgroup

---

[2]The term "lattice" has a precise mathematical meaning in terms of partially ordered sets.

which contains both of them, namely $\langle H, K \rangle$ (called the *join* of H and K), may be read off from the lattice as follows: trace paths upwards from H and K until a common subgroup A which contains H and K is reached (note that G itself always contains all subgroups so at least one such A exists). To ensure that $A = \langle H, K \rangle$ make sure there is no $A_1 \leq A$ (indicated by a downward path from A to A_1) with both H and K contained in A_1 (otherwise replace A with A_1 and repeat the process to see if $A_1 = \langle H, K \rangle$). By a symmetric process one can read off the largest subgroup of G which is contained in both H and K, namely their intersection (which is a subgroup by Proposition 8).

There are some limitations to this process, in particular it cannot be carried out per se for infinite groups. Even for finite groups of relatively small order, lattices can be quite complicated (see the book *Groups of Order 2^n, $n \leq 6$* by M. Hall and J. Senior, Macmillan, 1964, for some hair-raising examples). At the end of this section we shall describe how parts of a lattice may be drawn and used even for infinite groups.

Note that isomorphic groups have the same lattices (i.e., the same directed graphs). Nonisomorphic groups may also have identical lattices (this happens for two groups of order 16 — see the following exercises). Since the lattice of subgroups is only part of the data we shall carry in our descriptors of a group, this will not be a serious drawback (indeed, it might even be useful in seeing when two nonisomorphic groups have some common properties).

Examples

Except for the cyclic groups (Example 1) we have not proved that the following lattices are correct (e.g., contain all subgroups of the given group or have the right joins and intersections). For the moment we shall take these facts as given and, as we build up more theory in the course of the text, we shall assign as exercises the proofs that these are indeed correct.

(1) For $G = Z_n \cong \mathbb{Z}/n\mathbb{Z}$, by Theorem 7 the lattice of subgroups of G is the lattice of divisors of n (that is, the divisors of n are written on a page with n at the bottom, 1 at the top and paths upwards from a to b if $b \mid a$). Some specific examples for various values of n follow.

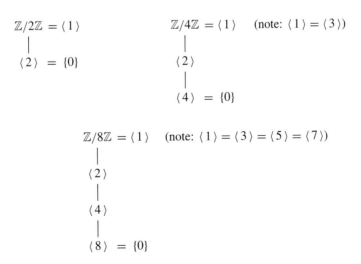

In general, if p is a prime, the lattice of $\mathbb{Z}/p^n\mathbb{Z}$ is

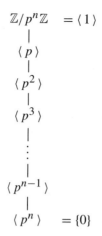

$$\mathbb{Z}/p^n\mathbb{Z} = \langle 1 \rangle$$
$$\langle p \rangle$$
$$\langle p^2 \rangle$$
$$\langle p^3 \rangle$$
$$\vdots$$
$$\langle p^{n-1} \rangle$$
$$\langle p^n \rangle = \{0\}$$

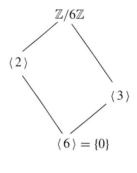

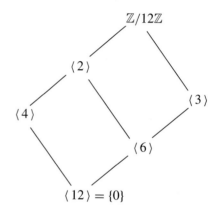

(2) The *Klein 4-group (Viergruppe)*, V_4, is the group of order 4 with multiplication table

·	1	a	b	c
1	1	a	b	c
a	a	1	c	b
b	b	c	1	a
c	c	b	a	1

and lattice

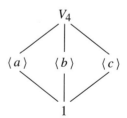

Note that V_4 is abelian and is not isomorphic to Z_4 (why?). We shall see that D_8 has an isomorphic copy of V_4 as a subgroup, so it will not be necessary to check that the associative law holds for the binary operation defined above.

(3) The lattice of S_3 is

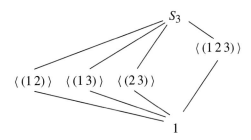

(4) Using our usual notation for $D_8 = \langle r, s \rangle$, the lattice of D_8 is

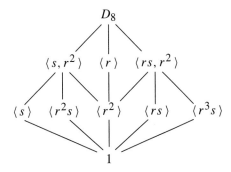

(5) The lattice of subgroups of Q_8 is

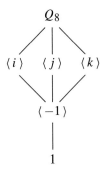

(6) The lattice of D_{16} is not a planar graph (cannot be drawn on a plane without lines crossing). One way of drawing it is

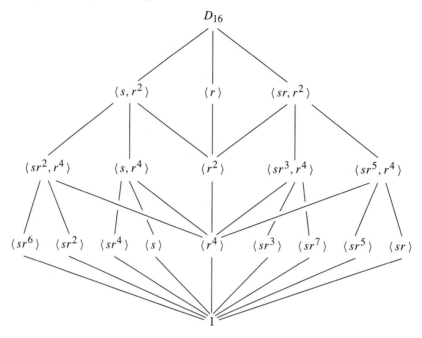

In many instances in both theoretical proofs and specific examples we shall be interested only in information concerning two (or some small number of) subgroups of a given group and their interrelationships. To depict these graphically we shall draw a *sublattice* of the entire group lattice which contains the relevant joins and intersections. An unbroken line in such a sublattice will not, in general, mean that there is no subgroup in between the endpoints of the line. These partial lattices for groups will also be used when we are dealing with infinite groups. For example, if we wished to discuss only the relationship between the subgroups $\langle sr^2, r^4 \rangle$ and $\langle r^2 \rangle$ of D_{16} we would draw the sublattice

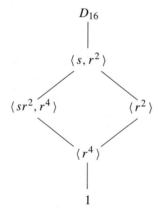

Note that $\langle s, r^2 \rangle$ and $\langle r^4 \rangle$ are precisely the join and intersection, respectively, of these two subgroups in D_{16}.

Finally, given the lattice of subgroups of a group, it is relatively easy to compute normalizers and centralizers. For example, in D_8 we can see that $C_{D_8}(s) = \langle s, r^2 \rangle$ because we first calculate that $r^2 \in C_{D_8}(s)$ (see Section 2). This proves $\langle s, r^2 \rangle \leq C_{D_8}(s)$ (note that an element always belongs to its own centralizer). The only subgroups which contain $\langle s, r^2 \rangle$ are that subgroup itself and all of D_8. We cannot have $C_{D_8}(s) = D_8$ because r does not commute with s (i.e., $r \notin C_{D_8}(s)$). This leaves only the claimed possibility for $C_{D_8}(s)$.

EXERCISES

1. Let H and K be subgroups of G. Exhibit all possible sublattices which show only G, 1, H, K and their joins and intersections. What distinguishes the different drawings?

2. In each of (a) to (d) list all subgroups of D_{16} that satisfy the given condition.
   (a) Subgroups that are contained in $\langle sr^2, r^4 \rangle$
   (b) Subgroups that are contained in $\langle sr^7, r^4 \rangle$
   (c) Subgroups that contain $\langle r^4 \rangle$
   (d) Subgroups that contain $\langle s \rangle$.

3. Show that the subgroup $\langle s, r^2 \rangle$ of D_8 is isomorphic to V_4.

4. Use the given lattice to find all pairs of elements that generate D_8 (there are 12 pairs).

5. Use the given lattice to find all elements $x \in D_{16}$ such that $D_{16} = \langle x, s \rangle$ (there are 8 such elements x).

6. Use the given lattices to help find the centralizers of every element in the following groups:
   (a) D_8 (b) Q_8 (c) S_3 (d) D_{16}.

7. Find the center of D_{16}.

8. In each of the following groups find the normalizer of each subgroup:
   (a) S_3 (b) Q_8.

9. Draw the lattices of subgroups of the following groups:
   (a) $\mathbb{Z}/16\mathbb{Z}$ (b) $\mathbb{Z}/24\mathbb{Z}$ (c) $\mathbb{Z}/48\mathbb{Z}$. [See Exercise 6 in Section 3.]

10. Classify groups of order 4 by proving that if $|G| = 4$ then $G \cong Z_4$ or $G \cong V_4$. [See Exercise 36, Section 1.1.]

11. Consider the group of order 16 with the following presentation:

$$QD_{16} = \langle \sigma, \tau \mid \sigma^8 = \tau^2 = 1, \ \sigma\tau = \tau\sigma^3 \rangle$$

(called the *quasidihedral* or *semidihedral* group of order 16). This group has three subgroups of order 8: $\langle \tau, \sigma^2 \rangle \cong D_8$, $\langle \sigma \rangle \cong Z_8$ and $\langle \sigma^2, \sigma\tau \rangle \cong Q_8$ and every proper subgroup is contained in one of these three subgroups. Fill in the missing subgroups in the lattice of all subgroups of the quasidihedral group on the following page, exhibiting each subgroup with at most two generators. (This is another example of a nonplanar lattice.)

The next three examples lead to two nonisomorphic groups that have the same lattice of subgroups.

12. The group $A = Z_2 \times Z_4 = \langle a, b \mid a^2 = b^4 = 1, \ ab = ba \rangle$ has order 8 and has three subgroups of order 4: $\langle a, b^2 \rangle \cong V_4$, $\langle b \rangle \cong Z_4$ and $\langle ab \rangle \cong Z_4$ and every proper

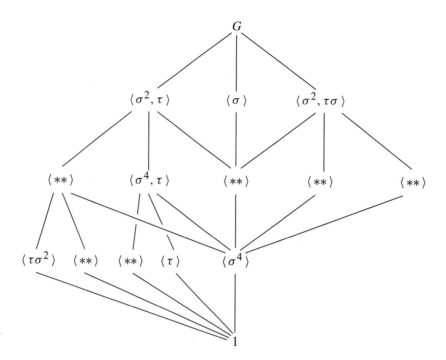

subgroup is contained in one of these three. Draw the lattice of all subgroups of A, giving each subgroup in terms of at most two generators.

13. The group $G = Z_2 \times Z_8 = \langle x, y \mid x^2 = y^8 = 1, \ xy = yx \rangle$ has order 16 and has three subgroups of order 8: $\langle x, y^2 \rangle \cong Z_2 \times Z_4$, $\langle y \rangle \cong Z_8$ and $\langle xy \rangle \cong Z_8$ and every proper subgroup is contained in one of these three. Draw the lattice of all subgroups of G, giving each subgroup in terms of at most two generators (cf. Exercise 12).

14. Let M be the group of order 16 with the following presentation:

$$\langle u, v \mid u^2 = v^8 = 1, \ vu = uv^5 \rangle$$

(sometimes called the *modular* group of order 16). It has three subgroups of order 8: $\langle u, v^2 \rangle$, $\langle v \rangle$ and $\langle uv \rangle$ and every proper subgroup is contained in one of these three. Prove that $\langle u, v^2 \rangle \cong Z_2 \times Z_4$, $\langle v \rangle \cong Z_8$ and $\langle uv \rangle \cong Z_8$. Show that the lattice of subgroups of M is the same as the lattice of subgroups of $Z_2 \times Z_8$ (cf. Exercise 13) but that these two groups are not isomorphic.

15. Describe the isomorphism type of each of the three subgroups of D_{16} of order 8.

16. Use the lattice of subgroups of the quasidihedral group of order 16 to show that every element of order 2 is contained in the proper subgroup $\langle \tau, \sigma^2 \rangle$ (cf. Exercise 11).

17. Use the lattice of subgroups of the modular group M of order 16 to show that the set $\{x \in M \mid x^2 = 1\}$ is a subgroup of M isomorphic to the Klein 4-group (cf. Exercise 14).

18. Use the lattice to help find the centralizer of every element of QD_{16} (cf. Exercise 11).

19. Use the lattice to help find $N_{D_{16}}(\langle s, r^4 \rangle)$.

20. Use the lattice of subgroups of QD_{16} (cf. Exercise 11) to help find the normalizers
    (a) $N_{QD_{16}}(\langle \tau\sigma \rangle)$ (b) $N_{QD_{16}}(\langle \tau, \sigma^4 \rangle)$.

72

CHAPTER 3

Quotient Groups and Homomorphisms

3.1 DEFINITIONS AND EXAMPLES

In this chapter we introduce the notion of a *quotient* group of a group G, which is another way of obtaining a "smaller" group from the group G and, as we did with subgroups, we shall use quotient groups to study the structure of G. The structure of the group G is reflected in the structure of the quotient groups and the subgroups of G. For example, we shall see that the lattice of subgroups for a *quotient* of G is reflected at the "top" (in a precise sense) of the lattice for G whereas the lattice for a *subgroup* of G occurs naturally at the "bottom." One can therefore obtain information about the group G by combining this information and we shall indicate how some classification theorems arise in this way.

The study of the quotient groups of G is essentially equivalent to the study of the homomorphisms of G, i.e., the maps of the group G to another group which respect the group structures. If φ is a homomorphism from G to a group H recall that the *fibers* of φ are the sets of elements of G projecting to single elements of H, which we can represent pictorially in Figure 1, where the vertical line in the box above a point a represents the fiber of φ over a.

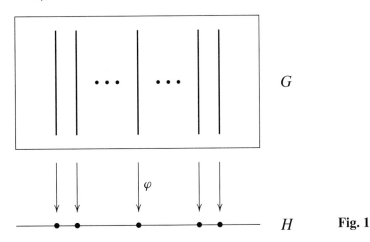

Fig. 1

The group operation in H provides a way to multiply two elements in the image of φ (i.e., two elements on the horizontal line in Figure 1). This suggests a natural multiplication of the *fibers* lying above these two points making *the set of fibers into a group*: if X_a is the fiber above a and X_b is the fiber above b then the product of X_a with X_b is defined to be the fiber X_{ab} above the product ab, i.e., $X_a X_b = X_{ab}$. This multiplication is associative since multiplication is associative in H, the identity is the fiber over the identity of H, and the inverse of the fiber over a is the fiber over a^{-1}, as is easily checked from the definition. For example, the associativity is proved as follows: $(X_a X_b)X_c = (X_{ab})X_c = X_{(ab)c}$ and $X_a(X_b X_c) = X_a(X_{bc}) = X_{a(bc)}$. Since $(ab)c = a(bc)$ in H, $(X_a X_b)X_c = X_a(X_b X_c)$. Roughly speaking, the group G is partitioned into pieces (the fibers) and these pieces themselves have the structure of a group, called a *quotient* group of G (a formal definition follows the example below).

Since the multiplication of fibers is defined from the multiplication in H, by construction the quotient group with this multiplication is naturally isomorphic to the image of G under the homomorphism φ (fiber X_a is identified with its image a in H).

Example

Let $G = \mathbb{Z}$, let $H = Z_n = \langle x \rangle$ be the cyclic group of order n and define $\varphi : \mathbb{Z} \to Z_n$ by $\varphi(a) = x^a$. Since

$$\varphi(a + b) = x^{a+b} = x^a x^b = \varphi(a)\varphi(b)$$

it follows that φ is a homomorphism (note that the operation in $\mathbb{Z}$ is addition and the operation in Z_n is multiplication). Note also that φ is surjective. The fiber of φ over x^a is then

$$\varphi^{-1}(x^a) = \{m \in \mathbb{Z} \mid x^m = x^a\} = \{m \in \mathbb{Z} \mid x^{m-a} = 1\}$$
$$= \{m \in \mathbb{Z} \mid n \text{ divides } m - a\} \qquad \text{(by Proposition 2.3)}$$
$$= \{m \in \mathbb{Z} \mid m \equiv a \pmod{n}\} = \bar{a},$$

i.e., the fibers of φ are precisely the residue classes modulo n. Figure 1 here becomes:

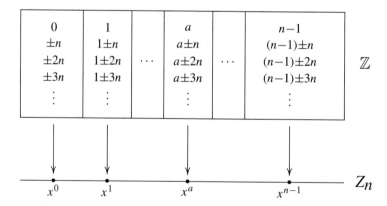

Fig. 2

The multiplication in Z_n is just $x^a x^b = x^{a+b}$. The corresponding fibers are $\bar{a}$, $\bar{b}$, and $\overline{a+b}$, so the corresponding group operation for the fibers is $\bar{a} \cdot \bar{b} = \overline{a+b}$. This is just the group $\mathbb{Z}/n\mathbb{Z}$ under addition, a group isomorphic to the image of φ (all of Z_n).

The identity of this group (the fiber above the identity in Z_n) consists of all the multiples of n in $\mathbb{Z}$, namely $n\mathbb{Z}$, a *subgroup* of $\mathbb{Z}$, and the remaining fibers are just translates, $a + n\mathbb{Z}$, of this subgroup. The group operation can also be defined directly by taking *representatives* from these fibers, adding these representatives in $\mathbb{Z}$ and taking the fiber containing this sum (this was the original definition of the group $\mathbb{Z}/n\mathbb{Z}$). From a computational point of view computing the product of $\bar{a}$ and $\bar{b}$ by simply adding representatives a and b is much easier than first computing the image of these fibers under φ (namely, x^a and x^b), multiplying these in H (obtaining x^{a+b}) and then taking the fiber over this product.

We first consider some basic properties of homomorphisms and their fibers. The fiber of a homomorphism $\varphi : G \to H$ lying above the identity of H is given a name:

Definition. If φ is a homomorphism $\varphi : G \to H$, the *kernel* of φ is the set

$$\{g \in G \mid \varphi(g) = 1\}$$

and will be denoted by ker φ (here 1 is the identity of H).

Proposition 1. Let G and H be groups and let $\varphi : G \to H$ be a homomorphism.
  (1) $\varphi(1_G) = 1_H$, where 1_G and 1_H are the identities of G and H, respectively.
  (2) $\varphi(g^{-1}) = \varphi(g)^{-1}$ for all $g \in G$.
  (3) $\varphi(g^n) = \varphi(g)^n$ for all $n \in \mathbb{Z}$.
  (4) ker φ is a subgroup of G.
  (5) im (φ), the image of G under φ, is a subgroup of H.

Proof: (1) Since $\varphi(1_G) = \varphi(1_G 1_G) = \varphi(1_G)\varphi(1_G)$, the cancellation laws show that (1) holds.
  (2) $\varphi(1_G) = \varphi(gg^{-1}) = \varphi(g)\varphi(g^{-1})$ and, by part (1), $\varphi(1_G) = 1_H$, hence

$$1_H = \varphi(g)\varphi(g^{-1}).$$

Multiplying both sides on the left by $\varphi(g)^{-1}$ and simplifying gives (2).
  (3) This is an easy exercise in induction for $n \in \mathbb{Z}^+$. By part (2), conclusion (3) holds for negative values of n as well.
  (4) Since $1_G \in$ ker φ, the kernel of φ is not empty. Let $x, y \in$ ker φ, that is $\varphi(x) = \varphi(y) = 1_H$. Then

$$\varphi(xy^{-1}) = \varphi(x)\varphi(y^{-1}) = \varphi(x)\varphi(y)^{-1} = 1_H 1_H^{-1} = 1_H$$

that is, $xy^{-1} \in$ ker φ. By the subgroup criterion, ker $\varphi \leq G$.
  (5) Since $\varphi(1_G) = 1_H$, the identity of H lies in the image of φ, so im(φ) is nonempty. If x and y are in im(φ), say $x = \varphi(a)$, $y = \varphi(b)$, then $y^{-1} = \varphi(b^{-1})$ by (2) so that $xy^{-1} = \varphi(a)\varphi(b^{-1}) = \varphi(ab^{-1})$ since φ is a homomorphism. Hence also xy^{-1} is in the image of φ, so im(φ) is a subgroup of H by the subgroup criterion.

We can now define some terminology associated with quotient groups.

Definition. Let $\varphi : G \to H$ be a homomorphism with kernel K. The *quotient group* or *factor group*, G/K (read G *modulo* K or simply G *mod* K), is the group whose elements are the fibers of φ with group operation defined above: namely if X is the fiber above a and Y is the fiber above b then the product of X with Y is defined to be the fiber above the product ab.

The notation emphasizes the fact that the kernel K is a *single element* in the group G/K and we shall see below (Proposition 2) that, as in the case of $\mathbb{Z}/n\mathbb{Z}$ above, the other elements of G/K are just the "translates" of the kernel K. Hence we may think of G/K as being obtained by collapsing or "dividing out" by K (or more precisely, by equivalence modulo K). This explains why G/K is referred to as a "quotient" group.

The definition of the quotient group G/K above requires the map φ explicitly, since the multiplication of the fibers is performed by first projecting the fibers to H via φ, multiplying in H and then determining the fiber over this product. Just as for $\mathbb{Z}/n\mathbb{Z}$ above, it is also possible to define the multiplication of fibers directly in terms of *representatives* from the fibers. This is computationally simpler and the map φ does not enter explicitly. We first show that the fibers of a homomorphism can be expressed in terms of the kernel of the homomorphism just as in the example above (where the kernel was $n\mathbb{Z}$ and the fibers were translates of the form $a + n\mathbb{Z}$).

Proposition 2. Let $\varphi : G \to H$ be a homomorphism of groups with kernel K. Let $X \in G/K$ be the fiber above a, i.e., $X = \varphi^{-1}(a)$. Then
  (1) For any $u \in X$, $X = \{uk \mid k \in K\}$
  (2) For any $u \in X$, $X = \{ku \mid k \in K\}$.

Proof: We prove (1) and leave the proof of (2) as an exercise. Let $u \in X$ so, by definition of X, $\varphi(u) = a$. Let

$$uK = \{uk \mid k \in K\}.$$

We first prove $uK \subseteq X$. For any $k \in K$,

$$\varphi(uk) = \varphi(u)\varphi(k) \qquad \text{(since } \varphi \text{ is a homomorphism)}$$
$$= \varphi(u)1 \qquad \text{(since } k \in \ker \varphi)$$
$$= a,$$

that is, $uk \in X$. This proves $uK \subseteq X$. To establish the reverse inclusion suppose $g \in X$ and let $k = u^{-1}g$. Then

$$\varphi(k) = \varphi(u^{-1})\varphi(g) = \varphi(u)^{-1}\varphi(g) \qquad \text{(by Proposition 1)}$$
$$= a^{-1}a = 1.$$

Thus $k \in \ker \varphi$. Since $k = u^{-1}g$, $g = uk \in uK$, establishing the inclusion $X \subseteq uK$. This proves (1).

The sets arising in Proposition 2 to describe the fibers of a homomorphism φ are defined for *any* subgroup K of G, not necessarily the kernel of some homomorphism (we shall determine necessary and sufficient conditions for a subgroup to be such a kernel shortly) and are given a name:

76

Definition. For any $N \leq G$ and any $g \in G$ let

$$gN = \{gn \mid n \in N\} \quad \text{and} \quad Ng = \{ng \mid n \in N\}$$

called respectively a *left coset* and a *right coset* of N in G. Any element of a coset is called a *representative* for the coset.

We have already seen in Proposition 2 that if N is the kernel of a homomorphism and g_1 is any representative for the coset gN then $g_1 N = gN$ (and if $g_1 \in Ng$ then $Ng_1 = Ng$). We shall see that this fact is valid for arbitrary subgroups N in Proposition 4 below, which explains the terminology of a *representative*.

If G is an additive group we shall write $g + N$ and $N + g$ for the left and right cosets of N in G with representative g, respectively. In general we can think of the left coset, gN, of N in G as the left translate of N by g. (The reader may wish to review Exercise 18 of Section 1.7 which proves that the right cosets of N in G are precisely the orbits of N acting on G by left multiplication.)

In terms of this definition, Proposition 2 shows that the fibers of a homomorphism are the left cosets of the kernel (and also the right cosets of the kernel), i.e., the elements of the quotient G/K are the left cosets gK, $g \in G$. In the example of $\mathbb{Z}/n\mathbb{Z}$ the multiplication in the quotient group could also be defined in terms of representatives for the cosets. The following result shows the same result is true for G/K in general (provided we know that K is the kernel of some homomorphism), namely that the product of two left cosets X and Y in G/K is computed by choosing any representative u of X, any representative v of Y, multiplying u and v in G and forming the coset $(uv)K$.

Theorem 3. Let G be a group and let K be the kernel of some homomorphism from G to another group. Then the set whose elements are the left cosets of K in G with operation defined by

$$uK \circ vK = (uv)K$$

forms a group, G/K. In particular, this operation is well defined in the sense that if u_1 is any element in uK and v_1 is any element in vK, then $u_1 v_1 \in uvK$, i.e., $u_1 v_1 K = uvK$ so that the multiplication does not depend on the choice of representatives for the cosets. The same statement is true with "right coset" in place of "left coset."

Proof: Let $X, Y \in G/K$ and let $Z = XY$ in G/K, so that by Proposition 2(1) X, Y and Z are (left) cosets of K. By assumption, K is the kernel of some homomorphism $\varphi : G \to H$ so $X = \varphi^{-1}(a)$ and $Y = \varphi^{-1}(b)$ for some $a, b \in H$. By definition of the operation in G/K, $Z = \varphi^{-1}(ab)$. Let u and v be arbitrary representatives of X, Y, respectively, so that $\varphi(u) = a$, $\varphi(v) = b$ and $X = uK$, $Y = vK$. We must show $uv \in Z$. Now

$$uv \in Z \quad \Leftrightarrow \quad uv \in \varphi^{-1}(ab)$$
$$\Leftrightarrow \quad \varphi(uv) = ab$$
$$\Leftrightarrow \quad \varphi(u)\varphi(v) = ab.$$

Since the latter equality does hold, $uv \in Z$ hence Z is the (left) coset uvK. (Exercise 2 below shows conversely that every $z \in Z$ can be written as uv, for some $u \in X$ and $v \in Y$.) This proves that the product of X with Y is the coset uvK for any choice of representatives $u \in X$, $v \in Y$ completing the proof of the first statements of the theorem. The last statement in the theorem follows immediately since, by Proposition 2, $uK = Ku$ and $vK = Kv$ for all u and v in G.

In terms of Figure 1, the multiplication in G/K via representatives can be pictured as in the following Figure 3.

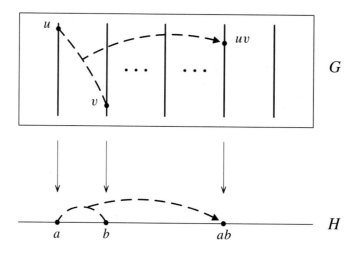

Fig. 3

We emphasize the fact that *the multiplication is independent of the particular representatives chosen*. Namely, the product (or sum, if the group is written additively) of two cosets X and Y is the coset uvK containing the product uv where u and v are *any* representatives for the cosets X and Y, respectively. This process of considering only the coset containing an element, or "reducing mod K" is the same as what we have been doing, in particular, in $\mathbb{Z}/n\mathbb{Z}$. A useful notation for denoting the coset uK containing a representative u is $\bar{u}$. With this notation (which we introduced in the Preliminaries in dealing with $\mathbb{Z}/n\mathbb{Z}$), the quotient group G/K is denoted $\overline{G}$ and the product of elements $\bar{u}$ and $\bar{v}$ is simply the coset containing uv, i.e., $\overline{uv}$. This notation also reinforces the fact that the cosets uK in G/K are *elements* $\bar{u}$ in G/K.

Examples

    (1) The first example in this chapter of the homomorphism φ from $\mathbb{Z}$ to Z_n has fibers the left (and also the right) cosets $a + n\mathbb{Z}$ of the kernel $n\mathbb{Z}$. Theorem 3 proves that these cosets form a group under addition of representatives, namely $\mathbb{Z}/n\mathbb{Z}$, which explains the notation for this group. The group is naturally isomorphic to its image under φ, so we recover the isomorphism $\mathbb{Z}/n\mathbb{Z} \cong Z_n$ of Chapter 2.

    (2) If $\varphi : G \to H$ is an *isomorphism*, then $K = 1$, the fibers of φ are the singleton subsets of G and so $G/1 \cong G$.

(3) Let G be any group, let $H = 1$ be the group of order 1 and define $\varphi : G \to H$ by $\varphi(g) = 1$, for all $g \in G$. It is immediate that φ is a homomorphism. This map is called the *trivial homomorphism*. Note that in this case $\ker \varphi = G$ and G/G is a group with the single element, G, i.e., $G/G \cong Z_1 = \{1\}$.

(4) Let $G = \mathbb{R}^2$ (operation vector addition), let $H = \mathbb{R}$ (operation addition) and define $\varphi : \mathbb{R}^2 \to \mathbb{R}$ by $\varphi((x, y)) = x$. Thus φ is projection onto the x-axis. We show φ is a homomorphism:

$$\varphi((x_1, y_1) + (x_2, y_2)) = \varphi((x_1 + x_2, y_1 + y_2))$$
$$= x_1 + x_2 = \varphi((x_1, y_1)) + \varphi((x_2, y_2)).$$

Now

$$\ker \varphi = \{(x, y) \mid \varphi((x, y)) = 0\}$$
$$= \{(x, y) \mid x = 0\} = \text{the } y\text{-axis.}$$

Note that $\ker \varphi$ is indeed a subgroup of $\mathbb{R}^2$ and that the fiber of φ over $a \in \mathbb{R}$ is the translate of the y-axis by a, i.e., the line $x = a$. This is also the left (and the right) coset of the kernel with representative $(a, 0)$ (or any other representative point projecting to a):

$$\overline{(a, 0)} = (a, 0) + y\text{-axis.}$$

Hence Figure 1 in this example becomes

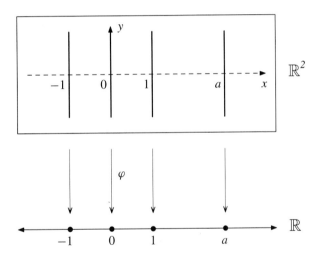

Fig. 4

The group operation (written additively here) can be described either by using the map φ: the sum of the line $(x = a)$ and the line $(x = b)$ is the line $(x = a + b)$; or directly in terms of coset representatives: the sum of the vertical line containing the point (a, y_1) and the vertical line containing the point (b, y_2) is the vertical line containing the point $(a + b, y_1 + y_2)$. Note in particular that the choice of representatives of these vertical lines is not important (i.e., the y-coordinates are not important).

(5) (An example where the group G is non-abelian.) Let $G = Q_8$ and let $H = V_4$ be the Klein 4-group (Section 2.5, Example 2). Define $\varphi : Q_8 \to V_4$ by

$$\varphi(\pm 1) = 1, \qquad \varphi(\pm i) = a, \qquad \varphi(\pm j) = b, \qquad \varphi(\pm k) = c.$$

The check that φ is a homomorphism is left as an exercise — relying on symmetry minimizes the work in showing $\varphi(xy) = \varphi(x)\varphi(y)$ for all x and y in Q_8. It is clear that φ is surjective and that ker $\varphi = \{\pm 1\}$. One might think of φ as an "absolute value" function on Q_8 so the fibers of φ are the sets $E = \{\pm 1\}$, $A = \{\pm i\}$, $B = \{\pm j\}$ and $C = \{\pm k\}$, which are collapsed to 1, a, b, and c respectively in $Q_8/\langle \pm 1 \rangle$ and these are the left (and also the right) cosets of ker φ (for example, $A = i \cdot \text{ker } \varphi = \{i, -i\} = \text{ker } \varphi \cdot i$).

By Theorem 3, if we are given a subgroup K of a group G which we know is the kernel of some homomorphism, we may define the quotient G/K without recourse to the homomorphism by the multiplication $uKvK = uvK$. This raises the question of whether it is possible to define the quotient group G/N similarly for *any* subgroup N of G. The answer is no in general since this multiplication is not in general well defined (cf. Proposition 5 later). In fact we shall see that it is possible to define the structure of a group on the cosets of N *if and only if* N is the kernel of some homomorphism (Proposition 7). We shall also give a criterion to determine when a subgroup N is such a kernel — this is the notion of a *normal* subgroup and we shall consider non-normal subgroups in subsequent sections.

We first show that the cosets of an arbitrary subgroup of G partition G (i.e., their union is all of G and distinct cosets have trivial intersection).

Proposition 4. Let N be any subgroup of the group G. The set of left cosets of N in G form a partition of G. Furthermore, for all u, $v \in G$, $uN = vN$ if and only if $v^{-1}u \in N$ and in particular, $uN = vN$ if and only if u and v are representatives of the same coset.

Proof: First of all note that since N is a subgroup of G, $1 \in N$. Thus $g = g \cdot 1 \in gN$ for all $g \in G$, i.e.,

$$G = \bigcup_{g \in G} gN.$$

To show that distinct left cosets have empty intersection, suppose $uN \cap vN \neq \emptyset$. We show $uN = vN$. Let $x \in uN \cap vN$. Write

$$x = un = vm, \qquad \text{for some } n, m \in N.$$

In the latter equality multiply both sides on the right by n^{-1} to get

$$u = vmn^{-1} = vm_1, \qquad \text{where } m_1 = mn^{-1} \in N.$$

Now for any element ut of uN $(t \in N)$,

$$ut = (vm_1)t = v(m_1 t) \in vN.$$

This proves $uN \subseteq vN$. By interchanging the roles of u and v one obtains similarly that $vN \subseteq uN$. Thus two cosets with nonempty intersection coincide.

By the first part of the proposition, $uN = vN$ if and only if $u \in vN$ if and only if $u = vn$, for some $n \in N$ if and only if $v^{-1}u \in N$, as claimed. Finally, $v \in uN$ is equivalent to saying v is a representative for uN, hence $uN = vN$ if and only if u and v are representatives for the same coset (namely the coset $uN = vN$).

80

Proposition 5. Let G be a group and let N be a subgroup of G.

(1) The operation on the set of left cosets of N in G described by

$$uN \cdot vN = (uv)N$$

is well defined if and only if $gng^{-1} \in N$ for all $g \in G$ and all $n \in N$.

(2) If the above operation is well defined, then it makes the set of left cosets of N in G into a group. In particular the identity of this group is the coset $1N$ and the inverse of gN is the coset $g^{-1}N$ i.e., $(gN)^{-1} = g^{-1}N$.

Proof: (1) Assume first that this operation is well defined, that is, for all $u, v \in G$,

$$\text{if } u, u_1 \in uN \text{ and } v, v_1 \in vN \quad \text{then} \quad uvN = u_1 v_1 N.$$

Let g be an arbitrary element of G and let n be an arbitrary element of N. Letting $u = 1, u_1 = n$ and $v = v_1 = g^{-1}$ and applying the assumption above we deduce that

$$1g^{-1}N = ng^{-1}N \quad \text{i.e.,} \quad g^{-1}N = ng^{-1}N.$$

Since $1 \in N$, $ng^{-1} \cdot 1 \in ng^{-1}N$. Thus $ng^{-1} \in g^{-1}N$, hence $ng^{-1} = g^{-1}n_1$, for some $n_1 \in N$. Multiplying both sides on the left by g gives $gng^{-1} = n_1 \in N$, as claimed.

Conversely, assume $gng^{-1} \in N$ for all $g \in G$ and all $n \in N$. To prove the operation stated above is well defined let $u, u_1 \in uN$ and $v, v_1 \in vN$. We may write

$$u_1 = un \text{ and } v_1 = vm, \quad \text{for some } n, m \in N.$$

We must prove that $u_1 v_1 \in uvN$:

$$u_1 v_1 = (un)(vm) = u(vv^{-1})nvm$$

$$= (uv)(v^{-1}nv)m = (uv)(n_1 m),$$

where $n_1 = v^{-1}nv = (v^{-1})n(v^{-1})^{-1}$ is an element of N by assumption. Now N is closed under products, so $n_1 m \in N$. Thus

$$u_1 v_1 = (uv)n_2, \quad \text{for some } n_2 \in N.$$

Thus the left cosets uvN and $u_1 v_1 N$ contain the common element $u_1 v_1$. By the preceding proposition they are equal. This proves that the operation is well defined.

(2) If the operation on cosets is well defined the group axioms are easy to check and are induced by their validity in G. For example, the associative law holds because for all $u, v, w \in G$,

$$(uN)(vNwN) = uN(vwN)$$

$$= u(vw)N$$

$$= (uv)wN = (uNvN)(wN),$$

since $u(vw) = (uv)w$ in G. The identity in G/N is the coset $1N$ and the inverse of gN is $g^{-1}N$ as is immediate from the definition of the multiplication.

As indicated before, the subgroups N satisfying the condition in Proposition 5 for which there is a natural group structure on the quotient G/N are given a name:

Definition. The element gng^{-1} is called the *conjugate* of $n \in N$ by g. The set $gNg^{-1} = \{gng^{-1} \mid n \in N\}$ is called the *conjugate* of N by g. The element g is said to *normalize N* if $gNg^{-1} = N$. A subgroup N of a group G is called *normal* if every element of G normalizes N, i.e., if $gNg^{-1} = N$ for all $g \in G$. If N is a normal subgroup of G we shall write $N \trianglelefteq G$.

Note that the structure of G is reflected in the structure of the quotient G/N when N is a normal subgroup (for example, the associativity of the multiplication in G/N is induced from the associativity in G and inverses in G/N are induced from inverses in G). We shall see more of the relationship of G to its quotient G/N when we consider the Isomorphism Theorems later in Section 3.

We summarize our results above as Theorem 6.

Theorem 6. Let N be a subgroup of the group G. The following are equivalent:
    (1) $N \trianglelefteq G$
    (2) $N_G(N) = G$ (recall $N_G(N)$ is the normalizer in G of N)
    (3) $gN = Ng$ for all $g \in G$
    (4) the operation on left cosets of N in G described in Proposition 5 makes the set of left cosets into a group
    (5) $gNg^{-1} \subseteq N$ for all $g \in G$.

Proof: We have already done the hard equivalences; the others are left as exercises.

As a practical matter, one tries to minimize the computations necessary to determine whether a given subgroup N is normal in a group G. In particular, one tries to avoid as much as possible the computation of all the conjugates gng^{-1} for $n \in N$ and $g \in G$. For example, the elements of N itself normalize N since N is a subgroup. Also, if one has a set of *generators* for N, it suffices to check that all conjugates of these generators lie in N to prove that N is a normal subgroup (this is because the conjugate of a product is the product of the conjugates and the conjugate of the inverse is the inverse of the conjugate) — this is Exercise 26 later. Similarly, if generators for G are also known, then it suffices to check that these generators for G normalize N. In particular, if generators for *both* N and G are known, this reduces the calculations to a small number of conjugations to check. If N is a *finite* group then it suffices to check that the conjugates of a set of generators for N by a set of generators for G are again elements of N (Exercise 29). Finally, it is often possible to prove directly that $N_G(N) = G$ without excessive computations (some examples appear in the next section), again proving that N is a normal subgroup of G without mindlessly computing all possible conjugates gng^{-1}.

We now prove that the normal subgroups are precisely the same as the kernels of homomorphisms considered earlier.

Proposition 7. A subgroup N of the group G is normal if and only if it is the kernel of some homomorphism.

Proof: If N is the kernel of the homomorphism φ, then Proposition 2 shows that the left cosets of N are the same as the right cosets of N (and both are the fibers of the map φ). By (3) of Theorem 6, N is then a normal subgroup. (Another direct proof of this from the definition of normality for N is given in the exercises).

Conversely, if $N \trianglelefteq G$, let $H = G/N$ and define $\pi : G \to G/N$ by

$$\pi(g) = gN \qquad \text{for all } g \in G.$$

By definition of the operation in G/N,

$$\pi(g_1 g_2) = (g_1 g_2)N = g_1 N g_2 N = \pi(g_1)\pi(g_2).$$

This proves π is a homomorphism. Now

$$\begin{aligned}
\ker \pi &= \{g \in G \mid \pi(g) = 1N\} \\
&= \{g \in G \mid gN = 1N\} \\
&= \{g \in G \mid g \in N\} = N.
\end{aligned}$$

Thus N is the kernel of the homomorphism π.

The homomorphism π constructed above demonstrating the normal subgroup N as the kernel of a homomorphism is given a name:

Definition. Let $N \trianglelefteq G$. The homomorphism $\pi : G \to G/N$ defined by $\pi(g) = gN$ is called the *natural projection (homomorphism)*[1] of G onto G/N. If $\overline{H} \le G/N$ is a subgroup of G/N, the *complete preimage* of $\overline{H}$ in G is the preimage of $\overline{H}$ under the natural projection homomorphism.

The complete preimage of a subgroup of G/N is a subgroup of G (cf. Exercise 1) which contains the subgroup N since these are the elements which map to the identity $\overline{1} \in \overline{H}$. We shall see in the Isomorphism Theorems in Section 3 that there is a natural correspondence between the subgroups of G that contain N and the subgroups of the quotient G/N.

We now have an "internal" criterion which determines precisely when a subgroup N of a given group G is the kernel of some homomorphism, namely,

$$N_G(N) = G.$$

We may thus think of the normalizer of a subgroup N of G as being a measure of "how close" N is to being a normal subgroup (this explains the choice of name for this subgroup). Keep in mind that the property of being normal is an *embedding* property, that is, it depends on the relation of N to G, not on the internal structure of N itself (the same group N may be a normal subgroup of G but not be normal in a larger group containing G).

We began the discussion of quotient groups with the existence of a homomorphism φ of G to H and showed the kernel of this homomorphism is a normal subgroup N of G and the quotient G/N (defined in terms of fibers originally) is naturally isomorphic

---

[1]The word "natural" has a precise mathematical meaning in the theory of categories; for our purposes we use the term to indicate that the definition of this homomorphism is a "coordinate free" projection i.e., is described only in terms of the elements themselves, not in terms of generators for G or N (cf. Appendix II).

to the image of G under φ in H. Conversely, if $N \trianglelefteq G$, we can find a group H (namely, G/N) and a homomorphism $\pi : G \to H$ such that $\ker \pi = N$ (namely, the natural projection). The study of homomorphic images of G (i.e., the images of homomorphisms from G into other groups) is thus equivalent to the study of quotient groups of G and we shall use homomorphisms to produce normal subgroups and vice versa.

We developed the theory of quotient groups by way of homomorphisms rather than simply defining the notion of a normal subgroup and its associated quotient group to emphasize the fact that the *elements* of the quotient are *subsets* (the fibers or cosets of the kernel N) of the original group G. The visualization in Figure 1 also emphasizes that N (and its cosets) are projected (or collapsed) onto single elements in the quotient G/N. Computations in the quotient group G/N are performed by taking *representatives* from the various cosets involved.

Some examples of normal subgroups and their associated quotients follow.

Examples

Let G be a group.
(1) The subgroups 1 and G are always normal in G; $G/1 \cong G$ and $G/G \cong 1$.
(2) If G is an *abelian* group, *any* subgroup N of G is normal because for all $g \in G$ and all $n \in N$,
$$gng^{-1} = gg^{-1}n = n \in N.$$

Note that it is important that G be abelian, not just that N be abelian. The structure of G/N may vary as we take different subgroups N of G. For instance, if $G = \mathbb{Z}$, then every subgroup N of G is cyclic:
$$N = \langle n \rangle = \langle -n \rangle = n\mathbb{Z}, \qquad \text{for some } n \in \mathbb{Z}$$
and $G/N = \mathbb{Z}/n\mathbb{Z}$ is a cyclic group with generator $\bar{1} = 1 + n\mathbb{Z}$ (note that 1 is a generator for G).

Suppose now that $G = Z_k$ is the cyclic group of order k. Let x be a generator of G and let $N \leq G$. By Theorem 2.7(1), $N = \langle x^d \rangle$, where d is the smallest power of x which lies in N. Now
$$G/N = \{gN \mid g \in G\} = \{x^\alpha N \mid \alpha \in \mathbb{Z}\}$$
and since $x^\alpha N = (xN)^\alpha$ (see Exercise 4 below), it follows that
$$G/N = \langle xN \rangle \qquad \text{i.e., } G/N \text{ is cyclic with } xN \text{ as a generator.}$$

By Exercise 5 below, the order of xN in G/N equals d. By Theorem 2.7(3), $d = \dfrac{|G|}{|N|}$. In summary,

quotient groups of a cyclic group are cyclic

and the image of a generator g for G is a generator $\bar{g}$ for the quotient. If in addition G is a *finite* cyclic group and $N \leq G$, then $|G/N| = \dfrac{|G|}{|N|}$ gives a formula for the order of the quotient group.

(3) If $N \leq Z(G)$, then $N \trianglelefteq G$ because for all $g \in G$ and all $n \in N$, $gng^{-1} = n \in N$, generalizing the previous example (where the center $Z(G)$ is all of G). Thus, in particular, $Z(G) \trianglelefteq G$. The subgroup $\langle -1 \rangle$ of Q_8 was previously seen to be the kernel of a homomorphism but since $\langle -1 \rangle = Z(Q_8)$ we obtain normality of this subgroup

now in another fashion. We already saw that $Q_8/\langle -1 \rangle \cong V_4$. The discussion for D_8 in the next paragraph could be applied equally well to Q_8 to give an independent identification of the isomorphism type of the quotient.

Let $G = D_8$ and let $Z = \langle r^2 \rangle = Z(D_8)$. Since $Z = \{1, r^2\}$, each coset, gZ, consists of the two element set $\{g, gr^2\}$. Since these cosets partition the 8 elements of D_8 into pairs, there must be 4 (disjoint) left cosets of Z in D_8:

$$\bar{1} = 1Z, \quad \bar{r} = rZ, \quad \bar{s} = sZ, \quad \text{and} \quad \overline{rs} = rsZ.$$

Now by the classification of groups of order 4 (Exercise 10, Section 2.5) we know that $D_8/Z(D_8) \cong Z_4$ or V_4. To determine which of these two is correct (i.e., determine the isomorphism type of the quotient) simply observe that

$$(\bar{r})^2 = r^2 Z = 1Z = \bar{1}$$
$$(\bar{s})^2 = s^2 Z = 1Z = \bar{1}$$
$$(\overline{rs})^2 = (rs)^2 Z = 1Z = \bar{1}$$

so every nonidentity element in D_8/Z has order 2. In particular there is no element of order 4 in the quotient, hence D_8/Z is not cyclic so $D_8/Z(D_8) \cong V_4$.

EXERCISES

Let G and H be groups.

1. Let $\varphi : G \to H$ be a homomorphism and let E be a subgroup of H. Prove that $\varphi^{-1}(E) \le G$ (i.e., the preimage or pullback of a subgroup under a homomorphism is a subgroup). If $E \trianglelefteq H$ prove that $\varphi^{-1}(E) \trianglelefteq G$. Deduce that $\ker \varphi \trianglelefteq G$.

2. Let $\varphi : G \to H$ be a homomorphism of groups with kernel K and let $a, b \in \varphi(G)$. Let $X \in G/K$ be the fiber above a and let Y be the fiber above b, i.e., $X = \varphi^{-1}(a)$, $Y = \varphi^{-1}(b)$. Fix an element u of X (so $\varphi(u) = a$). Prove that if $XY = Z$ in the quotient group G/K and w is any member of Z, then there is some $v \in Y$ such that $uv = w$. [Show $u^{-1}w \in Y$.]

3. Let A be an abelian group and let B be a subgroup of A. Prove that A/B is abelian. Give an example of a non-abelian group G containing a proper normal subgroup N such that G/N is abelian.

4. Prove that in the quotient group G/N, $(gN)^\alpha = g^\alpha N$ for all $\alpha \in \mathbb{Z}$.

5. Use the preceding exercise to prove that the order of the element gN in G/N is n, where n is the smallest positive integer such that $g^n \in N$ (and gN has infinite order if no such positive integer exists). Give an example to show that the order of gN in G/N may be strictly smaller than the order of g in G.

6. Define $\varphi : \mathbb{R}^\times \to \{\pm 1\}$ by letting $\varphi(x)$ be x divided by the absolute value of x. Describe the fibers of φ and prove that φ is a homomorphism.

7. Define $\pi : \mathbb{R}^2 \to \mathbb{R}$ by $\pi((x, y)) = x + y$. Prove that π is a surjective homomorphism and describe the kernel and fibers of π geometrically.

8. Let $\varphi : \mathbb{R}^\times \to \mathbb{R}^\times$ be the map sending x to the absolute value of x. Prove that φ is a homomorphism and find the image of φ. Describe the kernel and the fibers of φ.

9. Define $\varphi : \mathbb{C}^\times \to \mathbb{R}^\times$ by $\varphi(a + bi) = a^2 + b^2$. Prove that φ is a homomorphism and find the image of φ. Describe the kernel and the fibers of φ geometrically (as subsets of the plane).

10. Let $\varphi : \mathbb{Z}/8\mathbb{Z} \to \mathbb{Z}/4\mathbb{Z}$ by $\varphi(\bar{a}) = \bar{a}$. Show that this is a well defined, surjective homomorphism and describe its fibers and kernel explicitly (showing that φ is well defined involves the fact that $\bar{a}$ has a different meaning in the domain and range of φ).

11. Let F be a field and let $G = \{\begin{pmatrix} a & b \\ 0 & c \end{pmatrix} \mid a, b, c \in F,\ ac \neq 0\} \leq GL_2(F)$.

   (a) Prove that the map $\varphi : \begin{pmatrix} a & b \\ 0 & c \end{pmatrix} \mapsto a$ is a surjective homomorphism from G onto $F^\times$ (recall that $F^\times$ is the multiplicative group of nonzero elements in F). Describe the fibers and kernel of φ.

   (b) Prove that the map $\psi : \begin{pmatrix} a & b \\ 0 & c \end{pmatrix} \mapsto (a, c)$ is a surjective homomorphism from G onto $F^\times \times F^\times$. Describe the fibers and kernel of ψ.

   (c) Let $H = \{\begin{pmatrix} 1 & b \\ 0 & 1 \end{pmatrix} \mid b \in F\}$. Prove that H is isomorphic to the additive group F.

12. Let G be the additive group of real numbers, let H be the multiplicative group of complex numbers of absolute value 1 (the unit circle S^1 in the complex plane) and let $\varphi : G \to H$ be the homomorphism $\varphi : r \mapsto e^{2\pi i r}$. Draw the points on a real line which lie in the kernel of φ. Describe similarly the elements in the fibers of φ above the points -1, i, and $e^{4\pi i/3}$ of H. (Figure 1 of the text for this homomorphism φ is usually depicted using the following diagram.)

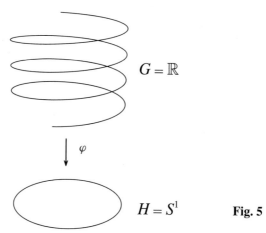

$G = \mathbb{R}$

φ

$H = S^1$

Fig. 5

13. Repeat the preceding exercise with the map φ replaced by the map $\varphi : r \mapsto e^{4\pi i r}$.

14. Consider the additive quotient group $\mathbb{Q}/\mathbb{Z}$.
   (a) Show that every coset of $\mathbb{Z}$ in $\mathbb{Q}$ contains exactly one representative $q \in \mathbb{Q}$ in the range $0 \leq q < 1$.
   (b) Show that every element of $\mathbb{Q}/\mathbb{Z}$ has finite order but that there are elements of arbitrarily large order.
   (c) Show that $\mathbb{Q}/\mathbb{Z}$ is the torsion subgroup of $\mathbb{R}/\mathbb{Z}$ (cf. Exercise 6, Section 2.1).
   (d) Prove that $\mathbb{Q}/\mathbb{Z}$ is isomorphic to the multiplicative group of root of unity in $\mathbb{C}^\times$.

15. Prove that a quotient of a divisible abelian group by any proper subgroup is also divisible. Deduce that $\mathbb{Q}/\mathbb{Z}$ is divisible (cf. Exercise 19, Section 2.4).

16. Let G be a group, let N be a normal subgroup of G and let $\overline{G} = G/N$. Prove that if

86

$G = \langle x, y \rangle$ then $\overline{G} = \langle \overline{x}, \overline{y} \rangle$. Prove more generally that if $G = \langle S \rangle$ for any subset S of G, then $\overline{G} = \langle \overline{S} \rangle$.

17. Let G be the dihedral group of order 16 (whose lattice appears in Section 2.5):
$$G = \langle r, s \mid r^8 = s^2 = 1, \ rs = sr^{-1} \rangle$$
and let $\overline{G} = G/\langle r^4 \rangle$ be the quotient of G by the subgroup generated by r^4 (this subgroup is the center of G, hence is normal).
   (a) Show that the order of $\overline{G}$ is 8.
   (b) Exhibit each element of $\overline{G}$ in the form $\overline{s}^a \overline{r}^b$, for some integers a and b.
   (c) Find the order of each of the elements of $\overline{G}$ exhibited in (b).
   (d) Write each of the following elements of $\overline{G}$ in the form $\overline{s}^a \overline{r}^b$, for some integers a and b as in (b): $\quad \overline{rs}, \quad \overline{sr^{-2}s}, \quad \overline{s^{-1}r^{-1}sr}$.
   (e) Prove that $\overline{H} = \langle \overline{s}, \overline{r}^2 \rangle$ is a normal subgroup of $\overline{G}$ and $\overline{H}$ is isomorphic to the Klein 4-group. Describe the isomorphism type of the complete preimage of $\overline{H}$ in G.
   (f) Find the center of $\overline{G}$ and describe the isomorphism type of $\overline{G}/Z(\overline{G})$.

18. Let G be the quasidihedral group of order 16 (whose lattice was computed in Exercise 11 of Section 2.5):
$$G = \langle \sigma, \tau \mid \sigma^8 = \tau^2 = 1, \ \sigma\tau = \tau\sigma^3 \rangle$$
and let $\overline{G} = G/\langle \sigma^4 \rangle$ be the quotient of G by the subgroup generated by σ^4 (this subgroup is the center of G, hence is normal).
   (a) Show that the order of $\overline{G}$ is 8.
   (b) Exhibit each element of $\overline{G}$ in the form $\overline{\tau}^a \overline{\sigma}^b$, for some integers a and b.
   (c) Find the order of each of the elements of $\overline{G}$ exhibited in (b).
   (d) Write each of the following elements of $\overline{G}$ in the form $\overline{\tau}^a \overline{\sigma}^b$, for some integers a and b as in (b): $\quad \overline{\sigma\tau}, \quad \overline{\tau\sigma^{-2}\tau}, \quad \overline{\tau^{-1}\sigma^{-1}\tau\sigma}$.
   (e) Prove that $\overline{G} \cong D_8$.

19. Let G be the modular group of order 16 (whose lattice was computed in Exercise 14 of Section 2.5):
$$G = \langle u, v \mid u^2 = v^8 = 1, \ vu = uv^5 \rangle$$
and let $\overline{G} = G/\langle v^4 \rangle$ be the quotient of G by the subgroup generated by v^4 (this subgroup is contained in the center of G, hence is normal).
   (a) Show that the order of $\overline{G}$ is 8.
   (b) Exhibit each element of $\overline{G}$ in the form $\overline{u}^a \overline{v}^b$, for some integers a and b.
   (c) Find the order of each of the elements of $\overline{G}$ exhibited in (b).
   (d) Write each of the following elements of $\overline{G}$ in the form $\overline{u}^a \overline{v}^b$, for some integers a and b as in (b): $\quad \overline{vu}, \quad \overline{uv^{-2}u}, \quad \overline{u^{-1}v^{-1}uv}$.
   (e) Prove that $\overline{G}$ is abelian and is isomorphic to $Z_2 \times Z_4$.

20. Let $G = \mathbb{Z}/24\mathbb{Z}$ and let $\widetilde{G} = G/\langle \overline{12} \rangle$, where for each integer a we simplify notation by writing $\overline{\overline{a}}$ as $\widetilde{a}$.
   (a) Show that $\widetilde{G} = \{\widetilde{0}, \widetilde{1}, \dots, \widetilde{11}\}$.
   (b) Find the order of each element of $\widetilde{G}$.
   (c) Prove that $\widetilde{G} \cong \mathbb{Z}/12\mathbb{Z}$. (Thus $(\mathbb{Z}/24\mathbb{Z})/(12\mathbb{Z}/24\mathbb{Z}) \cong \mathbb{Z}/12\mathbb{Z}$, just as if we inverted and cancelled the $24\mathbb{Z}$'s.)

21. Let $G = Z_4 \times Z_4$ be given in terms of the following generators and relations:
$$G = \langle x, y \mid x^4 = y^4 = 1, \ xy = yx \rangle.$$

Let $\overline{G} = G/\langle x^2 y^2 \rangle$ (note that every subgroup of the abelian group G is normal).

(a) Show that the order of $\overline{G}$ is 8.

(b) Exhibit each element of $\overline{G}$ in the form $\overline{x}^a \overline{y}^b$, for some integers a and b.

(c) Find the order of each of the elements of $\overline{G}$ exhibited in (b).

(d) Prove that $\overline{G} \cong Z_4 \times Z_2$.

22. (a) Prove that if H and K are normal subgroups of a group G then their intersection $H \cap K$ is also a normal subgroup of G.

(b) Prove that the intersection of an arbitrary nonempty collection of normal subgroups of a group is a normal subgroup (do not assume the collection is countable).

23. Prove that the join (cf. Section 2.5) of any nonempty collection of normal subgroups of a group is a normal subgroup.

24. Prove that if $N \trianglelefteq G$ and H is any subgroup of G then $N \cap H \trianglelefteq H$.

25. (a) Prove that a subgroup N of G is normal if and only if $gNg^{-1} \subseteq N$ for *all* $g \in G$.

(b) Let $G = GL_2(\mathbb{Q})$, let N be the subgroup of upper triangular matrices with integer entries and 1's on the diagonal, and let g be the diagonal matrix with entries $2, 1$. Show that $gNg^{-1} \subseteq N$ but g does *not* normalize N.

26. Let $a, b \in G$.

(a) Prove that the conjugate of the product of a and b is the product of the conjugate of a and the conjugate of b. Prove that the order of a and the order of any conjugate of a are the same.

(b) Prove that the conjugate of a^{-1} is the inverse of the conjugate of a.

(c) Let $N = \langle S \rangle$ for some subset S of G. Prove that $N \trianglelefteq G$ if $gSg^{-1} \subseteq N$ for all $g \in G$.

(d) Deduce that if N is the cyclic group $\langle x \rangle$, then N is normal in G if and only if for each $g \in G$, $gxg^{-1} = x^k$ for some $k \in \mathbb{Z}$.

(e) Let n be a positive integer. Prove that the subgroup N of G generated by all the elements of G of order n is a normal subgroup of G.

27. Let N be a *finite* subgroup of a group G. Show that $gNg^{-1} \subseteq N$ if and only if $gNg^{-1} = N$. Deduce that $N_G(N) = \{g \in G \mid gNg^{-1} \subseteq N\}$.

28. Let N be a *finite* subgroup of a group G and assume $N = \langle S \rangle$ for some subset S of G. Prove that an element $g \in G$ normalizes N if and only if $gSg^{-1} \subseteq N$.

29. Let N be a *finite* subgroup of G and suppose $G = \langle T \rangle$ and $N = \langle S \rangle$ for some subsets S and T of G. Prove that N is normal in G if and only if $tSt^{-1} \subseteq N$ for all $t \in T$.

30. Let $N \le G$ and let $g \in G$. Prove that $gN = Ng$ if and only if $g \in N_G(N)$.

31. Prove that if $H \le G$ and N is a normal subgroup of H then $H \le N_G(N)$. Deduce that $N_G(N)$ is the largest subgroup of G in which N is normal (i.e., is the join of all subgroups H for which $N \trianglelefteq H$).

32. Prove that every subgroup of Q_8 is normal. For each subgroup find the isomorphism type of its corresponding quotient. [You may use the lattice of subgroups for Q_8 in Section 2.5.]

33. Find all normal subgroups of D_8 and for each of these find the isomorphism type of its corresponding quotient. [You may use the lattice of subgroups for D_8 in Section 2.5.]

34. Let $D_{2n} = \langle r, s \mid r^n = s^2 = 1, \ rs = sr^{-1} \rangle$ be the usual presentation of the dihedral group of order $2n$ and let k be a positive integer dividing n.

(a) Prove that $\langle r^k \rangle$ is a normal subgroup of D_{2n}.

(b) Prove that $D_{2n}/\langle r^k \rangle \cong D_{2k}$.

35. Prove that $SL_n(F) \trianglelefteq GL_n(F)$ and describe the isomorphism type of the quotient group (cf. Exercise 9, Section 2.1).

36. Prove that if $G/Z(G)$ is cyclic then G is abelian. [If $G/Z(G)$ is cyclic with generator $xZ(G)$, show that every element of G can be written in the form $x^a z$ for some integer $a \in \mathbb{Z}$ and some element $z \in Z(G)$.]

37. Let A and B be groups. Show that $\{(a, 1) \mid a \in A\}$ is a normal subgroup of $A \times B$ and the quotient of $A \times B$ by this subgroup is isomorphic to B.

38. Let A be an abelian group and let D be the (diagonal) subgroup $\{(a, a) \mid a \in A\}$ of $A \times A$. Prove that D is a normal subgroup of $A \times A$ and $(A \times A)/D \cong A$.

39. Suppose A is the non-abelian group S_3 and D is the diagonal subgroup $\{(a, a) \mid a \in A\}$ of $A \times A$. Prove that D is not normal in $A \times A$.

40. Let G be a group, let N be a normal subgroup of G and let $\overline{G} = G/N$. Prove that $\overline{x}$ and $\overline{y}$ commute in $\overline{G}$ if and only if $x^{-1}y^{-1}xy \in N$. (The element $x^{-1}y^{-1}xy$ is called the *commutator* of x and y and is denoted by $[x, y]$.)

41. Let G be a group. Prove that $N = \langle x^{-1}y^{-1}xy \mid x, y \in G \rangle$ is a normal subgroup of G and G/N is abelian (N is called the *commutator subgroup* of G).

42. Assume both H and K are normal subgroups of G with $H \cap K = 1$. Prove that $xy = yx$ for all $x \in H$ and $y \in K$. [Show $x^{-1}y^{-1}xy \in H \cap K$.]

43. Assume $\mathcal{P} = \{A_i \mid i \in I\}$ is any partition of G with the property that $\mathcal{P}$ is a group under the "quotient operation" defined as follows: to compute the product of A_i with A_j take any element a_i of A_i and any element a_j of A_j and let $A_i A_j$ be the element of $\mathcal{P}$ containing $a_i a_j$ (this operation is assumed to be well defined). Prove that the element of $\mathcal{P}$ that contains the identity of G is a normal subgroup of G and the elements of $\mathcal{P}$ are the cosets of this subgroup (so $\mathcal{P}$ is just a quotient group of G in the usual sense).

3.2 MORE ON COSETS AND LAGRANGE'S THEOREM

In this section we continue the study of quotient groups. Since for finite groups one of the most important invariants of a group is its order we first prove that the order of a quotient group of a finite group can be readily computed: $|G/N| = \dfrac{|G|}{|N|}$. In fact we derive this as a consequence of a more general result, Lagrange's Theorem (see Exercise 19, Section 1.7). This theorem is one of the most important combinatorial results in finite group theory and will be used repeatedly. After indicating some easy consequences of Lagrange's Theorem we study more subtle questions concerning cosets of non-normal subgroups.

The proof of Lagrange's Theorem is straightforward and important. It is the same line of reasoning we used in Example 3 of the preceding section to compute $|D_8/Z(D_8)|$.

Theorem 8. *(Lagrange's Theorem)* If G is a finite group and H is a subgroup of G, then the order of H divides the order of G (i.e., $|H| \mid |G|$) and the number of left cosets of H in G equals $\dfrac{|G|}{|H|}$.

Proof: Let $|H| = n$ and let the number of left cosets of H in G equal k. By

Proposition 4 the set of left cosets of H in G partition G. By definition of a left coset the map:

$$H \to gH \qquad \text{defined by} \qquad h \mapsto gh$$

is a surjection from H to the left coset gH. The left cancellation law implies this map is injective since $gh_1 = gh_2$ implies $h_1 = h_2$. This proves that H and gH have the same order:

$$|gH| = |H| = n.$$

Since G is partitioned into k disjoint subsets each of which has cardinality n, $|G| = kn$. Thus $k = \dfrac{|G|}{n} = \dfrac{|G|}{|H|}$, completing the proof.

Definition. If G is a group (possibly infinite) and $H \leq G$, the number of left cosets of H in G is called the *index* of H in G and is denoted by $|G : H|$.

In the case of finite groups the index of H in G is $\dfrac{|G|}{|H|}$. For G an infinite group the quotient $\dfrac{|G|}{|H|}$ does not make sense. Infinite groups may have subgroups of finite or infinite index (e.g., $\{0\}$ is of infinite index in $\mathbb{Z}$ and $\langle n \rangle$ is of index n in $\mathbb{Z}$ for every $n > 0$).

We now derive some easy consequences of Lagrange's Theorem.

Corollary 9. If G is a finite group and $x \in G$, then the order of x divides the order of G. In particular $x^{|G|} = 1$ for all x in G.

Proof: By Proposition 2.2, $|x| = |\langle x \rangle|$. The first part of the corollary follows from Lagrange's Theorem applied to $H = \langle x \rangle$. The second statement is clear since now $|G|$ is a multiple of the order of x.

Corollary 10. If G is a group of prime order p, then G is cyclic, hence $G \cong Z_p$.

Proof: Let $x \in G$, $x \neq 1$. Thus $|\langle x \rangle| > 1$ and $|\langle x \rangle|$ divides $|G|$. Since $|G|$ is prime we must have $|\langle x \rangle| = |G|$, hence $G = \langle x \rangle$ is cyclic (with any nonidentity element x as generator). Theorem 2.4 completes the proof.

With Lagrange's Theorem in hand we examine some additional examples of normal subgroups.

Examples

(1) Let $H = \langle (1\ 2\ 3) \rangle \leq S_3$ and let $G = S_3$. We show $H \trianglelefteq S_3$. As noted in Section 2.2,

$$H \leq N_G(H) \leq G.$$

By Lagrange's Theorem, the order of H divides the order of $N_G(H)$ and the order of $N_G(H)$ divides the order of G. Since G has order 6 and H has order 3, the only possibilities for $N_G(H)$ are H or G. A direct computation gives

$$(1\ 2)(1\ 2\ 3)(1\ 2) = (1\ 3\ 2) = (1\ 2\ 3)^{-1}.$$

Since $(1\ 2) = (1\ 2)^{-1}$, this calculation shows that $(1\ 2)$ conjugates a generator of H to another generator of H. By Exercise 24 of Section 2.3 this is sufficient to prove that $(1\ 2) \in N_G(H)$. Thus $N_G(H) \neq H$ so $N_G(H) = G$, i.e., $H \trianglelefteq S_3$, as claimed. This argument illustrates that checking normality of a subgroup can often be reduced to a small number of calculations. A generalization of this example is given in the next example.

(2) Let G be any group containing a subgroup H of index 2. We prove $H \trianglelefteq G$. Let $g \in G - H$ so, by hypothesis, the two left cosets of H in G are $1H$ and gH. Since $1H = H$ and the cosets partition G, we must have $gH = G - H$. Now the two right cosets of H in G are $H1$ and Hg. Since $H1 = H$, we again must have $Hg = G - H$. Combining these gives $gH = Hg$, so every left coset of H in G is a right coset. By Theorem 6, $H \trianglelefteq G$. By definition of index, $|G/H| = 2$, so that $G/H \cong Z_2$. One must be careful to appreciate that the reason H is normal in this case is not because we can choose the same coset representatives 1 and g for both the left and right cosets of H but that there is a type of pigeon-hole principle at work: since $1H = H = H1$ for any subgroup H of any group G, the index assumption forces the remaining elements to comprise the remaining coset (either left or right). We shall see that this result is itself a special case of a result we shall prove in the next chapter.

Note that this result proves that $\langle i \rangle$, $\langle j \rangle$ and $\langle k \rangle$ are normal subgroups of Q_8 and that $\langle s, r^2 \rangle$, $\langle r \rangle$ and $\langle sr, r^2 \rangle$ are normal subgroups of D_8.

(3) The property "is a normal subgroup of" is not transitive. For example,

$$\langle s \rangle \trianglelefteq \langle s, r^2 \rangle \trianglelefteq D_8$$

(each subgroup is of index 2 in the next), however, $\langle s \rangle$ is not normal in D_8 because $rsr^{-1} = sr^2 \notin \langle s \rangle$.

We now examine some examples of non-normal subgroups. Although in abelian groups every subgroup is normal, this is not the case in non-abelian groups (in some sense Q_8 is the unique exception to this). In fact, there are groups G in which the only normal subgroups are the trivial ones: 1 and G. Such groups are called *simple groups* (simple does not mean easy, however). Simple groups play an important role in the study of general groups and this role will be described in Section 4. For now we emphasize that not every subgroup of a group G is normal in G; indeed, normal subgroups may be quite rare in G. The search for normal subgroups of a given group is in general a highly nontrivial problem.

Examples

(1) Let $H = \langle (1\ 2) \rangle \leq S_3$. Since H is of prime index 3 in S_3, by Lagrange's Theorem the only possibilities for $N_{S_3}(H)$ are H or S_3. Direct computation shows

$$(1\ 3)(1\ 2)(1\ 3)^{-1} = (1\ 3)(1\ 2)(1\ 3) = (2\ 3) \notin H$$

so $N_{S_3}(H) \neq S_3$, that is, H is not a normal subgroup of S_3. One can also see this by considering the left and right cosets of H; for instance

$$(1\ 3)H = \{(1\ 3), (1\ 2\ 3)\} \qquad \text{and} \qquad H(1\ 3) = \{(1\ 3), (1\ 3\ 2)\}.$$

Since the left coset $(1\ 3)H$ is the unique left coset of H containing $(1\ 3)$, the right coset $H(1\ 3)$ cannot be a left coset (see also Exercise 6). Note also that the "group operation" on the left cosets of H in S_3 defined by multiplying representatives is not even well defined. For example, consider the product of the two left cosets $1H$ and $(1\ 3)H$. The elements 1 and $(1\ 2)$ are both representatives for the coset $1H$, yet $1 \cdot (1\ 3) = (1\ 3)$ and $(1\ 2) \cdot (1\ 3) = (1\ 3\ 2)$ are not both elements of the same left coset as they should be if the product of these cosets were independent of the particular representatives chosen. This is an example of Theorem 6 which states that the cosets of a subgroup form a group *only* when the subgroup is a normal subgroup.

(2) Let $G = S_n$ for some $n \in \mathbb{Z}^+$ and fix some $i \in \{1, 2, \ldots, n\}$. As in Section 2.2 let

$$G_i = \{\sigma \in G \mid \sigma(i) = i\}$$

be the stabilizer of the point i. Suppose $\tau \in G$ and $\tau(i) = j$. It follows directly from the definition of G_i that for all $\sigma \in G_i$, $\tau\sigma(i) = j$. Furthermore, if $\mu \in G$ and $\mu(i) = j$, then $\tau^{-1}\mu(i) = i$, that is, $\tau^{-1}\mu \in G_i$, so $\mu \in \tau G_i$. This proves that

$$\tau G_i = \{\mu \in G \mid \mu(i) = j\},$$

i.e., the left coset τG_i consists of the permutations in S_n which take i to j. We can clearly see that distinct left cosets have empty intersection and that the number of distinct left cosets equals the number of distinct images of the integer i under the action of G, namely there are n distinct left cosets. Thus $|G : G_i| = n$. Using the same notation let $k = \tau^{-1}(i)$, so that $\tau(k) = i$. By similar reasoning we see that

$$G_i \tau = \{\lambda \in G \mid \lambda(k) = i\},$$

i.e., the right coset $G_i\tau$ consists of the permutations in S_n which take k to i. If $n > 2$, for some nonidentity element τ we have $\tau G_i \neq G_i\tau$ since there are certainly permutations which take i to j but do not take k to i. Thus G_i is not a normal subgroup. In fact $N_G(G_i) = G_i$ by Exercise 30 of Section 1, so G_i is in some sense far from being normal in S_n. This example generalizes the preceding one.

(3) In D_8 the only subgroup of order 2 which is normal is the center $\langle r^2 \rangle$.

We shall see many more examples of non-normal subgroups as we develop the theory.

The *full converse* to Lagrange's Theorem is *not* true: namely, if G is a finite group and n divides $|G|$, then G need not have a subgroup of order n. For example, let A be the group of symmetries of a regular tetrahedron. By Exercise 9 of Section 1.2, $|A| = 12$. Suppose A had a subgroup H of order 6. Since $\dfrac{|A|}{|H|} = 2$, H would be of index 2 in A, hence $H \trianglelefteq A$ and $A/H \cong Z_2$. Since the quotient group has order 2, the square of every element in the quotient is the identity, so for all $g \in A$, $(gH)^2 = 1H$, that is, for all $g \in A$, $g^2 \in H$. If g is an element of A of order 3, we obtain $g = (g^2)^2 \in H$, that is, H must contain all elements of A of order 3. This is a contradiction since $|H| = 6$ but one can easily exhibit 8 rotations of a tetrahedron of order 3.

There are some partial converses to Lagrange's Theorem. For finite *abelian* groups the full converse of Lagrange is true, namely an abelian group has a subgroup of order n for each divisor n of $|G|$ (in fact, this holds under weaker assumptions than "abelian"; we shall see this in Chapter 6). A partial converse which holds for arbitrary finite groups is the following result:

Theorem 11. *(Cauchy's Theorem)* If G is a finite group and p is a prime dividing $|G|$, then G has an element of order p.

Proof: We shall give a proof of this in the next chapter and another elegant proof is outlined in Exercise 9.

The strongest converse to Lagrange's Theorem which applies to *arbitrary* finite groups is the following:

Theorem 12. (Sylow) If G is a finite group of order $p^\alpha m$, where p is a prime and p does not divide m, then G has a subgroup of order p^α.

We shall prove this theorem in the next chapter and derive more information on the number of subgroups of order p^α.

We conclude this section with some useful results involving cosets.

Definition. Let H and K be subgroups of a group and define

$$HK = \{hk \mid h \in H, \ k \in K\}.$$

Proposition 13. If H and K are finite subgroups of a group then

$$|HK| = \frac{|H||K|}{|H \cap K|}.$$

Proof: Notice that HK is a union of left cosets of K, namely,

$$HK = \bigcup_{h \in H} hK.$$

Since each coset of K has $|K|$ elements it suffices to find the number of *distinct* left cosets of the form hK, $h \in H$. But $h_1 K = h_2 K$ for $h_1, h_2 \in H$ if and only if $h_2^{-1} h_1 \in K$. Thus

$$h_1 K = h_2 K \quad \Leftrightarrow \quad h_2^{-1} h_1 \in H \cap K \quad \Leftrightarrow \quad h_1 (H \cap K) = h_2 (H \cap K).$$

Thus the number of distinct cosets of the form hK, for $h \in H$ is the number of distinct cosets $h(H \cap K)$, for $h \in H$. The latter number, by Lagrange's Theorem, equals $\dfrac{|H|}{|H \cap K|}$. Thus HK consists of $\dfrac{|H|}{|H \cap K|}$ distinct cosets of K (each of which has $|K|$ elements) which gives the formula above.

Notice that there was no assumption that HK be a subgroup in Proposition 13. For example, if $G = S_3$, $H = \langle (1\,2) \rangle$ and $K = \langle (2\,3) \rangle$, then $|H| = |K| = 2$ and $|H \cap K| = 1$, so $|HK| = 4$. By Lagrange's Theorem HK cannot be a subgroup. As a consequence, we must have $S_3 = \langle (1\,2), (2\,3) \rangle$.

Proposition 14. If H and K are subgroups of a group, HK is a subgroup if and only if $HK = KH$.

Proof: Assume first that $HK = KH$ and let $a, b \in HK$. We prove $ab^{-1} \in HK$ so HK is a subgroup by the subgroup criterion. Let

$$a = h_1 k_1 \quad \text{and} \quad b = h_2 k_2,$$

for some $h_1, h_2 \in H$ and $k_1, k_2 \in K$. Thus $b^{-1} = k_2^{-1} h_2^{-1}$, so $ab^{-1} = h_1 k_1 k_2^{-1} h_2^{-1}$. Let $k_3 = k_1 k_2^{-1} \in K$ and $h_3 = h_2^{-1}$. Thus $ab^{-1} = h_1 k_3 h_3$. Since $HK = KH$,

$$k_3 h_3 = h_4 k_4, \quad \text{for some } h_4 \in H, \quad k_4 \in K.$$

Thus $ab^{-1} = h_1 h_4 k_4$, and since $h_1 h_4 \in H, k_4 \in K$, we obtain $ab^{-1} \in HK$, as desired.

Conversely, assume that HK is a subgroup of G. Since $K \leq HK$ and $H \leq HK$, by the closure property of subgroups, $KH \subseteq HK$. To show the reverse containment let $hk \in HK$. Since HK is assumed to be a subgroup, write $hk = a^{-1}$, for some $a \in HK$. If $a = h_1 k_1$, then

$$hk = (h_1 k_1)^{-1} = k_1^{-1} h_1^{-1} \in KH,$$

completing the proof.

Note that $HK = KH$ does *not* imply that the elements of H commute with those of K (contrary to what the notation may suggest) but rather that every product hk is of the form $k'h'$ (h need not be h' nor k be k') and conversely. For example, if $G = D_{2n}$, $H = \langle r \rangle$ and $K = \langle s \rangle$, then $G = HK = KH$ so that HK is a subgroup and $rs = sr^{-1}$ so the elements of H do not commute with the elements of K. This is an example of the following sufficient condition for HK to be a subgroup:

Corollary 15. If H and K are subgroups of G and $H \leq N_G(K)$, then HK is a subgroup of G. In particular, if $K \trianglelefteq G$ then $HK \leq G$ for any $H \leq G$.

Proof: We prove $HK = KH$. Let $h \in H, k \in K$. By assumption, $hkh^{-1} \in K$, hence

$$hk = (hkh^{-1})h \in KH.$$

This proves $HK \subseteq KH$. Similarly, $kh = h(h^{-1}kh) \in HK$, proving the reverse containment. The corollary follows now from the preceding proposition.

Definition. If A is any subset of $N_G(K)$ (or $C_G(K)$), we shall say A *normalizes K* (*centralizes K*, respectively).

With this terminology, Corollary 15 states that HK *is a subgroup if H normalizes K* (similarly, HK *is a subgroup if K normalizes H*).

In some instances one can prove that a finite group is a product of two of its subgroups by simply using the order formula in Proposition 13. For example, let $G = S_4$, $H = D_8$ and let $K = \langle (1\,2\,3) \rangle$, where we consider D_8 as a subgroup of S_4 by identifying each symmetry with its permutation on the 4 vertices of a square (under some fixed labelling). By Lagrange's Theorem, $H \cap K = 1$ (see Exercise 8). Proposition 13 then shows $|HK| = 24$ hence we must have $HK = S_4$. Since HK is a group, $HK = KH$. We leave as an exercise the verification that neither H nor K normalizes the other (so Corollary 15 could not have been used to give $HK = KH$).

Finally, throughout this chapter we have worked with left cosets of a subgroup. The same combinatorial results could equally well have been proved using right cosets. For normal subgroups this is trivial since left and right cosets are the same, but for non-normal subgroups some left cosets are not right cosets (for any choice of representative) so some (simple) verifications are necessary. For example, Lagrange's Theorem gives that in a finite group G

$$\text{the number of right cosets of the subgroup } H \text{ is } \frac{|G|}{|H|}.$$

Thus in a finite group the *number* of left cosets of H in G equals the *number* of right cosets even though the left cosets are not right cosets in general. This is also true for infinite groups as Exercise 12 below shows. Thus for purely combinatorial purposes one may use either left or right cosets (but not a mixture when a partition of G is needed). Our consistent use of left cosets is somewhat arbitrary although it will have some benefits when we discuss actions on cosets in the next chapter. Readers may encounter in some works the notation $H \setminus G$ to denote the set of right cosets of H in G.

In some papers one may also see the notation G/H used to denote the set of left cosets of H in G even when H is not normal in G (in which case G/H is called the *coset space* of left cosets of H in G). We shall not use this notation.

EXERCISES

Let G be a group.

1. Which of the following are permissible orders for subgroups of a group of order 120: 1, 2, 5, 7, 9, 15, 60, 240? For each permissible order give the corresponding index.

2. Prove that the lattice of subgroups of S_3 in Section 2.5 is correct (i.e., prove that it contains all subgroups of S_3 and that their pairwise joins and intersections are correctly drawn).

3. Prove that the lattice of subgroups of Q_8 in Section 2.5 is correct.

4. Show that if $|G| = pq$ for some primes p and q (not necessarily distinct) then either G is abelian or $Z(G) = 1$. [See Exercise 36 in Section 1.]

5. Let H be a subgroup of G and fix some element $g \in G$.
   (a) Prove that gHg^{-1} is a subgroup of G of the same order as H.
   (b) Deduce that if $n \in \mathbb{Z}^+$ and H is the unique subgroup of G of order n then $H \trianglelefteq G$.

6. Let $H \le G$ and let $g \in G$. Prove that if the right coset Hg equals *some* left coset of H in G then it equals the left coset gH and g must be in $N_G(H)$.

7. Let $H \le G$ and define a relation $\sim$ on G by $a \sim b$ if and only if $b^{-1}a \in H$. Prove that $\sim$ is an equivalence relation and describe the equivalence class of each $a \in G$. Use this to prove Proposition 4.

8. Prove that if H and K are finite subgroups of G whose orders are relatively prime then $H \cap K = 1$.

9. This exercise outlines a proof of Cauchy's Theorem due to James McKay (*Another proof of Cauchy's group theorem*, Amer. Math. Monthly, 66(1959), p. 119). Let G be a finite group and let p be a prime dividing $|G|$. Let $\mathcal{S}$ denote the set of p-tuples of elements of G the product of whose coordinates is 1:

$$\mathcal{S} = \{(x_1, x_2, \ldots, x_p) \mid x_i \in G \text{ and } x_1 x_2 \cdots x_p = 1\}.$$

(a) Show that $\mathcal{S}$ has $|G|^{p-1}$ elements, hence has order divisible by p.

Define the relation $\sim$ on $\mathcal{S}$ by letting $\alpha \sim \beta$ if β is a cyclic permutation of α.

(b) Show that a cyclic permutation of an element of $\mathcal{S}$ is again an element of $\mathcal{S}$.
(c) Prove that $\sim$ is an equivalence relation on $\mathcal{S}$.
(d) Prove that an equivalence class contains a single element if and only if it is of the form $(x, x, \ldots, x)$ with $x^p = 1$.
(e) Prove that every equivalence class has order 1 or p (this uses the fact that p is a *prime*). Deduce that $|G|^{p-1} = k + pd$, where k is the number of classes of size 1 and d is the number of classes of size p.
(f) Since $\{(1, 1, \ldots, 1)\}$ is an equivalence class of size 1, conclude from (e) that there must be a nonidentity element x in G with $x^p = 1$, i.e., G contains an element of order p. [Show $p \mid k$ and so $k > 1$.]

10. Suppose H and K are subgroups of finite index in the (possibly infinite) group G with $|G : H| = m$ and $|G : K| = n$. Prove that l.c.m.$(m, n) \le |G : H \cap K| \le mn$. Deduce that if m and n are relatively prime then $|G : H \cap K| = |G : H| \cdot |G : K|$.

11. Let $H \le K \le G$. Prove that $|G : H| = |G : K| \cdot |K : H|$ (do not assume G is finite).

12. Let $H \le G$. Prove that the map $x \mapsto x^{-1}$ sends each left coset of H in G onto a right coset of H and gives a bijection between the set of left cosets and the set of right cosets of H in G (hence the number of left cosets of H in G equals the number of right cosets).

13. Fix any labelling of the vertices of a square and use this to identify D_8 as a subgroup of S_4. Prove that the elements of D_8 and $\langle (1\,2\,3) \rangle$ do not commute in S_4.

14. Prove that S_4 does not have a normal subgroup of order 8 or a normal subgroup of order 3.

15. Let $G = S_n$ and for fixed $i \in \{1, 2, \ldots, n\}$ let G_i be the stabilizer of i. Prove that $G_i \cong S_{n-1}$.

16. Use Lagrange's Theorem in the multiplicative group $(\mathbb{Z}/p\mathbb{Z})^\times$ to prove *Fermat's Little Theorem*: if p is a prime then $a^p \equiv a \pmod{p}$ for all $a \in \mathbb{Z}$.

17. Let p be a prime and let n be a positive integer. Find the order of $\bar{p}$ in $(\mathbb{Z}/(p^n - 1)\mathbb{Z})^\times$ and deduce that $n \mid \varphi(p^n - 1)$ (here φ is Euler's function).

18. Let G be a finite group, let H be a subgroup of G and let $N \trianglelefteq G$. Prove that if $|H|$ and $|G : N|$ are relatively prime then $H \le N$.

19. Prove that if N is a normal subgroup of the finite group G and $(|N|, |G : N|) = 1$ then N is the unique subgroup of G of order $|N|$.

20. If A is an abelian group with $A \trianglelefteq G$ and B is any subgroup of G prove that $A \cap B \trianglelefteq AB$.

21. Prove that $\mathbb{Q}$ has no proper subgroups of finite index. Deduce that $\mathbb{Q}/\mathbb{Z}$ has no proper subgroups of finite index. [Recall Exercise 21, Section 1.6 and Exercise 15, Section 1.]

22. Use Lagrange's Theorem in the multiplicative group $(\mathbb{Z}/n\mathbb{Z})^\times$ to prove *Euler's Theorem*: $a^{\varphi(n)} \equiv 1 \bmod n$ for every integer a relatively prime to n, where φ denotes Euler's φ-function.

23. Determine the last two digits of $3^{3^{100}}$. [Determine $3^{100} \bmod \varphi(100)$ and use the previous exercise.]

3.3 THE ISOMORPHISM THEOREMS

In this section we derive some straightforward consequences of the relations between quotient groups and homomorphisms which were discussed in Section 1. In particular we consider the relation between the lattice of subgroups of a quotient group, G/N, and the lattice of subgroups of the group G. The first result restates our observations in Section 1 on the relation of the image of a homomorphism to the quotient by the kernel (sometimes called the Fundamental Theorem of Homomorphisms):

Theorem 16. *(The First Isomorphism Theorem)* If $\varphi : G \to H$ is a homomorphism of groups, then $\ker \varphi \trianglelefteq G$ and $G/\ker \varphi \cong \varphi(G)$.

Corollary 17. Let $\varphi : G \to H$ be a homomorphism of groups.
  (1) φ is injective if and only if $\ker \varphi = 1$.
  (2) $|G : \ker \varphi| = |\varphi(G)|$.

Proof: Exercise.

When we consider abstract vector spaces we shall see that Corollary 17(2) gives a formula possibly already familiar from the theory of linear transformations: if $\varphi : V \to W$ is a linear transformation of vector spaces, then $\dim V = \operatorname{rank} \varphi + \operatorname{nullity} \varphi$.

Theorem 18. *(The Second or Diamond Isomorphism Theorem)* Let G be a group, let A and B be subgroups of G and assume $A \leq N_G(B)$. Then AB is a subgroup of G, $B \trianglelefteq AB$, $A \cap B \trianglelefteq A$ and $AB/B \cong A/A \cap B$.

Proof: By Corollary 15, AB is a subgroup of G. Since $A \leq N_G(B)$ by assumption and $B \leq N_G(B)$ trivially, it follows that $AB \leq N_G(B)$, i.e., B is a normal subgroup of the subgroup AB.

Since B is normal in AB, the quotient group AB/B is well defined. Define the map $\varphi : A \to AB/B$ by $\varphi(a) = aB$. Since the group operation in AB/B is well defined it is easy to see that φ is a homomorphism:

$$\varphi(a_1 a_2) = (a_1 a_2)B = a_1 B \cdot a_2 B = \varphi(a_1)\varphi(a_2).$$

Alternatively, the map φ is just the restriction to the subgroup A of the natural projection homomorphism $\pi : AB \to AB/B$, so is also a homomorphism. It is clear from the definition of AB that φ is surjective. The identity in AB/B is the coset $1B$, so the kernel of φ consists of the elements $a \in A$ with $aB = 1B$, which by Proposition 4 are the elements $a \in B$, i.e., $\ker \varphi = A \cap B$. By the First Isomorphism Theorem, $A \cap B \trianglelefteq A$ and $A/A \cap B \cong AB/B$, completing the proof.

Note that this gives a new proof of the order formula in Proposition 13 in the special case that $A \leq N_G(B)$. The reason this theorem is called the Diamond Isomorphism is because of the portion of the lattice of subgroups of G involved (see Figure 6). The markings in the lattice lines indicate which quotients are isomorphic. The "quotient"

AB/A need not be a group (i.e., A need not be normal in AB), however we still have $|AB : A| = |B : A \cap B|$.

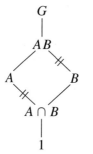

Fig. 6

The third Isomorphism Theorem considers the question of taking quotient groups of quotient groups.

Theorem 19. *(The Third Isomorphism Theorem)* Let G be a group and let H and K be normal subgroups of G with $H \leq K$. Then $K/H \trianglelefteq G/H$ and

$$(G/H)/(K/H) \cong G/K.$$

If we denote the quotient by H with a bar, this can be written

$$\overline{G}/\overline{K} \cong G/K.$$

Proof: We leave as an easy exercise the verification that $K/H \trianglelefteq G/H$. Define

$$\varphi : G/H \to G/K$$

$$(gH) \mapsto gK.$$

To show φ is well defined suppose $g_1 H = g_2 H$. Then $g_1 = g_2 h$, for some $h \in H$. Because $H \leq K$, the element h is also an element of K, hence $g_1 K = g_2 K$ i.e., $\varphi(g_1 H) = \varphi(g_2 H)$, which shows φ is well defined. Since g may be chosen arbitrarily in G, φ is a surjective homomorphism. Finally,

$$\ker \varphi = \{gH \in G/H \mid \varphi(gH) = 1K\}$$

$$= \{gH \in G/H \mid gK = 1K\}$$

$$= \{gH \in G/H \mid g \in K\} = K/H.$$

By the First Isomorphism Theorem, $(G/H)/(K/H) \cong G/K$.

An easy aid for remembering the Third Isomorphism Theorem is: "invert and cancel" (as one would for fractions). This theorem shows that we gain no new structural information from taking quotients of a quotient group.

The final isomorphism theorem describes the relation between the lattice of subgroups of the quotient group G/N and the lattice of subgroups of G. The lattice for G/N can be read immediately from the lattice for G by collapsing the group N to the identity. More precisely, there is a one-to-one correspondence between the subgroups of G containing N and the subgroups of G/N, so that the lattice for G/N (or rather, an isomorphic copy) appears in the lattice for G as the collection of subgroups of G between N and G. In particular, the lattice for G/N appears at the "top" of the lattice for G, a result we mentioned at the beginning of the chapter.

Theorem 20. *(The Fourth or Lattice Isomorphism Theorem)* Let G be a group and let N be a normal subgroup of G. Then there is a bijection from the set of subgroups A of G which contain N onto the set of subgroups $\overline{A} = A/N$ of G/N. In particular, every subgroup of $\overline{G}$ is of the form A/N for some subgroup A of G containing N (namely, its preimage in G under the natural projection homomorphism from G to G/N). This bijection has the following properties: for all $A, B \le G$ with $N \le A$ and $N \le B$,

    (1) $A \le B$ if and only if $\overline{A} \le \overline{B}$,
    (2) if $A \le B$, then $|B : A| = |\overline{B} : \overline{A}|$,
    (3) $\overline{\langle A, B \rangle} = \langle \overline{A}, \overline{B} \rangle$,
    (4) $\overline{A \cap B} = \overline{A} \cap \overline{B}$, and
    (5) $A \trianglelefteq G$ if and only if $\overline{A} \trianglelefteq \overline{G}$.

Proof: The complete preimage of a subgroup in G/N is a subgroup of G by Exercise 1 of Section 1. The numerous details of the theorem to check are all completely straightforward. We therefore leave the proof of this theorem to the exercises.

Examples

    (1) Let $G = Q_8$ and let N be the normal subgroup $\langle -1 \rangle$. The (isomorphic copy of the) lattice of G/N consists of the double lines in the lattice of G below. Note that we previously proved that $Q_8/\langle -1 \rangle \cong V_4$ and the two lattices do indeed coincide (see Section 2.5 for the lattices of Q_8 and V_4).

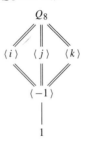

    (2) The same process gives us the lattice of $D_8/\langle r^2 \rangle$ (the double lines) in the lattice of D_8:

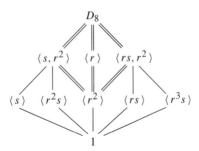

Note that in the second example above there are subgroups of G which do not directly correspond to subgroups in the quotient group G/N, namely the subgroups of G which do not contain the normal subgroup N. This is because the subgroup N projects to a point in G/N and so several subgroups of G can project to the same subgroup in the quotient. The image of the subgroup H of G under the natural projection homomorphism from G to G/N is the same as the image of the subgroup HN of G, and the subgroup HN of G contains N. Conversely, the preimage of a subgroup $\overline{H}$ of G/N contains N and is the unique subgroup of G containing N whose image in G/N is $\overline{H}$. It is the subgroups of G containing N which appear explicitly in the lattice for G/N.

The two lattices of groups of order 8 above emphasize the fact that the isomorphism type of a group cannot in general be determined from the knowledge of the isomorphism types of G/N and N, since $Q_8/\langle -1 \rangle \cong D_8/\langle r^2 \rangle$ and $\langle -1 \rangle \cong \langle r^2 \rangle$ yet Q_8 and D_8 are not isomorphic. We shall discuss this question further in the next section.

We shall often indicate the index of one subgroup in another in the lattice of subgroups, as follows:

where the integer n equals $|A : B|$. For example, all the unbroken edges in the lattices of Q_8 and D_8 would be labelled with 2. Thus the order of any subgroup, A, is the product of all integers which label any path upward from the identity to A. Also, by Theorem 20(2) these indices remain unchanged in quotients of G by normal subgroups of G contained in B, i.e., the portion of the lattice for G corresponding to the lattice of the quotient group has the correct indices for the quotient as well.

Finally we include a remark concerning the definition of homomorphisms on quotient groups. We have, in the course of the proof of the isomorphism theorems, encountered situations where a homomorphism φ on the quotient group G/N is specified by giving the value of φ on the coset gN in terms of the representative g alone. In each instance we then had to prove φ was well defined, i.e., was independent of the choice of g. In effect we are defining a homomorphism, Φ, on G itself by specifying the value of φ at g. Then independence of g is equivalent to requiring that Φ be trivial on N, so that

φ *is well defined on* G/N *if and only if* $N \leq \ker \Phi$.

This gives a simple criterion for defining homomorphisms on quotients (namely, define a homomorphism on G and check that N is contained in its kernel). In this situation we shall say the homomorphism Φ *factors through* N and φ is the *induced* homomorphism on G/N. This can be denoted pictorially as in Figure 7, where the diagram indicates that $\Phi = \varphi \circ \pi$, i.e., the image in H of an element in G does not depend on which path one takes in the diagram. If this is the case, then the diagram is said to *commute*.

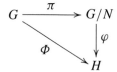

Fig. 7

At this point we have developed all the background material so that Section 6.3 on free groups and presentations may now be read.

EXERCISES

Let G be a group.

1. Let F be a finite field of order q and let $n \in \mathbb{Z}^+$. Prove that $|GL_n(F) : SL_n(F)| = q - 1$. [See Exercise 35, Section 1.]

2. Prove all parts of the Lattice Isomorphism Theorem.

3. Prove that if H is a normal subgroup of G of prime index p then for all $K \leq G$ either
   (i) $K \leq H$ or
   (ii) $G = HK$ and $|K : K \cap H| = p$.

4. Let C be a normal subgroup of the group A and let D be a normal subgroup of the group B. Prove that $(C \times D) \trianglelefteq (A \times B)$ and $(A \times B)/(C \times D) \cong (A/C) \times (B/D)$.

5. Let $QD_{16} = \langle \sigma, \tau \rangle$ be the quasidihedral group described in Exercise 11 of Section 2.5. Prove that $\langle \sigma^4 \rangle$ is normal in QD_{16} and use the Lattice Isomorphism Theorem to draw the lattice of subgroups of $QD_{16}/\langle \sigma^4 \rangle$. Which group of order 8 has the same lattice as this quotient? Use generators and relations for $QD_{16}/\langle \sigma^4 \rangle$ to decide the isomorphism type of this group.

6. Let $M = \langle v, u \rangle$ be the modular group of order 16 described in Exercise 14 of Section 2.5. Prove that $\langle v^4 \rangle$ is normal in M and use the Lattice Isomorphism Theorem to draw the lattice of subgroups of $M/\langle v^4 \rangle$. Which group of order 8 has the same lattice as this quotient? Use generators and relations for $M/\langle v^4 \rangle$ to decide the isomorphism type of this group.

7. Let M and N be normal subgroups of G such that $G = MN$. Prove that $G/(M \cap N) \cong (G/M) \times (G/N)$. [Draw the lattice.]

8. Let p be a prime and let G be the group of p-power roots of 1 in $\mathbb{C}$ (cf. Exercise 18, Section 2.4). Prove that the map $z \mapsto z^p$ is a surjective homomorphism. Deduce that G is isomorphic to a proper quotient of itself.

9. Let p be a prime and let G be a group of order $p^a m$, where p does not divide m. Assume P is a subgroup of G of order p^a and N is a normal subgroup of G of order $p^b n$, where p does not divide n. Prove that $|P \cap N| = p^b$ and $|PN/N| = p^{a-b}$. (The subgroup P of G is called a *Sylow p-subgroup* of G. This exercise shows that the intersection of any Sylow p-subgroup of G with a normal subgroup N is a Sylow p-subgroup of N.)

10. Generalize the preceding exercise as follows. A subgroup H of a finite group G is called a *Hall subgroup* of G if its index in G is relatively prime to its order: $(|G : H|, |H|) = 1$. Prove that if H is a Hall subgroup of G and $N \trianglelefteq G$, then $H \cap N$ is a Hall subgroup of N and HN/N is a Hall subgroup of G/N.

3.4 COMPOSITION SERIES AND THE HÖLDER PROGRAM

The remarks in the preceding section on lattices leave us with the intuitive picture that a quotient group G/N is the group whose structure (e.g., lattice) describes the structure of G "above" the normal subgroup N. Although this is somewhat vague, it gives at least some notion of the driving force behind one of the most powerful techniques in finite group theory (and even some branches of infinite group theory): the use of induction. In many instances the application of an inductive procedure follows a pattern similar to the following proof of a special case of Cauchy's Theorem. Although Cauchy's Theorem is valid for arbitrary groups (cf. Exercise 9 of Section 2), the following is a good example of the use of information on a normal subgroup N and on the quotient G/N to determine information about G, and we shall need this particular result in Chapter 4.

Proposition 21. If G is a finite abelian group and p is a prime dividing $|G|$, then G contains an element of order p.

Proof: The proof proceeds by induction on $|G|$, namely, we assume the result is valid for every group whose order is strictly smaller than the order of G and then prove the result valid for G (this is sometimes referred to as *complete* induction). Since $|G| > 1$, there is an element $x \in G$ with $x \neq 1$. If $|G| = p$ then x has order p by Lagrange's Theorem and we are done. We may therefore assume $|G| > p$.

Suppose p divides $|x|$ and write $|x| = pn$. By Proposition 2.5(3), $|x^n| = p$, and again we have an element of order p. We may therefore assume p does not divide $|x|$.

Let $N = \langle x \rangle$. Since G is abelian, $N \trianglelefteq G$. By Lagrange's Theorem, $|G/N| = \dfrac{|G|}{|N|}$ and since $N \neq 1$, $|G/N| < |G|$. Since p does not divide $|N|$, we must have $p \mid |G/N|$. We can now apply the induction assumption to the smaller group G/N to conclude it contains an element, $\bar{y} = yN$, of order p. Since $y \notin N$ ($\bar{y} \neq \bar{1}$) but $y^p \in N$ ($\bar{y}^p = \bar{1}$), we must have $\langle y^p \rangle \neq \langle y \rangle$, that is, $|y^p| < |y|$. Proposition 2.5(2) implies $p \mid |y|$. We are now in the situation described in the preceding paragraph, so that argument again produces an element of order p. The induction is complete.

The philosophy behind this method of proof is that if we have a sufficient amount of information about some normal subgroup, N, of a group G and sufficient information on G/N, then somehow we can piece this information together to force G itself to have some desired property. The induction comes into play because both N and G/N have smaller order than G. In general, just how much data are required is a delicate matter since, as we have already seen, the full isomorphism type of G cannot be determined from the isomorphism types of N and G/N alone.

Clearly a basic obstruction to this approach is the necessity of producing a normal subgroup, N, of G with $N \neq 1$ or G. In the preceding argument this was easy since G was abelian. Groups with no nontrivial proper normal subgroups are fundamental obstructions to this method of proof.

Definition. A (finite or infinite) group G is called *simple* if $|G| > 1$ and the only normal subgroups of G are 1 and G.

By Lagrange's Theorem if $|G|$ is a prime, its only subgroups (let alone normal ones) are 1 and G, so G is simple. In fact, every abelian simple group is isomorphic to Z_p, for some prime p (cf. Exercise 1). There are non-abelian simple groups (of both finite and infinite order), the smallest of which has order 60 (we shall introduce this group as a member of an infinite family of simple groups in the next section).

Simple groups, by definition, cannot be "factored" into pieces like N and G/N and as a result they play a role analogous to that of the primes in the arithmetic of $\mathbb{Z}$. This analogy is supported by a "unique factorization theorem" (for finite groups) which we now describe.

Definition. In a group G a sequence of subgroups

$$1 = N_0 \leq N_1 \leq N_2 \leq \cdots \leq N_{k-1} \leq N_k = G$$

is called a *composition series* if $N_i \trianglelefteq N_{i+1}$ and N_{i+1}/N_i is a simple group, $0 \leq i \leq k-1$. If the above sequence is a composition series, the quotient groups N_{i+1}/N_i are called *composition factors* of G.

Keep in mind that it is not assumed that each $N_i \trianglelefteq G$, only that $N_i \trianglelefteq N_{i+1}$. Thus

$$1 \trianglelefteq \langle s \rangle \trianglelefteq \langle s, r^2 \rangle \trianglelefteq D_8 \quad \text{and} \quad 1 \trianglelefteq \langle r^2 \rangle \trianglelefteq \langle r \rangle \trianglelefteq D_8$$

are two composition series for D_8 and in each series there are 3 composition factors, each of which is isomorphic to (the simple group) Z_2.

Theorem 22. (Jordan–Hölder) Let G be a finite group with $G \neq 1$. Then
  (1) G has a composition series and
  (2) The composition factors in a composition series are unique, namely, if
  $1 = N_0 \leq N_1 \leq \cdots \leq N_r = G$ and $1 = M_0 \leq M_1 \leq \cdots \leq M_s = G$ are
  two composition series for G, then $r = s$ and there is some permutation, π, of
  $\{1, 2, \ldots, r\}$ such that

$$M_{\pi(i)}/M_{\pi(i)-1} \cong N_i/N_{i-1}, \qquad 1 \leq i \leq r.$$

Proof: This is fairly straightforward. Since we shall not explicitly use this theorem to prove others in the text we outline the proof in a series of exercises at the end of this section.

Thus every finite group has a "factorization" (i.e., composition series) and although the series itself need not be unique (as D_8 shows) the number of composition factors and their isomorphism types are uniquely determined. Furthermore, nonisomorphic groups may have the same (up to isomorphism) list of composition factors (see Exercise 2). This motivates a two-part program for classifying all finite groups up to isomorphism:

The Hölder Program

(1) Classify all finite simple groups.
(2) Find all ways of "putting simple groups together" to form other groups.

These two problems form part of an underlying motivation for much of the development of group theory. Analogues of these problems may also be found as recurring themes throughout mathematics. We include a few more comments on the current status of progress on these problems.

The classification of finite simple groups (part (1) of the Hölder Program) was completed in 1980, about 100 years after the formulation of the Hölder Program. Efforts by over 100 mathematicians covering between 5,000 and 10,000 journal pages (spread over some 300 to 500 individual papers) have resulted in the proof of the following result:

Theorem. There is a list consisting of 18 (infinite) families of simple groups and 26 simple groups not belonging to these families (the *sporadic* simple groups) such that every finite simple group is isomorphic to one of the groups in this list.

One example of a family of simple groups is $\{Z_p \mid p \text{ a prime}\}$. A second infinite family in the list of finite simple groups is:

$$\{SL_n(\mathbb{F})/Z(SL_n(\mathbb{F})) \mid n \in \mathbb{Z}^+, n \geq 2 \text{ and } \mathbb{F} \text{ a finite field }\}.$$

These groups are all simple except for $SL_2(\mathbb{F}_2)$ and $SL_2(\mathbb{F}_3)$ where $\mathbb{F}_2$ is the finite field with 2 elements and $\mathbb{F}_3$ is the finite field with 3 elements. This is a 2-parameter family (n and $\mathbb{F}$ being independent parameters). We shall not prove these groups are simple (although it is not technically beyond the scope of the text) but rather refer the reader to the book *Finite Group Theory* (by M. Aschbacher, Cambridge University Press, 1986) for proofs and an extensive discussion of the simple group problem. A third family of finite simple groups, the alternating groups, is discussed in the next section; we shall prove these groups are simple in the next chapter.

To gain some idea of the complexity of the classification of finite simple groups the reader may wish to peruse the proof of one of the cornerstones of the entire classification:

Theorem. (Feit–Thompson) If G is a simple group of odd order, then $G \cong Z_p$ for some prime p.

This proof takes 255 pages of hard mathematics.[2]

Part (2) of the Hölder Program, sometimes called the *extension problem*, was rather vaguely formulated. A more precise description of "putting two groups together" is: given groups A and B, describe how to obtain all groups G containing a normal subgroup N such that $N \cong B$ and $G/N \cong A$. For instance, if $A = B = Z_2$, there are precisely two possibilities for G, namely, Z_4 and V_4 (see Exercise 10 of Section 2.5) and the Hölder program seeks to describe how the two groups of order 4 could have been built from two Z_2's without a priori knowledge of the existence of the groups of order 4. This part of the Hölder Program is extremely difficult, even when the subgroups involved are of small order. For example, all composition factors of a group G have order 2 if and only if $|G| = 2^n$, for some n (one implication is easy and we shall prove both implications in Chapter 6). It is known, however, that the number of nonisomorphic groups of order 2^n grows (exponentially) as a function of 2^n, so the number of ways of putting groups of 2-power order together is not bounded. Nonetheless, there are a wealth of interesting and powerful techniques in this subtle area which serve to unravel the structure of large classes of groups. We shall discuss only a couple of ways of building larger groups from smaller ones (in the sense above) but even from this limited excursion into the area of group extensions we shall construct numerous new examples of groups and prove some classification theorems.

One class of groups which figures prominently in the theory of polynomial equations is the class of *solvable* groups:

---

[2] *Solvability of groups of odd order*, Pacific Journal of Mathematics, 13(1963), pp. 775–1029.

Definition. A group G is *solvable* if there is a chain of subgroups

$$1 = G_0 \trianglelefteq G_1 \trianglelefteq G_2 \trianglelefteq \ldots \trianglelefteq G_s = G$$

such that G_{i+1}/G_i is abelian for $i = 0, 1, \ldots, s-1$.

The terminology comes from the correspondence in Galois Theory between these groups and polynomials which can be solved by radicals (which essentially means there is an algebraic formula for the roots). Exercise 8 shows that finite solvable groups are precisely those groups whose composition factors are all of prime order.

One remarkable property of finite solvable groups is the following generalization of Sylow's Theorem due to Philip Hall (cf. Theorem 6.11 and Theorem 19.8).

Theorem. The finite group G is solvable if and only if for every divisor n of $|G|$ such that $(n, \dfrac{|G|}{n}) = 1$, G has a subgroup of order n.

As another illustration of how properties of a group G can be deduced from combined information from a normal subgroup N and the quotient group G/N we prove

if N and G/N are solvable, then G is solvable.

To see this let $\overline{G} = G/N$, let $1 = N_0 \trianglelefteq N_1 \trianglelefteq \ldots \trianglelefteq N_n = N$ be a chain of subgroups of N such that N_{i+1}/N_i is abelian, $0 \le i < n$ and let $\overline{1} = \overline{G_0} \trianglelefteq \overline{G_1} \trianglelefteq \ldots \trianglelefteq \overline{G_m} = \overline{G}$ be a chain of subgroups of $\overline{G}$ such that $\overline{G_{i+1}}/\overline{G_i}$ is abelian, $0 \le i < m$. By the Lattice Isomorphism Theorem there are subgroups G_i of G with $N \le G_i$ such that $G_i/N = \overline{G_i}$ and $G_i \trianglelefteq G_{i+1}, 0 \le i < m$. By the Third Isomorphism Theorem

$$\overline{G_{i+1}}/\overline{G_i} = (G_{i+1}/N)/(G_i/N) \cong G_{i+1}/G_i.$$

Thus

$$1 = N_0 \trianglelefteq N_1 \trianglelefteq \ldots \trianglelefteq N_n = N = G_0 \trianglelefteq G_1 \trianglelefteq \ldots \trianglelefteq G_m = G$$

is a chain of subgroups of G all of whose successive quotient groups are abelian. This proves G is solvable.

It is inaccurate to say that finite group theory is concerned *only* with the Hölder Program. It *is* accurate to say that the Hölder Program suggests a large number of problems and motivates a number of algebraic techniques. For example, in the study of the extension problem where we are given groups A and B and wish to find G and $N \trianglelefteq G$ with $N \cong B$ and $G/N \cong A$, we shall see that (under certain conditions) we are led to an *action* of the group A on the set B. Such actions form the crux of the next chapter (and will result in information both about simple and non-simple groups) and this notion is a powerful one in mathematics not restricted to the theory of groups.

The final section of this chapter introduces another family of groups and although in line with our interest in simple groups, it will be of independent importance throughout the text, particularly in our study later of determinants and the solvability of polynomial equations.

EXERCISES

1. Prove that if G is an abelian simple group then $G \cong Z_p$ for some prime p (do not assume G is a finite group).

2. Exhibit all 3 composition series for Q_8 and all 7 composition series for D_8. List the composition factors in each case.

3. Find a composition series for the quasidihedral group of order 16 (cf. Exercise 11, Section 2.5). Deduce that QD_{16} is solvable.

4. Use Cauchy's Theorem and induction to show that a finite abelian group has a subgroup of order n for each positive divisor n of its order.

5. Prove that subgroups and quotient groups of a solvable group are solvable.

6. Prove part (1) of the Jordan–Hölder Theorem by induction on $|G|$.

7. If G is a finite group and $H \trianglelefteq G$ prove that there is a composition series of G, one of whose terms is H.

8. Let G be a *finite* group. Prove that the following are equivalent:
   (i) G is solvable
   (ii) G has a chain of subgroups: $1 = H_0 \trianglelefteq H_1 \trianglelefteq H_2 \trianglelefteq \ldots \trianglelefteq H_s = G$ such that H_{i+1}/H_i is cyclic, $0 \le i \le s-1$
   (iii) all composition factors of G are of prime order
   (iv) G has a chain of subgroups: $1 = N_0 \trianglelefteq N_1 \trianglelefteq N_2 \trianglelefteq \ldots \trianglelefteq N_t = G$ such that each N_i is a normal subgroup of G and N_{i+1}/N_i is abelian, $0 \le i \le t-1$.

   [For (iv), prove that a minimal nontrivial normal subgroup M of G is necessarily abelian and then use induction. To see that M is abelian, let $N \trianglelefteq M$ be of prime index (by (iii)) and show that $x^{-1}y^{-1}xy \in N$ for all $x, y \in M$ (cf. Exercise 40, Section 1). Apply the same argument to gNg^{-1} to show that $x^{-1}y^{-1}xy$ lies in the intersection of all G-conjugates of N, and use the minimality of M to conclude that $x^{-1}y^{-1}xy = 1$.]

9. Prove the following special case of part (2) of the Jordan–Hölder Theorem: assume the finite group G has two composition series

$$1 = N_0 \trianglelefteq N_1 \trianglelefteq \ldots \trianglelefteq N_r = G \qquad \text{and} \qquad 1 = M_0 \trianglelefteq M_1 \trianglelefteq M_2 = G.$$

   Show that $r = 2$ and that the list of composition factors is the same. [Use the Second Isomorphism Theorem.]

10. Prove part (2) of the Jordan–Hölder Theorem by induction on $\min\{r, s\}$. [Apply the inductive hypothesis to $H = N_{r-1} \cap M_{s-1}$ and use the preceding exercises.]

11. Prove that if H is a nontrivial normal subgroup of the solvable group G then there is a nontrivial subgroup A of H with $A \trianglelefteq G$ and A abelian.

12. Prove (without using the Feit–Thompson Theorem) that the following are equivalent:
    (i) every group of odd order is solvable
    (ii) the only simple groups of odd order are those of prime order.

3.5 TRANSPOSITIONS AND THE ALTERNATING GROUP

Transpositions and Generation of S_n

As we saw in Section 1.3 (and will prove in the next chapter) every element of S_n can be written as a product of disjoint cycles in an essentially unique fashion. In contrast,

106

every element of S_n can be written in many different ways as a (nondisjoint) product of cycles. For example, even in S_3 the element $\sigma = (1\,2\,3)$ may be written

$$\sigma = (1\,2\,3) = (1\,3)(1\,2) = (1\,2)(1\,3)(1\,2)(1\,3) = (1\,2)(2\,3)$$

and, in fact, there are an infinite number of different ways to write σ. Not requiring the cycles to be disjoint totally destroys the uniqueness of a representation of a permutation as a product of cycles. We can, however, obtain a sort of "parity check" from writing permutations (nonuniquely) as products of 2-cycles.

Definition. A 2-cycle is called a *transposition*.

Intuitively, every permutation of $\{1, 2, \ldots, n\}$ can be realized by a succession of transpositions or simple interchanges of pairs of elements (try this on a small deck of cards sometime!). We illustrate how this may be done. First observe that

$$(a_1\, a_2 \ldots a_m) = (a_1\, a_m)(a_1\, a_{m-1})(a_1\, a_{m-2}) \ldots (a_1\, a_2)$$

for any m-cycle. Now any permutation in S_n may be written as a product of cycles (for instance, its cycle decomposition). Writing each of these cycles in turn as a product of transpositions by the above procedure we see that

every element of S_n may be written as a product of transpositions

or, equivalently,

$$S_n = \langle\, T\, \rangle \quad \text{where} \quad T = \{(i\,\,j) \mid 1 \le i < j \le n\}.$$

For example, the permutation σ in Section 1.3 may be written

$$\sigma = (1\,12\,8\,10\,4)(2\,13)(5\,11\,7)(6\,9)$$
$$= (1\,4)(1\,10)(1\,8)(1\,12)(2\,13)(5\,7)(5\,11)(6\,9).$$

The Alternating Group

Again we emphasize that for any $\sigma \in S_n$ there may be many ways of writing σ as a product of transpositions. For fixed σ we now show that the parity (i.e., an odd or even number of terms) is the same for any product of transpositions equaling σ.

Let $x_1, \ldots, x_n$ be independent variables and let Δ be the polynomial

$$\Delta = \prod_{1 \le i < j \le n} (x_i - x_j),$$

i.e., the product of all the terms $x_i - x_j$ for $i < j$. For example, when $n = 4$,

$$\Delta = (x_1 - x_2)(x_1 - x_3)(x_1 - x_4)(x_2 - x_3)(x_2 - x_4)(x_3 - x_4).$$

For each $\sigma \in S_n$ let σ act on Δ by permuting the variables in the same way it permutes their indices:

$$\sigma(\Delta) = \prod_{1 \le i < j \le n} (x_{\sigma(i)} - x_{\sigma(j)}).$$

For example, if $n = 4$ and $\sigma = (1\ 2\ 3\ 4)$ then

$$\sigma(\Delta) = (x_2 - x_3)(x_2 - x_4)(x_2 - x_1)(x_3 - x_4)(x_3 - x_1)(x_4 - x_1)$$

(we have written the factors in the same order as above and applied σ to each factor to get $\sigma(\Delta)$). Note (in general) that Δ contains one factor $x_i - x_j$ for all $i < j$, and since σ is a bijection of the indices, $\sigma(\Delta)$ must contain either $x_i - x_j$ or $x_j - x_i$, but not both (and certainly no $x_i - x_i$ terms), for all $i < j$. If $\sigma(\Delta)$ has a factor $x_j - x_i$ where $j > i$, write this term as $-(x_i - x_j)$. Collecting all the changes in sign together we see that Δ and $\sigma(\Delta)$ have the same factors up to a product of -1's, i.e.,

$$\sigma(\Delta) = \pm\Delta, \qquad \text{for all } \sigma \in S_n.$$

For each $\sigma \in S_n$ let

$$\epsilon(\sigma) = \begin{cases} +1, & \text{if } \sigma(\Delta) = \Delta \\ -1, & \text{if } \sigma(\Delta) = -\Delta. \end{cases}$$

In the example above with $n = 4$ and $\sigma = (1\ 2\ 3\ 4)$, there are exactly 3 factors of the form $x_j - x_i$ where $j > i$ in $\sigma(\Delta)$, each of which contributes a factor of -1. Hence

$$(1\ 2\ 3\ 4)(\Delta) = (-1)^3(\Delta) = -\Delta,$$

so

$$\epsilon((1\ 2\ 3\ 4)) = -1.$$

Definition.
   (1) $\epsilon(\sigma)$ is called the *sign* of σ.
   (2) σ is called an *even permutation* if $\epsilon(\sigma) = 1$ and an *odd permutation* if $\epsilon(\sigma) = -1$.

The next result shows that the sign of a permutation defines a homomorphism.

Proposition 23. The map $\epsilon : S_n \to \{\pm 1\}$ is a homomorphism (where $\{\pm 1\}$ is a multiplicative version of the cyclic group of order 2).

Proof: By definition,

$$(\tau\sigma)(\Delta) = \prod_{1 \le i < j \le n} (x_{\tau\sigma(i)} - x_{\tau\sigma(j)}).$$

Suppose that $\sigma(\Delta)$ has exactly k factors of the form $x_j - x_i$ with $j > i$, that is $\epsilon(\sigma) = (-1)^k$. When calculating $(\tau\sigma)(\Delta)$, after first applying σ to the indices we see that $(\tau\sigma)(\Delta)$ has exactly k factors of the form $x_{\tau(j)} - x_{\tau(i)}$ with $j > i$. Interchanging the order of the terms in these k factors introduces the sign change $(-1)^k = \epsilon(\sigma)$, and now all factors of $(\tau\sigma)(\Delta)$ are of the form $x_{\tau(p)} - x_{\tau(q)}$, with $p < q$. Thus

$$(\tau\sigma)(\Delta) = \epsilon(\sigma) \prod_{1 \le p < q \le n} (x_{\tau(p)} - x_{\tau(q)}).$$

Since by definition of ϵ

$$\prod_{1 \le p < q \le n} (x_{\tau(p)} - x_{\tau(q)}) = \epsilon(\tau)\Delta$$

we have $(\tau\sigma)(\Delta) = \epsilon(\sigma)\epsilon(\tau)\Delta$. Thus $\epsilon(\tau\sigma) = \epsilon(\sigma)\epsilon(\tau) = \epsilon(\tau)\epsilon(\sigma)$, as claimed.

To see the proof in action, let $n = 4$, $\sigma = (1\ 2\ 3\ 4)$, $\tau = (4\ 2\ 3)$ so $\tau\sigma = (1\ 3\ 2\ 4)$. By definition (using the explicit Δ in this case),

$$(\tau\sigma)(\Delta) = (1\ 3\ 2\ 4)(\Delta)$$
$$= (x_3 - x_4)(x_3 - x_2)(x_3 - x_1)(x_4 - x_2)(x_4 - x_1)(x_2 - x_1)$$
$$= (-1)^5 \Delta$$

where all factors except the first one are flipped to recover Δ. This shows $\epsilon(\tau\sigma) = -1$. On the other hand, since we already computed $\sigma(\Delta)$

$$(\tau\sigma)(\Delta) = \tau(\sigma(\Delta))$$
$$= (x_{\tau(2)} - x_{\tau(3)})(x_{\tau(2)} - x_{\tau(4)})(x_{\tau(2)} - x_{\tau(1)})(x_{\tau(3)} - x_{\tau(4)}) \times$$
$$\times (x_{\tau(3)} - x_{\tau(1)})(x_{\tau(4)} - x_{\tau(1)})$$
$$= (-1)^3 \prod_{1 \le p < q \le 4} (x_{\tau(p)} - x_{\tau(q)}) = (-1)^3 \tau(\Delta)$$

where here the third, fifth, and sixth factors need to have their terms interchanged in order to put all factors in the form $x_{\tau(p)} - x_{\tau(q)}$ with $p < q$. We already calculated that $\epsilon(\sigma) = (-1)^3 = -1$ and, by the same method, it is easy to see that $\epsilon(\tau) = (-1)^2 = 1$ so $\epsilon(\tau\sigma) = -1 = \epsilon(\tau)\epsilon(\sigma)$.

The next step is to compute $\epsilon((i\ j))$, for any transposition $(i\ j)$. Rather than compute this directly for arbitrary i and j we do it first for $i = 1$ and $j = 2$ and reduce the general case to this. It is clear that applying $(1\ 2)$ to Δ (regardless of what n is) will flip exactly one factor, namely $x_1 - x_2$; thus $\epsilon((1\ 2)) = -1$. Now for any transposition $(i\ j)$ let λ be the permutation which interchanges 1 and i, interchanges 2 and j, and leaves all other numbers fixed (if $i = 1$ or $j = 2$, λ fixes i or j, respectively). Then it is easy to see that $(i\ j) = \lambda(1\ 2)\lambda$ (compute what the right hand side does to any $k \in \{1, 2, \ldots, n\}$). Since ϵ is a homomorphism we obtain

$$\epsilon((i\ j)) = \epsilon(\lambda(1\ 2)\lambda)$$
$$= \epsilon(\lambda)\epsilon((1\ 2))\epsilon(\lambda)$$
$$= (-1)\epsilon(\lambda)^2$$
$$= -1.$$

This proves

Proposition 24. Transpositions are all odd permutations and ϵ is a surjective homomorphism.

Definition. The *alternating group of degree n*, denoted by A_n, is the kernel of the homomorphism ϵ (i.e., the set of even permutations).

Note that by the First Isomorphism Theorem $S_n/A_n \cong \epsilon(S_n) = \{\pm 1\}$, so that the order of A_n is easily determined: $|A_n| = \frac{1}{2}|S_n| = \frac{1}{2}(n!)$. Also, $S_n - A_n$ is the coset of

A_n which is not the identity coset and this is the set of all odd permutations. The signs of permutations obey the usual $\mathbb{Z}/2\mathbb{Z}$ laws:

$$(even)(even) = (odd)(odd) = even$$
$$(even)(odd) = (odd)(even) = odd.$$

Moreover, since ϵ is a homomorphism and every $\sigma \in S_n$ is a product of transpositions, say $\sigma = \tau_1\tau_2 \cdots \tau_k$, then $\epsilon(\sigma) = \epsilon(\tau_1)\cdots\epsilon(\tau_k)$; since $\epsilon(\tau_i) = -1$, for $i = 1, 2, \ldots, k$, $\epsilon(\sigma) = (-1)^k$. Thus the class of k (mod 2), i.e., the parity of the number of transpositions in the product, is the same no matter how we write σ as a product of transpositions:

$$\epsilon(\sigma) = \begin{cases} +1, & \text{if } \sigma \text{ is a product of an even number of transpositions} \\ -1, & \text{if } \sigma \text{ is a product of an odd number of transpositions.} \end{cases}$$

Finally we give a quick way of computing $\epsilon(\sigma)$ from the cycle decomposition of σ. Recall that an m-cycle may be written as a product of $m - 1$ transpositions. Thus

an m-cycle is an odd permutation if and only if m is even.

For any permutation σ let $\alpha_1\alpha_2 \cdots \alpha_k$ be its cycle decomposition. Then $\epsilon(\sigma)$ is given by $\epsilon(\alpha_1) \cdots \epsilon(\alpha_k)$ and $\epsilon(\alpha_i) = -1$ if and only if the length of α_i is even. It follows that for $\epsilon(\sigma)$ to be -1 the product of the $\epsilon(\alpha_i)$'s must contain an odd number of factors of (-1). We summarize this in the following proposition:

Proposition 25. The permutation σ is odd if and only if the number of cycles of even length in its cycle decomposition is odd.

For example, $\sigma = (1\,2\,3\,4\,5\,6)(7\,8\,9)(10\,11)(12\,13\,14\,15)(16\,17\,18)$ has 3 cycles of even length, so $\epsilon(\sigma) = -1$. On the other hand, $\tau = (1\,12\,8\,10\,4)(2\,13)(5\,11\,7)(6\,9)$ has exactly 2 cycles of even length, hence $\epsilon(\tau) = 1$.

Be careful not to confuse the terms "odd" and "even" for a permutation σ with the parity of the order of σ. In fact, if σ is of odd order, all cycles in the cycle decomposition of σ have odd length so σ has an even (in this case 0) number of cycles of even length and hence is an even permutation. If $|\sigma|$ is even, σ may be either an even or an odd permutation; e.g., $(1\,2)$ is odd, $(1\,2)(3\,4)$ is even but both have order 2.

As we mentioned in the preceding section, the alternating groups A_n will be important in the study of solvability of polynomials. In the next chapter we shall prove:

A_n is a non-abelian simple group for all $n \geq 5$.

For small values of n, A_n is already familiar to us: A_1 and A_2 are both the trivial group and $|A_3| = 3$ (so $A_3 = \langle (1\,2\,3) \rangle \cong Z_3$). The group A_4 has order 12. Exercise 7 shows A_4 is isomorphic to the group of symmetries of a regular tetrahedron. The lattice of subgroups of A_4 appears in Figure 8 (Exercise 8 asserts that this is its complete lattice of subgroups). One of the nicer aspects of this lattice is that (unlike "virtually all groups") it is a planar graph (there are no crossing lines except at the vertices; see the lattice of D_{16} for a nonplanar lattice).

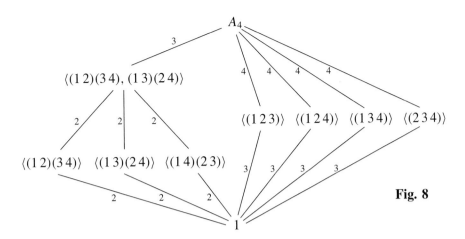

Fig. 8

EXERCISES

1. In Exercises 1 and 2 of Section 1.3 you were asked to find the cycle decomposition of some permutations. Write each of these permutations as a product of transpositions. Determine which of these is an even permutation and which is an odd permutation.

2. Prove that σ^2 is an even permutation for every permutation σ.

3. Prove that S_n is generated by $\{(i \ \ i+1) \mid 1 \le i \le n-1\}$. [Consider conjugates, viz. $(2\ 3)(1\ 2)(2\ 3)^{-1}$.]

4. Show that $S_n = \langle\,(1\ 2),\ (1\ 2\ 3\ldots n)\,\rangle$ for all $n \ge 2$.

5. Show that if p is prime, $S_p = \langle\,\sigma, \tau\,\rangle$ where σ is any transposition and τ is any p-cycle.

6. Show that $\langle\,(1\ 3),\ (1\ 2\ 3\ 4)\,\rangle$ is a proper subgroup of S_4. What is the isomorphism type of this subgroup?

7. Prove that the group of rigid motions of a tetrahedron is isomorphic to A_4. [Recall Exercise 20 in Section 1.7.]

8. Prove the lattice of subgroups of A_4 given in the text is correct. [By the preceding exercise and the comments following Lagrange's Theorem, A_4 has no subgroup of order 6.]

9. Prove that the (unique) subgroup of order 4 in A_4 is normal and is isomorphic to V_4.

10. Find a composition series for A_4. Deduce that A_4 is solvable.

11. Prove that S_4 has no subgroup isomorphic to Q_8.

12. Prove that A_n contains a subgroup isomorphic to S_{n-2} for each $n \ge 3$.

13. Prove that every element of order 2 in A_n is the square of an element of order 4 in S_n. [An element of order 2 in A_n is a product of $2k$ commuting transpositions.]

14. Prove that the subgroup of A_4 generated by any element of order 2 and any element of order 3 is all of A_4.

15. Prove that if x and y are distinct 3-cycles in S_4 with $x \ne y^{-1}$, then the subgroup of S_4 generated by x and y is A_4.

16. Let x and y be distinct 3-cycles in S_5 with $x \ne y^{-1}$.
    (a) Prove that if x and y fix a common element of $\{1, \ldots, 5\}$, then $\langle x, y \rangle \cong A_4$.
    (b) Prove that if x and y do not fix a common element of $\{1, \ldots, 5\}$, then $\langle x, y \rangle = A_5$.

17. If x and y are 3-cycles in S_n, prove that $\langle x, y \rangle$ is isomorphic to Z_3, A_4, A_5 or $Z_3 \times Z_3$.

CHAPTER 4

Group Actions

In this chapter we consider some of the consequences of a group acting on a set. It is an important and recurring idea in mathematics that when one object acts on another then much information can be obtained on both. As more structure is added to the set on which the group acts (for example, groups acting on groups or groups acting on vector spaces (considered in Chapter 18)), more information on the structure of the group becomes available. This study of group actions culminates here in the proof of Sylow's Theorem and the examples and classifications which accrue from it.

The concept of an action will recur as we study modules, vector spaces, canonical forms for matrices and Galois Theory, and is one of the fundamental unifying themes in the text.

4.1 GROUP ACTIONS AND PERMUTATION REPRESENTATIONS

In this section we give the basic theory of group actions and then apply this theory to subgroups of S_n acting on $\{1, 2, \ldots, n\}$ to prove that every element of S_n has a unique cycle decomposition. In Sections 2 and 3 we apply the general theory to two other specific group actions to derive some important results.

Let G be a group acting on a nonempty set A. Recall from Section 1.7 that for each $g \in G$ the map

$$\sigma_g : A \to A \qquad \text{defined by} \qquad \sigma_g : a \mapsto g \cdot a$$

is a permutation of A. We also saw in Section 1.7 that there is a homomorphism associated to an action of G on A:

$$\varphi : G \to S_A \qquad \text{defined by} \qquad \varphi(g) = \sigma_g,$$

called the *permutation representation* associated to the given action. Recall some additional terminology associated to group actions introduced in Sections 1.7 and 2.2.

Definition.
  (1) The *kernel* of the action is the set of elements of G that act trivially on every element of A: $\{g \in G \mid g \cdot a = a \text{ for all } a \in A\}$.
  (2) For each $a \in A$ the *stabilizer* of a in G is the set of elements of G that fix the element a: $\{g \in G \mid g \cdot a = a\}$ and is denoted by G_a.
  (3) An action is *faithful* if its kernel is the identity.

Note that the kernel of an action is precisely the same as the kernel of the associated permutation representation; in particular, the kernel is a normal subgroup of G. Two group elements induce the same permutation on A if and only if they are in the same coset of the kernel (if and only if they are in the same fiber of the permutation representation φ). In particular an action of G on A may also be viewed as a faithful action of the quotient group $G/\ker\varphi$ on A. Recall from Section 2.2 that the stabilizer in G of an element a of A is a subgroup of G. If a is a fixed element of A, then the kernel of the action is contained in the stabilizer G_a since the kernel of the action is the set of elements of G that stabilize every point, namely $\cap_{a\in A} G_a$.

Examples

(1) Let n be a positive integer. The group $G = S_n$ acts on the set $A = \{1, 2, \ldots, n\}$ by $\sigma \cdot i = \sigma(i)$ for all $i \in \{1, \ldots, n\}$. The permutation representation associated to this action is the identity map $\varphi : S_n \to S_n$. This action is faithful and for each $i \in \{1, \ldots, n\}$ the stabilizer G_i (the subgroup of all permutations fixing i) is isomorphic to S_{n-1} (cf. Exercise 15, Section 3.2).

(2) Let $G = D_8$ act on the set A consisting of the four vertices of a square. Label these vertices 1,2,3,4 in a clockwise fashion as in Figure 2 of Section 1.2. Let r be the rotation of the square clockwise by $\pi/2$ radians and let s be the reflection in the line which passes through vertices 1 and 3. Then the permutations of the vertices given by r and s are

$$\sigma_r = (1\ 2\ 3\ 4) \qquad \text{and} \qquad \sigma_s = (2\ 4).$$

Note that since the permutation representation is a homomorphism, the permutation of the four vertices corresponding to sr is $\sigma_{sr} = \sigma_s \sigma_r = (1\ 4)(2\ 3)$. The action of D_8 on the four vertices of a square is faithful since only the identity symmetry fixes all four vertices. The stabilizer of any vertex a is the subgroup of D_8 of order 2 generated by the reflection about the line passing through a and the center of the square (so, for example, the stabilizer of vertex 1 is $\langle s \rangle$).

(3) Label the four vertices of a square as in the preceding example and now let A be the set whose elements consist of unordered pairs of opposite vertices: $A = \{ \{1, 3\}, \{2, 4\} \}$. Then D_8 also acts on this set A since each symmetry of the square sends a pair of opposite vertices to a pair of opposite vertices. The rotation r interchanges the pairs $\{1, 3\}$ and $\{2, 4\}$; the reflection s fixes both unordered pairs of opposite vertices. Thus if we label the pairs $\{1, 3\}$ and $\{2, 4\}$ as **1** and **2**, respectively, then the permutations of A given by r and s are

$$\sigma_r = (\mathbf{1}\ \mathbf{2}) \qquad \text{and} \qquad \sigma_s = \text{the identity permutation.}$$

This action of D_8 is not faithful: its kernel is $\langle s, r^2 \rangle$. Moreover, for each $a \in A$ the stabilizer in D_8 of a is the same as the kernel of the action.

(4) Label the four vertices of a square as in Example 2 and now let A be the following set of unordered pairs of vertices: $\{ \{1, 2\}, \{3, 4\} \}$. The group D_8 does *not* act on this set A because $\{1, 2\} \in A$ but $r \cdot \{1, 2\} = \{2, 3\} \notin A$.

The relation between actions and homomorphisms into symmetric groups may be reversed. Namely, given any nonempty set A and any homomorphism φ of the group G into S_A we obtain an action of G on A by defining

$$g \cdot a = \varphi(g)(a)$$

for all $g \in G$ and all $a \in A$. The kernel of this action is the same as ker φ. The permutation representation associated to this action is precisely the given homomorphism φ. This proves the following result.

Proposition 1. For any group G and any nonempty set A there is a bijection between the actions of G on A and the homomorphisms of G into S_A.

In view of Proposition 1 the definition of a permutation representation may be rephrased.

Definition. If G is a group, a *permutation representation* of G is any homomorphism of G into the symmetric group S_A for some nonempty set A. We shall say a given action of G on A *affords* or *induces* the associated permutation representation of G.

We can think of a permutation representation as an analogue of the matrix representation of a linear transformation. In the case where A is a finite set of n elements we have $S_A \cong S_n$ (cf. Section 1.6), so by fixing a labelling of the elements of A we may consider our permutations as elements of the group S_n (which is exactly what we did in Examples 2 and 3 above), in the same way that fixing a basis for a vector space allows us to view a linear transformation as a matrix.

We now prove a combinatorial result about group actions which will have important consequences when we apply it to specific actions in subsequent sections.

Proposition 2. Let G be a group acting on the nonempty set A. The relation on A defined by

$$a \sim b \quad \text{if and only if} \quad a = g \cdot b \text{ for some } g \in G$$

is an equivalence relation. For each $a \in A$, the number of elements in the equivalence class containing a is $|G : G_a|$, the index of the stabilizer of a.

Proof: We first prove $\sim$ is an equivalence relation. By axiom 2 of an action, $a = 1 \cdot a$ for all $a \in A$, i.e., $a \sim a$ and the relation is reflexive. If $a \sim b$, then $a = g \cdot b$ for some $g \in G$ so that

$$g^{-1} \cdot a = g^{-1} \cdot (g \cdot b) = (g^{-1}g) \cdot b = 1 \cdot b = b$$

that is, $b \sim a$ and the relation is symmetric. Finally, if $a \sim b$ and $b \sim c$, then $a = g \cdot b$ and $b = h \cdot c$, for some $g, h \in G$ so

$$a = g \cdot b = g \cdot (h \cdot c) = (gh) \cdot c$$

hence $a \sim c$, and the relation is transitive.

To prove the last statement of the proposition we exhibit a bijection between the left cosets of G_a in G and the elements of the equivalence class of a. Let C_a be the class of a, so

$$C_a = \{g \cdot a \mid g \in G\}.$$

Suppose $b = g \cdot a \in C_a$. Then gG_a is a left coset of G_a in G. The map

$$b = g \cdot a \mapsto gG_a$$

is a map from $\mathcal{C}_a$ to the set of left cosets of G_a in G. This map is surjective since for any $g \in G$ the element $g \cdot a$ is an element of $\mathcal{C}_a$. Since $g \cdot a = h \cdot a$ if and only if $h^{-1}g \in G_a$ if and only if $gG_a = hG_a$, the map is also injective, hence is a bijection. This completes the proof.

By Proposition 2 a group G acting on the set A partitions A into disjoint equivalence classes under the action of G. These classes are given a name:

Definition. Let G be a group acting on the nonempty set A.
   (1) The equivalence class $\{g \cdot a \mid g \in G\}$ is called the *orbit* of G containing a.
   (2) The action of G on A is called *transitive* if there is only one orbit, i.e., given any two elements $a, b \in A$ there is some $g \in G$ such that $a = g \cdot b$.

Examples

   Let G be a group acting on the set A.
   (1) If G acts trivially on A then $G_a = G$ for all $a \in A$ and the orbits are the elements of A. This action is transitive if and only if $|A| = 1$.
   (2) The symmetric group $G = S_n$ acts transitively in its usual action as permutations on $A = \{1, 2, \dots, n\}$. Note that the stabilizer in G of any point i has index $n = |A|$ in S_n.
   (3) When the group G acts on the set A, any subgroup of G also acts on A. If G is transitive on A a subgroup of G need not be transitive on A. For example, if $G = \langle (1\ 2), (3\ 4) \rangle \le S_4$ then the orbits of G on $\{1, 2, 3, 4\}$ are $\{1, 2\}$ and $\{3, 4\}$ and there is no element of G that sends 2 to 3. The discussion below on cycle decompositions shows that when $\langle \sigma \rangle$ is any cyclic subgroup of S_n then the orbits of $\langle \sigma \rangle$ consist of the sets of numbers that appear in the individual cycles in the cycle decomposition of σ (for example, the orbits of $\langle (1\ 2)(3\ 4\ 5) \rangle$ are $\{1, 2\}$ and $\{3, 4, 5\}$).
   (4) The group D_8 acts transitively on the four vertices of the square and the stabilizer of any vertex is the subgroup of order 2 (and index 4) generated by the reflection about the line of symmetry passing through that point.
   (5) The group D_8 also acts transitively on the set of two pairs of opposite vertices. In this action the stabilizer of any point is $\langle s, r^2 \rangle$ (which is of index 2).

Cycle Decompositions

We now prove that every element of the symmetric group S_n has the unique cycle decomposition described in Section 1.3. Let $A = \{1, 2, \dots, n\}$, let σ be an element of S_n and let $G = \langle \sigma \rangle$. Then $\langle \sigma \rangle$ acts on A and so, by Proposition 2, it partitions $\{1, 2, \dots, n\}$ into a unique set of (disjoint) orbits. Let $\mathcal{O}$ be one of these orbits and let $x \in \mathcal{O}$. By (the proof of) Proposition 2 applied to $A = \mathcal{O}$ we see that there is a bijection between the left cosets of G_x in G and the elements of $\mathcal{O}$, given explicitly by

$$\sigma^i x \mapsto \sigma^i G_x.$$

Since G is a cyclic group, $G_x \trianglelefteq G$ and G/G_x is cyclic of order d, where d is the smallest positive integer for which $\sigma^d \in G_x$ (cf. Example 2 following Proposition 7 in Section 3.1). Also, $d = |G : G_x| = |\mathcal{O}|$. Thus the distinct cosets of G_x in G are

$$1G_x, \ \sigma G_x, \ \sigma^2 G_x, \ \dots, \ \sigma^{d-1}G_x.$$

This shows that the distinct elements of $\mathcal{O}$ are
$$x, \ \sigma(x), \ \sigma^2(x), \ \ldots, \ \sigma^{d-1}(x).$$
Ordering the elements of $\mathcal{O}$ in this manner shows that σ cycles the elements of $\mathcal{O}$, that is, on an orbit of size d, σ acts as a d-cycle. This proves the existence of a cycle decomposition for each $\sigma \in S_n$.

The orbits of $\langle \sigma \rangle$ are uniquely determined by σ. The only latitude is in which order the orbits are listed. Within each orbit, $\mathcal{O}$, we may begin with any element as a representative. Choosing $\sigma^i(x)$ instead of x as the initial representative simply produces the elements of $\mathcal{O}$ in the order
$$\sigma^i(x), \ \sigma^{i+1}(x), \ \ldots, \ \sigma^{d-1}(x), \ x, \ \sigma(x), \ \ldots, \ \sigma^{i-1}(x),$$
which is a cyclic permutation (forward $i - 1$ terms) of the original list. It follows that the cycle decomposition above is unique up to a rearrangement of the cycles and up to a cyclic permutation of the integers within each cycle.

Subgroups of symmetric groups are called *permutation groups*. For any subgroup G of S_n the orbits of G will refer to its orbits on $\{1, 2, \ldots, n\}$. The orbits of an element σ in S_n will mean the orbits of the group $\langle \sigma \rangle$ (namely the sets of integers comprising the cycles in its cycle decomposition).

The exercises below further illustrate how group theoretic information can be obtained from permutation representations.

EXERCISES

Let G be a group and let A be a nonempty set.

1. Let G act on the set A. Prove that if $a, b \in A$ and $b = g \cdot a$ for some $g \in G$, then $G_b = gG_a g^{-1}$ (G_a is the stabilizer of a). Deduce that if G acts transitively on A then the kernel of the action is $\cap_{g \in G} \, gG_a g^{-1}$.

2. Let G be a *permutation group* on the set A (i.e., $G \leq S_A$), let $\sigma \in G$ and let $a \in A$. Prove that $\sigma G_a \sigma^{-1} = G_{\sigma(a)}$. Deduce that if G acts transitively on A then
$$\bigcap_{\sigma \in G} \sigma G_a \sigma^{-1} = 1.$$

3. Assume that G is an abelian, transitive subgroup of S_A. Show that $\sigma(a) \neq a$ for all $\sigma \in G - \{1\}$ and all $a \in A$. Deduce that $|G| = |A|$. [Use the preceding exercise.]

4. Let S_3 act on the set Ω of ordered pairs: $\{(i, j) \mid 1 \leq i, j \leq 3\}$ by $\sigma((i, j)) = (\sigma(i), \sigma(j))$. Find the orbits of S_3 on Ω. For each $\sigma \in S_3$ find the cycle decomposition of σ under this action (i.e., find its cycle decomposition when σ is considered as an element of S_9 — first fix a labelling of these nine ordered pairs). For each orbit $\mathcal{O}$ of S_3 acting on these nine points pick some $a \in \mathcal{O}$ and find the stabilizer of a in S_3.

5. For each of parts (a) and (b) repeat the preceding exercise but with S_3 acting on the specified set:
   (a) the set of 27 triples $\{(i, j, k) \mid 1 \leq i, j, k \leq 3\}$
   (b) the set $\mathcal{P}(\{1, 2, 3\}) - \{\emptyset\}$ of all 7 nonempty subsets of $\{1, 2, 3\}$.

6. As in Exercise 12 of Section 2.2 let R be the set of all polynomials with integer coefficients in the independent variables x_1, x_2, x_3, x_4 and let S_4 act on R by permuting the indices of

the four variables:

$$\sigma \cdot p(x_1, x_2, x_3, x_4) = p(x_{\sigma(1)}, x_{\sigma(2)}, x_{\sigma(3)}, x_{\sigma(4)})$$

for all $\sigma \in S_4$.

   (a) Find the polynomials in the orbit of S_4 on R containing $x_1 + x_2$ (recall from Exercise 12 in Section 2.2 that the stabilizer of this polynomial has order 4).

   (b) Find the polynomials in the orbit of S_4 on R containing $x_1 x_2 + x_3 x_4$ (recall from Exercise 12 in Section 2.2 that the stabilizer of this polynomial has order 8).

   (c) Find the polynomials in the orbit of S_4 on R containing $(x_1 + x_2)(x_3 + x_4)$.

7. Let G be a transitive permutation group on the finite set A. A *block* is a nonempty subset B of A such that for all $\sigma \in G$ either $\sigma(B) = B$ or $\sigma(B) \cap B = \emptyset$ (here $\sigma(B)$ is the set $\{\sigma(b) \mid b \in B\}$).

   (a) Prove that if B is a block containing the element a of A, then the set G_B defined by $G_B = \{\sigma \in G \mid \sigma(B) = B\}$ is a subgroup of G containing G_a.

   (b) Show that if B is a block and $\sigma_1(B), \sigma_2(B), \ldots, \sigma_n(B)$ are all the distinct images of B under the elements of G, then these form a partition of A.

   (c) A (transitive) group G on a set A is said to be *primitive* if the only blocks in A are the trivial ones: the sets of size 1 and A itself. Show that S_4 is primitive on $A = \{1, 2, 3, 4\}$. Show that D_8 is not primitive as a permutation group on the four vertices of a square.

   (d) Prove that the transitive group G is primitive on A if and only if for each $a \in A$, the only subgroups of G containing G_a are G_a and G (i.e., G_a is a *maximal* subgroup of G, cf. Exercise 16, Section 2.4). [Use part (a).]

8. A transitive permutation group G on a set A is called *doubly transitive* if for any (hence all) $a \in A$ the subgroup G_a is transitive on the set $A - \{a\}$.

   (a) Prove that S_n is doubly transitive on $\{1, 2, \ldots, n\}$ for all $n \geq 2$.

   (b) Prove that a doubly transitive group is primitive. Deduce that D_8 is not doubly transitive in its action on the 4 vertices of a square.

9. Assume G acts transitively on the finite set A and let H be a normal subgroup of G. Let $\mathcal{O}_1, \mathcal{O}_2, \ldots, \mathcal{O}_r$ be the distinct orbits of H on A.

   (a) Prove that G permutes the sets $\mathcal{O}_1, \mathcal{O}_2, \ldots, \mathcal{O}_r$ in the sense that for each $g \in G$ and each $i \in \{1, \ldots, r\}$ there is a j such that $g\mathcal{O}_i = \mathcal{O}_j$, where $g\mathcal{O} = \{g \cdot a \mid a \in \mathcal{O}\}$ (i.e., in the notation of Exercise 7 the sets $\mathcal{O}_1, \ldots, \mathcal{O}_r$ are blocks). Prove that G is transitive on $\{\mathcal{O}_1, \ldots, \mathcal{O}_r\}$. Deduce that all orbits of H on A have the same cardinality.

   (b) Prove that if $a \in \mathcal{O}_1$ then $|\mathcal{O}_1| = |H : H \cap G_a|$ and prove that $r = |G : HG_a|$. [Draw the sublattice describing the Second Isomorphism Theorem for the subgroups H and G_a of G. Note that $H \cap G_a = H_a$.]

10. Let H and K be subgroups of the group G. For each $x \in G$ define the HK *double coset* of x in G to be the set

$$HxK = \{hxk \mid h \in H, \ k \in K\}.$$

   (a) Prove that HxK is the union of the left cosets $x_1 K, \ldots, x_n K$ where $\{x_1 K, \ldots, x_n K\}$ is the orbit containing xK of H acting by left multiplication on the set of left cosets of K.

   (b) Prove that HxK is a union of right cosets of H.

   (c) Show that HxK and HyK are either the same set or are disjoint for all $x, y \in G$. Show that the set of HK double cosets partitions G.

   (d) Prove that $|HxK| = |K| \cdot |H : H \cap xKx^{-1}|$.

   (e) Prove that $|HxK| = |H| \cdot |K : K \cap x^{-1}Hx|$.

4.2 GROUPS ACTING ON THEMSELVES BY LEFT MULTIPLICATION — CAYLEY'S THEOREM

In this section G is any group and we first consider G *acting on itself* (i.e., $A = G$) by *left multiplication*:

$$g \cdot a = ga \qquad \text{for all } g \in G, \ a \in G$$

where ga denotes the product of the two group elements g and a in G (if G is written additively, the action will be written $g \cdot a = g + a$ and called left translation). We saw in Section 1.7 that this satisfies the two axioms of a group action.

When G is a finite group of order n it is convenient to label the elements of G with the integers $1, 2, \ldots, n$ in order to describe the permutation representation afforded by this action. In this way the elements of G are listed as $g_1, g_2, \ldots, g_n$ and for each $g \in G$ the permutation σ_g may be described as a permutation of the indices $1, 2, \ldots, n$ as follows:

$$\sigma_g(i) = j \qquad \text{if and only if} \qquad gg_i = g_j.$$

A different labelling of the group elements will give a different description of σ_g as a permutation of $\{1, 2, \ldots, n\}$ (cf. the exercises).

Example

Let $G = \{1, a, b, c\}$ be the Klein 4-group whose group table is written out in Section 2.5. Label the group elements $1, a, b, c$ with the integers 1,2,3,4, respectively. Under this labelling we compute the permutation σ_a induced by the action of left multiplication by the group element a:

$a \cdot 1 = a1 = a$ and so $\sigma_a(1) = 2$
$a \cdot a = aa = 1$ and so $\sigma_a(2) = 1$
$a \cdot b = ab = c$ and so $\sigma_a(3) = 4$ and
$a \cdot c = ac = b$ and so $\sigma_a(4) = 3$.

With this labelling of the elements of G we see that $\sigma_a = (1\ 2)(3\ 4)$. In the permutation representation associated to the action of the Klein 4-group on itself by left multiplication one similarly computes that

$$a \mapsto \sigma_a = (1\ 2)(3\ 4) \qquad b \mapsto \sigma_b = (1\ 3)(2\ 4) \qquad c \mapsto \sigma_c = (1\ 4)(2\ 3),$$

which explicitly gives the permutation representation $G \to S_4$ associated to this action under this labelling.

It is easy to see (and we shall prove this shortly in a more general setting) that the action of a group on itself by left multiplication is always transitive and faithful, and that the stabilizer of any point is the identity subgroup (these facts can be checked by inspection for the above example).

We now consider a generalization of the action of a group by left multiplication on the set of its elements. Let H be any subgroup of G and let A be the set of all left cosets of H in G. Define an action of G on A by

$$g \cdot aH = gaH \qquad \text{for all } g \in G, \ aH \in A$$

where gaH is the left coset with representative ga. One easily checks that this satisfies the two axioms for a group action, i.e., that G does act on the set of left cosets of H

by left multiplication. In the special case when H is the identity subgroup of G the coset aH is just $\{a\}$ and if we identify the element a with the set $\{a\}$, this action by left multiplication on left cosets of the identity subgroup is the same as the action of G on itself by left multiplication.

When H is of finite index m in G it is convenient to label the left cosets of H with the integers $1, 2, \ldots, m$ in order to describe the permutation representation afforded by this action. In this way the distinct left cosets of H in G are listed as $a_1 H, a_2 H, \ldots, a_m H$ and for each $g \in G$ the permutation σ_g may be described as a permutation of the indices $1, 2, \ldots, m$ as follows:

$$\sigma_g(i) = j \qquad \text{if and only if} \qquad ga_i H = a_j H.$$

A different labelling of the group elements will give a different description of σ_g as a permutation of $\{1, 2, \ldots, m\}$ (cf. the exercises).

Example

Let $G = D_8$ and let $H = \langle s \rangle$. Label the distinct left cosets $1H, rH, r^2 H, r^3 H$ with the integers 1,2,3,4 respectively. Under this labelling we compute the permutation σ_s induced by the action of left multiplication by the group element s on the left cosets of H:

$$s \cdot 1H = sH = 1H \text{ and so } \sigma_s(1) = 1$$
$$s \cdot rH = srH = r^3 H \text{ and so } \sigma_s(2) = 4$$
$$s \cdot r^2 H = sr^2 H = r^2 H \text{ and so } \sigma_s(3) = 3$$
$$s \cdot r^3 H = sr^3 H = rH \text{ and so } \sigma_s(4) = 2.$$

With this labelling of the left cosets of H we obtain $\sigma_s = (2\ 4)$. In the permutation representation associated to the action of D_8 on the left cosets of $\langle s \rangle$ by left multiplication one similarly computes that $\sigma_r = (1\ 2\ 3\ 4)$. Note that the permutation representation is a homomorphism, so once its value has been determined on generators for D_8 its value on any other element can be determined (e.g., $\sigma_{sr^2} = \sigma_s \sigma_r^2$).

Theorem 3. Let G be a group, let H be a subgroup of G and let G act by left multiplication on the set A of left cosets of H in G. Let π_H be the associated permutation representation afforded by this action. Then
  (1) G acts transitively on A
  (2) the stabilizer in G of the point $1H \in A$ is the subgroup H
  (3) the kernel of the action (i.e., the kernel of π_H) is $\cap_{x \in G}\, xHx^{-1}$, and $\ker \pi_H$ is the largest normal subgroup of G contained in H.

Proof: To see that G acts transitively on A, let aH and bH be any two elements of A, and let $g = ba^{-1}$. Then $g \cdot aH = (ba^{-1})aH = bH$, and so the two arbitrary elements aH and bH of A lie in the same orbit, which proves (1). For (2), the stabilizer of the point $1H$ is, by definition, $\{g \in G \mid g \cdot 1H = 1H\}$, i.e., $\{g \in G \mid gH = H\} = H$.

By definition of π_H we have

$$\ker \pi_H = \{g \in G \mid gxH = xH \text{ for all } x \in G\}$$
$$= \{g \in G \mid (x^{-1}gx)H = H \text{ for all } x \in G\}$$
$$= \{g \in G \mid x^{-1}gx \in H \text{ for all } x \in G\}$$
$$= \{g \in G \mid g \in xHx^{-1} \text{ for all } x \in G\} = \bigcap_{x \in G} xHx^{-1},$$

which proves the first assertion of (3). The second assertion of (3) comes from observing first that $\ker \pi_H \trianglelefteq G$ and $\ker \pi_H \leq H$. If now N is any normal subgroup of G contained in H then we have $N = xNx^{-1} \leq xHx^{-1}$ for all $x \in G$ so that

$$N \leq \bigcap_{x \in G} xHx^{-1} = \ker \pi_H.$$

This shows that $\ker \pi_H$ is the largest normal subgroup of G contained in H.

Corollary 4. *(Cayley's Theorem)* Every group is isomorphic to a subgroup of some symmetric group. If G is a group of order n, then G is isomorphic to a subgroup of S_n.

Proof: Let $H = 1$ and apply the preceding theorem to obtain a homomorphism of G into S_G (here we are identifying the cosets of the identity subgroup with the elements of G). Since the kernel of this homomorphism is contained in $H = 1$, G is isomorphic to its image in S_G.

Note that G is isomorphic to a *subgroup* of a symmetric group, not to the full symmetric group itself. For example, we exhibited an isomorphism of the Klein 4-group with the subgroup $\langle (1\ 2)(3\ 4), (1\ 3)(2\ 4) \rangle$ of S_4. Recall that subgroups of symmetric groups are called *permutation groups* so Cayley's Theorem states that every group is isomorphic to a permutation group. The permutation representation afforded by left multiplication on the elements of G (cosets of $H = 1$) is called the *left regular representation* of G. One might think that we could study all groups more effectively by simply studying subgroups of symmetric groups (and all finite groups by studying subgroups of S_n, for all n). This approach alone is neither computationally nor theoretically practical, since to study groups of order n we would have to work in the much larger group S_n (cf. Exercise 7, for example).

Historically, finite groups were first studied not in an axiomatic setting as we have developed but as subgroups of S_n. Thus Cayley's Theorem proves that the historical notion of a group and the modern (axiomatic) one are equivalent. One advantage of the modern approach is that we are not, in our study of a given group, restricted to considering that group as a subgroup of some *particular* symmetric group (so in some sense our groups are "coordinate free").

The next result generalizes our result on the normality of subgroups of index 2.

Corollary 5. If G is a finite group of order n and p is the smallest prime dividing $|G|$, then any subgroup of index p is normal.

Remark: In general, a group of order n need not have a subgroup of index p (for example, A_4 has no subgroup of index 2).

Proof: Suppose $H \leq G$ and $|G : H| = p$. Let π_H be the permutation representation afforded by multiplication on the set of left cosets of H in G, let $K = \ker \pi_H$ and let $|H : K| = k$. Then $|G : K| = |G : H||H : K| = pk$. Since H has p left cosets, G/K is isomorphic to a subgroup of S_p (namely, the image of G under π_H) by the First Isomorphism Theorem. By Lagrange's Theorem, $pk = |G/K|$ divides $p!$.

120

Thus $k \mid \dfrac{p!}{p} = (p-1)!$. But all prime divisors of $(p-1)!$ are less than p and by the minimality of p, every prime divisor of k is greater than or equal to p. This forces $k = 1$, so $H = K \trianglelefteq G$, completing the proof.

EXERCISES

Let G be a group and let H be a subgroup of G.

1. Let $G = \{1, a, b, c\}$ be the Klein 4-group whose group table is written out in Section 2.5.
   (a) Label $1, a, b, c$ with the integers $1,2,4,3$, respectively, and prove that under the left regular representation of G into S_4 the nonidentity elements are mapped as follows:

   $$a \mapsto (1\ 2)(3\ 4) \qquad b \mapsto (1\ 4)(2\ 3) \qquad c \mapsto (1\ 3)(2\ 4).$$

   (b) Relabel $1, a, b, c$ as $1,4,2,3$, respectively, and compute the image of each element of G under the left regular representation of G into S_4. Show that the image of G in S_4 under this labelling is the same *subgroup* as the image of G in part (a) (even though the nonidentity elements individually map to different permutations under the two different labellings).

2. List the elements of S_3 as 1, $(1\ 2)$, $(2\ 3)$, $(1\ 3)$, $(1\ 2\ 3)$, $(1\ 3\ 2)$ and label these with the integers $1,2,3,4,5,6$ respectively. Exhibit the image of each element of S_3 under the left regular representation of S_3 into S_6.

3. Let r and s be the usual generators for the dihedral group of order 8.
   (a) List the elements of D_8 as $1, r, r^2, r^3, s, sr, sr^2, sr^3$ and label these with the integers $1, 2, \ldots, 8$ respectively. Exhibit the image of each element of D_8 under the left regular representation of D_8 into S_8.
   (b) Relabel this same list of elements of D_8 with the integers $1, 3, 5, 7, 2, 4, 6, 8$ respectively and recompute the image of each element of D_8 under the left regular representation with respect to this new labelling. Show that the two subgroups of S_8 obtained in parts (a) and (b) are different.

4. Use the left regular representation of Q_8 to produce two elements of S_8 which generate a subgroup of S_8 isomorphic to the quaternion group Q_8.

5. Let r and s be the usual generators for the dihedral group of order 8 and let $H = \langle s \rangle$. List the left cosets of H in D_8 as $1H, rH, r^2H$ and r^3H.
   (a) Label these cosets with the integers $1,2,3,4$, respectively. Exhibit the image of each element of D_8 under the representation π_H of D_8 into S_4 obtained from the action of D_8 by left multiplication on the set of 4 left cosets of H in D_8. Deduce that this representation is faithful (i.e., the elements of S_4 obtained form a subgroup isomorphic to D_8).
   (b) Repeat part (a) with the list of cosets relabelled by the integers $1,3,2,4$, respectively. Show that the permutations obtained from this labelling form a subgroup of S_4 that is different from the subgroup obtained in part (a).
   (c) Let $K = \langle sr \rangle$, list the cosets of K in D_8 as $1K, rK, r^2K$ and r^3K, and label these with the integers $1,2,3,4$. Prove that, with respect to this labelling, the image of D_8 under the representation π_K obtained from left multiplication on the cosets of K is the same *subgroup* of S_4 as in part (a) (even though the subgroups H and K are different and some of the elements of D_8 map to different permutations under the two homomorphisms).

6. Let r and s be the usual generators for the dihedral group of order 8 and let $N = \langle r^2 \rangle$. List the left cosets of N in D_8 as $1N, rN, sN$ and srN. Label these cosets with the integers 1,2,3,4 respectively. Exhibit the image of each element of D_8 under the representation π_N of D_8 into S_4 obtained from the action of D_8 by left multiplication on the set of 4 left cosets of N in D_8. Deduce that this representation is not faithful and prove that $\pi_N(D_8)$ is isomorphic to the Klein 4-group.

7. Let Q_8 be the quaternion group of order 8.
   (a) Prove that Q_8 is isomorphic to a subgroup of S_8.
   (b) Prove that Q_8 is not isomorphic to a subgroup of S_n for any $n \leq 7$. [If Q_8 acts on any set A of order ≤ 7 show that the stabilizer of any point $a \in A$ must contain the subgroup $\langle -1 \rangle$.]

8. Prove that if H has finite index n then there is a normal subgroup K of G with $K \leq H$ and $|G : K| \leq n!$.

9. Prove that if p is a prime and G is a group of order p^α for some $\alpha \in \mathbb{Z}^+$, then every subgroup of index p is normal in G. Deduce that every group of order p^2 has a normal subgroup of order p.

10. Prove that every non-abelian group of order 6 has a nonnormal subgroup of order 2. Use this to classify groups of order 6. [Produce an injective homomorphism into S_3.]

11. Let G be a finite group and let $\pi : G \to S_G$ be the left regular representation. Prove that if x is an element of G of order n and $|G| = mn$, then $\pi(x)$ is a product of m n-cycles.
Deduce that $\pi(x)$ is an odd permutation if and only if $|x|$ is even and $\dfrac{|G|}{|x|}$ is odd.

12. Let G and π be as in the preceding exercise. Prove that if $\pi(G)$ contains an odd permutation then G has a subgroup of index 2. [Use Exercise 3 in Section 3.3.]

13. Prove that if $|G| = 2k$ where k is odd then G has a subgroup of index 2. [Use Cauchy's Theorem to produce an element of order 2 and then use the preceding two exercises.]

14. Let G be a finite group of composite order n with the property that G has a subgroup of order k for each positive integer k dividing n. Prove that G is not simple.

4.3 GROUPS ACTING ON THEMSELVES BY CONJUGATION —THE CLASS EQUATION

In this section G is any group and we first consider G *acting on itself* (i.e., $A = G$) by *conjugation*:

$$g \cdot a = gag^{-1} \qquad \text{for all } g \in G, \ a \in G$$

where gag^{-1} is computed in the group G as usual. This definition satisfies the two axioms for a group action because

$$g_1 \cdot (g_2 \cdot a) = g_1 \cdot (g_2 a g_2^{-1}) = g_1(g_2 a g_2^{-1})g_1^{-1} = (g_1 g_2)a(g_1 g_2)^{-1} = (g_1 g_2) \cdot a$$

and

$$1 \cdot a = 1a1^{-1} = a$$

for all $g_1, g_2 \in G$ and all $a \in G$.

Definition. Two elements a and b of G are said to be *conjugate in G* if there is some $g \in G$ such that $b = gag^{-1}$ (i.e., if and only if they are in the same orbit of G acting on itself by conjugation). The orbits of G acting on itself by conjugation are called the *conjugacy classes of G.*

Examples

(1) If G is an abelian group then the action of G on itself by conjugation is the trivial action: $g \cdot a = a$, for all $g, a \in G$, and for each $a \in G$ the conjugacy class of a is $\{a\}$.

(2) If $|G| > 1$ then, unlike the action by left multiplication, G does *not* act transitively on itself by conjugation because $\{1\}$ is always a conjugacy class (i.e., an orbit for this action). More generally, the one element subset $\{a\}$ is a conjugacy class if and only if $gag^{-1} = a$ for all $g \in G$ if and only if a is in the center of G.

(3) In S_3 one can compute directly that the conjugacy classes are $\{1\}$, $\{(1\ 2), (1\ 3), (2\ 3)\}$ and $\{(1\ 2\ 3), (1\ 3\ 2)\}$. We shall shortly develop techniques for computing conjugacy classes more easily, particularly in symmetric groups.

As in the case of a group acting on itself by left multiplication, the action by conjugation can be generalized. If S is any subset of G, define

$$gSg^{-1} = \{gsg^{-1} \mid s \in S\}.$$

A group G acts on the set $\mathcal{P}(G)$ of all subsets of itself by defining $g \cdot S = gSg^{-1}$ for any $g \in G$ and $S \in \mathcal{P}(G)$. As above, this defines a group action of G on $\mathcal{P}(G)$. Note that if S is the one element set $\{s\}$ then $g \cdot S$ is the one element set $\{gsg^{-1}\}$ and so this action of G on all subsets of G may be considered as an extension of the action of G on itself by conjugation.

Definition. Two subsets S and T of G are said to be *conjugate in G* if there is some $g \in G$ such that $T = gSg^{-1}$ (i.e., if and only if they are in the same orbit of G acting on its subsets by conjugation).

We now apply Proposition 2 to the action of G by conjugation. Proposition 2 proves that if S is a subset of G, then the number of conjugates of S equals the index $|G : G_S|$ of the stabilizer G_S of S. For action by conjugation

$$G_S = \{g \in G \mid gSg^{-1} = S\} = N_G(S)$$

is the normalizer of S in G. We summarize this as

Proposition 6. The number of conjugates of a subset S in a group G is the index of the normalizer of S, $|G : N_G(S)|$. In particular, the number of conjugates of an element s of G is the index of the centralizer of s, $|G : C_G(s)|$.

Proof: The second assertion of the proposition follows from the observation that $N_G(\{s\}) = C_G(s)$.

The action of G on itself by conjugation partitions G into the conjugacy classes of G, whose orders can be computed by Proposition 6. Since the sum of the orders of these conjugacy classes is the order of G, we obtain the following important relation among these orders.

Theorem 7. *(The Class Equation)* Let G be a finite group and let $g_1, g_2, ..., g_r$ be representatives of the distinct conjugacy classes of G not contained in the center $Z(G)$ of G. Then

$$|G| = |Z(G)| + \sum_{i=1}^{r} |G : C_G(g_i)|.$$

Proof: As noted in Example 2 above the element $\{x\}$ is a conjugacy class of size 1 if and only if $x \in Z(G)$, since then $gxg^{-1} = x$ for all $g \in G$. Let $Z(G) = \{1, z_2, ..., z_m\}$, let $\mathcal{K}_1, \mathcal{K}_2, \ldots, \mathcal{K}_r$ be the conjugacy classes of G not contained in the center, and let g_i be a representative of $\mathcal{K}_i$ for each i. Then the full set of conjugacy classes of G is given by

$$\{1\}, \{z_2\}, \ldots, \{z_m\}, \mathcal{K}_1, \mathcal{K}_2, \ldots, \mathcal{K}_r.$$

Since these partition G we have

$$|G| = \sum_{i=1}^{m} 1 + \sum_{i=1}^{r} |\mathcal{K}_i|$$

$$= |Z(G)| + \sum_{i=1}^{r} |G : C_G(g_i)|,$$

where $|\mathcal{K}_i|$ is given by Proposition 6. This proves the class equation.

Note in particular that all the summands on the right hand side of the class equation are divisors of the group order since they are indices of subgroups of G. This restricts their possible values (cf. Exercise 6, for example).

Examples

(1) The class equation gives no information in an abelian group since conjugation is the trivial action and all conjugacy classes have size 1.

(2) In any group G we have $\langle g \rangle \leq C_G(g)$; this observation helps to minimize computations of conjugacy classes. For example, in the quaternion group Q_8 we see that $\langle i \rangle \leq C_{Q_8}(i) \leq Q_8$. Since $i \notin Z(Q_8)$ and $|Q_8 : \langle i \rangle| = 2$, we must have $C_{Q_8}(i) = \langle i \rangle$. Thus i has precisely 2 conjugates in Q_8, namely i and $-i = kik^{-1}$. The other conjugacy classes in Q_8 are determined similarly and are

$$\{1\}, \quad \{-1\}, \quad \{\pm i\}, \quad \{\pm j\}, \quad \{\pm k\}.$$

The first two classes form $Z(Q_8)$ and the class equation for this group is

$$|Q_8| = 2 + 2 + 2 + 2.$$

(3) In D_8 we may also use the fact that the three subgroups of index 2 are abelian to quickly see that if $x \notin Z(D_8)$, then $|C_{D_8}(x)| = 4$. The conjugacy classes of D_8 are

$$\{1\}, \quad \{r^2\}, \quad \{r, r^3\}, \quad \{s, sr^2\}, \quad \{sr, sr^3\}.$$

The first two classes form $Z(D_8)$ and the class equation for this group is

$$|D_8| = 2 + 2 + 2 + 2.$$

Before discussing more examples of conjugacy we give two important consequences of the class equation. The first application of the class equation is to show that groups of prime power order have nontrivial centers, which is the starting point for the study of groups of prime power order (to which we return in Chapter 6).

Theorem 8. If p is a prime and P is a group of prime power order p^α for some $\alpha \geq 1$, then P has a nontrivial center: $Z(P) \neq 1$.

Proof: By the class equation

$$|P| = |Z(P)| + \sum_{i=1}^{r} |P : C_P(g_i)|$$

where $g_1, \ldots, g_r$ are representatives of the distinct non-central conjugacy classes. By definition, $C_P(g_i) \neq P$ for $i = 1, 2, \ldots, r$ so p divides $|P : C_P(g_i)|$. Since p also divides $|P|$ it follows that p divides $|Z(P)|$, hence the center must be nontrivial.

Corollary 9. If $|P| = p^2$ for some prime p, then P is abelian. More precisely, P is isomorphic to either Z_{p^2} or $Z_p \times Z_p$.

Proof: Since $Z(P) \neq 1$ by the theorem, it follows that $P/Z(P)$ is cyclic. By Exercise 36, Section 3.1, P is abelian. If P has an element of order p^2, then P is cyclic. Assume therefore that every nonidentity element of P has order p. Let x be any nonidentity element of P and let $y \in P - \langle x \rangle$. Since $|\langle x, y \rangle| > |\langle x \rangle| = p$, we must have that $P = \langle x, y \rangle$. Both x and y have order p so $\langle x \rangle \times \langle y \rangle = Z_p \times Z_p$. It now follows directly that the map $(x^a, y^b) \mapsto x^a y^b$ is an isomorphism from $\langle x \rangle \times \langle y \rangle$ onto P. This completes the proof.

Conjugacy in S_n

We next consider conjugation in symmetric groups. Readers familiar with linear algebra will recognize that in the matrix group $GL_n(F)$, conjugation is the same as "change of basis": $A \mapsto PAP^{-1}$. The situation in S_n is analogous:

Proposition 10. Let σ, τ be elements of the symmetric group S_n and suppose σ has cycle decomposition

$$(a_1 \, a_2 \, \ldots \, a_{k_1}) \, (b_1 \, b_2 \, \ldots \, b_{k_2}) \ldots .$$

Then $\tau \sigma \tau^{-1}$ has cycle decomposition

$$(\tau(a_1) \, \tau(a_2) \, \ldots \, \tau(a_{k_1})) \, (\tau(b_1) \, \tau(b_2) \, \ldots \, \tau(b_{k_2})) \ldots,$$

that is, $\tau \sigma \tau^{-1}$ is obtained from σ by replacing each entry i in the cycle decomposition for σ by the entry $\tau(i)$.

Proof: Observe that if $\sigma(i) = j$, then

$$\tau \sigma \tau^{-1}(\tau(i)) = \tau(j).$$

Thus, if the ordered pair i, j appears in the cycle decomposition of σ, then the ordered pair $\tau(i), \tau(j)$ appears in the cycle decomposition of $\tau \sigma \tau^{-1}$. This completes the proof.

Example

Let $\sigma = (1\,2)(3\,4\,5)(6\,7\,8\,9)$ and let $\tau = (1\,3\,5\,7)(2\,4\,6\,8)$. Then

$$\tau\sigma\tau^{-1} = (3\,4)(5\,6\,7)(8\,1\,2\,9).$$

Definition.

(1) If $\sigma \in S_n$ is the product of disjoint cycles of lengths $n_1, n_2, \ldots, n_r$ with $n_1 \le n_2 \le \cdots \le n_r$ (including its 1-cycles) then the integers $n_1, n_2, \ldots, n_r$ are called the *cycle type* of σ.

(2) If $n \in \mathbb{Z}^+$, a *partition* of n is any nondecreasing sequence of positive integers whose sum is n.

Note that by the results of the preceding section the cycle type of a permutation is unique. For example, the cycle type of an m-cycle in S_n is $1, 1, \ldots, 1, m$, where the m is preceded by $n - m$ ones.

Proposition 11. Two elements of S_n are conjugate in S_n if and only if they have the same cycle type. The number of conjugacy classes of S_n equals the number of partitions of n.

Proof: By Proposition 10, conjugate permutations have the same cycle type. Conversely, suppose the permutations σ_1 and σ_2 have the same cycle type. Order the cycles in nondecreasing length, including 1-cycles (if several cycles of σ_1 and σ_2 have the same length then there are several ways of doing this). Ignoring parentheses, each cycle decomposition is a list in which all the integers from 1 to n appear exactly once. Define τ to be the function which maps the ith integer in the list for σ_1 to the ith integer in the list for σ_2. Thus τ is a permutation and since the parentheses which delineate the cycle decompositions appear at the same positions in each list, Proposition 10 ensures that $\tau\sigma_1\tau^{-1} = \sigma_2$, so that σ_1 and σ_2 are conjugate.

Since there is a bijection between the conjugacy classes of S_n and the permissible cycle types and each cycle type for a permutation in S_n is a partition of n, the second assertion of the proposition follows, completing the proof.

Examples

(1) Let $\sigma_1 = (1)(3\,5)(8\,9)(2\,4\,7\,6)$ and let $\sigma_2 = (3)(4\,7)(8\,1)(5\,2\,6\,9)$. Then define τ by $\tau(1) = 3, \tau(3) = 4, \tau(5) = 7, \tau(8) = 8$, etc. Then

$$\tau = (1\,3\,4\,2\,5\,7\,6\,9)(8)$$

and $\tau\sigma_1\tau^{-1} = \sigma_2$.

(2) If in the previous example we had reordered σ_2 as $\sigma_2 = (3)(8\,1)(4\,7)(5\,2\,6\,9)$ by interchanging the two cycles of length 2, then the corresponding τ described above is defined by $\tau(1) = 3, \tau(3) = 8, \tau(5) = 1, \tau(8) = 4$, etc., which gives the permutation

$$\tau = (1\,3\,8\,4\,2\,5)(6\,9\,7)$$

again with $\tau\sigma_1\tau^{-1} = \sigma_2$, which shows that there are many elements conjugating σ_1 into σ_2.

(3) If $n = 5$, the partitions of 5 and corresponding representatives of the conjugacy classes (with 1-cycles not written) are as given in the following table:

Partition of 5	Representative of Conjugacy Class
1, 1, 1, 1, 1	1
1, 1, 1, 2	(1 2)
1, 1, 3	(1 2 3)
1, 4	(1 2 3 4)
5	(1 2 3 4 5)
1, 2, 2	(1 2)(3 4)
2, 3	(1 2)(3 4 5)

Proposition 11 and Proposition 6 can be used to exhibit the centralizers of some elements in S_n. For example, if σ is an m-cycle in S_n, then the number of conjugates of σ (i.e., the number of m-cycles) is

$$\frac{n \cdot (n-1) \cdots (n-m+1)}{m}.$$

By Proposition 6 this is the index of the centralizer of σ: $\dfrac{|S_n|}{|C_{S_n}(\sigma)|}$. Since $|S_n| = n!$ we obtain

$$|C_{S_n}(\sigma)| = m \cdot (n-m)!.$$

The element σ certainly commutes with $1, \sigma, \sigma^2, \ldots, \sigma^{m-1}$. It also commutes with any permutation in S_n whose cycles are disjoint from σ and there are $(n-m)!$ permutations of this type (the full symmetric group on the numbers not appearing in σ). The product of elements of these two types already accounts for $m \cdot (n-m)!$ elements commuting with σ. By the order computation above, this is the full centralizer of σ in S_n. Explicitly,

if σ is an m-cycle in S_n, then $C_{S_n}(\sigma) = \{\sigma^i \tau \mid 0 \le i \le m-1, \ \tau \in S_{n-m}\}$

where S_{n-m} denotes the subgroup of S_n which fixes all integers appearing in the m-cycle σ (and is the identity subgroup if $m = n$ or $m = n-1$).

For example, the centralizer of $\sigma = (1\ 3\ 5)$ in S_7 is the subgroup

$$\{(1\ 3\ 5)^i \tau \mid i = 0, 1 \text{ or } 2, \text{ and } \tau \text{ fixes } 1, 3 \text{ and } 5\}.$$

Note that $\tau \in S_A$ where $A = \{2, 4, 6, 7\}$, so there are 4! choices for τ and the centralizer has order $3 \cdot 4! = 72$.

We shall discuss centralizers of other elements of S_n in the next exercises and in Chapter 5.

We can use this discussion of the conjugacy classes in S_n to give a combinatorial proof of the simplicity of A_5. We first observe that normal subgroups of a group G are the union of conjugacy classes of G, i.e.,

if $H \trianglelefteq G$, then for every conjugacy class $\mathcal{K}$ of G either $\mathcal{K} \subseteq H$ or $\mathcal{K} \cap H = \emptyset$.

This is because if $x \in \mathcal{K} \cap H$, then $gxg^{-1} \in gHg^{-1}$ for all $g \in G$. Since H is normal, $gHg^{-1} = H$, so that H contains all the conjugates of x, i.e., $\mathcal{K} \subseteq H$.

Theorem 12. A_5 is a simple group.

Proof: We first work out the conjugacy classes of A_5 and their orders. Proposition 11 does not apply directly since two elements of the same cycle type (which are conjugate in S_5) need *not* be conjugate in A_5. Exercises 19 to 22 analyze the relation of classes in S_n to classes in A_n in detail.

We have already seen that representatives of the cycle types of even permutations can be taken to be

$$1, \quad (1\ 2\ 3), \quad (1\ 2\ 3\ 4\ 5) \quad \text{and} \quad (1\ 2)(3\ 4).$$

The centralizers of 3-cycles and 5-cycles in S_5 were determined above, and checking which of these elements are contained in A_5 we see that

$$C_{A_5}((1\ 2\ 3)) = \langle (1\ 2\ 3) \rangle \quad \text{and} \quad C_{A_5}((1\ 2\ 3\ 4\ 5)) = \langle (1\ 2\ 3\ 4\ 5) \rangle.$$

These groups have orders 3 and 5 (index 20 and 12), respectively, so there are 20 distinct conjugates of $(1\ 2\ 3)$ and 12 distinct conjugates of $(1\ 2\ 3\ 4\ 5)$ in A_5. Since there are a total of twenty 3-cycles in S_5 (Exercise 16, Section 1.3) and all of these lie in A_5, we see that

$$\text{all twenty 3-cycles are conjugate in } A_5.$$

There are a total of twenty-four 5-cycles in A_5 but only 12 distinct conjugates of the 5-cycle $(1\ 2\ 3\ 4\ 5)$. Thus some 5-cycle, σ, is *not* conjugate to $(1\ 2\ 3\ 4\ 5)$ in A_5 (in fact, $(1\ 3\ 5\ 2\ 4)$ is not conjugate in A_5 to $(1\ 2\ 3\ 4\ 5)$ since the method of proof in Proposition 11 shows that any element of S_5 conjugating $(1\ 2\ 3\ 4\ 5)$ into $(1\ 3\ 5\ 2\ 4)$ must be an odd permutation). As above we see that σ also has 12 distinct conjugates in A_5, hence

$$\text{the 5-cycles lie in two conjugacy classes in } A_5, \text{ each of which has 12 elements.}$$

Since the 3-cycles and 5-cycles account for all the nonidentity elements of odd order, the 15 remaining nonidentity elements of A_5 must have order 2 and therefore have cycle type (2,2). It is easy to see that $(1\ 2)(3\ 4)$ commutes with $(1\ 3)(2\ 4)$ but does not commute with any element of odd order in A_5. It follows that $|C_{A_5}((12)(34))| = 4$. Thus $(1\ 2)(3\ 4)$ has 15 distinct conjugates in A_5, hence

$$\text{all 15 elements of order 2 in } A_5 \text{ are conjugate to } (1\ 2)(3\ 4).$$

In summary, the conjugacy classes of A_5 have orders 1, 15, 20, 12 and 12.

Now, suppose H were a normal subgroup of A_5. Then as we observed above, H would be the union of conjugacy classes of A_5. Then the order of H would be both a divisor of 60 (the order of A_5) and be the sum of some collection of the integers $\{1, 12, 12, 15, 20\}$ (the sizes of the conjugacy classes in A_5). A quick check shows the only possibilities are $|H| = 1$ or $|H| = 60$, so that A_5 has no proper, nontrivial normal subgroups.

Right Group Actions

As noted in Section 1.7, in the definition of an action the group elements appear to the left of the set elements and so our notion of an action might more precisely be termed a *left group action*. One can analogously define the notion of a *right group action* of the group G on the nonempty set A as a map from $A \times G$ to A, denoted by $a \cdot g$ for $a \in A$ and $g \in G$, that satisfies the axioms:

(1) $(a \cdot g_1) \cdot g_2 = a \cdot (g_1 g_2)$ for all $a \in A$, and $g_1, g_2 \in G$, and
(2) $a \cdot 1 = a$ for all $a \in A$.

In much of the literature on group theory, conjugation is written as a right group action using the following notation:

$$a^g = g^{-1} a g \qquad \text{for all } g, a \in G.$$

Similarly, for subsets S of G one defines $S^g = g^{-1} S g$. In this notation the two axioms for a right action are verified as follows:

$$(a^{g_1})^{g_2} = g_2^{-1}(g_1^{-1} a g_1) g_2 = (g_1 g_2)^{-1} a (g_1 g_2) = a^{(g_1 g_2)}$$

and

$$a^1 = 1^{-1} a 1 = a$$

for all $g_1, g_2, a \in G$. Thus the two axioms for this right action of a group on itself take the form of the familiar "laws of exponentiation." (Note that the integer power a^n of a group element a is easily distinguished from the conjugate a^g of a by the nature of the exponent: $n \in \mathbb{Z}$ but $g \in G$.) Because conjugation is so ubiquitous in the theory of groups, this notation is a useful and efficient shorthand (as opposed to always writing gag^{-1} or $g \cdot a$ for action on the left by conjugation).

For arbitrary group actions it is an easy exercise to check that if we are given a left group action of G on A then the map $A \times G \to A$ defined by $a \cdot g = g^{-1} \cdot a$ is a right group action. Conversely, given a right group action of G on A we can form a left group action by $g \cdot a = a \cdot g^{-1}$. Call these pairs *corresponding group actions*. Put another way, for corresponding group actions, g acts on the left in the same way that g^{-1} acts on the right. This is particularly transparent for the action of conjugation because the "left conjugate of a by g," namely gag^{-1}, is the same group element as the "right conjugate of a by g^{-1}," namely $a^{g^{-1}}$. Thus two elements or subsets of a group are "left conjugate" if and only if they are "right conjugate," and so the relation "conjugacy" is the same for the left and right corresponding actions. More generally, it is also an exercise (Exercise 1) to see that for any corresponding left and right actions the orbits are the same.

We have consistently used left actions since they are compatible with the notation of applying functions on the left (i.e., with the notation $\varphi(g)$); in this way left multiplication on the left cosets of a subgroup is a left action. Similarly, right multiplication on the right cosets of a subgroup is a right action and the associated permutation representation φ is a homomorphism provided the function $\varphi : G \to S_A$ is written on the right as $(g_1 g_2)\varphi$ (and also provided permutations in S_A are written on the right as functions from A to itself). There are instances where a set admits two actions by a group G: one naturally on the left and the other on the right, so that it is useful to be comfortable with both types of actions.

EXERCISES

Let G be a group.

1. Suppose G has a left action on a set A, denoted by $g \cdot a$ for all $g \in G$ and $a \in A$. Denote the corresponding right action on A by $a \cdot g$. Prove that the (equivalence) relations $\sim$ and $\sim'$ defined by

$$a \sim b \qquad \text{if and only if} \qquad a = g \cdot b \quad \text{for some } g \in G$$

and

$$a \sim' b \qquad \text{if and only if} \qquad a = b \cdot g \quad \text{for some } g \in G$$

are the same relation (i.e., $a \sim b$ if and only if $a \sim' b$).

2. Find all conjugacy classes and their sizes in the following groups:
   (a) D_8 (b) Q_8 (c) A_4.

3. Find all the conjugacy classes and their sizes in the following groups:
   (a) $Z_2 \times S_3$ (b) $S_3 \times S_3$ (c) $Z_3 \times A_4$.

4. Prove that if $S \subseteq G$ and $g \in G$ then $g N_G(S) g^{-1} = N_G(g S g^{-1})$ and $g C_G(S) g^{-1} = C_G(g S g^{-1})$.

5. If the center of G is of index n, prove that every conjugacy class has at most n elements.

6. Assume G is a non-abelian group of order 15. Prove that $Z(G) = 1$. Use the fact that $\langle g \rangle \le C_G(g)$ for all $g \in G$ to show that there is at most one possible class equation for G. [Use Exercise 36, Section 3.1.]

7. For $n = 3, 4, 6$ and 7 make lists of the partitions of n and give representatives for the corresponding conjugacy classes of S_n.

8. Prove that $Z(S_n) = 1$ for all $n \ge 3$.

9. Show that $|C_{S_n}((1\,2)(3\,4))| = 8 \cdot (n-4)!$ for all $n \ge 4$. Determine the elements in this centralizer explicitly.

10. Let σ be the 5-cycle $(1\,2\,3\,4\,5)$ in S_5. In each of (a) to (c) find an explicit element $\tau \in S_5$ which accomplishes the specified conjugation:
    (a) $\tau \sigma \tau^{-1} = \sigma^2$
    (b) $\tau \sigma \tau^{-1} = \sigma^{-1}$
    (c) $\tau \sigma \tau^{-1} = \sigma^{-2}$.

11. In each of (a) – (d) determine whether σ_1 and σ_2 are conjugate. If they are, give an explicit permutation τ such that $\tau \sigma_1 \tau^{-1} = \sigma_2$.
    (a) $\sigma_1 = (1\,2)(3\,4\,5)$ and $\sigma_2 = (1\,2\,3)(4\,5)$
    (b) $\sigma_1 = (1\,5)(3\,7\,2)(10\,6\,8\,11)$ and $\sigma_2 = (3\,7\,5\,10)(4\,9)(13\,11\,2)$
    (c) $\sigma_1 = (1\,5)(3\,7\,2)(10\,6\,8\,11)$ and $\sigma_2 = \sigma_1^3$
    (d) $\sigma_1 = (1\,3)(2\,4\,6)$ and $\sigma_2 = (3\,5)(2\,4)(5\,6)$.

12. Find a representative for each conjugacy class of elements of order 4 in S_8 and in S_{12}.

13. Find all finite groups which have exactly two conjugacy classes.

14. In Exercise 1 of Section 2 two labellings of the elements $\{1, a, b, c\}$ of the Klein 4-group V were chosen to give two versions of the left regular representation of V into S_4. Let π_1 be the version of regular representation obtained in part (a) of that exercise and let π_2 be the version obtained via the labelling in part (b). Let $\tau = (2\,4)$. Show that $\tau \circ \pi_1(g) \circ \tau^{-1} = \pi_2(g)$ for each $g \in V$ (i.e., conjugation by τ sends the image of π_1 to the image of π_2 elementwise).

15. Find an element of S_8 which conjugates the subgroup of S_8 obtained in part (a) of Exercise 3, Section 2 to the subgroup of S_8 obtained in part (b) of that same exercise (both of these subgroups are isomorphic to D_8).

16. Find an element of S_4 which conjugates the subgroup of S_4 obtained in part (a) of Exercise 5, Section 2 to the subgroup of S_4 obtained in part (b) of that same exercise (both of these subgroups are isomorphic to D_8).

17. Let A be a nonempty set and let X be any subset of S_A. Let

$$F(X) = \{a \in A \mid \sigma(a) = a \text{ for all } \sigma \in X\} \qquad \text{— the } \textit{fixed set} \text{ of } X.$$

Let $M(X) = A - F(X)$ be the elements which are *moved* by some element of X. Let $D = \{\sigma \in S_A \mid |M(\sigma)| < \infty\}$. Prove that D is a normal subgroup of S_A.

18. Let A be a set, let H be a subgroup of S_A and let $F(H)$ be the fixed points of H on A as defined in the preceding exercise. Prove that if $\tau \in N_{S_A}(H)$ then τ stabilizes the set $F(H)$ and its complement $A - F(H)$.

19. Assume H is a normal subgroup of G, $\mathcal{K}$ is a conjugacy class of G contained in H and $x \in \mathcal{K}$. Prove that $\mathcal{K}$ is a union of k conjugacy classes of equal size in H, where $k = |G : HC_G(x)|$. Deduce that a conjugacy class in S_n which consists of even permutations is either a single conjugacy class under the action of A_n or is a union of two classes of the same size in A_n. [Let $A = C_G(x)$ and $B = H$ so $A \cap B = C_H(x)$. Draw the lattice diagram associated to the Second Isomorphism Theorem and interpret the appropriate indices. See also Exercise 9, Section 1.]

20. Let $\sigma \in A_n$. Show that all elements in the conjugacy class of σ in S_n (i.e., all elements of the same cycle type as σ) are conjugate in A_n if and only if σ commutes with an odd permutation. [Use the preceding exercise.]

21. Let $\mathcal{K}$ be a conjugacy class in S_n and assume that $\mathcal{K} \subseteq A_n$. Show $\sigma \in S_n$ does *not* commute with any odd permutation if and only if the cycle type of σ consists of distinct odd integers. Deduce that $\mathcal{K}$ consists of two conjugacy classes in A_n if and only if the cycle type of an element of $\mathcal{K}$ consists of distinct odd integers. [Assume first that $\sigma \in \mathcal{K}$ does not commute with any odd permutation. Observe that σ commutes with each individual cycle in its cycle decomposition — use this to show that all its cycles must be of odd length. If two cycles have the same odd length, k, find a product of k transpositions which interchanges them and commutes with σ. Conversely, if the cycle type of σ consists of distinct integers, prove that σ commutes *only* with the group generated by the cycles in its cycle decomposition.]

22. Show that if n is odd then the set of all n-cycles consists of two conjugacy classes of equal size in A_n.

23. Recall (cf. Exercise 16, Section 2.4) that a proper subgroup M of G is called *maximal* if whenever $M \le H \le G$, either $H = M$ or $H = G$. Prove that if M is a maximal subgroup of G then either $N_G(M) = M$ or $N_G(M) = G$. Deduce that if M is a maximal subgroup of G that is not normal in G then the number of nonidentity elements of G that are contained in conjugates of M is at most $(|M| - 1)|G : M|$.

24. Assume H is a proper subgroup of the finite group G. Prove $G \ne \cup_{g \in G} gHg^{-1}$, i.e., G is not the union of the conjugates of any proper subgroup. [Put H in some maximal subgroup and use the preceding exercise.]

25. Let $G = GL_2(\mathbb{C})$ and let $H = \{\begin{pmatrix} a & b \\ 0 & c \end{pmatrix} \mid a, b, c \in \mathbb{C}, \ ac \ne 0\}$. Prove that every element of G is conjugate to some element of the subgroup H and deduce that G is the union of conjugates of H. [Show that every element of $GL_2(\mathbb{C})$ has an eigenvector.]

26. Let G be a transitive permutation group on the finite set A with $|A| > 1$. Show that there is some $\sigma \in G$ such that $\sigma(a) \neq a$ for all $a \in A$ (such an element σ is called *fixed point free*).

27. Let $g_1, g_2, \ldots, g_r$ be representatives of the conjugacy classes of the finite group G and assume these elements pairwise commute. Prove that G is abelian.

28. Let p and q be primes with $p < q$. Prove that a non-abelian group G of order pq has a nonnormal subgroup of index q, so that there exists an injective homomorphism into S_q. Deduce that G is isomorphic to a subgroup of the normalizer in S_q of the cyclic group generated by the q-cycle $(1\,2\ldots q)$.

29. Let p be a prime and let G be a group of order p^α. Prove that G has a subgroup of order p^β, for every β with $0 \leq \beta \leq \alpha$. [Use Theorem 8 and induction on α.]

30. If G is a group of odd order, prove for any nonidentity element $x \in G$ that x and x^{-1} are not conjugate in G.

31. Using the usual generators and relations for the dihedral group D_{2n} (cf. Section 1.2) show that for $n = 2k$ an even integer the conjugacy classes in D_{2n} are the following: $\{1\}$, $\{r^k\}$, $\{r^{\pm 1}\}$, $\{r^{\pm 2}\}$, $\ldots$, $\{r^{\pm(k-1)}\}$, $\{sr^{2b} \mid b = 1, \ldots, k\}$ and $\{sr^{2b-1} \mid b = 1, \ldots, k\}$. Give the class equation for D_{2n}.

32. For $n = 2k + 1$ an odd integer show that the conjugacy classes in D_{2n} are $\{1\}$, $\{r^{\pm 1}\}$, $\{r^{\pm 2}\}$, $\ldots$, $\{r^{\pm k}\}$, $\{sr^b \mid b = 1, \ldots, n\}$. Give the class equation for D_{2n}.

33. This exercise gives a formula for the size of each conjugacy class in S_n. Let σ be a permutation in S_n and let $m_1, m_2, \ldots, m_s$ be the *distinct* integers which appear in the cycle type of σ (including 1-cycles). For each $i \in \{1, 2, \ldots, s\}$ assume σ has k_i cycles of length m_i (so that $\Sigma_{i=1}^{s} k_i m_i = n$). Prove that the number of conjugates of σ is

$$\frac{n!}{(k_1!m_1^{k_1})(k_2!m_2^{k_2})\ldots(k_s!m_s^{k_s})}.$$

[See Exercises 16 and 17 in Section 1.3 where this formula was given in some special cases.]

34. Prove that if p is a prime and P is a subgroup of S_p of order p, then $|N_{S_p}(P)| = p(p-1)$. [Argue that every conjugate of P contains exactly $p - 1$ p-cycles and use the formula for the number of p-cycles to compute the index of $N_{S_p}(P)$ in S_p.]

35. Let p be a prime. Find a formula for the number of conjugacy classes of elements of order p in S_n (using the greatest integer function).

36. Let $\pi : G \to S_G$ be the left regular representation afforded by the action of G on itself by left multiplication. For each $g \in G$ denote the permutation $\pi(g)$ by σ_g, so that $\sigma_g(x) = gx$ for all $x \in G$. Let $\lambda : G \to S_G$ be the permutation representation afforded by the corresponding right action of G on itself, and for each $h \in G$ denote the permutation $\lambda(h)$ by τ_h. Thus $\tau_h(x) = xh^{-1}$ for all $x \in G$ (λ is called the *right regular representation* of G).

  (a) Prove that σ_g and τ_h commute for all $g, h \in G$. (Thus the centralizer in S_G of $\pi(G)$ contains the subgroup $\lambda(G)$, which is isomorphic to G).

  (b) Prove that $\sigma_g = \tau_g$ if and only if g is an element of order 1 or 2 in the center of G.

  (c) Prove that $\sigma_g = \tau_h$ if and only if g and h lie in the center of G and $g = h^{-1}$. Deduce that $\pi(G) \cap \lambda(G) = \pi(Z(G)) = \lambda(Z(G))$.

4.4 AUTOMORPHISMS

Definition. Let G be a group. An isomorphism from G onto itself is called an *automorphism* of G. The set of all automorphisms of G is denoted by $\text{Aut}(G)$.

We leave as an exercise the simple verification that $\text{Aut}(G)$ is a group under composition of automorphisms, the *automorphism group* of G (composition of automorphisms is defined since the domain and range of each automorphism is the same). Notice that automorphisms of a group G are, in particular, permutations of the set G so $\text{Aut}(G)$ is a subgroup of S_G.

One of the most important examples of an automorphism of a group G is provided by conjugation by a fixed element in G. The next result discusses this in a slightly more general context.

Proposition 13. Let H be a normal subgroup of the group G. Then G acts by conjugation on H as automorphisms of H. More specifically, the action of G on H by conjugation is defined for each $g \in G$ by

$$h \mapsto ghg^{-1} \qquad \text{for each } h \in H.$$

For each $g \in G$, conjugation by g is an automorphism of H. The permutation representation afforded by this action is a homomorphism of G into $\text{Aut}(H)$ with kernel $C_G(H)$. In particular, $G/C_G(H)$ is isomorphic to a subgroup of $\text{Aut}(H)$.

Proof: (cf. Exercise 17, Section 1.7) Let φ_g be conjugation by g. Note that because g normalizes H, φ_g maps H to itself. Since we have already seen that conjugation defines an action, it follows that $\varphi_1 = 1$ (the identity map on H) and $\varphi_a \circ \varphi_b = \varphi_{ab}$ for all $a, b \in G$. Thus each φ_g gives a bijection from H to itself since it has a 2-sided inverse $\varphi_{g^{-1}}$. Each φ_g is a homomorphism from H to H because

$$\varphi_g(hk) = g(hk)g^{-1} = gh(gg^{-1})kg^{-1} = (ghg^{-1})(gkg^{-1}) = \varphi_g(h)\varphi_g(k)$$

for all $h, k \in H$. This proves that conjugation by any fixed element of G defines an automorphism of H.

By the preceding remark, the permutation representation $\psi : G \to S_H$ defined by $\psi(g) = \varphi_g$ (which we have already proved is a homomorphism) has image contained in the subgroup $\text{Aut}(H)$ of S_H. Finally,

$$\ker \psi = \{g \in G \mid \varphi_g = \text{id}\}$$
$$= \{g \in G \mid ghg^{-1} = h \text{ for all } h \in H\}$$
$$= C_G(H).$$

The First Isomorphism Theorem implies the final statement of the proposition.

Proposition 13 shows that a group acts by conjugation on a normal subgroup as *structure preserving* permutations, i.e., as automorphisms. In particular, this action must send subgroups to subgroups, elements of order n to elements of order n, etc. Two specific applications of this proposition are described in the next two corollaries.

Corollary 14. If K is any subgroup of the group G and $g \in G$, then $K \cong gKg^{-1}$. Conjugate elements and conjugate subgroups have the same order.

Proof: Letting $G = H$ in the proposition shows that conjugation by $g \in G$ is an automorphism of G, from which the corollary follows.

Corollary 15. For any subgroup H of a group G, the quotient group $N_G(H)/C_G(H)$ is isomorphic to a subgroup of $\text{Aut}(H)$. In particular, $G/Z(G)$ is isomorphic to a subgroup of $\text{Aut}(G)$.

Proof: Since H is a normal subgroup of the group $N_G(H)$, Proposition 13 (applied with $N_G(H)$ playing the role of G) implies the first assertion. The second assertion is the special case when $H = G$, in which case $N_G(G) = G$ and $C_G(G) = Z(G)$.

Definition. Let G be a group and let $g \in G$. Conjugation by g is called an *inner automorphism* of G and the subgroup of $\text{Aut}(G)$ consisting of all inner automorphisms is denoted by $\text{Inn}(G)$.

Note that the collection of inner automorphisms of G is in fact a subgroup of $\text{Aut}(G)$ and that by Corollary 15, $\text{Inn}(G) \cong G/Z(G)$. Note also that if H is a normal subgroup of G, conjugation by an element of G when restricted to H is an automorphism of H but need not be an inner automorphism of H (as we shall see).

Examples

    (1) A group G is abelian if and only if every inner automorphism is trivial. If H is an abelian normal subgroup of G and H is not contained in $Z(G)$, then there is some $g \in G$ such that conjugation by g restricted to H is not an inner automorphism of H. An explicit example of this is $G = A_4$, H is the Klein 4-group in G and g is any 3-cycle.

    (2) Since $Z(Q_8) = \langle -1 \rangle$ we have $\text{Inn}(Q_8) \cong V_4$.

    (3) Since $Z(D_8) = \langle r^2 \rangle$ we have $\text{Inn}(D_8) \cong V_4$.

    (4) Since for all $n \geq 3$, $Z(S_n) = 1$ we have $\text{Inn}(S_n) \cong S_n$.

Corollary 15 shows that any information we have about the automorphism group of a subgroup H of a group G translates into information about $N_G(H)/C_G(H)$. For example, if $H \cong Z_2$, then since H has unique elements of orders 1 and 2, Corollary 14 forces $\text{Aut}(H) = 1$. Thus if $H \cong Z_2$, $N_G(H) = C_G(H)$; if in addition H is a normal subgroup of G, then $H \leq Z(G)$ (cf. Exercise 10, Section 2.2).

Although the preceding example was fairly trivial, it illustrates that the action of G by conjugation on a *normal* subgroup H can be restricted by knowledge of the automorphism group of H. This in turn can be used to investigate the structure of G and will lead to some classification theorems when we consider semidirect products in Section 5.5.

A notion which will be used in later sections most naturally warrants introduction here:

Definition. A subgroup H of a group G is called *characteristic* in G, denoted H char G, if every automorphism of G maps H to itself, i.e., $\sigma(H) = H$ for all $\sigma \in \text{Aut}(G)$.

Results concerning characteristic subgroups which we shall use later (and whose proofs are relegated to the exercises) are

(1) characteristic subgroups are normal,
(2) if H is the unique subgroup of G of a given order, then H is characteristic in G, and
(3) if K char H and $H \trianglelefteq G$, then $K \trianglelefteq G$ (so although "normality" is not a transitive property (i.e., a normal subgroup of a normal subgroup need not be normal), a characteristic subgroup of a normal subgroup is normal).

Thus we may think of characteristic subgroups as "strongly normal" subgroups. For example, property (2) and Theorem 2.7 imply that every subgroup of a cyclic group is characteristic.

We close this section with some results on automorphism groups of specific groups.

Proposition 16. The automorphism group of the cyclic group of order n is isomorphic to $(\mathbb{Z}/n\mathbb{Z})^{\times}$, an abelian group of order $\varphi(n)$ (where φ is Euler's function).

Proof: Let x be a generator of the cyclic group Z_n. If $\psi \in \text{Aut}(Z_n)$, then $\psi(x) = x^a$ for some $a \in \mathbb{Z}$ and the integer a uniquely determines ψ. Denote this automorphism by ψ_a. As usual, since $|x| = n$, the integer a is only defined mod n. Since ψ_a is an automorphism, x and x^a must have the same order, hence $(a, n) = 1$. Furthermore, for every a relatively prime to n, the map $x \mapsto x^a$ is an automorphism of Z_n. Hence we have a surjective map

$$\Psi : \text{Aut}(Z_n) \to (\mathbb{Z}/n\mathbb{Z})^{\times}$$
$$\psi_a \mapsto a \ (\text{mod } n).$$

The map Ψ is a homomorphism because

$$\psi_a \circ \psi_b(x) = \psi_a(x^b) = (x^b)^a = x^{ab} = \psi_{ab}(x)$$

for all $\psi_a, \psi_b \in \text{Aut}(Z_n)$, so that

$$\Psi(\psi_a \circ \psi_b) = \Psi(\psi_{ab}) = ab \ (\text{mod } n) = \Psi(\psi_a)\Psi(\psi_b).$$

Finally, Ψ is clearly injective, hence is an isomorphism.

A complete description of the isomorphism type of $\text{Aut}(Z_n)$ is given at the end of Section 9.5.

Example

Assume G is a group of order pq, where p and q are primes (not necessarily distinct) with $p \leq q$. If $p \nmid q - 1$, we prove G is abelian.

If $Z(G) \neq 1$, Lagrange's Theorem forces $G/Z(G)$ to be cyclic, hence G is abelian by Exercise 36, Section 3.1. Hence we may assume $Z(G) = 1$.

If every nonidentity element of G has order p, then the centralizer of every nonidentity element has index q, so the class equation for G reads

$$pq = 1 + kq.$$

This is impossible since q divides pq and kq but not 1. Thus G contains an element, x, of order q.

Let $H = \langle x \rangle$. Since H has index p and p is the smallest prime dividing $|G|$, the subgroup H is normal in G by Corollary 5. Since $Z(G) = 1$, we must have $C_G(H) = H$. Thus $G/H = N_G(H)/C_G(H)$ is a group of order p isomorphic to a subgroup of Aut(H) by Corollary 15. But by Proposition 16, Aut(H) has order $\varphi(q) = q - 1$, which by Lagrange's Theorem would imply $p \mid q - 1$, contrary to assumption. This shows that G must be abelian.

One can check that every group of order pq, where p and q are distinct primes with $p < q$ and $p \nmid q - 1$ is *cyclic* (see the exercises). This is the first instance where there is a unique isomorphism type of group whose order is *composite*. For instance, every group of order 15 is cyclic.

The next proposition summarizes some results on automorphism groups of known groups and will be proved later. Part 3 of this proposition illustrates how the theory of vector spaces comes into play in group theory.

Proposition 17.
(1) If p is an odd prime and $n \in \mathbb{Z}^+$, then the automorphism group of the cyclic group of order p is cyclic of order $p - 1$. More generally, the automorphism group of the cyclic group of order p^n is cyclic of order $p^{n-1}(p-1)$ (cf. Corollary 20, Section 9.5).

(2) For all $n \geq 3$ the automorphism group of the cyclic group of order 2^n is isomorphic to $Z_2 \times Z_{2^{n-2}}$, and in particular is not cyclic but has a cyclic subgroup of index 2 (cf. Corollary 20, Section 9.5).

(3) Let p be a prime and let V be an abelian group (written additively) with the property that $pv = 0$ for all $v \in V$. If $|V| = p^n$, then V is an n-dimensional vector space over the field $\mathbb{F}_p = \mathbb{Z}/p\mathbb{Z}$. The automorphisms of V are precisely the nonsingular linear transformations from V to itself, that is

$$\text{Aut}(V) \cong GL(V) \cong GL_n(\mathbb{F}_p).$$

In particular, the order of Aut(V) is as given in Section 1.4 (cf. the examples in Sections 10.2 and 11.1).

(4) For all $n \neq 6$ we have Aut$(S_n) = \text{Inn}(S_n) \cong S_n$ (cf. Exercise 18). For $n = 6$ we have $|\text{Aut}(S_6) : \text{Inn}(S_6)| = 2$ (cf. the following Exercise 19 and also Exercise 10 in Section 6.3).

(5) Aut$(D_8) \cong D_8$ and Aut$(Q_8) \cong S_4$ (cf. the following Exercises 4 and 5 and also Exercise 9 in Section 6.3).

The group V described in Part 3 of the proposition is called the *elementary abelian group* of order p^n (we shall see in Chapter 5 that it is uniquely determined up to isomorphism by p and n). The Klein 4-group, V_4, is the elementary abelian group of order 4. This proposition asserts that

$$\text{Aut}(V_4) \cong GL_2(\mathbb{F}_2).$$

By the exercises in Section 1.4, the latter group has order 6. But $\text{Aut}(V_4)$ permutes the 3 nonidentity elements of V_4, and this action of $\text{Aut}(V_4)$ on $V_4 - \{1\}$ gives an injective permutation representation of $\text{Aut}(V_4)$ into S_3. By order considerations, the homomorphism is onto, so

$$\text{Aut}(V_4) \cong GL_2(\mathbb{F}_2) \cong S_3.$$

Note that V_4 is abelian, so $\text{Inn}(V_4) = 1$.

For any prime p, the elementary abelian group of order p^2 is $Z_p \times Z_p$. Its automorphism group, $GL_2(\mathbb{F}_p)$, has order $p(p-1)^2(p+1)$. Thus Corollary 9 implies that for p a prime

$$\text{if } |P| = p^2, \quad |\text{Aut}(P)| = p(p-1) \text{ or } p(p-1)^2(p+1)$$

according to whether P is cyclic or elementary abelian, respectively.

Example

Suppose G is a group of order $45 = 3^2 5$ with a normal subgroup P of order 3^2. We show that G is necessarily abelian.

The quotient $G/C_G(P)$ is isomorphic to a subgroup of $\text{Aut}(P)$ by Corollary 15, and $\text{Aut}(P)$ has order 6 or 48 (according to whether P is cyclic or elementary abelian, respectively) by the preceding paragraph. On the other hand, since the order of P is the square of a prime, P is an abelian group, hence $P \leq C_G(P)$. It follows that $|C_G(P)|$ is divisible by 9, which implies $|G/C_G(P)|$ is 1 or 5. Together these imply $|G/C_G(P)| = 1$, i.e., $C_G(P) = G$ and $P \leq Z(G)$. Since then $G/Z(G)$ is cyclic, G must be an abelian group.

EXERCISES

Let G be a group.

1. If $\sigma \in \text{Aut}(G)$ and φ_g is conjugation by g prove $\sigma \varphi_g \sigma^{-1} = \varphi_{\sigma(g)}$. Deduce that $\text{Inn}(G) \trianglelefteq \text{Aut}(G)$. (The group $\text{Aut}(G)/\text{Inn}(G)$ is called the *outer automorphism group* of G.)

2. Prove that if G is an abelian group of order pq, where p and q are distinct primes, then G is cyclic. [Use Cauchy's Theorem to produce elements of order p and q and consider the order of their product.]

3. Prove that under any automorphism of D_8, r has at most 2 possible images and s has at most 4 possible images (r and s are the usual generators — cf. Section 1.2). Deduce that $|\text{Aut}(D_8)| \leq 8$.

4. Use arguments similar to those in the preceding exercise to show $|\text{Aut}(Q_8)| \leq 24$.

5. Use the fact that $D_8 \trianglelefteq D_{16}$ to prove that $\text{Aut}(D_8) \cong D_8$.

6. Prove that characteristic subgroups are normal. Give an example of a normal subgroup that is not characteristic.

7. If H is the unique subgroup of a given order in a group G prove H is characteristic in G.

8. Let G be a group with subgroups H and K with $H \leq K$.
   (a) Prove that if H is characteristic in K and K is normal in G then H is normal in G.
   (b) Prove that if H is characteristic in K and K is characteristic in G then H is characteristic in G. Use this to prove that the Klein 4-group V_4 is characteristic in S_4.
   (c) Give an example to show that if H is normal in K and K is characteristic in G then H need not be normal in G.

9. If r, s are the usual generators for the dihedral group D_{2n}, use the preceding two exercises to deduce that every subgroup of $\langle r \rangle$ is normal in D_{2n}.

10. Let G be a group, let A be an abelian normal subgroup of G, and write $\overline{G} = G/A$. Show that $\overline{G}$ acts (on the left) by conjugation on A by $\overline{g} \cdot a = gag^{-1}$, where g is *any* representative of the coset $\overline{g}$ (in particular, show that this action is well defined). Give an explicit example to show that this action is not well defined if A is non-abelian.

11. If p is a prime and P is a subgroup of S_p of order p, prove $N_{S_p}(P)/C_{S_p}(P) \cong \text{Aut}(P)$. [Use Exercise 34, Section 3.]

12. Let G be a group of order 3825. Prove that if H is a normal subgroup of order 17 in G then $H \leq Z(G)$.

13. Let G be a group of order 203. Prove that if H is a normal subgroup of order 7 in G then $H \leq Z(G)$. Deduce that G is abelian in this case.

14. Let G be a group of order 1575. Prove that if H is a normal subgroup of order 9 in G then $H \leq Z(G)$.

15. Prove that each of the following (multiplicative) groups is cyclic: $(\mathbb{Z}/5\mathbb{Z})^{\times}$, $(\mathbb{Z}/9\mathbb{Z})^{\times}$ and $(\mathbb{Z}/18\mathbb{Z})^{\times}$.

16. Prove that $(\mathbb{Z}/24\mathbb{Z})^{\times}$ is an elementary abelian group of order 8. (We shall see later that $(\mathbb{Z}/n\mathbb{Z})^{\times}$ is an elementary abelian group if and only if $n \mid 24$.)

17. Let $G = \langle x \rangle$ be a cyclic group of order n. For $n = 2, 3, 4, 5, 6$ write out the elements of $\text{Aut}(G)$ explicitly (by Proposition 16 above we know $\text{Aut}(G) \cong (\mathbb{Z}/n\mathbb{Z})^{\times}$, so for each element $a \in (\mathbb{Z}/n\mathbb{Z})^{\times}$, write out explicitly what the automorphism ψ_a does to the elements $\{1, x, x^2, \ldots, x^{n-1}\}$ of G).

18. This exercise shows that for $n \neq 6$ every automorphism of S_n is inner. Fix an integer $n \geq 2$ with $n \neq 6$.
  (a) Prove that the automorphism group of a group G permutes the conjugacy classes of G, i.e., for each $\sigma \in \text{Aut}(G)$ and each conjugacy class $\mathcal{K}$ of G the set $\sigma(\mathcal{K})$ is also a conjugacy class of G.
  (b) Let $\mathcal{K}$ be the conjugacy class of transpositions in S_n and let $\mathcal{K}'$ be the conjugacy class of any element of order 2 in S_n that is not a transposition. Prove that $|\mathcal{K}| \neq |\mathcal{K}'|$. Deduce that any automorphism of S_n sends transpositions to transpositions. [See Exercise 33 in Section 3.]
  (c) Prove that for each $\sigma \in \text{Aut}(S_n)$

$$\sigma : (1\ 2) \mapsto (a\ b_2), \qquad \sigma : (1\ 3) \mapsto (a\ b_3), \qquad \ldots, \qquad \sigma : (1\ n) \mapsto (a\ b_n)$$

  for some distinct integers $a, b_2, b_3, \ldots, b_n \in \{1, 2, \ldots, n\}$.
  (d) Show that $(1\ 2), (1\ 3), \ldots, (1\ n)$ generate S_n and deduce that any automorphism of S_n is uniquely determined by its action on these elements. Use (c) to show that S_n has at most $n!$ automorphisms and conclude that $\text{Aut}(S_n) = \text{Inn}(S_n)$ for $n \neq 6$.

19. This exercise shows that $|\text{Aut}(S_6) : \text{Inn}(S_6)| \leq 2$ (Exercise 10 in Section 6.3 shows that equality holds by exhibiting an automorphism of S_6 that is not inner).
  (a) Let $\mathcal{K}$ be the conjugacy class of transpositions in S_6 and let $\mathcal{K}'$ be the conjugacy class of any element of order 2 in S_6 that is not a transposition. Prove that $|\mathcal{K}| \neq |\mathcal{K}'|$ unless $\mathcal{K}'$ is the conjugacy class of products of three disjoint transpositions. Deduce that $\text{Aut}(S_6)$ has a subgroup of index at most 2 which sends transpositions to transpositions.
  (b) Prove that $|\text{Aut}(S_6) : \text{Inn}(S_6)| \leq 2$. [Follow the same steps as in (c) and (d) of the preceding exercise to show that any automorphism that sends transpositions to transpositions is inner.]

The next exercise introduces a subgroup, $J(P)$, which (like the center of P) is defined for an arbitrary finite group P (although in most applications P is a group whose order is a power of a prime). This subgroup was defined by J. Thompson in 1964 and it now plays a pivotal role in the study of finite groups, in particular, in the classification of finite simple groups.

20. For any finite group P let $d(P)$ be the minimum number of generators of P (so, for example, $d(P) = 1$ if and only if P is a nontrivial cyclic group and $d(Q_8) = 2$). Let $m(P)$ be the maximum of the integers $d(A)$ as A runs over all *abelian* subgroups of P (so, for example, $m(Q_8) = 1$ and $m(D_8) = 2$). Define

$$J(P) = \langle\, A \mid A \text{ is an abelian subgroup of } P \text{ with } d(A) = m(P) \,\rangle.$$

($J(P)$ is called the *Thompson subgroup* of P.)
  (a) Prove that $J(P)$ is a characteristic subgroup of P.
  (b) For each of the following groups P list all abelian subgroups A of P that satisfy $d(A) = m(P)$: Q_8, D_8, D_{16} and QD_{16} (where QD_{16} is the quasidihedral group of order 16 defined in Exercise 11 of Section 2.5). [Use the lattices of subgroups for these groups in Section 2.5.]
  (c) Show that $J(Q_8) = Q_8$, $J(D_8) = D_8$, $J(D_{16}) = D_{16}$ and $J(QD_{16})$ is a dihedral subgroup of order 8 in QD_{16}.
  (d) Prove that if $Q \le P$ and $J(P)$ is a subgroup of Q, then $J(P) = J(Q)$. Deduce that if P is a subgroup (not necessarily normal) of the finite group G and $J(P)$ is contained in some subgroup Q of P such that $Q \trianglelefteq G$, then $J(P) \trianglelefteq G$.

4.5 SYLOW'S THEOREM

In this section we prove a partial converse to Lagrange's Theorem and derive numerous consequences, some of which will lead to classification theorems in the next chapter.

Definition.   Let G be a group and let p be a prime.
  (1) A group of order p^α for some $\alpha \ge 0$ is called a *p-group*. Subgroups of G which are p-groups are called *p-subgroups*.
  (2) If G is a group of order $p^\alpha m$, where $p \nmid m$, then a subgroup of order p^α is called a *Sylow p-subgroup* of G.
  (3) The set of Sylow p-subgroups of G will be denoted by $Syl_p(G)$ and the number of Sylow p-subgroups of G will be denoted by $n_p(G)$ (or just n_p when G is clear from the context).

Theorem 18. *(Sylow's Theorem)* Let G be a group of order $p^\alpha m$, where p is a prime not dividing m.
  (1) Sylow p-subgroups of G exist, i.e., $Syl_p(G) \ne \emptyset$.
  (2) If P is a Sylow p-subgroup of G and Q is any p-subgroup of G, then there exists $g \in G$ such that $Q \le gPg^{-1}$, i.e., Q is contained in some conjugate of P. In particular, any two Sylow p-subgroups of G are conjugate in G.
  (3) The number of Sylow p-subgroups of G is of the form $1 + kp$, i.e.,

$$n_p \equiv 1 \pmod{p}.$$

Further, n_p is the index in G of the normalizer $N_G(P)$ for any Sylow p-subgroup P, hence n_p divides m.

We first prove the following lemma:

Lemma 19. Let $P \in Syl_p(G)$. If Q is any p-subgroup of G, then $Q \cap N_G(P) = Q \cap P$.

Proof: Let $H = N_G(P) \cap Q$. Since $P \leq N_G(P)$ it is clear that $P \cap Q \leq H$, so we must prove the reverse inclusion. Since by definition $H \leq Q$, this is equivalent to showing $H \leq P$. We do this by demonstrating that PH is a p-subgroup of G containing both P and H; but P is a p-subgroup of G of largest possible order, so we must have $PH = P$, i.e., $H \leq P$.

Since $H \leq N_G(P)$, by Corollary 15 in Section 3.2, PH is a subgroup. By Proposition 13 in the same section

$$|PH| = \frac{|P||H|}{|P \cap H|}.$$

All the numbers in the above quotient are powers of p, so PH is a p-group. Moreover, P is a subgroup of PH so the order of PH is divisible by p^α, the largest power of p which divides $|G|$. These two facts force $|PH| = p^\alpha = |P|$. This in turn implies $P = PH$ and $H \leq P$. This establishes the lemma.

Proof of Sylow's Theorem (1) Proceed by induction on $|G|$. If $|G| = 1$, there is nothing to prove. Assume inductively the existence of Sylow p-subgroups for all groups of order less than $|G|$.

If p divides $|Z(G)|$, then by Cauchy's Theorem for abelian groups (Proposition 21, Section 3.4) $Z(G)$ has a subgroup, N, of order p. Let $\overline{G} = G/N$, so that $|\overline{G}| = p^{\alpha-1}m$. By induction, $\overline{G}$ has a subgroup $\overline{P}$ of order $p^{\alpha-1}$. If we let P be the subgroup of G containing N such that $P/N = \overline{P}$ then $|P| = |P/N| \cdot |N| = p^\alpha$ and P is a Sylow p-subgroup of G. We are reduced to the case when p does not divide $|Z(G)|$.

Let $g_1, g_2, \ldots, g_r$ be representatives of the distinct non-central conjugacy classes of G. The class equation for G is

$$|G| = |Z(G)| + \sum_{i=1}^{r} |G : C_G(g_i)|.$$

If $p \mid |G : C_G(g_i)|$ for all i, then since $p \mid |G|$, we would also have $p \mid |Z(G)|$, a contradiction. Thus for some i, p does not divide $|G : C_G(g_i)|$. For this i let $H = C_G(g_i)$ so that

$$|H| = p^\alpha k, \quad \text{where } p \nmid k.$$

Since $g_i \notin Z(G)$, $|H| < |G|$. By induction, H has a Sylow p-subgroup, P, which of course is also a subgroup of G. Since $|P| = p^\alpha$, P is a Sylow p-subgroup of G. This completes the induction and establishes (1).

Before proving (2) and (3) we make some calculations. By (1) there exists a Sylow p-subgroup, P, of G. Let

$$\{P_1, P_2, \ldots, P_r\} = S$$

be the set of all conjugates of P (i.e., $S = \{gPg^{-1} \mid g \in G\}$) and let Q be *any* p-subgroup of G. By definition of S, G, hence also Q, acts by conjugation on S. Write S as a disjoint union of orbits under this action by Q:

$$S = \mathcal{O}_1 \cup \mathcal{O}_2 \cup \cdots \cup \mathcal{O}_s$$

where $r = |\mathcal{O}_1| + \cdots + |\mathcal{O}_s|$. Keep in mind that r does not depend on Q but the number of Q-orbits s does (note that by definition, G has only one orbit on $\mathcal{S}$ but a subgroup Q of G may have more than one orbit). Renumber the elements of $\mathcal{S}$ if necessary so that the first s elements of $\mathcal{S}$ are representatives of the Q-orbits: $P_i \in \mathcal{O}_i$, $1 \le i \le s$. It follows from Proposition 2 that $|\mathcal{O}_i| = |Q : N_Q(P_i)|$. By definition, $N_Q(P_i) = N_G(P_i) \cap Q$ and by Lemma 19, $N_G(P_i) \cap Q = P_i \cap Q$. Combining these two facts gives

$$|\mathcal{O}_i| = |Q : P_i \cap Q|, \qquad 1 \le i \le s. \tag{4.1}$$

We are now in a position to prove that $r \equiv 1 \pmod{p}$. Since Q was arbitrary we may take $Q = P_1$ above, so that (1) gives

$$|\mathcal{O}_1| = 1.$$

Now, for all $i > 1$, $P_1 \ne P_i$, so $P_1 \cap P_i < P_1$. By (1)

$$|\mathcal{O}_i| = |P_1 : P_1 \cap P_i| > 1, \qquad 2 \le i \le s.$$

Since P_1 is a p-group, $|P_1 : P_1 \cap P_i|$ must be a power of p, so that

$$p \mid |\mathcal{O}_i|, \qquad 2 \le i \le s.$$

Thus

$$r = |\mathcal{O}_1| + (|\mathcal{O}_2| + \ldots + |\mathcal{O}_s|) \equiv 1 \pmod{p}.$$

We now prove parts (2) and (3). Let Q be any p-subgroup of G. Suppose Q is not contained in P_i for any $i \in \{1, 2, \ldots, r\}$ (i.e., $Q \nleq gPg^{-1}$ for any $g \in G$). In this situation, $Q \cap P_i < Q$ for all i, so by (1)

$$|\mathcal{O}_i| = |Q : Q \cap P_i| > 1, \qquad 1 \le i \le s.$$

Thus $p \mid |\mathcal{O}_i|$ for all i, so p divides $|\mathcal{O}_1| + \ldots + |\mathcal{O}_s| = r$. This contradicts the fact that $r \equiv 1 \pmod{p}$ (remember, r does not depend on the choice of Q). This contradiction proves $Q \le gPg^{-1}$ for some $g \in G$.

To see that all Sylow p-subgroups of G are conjugate, let Q be any Sylow p-subgroup of G. By the preceding argument, $Q \le gPg^{-1}$ for some $g \in G$. Since $|gPg^{-1}| = |Q| = p^\alpha$, we must have $gPg^{-1} = Q$. This establishes part (2) of the theorem. In particular, $\mathcal{S} = Syl_p(G)$ since *every* Sylow p-subgroup of G is conjugate to P, and so $n_p = r \equiv 1 \pmod{p}$, which is the first part of (3).

Finally, since all Sylow p-subgroups are conjugate, Proposition 6 shows that

$$n_p = |G : N_G(P)| \quad \text{for any } P \in Syl_p(G),$$

completing the proof of Sylow's Theorem.

Note that the conjugacy part of Sylow's Theorem together with Corollary 14 shows that *any two Sylow p-subgroups of a group (for the same prime p) are isomorphic.*

Corollary 20. Let P be a Sylow p-subgroup of G. Then the following are equivalent:

   (1) P is the unique Sylow p-subgroup of G, i.e., $n_p = 1$

   (2) P is normal in G

   (3) P is characteristic in G

   (4) All subgroups generated by elements of p-power order are p-groups, i.e., if X is any subset of G such that $|x|$ is a power of p for all $x \in X$, then $\langle X \rangle$ is a p-group.

Proof: If (1) holds, then $gPg^{-1} = P$ for all $g \in G$ since $gPg^{-1} \in Syl_p(G)$, i.e., P is normal in G. Hence (1) implies (2). Conversely, if $P \trianglelefteq G$ and $Q \in Syl_p(G)$, then by Sylow's Theorem there exists $g \in G$ such that $Q = gPg^{-1} = P$. Thus $Syl_p(G) = \{P\}$ and (2) implies (1).

Since characteristic subgroups are normal, (3) implies (2). Conversely, if $P \trianglelefteq G$, we just proved P is the unique subgroup of G of order p^α, hence P char G. Thus (2) and (3) are equivalent.

Finally, assume (1) holds and suppose X is a subset of G such that $|x|$ is a power of p for all $x \in X$. By the conjugacy part of Sylow's Theorem, for each $x \in X$ there is some $g \in G$ such that $x \in gPg^{-1} = P$. Thus $X \subseteq P$, and so $\langle X \rangle \le P$, and $\langle X \rangle$ is a p-group. Conversely, if (4) holds, let X be the union of all Sylow p-subgroups of G. If P is any Sylow p-subgroup, P is a subgroup of the p-group $\langle X \rangle$. Since P is a p-subgroup of G of maximal order, we must have $P = \langle X \rangle$, so (1) holds.

Examples

Let G be a finite group and let p be a prime.

   (1) If p does not divide the order of G, the Sylow p-subgroup of G is the trivial group (and all parts of Sylow's Theorem hold trivially). If $|G| = p^\alpha$, G is the unique Sylow p-subgroup of G.

   (2) A finite abelian group has a unique Sylow p-subgroup for each prime p. This subgroup consists of all elements x whose order is a power of p. This is sometimes called the *p-primary component* of the abelian group.

   (3) S_3 has three Sylow 2-subgroups: $\langle (1\,2) \rangle$, $\langle (2\,3) \rangle$ and $\langle (1\,3) \rangle$. It has a unique (hence normal) Sylow 3-subgroup: $\langle (1\,2\,3) \rangle = A_3$. Note that $3 \equiv 1 \pmod 2$.

   (4) A_4 has a unique Sylow 2-subgroup: $\langle (1\,2)(3\,4), (1\,3)(2\,4) \rangle \cong V_4$. It has four Sylow 3-subgroups: $\langle (1\,2\,3) \rangle$, $\langle (1\,2\,4) \rangle$, $\langle (1\,3\,4) \rangle$ and $\langle (2\,3\,4) \rangle$. Note that $4 \equiv 1 \pmod 3$.

   (5) S_4 has $n_2 = 3$ and $n_3 = 4$. Since S_4 contains a subgroup isomorphic to D_8, every Sylow 2-subgroup of S_4 is isomorphic to D_8.

Applications of Sylow's Theorem

We now give some applications of Sylow's Theorem. Most of the examples use Sylow's Theorem to prove that a group of a particular order is not simple. After discussing methods of constructing larger groups from smaller ones (for example, the formation of semidirect products) we shall be able to use these results to classify groups of some specific orders n (as we already did for $n = 15$).

Since Sylow's Theorem ensures the existence of p-subgroups of a finite group, it is worthwhile to study groups of prime power order more closely. This will be done in Chapter 6 and many more applications of Sylow's Theorem will be discussed there.

For groups of small order, the congruence condition of Sylow's Theorem alone is often sufficient to force the existence of a *normal* subgroup. The first step in any numerical application of Sylow's Theorem is to factor the group order into prime powers. The largest prime divisors of the group order tend to give the fewest possible values for n_p (for example, the congruence condition on n_2 gives no restriction whatsoever), which limits the structure of the group G. In the following examples we shall see situations where Sylow's Theorem alone does not force the existence of a normal subgroup, however some additional argument (often involving studying the elements of order p for a number of different primes p) proves the existence of a normal Sylow subgroup.

Example: (Groups of order pq, p and q primes with $p < q$)

Suppose $|G| = pq$ for primes p and q with $p < q$. Let $P \in Syl_p(G)$ and let $Q \in Syl_q(G)$. We show that Q is normal in G and if P is also normal in G, then G is cyclic.

Now the three conditions: $n_q = 1 + kq$ for some $k \geq 0$, n_q divides p and $p < q$, together force $k = 0$. Since $n_q = 1$, $Q \trianglelefteq G$.

Since n_p divides the prime q, the only possibilities are $n_p = 1$ or q. In particular, if $p \nmid q - 1$, (that is, if $q \not\equiv 1 \pmod{p}$), then n_p cannot equal q, so $P \trianglelefteq G$.

Let $P = \langle x \rangle$ and $Q = \langle y \rangle$. If $P \trianglelefteq G$, then since $G/C_G(P)$ is isomorphic to a subgroup of $\mathrm{Aut}(Z_p)$ and the latter group has order $p - 1$, Lagrange's Theorem together with the observation that neither p nor q can divide $p - 1$ implies that $G = C_G(P)$. In this case $x \in P \leq Z(G)$ so x and y commute. (Alternatively, this follows immediately from Exercise 42 of Section 3.1.) This means $|xy| = pq$ (cf. the exercises in Section 2.3), hence in this case G is cyclic: $G \cong Z_{pq}$.

If $p \mid q - 1$, we shall see in Chapter 5 that there is a unique non-abelian group of order pq (in which, necessarily, $n_p = q$). We can prove the existence of this group now. Let Q be a Sylow q-subgroup of the symmetric group of degree q, S_q. By Exercise 34 in Section 3, $|N_{S_q}(Q)| = q(q - 1)$. By assumption, $p \mid q - 1$ so by Cauchy's Theorem $N_{S_q}(Q)$ has a subgroup, P, of order p. By Corollary 15 in Section 3.2, PQ is a group of order pq. Since $C_{S_q}(Q) = Q$ (Example 2, Section 3), PQ is a non-abelian group. The essential ingredient in the uniqueness proof of PQ is Proposition 17 on the cyclicity of $\mathrm{Aut}(Z_q)$.

Example: (Groups of order 30)

Let G be a group of order 30. We show that G has a normal subgroup isomorphic to Z_{15}. We shall use this information to classify groups of order 30 in the next chapter. Note that any subgroup of order 15 is necessarily normal (since it is of index 2) and cyclic (by the preceding result) so it is only necessary to show there exists a subgroup of order 15. The quickest way of doing this is to quote Exercise 13 in Section 2. We give an alternate argument which illustrates how Sylow's Theorem can be used in conjunction with a counting of elements of prime order to produce a normal subgroup.

Let $P \in Syl_5(G)$ and let $Q \in Syl_3(G)$. If either P or Q is normal in G, by Corollary 15, Chapter 3, PQ is a group of order 15. Note also that if either P or Q is normal, then both P and Q are characteristic subgroups of PQ, and since $PQ \trianglelefteq G$, *both* P and Q are normal in G (Exercise 8(a), Section 4). Assume therefore that neither Sylow subgroup is normal. The only possibilities by Part 3 of Sylow's Theorem are $n_5 = 6$ and $n_3 = 10$. Each element of order 5 lies in a Sylow 5-subgroup, each Sylow 5-subgroup contains 4 nonidentity elements and, by Lagrange's Theorem, distinct Sylow 5-subgroups intersect in the identity. Thus the number of elements of order 5 in G is the number of nonidentity elements in one Sylow 5-subgroup times the number of Sylow 5-subgroups. This would be $4 \cdot 6 = 24$ elements of order 5. By similar reasoning, the number of elements of order 3 would be $2 \cdot 10 = 20$. This is absurd since a group of order 30 cannot contain $24 + 20 = 44$ distinct elements. One of P or Q (hence both) must be normal in G.

This sort of counting technique is frequently useful (cf. also Section 6.2) and works particularly well when the Sylow p-subgroups have order p (as in this example), since then the intersection of two distinct Sylow p-subgroups must be the identity. If the order of the Sylow p-subgroup is p^{α} with $\alpha \geq 2$, greater care is required in counting elements, since in this case distinct Sylow p-subgroups may have many more elements in common, i.e., the intersection may be nontrivial.

Example: (Groups of order 12)

Let G be a group of order 12. We show that either G has a normal Sylow 3-subgroup or $G \cong A_4$ (in the latter case G has a normal Sylow 2-subgroup). We shall use this information to classify groups of order 12 in the next chapter.

Suppose $n_3 \neq 1$ and let $P \in Syl_3(G)$. Since $n_3 \mid 4$ and $n_3 \equiv 1 \pmod{3}$, it follows that $n_3 = 4$. Since distinct Sylow 3-subgroups intersect in the identity and each contains two elements of order 3, G contains $2 \cdot 4 = 8$ elements of order 3. Since $|G : N_G(P)| = n_3 = 4$, $N_G(P) = P$. Now G acts by conjugation on its four Sylow 3-subgroups, so this action affords a permutation representation

$$\varphi : G \to S_4$$

(note that we could also act by left multiplication on the left cosets of P and use Theorem 3). The kernel K of this action is the subgroup of G which normalizes all Sylow 3-subgroups of G. In particular, $K \leq N_G(P) = P$. Since P is not normal in G by assumption, $K = 1$, i.e., φ is injective and

$$G \cong \varphi(G) \leq S_4.$$

Since G contains 8 elements of order 3 and there are precisely 8 elements of order 3 in S_4, all contained in A_4, it follows that $\varphi(G)$ intersects A_4 in a subgroup of order at least 8. Since both groups have order 12 it follows that $\varphi(G) = A_4$, so that $G \cong A_4$.

Note that A_4 does indeed have 4 Sylow 3-subgroups (see Example 4 following Corollary 20), so that such a group G does exist. Also, let V be a Sylow 2-subgroup of A_4. Since $|V| = 4$, it contains all of the remaining elements of A_4. In particular, there cannot be another Sylow 2-subgroup. Thus $n_2(A_4) = 1$, i.e., $V \trianglelefteq A_4$ (which one can also see directly because V is the identity together with the three elements of S_4 which are products of two disjoint transpositions, that is, V is a union of conjugacy classes).

Example: (Groups of order p^2q, p and q distinct primes)

Let G be a group of order p^2q. We show that G has a normal Sylow subgroup (for either p or q). We shall use this information to classify some groups of this order in the next chapter (cf. Exercises 8 to 12 of Section 5.5). Let $P \in Syl_p(G)$ and let $Q \in Syl_q(G)$.

Consider first when $p > q$. Since $n_p \mid q$ and $n_p = 1 + kp$, we must have $n_p = 1$. Thus $P \trianglelefteq G$.

Consider now the case $p < q$. If $n_q = 1$, Q is normal in G. Assume therefore that $n_q > 1$, i.e., $n_q = 1 + tq$, for some $t > 0$. Now n_q divides p^2 so $n_q = p$ or p^2. Since $q > p$ we cannot have $n_q = p$, hence $n_q = p^2$. Thus

$$tq = p^2 - 1 = (p-1)(p+1).$$

Since q is prime, either $q \mid p - 1$ or $q \mid p + 1$. The former is impossible since $q > p$ so the latter holds. Since $q > p$ but $q \mid p + 1$, we must have $q = p + 1$. This forces $p = 2$, $q = 3$ and $|G| = 12$. The result now follows from the preceding example.

Groups of Order 60

We illustrate how Sylow's Theorems can be used to unravel the structure of groups of a given order even if some groups of that order may be simple. Note the technique of changing from one prime to another and the inductive process where we use results on groups of order < 60 to study groups of order 60.

Proposition 21. If $|G| = 60$ and G has more than one Sylow 5-subgroup, then G is simple.

Proof: Suppose by way of contradiction that $|G| = 60$ and $n_5 > 1$ but that there exists H a normal subgroup of G with $H \neq 1$ or G. By Sylow's Theorem the only possibility for n_5 is 6. Let $P \in Syl_5(G)$, so that $|N_G(P)| = 10$ since its index is n_5.

If $5 \mid |H|$ then H contains a Sylow 5-subgroup of G and since H is normal, it contains all 6 conjugates of this subgroup. In particular, $|H| \geq 1 + 6 \cdot 4 = 25$, and the only possibility is $|H| = 30$. This leads to a contradiction since a previous example proved that any group of order 30 has a normal (hence unique) Sylow 5-subgroup. This argument shows 5 does not divide $|H|$ for any proper normal subgroup H of G.

If $|H| = 6$ or 12, H has a normal, hence characteristic, Sylow subgroup, which is therefore also normal in G. Replacing H by this subgroup if necessary, we may assume $|H| = 2, 3$ or 4. Let $\overline{G} = G/H$, so $|\overline{G}| = 30, 20$ or 15. In each case, $\overline{G}$ has a normal subgroup $\overline{P}$ of order 5 by previous results. If we let H_1 be the complete preimage of $\overline{P}$ in G, then $H_1 \trianglelefteq G$, $H_1 \neq G$ and $5 \mid |H_1|$. This contradicts the preceding paragraph and so completes the proof.

Corollary 22. A_5 is simple.

Proof: The subgroups $\langle (1\,2\,3\,4\,5) \rangle$ and $\langle (1\,3\,2\,4\,5) \rangle$ are distinct Sylow 5-subgroups of A_5 so the result follows immediately from the proposition.

The next proposition shows that there is a unique simple group of order 60.

Proposition 23. If G is a simple group of order 60, then $G \cong A_5$.

Proof: Let G be a simple group of order 60, so $n_2 = 3, 5$ or 15. Let $P \in Syl_2(G)$ and let $N = N_G(P)$, so $|G : N| = n_2$.

First observe that G has no proper subgroup H of index less that 5, as follows: if H were a subgroup of G of index 4, 3 or 2, then, by Theorem 3, G would have a normal subgroup K contained in H with G/K isomorphic to a subgroup of S_4, S_3 or S_2. Since $K \neq G$, simplicity forces $K = 1$. This is impossible since $60 \, (= |G|)$ does not divide $4!$. This argument shows, in particular, that $n_2 \neq 3$.

If $n_2 = 5$, then N has index 5 in G so the action of G by left multiplication on the set of left cosets of N gives a permutation representation of G into S_5. Since (as above) the kernel of this representation is a proper normal subgroup and G is simple, the kernel is 1 and G is isomorphic to a subgroup of S_5. Identify G with this isomorphic copy so that we may assume $G \leq S_5$. If G is not contained in A_5, then $S_5 = GA_5$ and, by the Second Isomorphism Theorem, $A_5 \cap G$ is of index 2 in G. Since G has no (normal) subgroup of index 2, this is a contradiction. This argument proves $G \leq A_5$. Since $|G| = |A_5|$, the isomorphic copy of G in S_5 coincides with A_5, as desired.

Finally, assume $n_2 = 15$. If for every pair of distinct Sylow 2-subgroups P and Q of G, $P \cap Q = 1$, then the number of nonidentity elements in Sylow 2-subgroups of G would be $(4-1) \cdot 15 = 45$. But $n_5 = 6$ so the number of elements of order 5 in G is $(5-1) \cdot 6 = 24$, accounting for 69 elements. This contradiction proves that there exist distinct Sylow 2-subgroups P and Q with $|P \cap Q| = 2$. Let $M = N_G(P \cap Q)$. Since P and Q are abelian (being groups of order 4), P and Q are subgroups of M and since G is simple, $M \neq G$. Thus 4 divides $|M|$ and $|M| > 4$ (otherwise, $P = M = Q$). The only possibility is $|M| = 12$, i.e., M has index 5 in G (recall M cannot have index 3 or 1). But now the argument of the preceding paragraph applied to M in place of N gives $G \cong A_5$. This leads to a contradiction in this case because $n_2(A_5) = 5$ (cf. the exercises). The proof is complete.

EXERCISES

Let G be a finite group and let p be a prime.

1. Prove that if $P \in Syl_p(G)$ and H is a subgroup of G containing P then $P \in Syl_p(H)$. Give an example to show that, in general, a Sylow p-subgroup of a subgroup of G need not be a Sylow p-subgroup of G.

2. Prove that if H is a subgroup of G and $Q \in Syl_p(H)$ then $gQg^{-1} \in Syl_p(gHg^{-1})$ for all $g \in G$.

3. Use Sylow's Theorem to prove Cauchy's Theorem. (Note that we only used Cauchy's Theorem for abelian groups — Proposition 3.21 — in the proof of Sylow's Theorem so this line of reasoning is not circular.)

4. Exhibit all Sylow 2-subgroups and Sylow 3-subgroups of D_{12} and $S_3 \times S_3$.

5. Show that a Sylow p-subgroup of D_{2n} is cyclic and normal for every odd prime p.

6. Exhibit all Sylow 3-subgroups of A_4 and all Sylow 3-subgroups of S_4.

7. Exhibit all Sylow 2-subgroups of S_4 and find elements of S_4 which conjugate one of these into each of the others.

8. Exhibit two distinct Sylow 2-subgroups of S_5 and an element of S_5 that conjugates one into the other.

9. Exhibit all Sylow 3-subgroups of $SL_2(\mathbb{F}_3)$ (cf. Exercise 9, Section 2.1).

10. Prove that the subgroup of $SL_2(\mathbb{F}_3)$ generated by $\begin{pmatrix} 0 & -1 \\ 1 & 0 \end{pmatrix}$ and $\begin{pmatrix} 1 & 1 \\ 1 & -1 \end{pmatrix}$ is the unique Sylow 2-subgroup of $SL_2(\mathbb{F}_3)$ (cf. Exercise 10, Section 2.4).

11. Show that the center of $SL_2(\mathbb{F}_3)$ is the group of order 2 consisting of $\pm I$, where I is the identity matrix. Prove that $SL_2(\mathbb{F}_3)/Z(SL_2(\mathbb{F}_3)) \cong A_4$. [Use facts about groups of order 12.]

12. Let $2n = 2^a k$ where k is odd. Prove that the number of Sylow 2-subgroups of D_{2n} is k. [Prove that if $P \in Syl_2(D_{2n})$ then $N_{D_{2n}}(P) = P$.]

13. Prove that a group of order 56 has a normal Sylow p-subgroup for some prime p dividing its order.

14. Prove that a group of order 312 has a normal Sylow p-subgroup for some prime p dividing its order.

15. Prove that a group of order 351 has a normal Sylow p-subgroup for some prime p dividing its order.

16. Let $|G| = pqr$, where p, q and r are primes with $p < q < r$. Prove that G has a normal Sylow subgroup for either p, q or r.

17. Prove that if $|G| = 105$ then G has a normal Sylow 5-subgroup and a normal Sylow 7-subgroup.

18. Prove that a group of order 200 has a normal Sylow 5-subgroup.

19. Prove that if $|G| = 6545$ then G is not simple.

20. Prove that if $|G| = 1365$ then G is not simple.

21. Prove that if $|G| = 2907$ then G is not simple.

22. Prove that if $|G| = 132$ then G is not simple.

23. Prove that if $|G| = 462$ then G is not simple.

24. Prove that if G is a group of order 231 then $Z(G)$ contains a Sylow 11-subgroup of G and a Sylow 7-subgroup is normal in G.

25. Prove that if G is a group of order 385 then $Z(G)$ contains a Sylow 7-subgroup of G and a Sylow 11-subgroup is normal in G.

26. Let G be a group of order 105. Prove that if a Sylow 3-subgroup of G is normal then G is abelian.

27. Let G be a group of order 315 which has a normal Sylow 3-subgroup. Prove that $Z(G)$ contains a Sylow 3-subgroup of G and deduce that G is abelian.

28. Let G be a group of order 1575. Prove that if a Sylow 3-subgroup of G is normal then a Sylow 5-subgroup and a Sylow 7-subgroup are normal. In this situation prove that G is abelian.

29. If G is a non-abelian simple group of order < 100, prove that $G \cong A_5$. [Eliminate all orders but 60.]

30. How many elements of order 7 must there be in a simple group of order 168?

31. For $p = 2$, 3 and 5 find $n_p(A_5)$ and $n_p(S_5)$. [Note that $A_4 \leq A_5$.]

32. Let P be a Sylow p-subgroup of H and let H be a subgroup of K. If $P \trianglelefteq H$ and $H \trianglelefteq K$, prove that P is normal in K. Deduce that if $P \in Syl_p(G)$ and $H = N_G(P)$, then $N_G(H) = H$ (in words: *normalizers of Sylow p-subgroups are self-normalizing*).

33. Let P be a normal Sylow p-subgroup of G and let H be any subgroup of G. Prove that $P \cap H$ is the unique Sylow p-subgroup of H.

34. Let $P \in Syl_p(G)$ and assume $N \trianglelefteq G$. Use the conjugacy part of Sylow's Theorem to prove that $P \cap N$ is a Sylow p-subgroup of N. Deduce that PN/N is a Sylow p-subgroup of G/N (note that this may also be done by the Second Isomorphism Theorem — cf. Exercise 9, Section 3.3).

35. Let $P \in Syl_p(G)$ and let $H \leq G$. Prove that $gPg^{-1} \cap H$ is a Sylow p-subgroup of H for some $g \in G$. Give an explicit example showing that $hPh^{-1} \cap H$ is not necessarily a Sylow p-subgroup of H for any $h \in H$ (in particular, we cannot always take $g = 1$ in the first part of this problem, as we could when H was normal in G).

36. Prove that if N is a normal subgroup of G then $n_p(G/N) \leq n_p(G)$.

37. Let R be a normal p-subgroup of G (not necessarily a Sylow subgroup).

    (a) Prove that R is contained in every Sylow p-subgroup of G.

    (b) If S is another normal p-subgroup of G, prove that RS is also a normal p-subgroup of G.

    (c) The subgroup $O_p(G)$ is defined to be the group generated by all normal p-subgroups of G. Prove that $O_p(G)$ is the unique largest normal p-subgroup of G and $O_p(G)$ equals the intersection of all Sylow p-subgroups of G.

    (d) Let $\overline{G} = G/O_p(G)$. Prove that $O_p(\overline{G}) = \overline{1}$ (i.e., $\overline{G}$ has no nontrivial normal p-subgroup).

38. Use the method of proof in Sylow's Theorem to show that if n_p is not congruent to $1 \pmod{p^2}$ then there are distinct Sylow p-subgroups P and Q of G such that $|P : P \cap Q| = |Q : P \cap Q| = p$.

39. Show that the subgroup of strictly upper triangular matrices in $GL_n(\mathbb{F}_p)$ (cf. Exercise 17, Section 2.1) is a Sylow p-subgroup of this finite group. [Use the order formula in Section 1.4 to find the order of a Sylow p-subgroup of $GL_n(\mathbb{F}_p)$.]

40. Prove that the number of Sylow p-subgroups of $GL_2(\mathbb{F}_p)$ is $p + 1$. [Exhibit two distinct Sylow p-subgroups.]

41. Prove that $SL_2(\mathbb{F}_4) \cong A_5$ (cf. Exercise 9, Section 2.1 for the definition of $SL_2(\mathbb{F}_4)$).

42. Prove that the group of rigid motions in $\mathbb{R}^3$ of an icosahedron is isomorphic to A_5. [Recall that the order of this group is 60: Exercise 13, Section 1.2.]

43. Prove that the group of rigid motions in $\mathbb{R}^3$ of a dodecahedron is isomorphic to A_5. (As with the cube and the tetrahedron, the icosahedron and the dodecahedron are dual solids.) [Recall that the order of this group is 60: Exercise 12, Section 1.2.]

44. Let p be the smallest prime dividing the order of the finite group G. If $P \in Syl_p(G)$ and P is cyclic prove that $N_G(P) = C_G(P)$.

45. Find generators for a Sylow p-subgroup of S_{2p}, where p is an odd prime. Show that this is an abelian group of order p^2.

46. Find generators for a Sylow p-subgroup of S_{p^2}, where p is a prime. Show that this is a non-abelian group of order p^{p+1}.

47. Write and execute a computer program which

    (i) gives each odd number $n < 10,000$ that is not a power of a prime and with the property that for each prime divisor p of n the corresponding n_p is not forced to be 1 for all groups of order n by the congruence condition of Sylow's Theorem, and

    (ii) gives for each n in (i) the factorization of n into prime powers and gives the list of all permissible values of n_p for all primes p dividing n (i.e., those values not ruled out by Part 3 of Sylow's Theorem).

48. Carry out the same process as in the preceding exercise for all even numbers less than 1000. Explain the relative lengths of the lists versus the number of integers tested.

49. Prove that if $|G| = 2^n m$ where m is odd and G has a cyclic Sylow 2-subgroup then G has a normal subgroup of order m. [Use induction and Exercises 11 and 12 in Section 2.]

50. Prove that if U and W are normal subsets of a Sylow p-subgroup P of G then U is conjugate to W in G if and only if U is conjugate to W in $N_G(P)$. Deduce that two elements in the center of P are conjugate in G if and only if they are conjugate in $N_G(P)$. (A subset U of P is normal in P if $N_P(U) = P$.)

51. Let P be a Sylow p-subgroup of G and let M be any subgroup of G which contains $N_G(P)$. Prove that $|G : M| \equiv 1 \pmod{p}$.

The following sequence of exercises leads to the classification of all numbers n with the property that every group of order n is cyclic (for example, $n = 15$ is such an integer). These arguments are a vastly simplified prototype for the proof that every group of odd order is solvable in the sense that they use the *structure* (commutativity) of the proper subgroups and their *embedding* in the whole group (we shall see that distinct maximal subgroups intersect in the identity) to obtain a contradiction by counting arguments. In the proof that groups of odd order are solvable one uses induction to reduce to the situation in which a minimal counterexample is a simple group — but here every proper subgroup is solvable (not abelian as in our situation). The analysis of the structure and embedding of the maximal subgroups in this situation is much more complicated and the counting arguments are (roughly speaking) replaced by character theory arguments (as will be discussed in Part VI).

52. Suppose G is a finite simple group in which every proper subgroup is abelian. If M and N are distinct maximal subgroups of G prove $M \cap N = 1$. [See Exercise 23 in Section 3.]

53. Use the preceding exercise to prove that if G is any non-abelian group in which every proper subgroup is abelian then G is not simple. [Let G be a counterexample to this assertion and use Exercise 24 in Section 3 to show that G has more than one conjugacy class of maximal subgroups. Use the method of Exercise 23 in Section 3 to count the elements which lie in all conjugates of M and N, where M and N are nonconjugate maximal subgroups of G; show that this gives more than $|G|$ elements.]

54. Prove the following classification: if G is a finite group of order $p_1 p_2 \ldots p_r$ where the p_i's are distinct primes such that p_i does not divide $p_j - 1$ for all i and j, then G is cyclic. [By induction, every proper subgroup of G is cyclic, so G is not simple by the preceding exercise. If N is a nontrivial proper normal subgroup, N is cyclic and $G/C_G(N)$ acts as automorphisms of N. Use Proposition 16 to show that $N \leq Z(G)$ and use induction to show $G/Z(G)$ is cyclic, hence G is abelian by Exercise 36 of Section 3.1.]

55. Prove the converse to the preceding exercise: if $n \geq 2$ is an integer such that every group of order n is cyclic, then $n = p_1 p_2 \ldots p_r$ is a product of distinct primes and p_i does not divide $p_j - 1$ for all i, j. [If n is not of this form, construct noncyclic groups of order n using direct products of noncyclic groups of order p^2 and pq, where $p \mid q - 1$.]

56. If G is a finite group in which every proper subgroup is abelian, show that G is solvable.

4.6 THE SIMPLICITY OF A_n

There are a number of proofs of the simplicity of A_n, $n \geq 5$. The most elementary involves showing A_n is generated by 3-cycles. Then one shows that a normal subgroup must contain one 3-cycle hence must contain all the 3-cycles so cannot be a proper subgroup. We include a less computational approach.

Note that A_3 is an abelian simple group and that A_4 is not simple ($n_2(A_4) = 1$).

Theorem 24. A_n is simple for all $n \geq 5$.

Proof: By induction on n. The result has already been established for $n = 5$, so assume $n \geq 6$ and let $G = A_n$. Assume there exists $H \trianglelefteq G$ with $H \neq 1$ or G.

For each $i \in \{1, 2, \ldots, n\}$ let G_i be the stabilizer of i in the natural action of G on $i \in \{1, 2, \ldots, n\}$. Thus $G_i \leq G$ and $G_i \cong A_{n-1}$. By induction, G_i is simple for $1 \leq i \leq n$.

Suppose first that there is some $\tau \in H$ with $\tau \neq 1$ but $\tau(i) = i$ for some $i \in \{1, 2, \ldots, n\}$. Since $\tau \in H \cap G_i$ and $H \cap G_i \trianglelefteq G_i$, by the simplicity of G_i we must have $H \cap G_i = G_i$, that is

$$G_i \leq H.$$

By Exercise 2 of Section 1, $\sigma G_i \sigma^{-1} = G_{\sigma(i)}$, so for all i, $\sigma G_i \sigma^{-1} \leq \sigma H \sigma^{-1} = H$. Thus

$$G_j \leq H, \quad \text{for all } j \in \{1, 2, \ldots, n\}.$$

Any $\lambda \in A_n$ may be written as a product of an even number, $2t$, of transpositions, so

$$\lambda = \lambda_1 \lambda_2 \cdots \lambda_t,$$

where λ_k is a product of two transpositions. Since $n > 4$ each $\lambda_k \in G_j$, for some j, hence

$$G = \langle G_1, G_2, \ldots, G_n \rangle \leq H,$$

which is a contradiction. Therefore if $\tau \neq 1$ is an element of H then $\tau(i) \neq i$ for all $i \in \{1, 2, \ldots, n\}$, i.e., no nonidentity element of H fixes any element of $\{1, 2, \ldots, n\}$.

It follows that if τ_1, τ_2 are elements of H with

$$\tau_1(i) = \tau_2(i) \text{ for some } i, \text{ then } \tau_1 = \tau_2 \tag{4.2}$$

since then $\tau_2^{-1} \tau_1(i) = i$.

Suppose there exists a $\tau \in H$ such that the cycle decomposition of τ contains a cycle of length ≥ 3, say

$$\tau = (a_1 \, a_2 \, a_3 \ldots)(b_1 \, b_2 \ldots) \ldots .$$

Let $\sigma \in G$ be an element with $\sigma(a_1) = a_1$, $\sigma(a_2) = a_2$ but $\sigma(a_3) \neq a_3$ (note that such a σ exists in A_n since $n \geq 5$). By Proposition 10

$$\tau_1 = \sigma \tau \sigma^{-1} = (a_1 \, a_2 \, \sigma(a_3) \ldots)(\sigma(b_1) \, \sigma(b_2) \ldots) \ldots$$

so τ and τ_1 are distinct elements of H with $\tau(a_1) = \tau_1(a_1) = a_2$, contrary to (2). This proves that only 2-cycles can appear in the cycle decomposition of nonidentity elements of H.

Let $\tau \in H$ with $\tau \neq 1$, so that

$$\tau = (a_1 \, a_2)(a_3 \, a_4)(a_5 \, a_6) \ldots$$

(note that $n \geq 6$ is used here). Let $\sigma = (a_1 \, a_2)(a_3 \, a_5) \in G$. Then

$$\tau_1 = \sigma \tau \sigma^{-1} = (a_1 \, a_2)(a_5 \, a_4)(a_3 \, a_6) \ldots,$$

hence τ and τ_1 are distinct elements of H with $\tau(a_1) = \tau_1(a_1) = a_2$, again contrary to (2). This completes the proof of the simplicity of A_n.

EXERCISES

Let G be a group and let Ω be an infinite set.

1. Prove that A_n does not have a proper subgroup of index $< n$ for all $n \geq 5$.

2. Find all normal subgroups of S_n for all $n \geq 5$.

3. Prove that A_n is the only proper subgroup of index $< n$ in S_n for all $n \geq 5$.

4. Prove that A_n is generated by the set of all 3-cycles for each $n \geq 3$.

5. Prove that if there exists a chain of subgroups $G_1 \leq G_2 \leq \ldots \leq G$ such that $G = \cup_{i=1}^{\infty} G_i$ and each G_i is simple then G is simple.

6. Let D be the subgroup of S_Ω consisting of permutations which move only a finite number of elements of Ω (described in Exercise 17 in Section 3) and let A be the set of all elements $\sigma \in D$ such that σ acts as an even permutation on the (finite) set of points it moves. Prove that A is an infinite simple group. [Show that every pair of elements of A lie in a finite simple subgroup of A.]

7. Under the notation of the preceding, exercise prove that if $H \trianglelefteq S_\Omega$ and $H \neq 1$ then $A \leq H$, i.e., A is the unique (nontrivial) minimal normal subgroup of S_Ω.

8. Under the notation of the preceding two exercises prove that $|D| = |A| = |\Omega|$. Deduce that
$$\text{if } S_\Omega \cong S_\Delta \text{ then } |\Omega| = |\Delta|.$$

[Use the fact that D is generated by transpositions. You may assume that countable unions and finite direct products of sets of cardinality $|\Omega|$ also have cardinality $|\Omega|$.]

CHAPTER 5

Direct and Semidirect Products and Abelian Groups

In this chapter we consider two of the easier methods for constructing larger groups from smaller ones, namely the notions of direct and semidirect products. This allows us to state the Fundamental Theorem on Finitely Generated Abelian Groups, which in particular completely classifies all finite abelian groups.

5.1 DIRECT PRODUCTS

We begin with the definition of the direct product of a finite and of a countable number of groups (the direct product of an arbitrary collection of groups is considered in the exercises).

Definition.

(1) The *direct product* $G_1 \times G_2 \times \cdots \times G_n$ of the groups $G_1, G_2, \ldots, G_n$ with operations $\star_1, \star_2, \ldots, \star_n$, respectively, is the set of n-tuples $(g_1, g_2, \ldots, g_n)$ where $g_i \in G_i$ with operation defined componentwise:

$$(g_1, g_2, \ldots, g_n) \star (h_1, h_2, \ldots, h_n) = (g_1 \star_1 h_1, g_2 \star_2 h_2, \ldots, g_n \star_n h_n).$$

(2) Similarly, the *direct product* $G_1 \times G_2 \times \cdots$ of the groups $G_1, G_2, \ldots$ with operations $\star_1, \star_2, \ldots$, respectively, is the set of sequences $(g_1, g_2, \ldots)$ where $g_i \in G_i$ with operation defined componentwise:

$$(g_1, g_2, \ldots) \star (h_1, h_2, \ldots) = (g_1 \star_1 h_1, g_2 \star_2 h_2, \ldots).$$

Although the operations may be different in each of the factors of a direct product, we shall, as usual, write all abstract groups multiplicatively, so that the operation in (1) above, for example, becomes simply

$$(g_1, g_2, \ldots, g_n)(h_1, h_2, \ldots, h_n) = (g_1 h_1, g_2 h_2, \ldots, g_n h_n).$$

Examples

(1) Suppose $G_i = \mathbb{R}$ (operation addition) for $i = 1, 2, \ldots, n$. Then $\mathbb{R} \times \mathbb{R} \times \cdots \times \mathbb{R}$ (n-factors) is the familiar Euclidean n-space $\mathbb{R}^n$ with usual vector addition:

$$(a_1, a_2, \ldots, a_n) + (b_1, b_2, \ldots, b_n) = (a_1 + b_1, a_2 + b_2, \ldots, a_n + b_n).$$

(2) To illustrate that groups forming the direct product (and corresponding operations) may be completely general, let $G_1 = \mathbb{Z}$, let $G_2 = S_3$ and let $G_3 = GL_2(\mathbb{R})$, where the group operations are addition, composition, and matrix multiplication, respectively. Then the operation in $G_1 \times G_2 \times G_3$ is defined by

$$(n, \sigma, \begin{pmatrix} a & b \\ c & d \end{pmatrix}) \, (m, \tau, \begin{pmatrix} p & q \\ r & s \end{pmatrix}) = (n + m, \sigma \circ \tau, \begin{pmatrix} ap + br & aq + bs \\ cp + dr & cq + ds \end{pmatrix}).$$

Proposition 1. If $G_1, \ldots, G_n$ are groups, their direct product is a group of order $|G_1| \, |G_2| \cdots |G_n|$ (if any G_i is infinite, so is the direct product).

Proof: Let $G = G_1 \times G_2 \times \cdots \times G_n$. The proof that the group axioms hold for G is straightforward since each axiom is a consequence of the fact that the same axiom holds in each factor, G_i, and the operation on G is defined componentwise. For example, the associative law is verified as follows:

Let $(a_1, a_2, \ldots, a_n)$, $(b_1, b_2, \ldots, b_n)$, and $(c_1, c_2, \ldots, c_n) \in G$. Then

$$(a_1, a_2, \ldots, a_n)\big[(b_1, b_2, \ldots, b_n)(c_1, c_2, \ldots, c_n)\big]$$
$$= (a_1, a_2, \ldots, a_n)(b_1 c_1, b_2 c_2, \ldots, b_n c_n)$$
$$= (a_1(b_1 c_1), a_2(b_2 c_2), \ldots, a_n(b_n c_n))$$
$$= ((a_1 b_1)c_1, (a_2 b_2)c_2, \ldots, (a_n b_n)c_n)$$
$$= \big[(a_1, a_2, \cdots, a_n)(b_1, b_2, \ldots, b_n)\big](c_1, c_2, \ldots, c_n),$$

where in the third step we have used the associative law in each component. The remaining verification that the direct product is a group is similar: the identity of G is the n-tuple $(1_1, 1_2, \ldots, 1_n)$, where 1_i is the identity of G_i and the inverse of $(g_1, g_2, \ldots, g_n)$ is $(g_1^{-1}, g_2^{-1}, \ldots, g_n^{-1})$, where g_i^{-1} is the inverse of g_i in G_i.

The formula for the order of G is clear.

If the factors of the direct product are rearranged, the resulting direct product is isomorphic to the original one (cf. Exercise 7).

The next proposition shows that a direct product, $G_1 \times G_2 \times \cdots \times G_n$, contains an isomorphic copy of each G_i. One can think of these specific copies as the "coordinate axes" of the direct product since, in the case of $\mathbb{R} \times \mathbb{R}$, they coincide with the x and y axes. One should be careful, however, not to think of these "coordinate axes" as the *only* copies of the groups G_i in the direct product. For example in $\mathbb{R} \times \mathbb{R}$ any line through the origin is a subgroup of $\mathbb{R} \times \mathbb{R}$ isomorphic to $\mathbb{R}$ (and $\mathbb{R} \times \mathbb{R}$ has infinitely many pairs of lines which are coordinate axes, viz. any rotation of a given coordinate system). The second part of the proposition shows that there are *projection homomorphisms* onto each of the components.

Proposition 2. Let $G_1, G_2, \ldots, G_n$ be groups and let $G = G_1 \times \cdots \times G_n$ be their direct product.

(1) For each fixed i the set of elements of G which have the identity of G_j in the jth position for all $j \neq i$ and arbitrary elements of G_i in position i is a subgroup of G isomorphic to G_i:

$$G_i \cong \{(1, 1, \ldots, 1, g_i, 1, \ldots, 1) \mid g_i \in G_i\},$$

(here g_i appears in the ith position). If we identify G_i with this subgroup, then $G_i \trianglelefteq G$ and

$$G/G_i \cong G_1 \times \cdots \times G_{i-1} \times G_{i+1} \times \cdots \times G_n.$$

(2) For each fixed i define $\pi_i : G \to G_i$ by

$$\pi_i((g_1, g_2, \ldots, g_n)) = g_i.$$

Then π_i is a surjective homomorphism with

$$\ker \pi_i = \{(g_1, \ldots, g_{i-1}, 1, g_{i+1}, \ldots, g_n) \mid g_j \in G_j \text{ for all } j \neq i\}$$
$$\cong G_1 \times \cdots \times G_{i-1} \times G_{i+1} \times \cdots \times G_n$$

(here the 1 appears in position i).

(3) Under the identifications in part (1), if $x \in G_i$ and $y \in G_j$ for some $i \neq j$, then $xy = yx$.

Proof: (1) Since the operation in G is defined componentwise, it follows easily from the subgroup criterion that $\{(1, 1, \ldots, 1, g_i, 1, \ldots, 1) \mid g_i \in G_i\}$ is a subgroup of G. Furthermore, the map $g_i \mapsto (1, 1, \ldots, 1, g_i, 1, \ldots, 1)$ is seen to be an isomorphism of G_i with this subgroup. Identify G_i with this isomorphic copy in G.

To prove the remaining parts of (1) consider the map

$$\varphi : G \longrightarrow G_1 \times \cdots \times G_{i-1} \times G_{i+1} \times \cdots \times G_n$$

defined by

$$\varphi(g_1, g_2, \ldots, g_n) = (g_1, \ldots, g_{i-1}, g_{i+1}, \ldots, g_n)$$

(i.e., φ erases the ith component of G). The map φ is a homomorphism since

$$\varphi((g_1, \ldots, g_n)(h_1, \ldots, h_n)) = \varphi((g_1 h_1, \ldots, g_n h_n))$$
$$= (g_1 h_1, \ldots, g_{i-1} h_{i-1}, g_{i+1} h_{i+1}, \ldots, g_n h_n)$$
$$= (g_1, \ldots, g_{i-1}, g_{i+1}, \ldots, g_n)(h_1, \ldots, h_{i-1}, h_{i+1}, \ldots, h_n)$$
$$= \varphi((g_1, \ldots, g_n))\varphi((h_1, \ldots, h_n)).$$

Since the entries in position j are arbitrary elements of G_j for all j, φ is surjective. Furthermore,

$$\ker \varphi = \{(g_1, \ldots, g_n) \mid g_j = 1 \text{ for all } j \neq i\} = G_i.$$

This proves that G_i is a normal subgroup of G (in particular, it again proves this copy of G_i is a subgroup) and the First Isomorphism Theorem gives the final assertion of part (1).

In (2) the argument that π_i is a surjective homomorphism and the kernel is the subgroup described is very similar to that in part (1), so the details are left to the reader.

In part (3) if $x = (1, \ldots, 1, g_i, 1, \ldots, 1)$ and $y = (1, \ldots, 1, g_j, 1, \ldots, 1)$, where the indicated entries appear in positions i, j respectively, then

$$xy = (1, \ldots, 1, g_i, 1, \ldots, 1, g_j, 1, \ldots, 1) = yx$$

(where the notation is chosen so that $i < j$). This completes the proof.

A generalization of this proposition appears as Exercise 2.

We shall continue to identify the "coordinate axis" subgroups described in part (1) of the proposition with their isomorphic copies, the G_i's. The ith such subgroup is often called the ith *component* or i^{th} *factor* of G. For instance, when we wish to calculate in $Z_n \times Z_m$ we can let x be a generator of the first factor, let y be a generator of the second factor and write the elements of $Z_n \times Z_m$ in the form $x^a y^b$. This replaces the formal ordered pairs $(x, 1)$ and $(1, y)$ with x and y (so $x^a y^b$ replaces (x^a, y^b)).

Examples

(1) Under the notation of Proposition 2 it follows from part (3) that if $x_i \in G_i$, $1 \le i \le n$, then for all $k \in \mathbb{Z}$

$$(x_1 x_2 \ldots x_n)^k = x_1^k x_2^k \ldots x_n^k.$$

Since the order of $x_1 x_2 \ldots x_n$ is the smallest positive integer k such that $x_i^k = 1$ for all i, we see that

$$|x_1 x_2 \ldots x_n| = \text{l.c.m.}(|x_1|, |x_2|, \ldots, |x_n|)$$

(where this order is infinite if and only if one of the x_i's has infinite order).

(2) Let p be a prime and for $n \in \mathbb{Z}^+$ consider

$$E_{p^n} = Z_p \times Z_p \times \cdots \times Z_p \qquad (n \text{ factors}).$$

Then E_{p^n} is an abelian group of order p^n with the property that $x^p = 1$ for all $x \in E_{p^n}$. This group is the *elementary abelian* group of order p^n described in Section 4.4.

(3) For p a prime, we show that the elementary abelian group of order p^2 has exactly $p + 1$ subgroups of order p (in particular, there are more than the two obvious ones). Let $E = E_{p^2}$. Since each nonidentity element of E has order p, each of these generates a cyclic subgroup of E of order p. By Lagrange's Theorem distinct subgroups of order p intersect trivially. Thus the $p^2 - 1$ nonidentity elements of E are partitioned into subsets of size $p - 1$ (i.e., each of these subsets consists of the nonidentity elements of some subgroup of order p). There must therefore be

$$\frac{p^2 - 1}{p - 1} = p + 1$$

subgroups of order p. When $p = 2$, E is the Klein 4-group which we have already seen has 3 subgroups of order 2 (cf. also Exercises 10 and 11).

EXERCISES

1. Show that the center of a direct product is the direct product of the centers:
$$Z(G_1 \times G_2 \times \cdots \times G_n) = Z(G_1) \times Z(G_2) \times \cdots \times Z(G_n).$$
Deduce that a direct product of groups is abelian if and only if each of the factors is abelian.

2. Let $G_1, G_2, \ldots, G_n$ be groups and let $G = G_1 \times \cdots \times G_n$. Let I be a proper, nonempty subset of $\{1, \ldots, n\}$ and let $J = \{1, \ldots, n\} - I$. Define G_I to be the set of elements of G that have the identity of G_j in position j for all $j \in J$.
   (a) Prove that G_I is isomorphic to the direct product of the groups $G_i, i \in I$.
   (b) Prove that G_I is a normal subgroup of G and $G/G_I \cong G_J$.
   (c) Prove that $G \cong G_I \times G_J$.

3. Under the notation of the preceding exercise let I and K be any disjoint nonempty subsets of $\{1, 2, \ldots, n\}$ and let G_I and G_K be the subgroups of G defined above. Prove that $xy = yx$ for all $x \in G_I$ and all $y \in G_K$.

4. Let A and B be finite groups and let p be a prime. Prove that any Sylow p-subgroup of $A \times B$ is of the form $P \times Q$, where $P \in Syl_p(A)$ and $Q \in Syl_p(B)$. Prove that $n_p(A \times B) = n_p(A)n_p(B)$. Generalize both of these results to a direct product of any finite number of finite groups (so that the number of Sylow p-subgroups of a direct product is the product of the numbers of Sylow p-subgroups of the factors).

5. Exhibit a nonnormal subgroup of $Q_8 \times Z_4$ (note that every subgroup of each factor is normal).

6. Show that all subgroups of $Q_8 \times E_{2^n}$ are normal.

7. Let $G_1, G_2, \ldots, G_n$ be groups and let π be a fixed element of S_n. Prove that the map
$$\varphi_\pi : G_1 \times G_2 \times \cdots \times G_n \to G_{\pi^{-1}(1)} \times G_{\pi^{-1}(2)} \times \cdots \times G_{\pi^{-1}(n)}$$
defined by
$$\varphi_\pi(g_1, g_2, \ldots, g_n) = (g_{\pi^{-1}(1)}, g_{\pi^{-1}(2)}, \ldots, g_{\pi^{-1}(n)})$$
is an isomorphism (so that changing the order of the factors in a direct product does not change the isomorphism type).

8. Let $G_1 = G_2 = \cdots = G_n$ and let $G = G_1 \times \cdots \times G_n$. Under the notation of the preceding exercise show that $\varphi_\pi \in \text{Aut}(G)$. Show also that the map $\pi \mapsto \varphi_\pi$ is an injective homomorphism of S_n into $\text{Aut}(G)$. (In particular, $\varphi_{\pi_1} \circ \varphi_{\pi_2} = \varphi_{\pi_1\pi_2}$. It is at this point that the π^{-1}'s in the definition of φ_π are needed. The underlying reason for this is because if e_i is the n-tuple with 1 in position i and zeros elsewhere, $1 \le i \le n$, then S_n acts on $\{e_1, \ldots, e_n\}$ by $\pi \cdot e_i = e_{\pi(i)}$; this is a left group action. If the n-tuple $(g_1, \ldots, g_n)$ is represented by $g_1e_1 + \cdots + g_ne_n$, then this left group action on $\{e_1, \ldots, e_n\}$ extends to a left group action on sums by
$$\pi \cdot (g_1e_1 + g_2e_2 + \cdots + g_ne_n) = g_1e_{\pi(1)} + g_2e_{\pi(2)} + \cdots + g_ne_{\pi(n)}.$$
The coefficient of $e_{\pi(i)}$ on the right hand side is g_i, so the coefficient of e_i is $g_{\pi^{-1}(i)}$. Thus the right hand side may be rewritten as $g_{\pi^{-1}(1)}e_1 + g_{\pi^{-1}(2)}e_2 + \cdots + g_{\pi^{-1}(n)}e_n$, which is precisely the sum attached to the n-tuple $(g_{\pi^{-1}(1)}, g_{\pi^{-1}(2)}, \ldots, g_{\pi^{-1}(n)})$. In other words, any permutation of the "position vectors" $e_1, \ldots, e_n$ (which fixes their coefficients) is the same as the inverse permutation on the coefficients (fixing the e_i's). If one uses π's in place of π^{-1}'s in the definition of φ_π then the map $\pi \mapsto \varphi_\pi$ is not necessarily a homomorphism — it corresponds to a *right* group action.)

9. Let G_i be a field F for all i and use the preceding exercise to show that the set of $n \times n$ matrices with one 1 in each row and each column is a subgroup of $GL_n(F)$ isomorphic to S_n (these matrices are called *permutation matrices* since they simply permute the standard basis $e_1, \ldots, e_n$ (as above) of the n-dimensional vector space $F \times F \times \cdots \times F$).

10. Let p be a prime. Let A and B be two cyclic groups of order p with generators x and y, respectively. Set $E = A \times B$ so that E is the elementary abelian group of order p^2: E_{p^2}. Prove that the distinct subgroups of E of order p are

$$\langle x \rangle, \quad \langle xy \rangle, \quad \langle xy^2 \rangle, \quad \ldots, \quad \langle xy^{p-1} \rangle, \quad \langle y \rangle$$

(note that there are $p + 1$ of them).

11. Let p be a prime and let $n \in \mathbb{Z}^+$. Find a formula for the number of subgroups of order p in the elementary abelian group E_{p^n}.

12. Let A and B be groups. Assume $Z(A)$ contains a subgroup Z_1 and $Z(B)$ contains a subgroup Z_2 with $Z_1 \cong Z_2$. Let this isomorphism be given by the map $x_i \mapsto y_i$ for all $x_i \in Z_1$. A *central product* of A and B is a quotient

$$(A \times B)/Z \quad \text{where} \quad Z = \{(x_i, y_i^{-1}) \mid x_i \in Z_1\}$$

and is denoted by $A * B$ — it is not unique since it depends on Z_1, Z_2 and the isomorphism between them. (Think of $A * B$ as the direct product of A and B "collapsed" by identifying each element $x_i \in Z_1$ with its corresponding element $y_i \in Z_2$.)

   (a) Prove that the images of A and B in the quotient group $A * B$ are isomorphic to A and B, respectively, and that these images intersect in a central subgroup isomorphic to Z_1. Find $|A * B|$.

   (b) Let $Z_4 = \langle x \rangle$. Let $D_8 = \langle r, s \rangle$ and $Q_8 = \langle i, j \rangle$ be given by their usual generators and relations. Let $Z_4 * D_8$ be the central product of Z_4 and D_8 which identifies x^2 and r^2 (i.e., $Z_1 = \langle x^2 \rangle$, $Z_2 = \langle r^2 \rangle$ and the isomorphism is $x^2 \mapsto r^2$) and let $Z_4 * Q_8$ be the central product of Z_4 and Q_8 which identifies x^2 and -1. Prove that $Z_4 * D_8 \cong Z_4 * Q_8$.

13. Give presentations for the groups $Z_4 * D_8$ and $Z_4 * Q_8$ constructed in the preceding exercise.

14. Let $G = A_1 \times A_2 \times \cdots \times A_n$ and for each i let B_i be a normal subgroup of A_i. Prove that $B_1 \times B_2 \times \cdots \times B_n \trianglelefteq G$ and that

$$(A_1 \times A_2 \times \cdots \times A_n)/(B_1 \times B_2 \times \cdots \times B_n) \cong (A_1/B_1) \times (A_2/B_2) \times \cdots \times (A_n/B_n).$$

The following exercise describes the direct product of an arbitrary collection of groups. The terminology for the Cartesian product of an arbitrary collection of sets may be found in the Appendix.

15. Let I be any nonempty index set and let $(G_i, \star_i)$ be a group for each $i \in I$. The *direct product* of the groups $G_i, i \in I$ is the set $G = \prod_{i \in I} G_i$ (the Cartesian product of the G_i's) with a binary operation defined as follows: if $\prod a_i$ and $\prod b_i$ are elements of G, then their product in G is given by

$$\left(\prod_{i \in I} a_i \right) \left(\prod_{i \in I} b_i \right) = \prod_{i \in I} (a_i \star_i b_i)$$

(i.e., the group operation in the direct product is defined componentwise).

   (a) Show that this binary operation is well defined and associative.

   (b) Show that the element $\prod 1_i$ satisfies the axiom for the identity of G, where 1_i is the identity of G_i for all i.

(c) Show that the element $\prod a_i^{-1}$ is the inverse of $\prod a_i$, where the inverse of each component element a_i is taken in the group G_i.

Conclude that the direct product is a group.
(Note that if $I = \{1, 2, \ldots, n\}$, this definition of the direct product is the same as the n-tuple definition in the text.)

16. State and prove the generalization of Proposition 2 to arbitrary direct products.

17. Let I be any nonempty index set and let G_i be a group for each $i \in I$. The *restricted direct product* or *direct sum* of the groups G_i is the set of elements of the direct product which are the identity in all but finitely many components, that is, the set of all elements $\prod a_i \in \prod_{i \in I} G_i$ such that $a_i = 1_i$ for all but a finite number of $i \in I$.
  (a) Prove that the restricted direct product is a subgroup of the direct product.
  (b) Prove that the restricted direct product is normal in the direct product.
  (c) Let $I = \mathbb{Z}^+$ and let p_i be the ith integer prime. Show that if $G_i = \mathbb{Z}/p_i\mathbb{Z}$ for all $i \in \mathbb{Z}^+$, then every element of the restricted direct product of the G_i's has finite order but $\prod_{i \in \mathbb{Z}^+} G_i$ has elements of infinite order. Show that in this example the restricted direct product is the torsion subgroup of the direct product (cf. Exercise 6, Section 2.1).

18. In each of (a) to (e) give an example of a group with the specified properties:
  (a) an infinite group in which every element has order 1 or 2
  (b) an infinite group in which every element has finite order but for each positive integer n there is an element of order n
  (c) a group with an element of infinite order and an element of order 2
  (d) a group G such that every finite group is isomorphic to some subgroup of G
  (e) a nontrivial group G such that $G \cong G \times G$.

5.2 THE FUNDAMENTAL THEOREM OF FINITELY GENERATED ABELIAN GROUPS

Definition.
  (1) A group G is *finitely generated* if there is a finite subset A of G such that $G = \langle A \rangle$.
  (2) For each $r \in \mathbb{Z}$ with $r \geq 0$, let $\mathbb{Z}^r = \mathbb{Z} \times \mathbb{Z} \times \cdots \times \mathbb{Z}$ be the direct product of r copies of the group $\mathbb{Z}$, where $\mathbb{Z}^0 = 1$. The group $\mathbb{Z}^r$ is called the *free abelian group of rank r*.

Note that any finite group G is, a fortiori, finitely generated: simply take $A = G$ as a set of generators. Also, $\mathbb{Z}^r$ is finitely generated by $e_1, e_2, \ldots, e_r$, where e_i is the r-tuple with 1 in position i and zeros elsewhere. We can now state the fundamental classification theorem for (finitely generated) abelian groups.

Theorem 3. *(Fundamental Theorem of Finitely Generated Abelian Groups)* Let G be a finitely generated abelian group. Then
  (1)
$$G \cong \mathbb{Z}^r \times Z_{n_1} \times Z_{n_2} \times \cdots \times Z_{n_s},$$

for some integers $r, n_1, n_2, \ldots, n_s$ satisfying the following conditions:

(a) $r \geq 0$ and $n_j \geq 2$ for all j, and

(b) $n_{i+1} \mid n_i$ for $1 \leq i \leq s - 1$

(2) the expression in (1) is unique: if $G \cong \mathbb{Z}^t \times Z_{m_1} \times Z_{m_2} \times \cdots \times Z_{m_u}$, where t and $m_1, m_2, \ldots, m_u$ satisfy (a) and (b) (i.e., $t \geq 0$, $m_j \geq 2$ for all j and $m_{i+1} \mid m_i$ for $1 \leq i \leq u - 1$), then $t = r$, $u = s$ and $m_i = n_i$ for all i.

Proof: We shall derive this theorem in Section 12.1 as a consequence of a more general classification theorem. For finite groups we shall give an alternate proof at the end of Section 6.1.

Definition. The integer r in Theorem 3 is called the *free rank* or *Betti number* of G and the integers $n_1, n_2, \ldots, n_s$ are called the *invariant factors* of G. The description of G in Theorem 3(1) is called the *invariant factor decomposition* of G.

Theorem 3 asserts that the free rank and (ordered) list of invariant factors of an abelian group are uniquely determined, so that two finitely generated abelian groups are isomorphic if and only if they have the same free rank and the same list of invariant factors. Observe that a finitely generated abelian group is a finite group if and only if its free rank is zero.

The order of a finite abelian group is just the product of its invariant factors (by Proposition 1). If G is a finite abelian group with invariant factors $n_1, n_2, \ldots, n_s$, where $n_{i+1} \mid n_i$, $1 \leq i \leq s - 1$, then G is said to be of *type* $(n_1, n_2, \ldots, n_s)$.

Theorem 3 gives an effective way of listing *all* finite abelian groups of a given order. Namely, to find (up to isomorphism) all abelian groups of a given order n one must find all finite sequences of integers $n_1, n_2, \ldots, n_s$ such that

(1) $n_j \geq 2$ for all $j \in \{1, 2, \ldots, s\}$,

(2) $n_{i+1} \mid n_i$, $1 \leq i \leq s - 1$, and

(3) $n_1 n_2 \cdots n_s = n$.

Theorem 3 states that there is a bijection between the set of such sequences and the set of isomorphism classes of finite abelian groups of order n (where each sequence corresponds to the list of invariant factors of a finite abelian group).

Before illustrating how to find all such sequences for a specific value of n we make some general comments. First note that $n_1 \geq n_2 \geq \cdots \geq n_s$, so n_1 is the largest invariant factor. Also, by property (3) each n_i divides n. If p is any prime divisor of n then by (3) we see that p must divide n_i for some i. Then, by (2), p also divides n_j for all $j \leq i$. It follows that

every prime divisor of n must divide the first invariant factor n_1.

In particular, if n is the product of distinct primes (all to the first power)[1] we see that $n \mid n_1$, hence $n = n_1$. This proves that if n is squarefree, there is only one possible list of invariant factors for an abelian group of order n (namely, the list $n_1 = n$):

---

[1] Such integers are called *squarefree* since they are not divisible by any square > 1.

Corollary 4. If n is the product of distinct primes, then up to isomorphism the only abelian group of order n is the cyclic group of order n, Z_n.

The factorization of n into prime powers is the first step in determining all possible lists of invariant factors for abelian groups of order n.

Example

Suppose $n = 180 = 2^2 \cdot 3^2 \cdot 5$. As noted above we must have $2 \cdot 3 \cdot 5 \mid n_1$, so possible values of n_1 are

$$n_1 = 2^2 \cdot 3^2 \cdot 5, \quad 2^2 \cdot 3 \cdot 5, \quad 2 \cdot 3^2 \cdot 5, \quad \text{or} \quad 2 \cdot 3 \cdot 5.$$

For each of these one must work out all possible n_2's (subject to $n_2 \mid n_1$ and $n_1 n_2 \mid n$). For each resulting pair n_1, n_2 one must work out all possible n_3's etc. until all lists satisfying (1) to (3) are obtained.

For instance, if $n_1 = 2 \cdot 3^2 \cdot 5$, the only number n_2 dividing n_1 with $n_1 n_2$ dividing n is $n_2 = 2$. In this case $n_1 n_2 = n$, so this list is complete: $2 \cdot 3^2 \cdot 5$, 2. The abelian group corresponding to this list is $Z_{90} \times Z_2$.

If $n_1 = 2 \cdot 3 \cdot 5$, the only candidates for n_2 are $n_2 = 2, 3$ or 6. If $n_2 = 2$ or 3, then since $n_3 \mid n_2$ we would necessarily have $n_3 = n_2$ (and there must be a third term in the list by property (3)). This leads to a contradiction because $n_1 n_2 n_3$ would be divisible by 2^3 or 3^3 respectively, but n is not divisible by either of these numbers. Thus the only list of invariant factors whose first term is $2 \cdot 3 \cdot 5$ is $2 \cdot 3 \cdot 5$, $2 \cdot 3$. The corresponding abelian group is $Z_{30} \times Z_6$.

Similarly, all permissible lists of invariant factors and the corresponding abelian groups of order 180 are easily seen to be the following:

Invariant Factors	Abelian Groups
$2^2 \cdot 3^2 \cdot 5$	Z_{180}
$2 \cdot 3^2 \cdot 5$, 2	$Z_{90} \times Z_2$
$2^2 \cdot 3 \cdot 5$, 3	$Z_{60} \times Z_3$
$2 \cdot 3 \cdot 5$, $2 \cdot 3$	$Z_{30} \times Z_6$

The process we carried out above was somewhat *ad hoc*, however it indicates that the determination of lists of invariant factors of all abelian groups of a given order n relies strongly on the factorization of n. The following theorem (which we shall see is equivalent to the Fundamental Theorem in the case of finite abelian groups) gives a more systematic and computationally much faster way of determining all finite abelian groups of a given order. More specifically, if the factorization of n is

$$n = p_1^{\alpha_1} p_2^{\alpha_2} \cdots p_k^{\alpha_k},$$

it shows that all permissible lists of invariant factors for abelian groups of order n may be determined by finding permissible lists for groups of order $p_i^{\alpha_i}$ for each i. For a prime power, p^α, we shall see that the problem of determining all permissible lists is equivalent to the determination of all partitions of α (and does not depend on p).

Theorem 5. Let G be an abelian group of order $n > 1$ and let the unique factorization of n into distinct prime powers be

$$n = p_1^{\alpha_1} p_2^{\alpha_2} \cdots p_k^{\alpha_k}.$$

Then
   (1) $G \cong A_1 \times A_2 \times \cdots \times A_k$, where $|A_i| = p_i^{\alpha_i}$
   (2) for each $A \in \{A_1, A_2, \ldots, A_k\}$ with $|A| = p^\alpha$,

$$A \cong Z_{p^{\beta_1}} \times Z_{p^{\beta_2}} \times \cdots \times Z_{p^{\beta_t}}$$

   with $\beta_1 \geq \beta_2 \geq \cdots \geq \beta_t \geq 1$ and $\beta_1 + \beta_2 + \cdots + \beta_t = \alpha$ (where t and $\beta_1, \ldots, \beta_t$ depend on i)
   (3) the decomposition in (1) and (2) is unique, i.e., if $G \cong B_1 \times B_2 \times \cdots \times B_m$, with $|B_i| = p_i^{\alpha_i}$ for all i, then $B_i \cong A_i$ and B_i and A_i have the same invariant factors.

Definition. The integers p^{β_j} described in the preceding theorem are called the *elementary divisors* of G. The description of G in Theorem 5(1) and 5(2) is called the *elementary divisor decomposition* of G.

The subgroups A_i described in part (1) of the theorem are the Sylow p_i-subgroups of G. Thus (1) says that G is isomorphic to the direct product of its Sylow subgroups (note that they are normal — since G is abelian — hence unique). Part 1 is often referred to as *The Primary Decomposition Theorem* for finite abelian groups.[2] As with Theorem 3, we shall prove this theorem later.

Note that for p a prime, $p^\beta \mid p^\gamma$ if and only if $\beta \leq \gamma$. Furthermore, $p^{\beta_1} \cdots p^{\beta_t} = p^\alpha$ if and only if $\beta_1 + \cdots + \beta_t = \alpha$. Thus the decomposition of A appearing in part (2) of Theorem 5 is the invariant factor decomposition of A with the "divisibility" conditions on the integers p^{β_j} translated into "additive" conditions on their exponents. The *elementary divisors* of G are now seen to be the *invariant factors of the Sylow p-subgroups* as p runs over all prime divisors of G.

By Theorem 5, in order to find all abelian groups of order $n = p_1^{\alpha_1} p_2^{\alpha_2} \cdots p_k^{\alpha_k}$ one must find for each i, $1 \leq i \leq k$, all possible lists of invariant factors for groups of order $p_i^{\alpha_i}$. The set of elementary divisors of each abelian group is then obtained by taking one set of invariant factors from each of the k lists. The abelian groups are the direct products of the cyclic groups whose orders are the elementary divisors (and distinct lists of elementary divisors give nonisomorphic groups). The advantage of this process over the one described following Theorem 2 is that it is easier to systematize how to obtain all possible lists of invariant factors, $p^{\beta_1}, p^{\beta_2}, \ldots, p^{\beta_t}$, for a group of prime power order p^β. Conditions (1) to (3) for invariant factors described earlier then become

   (1) $\beta_j \geq 1$ for all $j \in \{1, 2, \ldots, t\}$,
   (2) $\beta_i \geq \beta_{i+1}$ for all i, and
   (3) $\beta_1 + \beta_2 + \cdots + \beta_t = \beta$.

---

[2]Recall that for abelian groups the Sylow p-subgroups are sometimes called the p-primary components.

Hence, each list of invariant factors in this case is simply a *partition* of β (ordered in descending order). In particular, the number of nonisomorphic abelian groups of order p^β (= the number of distinct lists) equals the number of partitions of β. This number is independent of the prime p. For example the number of abelian groups of order p^5 is obtained from the list of partitions of 5:

Invariant Factors	Abelian Groups
5	Z_{p^5}
4, 1	$Z_{p^4} \times Z_p$
3, 2	$Z_{p^3} \times Z_{p^2}$
3, 1, 1	$Z_{p^3} \times Z_p \times Z_p$
2, 2, 1	$Z_{p^2} \times Z_{p^2} \times Z_p$
2, 1, 1, 1	$Z_{p^2} \times Z_p \times Z_p \times Z_p$
1, 1, 1, 1, 1	$Z_p \times Z_p \times Z_p \times Z_p \times Z_p$

Thus there are precisely 7 nonisomorphic groups of order p^5, the first in the list being the cyclic group, Z_{p^5}, and the last in the list being the elementary abelian group, E_{p^5}.

If $n = p_1^{\alpha_1} p_2^{\alpha_2} \cdots p_k^{\alpha_k}$ and q_i is the number of partitions of α_i, we see that the number of (distinct, nonisomorphic) abelian groups of order n equals $q_1 q_2 \cdots q_k$.

Example

If $n = 1800 = 2^3 3^2 5^2$ we list the abelian groups of this order as follows:

Order p^β	Partitions of β	Abelian Groups
2^3	3; 2, 1; 1, 1, 1	$Z_8,\ Z_4 \times Z_2,\ Z_2 \times Z_2 \times Z_2$
3^2	2; 1, 1	$Z_9,\ Z_3 \times Z_3$
5^2	2; 1, 1	$Z_{25},\ Z_5 \times Z_5$

We obtain the abelian groups of order 1800 by taking one abelian group from each of the three lists (right hand column above) and taking their direct product. Doing this in all possible ways gives all isomorphism types:

$Z_8 \times Z_9 \times Z_{25}$ $Z_4 \times Z_2 \times Z_3 \times Z_3 \times Z_{25}$

$Z_8 \times Z_9 \times Z_5 \times Z_5$ $Z_4 \times Z_2 \times Z_3 \times Z_3 \times Z_5 \times Z_5$

$Z_8 \times Z_3 \times Z_3 \times Z_{25}$ $Z_2 \times Z_2 \times Z_2 \times Z_9 \times Z_{25}$

$Z_8 \times Z_3 \times Z_3 \times Z_5 \times Z_5$ $Z_2 \times Z_2 \times Z_2 \times Z_9 \times Z_5 \times Z_5$

$Z_4 \times Z_2 \times Z_9 \times Z_{25}$ $Z_2 \times Z_2 \times Z_2 \times Z_3 \times Z_3 \times Z_{25}$

$Z_4 \times Z_2 \times Z_9 \times Z_5 \times Z_5$ $Z_2 \times Z_2 \times Z_2 \times Z_3 \times Z_3 \times Z_5 \times Z_5$.

By the Fundamental Theorems above, this is a *complete list* of all abelian groups of order 1800 — every abelian group of this order is isomorphic to precisely one of the groups above and no two of the groups in this list are isomorphic.

We emphasize that the elementary divisors of G are not invariant factors of G (but invariant factors of *subgroups* of G). For instance, in case 1 above the elementary divisors 8, 9, 25 do not satisfy the divisibility criterion of a list of invariant factors.

Our next aim is to illustrate how to pass from a list of invariant factors of a finite abelian group to its list of elementary divisors and vice versa. We show how to determine these invariants of the group no matter how it is given as a direct product of cyclic groups. We need the following proposition.

Proposition 6. Let $m, n \in \mathbb{Z}^+$.
  (1) $Z_m \times Z_n \cong Z_{mn}$ if and only if $(m, n) = 1$.
  (2) If $n = p_1^{\alpha_1} p_2^{\alpha_2} \cdots p_k^{\alpha_k}$ then $Z_n \cong Z_{p_1^{\alpha_1}} \times Z_{p_2^{\alpha_2}} \times \cdots \times Z_{p_k^{\alpha_k}}$.

Proof: Since (2) is an easy exercise using (1) and induction on k, we concentrate on proving (1). Let $Z_m = \langle x \rangle$, $Z_n = \langle y \rangle$ and let $l = \text{l.c.m.}(m, n)$. Note that $l = mn$ if and only if $(m, n) = 1$. Let $x^a y^b$ be a typical element of $Z_m \times Z_n$. Then (as noted in Example 1, Section 1)

$$(x^a y^b)^l = x^{la} y^{lb}$$
$$= 1^a 1^b = 1 \qquad \text{(because } m \mid l \text{ and } n \mid l\text{).}$$

If $(m, n) \neq 1$, every element of $Z_m \times Z_n$ has order at most l, hence has order strictly less than mn, so $Z_m \times Z_n$ cannot be isomorphic to Z_{mn}.

Conversely, if $(m, n) = 1$, then $|xy| = \text{l.c.m.}(|x|, |y|) = mn$. Thus, by order considerations, $Z_m \times Z_n = \langle xy \rangle$ is cyclic, completing the proof.

Obtaining Elementary Divisors from Invariant Factors

Suppose G is given as an abelian group of type $(n_1, n_2, \ldots, n_s)$, that is

$$G \cong Z_{n_1} \times Z_{n_2} \times \cdots \times Z_{n_s}.$$

Let $n = p_1^{\alpha_1} p_2^{\alpha_2} \cdots p_k^{\alpha_k} = n_1 n_2 \cdots n_s$. Factor each n_i as

$$n_i = p_1^{\beta_{i1}} p_2^{\beta_{i2}} \cdots p_k^{\beta_{ik}}, \quad \text{where } \beta_{ij} \geq 0.$$

By the proposition above,

$$Z_{n_i} \cong Z_{p_1^{\beta_{i1}}} \times \cdots \times Z_{p_k^{\beta_{ik}}},$$

for each i. If $\beta_{ij} = 0$, $Z_{p_j^{\beta_{ij}}} = 1$ and this factor may be deleted from the direct product without changing the isomorphism type. Then the elementary divisors of G are precisely the integers

$$p_j^{\beta_{ij}}, \qquad 1 \leq j \leq k, \quad 1 \leq i \leq s \text{ such that } \beta_{ij} \neq 0.$$

For example, if $|G| = 2^3 \cdot 3^2 \cdot 5^2$ and G is of type $(30, 30, 2)$, then

$$G \cong Z_{30} \times Z_{30} \times Z_2.$$

Since $Z_{30} \cong Z_2 \times Z_3 \times Z_5$, $G \cong Z_2 \times Z_3 \times Z_5 \times Z_2 \times Z_3 \times Z_5 \times Z_2$. The elementary divisors of G are therefore $2, 3, 5, 2, 3, 5, 2$, or, grouping like primes together (note that rearranging the order of the factors in a direct product does not affect the isomorphism type (Exercise 7 of Section 1)), $2, 2, 2, \quad 3, 3, \quad 5, 5$. In particular, G is isomorphic to the last group in the list in the example above.

If for each j one collects all the factors $Z_{p_j^{\beta_{ij}}}$ together, the resulting direct product forms the Sylow p_j-subgroup, A_j, of G. Thus the Sylow 2-subgroup of the group in the preceding paragraph is isomorphic to $Z_2 \times Z_2 \times Z_2$ (i.e., the elementary abelian group of order 8).

Obtaining Elementary Divisors from any cyclic decomposition

The same process described above will give the elementary divisors of a finite abelian group G whenever G is given as a direct product of cyclic groups (not just when the orders of the cyclic components are the invariant factors). For example, if $G = Z_6 \times Z_{15}$, the list 6, 15 is neither that of the invariant factors (the divisibility condition fails) nor that of elementary divisors (they are not prime powers). To find the elementary divisors, factor $6 = 2 \cdot 3$ and $15 = 3 \cdot 5$. Then the prime powers 2, 3, 3, 5 are the elementary divisors and

$$G \cong Z_2 \times Z_3 \times Z_3 \times Z_5.$$

Obtaining Invariant Factors from Elementary Divisors

Suppose G is an abelian group of order n, where $n = p_1^{\alpha_1} p_2^{\alpha_2} \cdots p_k^{\alpha_k}$ and we are given the elementary divisors of G. The invariant factors of G are obtained by following these steps:

(1) First group all elementary divisors which are powers of the same prime together. In this way we obtain k lists of integers (one for each p_j).

(2) In each of these k lists arrange the integers in nonincreasing order.

(3) Among these k lists suppose that the longest (i.e., the one with the most terms) consists of t integers. Make each of the k lists of length t by appending an appropriate number of 1's at the end of each list.

(4) For each $i \in \{1, 2, \dots, t\}$ the ith invariant factor, n_i, is obtained by taking the product of the ith integer in each of the t (ordered) lists.

The point of ordering the lists in this way is to ensure that we have the divisibility condition $n_{i+1} \mid n_i$.

Suppose, for example, that the elementary divisors of G are given as 2, 3, 2, 25, 3, 2 (so $|G| = 2^3 \cdot 3^2 \cdot 5^2$). Regrouping and increasing each list to have 3 ($= t$) members gives:

$p = 2$	$p = 3$	$p = 5$
2	3	25
2	3	1
2	1	1

so the invariant factors of G are $2 \cdot 3 \cdot 25$, $2 \cdot 3 \cdot 1$, $2 \cdot 1 \cdot 1$ and

$$G \cong Z_{150} \times Z_6 \times Z_2.$$

Note that this is the penultimate group in the list classifying abelian groups of order 1800 computed above.

The invariant factor decompositions of the abelian groups of order 1800 are as follows, where the ith group in this list is isomorphic to the ith group computed in the previous list:

$$Z_{1800}$$

$$Z_{360} \times Z_5 \qquad\qquad Z_{300} \times Z_6$$
$$Z_{600} \times Z_3 \qquad\qquad Z_{60} \times Z_{30}$$
$$Z_{120} \times Z_{15} \qquad\quad Z_{450} \times Z_2 \times Z_2$$
$$Z_{900} \times Z_2 \qquad\quad Z_{90} \times Z_{10} \times Z_2$$
$$Z_{180} \times Z_{10} \qquad\quad Z_{150} \times Z_6 \times Z_2$$
$$Z_{30} \times Z_{30} \times Z_2.$$

Using the uniqueness statements of the Fundamental Theorems 3 and 5, we can use these processes to determine whether any two direct products of finite cyclic groups are isomorphic. For instance, if one wanted to know whether $Z_6 \times Z_{15} \cong Z_{10} \times Z_9$, first determine whether they have the same order (both are of order 90) and then (the easiest way in general) determine whether they have the same elementary divisors:

$Z_6 \times Z_{15}$ has elementary divisors 2, 3, 3, 5 and is isomorphic to $Z_2 \times Z_3 \times Z_3 \times Z_5$

$Z_{10} \times Z_9$ has elementary divisors 2, 5, 9 and is isomorphic to $Z_2 \times Z_5 \times Z_9$.

The lists of elementary divisors are different so (by Theorem 5) they are not isomorphic. Note that $Z_6 \times Z_{15}$ has no element of order 9 whereas $Z_{10} \times Z_9$ does (cf. Exercise 5).

The processes we described above (with some elaboration) form a proof (via Proposition 6) that for finite abelian groups Theorems 3 and 5 are equivalent (i.e., one implies the other). We leave the details to the reader.

One can now better understand some of the power and some of the limitations of classification theorems. On one hand, given any positive integer n one can explicitly describe all abelian groups of order n, a significant achievement. On the other hand, the amount of information necessary to determine which of the isomorphism types of groups of order n a particular group belongs to may be considerable (and is large if n is divisible by large powers of primes).

We close this section with some terminology which will be useful in later sections.

Definition.
   (1) If G is a finite abelian group of type $(n_1, n_2, \ldots, n_t)$, the integer t is called the *rank* of G (the free rank of G is 0 so there will be no confusion).
   (2) If G is any group, the *exponent* of G is the smallest positive integer n such that $x^n = 1$ for all $x \in G$ (if no such integer exists the exponent of G is ∞).

EXERCISES

1. In each of parts (a) to (e) give the number of nonisomorphic abelian groups of the specified order — do not list the groups: **(a)** order 100, **(b)** order 576, **(c)** order 1155, **(d)** order 42875, **(e)** order 2704.

2. In each of parts (a) to (e) give the lists of invariant factors for all abelian groups of the specified order:
   (a) order 270, **(b)** order 9801, **(c)** order 320, **(d)** order 105, **(e)** order 44100.

3. In each of parts (a) to (e) give the lists of elementary divisors for all abelian groups of the specified order and then match each list with the corresponding list of invariant factors

found in the preceding exercise:

(a) order 270, (b) order 9801, (c) order 320, (d) order 105, (e) order 44100.

4. In each of parts (a) to (d) determine which pairs of abelian groups listed are isomorphic (here the expression $\{a_1, a_2, \ldots, a_k\}$ denotes the abelian group $Z_{a_1} \times Z_{a_2} \times \cdots \times Z_{a_k}$).
   (a) $\{4, 9\}$, $\{6, 6\}$, $\{8, 3\}$, $\{9, 4\}$, $\{6, 4\}$, $\{64\}$.
   (b) $\{2^2, 2 \cdot 3^2\}$, $\{2^2 \cdot 3, 2 \cdot 3\}$, $\{2^3 \cdot 3^2\}$, $\{2^2 \cdot 3^2, 2\}$.
   (c) $\{5^2 \cdot 7^2, 3^2 \cdot 5 \cdot 7\}$, $\{3^2 \cdot 5^2 \cdot 7, 5 \cdot 7^2\}$, $\{3 \cdot 5^2, 7^2, 3 \cdot 5 \cdot 7\}$,
       $\{5^2 \cdot 7, 3^2 \cdot 5, 7^2\}$.
   (d) $\{2^2 \cdot 5 \cdot 7, 2^3 \cdot 5^3, 2 \cdot 5^2\}$, $\{2^3 \cdot 5^3 \cdot 7, 2^3 \cdot 5^3\}$, $\{2^2, 2 \cdot 7, 2^3, 5^3, 5^3\}$,
       $\{2 \cdot 5^3, 2^2 \cdot 5^3, 2^3, 7\}$.

5. Let G be a finite abelian group of type $(n_1, n_2, \ldots, n_t)$. Prove that G contains an element of order m if and only if $m \mid n_1$. Deduce that G is of exponent n_1.

6. Prove that any finite group has a finite exponent. Give an example of an infinite group with finite exponent. Does a finite group of exponent m always contain an element of order m?

7. Let p be a prime and let $A = \langle x_1 \rangle \times \langle x_2 \rangle \times \cdots \times \langle x_n \rangle$ be an abelian p-group, where $|x_i| = p^{\alpha_i} > 1$ for all i. Define the pth-power map

$$\varphi : A \to A \quad \text{by} \quad \varphi : x \mapsto x^p.$$

   (a) Prove that φ is a homomorphism.
   (b) Describe the image and kernel of φ in terms of the given generators.
   (c) Prove both $\ker \varphi$ and $A/\operatorname{im} \varphi$ have rank n (i.e., have the same rank as A) and prove these groups are both isomorphic to the elementary abelian group, E_{p^n}, of order p^n.

8. Let A be a finite abelian group (written multiplicatively) and let p be a prime. Let

$$A^p = \{a^p \mid a \in A\} \quad \text{and} \quad A_p = \{x \mid x^p = 1\}$$

   (so A^p and A_p are the image and kernel of the pth-power map, respectively).
   (a) Prove that $A/A^p \cong A_p$. [Show that they are both elementary abelian and they have the same order.]
   (b) Prove that the number of subgroups of A of order p equals the number of subgroups of A of index p. [Reduce to the case where A is an elementary abelian p-group.]

9. Let $A = Z_{60} \times Z_{45} \times Z_{12} \times Z_{36}$. Find the number of elements of order 2 and the number of subgroups of index 2 in A.

10. Let n and k be positive integers and let A be the free abelian group of rank n (written additively). Prove that A/kA is isomorphic to the direct product of n copies of $\mathbb{Z}/k\mathbb{Z}$ (here $kA = \{ka \mid a \in A\}$). [See Exercise 14, Section 1.]

11. Let G be a nontrivial finite abelian group of rank t.
   (a) Prove that the rank of G equals the maximum of the ranks of its Sylow subgroups.
   (b) Prove that G can be generated by t elements but no subset with fewer than t elements generates G. [One way of doing this is by using part (a) together with Exercise 7.]

12. Let n and m be positive integers with $d = (n, m)$. Let $Z_n = \langle x \rangle$ and $Z_m = \langle y \rangle$. Let A be the central product of $\langle x \rangle$ and $\langle y \rangle$ with an element of order d identified, which has presentation $\langle x, y \mid x^n = y^m = 1, \ xy = yx, \ x^{\frac{n}{d}} = y^{\frac{m}{d}} \rangle$. Describe A as a direct product of two cyclic groups.

13. Let $A = \langle x_1 \rangle \times \cdots \times \langle x_r \rangle$ be a finite abelian group with $|x_i| = n_i$ for $1 \le i \le r$. Find a presentation for A. Prove that if G is any group containing commuting elements $g_1, \ldots, g_r$ such that $g_i^{n_i} = 1$ for $1 \le i \le r$, then there is a unique homomorphism from A to G which sends x_i to g_i for all i.

166

14. For any group G define the *dual group* of G (denoted $\widehat{G}$) to be the set of all homomorphisms from G into the multiplicative group of roots of unity in $\mathbb{C}$. Define a group operation in $\widehat{G}$ by pointwise multiplication of functions: if χ, ψ are homomorphisms from G into the group of roots of unity then $\chi\psi$ is the homomorphism given by $(\chi\psi)(g) = \chi(g)\psi(g)$ for all $g \in G$, where the latter multiplication takes place in $\mathbb{C}$.

   (a) Show that this operation on $\widehat{G}$ makes $\widehat{G}$ into an abelian group. [Show that the identity is the map $g \mapsto 1$ for all $g \in G$ and the inverse of $\chi \in \widehat{G}$ is the map $g \mapsto \chi(g)^{-1}$.]

   (b) If G is a finite abelian group, prove that $\widehat{G} \cong G$. [Write G as $\langle x_1 \rangle \times \cdots \times \langle x_r \rangle$ and if $n_i = |x_i|$ define χ_i to be the homomorphism which sends x_i to $e^{2\pi i/n_i}$ and sends x_j to 1, for all $j \neq i$. Prove χ_i has order n_i in $\widehat{G}$ and $\widehat{G} = \langle \chi_1 \rangle \times \cdots \times \langle \chi_r \rangle$.]

   (This result is often phrased: a finite abelian group is self-dual. It implies that the lattice diagram of a finite abelian group is the same when it is turned upside down. Note however that there is no *natural* isomorphism between G and its dual (the isomorphism depends on a choice of a set of generators for G). This is frequently stated in the form: a finite abelian group is *noncanonically* isomorphic to its dual.)

15. Let $G = \langle x \rangle \times \langle y \rangle$ where $|x| = 8$ and $|y| = 4$.

   (a) Find all pairs a, b in G such that $G = \langle a \rangle \times \langle b \rangle$ (where a and b are expressed in terms of x and y).

   (b) Let $H = \langle x^2 y, y^2 \rangle \cong Z_4 \times Z_2$. Prove that there are no elements a, b of G such that $G = \langle a \rangle \times \langle b \rangle$ and $H = \langle a^2 \rangle \times \langle b^2 \rangle$ (i.e., one cannot pick direct product generators for G in such a way that some powers of these are direct product generators for H).

16. Prove that no finitely generated abelian group is divisible (cf. Exercise 19, Section 2.4).

5.3 TABLE OF GROUPS OF SMALL ORDER

At this point we can give a table of the isomorphism types for most of the groups of small order.

Each of the unfamiliar non-abelian groups in the table on the following page will be constructed in Section 5 on semidirect products (which will also explain the notation used for them). For the present we give generators and relations for each of them (i.e., presentations of them).

The group $Z_3 \rtimes Z_4$ of order 12 can be described by the generators and relations:

$$\langle x, y \mid x^4 = y^3 = 1, \ x^{-1}yx = y^{-1} \rangle,$$

namely, it has a normal Sylow 3-subgroup ($\langle y \rangle$) which is inverted by an element of order 4 (x) acting by conjugation (x^2 centralizes y).

The group $(Z_3 \times Z_3) \rtimes Z_2$ has generators and relations:

$$\langle x, y, z \mid x^2 = y^3 = z^3 = 1, \ yz = zy, \ x^{-1}yx = y^{-1}, \ x^{-1}zx = z^{-1} \rangle,$$

namely, it has a normal Sylow 3-subgroup isomorphic to $Z_3 \times Z_3$ ($\langle y, z \rangle$) inverted by an element of order 2 (x) acting by conjugation.

The group $Z_5 \rtimes Z_4$ of order 20 has generators and relations:

$$\langle x, y \mid x^4 = y^5 = 1, \ x^{-1}yx = y^{-1} \rangle,$$

namely, it has a normal Sylow 5-subgroup ($\langle y \rangle$) which is inverted by an element of order 4 (x) acting by conjugation (x^2 centralizes y).

Order	No. of Isomorphism Types	Abelian Groups	Non-abelian Groups
1	1	Z_1	none
2	1	Z_2	none
3	1	Z_3	none
4	2	$Z_4, Z_2 \times Z_2$	none
5	1	Z_5	none
6	2	Z_6	S_3
7	1	Z_7	none
8	5	$Z_8, Z_4 \times Z_2,$ $Z_2 \times Z_2 \times Z_2$	D_8, Q_8
9	2	$Z_9, Z_3 \times Z_3$	none
10	2	Z_{10}	D_{10}
11	1	Z_{11}	none
12	5	$Z_{12}, Z_6 \times Z_2$	$A_4, D_{12}, Z_3 \rtimes Z_4$
13	1	Z_{13}	none
14	2	Z_{14}	D_{14}
15	1	Z_{15}	none
16	14	$Z_{16}, Z_8 \times Z_2,$ $Z_4 \times Z_4, Z_4 \times Z_2 \times Z_2,$ $Z_2 \times Z_2 \times Z_2 \times Z_2$	not listed
17	1	Z_{17}	none
18	5	$Z_{18}, Z_6 \times Z_3$	$D_{18}, S_3 \times Z_3,$ $(Z_3 \times Z_3) \rtimes Z_2$
19	1	Z_{19}	none
20	5	$Z_{20}, Z_{10} \times Z_2$	D_{20} $Z_5 \rtimes Z_4, F_{20}$

The group F_{20} of order 20 has generators and relations:

$$\langle x, y \mid x^4 = y^5 = 1, \ xyx^{-1} = y^2 \rangle,$$

namely, it has a normal Sylow 5-subgroup ($\langle y \rangle$) which is squared by an element of order 4 (x) acting by conjugation. One can check that this group occurs as the normalizer of a Sylow 5-subgroup in S_5, e.g.,

$$F_{20} = \langle (2\,3\,5\,4), \ (1\,2\,3\,4\,5) \rangle.$$

This group is called the *Frobenius group* of order 20.

1. Prove that D_{16}, $Z_2 \times D_8$, $Z_2 \times Q_8$, $Z_4 * D_8$, QD_{16} and M are nonisomorphic non-abelian groups of order 16 (where $Z_4 * D_8$ is described in Exercise 12, Section 1 and QD_{16} and M are described in the exercises in Section 2.5).

5.4 RECOGNIZING DIRECT PRODUCTS

So far we have seen that direct products may be used to both construct "larger" groups from "smaller" ones and to decompose finitely generated abelian groups into cyclic factors. Even certain non-abelian groups, which may be given in some other form, may be decomposed as direct products of smaller groups. The purpose of this section is to indicate a criterion to recognize when a group is the direct product of some of its subgroups and to illustrate the criterion with some examples.

Before doing so we introduce some standard notation and elementary results on commutators which will streamline the presentation and which will be used again in Chapter 6 when we consider nilpotent groups.

Definition. Let G be a group, let x, $y \in G$ and let A, B be nonempty subsets of G.
  (1) Define $[x, y] = x^{-1}y^{-1}xy$, called the *commutator* of x and y.
  (2) Define $[A, B] = \langle [a, b] \mid a \in A, \ b \in B \rangle$, the group generated by commutators of elements from A and from B.
  (3) Define $G' = \langle [x, y] \mid x, y \in G \rangle$, the subgroup of G generated by commutators of elements from G, called the *commutator subgroup* of G.

The commutator of x and y is 1 if and only if x and y commute, which explains the terminology. The following proposition shows how commutators measure the "difference" in G between xy and yx.

Proposition 7. Let G be a group, let x, $y \in G$ and let $H \leq G$. Then
  (1) $xy = yx[x, y]$ (in particular, $xy = yx$ if and only if $[x, y] = 1$).
  (2) $H \trianglelefteq G$ if and only if $[H, G] \leq H$.
  (3) $\sigma[x, y] = [\sigma(x), \sigma(y)]$ for any automorphism σ of G, G' char G and G/G' is abelian.
  (4) G/G' is the largest abelian quotient of G in the sense that if $H \trianglelefteq G$ and G/H is abelian, then $G' \leq H$. Conversely, if $G' \leq H$, then $H \trianglelefteq G$ and G/H is abelian.
  (5) If $\varphi : G \to A$ is any homomorphism of G into an abelian group A, then φ factors through G' i.e., $G' \leq \ker \varphi$ and the following diagram commutes:

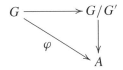

Proof: (1) This is immediate from the definition of $[x, y]$.

(2) By definition, $H \trianglelefteq G$ if and only if $g^{-1}hg \in H$ for all $g \in G$ and all $h \in H$. For $h \in H$, $g^{-1}hg \in H$ if and only if $h^{-1}g^{-1}hg \in H$, so that $H \trianglelefteq G$ if and only if $[h, g] \in H$ for all $h \in H$ and all $g \in G$. Thus $H \trianglelefteq G$ if and only if $[H, G] \leq H$, which is (2).

(3) Let $\sigma \in \mathrm{Aut}(G)$ be an automorphism of G and let $x, y \in G$. Then

$$\sigma([x, y]) = \sigma(x^{-1}y^{-1}xy)$$
$$= \sigma(x)^{-1}\sigma(y)^{-1}\sigma(x)\sigma(y)$$
$$= [\sigma(x), \sigma(y)].$$

Thus for every commutator $[x, y]$ of G', $\sigma([x, y])$ is again a commutator. Since σ has a 2-sided inverse, it follows that it maps the set of commutators bijectively onto itself. Since the commutators are a generating set for G', $\sigma(G') = G'$, that is, G' char G.

To see that G/G' is abelian, let xG' and yG' be arbitrary elements of G/G'. By definition of the group operation in G/G' and since $[x, y] \in G'$ we have

$$(xG')(yG') = (xy)G'$$
$$= (yx[x, y])G'$$
$$= (yx)G' = (yG')(xG'),$$

which completes the proof of (3).

(4) Suppose $H \trianglelefteq G$ and G/H is abelian. Then for all $x, y \in G$ we have $(xH)(yH) = (yH)(xH)$, so

$$1H = (xH)^{-1}(yH)^{-1}(xH)(yH)$$
$$= x^{-1}y^{-1}xyH$$
$$= [x, y]H.$$

Thus $[x, y] \in H$ for all $x, y \in G$, so that $G' \leq H$.

Conversely, if $G' \leq H$, then since G/G' is abelian by (3), every subgroup of G/G' is normal. In particular, $H/G' \trianglelefteq G/G'$. By the Lattice Isomorphism Theorem $H \trianglelefteq G$. By the Third Isomorphism Theorem

$$G/H \cong (G/G')/(H/G')$$

hence G/H is abelian (being isomorphic to a quotient of the abelian group G/G'). This proves (4).

(5) This is (4) phrased in terms of homomorphisms.

Passing to the quotient by the commutator subgroup of G collapses all commutators to the identity so that all elements in the quotient group commute. As (4) indicates, a strong converse to this also holds: a quotient of G by H is abelian if and only if the commutator subgroup is contained in H (i.e., if and only if G' is mapped to the identity in the quotient G/H).

We shall exhibit a group (of order 96) in the next section with the property that one of the elements of its commutator subgroup *cannot* be written as a single commutator $[x, y]$ for any x and y. Thus G' does not necessarily consist only of the set of (single) commutators (but is the group *generated* by these elements).

Examples

(1) A group G is abelian if and only if $G' = 1$.

(2) Sometimes it is possible to compute the commutator subgroup of a group without actually calculating commutators explicitly. For instance, if $G = D_8$, then since $Z(D_8) = \langle r^2 \rangle \trianglelefteq D_8$ and $D_8/Z(D_8)$ is abelian (the Klein 4-group), the commutator subgroup D_8' is a subgroup of $Z(D_8)$. Since D_8 is not itself abelian its commutator subgroup is nontrivial. The only possibility is that $D_8' = Z(D_8)$. By a similar argument, $Q_8' = Z(Q_8) = \langle -1 \rangle$. More generally, if G is any non-abelian group of order p^3, where p is a prime, $G' = Z(G)$ and $|G'| = p$ (Exercise 7).

(3) Let $D_{2n} = \langle r, s \mid r^n = s^2 = 1, s^{-1}rs = r^{-1} \rangle$. Since $[r, s] = r^{-2}$, we have $\langle r^{-2} \rangle = \langle r^2 \rangle \leq D_{2n}'$. Furthermore, $\langle r^2 \rangle \trianglelefteq D_{2n}$ and the images of r and s in $D_{2n}/\langle r^2 \rangle$ generate this quotient. They are commuting elements of order ≤ 2, so the quotient is abelian and $D_{2n}' \leq \langle r^2 \rangle$. Thus $D_{2n}' = \langle r^2 \rangle$. Finally, note that if $n \, (= |r|)$ is odd, $\langle r^2 \rangle = \langle r \rangle$ whereas if n is even, $\langle r^2 \rangle$ is of index 2 in $\langle r \rangle$. Hence D_{2n}' is of index 2 or 4 in D_{2n} according to whether n is odd or even, respectively.

(4) Since conjugation by $g \in G$ is an automorphism of G, $[a^g, b^g] = [a, b]^g$ for all $a, b \in G$ by (3) of the proposition, i.e., conjugates of commutators are also commutators. For example, once we exhibit an element of one cycle type in S_n as a commutator, every element of the same cycle type is also a commutator (cf. Section 4.3). For example, every 5-cycle is a commutator in S_5 as follows: labelling the vertices of a pentagon as $1, \ldots, 5$ we see that $D_{10} \leq S_5$ (a subgroup of A_5 in fact). By the preceding example an element of order 5 is a commutator in D_{10}, hence also in S_5. Explicitly, $(1\,4\,2\,5\,3) = [(1\,2\,3\,4\,5), (2\,5)(4\,3)]$.

The next result actually follows from the proof of Proposition 3.13 but we isolate it explicitly for reference:

Proposition 8. Let H and K be subgroups of the group G. The number of distinct ways of writing each element of the set HK in the form hk, for some $h \in H$ and $k \in K$ is $|H \cap K|$. In particular, if $H \cap K = 1$, then each element of HK can be written uniquely as a product hk, for some $h \in H$ and $k \in K$.

Proof: Exercise.

The main result of this section is the following *recognition theorem*.

Theorem 9. Suppose G is a group with subgroups H and K such that

(1) H and K are normal in G, and
(2) $H \cap K = 1$.

Then $HK \cong H \times K$.

Proof: Observe that by hypothesis (1), HK is a subgroup of G (see Corollary 3.15). Let $h \in H$ and let $k \in K$. Since $H \trianglelefteq G$, $k^{-1}hk \in H$, so that $h^{-1}(k^{-1}hk) \in H$. Similarly, $(h^{-1}k^{-1}h)k \in K$. Since $H \cap K = 1$ it follows that $h^{-1}k^{-1}hk = 1$, i.e., $hk = kh$ so that every element of H commutes with every element of K.

By the preceding proposition each element of HK can be written uniquely as a product hk, with $h \in H$ and $k \in K$. Thus the map

$$\varphi : HK \to H \times K$$
$$hk \mapsto (h, k)$$

is well defined. To see that φ is a homomorphism note that if $h_1, h_2 \in H$ and $k_1, k_2 \in K$, then we have seen that h_2 and k_1 commute. Thus

$$(h_1 k_1)(h_2 k_2) = (h_1 h_2)(k_1 k_2)$$

and the latter product is the unique way of writing $(h_1 k_1)(h_2 k_2)$ in the form hk with $h \in H$ and $k \in K$. This shows that

$$\varphi(h_1 k_1 h_2 k_2) = \varphi(h_1 h_2 k_1 k_2)$$
$$= (h_1 h_2, k_1 k_2)$$
$$= (h_1, k_1)(h_2, k_2) = \varphi(h_1 k_1)\varphi(h_2 k_2)$$

so that φ is a homomorphism. The homomorphism φ is a bijection since the representation of each element of HK as a product of the form hk is unique, which proves that φ is an isomorphism.

Definition. If G is a group and H and K are normal subgroups of G with $H \cap K = 1$, we call HK the *internal direct product* of H and K. We shall (when emphasis is called for) call $H \times K$ the *external direct product* of H and K.

The distinction between internal and external direct product is (by Theorem 9) purely notational: the elements of the internal direct product are written in the form hk, whereas those of the external direct product are written as ordered pairs (h, k). We have in previous instances passed between these. For example, when $Z_n = \langle a \rangle$ and $Z_m = \langle b \rangle$ we wrote $x = (a, 1)$ and $y = (1, b)$ so that every element of $Z_n \times Z_m$ was written in the form $x^r y^s$.

Examples

(1) If n is a positive odd integer, we show $D_{4n} \cong D_{2n} \times Z_2$. To see this let

$$D_{4n} = \langle r, s \mid r^{2n} = s^2 = 1, \ srs = r^{-1} \rangle$$

be the usual presentation of D_{4n}. Let $H = \langle s, r^2 \rangle$ and let $K = \langle r^n \rangle$. Geometrically, if D_{4n} is the group of symmetries of a regular $2n$-gon, H is the group of symmetries of the regular n-gon inscribed in the $2n$-gon by joining vertex $2i$ to vertex $2i + 2$, for all $i \bmod 2n$ (and if one lets $r_1 = r^2$, H has the usual presentation of the dihedral group of order $2n$ with generators r_1 and s). Note that $H \trianglelefteq D_{4n}$ (it has index 2). Since $|r| = 2n$, $|r^n| = 2$. Since $srs = r^{-1}$, we have $sr^n s = r^{-n} = r^n$, that is, s centralizes r^n. Since clearly r centralizes r^n, $K \leq Z(D_{4n})$. Thus $K \trianglelefteq D_{4n}$. Finally, $K \not\leq H$ since r^2 has odd order (or because r^n sends vertex i into vertex $i + n$, hence does not preserve the set of even vertices of the $2n$-gon). Thus $H \cap K = 1$ by Lagrange. Theorem 9 now completes the proof.

(2) Let I be a subset of $\{1, 2, \ldots, n\}$ and let G be the setwise stabilizer of I in S_n, i.e.,

$$G = \{\sigma \in S_n \mid \sigma(i) \in I \text{ for all } i \in I\}.$$

Let $J = \{1, 2, \ldots, n\} - I$ be the complement of I and note that G is also the setwise stabilizer of J. Let H be the *pointwise* stabilizer of I and let K be the *pointwise* stabilizer of $\{1, 2, \ldots, n\} - I$, i.e.,

$$H = \{\sigma \in G \mid \sigma(i) = i \text{ for all } i \in I\}$$
$$K = \{\tau \in G \mid \tau(j) = j \text{ for all } j \in J\}.$$

It is easy to see that H and K are normal subgroups of G (in fact they are kernels of the actions of G on I and J, respectively). Since any element of $H \cap K$ fixes all of $\{1, 2, \ldots, n\}$, we have $H \cap K = 1$. Finally, since every element σ of G stabilizes the sets I and J, each cycle in the cycle decomposition of σ involves only elements of I or only elements of J. Thus σ may be written as a product $\sigma_I \sigma_J$, where $\sigma_I \in H$ and $\sigma_J \in K$. This proves $G = HK$. By Theorem 9, $G \cong H \times K$. Now any permutation of J can be extended to a permutation in S_n by letting it act as the identity on I. These are precisely the permutations in H (and similarly the permutations in K are the permutations of I which are the identity on J), so

$$H \cong S_J \quad K \cong S_I \quad \text{and} \quad G \cong S_m \times S_{n-m},$$

where $m = |J|$ (and, by convention, $S_\emptyset = 1$).

(3) Let $\sigma \in S_n$ and let I be the subset of $\{1, 2, \ldots, n\}$ fixed pointwise by σ:

$$I = \{i \in \{1, 2, \ldots, n\} \mid \sigma(i) = i\}.$$

If $C = C_{S_n}(\sigma)$, then by Exercise 18 of Section 4.3, C stabilizes the set I and its complement J. By the preceding example, C is isomorphic to a subgroup of $H \times K$, where H is the subgroup of all permutations in S_n fixing I pointwise and K is the set of all permutations fixing J pointwise. Note that $\sigma \in H$. Thus each element, α, of C can be written (uniquely) as $\alpha = \alpha_I \alpha_J$, for some $\alpha_I \in H$ and $\alpha_J \in K$. Note further that if τ is any permutation of $\{1, 2, \ldots, n\}$ which fixes each $j \in J$ (i.e., any element of K), then σ and τ commute (since they move no common integers). Thus C contains all such τ, i.e., C contains the subgroup K. This proves that the group C consists of all elements $\alpha_I \alpha_J \in H \times K$ such that α_J is arbitrary in K and α_I commutes with σ in H:

$$C_{S_n}(\sigma) = C_H(\sigma) \times K$$
$$\cong C_{S_J}(\sigma) \times S_I.$$

In particular, if σ is an m-cycle in S_n,

$$C_{S_n}(\sigma) = \langle \sigma \rangle \times S_{n-m}.$$

The latter group has order $m(n - m)!$, as computed in Section 4.3.

EXERCISES

Let G be a group.

1. Prove that if $x, y \in G$ then $[y, x] = [x, y]^{-1}$. Deduce that for any subsets A and B of G, $[A, B] = [B, A]$ (recall that $[A, B]$ is the *subgroup* of G generated by the commutators $[a, b]$).

2. Prove that a subgroup H of G is normal if and only if $[G, H] \leq H$.

3. Let $a, b, c \in G$. Prove that
   (a) $[a, bc] = [a, c](c^{-1}[a, b]c)$

(b) $[ab, c] = (b^{-1}[a, c]b)[b, c]$.

4. Find the commutator subgroups of S_4 and A_4.

5. Prove that A_n is the commutator subgroup of S_n for all $n \geq 5$.

6. Exhibit a representative of each cycle type of A_5 as a commutator in S_5.

7. Prove that if p is a prime and P is a non-abelian group of order p^3 then $P' = Z(P)$.

8. Assume $x, y \in G$ and both x and y commute with $[x, y]$. Prove that for all $n \in \mathbb{Z}^+$, $(xy)^n = x^n y^n [y, x]^{\frac{n(n-1)}{2}}$.

9. Prove that if p is an odd prime and P is a group of order p^3 then the pth power map $x \mapsto x^p$ is a homomorphism of P into $Z(P)$. If P is not cyclic, show that the kernel of the pth power map has order p^2 or p^3. Is the squaring map a homomorphism in non-abelian groups of order 8? Where is the oddness of p needed in the above proof? [Use Exercise 8.]

10. Prove that a finite abelian group is the direct product of its Sylow subgroups.

11. Prove that if $G = HK$ where H and K are characteristic subgroups of G with $H \cap K = 1$ then $\text{Aut}(G) \cong \text{Aut}(H) \times \text{Aut}(K)$. Deduce that if G is an abelian group of finite order then $\text{Aut}(G)$ is isomorphic to the direct product of the automorphism groups of its Sylow subgroups.

12. Use Theorem 4.17 to describe the automorphism group of a finite cyclic group.

13. Prove that D_{8n} is not isomorphic to $D_{4n} \times Z_2$.

14. Let $G = \{(a_{ij}) \in GL_n(F) \mid a_{ij} = 0 \text{ if } i > j, \text{ and } a_{11} = a_{22} = \cdots = a_{nn}\}$, where F is a field, be the group of upper triangular matrices all of whose diagonal entries are equal. Prove that $G \cong D \times U$, where D is the group of all nonzero multiples of the identity matrix and U is the group of upper triangular matrices with 1's down the diagonal.

15. If A and B are normal subgroups of G such that G/A and G/B are both abelian, prove that $G/(A \cap B)$ is abelian.

16. Prove that if K is a normal subgroup of G then $K' \trianglelefteq G$.

17. If K is a normal subgroup of G and K is cyclic, prove that $G' \leq C_G(K)$. [Recall that the automorphism group of a cyclic group is abelian.]

18. Let $K_1, K_2, \ldots, K_n$ be non-abelian simple groups and let $G = K_1 \times K_2 \times \cdots \times K_n$. Prove that every normal subgroup of G is of the form G_I for some subset I of $\{1, 2, \ldots, n\}$ (where G_I is defined in Exercise 2 of Section 1). [If $N \trianglelefteq G$ and $x = (a_1, \ldots, a_n) \in N$ with some $a_i \neq 1$, then show that there is some $g_i \in G_i$ not commuting with a_i. Show $[(1, \ldots, g_i, \ldots, 1), x] \in K_i \cap N$ and deduce $K_i \leq N$.]

19. A group H is called *perfect* if $H' = H$ (i.e., H equals its own commutator subgroup).
  (a) Prove that every non-abelian simple group is perfect.
  (b) Prove that if H and K are perfect subgroups of a group G then $\langle H, K \rangle$ is also perfect. Extend this to show that the subgroup of G generated by any collection of perfect subgroups is perfect.
  (c) Prove that any conjugate of a perfect subgroup is perfect.
  (d) Prove that any group G has a unique maximal perfect subgroup and that this subgroup is normal.

20. Let $H(F)$ be the Heisenberg group over the field F, cf. Exercise 11 of Section 1.4. Find an explicit formula for the commutator $[X, Y]$, where $X, Y \in H(F)$, and show that the commutator subgroup of $H(F)$ equals the center of $H(F)$ (cf. Section 2.2, Exercise 14).

5.5 SEMIDIRECT PRODUCTS

In this section we study the "semidirect product" of two groups H and K, which is a generalization of the notion of the direct product of H and K obtained by relaxing the requirement that both H and K be normal. This construction will enable us (in certain circumstances) to build a "larger" group from the groups H and K in such a way that G contains subgroups isomorphic to H and K, respectively, as in the case of direct products. In this case the subgroup H will be normal in G but the subgroup K will not necessarily be normal (as it is for direct products). Thus, for instance, we shall be able to construct non-abelian groups even if H and K are abelian. This construction will allow us to enlarge considerably the set of examples of groups at our disposal. As in the preceding section, we shall then prove a recognition theorem that will enable us to decompose some familiar groups into smaller "factors," from which we shall be able to derive some classification theorems.

By way of motivation suppose we already have a group G containing subgroups H and K such that

(a) $H \trianglelefteq G$ (but K is not necessarily normal in G), and
(b) $H \cap K = 1$.

It is still true that HK is a subgroup of G (Corollary 3.15) and, by Proposition 8, every element of HK can be written uniquely as a product hk, for some $h \in H$ and $k \in K$, i.e., there is a bijection between HK and the collection of ordered pairs (h, k), given by $hk \mapsto (h, k)$ (so the group H appears as the set of elements $(h, 1)$ and K appears as the set of elements $(1, k)$). Given two elements $h_1 k_1$ and $h_2 k_2$ of HK, we first see how to write their product (in G) in the same form:

$$
\begin{aligned}
(h_1 k_1)(h_2 k_2) &= h_1 k_1 h_2 (k_1^{-1} k_1) k_2 \\
&= h_1 (k_1 h_2 k_1^{-1}) k_1 k_2 \\
&= h_3 k_3,
\end{aligned}
\tag{5.1}
$$

where $h_3 = h_1(k_1 h_2 k_1^{-1})$ and $k_3 = k_1 k_2$. Note that since $H \trianglelefteq G$, $k_1 h_2 k_1^{-1} \in H$, so $h_3 \in H$ and $k_3 \in K$.

These calculations were predicated on the assumption that there *already existed* a group G containing subgroups H and K with $H \trianglelefteq G$ and $H \cap K = 1$. The basic idea of the semidirect product is to turn this construction around, namely start with two (abstract) groups H and K and try to *define* a group containing (an isomorphic copy of) them in such a way that (a) and (b) above hold. To do this, we write equation (1), which defines the multiplication of elements in our group, in a way that makes sense even if we do not already know there is a group containing H and K as above. The point is that k_3 in equation (1) is obtained only from multiplication in K (namely $k_1 k_2$) and h_3 is obtained from multiplying h_1 and $k_1 h_2 k_1^{-1}$ in H. If we can understand where the element $k_1 h_2 k_1^{-1}$ arises (in terms of H and K and without reference to G), then the group HK will have been described entirely in terms of H and K. We can then use this description to *define* the group HK using equation (1) to define the multiplication.

Since H is normal in G, the group K *acts* on H by conjugation:

$$k \cdot h = khk^{-1} \qquad \text{for } h \in H, k \in K$$

(we use the symbol $\cdot$ to emphasize the action) so that (1) can be written

$$(h_1 k_1)(h_2 k_2) = (h_1 \, k_1 \cdot h_2)(k_1 k_2). \qquad (5.2)$$

The action of K on H by conjugation gives a homomorphism φ of K into $\text{Aut}(H)$, so (2) shows that the multiplication in HK depends only on the multiplication in H, the multiplication in K and the homomorphism φ, hence is defined intrinsically in terms of H and K.

We now use this interpretation to define a group given two groups H and K and a homomorphism φ from K to $\text{Aut}(H)$ (which will turn out to define conjugation in the resulting group).

Theorem 10. Let H and K be groups and let φ be a homomorphism from K into $\text{Aut}(H)$. Let $\cdot$ denote the (left) action of K on H determined by φ. Let G be the set of ordered pairs (h, k) with $h \in H$ and $k \in K$ and define the following multiplication on G:

$$(h_1, k_1)(h_2, k_2) = (h_1 \, k_1 \cdot h_2 \, , \, k_1 k_2).$$

(1) This multiplication makes G into a group of order $|G| = |H||K|$.
(2) The sets $\{(h, 1) \mid h \in H\}$ and $\{(1, k) \mid k \in K\}$ are subgroups of G and the maps $h \mapsto (h, 1)$ for $h \in H$ and $k \mapsto (1, k)$ for $k \in K$ are isomorphisms of these subgroups with the groups H and K respectively:

$$H \cong \{(h, 1) \mid h \in H\} \quad \text{and} \quad K \cong \{(1, k) \mid k \in K\}.$$

Identifying H and K with their isomorphic copies in G described in (2) we have
(3) $H \trianglelefteq G$
(4) $H \cap K = 1$
(5) for all $h \in H$ and $k \in K$, $khk^{-1} = k \cdot h = \varphi(k)(h)$.

Proof: It is straightforward to check that G is a group under this multiplication using the fact that $\cdot$ is an action of K on H. For example, the associative law is verified as follows:

$$\begin{aligned}
\big((a, x)(b, y)\big)(c, z) &= (a \, x \cdot b \, , \, xy)(c, z) \\
&= (a \, x \cdot b \, (xy) \cdot c \, , \, xyz) \\
&= (a \, x \cdot b \, x \cdot (y \cdot c) \, , \, xyz) \\
&= (a \, x \cdot (b \, y \cdot c) \, , \, xyz) \\
&= (a, x)(b \, y \cdot c \, , \, yz) \\
&= (a, x)\big((b, y)(c, z)\big)
\end{aligned}$$

for all $(a, x), (b, y), (c, z) \in G$. We leave as an exercise the verification that $(1,1)$ is the identity of G and that

$$(h, k)^{-1} = (k^{-1} \cdot h^{-1} \, , \, k^{-1})$$

for each $(h, k) \in G$. The order of the group G is clearly the product of the orders of H and K, which proves (1).

Let $\widetilde{H} = \{(h, 1) \mid h \in H\}$ and $\widetilde{K} = \{(1, k) \mid k \in K\}$. We have

$$(a, 1)(b, 1) = (a \; 1 \cdot b, \; 1) = (ab, 1)$$

for all $a, b \in H$ and

$$(1, x)(1, y) = (1, xy)$$

for all $x, y \in K$, which show that $\widetilde{H}$ and $\widetilde{K}$ are subgroups of G and that the maps in (2) are isomorphisms.

It is clear that $\widetilde{H} \cap \widetilde{K} = 1$, which is (4). Now,

$$\begin{aligned}
(1, k)(h, 1)(1, k)^{-1} &= \big((1, k)(h, 1)\big)(1, k^{-1}) \\
&= (k \cdot h, \; k)(1, k^{-1}) \\
&= (k \cdot h \; k \cdot 1, \; kk^{-1}) \\
&= (k \cdot h, 1)
\end{aligned}$$

so that identifying $(h, 1)$ with h and $(1, k)$ with k by the isomorphisms in (2) we have $khk^{-1} = k \cdot h$, which is (5).

Finally, we have just seen that (under the identifications in (2)) $K \leq N_G(H)$. Since $G = HK$ and certainly $H \leq N_G(H)$, we have $N_G(H) = G$, i.e., $H \trianglelefteq G$, which proves (3) and completes the proof.

Definition. Let H and K be groups and let φ be a homomorphism from K into $\text{Aut}(H)$. The group described in Theorem 10 is called the *semidirect product* of H and K with respect to φ and will be denoted by $H \rtimes_{\varphi} K$ (when there is no danger of confusion we shall simply write $H \rtimes K$).

The notation is chosen to remind us that the copy of H in $H \rtimes K$ is the normal "factor" and that the construction of a semidirect product is not symmetric in H and K (unlike that of a direct product). Before giving some examples we clarify exactly when the semidirect product of H and K is their direct product (in particular, we see that direct products are a special case of semidirect products). See also Exercise 1.

Proposition 11. Let H and K be groups and let $\varphi : K \to \text{Aut}(H)$ be a homomorphism. Then the following are equivalent:
   (1) the identity (set) map between $H \rtimes K$ and $H \times K$ is a group homomorphism (hence an isomorphism)
   (2) φ is the trivial homomorphism from K into $\text{Aut}(H)$
   (3) $K \trianglelefteq H \rtimes K$.

Proof: $(1) \Rightarrow (2)$ By definition of the group operation in $H \rtimes K$

$$(h_1, k_1)(h_2, k_2) = (h_1 \; k_1 \cdot h_2, \; k_1 k_2)$$

for all $h_1, h_2 \in H$ and $k_1, k_2 \in K$. By assumption (1), $(h_1, k_1)(h_2, k_2) = (h_1 h_2, k_1 k_2)$. Equating the first factors of these ordered pairs gives $k_1 \cdot h_2 = h_2$ for all $h_2 \in H$ and all $k_1 \in K$, i.e., K acts trivially on H. This is (2).

$(2) \Rightarrow (3)$ If φ is trivial, then the action of K on H is trivial, so that the elements of H commute with those of K by Theorem 10(5). In particular, H normalizes K. Since K normalizes itself, $G = HK$ normalizes K, which is (3).

$(3) \Rightarrow (1)$ If K is normal in $H \rtimes K$ then (as in the proof of Theorem 9) for all $h \in H$ and $k \in K$, $[h, k] \in H \cap K = 1$. Thus $hk = kh$ and the action of K on H is trivial. The multiplication in the semidirect product is then the same as that in the direct product:

$$(h_1, k_1)(h_2, k_2) = (h_1 h_2, k_1 k_2)$$

for all $h_1, h_2 \in H$ and $k_1, k_2 \in K$. This gives (1) and completes the proof.

Examples

In all examples H and K are groups and φ is a homomorphism from K into Aut (H) with associated action of K on H denoted by a dot. Let $G = H \rtimes K$ and as in Theorem 10 we identify H and K as subgroups of G. We shall use Propositions 4.16 and 4.17 to determine homomorphisms φ for some specific groups H. In each of the following examples the proof that φ is a homomorphism is easy (since K will often be cyclic) so the details are omitted.

(1) Let H be any abelian group (even of infinite order) and let $K = \langle x \rangle \cong Z_2$ be the group of order 2. Define $\varphi : K \to \text{Aut}(H)$ by mapping x to the automorphism of inversion on H so that the associated action is $x \cdot h = h^{-1}$, for all $h \in H$. Then G contains the subgroup H of index 2 and

$$xhx^{-1} = h^{-1} \qquad \text{for all } h \in H.$$

Of particular interest is the case when H is cyclic: if $H = Z_n$, one recognizes G as D_{2n} and if $H = \mathbb{Z}$ we denote G by D_∞.

(2) We can generalize the preceding example in a number of ways. One way is to let H be any abelian group and to let $K = \langle x \rangle \cong Z_{2n}$ be cyclic of order $2n$. Define φ again by mapping x to inversion, so that x^2 acts as the identity on H. In G, $xhx^{-1} = h^{-1}$ and $x^2 h x^{-2} = h$ for all $h \in H$. Thus $x^2 \in Z(G)$. In particular, if $H = Z_3$ and $K = Z_4$, G is a non-abelian group of order 12 which is not isomorphic to A_4 or D_{12} (since its Sylow 2-subgroup, K, is *cyclic* of order 4).

(3) Following up on the preceding example let $H = \langle h \rangle \cong Z_{2^n}$ for $n \geq 2$ and let $K = \langle x \rangle \cong Z_4$ with $xhx^{-1} = h^{-1}$ in G. As noted above, $x^2 \in Z(G)$. Since x inverts h (i.e., inverts H), x inverts the unique subgroup $\langle z \rangle$ of order 2 in H, where $z = h^{2^{n-1}}$. Thus $xzx^{-1} = z^{-1} = z$, so x centralizes z. It follows that $z \in Z(G)$. Thus $x^2 z \in Z(G)$ hence $\langle x^2 z \rangle \trianglelefteq G$. Let $\overline{G} = G/\langle x^2 z \rangle$. Since x^2 and z are distinct commuting elements of order 2, the order of $x^2 z$ is 2, so $|\overline{G}| = \frac{1}{2}|G| = 2^{n+1}$. By factoring out the product $x^2 z$ to form $\overline{G}$ we identify x^2 and $h^{2^{n-1}}$ in the quotient. In particular, when $n = 2$, both $\bar{x}$ and $\bar{h}$ have order 4, $\bar{x}$ inverts $\bar{h}$ and $\bar{h}^2 = \bar{x}^2$. It follows that $\overline{G} \cong Q_8$ in this case. In general, one can check that $\overline{G}$ has a unique subgroup of order 2 (namely $\langle \bar{x}^2 \rangle$) which equals the center of $\overline{G}$. The group $\overline{G}$ is called the *generalized quaternion group* of order 2^{n+1} and is denoted by $Q_{2^{n+1}}$:

$$Q_{2^{n+1}} = \langle h, x \mid h^{2^n} = x^4 = 1, \; xhx^{-1} = h^{-1}, \; h^{2^{n-1}} = x^2 \rangle.$$

(4) Let $H = \mathbb{Q}$ (under addition) and let $K = \langle x \rangle \cong \mathbb{Z}$. Define φ by mapping x to the map "multiplication by 2" on H, so that x acts on $h \in H$ by $x \cdot h = 2h$. Note that multiplication by 2 is an automorphism of H because it has a 2-sided inverse, namely

multiplication by $\frac{1}{2}$. In the group G, $\mathbb{Z} \le \mathbb{Q}$ and the conjugate $x\mathbb{Z}x^{-1}$ of $\mathbb{Z}$ is a *proper* subgroup of $\mathbb{Z}$ (namely $2\mathbb{Z}$). Thus $x \notin N_G(\mathbb{Z})$ even though $x\mathbb{Z}x^{-1} \le \mathbb{Z}$ (note that $x^{-1}\mathbb{Z}x$ is not contained in $\mathbb{Z}$). This shows that in order to prove an element g normalizes a subgroup A in an *infinite* group it is not sufficient in general to show that the conjugate of A by g is just *contained* in A (which is sufficient for finite groups).

(5) For H any group let $K = \mathrm{Aut}(H)$ with φ the identity map from K to $\mathrm{Aut}(H)$. The semidirect product $H \rtimes \mathrm{Aut}(H)$ is called the *holomorph* of H and will be denoted by $\mathrm{Hol}(H)$. Some holomorphs are described below; verifications of these isomorphisms are given as exercises at the end of this chapter.

(a) $\mathrm{Hol}(Z_2 \times Z_2) \cong S_4$.
(b) If $|G| = n$ and $\pi : G \to S_n$ is the left regular representation (Section 4.2), then $N_{S_n}(\pi(G)) \cong \mathrm{Hol}(G)$. In particular, since the left regular representation of a generator of Z_n is an n-cycle in S_n we obtain that for any n-cycle $(1\,2\,\dots\,n)$:

$$N_{S_n}(\langle\, (1\,2\,\dots\,n)\,\rangle) \cong \mathrm{Hol}(Z_n) = Z_n \rtimes \mathrm{Aut}(Z_n).$$

Note that the latter group has order $n\varphi(n)$.

(6) Let p and q be primes with $p < q$, let $H = Z_q$ and let $K = Z_p$. We have already seen that if p does not divide $q - 1$ then every group of order pq is cyclic (see the example following Proposition 4.16). This is consistent with the fact that if p does not divide $q - 1$, there is no nontrivial homomorphism from Z_p into $\mathrm{Aut}(Z_q)$ (the latter group is cyclic of order $q - 1$ by Proposition 4.17). Assume now that $p \mid q - 1$. By Cauchy's Theorem, $\mathrm{Aut}(Z_q)$ contains a subgroup of order p (which is unique because $\mathrm{Aut}(Z_q)$ is cyclic). Thus there is a nontrivial homomorphism, φ, from K into $\mathrm{Aut}(H)$. The associated group $G = H \rtimes K$ has order pq and K is not normal in G (Proposition 11). In particular, G is non-abelian. We shall prove shortly that G is (up to isomorphism) the unique non-abelian group of order pq. If $p = 2$, G must be isomorphic to D_{2q}.

(7) Let p be an odd prime. We construct two nonisomorphic non-abelian groups of order p^3 (we shall later prove that any non-abelian group of order p^3 is isomorphic to one of these two).

Let $H = Z_p \times Z_p$ and let $K = Z_p$. By Proposition 4.17, $\mathrm{Aut}(H) \cong GL_2(\mathbb{F}_p)$ and $|GL_2(\mathbb{F}_p)| = (p^2 - 1)(p^2 - p)$. Since $p \mid |\mathrm{Aut}(H)|$, by Cauchy's Theorem H has an automorphism of order p. Thus there is a nontrivial homomorphism, φ, from K into $\mathrm{Aut}(H)$ and so the associated group $H \rtimes K$ is a non-abelian group of order p^3. More explicitly, if $H = \langle a \rangle \times \langle b \rangle$, and x is a generator for K then x acts on a and b by

$$x \cdot a = ab \quad \text{and} \quad x \cdot b = b$$

which defines the action of x on all of H. With respect to the $\mathbb{F}_p$-basis a, b of the 2-dimensional vector space H the action of x (which can be considered in additive notation as a nonsingular linear transformation) has matrix

$$\begin{pmatrix} 1 & 0 \\ 1 & 1 \end{pmatrix} \in GL_2(\mathbb{F}_p).$$

The resulting semidirect product has the presentation

$$\langle x, a, b \mid x^p = a^p = b^p = 1, \ ab = ba, \ xax^{-1} = ab, \ xbx^{-1} = b \rangle$$

(in fact, this group is generated by $\{x, a\}$, and is called the *Heisenberg group* over $\mathbb{Z}/p\mathbb{Z}$, cf. Exercise 25).

Next let $H = Z_{p^2}$ and $K = Z_p$. Again by Proposition 4.17, $\mathrm{Aut}(H) \cong Z_{p(p-1)}$, so H admits an automorphism of order p. Thus there is a nontrivial homomorphism,

φ, from K into $\mathrm{Aut}(H)$ and so the group $H \rtimes K$ is non-abelian and of order p^3. More explicitly, if $H = \langle y \rangle$, and x is a generator for K then x acts on y by

$$x \cdot y = y^{1+p}.$$

The resulting semidirect product has the presentation

$$\langle x, y \mid x^p = y^{p^2} = 1, \ xyx^{-1} = y^{1+p} \rangle.$$

These two groups are not isomorphic (the former contains no element of order p^2, cf. Exercise 25, and the latter clearly does, namely y).

(8) Let $H = Q_8 \times (Z_2 \times Z_2) = \langle i, j \rangle \times (\langle a \rangle \times \langle b \rangle)$ and let $K = \langle y \rangle \cong Z_3$. The map defined by

$$i \mapsto j \qquad j \mapsto k = ij \qquad a \mapsto b \qquad b \mapsto ab$$

is easily seen to give an automorphism of H of order 3. Let φ be the homomorphism from K to $\mathrm{Aut}(H)$ defined by mapping y to this automorphism, and let G be the associated semidirect product, so that $y \in G$ acts by

$$y \cdot i = j \qquad y \cdot j = k \qquad y \cdot a = b \qquad y \cdot b = ab.$$

The group $G = H \rtimes K$ is a non-abelian group of order 96 with the property that the element $i^2 a \in G'$ but $i^2 a$ cannot be expressed as a single commutator $[x, y]$, for any $x, y \in G$ (checking the latter assertion is an elementary calculation).

As in the case of direct products we now prove a recognition theorem for semidirect products. This theorem will enable us to "break down" or "factor" all groups of certain orders and, as a result, classify groups of those orders. The strategy is discussed in greater detail following this theorem.

Theorem 12. Suppose G is a group with subgroups H and K such that
    (1) $H \trianglelefteq G$, and
    (2) $H \cap K = 1$.

Let $\varphi : K \rightarrow \mathrm{Aut}(H)$ be the homomorphism defined by mapping $k \in K$ to the automorphism of left conjugation by k on H. Then $HK \cong H \rtimes K$. In particular, if $G = HK$ with H and K satisfying (1) and (2), then G is the semidirect product of H and K.

Proof: Note that since $H \trianglelefteq G$, HK is a subgroup of G. By Proposition 8 every element of HK can be written uniquely in the form hk, for some $h \in H$ and $k \in K$. Thus the map $hk \mapsto (h, k)$ is a *set* bijection from HK onto $H \rtimes K$. The fact that this map is a homomorphism is the computation at the beginning of this section which led us to the formulation of the definition of the semidirect product.

Definition. Let H be a subgroup of the group G. A subgroup K of G is called a *complement* for H in G if $G = HK$ and $H \cap K = 1$.

With this terminology, the criterion for recognizing a semidirect product is simply that there must exist a complement for some proper *normal* subgroup of G. Not every group is the semidirect product of two of its proper subgroups (for example, if the group is simple), but as we have seen, the notion of a semidirect product greatly increases our list of known groups.

Some Classifications

We now apply Theorem 12 to classify groups of order n for certain values of n. The basic idea in each of the following arguments is to

(a) show every group of order n has proper subgroups H and K satisfying the hypothesis of Theorem 12 with $G = HK$

(b) find all possible isomorphism types for H and K

(c) for each pair H, K found in (b) find all possible homomorphisms $\varphi : K \to \text{Aut}(H)$

(d) for each triple H, K, φ found in (c) form the semidirect product $H \rtimes K$ (so any group G of order n is isomorphic to one of these explicitly constructed groups) and among all these semidirect products determine which pairs are isomorphic. This results in a list of the distinct isomorphism types of groups of order n.

In order to start this process we must first find subgroups H and K (of an arbitrary group G of order n) satisfying the above conditions. In the case of "small" values of n we can often do this by Sylow's Theorem. To show *normality* of H we use the conjugacy part of Sylow's Theorem or other normality criteria established in Chapter 4 (e.g., Corollary 4.5). Some of this work has already been done in the examples in Section 4.5. In many of the examples that follow, $|H|$ and $|K|$ are relatively prime, so $H \cap K = 1$ holds by Lagrange's Theorem.

Since H and K are proper subgroups of G one should think of the determination of H and K as being achieved inductively. In the examples we discuss, H and K will have sufficiently small order that we shall know all possible isomorphism types from previous results. For example, in most instances H and K will be of prime or prime squared order.

There will be relatively few possible homomorphisms $\varphi : K \to \text{Aut}(H)$ in our examples, particularly after we take into account certain symmetries (such as replacing one generator of K by another when K is cyclic).

Finally, the semidirect products which emerge from this process will, in our examples, be small in number and we shall find that, for the most part, they are (pairwise) *not* isomorphic. In general, this can be a more delicate problem, as Exercise 4 indicates.

We emphasize that this approach to "factoring" every group of some given order n as a semidirect product does not work for arbitrary n. For example, Q_8 is not a semidirect product since no proper subgroup has a complement (although we saw that it is a *quotient* of a semidirect product). Empirically, this process generally works well when the group order n is not divisible by a large power of any prime. At the other extreme, only a small percentage of the groups of order p^α for large α (p a prime) are nontrivial semidirect products.

Example: (Groups of Order pq, p and q primes with $p < q$)

Let G be any group of order pq, let $P \in Syl_p(G)$ and let $Q \in Syl_q(G)$. In Example 1 of the applications of Sylow's Theorems we proved that $G \cong Q \rtimes P$, for some $\varphi : P \to \text{Aut}(Q)$. Since P and Q are of prime order, they are cyclic. The group $\text{Aut}(Q)$ is cyclic of order $q - 1$. If p does not divide $q - 1$, the only homomorphism from P to $\text{Aut}(Q)$ is the trivial homomorphism, hence the only semidirect product in this case is the direct product, i.e., G is cyclic.

Consider now the case when $p \mid q - 1$ and let $P = \langle y \rangle$. Since $\text{Aut}(Q)$ is cyclic it contains a unique subgroup of order p, say $\langle \gamma \rangle$, and any homomorphism $\varphi : P \to \text{Aut}(Q)$

must map y to a power of γ. There are therefore p homomorphisms $\varphi_i : P \to \mathrm{Aut}(Q)$ given by $\varphi_i(y) = \gamma^i, 0 \le i \le p - 1$. Since φ_0 is the trivial homomorphism, $Q \rtimes_{\varphi_0} P \cong Q \times P$ as before. Each φ_i for $i \ne 0$ gives rise to a non-abelian group, G_i, of order pq. It is straightforward to check that these groups are all isomorphic because for each $\varphi_i, i > 0$, there is some generator y_i of P such that $\varphi_i(y_i) = \gamma$. Thus, up to a choice for the (arbitrary) generator of P, these semidirect products are all the same (see Exercise 6. See also Exercise 28 of Section 4.3).

Example: (Groups of Order 30)

By the examples following Sylow's Theorem every group G of order 30 contains a subgroup H of order 15. By the preceding example H is cyclic and H is normal in G (index 2). By Sylow's Theorem there is a subgroup K of G of order 2. Thus $G = HK$ and $H \cap K = 1$ so $G \cong H \rtimes K$, for some $\varphi : K \to \mathrm{Aut}(H)$. By Proposition 4.16,

$$\mathrm{Aut}(Z_{15}) \cong (\mathbb{Z}/15\mathbb{Z})^\times \cong Z_4 \times Z_2.$$

The latter isomorphism can be computed directly, or one can use Exercise 11 of the preceding section: writing H as $\langle a \rangle \times \langle b \rangle \cong Z_5 \times Z_3$, we have (since these two subgroups are characteristic in H)

$$\mathrm{Aut}(H) \cong \mathrm{Aut}(Z_5) \times \mathrm{Aut}(Z_3).$$

In particular, $\mathrm{Aut}(H)$ contains precisely three elements of order 2, whose actions on the group $H = \langle a \rangle \times \langle b \rangle$ are the following:

$$\begin{Bmatrix} a & \mapsto & a \\ b & \mapsto & b^{-1} \end{Bmatrix} \qquad \begin{Bmatrix} a & \mapsto & a^{-1} \\ b & \mapsto & b \end{Bmatrix} \qquad \begin{Bmatrix} a & \mapsto & a^{-1} \\ b & \mapsto & b^{-1} \end{Bmatrix}.$$

Thus there are three nontrivial homomorphisms from K into $\mathrm{Aut}(H)$ given by sending the generator of K into one of these three elements of order 2 (as usual, the trivial homomorphism gives the direct product: $H \times K \cong Z_{30}$).

Let $K = \langle k \rangle$. If the homomorphism $\varphi_1 : K \to \mathrm{Aut}(H)$ is defined by mapping k to the first automorphism above (so that $k \cdot a = a$ and $k \cdot b = b^{-1}$ gives the action of k on H) then $G_1 = H \rtimes_{\varphi_1} K$ is easily seen to be isomorphic to $Z_5 \times D_6$ (note that in this semidirect product k centralizes the element a of H of order 5, so the factorization as a direct product is $\langle a \rangle \times \langle b, k \rangle$).

If φ_2 is defined by mapping k to the second automorphism above, then $G_2 = H \rtimes_{\varphi_2} K$ is easily seen to be isomorphic to $Z_3 \times D_{10}$ (note that in this semidirect product k centralizes the element b of H of order 3, so the factorization as a direct product is $\langle b \rangle \times \langle a, k \rangle$).

If φ_3 is defined by mapping k to the third automorphism above then $G_3 = H \rtimes_{\varphi_3} K$ is easily seen to be isomorphic to D_{30}.

Note that these groups are all nonisomorphic since their centers have orders 30 (in the abelian case), 5 (for G_1), 3 (for G_2), and 1 (for G_3).

We emphasize that although (in hindsight) this procedure does not give rise to any groups we could not already have constructed using only direct products, the argument proves that this is the *complete* list of isomorphism types of groups of order 30.

Example: (Groups of Order 12)

Let G be a group of order 12, let $V \in Syl_2(G)$ and let $T \in Syl_3(G)$. By the discussion of groups of order 12 in Section 4.5 we know that either V or T is normal in G (for purposes of illustration we shall not invoke the full force of our results from Chapter 4, namely that either $T \trianglelefteq G$ or $G \cong A_4$). By Lagrange's Theorem $V \cap T = 1$. Thus G is a semidirect product. Note that $V \cong Z_4$ or $Z_2 \times Z_2$ and $T \cong Z_3$.

Case 1: $V \trianglelefteq G$

We must determine all possible homomorphisms from T into $\mathrm{Aut}(V)$. If $V \cong Z_4$, then $\mathrm{Aut}(V) \cong Z_2$ and there are no nontrivial homomorphisms from T into $\mathrm{Aut}(V)$. Thus the only group of order 12 with a normal cyclic Sylow 2-subgroup is Z_{12}.

Assume therefore that $V \cong Z_2 \times Z_2$. In this case $\mathrm{Aut}(V) \cong S_3$ and there is a unique subgroup of $\mathrm{Aut}(V)$ of order 3, say $\langle \gamma \rangle$. Thus if $T = \langle y \rangle$, there are three possible homomorphisms from T into $\mathrm{Aut}(V)$:

$$\varphi_i : T \to \mathrm{Aut}(V) \quad \text{defined by} \quad \varphi_i(y) = \gamma^i, \quad i = 0, 1, 2.$$

As usual, φ_0 is the trivial homomorphism, which gives rise to the direct product $Z_2 \times Z_2 \times Z_3$. Homomorphisms φ_1 and φ_2 give rise to isomorphic semidirect products because they differ only in the choice of a generator for T (i.e., $\varphi_1(y) = \gamma$ and $\varphi_2(y') = \gamma$, where $y' = y^2$ and y' is another choice of generator for T — see also Exercise 6). The unique non-abelian group in this case is A_4.

Case 2: $T \trianglelefteq G$

We must determine all possible homomorphisms from V into $\mathrm{Aut}(T)$. Note that $\mathrm{Aut}(T) = \langle \lambda \rangle \cong Z_2$, where λ inverts T. If $V = \langle x \rangle \cong Z_4$, there are precisely two homomorphisms from V into $\mathrm{Aut}(T)$: the trivial homomorphism and the homomorphism which sends x to λ. As usual, the trivial homomorphism gives rise to the direct product: $Z_3 \times Z_4 \cong Z_{12}$. The nontrivial homomorphism gives the semidirect product which was discussed in Example 2 following Proposition 11 of this section.

Finally, assume $V = \langle a \rangle \times \langle b \rangle \cong Z_2 \times Z_2$. There are precisely three nontrivial homomorphisms from V into $\mathrm{Aut}(T)$ determined by specifying their kernels as one of the three subgroups of order 2 in V. For example, $\varphi_1(a) = \lambda$ and $\varphi_1(b) = \lambda$ has kernel $\langle ab \rangle$, that is, in this semidirect product both a and b act by inverting T and ab centralizes T. If φ_2 and φ_3 have kernels $\langle a \rangle$ and $\langle b \rangle$, respectively, then one easily checks that the resulting three semidirect products are all isomorphic to $S_3 \times Z_2$, where the Z_2 direct factor is the kernel of φ_i. For example,

$$T \rtimes_{\varphi_1} V = \langle a, T \rangle \times \langle ab \rangle.$$

In summary, there are precisely 5 groups of order 12, three of which are non-abelian.

Example: (Groups of Order p^3, p an odd prime)

Let G be a group of order p^3, p an odd prime, and assume G is not cyclic. By Exercise 9 of the previous section the map $x \mapsto x^p$ is a homomorphism from G into $Z(G)$ and the kernel of this homomorphism has order p^2 or p^3. In the former case G must contain an element of order p^2 and in the latter case every nonidentity element of G has order p.

Case 1: G has an element of order p^2

Let x be an element of order p^2 and let $H = \langle x \rangle$. Note that since H has index p, H is normal in G by Corollary 4.5. If E is the kernel of the pth power map, then in this case $E \cong Z_p \times Z_p$ and $E \cap H = \langle x^p \rangle$. Let y be any element of $E - H$ and let $K = \langle y \rangle$. By construction, $H \cap K = 1$ and so G is isomorphic to $Z_{p^2} \rtimes Z_p$, for some $\varphi : K \to \mathrm{Aut}(H)$. If φ is the trivial homomorphism, $G \cong Z_{p^2} \times Z_p$, so we need only consider the nontrivial homomorphisms. By Proposition 4.17 $\mathrm{Aut}(H) \cong Z_{p(p-1)}$ is cyclic and so contains a unique subgroup of order p, explicitly given by $\langle \gamma \rangle$ where

$$\gamma(x) = x^{1+p}.$$

As usual, up to choice of a generator for the cyclic group K, there is only one nontrivial homomorphism, φ, from K into $\mathrm{Aut}(H)$, given by $\varphi(y) = \gamma$; hence up to isomorphism there is a unique non-abelian group $H \rtimes K$ in this case. This group is described in Example 7 above.

Case 2: every nonidentity element of G has order p

In this case let H be any subgroup of G of order p^2 (see Exercise 29, Section 4.3). Necessarily $H \cong Z_p \times Z_p$. Let $K = \langle y \rangle$ for any element y of $G - H$. Since H has index p, $H \trianglelefteq G$ and since K has order p but is not contained in H, $H \cap K = 1$. Then G is isomorphic to $(Z_p \times Z_p) \rtimes Z_p$, for some $\varphi : K \to \text{Aut}(H)$. If φ is trivial, $G \cong Z_p \times Z_p \times Z_p$ (the elementary abelian group), so we may assume φ is nontrivial. By Proposition 4.17,

$$\text{Aut}(H) \cong GL_2(\mathbb{F}_p)$$

so $|\text{Aut}(H)| = (p^2 - 1)(p^2 - p)$. Note that a Sylow p-subgroup of $\text{Aut}(H)$ has order p so all subgroups of order p in $\text{Aut}(H)$ are conjugate in $\text{Aut}(H)$ by Sylow's Theorem. Explicitly, (as discussed in Example 7 above) every subgroup of order p in $\text{Aut}(H)$ is conjugate to $\langle \gamma \rangle$, where if $H = \langle a \rangle \times \langle b \rangle$, the automorphism γ is defined by

$$\gamma(a) = ab \quad \text{and} \quad \gamma(b) = b.$$

With respect to the $\mathbb{F}_p$-basis a, b of the 2-dimensional vector space H the automorphism has matrix

$$\begin{pmatrix} 1 & 0 \\ 1 & 1 \end{pmatrix} \in GL_2(\mathbb{F}_p).$$

Thus (again quoting Exercise 6) there is a unique isomorphism type of semidirect product in this case.

Finally, since the two non-abelian groups have different orders for the kernels of the pth power maps, they are not isomorphic. A presentation for this group is also given in Example 7 above.

EXERCISES

Let H and K be groups, let φ be a homomorphism from K into $\text{Aut}(H)$ and, as usual, identify H and K as subgroups of $G = H \rtimes_\varphi K$.

1. Prove that $C_K(H) = \ker \varphi$ (recall that $C_K(H) = C_G(H) \cap K$).

2. Prove that $C_H(K) = N_H(K)$.

3. In Example 1 following the proof of Proposition 11 prove that every element of $G - H$ has order 2. Prove that G is abelian if and only if $h^2 = 1$ for all $h \in H$.

4. Let $p = 2$ and check that the construction of the two non-abelian groups of order p^3 is valid in this case. Prove that *both* resulting groups are isomorphic to D_8.

5. Let $G = \text{Hol}(Z_2 \times Z_2)$.
 (a) Prove that $G = H \rtimes K$ where $H = Z_2 \times Z_2$ and $K \cong S_3$. Deduce that $|G| = 24$.
 (b) Prove that G is isomorphic to S_4. [Obtain a homomorphism from G into S_4 by letting G act on the left cosets of K. Use Exercise 1 to show this representation is faithful.]

6. Assume that K is a cyclic group, H is an arbitrary group and φ_1 and φ_2 are homomorphisms from K into $\text{Aut}(H)$ such that $\varphi_1(K)$ and $\varphi_2(K)$ are conjugate subgroups of $\text{Aut}(H)$. If K is infinite assume φ_1 and φ_2 are injective. Prove by constructing an explicit isomorphism that $H \rtimes_{\varphi_1} K \cong H \rtimes_{\varphi_2} K$ (in particular, if the subgroups $\varphi_1(K)$ and $\varphi_2(K)$ are equal in $\text{Aut}(H)$, then the resulting semidirect products are isomorphic). [Suppose $\sigma \varphi_1(K) \sigma^{-1} = \varphi_2(K)$ so that for some $a \in \mathbb{Z}$ we have $\sigma \varphi_1(k) \sigma^{-1} = \varphi_2(k)^a$ for all $k \in K$. Show that the map

$\psi : H \rtimes_{\varphi_1} K \rightarrow H \rtimes_{\varphi_2} K$ defined by $\psi((h, k)) = (\sigma(h), k^a)$ is a homomorphism. Show ψ is bijective by constructing a 2-sided inverse.]

7. This exercise describes thirteen isomorphism types of groups of order 56. (It is not too difficult to show that every group of order 56 is isomorphic to one of these.)
   (a) Prove that there are three abelian groups of order 56.
   (b) Prove that every group of order 56 has either a normal Sylow 2-subgroup or a normal Sylow 7-subgroup.
   (c) Construct the following non-abelian groups of order 56 which have a normal Sylow 7-subgroup and whose Sylow 2-subgroup S is as specified:
   > one group when $S \cong Z_2 \times Z_2 \times Z_2$
   > two nonisomorphic groups when $S \cong Z_4 \times Z_2$
   > one group when $S \cong Z_8$
   > two nonisomorphic groups when $S \cong Q_8$
   > three nonisomorphic groups when $S \cong D_8$.
   [For a particular S, two groups are not isomorphic if the kernels of the maps from S into $\text{Aut}(Z_7)$ are not isomorphic.]
   (d) Let G be a group of order 56 with a nonnormal Sylow 7-subgroup. Prove that if S is the Sylow 2-subgroup of G then $S \cong Z_2 \times Z_2 \times Z_2$. [Let an element of order 7 act by conjugation on the seven nonidentity elements of S and deduce that they all have the same order.]
   (e) Prove that there is a unique group of order 56 with a nonnormal Sylow 7-subgroup. [For existence use the fact that $|GL_3(\mathbb{F}_2)| = 168$; for uniqueness use Exercise 6.]

8. Construct a non-abelian group of order 75. Classify all groups of order 75 (there are three of them). [Use Exercise 6 to show that the non-abelian group is unique.] (The classification of groups of order pq^2, where p and q are primes with $p < q$ and p not dividing $q - 1$, is quite similar.)

9. Show that the matrix $\begin{pmatrix} 0 & -1 \\ 1 & 4 \end{pmatrix}$ is an element of order 5 in $GL_2(\mathbb{F}_{19})$. Use this matrix to construct a non-abelian group of order 1805 and give a presentation of this group. Classify groups of order 1805 (there are three isomorphism types). [Use Exercise 6 to prove uniqueness of the non-abelian group.] (A general method for finding elements of prime order in $GL_n(\mathbb{F}_p)$ is described in the exercises in Section 12.2; this particular matrix of order 5 in $GL_2(\mathbb{F}_{19})$ appears in Exercise 16 of that section as an illustration of the method.)

10. This exercise classifies the groups of order 147 (there are six isomorphism types).
   (a) Prove that there are two abelian groups of order 147.
   (b) Prove that every group of order 147 has a normal Sylow 7-subgroup.
   (c) Prove that there is a unique non-abelian group whose Sylow 7-subgroup is cyclic.
   (d) Let $t_1 = \begin{pmatrix} 2 & 0 \\ 0 & 1 \end{pmatrix}$ and $t_2 = \begin{pmatrix} 1 & 0 \\ 0 & 2 \end{pmatrix}$ be elements of $GL_2(\mathbb{F}_7)$. Prove $P = \langle t_1, t_2 \rangle$ is a Sylow 3-subgroup of $GL_2(\mathbb{F}_7)$ and that $P \cong Z_3 \times Z_3$. Deduce that every subgroup of $GL_2(\mathbb{F}_7)$ of order 3 is conjugate in $GL_2(\mathbb{F}_7)$ to a subgroup of P.
   (e) By Example 3 in Section 1 the group P has four subgroups of order 3 and these are: $P_1 = \langle t_1 \rangle$, $P_2 = \langle t_2 \rangle$, $P_3 = \langle t_1 t_2 \rangle$, and $P_4 = \langle t_1 t_2^2 \rangle$. For $i = 1, 2, 3, 4$ let $G_i = (Z_7 \times Z_7) \rtimes_{\varphi_i} Z_3$, where φ_i is an isomorphism of Z_3 with the subgroup P_i of $\text{Aut}(Z_7 \times Z_7)$. For each i describe G_i in terms of generators and relations. Deduce that $G_1 \cong G_2$.
   (f) Prove that G_1 is not isomorphic to either G_3 or G_4. [Show that the center of G_1 has

order 7 whereas the centers of G_3 and G_4 are trivial.]

    (g) Prove that G_3 is not isomorphic to G_4. [Show that every subgroup of order 7 in G_3 is normal in G_3 but that G_4 has subgroups of order 7 that are not normal.]

    (h) Classify the groups of order 147 by showing that the six nonisomorphic groups described above (two from part (a), one from part (c) and G_1, G_3, and G_4) are all the groups of order 147. [Use Exercise 6 and part (d).] (The classification of groups of order pq^2, where p and q are primes with $p < q$ and $p \mid q - 1$, is quite similar.)

11. Classify groups of order 28 (there are four isomorphism types).

12. Classify the groups of order 20 (there are five isomorphism types).

13. Classify groups of order $4p$, where p is a prime greater than 3. [There are four isomorphism types when $p \equiv 3 \pmod 4$ and five isomorphism types when $p \equiv 1 \pmod 4$.]

14. This exercise classifies the groups of order 60 (there are thirteen isomorphism types). Let G be a group of order 60, let P be a Sylow 5-subgroup of G and let Q be a Sylow 3-subgroup of G.

    (a) Prove that if P is not normal in G then $G \cong A_5$. [See Section 4.5.]

    (b) Prove that if $P \trianglelefteq G$ but Q is not normal in G then $G \cong A_4 \times Z_5$. [Show in this case that $P \le Z(G)$, $G/P \cong A_4$, a Sylow 2-subgroup T of G is normal and $TQ \cong A_4$.]

    (c) Prove that if both P and Q are normal in G then $G \cong Z_{15} \rtimes T$ where $T \cong Z_4$ or $Z_2 \times Z_2$. Show in this case that there are six isomorphism types when T is cyclic (one abelian) and there are five isomorphism types when T is the Klein 4-group (one abelian). [Use the same ideas as in the classifications of groups of orders 30 and 20.]

15. Let p be an odd prime. Prove that every element of order 2 in $GL_2(\mathbb{F}_p)$ is conjugate to a diagonal matrix with ± 1's on the diagonal. Classify the groups of order $2p^2$. [If A is a 2×2 matrix with $A^2 = I$ and v_1, v_2 is a basis for the underlying vector space, look at A acting on the vectors $w_1 = v_1 + v_2$ and $w_2 = v_1 - v_2$.]

16. Show that there are exactly 4 distinct homomorphisms from Z_2 into $\text{Aut}(Z_8)$. Prove that the resulting semidirect products are the groups: $Z_8 \times Z_2$, D_{16}, the quasidihedral group QD_{16} and the modular group M (cf. the exercises in Section 2.5).

17. Show that for any $n \ge 3$ there are exactly 4 distinct homomorphisms from Z_2 into $\text{Aut}(Z_{2^n})$. Prove that the resulting semidirect products give 4 nonisomorphic groups of order 2^{n+1}. [Recall Exercises 21 to 23 in Section 2.3.] (These four groups together with the cyclic group and the generalized quaternion group, $Q_{2^{n+1}}$, are all the groups of order 2^{n+1} which possess a cyclic subgroup of index 2.)

18. Show that if H is any group then there is a group G that contains H as a normal subgroup with the property that for every automorphism σ of H there is an element $g \in G$ such that conjugation by g when restricted to H is the given automorphism σ, i.e., every automorphism of H is obtained as an inner automorphism of G restricted to H.

19. Let H be a group of order n, let $K = \text{Aut}(H)$ and form $G = \text{Hol}(H) = H \rtimes K$ (where φ is the identity homomorphism). Let G act by left multiplication on the left cosets of K in G and let π be the associated permutation representation $\pi : G \to S_n$.

    (a) Prove the elements of H are coset representatives for the left cosets of K in G and with this choice of coset representatives π restricted to H is the regular representation of H.

    (b) Prove $\pi(G)$ is the normalizer in S_n of $\pi(H)$. Deduce that under the regular representation of any finite group H of order n, the normalizer in S_n of the image of H is isomorphic to $\text{Hol}(H)$. [Show $|G| = |N_{S_n}(\pi(H))|$ using Exercises 1 and 2 above.]

    (c) Deduce that the normalizer of the group generated by an n-cycle in S_n is isomorphic to $\text{Hol}(Z_n)$ and has order $n\varphi(n)$.

20. Let p be an odd prime. Prove that if P is a non-cyclic p-group then P contains a normal subgroup U with $U \cong Z_p \times Z_p$. Deduce that for odd primes p a p-group that contains a unique subgroup of order p is cyclic. (For $p = 2$ it is a theorem that the generalized quaternion groups Q_{2^n} are the only non-cyclic 2-groups which contain a unique subgroup of order 2). [Proceed by induction on $|P|$. Let Z be a subgroup of order p in $Z(P)$ and let $\overline{P} = P/Z$. If $\overline{P}$ is cyclic then P is abelian by Exercise 36 in Section 3.1 — show the result is true for abelian groups. When $\overline{P}$ is not cyclic use induction to produce a normal subgroup $\overline{H}$ of $\overline{P}$ with $\overline{H} \cong Z_p \times Z_p$. Let H be the complete preimage of $\overline{H}$ in P, so $|H| = p^3$. Let $H_0 = \{x \in H \mid x^p = 1\}$ so that H_0 is a characteristic subgroup of H of order p^2 or p^3 by Exercise 9 in Section 4. Show that a suitable subgroup of H_0 gives the desired normal subgroup U.]

21. Let p be an odd prime and let P be a p-group. Prove that if every subgroup of P is normal then P is abelian. (Note that Q_8 is a non-abelian 2-group with this property, so the result is false for $p = 2$.) [Use the preceding exercises and Exercise 15 of Section 4.]

22. Let F be a field let n be a positive integer and let G be the group of upper triangular matrices in $GL_n(F)$ (cf. Exercise 16, Section 2.1)
   (a) Prove that G is the semidirect product $U \rtimes D$ where U is the set of upper triangular matrices with 1's down the diagonal (cf. Exercise 17, Section 2.1) and D is the set of diagonal matrices in $GL_n(F)$.
   (b) Let $n=2$. Recall that $U \cong F$ and $D \cong F^\times \times F^\times$ (cf. Exercise 11 in Section 3.1). Describe the homomorphism from D into $\text{Aut}(U)$ explicitly in terms of these isomorphisms (i.e., show how each element of $F^\times \times F^\times$ acts as an automorphism on F).

23. Let K and L be groups, let n be a positive integer, let $\rho : K \to S_n$ be a homomorphism and let H be the direct product of n copies of L. In Exercise 8 of Section 1 an injective homomorphism ψ from S_n into $\text{Aut}(H)$ was constructed by letting the elements of S_n permute the n factors of H. The composition $\psi \circ \rho$ is a homomorphism from K into $\text{Aut}(H)$. The *wreath product* of L by K is the semidirect product $H \rtimes K$ with respect to this homomorphism and is denoted by $L \wr K$ (this wreath product depends on the choice of permutation representation ρ of K — if none is given explicitly, ρ is assumed to be the left regular representation of K).
   (a) Assume K and L are finite groups and ρ is the left regular representation of K. Find $|L \wr K|$ in terms of $|K|$ and $|L|$.
   (b) Let p be a prime, let $K = L = Z_p$ and let ρ be the left regular representation of K. Prove that $Z_p \wr Z_p$ is a non-abelian group of order p^{p+1} and is isomorphic to a Sylow p-subgroup of S_{p^2}. [The p copies of Z_p whose direct product makes up H may be represented by p disjoint p-cycles; these are cyclically permuted by K.]

24. Let n be an integer > 1. Prove the following classification: every group of order n is abelian if and only if $n = p_1^{\alpha_1} p_2^{\alpha_2} \dots p_r^{\alpha_r}$, where $p_1, \dots, p_r$ are distinct primes, $\alpha_i = 1$ or 2 for all $i \in \{1, \dots, r\}$ and p_i does not divide $p_j^{\alpha_j} - 1$ for all i and j. [See Exercise 56 in Section 4.5.]

25. Let $H(\mathbb{F}_p)$ be the Heisenberg group over the finite field $\mathbb{F}_p = \mathbb{Z}/p\mathbb{Z}$ (cf. Exercise 20 in Section 4). Prove that $H(\mathbb{F}_2) \cong D_8$, and that $H(\mathbb{F}_p)$ has exponent p and is isomorphic to the first non-abelian group in Example 7.

CHAPTER 6

Further Topics in Group Theory

6.1 p-GROUPS, NILPOTENT GROUPS, AND SOLVABLE GROUPS

Let p be a prime and let G be a finite group of order $p^a n$, where p does not divide n. Recall that a (finite) p-group is any group whose order is a power of p. Sylow's Theorem shows that p-groups abound as subgroups of G and in order to exploit this phenomenon to unravel the structure of finite groups it will be necessary to establish some basic properties of p-groups. In the next section we shall apply these results in many specific instances.

Before giving the results on p-groups we first recall a definition that has appeared in some earlier exercises.

Definition. A *maximal subgroup* of a group G is a proper subgroup M of G such that there are no subgroups H of G with $M < H < G$.

By order considerations every proper subgroup of a finite group is contained in some maximal subgroup. In contrast, infinite groups may or may not have maximal subgroups. For example, $p\mathbb{Z}$ is a maximal subgroup of $\mathbb{Z}$ whereas $\mathbb{Q}$ (under +) has no maximal subgroups (cf. Exercise 16 at the end of this section).

We now collect all the properties of p-groups we shall need into an omnibus theorem:

Theorem 1. Let p be a prime and let P be a group of order p^a, $a \geq 1$. Then
   (1) The center of P is nontrivial: $Z(P) \neq 1$.
   (2) If H is a nontrivial normal subgroup of P then H intersects the center non-trivially: $H \cap Z(P) \neq 1$. In particular, every normal subgroup of order p is contained in the center.
   (3) If H is a normal subgroup of P then H contains a subgroup of order p^b that is normal in P for each divisor p^b of $|H|$. In particular, P has a normal subgroup of order p^b for every $b \in \{0, 1, \ldots, a\}$.
   (4) If $H < P$ then $H < N_P(H)$ (i.e., every proper subgroup of P is a proper subgroup of its normalizer in P).
   (5) Every maximal subgroup of P is of index p and is normal in P.

Proof: These results rely ultimately on the class equation and it may be useful for the reader to review Section 4.3.

Part 1 is Theorem 8 of Chapter 4 and is also the special case of part 2 when $H = P$. We therefore begin by proving (2); we shall not quote Theorem 8 of Chapter 4 although the argument that follows is only a slight generalization of the one in Chapter 4. Let H be a nontrivial normal subgroup of P. Recall that for each conjugacy class $\mathcal{C}$ of P, either $\mathcal{C} \subseteq H$ or $\mathcal{C} \cap H = \emptyset$ because H is normal (this easy fact was shown in a remark preceding Theorem 4.12). Pick representatives of the conjugacy classes of P:

$$a_1, a_2, \ldots, a_r$$

with $a_1, \ldots, a_k \in H$ and $a_{k+1}, \ldots, a_r \notin H$. Let $\mathcal{C}_i$ be the conjugacy class of a_i in P, for all i. Thus

$$\mathcal{C}_i \subseteq H, \quad 1 \le i \le k \qquad \text{and} \qquad \mathcal{C}_i \cap H = \emptyset, \quad k+1 \le i \le r.$$

By renumbering $a_1, \ldots, a_k$ if necessary we may assume $a_1, \ldots, a_s$ represent classes of size 1 (i.e., are in the center of P) and $a_{s+1}, \ldots, a_k$ represent classes of size > 1. Since H is the disjoint union of these we have

$$|H| = |H \cap Z(P)| + \sum_{i=s+1}^{k} \frac{|P|}{|C_P(a_i)|}.$$

Now p divides $|H|$ and p divides each term in the sum $\sum_{i=s+1}^{k} |P : C_P(a_i)|$ so p divides their difference: $|H \cap Z(P)|$. This proves $H \cap Z(P) \ne 1$. If $|H| = p$, since $H \cap Z(P) \ne 1$ we must have $H \le Z(P)$. This completes the proof of (2).

Next we prove (3) by induction on a. If $a \le 1$ or $H = 1$, the result is trivial. Assume therefore that $a > 1$ and $H \ne 1$. By part 2, $H \cap Z(P) \ne 1$ so by Cauchy's Theorem $H \cap Z(P)$ contains a (normal) subgroup Z of order p. Use bar notation to denote passage to the quotient group P/Z. This quotient has order p^{a-1} and $\overline{H} \trianglelefteq \overline{P}$. By induction, for every nonnegative integer b such that p^b divides $|\overline{H}|$ there is a subgroup $\overline{K}$ of $\overline{H}$ of order p^b that is normal in $\overline{P}$. If K is the complete preimage of $\overline{K}$ in P then $|K| = p^{b+1}$. The set of all subgroups of H obtained by this process together with the identity subgroup provides a subgroup of H that is normal in P for each divisor of $|H|$. The second assertion of part 3 is the special case $H = P$. This establishes part 3.

We prove (4) also by induction on $|P|$. If P is abelian then all subgroups of P are normal in P and the result is trivial. We may therefore assume $|P| > p$ (in fact, $|P| > p^2$ by Corollary 4.9). Let H be a proper subgroup of P. Since all elements of $Z(P)$ commute with all elements of P, $Z(P)$ normalizes every subgroup of P. By part 1 we have that $Z(P) \ne 1$. If $Z(P)$ is not contained in H, then H is properly contained in $\langle H, Z(P) \rangle$ and the latter subgroup is contained in $N_P(H)$ so (4) holds. We may therefore assume $Z(P) \le H$. Use bar notation to denote passage to the quotient $P/Z(P)$. Since $\overline{P}$ has smaller order than P by (1), by induction $\overline{H}$ is properly contained in $N_{\overline{P}}(\overline{H})$. It follows directly from the Lattice Isomorphism Theorem that $N_P(H)$ is the complete preimage in P of $N_{\overline{P}}(\overline{H})$, hence we obtain proper containment of H in its normalizer in this case as well. This completes the induction.

To prove (5) let M be a maximal subgroup of P. By definition, $M < P$ so by part 4, $M < N_P(M)$. By definition of maximality we must therefore have $N_P(M) = P$, i.e., $M \trianglelefteq P$. The Lattice Isomorphism Theorem shows that P/M is a p-group with no proper nontrivial subgroups because M is a maximal subgroup. By part 3, however,

P/M has subgroups of every order dividing $|P/M|$. The only possibility is $|P/M| = p$. This proves (5) and completes the proof of the theorem.

Definition.
  (1) For any (finite or infinite) group G define the following subgroups inductively:

$$Z_0(G) = 1, \qquad Z_1(G) = Z(G)$$

  and $Z_{i+1}(G)$ is the subgroup of G containing $Z_i(G)$ such that

$$Z_{i+1}(G)/Z_i(G) = Z(G/Z_i(G))$$

  (i.e., $Z_{i+1}(G)$ is the complete preimage in G of the center of $G/Z_i(G)$ under the natural projection). The chain of subgroups

$$Z_0(G) \le Z_1(G) \le Z_2(G) \le \cdots$$

  is called the *upper central series of G*. (The use of the term "upper" indicates that $Z_i(G) \le Z_{i+1}(G)$.)
  (2) A group G is called *nilpotent* if $Z_c(G) = G$ for some $c \in \mathbb{Z}$. The smallest such c is called the *nilpotence class of G*.

One of the exercises at the end of this section shows that $Z_i(G)$ is a characteristic (hence normal) subgroup of G for all i. We use this fact freely from now on.

Remarks:
(1) If G is abelian then G is nilpotent (of class 1, provided $|G| > 1$), since in this case $G = Z(G) = Z_1(G)$. One should think of nilpotent groups as lying between abelian and solvable groups in the hierarchy of structure (recall that solvable groups were introduced in Section 3.4; we shall discuss solvable groups further at the end of this section):

cyclic groups $\subset$ abelian groups $\subset$ nilpotent groups $\subset$ solvable groups $\subset$ all groups

(all of the above containments are proper, as we shall verify shortly).
(2) For any finite group there must, by order considerations, be an integer n such that

$$Z_n(G) = Z_{n+1}(G) = Z_{n+2}(G) = \cdots .$$

For example, $Z_n(S_3) = 1$ for all $n \in \mathbb{Z}^+$. Once two terms in the upper central series are the same, the chain stabilizes at that point (i.e., all terms thereafter are equal to these two). For example, if $G = Z_2 \times S_3$,

$$Z(G) = Z_1(G) = Z_2(G) = Z_n(G) \quad \text{has order 2 for all } n.$$

By definition, $Z_n(G)$ is a proper subgroup of G for all n for non-nilpotent groups.
(3) For infinite groups G it may happen that all $Z_i(G)$ are proper subgroups of G (so G is not nilpotent) but

$$G = \bigcup_{i=0}^{\infty} Z_i(G).$$

Groups for which this hold are called *hypernilpotent* — they enjoy some (but not all) of the properties of nilpotent groups. While we shall be dealing mainly with finite nilpotent groups, results that do not involve the notion of order, Sylow subgroups etc. also hold for infinite groups. Even for infinite groups one of the main techniques for dealing with nilpotent groups is induction on the nilpotence class.

Proposition 2. Let p be a prime and let P be a group of order p^a. Then P is nilpotent of nilpotence class at most $a - 1$ for $a \geq 2$ (and class equal to a when $a = 0$ or 1).

Proof: For each $i \geq 0$, $P/Z_i(P)$ is a p-group, so

$$\text{if } |P/Z_i(P)| > 1 \text{ then } Z(P/Z_i(P)) \neq 1$$

by Theorem 1(1). Thus if $Z_i(P) \neq P$ then $|Z_{i+1}(P)| \geq p|Z_i(P)|$ and so $|Z_{i+1}(P)| \geq p^{i+1}$. In particular, $|Z_a(P)| \geq p^a$, so $P = Z_a(P)$. Thus P is nilpotent of class $\leq a$. The only way P could be of nilpotence class exactly equal to a would be if $|Z_i(P)| = p^i$ for all i. In this case, however, $Z_{a-2}(P)$ would have index p^2 in P, so $P/Z_{a-2}(P)$ would be abelian (by Corollary 4.9). But then $P/Z_{a-2}(P)$ would equal its center and so $Z_{a-1}(P)$ would equal P, a contradiction. This proves that the class of P is $\leq a - 1$.

Example

Both D_8 and Q_8 are nilpotent of class 2. More generally, D_{2^n} is nilpotent of class $n - 1$. This can be proved inductively by showing that $|Z(D_{2^n})| = 2$ and $D_{2^n}/Z(D_{2^n}) \cong D_{2^{n-1}}$ for $n \geq 3$ (the details are left as an exercise). If n is not a power of 2, D_{2n} is not nilpotent (cf. Exercise 10).

We now give some equivalent (and often more workable) characterizations of nilpotence for *finite* groups:

Theorem 3. Let G be a finite group, let $p_1, p_2, \ldots, p_s$ be the distinct primes dividing its order and let $P_i \in Syl_{p_i}(G)$, $1 \leq i \leq s$. Then the following are equivalent:
  (1) G is nilpotent
  (2) if $H < G$ then $H < N_G(H)$, i.e., every proper subgroup of G is a proper subgroup of its normalizer in G
  (3) $P_i \trianglelefteq G$ for $1 \leq i \leq s$, i.e., every Sylow subgroup is normal in G
  (4) $G \cong P_1 \times P_2 \times \cdots \times P_s$.

Proof: The proof that (1) implies (2) is the same argument as for p-groups — the only fact we needed was if G is nilpotent then so is $G/Z(G)$ — so the details are omitted (cf. the exercises).

To show that (2) implies (3) let $P = P_i$ for some i and let $N = N_G(P)$. Since $P \trianglelefteq N$, Corollary 4.20 gives that P is characteristic in N. Since P char $N \trianglelefteq N_G(N)$ we get that $P \trianglelefteq N_G(N)$. This means $N_G(N) \leq N$ and hence $N_G(N) = N$. By (2) we must therefore have $N = G$, which gives (3).

Next we prove (3) implies (4). For any t, $1 \leq t \leq s$ we show inductively that

$$P_1 P_2 \cdots P_t \cong P_1 \times P_2 \times \cdots \times P_t.$$

Note first that each P_i is normal in G so $P_1 \cdots P_t$ is a subgroup of G. Let H be the product $P_1 \cdots P_{t-1}$ and let $K = P_t$, so by induction $H \cong P_1 \times \cdots \times P_{t-1}$. In particular, $|H| = |P_1| \cdot |P_2| \cdots |P_{t-1}|$. Since $|K| = |P_t|$, the orders of H and K are relatively prime. Lagrange's Theorem implies $H \cap K = 1$. By definition, $P_1 \cdots P_t = HK$, hence Theorem 5.9 gives

$$HK \cong H \times K = (P_1 \times \cdots \times P_{t-1}) \times P_t \cong P_1 \times \cdots \times P_t$$

which completes the induction. Now take $t = s$ to obtain (4).

Finally, to prove (4) implies (1) use Exercise 1 of Section 5.1 to obtain

$$Z(P_1 \times \cdots \times P_s) \cong Z(P_1) \times \cdots \times Z(P_s).$$

By Exercise 14 in Section 5.1,

$$G/Z(G) = (P_1/Z(P_1)) \times \cdots \times (P_s/Z(P_s)).$$

Thus the hypotheses of (4) also hold for $G/Z(G)$. By Theorem 1, if $P_i \neq 1$ then $Z(P_i) \neq 1$, so if $G \neq 1$, $|G/Z(G)| < |G|$. By induction, $G/Z(G)$ is nilpotent, so by Exercise 6, G is nilpotent. This completes the proof.

Note that the first part of the Fundamental Theorem of Finite Abelian Groups (Theorem 5 in Section 5.2) follows immediately from the above theorem (we shall give another proof later as a consequence of the Chinese Remainder Theorem):

Corollary 4. A finite abelian group is the direct product of its Sylow subgroups.

Next we prove a proposition which will be used later to show that the multiplicative group of a finite field is cyclic (without using the Fundamental Theorem of Finite Abelian Groups).

Proposition 5. If G is a finite group such that for all positive integers n dividing its order, G contains at most n elements x satisfying $x^n = 1$, then G is cyclic.

Proof: Let $|G| = p_1^{\alpha_1} \cdots p_s^{\alpha_s}$ and let P_i be a Sylow p_i-subgroup of G for $i = 1, 2, \ldots, s$. Since $p_i^{\alpha_i} \mid |G|$ and the $p_i^{\alpha_i}$ elements of P_i are solutions of $x^{p_i^{\alpha_i}} = 1$, by hypothesis P_i must contain *all* solutions to this equation in G. It follows that P_i is the unique (hence normal) Sylow p_i-subgroup of G. By Theorem 3, G is the direct product of its Sylow subgroups. By Theorem 1, each P_i possesses a normal subgroup M_i of index p_i. Since $|M_i| = p_i^{\alpha_i-1}$ and G has at most $p_i^{\alpha_i-1}$ solutions to $x^{p_i^{\alpha_i-1}} = 1$, by Lagrange's Theorem (Corollary 9, Section 3.2) M contains all elements x of G satisfying $x^{p_i^{\alpha_i-1}} = 1$. Thus any element of P_i not contained in M_i satisfies $x^{p_i^{\alpha_i}} = 1$ but $x^{p_i^{\alpha_i-1}} \neq 1$, i.e., x is an element of order $p_i^{\alpha_i}$. This proves P_i is cyclic for all i, so G is the direct product of cyclic groups of relatively prime order, hence is cyclic.

The next proposition is called Frattini's Argument. We shall apply it to give another characterization of finite nilpotent groups. It will also be a valuable tool in the next section.

Proposition 6. *(Frattini's Argument)* Let G be a finite group, let H be a normal subgroup of G and let P be a Sylow p-subgroup of H. Then $G = HN_G(P)$ and $|G : H|$ divides $|N_G(P)|$.

Proof: By Corollary 3.15, $HN_G(P)$ is a subgroup of G and $HN_G(P) = N_G(P)H$ since H is a normal subgroup of G. Let $g \in G$. Since $P^g \leq H^g = H$, both P and P^g are Sylow p-subgroups of H. By Sylow's Theorem applied in H, there exists $x \in H$ such that $P^g = P^x$. Thus $gx^{-1} \in N_G(P)$ and so $g \in N_G(P)x$. Since g was an arbitrary element of G, this proves $G = N_G(P)H$.

Apply the Second Isomorphism Theorem to $G = N_G(P)H$ to conclude that

$$|G : H| = |N_G(P) : N_G(P) \cap H|$$

so $|G : H|$ divides $|N_G(P)|$, completing the proof.

Proposition 7. A finite group is nilpotent if and only if every maximal subgroup is normal.

Proof: Let G be a finite nilpotent group and let M be a maximal subgroup of G. As in the proof of Theorem 1, since $M < N_G(M)$ (by Theorem 3(2)) maximality of M forces $N_G(M) = G$, i.e., $M \trianglelefteq G$.

Conversely, assume every maximal subgroup of the finite group G is normal. Let P be a Sylow p-subgroup of G. We prove $P \trianglelefteq G$ and conclude that G is nilpotent by Theorem 3(3). If P is not normal in G let M be a maximal subgroup of G containing $N_G(P)$. By hypothesis, $M \trianglelefteq G$ hence by Frattini's Argument $G = MN_G(P)$. Since $N_G(P) \leq M$ we have $MN_G(P) = M$, a contradiction. This establishes the converse.

Commutators and the Lower Central Series

For the sake of completeness we include the definition of the *lower central series* of a group and state its relation to the upper central series. Since we shall not be using these results in the future, the proofs are left as (straightforward) exercises.

Recall that the commutator of two elements x, y in a group G is defined as

$$[x, y] = x^{-1}y^{-1}xy,$$

and the commutator of two subgroups H and K of G is

$$[H, K] = \langle [h, k] \mid h \in H, \ k \in K \rangle.$$

Basic properties of commutators and the commutator subgroup were established in Section 5.4.

Definition. For any (finite or infinite) group G define the following subgroups inductively:

$$G^0 = G, \qquad G^1 = [G, G] \quad \text{and} \quad G^{i+1} = [G, G^i].$$

The chain of groups

$$G^0 \geq G^1 \geq G^2 \geq \cdots$$

is called the *lower central series of G*. (The term "lower" indicates that $G^i \geq G^{i+1}$.)

As with the upper central series we include in the exercises at the end of this section the verification that G^i is a characteristic subgroup of G for all i. The next theorem shows the relation between the upper and lower central series of a group.

Theorem 8. A group G is nilpotent if and only if $G^n = 1$ for some $n \geq 0$. More precisely, G is nilpotent of class c if and only if c is the smallest nonnegative integer such that $G^c = 1$. If G is nilpotent of class c then

$$G^{c-i} \leq Z_i(G) \quad \text{for all } i \in \{0, 1, \ldots, c\}.$$

Proof: This is proved by a straightforward induction on the length of either the upper or lower central series.

The terms of the upper and lower central series do not necessarily coincide in general although in some groups this does occur.

Remarks:
(1) If G is abelian, we have already seen that $G' = G^1 = 1$ so the lower central series terminates in the identity after one term.
(2) As with the upper central series, for any finite group there must, by order considerations, be an integer n such that

$$G^n = G^{n+1} = G^{n+2} = \cdots.$$

For non-nilpotent groups, G^n is a nontrivial subgroup of G. For example, in Section 5.4 we showed that $S_3' = S_3^1 = A_3$. Since S_3 is not nilpotent, we must have $S_3^2 = A_3$. In fact

$$(123) = [(12), (132)] \in [S_3, S_3^1] = S_3^2.$$

Once two terms in the lower central series are the same, the chain stabilizes at that point i.e., all terms thereafter are equal to these two. Thus $S_3^i = A_3$ for all $i \geq 2$. Note that S_3 is an example where the lower central series has two distinct terms whereas all terms in the upper central series are equal to the identity (in particular, for non-nilpotent groups these series need not have the same length).

Solvable Groups and the Derived Series

Recall that in Section 3.4 a solvable group was defined as one possessing a series:

$$1 = H_0 \trianglelefteq H_1 \trianglelefteq \cdots \trianglelefteq H_s = G$$

such that each factor H_{i+1}/H_i is abelian. We now give another characterization of solvability in terms of a descending series of characteristic subgroups.

Definition. For any group G define the following sequence of subgroups inductively:

$$G^{(0)} = G, \qquad G^{(1)} = [G, G] \quad \text{and} \quad G^{(i+1)} = [G^{(i)}, G^{(i)}] \quad \text{for all } i \geq 1.$$

This series of subgroups is called the *derived* or *commutator* series of G.

The terms of this series are also often written as: $G^{(1)} = G'$, $G^{(2)} = G''$, etc. Again it is left as an exercise to show that each $G^{(i)}$ is characteristic in G for all i.

It is important to note that although $G^{(0)} = G^0$ and $G^{(1)} = G^1$, it is not in general true that $G^{(i)} = G^i$. The difference is that the definition of the $i+1^{\text{st}}$ term in the lower central series is the commutator of the ith term with the *whole* group G whereas the $i+1^{\text{st}}$ term in the derived series is the commutator of the ith term with itself. Hence

$$G^{(i)} \leq G^i \quad \text{for all } i$$

and the containment can be proper. For example, in $G = S_3$ we have already seen that $G^1 = G' = A_3$ and $G^2 = [S_3, A_3] = A_3$, whereas $G^{(2)} = [A_3, A_3] = 1$ (A_3 being abelian).

Theorem 9. A group G is solvable if and only if $G^{(n)} = 1$ for some $n \geq 0$.

Proof: Assume first that G is solvable and so possesses a series

$$1 = H_0 \trianglelefteq H_1 \trianglelefteq \cdots \trianglelefteq H_s = G$$

such that each factor H_{i+1}/H_i is abelian. We prove by induction that $G^{(i)} \leq H_{s-i}$. This is true for $i = 0$, so assume $G^{(i)} \leq H_{s-i}$. Then

$$G^{(i+1)} = [G^{(i)}, G^{(i)}] \leq [H_{s-i}, H_{s-i}].$$

Since H_{s-i}/H_{s-i-1} is abelian, by Proposition 5.7(4), $[H_{s-i}, H_{s-i}] \leq H_{s-i-1}$. Thus $G^{(i+1)} \leq H_{s-i-1}$, which completes the induction. Since $H_0 = 1$ we have $G^{(s)} = 1$.

Conversely, if $G^{(n)} = 1$ for some $n \geq 0$, Proposition 5.7(4) shows that if we take H_i to be $G^{(n-i)}$ then H_i is a normal subgroup of H_{i+1} with abelian quotient, so the derived series itself satisfies the defining condition for solvability of G. This completes the proof.

If G is solvable, the smallest nonnegative n for which $G^{(n)} = 1$ is called the *solvable length* of G. The derived series is a series of shortest length whose successive quotients are abelian and it has the additional property that it consists of subgroups that are characteristic in the *whole* group (as opposed to each just being normal in the *next* in the initial definition of solvability). Its "intrinsic" definition also makes it easier to work with in many instances, as the following proposition (which reproves some results and exercises from Section 3.4) illustrates.

Proposition 10. Let G and K be groups, let H be a subgroup of G and let $\varphi : G \to K$ be a surjective homomorphism.

  (1) $H^{(i)} \leq G^{(i)}$ for all $i \geq 0$. In particular, if G is solvable, then so is H, i.e., subgroups of solvable groups are solvable (and the solvable length of H is less than or equal to the solvable length of G).

(2) $\varphi(G^{(i)}) = K^{(i)}$. In particular, homomorphic images and quotient groups of solvable groups are solvable (of solvable length less than or equal to that of the domain group).

(3) If N is normal in G and both N and G/N are solvable then so is G.

Proof: Part 1 follows from the observation that since $H \leq G$, by definition of commutator subgroups, $[H, H] \leq [G, G]$, i.e., $H^{(1)} \leq G^{(1)}$. Then, by induction,

$$H^{(i)} \leq G^{(i)} \quad \text{for all } i \in \mathbb{Z}^+.$$

In particular, if $G^{(n)} = 1$ for some n, then also $H^{(n)} = 1$. This establishes (1).

To prove (2) note that by definition of commutators,

$$\varphi([x, y]) = [\varphi(x), \varphi(y)]$$

so by induction $\varphi(G^{(i)}) \leq K^{(i)}$. Since φ is surjective, every commutator in K is the image of a commutator in G, hence again by induction we obtain equality for all i. Again, if $G^{(n)} = 1$ for some n then $K^{(n)} = 1$. This proves (2).

Finally, if G/N and N are solvable, of lengths n and m respectively then by (2) applied to the natural projection $\varphi : G \rightarrow G/N$ we obtain

$$\varphi(G^{(n)}) = (G/N)^{(n)} = 1N$$

i.e., $G^{(n)} \leq N$. Thus $G^{(n+m)} = (G^{(n)})^{(m)} \leq N^{(m)} = 1$. Theorem 9 shows that G is solvable, which completes the proof.

Some additional conditions under which finite groups are solvable are the following:

Theorem 11. Let G be a finite group.
  (1) (Burnside) If $|G| = p^a q^b$ for some primes p and q, then G is solvable.
  (2) (Philip Hall) If for every prime p dividing $|G|$ we factor the order of G as $|G| = p^a m$ where $(p, m) = 1$, and G has a subgroup of order m, then G is solvable (i.e., if for all primes p, G has a subgroup whose index equals the order of a Sylow p-subgroup, then G is solvable — such subgroups are called Sylow p-complements).
  (3) (Feit–Thompson) If $|G|$ is odd then G is solvable.
  (4) (Thompson) If for every pair of elements $x, y \in G$, $\langle x, y \rangle$ is a solvable group, then G is solvable.

We shall prove Burnside's Theorem in Chapter 19 and deduce Philip Hall's generalization of it. As mentioned in Section 3.5, the proof of the Feit–Thompson Theorem takes 255 pages. Thompson's Theorem was first proved as a consequence of a 475 page paper (that in turn relies ultimately on the Feit–Thompson Theorem).

A Proof of the Fundamental Theorem of Finite Abelian Groups

We sketch a group-theoretic proof of the result that every finite abelian group is a direct product of cyclic groups (i.e., Parts 1 and 2 of Theorem 5, Section 5.2) — the Classification of Finitely Generated Abelian Groups (Theorem 3, Section 5.2) will be derived as a consequence of a more general theorem in Chapter 12.

By Corollary 4 it suffices to prove that for p a prime, any abelian p-group is a direct product of cyclic groups (the divisibility condition in Theorem 5.5 is trivially achieved by reordering factors). Let A be an abelian p-group. We proceed by induction on $|A|$.

If E is an elementary abelian p-group (i.e., $x^p = 1$ for all $x \in E$), we first prove the following result:

for any $x \in E$, there exists $M \le E$ with $E = M \times \langle x \rangle$.

If $x = 1$, let $M = E$. Otherwise let M be a subgroup of E of maximal order subject to the condition that x not be an element of M. If M is not of index p in E, let $\overline{E} = E/M$. Then $\overline{E}$ is elementary abelian and there exists $\overline{y} \in \overline{E} - \langle \overline{x} \rangle$. Since $\overline{y}$ has order p, we also have $\overline{x} \notin \langle \overline{y} \rangle$. The complete preimage of $\langle \overline{y} \rangle$ in E is a subgroup of E that does not contain x and whose order is larger than the order of M, contrary to the choice of M. This proves $|E : M| = p$, hence

$$E = M\langle x \rangle \quad \text{and} \quad M \cap \langle x \rangle = 1.$$

By the recognition theorem for direct products, Theorem 5.9, $E = M \times \langle x \rangle$, as asserted.

Now let $\varphi : A \to A$ be defined by $\varphi(x) = x^p$ (see Exercise 7, Section 5.2). Then φ is a homomorphism since A is abelian. Denote the kernel of φ by K and denote the image of φ by H. By definition $K = \{x \in A \mid x^p = 1\}$ and H is the subgroup of A consisting of pth powers. Note that both K and A/H are elementary abelian. By the First Isomorphism Theorem

$$|A : H| = |K|.$$

By induction,

$$H = \langle h_1 \rangle \times \cdots \times \langle h_r \rangle$$
$$\cong Z_{p^{\alpha_1}} \times \cdots \times Z_{p^{\alpha_r}} \quad \alpha_i \ge 1, \ i = 1, 2, \dots, r.$$

By definition of φ, there exist elements $g_i \in A$ such that $g_i^p = h_i$, $1 \le i \le r$. Let $A_0 = \langle g_1, \cdots, g_r \rangle$. It is an exercise to see that

(a) $A_0 = \langle g_1 \rangle \times \cdots \times \langle g_r \rangle$,

(b) $A_0/H = \langle g_1 H \rangle \times \cdots \times \langle g_r H \rangle$ is elementary abelian of order p^r, and

(c) $H \cap K = \langle h_1^{p^{\alpha_1 - 1}} \rangle \times \cdots \times \langle h_r^{p^{\alpha_r - 1}} \rangle$ is elementary abelian of order p^r.

If K is contained in H, then $|K| = |K \cap H| = p^r = |A_0 : H|$. In this case by comparing orders we see that $A_0 = A$ and the theorem is proved. Assume therefore that K is not a subgroup of H and use the bar notation to denote passage to the quotient group A/H. Let $x \in K - H$, so $|\overline{x}| = |x| = p$. By the initial remark of the proof applied to the elementary abelian p-group $E = \overline{A}$, there is a subgroup $\overline{M}$ of $\overline{A}$ such that

$$\overline{A} = \overline{M} \times \langle \overline{x} \rangle.$$

If M is the complete preimage in A of $\overline{M}$, then since x has order p and $x \notin M$, we have $\langle x \rangle \cap M = 1$. By the recognition theorem for direct products,

$$A = M \times \langle x \rangle.$$

By induction, M is a direct product of cyclic groups, hence so is A. This completes the proof.

The uniqueness of the decomposition of a finite abelian group into a direct product of cyclic groups (Part 3 of Theorem 5.5) can also be proved by induction using the pth-power map (i.e., using Exercise 7, Section 5.2). This is essentially the procedure we follow in Section 12.1 for the uniqueness part of the proof of the Fundamental Theorem of Finitely Generated Abelian Groups.

EXERCISES

1. Prove that $Z_i(G)$ is a characteristic subgroup of G for all i.

2. Prove Parts 2 and 4 of Theorem 1 for G a finite nilpotent group, not necessarily a p-group.

3. If G is finite prove that G is nilpotent if and only if it has a normal subgroup of each order dividing $|G|$, and is cyclic if and only if it has a unique subgroup of each order dividing $|G|$.

4. Prove that a maximal subgroup of a finite nilpotent group has prime index.

5. Prove Parts 2 and 4 of Theorem 1 for G an infinite nilpotent group.

6. Show that if $G/Z(G)$ is nilpotent then G is nilpotent.

7. Prove that subgroups and quotient groups of nilpotent groups are nilpotent (your proof should work for infinite groups). Give an explicit example of a group G which possesses a normal subgroup H such that both H and G/H are nilpotent but G is not nilpotent.

8. Prove that if p is a prime and P is a non-abelian group of order p^3 then $|Z(P)| = p$ and $P/Z(P) \cong Z_p \times Z_p$.

9. Prove that a finite group G is nilpotent if and only if whenever $a,\ b \in G$ with $(|a|, |b|) = 1$ then $ab = ba$. [Use Part 4 of Theorem 3.]

10. Prove that D_{2n} is nilpotent if and only if n is a power of 2. [Use Exercise 9.]

11. Give another proof of Proposition 5 under the additional assumption that G is abelian by invoking the Fundamental Theorem of Finite Abelian Groups.

12. Find the upper and lower central series for A_4 and S_4.

13. Find the upper and lower central series for A_n and S_n, $n \geq 5$.

14. Prove that G^i is a characteristic subgroup of G for all i.

15. Prove that $Z_i(D_{2^n}) = D_{2^n}^{n-1-i}$.

16. Prove that $\mathbb{Q}$ has no maximal subgroups. [Recall Exercise 21, Section 3.2.]

17. Prove that $G^{(i)}$ is a characteristic subgroup of G for all i.

18. Show that if G'/G'' and G''/G''' are both cyclic then then $G'' = G'''$. [You may assume $G''' = 1$. Then G/G'' acts by conjugation on the cyclic group G''.]

19. Show that there is no group whose commutator subgroup is isomorphic to S_4. [Use the preceding exercise.]

20. Let p be a prime, let P be a p-subgroup of the finite group G, let N be a normal subgroup of G whose order is relatively prime to p and let $\overline{G} = G/N$. Prove the following:
    (a) $N_{\overline{G}}(\overline{P}) = \overline{N_G(P)}$ [Use Frattini's Argument.]
    (b) $C_{\overline{G}}(\overline{P}) = \overline{C_G(P)}$. [Use part (a).]

For any group G the *Frattini subgroup* of G (denoted by $\Phi(G)$) is defined to be the intersection of all the maximal subgroups of G (if G has no maximal subgroups, set $\Phi(G) = G$). The next few exercises deal with this important subgroup.

21. Prove that $\Phi(G)$ is a characteristic subgroup of G.

22. Prove that if $N \trianglelefteq G$ then $\Phi(N) \leq \Phi(G)$. Give an explicit example where this containment does not hold if N is not normal in G.

23. Compute $\Phi(S_3)$, $\Phi(A_4)$, $\Phi(S_4)$, $\Phi(A_5)$ and $\Phi(S_5)$.

24. Say an element x of G is a *nongenerator* if for every proper subgroup H of G, $\langle x, H \rangle$ is also a proper subgroup of G. Prove that $\Phi(G)$ is the set of nongenerators of G (here $|G| > 1$).

25. Let G be a finite group. Prove that $\Phi(G)$ is nilpotent. [Use Frattini's Argument to prove that every Sylow subgroup of $\Phi(G)$ is normal in G.]

26. Let p be a prime, let P be a finite p-group and let $\overline{P} = P/\Phi(P)$.
    (a) Prove that $\overline{P}$ is an elementary abelian p-group. [Show that $P' \leq \Phi(P)$ and that $x^p \in \Phi(P)$ for all $x \in P$.]
    (b) Prove that if N is any normal subgroup of P such that P/N is elementary abelian then $\Phi(P) \leq N$. State this (universal) property in terms of homomorphisms and commutative diagrams.
    (c) Let $\overline{P}$ be elementary abelian of order p^r (by (a)). Deduce from Exercise 24 that if $\overline{x_1}, \overline{x_2}, \ldots, \overline{x_r}$ are any basis for the r-dimensional vector space $\overline{P}$ over $\mathbb{F}_p$ and if x_i is any element of the coset $\overline{x_i}$, then $P = \langle x_1, x_2, \ldots, x_r \rangle$. Show conversely that if $y_1, y_2, \ldots, y_s$ is any set of generators for P, then $s \geq r$ (you may assume that every minimal generating set for an r-dimensional vector space has r elements, i.e., every basis has r elements). Deduce *Burnside's Basis Theorem*: a set $y_1, \ldots, y_s$ is a minimal generating set for P if and only if $\overline{y_1}, \ldots, \overline{y_s}$ is a basis of $\overline{P} = P/\Phi(P)$. Deduce that any minimal generating set for P has r elements.
    (d) Prove that if $P/\Phi(P)$ is cyclic then P is cyclic. Deduce that if P/P' is cyclic then so is P.
    (e) Let σ be any automorphism of P of prime order q with $q \neq p$. Show that if σ fixes the coset $x\Phi(P)$ then σ fixes some element of this coset (note that since $\Phi(P)$ is characteristic in P every automorphism of P induces an automorphism of $P/\Phi(P)$). [Use the observation that σ acts a permutation of order 1 or q on the p^a elements in the coset $x\Phi(P)$.]
    (f) Use parts (e) and (c) to deduce that every nontrivial automorphism of P of order prime to p induces a nontrivial automorphism on $P/\Phi(P)$. Deduce that any group of automorphisms of P which has order prime to p is isomorphic to a subgroup of $\text{Aut}(\overline{P}) = GL_r(\mathbb{F}_p)$.

27. Generalize part (d) of the preceding exercise as follows: let p be a prime, let P be a p-group and let $\overline{P} = P/\Phi(P)$ be elementary abelian of order p^r. Prove that P has exactly $\dfrac{p^r - 1}{p - 1}$ maximal subgroups. [Since every maximal subgroup of P contains $\Phi(P)$, the maximal subgroups of P are, by the Lattice Isomorphism Theorem, in bijective correspondence with the maximal subgroups of the elementary abelian group $\overline{P}$. It therefore suffices to show that the number of maximal subgroups of an elementary abelian p-group of order p^r is as stated above. One way of doing this is to use the result that an abelian group is isomorphic to its dual group (cf. Exercise 14 in Section 5.2) so the number of subgroups of *index* p equals the number of subgroups of *order* p.]

28. Prove that if p is a prime and $P = Z_p \times Z_{p^2}$ then $|\Phi(P)| = p$ and $P/\Phi(P) \cong Z_p \times Z_p$. Deduce that P has $p + 1$ maximal subgroups.

29. Prove that if p is a prime and P is a non-abelian group of order p^3 then $\Phi(P) = Z(P)$ and $P/\Phi(P) \cong Z_p \times Z_p$. Deduce that P has $p + 1$ maximal subgroups.

30. Let p be an odd prime, let $P_1 = Z_p \times Z_{p^2}$ and let P_2 be the non-abelian group of order p^3 which has an element of order p^2. Prove that P_1 and P_2 have the same lattice of subgroups.

31. For any group G a *minimal normal subgroup* is a normal subgroup M of G such that the only normal subgroups of G which are contained in M are 1 and M. Prove that every minimal normal subgroup of a finite solvable group is an elementary abelian p-group for some prime p. [If M is a minimal normal subgroup of G, consider its characteristic subgroups: M' and $\langle x^p \mid x \in M \rangle$.]

32. Prove that every maximal subgroup of a finite solvable group has prime power index. [Let H be a maximal subgroup of G and let M be a minimal normal subgroup of G — cf. the preceding exercise. Apply induction to G/M and consider separately the two cases: $M \le H$ and $M \not\le H$.]

33. Let π be any set of primes. A subgroup H of a finite group is called a *Hall π-subgroup* of G if the only primes dividing $|H|$ are in the set π and $|H|$ is relatively prime to $|G : H|$. (Note that if $\pi = \{p\}$, Hall π-subgroups are the same as Sylow p-subgroups. Hall subgroups were introduced in Exercise 10 of Section 3.3). Prove the following generalization of Sylow's Theorem for solvable groups: if G is a finite solvable group then for every set π of primes, G has a Hall π-subgroup and any two Hall π-subgroups (for the same set π) are conjugate in G. [Fix π and proceed by induction on $|G|$, proving both existence and conjugacy at once. Let M be a minimal normal subgroup of G, so M is a p-group for some prime p. If $p \in \pi$, apply induction to G/M. If $p \notin \pi$, reduce to the case $|G| = p^\alpha n$, where $p^\alpha = |M|$ and n is the order of a Hall π-subgroup of G. In this case let N/M be a minimal normal subgroup of G/M, so N/M is a q-group for some prime $q \ne p$. Let $Q \in Syl_q(N)$. If $Q \trianglelefteq G$ argue as before with Q in place of M. If Q is not normal in G, use Frattini's Argument to show $N_G(Q)$ is a Hall π-subgroup of G and establish conjugacy in this case too.]

The following result shows how to produce normal p-subgroups of some groups on which the elements of order prime to p act faithfully by conjugation. Exercise 26(f) then applies to restrict these actions and give some information about the structure of the group.

34. Let p be a prime dividing the order of the finite solvable group G. Assume G has no nontrivial normal subgroups of order prime to p. Let P be the largest normal p-subgroup of G (cf. Exercise 37, Section 4.5). Note that Exercise 31 above shows that $P \ne 1$. Prove that $C_G(P) \le P$, i.e., $C_G(P) = Z(P)$. [Let $N = C_G(P)$ and use the preceding exercise to show $N = Z(P) \times H$ for some Hall π-subgroup H of N — here π is the set of all prime divisors of $|N|$ except for p. Show $H \trianglelefteq G$ to obtain the desired conclusion: $H = 1$.]

35. Prove that if G is a finite group in which every proper subgroup is nilpotent, then G is solvable. [Show that a minimal counterexample is simple. Let M and N be distinct maximal subgroups chosen with $|M \cap N|$ as large as possible and apply Part 2 of Theorem 3 to show that $M \cap N = 1$. Now apply the methods of Exercise 53 in Section 4.5.]

36. Let p be a prime, let V be a nonzero finite dimensional vector space over the field of p elements and let φ be an element of $GL(V)$ of order a power of p (i.e., V is a nontrivial elementary abelian p-group and φ is an automorphism of V of p-power order). Prove that there is some nonzero element $v \in V$ such that $\varphi(v) = v$, i.e., φ has a nonzero fixed point on V.

37. Let V be a finite dimensional vector space over the field of 2 elements and let φ be an element of $GL(V)$ of order 2. (i.e., V is a nontrivial elementary abelian 2-group and φ is an

automorphism of V of order 2). Prove that the map $v \mapsto v + \varphi(v)$ is a homomorphism from V to itself. Show that every element in the image of this map is fixed by φ. Deduce that the subspace of elements of V which are fixed by φ has dimension $\geq \frac{1}{2}$(dimension V). (Note that if G is the semidirect product of V with $\langle \varphi \rangle$, where $V \trianglelefteq G$ and φ acts by conjugation on V by sending each $v \in V$ to $\varphi(v)$, then the fixed points of φ on V are $C_V(\varphi)$ and the above map is simply the commutator map: $v \mapsto [v, \varphi]$. In this terminology the problem is to show that $|C_V(\varphi)|^2 \geq |V|$.)

38. Use the preceding exercise to prove that if P is a 2-group which has a cyclic center and M is a subgroup of index 2 in P, then the center of M has rank ≤ 2. [The group G/M of order 2 acts by conjugation on the $\mathbb{F}_2$ vector space: $\{z \in Z(M) \mid z^2 = 1\}$ and the fixed points of this action are in the center of P.]

6.2 APPLICATIONS IN GROUPS OF MEDIUM ORDER

The purpose of this section is to work through a number of examples which illustrate many of the techniques we have developed. These examples use Sylow's Theorems extensively and demonstrate how they are applied in the study of finite groups. Motivated by the Hölder Program we address primarily the problem of showing that for certain n every group of order n has a proper, nontrivial normal subgroup (i.e., there are no simple groups of order n). In most cases we shall stop once this has been accomplished. However readers should be aware that in the process of achieving this result we shall already have determined a great deal of information about arbitrary groups of given order n for the n that we consider. This information could be built upon to classify groups of these orders (but in general this requires techniques beyond the simple use of semidirect products to construct groups).

Since for p a prime we have already proved that there are no simple p-groups (other than the cyclic group of order p, Z_p) and since the structure of p-groups can be very complicated (recall the table in Section 5.3), we shall not study the structure of p-groups explicitly. Rather, the theory of p-groups developed in the preceding section will be applied to subgroups of groups of non-prime-power order.

Finally, for certain n (e.g., 60, 168, 360, 504,...) there do exist simple groups of order n so, of course, we cannot force every group of these orders to be nonsimple. As in Section 4.5 we can, in certain cases, prove there is a unique simple group of order n and unravel some of its internal structure (Sylow numbers, etc.). We shall study simple groups of order 168 as an additional test case. Thus the Sylow Theorems will be applied in a number of different contexts to show how groups of a given order may be manipulated.

We shall end this section with some comments on the existence problem for groups, particularly for finite simple groups.

For $n < 10000$ there are 60 odd, non-prime-power numbers for which the congruence conditions of Sylow's Theorems do *not* force at least one of the Sylow subgroups to be normal i.e., n_p can be > 1 for all primes $p \mid n$ (recall that n_p denotes the number of Sylow p-subgroups). For example, no numbers of the form pq, where p and q are distinct primes occur in our list by results of Section 4.5. In contrast, for even numbers < 500 there are already 46 candidates for orders of simple groups (the congruence conditions allow many more possibilities). Many of our numerical examples arise from these lists of numbers and we often use odd numbers because the Sylow congruence conditions allow fewer values for n_p. The purpose of these examples is to illustrate the use of the results we have proved. Many of these examples can be dealt with by more advanced techniques (for example, the Feit–Thompson Theorem proves that there are *no* simple groups of odd composite order).

As we saw in the case $n = 30$ in Section 4.5, even though Sylow's Theorem permitted $n_5 = 6$ and $n_3 = 10$, further examination showed that any group of order 30 must have both $n_5 = 1$ and $n_3 = 1$. Thus the congruence part of Sylow's Theorem is a sufficient but by no means necessary condition for normality of a Sylow subgroup. For many n (e.g., $n = 120$) we can prove that there are no simple groups of order n, so there is a nontrivial normal subgroup but this subgroup may not be a Sylow subgroup. For example, S_5 and $SL_2(\mathbb{F}_5)$ both have order 120. The group S_5 has a unique nontrivial proper normal subgroup of order 60 (A_5) and $SL_2(\mathbb{F}_5)$ has a unique nontrivial proper normal subgroup of order 2 ($Z(SL_2(\mathbb{F}_5)) \cong Z_2$), neither of which is a Sylow subgroup. Our techniques for producing normal subgroups must be flexible enough to cover such diverse possibilities. In this section we shall examine Sylow subgroups for different primes dividing n, intersections of Sylow subgroups, normalizers of p-subgroups and many other less obvious subgroups. The elementary methods we outline are by no means exhaustive, even for groups of "medium" order.

Some Techniques

Before listing some techniques for producing normal subgroups in groups of a given ("medium") order we note that in all the problems where one deals with groups of order n, for some specific n, it is first necessary to factor n into prime powers and then to compute the permissible values of n_p, for all primes p dividing n. We emphasize the need to be comfortable computing mod p when carrying out the last step. The techniques we describe may be listed as follows:

(1) Counting elements.
(2) Exploiting subgroups of small index.
(3) Permutation representations.
(4) Playing p-subgroups off against each other for different primes p.
(5) Studying normalizers of intersections of Sylow p-subgroups.

Counting Elements

Let G be a group of order n, let p be a prime dividing n and let $P \in Syl_p(G)$. If $|P| = p$, then every nonidentity element of P has order p and every element of G of order p lies in some conjugate of P. By Lagrange's Theorem distinct conjugates of P intersect in the identity, hence in this case the number of elements of G of order p is $n_p(p - 1)$.

If Sylow p-subgroups for different primes p have prime order and we assume none of these is normal, we can sometimes show that the number of elements of prime order is $> |G|$. This contradiction would show that at least one of the n_p's must be 1 (i.e., some Sylow subgroup is normal in G).

This is the argument we used (in Section 4.5) to prove that there are no simple groups of order 30. For another example, suppose $|G| = 105 = 3 \cdot 5 \cdot 7$. If G were simple, we must have $n_3 = 7, n_5 = 21$ and $n_7 = 15$. Thus

the number of elements of order 3 is $7 \cdot 2$	$=$	14
the number of elements of order 5 is $21 \cdot 4$	$=$	84
the number of elements of order 7 is $15 \cdot 6$	$=$	90
the number of elements of prime order is 188	$>$	$|G|$.

Sometimes counting elements of prime order does not lead to too many elements. However, there may be so few elements remaining that there must be a normal subgroup involving these elements. This was (in essence) the technique used in Section 4.5 to show that in a group of order 12 either $n_2 = 1$ or $n_3 = 1$. This technique works particularly well when G has a Sylow p-subgroup P of order p such that $N_G(P) = P$. For example, let $|G| = 56$. If G were simple, the only possibility for the number of Sylow 7-subgroups is 8, so

$$\text{the number of elements of order 7 is } 8 \cdot 6 = 48.$$

Thus there are $56 - 48 = 8$ elements remaining in G. Since a Sylow 2-subgroup contains 8 elements (none of which have order 7), there can be at most one Sylow 2-subgroup, hence G has a normal Sylow 2-subgroup.

Exploiting Subgroups of Small Index

Recall that the results of Section 4.2 show that if G has a subgroup H of index k, then there is a homomorphism from G into the symmetric group S_k whose kernel is contained in H. If $k > 1$, this kernel is a proper normal subgroup of G and if we are trying to prove that G is not simple, we may, by way of contradiction, assume that this kernel is the identity. Then, by the First Isomorphism Theorem, G is isomorphic to a subgroup of S_k. In particular, the order of G divides $k!$. This argument shows that if k is the smallest integer with $|G|$ dividing $k!$ for a finite simple group G then G contains no proper subgroups of index less than k. This smallest permissible index k should be calculated at the outset of the study of groups of a given order n. In the examples we consider this is usually quite easy: n will often factor as

$$p_1^{\alpha_1} p_2^{\alpha_2} \cdots p_s^{\alpha_s} \quad \text{with} \quad p_1 < p_2 < \cdots < p_s$$

and α_s is usually equal to 1 or 2 in our examples. In this case the minimal index of a proper subgroup will have to be at least p_s (respectively $2p_s$) and this is often its exact value.

For example, there is no simple group of order 3393, because if $n = 3393 = 3^2 \cdot 13 \cdot 29$, then the minimal index of a proper subgroup is 29 (n does not divide 28! because 29 does not divide 28!). However any simple group of order 3393 must have $n_3 = 13$, so for $P \in Syl_3(G)$, $N_G(P)$ has index 13, a contradiction.

Permutation Representations

This method is a refinement of the preceding one. As above, if G is a simple group of order n with a proper subgroup of index k, then G is isomorphic to a subgroup of S_k. We may identify G with this subgroup and so assume $G \leq S_k$. Rather than relying only on Lagrange's Theorem for our contradiction (this was what we did for the preceding technique) we can sometimes show by calculating within S_k that S_k contains no simple subgroup of order n. Two restrictions which may enable one to show such a result are

(1) if G contains an element or subgroup of a particular order, so must S_k, and

(2) if $P \in Syl_p(G)$ and if P is also a Sylow p-subgroup of S_k, then $|N_G(P)|$ must divide $|N_{S_k}(P)|$.

Condition (2) arises frequently when p is a prime, $k = p$ or $p + 1$ and G has a subgroup of index k. In this case p^2 does not divide k!, so Sylow p-subgroups of G are also Sylow p-subgroups of S_k. Since now Sylow p-subgroups of S_k are precisely the groups generated by a p-cycle, and distinct Sylow p-subgroups intersect in the identity,

$$\text{the no. of Sylow } p\text{-subgroups of } S_k = \frac{\text{the no. of } p\text{-cycles}}{\text{the no. of } p\text{-cycles in a Sylow } p\text{-subgroup}}$$

$$= \frac{k \cdot (k-1) \cdots (k-p+1)}{p(p-1)}.$$

This number gives the index in S_k of the normalizer of a Sylow p-subgroup of S_k. Thus for $k = p$ or $p + 1$

$$|N_{S_k}(P)| = p(p-1) \qquad (k = p \text{ or } k = p + 1)$$

(cf. also the corresponding discussion for centralizers of elements in symmetric groups in Section 4.3 and the last exercises in Section 4.3). This proves, under the above hypotheses, that $|N_G(P)|$ must divide $p(p-1)$.

For example, if G were a simple group of order $396 = 2^2 \cdot 3^2 \cdot 11$, we must have $n_{11} = 12$, so if $P \in Syl_{11}(G)$, $|G : N_G(P)| = 12$ and $|N_G(P)| = 33$. Since G has a subgroup of index 12, G is isomorphic to a subgroup of S_{12}. But then (considering G as actually contained in S_{12}) $P \in Syl_{11}(S_{12})$ and $|N_{S_{12}}(P)| = 110$. Since $N_G(P) \le N_{S_{12}}(P)$, this would imply $33 \mid 110$, clearly impossible, so we cannot have a simple group of order 396.

We can sometimes squeeze a little bit more out of this method by working in A_k rather than S_k. This slight improvement helps only occasionally and only for groups of even order. It is based on the following observations (the first of which we have made earlier in the text).

Proposition 12.
    (1) If G has no subgroup of index 2 and $G \le S_k$, then $G \le A_k$.
    (2) If $P \in Syl_p(S_k)$ for some odd prime p, then $P \in Syl_p(A_k)$ and $|N_{A_k}(P)| = \frac{1}{2}|N_{S_k}(P)|$.

Proof: The first assertion follows from the Second Isomorphism Theorem: if G is not contained in A_k, then $A_k < GA_k$ so we must have $GA_k = S_k$. But now

$$2 = |S_k : A_k| = |GA_k : A_k| = |G : G \cap A_k|$$

so G has a subgroup, $G \cap A_k$, of index 2.

To prove (2) note that if $P \in Syl_p(S_k)$, for some odd prime p, by (1) (or order considerations) $P \leq A_k$, hence $P \in Syl_p(A_k)$ as well. By Frattini's Argument (Proposition 6)

$$S_k = N_{S_k}(P)A_k$$

so, in particular, $N_{S_k}(P)$ is not contained in A_k. This forces $N_{S_k}(P) \cap A_k \ (= N_{A_k}(P))$ to be a subgroup of index 2 in $N_{S_k}(P)$.

For example, there is no simple group of order 264. Suppose G were a simple group of order $264 = 2^3 \cdot 3 \cdot 11$. We must have $n_{11} = 12$. As usual, G would be isomorphic to a subgroup of S_{12}. Since G is simple (hence contains no subgroup of index 2), $G \leq A_{12}$. Let $P \in Syl_{11}(G)$. Since $n_{11} = 12 = |G : N_G(P)|$, we have $|N_G(P)| = 22$. As above,

$$|N_{A_{12}}(P)| = \tfrac{1}{2}|N_{S_{12}}(P)| = \tfrac{1}{2}11(11-1) = 55;$$

however, 22 does not divide 55, a contradiction to $N_G(P) \leq N_{A_{12}}(P)$.

Finally, we emphasize that we have only barely touched upon the combinatorial information available from certain permutation representations. Whenever possible in the remaining examples we shall illustrate other applications of this technique.

Playing p-Subgroups Off Against Each Other for Different Primes p

Suppose p and q are distinct primes such that every group of order pq is cyclic. This is equivalent to $p \nmid q - 1$, where $p < q$. If G has a Sylow q-subgroup Q of order q and $p \mid |N_G(Q)|$, applying Cauchy's Theorem in $N_G(Q)$ gives a group P of order p normalizing Q (note that P need not be a Sylow p-subgroup of G). Thus PQ is a group and if PQ is abelian, we obtain

$$PQ \leq N_G(P) \quad \text{and so} \quad q \mid |N_G(P)|.$$

(A symmetric argument applies if Sylow p-subgroups of G have order p and q divides the order of a Sylow p-normalizer). This numerical information alone may be sufficient to force $N_G(P) = G$ (i.e., $P \trianglelefteq G$), or at least to force $N_G(P)$ to have index smaller than the minimal index permitted by permutation representations, giving a contradiction by a preceding technique.

For example, there are no simple groups of order 1785. If there were, let G be a simple group of order $1785 = 3 \cdot 5 \cdot 7 \cdot 17$. The only possible value for n_{17} is 35, so if Q is a Sylow 17-subgroup, $|G : N_G(Q)| = 35$. Thus $|N_G(Q)| = 3 \cdot 17$. Let P be a Sylow 3-subgroup of $N_G(Q)$. The group PQ is abelian since 3 does not divide $17 - 1$, so $Q \leq N_G(P)$ and $17 \mid |N_G(P)|$. In this case $P \in Syl_3(G)$. The permissible values of n_3 are 7, 85 and 595; however, since $17 \mid |N_G(P)|$, we cannot have $17 \mid |G : N_G(P)| = n_3$. Thus $n_3 = 7$. But G has no proper subgroup of index < 17 (the minimal index of a proper subgroup is 17 for this order), a contradiction. Alternatively, if $n_3 = 7$, then $|N_G(P)| = 3 \cdot 5 \cdot 17$, and by Sylow's Theorem applied in $N_G(P)$ we have $Q \trianglelefteq N_G(P)$. This contradicts the fact that $|N_G(Q)| = 3 \cdot 17$.

We can refine this method by not requiring P and Q to be of prime order. Namely, if p and q are distinct primes dividing $|G|$ such that $Q \in Syl_q(G)$ and $p \mid |N_G(Q)|$, let $P \in Syl_p(N_G(Q))$. We can then apply Sylow's Theorems in $N_G(Q)$ to see whether

$P \trianglelefteq N_G(Q)$, and if so, force $N_G(P)$ to be of small index. If P is a Sylow p-subgroup of the whole group G, we can use the congruence part of Sylow's Theorem to put further restrictions on $|N_G(P)|$ (as we did in the preceding example). If P is not a Sylow p-subgroup of G, then by the second part of Sylow's Theorem $P \le P^* \in Syl_p(G)$. In this case since $P < P^*$, Theorem 1(4) shows that $P < N_{P^*}(P)$. Thus $N_G(P)$ (which contains $N_{P^*}(P)$) has order divisible by a larger power of p than divides $|P|$ (as well as being divisible by $|Q|$).

For example, there are no simple groups of order 3675. If there were, let G be a simple group of order $3675 = 3 \cdot 5^2 \cdot 7^2$. The only possibility for n_7 is 15, so for $Q \in Syl_7(G)$, $|G : N_G(Q)| = 15$ and $|N_G(Q)| = 245 = 5 \cdot 7^2$. Let $N = N_G(Q)$ and let $P \in Syl_5(N)$. By the congruence conditions of Sylow's Theorem applied in N we get $P \trianglelefteq N$. Since $|P| = 5$, P is not itself a Sylow 5-subgroup of G so P is contained in some Sylow 5-subgroup P^* of G. Since P is of index 5 in the 5-group P^*, $P \trianglelefteq P^*$ by Theorem 1, that is $P^* \le N_G(P)$. This proves

$$\langle N, P^* \rangle \le N_G(P) \quad \text{so} \quad 7^2 \cdot 5^2 \mid |N_G(P)|.$$

Thus $|G : N_G(P)| \mid 3$, which is impossible since P is not normal and G has no subgroup of index 3.

Studying Normalizers of Intersections of Sylow p-Subgroups

One of the reasons the counting arguments in the first method above do not immediately generalize to Sylow subgroups which are not of prime order is because if $P \in Syl_p(G)$ for some prime p and $|P| = p^a$, $a \ge 2$, then it need not be the case that distinct conjugates of P intersect in the identity subgroup. If distinct conjugates of P *do* intersect in the identity, we can again count to find that the number of elements of p-power order is $n_p(|P| - 1)$.

Suppose, however, there exists $R \in Syl_p(G)$ with $R \ne P$ and $P \cap R \ne 1$. Let $P_0 = P \cap R$. Then $P_0 < P$ and $P_0 < R$, hence by Theorem 1

$$P_0 < N_P(P_0) \quad \text{and} \quad P_0 < N_R(P_0).$$

One can try to use this to prove that the normalizer in G of P_0 is sufficiently large (i.e., of sufficiently small index) to obtain a contradiction by previous methods (note that this normalizer is a proper subgroup since $P_0 \ne 1$).

One special case where this works particularly well is when $|P_0| = p^{a-1}$ i.e., the two Sylow p-subgroups R and P have large intersection. In this case set $N = N_G(P_0)$. Then by the above reasoning (i.e., since P_0 is a maximal subgroup of the p-groups P and R), $P_0 \trianglelefteq P$ and $P_0 \trianglelefteq R$, that is,

$$N \text{ has 2 distinct Sylow } p\text{-subgroups: } P \text{ and } R.$$

In particular, $|N| = p^a k$, where (by Sylow's Theorem) $k \ge p + 1$.

Recapitulating, if Sylow p-subgroups pairwise intersect in the identity, then counting elements of p-power order is possible; otherwise there is some intersection of Sylow p-subgroups whose normalizer is "large." Since for an arbitrary group order one cannot necessarily tell which of these two phenomena occurs, it may be necessary to split the nonsimplicity argument into two (mutually exclusive) cases and derive a contradiction in each. This process is especially amenable when the order of a Sylow p-subgroup is p^2 (for example, this line of reasoning was used to count elements of 2-power order in the proof that a simple group of order 60 is isomorphic to A_5 — Proposition 23, Section 4.5).

Before proceeding with an example we state a lemma which gives a sufficient condition to force a nontrivial Sylow intersection.

Lemma 13. In a finite group G if $n_p \not\equiv 1 \pmod{p^2}$, then there are distinct Sylow p-subgroups P and R of G such that $P \cap R$ is of index p in both P and R (hence is normal in each).

Proof: The argument is an easy refinement of the proof of the congruence part of Sylow's Theorem (cf. the exercises at the end of Section 4.5). Let P act by conjugation on the set $Syl_p(G)$. Let $\mathcal{O}_1, \ldots, \mathcal{O}_s$ be the orbits under this action with $\mathcal{O}_1 = \{P\}$. If p^2 divides $|P : P \cap R|$ for all Sylow p-subgroups R of G different from P, then each $\mathcal{O}_i$ has size divisible by p^2, $i = 2, 3, \ldots, s$. In this case, since n_p is the sum of the lengths of the orbits we would have $n_p = 1 + kp^2$, contrary to assumption. Thus for some $R \in Syl_p(G)$, $|P : P \cap R| = p$.

For example, there are no simple groups of order 1053. If there were, let G be a simple group of order $1053 = 3^4 \cdot 13$ and let $P \in Syl_3(G)$. We must have $n_3 = 13$. But $13 \not\equiv 1 \pmod{3^2}$ so there exist $P, R \in Syl_3(G)$ such that $|P \cap R| = 3^3$. Let $N = N_G(P \cap R)$, so by the above arguments $P, R \leq N$. Thus $3^4 \mid |N|$ and $|N| > 3^4$. The only possibility is $N = G$, i.e., $P \cap R \trianglelefteq G$, a contradiction.

Simple Groups of Order 168

We now show how many of our techniques can be used to unravel the structure of and then classify certain simple groups by classifying the simple groups of order 168. Because there are no nontrivial normal subgroups in simple groups, this process departs from the methods in Section 5.5, but the overall approach typifies methods used in the study of finite simple groups.

We begin by assuming there is a simple group G of order $168 = 2^3 \cdot 3 \cdot 7$. We first work out many of its properties: the number and structure of its Sylow subgroups, the conjugacy classes, etc. All of these calculations are based only on the order and simplicity of G. We use these results to first prove the uniqueness of G; and ultimately we prove the existence of the simple group of order 168.

Because $|G|$ does not divide 6! we have

(1)   *G has no proper subgroup of index less than 7,*

since otherwise the action of G on the cosets of the subgroup would give a (necessarily injective since G is simple) homomorphism from G into some S_n with $n \leq 6$.

The simplicity of G and Sylow's Theorem also immediately imply that

(2) $n_7 = 8$, *so the normalizer of a Sylow 7-subgroup has order 21. In particular, no element of order 2 normalizes a Sylow 7-subgroup and G has no elements of order 14.*

If G had an element of order 21 then the normalizer of a Sylow 3-subgroup of G would have order divisible by 7. Thus n_3 would be relatively prime to 7. Since then $n_3 \mid 8$ we would have $n_3 = 4$ contrary to (1). This proves:

(3)   *G has no elements of order 21.*

By Sylow's Theorem $n_3 = 7$ or 28; we next rule out the former possibility. Assume $n_3 = 7$, let $P \in Syl_3(G)$ and let T be a Sylow 2-subgroup of the group $N_G(P)$ of order 24. Each Sylow 3-subgroup normalizes some Sylow 7-subgroup of G so P normalizes a Sylow 7-subgroup R of G. For every $t \in T$ we also have that $P = tPt^{-1}$ normalizes tRt^{-1}. The subgroup T acts by conjugation on the set of eight Sylow 7-subgroups of G and since no element of order 2 in G normalizes a Sylow 7-subgroup by (2), it follows that T acts transitively, i.e., every Sylow 7-subgroup of G is one of the tRt^{-1}. Hence P normalizes every Sylow 7-subgroup of G, i.e., P is contained in the intersection of the normalizers of all Sylow 7-subgroups. But this intersection is a proper normal subgroup of G, so it must be trivial. This contradiction proves:

(4) $n_3 = 28$ *and the normalizer of a Sylow 3-subgroup has order 6.*

Since $n_2 = 7$ or 21, we have $n_2 \not\equiv 1 \bmod 8$, so by Exercise 21 there is a pair of distinct Sylow 2-subgroups that have nontrivial intersection; over all such pairs let T_1 and T_2 be chosen with $U = T_1 \cap T_2$ of maximal order. We next prove

(5)   *U is a Klein 4-group and* $N_G(U) \cong S_4$.

Let $N = N_G(U)$. Since $|U| = 2$ or 4 and N permutes the nonidentity elements of U by conjugation, a subgroup of order 7 in N would commute with some element of order 2 in U, contradicting (2). It follows that the order of N is not divisible by 7. By Exercise 13, N has more than one Sylow 2-subgroup, hence $|N| = 2^a \cdot 3$, where $a = 2$ or 3. Let $P \in Syl_3(N)$. Since P is a Sylow 3-subgroup of G, by (4) the group $N_N(P)$ has order 3 or 6 (with P as its unique subgroup of order 3). Thus by Sylow's Theorem N must have four Sylow 3-subgroups, and these are permuted transitively by N under conjugation. Since any group of order 12 must have either a normal Sylow 2-subgroup or a normal Sylow 3-subgroup (cf. Section 4.5), $|N| = 24$. Let K be the kernel of N acting by conjugation on its four Sylow 3-subgroups, so K is the intersection of the normalizers of the Sylow 3-subgroups of N. If $K = 1$ then $N \cong S_4$ as asserted; so consider when $K \neq 1$. Since $K \leq N_N(P)$, the group K has order dividing 6, and since P does not normalize another Sylow 3-subgroup, P is not contained in K. It follows that $|K| = 2$. But now N/K is a group of order 12 which is seen to have more than one Sylow 2-subgroup and four Sylow 3-subgroups, contrary to the property of groups of order 12 cited earlier. This proves $N \cong S_4$. Since S_4 has a unique nontrivial normal 2-subgroup, V_4, (5) holds. Since $N \cong S_4$, it follows that N contains a Sylow 2-subgroup of G and also that $N_N(P) \cong S_3$ (so also $N_G(P) \cong S_3$ by (4)). Hence we obtain

(6)   *Sylow 2-subgroups of G are isomorphic to* D_8, *and*

(7)   *the normalizer in G of a Sylow 3-subgroup is isomorphic to* S_3 *and so G has no elements of order 6.*

By (2) and (7), no element of order 2 commutes with an element of odd prime order. If $T \in Syl_2(G)$, then $T \cong D_8$ by (6), so $Z(T) = \langle z \rangle$ where z is an element of order 2. Then $T \leq C_G(z)$ and $|C_G(z)|$ has no odd prime factors by what was just said, so $C_G(z) = T$. Since any element normalizing T would normalize its center, hence commute with z, it follows that Sylow 2-subgroups of G are self-normalizing. This gives

(8) $n_2 = 21$ and $C_G(z) = T$, where $T \in Syl_2(G)$ and $Z(T) = \langle z \rangle$.

Since $|C_G(z)| = 8$, the element z in (8) has 21 conjugates. By (6), G has one conjugacy class of elements of order 4, which by (6) and (8) contains 42 elements. By (2) there are 48 elements of order 7, and by (4) there are 56 elements of order 3. These account for all 167 nonidentity elements of G, and so every element of order 2 must be conjugate to z, i.e.,

(9) G has a unique conjugacy class of elements of order 2.

Continuing with the same notation, let $T \in Syl_2(G)$ with $U \leq T$ and let W be the other Klein 4-group in T. It follows from Sylow's Theorem that U and W are not conjugate in G since they are not conjugate in $N_G(T) = T$ (cf. Exercise 50 in Section 4.5). We argue next that

(10) $N_G(W) \cong S_4$.

To see this let $W = \langle z, w \rangle$ where, as before, $\langle z \rangle = Z(T)$. Since w is conjugate in G to z, $C_G(w) = T_0$ is another Sylow 2-subgroup of G containing W but different from T. Thus $W = T \cap T_0$. Since U was an arbitrary maximal intersection of Sylow 2-subgroups of G, the argument giving (5) implies (10).

We now record results which we have proved or which are easy consequences of (1) to (10).

Proposition 14. If G is a simple group of order 168, then the following hold:
  (1) $n_2 = 21$, $n_3 = 28$ and $n_7 = 8$
  (2) Sylow 2-subgroups of G are dihedral, Sylow 3- and 7-subgroups are cyclic
  (3) G is isomorphic to a subgroup of A_7 and G has no subgroup of index ≤ 6
  (4) the conjugacy classes of G are the following: the identity; two classes of elements of order 7 each of which contains 24 elements (represented by any element of order 7 and its inverse); one class of elements of order 3 containing 56 elements; one class of elements of order 4 containing 42 elements; one class of elements of order 2 containing 21 elements
      (in particular, every element of G has order a power of a prime)
  (5) if $T \in Syl_2(G)$ and U, W are the two Klein 4-groups in T, then U and W are not conjugate in G and $N_G(U) \cong N_G(W) \cong S_4$
  (6) G has precisely three conjugacy classes of maximal subgroups, two of which are isomorphic to S_4 and one of which is isomorphic to the non-abelian group of order 21.

All of the calculations above were predicated on the assumption that there exists a simple group of order 168. The fact that none of these arguments leads to a contradiction does not *prove* the existence of such a group, but rather just gives strong evidence that there *may* be a simple group of this order. We next illustrate how the internal subgroup structure of G gives rise to a geometry on which G acts, and so leads to a proof that a simple group of order 168 is unique, if it exists (which we shall also show).

Continuing the above notation let $U_1, \ldots, U_7$ be the conjugates of U and let $W_1, \ldots, W_7$ be the conjugates of W. Call the U_i *points* and the W_j *lines*. Define an "incidence relation" by specifying that

> *the point U_i is on the line W_j if and only if U_i normalizes W_j.*

Note that U_i normalizes W_j if and only if $U_i W_j \cong D_8$, which in turn occurs if and only if W_j normalizes U_i. In each point or line stabilizer—which is isomorphic to S_4—there is a unique normal 4-group, V, and precisely three other (nonnormal) 4-groups A_1, A_2, A_3. The groups $V A_i$ are the three Sylow 2-subgroups of the S_4. We therefore have:

(11)   *each line contains exactly 3 points and each point lies on exactly 3 lines.*

Since any two nonnormal 4-groups in an S_4 generate the S_4, hence uniquely determine the other two Klein groups in that S_4, we obtain

(12)   *any 2 points on a line uniquely determine the line (and the third point on it).*

Since there are 7 points and 7 lines, elementary counting now shows that

(13)   *each pair of points lies on a unique line, and each pair of lines intersects in a unique point.*

(This configuration of points and lines thus satisfies axioms for what is termed a *projective plane*.) It is now straightforward to show that the incidence geometry is uniquely determined and may be represented by the graph in Figure 1, where points are vertices and lines are the six sides and medians of the triangle together with the inscribed circle—see Exercise 27. This incidence geometry is called the *projective plane of order 2* or the *Fano Plane*, and will be denoted by $\mathcal{F}$. (Generally, a projective plane of "order" N has $N^2 + N + 1$ points, and the same number of lines.) Note that at this point the projective plane $\mathcal{F}$ *does* exist—we have explicitly exhibited points and lines satisfying (11) to (13)—even though the group G is not yet known to exist.

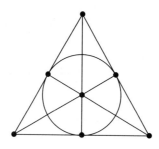

Figure 1

An *automorphism* of this plane is any permutation of points and lines that preserves the incidence relation. For example, any of the six symmetries of the triangle in Figure 1 give automorphisms of $\mathcal{F}$, but we shall see that $\mathcal{F}$ has many more automorphisms than these.

Each $g \in G$ acts by conjugation on the set of points and lines, and this action preserves the incidence relation. Only the identity element in G fixes all points and so via this action the group G would be isomorphic to a subgroup of the group of Aut($\mathcal{F}$), the group of all automorphisms of $\mathcal{F}$.

Any automorphism of $\mathcal{F}$ that fixes two points on a line as well as a third point not on that line is easily seen to fix all points. Thus any automorphism of $\mathcal{F}$ is uniquely determined by its action on any three noncollinear points. Since one easily computes that there are 168 such triples, $\mathcal{F}$ has at most 168 automorphisms. This proves

if the simple group G exists it is unique and $G \cong$ Aut($\mathcal{F}$).

Two steps in the classification process yet remain: to prove that $\mathcal{F}$ does have 168 automorphisms and to prove Aut($\mathcal{F}$) is indeed a simple group. Although one can do these graph-theoretically, we adopt an approach following ideas from the theory of "algebraic groups." Let V be a 3-dimensional vector space over the field of 2 elements, $\mathbb{F}_2$, so V is the elementary abelian 2-group $Z_2 \times Z_2 \times Z_2$ of order 8. By Proposition 17 in Section 4.4, Aut(V) $= GL(V) \cong GL_3(\mathbb{F}_2)$ has order 168. Call the seven 1-dimensional subspaces (i.e., the nontrivial cyclic subgroups) of V *points*, call the seven 2-dimensional subspaces (i.e., the subgroups of order 4) *lines*, and say the point p is *incident to* the line L if $p \subset L$. Then the points and lines are easily seen to satisfy the same axioms (11) to (13) above, hence to represent the Fano Plane. Since $GL(V)$ acts faithfully on these points and lines preserving incidence, Aut($\mathcal{F}$) has order at least 168. In light of the established upper bound for |Aut($\mathcal{F}$)| this proves

Aut($\mathcal{F}$) $\cong GL(V) \cong GL_3(\mathbb{F}_2)$ *and* Aut($\mathcal{F}$) *has order 168.*

Finally we prove that $GL(V)$ is a simple group. By way of contradiction assume H is a proper nontrivial normal subgroup of $GL(V)$. Let Ω be the 7 points and let N be the stabilizer in $GL(V)$ of some point in Ω. Since $GL(V)$ acts transitively on Ω, N has index 7. Since the intersection of all conjugates of N fixes all points, this intersection is the identity. Thus $H \not\leq N$, and so $GL(V) = HN$. Since $|H : H \cap N| = |HN : N|$ we have $7 \mid |H|$. Since $GL(V)$ is isomorphic to a subgroup of S_7 and since Sylow 7-subgroups of S_7 have normalizers of order 42, $GL(V)$ does not have a normal Sylow 7-subgroup, so by Sylow's Theorem $n_7(GL(V)) = 8$. A normal Sylow 7-subgroup of H would be characteristic in H, hence normal in $GL(V)$, so also H does not have a unique Sylow 7-subgroup. Since $n_7(H) \equiv 1 \bmod 7$ and $n_7(H) \leq n_7(GL(V)) = 8$ we must have $n_7(H) = 8$. This implies $|H|$ is divisible by 8, so $56 \mid |H|$, and since H is proper we must have $|H| = 56$. By usual counting arguments (cf. Exercise 7(b) of Section 5.5) H has a normal, hence characteristic, Sylow 2-subgroup, which is therefore normal in $GL(V)$. But then $GL(V)$ would have a unique Sylow 2-subgroup. Since the set of upper triangular matrices and the set of lower triangular matrices are two subgroups of $GL_3(\mathbb{F}_2)$ each of order 8, we have a contradiction. In summary we have now proven the following theorem.

Theorem 15. Up to isomorphism there is a unique simple group of order 168, $GL_3(\mathbb{F}_2)$, which is also the automorphism group of the projective plane $\mathcal{F}$.

Note that we might just as well have called the W_j points and the U_i lines. This "duality" between points and lines together with the uniqueness of a simple group of order 168 may be used to prove the existence of an outer automorphism of G that interchanges points and lines i.e., conjugates U to W.

Many families of finite simple groups can be classified by analogous methods. In more general settings geometric structures known as *buildings* play the role of the projective plane (which is a special case of a building of type $\mathcal{A}_2$). In this context the subgroups $N_G(U)$ and $N_G(W)$ are *parabolic subgroups* of G, and U, W are their *unipotent radicals* respectively. In particular, all the simple linear groups (cf. Section 3.4) are characterized by the structure and intersections of their parabolic subgroups, or equivalently, by their action on an associated building.

Remarks on the Existence Problem for Groups

As in other areas of mathematics (such as the theory of differential equations) one may hypothesize the existence of a mathematical system (e.g., solution to an equation) and derive a great deal of information about this proposed system. In general, if after considerable effort no contradiction is reached based on the initial hypothesis one begins to suspect that there actually is a system which does satisfy the conditions hypothesized. However, no amount of consistent data will *prove* existence. Suppose we carried out an analysis of a hypothetical simple group G of order $3^3 \cdot 7 \cdot 13 \cdot 409$ analogous to our analysis of a simple group of order 168 (which we showed to exist). After a certain amount of effort we could show that there are unique possible Sylow numbers:

$$n_3 = 7 \cdot 409 \qquad n_7 = 3^2 \cdot 13 \cdot 409 \qquad n_{13} = 3^2 \cdot 7 \cdot 409 \qquad n_{409} = 3^2 \cdot 7 \cdot 13.$$

We could further show that such a G would have no elements of order pq, p and q distinct primes, no elements of order 9, and that distinct Sylow subgroups would intersect in the identity. We could then count the elements in Sylow p-subgroups for all primes p and we would find that these would total to exactly $|G|$. At this point we would have the complete subgroup structure and class equation for G. We might then guess that there *is* a simple group of this order, but the Feit–Thompson Theorem asserts that there are *no* simple groups of odd composite order. (Note, however, that the configuration for a possible simple group of order $3^3 \cdot 7 \cdot 13 \cdot 409$ is among the cases that must be dealt with in the *proof* of the Feit-Thompson Theorem, so quoting this result in this instance is actually circular. We prove no simple group of this order exists in Section 19.3; see also Exercise 29.) The point is that even though we have as much data in this case as we had in the order 168 situation (i.e., Proposition 14), we cannot prove existence without some new techniques.

When we are dealing with nonsimple groups we have at least one method of building larger groups from smaller ones: semidirect products. Even though this method is fairly restrictive it conveys the notion that nonsimple groups may be built up from smaller groups in some constructive fashion. This process breaks down completely for simple groups; and so this demarcation of techniques reinforces our appreciation for the Hölder

Program: determining the simple groups, and finding how these groups are put together to form larger groups.

The study of simple groups, as illustrated in the preceding discussion of groups of order 168, uses many of the same tools as the study of nonsimple groups (to unravel their subgroup structures, etc.) but also requires other techniques for their construction. As we mentioned at the end of that discussion, these often involve algebraic or geometric methods which construct simple groups as automorphisms of mathematical structures that have intrinsic interest, and thereby link group theory to other areas of mathematics and science in fascinating ways. Thus while we have come a long way in the analysis of finite groups, there are a number of different areas in this branch of mathematics on which we have just touched.

The analysis of infinite groups generally involves quite different methods, and in the next section we introduce some of these.

EXERCISES

Counting elements:

1. Prove that for fixed $P \in Syl_p(G)$ if $P \cap R = 1$ for all $R \in Syl_p(G) - \{P\}$, then $P_1 \cap P_2 = 1$ whenever P_1 and P_2 are distinct Sylow p-subgroups of G. Deduce in this case that the number of nonidentity elements of p-power order in G is $(|P| - 1)|G : N_G(P)|$.

2. In the group $S_3 \times S_3$ exhibit a pair of Sylow 2-subgroups that intersect in the identity and exhibit another pair that intersect in a group of order 2.

3. Prove that if $|G| = 380$ then G is not simple. [Just count elements of odd prime order.]

4. Prove that there are no simple groups of order 80, 351, 3875 or 5313.

5. Let G be a solvable group of order pm, where p is a prime not dividing m, and let $P \in Syl_p(G)$. If $N_G(P) = P$, prove that G has a normal subgroup of order m. Where was the solvability of G needed in the proof? (This result is true for nonsolvable groups as well — it is a special case of *Burnside's N/C-Theorem.*)

Exploiting subgroups of small index:

6. Prove that there are no simple groups of order 2205, 4125, 5103, 6545 or 6435.

Permutation representations:

7. Prove that there are no simple groups of order 1755 or 5265. [Use Sylow 3-subgroups to show $G \leq S_{13}$ and look at the normalizer of a Sylow 13-subgroup.]

8. Prove that there are no simple groups of order 792 or 918.

9. Prove that there are no simple groups of order 336.

Playing p-subgroups off against each other:

10. Prove that there are no simple groups of order 4095, 4389, 5313 or 6669.

11. Prove that there are no simple groups of order 4851 or 5145.

12. Prove that there are no simple groups of order 9555. [Let $Q \in Syl_{13}(G)$ and let $P \in Syl_7(N_G(Q))$. Argue that $Q \trianglelefteq N_G(P)$ — why is this a contradiction?]

Normalizers of Sylow intersections:

13. Let G be a group with more than one Sylow p-subgroup. Over all pairs of distinct Sylow p-subgroups let P and Q be chosen so that $|P \cap Q|$ is maximal. Show that $N_G(P \cap Q)$

has more than one Sylow p-subgroup and that any two distinct Sylow p-subgroups of $N_G(P \cap Q)$ intersect in the subgroup $P \cap Q$. (Thus $|N_G(P \cap Q)|$ is divisible by $p \cdot |P \cap Q|$ and by some prime other than p. Note that Sylow p-subgroups of $N_G(P \cap Q)$ need not be Sylow in G.)

14. Prove that there are no simple groups of order 144, 525, 2025 or 3159.

General exercises:

15. Classify groups of order 105.

16. Prove that there are no non-abelian simple groups of odd order < 10000.

17. **(a)** Prove that there is no simple group of order 420.
    (b) Prove that there are no simple groups of even order < 500 except for orders 2, 60, 168 and 360.

18. Prove that if G is a group of order 36 then G has either a normal Sylow 2-subgroup or a normal Sylow 3-subgroup.

19. Show that a group of order 12 with no subgroup of order 6 is isomorphic to A_4.

20. Show that a group of order 24 with no element of order 6 is isomorphic to S_4.

21. Generalize Lemma 13 by proving that if $n_p \not\equiv 1 \pmod{p^k}$ then there are distinct Sylow p-subgroups P and R of G such that $P \cap R$ is of index $\leq p^{k-1}$ in both P and R.

22. Suppose over all pairs of distinct Sylow p-subgroups of G, P and R are chosen with $|P \cap R|$ maximal. Prove that $N_G(P \cap R)$ is not a p-group.

23. Let A and B be normal subsets of a Sylow p-subgroup P of G. Prove that if A and B are conjugate in G then they are conjugate in $N_G(P)$.

24. Let G be a group of order pqr where p, q and r are primes with $p < q < r$. Prove that a Sylow r-subgroup of G is normal.

25. Let G be a simple group of order p^2qr where p, q and r are primes. Prove that $|G| = 60$.

26. Prove or construct a counterexample to the assertion: if G is a group of order 168 with more than one Sylow 7-subgroup then G is simple.

27. Show that if $\mathcal{F}$ is any set of points and lines satisfying properties (11) to (13) in the subsection on simple groups of order 168 then the graph of incidences for $\mathcal{F}$ is uniquely determined and is the same as Figure 1 (up to relabeling points and lines). [Take a line and any point not on this line. Depict the line as the base of an equilateral triangle and the point as the vertex of this triangle not on the base. Use the axioms to show that the incidences of the remaining points and lines are then uniquely determined as in Figure 1.]

28. Let G be a simple group of order $3^3 \cdot 7 \cdot 13 \cdot 409$. Compute all permissible values of n_p for each $p \in \{3, 7, 13, 409\}$ and reduce to the case where there is a unique possible value for each n_p.

29. Given the information on the Sylow numbers for a hypothetical simple group of order $3^3 \cdot 7 \cdot 13 \cdot 409$, prove that there is no such group. [Work with the permutation representation of degree 819.]

30. Suppose G is a simple group of order 720. Find as many properties of G as you can (Sylow numbers, isomorphism type of Sylow subgroups, conjugacy classes, etc.). Is there such a group?

6.3 A WORD ON FREE GROUPS

In this section we introduce the basic theory of so-called free groups. This will enable us to make precise the notions of generators and relations which were used in earlier chapters. The results of this section rely only on the basic theory of homomorphisms.

The basic idea of a free group $F(S)$ generated by a set S is that there are no relations satisfied by any of the elements in S (S is "free" of relations). For example, if S is the set $\{a, b\}$ then the elements of the free group on the two generators a and b are of the form a, aa, ab, $abab$, bab, etc., called *words* in a and b, together with the inverses of these elements, and all these elements are considered distinct. If we group like terms together, then we obtain elements of the familiar form a, b^{-3}, $aba^{-1}b^2$ etc. Such elements are multiplied by concatenating their words (for example, the product of aba and $b^{-1}a^3b$ would simply be $abab^{-1}a^3b$). It is natural at the outset (even before we know S is contained in some group) to simply *define* $F(S)$ to be the set of all words in S, where two such expressions are multiplied in $F(S)$ by concatenating them. Although in essence this is what we do, it is necessary to be more formal in order to prove that this concatenation operation is well defined and associative. After all, even the familiar notation a^n for the product $a \cdot a \cdots a$ (n terms) is permissible only because we know that this product is independent of the way it is bracketed (cf. the generalized associative law in Section 1.1). The formal construction of $F(S)$ is carried out below for an arbitrary set S.

One important property reflecting the fact that there are no relations that must be satisfied by the generators in S is that any *map* from the *set S* to a group G can be uniquely extended to a *homomorphism* from the *group $F(S)$* to G (basically since we have specified where the generators must go and the images of all the other elements are uniquely determined by the homomorphism property — the fact that there are no relations to worry about means that we can specify the images of the generators *arbitrarily*). This is frequently referred to as the *universal* property of the free group and in fact characterizes the group $F(S)$.

The notion of "freeness" occurs in many algebraic systems and it may already be familiar (using a different terminology) from elementary vector space theory. When the algebraic systems are vector spaces, $F(S)$ is simply the vector space which has S as a basis. Every vector in this space is a unique linear combination of the elements of S (the analogue of a "word"). Any set map from the basis S to another vector space V extends uniquely to a linear transformation (i.e., vector space homomorphism) from $F(S)$ to V.

Before beginning the construction of $F(S)$ we mention that one often sees the universal property described in the language of commutative diagrams. In this form it reads (for groups) as follows: given any set map φ from the set S to a group G there is a unique homomorphism $\Phi : F(S) \to G$ such that $\Phi|_S = \varphi$ i.e., such that the following diagram commutes:

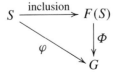

As mentioned above, the only difficulty with the construction of $F(S)$ is the verification that the concatenation operation on the words in $F(S)$ is well defined and associative. To prove the associative property for multiplication of words we return to the most basic level where all the exponents in the words of S are ± 1.

We first introduce inverses for elements of S and an identity.

Let S^{-1} be any set disjoint from S such that there is a bijection from S to S^{-1}. For each $s \in S$ denote its corresponding element in S^{-1} by s^{-1} and similarly for each $t \in S^{-1}$ let the corresponding element of S be denoted by t^{-1} (so $(s^{-1})^{-1} = s$). Take a singleton set not contained in $S \cup S^{-1}$ and call it $\{1\}$. Let $1^{-1} = 1$ and for any $x \in S \cup S^{-1} \cup \{1\}$ let $x^1 = x$.

Next we describe the elements of the free group on the set S. A *word* on S is by definition a sequence

$$(s_1, s_2, s_3, \dots) \quad \text{where } s_i \in S \cup S^{-1} \cup \{1\} \text{ and } s_i = 1 \text{ for all } i \text{ sufficiently large}$$

(that is, for each sequence there is an N such that $s_i = 1$ for all $i \geq N$). Thus we can think of a word as a finite product of elements of S and their inverses (where repetitions are allowed). Next, in order to assure uniqueness of expressions we consider only words which have no obvious "cancellations" between adjacent terms (such as $baa^{-1}b = bb$). The word $(s_1, s_2, s_3, \dots)$ is said to be *reduced* if

(1) $s_{i+1} \neq s_i^{-1}$ for all i with $s_i \neq 1$, and
(2) if $s_k = 1$ for some k, then $s_i = 1$ for all $i \geq k$.

The reduced word $(1, 1, 1, \dots)$ is called the *empty word* and is denoted by 1. We now simplify the notation by writing the reduced word $(s_1^{\epsilon_1}, s_2^{\epsilon_2}, \dots, s_n^{\epsilon_n}, 1, 1, 1, \dots)$, $s_i \in S$, $\epsilon_i = \pm 1$, as $s_1^{\epsilon_1} s_2^{\epsilon_2} \dots s_n^{\epsilon_n}$. Note that by definition, reduced words $r_1^{\delta_1} r_2^{\delta_2} \dots r_m^{\delta_m}$ and $s_1^{\epsilon_1} s_2^{\epsilon_2} \dots s_n^{\epsilon_n}$ are equal if and only if $n = m$, $r_i = s_i$ and $\delta_i = \epsilon_i$, $1 \leq i \leq n$. Let $F(S)$ be the set of reduced words on S and embed S into $F(S)$ by

$$s \mapsto (s, 1, 1, 1, \dots).$$

Under this set injection we identify S with its image and henceforth consider S as a subset of $F(S)$. Note that if $S = \emptyset$, $F(S) = \{1\}$.

We are now in a position to introduce the binary operation on $F(S)$. The principal technical difficulty is to ensure that the product of two reduced words is again a *reduced* word. Although the definition appears to be complicated it is simply the formal rule for "successive cancellation" of juxtaposed terms which are inverses of each other (e.g., $ab^{-1}a$ times $a^{-1}ba$ should reduce to aa). Let $r_1^{\delta_1} r_2^{\delta_2} \dots r_m^{\delta_m}$ and $s_1^{\epsilon_1} s_2^{\epsilon_2} \dots s_n^{\epsilon_n}$ be reduced words and assume first that $m \leq n$. Let k be the smallest integer in the range $1 \leq k \leq m+1$ such that $s_k^{\epsilon_k} \neq r_{m-k+1}^{-\delta_{m-k+1}}$. Then the product of these reduced words is defined to be:

$$(r_1^{\delta_1} r_2^{\delta_2} \dots r_m^{\delta_m})(s_1^{\epsilon_1} s_2^{\epsilon_2} \dots s_n^{\epsilon_n}) = \begin{cases} r_1^{\delta_1} \dots r_{m-k+1}^{\delta_{m-k+1}} s_k^{\epsilon_k} \dots s_n^{\epsilon_n}, & \text{if } k \leq m \\ s_{m+1}^{\epsilon_{m+1}} \dots s_n^{\epsilon_n}, & \text{if } k = m+1 \leq n \\ 1, & \text{if } k = m+1 \text{ and } m = n. \end{cases}$$

The product is defined similarly when $m \geq n$, so in either case it results in a reduced word.

Theorem 16. $F(S)$ is a group under the binary operation defined above.

Proof: One easily checks that 1 is an identity and that the inverse of the reduced word $s_1^{\epsilon_1} s_2^{\epsilon_2} \ldots s_n^{\epsilon_n}$ is the reduced word $s_n^{-\epsilon_n} s_{n-1}^{-\epsilon_{n-1}} \ldots s_1^{-\epsilon_1}$. The difficult part of the proof is the verification of the associative law. This can be done by induction on the "length" of the words involved and considering various cases or one can proceed as follows: For each $s \in S \cup S^{-1} \cup \{1\}$ define $\sigma_s : F(S) \to F(S)$ by

$$
\sigma_s(s_1^{\epsilon_1} s_2^{\epsilon_2} \ldots s_n^{\epsilon_n}) = \begin{cases} s \cdot s_1^{\epsilon_1} s_2^{\epsilon_2} \ldots s_n^{\epsilon_n}, & \text{if } s_1^{\epsilon_1} \neq s^{-1} \\ s_2^{\epsilon_2} s_3^{\epsilon_3} \ldots s_n^{\epsilon_n}, & \text{if } s_1^{\epsilon_1} = s^{-1}. \end{cases}
$$

Since $\sigma_{s^{-1}} \circ \sigma_s$ is the identity map of $F(S) \to F(S)$, σ_s is a permutation of $F(S)$. Let $A(F)$ be the subgroup of the symmetric group on the set $F(S)$ which is generated by $\{\sigma_s \mid s \in S\}$. It is easy to see that the map

$$
s_1^{\epsilon_1} s_2^{\epsilon_2} \ldots s_n^{\epsilon_n} \mapsto \sigma_{s_1}^{\epsilon_1} \circ \sigma_{s_2}^{\epsilon_2} \circ \ldots \circ \sigma_{s_n}^{\epsilon_n}
$$

is a (set) bijection between $F(S)$ and $A(S)$ which respects their binary operations. Since $A(S)$ is a group, hence associative, so is $F(S)$.

The universal property of free groups now follows easily.

Theorem 17. Let G be a group, S a set and $\varphi : S \to G$ a set map. Then there is a unique group homomorphism $\Phi : F(S) \to G$ such that the following diagram commutes:

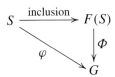

Proof: Such a map Φ must satisfy $\Phi(s_1^{\epsilon_1} s_2^{\epsilon_2} \ldots s_n^{\epsilon_n}) = \varphi(s_1)^{\epsilon_1} \varphi(s_2)^{\epsilon_2} \ldots \varphi(s_n)^{\epsilon_n}$ if it is to be a homomorphism (which proves uniqueness), and it is straightforward to check that this map is in fact a homomorphism (which proves existence).

Corollary 18. $F(S)$ is unique up to a unique isomorphism which is the identity map on the set S.

Proof: This follows from the universal property. Suppose $F(S)$ and $F'(S)$ are two free groups generated by S. Since S is contained in both $F(S)$ and $F'(S)$, we have natural injections $S \hookrightarrow F'(S)$ and $S \hookrightarrow F(S)$. By the universal property in the theorem, it follows that we have unique associated group homomorphisms $\Phi : F(S) \to F'(S)$ and $\Phi' : F'(S) \to F(S)$ which are both the identity on S. The composite $\Phi'\Phi$ is a homomorphism from $F(S)$ to $F(S)$ which is the identity on S, so by the uniqueness statement in the theorem, it must be the identity map. Similarly $\Phi\Phi'$ is the identity, so Φ is an isomorphism (with inverse Φ'), which proves the corollary.

Definition. The group $F(S)$ is called the *free group* on the set S. A group F is a *free group* if there is some set S such that $F = F(S)$ — in this case we call S a set of *free generators* (or a *free basis*) of F. The cardinality of S is called the *rank* of the free group.

One can now simplify expressions in a free group by using exponential notation, so we write $a^3 b^{-2}$ instead of the formal reduced word $aaab^{-1}b^{-1}$. Expressions like aba, however, cannot be simplified in the free group on $\{a, b\}$. We mention one important theorem in this area.

Theorem 19. (Schreier) Subgroups of a free group are free.

This is not trivial to prove and we do not include a proof. There is a nice proof of this result using covering spaces (cf. *Trees* by J.-P. Serre, Springer-Verlag, 1980).

Presentations

Let G be any group. Then G is a homomorphic image of a free group: take $S = G$ and φ as the identity map from G to G; then Theorem 16 produces a (surjective) homomorphism from $F(G)$ onto G. More generally, if S is any subset of G such that $G = \langle S \rangle$, then again there is a unique surjective homomorphism from $F(S)$ onto G which is the identity on S. (Note that we can now independently formulate the notion that a subset *generates* a group by noting that $G = \langle S \rangle$ if and only if the map $\pi : F(S) \to G$ which extends the identity map of S to G is surjective.)

Definition. Let S be a subset of a group G such that $G = \langle S \rangle$.

(1) A *presentation* for G is a pair (S, R), where R is a set of words in $F(S)$ such that the normal closure of $\langle R \rangle$ in $F(S)$ (the smallest normal subgroup containing $\langle R \rangle$) equals the kernel of the homomorphism $\pi : F(S) \to G$ (where π extends the identity map from S to S). The elements of S are called *generators* and those of R are called *relations* of G.

(2) We say G is *finitely generated* if there is a presentation (S, R) such that S is a finite set and we say G is *finitely presented* if there is a presentation (S, R) with both S and R finite sets.

Note that if (S, R) is a presentation, the kernel of the map $F(S) \to G$ is not $\langle R \rangle$ itself but rather the (much larger) group generated by R and *all conjugates* of elements in R. Note that even for a fixed set S a group will have many different presentations (we can always throw redundant relations into R, for example). If G is finitely presented with $S = \{s_1, s_2, \ldots, s_n\}$ and $R = \{w_1, w_2, \ldots, w_k\}$, we write (as we have in preceding chapters):

$$G = \langle s_1, s_2, \ldots, s_n \mid w_1 = w_2 = \cdots = w_k = 1 \rangle$$

and if w is the word $w_1 w_2^{-1}$, we shall write $w_1 = w_2$ instead of $w = 1$.

Examples

(1) Every finite group is finitely presented. To see this let $G = \{g_1, \ldots, g_n\}$ be a finite group. Let $S = G$ and let $\pi : F(S) \to G$ be the homomorphism extending the identity map of S. Let R_0 be the set of words $g_i g_j g_k^{-1}$, where $i, j = 1, \ldots, n$ and $g_i g_j = g_k$ in G. Clearly $R_0 \le \ker \pi$. If N is the normal closure of R_0 in $F(S)$ and $\widetilde{G} = F(S)/N$, then G is a homomorphic image of $\widetilde{G}$ (i.e., π factors through N). Moreover, the set of elements $\{\widetilde{g}_i \mid i = 1, \ldots, n\}$ is closed under multiplication. Since this set generates $\widetilde{G}$, it must equal $\widetilde{G}$. Thus $|\widetilde{G}| = |G|$ and so $N = \ker \pi$ and (S, R_0) is a presentation of G.

This illustrates a sufficient condition for (S, R) to be a presentation for a given finite group G:

   (i) S must be a generating set for G, and

   (ii) any group generated by S satisfying the relations in R must have order $\le |G|$.

(2) Abelian groups can be presented easily. For instance

$$\mathbb{Z} \cong F(\{a\}) = \langle\, a\,\rangle,$$

$$\mathbb{Z} \times \mathbb{Z} \cong \langle\, a, b \mid [a, b] = 1\,\rangle,$$

$$Z_n \times Z_m \cong \langle\, a, b \mid a^n = b^m = [a, b] = 1\,\rangle.$$

(Recall $[a, b] = a^{-1}b^{-1}ab$).

(3) Some familiar non-abelian groups introduced in earlier chapters have simple presentations:

$$D_{2n} = \langle\, r, s \mid r^n = s^2 = 1,\ s^{-1}rs = r^{-1}\,\rangle$$

$$Q_8 = \langle\, i, j \mid i^4 = 1,\ j^2 = i^2,\ j^{-1}ij = i^{-1}\,\rangle.$$

To check, for example, the presentation for D_{2n} note that the relations in the presentation $\langle\, r, s \mid r^n = s^2 = 1,\ s^{-1}rs = r^{-1}\,\rangle$ imply that this group has a normal subgroup (generated by r) of order $\le n$ whose quotient is generated by s (which has order ≤ 2). Thus any group with these generators and relations has order at most $2n$. Since we already know of the existence of the group D_{2n} of order $2n$ satisfying these conditions, the abstract presentation must equal D_{2n}.

(4) As mentioned in Section 1.2, in general it is extremely difficult even to determine if a given set of generators and relations is or is not the identity group (let alone determine whether it is some other nontrivial finite group). For example, in the following two presentations the first group is an *infinite* group and the second is the *identity* group (cf. *Trees*, Chapter 1):

$$\langle\, x_1, x_2, x_3, x_4 \mid x_2 x_1 x_2^{-1} = x_1^2,\ x_3 x_2 x_3^{-1} = x_2^2,\ x_4 x_3 x_4^{-1} = x_3^2,\ x_1 x_4 x_1^{-1} = x_4^2\,\rangle$$

$$\langle\, x_1, x_2, x_3, \mid x_2 x_1 x_2^{-1} = x_1^2,\ x_3 x_2 x_3^{-1} = x_2^2,\ x_1 x_3 x_1^{-1} = x_3^2\,\rangle.$$

(5) It is easy to see that S_n is generated by the transpositions $(1\,2), (2\,3), \ldots, (n-1\,n)$, and that these satisfy the relations

$$((i\,i+1)(i+1\,i+2))^3 = 1 \quad \text{and} \quad [(i\,i+1), (j\,j+1)] = 1, \text{ whenever } |i - j| \ge 2$$

(here $|i - j|$ denotes the absolute value of the integer $i - j$). One can prove by induction on n that these form a presentation of S_n:

$$S_n \cong \langle\, t_1, \ldots, t_{n-1} \mid t_i^2 = 1,\ (t_i t_{i+1})^3 = 1, \text{ and } [t_i, t_j] = 1$$
$$\text{whenever } |i - j| \ge 2,\ 1 \le i, j \le n - 1\,\rangle.$$

As mentioned in Section 1.6 we can use presentations of a group to find homomorphisms between groups or to find automorphisms of a group. We did this in classifying groups of order 6, for example, when we proved that any non-abelian group of order 6 was generated by an element of order 3 and an element of order 2 inverting it; thus there is a homomorphism from S_3 onto any non-abelian group of order 6 (hence an isomorphism, by computing orders). More generally, suppose G is presented by, say, generators a, b with relations $r_1, \ldots, r_k$. If a', b' are any elements of a group H satisfying these relations, there is a homomorphism from G into H. Namely, if $\pi : F(\{a, b\}) \to G$ is the presentation homomorphism, we can define $\pi' : F(\{a, b\}) \to H$ by $\pi'(a) = a'$ and $\pi'(b) = b'$. Then $\ker \pi \le \ker \pi'$ so π' factors through $\ker \pi$ and we obtain

$$G \cong F(\{a, b\})/\ker \pi \longrightarrow H.$$

In, particular, if $\langle a', b' \rangle = H = G$, this homomorphism is an automorphism of G. Conversely, any automorphism must send a set of generators to another set of generators satisfying the same relations. For example, $D_8 = \langle a, b \mid a^2 = b^4 = 1, \ aba = b^{-1} \rangle$ and any pair a', b' of elements, where a' is a noncentral element of order 2 and b' is of order 4, satisfies the same relations. Since there are four noncentral elements of order 2 and two elements of order 4, D_8 has 8 automorphisms.

Similarly, any pair of elements of order 4 in Q_8 which are not equal or inverses of each other necessarily generate Q_8 and satisfy the relations given in Example 3 above. It is easy to check that there are 24 such pairs, so

$$|\text{Aut}(Q_8)| = 24.$$

Free objects can be constructed in (many, but not all) other categories. For instance, a *monoid* is a set together with a binary operation satisfying all of the group axioms except the axiom specifying the existence of inverses. Free objects in the category of monoids play a fundamental role in theoretical computer science where they model the behavior of machines (Turing machines, etc.). We shall encounter free algebras (i.e., polynomial algebras) and free modules in later chapters.

EXERCISES

1. Let F_1 and F_2 be free groups of finite rank. Prove that $F_1 \cong F_2$ if and only if they have the same rank. What facts do you need in order to extend your proof to infinite ranks (where the result is also true)?

2. Prove that if $|S| > 1$ then $F(S)$ is non-abelian.

3. Prove that the commutator subgroup of the free group on 2 generators is not finitely generated (in particular, subgroups of finitely generated groups need not be finitely generated).

4. Prove that every nonidentity element of a free group is of infinite order.

5. Establish a finite presentation for A_4 using 2 generators.

6. Establish a finite presentation for S_4 using 2 generators.

7. Prove that the following is a presentation for the quaternion group of order 8:

$$Q_8 = \langle a, b \mid a^2 = b^2, \ a^{-1}ba = b^{-1} \rangle.$$

8. Use presentations to find the orders of the automorphism groups of the groups $Z_2 \times Z_4$ and $Z_4 \times Z_4$.

9. Prove that $\text{Aut}(Q_8) \cong S_4$.

10. This exercise exhibits an automorphism of S_6 that is not inner (hence, together with Exercise 19 in Section 4.4 it shows that $|\text{Aut}(S_6) : \text{Inn}(S_6)| = 2$). Let $t_1' = (1\ 2)(3\ 4)(5\ 6)$, $t_2' = (1\ 4)(2\ 5)(3\ 6)$, $t_3' = (1\ 3)(2\ 4)(5\ 6)$, $t_4' = (1\ 2)(3\ 6)(4\ 5)$, and $t_5' = (1\ 4)(2\ 3)(5\ 6)$. Show that $t_1', \ldots, t_5'$ satisfy the following relations:

$(t_i')^2 = 1$ for all i,

$(t_i' t_j')^2 = 1$ for all i and j with $|i - j| \geq 2$, and

$(t_i' t_{i+1}')^3 = 1$ for all $i \in \{1, 2, 3, 4\}$.

Deduce that $S_6 = \langle t_1', \ldots, t_5' \rangle$ and that the map

$$(1\ 2) \mapsto t_1', \quad (2\ 3) \mapsto t_2', \quad (3\ 4) \mapsto t_3', \quad (4\ 5) \mapsto t_4', \quad (5\ 6) \mapsto t_5'$$

extends to an automorphism of S_6 (which is clearly not inner since it does not send transpositions to transpositions). [Use the presentation for S_6 described in Example 5.]

11. Let S be a set. The group with presentation (S, R), where $R = \{[s, t] \mid s, t \in S\}$ is called the *free abelian* group on S — denote it by $A(S)$. Prove that $A(S)$ has the following universal property: if G is any abelian group and $\varphi : S \to G$ is any set map, then there is a unique group homomorphism $\Phi : A(S) \to G$ such that $\Phi|_S = \varphi$. Deduce that if A is a free abelian group on a set of cardinality n then

$$A \cong \mathbb{Z} \times \mathbb{Z} \times \cdots \times \mathbb{Z} \quad (n \text{ factors}).$$

12. Let S be a set and let c be a positive integer. Formulate the notion of a *free nilpotent group* on S of nilpotence class c and prove it has the appropriate universal property with respect to nilpotent groups of class $\leq c$.

13. Prove that there cannot be a nilpotent group N generated by two elements with the property that *every* nilpotent group which is generated by two elements is a homomorphic image of N (i.e., the specification of the class c in the preceding problem was necessary).

14. Prove that $G = \langle x, y \mid x^3 = y^3 = (xy)^3 = 1 \rangle$ is an infinite group as follows. Let p be a prime congruent to 1 mod 3 and let G_p be the non-abelian group of order $3p$. Let $a, b \in G_p$ with $|a| = p$ and $|b| = 3$. Prove that ab and ab^2 both have order 3. Deduce that G_p is a homomorphic image of G and from this conclude G is an infinite group, using the fact that there are infinitely many primes $p \equiv 1 \bmod 3$. [Note every nonidentity element of G_p has order 3 or p.]

Part II

RING THEORY

The theory of groups is concerned with general properties of certain objects having an algebraic structure defined by a single binary operation. The study of rings is concerned with objects possessing two binary operations (called addition and multiplication) related by the distributive laws. We first study analogues for the basic points of development in the structure theory of groups. In particular, we introduce subrings, quotient rings, ideals (which are the analogues of normal subgroups) and ring homomorphisms. We then focus on questions about general rings which arise naturally from the presence of two binary operations. Questions concerning multiplicative inverses lead to the notion of fields and eventually to the construction of some specific fields such as finite fields. The study of the arithmetic (divisibility, greatest common divisors, etc.) of rings such as the familiar ring of integers, $\mathbb{Z}$, leads to the notion of primes and unique factorizations in Chapter 8. The results of Chapters 7 and 8 are then applied to rings of polynomials in Chapter 9.

The basic theory of rings developed in Part II is the cornerstone for the remaining four parts of the book. The theory of ring actions (modules) comprises Part III of the book. There we shall see how the structure of rings is reflected in the structure of the objects on which they act and this will enable us to prove some powerful classification theorems. The structure theory of rings, in particular of polynomial rings, forms the basis in Part IV for the theory of fields and polynomial equations over fields. There the rich interplay among ring theory, field theory and group theory leads to many beautiful results on the structure of fields and the theory of roots of polynomials. Part V continues the study of rings and applications of ring theory to such topics as geometry and the theory of extensions. In Part VI the study of certain specific kinds of rings (group rings) and the objects (modules) on which they act again gives deep classification theorems whose consequences are then exploited to provide new results and insights into finite groups.

CHAPTER 7

Introduction to Rings

7.1 BASIC DEFINITIONS AND EXAMPLES

Definition.
(1) A *ring* R is a set together with two binary operations $+$ and $\times$ (called addition and multiplication) satisfying the following axioms:
    (i) $(R, +)$ is an *abelian* group,
    (ii) $\times$ is associative : $(a \times b) \times c = a \times (b \times c)$ for all $a, b, c \in R$,
    (iii) the *distributive laws* hold in R : for all $a, b, c \in R$

$$(a+b) \times c = (a \times c) + (b \times c) \quad \text{and} \quad a \times (b+c) = (a \times b) + (a \times c).$$

(2) The ring R is *commutative* if multiplication is commutative.
(3) The ring R is said to have an *identity* (or *contain a* 1) if there is an element $1 \in R$ with

$$1 \times a = a \times 1 = a \qquad \text{for all } a \in R.$$

We shall usually write simply ab rather than $a \times b$ for $a, b \in R$. The additive identity of R will always be denoted by 0 and the additive inverse of the ring element a will be denoted by $-a$.

The condition that R be a group under addition is a fairly natural one, but it may seem artificial to require that this group be *abelian*. One motivation for this is that if the ring R has a 1, the commutativity under addition is *forced* by the distributive laws. To see this, compute the product $(1+1)(a+b)$ in two different ways, using the distributive laws (but not assuming that addition is commutative). One obtains

$$(1 + 1)(a + b) = 1(a + b) + 1(a + b) = 1a + 1b + 1a + 1b = a + b + a + b$$

and

$$(1 + 1)(a + b) = (1 + 1)a + (1 + 1)b = 1a + 1a + 1b + 1b = a + a + b + b.$$

Since R is a group under addition, this implies $b + a = a + b$, i.e., that R under addition is necessarily commutative.

Fields are one of the most important examples of rings. Note that their definition below is just another formulation of the one given in Section 1.4.

223

Definition. A ring R with identity 1, where $1 \neq 0$, is called a *division ring* (or *skew field*) if every nonzero element $a \in R$ has a multiplicative inverse, i.e., there exists $b \in R$ such that $ab = ba = 1$. A commutative division ring is called a *field*.

More examples of rings follow.

Examples

(1) The simplest examples of rings are the *trivial rings* obtained by taking R to be any commutative group (denoting the group operation by $+$) and defining the multiplication $\times$ on R by $a \times b = 0$ for all $a, b \in R$. It is easy to see that this multiplication defines a commutative ring. In particular, if $R = \{0\}$ is the trivial group, the resulting ring R is called the *zero ring*, denoted $R = 0$. Except for the zero ring, a trivial ring does not contain an identity ($R = 0$ is the only ring where $1 = 0$; we shall often exclude this ring by imposing the condition $1 \neq 0$). Although trivial rings have two binary operations, multiplication adds no new structure to the additive group and the theory of rings gives no information which could not already be obtained from (abelian) group theory.

(2) The ring of integers, $\mathbb{Z}$, under the usual operations of addition and multiplication is a commutative ring with identity (the integer 1). The ring axioms (as with the additive group axioms) follow from the basic axioms for the system of natural numbers. Note that under *multiplication* $\mathbb{Z} - \{0\}$ is *not* a group (in fact, there are very few multiplicative inverses to elements in this ring). We shall come back to the question of these inverses shortly.

(3) Similarly, the rational numbers, $\mathbb{Q}$, the real numbers, $\mathbb{R}$, and the complex numbers, $\mathbb{C}$, are commutative rings with identity (in fact they are fields). The ring axioms for each of these follow ultimately from the ring axioms for $\mathbb{Z}$. We shall verify this when we construct $\mathbb{Q}$ from $\mathbb{Z}$ (Section 7.5) and $\mathbb{C}$ from $\mathbb{R}$ (Example 1, Section 13.1); both of these constructions will be special cases of more general processes. The construction of $\mathbb{R}$ from $\mathbb{Q}$ (and subsequent verification of the ring axioms) is carried out in basic analysis texts.

(4) The quotient group $\mathbb{Z}/n\mathbb{Z}$ is a commutative ring with identity (the element $\bar{1}$) under the operations of addition and multiplication of residue classes (frequently referred to as "modular arithmetic"). We saw that the additive abelian group axioms followed from the general principles of the theory of quotient groups (indeed this was the prototypical quotient group). We shall shortly prove that the remaining ring axioms (in particular, the fact that multiplication of residue classes is well defined) follow analogously from the general theory of quotient rings.

In all of the examples so far the rings have been commutative. Historically, one of the first noncommutative rings was discovered in 1843 by Sir William Rowan Hamilton (1805–1865). This ring, which is a division ring, was extremely influential in the subsequent development of mathematics and it continues to play an important role in certain areas of mathematics and physics.

(5) (The *(real) Hamilton Quaternions*) Let $\mathbb{H}$ be the collection of elements of the form $a + bi + cj + dk$ where $a, b, c, d \in \mathbb{R}$ are real numbers (loosely, "polynomials in $1, i, j, k$ with real coefficients") where addition is defined "componentwise" by

$$(a+bi+cj+dk) + (a'+b'i+c'j+d'k) = (a+a') + (b+b')i + (c+c')j + (d+d')k$$

and multiplication is defined by expanding $(a + bi + cj + dk)(a' + b'i + c'j + d'k)$ using the distributive law (being careful about the order of terms) and simplifying using the relations

$$i^2 = j^2 = k^2 = -1, \quad ij = -ji = k, \quad jk = -kj = i, \quad ki = -ik = j$$

(where the real number coefficients commute with i, j and k). For example,

$$(1+i+2j)(j+k) = 1(j+k) + i(j+k) + 2j(j+k) = j + k + ij + ik + 2j^2 + 2jk$$
$$= j + k + k + (-j) + 2(-1) + 2(i) = -2 + 2i + 2k.$$

The fact that $\mathbb{H}$ is a ring may be proved by a straightforward, albeit lengthy, check of the axioms (associativity of multiplication is particularly tedious). The Hamilton Quaternions are a noncommutative ring with identity ($1 = 1 + 0i + 0j + 0k$). Similarly, one can define the ring of *rational* Hamilton Quaternions by taking a, b, c, d to be rational numbers above. Both the real and rational Hamilton Quaternions are *division rings*, where inverses of nonzero elements are given by

$$(a + bi + cj + dk)^{-1} = \frac{a - bi - cj - dk}{a^2 + b^2 + c^2 + d^2}.$$

(6) One important class of rings is obtained by considering rings of functions. Let X be any nonempty set and let A be any ring. The collection, R, of all (set) functions $f : X \to A$ is a ring under the usual definition of pointwise addition and multiplication of functions: $(f + g)(x) = f(x) + g(x)$ and $(fg)(x) = f(x)g(x)$. Each ring axiom for R follows directly from the corresponding axiom for A. The ring R is commutative if and only if A is commutative and R has a 1 if and only if A has a 1 (in which case the 1 of R is necessarily the constant function 1 on X).

If X and A have more structure, we may form other rings of functions which respect those structures. For instance, if A is the ring of real numbers $\mathbb{R}$ and X is the closed interval $[0, 1]$ in $\mathbb{R}$ we may form the ring of all *continuous* functions from $[0, 1]$ to $\mathbb{R}$ (here we need basic limit theorems to guarantee that sums and products of continuous functions are continuous) — this is a commutative ring with 1.

(7) An example of a ring which does not have an identity is the ring $2\mathbb{Z}$ of even integers under usual addition and multiplication of integers (the sum and product of even integers is an even integer).

Another example which arises naturally in analysis is constructed as follows. A function $f : \mathbb{R} \to \mathbb{R}$ is said to have *compact support* if there are real numbers a, b (depending on f) such that $f(x) = 0$ for all $x \notin [a, b]$ (i.e., f is zero outside some bounded interval). The set of all functions $f : \mathbb{R} \to \mathbb{R}$ with compact support is a commutative ring without identity (since an identity could not have compact support). Similarly, the set of all continuous functions $f : \mathbb{R} \to \mathbb{R}$ with compact support is a commutative ring without identity.

In the next section we give three important ways of constructing "larger" rings from a given ring (analogous to Example 6 above) and thus greatly expand our list of examples. Before doing so we mention some basic properties of arbitrary rings. The ring $\mathbb{Z}$ is a good example to keep in mind, although this ring has a good deal more algebraic structure than a general ring (for example, it is commutative and has an identity). Nonetheless, its basic arithmetic holds for general rings as the following result shows.

Proposition 1. Let R be a ring. Then

   (1) $0a = a0 = 0$ for all $a \in R$.
   (2) $(-a)b = a(-b) = -(ab)$ for all $a, b \in R$ (recall $-a$ is the additive inverse of a).
   (3) $(-a)(-b) = ab$ for all $a, b \in R$.
   (4) if R has an identity 1, then the identity is unique and $-a = (-1)a$.

Proof: These all follow from the distributive laws and cancellation in the additive group R. For example, (1) follows from $0a = (0 + 0)a = 0a + 0a$. The equality $(-a)b = -(ab)$ in (2) follows from $ab + (-a)b = (a + (-a))b = 0b = 0$. The rest follow similarly and are left to the reader.

This proposition shows that because of the distributive laws the additive and multiplicative structures of a ring behave well with respect to one another, just as in the familiar example of the integers.

Unlike the integers, however, general rings may possess many elements that have multiplicative inverses or may have nonzero elements a and b whose product is zero. These two properties of elements, which relate to the multiplicative structure of a ring, are given special names.

Definition. Let R be a ring.

   (1) A nonzero element a of R is called a *zero divisor* if there is a nonzero element b in R such that either $ab = 0$ or $ba = 0$.
   (2) Assume R has an identity $1 \neq 0$. An element u of R is called a *unit* in R if there is some v in R such that $uv = vu = 1$. The set of units in R is denoted $R^{\times}$.

It is easy to see that the units in a ring R form a group under multiplication so $R^{\times}$ will be referred to as the *group of units* of R. In this terminology a *field is a commutative ring F with identity $1 \neq 0$ in which every nonzero element is a unit*, i.e., $F^{\times} = F - \{0\}$.

Observe that *a zero divisor can never be a unit*. Suppose for example that a is a unit in R and that $ab = 0$ for some nonzero b in R. Then $va = 1$ for some $v \in R$, so $b = 1b = (va)b = v(ab) = v0 = 0$, a contradiction. Similarly, if $ba = 0$ for some nonzero b then a cannot be a unit.

This shows in particular that fields contain no zero divisors.

Examples

   (1) The ring $\mathbb{Z}$ of integers has no zero divisors and its only units are ± 1, i.e., $\mathbb{Z}^{\times} = \{\pm 1\}$. Note that every nonzero integer has an inverse in the *larger ring* $\mathbb{Q}$, so the property of being a unit depends on the ring in which an element is viewed.
   (2) Let n be an integer ≥ 2. In the ring $\mathbb{Z}/n\mathbb{Z}$ the elements $\bar{u}$ for which u and n are relatively prime are units (we shall prove this in the next chapter). Thus our use of the notation $(\mathbb{Z}/n\mathbb{Z})^{\times}$ is consistent with the definition of the group of units in an arbitrary ring.

      If, on the other hand, a is a nonzero integer and a is not relatively prime to n then we show that $\bar{a}$ is a zero divisor in $\mathbb{Z}/n\mathbb{Z}$. To see this let d be the g.c.d. of a and n and let $b = \dfrac{n}{d}$. By assumption $d > 1$ so $0 < b < n$, i.e., $\bar{b} \neq \bar{0}$. But by construction n

divides ab, that is, $\overline{ab} = \overline{0}$ in $\mathbb{Z}/n\mathbb{Z}$. This shows that *every nonzero element of* $\mathbb{Z}/n\mathbb{Z}$ *is either a unit or a zero divisor*. Furthermore, every nonzero element is a unit if and only if every integer a in the range $0 < a < n$ is relatively prime to n. This happens if and only if n is a prime, i.e., $\mathbb{Z}/n\mathbb{Z}$ *is a field if and only if* n *is a prime*.

(3) If R is the ring of all functions from the closed interval $[0,1]$ to $\mathbb{R}$ then the units of R are the functions that are not zero at any point (for such f its inverse is the function $\frac{1}{f}$). If f is not a unit and not zero then f is a zero divisor because if we define

$$g(x) = \begin{cases} 0, & \text{if } f(x) \neq 0 \\ 1, & \text{if } f(x) = 0 \end{cases}$$

then g is not the zero function but $f(x)g(x) = 0$ for all x.

(4) If R is the ring of all *continuous* functions from the closed interval $[0,1]$ to $\mathbb{R}$ then the units of R are still the functions that are not zero at any point, but now there are functions that are neither units nor zero divisors. For instance, $f(x) = x - \frac{1}{2}$ has only one zero (at $x = \frac{1}{2}$) so f is not a unit. On the other hand, if $gf = 0$ then g must be zero for all $x \neq \frac{1}{2}$, and the only *continuous* function with this property is the zero function. Hence f is neither a unit nor a zero divisor. Similarly, no function with only a finite (or countable) number of zeros on $[0,1]$ is a zero divisor. This ring also contains many zero divisors. For instance let

$$f(x) = \begin{cases} 0, & 0 \leq x \leq \frac{1}{2} \\ x - \frac{1}{2}, & \frac{1}{2} \leq x \leq 1 \end{cases}$$

and let $g(x) = f(1 - x)$. Then f and g are nonzero continuous functions whose product is the zero function.

(5) Let D be a rational number that is not a perfect square in $\mathbb{Q}$ and define

$$\mathbb{Q}(\sqrt{D}) = \{a + b\sqrt{D} \mid a, b \in \mathbb{Q}\}$$

as a subset of $\mathbb{C}$. This set is clearly closed under subtraction, and the identity $(a + b\sqrt{D})(c + d\sqrt{D}) = (ac + bdD) + (ad + bc)\sqrt{D}$ shows that it is also closed under multiplication. Hence $\mathbb{Q}(\sqrt{D})$ is a subring of $\mathbb{C}$ (even a subring of $\mathbb{R}$ if $D > 0$), so in particular is a commutative ring with identity. It is easy to show that the assumption that D is not a square implies that every element of $\mathbb{Q}(\sqrt{D})$ may be written uniquely in the form $a + b\sqrt{D}$. This assumption also implies that if a and b are not both 0 then $a^2 - Db^2$ is nonzero, and since $(a + b\sqrt{D})(a - b\sqrt{D}) = a^2 - Db^2$ it follows that if $a + b\sqrt{D} \neq 0$ (i.e., one of a or b is nonzero) then $\dfrac{a - b\sqrt{D}}{a^2 - Db^2}$ is the inverse of $a + b\sqrt{D}$ in $\mathbb{Q}(\sqrt{D})$. This shows that every nonzero element in this commutative ring is a unit, i.e., $\mathbb{Q}(\sqrt{D})$ is a field (called a *quadratic field*, cf. Section 13.2).

The rational number D may be written $D = f^2 D'$ for some rational number f and a unique integer D' where D' is not divisible by the square of any integer greater than 1, i.e., D' is either -1 or ± 1 times the product of distinct primes in $\mathbb{Z}$ (for example, $8/5 = (2/5)^2 \cdot 10$). Call D' the *squarefree part* of D. Then $\sqrt{D} = f\sqrt{D'}$, and so $\mathbb{Q}(\sqrt{D}) = \mathbb{Q}(\sqrt{D'})$. Thus *there is no loss in assuming that* D *is a squarefree integer* (i.e., $f = 1$) *in the definition of the quadratic field* $\mathbb{Q}(\sqrt{D})$.

Rings having some of the same characteristics as the integers $\mathbb{Z}$ are given a name:

Definition. A commutative ring with identity $1 \neq 0$ is called an *integral domain* if it has no zero divisors.

The absence of zero divisors in integral domains give these rings a cancellation property:

Proposition 2. Assume a, b and c are elements of any ring with a not a zero divisor. If $ab = ac$, then either $a = 0$ or $b = c$ (i.e., if $a \neq 0$ we can cancel the a's). In particular, if a, b, c are any elements in an integral domain and $ab = ac$, then either $a = 0$ or $b = c$.

Proof: If $ab = ac$ then $a(b - c) = 0$ so either $a = 0$ or $b - c = 0$. The second statement follows from the first and the definition of an integral domain.

Corollary 3. Any finite integral domain is a field.

Proof: Let R be a finite integral domain and let a be a nonzero element of R. By the cancellation law the map $x \mapsto ax$ is an injective function. Since R is finite this map is also surjective. In particular, there is some $b \in R$ such that $ab = 1$, i.e., a is a unit in R. Since a was an arbitrary nonzero element, R is a field.

A remarkable result of Wedderburn is that a finite division ring is necessarily commutative, i.e., is a field. A proof of this theorem is outlined in the exercises at the end of Section 13.6.

In Section 5 we study the relation between zero divisors and units in greater detail. We shall see that every nonzero element of a commutative ring that is not a zero divisor has a multiplicative inverse in some larger ring. This gives another perspective on the cancellation law in Proposition 2.

Having defined the notion of a ring, there is a natural notion of a subring.

Definition. A *subring* of the ring R is a subgroup of R that is closed under multiplication.

In other words, a subset S of a ring R is a subring if the operations of addition and multiplication in R when restricted to S give S the structure of a ring. To show that a subset of a ring R is a subring it suffices to check that it is *nonempty* and *closed under subtraction and under multiplication*.

Examples

A number of the examples above were also subrings.
   (1) $\mathbb{Z}$ is a subring of $\mathbb{Q}$ and $\mathbb{Q}$ is a subring of $\mathbb{R}$. The property "is a subring of" is clearly transitive.
   (2) $2\mathbb{Z}$ is a subring of $\mathbb{Z}$, as is $n\mathbb{Z}$ for any integer n. The ring $\mathbb{Z}/n\mathbb{Z}$ is not a subring (or a subgroup) of $\mathbb{Z}$ for any $n \geq 2$.

(3) The ring of all continuous functions from $\mathbb{R}$ to $\mathbb{R}$ is a subring of the ring of all functions from $\mathbb{R}$ to $\mathbb{R}$. The ring of all differentiable functions from $\mathbb{R}$ to $\mathbb{R}$ is a subring of both.

(4) $S = \mathbb{Z} + \mathbb{Z}i + \mathbb{Z}j + \mathbb{Z}k$, the *'naive' integral* Quaternions, form a subring of either the real or the rational Quaternions – it is easy to check that multiplying two such quaternions together gives another quaternion with integer coefficients. The set of quaternions $\mathbb{Z}(1 + i + j + k)/2 + \mathbb{Z}i + \mathbb{Z}j + \mathbb{Z}k$ (i.e., S together with the odd integer multiples of $(1 + i + j + k)/2$), called the *Hurwitz* or *integral* Quaternions, is also a subring (again an easy check), containing S as a subring. These rings (which are not division rings) can be used to give proofs for a number of results in number theory.

(5) If R is a subring of a field F that contains the identity of F then R is an integral domain. The converse of this is also true, namely any integral domain is contained in a field (cf. Section 5).

Example: (Quadratic Integer Rings)

Let D be a squarefree integer. It is immediate from the addition and multiplication that the subset $\mathbb{Z}[\sqrt{D}] = \{a + b\sqrt{D} \mid a, b \in \mathbb{Z}\}$ forms a subring of the quadratic field $\mathbb{Q}(\sqrt{D})$ defined earlier. If $D \equiv 1 \bmod 4$ then the slightly larger subset

$$\mathbb{Z}[\frac{1 + \sqrt{D}}{2}] = \{a + b\frac{1 + \sqrt{D}}{2} \mid a, b \in \mathbb{Z}\}$$

is also a subring: closure under addition is immediate and $(a + b\frac{1+\sqrt{D}}{2})(c + d\frac{1+\sqrt{D}}{2}) = (ac + bd\frac{D-1}{4}) + (ad + bc + bd)\frac{1+\sqrt{D}}{2}$ together with the congruence on D shows closure under multiplication.

Define

$$\mathcal{O} = \mathcal{O}_{\mathbb{Q}(\sqrt{D})} = \mathbb{Z}[\omega] = \{a + b\omega \mid a, b \in \mathbb{Z}\},$$

where

$$\omega = \begin{cases} \sqrt{D}, & \text{if } D \equiv 2, 3 \bmod 4 \\ \dfrac{1 + \sqrt{D}}{2}, & \text{if } D \equiv 1 \bmod 4, \end{cases}$$

called the *ring of integers* in the quadratic field $\mathbb{Q}(\sqrt{D})$. The terminology comes from the fact that the elements of the subring $\mathcal{O}$ of the field $\mathbb{Q}(\sqrt{D})$ have many properties analogous to those of the subring of integers $\mathbb{Z}$ in the field of rational numbers $\mathbb{Q}$ (and are the *integral closure* of $\mathbb{Z}$ in $\mathbb{Q}(\sqrt{D})$ as explained in Section 15.3).

In the special case when $D = -1$ we obtain the ring $\mathbb{Z}[i]$ of *Gaussian integers*, which are the complex numbers $a + bi \in \mathbb{C}$ with a and b both *integers*. These numbers were originally introduced by Gauss around 1800 in order to state the biquadratic reciprocity law which deals with the beautiful relations that exist among fourth powers modulo primes. We shall shortly see another useful application of the algebraic structure of this ring to number theoretic questions.

Define the *field norm* $N : \mathbb{Q}(\sqrt{D}) \to \mathbb{Q}$ by

$$N(a + b\sqrt{D}) = (a + b\sqrt{D})(a - b\sqrt{D}) = a^2 - Db^2 \in \mathbb{Q},$$

which, as previously mentioned, is nonzero if $a + b\sqrt{D} \neq 0$. This norm gives a measure of "size" in the field $\mathbb{Q}(\sqrt{D})$. For instance when $D = -1$ the norm of $a + bi$ is $a^2 + b^2$, which is the square of the length of this complex number considered as a vector in the complex plane. We shall use the field norm in this and subsequent examples to establish many properties of the rings $\mathcal{O}$.

It is easy to check that N is *multiplicative*, i.e., that $N(\alpha\beta) = N(\alpha)N(\beta)$ for all $\alpha, \beta \in \mathbb{Q}(\sqrt{D})$. On the subring $\mathcal{O}$ it is also easy to see that the field norm is given by

$$N(a + b\omega) = (a + b\omega)(a + b\overline{\omega}) = \begin{cases} a^2 - Db^2, & \text{if } D \equiv 2, 3 \bmod 4 \\ a^2 + ab + \dfrac{1 - D}{4}b^2, & \text{if } D \equiv 1 \bmod 4 \end{cases}$$

where

$$\overline{\omega} = \begin{cases} -\sqrt{D}, & \text{if } D \equiv 2, 3 \bmod 4 \\ \dfrac{1 - \sqrt{D}}{2}, & \text{if } D \equiv 1 \bmod 4. \end{cases}$$

It follows that $N(\alpha)$ is in fact an *integer* for every $\alpha \in \mathcal{O}$.

We may use this norm to characterize the units in $\mathcal{O}$. If $\alpha \in \mathcal{O}$ has field norm $N(\alpha) = \pm 1$, the previous formula shows that $(a + b\omega)^{-1} = \pm(a + b\overline{\omega})$, which is again an element of $\mathcal{O}$ and so α is a unit in $\mathcal{O}$. Suppose conversely that α is a unit in $\mathcal{O}$, say $\alpha\beta = 1$ for some $\beta \in \mathcal{O}$. Then the multiplicative property of the field norm implies that $N(\alpha)N(\beta) = N(\alpha\beta) = N(1) = 1$. Since both $N(\alpha)$ and $N(\beta)$ are integers, each must be ± 1. Hence,

the element α is a unit in $\mathcal{O}$ if and only if $N(\alpha) = \pm 1$.

In particular the determination of the integer solutions to the equation $x^2 - Dy^2 = \pm 1$ (called *Pell's equation* in elementary number theory) is essentially equivalent to the determination of the units in the ring $\mathcal{O}$.

When $D = -1$, the units in the Gaussian integers $\mathbb{Z}[i]$ are the elements $a + bi$ with $a^2 + b^2 = \pm 1$, $a, b \in \mathbb{Z}$, so the group of units consists of $\{\pm 1, \pm i\}$. When $D = -3$, the units in $\mathbb{Z}[(1 + \sqrt{-3})/2]$ are determined by the integers a, b with $a^2 + ab + b^2 = \pm 1$, i.e., with $(2a + b)^2 + 3b^2 = \pm 4$, from which it is easy to see that the group of units is a group of order 6 given by $\{\pm 1, \pm\rho, \pm\rho^2\}$ where $\rho = (-1 + \sqrt{-3})/2$. For any other $D < 0$ it is similarly straightforward to see that the only units are $\{\pm 1\}$.

By contrast, when $D > 0$ it can be shown that the group of units $\mathcal{O}^\times$ is always infinite. For example, it is easy to check that $1 + \sqrt{2}$ is a unit in the ring $\mathcal{O} = \mathbb{Z}[\sqrt{2}]$ (with field norm -1) and that $\{\pm(1 + \sqrt{2})^n \mid n \in \mathbb{Z}\}$, is an infinite set of distinct units (in fact the full group of units in this case, but this is harder to prove).

EXERCISES

Let R be a ring with 1.

1. Show that $(-1)^2 = 1$ in R.

2. Prove that if u is a unit in R then so is $-u$.

3. Let R be a ring with identity and let S be a subring of R containing the identity. Prove that if u is a unit in S then u is a unit in R. Show by example that the converse is false.

4. Prove that the intersection of any nonempty collection of subrings of a ring is also a subring.

5. Decide which of the following (a) – (f) are subrings of $\mathbb{Q}$:
   (a) the set of all rational numbers with odd denominators (when written in lowest terms)
   (b) the set of all rational numbers with even denominators (when written in lowest terms)
   (c) the set of nonnegative rational numbers
   (d) the set of squares of rational numbers
   (e) the set of all rational numbers with odd numerators (when written in lowest terms)

(f) the set of all rational numbers with even numerators (when written in lowest terms).

6. Decide which of the following are subrings of the ring of all functions from the closed interval $[0,1]$ to $\mathbb{R}$:
   (a) the set of all functions $f(x)$ such that $f(q) = 0$ for all $q \in \mathbb{Q} \cap [0, 1]$
   (b) the set of all polynomial functions
   (c) the set of all functions which have only a finite number of zeros, together with the zero function
   (d) the set of all functions which have an infinite number of zeros
   (e) the set of all functions f such that $\lim_{x \to 1^-} f(x) = 0$
   (f) the set of all rational linear combinations of the functions $\sin nx$ and $\cos mx$, where $m, n \in \{0, 1, 2, \dots\}$.

7. The *center* of a ring R is $\{z \in R \mid zr = rz \text{ for all } r \in R\}$ (i.e., is the set of all elements which commute with every element of R). Prove that the center of a ring is a subring that contains the identity. Prove that the center of a division ring is a field.

8. Describe the center of the real Hamilton Quaternions $\mathbb{H}$. Prove that $\{a + bi \mid a, b \in \mathbb{R}\}$ is a subring of $\mathbb{H}$ which is a field but is not contained in the center of $\mathbb{H}$.

9. For a fixed element $a \in R$ define $C(a) = \{r \in R \mid ra = ar\}$. Prove that $C(a)$ is a subring of R containing a. Prove that the center of R is the intersection of the subrings $C(a)$ over all $a \in R$.

10. Prove that if D is a division ring then $C(a)$ is a division ring for all $a \in D$ (cf. the preceding exercise).

11. Prove that if R is an integral domain and $x^2 = 1$ for some $x \in R$ then $x = \pm 1$.

12. Prove that any subring of a field which contains the identity is an integral domain.

13. An element x in R is called *nilpotent* if $x^m = 0$ for some $m \in \mathbb{Z}^+$.
    (a) Show that if $n = a^k b$ for some integers a and b then $\overline{ab}$ is a nilpotent element of $\mathbb{Z}/n\mathbb{Z}$.
    (b) If $a \in \mathbb{Z}$ is an integer, show that the element $\bar{a} \in \mathbb{Z}/n\mathbb{Z}$ is nilpotent if and only if every prime divisor of n is also a divisor of a. In particular, determine the nilpotent elements of $\mathbb{Z}/72\mathbb{Z}$ explicitly.
    (c) Let R be the ring of functions from a nonempty set X to a field F. Prove that R contains no nonzero nilpotent elements.

14. Let x be a nilpotent element of the commutative ring R (cf. the preceding exercise).
    (a) Prove that x is either zero or a zero divisor.
    (b) Prove that rx is nilpotent for all $r \in R$.
    (c) Prove that $1 + x$ is a unit in R.
    (d) Deduce that the sum of a nilpotent element and a unit is a unit.

15. A ring R is called a *Boolean ring* if $a^2 = a$ for all $a \in R$. Prove that every Boolean ring is commutative.

16. Prove that the only Boolean ring that is an integral domain is $\mathbb{Z}/2\mathbb{Z}$.

17. Let R and S be rings. Prove that the direct product $R \times S$ is a ring under componentwise addition and multiplication. Prove that $R \times S$ is commutative if and only if both R and S are commutative. Prove that $R \times S$ has an identity if and only if both R and S have identities.

18. Prove that $\{(r, r) \mid r \in R\}$ is a subring of $R \times R$.

19. Let I be any nonempty index set and let R_i be a ring for each $i \in I$. Prove that the direct

product $\prod_{i \in I} R_i$ is a ring under componentwise addition and multiplication.

20. Let R be the collection of sequences $(a_1, a_2, a_3, \ldots)$ of integers $a_1, a_2, a_3, \ldots$ where all but finitely many of the a_i are 0 (called the *direct sum* of infinitely many copies of $\mathbb{Z}$). Prove that R is a ring under componentwise addition and multiplication which does not have an identity.

21. Let X be any nonempty set and let $\mathcal{P}(X)$ be the set of all subsets of X (the *power set* of X). Define addition and multiplication on $\mathcal{P}(X)$ by

$$A + B = (A - B) \cup (B - A) \qquad \text{and} \qquad A \times B = A \cap B$$

i.e., addition is symmetric difference and multiplication is intersection.
   (a) Prove that $\mathcal{P}(X)$ is a ring under these operations ($\mathcal{P}(X)$ and its subrings are often referred to as *rings of sets*).
   (b) Prove that this ring is commutative, has an identity and is a Boolean ring.

22. Give an example of an infinite Boolean ring.

23. Let D be a squarefree integer, and let $\mathcal{O}$ be the ring of integers in the quadratic field $\mathbb{Q}(\sqrt{D})$. For any positive integer f prove that the set $\mathcal{O}_f = \mathbb{Z}[f\omega] = \{a + bf\omega \mid a, b \in \mathbb{Z}\}$ is a subring of $\mathcal{O}$ containing the identity. Prove that $[\mathcal{O} : \mathcal{O}_f] = f$ (index as additive abelian groups). Prove conversely that a subring of $\mathcal{O}$ containing the identity and having finite index f in $\mathcal{O}$ (as additive abelian group) is equal to $\mathcal{O}_f$. (The ring $\mathcal{O}_f$ is called the *order of conductor* f in the field $\mathbb{Q}(\sqrt{D})$. The ring of integers $\mathcal{O}$ is called the *maximal order* in $\mathbb{Q}(\sqrt{D})$.)

24. Show for $D = 3, 5, 6$, and 7 that the group of units $\mathcal{O}^{\times}$ of the quadratic integer ring $\mathcal{O}$ is infinite by exhibiting an explicit unit of infinite (multiplicative) order in each ring.

25. Let $S = \mathbb{Z} + \mathbb{Z}i + \mathbb{Z}j + \mathbb{Z}k$ be the ring of naive integral Hamilton Quaternions and define

$$N : S \to \mathbb{Z} \qquad \text{by} \qquad N(a + bi + cj + dk) = a^2 + b^2 + c^2 + d^2$$

(the map N is called a *norm*).
   (a) Prove that $N(\alpha) = \alpha\bar{\alpha}$ for all $\alpha \in S$, where if $\alpha = a + bi + cj + dk$ then $\bar{\alpha} = a - bi - cj - dk$.
   (b) Prove that $N(\alpha\beta) = N(\alpha)N(\beta)$ for all $\alpha, \beta \in S$.
   (c) Prove that an element of S is a unit if and only if it has norm $+1$. Show that $S^{\times}$ is isomorphic to the quaternion group of order 8. [The inverse in the ring of rational quaternions of a nonzero element α is $\dfrac{\bar{\alpha}}{N(\alpha)}$.]

26. Let K be a field. A *discrete valuation* on K is a function $v : K^{\times} \to \mathbb{Z}$ satisfying
   (i) $v(ab) = v(a) + v(b)$ (i.e., v is a homomorphism from the multiplicative group of nonzero elements of K to $\mathbb{Z}$),
   (ii) v is surjective, and
   (iii) $v(x + y) \geq \min\{v(x), v(y)\}$ for all $x, y \in K^{\times}$ with $x + y \neq 0$.
   The set $R = \{x \in K^{\times} \mid v(x) \geq 0\} \cup \{0\}$ is called the *valuation ring* of v.
   (a) Prove that R is a subring of K which contains the identity. (In general, a ring R is called a *discrete valuation ring* if there is some field K and some discrete valuation v on K such that R is the valuation ring of v.)
   (b) Prove that for each nonzero element $x \in K$ either x or x^{-1} is in R.
   (c) Prove that an element x is a unit of R if and only if $v(x) = 0$.

27. A specific example of a discrete valuation ring (cf. the preceding exercise) is obtained

232

when p is a prime, $K = \mathbb{Q}$ and

$$\nu_p : \mathbb{Q}^\times \to \mathbb{Z} \quad \text{by} \quad \nu_p(\frac{a}{b}) = \alpha \quad \text{where} \quad \frac{a}{b} = p^\alpha \frac{c}{d}, \quad p \nmid c \text{ and } p \nmid d.$$

Prove that the corresponding valuation ring R is the ring of all rational numbers whose denominators are relatively prime to p. Describe the units of this valuation ring.

28. Let R be a ring with $1 \neq 0$. A nonzero element a is called a *left zero divisor* in R if there is a nonzero element $x \in R$ such that $ax = 0$. Symmetrically, $b \neq 0$ is a *right zero divisor* if there is a nonzero $y \in R$ such that $yb = 0$ (so a zero divisor is an element which is either a left or a right zero divisor). An element $u \in R$ has a *left inverse* in R if there is some $s \in R$ such that $su = 1$. Symmetrically, v has a *right inverse* if $vt = 1$ for some $t \in R$.
    (a) Prove that u is a unit if and only if it has both a right and a left inverse (i.e., u must have a two-sided inverse).
    (b) Prove that if u has a right inverse then u is not a right zero divisor.
    (c) Prove that if u has more than one right inverse then u is a left zero divisor.
    (d) Prove that if R is a finite ring then every element that has a right inverse is a unit (i.e., has a two-sided inverse).

29. Let A be any commutative ring with identity $1 \neq 0$. Let R be the set of all group homomorphisms of the additive group A to itself with addition defined as pointwise addition of functions and multiplication defined as function composition. Prove that these operations make R into a ring with identity. Prove that the units of R are the group automorphisms of A (cf. Exercise 20, Section 1.6).

30. Let $A = \mathbb{Z} \times \mathbb{Z} \times \mathbb{Z} \times \cdots$ be the direct product of copies of $\mathbb{Z}$ indexed by the positive integers (so A is a ring under componentwise addition and multiplication) and let R be the ring of all group homomorphisms from A to itself as described in the preceding exercise. Let φ be the element of R defined by $\varphi(a_1, a_2, a_3, \dots) = (a_2, a_3, \dots)$. Let ψ be the element of R defined by $\psi(a_1, a_2, a_3, \dots) = (0, a_1, a_2, a_3, \dots)$.
    (a) Prove that $\varphi\psi$ is the identity of R but $\psi\varphi$ is not the identity of R (i.e., ψ is a *right inverse* for φ but not a left inverse).
    (b) Exhibit infinitely many right inverses for φ.
    (c) Find a nonzero element π in R such that $\varphi\pi = 0$ but $\pi\varphi \neq 0$.
    (d) Prove that there is no nonzero element $\lambda \in R$ such that $\lambda\varphi = 0$ (i.e., φ is a left zero divisor but not a right zero divisor).

7.2 EXAMPLES: POLYNOMIAL RINGS, MATRIX RINGS, AND GROUP RINGS

We introduce here three important types of rings: polynomial rings, matrix rings, and group rings. We shall see in the course of the text that these three classes of rings are often related. For example, we shall see in Part VI that the group ring of a group G over the complex numbers $\mathbb{C}$ is a direct product of matrix rings over $\mathbb{C}$.

These rings also have many important applications, in addition to being interesting in their own right. In Part III we shall use polynomial rings to prove some classification theorems for matrices which, in particular, determine when a matrix is similar to a diagonal matrix. In Part VI we shall use group rings to study group actions and to prove some additional important classification theorems.

Polynomial Rings

Fix a commutative ring R with identity. We define the ring of polynomials in a form which may already be familiar, at least for polynomials with real coefficients. A definition in terms of Cartesian products is given in Appendix I. Let x be an indeterminate. The formal sum

$$a_n x^n + a_{n-1} x^{n-1} + \cdots + a_1 x + a_0$$

with $n \geq 0$ and each $a_i \in R$ is called a *polynomial* in x with coefficients a_i in R. If $a_n \neq 0$, then the polynomial is said to be of *degree n*, $a_n x^n$ is called the *leading term*, and a_n is called the *leading coefficient* (where the leading coefficient of the zero polynomial is taken to be 0). The polynomial is *monic* if $a_n = 1$. The set of all such polynomials is called the ring of *polynomials in the variable x with coefficients in R* and will be denoted $R[x]$.

The operations of addition and multiplication which make $R[x]$ into a ring are the same operations familiar from elementary algebra: addition is "componentwise"

$$(a_n x^n + a_{n-1} x^{n-1} + \cdots + a_1 x + a_0) + (b_n x^n + b_{n-1} x^{n-1} + \cdots + b_1 x + b_0)$$
$$= (a_n + b_n) x^n + (a_{n-1} + b_{n-1}) x^{n-1} + \cdots + (a_1 + b_1) x + (a_0 + b_0)$$

(here a_n or b_n may be zero in order for addition of polynomials of different degrees to be defined). Multiplication is performed by first defining $(ax^i)(bx^j) = abx^{i+j}$ for polynomials with only one nonzero term and then extending to all polynomials by the distributive laws (usually referred to as "expanding out and collecting like terms"):

$$(a_0 + a_1 x + a_2 x^2 + \dots) \times (b_0 + b_1 x + b_2 x^2 + \dots)$$
$$= a_0 b_0 + (a_0 b_1 + a_1 b_0) x + (a_0 b_2 + a_1 b_1 + a_2 b_0) x^2 + \dots$$

(in general, the coefficient of x^k in the product will be $\sum_{i=0}^{k} a_i b_{k-i}$). These operations make sense since R is a ring so the sums and products of the coefficients are defined. An easy verification proves that $R[x]$ is indeed a ring with these definitions of addition and multiplication.

The ring R appears in $R[x]$ as the *constant polynomials*. Note that by definition of the multiplication, $R[x]$ is a *commutative ring with identity* (the identity 1 from R).

The coefficient ring R above was assumed to be a commutative ring since that is the situation we shall be primarily interested in, but note that the definition of the addition and multiplication in $R[x]$ above would be valid even if R were not commutative or did not have an identity. If the coefficient ring R is the integers $\mathbb{Z}$ (respectively, the rationals $\mathbb{Q}$) the polynomial ring $\mathbb{Z}[x]$ (respectively, $\mathbb{Q}[x]$) is the ring of polynomials with integer (rational) coefficients familiar from elementary algebra.

Another example is the polynomial ring $\mathbb{Z}/3\mathbb{Z}[x]$ of polynomials in x with coefficients in $\mathbb{Z}/3\mathbb{Z}$. This ring consists of nonnegative powers of x with coefficients 0, 1, and 2 with calculations on the coefficients performed modulo 3. For example, if

$$p(x) = x^2 + 2x + 1 \quad \text{and} \quad q(x) = x^3 + x + 2$$

then

$$p(x) + q(x) = x^3 + x^2$$

and
$$p(x)q(x) = x^5 + 2x^4 + 2x^3 + x^2 + 2x + 2.$$

The ring in which the coefficients are taken makes a substantial difference in the behavior of polynomials. For example, the polynomial $x^2 + 1$ is not a perfect square in the polynomial ring $\mathbb{Z}[x]$, but *is* a perfect square in the polynomial ring $\mathbb{Z}/2\mathbb{Z}[x]$, since $(x + 1)^2 = x^2 + 2x + 1 = x^2 + 1$ in this ring.

Proposition 4. Let R be an integral domain and let $p(x), q(x)$ be nonzero elements of $R[x]$. Then
   (1) degree $p(x)q(x)$ = degree $p(x)$ + degree $q(x)$,
   (2) the units of $R[x]$ are just the units of R,
   (3) $R[x]$ is an integral domain.

Proof: If R has no zero divisors then neither does $R[x]$; if $p(x)$ and $q(x)$ are polynomials with leading terms $a_n x^n$ and $b_m x^m$, respectively, then the leading term of $p(x)q(x)$ is $a_n b_m x^{n+m}$, and $a_n b_m \neq 0$. This proves (3) and also verifies (1). If $p(x)$ is a unit, say $p(x)q(x) = 1$ in $R[x]$, then degree $p(x)$ + degree $q(x) = 0$, so both $p(x)$ and $q(x)$ are elements of R, hence are units in R since their product is 1. This proves (2).

If the ring R has zero divisors then so does $R[x]$, because $R \subset R[x]$. Also, if $f(x)$ is a zero divisor in $R[x]$ (i.e., $f(x)g(x) = 0$ for some nonzero $g(x) \in R[x]$) then in fact $cf(x) = 0$ for some nonzero $c \in R$ (cf. Exercise 2).

If S is a subring of R then $S[x]$ is a subring of $R[x]$. For instance, $\mathbb{Z}[x]$ is a subring of $\mathbb{Q}[x]$. Some other examples of subrings of $R[x]$ are the set of all polynomials in x^2 (i.e., in which only even powers of x appear) and the set of all polynomials with zero constant term (the latter subring does not have an identity).

Polynomial rings, particularly those over fields, will be studied extensively in Chapter 9.

Matrix Rings

Fix an arbitrary ring R and let n be a positive integer. Let $M_n(R)$ be the set of all $n \times n$ *matrices with entries from R*. The element (a_{ij}) of $M_n(R)$ is an $n \times n$ square array of elements of R whose entry in row i and column j is $a_{ij} \in R$. The set of matrices becomes a ring under the usual rules by which matrices of real numbers are added and multiplied. Addition is componentwise: the i, j entry of the matrix $(a_{ij}) + (b_{ij})$ is $a_{ij} + b_{ij}$. The i, j entry of the matrix product $(a_{ij}) \times (b_{ij})$ is $\sum_{k=1}^{n} a_{ik}b_{kj}$ (note that these matrices need to be square in order that multiplication of any two elements be defined). It is a straightforward calculation to check that these operations make $M_n(R)$ into a ring. When R is a field we shall prove that $M_n(R)$ is a ring by less computational means in Part III.

Note that if R is any nontrivial ring (even a commutative one) and $n \geq 2$ then $M_n(R)$ is *not commutative*: if $ab \neq 0$ in R let A be the matrix with a in position 1,1 and zeros elsewhere and let B be the matrix with b in position 1,2 and zeros elsewhere; then ab is the (nonzero) entry in position 1,2 of AB whereas BA is the zero matrix.

These two matrices also show that $M_n(R)$ has zero divisors for all nonzero rings R whenever $n \geq 2$.

An element (a_{ij}) of $M_n(R)$ is called a *scalar matrix* if for some $a \in R$, $a_{ii} = a$ for all $i \in \{1, \ldots, n\}$ and $a_{ij} = 0$ for all $i \neq j$ (i.e., all diagonal entries equal a and all off-diagonal entries are 0). The set of scalar matrices is a subring of $M_n(R)$. This subring is a copy of R (i.e., is "isomorphic" to R): if the matrix A has the element a along the main diagonal and the matrix B has the element b along the main diagonal then the matrix $A + B$ has $a + b$ along the diagonal and AB has ab along the diagonal (and all other entries 0). If R is commutative, the scalar matrices commute with all elements of $M_n(R)$. If R has a 1, then the scalar matrix with 1's down the diagonal (the $n \times n$ *identity matrix*) is the 1 of $M_n(R)$. In this case the units in $M_n(R)$ are the invertible $n \times n$ matrices and the group of units is denoted $GL_n(R)$, the *general linear group* of degree n over R.

If S is a subring of R then $M_n(S)$ is a subring of $M_n(R)$. For instance $M_n(\mathbb{Z})$ is a subring of $M_n(\mathbb{Q})$ and $M_n(2\mathbb{Z})$ is a subring of both of these. Another example of a subring of $M_n(R)$ is the set of *upper triangular* matrices: $\{(a_{ij}) \mid a_{pq} = 0 \text{ whenever } p > q\}$ (the set of matrices all of whose entries below the main diagonal are zero) — one easily checks that the sum and product of upper triangular matrices is upper triangular.

Group Rings

Fix a commutative ring R with identity $1 \neq 0$ and let $G = \{g_1, g_2, \ldots, g_n\}$ be any finite group with group operation written multiplicatively. Define the *group ring*, RG, of G with coefficients in R to be the set of all formal sums

$$a_1 g_1 + a_2 g_2 + \cdots + a_n g_n \qquad a_i \in R, \quad 1 \leq i \leq n.$$

If g_1 is the identity of G we shall write $a_1 g_1$ simply as a_1. Similarly, we shall write the element $1g$ for $g \in G$ simply as g.

Addition is defined "componentwise"

$$(a_1 g_1 + a_2 g_2 + \cdots + a_n g_n) + (b_1 g_1 + b_2 g_2 + \cdots + b_n g_n)$$
$$= (a_1 + b_1)g_1 + (a_2 + b_2)g_2 + \cdots + (a_n + b_n)g_n.$$

Multiplication is performed by first defining $(a g_i)(b g_j) = (ab)g_k$, where the product ab is taken in R and $g_i g_j = g_k$ is the product in the group G. This product is then extended to all formal sums by the distributive laws so that the coefficient of g_k in the product $(a_1 g_1 + \cdots + a_n g_n) \times (b_1 g_1 + \cdots + b_n g_n)$ is $\sum_{g_i g_j = g_k} a_i b_j$. It is straightforward to check that these operations make RG into a ring (again, commutativity of R is not needed). The associativity of multiplication follows from the associativity of the group operation in G. The ring RG is commutative if and only if G is a commutative group.

Example

Let $G = D_8$ be the dihedral group of order 8 with the usual generators r, s ($r^4 = s^2 = 1$ and $rs = sr^{-1}$) and let $R = \mathbb{Z}$. The elements $\alpha = r + r^2 - 2s$ and $\beta = -3r^2 + rs$ are typical members of $\mathbb{Z}D_8$. Their sum and product are then

$$\alpha + \beta = r - 2r^2 - 2s + rs$$

$$\alpha\beta = (r + r^2 - 2s)(-3r^2 + rs)$$

$$= r(-3r^2 + rs) + r^2(-3r^2 + rs) - 2s(-3r^2 + rs)$$

$$= -3r^3 + r^2 s - 3 + r^3 s + 6r^2 s - 2r^3$$

$$= -3 - 5r^3 + 7r^2 s + r^3 s.$$

The ring R appears in RG as the "constant" formal sums i.e., the R-multiples of the identity of G (note that the definition of the addition and multiplication in RG restricted to these elements is just the addition and multiplication in R). These elements of R commute with all elements of RG. The identity of R is the identity of RG.

The group G also appears in RG (the element g_i appears as $1g_i$ — for example, $r, s \in D_8$ are also elements of the group ring $\mathbb{Z}D_8$ above) — multiplication in the ring RG restricted to G is just the group operation. In particular, each element of G has a multiplicative inverse in the ring RG (namely, its inverse in G). This says that G is a *subgroup of the group of units of RG.*

If $|G| > 1$ then RG always has zero divisors. For example, let g be any element of G of order $m > 1$. Then

$$(1 - g)(1 + g + \cdots + g^{m-1}) = 1 - g^m = 1 - 1 = 0$$

so $1 - g$ is a zero divisor (note that by definition of RG neither of the formal sums in the above product is zero).

If S is a subring of R then SG is a subring of RG. For instance, $\mathbb{Z}G$ (called the *integral group ring* of G) is a subring of $\mathbb{Q}G$ (the *rational group ring* of G). Furthermore, if H is a subgroup of G then RH is a subring of RG. The set of all elements of RG whose coefficients sum to zero is a subring (without identity). If $|G| > 1$, the set of elements with zero "constant term" (i.e., the coefficient of the identity of G is zero) is *not* a subring (it is not closed under multiplication).

Note that the group ring $\mathbb{R}Q_8$ is *not* the same ring as the Hamilton Quaternions $\mathbb{H}$ even though the latter contains a copy of the quaternion group Q_8 (under multiplication). One difference is that the unique element of order 2 in Q_8 (usually denoted by -1) is not the additive inverse of 1 in $\mathbb{R}Q_8$. In other words, if we temporarily denote the identity of the group Q_8 by g_1 and the unique element of order 2 by g_2, then $g_1 + g_2$ is not zero in $\mathbb{R}Q_8$, whereas $1 + (-1)$ is zero in $\mathbb{H}$. Furthermore, as noted above, the group ring $\mathbb{R}Q_8$ contains zero divisors hence is not a division ring.

Group rings over fields will be studied extensively in Chapter 18.

EXERCISES

Let R be a commutative ring with 1.

1. Let $p(x) = 2x^3 - 3x^2 + 4x - 5$ and let $q(x) = 7x^3 + 33x - 4$. In each of parts (a), (b) and (c) compute $p(x) + q(x)$ and $p(x)q(x)$ under the assumption that the coefficients of the two given polynomials are taken from the specified ring (where the integer coefficients are taken mod n in parts (b) and (c)):
   (a) $R = \mathbb{Z}$,   (b) $R = \mathbb{Z}/2\mathbb{Z}$,   (c) $R = \mathbb{Z}/3\mathbb{Z}$.

2. Let $p(x) = a_n x^n + a_{n-1} x^{n-1} + \cdots + a_1 x + a_0$ be an element of the polynomial ring $R[x]$. Prove that $p(x)$ is a zero divisor in $R[x]$ if and only if there is a nonzero $b \in R$ such that $bp(x) = 0$. [Let $g(x) = b_m x^m + b_{m-1} x^{m-1} + \cdots + b_0$ be a nonzero polynomial of minimal degree such that $g(x)p(x) = 0$. Show that $b_m a_n = 0$ and so $a_n g(x)$ is a polynomial of degree less than m that also gives 0 when multiplied by $p(x)$. Conclude that $a_n g(x) = 0$. Apply a similar argument to show by induction on i that $a_{n-i} g(x) = 0$ for $i = 0, 1, \ldots, n$, and show that this implies $b_m p(x) = 0$.]

3. Define the set $R[[x]]$ of *formal power series* in the indeterminate x with coefficients from R to be all formal infinite sums

$$\sum_{n=0}^{\infty} a_n x^n = a_0 + a_1 x + a_2 x^2 + a_3 x^3 + \cdots .$$

Define addition and multiplication of power series in the same way as for power series with real or complex coefficients i.e., extend polynomial addition and multiplication to power series as though they were "polynomials of infinite degree":

$$\sum_{n=0}^{\infty} a_n x^n + \sum_{n=0}^{\infty} b_n x^n = \sum_{n=0}^{\infty} (a_n + b_n) x^n$$

$$\sum_{n=0}^{\infty} a_n x^n \times \sum_{n=0}^{\infty} b_n x^n = \sum_{n=0}^{\infty} \left(\sum_{k=0}^{n} a_k b_{n-k} \right) x^n .$$

(The term "formal" is used here to indicate that convergence is not considered, so that formal power series need not represent functions on R.)
   (a) Prove that $R[[x]]$ is a commutative ring with 1.
   (b) Show that $1 - x$ is a unit in $R[[x]]$ with inverse $1 + x + x^2 + \cdots$.
   (c) Prove that $\sum_{n=0}^{\infty} a_n x^n$ is a unit in $R[[x]]$ if and only if a_0 is a unit in R.

4. Prove that if R is an integral domain then the ring of formal power series $R[[x]]$ is also an integral domain.

5. Let F be a field and define the ring $F((x))$ of *formal Laurent series* with coefficients from F by

$$F((x)) = \{ \sum_{n \geq N} a_n x^n \mid a_n \in F \text{ and } N \in \mathbb{Z} \}.$$

(Every element of $F((x))$ is a power series in x plus a polynomial in $1/x$, i.e., each element of $F((x))$ has only a finite number of terms with negative powers of x.)
   (a) Prove that $F((x))$ is a field.
   (b) Define the map

$$v : F((x))^{\times} \to \mathbb{Z} \quad \text{by} \quad v\left(\sum_{n \geq N} a_n x^n \right) = N$$

where a_N is the first nonzero coefficient of the series (i.e., N is the "order of zero or pole of the series at 0"). Prove that v is a discrete valuation on $F((x))$ whose discrete valuation ring is $F[[x]]$, the ring of formal power series (cf. Exercise 26, Section 1).

6. Let S be a ring with identity $1 \neq 0$. Let $n \in \mathbb{Z}^+$ and let A be an $n \times n$ matrix with entries from S whose i, j entry is a_{ij}. Let E_{ij} be the element of $M_n(S)$ whose i, j entry is 1 and whose other entries are all 0.

(a) Prove that $E_{ij}A$ is the matrix whose i^{th} row equals the jth row of A and all other rows are zero.

(b) Prove that AE_{ij} is the matrix whose j^{th} column equals the ith column of A and all other columns are zero.

(c) Deduce that $E_{pq}AE_{rs}$ is the matrix whose p, s entry is a_{qr} and all other entries are zero.

7. Prove that the center of the ring $M_n(R)$ is the set of scalar matrices (cf. Exercise 7, Section 1). [Use the preceding exercise.]

8. Let S be any ring and let $n \geq 2$ be an integer. Prove that if A is any strictly upper triangular matrix in $M_n(S)$ then $A^n = 0$ (a strictly upper triangular matrix is one whose entries on and below the main diagonal are all zero).

9. Let $\alpha = r + r^2 - 2s$ and $\beta = -3r^2 + rs$ be the two elements of the integral group ring $\mathbb{Z}D_8$ described in this section. Compute the following elements of $\mathbb{Z}D_8$:
(a) $\beta\alpha$, (b) α^2, (c) $\alpha\beta - \beta\alpha$, (d) $\beta\alpha\beta$.

10. Consider the following elements of the integral group ring $\mathbb{Z}S_3$:

$$\alpha = 3(1\ 2) - 5(2\ 3) + 14(1\ 2\ 3) \qquad \text{and} \qquad \beta = 6(1) + 2(2\ 3) - 7(1\ 3\ 2)$$

(where (1) is the identity of S_3). Compute the following elements:
(a) $\alpha + \beta$, (b) $2\alpha - 3\beta$, (c) $\alpha\beta$, (d) $\beta\alpha$, (e) α^2.

11. Repeat the preceding exercise under the assumption that the coefficients of α and β are in $\mathbb{Z}/3\mathbb{Z}$ (i.e., $\alpha, \beta \in \mathbb{Z}/3\mathbb{Z}S_3$).

12. Let $G = \{g_1, \dots, g_n\}$ be a finite group. Prove that the element $N = g_1 + g_2 + \dots + g_n$ is in the center of the group ring RG (cf. Exercise 7, Section 1).

13. Let $\mathcal{K} = \{k_1, \dots, k_m\}$ be a conjugacy class in the finite group G.
(a) Prove that the element $K = k_1 + \dots + k_m$ is in the center of the group ring RG (cf. Exercise 7, Section 1). [Check that $g^{-1}Kg = K$ for all $g \in G$.]
(b) Let $\mathcal{K}_1, \dots, \mathcal{K}_r$ be the conjugacy classes of G and for each $\mathcal{K}_i$ let K_i be the element of RG that is the sum of the members of $\mathcal{K}_i$. Prove that an element $\alpha \in RG$ is in the center of RG if and only if $\alpha = a_1 K_1 + a_2 K_2 + \dots + a_r K_r$ for some $a_1, a_2, \dots, a_r \in R$.

7.3 RING HOMOMORPHISMS AND QUOTIENT RINGS

A ring homomorphism is a map from one ring to another that respects the additive and multiplicative structures:

Definition. Let R and S be rings.
(1) A *ring homomorphism* is a map $\varphi : R \to S$ satisfying
    (i) $\varphi(a + b) = \varphi(a) + \varphi(b)$ for all $a, b \in R$ (so φ is a group homomorphism on the additive groups) and
    (ii) $\varphi(ab) = \varphi(a)\varphi(b)$ for all $a, b \in R$.
(2) The *kernel* of the ring homomorphism φ, denoted ker φ, is the set of elements of R that map to 0 in S (i.e., the kernel of φ viewed as a homomorphism of additive groups).
(3) A bijective ring homomorphism is called an *isomorphism*.

If the context is clear we shall simply use the term "homomorphism" instead of "ring homomorphism." Similarly, if A and B are rings, $A \cong B$ will always mean an isomorphism of rings unless otherwise stated.

Examples

(1) The map $\varphi : \mathbb{Z} \to \mathbb{Z}/2\mathbb{Z}$ defined by sending an even integer to 0 and an odd integer to 1 is a ring homomorphism. The map is additive since the sum of two even or odd integers is even and the sum of an even integer and an odd integer is odd. The map is multiplicative since the product of two odd integers is odd and the product of an even integer with any integer is even. The kernel of φ (the fiber of φ above $0 \in \mathbb{Z}/2\mathbb{Z}$) is the set of even integers. The fiber of φ above $1 \in \mathbb{Z}/2\mathbb{Z}$ is the set of odd integers.

(2) For $n \in \mathbb{Z}$ the maps $\varphi_n : \mathbb{Z} \to \mathbb{Z}$ defined by $\varphi_n(x) = nx$ are *not* in general ring homomorphisms because $\varphi_n(xy) = nxy$ whereas $\varphi_n(x)\varphi_n(y) = nxny = n^2xy$. Hence φ_n is a ring homomorphism only when $n^2 = n$, i.e., $n = 0, 1$. Note however that φ_n is always a *group homomorphism* on the additive groups. Thus care should be exercised when dealing with rings to be sure to check that *both* ring operations are preserved. Note that φ_0 is the zero homomorphism and φ_1 is the identity homomorphism.

(3) Let $\varphi : \mathbb{Q}[x] \to \mathbb{Q}$ be the map from the ring of polynomials in x with rational coefficients to the rationals defined by $\varphi(p(x)) = p(0)$ (i.e., mapping the polynomial to its constant term). Then φ is a ring homomorphism since the constant term of the sum of two polynomials is the sum of their constant terms and the constant term of the product of two polynomials is the product of their constant terms. The fiber above $a \in \mathbb{Q}$ consists of the set of polynomials with a as constant term. In particular, the kernel of φ consists of the polynomials with constant term 0.

Proposition 5. Let R and S be rings and let $\varphi : R \to S$ be a homomorphism.

(1) The image of φ is a subring of S.

(2) The kernel of φ is a subring of R. Furthermore, if $\alpha \in \ker \varphi$ then $r\alpha$ and $\alpha r \in \ker \varphi$ for every $r \in R$, i.e., $\ker \varphi$ is closed under multiplication by elements from R.

Proof: (1) If $s_1, s_2 \in \text{im } \varphi$ then $s_1 = \varphi(r_1)$ and $s_2 = \varphi(r_2)$ for some $r_1, r_2 \in R$. Then $\varphi(r_1 - r_2) = s_1 - s_2$ and $\varphi(r_1 r_2) = s_1 s_2$. This shows $s_1 - s_2, s_1 s_2 \in \text{im } \varphi$, so the image of φ is closed under subtraction and under multiplication, hence is a subring of S.

(2) If $\alpha, \beta \in \ker \varphi$ then $\varphi(\alpha) = \varphi(\beta) = 0$. Hence $\varphi(\alpha - \beta) = 0$ and $\varphi(\alpha\beta) = 0$, so $\ker \varphi$ is closed under subtraction and under multiplication, so is a subring of R. Similarly, for any $r \in R$ we have $\varphi(r\alpha) = \varphi(r)\varphi(\alpha) = \varphi(r)\,0 = 0$, and also $\varphi(\alpha r) = \varphi(\alpha)\varphi(r) = 0\,\varphi(r) = 0$, so $r\alpha, \alpha r \in \ker \varphi$.

In the case of a homomorphism φ of groups we saw that the fibers of the homomorphism have the structure of a group naturally isomorphic to the image of φ, which led to the notion of a quotient group by a normal subgroup. An analogous result is true for a homomorphism of rings.

Let $\varphi : R \to S$ be a ring homomorphism with kernel I. Since R and S are in particular additive abelian groups, φ is in particular a homomorphism of abelian groups and the fibers of φ are the additive cosets $r + I$ of the kernel I (more precisely, if r is any element of R mapping to $a \in S$, $\varphi(r) = a$, then the fiber of φ over a is the coset $r + I$ of the kernel I). These fibers have the structure of a ring naturally isomorphic to the image of φ: if X is the fiber over $a \in S$ and Y is the fiber over $b \in S$, then $X + Y$ is the fiber over $a + b$ and XY is the fiber over ab. In terms of cosets of the kernel I this addition and multiplication is

$$(r + I) + (s + I) = (r + s) + I \tag{7.1}$$

$$(r + I) \times (s + I) = (rs) + I. \tag{7.2}$$

As in the case for groups, the verification that these operations define a ring structure on the collection of cosets of the kernel I ultimately rests on the corresponding ring properties of S. This ring of cosets is called the *quotient ring* of R by $I = \ker \varphi$ and is denoted R/I. Note that the additive structure of the ring R/I is just the additive quotient group of the additive abelian group R by the (necessarily normal) subgroup I. When I is the kernel of some homomorphism φ this additive abelian quotient group also has a multiplicative structure, defined by (7.2), which makes R/I into a ring.

As in the case for groups, we can also consider whether (1) and (2) can be used to define a ring structure on the collection of cosets of an *arbitrary* subgroup I of R. Note that since R is an abelian additive group, the subgroup I is necessarily normal so that the quotient R/I of cosets of I is automatically an additive abelian group. The question then is whether this quotient group also has a *multiplicative* structure induced from the multiplication in R, defined by (2). The answer is no in general (just as the answer is no in trying to form the quotient by an arbitrary subgroup of a group), which leads to the notion of an *ideal* in R (the analogue for rings of a normal subgroup of a group). We shall then see that the ideals of R are exactly the kernels of the ring homomorphisms of R (the analogue for rings of the characterization of normal subgroups as the kernels of group homomorphisms).

Let I be an arbitrary subgroup of the additive group R. We consider when the multiplication of cosets in (2) is well defined and makes the additive abelian group R/I into a ring. The statement that the multiplication in (2) is well defined is the statement that the multiplication is independent of the particular representatives r and s chosen, i.e., that we obtain the same coset on the right if instead we use the representatives $r + \alpha$ and $s + \beta$ for any $\alpha, \beta \in I$. In other words, we must have

$$(r + \alpha)(s + \beta) + I = rs + I \tag{$*$}$$

for all $r, s \in R$ and all $\alpha, \beta \in I$.

Letting $r = s = 0$, we see that I must be closed under multiplication, i.e., I must be a *subring* of R.

Next, by letting $s = 0$ and letting r be arbitrary, we see that we must have $r\beta \in I$ for every $r \in R$ and every $\beta \in I$, i.e., that I must be closed under multiplication on the left by elements from R. Letting $r = 0$ and letting s be arbitrary, we see similarly that I must be closed under multiplication on the right by elements from R.

Conversely, if I is closed under multiplication on the left and on the right by elements from R then the relation $(*)$ is satisfied for all $\alpha, \beta \in I$. Hence this is a necessary and sufficient condition for the multiplication in (2) to be well defined.

Finally, if the multiplication of cosets defined by (2) is well defined, then this multiplication makes the additive quotient group R/I into a ring. Each ring axiom in the quotient follows directly from the corresponding axiom in R. For example, one of the distributive laws is verified as follows:

$$
\begin{aligned}
(r + I)[(s + I) + (t + I)] &= (r + I)[(s + t) + I] \\
&= r(s + t) + I = (rs + rt) + I \\
&= (rs + I) + (rt + I) \\
&= [(r + I)(s + I)] + [(r + I)(t + I)].
\end{aligned}
$$

This shows that the quotient R/I of the ring R by a subgroup I has a natural ring structure if and only if I is also closed under multiplication on the left and on the right by elements from R (so in particular must be a subring of R since it is closed under multiplication). As mentioned, such subrings I are called the *ideals* of R:

Definition. Let R be a ring, let I be a subset of R and let $r \in R$.
   (1) $rI = \{ra \mid a \in I\}$ and $Ir = \{ar \mid a \in I\}$.
   (2) A subset I of R is a *left ideal* of R if
      (i) I is a subring of R, and
      (ii) I is closed under left multiplication by elements from R, i.e., $rI \subseteq I$
         for all $r \in R$.
   Similarly I is a *right ideal* if (i) holds and in place of (ii) one has
      (ii)′ I is closed under right multiplication by elements from R, i.e., $Ir \subseteq I$
         for all $r \in R$.
   (3) A subset I that is both a left ideal and a right ideal is called an *ideal* (or, for added emphasis, a *two-sided ideal*) of R.

For commutative rings the notions of left, right and two-sided ideal coincide. We emphasize that to prove a subset I of a ring R is an ideal it is necessary to prove that I is nonempty, closed under subtraction and closed under multiplication by all the elements of R (and not just by elements of I). If R has a 1 then $(-1)a = -a$ so in this case I is an ideal if it is nonempty, closed under addition and closed under multiplication by all the elements of R.

Note also that the last part of Proposition 5 proves that the kernel of any ring homomorphism is an ideal.

We summarize the preceding discussion in the following proposition.

Proposition 6. Let R be a ring and let I be an ideal of R. Then the (additive) quotient group R/I is a ring under the binary operations:

$$
(r + I) + (s + I) = (r + s) + I \quad \text{and} \quad (r + I) \times (s + I) = (rs) + I
$$

for all $r, s \in R$. Conversely, if I is any subgroup such that the above operations are well defined, then I is an ideal of R.

Definition. When I is an ideal of R the ring R/I with the operations in the previous proposition is called the *quotient ring* of R by I.

Theorem 7.

    (1) *(The First Isomorphism Theorem for Rings)* If $\varphi : R \to S$ is a homomorphism of rings, then the kernel of φ is an ideal of R, the image of φ is a subring of S and $R/\ker \varphi$ is isomorphic as a ring to $\varphi(R)$.

    (2) If I is any ideal of R, then the map

$$R \to R/I \qquad \text{defined by} \qquad r \mapsto r + I$$

is a surjective ring homomorphism with kernel I (this homomorphism is called the *natural projection* of R onto R/I). Thus every ideal is the kernel of a ring homomorphism and vice versa.

Proof: This is just a matter of collecting previous calculations. If I is the kernel of φ, then the cosets (under addition) of I are precisely the fibers of φ. In particular, the cosets $r + I$, $s + I$ and $rs + I$ are the fibers of φ over $\varphi(r)$, $\varphi(s)$ and $\varphi(rs)$, respectively. Since φ is a ring homomorphism $\varphi(r)\varphi(s) = \varphi(rs)$, hence $(r + I)(s + I) = rs + I$. Multiplication of cosets is well defined and so I is an ideal and R/I is a ring. The correspondence $r + I \mapsto \varphi(r)$ is a bijection between the rings R/I and $\varphi(R)$ which respects addition and multiplication, hence is a ring isomorphism.

If I is any ideal, then R/I is a ring (in particular is an abelian group) and the map $\pi : r \mapsto r + I$ is a group homomorphism with kernel I. It remains to check that π is a ring homomorphism. This is immediate from the definition of multiplication in R/I:

$$\pi : rs \mapsto rs + I = (r + I)(s + I) = \pi(r)\pi(s).$$

As with groups we shall often use the bar notation for reduction mod I: $\bar{r} = r + I$. With this notation the addition and multiplication in the quotient ring R/I become simply $\bar{r} + \bar{s} = \overline{r + s}$ and $\bar{r}\bar{s} = \overline{rs}$.

Examples

Let R be a ring.

    (1) The subrings R and $\{0\}$ are ideals. An ideal I is *proper* if $I \neq R$. The ideal $\{0\}$ is called the *trivial ideal* and is denoted by 0.

    (2) It is immediate that $n\mathbb{Z}$ is an ideal of $\mathbb{Z}$ for any $n \in \mathbb{Z}$ and these are the only ideals of $\mathbb{Z}$ since in particular these are the only subgroups of $\mathbb{Z}$. The associated quotient ring is $\mathbb{Z}/n\mathbb{Z}$ (which explains the choice of notation and which we have now proved is a ring), introduced in Chapter 0. For example, if $n = 15$ then the elements of $\mathbb{Z}/15\mathbb{Z}$ are the cosets $\bar{0}, \bar{1}, \ldots, \overline{13}, \overline{14}$. To add (or multiply) in the quotient, simply choose any representatives for the two cosets, add (multiply, respectively) these representatives in the integers $\mathbb{Z}$, and take the corresponding coset containing this sum (product, respectively). For example, $\bar{7} + \overline{11} = \overline{18}$ and $\overline{18} = \bar{3}$, so $\bar{7} + \overline{11} = \bar{3}$ in $\mathbb{Z}/15\mathbb{Z}$. Similarly, $\bar{7}\,\overline{11} = \overline{77} = \bar{2}$ in $\mathbb{Z}/15\mathbb{Z}$. We could also express this by writing $7 + 11 \equiv 3 \bmod 15$, $7(11) \equiv 2 \bmod 15$.

        The natural projection $\mathbb{Z} \to \mathbb{Z}/n\mathbb{Z}$ is called *reduction mod n* and will be discussed further at the end of these examples.

(3) Let $R = \mathbb{Z}[x]$ be the ring of polynomials in x with integer coefficients. Let I be the collection of polynomials whose terms are of degree at least 2 (i.e., having no terms of degree 0 or degree 1) together with the zero polynomial. Then I is an ideal: the sum of two such polynomials again has terms of degree at least 2 and the product of a polynomial whose terms are of degree at least 2 with *any* polynomial again only has terms of degree at least 2. Two polynomials $p(x), q(x)$ are in the same coset of I if and only if they differ by a polynomial whose terms are of degree at least 2, i.e., if and only if $p(x)$ and $q(x)$ have the same constant and first degree terms. For example, the polynomials $3 + 5x + x^3 + x^5$ and $3 + 5x - x^4$ are in the same coset of I. It follows easily that a complete set of representatives for the quotient R/I is given by the polynomials $a + bx$ of degree at most 1.

Addition and multiplication in the quotient are again performed by representatives. For example,

$$(\overline{1 + 3x}) + (\overline{-4 + 5x}) = \overline{-3 + 8x}$$

and

$$(\overline{1 + 3x})(\overline{-4 + 5x}) = \overline{(-4 - 7x + 15x^2)} = \overline{-4 - 7x}.$$

Note that in this quotient ring R/I we have $\bar{x}\,\bar{x} = \overline{x^2} = \bar{0}$, for example, so that R/I has zero divisors, even though $R = \mathbb{Z}[x]$ does not.

(4) Let A be a ring, let X be any nonempty set and let R be the ring of all functions from X to A. For each fixed $c \in X$ the map

$$E_c : R \to A \quad \text{defined by} \quad E_c(f) = f(c)$$

(called *evaluation at* c) is a ring homomorphism because the operations in R are pointwise addition and multiplication of functions. The kernel of E_c is given by $\{f \in R \mid f(c) = 0\}$ (the set of functions from X to A that vanish at c). Also, E_c is surjective: given any $a \in A$ the constant function $f(x) = a$ maps to a under evaluation at c. Thus $R/\ker E_c \cong A$.

Similarly, let X be the closed interval $[0,1]$ in $\mathbb{R}$ and let R be the ring of all continuous real valued functions on $[0,1]$. For each $c \in [0, 1]$, evaluation at c is a surjective ring homomorphism (since R contains the constant functions) and so $R/\ker E_c \cong \mathbb{R}$. The kernel of E_c is the ideal of all continuous functions whose graph crosses the x-axis at c. More generally, the fiber of E_c above the real number y_0 is the set of all continuous functions that pass through the point (c, y_0).

(5) The map from the polynomial ring $R[x]$ to R defined by $p(x) \mapsto p(0)$ (evaluation at 0) is a ring homomorphism whose kernel is the set of all polynomials whose constant term is zero, i.e., $p(0) = 0$. We can compose this homomorphism with any homomorphism from R to another ring S to obtain a ring homomorphism from $R[x]$ to S. For example, let $R = \mathbb{Z}$ and consider the homomorphism $\mathbb{Z}[x] \to \mathbb{Z}/2\mathbb{Z}$ defined by the composition $p(x) \mapsto p(0) \mapsto p(0) \bmod 2 \in \mathbb{Z}/2\mathbb{Z}$. The kernel of this composite map is given by $\{p(x) \in \mathbb{Z}[x] \mid p(0) \in 2\mathbb{Z}\}$, i.e., the set of all polynomials with integer coefficients whose constant term is even. The other fiber of this homomorphism is the coset of polynomials whose constant term is odd, as we determined earlier. Since the homomorphism is clearly surjective, the quotient ring is $\mathbb{Z}/2\mathbb{Z}$.

(6) Fix some $n \in \mathbb{Z}$ with $n \geq 2$ and consider the noncommutative ring $M_n(R)$. If J is any ideal of R then $M_n(J)$, the $n \times n$ matrices whose entries come from J, is a two-sided ideal of $M_n(R)$. This ideal is the kernel of the surjective homomorphism $M_n(R) \to M_n(R/J)$ which reduces each entry of a matrix mod J, i.e., which maps each entry a_{ij} to $\overline{a_{ij}}$ (here bar denotes passage to R/J). For instance, when $n = 3$ and $R = \mathbb{Z}$, the 3×3 matrices whose entries are all even is the two-sided ideal $M_3(2\mathbb{Z})$

of $M_3(\mathbb{Z})$ and the quotient $M_3(\mathbb{Z})/M_3(2\mathbb{Z})$ is isomorphic to $M_3(\mathbb{Z}/2\mathbb{Z})$. If the ring R has an identity then the exercises below show that every two-sided ideal of $M_n(R)$ is of the form $M_n(J)$ for some two-sided ideal J of R.

(7) Let R be a commutative ring with 1 and let $G = \{g_1, \ldots, g_n\}$ be a finite group. The map from the group ring RG to R defined by $\sum_{i=1}^n a_i g_i \mapsto \sum_{i=1}^n a_i$ is easily seen to be a homomorphism, called the *augmentation map* . The kernel of the augmentation map, the *augmentation ideal*, is the set of elements of RG whose coefficients sum to 0. For example, $g_i - g_j$ is an element of the augmentation ideal for all i, j. Since the augmentation map is surjective, the quotient ring is isomorphic to R.

Another ideal in RG is $\{\sum_{i=1}^n ag_i \mid a \in R\}$, i.e., the formal sums whose coefficients are all equal (equivalently, all R-multiples of the element $g_1 + \cdots + g_n$).

(8) Let R be a commutative ring with identity $1 \neq 0$ and let $n \in \mathbb{Z}$ with $n \geq 2$. We exhibit some one-sided ideals in the ring $M_n(R)$. For each $j \in \{1, 2, \ldots, n\}$ let L_j be the set of all $n \times n$ matrices in $M_n(R)$ with arbitrary entries in the jth column and zeros in all other columns. It is clear that L_j is closed under subtraction. It follows directly from the definition of matrix multiplication that for any matrix $T \in M_n(R)$ and any $A \in L_j$ the product TA has zero entries in the ith column for all $i \neq j$. This shows L_j is a *left ideal* of $M_n(R)$. Moreover, L_j is *not* a *right* ideal (hence is not a two-sided ideal). To see this, let E_{pq} be the matrix with 1 in the pth row and q^{th} column and zeros elsewhere ($p, q \in \{1, \ldots, n\}$). Then $E_{1j} \in L_j$ but $E_{1j}E_{ji} = E_{1i} \notin L_j$ if $i \neq j$, so L_j is not closed under right multiplication by arbitrary ring elements. An analogous argument shows that if R_j is the set of all $n \times n$ matrices in $M_n(R)$ with arbitrary entries in the jth row and zeros in all other rows, then R_j is a *right* ideal which is not a *left* ideal. These one-sided ideals will play an important role in Part VI.

Example: (The Reduction Homomorphism)

The canonical projection map from $\mathbb{Z}$ to $\mathbb{Z}/n\mathbb{Z}$ obtained by factoring out by the ideal $n\mathbb{Z}$ of $\mathbb{Z}$ is usually referred to as "reducing modulo n." The fact that this is a *ring homomorphism* has important consequences for elementary number theory. For example, suppose we are trying to solve the equation

$$x^2 + y^2 = 3z^2$$

in *integers x, y* and *z* (such problems are frequently referred to as *Diophantine equations* after Diophantus, who was one of the first to systematically examine the existence of *integer* solutions of equations). Suppose such integers exist. Observe first that we may assume x, y and z have no factors in common, since otherwise we could divide through this equation by the square of this common factor and obtain another set of integer solutions smaller than the initial ones. This equation simply states a relation between these elements in the *ring* $\mathbb{Z}$. As such, the same relation must also hold in any *quotient* ring as well. In particular, this relation must hold in $\mathbb{Z}/n\mathbb{Z}$ for any integer n. The choice $n = 4$ is particularly efficacious, for the following reason: the squares mod 4 are just $0^2, 1^2, 2^2, 3^2$, i.e., 0, 1 (mod 4). Reading the above equation mod 4 (that is, considering this equation in the quotient ring $\mathbb{Z}/4\mathbb{Z}$), we must have

$$\begin{Bmatrix} 0 \\ 1 \end{Bmatrix} + \begin{Bmatrix} 0 \\ 1 \end{Bmatrix} \equiv 3 \begin{Bmatrix} 0 \\ 1 \end{Bmatrix} \equiv \begin{Bmatrix} 0 \\ 3 \end{Bmatrix} \quad (\text{mod } 4)$$

where the $\begin{Bmatrix} 0 \\ 1 \end{Bmatrix}$, for example, indicates that either a 0 or a 1 may be taken. Checking the few possibilities shows that we must take the 0 each time. This means that each of x, y and z must be even integers (squares of the odd integers gave us 1 mod 4). But this contradicts the assumption of no common factors for these integers, and shows that this equation has *no solutions in nonzero integers*.

Note that even had solutions existed, this technique gives information about the possible residues of the solutions mod n (since we could just as well have examined the possibilities mod n as mod 4) and note that for each choice of n we have only a *finite* problem to solve because there are only finitely many residue classes mod n. Together with the Chinese Remainder Theorem (described in Section 6), we can then determine the possible solutions modulo very large integers, which greatly assists in finding them numerically (when they exist). We also observe that this technique has a number of limitations — for example, there are equations which have nontrivial solutions modulo every integer, but which do not have nontrivial integer solutions. An easy example (but extremely hard to verify that it does indeed have this property) is the equation

$$3x^3 + 4y^3 + 5z^3 = 0.$$

Another example is provided by $x^2 + y^2 + z^2 + w^2 = -1$, which clearly has no integer solutions (or even real solutions), but has solutions modulo every integer n (this follows easily given the result from elementary number theory that every positive integer is a sum of four squares, since then $n - 1$ is a sum of four squares).

As a final example of this technique, we mention that the map from the ring $\mathbb{Z}[x]$ of polynomials with integer coefficients to the ring $\mathbb{Z}/p\mathbb{Z}[x]$ of polynomials with coefficients in $\mathbb{Z}/p\mathbb{Z}$ for a prime p given by *reducing the coefficients modulo p* is a ring homomorphism. This example of reduction will be used in Chapter 9 in trying to determine whether polynomials can be factored.

The following theorem gives the remaining Isomorphism Theorems for rings. Each of these may be proved as follows: first use the corresponding theorem from group theory to obtain an isomorphism of *additive groups* (or correspondence of groups, in the case of the Fourth Isomorphism Theorem) and then check that this group isomorphism (or correspondence, respectively) is a multiplicative map, and so defines a *ring* isomorphism. In each case the verification is immediate from the definition of multiplication in quotient rings. For example, the map that gives the isomorphism in (2) below is defined by $\varphi : r + I \mapsto r + J$. This map is multiplicative since $(r_1 + I)(r_2 + I) = r_1 r_2 + I$ by the definition of the multiplication in the quotient ring R/I, and $r_1 r_2 + I \mapsto r_1 r_2 + J = (r_1 + J)(r_2 + J)$ by the definition of the multiplication in the quotient ring R/J, i.e., $\varphi(r_1 r_2) = \varphi(r_1)\varphi(r_2)$. The proofs for the other parts of the theorem are similar.

Theorem 8. Let R be a ring.
   (1) *(The Second Isomorphism Theorem for Rings)* Let A be a subring and let B be an ideal of R. Then $A + B = \{a + b \mid a \in A, \ b \in B\}$ is a subring of R, $A \cap B$ is an ideal of A and $(A + B)/B \cong A/(A \cap B)$.
   (2) *(The Third Isomorphism Theorem for Rings)* Let I and J be ideals of R with $I \subseteq J$. Then J/I is an ideal of R/I and $(R/I)/(J/I) \cong R/J$.
   (3) *(The Fourth or Lattice Isomorphism Theorem for Rings)* Let I be an ideal of R. The correspondence $A \leftrightarrow A/I$ is an inclusion preserving bijection between the set of subrings A of R that contain I and the set of subrings of R/I. Furthermore, A (a subring containing I) is an ideal of R if and only if A/I is an ideal of R/I.

Example

Let $R = \mathbb{Z}$ and let I be the ideal $12\mathbb{Z}$. The quotient ring $\overline{R} = R/I = \mathbb{Z}/12\mathbb{Z}$ has ideals $\overline{R}$, $2\mathbb{Z}/12\mathbb{Z}$, $3\mathbb{Z}/12\mathbb{Z}$, $4\mathbb{Z}/12\mathbb{Z}$, $6\mathbb{Z}/12\mathbb{Z}$, and $\overline{0} = 12\mathbb{Z}/12\mathbb{Z}$ corresponding to the ideals $R = \mathbb{Z}$, $2\mathbb{Z}$, $3\mathbb{Z}$, $4\mathbb{Z}$, $6\mathbb{Z}$ and $12\mathbb{Z} = I$ of R containing I, respectively.

If I and J are ideals in the ring R then the set of sums $a + b$ with $a \in I$ and $b \in J$ is not only a subring of R (as in the Second Isomorphism Theorem for Rings), but is an *ideal* in R (the set is clearly closed under sums and $r(a + b) = ra + rb \in I + J$ since $ra \in I$ and $rb \in J$). We can also define the product of two ideals:

Definition. Let I and J be ideals of R.

(1) Define the *sum* of I and J by $I + J = \{a + b \mid a \in I, \ b \in J\}$.

(2) Define the *product* of I and J, denoted by IJ, to be the set of all finite sums of elements of the form ab with $a \in I$ and $b \in J$.

(3) For any $n \geq 1$, define the nth *power* of I, denoted by I^n, to be the set consisting of all finite sums of elements of the form $a_1 a_2 \cdots a_n$ with $a_i \in I$ for all i. Equivalently, I^n is defined inductively by defining $I^1 = I$, and $I^n = II^{n-1}$ for $n = 2, 3, \ldots$.

It is easy to see that the sum $I + J$ of the ideals I and J is the smallest ideal of R containing both I and J and that the product IJ is an ideal contained in $I \cap J$ (but may be strictly smaller, cf. the exercises). Note also that the elements of the product ideal IJ are *finite sums* of products of elements ab from I and J. The set $\{ab \mid a \in I, \ b \in J\}$ consisting just of products of elements from I and J is in general not closed under addition, hence is not in general an ideal.

Examples

(1) Let $I = 6\mathbb{Z}$ and $J = 10\mathbb{Z}$ in $\mathbb{Z}$. Then $I + J$ consists of all integers of the form $6x + 10y$ with $x, y \in \mathbb{Z}$. Since every such integer is divisible by 2, the ideal $I + J$ is contained in $2\mathbb{Z}$. On the other hand, $2 = 6(2) + 10(-1)$ shows that the ideal $I + J$ contains the ideal $2\mathbb{Z}$, so that $6\mathbb{Z} + 10\mathbb{Z} = 2\mathbb{Z}$. In general, $m\mathbb{Z} + n\mathbb{Z} = d\mathbb{Z}$, where d is the greatest common divisor of m and n. The product IJ consists of all finite sums of elements of the form $(6x)(10y)$ with $x, y \in \mathbb{Z}$, which clearly gives the ideal $60\mathbb{Z}$.

(2) Let I be the ideal in $\mathbb{Z}[x]$ consisting of the polynomials with integer coefficients whose constant term is even (cf. Example 5). The two polynomials 2 and x are contained in I, so both $4 = 2 \cdot 2$ and $x^2 = x \cdot x$ are elements of the product ideal $I^2 = II$, as is their sum $x^2 + 4$. It is easy to check, however, that $x^2 + 4$ cannot be written as a single product $p(x)q(x)$ of two elements of I.

EXERCISES

Let R be a ring with identity $1 \neq 0$.

1. Prove that the rings $2\mathbb{Z}$ and $3\mathbb{Z}$ are not isomorphic.

2. Prove that the rings $\mathbb{Z}[x]$ and $\mathbb{Q}[x]$ are not isomorphic.

3. Find all homomorphic images of $\mathbb{Z}$.

4. Find all ring homomorphisms from $\mathbb{Z}$ to $\mathbb{Z}/30\mathbb{Z}$. In each case describe the kernel and the image.

5. Describe all ring homomorphisms from the ring $\mathbb{Z} \times \mathbb{Z}$ to $\mathbb{Z}$. In each case describe the kernel and the image.

6. Decide which of the following are ring homomorphisms from $M_2(\mathbb{Z})$ to $\mathbb{Z}$:

(a) $\begin{pmatrix} a & b \\ c & d \end{pmatrix} \mapsto a$ (projection onto the 1,1 entry)

(b) $\begin{pmatrix} a & b \\ c & d \end{pmatrix} \mapsto a + d$ (the *trace* of the matrix)

(c) $\begin{pmatrix} a & b \\ c & d \end{pmatrix} \mapsto ad - bc$ (the *determinant* of the matrix).

7. Let $R = \{ \begin{pmatrix} a & b \\ 0 & d \end{pmatrix} \mid a, b, d \in \mathbb{Z} \}$ be the subring of $M_2(\mathbb{Z})$ of upper triangular matrices. Prove that the map

$$\varphi : R \to \mathbb{Z} \times \mathbb{Z} \quad \text{defined by} \quad \varphi : \begin{pmatrix} a & b \\ 0 & d \end{pmatrix} \mapsto (a, d)$$

is a surjective homomorphism and describe its kernel.

8. Decide which of the following are ideals of the ring $\mathbb{Z} \times \mathbb{Z}$:
  (a) $\{(a, a) \mid a \in \mathbb{Z}\}$
  (b) $\{(2a, 2b) \mid a, b \in \mathbb{Z}\}$
  (c) $\{(2a, 0) \mid a \in \mathbb{Z}\}$
  (d) $\{(a, -a) \mid a \in \mathbb{Z}\}$.

9. Decide which of the sets in Exercise 6 of Section 1 are ideals of the ring of all functions from $[0,1]$ to $\mathbb{R}$.

10. Decide which of the following are ideals of the ring $\mathbb{Z}[x]$:
  (a) the set of all polynomials whose constant term is a multiple of 3
  (b) the set of all polynomials whose coefficient of x^2 is a multiple of 3
  (c) the set of all polynomials whose constant term, coefficient of x and coefficient of x^2 are zero
  (d) $\mathbb{Z}[x^2]$ (i.e., the polynomials in which only even powers of x appear)
  (e) the set of polynomials whose coefficients sum to zero
  (f) the set of polynomials $p(x)$ such that $p'(0) = 0$, where $p'(x)$ is the usual first derivative of $p(x)$ with respect to x.

11. Let R be the ring of all continuous real valued functions on the closed interval $[0, 1]$. Prove that the map $\varphi : R \to \mathbb{R}$ defined by $\varphi(f) = \int_0^1 f(t)dt$ is a homomorphism of additive groups but not a ring homomorphism.

12. Let D be an integer that is not a perfect square in $\mathbb{Z}$ and let $S = \{ \begin{pmatrix} a & b \\ Db & a \end{pmatrix} \mid a, b \in \mathbb{Z} \}$.
  (a) Prove that S is a subring of $M_2(\mathbb{Z})$.
  (b) If D is not a perfect square in $\mathbb{Z}$ prove that the map $\varphi : \mathbb{Z}[\sqrt{D}] \to S$ defined by
$$\varphi(a + b\sqrt{D}) = \begin{pmatrix} a & b \\ Db & a \end{pmatrix} \text{ is a ring isomorphism.}$$
  (c) If $D \equiv 1 \bmod 4$ is squarefree, prove that the set $\{ \begin{pmatrix} a & b \\ (D-1)b/4 & a+b \end{pmatrix} \mid a, b \in \mathbb{Z} \}$

is a subring of $M_2(\mathbb{Z})$ and is isomorphic to the quadratic integer ring $\mathcal{O}$.

248

13. Prove that the ring $M_2(\mathbb{R})$ contains a subring that is isomorphic to $\mathbb{C}$.

14. Prove that the ring $M_4(\mathbb{R})$ contains a subring that is isomorphic to the real Hamilton Quaternions, $\mathbb{H}$.

15. Let X be a nonempty set and let $\mathcal{P}(X)$ be the Boolean ring of all subsets of X defined in Exercise 21 of Section 1. Let R be the ring of all functions from X into $\mathbb{Z}/2\mathbb{Z}$. For each $A \in \mathcal{P}(X)$ define the function

$$\chi_A : X \to \mathbb{Z}/2\mathbb{Z} \qquad \text{by} \qquad \chi_A(x) = \begin{cases} 1 & \text{if } x \in A \\ 0 & \text{if } x \notin A \end{cases}$$

(χ_A is called the *characteristic function of* A with values in $\mathbb{Z}/2\mathbb{Z}$). Prove that the map $\mathcal{P}(X) \to R$ defined by $A \mapsto \chi_A$ is a ring isomorphism.

16. Let $\varphi : R \to S$ be a surjective homomorphism of rings. Prove that the image of the center of R is contained in the center of S (cf. Exercise 7 of Section 1).

17. Let R and S be nonzero rings with identity and denote their respective identities by 1_R and 1_S. Let $\varphi : R \to S$ be a nonzero homomorphism of rings.
   (a) Prove that if $\varphi(1_R) \neq 1_S$ then $\varphi(1_R)$ is a zero divisor in S. Deduce that if S is an integral domain then every ring homomorphism from R to S sends the identity of R to the identity of S.
   (b) Prove that if $\varphi(1_R) = 1_S$ then $\varphi(u)$ is a unit in S and that $\varphi(u^{-1}) = \varphi(u)^{-1}$ for each unit u of R.

18. **(a)** If I and J are ideals of R prove that their intersection $I \cap J$ is also an ideal of R.
   (b) Prove that the intersection of an arbitrary nonempty collection of ideals is again an ideal (do not assume the collection is countable).

19. Prove that if $I_1 \subseteq I_2 \subseteq \cdots$ are ideals of R then $\cup_{n=1}^{\infty} I_n$ is an ideal of R.

20. Let I be an ideal of R and let S be a subring of R. Prove that $I \cap S$ is an ideal of S. Show by example that not every ideal of a subring S of a ring R need be of the form $I \cap S$ for some ideal I of R.

21. Prove that every (two-sided) ideal of $M_n(R)$ is equal to $M_n(J)$ for some (two-sided) ideal J of R. [Use Exercise 6(c) of Section 2 to show first that the set of entries of matrices in an ideal of $M_n(R)$ form an ideal in R.]

22. Let a be an element of the ring R.
   (a) Prove that $\{x \in R \mid ax = 0\}$ is a right ideal and $\{y \in R \mid ya = 0\}$ is a left ideal (called respectively the right and left *annihilators* of a in R).
   (b) Prove that if L is a left ideal of R then $\{x \in R \mid xa = 0 \text{ for all } a \in L\}$ is a two-sided ideal (called the left *annihilator* of L in R).

23. Let S be a subring of R and let I be an ideal of R. Prove that if $S \cap I = 0$ then $\overline{S} \cong S$, where the bar denotes passage to R/I.

24. Let $\varphi : R \to S$ be a ring homomorphism.
   (a) Prove that if J is an ideal of S then $\varphi^{-1}(J)$ is an ideal of R. Apply this to the special case when R is a subring of S and φ is the inclusion homomorphism to deduce that if J is an ideal of S then $J \cap R$ is an ideal of R.
   (b) Prove that if φ is surjective and I is an ideal of R then $\varphi(I)$ is an ideal of S. Give an example where this fails if φ is not surjective.

25. Assume R is a commutative ring with 1. Prove that the Binomial Theorem

$$(a + b)^n = \sum_{k=0}^{n} \binom{n}{k} a^k b^{n-k}$$

holds in R, where the binomial coefficient $\binom{n}{k}$ is interpreted in R as the sum $1+1+\cdots+1$ of the identity 1 in R taken $\binom{n}{k}$ times.

26. The *characteristic* of a ring R is the smallest positive integer n such that $1+1+\cdots+1 = 0$ (n times) in R; if no such integer exists the characteristic of R is said to be 0. For example, $\mathbb{Z}/n\mathbb{Z}$ is a ring of characteristic n for each positive integer n and $\mathbb{Z}$ is a ring of characteristic 0.

  (a) Prove that the map $\mathbb{Z} \to R$ defined by

$$k \mapsto \begin{cases} 1 + 1 + \cdots + 1 \ (k \text{ times}) & \text{if } k > 0 \\ 0 & \text{if } k = 0 \\ -1 - 1 - \cdots - 1 \ (-k \text{ times}) & \text{if } k < 0 \end{cases}$$

  is a ring homomorphism whose kernel is $n\mathbb{Z}$, where n is the characteristic of R (this explains the use of the terminology "characteristic 0" instead of the archaic phrase "characteristic ∞" for rings in which no sum of 1's is zero).
  (b) Determine the characteristics of the rings $\mathbb{Q}$, $\mathbb{Z}[x]$, $\mathbb{Z}/n\mathbb{Z}[x]$.
  (c) Prove that if p is a prime and if R is a commutative ring of characteristic p, then $(a+b)^p = a^p + b^p$ for all $a, b \in R$.

27. Prove that a nonzero Boolean ring has characteristic 2 (cf. Exercise 15, Section 1).

28. Prove that an integral domain has characteristic p, where p is either a prime or 0 (cf. Exercise 26).

29. Let R be a commutative ring. Recall (cf. Exercise 13, Section 1) that an element $x \in R$ is nilpotent if $x^n = 0$ for some $n \in \mathbb{Z}^+$. Prove that the set of nilpotent elements form an ideal — called the *nilradical* of R and denoted by $\mathfrak{N}(R)$. [Use the Binomial Theorem to show $\mathfrak{N}(R)$ is closed under addition.]

30. Prove that if R is a commutative ring and $\mathfrak{N}(R)$ is its nilradical (cf. the preceding exercise) then zero is the only nilpotent element of $R/\mathfrak{N}(R)$ i.e., prove that $\mathfrak{N}(R/\mathfrak{N}(R)) = 0$.

31. Prove that the elements $\begin{pmatrix} 0 & 1 \\ 0 & 0 \end{pmatrix}$ and $\begin{pmatrix} 0 & 0 \\ 1 & 0 \end{pmatrix}$ are nilpotent elements of $M_2(\mathbb{Z})$ whose sum is not nilpotent (note that these two matrices do not commute). Deduce that the set of nilpotent elements in the noncommutative ring $M_2(\mathbb{Z})$ is not an ideal.

32. Let $\varphi : R \to S$ be a homomorphism of rings. Prove that if x is a nilpotent element of R then $\varphi(x)$ is nilpotent in S.

33. Assume R is commutative. Let $p(x) = a_n x^n + a_{n-1} x^{n-1} + \cdots + a_1 x + a_0$ be an element of the polynomial ring $R[x]$.
  (a) Prove that $p(x)$ is a unit in $R[x]$ if and only if a_0 is a unit and $a_1, a_2, \ldots, a_n$ are nilpotent in R. [See Exercise 14 of Section 1.]
  (b) Prove that $p(x)$ is nilpotent in $R[x]$ if and only if $a_0, a_1, \ldots, a_n$ are nilpotent elements of R.

34. Let I and J be ideals of R.
  (a) Prove that $I + J$ is the smallest ideal of R containing both I and J.
  (b) Prove that IJ is an ideal contained in $I \cap J$.
  (c) Give an example where $IJ \neq I \cap J$.
  (d) Prove that if R is commutative and if $I + J = R$ then $IJ = I \cap J$.

35. Let I, J, K be ideals of R.
  (a) Prove that $I(J + K) = IJ + IK$ and $(I + J)K = IK + JK$.
  (b) Prove that if $J \subseteq I$ then $I \cap (J + K) = J + (I \cap K)$.

36. Show that if I is the ideal of all polynomials in $\mathbb{Z}[x]$ with zero constant term then $I^n = \{a_n x^n + a_{n+1} x^{n+1} + \cdots + a_{n+m} x^{n+m} \mid a_i \in \mathbb{Z}, \ m \geq 0\}$ is the set of polynomials whose first nonzero term has degree at least n.

37. An ideal N is called *nilpotent* if N^n is the zero ideal for some $n \geq 1$. Prove that the ideal $p\mathbb{Z}/p^m\mathbb{Z}$ is a nilpotent ideal in the ring $\mathbb{Z}/p^m\mathbb{Z}$.

7.4 PROPERTIES OF IDEALS

Throughout this section R is a ring with identity $1 \neq 0$.

Definition. Let A be any subset of the ring R.

(1) Let (A) denote the smallest ideal of R containing A, called *the ideal generated by A*.

(2) Let RA denote the set of all finite sums of elements of the form ra with $r \in R$ and $a \in A$ i.e., $RA = \{r_1 a_1 + r_2 a_2 + \cdots + r_n a_n \mid r_i \in R, \ a_i \in A, \ n \in \mathbb{Z}^+\}$ (where the convention is $RA = 0$ if $A = \emptyset$).

Similarly, $AR = \{a_1 r_1 + a_2 r_2 + \cdots + a_n r_n \mid r_i \in R, \ a_i \in A, \ n \in \mathbb{Z}^+\}$ and $RAR = \{r_1 a_1 r_1' + r_2 a_2 r_2' + \cdots + r_n a_n r_n' \mid r_i, r_i' \in R, \ a_i \in A, \ n \in \mathbb{Z}^+\}$.

(3) An ideal generated by a single element is called a *principal ideal*.

(4) An ideal generated by a finite set is called a *finitely generated ideal*.

When $A = \{a\}$ or $\{a_1, a_2, \ldots\}$, etc., we shall drop the set brackets and simply write $(a), (a_1, a_2, \ldots)$ for (A), respectively.

The notion of ideals generated by subsets of a ring is analogous to that of subgroups generated by subsets of a group (Section 2.4). Since the intersection of any nonempty collection of ideals of R is also an ideal (cf. Exercise 18, Section 3) and A is always contained in at least one ideal (namely R), we have

$$(A) = \bigcap_{\substack{I \text{ an ideal} \\ A \subseteq I}} I \ ,$$

i.e., (A) is the intersection of all ideals of R that contain the set A.

The *left ideal generated by A* is the intersection of all left ideals of R that contain A. This left ideal is obtained from A by closing A under all the operations that define a left ideal. It is immediate from the definition that RA is closed under addition and under left multiplication by any ring element. Since R has an identity, RA contains A. Thus RA is a left ideal of R which contains A. Conversely, any left ideal which contains A must contain all finite sums of elements of the form ra, $r \in R$ and $a \in A$ and so must contain RA. Thus RA *is precisely the left ideal generated by A*. Similarly, AR *is the right ideal generated by A* and RAR *is the (two-sided) ideal generated by A*. In particular,

if R is commutative then $RA = AR = RAR = (A)$.

When R is a commutative ring and $a \in R$, the principal ideal (a) generated by a is just the set of all R-multiples of a. If R is not commutative, however, the set

$\{ras \mid r, s \in R\}$ is not necessarily the two-sided ideal generated by a since it need not be closed under addition (in this case the ideal generated by a is the ideal RaR, which consists of all *finite sums* of elements of the form ras, $r, s \in R$).

The formation of principal ideals in a commutative ring is a particularly simple way of creating ideals, similar to generating cyclic subgroups of a group. Notice that the element $b \in R$ belongs to the ideal (a) if and only if $b = ra$ for some $r \in R$, i.e., if and only if b is a *multiple of* a or, put another way, a *divides* b in R. Also, $b \in (a)$ if and only if $(b) \subseteq (a)$. Thus containment relations between ideals, in particular between principal ideals, is seen to capture some of the arithmetic of general commutative rings. Commutative rings in which all ideals are principal are among the easiest to study and these will play an important role in Chapters 8 and 9.

Examples

(1) The trivial ideal 0 and the ideal R are both principal: $0 = (0)$ and $R = (1)$.

(2) In $\mathbb{Z}$ we have $n\mathbb{Z} = \mathbb{Z}n = (n) = (-n)$ for all integers n. Thus our notation for aR is consistent with the definition of $n\mathbb{Z}$ we have been using. As noted in the preceding section, these are all the ideals of $\mathbb{Z}$ so *every* ideal of $\mathbb{Z}$ is principal. For positive integers n and m, $n\mathbb{Z} \subseteq m\mathbb{Z}$ if and only if m divides n in $\mathbb{Z}$, so the lattice of ideals containing $n\mathbb{Z}$ is the same as the lattice of divisors of n. Furthermore, the ideal generated by two nonzero integers n and m is the principal ideal generated by their greatest common divisor, d: $(n, m) = (d)$. The notation for (n, m) as the greatest common divisor of n and m is thus consistent with the same notation for the ideal generated by n and m (although a principal generator for the ideal generated by n and m is determined only up to a $\pm$ sign — we could make it unique by choosing a nonnegative generator). In particular, n and m are relatively prime if and only if $(n, m) = (1)$.

(3) We show that the ideal $(2, x)$ generated by 2 and x in $\mathbb{Z}[x]$ is *not* a principal ideal. Observe that $(2, x) = \{2p(x) + xq(x) \mid p(x), q(x) \in \mathbb{Z}[x]\}$ and so this ideal consists precisely of the polynomials with integer coefficients whose constant term is even (as discussed in Example 5 in the preceding section) — in particular, this is a proper ideal. Assume by way of contradiction that $(2, x) = (a(x))$ for some $a(x) \in \mathbb{Z}[x]$. Since $2 \in (a(x))$ there must be some $p(x)$ such that $2 = p(x)a(x)$. The degree of $p(x)a(x)$ equals degree $p(x) +$ degree $a(x)$, hence both $p(x)$ and $a(x)$ must be constant polynomials, i.e., integers. Since 2 is a prime number, $a(x), p(x) \in \{\pm 1, \pm 2\}$. If $a(x)$ were ± 1 then every polynomial would be a multiple of $a(x)$, contrary to $(a(x))$ being a proper ideal. The only possibility is $a(x) = \pm 2$. But now $x \in (a(x)) = (2) = (-2)$ and so $x = 2q(x)$ for some polynomial $q(x)$ with integer coefficients, clearly impossible. This contradiction proves that $(2, x)$ is not principal.

Note that the symbol (A) is ambiguous if the ring is not specified: the ideal generated by 2 and x in $\mathbb{Q}[x]$ is the entire ring (1) since it contains the element $\frac{1}{2}2 = 1$.

We shall see in Chapter 9 that for any *field* F, all ideals of $F[x]$ *are* principal.

(4) If R is the ring of all functions from the closed interval $[0,1]$ into $\mathbb{R}$ let M be the ideal $\{f \mid f(\frac{1}{2}) = 0\}$ (the kernel of evaluation at $\frac{1}{2}$). Let $g(x)$ be the function which is zero at $x = \frac{1}{2}$ and 1 at all other points. Then $f = fg$ for all $f \in M$ so M is a principal ideal with generator g. In fact, *any* function which is zero at $\frac{1}{2}$ and nonzero at all other points is another generator for the same ideal M.

On the other hand, if R is the ring of all *continuous* functions from $[0,1]$ to $\mathbb{R}$ then $\{f \mid f(\frac{1}{2}) = 0\}$ is *not* principal nor is it even finitely generated (cf. the exercises).

(5) If G is a finite group and R is a commutative ring with 1 then the augmentation ideal is generated by the set $\{g - 1 \mid g \in G\}$, although this need not be a minimal set of generators. For example, if G is a cyclic group with generator σ, then the augmentation ideal is a principal ideal with generator $\sigma - 1$.

Proposition 9. Let I be an ideal of R.
   (1) $I = R$ if and only if I contains a unit.
   (2) Assume R is commutative. Then R is a field if and only if its only ideals are 0 and R.

Proof: (1) If $I = R$ then I contains the unit 1. Conversely, if u is a unit in I with inverse v, then for any $r \in R$

$$r = r \cdot 1 = r(vu) = (rv)u \in I$$

hence $R = I$.

   (2) The ring R is a field if and only if every nonzero element is a unit. If R is a field every nonzero ideal contains a unit, so by the first part R is the only nonzero ideal. Conversely, if 0 and R are the only ideals of R let u be any nonzero element of R. By hypothesis $(u) = R$ and so $1 \in (u)$. Thus there is some $v \in R$ such that $1 = vu$, i.e., u is a unit. Every nonzero element of R is therefore a unit and so R is a field.

Corollary 10. If R is a field then any nonzero ring homomorphism from R into another ring is an injection.

Proof: The kernel of a ring homomorphism is an ideal. The kernel of a nonzero homomorphism is a proper ideal hence is 0 by the proposition.

These results show that the ideal structure of fields is trivial. Our approach to studying an algebraic structure through its homomorphisms will still play a fundamental role in field theory (Part IV) when we study injective homomorphisms (embeddings) of one field into another and automorphisms of fields (isomorphisms of a field to itself).

If D is a ring with identity $1 \neq 0$ in which the only left ideals and the only right ideals are 0 and D, then D is a division ring. Conversely, the only (left, right or two-sided) ideals in a division ring D are 0 and D, which gives an analogue of Proposition 9(2) if R is not commutative (see the exercises). However, if F is a field, then for any $n \geq 2$ the only two-sided ideals in the matrix ring $M_n(F)$ are 0 and $M_n(F)$, even though this is not a division ring (it does have proper, nontrivial, left and right ideals: cf. Section 3), which shows that Proposition 9(2) does not hold for noncommutative rings. Rings whose only two-sided ideals are 0 and the whole ring (which are called *simple rings*) will be studied in Chapter 18.

One important class of ideals are those which are not contained in any other proper ideal:

Definition. An ideal M in an arbitrary ring S is called a *maximal ideal* if $M \neq S$ and the only ideals containing M are M and S.

A general ring need not have maximal ideals. For example, take any abelian group which has no maximal subgroups (for example, $\mathbb{Q}$ — cf. Exercise 16, Section 6.1) and make it into a trivial ring by defining $ab = 0$ for all a, b. In such a ring the ideals are simply the subgroups and so there are no maximal ideals. The zero ring has no maximal ideals, hence any result involving maximal ideals forces a ring to be nonzero. The next proposition shows that rings with an identity $1 \neq 0$ always possess maximal ideals. Like many such general existence theorems (e.g., the result that a finitely generated group has maximal subgroups or that every vector space has a basis) the proof relies on Zorn's Lemma (see Appendix I). In many specific rings, however, the presence of maximal ideals is often obvious, independent of Zorn's Lemma.

Proposition 11. In a ring with identity every proper ideal is contained in a maximal ideal.

Proof: Let R be a ring with identity and let I be a proper ideal (so R cannot be the zero ring, i.e., $1 \neq 0$). Let $\mathcal{S}$ be the set of all proper ideals of R which contain I. Then $\mathcal{S}$ is nonempty ($I \in \mathcal{S}$) and is partially ordered by inclusion. If $\mathcal{C}$ is a chain in $\mathcal{S}$, define J to be the union of all ideals in $\mathcal{C}$:

$$J = \bigcup_{A \in \mathcal{C}} A.$$

We first show that J is an ideal. Certainly J is nonempty because $\mathcal{C}$ is nonempty — specifically, $0 \in J$ since 0 is in every ideal A. If $a, b \in J$, then there are ideals $A, B \in \mathcal{C}$ such that $a \in A$ and $b \in B$. By definition of a chain either $A \subseteq B$ or $B \subseteq A$. In either case $a - b \in J$, so J is closed under subtraction. Since each $A \in \mathcal{C}$ is closed under left and right multiplication by elements of R, so is J. This proves J is an ideal.

If J is not a proper ideal then $1 \in J$. In this case, by definition of J we must have $1 \in A$ for some $A \in \mathcal{C}$. This is a contradiction because each A is a proper ideal ($A \in \mathcal{C} \subseteq \mathcal{S}$). This proves that each chain has an upper bound in $\mathcal{S}$. By Zorn's Lemma $\mathcal{S}$ has a maximal element which is therefore a maximal (proper) ideal containing I.

For commutative rings the next result characterizes maximal ideals by the structure of their quotient rings.

Proposition 12. Assume R is commutative. The ideal M is a maximal ideal if and only if the quotient ring R/M is a field.

Proof: This follows from the Lattice Isomorphism Theorem together with Proposition 9(2). The ideal M is maximal if and only if there are no ideals I with $M \subset I \subset R$. By the Lattice Isomorphism Theorem the ideals of R containing M correspond bijectively with the ideals of R/M, so M is maximal if and only if the only ideals of R/M are 0 and R/M. By Proposition 9(2) we see that M is maximal if and only if R/M is a field.

The proposition above indicates how to *construct* some fields: take the quotient of any commutative ring R with identity by a maximal ideal in R. We shall use this in Part IV to construct all finite fields by taking quotients of the ring $\mathbb{Z}[x]$ by maximal ideals.

Examples

    (1) Let n be a nonnegative integer. The ideal $n\mathbb{Z}$ of $\mathbb{Z}$ is a maximal ideal if and only if $\mathbb{Z}/n\mathbb{Z}$ is a field. We saw in Section 3 that this is the case if and only if n is a prime number. This also follows directly from the containment of ideals of $\mathbb{Z}$ described in Example 2 above.

    (2) The ideal $(2, x)$ is a maximal ideal in $\mathbb{Z}[x]$ because its quotient ring is the field $\mathbb{Z}/2\mathbb{Z}$ — cf. Example 3 above and Example 5 at the end of Section 3.

    (3) The ideal (x) in $\mathbb{Z}[x]$ is not a maximal ideal because $(x) \subset (2, x) \subset \mathbb{Z}[x]$. The quotient ring $\mathbb{Z}[x]/(x)$ is isomorphic to $\mathbb{Z}$ (the ideal (x) in $\mathbb{Z}[x]$ is the kernel of the surjective ring homomorphism from $\mathbb{Z}[x]$ to $\mathbb{Z}$ given by evaluation at 0). Since $\mathbb{Z}$ is not a field, we see again that (x) is not a maximal ideal in $\mathbb{Z}[x]$.

    (4) Let R be the ring of all functions from $[0,1]$ to $\mathbb{R}$ and for each $a \in [0, 1]$ let M_a be the kernel of evaluation at a. Since evaluation is a surjective homomorphism from R to $\mathbb{R}$, we see that $R/M_a \cong \mathbb{R}$ and hence M_a is a maximal ideal. Similarly, the kernel of evaluation at any fixed point is a maximal ideal in the ring of continuous real valued functions on $[0, 1]$.

    (5) If F is a field and G is a finite group, then the augmentation ideal I is a maximal ideal of the group ring FG (cf. Example 7 at the end of the preceding section). The augmentation ideal is the kernel of the augmentation map which is a surjective homomorphism onto the field F (i.e., $FG/I \cong F$, a field). Note that Proposition 12 does not apply directly since FG need not be commutative, however, the implication in Proposition 12 that I is a maximal ideal if R/I is a field holds for arbitrary rings.

Definition. Assume R is commutative. An ideal P is called a *prime ideal* if $P \neq R$ and whenever the product ab of two elements $a, b \in R$ is an element of P, then at least one of a and b is an element of P.

    The notion of a maximal ideal is fairly intuitive but the definition of a prime ideal may seem a little strange. It is, however, a natural generalization of the notion of a "prime" in the integers $\mathbb{Z}$. Let n be a nonnegative integer. According to the above definition the ideal $n\mathbb{Z}$ is a *prime* ideal provided $n \neq 1$ (to ensure that the ideal is proper) and provided every time the product ab of two integers is an element of $n\mathbb{Z}$, at least one of a, b is an element of $n\mathbb{Z}$. Put another way, if $n \neq 0$, it must have the property that whenever n divides ab, n must divide a or divide b. This is equivalent to the usual definition that n is a prime number. Thus *the prime ideals of $\mathbb{Z}$ are just the ideals $p\mathbb{Z}$ of $\mathbb{Z}$ generated by prime numbers p together with the ideal 0.*

    For the integers $\mathbb{Z}$ there is no difference between the maximal ideals and the nonzero prime ideals. This is not true in general, but we shall see shortly that every maximal ideal is a prime ideal. First we translate the notion of prime ideals into properties of quotient rings as we did for maximal ideals in Proposition 12. Recall that an integral domain is a commutative ring with identity $1 \neq 0$ that has no zero divisors.

Proposition 13. Assume R is commutative. Then the ideal P is a prime ideal in R if and only if the quotient ring R/P is an integral domain.

    Proof: This proof is simply a matter of translating the definition of a prime ideal into the language of quotients. The ideal P is prime if and only if $P \neq R$ and whenever

$ab \in P$, then either $a \in P$ or $b \in P$. Use the bar notation for elements of R/P: $\bar{r} = r + P$. Note that $r \in P$ if and only if the element $\bar{r}$ is zero in the quotient ring R/P. Thus in the terminology of quotients P is a prime ideal if and only if $\bar{R} \neq \bar{0}$ and whenever $\overline{ab} = \bar{a}\bar{b} = \bar{0}$, then either $\bar{a} = \bar{0}$ or $\bar{b} = \bar{0}$, i.e., R/P is an integral domain.

It follows in particular that a commutative ring with identity is an integral domain if and only if 0 is a prime ideal.

Corollary 14. Assume R is commutative. Every maximal ideal of R is a prime ideal.

Proof: If M is a maximal ideal then R/M is a field by Proposition 12. A field is an integral domain so the corollary follows from Proposition 13.

Examples

    (1) The principal ideals generated by primes in $\mathbb{Z}$ are both prime and maximal ideals. The zero ideal in $\mathbb{Z}$ is prime but not maximal.

    (2) The ideal (x) is a prime ideal in $\mathbb{Z}[x]$ since $\mathbb{Z}[x]/(x) \cong \mathbb{Z}$. This ideal is not a maximal ideal. The ideal 0 is a prime ideal in $\mathbb{Z}[x]$, but is not a maximal ideal.

EXERCISES

Let R be a ring with identity $1 \neq 0$.

1. Let L_j be the left ideal of $M_n(R)$ consisting of arbitrary entries in the jth column and zero in all other entries and let E_{ij} be the element of $M_n(R)$ whose i, j entry is 1 and whose other entries are all 0. Prove that $L_j = M_n(R)E_{ij}$ for any i. [See Exercise 6, Section 2.]

2. Assume R is commutative. Prove that the augmentation ideal in the group ring RG is generated by $\{g - 1 \mid g \in G\}$. Prove that if $G = \langle \sigma \rangle$ is cyclic then the augmentation ideal is generated by $\sigma - 1$.

3. **(a)** Let p be a prime and let G be an abelian group of order p^n. Prove that the nilradical of the group ring $\mathbb{F}_p G$ is the augmentation ideal (cf. Exercise 29, Section 3). [Use the preceding exercise.]
   (b) Let $G = \{g_1, \ldots, g_n\}$ be a finite group and assume R is commutative. Prove that if r is any element of the augmentation ideal of RG then $r(g_1 + \cdots + g_n) = 0$. [Use the preceding exercise.]

4. Assume R is commutative. Prove that R is a field if and only if 0 is a maximal ideal.

5. Prove that if M is an ideal such that R/M is a field then M is a maximal ideal (do not assume R is commutative).

6. Prove that R is a division ring if and only if its only left ideals are (0) and R. (The analogous result holds when "left" is replaced by "right.")

7. Let R be a commutative ring with 1. Prove that the principal ideal generated by x in the polynomial ring $R[x]$ is a prime ideal if and only if R is an integral domain. Prove that (x) is a maximal ideal if and only if R is a field.

8. Let R be an integral domain. Prove that $(a) = (b)$ for some elements $a, b \in R$, if and only if $a = ub$ for some unit u of R.

9. Let R be the ring of all continuous functions on $[0, 1]$ and let I be the collection of functions $f(x)$ in R with $f(1/3) = f(1/2) = 0$. Prove that I is an ideal of R but is not a prime ideal.

10. Assume R is commutative. Prove that if P is a prime ideal of R and P contains no zero divisors then R is an integral domain.

11. Assume R is commutative. Let I and J be ideals of R and assume P is a prime ideal of R that contains IJ (for example, if P contains $I \cap J$). Prove either I or J is contained in P.

12. Assume R is commutative and suppose $I = (a_1, a_2, \ldots, a_n)$ and $J = (b_1, b_2, \ldots, b_m)$ are two finitely generated ideals in R. Prove that the product ideal IJ is finitely generated by the elements $a_i b_j$ for $i = 1, 2, \ldots, n$, and $j = 1, 2, \ldots, m$.

13. Let $\varphi : R \to S$ be a homomorphism of commutative rings.
    (a) Prove that if P is a prime ideal of S then either $\varphi^{-1}(P) = R$ or $\varphi^{-1}(P)$ is a prime ideal of R. Apply this to the special case when R is a subring of S and φ is the inclusion homomorphism to deduce that if P is a prime ideal of S then $P \cap R$ is either R or a prime ideal of R.
    (b) Prove that if M is a maximal ideal of S and φ is surjective then $\varphi^{-1}(M)$ is a maximal ideal of R. Give an example to show that this need not be the case if φ is not surjective.

14. Assume R is commutative. Let x be an indeterminate, let $f(x)$ be a monic polynomial in $R[x]$ of degree $n \geq 1$ and use the bar notation to denote passage to the quotient ring $R[x]/(f(x))$.
    (a) Show that every element of $R[x]/(f(x))$ is of the form $\overline{p(x)}$ for some polynomial $p(x) \in R[x]$ of degree less than n, i.e.,
    $$R[x]/(f(x)) = \{\overline{a_0 + a_1 x + \cdots + a_{n-1}x^{n-1}} \mid a_0, a_1, \ldots, a_{n-1} \in R\}.$$
    [If $f(x) = x^n + b_{n-1}x^{n-1} + \cdots + b_0$ then $\overline{x^n} = -(b_{n-1}x^{n-1} + \cdots + b_0)$. Use this to reduce powers of $\overline{x}$ in the quotient ring.]
    (b) Prove that if $p(x)$ and $q(x)$ are distinct polynomials in $R[x]$ which are both of degree less than n, then $\overline{p(x)} \neq \overline{q(x)}$. [Otherwise $p(x) - q(x)$ is an $R[x]$-multiple of the monic polynomial $f(x)$.]
    (c) If $f(x) = a(x)b(x)$ where both $a(x)$ and $b(x)$ have degree less than n, prove that $\overline{a(x)}$ is a zero divisor in $R[x]/(f(x))$.
    (d) If $f(x) = x^n - a$ for some nilpotent element $a \in R$, prove that $\overline{x}$ is nilpotent in $R[x]/(f(x))$.
    (e) Let p be a prime, assume $R = \mathbb{F}_p$ and $f(x) = x^p - a$ for some $a \in \mathbb{F}_p$. Prove that $\overline{x - a}$ is nilpotent in $R[x]/(f(x))$. [Use Exercise 26(c) of Section 3.]

15. Let $x^2 + x + 1$ be an element of the polynomial ring $E = \mathbb{F}_2[x]$ and use the bar notation to denote passage to the quotient ring $\mathbb{F}_2[x]/(x^2 + x + 1)$.
    (a) Prove that $\overline{E}$ has 4 elements: $\overline{0}, \overline{1}, \overline{x}$ and $\overline{x+1}$.
    (b) Write out the 4×4 addition table for $\overline{E}$ and deduce that the additive group $\overline{E}$ is isomorphic to the Klein 4-group.
    (c) Write out the 4×4 multiplication table for $\overline{E}$ and prove that $\overline{E}^{\times}$ is isomorphic to the cyclic group of order 3. Deduce that $\overline{E}$ is a field.

16. Let $x^4 - 16$ be an element of the polynomial ring $E = \mathbb{Z}[x]$ and use the bar notation to denote passage to the quotient ring $\mathbb{Z}[x]/(x^4 - 16)$.
    (a) Find a polynomial of degree ≤ 3 that is congruent to $7x^{13} - 11x^9 + 5x^5 - 2x^3 + 3$ modulo $(x^4 - 16)$.
    (b) Prove that $\overline{x - 2}$ and $\overline{x + 2}$ are zero divisors in $\overline{E}$.

17. Let $x^3 - 2x + 1$ be an element of the polynomial ring $E = \mathbb{Z}[x]$ and use the bar notation to denote passage to the quotient ring $\mathbb{Z}[x]/(x^3 - 2x + 1)$. Let $p(x) = 2x^7 - 7x^5 + 4x^3 - 9x + 1$ and let $q(x) = (x - 1)^4$.

(a) Express each of the following elements of $\overline{E}$ in the form $\overline{f(x)}$ for some polynomial $f(x)$ of degree ≤ 2: $\overline{p(x)}, \overline{q(x)}, \overline{p(x) + q(x)}$ and $\overline{p(x)q(x)}$.
(b) Prove that $\overline{E}$ is not an integral domain.
(c) Prove that $\overline{x}$ is a unit in $\overline{E}$.

18. Prove that if R is an integral domain and $R[[x]]$ is the ring of formal power series in the indeterminate x then the principal ideal generated by x is a prime ideal (cf. Exercise 3, Section 2). Prove that the principal ideal generated by x is a maximal ideal if and only if R is a field.

19. Let R be a finite commutative ring with identity. Prove that every prime ideal of R is a maximal ideal.

20. Prove that a nonzero finite commutative ring that has no zero divisors is a field (if the ring has an identity, this is Corollary 3, so do not assume the ring has a 1).

21. Prove that a finite ring with identity $1 \neq 0$ that has no zero divisors is a field (you may quote Wedderburn's Theorem).

22. Let $p \in \mathbb{Z}^+$ be a prime and let the $\mathbb{F}_p$ Quaternions be defined by

$$a + bi + cj + dk \qquad a, b, c, d \in \mathbb{Z}/p\mathbb{Z}$$

where addition is componentwise and multiplication is defined using the same relations on i, j, k as for the real Quaternions.
(a) Prove that the $\mathbb{F}_p$ Quaternions are a homomorphic image of the naive integral Quaternions $\mathbb{Z} + \mathbb{Z}i + \mathbb{Z}j + \mathbb{Z}k$ (cf. Section 1).
(b) Prove that the $\mathbb{F}_p$ Quaternions contain zero divisors (and so they cannot be a division ring). [Use the preceding exercise.]

23. Prove that in a Boolean ring (cf. Exercise 15, Section 1) every prime ideal is a maximal ideal.

24. Prove that in a Boolean ring every finitely generated ideal is principal.

25. Assume R is commutative and for each $a \in R$ there is an integer $n > 1$ (depending on a) such that $a^n = a$. Prove that every prime ideal of R is a maximal ideal.

26. Prove that a prime ideal in a commutative ring R contains every nilpotent element (cf. Exercise 13, Section 1). Deduce that the nilradical of R (cf. Exercise 29, Section 3) is contained in the intersection of all the prime ideals of R. (It is shown in Section 15.2 that the nilradical of R is equal to the intersection of all prime ideals of R.)

27. Let R be a commutative ring with $1 \neq 0$. Prove that if a is a nilpotent element of R then $1 - ab$ is a unit for all $b \in R$.

28. Prove that if R is a commutative ring and $N = (a_1, a_2, \ldots, a_m)$ where each a_i is a nilpotent element, then N is a nilpotent ideal (cf. Exercise 37, Section 3). Deduce that if the nilradical of R is finitely generated then it is a nilpotent ideal.

29. Let p be a prime and let G be a finite group of order a power of p (i.e., a p-group). Prove that the augmentation ideal in the group ring $\mathbb{Z}/p\mathbb{Z}G$ is a nilpotent ideal. (Note that this ring may be noncommutative.) [Use Exercise 2.]

30. Let I be an ideal of the commutative ring R and define

$$\text{rad } I = \{r \in R \mid r^n \in I \text{ for some } n \in \mathbb{Z}^+\}$$

called the *radical* of I. Prove that rad I is an ideal containing I and that $(\text{rad } I)/I$ is the nilradical of the quotient ring R/I, i.e., $(\text{rad } I)/I = \mathfrak{N}(R/I)$ (cf. Exercise 29, Section 3).

31. An ideal I of the commutative ring R is called a *radical ideal* if rad $I = I$.

258

(a) Prove that every prime ideal of R is a radical ideal.

(b) Let $n > 1$ be an integer. Prove that 0 is a radical ideal in $\mathbb{Z}/n\mathbb{Z}$ if and only if n is a product of distinct primes to the first power (i.e., n is square free). Deduce that (n) is a radical ideal of $\mathbb{Z}$ if and only if n is a product of distinct primes in $\mathbb{Z}$.

32. Let I be an ideal of the commutative ring R and define

$$\text{Jac } I \text{ to be the intersection of all maximal ideals of } R \text{ that contain } I$$

where the convention is that $\text{Jac } R = R$. (If I is the zero ideal, $\text{Jac } 0$ is called the *Jacobson radical* of the ring R, so $\text{Jac } I$ is the preimage in R of the Jacobson radical of R/I.)

(a) Prove that $\text{Jac } I$ is an ideal of R containing I.

(b) Prove that $\text{rad } I \subseteq \text{Jac } I$, where $\text{rad } I$ is the radical of I defined in Exercise 30.

(c) Let $n > 1$ be an integer. Describe $\text{Jac } n\mathbb{Z}$ in terms of the prime factorization of n.

33. Let R be the ring of all continuous functions from the closed interval $[0,1]$ to $\mathbb{R}$ and for each $c \in [0, 1]$ let $M_c = \{f \in R \mid f(c) = 0\}$ (recall that M_c was shown to be a maximal ideal of R).

(a) Prove that if M is *any* maximal ideal of R then there is a real number $c \in [0, 1]$ such that $M = M_c$.

(b) Prove that if b and c are distinct points in $[0,1]$ then $M_b \neq M_c$.

(c) Prove that M_c is not equal to the principal ideal generated by $x - c$.

(d) Prove that M_c is not a finitely generated ideal.

The preceding exercise shows that there is a bijection between the *points* of the closed interval $[0,1]$ and the set of *maximal ideals* in the ring R of all of continuous functions on $[0,1]$ given by $c \leftrightarrow M_c$. For any subset X of $\mathbb{R}$ or, more generally, for any completely regular topological space X, the map $c \mapsto M_c$ is an *injection* from X to the set of maximal ideals of R, where R is the ring of all bounded continuous real valued functions on X and M_c is the maximal ideal of functions that vanish at c. Let $\beta(X)$ be the set of maximal ideals of R. One can put a topology on $\beta(X)$ in such a way that if we identify X with its image in $\beta(X)$ then X (in its given topology) becomes a subspace of $\beta(X)$. Moreover, $\beta(X)$ is a compact space under this topology and is called the *Stone-Čech compactification* of X.

34. Let R be the ring of all continuous functions from $\mathbb{R}$ to $\mathbb{R}$ and for each $c \in \mathbb{R}$ let M_c be the maximal ideal $\{f \in R \mid f(c) = 0\}$.

(a) Let I be the collection of functions $f(x)$ in R with *compact support* (i.e., $f(x) = 0$ for $|x|$ sufficiently large). Prove that I is an ideal of R that is not a prime ideal.

(b) Let M be a maximal ideal of R containing I (properly, by (a)). Prove that $M \neq M_c$ for any $c \in \mathbb{R}$ (cf. the preceding exercise).

35. Let $A = (a_1, a_2, \ldots, a_n)$ be a nonzero finitely generated ideal of R. Prove that there is an ideal B which is maximal with respect to the property that it does not contain A. [Use Zorn's Lemma.]

36. Assume R is commutative. Prove that the set of prime ideals in R has a minimal element with respect to inclusion (possibly the zero ideal). [Use Zorn's Lemma.]

37. A commutative ring R is called a *local ring* if it has a unique maximal ideal. Prove that if R is a local ring with maximal ideal M then every element of $R - M$ is a unit. Prove conversely that if R is a commutative ring with 1 in which the set of nonunits forms an ideal M, then R is a local ring with unique maximal ideal M.

38. Prove that the ring of all rational numbers whose denominators is odd is a local ring whose unique maximal ideal is the principal ideal generated by 2.

39. Following the notation of Exercise 26 in Section 1, let K be a field, let ν be a discrete

valuation on K and let R be the valuation ring of v. For each integer $k \geq 0$ define $A_k = \{r \in R \mid v(r) \geq k\} \cup \{0\}$.
  (a) Prove that A_k is a principal ideal and that $A_0 \supseteq A_1 \supseteq A_2 \supseteq \cdots$.
  (b) Prove that if I is any nonzero ideal of R, then $I = A_k$ for some $k \geq 0$. Deduce that R is a local ring with unique maximal ideal A_1.

40. Assume R is commutative. Prove that the following are equivalent: (see also Exercises 13 and 14 in Section 1)
  (i) R has exactly one prime ideal
  (ii) every element of R is either nilpotent or a unit
  (iii) $R/\mathfrak{N}(R)$ is a field (cf. Exercise 29, Section 3).

41. A proper ideal Q of the commutative ring R is called *primary* if whenever $ab \in Q$ and $a \notin Q$ then $b^n \in Q$ for some positive integer n. (Note that the symmetry between a and b in this definition implies that if Q is a primary ideal and $ab \in Q$ with *neither* a nor b in Q, then a positive power of a and a positive power of b both lie in Q.) Establish the following facts about primary ideals.
  (a) The primary ideals of $\mathbb{Z}$ are 0 and (p^n), where p is a prime and n is a positive integer.
  (b) Every prime ideal of R is a primary ideal.
  (c) An ideal Q of R is primary if and only if every zero divisor in R/Q is a nilpotent element of R/Q.
  (d) If Q is a primary ideal then rad(Q) is a prime ideal (cf. Exercise 30).

7.5 RINGS OF FRACTIONS

Throughout this section R is a commutative ring. Proposition 2 shows that if a is not zero nor a zero divisor and $ab = ac$ in R then $b = c$. Thus a nonzero element that is not a zero divisor enjoys some of the properties of a unit without necessarily possessing a multiplicative inverse in R. On the other hand, we saw in Section 1 that a zero divisor a cannot be a unit in R and, by definition, if a is a zero divisor we cannot always cancel the a's in the equation $ab = ac$ to obtain $b = c$ (take $c = 0$ for example). The aim of this section is to prove that a commutative ring R is always a subring of a larger ring Q in which every nonzero element of R that is not a zero divisor is a unit in Q. The principal application of this will be to integral domains, in which case this ring Q will be a field — called its *field of fractions* or *quotient field*. Indeed, the paradigm for the construction of Q from R is the one offered by the construction of the field of rational numbers from the integral domain $\mathbb{Z}$.

  In order to see the essential features of the construction of the field $\mathbb{Q}$ from the integral domain $\mathbb{Z}$ we review the basic properties of fractions. Each rational number may be represented in many different ways as the quotient of two integers (for example, $\frac{1}{2} = \frac{2}{4} = \frac{3}{6} = \ldots$, etc.). These representations are related by

$$\frac{a}{b} = \frac{c}{d} \qquad \text{if and only if} \qquad ad = bc.$$

In more precise terms, the fraction $\frac{a}{b}$ is the equivalence class of ordered pairs (a, b) of integers with $b \neq 0$ under the equivalence relation: $(a, b) \sim (c, d)$ if and only if

$ad = bc$. The arithmetic operations on fractions are given by

$$\frac{a}{b} + \frac{c}{d} = \frac{ad + bc}{bd} \quad \text{and} \quad \frac{a}{b} \times \frac{c}{d} = \frac{ac}{bd}.$$

These are well defined (independent of choice of representatives of the equivalence classes) and make the set of fractions into a commutative ring (in fact, a field), $\mathbb{Q}$. The integers $\mathbb{Z}$ are identified with the subring $\{\frac{a}{1} \mid a \in \mathbb{Z}\}$ of $\mathbb{Q}$ and every nonzero integer a has an inverse $\frac{1}{a}$ in $\mathbb{Q}$.

It seems reasonable to attempt to follow the same steps for any commutative ring R, allowing arbitrary denominators. If, however, b is zero or a zero divisor in R, say $bd = 0$, and if we allow b as a denominator, then we should expect to have

$$d = \frac{d}{1} = \frac{bd}{b} = \frac{0}{b} = 0$$

in the "ring of fractions" (where, for convenience, we have assumed R has a 1). Thus if we allow zero or zero divisors as denominators there must be some collapsing in the sense that we cannot expect R to appear naturally as a subring of this "ring of fractions." A second restriction is more obviously imposed by the laws of addition and multiplication: if ring elements b and d are allowed as denominators, then bd must also be a denominator, i.e., the set of denominators must be closed under multiplication in R. The main result of this section shows that these two restrictions are sufficient to construct a ring of fractions for R. Note that this theorem includes the construction of $\mathbb{Q}$ from $\mathbb{Z}$ as a special case.

Theorem 15. Let R be a commutative ring. Let D be any nonempty subset of R that does not contain 0, does not contain any zero divisors and is closed under multiplication (i.e., $ab \in D$ for all $a, b \in D$). Then there is a commutative ring Q with 1 such that Q contains R as a subring and every element of D is a unit in Q. The ring Q has the following additional properties.

(1) every element of Q is of the form rd^{-1} for some $r \in R$ and $d \in D$. In particular, if $D = R - \{0\}$ then Q is a field.

(2) (uniqueness of Q) The ring Q is the *"smallest"* ring containing R in which all elements of D become units, in the following sense. Let S be any commutative ring with identity and let $\varphi : R \to S$ be any injective ring homomorphism such that $\varphi(d)$ is a unit in S for every $d \in D$. Then there is an injective homomorphism $\Phi : Q \to S$ such that $\Phi|_R = \varphi$. In other words, any ring containing an isomorphic copy of R in which all the elements of D become units must also contain an isomorphic copy of Q.

Remark: In Section 15.4 a more general construction is given. The proof of the general result is more technical but relies on the same basic rationale and steps as the proof of Theorem 15. Readers wishing greater generality may read the proof below and the beginning of Section 15.4 in concert.

Proof: Let $\mathcal{F} = \{(r, d) \mid r \in R, \ d \in D\}$ and define the relation $\sim$ on $\mathcal{F}$ by

$$(r, d) \sim (s, e) \qquad \text{if and only if} \qquad re = sd.$$

It is immediate that this relation is reflexive and symmetric. Suppose $(r, d) \sim (s, e)$ and $(s, e) \sim (t, f)$. Then $re - sd = 0$ and $sf - te = 0$. Multiplying the first of these equations by f and the second by d and adding them gives $(rf - td)e = 0$. Since $e \in D$ is neither zero nor a zero divisor we must have $rf - td = 0$, i.e., $(r, d) \sim (t, f)$. This proves $\sim$ is transitive, hence an equivalence relation. Denote the equivalence class of (r, d) by $\dfrac{r}{d}$:

$$\frac{r}{d} = \{(a, b) \mid a \in R, \ b \in D \text{ and } rb = ad\}.$$

Let Q be the set of equivalence classes under $\sim$. Note that $\dfrac{r}{d} = \dfrac{re}{de}$ in Q for all $e \in D$, since D is closed under multiplication.

We now define an additive and multiplicative structure on Q:

$$\frac{a}{b} + \frac{c}{d} = \frac{ad + bc}{bd} \qquad \text{and} \qquad \frac{a}{b} \times \frac{c}{d} = \frac{ac}{bd}.$$

In order to prove that Q is a commutative ring with identity there are a number of things to check:

(1) these operations are well defined (i.e., do not depend on the choice of representatives for the equivalence classes),

(2) Q is an abelian group under addition, where the additive identity is $\dfrac{0}{d}$ for any $d \in D$ and the additive inverse of $\dfrac{a}{d}$ is $\dfrac{-a}{d}$,

(3) multiplication is associative, distributive and commutative, and

(4) Q has an identity $\left(= \dfrac{d}{d} \text{ for any } d \in D\right)$.

These are all completely straightforward calculations involving only arithmetic in R and the definition of $\sim$. Again we need D to be closed under multiplication for addition and multiplication to be defined.

For example, to check that addition is well defined assume $\dfrac{a}{b} = \dfrac{a'}{b'}$ (i.e., $ab' = a'b$) and $\dfrac{c}{d} = \dfrac{c'}{d'}$ (i.e., $cd' = c'd$). We must show that $\dfrac{ad + bc}{bd} = \dfrac{a'd' + b'c'}{b'd'}$, i.e.,

$$(ad + bc)(b'd') = (a'd' + b'c')(bd).$$

The left hand side of this equation is $ab'dd' + cd'bb'$ substituting $a'b$ for ab' and $c'd$ for cd' gives $a'bdd' + c'dbb'$, which is the right hand side. Hence addition of fractions is well defined. Checking the details in the other parts of (1) to (4) involves even easier manipulations and so is left as an exercise.

Next we embed R into Q by defining

$$\iota : R \to Q \qquad \text{by} \qquad \iota : r \mapsto \frac{rd}{d} \qquad \text{where } d \text{ is any element of } D.$$

Since $\dfrac{rd}{d} = \dfrac{re}{e}$ for all $d, e \in D$, $\iota(r)$ does not depend on the choice of $d \in D$. Since D is closed under multiplication, one checks directly that ι is a ring homomorphism.

Furthermore, ι is injective because

$$\iota(r) = 0 \Leftrightarrow \frac{rd}{d} = \frac{0}{d} \Leftrightarrow rd^2 = 0 \Leftrightarrow r = 0$$

because d (hence also d^2) is neither zero nor a zero divisor. The subring $\iota(R)$ of Q is therefore isomorphic to R. We henceforth identify each $r \in R$ with $\iota(r)$ and so consider R as a subring of Q.

Next note that each $d \in D$ has a multiplicative inverse in Q: namely, if d is represented by the fraction $\dfrac{de}{e}$ then its multiplicative inverse is $\dfrac{e}{de}$. One then sees that every element of Q may be written as $r \cdot d^{-1}$ for some $r \in R$ and some $d \in D$. In particular, if $D = R - \{0\}$, every nonzero element of Q has a multiplicative inverse and Q is a field.

It remains to establish the uniqueness property of Q. Assume $\varphi : R \to S$ is an injective ring homomorphism such that $\varphi(d)$ is a unit in S for all $d \in D$. Extend φ to a map $\Phi : Q \to S$ by defining $\Phi(rd^{-1}) = \varphi(r)\varphi(d)^{-1}$ for all $r \in R, d \in D$. This map is well defined, since $rd^{-1} = se^{-1}$ implies $re = sd$, so $\varphi(r)\varphi(e) = \varphi(s)\varphi(d)$, and then

$$\Phi(rd^{-1}) = \varphi(r)\varphi(d)^{-1} = \varphi(s)\varphi(e)^{-1} = \Phi(se^{-1}).$$

It is straightforward to check that Φ is a ring homomorphism — the details are left as an exercise. Finally, Φ is injective because $rd^{-1} \in \ker \Phi$ implies $r \in \ker \Phi \cap R = \ker \varphi$; since φ is injective this forces r and hence also rd^{-1} to be zero. This completes the proof.

Definition. Let R, D and Q be as in Theorem 15.
   (1) The ring Q is called the *ring of fractions* of D with respect to R and is denoted $D^{-1}R$.
   (2) If R is an integral domain and $D = R - \{0\}$, Q is called the *field of fractions* or *quotient field* of R.

If A is a subset of a field F (for example, if A is a subring of F), then the intersection of all the subfields of F containing A is a subfield of F and is called the subfield *generated* by A. This subfield is the smallest subfield of F containing A (namely, any subfield of F containing A contains the subfield generated by A).

The next corollary shows that the smallest field containing an integral domain R is its field of fractions.

Corollary 16. Let R be an integral domain and let Q be the field of fractions of R. If a field F contains a subring R' isomorphic to R then the subfield of F generated by R' is isomorphic to Q.

Proof: Let $\varphi : R \cong R' \subseteq F$ be a (ring) isomorphism of R to R'. In particular, $\varphi : R \to F$ is an injective homomorphism from R into the field F. Let $\Phi : Q \to F$ be the extension of φ to Q as in the theorem. By Theorem 15, Φ is injective, so $\Phi(Q)$ is an isomorphic copy of Q in F containing $\varphi(R) = R'$. Now, any subfield of F containing $R' = \varphi(R)$ contains the elements $\varphi(r_1)\varphi(r_2)^{-1} = \varphi(r_1 r_2^{-1})$ for all $r_1, r_2 \in R$. Since every element of Q is of the form $r_1 r_2^{-1}$ for some $r_1, r_2 \in R$, it follows that any subfield of F containing R' contains the field $\Phi(Q)$, so that $\Phi(Q)$ is the subfield of F generated by R', proving the corollary.

Examples

(1) If R is a field then its field of fractions is just R itself.

(2) The integers $\mathbb{Z}$ are an integral domain whose field of fractions is the field $\mathbb{Q}$ of rational numbers. The quadratic integer ring $\mathcal{O}$ of Section 1 is an integral domain whose field of fractions is the quadratic field $\mathbb{Q}(\sqrt{D})$.

(3) The subring $2\mathbb{Z}$ of $\mathbb{Z}$ also has no zero divisors (but has no identity). Its field of fractions is also $\mathbb{Q}$. Note how an identity "appears" in the field of fractions.

(4) If R is any integral domain, then the polynomial ring $R[x]$ is also an integral domain. The associated field of fractions is the field of *rational functions* in the variable x over R. The elements of this field are of the form $\dfrac{p(x)}{q(x)}$, where $p(x)$ and $q(x)$ are polynomials with coefficients in R with $q(x)$ not the zero polynomial. In particular, $p(x)$ and $q(x)$ may both be constant polynomials, so the field of rational functions contains the field of fractions of R: elements of the form $\dfrac{a}{b}$ such that $a, b \in R$ and $b \neq 0$. If F is a field, we shall denote the field of rational functions by $F(x)$. Thus if F is the field of fractions of the integral domain R then the field of rational functions over R is the same as the field of rational functions over F, namely $F(x)$.

For example, suppose $R = \mathbb{Z}$, so $F = \mathbb{Q}$. If $p(x), q(x)$ are polynomials in $\mathbb{Q}[x]$ then for some integer N, $Np(x), Nq(x)$ have integer coefficients (let N be a common denominator for all the coefficients in $p(x)$ and $q(x)$, for example). Then $\dfrac{p(x)}{q(x)} = \dfrac{Np(x)}{Nq(x)}$ can be written as the quotient of two polynomials with integer coefficients, so the field of fractions of $\mathbb{Q}[x]$ is the same as the field of fractions of $\mathbb{Z}[x]$.

(5) If R is any commutative ring with identity and d is neither zero nor a zero divisor in R we may form the ring $R[1/d]$ by setting $D = \{1, d, d^2, d^3, \dots\}$ and defining $R[1/d]$ to be the ring of fractions $D^{-1}R$. Note that R is the subring of elements of the form $\dfrac{r}{1}$. In this way any nonzero element of R that is not a zero divisor can be inverted in a larger ring containing R. Note that the elements of $R[1/d]$ look like polynomials in $1/d$ with coefficients in R, which explains the notation.

EXERCISES

Let R be a commutative ring with identity $1 \neq 0$.

1. Fill in all the details in the proof of Theorem 15.

2. Let R be an integral domain and let D be a nonempty subset of R that is closed under multiplication. Prove that the ring of fractions $D^{-1}R$ is isomorphic to a subring of the quotient field of R (hence is also an integral domain).

3. Let F be a field. Prove that F contains a unique smallest subfield F_0 and that F_0 is isomorphic to either $\mathbb{Q}$ or $\mathbb{Z}/p\mathbb{Z}$ for some prime p (F_0 is called the *prime subfield* of F). [See Exercise 26, Section 3.]

4. Prove that any subfield of $\mathbb{R}$ must contain $\mathbb{Q}$.

5. If F is a field, prove that the field of fractions of $F[[x]]$ (the ring of formal power series in the indeterminate x with coefficients in F) is the ring $F((x))$ of formal Laurent series (cf. Exercises 3 and 5 of Section 2). Show the field of fractions of the power series ring $\mathbb{Z}[[x]]$ is *properly* contained in the field of Laurent series $\mathbb{Q}((x))$. [Consider the series for e^x.]

6. Prove that the real numbers, $\mathbb{R}$, contain a subring A with $1 \in A$ and A maximal (under inclusion) with respect to the property that $\frac{1}{2} \notin A$. [Use Zorn's Lemma.] (Exercise 13 in Section 15.3 shows $\mathbb{R}$ is the quotient field of A, so $\mathbb{R}$ is the quotient field of a proper subring.)

7.6 THE CHINESE REMAINDER THEOREM

Throughout this section we shall assume unless otherwise stated that all rings are commutative with an identity $1 \neq 0$.

Given an arbitrary collection of rings (not necessarily satisfying the conventions above), their *(ring) direct product* is defined to be their direct product as (abelian) groups made into a ring by defining multiplication componentwise. In particular, if R_1 and R_2 are two rings, we shall denote by $R_1 \times R_2$ their direct product (as rings), that is, the set of ordered pairs (r_1, r_2) with $r_1 \in R_1$ and $r_2 \in R_2$ where addition and multiplication are performed componentwise:

$$(r_1, r_2) + (s_1, s_2) = (r_1 + s_1, r_2 + s_2) \quad \text{and} \quad (r_1, r_2)(s_1, s_2) = (r_1 s_1, r_2 s_2).$$

We note that a map φ from a ring R into a direct product ring is a homomorphism if and only if the induced maps into each of the components are homomorphisms.

There is a generalization to arbitrary rings of the notion in $\mathbb{Z}$ of two integers n and m being relatively prime (even to rings where the notion of greatest common divisor is not defined). In $\mathbb{Z}$ this is equivalent to being able to solve the equation $nx + my = 1$ in integers x and y (this fact was stated in Chapter 0 and will be proved in Chapter 8). This in turn is equivalent to $n\mathbb{Z} + m\mathbb{Z} = \mathbb{Z}$ as ideals (in general, $n\mathbb{Z} + m\mathbb{Z} = (m, n)\mathbb{Z}$). This motivates the following definition:

Definition. The ideals A and B of the ring R are said to be *comaximal* if $A + B = R$.

Recall that the *product*, AB, of the ideals A and B of R is the ideal consisting of all finite sums of elements of the form xy, $x \in A$ and $y \in B$ (cf. Exercise 34, Section 3). If $A = (a)$ and $B = (b)$, then $AB = (ab)$. More generally, the product of the ideals $A_1, A_2, \ldots, A_k$ is the ideal of all finite sums of elements of the form $x_1 x_2 \cdots x_k$ such that $x_i \in A_i$ for all i. If $A_i = (a_i)$, then $A_1 \cdots A_k = (a_1 \cdots a_k)$.

Theorem 17. *(Chinese Remainder Theorem)* Let $A_1, A_2, \ldots, A_k$ be ideals in R. The map

$$R \to R/A_1 \times R/A_2 \times \cdots \times R/A_k \qquad \text{defined by} \qquad r \mapsto (r + A_1, r + A_2, \ldots, r + A_k)$$

is a ring homomorphism with kernel $A_1 \cap A_2 \cap \cdots \cap A_k$. If for each $i, j \in \{1, 2, \ldots, k\}$ with $i \neq j$ the ideals A_i and A_j are comaximal, then this map is surjective and $A_1 \cap A_2 \cap \cdots \cap A_k = A_1 A_2 \cdots A_k$, so

$$R/(A_1 A_2 \cdots A_k) = R/(A_1 \cap A_2 \cap \cdots \cap A_k) \cong R/A_1 \times R/A_2 \times \cdots \times R/A_k.$$

Proof: We first prove this for $k = 2$; the general case will follow by induction. Let $A = A_1$ and $B = A_2$. Consider the map $\varphi : R \rightarrow R/A \times R/B$ defined by $\varphi(r) = (r \bmod A, r \bmod B)$, where mod A means the class in R/A containing r (that is, $r + A$). This map is a ring homomorphism because φ is just the natural projection of R into R/A and R/B for the two components. The kernel of φ consists of all the elements $r \in R$ that are in A and in B, i.e., $A \cap B$. To complete the proof in this case it remains to show that when A and B are comaximal, φ is surjective and $A \cap B = AB$. Since $A + B = R$, there are elements $x \in A$ and $y \in B$ such that $x + y = 1$. This equation shows that $\varphi(x) = (0, 1)$ and $\varphi(y) = (1, 0)$ since, for example, x is an element of A and $x = 1 - y \in 1 + B$. If now $(r_1 \bmod A, r_2 \bmod B)$ is an arbitrary element in $R/A \times R/B$, then the element $r_2 x + r_1 y$ maps to this element since

$$
\begin{aligned}
\varphi(r_2 x + r_1 y) &= \varphi(r_2)\varphi(x) + \varphi(r_1)\varphi(y) \\
&= (r_2 \bmod A, r_2 \bmod B)(0, 1) + (r_1 \bmod A, r_1 \bmod B)(1, 0) \\
&= (0, r_2 \bmod B) + (r_1 \bmod A, 0) \\
&= (r_1 \bmod A, r_2 \bmod B).
\end{aligned}
$$

This shows that φ is indeed surjective. Finally, the ideal AB is always contained in $A \cap B$. If A and B are comaximal and x and y are as above, then for any $c \in A \cap B$, $c = c1 = cx + cy \in AB$. This establishes the reverse inclusion $A \cap B \subseteq AB$ and completes the proof when $k = 2$.

The general case follows easily by induction from the case of two ideals using $A = A_1$ and $B = A_2 \cdots A_k$ once we show that A_1 and $A_2 \cdots A_k$ are comaximal. By hypothesis, for each $i \in \{2, 3, \ldots, k\}$ there are elements $x_i \in A_1$ and $y_i \in A_i$ such that $x_i + y_i = 1$. Since $x_i + y_i \equiv y_i \bmod A_1$, it follows that $1 = (x_2 + y_2) \cdots (x_k + y_k)$ is an element in $A_1 + (A_2 \cdots A_k)$. This completes the proof.

This theorem obtained its name from the special case $\mathbb{Z}/mn\mathbb{Z} \cong (\mathbb{Z}/m\mathbb{Z}) \times (\mathbb{Z}/n\mathbb{Z})$ *as rings* when m and n are relatively prime integers. We proved this isomorphism just for the additive groups earlier. This isomorphism, phrased in number-theoretic terms, relates to simultaneously solving two congruences modulo relatively prime integers (and states that such congruences can always be solved, and uniquely). Such problems were considered by the ancient Chinese, hence the name. Some examples are provided in the exercises.

Since the isomorphism in the Chinese Remainder Theorem is an isomorphism of *rings*, in particular the groups of *units* on both sides must be isomorphic. It is easy to see that the units in any direct product of rings are the elements that have units in each of the coordinates. In the case of $\mathbb{Z}/mn\mathbb{Z}$ the Chinese Remainder Theorem gives the following isomorphism on the groups of units:

$$(\mathbb{Z}/mn\mathbb{Z})^\times \cong (\mathbb{Z}/m\mathbb{Z})^\times \times (\mathbb{Z}/n\mathbb{Z})^\times.$$

More generally we have the following result.

Corollary 18. Let n be a positive integer and let $p_1^{\alpha_1} p_2^{\alpha_2} \ldots p_k^{\alpha_k}$ be its factorization into powers of distinct primes. Then

$$\mathbb{Z}/n\mathbb{Z} \cong (\mathbb{Z}/p_1^{\alpha_1}\mathbb{Z}) \times (\mathbb{Z}/p_2^{\alpha_2}\mathbb{Z}) \times \cdots \times (\mathbb{Z}/p_k^{\alpha_k}\mathbb{Z}),$$

as rings, so in particular we have the following isomorphism of multiplicative groups:

$$(\mathbb{Z}/n\mathbb{Z})^\times \cong (\mathbb{Z}/p_1^{\alpha_1}\mathbb{Z})^\times \times (\mathbb{Z}/p_2^{\alpha_2}\mathbb{Z})^\times \times \cdots \times (\mathbb{Z}/p_k^{\alpha_k}\mathbb{Z})^\times.$$

If we compare orders on the two sides of this last isomorphism, we obtain the formula

$$\varphi(n) = \varphi(p_1^{\alpha_1})\varphi(p_2^{\alpha_2}) \ldots \varphi(p_k^{\alpha_k})$$

for the Euler φ-function. This in turn implies that φ is what in elementary number theory is termed a *multiplicative function*, namely that $\varphi(ab) = \varphi(a)\varphi(b)$ whenever a and b are relatively prime positive integers. The value of φ on prime powers p^α is easily seen to be $\varphi(p^\alpha) = p^{\alpha-1}(p - 1)$ (cf. Chapter 0). From this and the multiplicativity of φ we obtain its value on all positive integers.

Corollary 18 is also a step toward a determination of the decomposition of the abelian group $(\mathbb{Z}/n\mathbb{Z})^\times$ into a direct product of cyclic groups. The complete structure is derived at the end of Section 9.5.

EXERCISES

Let R be a ring with identity $1 \neq 0$.

1. An element $e \in R$ is called an *idempotent* if $e^2 = e$. Assume e is an idempotent in R and $er = re$ for all $r \in R$. Prove that Re and $R(1 - e)$ are two-sided ideals of R and that $R \cong Re \times R(1 - e)$. Show that e and $1 - e$ are identities for the subrings Re and $R(1 - e)$ respectively.

2. Let R be a finite Boolean ring with identity $1 \neq 0$ (cf. Exercise 15 of Section 1). Prove that $R \cong \mathbb{Z}/2\mathbb{Z} \times \cdots \times \mathbb{Z}/2\mathbb{Z}$. [Use the preceding exercise.]

3. Let R and S be rings with identities. Prove that every ideal of $R \times S$ is of the form $I \times J$ where I is an ideal of R and J is an ideal of S.

4. Prove that if R and S are nonzero rings then $R \times S$ is never a field.

5. Let $n_1, n_2, \ldots, n_k$ be integers which are relatively prime in pairs: $(n_i, n_j) = 1$ for all $i \neq j$.
   (a) Show that the Chinese Remainder Theorem implies that for any $a_1, \ldots, a_k \in \mathbb{Z}$ there is a solution $x \in \mathbb{Z}$ to the simultaneous congruences

   $$x \equiv a_1 \bmod n_1, \qquad x \equiv a_2 \bmod n_2, \qquad \ldots, \qquad x \equiv a_k \bmod n_k$$

   and that the solution x is unique mod $n = n_1 n_2 \ldots n_k$.
   (b) Let $n_i' = n/n_i$ be the quotient of n by n_i, which is relatively prime to n_i by assumption. Let t_i be the inverse of $n_i' \bmod n_i$. Prove that the solution x in (a) is given by

   $$x = a_1 t_1 n_1' + a_2 t_2 n_2' + \cdots + a_k t_k n_k' \bmod n.$$

   Note that the elements t_i can be quickly found by the Euclidean Algorithm as described in Section 2 of the Preliminaries chapter (writing $an_i + bn_i' = (n_i, n_i') = 1$ gives $t_i = b$) and that these then quickly give the solutions to the system of congruences above for any choice of $a_1, a_2, \ldots, a_k$.

(c) Solve the simultaneous system of congruences

$$x \equiv 1 \bmod 8, \qquad x \equiv 2 \bmod 25, \qquad \text{and} \quad x \equiv 3 \bmod 81$$

and the simultaneous system

$$y \equiv 5 \bmod 8, \qquad y \equiv 12 \bmod 25, \qquad \text{and} \quad y \equiv 47 \bmod 81.$$

6. Let $f_1(x), f_2(x), \ldots, f_k(x)$ be polynomials with integer coefficients of the same degree d. Let $n_1, n_2, \ldots, n_k$ be integers which are relatively prime in pairs (i.e., $(n_i, n_j) = 1$ for all $i \neq j$). Use the Chinese Remainder Theorem to prove there exists a polynomial $f(x)$ with integer coefficients and of degree d with

$$f(x) \equiv f_1(x) \bmod n_1, \qquad f(x) \equiv f_2(x) \bmod n_2, \qquad \ldots, \qquad f(x) \equiv f_k(x) \bmod n_k$$

i.e., the coefficients of $f(x)$ agree with the coefficients of $f_i(x) \bmod n_i$. Show that if all the $f_i(x)$ are monic, then $f(x)$ may also be chosen monic. [Apply the Chinese Remainder Theorem in $\mathbb{Z}$ to each of the coefficients separately.]

7. Let m and n be positive integers with n dividing m. Prove that the natural surjective ring projection $\mathbb{Z}/m\mathbb{Z} \to \mathbb{Z}/n\mathbb{Z}$ is also surjective on the units: $(\mathbb{Z}/m\mathbb{Z})^\times \to (\mathbb{Z}/n\mathbb{Z})^\times$.

The next four exercises develop the concept of *direct limits* and the "dual" notion of *inverse limits*. In these exercises I is a nonempty index set with a partial order $\leq$ (cf. Appendix I). For each $i \in I$ let A_i be an additive abelian group. In Exercise 8 assume also that I is a *directed set*: for every $i, j \in I$ there is some $k \in I$ with $i \leq k$ and $j \leq k$.

8. Suppose for every pair of indices i, j with $i \leq j$ there is a map $\rho_{ij} : A_i \to A_j$ such that the following hold:

i. $\rho_{jk} \circ \rho_{ij} = \rho_{ik}$ whenever $i \leq j \leq k$, and
ii. $\rho_{ii} = 1$ for all $i \in I$.

Let B be the disjoint union of all the A_i. Define a relation $\sim$ on B by

$$a \sim b \text{ if and only if there exists } k \text{ with } i, j \leq k \text{ and } \rho_{ik}(a) = \rho_{jk}(b),$$

for $a \in A_i$ and $b \in A_j$.
  (a) Show that $\sim$ is an equivalence relation on B. (The set of equivalence classes is called the *direct* or *inductive limit* of the directed system $\{A_i\}$, and is denoted $\varinjlim A_i$. In the remaining parts of this exercise let $A = \varinjlim A_i$.)
  (b) Let $\overline{x}$ denote the class of x in A and define $\rho_i : A_i \to A$ by $\rho_i(a) = \overline{a}$. Show that if each ρ_{ij} is injective, then so is ρ_i for all i (so we may then identify each A_i as a subset of A).
  (c) Assume all ρ_{ij} are group homomorphisms. For $a \in A_i, b \in A_j$ show that the operation

$$\overline{a} + \overline{b} = \overline{\rho_{ik}(a) + \rho_{jk}(b)}$$

where k is any index with $i, j \leq k$, is well defined and makes A into an abelian group. Deduce that the maps ρ_i in (b) are group homomorphisms from A_i to A.
  (d) Show that if all A_i are commutative rings with 1 and all ρ_{ij} are ring homomorphisms that send 1 to 1, then A may likewise be given the structure of a commutative ring with 1 such that all ρ_i are ring homomorphisms.
  (e) Under the hypotheses in (c) prove that the direct limit has the following *universal property*: if C is any abelian group such that for each $i \in I$ there is a homomorphism $\varphi_i : A_i \to C$ with $\varphi_i = \varphi_j \circ \rho_{ij}$ whenever $i \leq j$, then there is a unique homomorphism $\varphi : A \to C$ such that $\varphi \circ \rho_i = \varphi_i$ for all i.

9. Let I be the collection of open intervals $U = (a, b)$ on the real line containing a fixed real number p. Order these by reverse inclusion: $U \le V$ if $V \subseteq U$ (note that I is a directed set). For each U let A_U be the ring of continuous real valued functions on U. For $V \subseteq U$ define the *restriction maps* $\rho_{UV} : A_U \to A_V$ by $f \mapsto f|_V$, the usual restriction of a function on U to a function on the subset V (which is easily seen to be a ring homomorphism). Let $A = \varinjlim A_U$ be the direct limit. In the notation of the preceding exercise, show that the maps $\rho_U : A_U \to A$ are *not* injective but are all surjective (A is called the ring of *germs of continuous functions* at p).

We now develop the notion of *inverse limits*. Continue to assume I is a partially ordered set (but not necessarily directed), and A_i is a group for all $i \in I$.

10. Suppose for every pair of indices i, j with $i \le j$ there is a map $\mu_{ji} : A_j \to A_i$ such that the following hold:

   i. $\mu_{ji} \circ \mu_{kj} = \mu_{ki}$ whenever $i \le j \le k$, and
   ii. $\mu_{ii} = 1$ for all $i \in I$.

   Let P be the subset of elements $(a_i)_{i \in I}$ in the direct product $\prod_{i \in I} A_i$ such that $\mu_{ji}(a_j) = a_i$ whenever $i \le j$ (here a_i and a_j are the ith and j^{th} components respectively of the element in the direct product). The set P is called the *inverse* or *projective limit* of the system $\{A_i\}$, and is denoted $\varprojlim A_i$.)

   (a) Assume all μ_{ji} are group homomorphisms. Show that P is a subgroup of the direct product group (cf. Exercise 15, Section 5.1).

   (b) Assume the hypotheses in (a), and let $I = \mathbb{Z}^+$ (usual ordering). For each $i \in I$ let $\mu_i : P \to A_i$ be the projection of P onto its ith component. Show that if each μ_{ji} is surjective, then so is μ_i for all i (so each A_i is a quotient group of P).

   (c) Show that if all A_i are commutative rings with 1 and all μ_{ji} are ring homomorphisms that send 1 to 1, then P may likewise be given the structure of a commutative ring with 1 such that all μ_i are ring homomorphisms.

   (d) Under the hypotheses in (a) prove that the inverse limit has the following *universal property*: if D is any group such that for each $i \in I$ there is a homomorphism $\pi_i : D \to A_i$ with $\pi_i = \mu_{ji} \circ \pi_j$ whenever $i \le j$, then there is a unique homomorphism $\pi : D \to P$ such that $\mu_i \circ \pi = \pi_i$ for all i.

11. Let p be a prime let $I = \mathbb{Z}^+$, let $A_i = \mathbb{Z}/p^i \mathbb{Z}$ and let μ_{ji} be the natural projection maps

$$\mu_{ji} : a \ (\text{mod } p^j) \longmapsto a \ (\text{mod } p^i).$$

   The inverse limit $\varprojlim \mathbb{Z}/p^i \mathbb{Z}$ is called the ring of *p-adic integers*, and is denoted by $\mathbb{Z}_p$.

   (a) Show that every element of $\mathbb{Z}_p$ may be written uniquely as an infinite formal sum $b_0 + b_1 p + b_2 p^2 + b_3 p^3 + \cdots$ with each $b_i \in \{0, 1, \ldots, p - 1\}$. Describe the rules for adding and multiplying such formal sums corresponding to addition and multiplication in the ring $\mathbb{Z}_p$. [Write a least residue in each $\mathbb{Z}/p^i \mathbb{Z}$ in its base p expansion and then describe the maps μ_{ji}.] (Note in particular that $\mathbb{Z}_p$ is uncountable.)

   (b) Prove that $\mathbb{Z}_p$ is an integral domain that contains a copy of the integers.

   (c) Prove that $b_0 + b_1 p + b_2 p^2 + b_3 p^3 + \cdots$ as in (a) is a unit in $\mathbb{Z}_p$ if and only if $b_0 \ne 0$.

   (d) Prove that $p\mathbb{Z}_p$ is the unique maximal ideal of $\mathbb{Z}_p$ and $\mathbb{Z}_p/p\mathbb{Z}_p \cong \mathbb{Z}/p\mathbb{Z}$ (where $p = 0 + 1p + 0p^2 + 0p^3 + \cdots$). Prove that every nonzero ideal of $\mathbb{Z}_p$ is of the form $p^n \mathbb{Z}_p$ for some integer $n \ge 0$.

   (e) Show that if $a_1 \not\equiv 0 \ (\text{mod } p)$ then there is an element $a = (a_i)$ in the inverse limit $\mathbb{Z}_p$ satisfying $a_j^{p-1} \equiv 1 \ (\text{mod } p^j)$ and $\mu_{j1}(a_j) = a_1$ for all j. Deduce that $\mathbb{Z}_p$ contains $p - 1$ distinct $(p - 1)^{\text{st}}$ roots of 1.

CHAPTER 8

Euclidean Domains, Principal Ideal Domains, and Unique Factorization Domains

There are a number of classes of rings with more algebraic structure than generic rings. Those considered in this chapter are rings with a division algorithm (Euclidean Domains), rings in which every ideal is principal (Principal Ideal Domains) and rings in which elements have factorizations into primes (Unique Factorization Domains). The principal examples of such rings are the ring $\mathbb{Z}$ of integers and polynomial rings $F[x]$ with coefficients in some field F. We prove here all the theorems on the integers $\mathbb{Z}$ stated in the Preliminaries chapter as special cases of results valid for more general rings. These results will be applied to the special case of the ring $F[x]$ in the next chapter.

All rings in this chapter are commutative.

8.1 EUCLIDEAN DOMAINS

We first define the notion of a *norm* on an integral domain R. This is essentially no more than a measure of "size" in R.

Definition. Any function $N : R \to \mathbb{Z}^+ \cup \{0\}$ with $N(0) = 0$ is called a *norm* on the integral domain R. If $N(a) > 0$ for $a \neq 0$ define N to be a *positive norm*.

We observe that this notion of a norm is fairly weak and that it is possible for the same integral domain R to possess several different norms.

Definition. The integral domain R is said to be a *Euclidean Domain* (or possess a *Division Algorithm*) if there is a norm N on R such that for any two elements a and b of R with $b \neq 0$ there exist elements q and r in R with

$$a = qb + r \qquad \text{with } r = 0 \text{ or } N(r) < N(b).$$

The element q is called the *quotient* and the element r the *remainder* of the division.

The importance of the existence of a Division Algorithm on an integral domain R is that it allows a *Euclidean Algorithm* for two elements a and b of R: by successive "divisions" (these actually *are* divisions in the field of fractions of R) we can write

$$a = q_0 b + r_0 \tag{0}$$

$$b = q_1 r_0 + r_1 \tag{1}$$

$$r_0 = q_2 r_1 + r_2 \tag{2}$$

$$\vdots$$

$$r_{n-2} = q_n r_{n-1} + r_n \tag{n}$$

$$r_{n-1} = q_{n+1} r_n \tag{n+1}$$

where r_n is the last nonzero remainder. Such an r_n exists since $N(b) > N(r_0) > N(r_1) > \cdots > N(r_n)$ is a decreasing sequence of nonnegative integers if the remainders are nonzero, and such a sequence cannot continue indefinitely. Note also that there is no guarantee that these elements are *unique*.

Examples

(0) Fields are trivial examples of Euclidean Domains where any norm will satisfy the defining condition (e.g., $N(a) = 0$ for all a). This is because for every a, b with $b \neq 0$ we have $a = qb + 0$, where $q = ab^{-1}$.

(1) The integers $\mathbb{Z}$ are a Euclidean Domain with norm given by $N(a) = |a|$, the usual absolute value. The existence of a Division Algorithm in $\mathbb{Z}$ (the familiar "long division" of elementary arithmetic) is verified as follows. Let a and b be two nonzero integers and suppose first that $b > 0$. The half open intervals $[nb, (n+1)b)$, $n \in \mathbb{Z}$ partition the real line and so a is in one of them, say $a \in [kb, (k+1)b)$. For $q = k$ we have $a - qb = r \in [0, |b|)$ as needed. If $b < 0$ (so $-b > 0$), by what we have just seen there is an integer q such that $a = q(-b) + r$ with either $r = 0$ or $|r| < |-b|$; then $a = (-q)b + r$ satisfies the requirements of the Division Algorithm for a and b. This argument can be made more formal by using induction on $|a|$.

Note that if a is not a multiple of b there are always two possibilities for the pair q, r: the proof above always produced a positive remainder r. If for example $b > 0$ and q, r are as above with $r > 0$, then $a = q'b + r'$ with $q' = q + 1$ and $r' = r - b$ also satisfy the conditions of the Division Algorithm applied to a, b. Thus $5 = 2 \cdot 2 + 1 = 3 \cdot 2 - 1$ are the two ways of applying the Division Algorithm in $\mathbb{Z}$ to $a = 5$ and $b = 2$. The quotient and remainder are unique if we require the remainder to be nonnegative.

(2) If F is a field, then the polynomial ring $F[x]$ is a Euclidean Domain with norm given by $N(p(x)) =$ the degree of $p(x)$. The Division Algorithm for polynomials is simply "long division" of polynomials which may be familiar for polynomials with real coefficients. The proof is very similar to that for $\mathbb{Z}$ and is given in the next chapter (although for polynomials the quotient and remainder are shown to be unique). In order for a polynomial ring to be a Euclidean Domain the coefficients must come from a field since the division algorithm ultimately rests on being able to divide arbitrary nonzero coefficients. We shall prove in Section 2 that $R[x]$ is not a Euclidean Domain if R is not a field.

(3) The quadratic integer rings $\mathcal{O}$ in Section 7.1 are integral domains with a norm defined by the absolute value of the field norm (to ensure the values taken are nonnegative;

when $D < 0$ the field norm is itself a norm), but in general $\mathcal{O}$ is not Euclidean with respect to this norm (or any other norm). The Gaussian integers $\mathbb{Z}[i]$ (where $D = -1$), however, are a Euclidean Domain with respect to the norm $N(a + bi) = a^2 + b^2$, as we now show (cf. also the end of Section 3).

Let $\alpha = a + bi$, $\beta = c + di$ be two elements of $\mathbb{Z}[i]$ with $\beta \neq 0$. Then in the field $\mathbb{Q}(i)$ we have $\dfrac{\alpha}{\beta} = r + si$ where $r = (ac + bd)/(c^2 + d^2)$ and $s = (bc - ad)/(c^2 + d^2)$ are rational numbers. Let p be an integer closest to the rational number r and let q be an integer closest to the rational number s, so that both $|r - p|$ and $|s - q|$ are at most $1/2$. The Division Algorithm follows immediately once we show

$$\alpha = (p + qi)\beta + \gamma \qquad \text{for some } \gamma \in \mathbb{Z}[i] \text{ with} \quad N(\gamma) \leq \frac{1}{2}N(\beta)$$

which is even stronger than necessary. Let $\theta = (r - p) + (s - q)i$ and set $\gamma = \beta\theta$. Then $\gamma = \alpha - (p + qi)\beta$, so that $\gamma \in \mathbb{Z}[i]$ is a Gaussian integer and $\alpha = (p + qi)\beta + \gamma$. Since $N(\theta) = (r - p)^2 + (s - q)^2$ is at most $1/4 + 1/4 = 1/2$, the multiplicativity of the norm N implies that $N(\gamma) = N(\theta)N(\beta) \leq \dfrac{1}{2}N(\beta)$ as claimed.

Note that the algorithm is quite explicit since a quotient $p + qi$ is quickly determined from the rational numbers r and s, and then the remainder $\gamma = \alpha - (p + qi)\beta$ is easily computed. Note also that the quotient need not be unique: if r (or s) is half of an odd integer then there are two choices for p (or for q, respectively).

This proof that $\mathbb{Z}[i]$ is a Euclidean Domain can also be used to show that $\mathcal{O}$ is a Euclidean Domain (with respect to the field norm defined in Section 7.1) for $D = -2, -3, -7, -11$ (cf. the exercises). We shall see shortly that $\mathbb{Z}[\sqrt{-5}]$ is not Euclidean with respect to any norm, and a proof that $\mathbb{Z}[(1 + \sqrt{-19})/2]$ is not a Euclidean Domain with respect to any norm appears at the end of this section.

(4) Recall (cf. Exercise 26 in Section 7.1) that a *discrete valuation ring* is obtained as follows. Let K be a field. A *discrete valuation* on K is a function $\nu : K^\times \to \mathbb{Z}$ satisfying

(i) $\nu(ab) = \nu(a) + \nu(b)$ (i.e., ν is a homomorphism from the multiplicative group of nonzero elements of K to $\mathbb{Z}$),

(ii) ν is surjective, and

(iii) $\nu(x + y) \geq \min\{\nu(x), \nu(y)\}$ for all $x, y \in K^\times$ with $x + y \neq 0$.

The set $\{x \in K^\times \mid \nu(x) \geq 0\} \cup \{0\}$ is a subring of K called the valuation ring of ν. An integral domain R is called a discrete valuation ring if there is a valuation ν on its field of fractions such that R is the valuation ring of ν.

For example the ring R of all rational numbers whose denominators are relatively prime to the fixed prime $p \in \mathbb{Z}$ is a discrete valuation ring contained in $\mathbb{Q}$.

A discrete valuation ring is easily seen to be a Euclidean Domain with respect to the norm defined by $N(0) = 0$ and $N = \nu$ on the nonzero elements of R. This is because for $a, b \in R$ with $b \neq 0$

(a) if $N(a) < N(b)$ then $a = 0 \cdot b + a$, and

(b) if $N(a) \geq N(b)$ then it follows from property (i) of a discrete valuation that $q = ab^{-1} \in R$, so $a = qb + 0$.

The first implication of a Division Algorithm for the integral domain R is that it forces every ideal of R to be *principal*.

Proposition 1. Every ideal in a Euclidean Domain is principal. More precisely, if I is any nonzero ideal in the Euclidean Domain R then $I = (d)$, where d is any nonzero element of I of minimum norm.

Proof: If I is the zero ideal, there is nothing to prove. Otherwise let d be any nonzero element of I of minimum norm (such a d exists since the set $\{N(a) \mid a \in I\}$ has a minimum element by the Well Ordering of $\mathbb{Z}$). Clearly $(d) \subseteq I$ since d is an element of I. To show the reverse inclusion let a be any element of I and use the Division Algorithm to write $a = qd + r$ with $r = 0$ or $N(r) < N(d)$. Then $r = a - qd$ and both a and qd are in I, so r is also an element of I. By the minimality of the norm of d, we see that r must be 0. Thus $a = qd \in (d)$ showing $I = (d)$.

Proposition 1 shows that every ideal of $\mathbb{Z}$ is principal. This fundamental property of $\mathbb{Z}$ was previously determined (in Section 7.3) from the (additive) group structure of $\mathbb{Z}$, using the classification of the subgroups of cyclic groups in Section 2.3. Note that these are really the same proof, since the results in Section 2.3 ultimately relied on the Euclidean Algorithm in $\mathbb{Z}$.

Proposition 1 can also be used to prove that some integral domains R are *not* Euclidean Domains (with respect to *any* norm) by proving the existence of ideals of R that are not principal.

Examples

(1) Let $R = \mathbb{Z}[x]$. Since the ideal $(2, x)$ is not principal (cf. Example 3 at the beginning of Section 7.4), it follows that the ring $\mathbb{Z}[x]$ of polynomials with *integer* coefficients is *not* a Euclidean Domain (for any choice of norm), even though the ring $\mathbb{Q}[x]$ of polynomials with *rational* coefficients is a Euclidean Domain.

(2) Let R be the quadratic integer ring $\mathbb{Z}[\sqrt{-5}]$, let N be the associated field norm $N(a + b\sqrt{-5}) = a^2 + 5b^2$ and consider the ideal $I = (3, 2 + \sqrt{-5})$ generated by 3 and $2 + \sqrt{-5}$. Suppose $I = (a + b\sqrt{-5})$, $a, b \in \mathbb{Z}$, were principal, i.e., $3 = \alpha(a + b\sqrt{-5})$ and $2 + \sqrt{-5} = \beta(a + b\sqrt{-5})$ for some $\alpha, \beta \in R$. Taking norms in the first equation gives $9 = N(\alpha)(a^2 + 5b^2)$ and since $a^2 + 5b^2$ is a positive integer it must be 1, 3 or 9. If the value is 9 then $N(\alpha) = 1$ and $\alpha = \pm 1$, so $a + b\sqrt{-5} = \pm 3$, which is impossible by the second equation since the coefficients of $2 + \sqrt{-5}$ are not divisible by 3. The value cannot be 3 since there are no integer solutions to $a^2 + 5b^2 = 3$. If the value is 1, then $a + b\sqrt{-5} = \pm 1$ and the ideal I would be the entire ring R. But then 1 would be an element of I, so $3\gamma + (2 + \sqrt{-5})\delta = 1$ for some $\gamma, \delta \in R$. Multiplying both sides by $2 - \sqrt{-5}$ would then imply that $2 - \sqrt{-5}$ is a multiple of 3 in R, a contradiction. It follows that I is not a principal ideal and so R is not a Euclidean Domain (with respect to any norm).

One of the fundamental consequences of the Euclidean Algorithm in $\mathbb{Z}$ is that it produces a greatest common divisor of two nonzero elements. This is true in any Euclidean Domain. The notion of a greatest common divisor of two elements (if it exists) can be made precise in general rings.

Definition. Let R be a commutative ring and let $a, b \in R$ with $b \neq 0$.

> **(1)** a is said to be a *multiple* of b if there exists an element $x \in R$ with $a = bx$. In this case b is said to *divide* a or be a *divisor* of a, written $b \mid a$.
>
> **(2)** A *greatest common divisor* of a and b is a nonzero element d such that
>
>> **(i)** $d \mid a$ and $d \mid b$, and
>>
>> **(ii)** if $d' \mid a$ and $d' \mid b$ then $d' \mid d$.
>
> A greatest common divisor of a and b will be denoted by g.c.d.(a, b), or (abusing the notation) simply (a, b).

Note that $b \mid a$ in a ring R if and only if $a \in (b)$ if and only if $(a) \subseteq (b)$. In particular, if d is any divisor of both a and b then (d) must contain both a and b and hence must contain the ideal generated by a and b. The defining properties (i) and (ii) of a greatest common divisor of a and b translated into the language of ideals therefore become (respectively):

if I is the ideal of R generated by a and b, then d is a greatest common divisor of a and b if

(i) I is contained in the principal ideal (d), and

(ii) if (d') is any principal ideal containing I then $(d) \subseteq (d')$.

Thus a greatest common divisor of a and b (if such exists) is a generator for the unique smallest principal ideal containing a and b. There are rings in which greatest common divisors do not exist.

This discussion immediately gives the following *sufficient* condition for the existence of a greatest common divisor.

Proposition 2. If a and b are nonzero elements in the commutative ring R such that the ideal generated by a and b is a principal ideal (d), then d is a greatest common divisor of a and b.

This explains why the symbol (a, b) is often used to denote both the ideal generated by a and b and a greatest common divisor of a and b. An integral domain in which every ideal (a, b) generated by two elements is principal is called a *Bezout Domain*. The exercises in this and subsequent sections explore these rings and show that there are Bezout Domains containing nonprincipal (necessarily infinitely generated) ideals.

Note that the condition in Proposition 2 is *not* a *necessary* condition. For example, in the ring $R = \mathbb{Z}[x]$ the elements 2 and x generate a maximal, nonprincipal ideal (cf. the examples in Section 7.4). Thus $R = (1)$ is the unique principal ideal containing both 2 and x, so 1 is a greatest common divisor of 2 and x. We shall see other examples along these lines in Section 3.

Before returning to Euclidean Domains we examine the uniqueness of greatest common divisors.

Proposition 3. Let R be an integral domain. If two elements d and d' of R generate the same principal ideal, i.e., $(d) = (d')$, then $d' = ud$ for some unit u in R. In particular, if d and d' are both greatest common divisors of a and b, then $d' = ud$ for some unit u.

Proof: This is clear if either d or d' is zero so we may assume d and d' are nonzero. Since $d \in (d')$ there is some $x \in R$ such that $d = xd'$. Since $d' \in (d)$ there is some $y \in R$ such that $d' = yd$. Thus $d = xyd$ and so $d(1 - xy) = 0$. Since $d \neq 0$, $xy = 1$, that is, both x and y are units. This proves the first assertion. The second assertion follows from the first since any two greatest common divisors of a and b generate the same principal ideal (they divide each other).

One of the most important properties of Euclidean Domains is that greatest common divisors always exist and *can be computed algorithmically.*

Theorem 4. Let R be a Euclidean Domain and let a and b be nonzero elements of R. Let $d = r_n$ be the last nonzero remainder in the Euclidean Algorithm for a and b described at the beginning of this chapter. Then
  (1) d is a greatest common divisor of a and b, and
  (2) the principal ideal (d) is the ideal generated by a and b. In particular, d can be written as an *R-linear combination* of a and b, i.e., there are elements x and y in R such that
$$d = ax + by.$$

Proof: By Proposition 1, the ideal generated by a and b is principal so a, b do have a greatest common divisor, namely any element which generates the (principal) ideal (a, b). Both parts of the theorem will follow therefore once we show $d = r_n$ generates this ideal, i.e., once we show that

(i) $d \mid a$ and $d \mid b$ (so $(a, b) \subseteq (d)$)
(ii) d is an R-linear combination of a and b (so $(d) \subseteq (a, b)$).

To prove that d divides both a and b simply keep track of the divisibilities in the Euclidean Algorithm. Starting from the $(n+1)^{\text{st}}$ equation, $r_{n-1} = q_{n+1} r_n$, we see that $r_n \mid r_{n-1}$. Clearly $r_n \mid r_n$. By induction (proceeding from index n downwards to index 0) assume r_n divides r_{k+1} and r_k. By the $(k+1)^{\text{st}}$ equation, $r_{k-1} = q_{k+1} r_k + r_{k+1}$, and since r_n divides both terms on the right hand side we see that r_n also divides r_{k-1}. From the 1st equation in the Euclidean Algorithm we obtain that r_n divides b and then from the 0th equation we get that r_n divides a. Thus (i) holds.

To prove that r_n is in the ideal (a, b) generated by a and b proceed similarly by induction proceeding from equation (0) to equation (n). It follows from equation (0) that $r_0 \in (a, b)$ and by equation (1) that $r_1 = b - q_1 r_0 \in (b, r_0) \subseteq (a, b)$. By induction assume $r_{k-1}, r_k \in (a, b)$. Then by the $(k+1)^{\text{st}}$ equation
$$r_{k+1} = r_{k-1} - q_{k+1} r_k \in (r_{k-1}, r_k) \subseteq (a, b).$$
This induction shows $r_n \in (a, b)$, which completes the proof.

Much of the material above may be familiar from elementary arithmetic in the case of the integers $\mathbb{Z}$, except possibly for the translation into the language of ideals. For example, if $a = 2210$ and $b = 1131$ then the smallest ideal of $\mathbb{Z}$ that contains both a and b (the ideal generated by a and b) is $13\mathbb{Z}$, since 13 is the greatest common divisor of 2210 and 1131. This fact follows quickly from the Euclidean Algorithm:
$$2210 = 1 \cdot 1131 + 1079$$

$$1131 = 1 \cdot 1079 + 52$$
$$1079 = 20 \cdot 52 + 39$$
$$52 = 1 \cdot 39 + 13$$
$$39 = 3 \cdot 13$$

so that $13 = (2210, 1131)$ is the last nonzero remainder. Using the procedure of Theorem 4 we can also write 13 as a linear combination of 2210 and 1131 by first solving the next to last equation above for $13 = 52 - 1 \cdot 39$, then using previous equations to solve for 39 and 52, etc., finally writing 13 entirely in terms of 2210 and 1131. The answer in this case is

$$13 = (-22) \cdot 2210 + 43 \cdot 1131.$$

The Euclidean Algorithm in the integers $\mathbb{Z}$ is extremely fast. It is a theorem that the number of steps required to determine the greatest common divisor of two integers a and b is at worst 5 times the number of digits of the smaller of the two numbers. Put another way, this algorithm is *logarithmic* in the size of the integers. To obtain an appreciation of the speed implied here, notice that for the example above we would have expected at worst $5 \cdot 4 = 20$ divisions (the example required far fewer). If we had started with integers on the order of 10^{100} (large numbers by physical standards), we would have expected at worst only 500 divisions.

There is no uniqueness statement for the integers x and y in $(a, b) = ax + by$. Indeed, $x' = x + b$ and $y' = y - a$ satisfy $(a, b) = ax' + by'$. This is essentially the only possibility — one can prove that if x_0 and y_0 are solutions to the equation $ax + by = N$, then any other solutions x and y to this equation are of the form

$$x = x_0 + m \frac{b}{(a, b)}$$
$$y = y_0 - m \frac{a}{(a, b)}$$

for some integer m (positive or negative).

This latter theorem (a proof of which is outlined in the exercises) provides a complete solution of the *First Order Diophantine Equation* $ax + by = N$ provided we know there is *at least one* solution to this equation. But the equation $ax + by = N$ is simply another way of stating that N is an element of the ideal generated by a and b. Since we know this ideal is just (d), the principal ideal generated by the greatest common divisor d of a and b, this is the same as saying $N \in (d)$, i.e., N is divisible by d. Hence, *the equation $ax + by = N$ is solvable in integers x and y if and only if N is divisible by the g.c.d. of a and b* (and then the result quoted above gives a full set of solutions of this equation).

We end this section with another criterion that can sometimes be used to prove that a given integral domain is not a Euclidean Domain.[1] For any integral domain let

---

[1] The material here and in some of the following section follows the exposition by J.C. Wilson in *A principal ideal ring that is not a Euclidean ring*, Math. Mag., 46(1973), pp. 34–38, of ideas of Th. Motzkin, and use a simplification by Kenneth S. Williams in *Note on non-Euclidean Principal Ideal Domains*, Math. Mag., 48(1975), pp. 176–177.

$\widetilde{R} = R^{\times} \cup \{0\}$ denote the collection of units of R together with 0. An element $u \in R - \widetilde{R}$ is called a *universal side divisor* if for every $x \in R$ there is some $z \in \widetilde{R}$ such that u divides $x - z$ in R, i.e., there is a type of "division algorithm" for u: every x may be written $x = qu + z$ where z is either zero or a unit. The existence of universal side divisors is a weakening of the Euclidean condition:

Proposition 5. Let R be an integral domain that is not a field. If R is a Euclidean Domain then there are universal side divisors in R.

Proof: Suppose R is Euclidean with respect to some norm N and let u be an element of $R - \widetilde{R}$ (which is nonempty since R is not a field) of minimal norm. For any $x \in R$, write $x = qu + r$ where r is either 0 or $N(r) < N(u)$. In either case the minimality of u implies $r \in \widetilde{R}$. Hence u is a universal side divisor in R.

Example

We can use Proposition 5 to prove that the quadratic integer ring $R = \mathbb{Z}[(1 + \sqrt{-19})/2]$ is not a Euclidean Domain with respect to any norm by showing that R contains no universal side divisors (we shall see in the next section that all of the ideals in R are principal, so the technique in the examples following Proposition 1 do not apply to this ring). We have already determined that ± 1 are the only units in R and so $\widetilde{R} = \{0, \pm 1\}$. Suppose $u \in R$ is a universal side divisor and let $N(a + b(1 + \sqrt{-19})/2) = a^2 + ab + 5b^2$ denote the field norm on R as in Section 7.1. Note that if $a, b \in \mathbb{Z}$ and $b \neq 0$ then $a^2 + ab + 5b^2 = (a + b/2)^2 + 19/4b^2 \geq 5$ and so the smallest nonzero values of N on R are 1 (for the units ± 1) and 4 (for ± 2). Taking $x = 2$ in the definition of a universal side divisor it follows that u must divide one of $2 - 0$ or 2 ± 1 in R, i.e., u is a nonunit divisor of 2 or 3 in R. If $2 = \alpha\beta$ then $4 = N(\alpha)N(\beta)$ and by the remark above it follows that one of α or β has norm 1, i.e., equals ± 1. Hence the only divisors of 2 in R are $\{\pm 1, \pm 2\}$. Similarly, the only divisors of 3 in R are $\{\pm 1, \pm 3\}$, so the only possible values for u are ± 2 or ± 3. Taking $x = (1 + \sqrt{-19})/2$ it is easy to check that none of $x, x \pm 1$ are divisible by ± 2 or ± 3 in R, so none of these is a universal side divisor.

EXERCISES

1. For each of the following five pairs of integers a and b, determine their greatest common divisor d and write d as a linear combination $ax + by$ of a and b.
   (a) $a = 20, b = 13$.
   (b) $a = 69, b = 372$.
   (c) $a = 11391, b = 5673$.
   (d) $a = 507885, b = 60808$.
   (e) $a = 91442056588823, b = 779086434385541$ (the Euclidean Algorithm requires only 7 steps for these integers).

2. For each of the following pairs of integers a and n, show that a is relatively prime to n and determine the inverse of $a \mod n$ (cf. Section 3 of the Preliminaries chapter).
   (a) $a = 13, n = 20$.
   (b) $a = 69, n = 89$.
   (c) $a = 1891, n = 3797$.

(d) $a = 6003722857$, $n = 77695236973$ (the Euclidean Algorithm requires only 3 steps for these integers).

3. Let R be a Euclidean Domain. Let m be the minimum integer in the set of norms of nonzero elements of R. Prove that every nonzero element of R of norm m is a unit. Deduce that a nonzero element of norm zero (if such an element exists) is a unit.

4. Let R be a Euclidean Domain.
   (a) Prove that if $(a, b) = 1$ and a divides bc, then a divides c. More generally, show that if a divides bc with nonzero a, b then $\dfrac{a}{(a, b)}$ divides c.
   (b) Consider the Diophantine Equation $ax + by = N$ where a, b and N are integers and a, b are nonzero. Suppose x_0, y_0 is a solution: $ax_0 + by_0 = N$. Prove that the full set of solutions to this equation is given by
   $$x = x_0 + m\frac{b}{(a, b)}, \qquad y = y_0 - m\frac{a}{(a, b)}$$
   as m ranges over the integers. [If x, y is a solution to $ax + by = N$, show that $a(x - x_0) = b(y_0 - y)$ and use (a).]

5. Determine all integer solutions of the following equations:
   (a) $2x + 4y = 5$
   (b) $17x + 29y = 31$
   (c) $85x + 145y = 505$.

6. (*The Postage Stamp Problem*) Let a and b be two relatively prime positive integers. Prove that every sufficiently large positive integer N can be written as a linear combination $ax + by$ of a and b where x and y are both *nonnegative*, i.e., there is an integer N_0 such that for all $N \geq N_0$ the equation $ax + by = N$ can be solved with both x and y nonnegative integers. Prove in fact that the integer $ab - a - b$ cannot be written as a positive linear combination of a and b but that every integer greater than $ab - a - b$ is a positive linear combination of a and b (so every "postage" greater than $ab - a - b$ can be obtained using only stamps in denominations a and b).

7. Find a generator for the ideal $(85, 1+13i)$ in $\mathbb{Z}[i]$, i.e., a greatest common divisor for 85 and $1+13i$, by the Euclidean Algorithm. Do the same for the ideal $(47 - 13i, 53 + 56i)$.

It is known (but not so easy to prove) that $D = -1, -2, -3, -7, -11, -19, -43, -67$, and -163 are the only negative values of D for which every ideal in $\mathcal{O}$ is principal (i.e., $\mathcal{O}$ is a P.I.D. in the terminology of the next section). The results of the next exercise determine precisely which quadratic integer rings with $D < 0$ are Euclidean.

8. Let $F = \mathbb{Q}(\sqrt{D})$ be a quadratic field with associated quadratic integer ring $\mathcal{O}$ and field norm N as in Section 7.1.
   (a) Suppose D is $-1, -2, -3, -7$ or -11. Prove that $\mathcal{O}$ is a Euclidean Domain with respect to N. [Modify the proof for $\mathbb{Z}[i]$ ($D = -1$) in the text. For $D = -3, -7, -11$ prove that every element of F differs from an element in $\mathcal{O}$ by an element whose norm is at most $(1 + |D|)^2/(16|D|)$, which is less than 1 for these values of D. Plotting the points of $\mathcal{O}$ in $\mathbb{C}$ may be helpful.]
   (b) Suppose that $D = -43, -67$, or -163. Prove that $\mathcal{O}$ is not a Euclidean Domain with respect to any norm. [Apply the same proof as for $D = -19$ in the text.]

9. Prove that the ring of integers $\mathcal{O}$ in the quadratic integer ring $\mathbb{Q}(\sqrt{2})$ is a Euclidean Domain with respect to the norm given by the absolute value of the field norm N in Section 7.1.

10. Prove that the quotient ring $\mathbb{Z}[i]/I$ is finite for any nonzero ideal I of $\mathbb{Z}[i]$. [Use the fact

that $I = (\alpha)$ for some nonzero α and then use the Division Algorithm in this Euclidean Domain to see that every coset of I is represented by an element of norm less than $N(\alpha)$.]

11. Let R be a commutative ring with 1 and let a and b be nonzero elements of R. A *least common multiple* of a and b is an element e of R such that

   (i) $a \mid e$ and $b \mid e$, and
   (ii) if $a \mid e'$ and $b \mid e'$ then $e \mid e'$.

(a) Prove that a least common multiple of a and b (if such exists) is a generator for the unique largest principal ideal contained in $(a) \cap (b)$.

(b) Deduce that any two nonzero elements in a Euclidean Domain have a least common multiple which is unique up to multiplication by a unit.

(c) Prove that in a Euclidean Domain the least common multiple of a and b is $\dfrac{ab}{(a, b)}$, where (a, b) is the greatest common divisor of a and b.

12. (*A Public Key Code*) Let N be a positive integer. Let M be an integer relatively prime to N and let d be an integer relatively prime to $\varphi(N)$, where φ denotes Euler's φ-function. Prove that if $M_1 \equiv M^d \pmod{N}$ then $M \equiv M_1^{d'} \pmod{N}$ where d' is the inverse of d mod $\varphi(N)$: $dd' \equiv 1 \pmod{\varphi(N)}$.

Remark: This result is the basis for a standard *Public Key Code*. Suppose $N = pq$ is the product of two distinct large primes (each on the order of 100 digits, for example). If M is a message, then $M_1 \equiv M^d \pmod{N}$ is a scrambled (*encoded*) version of M, which can be unscrambled (*decoded*) by computing $M_1^{d'} \pmod{N}$ (these powers can be computed quite easily even for large values of M and N by successive squarings). The values of N and d (but not p and q) are made publicly known (hence the name) and then anyone with a message M can send their encoded message $M^d \pmod{N}$. To decode the message it seems necessary to determine d', which requires the determination of the value $\varphi(N) = \varphi(pq) = (p-1)(q-1)$ (no one has as yet *proved* that there is no other decoding scheme, however). The success of this method as a code rests on the necessity of determining the *factorization* of N into primes, for which no sufficiently efficient algorithm exists (for example, the most naive method of checking all factors up to $\sqrt{N}$ would here require on the order of 10^{100} computations, or approximately 300 years even at 10 billion computations per second, and of course one can always increase the size of p and q).

8.2 PRINCIPAL IDEAL DOMAINS (P.I.D.s)

Definition.   A *Principal Ideal Domain* (P.I.D.) is an integral domain in which every ideal is principal.

Proposition 1 proved that *every Euclidean Domain is a Principal Ideal Domain* so that every result about Principal Ideal Domains automatically holds for Euclidean Domains.

Examples

   (1) As mentioned after Proposition 1, the integers $\mathbb{Z}$ are a P.I.D. We saw in Section 7.4 that the polynomial ring $\mathbb{Z}[x]$ contains nonprincipal ideals, hence is not a P.I.D.
   (2) Example 2 following Proposition 1 showed that the quadratic integer ring $\mathbb{Z}[\sqrt{-5}]$ is not a P.I.D., in fact the ideal $(3, 1 + \sqrt{-5})$ is a nonprincipal ideal. It is possible

for the product IJ of two nonprincipal ideals I and J to be principal, for example the ideals $(3, 1 + \sqrt{-5})$ and $(3, 1 - \sqrt{-5})$ are both nonprincipal and their product is the principal ideal generated by 3, i.e., $(3, 1 + \sqrt{-5})(3, 1 - \sqrt{-5}) = (3)$ (cf. Exercise 5 and the example preceding Proposition 12 below).

It is not true that every Principal Ideal Domain is a Euclidean Domain. We shall prove below that the quadratic integer ring $\mathbb{Z}[(1 + \sqrt{-19})/2]$, which was shown not to be a Euclidean Domain in the previous section, nevertheless is a P.I.D.

From an ideal-theoretic point of view Principal Ideal Domains are a natural class of rings to study beyond rings which are fields (where the ideals are just the trivial ones: (0) and (1)). Many of the properties enjoyed by Euclidean Domains are also satisfied by Principal Ideal Domains. A significant advantage of Euclidean Domains over Principal Ideal Domains, however, is that although greatest common divisors exist in both settings, in Euclidean Domains one has an *algorithm* for computing them. Thus (as we shall see in Chapter 12 in particular) results which depend on the existence of greatest common divisors may often be proved in the larger class of Principal Ideal Domains although computation of examples (i.e., concrete applications of these results) are more effectively carried out using a Euclidean Algorithm (if one is available).

We collect some facts about greatest common divisors proved in the preceding section.

Proposition 6. Let R be a Principal Ideal Domain and let a and b be nonzero elements of R. Let d be a generator for the principal ideal generated by a and b. Then
   (1) d is a greatest common divisor of a and b
   (2) d can be written as an R-*linear combination* of a and b, i.e., there are elements x and y in R with
$$d = ax + by$$

   (3) d is unique up to multiplication by a unit of R.

Proof: This is just Propositions 2 and 3.

Recall that maximal ideals are always prime ideals but the converse is not true in general. We observed in Section 7.4, however, that every nonzero prime ideal of $\mathbb{Z}$ is a maximal ideal. This useful fact is true in an arbitrary Principal Ideal Domain, as the following proposition shows.

Proposition 7. Every nonzero prime ideal in a Principal Ideal Domain is a maximal ideal.

Proof: Let (p) be a nonzero prime ideal in the Principal Ideal Domain R and let $I = (m)$ be any ideal containing (p). We must show that $I = (p)$ or $I = R$. Now $p \in (m)$ so $p = rm$ for some $r \in R$. Since (p) is a prime ideal and $rm \in (p)$, either r or m must lie in (p). If $m \in (p)$ then $(p) = (m) = I$. If, on the other hand, $r \in (p)$ write $r = ps$. In this case $p = rm = psm$, so $sm = 1$ (recall that R is an integral domain) and m is a unit so $I = R$.

As we have already mentioned, if F is a field, then the polynomial ring $F[x]$ is a Euclidean Domain, hence also a Principal Ideal Domain (this will be proved in the next chapter). The converse to this is also true. Intuitively, if I is an ideal in R (such as the ideal (2) in $\mathbb{Z}$) then the ideal (I, x) in $R[x]$ (such as the ideal $(2, x)$ in $\mathbb{Z}[x]$) requires one more generator than does I, hence in general is not principal.

Corollary 8. If R is any commutative ring such that the polynomial ring $R[x]$ is a Principal Ideal Domain (or a Euclidean Domain), then R is necessarily a field.

Proof: Assume $R[x]$ is a Principal Ideal Domain. Since R is a subring of $R[x]$ then R must be an integral domain (recall that $R[x]$ has an identity if and only if R does). The ideal (x) is a nonzero prime ideal in $R[x]$ because $R[x]/(x)$ is isomorphic to the integral domain R. By Proposition 7, (x) is a maximal ideal, hence the quotient R is a field by Proposition 12 in Section 7.4.

The last result in this section will be used to prove that not every P.I.D. is a Euclidean Domain and relates the principal ideal property with another weakening of the Euclidean condition.

Definition. Define N to be a *Dedekind–Hasse norm* if N is a positive norm and for every nonzero $a, b \in R$ either a is an element of the ideal (b) or there is a nonzero element in the ideal (a, b) of norm strictly smaller than the norm of b (i.e., either b divides a in R or there exist $s, t \in R$ with $0 < N(sa - tb) < N(b)$).

Note that R is Euclidean with respect to a positive norm N if it is always possible to satisfy the Dedekind–Hasse condition with $s = 1$, so this is indeed a weakening of the Euclidean condition.

Proposition 9. The integral domain R is a P.I.D. if and only if R has a Dedekind–Hasse norm.[2]

Proof: Let I be any nonzero ideal in R and let b be a nonzero element of I with $N(b)$ minimal. Suppose a is any nonzero element in I, so that the ideal (a, b) is contained in I. Then the Dedekind–Hasse condition on N and the minimality of b implies that $a \in (b)$, so $I = (b)$ is principal. The converse will be proved in the next section (Corollary 16).

---

[2]That a Dedekind–Hasse norm on R implies that R is a P.I.D. (and is equivalent when R is a ring of algebraic integers) is the classical *Criterion of Dedekind and Hasse*, cf. *Über eindeutige Zerlegung in Primelemente oder in Primhauptideale in Integritätsbereichen*, Jour. für die Reine und Angew. Math., 159(1928), pp. 3–12. The observation that the converse holds generally is more recent and due to John Greene, *Principal Ideal Domains are almost Euclidean*, Amer. Math. Monthly, 104(1997), pp. 154–156.

Example

Let $R = \mathbb{Z}[(1+\sqrt{-19})/2]$ be the quadratic integer ring considered at the end of the previous section. We show that the positive field norm $N(a + b(1 + \sqrt{-19})/2) = a^2 + ab + 5b^2$ defined on R is a Dedekind–Hasse norm, which by Proposition 9 and the results of the previous section will prove that R is a P.I.D. but not a Euclidean Domain.

Suppose α, β are nonzero elements of R and $\alpha/\beta \notin R$. We must show that there are elements $s, t \in R$ with $0 < N(s\alpha - t\beta) < N(\beta)$, which by the multiplicativity of the field norm is equivalent to

$$0 < N(\frac{\alpha}{\beta}s - t) < 1. \tag{$*$}$$

Write $\dfrac{\alpha}{\beta} = \dfrac{a + b\sqrt{-19}}{c} \in \mathbb{Q}[\sqrt{-19}]$ with integers a, b, c having no common divisor and with $c > 1$ (since β is assumed not to divide α). Since a, b, c have no common divisor there are integers x, y, z with $ax + by + cz = 1$. Write $ay - 19bx = cq + r$ for some quotient q and remainder r with $|r| \le c/2$ and let $s = y + x\sqrt{-19}$ and $t = q - z\sqrt{-19}$. Then a quick computation shows that

$$0 < N(\frac{\alpha}{\beta}s - t) = \frac{(ay - 19bx - cq)^2 + 19(ax + by + cz)^2}{c^2} = \frac{r^2 + 19}{c^2} \le \frac{1}{4} + \frac{19}{c^2}$$

and so $(*)$ is satisfied with this s and t provided $c \ge 5$ (note $r^2 + 19 \le 23$ when $c = 5$).

Suppose that $c = 2$. Then one of a, b is even and the other is odd (otherwise $\alpha/\beta \in R$), and then a quick check shows that $s = 1$ and $t = \dfrac{(a - 1) + b\sqrt{-19}}{2}$ are elements of R satisfying $(*)$.

Suppose that $c = 3$. The integer $a^2 + 19b^2$ is not divisible by 3 (modulo 3 this is $a^2 + b^2$ which is easily seen to be 0 modulo 3 if and only if a and b are both 0 modulo 3; but then a, b, c have a common factor). Write $a^2 + 19b^2 = 3q + r$ with $r = 1$ or 2. Then again a quick check shows that $s = a - b\sqrt{-19}, t = q$ are elements of R satisfying $(*)$.

Finally, suppose that $c = 4$, so a and b are not both even. If one of a, b is even and the other odd, then $a^2 + 19b^2$ is odd, so we can write $a^2 + 19b^2 = 4q + r$ for some $q, r \in \mathbb{Z}$ and $0 < r < 4$. Then $s = a - b\sqrt{-19}$ and $t = q$ satisfy $(*)$. If a and b are both odd, then $a^2 + 19b^2 \equiv 1 + 3 \bmod 8$, so we can write $a^2 + 19b^2 = 8q + 4$ for some $q \in \mathbb{Z}$. Then $s = \dfrac{a - b\sqrt{-19}}{2}$ and $t = q$ are elements of R that satisfy $(*)$.

EXERCISES

1. Prove that in a Principal Ideal Domain two ideals (a) and (b) are comaximal (cf. Section 7.6) if and only if a greatest common divisor of a and b is 1 (in which case a and b are said to be *coprime* or *relatively prime*).

2. Prove that any two nonzero elements of a P.I.D. have a least common multiple (cf. Exercise 11, Section 1).

3. Prove that a quotient of a P.I.D. by a prime ideal is again a P.I.D.

4. Let R be an integral domain. Prove that if the following two conditions hold then R is a Principal Ideal Domain:
   (i) any two nonzero elements a and b in R have a greatest common divisor which can be written in the form $ra + sb$ for some $r, s \in R$, and

(ii) if $a_1, a_2, a_3, \ldots$ are nonzero elements of R such that $a_{i+1} \mid a_i$ for all i, then there is a positive integer N such that a_n is a unit times a_N for all $n \geq N$.

5. Let R be the quadratic integer ring $\mathbb{Z}[\sqrt{-5}\,]$. Define the ideals $I_2 = (2, 1 + \sqrt{-5}\,)$, $I_3 = (3, 2 + \sqrt{-5}\,)$, and $I_3' = (3, 2 - \sqrt{-5}\,)$.

   (a) Prove that I_2, I_3, and I_3' are nonprincipal ideals in R. [Note that Example 2 following Proposition 1 proves this for I_3.]

   (b) Prove that the product of two nonprincipal ideals can be principal by showing that I_2^2 is the principal ideal generated by 2, i.e., $I_2^2 = (2)$.

   (c) Prove similarly that $I_2 I_3 = (1 - \sqrt{-5}\,)$ and $I_2 I_3' = (1 + \sqrt{-5}\,)$ are principal. Conclude that the principal ideal (6) is the product of 4 ideals: $(6) = I_2^2 I_3 I_3'$.

6. Let R be an integral domain and suppose that every *prime* ideal in R is principal. This exercise proves that every ideal of R is principal, i.e., R is a P.I.D.

   (a) Assume that the set of ideals of R that are not principal is nonempty and prove that this set has a maximal element under inclusion (which, by hypothesis, is not prime). [Use Zorn's Lemma.]

   (b) Let I be an ideal which is maximal with respect to being nonprincipal, and let $a, b \in R$ with $ab \in I$ but $a \notin I$ and $b \notin I$. Let $I_a = (I, a)$ be the ideal generated by I and a, let $I_b = (I, b)$ be the ideal generated by I and b, and define $J = \{r \in R \mid rI_a \subseteq I\}$. Prove that $I_a = (\alpha)$ and $J = (\beta)$ are principal ideals in R with $I \subsetneq I_b \subseteq J$ and $I_a J = (\alpha\beta) \subseteq I$.

   (c) If $x \in I$ show that $x = s\alpha$ for some $s \in J$. Deduce that $I = I_a J$ is principal, a contradiction, and conclude that R is a P.I.D.

7. An integral domain R in which every ideal generated by two elements is principal (i.e., for every $a, b \in R$, $(a, b) = (d)$ for some $d \in R$) is called a *Bezout Domain*. [cf. also Exercise 11 in Section 3.]

   (a) Prove that the integral domain R is a Bezout Domain if and only if every pair of elements a, b of R has a g.c.d. d in R that can be written as an R-linear combination of a and b, i.e., $d = ax + by$ for some $x, y \in R$.

   (b) Prove that every finitely generated ideal of a Bezout Domain is principal. [cf. the exercises in Sections 9.2 and 9.3 for Bezout Domains in which not every ideal is principal.]

   (c) Let F be the fraction field of the Bezout Domain R. Prove that every element of F can be written in the form a/b with $a, b \in R$ and a and b relatively prime (cf. Exercise 1).

8. Prove that if R is a Principal Ideal Domain and D is a multiplicatively closed subset of R, then $D^{-1}R$ is also a P.I.D. (cf. Section 7.5).

8.3 UNIQUE FACTORIZATION DOMAINS (U.F.D.s)

In the case of the integers $\mathbb{Z}$, there is another method for determining the greatest common divisor of two elements a and b familiar from elementary arithmetic, namely the notion of "factorization into primes" for a and b, from which the greatest common divisor can easily be determined. This can also be extended to a larger class of rings called Unique Factorization Domains (U.F.D.s) — these will be defined shortly. We shall then prove that

every Principal Ideal Domain is a Unique Factorization Domain so that every result about Unique Factorization Domains will automatically hold for both Euclidean Domains and Principal Ideal Domains.

We first introduce some terminology.

Definition. Let R be an integral domain.

(1) Suppose $r \in R$ is nonzero and is not a unit. Then r is called *irreducible* in R if whenever $r = ab$ with $a, b \in R$, at least one of a or b must be a unit in R. Otherwise r is said to be *reducible*.

(2) The nonzero element $p \in R$ is called *prime* in R if the ideal (p) generated by p is a prime ideal. In other words, a nonzero element p is a prime if it is not a unit and whenever $p \mid ab$ for any $a, b \in R$, then either $p \mid a$ or $p \mid b$.

(3) Two elements a and b of R differing by a unit are said to be *associate* in R (i.e., $a = ub$ for some unit u in R).

Proposition 10. In an integral domain a prime element is always irreducible.

Proof: Suppose (p) is a nonzero prime ideal and $p = ab$. Then $ab = p \in (p)$, so by definition of prime ideal one of a or b, say a, is in (p). Thus $a = pr$ for some r. This implies $p = ab = prb$ so $rb = 1$ and b is a unit. This shows that p is irreducible.

It is not true in general that an irreducible element is necessarily prime. For example, consider the element 3 in the quadratic integer ring $R = \mathbb{Z}[\sqrt{-5}]$. The computations in Section 1 show that 3 is irreducible in R, but 3 is not a prime since $(2+\sqrt{-5})(2-\sqrt{-5}) = 3^2$ is divisible by 3, but neither $2+\sqrt{-5}$ nor $2-\sqrt{-5}$ is divisible by 3 in R.

If R is a Principal Ideal Domain however, the notions of prime and irreducible elements are the same. In particular these notions coincide in $\mathbb{Z}$ and in $F[x]$ (where F is a field).

Proposition 11. In a Principal Ideal Domain a nonzero element is a prime if and only if it is irreducible.

Proof: We have shown above that prime implies irreducible. We must show conversely that if p is irreducible, then p is a prime, i.e., the ideal (p) is a prime ideal. If M is any ideal containing (p) then by hypothesis $M = (m)$ is a principal ideal. Since $p \in (m)$, $p = rm$ for some r. But p is irreducible so by definition either r or m is a unit. This means either $(p) = (m)$ or $(m) = (1)$, respectively. Thus the only ideals containing (p) are (p) or (1), i.e., (p) is a maximal ideal. Since maximal ideals are prime ideals, the proof is complete.

Example

Proposition 11 gives another proof that the quadratic integer ring $\mathbb{Z}[\sqrt{-5}]$ is not a P.I.D. since 3 is irreducible but not prime in this ring.

The irreducible elements in the integers $\mathbb{Z}$ are the prime numbers (and their negatives) familiar from elementary arithmetic, and two integers a and b are associates of one another if and only if $a = \pm b$.

In the integers $\mathbb{Z}$ any integer n can be written as a product of primes (not necessarily distinct), as follows. If n is not itself a prime then by definition it is possible to write $n = n_1 n_2$ for two other integers n_1 and n_2 neither of which is a unit, i.e., neither of which is ± 1. Both n_1 and n_2 must be smaller in absolute value than n itself. If they are both primes, we have already written n as a product of primes. If one of n_1 or n_2 is not prime, then it in turn can be factored into two (smaller) integers. Since integers cannot decrease in absolute value indefinitely, we must at some point be left only with prime integer factors, and so we have written n as a product of primes.

For example, if $n = 2210$, the algorithm above proceeds as follows: n is not itself prime, since we can write $n = 2 \cdot 1105$. The integer 2 is a prime, but 1105 is not: $1105 = 5 \cdot 221$. The integer 5 is prime, but 221 is not: $221 = 13 \cdot 17$. Here the algorithm terminates, since both 13 and 17 are primes. This gives the *prime factorization* of 2210 as $2210 = 2 \cdot 5 \cdot 13 \cdot 17$. Similarly, we find $1131 = 3 \cdot 13 \cdot 29$. In these examples each prime occurs only to the first power, but of course this need not be the case generally.

In the ring $\mathbb{Z}$ not only is it true that every integer n can be written as a product of primes, but in fact this decomposition is *unique* in the sense that any two prime factorizations of the same positive integer n differ only in the order in which the positive prime factors are written. The restriction to positive integers is to avoid considering the factorizations $(3)(5)$ and $(-3)(-5)$ of 15 as essentially distinct. This *unique factorization* property of $\mathbb{Z}$ (which we shall prove very shortly) is extremely useful for the arithmetic of the integers. General rings with the analogous property are given a name.

Definition. A *Unique Factorization Domain* (U.F.D.) is an integral domain R in which every nonzero element $r \in R$ which is not a unit has the following two properties:

> **(i)** r can be written as a finite product of irreducibles p_i of R (not necessarily distinct): $r = p_1 p_2 \cdots p_n$ and
>
> **(ii)** the decomposition in (i) is *unique up to associates*: namely, if $r = q_1 q_2 \cdots q_m$ is another factorization of r into irreducibles, then $m = n$ and there is some renumbering of the factors so that p_i is associate to q_i for $i = 1, 2, \ldots, n$.

Examples

> **(1)** A field F is trivially a Unique Factorization Domain since every nonzero element is a unit, so there are no elements for which properties (i) and (ii) must be verified.
>
> **(2)** As indicated above, we shall prove shortly that every Principal Ideal Domain is a Unique Factorization Domain (so, in particular, $\mathbb{Z}$ and $F[x]$ where F is a field are both Unique Factorization Domains).
>
> **(3)** We shall also prove in the next chapter that the ring $R[x]$ of polynomials is a Unique Factorization Domain whenever R itself is a Unique Factorization Domain (in contrast to the properties of being a Principal Ideal Domain or being a Euclidean Domain, which do not carry over from a ring R to the polynomial ring $R[x]$). This result together with the preceding example will show that $\mathbb{Z}[x]$ is a Unique Factorization Domain.
>
> **(4)** The subring of the Gaussian integers $R = \mathbb{Z}[2i] = \{a + 2bi \mid a, b \in \mathbb{Z}\}$, where $i^2 = -1$, is an integral domain but not a Unique Factorization Domain (rings of this nature were introduced in Exercise 23 of Section 7.1). The elements 2 and $2i$ are

irreducibles which are not associates in R since $i \notin R$, and $4 = 2 \cdot 2 = (-2i) \cdot (2i)$ has two distinct factorizations in R. One may also check directly that $2i$ is irreducible but not prime in R (since $R/(2i) \cong \mathbb{Z}/4\mathbb{Z}$). In the larger ring of Gaussian integers, $\mathbb{Z}[i]$, (which is a Unique Factorization Domain) 2 and $2i$ *are* associates since i is a unit in this larger ring. We shall give a slightly different proof that $\mathbb{Z}[2i]$ is not a Unique Factorization Domain at the end of Section 9.3 (one in which we do not have to check that 2 and $2i$ are irreducibles).

(5) The quadratic integer ring $\mathbb{Z}[\sqrt{-5}]$ is another example of an integral domain that is not a Unique Factorization Domain, since $6 = 2 \cdot 3 = (1 + \sqrt{-5})(1 - \sqrt{-5})$ gives two distinct factorizations of 6 into irreducibles. The principal ideal (6) in $\mathbb{Z}[\sqrt{-5}]$ can be written as a product of 4 nonprincipal prime ideals: $(6) = P_2^2 P_3 P_3'$ and the two distinct factorizations of the element 6 in $\mathbb{Z}[\sqrt{-5}]$ can be interpreted as arising from two rearrangements of this product of ideals into products of principal ideals: the product of $P_2^2 = (2)$ with $P_3 P_3' = (3)$, and the product of $P_2 P_3 = (1 + \sqrt{-5})$ with $P_2 P_3' = (1 - \sqrt{-5})$ (cf. Exercise 8).

While the *elements* of the quadratic integer ring $\mathcal{O}$ need not have unique factorization, it is a theorem (Corollary 16.16) that every *ideal* in $\mathcal{O}$ can be written uniquely as a product of prime *ideals*. The unique factorization of ideals into the product of prime ideals holds in general for rings of integers of algebraic number fields (examples of which are the quadratic integer rings) and leads to the notion of a Dedekind Domain considered in Chapter 16. It was the failure to have unique factorization into irreducibles for elements in algebraic integer rings in number theory that originally led to the definition of an ideal. The resulting uniqueness of the decomposition into prime ideals in these rings gave the elements of the ideals an "ideal" (in the sense of "perfect" or "desirable") behavior that is the basis for the choice of terminology for these (now fundamental) algebraic objects.

The first property of irreducible elements in a Unique Factorization Domain is that they are also primes. One might think that we could deduce Proposition 11 from this proposition together with the previously mentioned theorem (that we shall prove shortly) that every Principal Ideal Domain is a Unique Factorization Domain, however Proposition 11 will be used in the proof of the latter theorem.

Proposition 12. In a Unique Factorization Domain a nonzero element is a prime if and only if it is irreducible.

Proof: Let R be a Unique Factorization Domain. Since by Proposition 10, primes of R are irreducible it remains to prove that each irreducible element is a prime. Let p be an irreducible in R and assume $p \mid ab$ for some $a, b \in R$; we must show that p divides either a or b. To say that p divides ab is to say $ab = pc$ for some c in R. Writing a and b as a product of irreducibles, we see from this last equation and from the *uniqueness* of the decomposition into irreducibles of ab that the irreducible element p must be *associate* to one of the irreducibles occurring either in the factorization of a or in the factorization of b. We may assume that p is associate to one of the irreducibles in the factorization of a, i.e., that a can be written as a product $a = (up)p_2 \cdots p_n$ for u a unit and some (possibly empty set of) irreducibles $p_2, \ldots, p_n$. But then p divides a, since $a = pd$ with $d = up_2 \cdots p_n$, completing the proof.

In a Unique Factorization Domain we shall now use the terms "prime" and "irreducible" interchangeably although we shall usually refer to the "primes" in $\mathbb{Z}$ and the "irreducibles" in $F[x]$.

We shall use the preceding proposition to show that in a Unique Factorization Domain any two nonzero elements a and b have a greatest common divisor:

Proposition 13. Let a and b be two nonzero elements of the Unique Factorization Domain R and suppose

$$a = u p_1^{e_1} p_2^{e_2} \cdots p_n^{e_n} \qquad \text{and} \qquad b = v p_1^{f_1} p_2^{f_2} \cdots p_n^{f_n}$$

are prime factorizations for a and b, where u and v are units, the primes $p_1, p_2, \ldots, p_n$ are *distinct* and the exponents e_i and f_i are ≥ 0. Then the element

$$d = p_1^{\min (e_1, f_1)} p_2^{\min (e_2, f_2)} \cdots p_n^{\min (e_n, f_n)}$$

(where $d = 1$ if all the exponents are 0) is a greatest common divisor of a and b.

Proof: Since the exponents of each of the primes occurring in d are no larger than the exponents occurring in the factorizations of both a and b, d divides both a and b. To show that d is a greatest common divisor, let c be any common divisor of a and b and let $c = q_1^{g_1} q_2^{g_2} \cdots q_m^{g_m}$ be the prime factorization of c. Since each q_i divides c, hence divides a and b, we see from the preceding proposition that q_i must divide one of the primes p_j. In particular, up to associates (so up to multiplication by a unit) the primes occurring in c must be a subset of the primes occurring in a and b : $\{q_1, q_2, \ldots, q_m\} \subseteq \{p_1, p_2, \ldots, p_n\}$. Similarly, the exponents for the primes occurring in c must be no larger than those occurring in d. This implies that c divides d, completing the proof.

Example

In the example above, where $a = 2210$ and $b = 1131$, we find immediately from their prime factorizations that $(a, b) = 13$. Note that if the prime factorizations for a and b are known, the proposition above gives their greatest common divisor instantly, but that finding these prime factorizations is extremely time-consuming computationally. The Euclidean Algorithm is the fastest method for determining the g.c.d. of two integers but unfortunately it gives almost no information on the prime factorizations of the integers.

We now come to one of the principal results relating some of the rings introduced in this chapter.

Theorem 14. Every Principal Ideal Domain is a Unique Factorization Domain. In particular, every Euclidean Domain is a Unique Factorization Domain.

Proof: Note that the second assertion follows from the first since Euclidean Domains are Principal Ideal Domains. To prove the first assertion let R be a Principal Ideal Domain and let r be a nonzero element of R which is not a unit. We must show first that r can be written as a finite product of irreducible elements of R and then we must verify that this decomposition is unique up to units.

The method of proof of the first part is precisely analogous to the determination of the prime factor decomposition of an integer. Assume r is nonzero and is not a unit. If r is itself irreducible, then we are done. If not, then by definition r can be written as a product $r = r_1 r_2$ where neither r_1 nor r_2 is a unit. If both these elements are irreducibles, then again we are done, having written r as a product of irreducible elements. Otherwise, at least one of the two elements, say r_1 is reducible, hence can be written as a product of two nonunit elements $r_1 = r_{11} r_{12}$, and so forth. What we must verify is that this process *terminates*, i.e., that we must necessarily reach a point where all of the elements obtained as factors of r are irreducible. Suppose this is not the case. From the factorization $r = r_1 r_2$ we obtain a *proper* inclusion of ideals: $(r) \subset (r_1) \subset R$. The first inclusion is proper since r_2 is not a unit, and the last inclusion is proper since r_1 is not a unit. From the factorization of r_1 we similarly obtain $(r) \subset (r_1) \subset (r_{11}) \subset R$. If this process of factorization did not terminate after a finite number of steps, then we would obtain an *infinite ascending chain* of ideals:

$$(r) \subset (r_1) \subset (r_{11}) \subset \cdots \subset R$$

where all containments are proper, and the Axiom of Choice ensures that an infinite chain exists (cf. Appendix I).

We now show that any ascending chain $I_1 \subseteq I_2 \subseteq \cdots \subseteq R$ of ideals in a Principal Ideal Domain eventually becomes stationary, i.e., there is some positive integer n such that $I_k = I_n$ for all $k \geq n$.[3] In particular, it is not possible to have an infinite ascending chain of ideals where all containments are proper. Let $I = \cup_{i=1}^{\infty} I_i$. It follows easily (as in the proof of Proposition 11 in Section 7.4) that I is an ideal. Since R is a Principal Ideal Domain it is principally generated, say $I = (a)$. Since I is the union of the ideals above, a must be an element of one of the ideals in the chain, say $a \in I_n$. But then we have $I_n \subseteq I = (a) \subseteq I_n$ and so $I = I_n$ and the chain becomes stationary at I_n. This proves that every nonzero element of R which is not a unit has some factorization into irreducibles in R.

It remains to prove that the above decomposition is essentially unique. We proceed by induction on the number, n, of irreducible factors in some factorization of the element r. If $n = 0$, then r is a unit. If we had $r = qc$ (some other factorization) for some irreducible q, then q would divide a unit, hence would itself be a unit, a contradiction. Suppose now that n is at least 1 and that we have two products

$$r = p_1 p_2 \cdots p_n = q_1 q_2 \cdots q_m \qquad m \geq n$$

for r where the p_i and q_j are (not necessarily distinct) irreducibles. Since then p_1 divides the product on the right, we see by Proposition 11 that p_1 must divide one of the factors. Renumbering if necessary, we may assume p_1 divides q_1. But then $q_1 = p_1 u$ for some element u of R which must in fact be a unit since q_1 is irreducible. Thus p_1 and q_1 are associates. Cancelling p_1 (recall we are in an integral domain, so this is legitimate), we obtain the equation

$$p_2 \cdots p_n = u q_2 q_3 \cdots q_m = q_2' q_3 \cdots q_m \qquad m \geq n.$$

---

[3]This same argument can be used to prove the more general statement: an ascending chain of ideals becomes stationary in any commutative ring where all the ideals are *finitely generated*. This result will be needed in Chapter 12 where the details will be repeated.

where $q_2' = uq_2$ is again an irreducible (associate to q_2). By induction on n, we conclude that each of the factors on the left matches bijectively (up to associates) with the factors on the far right, hence with the factors in the middle (which are the same, up to associates). Since p_1 and q_1 (after the initial renumbering) have already been shown to be associate, this completes the induction step and the proof of the theorem.

Corollary 15. *(Fundamental Theorem of Arithmetic)* The integers $\mathbb{Z}$ are a Unique Factorization Domain.

Proof: The integers $\mathbb{Z}$ are a Euclidean Domain, hence are a Unique Factorization Domain by the theorem.

We can now complete the equivalence (Proposition 9) between the existence of a Dedekind–Hasse norm on the integral domain R and whether R is a P.I.D.

Corollary 16. Let R be a P.I.D. Then there exists a multiplicative Dedekind–Hasse norm on R.

Proof: If R is a P.I.D. then R is a U.F.D. Define the norm N by setting $N(0) = 0$, $N(u) = 1$ if u is a unit, and $N(a) = 2^n$ if $a = p_1 p_2 \cdots p_n$ where the p_i are irreducibles in R (well defined since the number of irreducible factors of a is unique). Clearly $N(ab) = N(a)N(b)$ so N is positive and multiplicative. To show that N is a Dedekind–Hasse norm, suppose that a, b are nonzero elements of R. Then the ideal generated by a and b is principal by assumption, say $(a, b) = (r)$. If a is not contained in the ideal (b) then also r is not contained in (b), i.e., r is not divisible by b. Since $b = xr$ for some $x \in R$, it follows that x is not a unit in R and so $N(b) = N(x)N(r) > N(r)$. Hence (a, b) contains a nonzero element with norm strictly smaller than the norm of b, completing the proof.

Factorization in the Gaussian Integers

We end our discussion of Unique Factorization Domains by describing the irreducible elements in the Gaussian integers $\mathbb{Z}[i]$ and the corresponding application to a famous theorem of Fermat in elementary number theory. This is particularly appropriate since the classical study of $\mathbb{Z}[i]$ initiated the algebraic study of rings.

In general, let $\mathcal{O}$ be a quadratic integer ring and let N be the associated field norm introduced in Section 7.1. Suppose $\alpha \in \mathcal{O}$ is an element whose norm is a prime p in $\mathbb{Z}$. If $\alpha = \beta\gamma$ for some $\beta, \gamma \in \mathcal{O}$ then $p = N(\alpha) = N(\beta)N(\gamma)$ so that one of $N(\beta)$ or $N(\gamma)$ is ± 1 and the other is $\pm p$. Since we have seen that an element of $\mathcal{O}$ has norm ± 1 if and only if it is a unit in $\mathcal{O}$, one of the factors of α is a unit. It follows that

$$\text{if } N(\alpha) \text{ is } \pm \text{ a prime (in } \mathbb{Z} \text{), then } \alpha \text{ is irreducible in } \mathcal{O}.$$

Suppose that π is a prime element in $\mathcal{O}$ and let (π) be the ideal generated by π in $\mathcal{O}$. Since (π) is a prime ideal in $\mathcal{O}$ it is easy to check that $(\pi) \cap \mathbb{Z}$ is a prime ideal in $\mathbb{Z}$ (if a and b are integers with $ab \in (\pi)$ then either a or b is an element of (π), so a or b is in $(\pi) \cap \mathbb{Z}$). Since $N(\pi)$ is a nonzero integer in (π) we have $(\pi) \cap \mathbb{Z} = p\mathbb{Z}$ for some integer prime p. It follows from $p \in (\pi)$ that π is a divisor in $\mathcal{O}$ of the integer prime p, and so the prime elements in $\mathcal{O}$ can be found by determining how the primes in $\mathbb{Z}$ factor in the larger ring $\mathcal{O}$. Suppose π divides the prime p in $\mathcal{O}$, say $p = \pi\pi'$. Then $N(\pi)N(\pi') = N(p) = p^2$, so since π is not a unit there are only two possibilities: either $N(\pi) = \pm p^2$ or $N(\pi) = \pm p$. In the former case $N(\pi') = \pm 1$, hence π' is a unit and $p = \pi$ (up to associates) is irreducible in $\mathcal{O}$. In the latter case $N(\pi) = N(\pi') = \pm p$, hence π' is also irreducible and $p = \pi\pi'$ is the product of precisely two irreducibles.

Consider now the special case $D = -1$ of the Gaussian integers $\mathbb{Z}[i]$. We have seen that the units in $\mathbb{Z}[i]$ are the elements ± 1 and $\pm i$. We proved in Section 1 that $\mathbb{Z}[i]$ is a Euclidean Domain, hence is also a Principal Ideal Domain and a Unique Factorization Domain, so the irreducible elements are the same as the prime elements, and can be determined by seeing how the primes in $\mathbb{Z}$ factor in the larger ring $\mathbb{Z}[i]$.

In this case $\alpha = a + bi$ has $N(\alpha) = \alpha\overline{\alpha} = a^2 + b^2$, where $\overline{\alpha} = a - bi$ is the complex conjugate of α. It follows by what we just saw that *p factors in $\mathbb{Z}[i]$ into precisely two irreducibles if and only if $p = a^2 + b^2$ is the sum of two integer squares (otherwise p remains irreducible in $\mathbb{Z}[i]$).* If $p = a^2 + b^2$ then the corresponding irreducible elements in $\mathbb{Z}[i]$ are $a \pm bi$.

Clearly $2 = 1^2 + 1^2$ is the sum of two squares, giving the factorization $2 = (1+i)(1-i) = -i(1+i)^2$. The irreducibles $1 + i$ and $1 - i = -i(1+i)$ are associates and it is easy to check that this is the only situation in which conjugate irreducibles $a + bi$ and $a - bi$ can be associates.

Since the square of any integer is congruent to either 0 or 1 modulo 4, an odd prime in $\mathbb{Z}$ that is the sum of two squares must be congruent to 1 modulo 4. Thus if p is a prime of $\mathbb{Z}$ with $p \equiv 3 \bmod 4$ then p is not the sum of two squares and p remains irreducible in $\mathbb{Z}[i]$.

Suppose now that p is a prime of $\mathbb{Z}$ with $p \equiv 1 \bmod 4$. We shall prove that p cannot be irreducible in $\mathbb{Z}[i]$ which will show that $p = (a + bi)(a - bi)$ factors as the product of two distinct irreducibles in $\mathbb{Z}[i]$ or, equivalently, that $p = a^2 + b^2$ is the sum of two squares. We first prove the following result from elementary number theory:

Lemma 17. The prime number $p \in \mathbb{Z}$ divides an integer of the form $n^2 + 1$ if and only if p is either 2 or is an odd prime congruent to 1 modulo 4.

Proof: The statement for $p = 2$ is trivial since $2 \mid 1^2 + 1$. If p is an odd prime, note that $p \mid n^2 + 1$ is equivalent to $n^2 = -1$ in $\mathbb{Z}/p\mathbb{Z}$. This in turn is equivalent to saying the residue class of n is of order 4 in the multiplicative group $(\mathbb{Z}/p\mathbb{Z})^\times$. Thus p divides an integer of the form $n^2 + 1$ if and only if $(\mathbb{Z}/p\mathbb{Z})^\times$ contains an element of order 4. By Lagrange's Theorem, if $(\mathbb{Z}/p\mathbb{Z})^\times$ contains an element of order 4 then $|(\mathbb{Z}/p\mathbb{Z})^\times| = p - 1$ is divisible by 4, i.e., p is congruent to 1 modulo 4.

Conversely, suppose $p - 1$ is divisible by 4. We first argue that $(\mathbb{Z}/p\mathbb{Z})^\times$ contains a unique element of order 2. If $m^2 \equiv 1 \bmod p$ then p divides $m^2 - 1 = (m-1)(m+1)$. Thus p divides either $m - 1$ (i.e., $m \equiv 1 \bmod p$) or $m + 1$ (i.e., $m \equiv -1 \bmod p$), so -1 is the unique residue class of order 2 in $(\mathbb{Z}/p\mathbb{Z})^\times$. Now the abelian group $(\mathbb{Z}/p\mathbb{Z})^\times$ contains a subgroup H of order 4 (for example, the quotient by the subgroup $\{\pm 1\}$ contains a subgroup of order 2 whose preimage is a subgroup of order 4 in $(\mathbb{Z}/p\mathbb{Z})^\times$).

Since the Klein 4-group has three elements of order 2 whereas $(\mathbb{Z}/p\mathbb{Z})^\times$ — hence also H — has a unique element of order 2, H must be the cyclic group of order 4. Thus $(\mathbb{Z}/p\mathbb{Z})^\times$ contains an element of order 4, namely a generator for H.

Remark: We shall prove later (Corollary 19 in Section 9.5) that $(\mathbb{Z}/p\mathbb{Z})^\times$ is a cyclic group, from which it is immediate that there is an element of order 4 if and only if $p - 1$ is divisible by 4.

By Lemma 17, if $p \equiv 1 \bmod 4$ is a prime then p divides $n^2 + 1$ in $\mathbb{Z}$ for some $n \in \mathbb{Z}$, so certainly p divides $n^2 + 1 = (n + i)(n - i)$ in $\mathbb{Z}[i]$. If p were irreducible in $\mathbb{Z}[i]$ then p would divide either $n + i$ or $n - i$ in $\mathbb{Z}[i]$. In this situation, since p is a real number, it would follow that p divides both $n + i$ and its complex conjugate $n - i$; hence p would divide their difference, $2i$. This is clearly not the case. We have proved the following result:

Proposition 18.
    (1) *(Fermat's Theorem on sums of squares)* The prime p is the sum of two integer squares, $p = a^2 + b^2$, $a, b \in \mathbb{Z}$, if and only if $p = 2$ or $p \equiv 1 \bmod 4$. Except for interchanging a and b or changing the signs of a and b, the representation of p as a sum of two squares is unique.
    (2) The irreducible elements in the Gaussian integers $\mathbb{Z}[i]$ are as follows:
        (a) $1 + i$ (which has norm 2),
        (b) the primes $p \in \mathbb{Z}$ with $p \equiv 3 \bmod 4$ (which have norm p^2), and
        (c) $a + bi$, $a - bi$, the distinct irreducible factors of $p = a^2 + b^2 = (a + bi)(a - bi)$ for the primes $p \in \mathbb{Z}$ with $p \equiv 1 \bmod 4$ (both of which have norm p).

The first part of Proposition 18 is a famous theorem of Fermat in elementary number theory, for which a number of alternate proofs can be given.

More generally, the question of whether the integer $n \in \mathbb{Z}$ can be written as a sum of two integer squares, $n = A^2 + B^2$, is equivalent to the question of whether n is the norm of an element $A + Bi$ in the Gaussian integers, i.e., $n = A^2 + B^2 = N(A + Bi)$. Writing $A + Bi = \pi_1 \pi_2 \cdots \pi_k$ as a product of irreducibles (uniquely up to units) it follows from the explicit description of the irreducibles in $\mathbb{Z}[i]$ in Proposition 18 that n is a norm if and only if the prime divisors of n that are congruent to 3 mod 4 occur to even exponents. Further, if this condition on n is satisfied, then the uniqueness of the factorization of $A + Bi$ in $\mathbb{Z}[i]$ allows us to count the number of representations of n as a sum of two squares, as in the following corollary.

Corollary 19. Let n be a positive integer and write

$$n = 2^k p_1^{a_1} \cdots p_r^{a_r} q_1^{b_1} \cdots q_s^{b_s}$$

where $p_1, \dots, p_r$ are distinct primes congruent to 1 modulo 4 and $q_1, \dots, q_s$ are distinct primes congruent to 3 modulo 4. Then n can be written as a sum of two squares in $\mathbb{Z}$, i.e., $n = A^2 + B^2$ with $A, B \in \mathbb{Z}$, if and only if each b_i is even. Further, if this condition on n is satisfied, then the number of representations of n as a sum of two squares is $4(a_1 + 1)(a_2 + 1) \cdots (a_r + 1)$.

Proof: The first statement in the corollary was proved above. Assume now that $b_1, \ldots, b_s$ are all even. For each prime p_i congruent to 1 modulo 4 write $p_i = \pi_i \overline{\pi_i}$ for $i = 1, 2, \ldots, r$, where π_i and $\overline{\pi_i}$ are irreducibles as in (2)(c) of Proposition 18. If $N(A + Bi) = n$ then examining norms we see that, up to units, the factorization of $A + Bi$ into irreducibles in $\mathbb{Z}[i]$ is given by

$$A + Bi = (1 + i)^k (\pi_1^{a_{1,1}} \overline{\pi_1}^{a_{1,2}}) \ldots (\pi_r^{a_{r,1}} \overline{\pi_r}^{a_{r,2}}) q_1^{b_1/2} \ldots q_s^{b_s/2}$$

with nonnegative integers $a_{i,1}, a_{i,2}$ satisfying $a_{i,1} + a_{i,2} = a_i$ for $i = 1, 2, \ldots, r$. Since $a_{i,1}$ can have the values $0, 1, \ldots, a_i$ (and then $a_{i,2}$ is determined), there are a total of $(a_1 + 1)(a_2 + 1) \cdots (a_r + 1)$ distinct elements $A + Bi$ in $\mathbb{Z}[i]$ of norm n, up to units. Finally, since there are four units in $\mathbb{Z}[i]$, the second statement in the corollary follows.

Example

Since $493 = 17 \cdot 29$ and both primes are congruent to 1 modulo 4, $493 = A^2 + B^2$ is the sum of two integer squares. Since $17 = (4 + i)(4 - i)$ and $29 = (5 + 2i)(5 - 2i)$ the possible factorizations of $A + Bi$ in $\mathbb{Z}[i]$ up to units are $(4 + i)(5 + 2i) = 18 + 13i$, $(4 + i)(5 - 2i) = 22 - 3i$, $(4 - i)(5 - 2i) = 22 + 3i$, and $(4 - i)(5 - 2i) = 18 - 13i$. Multiplying by -1 reverses both signs and multiplication by i interchanges the A and B and introduces one sign change. Then $493 = (\pm 18)^2 + (\pm 13)^2 = (\pm 22)^2 + (\pm 3)^2$ with all possible choices of signs give 8 of the 16 possible representations of 493 as the sum of two squares; the remaining 8 are obtained by interchanging the two summands.

Similarly, the integer $58000957 = 7^6 \cdot 17 \cdot 29$ can be written as a sum of two squares in precisely 16 ways, obtained by multiplying each of the integers A, B in $493 = A^2 + B^2$ above by 7^3.

Summary

In summary, we have the following inclusions among classes of commutative rings with identity:

$$\textit{fields} \subset \textit{Euclidean Domains} \subset \textit{P.I.D.s} \subset \textit{U.F.D.s} \subset \textit{integral domains}$$

with all containments being proper. Recall that $\mathbb{Z}$ is a Euclidean Domain that is not a field, the quadratic integer ring $\mathbb{Z}[(1 + \sqrt{-19})/2]$ is a Principal Ideal Domain that is not a Euclidean Domain, $\mathbb{Z}[x]$ is a Unique Factorization Domain (Theorem 7 in Chapter 9) that is not a Principal Ideal Domain and $\mathbb{Z}[\sqrt{-5}]$ is an integral domain that is not a Unique Factorization Domain.

EXERCISES

1. Let $G = \mathbb{Q}^\times$ be the multiplicative group of nonzero rational numbers. If $\alpha = p/q \in G$, where p and q are relatively prime integers, let $\varphi : G \to G$ be the map which interchanges the primes 2 and 3 in the prime power factorizations of p and q (so, for example, $\varphi(2^4 3^{11} 5^1 13^2) = 3^4 2^{11} 5^1 13^2$, $\varphi(3/16) = \varphi(3/2^4) = 2/3^4 = 2/81$, and φ is the identity on all rational numbers with numerators and denominators relatively prime to 2 and to 3).
   (a) Prove that φ is a group isomorphism.
   (b) Prove that there are infinitely many isomorphisms of the group G to itself.

(c) Prove that none of the isomorphisms above can be extended to an isomorphism of the *ring* $\mathbb{Q}$ to itself. In fact prove that the identity map is the only ring isomorphism of $\mathbb{Q}$.

2. Let a and b be nonzero elements of the Unique Factorization Domain R. Prove that a and b have a least common multiple (cf. Exercise 11 of Section 1) and describe it in terms of the prime factorizations of a and b in the same fashion that Proposition 13 describes their greatest common divisor.

3. Determine all the representations of the integer $2130797 = 17^2 \cdot 73 \cdot 101$ as a sum of two squares.

4. Prove that if an integer is the sum of two rational squares, then it is the sum of two integer squares (for example, $13 = (1/5)^2 + (18/5)^2 = 2^2 + 3^2$).

5. Let $R = \mathbb{Z}[\sqrt{-n}\,]$ where n is a squarefree integer greater than 3.
   (a) Prove that 2, $\sqrt{-n}$ and $1 + \sqrt{-n}$ are irreducibles in R.
   (b) Prove that R is not a U.F.D. Conclude that the quadratic integer ring $\mathcal{O}$ is not a U.F.D. for $D \equiv 2, 3 \bmod 4$, $D < -3$ (so also not Euclidean and not a P.I.D.). [Show that either $\sqrt{-n}$ or $1 + \sqrt{-n}$ is not prime.]
   (c) Give an explicit ideal in R that is not principal. [Using (b) consider a maximal ideal containing the nonprime ideal $(\sqrt{-n}\,)$ or $(1 + \sqrt{-n}\,)$.]

6. (a) Prove that the quotient ring $\mathbb{Z}[i]/(1 + i)$ is a field of order 2.
   (b) Let $q \in \mathbb{Z}$ be a prime with $q \equiv 3 \bmod 4$. Prove that the quotient ring $\mathbb{Z}[i]/(q)$ is a field with q^2 elements.
   (c) Let $p \in \mathbb{Z}$ be a prime with $p \equiv 1 \bmod 4$ and write $p = \pi\bar{\pi}$ as in Proposition 18. Show that the hypotheses for the Chinese Remainder Theorem (Theorem 17 in Section 7.6) are satisfied and that $\mathbb{Z}[i]/(p) \cong \mathbb{Z}[i]/(\pi) \times \mathbb{Z}[i]/(\bar{\pi})$ as rings. Show that the quotient ring $\mathbb{Z}[i]/(p)$ has order p^2 and conclude that $\mathbb{Z}[i]/(\pi)$ and $\mathbb{Z}[i]/(\bar{\pi})$ are both fields of order p.

7. Let π be an irreducible element in $\mathbb{Z}[i]$.
   (a) For any integer $n \geq 0$, prove that $(\pi^{n+1}) = \pi^{n+1}\mathbb{Z}[i]$ is an ideal in $(\pi^n) = \pi^n\mathbb{Z}[i]$ and that multiplication by π^n induces an isomorphism $\mathbb{Z}[i]/(\pi) \cong (\pi^n)/(\pi^{n+1})$ as additive abelian groups.
   (b) Prove that $|\mathbb{Z}[i]/(\pi^n)| = |\mathbb{Z}[i]/(\pi)|^n$.
   (c) Prove for any nonzero α in $\mathbb{Z}[i]$ that the quotient ring $\mathbb{Z}[i]/(\alpha)$ has order equal to $N(\alpha)$. [Use (b) together with the Chinese Remainder Theorem and the results of the previous exercise.]

8. Let R be the quadratic integer ring $\mathbb{Z}[\sqrt{-5}\,]$ and define the ideals $I_2 = (2, 1 + \sqrt{-5}\,)$, $I_3 = (3, 2 + \sqrt{-5}\,)$, and $I_3' = (3, 2 - \sqrt{-5}\,)$.
   (a) Prove that 2, 3, $1 + \sqrt{-5}$ and $1 - \sqrt{-5}$ are irreducibles in R, no two of which are associate in R, and that $6 = 2 \cdot 3 = (1 + \sqrt{-5}\,) \cdot (1 - \sqrt{-5}\,)$ are two distinct factorizations of 6 into irreducibles in R.
   (b) Prove that I_2, I_3, and I_3' are prime ideals in R. [One approach: for I_3, observe that $R/I_3 \cong (R/(3))/(I_3/(3))$ by the Third Isomorphism Theorem for Rings. Show that $R/(3)$ has 9 elements, $(I_3/(3))$ has 3 elements, and that $R/I_3 \cong \mathbb{Z}/3\mathbb{Z}$ as an additive abelian group. Conclude that I_3 is a maximal (hence prime) ideal and that $R/I_3 \cong \mathbb{Z}/3\mathbb{Z}$ as rings.]
   (c) Show that the factorizations in (a) imply the equality of ideals $(6) = (2)(3)$ and $(6) = (1 + \sqrt{-5}\,)(1 - \sqrt{-5}\,)$. Show that these two ideal factorizations give the same factorization of the ideal (6) as the product of prime ideals (cf. Exercise 5 in Section 2).

9. Suppose that the quadratic integer ring $\mathcal{O}$ is a P.I.D. Prove that the absolute value of the field norm N on $\mathcal{O}$ (cf. Section 7.1) is a Dedekind–Hasse norm on $\mathcal{O}$. Conclude that if the quadratic integer ring $\mathcal{O}$ possesses *any* Dedekind–Hasse norm, then in fact the absolute value of the field norm on $\mathcal{O}$ already provides a Dedekind–Hasse norm on $\mathcal{O}$. [If $\alpha, \beta \in \mathcal{O}$ then $(\alpha, \beta) = (\gamma)$ for some $\gamma \in \mathcal{O}$. Show that if β does not divide α then $0 < |N(\gamma)| < |N(\beta)|$ — use the fact that the units in $\mathcal{O}$ are precisely the elements whose norm is ± 1.]

Remark: If $\mathcal{O}$ is a Euclidean Domain with respect to some norm it is not necessarily true that it is a Euclidean Domain with respect to the absolute value of the field norm (although this is true for $D < 0$, cf. Exercise 8 in Section 1). An example is $D = 69$ (cf. D. Clark, *A quadratic field which is Euclidean but not norm-Euclidean*, Manuscripta Math., 83(1994), pp. 327–330).

10. (*k-stage Euclidean Domains*) Let R be an integral domain and let $N : R \to \mathbb{Z}^+ \cup \{0\}$ be a norm on R. The ring R is Euclidean with respect to N if for any $a, b \in R$ with $b \neq 0$, there exist elements q and r in R with

$$a = qb + r \qquad \text{with } r = 0 \text{ or } N(r) < N(b).$$

Suppose now that this condition is weakened, namely that for any $a, b \in R$ with $b \neq 0$, there exist elements q, q' and r, r' in R with

$$a = qb + r, \quad b = q'r + r' \qquad \text{with } r' = 0 \text{ or } N(r') < N(b),$$

i.e., the remainder after two divisions is smaller. Call such a domain a *2-stage Euclidean Domain*.

(a) Prove that iterating the divisions in a 2-stage Euclidean Domain produces a greatest common divisor of a and b which is a linear combination of a and b. Conclude that every *finitely generated* ideal of a 2-stage Euclidean Domain is principal. (There are 2-stage Euclidean Domains that are *not* P.I.D.s, however.) [Imitate the proof of Theorem 4.]

(b) Prove that a 2-stage Euclidean Domain in which every nonzero nonunit can be factored into a finite number of irreducibles is a Unique Factorization Domain. [Prove first that irreducible elements are prime, as follows. If p is irreducible and $p \mid ab$ with p not dividing a, use part (a) to write $px + ay = 1$ for some x, y. Multiply through by b to conclude that $p \mid b$, so p is prime. Now follow the proof of uniqueness in Theorem 14.]

(c) Make the obvious generalization to define the notion of a k-stage Euclidean Domain for any integer $k \geq 1$. Prove that statements (a) and (b) remain valid if "2-stage Euclidean" is replaced by "k-stage Euclidean."

Remarks: There are examples of rings which are 2-stage Euclidean but are not Euclidean. There are also examples of rings which are not Euclidean with respect to a given norm but which are k-stage Euclidean with respect to the norm (for example, the ring $\mathbb{Z}[\sqrt{14}]$ is not Euclidean with respect to the usual norm $N(a + b\sqrt{14}) = |a^2 - 14b^2|$, but is 2-stage Euclidean with respect to this norm). The k-stage Euclidean condition is also related to the question of whether the group $GL_n(R)$ of invertible $n \times n$ matrices with entries from R is generated by elementary matrices (matrices with 1's along the main diagonal, a single nonzero element from R somewhere off the main diagonal, and 0's elsewhere), together with diagonal and permutation matrices.

11. (*Characterization of P.I.D.s*) Prove that R is a P.I.D. if and only if R is a U.F.D. that is also a Bezout Domain (cf. Exercise 7 in Section 2). [One direction is given by Theorem 14. For the converse, let a be a nonzero element of the ideal I with a minimal number of irreducible factors. Prove that $I = (a)$ by showing that if there is an element $b \in I$ that is not in (a) then $(a, b) = (d)$ leads to a contradiction.]

CHAPTER 9

Polynomial Rings

We begin this chapter on polynomial rings with a summary of facts from the preceding two chapters (with references where needed). The basic definitions were given in slightly greater detail in Section 7.2. For convenience, the ring R will always be a commutative ring with identity $1 \neq 0$.

9.1 DEFINITIONS AND BASIC PROPERTIES

The polynomial ring $R[x]$ in the indeterminate x with coefficients from R is the set of all formal sums $a_n x^n + a_{n-1} x^{n-1} + \cdots + a_1 x + a_0$ with $n \geq 0$ and each $a_i \in R$. If $a_n \neq 0$ then the polynomial is of degree n, $a_n x^n$ is the leading term, and a_n is the leading coefficient (where the leading coefficient of the zero polynomial is defined to be 0). The polynomial is monic if $a_n = 1$. Addition of polynomials is "componentwise":

$$\sum_{i=0}^{n} a_i x^i + \sum_{i=0}^{n} b_i x^i = \sum_{i=0}^{n} (a_i + b_i) x^i$$

(here a_n or b_n may be zero in order for addition of polynomials of different degrees to be defined). Multiplication is performed by first defining $(ax^i)(bx^j) = ab x^{i+j}$ and then extending to all polynomials by the distributive laws so that in general

$$\left(\sum_{i=0}^{n} a_i x^i \right) \times \left(\sum_{i=0}^{m} b_i x^i \right) = \sum_{k=0}^{n+m} \left(\sum_{i=0}^{k} a_i b_{k-i} \right) x^k .$$

In this way $R[x]$ is a commutative ring with identity (the identity 1 from R) in which we identify R with the subring of constant polynomials.

We have already noted that if R is an integral domain then the leading term of a product of polynomials is the product of the leading terms of the factors. The following is Proposition 4 of Section 7.2 which we record here for completeness.

Proposition 1. Let R be an integral domain. Then
    (1) degree $p(x)q(x) = $ degree $p(x) + $ degree $q(x)$ if $p(x), q(x)$ are nonzero
    (2) the units of $R[x]$ are just the units of R
    (3) $R[x]$ is an integral domain.

Recall also that if R is an integral domain, the quotient field of $R[x]$ consists of all quotients $\dfrac{p(x)}{q(x)}$ where $q(x)$ is not the zero polynomial (and is called the field of rational functions in x with coefficients in R).

The next result describes a relation between the ideals of R and those of $R[x]$.

Proposition 2. Let I be an ideal of the ring R and let $(I) = I[x]$ denote the ideal of $R[x]$ generated by I (the set of polynomials with coefficients in I). Then

$$R[x]/(I) \cong (R/I)[x].$$

In particular, if I is a prime ideal of R then (I) is a prime ideal of $R[x]$.

Proof: There is a natural map $\varphi : R[x] \to (R/I)[x]$ given by reducing each of the coefficients of a polynomial modulo I. The definition of addition and multiplication in these two rings shows that φ is a ring homomorphism. The kernel is precisely the set of polynomials each of whose coefficients is an element of I, which is to say that $\ker \varphi = I[x] = (I)$, proving the first part of the proposition. The last statement follows from Proposition 1, since if I is a prime ideal in R, then R/I is an integral domain, hence also $(R/I)[x]$ is an integral domain. This shows if I is a prime ideal of R, then (I) is a prime ideal of $R[x]$.

Note that it is not true that if I is a maximal ideal of R then (I) is a maximal ideal of $R[x]$. However, if I is maximal in R then the ideal of $R[x]$ generated by I and x is maximal in $R[x]$.

We now give an example of the "reduction homomorphism" of Proposition 2 which will be useful on a number of occasions later ("reduction homomorphisms" were also discussed at the end of Section 7.3 with reference to reducing the integers mod n).

Example

Let $R = \mathbb{Z}$ and consider the ideal $n\mathbb{Z}$ of $\mathbb{Z}$. Then the isomorphism above can be written

$$\mathbb{Z}[x]/n\mathbb{Z}[x] \cong \mathbb{Z}/n\mathbb{Z}[x]$$

and the natural projection map of $\mathbb{Z}[x]$ to $\mathbb{Z}/n\mathbb{Z}[x]$ by reducing the coefficients modulo n is a ring homomorphism. If n is composite, then the quotient ring is not an integral domain. If, however, n is a prime p, then $\mathbb{Z}/p\mathbb{Z}$ is a field and so $\mathbb{Z}/p\mathbb{Z}[x]$ is an integral domain (in fact, a Euclidean Domain, as we shall see shortly). We also see that the set of polynomials whose coefficients are divisible by p is a prime ideal in $\mathbb{Z}[x]$.

We close this section with a description of the natural extension to polynomial rings in *several* variables.

Definition. The *polynomial ring in the variables $x_1, x_2, \ldots, x_n$ with coefficients in R,* denoted $R[x_1, x_2, \ldots, x_n]$, is defined inductively by

$$R[x_1, x_2, \ldots, x_n] = R[x_1, x_2, \ldots, x_{n-1}][x_n]$$

This definition means that we can consider polynomials in n variables with coefficients in R simply as polynomials in *one* variable (say x_n) but now with coefficients that

296

are themselves *polynomials in n − 1 variables*. In a slightly more concrete formulation, a nonzero polynomial in $x_1, x_2, \ldots, x_n$ with coefficients in R is a finite sum of nonzero *monomial terms*, i.e., a finite sum of elements of the form

$$ax_1^{d_1} x_2^{d_2} \ldots x_n^{d_n}$$

where $a \in R$ (the *coefficient* of the term) and the d_i are nonnegative integers. A monic term $x_1^{d_1} x_2^{d_2} \ldots x_n^{d_n}$ is called simply a *monomial* and is the *monomial part* of the term $ax_1^{d_1} x_2^{d_2} \ldots x_n^{d_n}$. The exponent d_i is called the *degree in x_i* of the term and the sum

$$d = d_1 + d_2 + \cdots + d_n$$

is called the *degree* of the term. The ordered n-tuple $(d_1, d_2, \ldots, d_n)$ is the *multidegree* of the term. The *degree* of a nonzero polynomial is the largest degree of any of its monomial terms. A polynomial is called *homogeneous* or a *form* if all its terms have the same degree. If f is a nonzero polynomial in n variables, the sum of all the monomial terms in f of degree k is called the *homogeneous component of f of degree k*. If f has degree d then f may be written uniquely as the sum $f_0 + f_1 + \cdots + f_d$ where f_k is the homogeneous component of f of degree k, for $0 \le k \le d$ (where some f_k may be zero).

Finally, to define a polynomial ring in an *arbitrary* number of variables with coefficients in R we take finite sums of monomial terms of the type above (but where the variables are not restricted to just $x_1, \ldots, x_n$), with the natural addition and multiplication. Alternatively, we could define this ring as the *union* of *all* the polynomial rings in a *finite* number of the variables being considered.

Example

The polynomial ring $\mathbb{Z}[x, y]$ in two variables x and y with integer coefficients consists of all finite sums of monomial terms of the form $ax^i y^j$ (of degree $i + j$). For example,

$$p(x, y) = 2x^3 + xy - y^2$$

and

$$q(x, y) = -3xy + 2y^2 + x^2 y^3$$

are both elements of $\mathbb{Z}[x, y]$, of degrees 3 and 5, respectively. We have

$$p(x, y) + q(x, y) = 2x^3 - 2xy + y^2 + x^2 y^3$$

and

$$p(x, y)q(x, y) = -6x^4 y + 4x^3 y^2 + 2x^5 y^3 - 3x^2 y^2 + 5xy^3 + x^3 y^4 - 2y^4 - x^2 y^5,$$

a polynomial of degree 8. To view this last polynomial, say, as a polynomial in y with coefficients in $\mathbb{Z}[x]$ as in the definition of several variable polynomial rings above, we would write the polynomial in the form

$$(-6x^4)y + (4x^3 - 3x^2)y^2 + (2x^5 + 5x)y^3 + (x^3 - 2)y^4 - (x^2)y^5.$$

The nonzero homogeneous components of $f = f(x, y) = p(x, y)q(x, y)$ are the polynomials $f_4 = -3x^2 y^2 + 5xy^3 - 2y^4$ (degree 4), $f_5 = -6x^4 y + 4x^3 y^2$ (degree 5), $f_7 = x^3 y^4 - x^2 y^5$ (degree 7), and $f_8 = 2x^5 y^3$ (degree 8).

Each of the statements in Proposition 1 is true for polynomial rings with an arbitrary number of variables. This follows by induction for finitely many variables and from the definition in terms of unions in the case of polynomial rings in arbitrarily many variables.

EXERCISES

1. Let $p(x, y, z) = 2x^2y - 3xy^3z + 4y^2z^5$ and $q(x, y, z) = 7x^2 + 5x^2y^3z^4 - 3x^2z^3$ be polynomials in $\mathbb{Z}[x, y, z]$.
   (a) Write each of p and q as a polynomial in x with coefficients in $\mathbb{Z}[y, z]$.
   (b) Find the degree of each of p and q.
   (c) Find the degree of p and q in each of the three variables x, y and z.
   (d) Compute pq and find the degree of pq in each of the three variables x, y and z.
   (e) Write pq as a polynomial in the variable z with coefficients in $\mathbb{Z}[x, y]$.

2. Repeat the preceding exercise under the assumption that the coefficients of p and q are in $\mathbb{Z}/3\mathbb{Z}$.

3. If R is a commutative ring and $x_1, x_2, \ldots, x_n$ are independent variables over R, prove that $R[x_{\pi(1)}, x_{\pi(2)}, \ldots, x_{\pi(n)}]$ is isomorphic to $R[x_1, x_2, \ldots, x_n]$ for any permutation π of $\{1, 2, \ldots, n\}$.

4. Prove that the ideals (x) and (x, y) are prime ideals in $\mathbb{Q}[x, y]$ but only the latter ideal is a maximal ideal.

5. Prove that (x, y) and $(2, x, y)$ are prime ideals in $\mathbb{Z}[x, y]$ but only the latter ideal is a maximal ideal.

6. Prove that (x, y) is not a principal ideal in $\mathbb{Q}[x, y]$.

7. Let R be a commutative ring with 1. Prove that a polynomial ring in more than one variable over R is not a Principal Ideal Domain.

8. Let F be a field and let $R = F[x, x^2y, x^3y^2, \ldots, x^ny^{n-1}, \ldots]$ be a subring of the polynomial ring $F[x, y]$.
   (a) Prove that the fields of fractions of R and $F[x, y]$ are the same.
   (b) Prove that R contains an ideal that is not finitely generated.

9. Prove that a polynomial ring in infinitely many variables with coefficients in any commutative ring contains ideals that are not finitely generated.

10. Prove that the ring $\mathbb{Z}[x_1, x_2, x_3, \ldots]/(x_1x_2, x_3x_4, x_5x_6, \ldots)$ contains infinitely many minimal prime ideals (cf. Exercise 36 of Section 7.4).

11. Show that the radical of the ideal $I = (x, y^2)$ in $\mathbb{Q}[x, y]$ is (x, y) (cf. Exercise 30, Section 7.4). Deduce that I is a primary ideal that is not a power of a prime ideal (cf. Exercise 41, Section 7.4).

12. Let $R = \mathbb{Q}[x, y, z]$ and let bars denote passage to $\mathbb{Q}[x, y, z]/(xy - z^2)$. Prove that $\overline{P} = (\overline{x}, \overline{z})$ is a prime ideal. Show that $\overline{xy} \in \overline{P}^2$ but that no power of $\overline{y}$ lies in $\overline{P}^2$. (This shows $\overline{P}$ is a prime ideal whose square is *not* a primary ideal — cf. Exercise 41, Section 7.4).

13. Prove that the rings $F[x, y]/(y^2 - x)$ and $F[x, y]/(y^2 - x^2)$ are not isomorphic for any field F.

14. Let R be an integral domain and let i, j be relatively prime integers. Prove that the ideal $(x^i - y^j)$ is a prime ideal in $R[x, y]$. [Consider the ring homomorphism φ from $R[x, y]$ to $R[t]$ defined by mapping x to t^j and mapping y to t^i. Show that an element of $R[x, y]$

differs from an element in $(x^i - y^j)$ by a polynomial $f(x)$ of degree at most $j - 1$ in y and observe that the exponents of $\varphi(x^r y^s)$ are distinct for $0 \leq s < j$.]

15. Let $p(x_1, x_2, \ldots, x_n)$ be a homogeneous polynomial of degree k in $R[x_1, \ldots, x_n]$. Prove that for all $\lambda \in R$ we have $p(\lambda x_1, \lambda x_2, \ldots, \lambda x_n) = \lambda^k p(x_1, x_2, \ldots, x_n)$.

16. Prove that the product of two homogeneous polynomials is again homogeneous.

17. An ideal I in $R[x_1, \ldots, x_n]$ is called a *homogeneous ideal* if whenever $p \in I$ then each homogeneous component of p is also in I. Prove that an ideal is a homogeneous ideal if and only if it may be generated by homogeneous polynomials. [Use induction on degrees to show the "if" implication.]

The following exercise shows that some care must be taken when working with polynomials over noncommutative rings R (the ring operations in $R[x]$ are defined in the same way as for commutative rings R), in particular when considering polynomials as functions.

18. Let R be an arbitrary ring and let Func(R) be the ring of all functions from R to itself. If $p(x) \in R[x]$ is a polynomial, let $f_p \in$ Func(R) be the function on R defined by $f_p(r) = p(r)$ (the usual way of viewing a polynomial in $R[x]$ as defining a function on R by "evaluating at r").
  (a) For fixed $a \in R$, prove that "evaluation at a" is a ring homomorphism from Func(R) to R (cf. Example 4 following Theorem 7 in Section 7.3).
  (b) Prove that the map $\varphi : R[x] \to$ Func(R) defined by $\varphi(p(x)) = f_p$ is not a ring homomorphism in general. Deduce that polynomial identities need not give corresponding identities when the polynomials are viewed as functions. [If $R = \mathbb{H}$ is the ring of real Hamilton Quaternions show that $p(x) = x^2 + 1$ factors as $(x + i)(x - i)$, but that $p(j) = 0$ while $(j + i)(j - i) \neq 0$.]
  (c) For fixed $a \in R$, prove that the composite "evaluation at a" of the maps in (a) and (b) mapping $R[x]$ to R is a ring homomorphism if and only if a is in the center of R.

9.2 POLYNOMIAL RINGS OVER FIELDS I

We now consider more carefully the situation where the coefficient ring is a *field* F. We can define a *norm* on $F[x]$ by defining $N(p(x)) =$ degree of $p(x)$ (where we set $N(0) = 0$). From elementary algebra we know that we can divide one polynomial with, say, rational coefficients by another (nonzero) polynomial with rational coefficients to obtain a quotient and remainder. The same is true over any field.

Theorem 3. Let F be a field. The polynomial ring $F[x]$ is a Euclidean Domain. Specifically, if $a(x)$ and $b(x)$ are two polynomials in $F[x]$ with $b(x)$ nonzero, then there are *unique* $q(x)$ and $r(x)$ in $F[x]$ such that

$$a(x) = q(x)b(x) + r(x) \qquad \text{with } r(x) = 0 \text{ or degree } r(x) < \text{degree } b(x) \,.$$

Proof: If $a(x)$ is the zero polynomial then take $q(x) = r(x) = 0$. We may therefore assume $a(x) \neq 0$ and prove the existence of $q(x)$ and $r(x)$ by induction on $n =$ degree $a(x)$. Let $b(x)$ have degree m. If $n < m$ take $q(x) = 0$ and $r(x) = a(x)$. Otherwise $n \geq m$. Write

$$a(x) = a_n x^n + a_{n-1} x^{n-1} + \cdots + a_1 x + a_0$$

and

$$b(x) = b_m x^m + b_{m-1} x^{m-1} + \cdots + b_1 x + b_0.$$

Then the polynomial $a'(x) = a(x) - \dfrac{a_n}{b_m} x^{n-m} b(x)$ is of degree less than n (we have arranged to subtract the leading term from $a(x)$). Note that this polynomial is well defined because the coefficients are taken from a *field* and $b_m \neq 0$. By induction then, there exist polynomials $q'(x)$ and $r(x)$ with

$$a'(x) = q'(x)b(x) + r(x) \qquad \text{with } r(x) = 0 \text{ or degree } r(x) < \text{degree } b(x).$$

Then, letting $q(x) = q'(x) + \dfrac{a_n}{b_m} x^{n-m}$ we have

$$a(x) = q(x)b(x) + r(x) \qquad \text{with } r(x) = 0 \text{ or degree } r(x) < \text{degree } b(x)$$

completing the induction step.

As for the uniqueness, suppose $q_1(x)$ and $r_1(x)$ also satisfied the conditions of the theorem. Then both $a(x) - q(x)b(x)$ and $a(x) - q_1(x)b(x)$ are of degree less than $m = \text{degree } b(x)$. The difference of these two polynomials, i.e., $b(x)(q(x) - q_1(x))$ is also of degree less than m. But the degree of the product of two nonzero polynomials is the sum of their degrees (since F is an integral domain), hence $q(x) - q_1(x)$ must be 0, that is, $q(x) = q_1(x)$. This implies $r(x) = r_1(x)$, completing the proof.

Corollary 4. If F is a field, then $F[x]$ is a Principal Ideal Domain and a Unique Factorization Domain.

Proof: This is immediate from the results of the last chapter.

Recall also from Corollary 8 in Section 8.2 that if R is any commutative ring such that $R[x]$ is a Principal Ideal Domain (or Euclidean Domain) then R must be a field. We shall see in the next section, however, that $R[x]$ is a Unique Factorization Domain whenever R itself is a Unique Factorization Domain.

Examples

(1) By the above remarks the ring $\mathbb{Z}[x]$ is not a Principal Ideal Domain. As we have already seen (Example 3 beginning of Section 7.4) the ideal $(2, x)$ is not principal in this ring.

(2) $\mathbb{Q}[x]$ is a Principal Ideal Domain since the coefficients lie in the field $\mathbb{Q}$. The ideal generated in $\mathbb{Z}[x]$ by 2 and x is not principal in the subring $\mathbb{Z}[x]$ of $\mathbb{Q}[x]$. However, the ideal generated in $\mathbb{Q}[x]$ is principal; in fact it is the entire ring (so has 1 as a generator) since 2 is a unit in $\mathbb{Q}[x]$.

(3) If p is a prime, the ring $\mathbb{Z}/p\mathbb{Z}[x]$ obtained by reducing $\mathbb{Z}[x]$ modulo the prime ideal (p) is a Principal Ideal Domain, since the coefficients lie in the field $\mathbb{Z}/p\mathbb{Z}$. This example shows that the quotient of a ring which is not a Principal Ideal Domain *may* be a Principal Ideal Domain. To follow the ideal $(2, x)$ above in this example, note that if $p = 2$, then the ideal $(2, x)$ reduces to the ideal (x) in the quotient $\mathbb{Z}/2\mathbb{Z}[x]$, which is a proper (maximal) ideal. If $p \neq 2$, then 2 is a unit in the quotient, so the ideal $(2, x)$ reduces to the entire ring $\mathbb{Z}/p\mathbb{Z}[x]$.

(4) $\mathbb{Q}[x, y]$, the ring of polynomials in two variables with rational coefficients, is *not* a Principal Ideal Domain since this ring is $\mathbb{Q}[x][y]$ and $\mathbb{Q}[x]$ is not a field (any element

300

of positive degree is not invertible). It is an exercise to see that the ideal (x, y) is not a principal ideal in this ring. We shall see shortly that $\mathbb{Q}[x, y]$ *is* a Unique Factorization Domain.

We note that the quotient and remainder in the Division Algorithm applied to $a(x), b(x) \in F[x]$ are *independent of field extensions* in the following sense. Suppose the field F is contained in the field E and $a(x) = Q(x)b(x) + R(x)$ for some $Q(x)$, $R(x)$ satisfying the conditions of Theorem 3 in $E[x]$. Write $a(x) = q(x)b(x) + r(x)$ for some $q(x), r(x) \in F[x]$ and apply the uniqueness condition of Theorem 3 in the ring $E[x]$ to deduce that $Q(x) = q(x)$ and $R(x) = r(x)$. In particular, $b(x)$ divides $a(x)$ in the ring $E[x]$ if and only if $b(x)$ divides $a(x)$ in $F[x]$. Also, the greatest common divisor of $a(x)$ and $b(x)$ (which can be obtained from the Euclidean Algorithm) is the same, once we make it unique by specifying it to be monic, whether these elements are viewed in $F[x]$ or in $E[x]$.

EXERCISES

Let F be a field and let x be an indeterminate over F.

1. Let $f(x) \in F[x]$ be a polynomial of degree $n \geq 1$ and let bars denote passage to the quotient $F[x]/(f(x))$. Prove that for each $\overline{g(x)}$ there is a unique polynomial $g_0(x)$ of degree $\leq n - 1$ such that $\overline{g(x)} = \overline{g_0(x)}$ (equivalently, the elements $\overline{1}, \overline{x}, \ldots, x^{n-1}$ are a *basis* of the vector space $F[x]/(f(x))$ over F — in particular, the dimension of this space is n). [Use the Division Algorithm.]

2. Let F be a finite field of order q and let $f(x)$ be a polynomial in $F[x]$ of degree $n \geq 1$. Prove that $F[x]/(f(x))$ has q^n elements. [Use the preceding exercise.]

3. Let $f(x)$ be a polynomial in $F[x]$. Prove that $F[x]/(f(x))$ is a field if and only if $f(x)$ is irreducible. [Use Proposition 7, Section 8.2.]

4. Let F be a finite field. Prove that $F[x]$ contains infinitely many primes. (Note that over an infinite field the polynomials of degree 1 are an infinite set of primes in the ring of polynomials).

5. Exhibit *all* the ideals in the ring $F[x]/(p(x))$, where F is a field and $p(x)$ is a polynomial in $F[x]$ (describe them in terms of the factorization of $p(x)$).

6. Describe (briefly) the ring structure of the following rings:
   (a) $\mathbb{Z}[x]/(2)$, **(b)** $\mathbb{Z}[x]/(x)$, **(c)** $\mathbb{Z}[x]/(x^2)$, **(d)** $\mathbb{Z}[x, y]/(x^2, y^2, 2)$.
   Show that $\alpha^2 = 0$ or 1 for every α in the last ring and determine those elements with $\alpha^2 = 0$. Determine the characteristics of each of these rings (cf. Exercise 26, Section 7.3).

7. Determine all the ideals of the ring $\mathbb{Z}[x]/(2, x^3 + 1)$.

8. Determine the greatest common divisor of $a(x) = x^3 - 2$ and $b(x) = x + 1$ in $\mathbb{Q}[x]$ and write it as a linear combination (in $\mathbb{Q}[x]$) of $a(x)$ and $b(x)$.

9. Determine the greatest common divisor of $a(x) = x^5 + 2x^3 + x^2 + x + 1$ and the polynomial $b(x) = x^5 + x^4 + 2x^3 + 2x^2 + 2x + 1$ in $\mathbb{Q}[x]$ and write it as a linear combination (in $\mathbb{Q}[x]$) of $a(x)$ and $b(x)$.

10. Determine the greatest common divisor of $a(x) = x^3 + 4x^2 + x - 6$ and $b(x) = x^5 - 6x + 5$ in $\mathbb{Q}[x]$ and write it as a linear combination (in $\mathbb{Q}[x]$) of $a(x)$ and $b(x)$.

11. Suppose $f(x)$ and $g(x)$ are two nonzero polynomials in $\mathbb{Q}[x]$ with greatest common divisor $d(x)$.

(a) Given $h(x) \in \mathbb{Q}[x]$, show that there are polynomials $a(x), b(x) \in \mathbb{Q}[x]$ satisfying the equation $a(x)f(x) + b(x)g(x) = h(x)$ if and only if $h(x)$ is divisible by $d(x)$.

(b) If $a_0(x), b_0(x) \in \mathbb{Q}[x]$ are particular solutions to the equation in (a), show that the full set of solutions to this equation is given by

$$a(x) = a_0(x) + m(x)\frac{g(x)}{d(x)}$$

$$b(x) = b_0(x) - m(x)\frac{f(x)}{d(x)}$$

as $m(x)$ ranges over the polynomials in $\mathbb{Q}[x]$. [cf. Exercise 4 in Section 8.1]

12. Let $F[x, y_1, y_2, \dots]$ be the polynomial ring in the infinite set of variables $x, y_1, y_2, \dots$ over the field F, and let I be the ideal $(x - y_1^2, y_1 - y_2^2, \dots, y_i - y_{i+1}^2, \dots)$ in this ring. Define R to be the ring $F[x, y_1, y_2, \dots]/I$, so that in R the square of each y_{i+1} is y_i and $y_1^2 = x$ modulo I, i.e., x has a 2^i th root, for every i. Denote the image of y_i in R as $x^{1/2^i}$. Let R_n be the subring of R generated by F and $x^{1/2^n}$.

(a) Prove that $R_1 \subseteq R_2 \subseteq \cdots$ and that R is the union of all R_n, i.e., $R = \cup_{n=1}^{\infty} R_n$.

(b) Prove that R_n is isomorphic to a polynomial ring in one variable over F, so that R_n is a P.I.D. Deduce that R is a Bezout Domain (cf. Exercise 7 in Section 8.2). [First show that the ring $S_n = F[x, y_1, \dots, y_n]/(x - y_1^2, y_1 - y_2^2, \dots, y_{n-1} - y_n^2)$ is isomorphic to the polynomial ring $F[y_n]$. Then show any polynomial relation y_n satisfies in R_n gives a corresponding relation in S_N for some $N \geq n$.]

(c) Prove that the ideal generated by $x, x^{1/2}, x^{1/4}, \dots$ in R is not finitely generated (so R is not a P.I.D.).

13. This exercise introduces a noncommutative ring which is a "right" Euclidean Domain (and a "left" Principal Ideal Domain) but is not a "left" Euclidean Domain (and not a "right" Principal Ideal Domain). Let F be a field of characteristic p in which not every element is a pth power: $F \neq F^p$ (for example the field $F = \mathbb{F}_p(t)$ of rational functions in the variable t with coefficients in $\mathbb{F}_p$ is such a field). Let $R = F\{x\}$ be the "twisted" polynomial ring of polynomials $\sum_{i=0}^{n} a_i x^i$ in x with coefficients in F with the usual (termwise) addition

$$\sum_{i=0}^{n} a_i x^i + \sum_{i=0}^{n} b_i x^i = \sum_{i=0}^{n} (a_i + b_i) x^i$$

but with a noncommutative multiplication defined by

$$\left(\sum_{i=0}^{n} a_i x^i\right)\left(\sum_{j=0}^{m} b_j x^j\right) = \sum_{k=0}^{n+m}\left(\sum_{i+j=k} a_i b_j^{p^i}\right) x^k \, .$$

This multiplication arises from defining $xa = a^p x$ for every $a \in F$ (so the powers of x do not commute with the coefficients) and extending in a natural way. Let N be the norm defined by taking the degree of a polynomial in R: $N(f) = \deg(f)$.

(a) Show that $x^k a = a^{p^k} x^k$ for every $a \in F$ and every integer $k \geq 0$ and that R is a ring with this definition of multiplication. [Use the fact that $(a+b)^p = a^p + b^p$ for every $a, b \in F$ since F has characteristic p, so also $(a+b)^{p^k} = a^{p^k} + b^{p^k}$ for every $a, b \in F$.]

(b) Prove that the degree of a product of two elements of R is the sum of the degrees of the elements. Prove that R has no zero divisors.

(c) Prove that R is "right Euclidean" with respect to N, i.e., for any polynomials $f, g \in R$ with $g \neq 0$, there exist polynomials q and r in R with

$$f = qg + r \qquad \text{with } r = 0 \text{ or } \deg(r) < \deg(g).$$

Use this to prove that every *left* ideal of R is principal.

(d) Let $f = \theta x$ for some $\theta \in F$, $\theta \notin F^p$ and let $g = x$. Prove that there are no polynomials q and r in R with

$$f = gq + r \qquad \text{with } r = 0 \text{ or } \deg(r) < \deg(g),$$

so in particular R is not "left Euclidean" with respect to N. Prove that the right ideal of R generated by x and θx is not principal. Conclude that R is not "left Euclidean" with respect to *any* norm.

9.3 POLYNOMIAL RINGS THAT ARE UNIQUE FACTORIZATION DOMAINS

We have seen in Proposition 1 that if R is an integral domain then $R[x]$ is also an integral domain. Also, such an R can be embedded in its field of fractions F (Theorem 15, Section 7.5), so that $R[x] \subseteq F[x]$ is a subring, and $F[x]$ is a Euclidean Domain (hence a Principal Ideal Domain and a Unique Factorization Domain). Many computations for $R[x]$ may be accomplished in $F[x]$ at the expense of allowing fractional coefficients. This raises the immediate question of how computations (such as factorizations of polynomials) in $F[x]$ can be used to give information in $R[x]$.

For instance, suppose $p(x)$ is a polynomial in $R[x]$. Since $F[x]$ is a Unique Factorization Domain we can factor $p(x)$ uniquely into a product of irreducibles in $F[x]$. It is natural to ask whether we can do the same in $R[x]$, i.e., is $R[x]$ a Unique Factorization Domain? In general the answer is no because if $R[x]$ were a Unique Factorization Domain, the constant polynomials would have to be uniquely factored into irreducible elements of $R[x]$, necessarily of degree 0 since the degrees of products add, that is, R would itself have to be a Unique Factorization Domain. Thus if R is an integral domain which is not a Unique Factorization Domain, $R[x]$ cannot be a Unique Factorization Domain. On the other hand, it turns out that if R is a Unique Factorization Domain, then $R[x]$ is also a Unique Factorization Domain. The method of proving this is to first factor uniquely in $F[x]$ and then "clear denominators" to obtain a unique factorization in $R[x]$. The first step in making this precise is to compare the factorization of a polynomial in $F[x]$ to a factorization in $R[x]$.

Proposition 5. *(Gauss' Lemma)* Let R be a Unique Factorization Domain with field of fractions F and let $p(x) \in R[x]$. If $p(x)$ is reducible in $F[x]$ then $p(x)$ is reducible in $R[x]$. More precisely, if $p(x) = A(x)B(x)$ for some nonconstant polynomials $A(x), B(x) \in F[x]$, then there are nonzero elements $r, s \in F$ such that $rA(x) = a(x)$ and $sB(x) = b(x)$ both lie in $R[x]$ and $p(x) = a(x)b(x)$ is a factorization in $R[x]$.

Proof: The coefficients of the polynomials on the right hand side of the equation $p(x) = A(x)B(x)$ are elements in the field F, hence are quotients of elements from the Unique Factorization Domain R. Multiplying through by a common denominator for all these coefficients, we obtain an equation $dp(x) = a'(x)b'(x)$ where now $a'(x)$ and $b'(x)$ are elements of $R[x]$ and d is a nonzero element of R. If d is a unit in R, the proposition is true with $a(x) = d^{-1}a'(x)$ and $b(x) = b'(x)$. Assume d is not a unit and write d as a product of irreducibles in R, say $d = p_1 \cdots p_n$. Since p_1 is irreducible in R, the ideal (p_1) is prime (cf. Proposition 12, Section 8.3), so by Proposition 2 above, the ideal $p_1 R[x]$ is prime in $R[x]$ and $(R/p_1 R)[x]$ is an integral domain. Reducing the equation $dp(x) = a'(x)b'(x)$ modulo p_1, we obtain the equation $0 = \overline{a'(x)}\,\overline{b'(x)}$ in this integral domain (the bars denote the images of these polynomials in the quotient ring), hence one of the two factors, say $\overline{a'(x)}$ must be 0. But this means all the coefficients of $a'(x)$ are divisible by p_1, so that $\frac{1}{p_1}a'(x)$ also has coefficients in R. In other words, in the equation $dp(x) = a'(x)b'(x)$ we can cancel a factor of p_1 from d (on the left) and from either $a'(x)$ or $b'(x)$ (on the right) and still have an equation in $R[x]$. But now the factor d on the left hand side has one fewer irreducible factors. Proceeding in the same fashion with each of the remaining factors of d, we can cancel all of the factors of d into the two polynomials on the right hand side, leaving an equation $p(x) = a(x)b(x)$ with $a(x), b(x) \in R[x]$ and with $a(x), b(x)$ being F-multiples of $A(x)$, $B(x)$, respectively. This completes the proof.

Note that we cannot prove that $a(x)$ and $b(x)$ are necessarily R-multiples of $A(x)$, $B(x)$, respectively, because, for example, we could factor x^2 in $\mathbb{Q}[x]$ with $A(x) = 2x$ and $B(x) = \frac{1}{2}x$ but no *integer* multiples of $A(x)$ and $B(x)$ give a factorization of x^2 in $\mathbb{Z}[x]$.

The elements of the ring R become *units* in the Unique Factorization Domain $F[x]$ (the units in $F[x]$ being the nonzero elements of F). For example, $7x$ factors in $\mathbb{Z}[x]$ into a product of two irreducibles: 7 and x (so $7x$ is not irreducible in $\mathbb{Z}[x]$), whereas $7x$ is the unit 7 times the irreducible x in $\mathbb{Q}[x]$ (so $7x$ is irreducible in $\mathbb{Q}[x]$). The following corollary shows that this is essentially the *only* difference between the irreducible elements in $R[x]$ and those in $F[x]$.

Corollary 6. Let R be a Unique Factorization Domain, let F be its field of fractions and let $p(x) \in R[x]$. Suppose the greatest common divisor of the coefficients of $p(x)$ is 1. Then $p(x)$ is irreducible in $R[x]$ if and only if it is irreducible in $F[x]$. In particular, if $p(x)$ is a monic polynomial that is irreducible in $R[x]$, then $p(x)$ is irreducible in $F[x]$.

Proof: By Gauss' Lemma above, if $p(x)$ is reducible in $F[x]$, then it is reducible in $R[x]$. Conversely, the assumption on the greatest common divisor of the coefficients of $p(x)$ implies that if it is reducible in $R[x]$, then $p(x) = a(x)b(x)$ where neither $a(x)$ nor $b(x)$ are constant polynomials in $R[x]$. This same factorization shows that $p(x)$ is reducible in $F[x]$, completing the proof.

Theorem 7. R is a Unique Factorization Domain if and only if $R[x]$ is a Unique Factorization Domain.

Proof: We have indicated above that $R[x]$ a Unique Factorization Domain forces R to be a Unique Factorization Domain. Suppose conversely that R is a Unique Factorization Domain, F is its field of fractions and $p(x)$ is a nonzero element of $R[x]$. Let d be the greatest common divisor of the coefficients of $p(x)$, so that $p(x) = dp'(x)$, where the g.c.d. of the coefficients of $p'(x)$ is 1. Such a factorization of $p(x)$ is unique up to a change in d (so up to a unit in R), and since d can be factored uniquely into irreducibles in R (and these are also irreducibles in the larger ring $R[x]$), it suffices to prove that $p'(x)$ can be factored uniquely into irreducibles in $R[x]$. Thus we may assume that the greatest common divisor of the coefficients of $p(x)$ is 1. We may further assume $p(x)$ is not a unit in $R[x]$, i.e., degree $p(x) > 0$.

Since $F[x]$ is a Unique Factorization Domain, $p(x)$ can be factored uniquely into irreducibles in $F[x]$. By Gauss' Lemma, such a factorization implies there is a factorization of $p(x)$ in $R[x]$ whose factors are F-multiples of the factors in $F[x]$. Since the greatest common divisor of the coefficients of $p(x)$ is 1, the g.c.d. of the coefficients in each of these factors in $R[x]$ must be 1. By Corollary 6, each of these factors is an irreducible in $R[x]$. This shows that $p(x)$ can be written as a finite product of irreducibles in $R[x]$.

The uniqueness of the factorization of $p(x)$ follows from the uniqueness in $F[x]$. Suppose

$$p(x) = q_1(x) \cdots q_r(x) = q_1'(x) \cdots q_s'(x)$$

are two factorizations of $p(x)$ into irreducibles in $R[x]$. Since the g.c.d. of the coefficients of $p(x)$ is 1, the same is true for each of the irreducible factors above — in particular, each has positive degree. By Corollary 6, each $q_i(x)$ and $q_j'(x)$ is an irreducible in $F[x]$. By unique factorization in $F[x]$, $r = s$ and, possibly after rearrangement, $q_i(x)$ and $q_i'(x)$ are associates in $F[x]$ for all $i \in \{1, \ldots, r\}$. It remains to show they are associates in $R[x]$. Since the units of $F[x]$ are precisely the elements of $F^\times$ we need to consider when $q(x) = \frac{a}{b} q'(x)$ for some $q(x), q'(x) \in R[x]$ and nonzero elements a, b of R, where the greatest common divisor of the coefficients of each of $q(x)$ and $q'(x)$ is 1. In this case $bq(x) = aq'(x)$; the g.c.d. of the coefficients on the left hand side is b and on the right hand side is a. Since in a Unique Factorization Domain the g.c.d. of the coefficients of a nonzero polynomial is unique up to units, $a = ub$ for some unit u in R. Thus $q(x) = uq'(x)$ and so $q(x)$ and $q'(x)$ are associates in R as well. This completes the proof.

Corollary 8. If R is a Unique Factorization Domain, then a polynomial ring in an arbitrary number of variables with coefficients in R is also a Unique Factorization Domain.

Proof: For finitely many variables, this follows by induction from Theorem 7, since a polynomial ring in n variables can be considered as a polynomial ring in one variable with coefficients in a polynomial ring in $n - 1$ variables. The general case follows from the definition of a polynomial ring in an arbitrary number of variables as the union of polynomial rings in finitely many variables.

Examples

(1) $\mathbb{Z}[x]$, $\mathbb{Z}[x, y]$, etc. are Unique Factorization Domains. The ring $\mathbb{Z}[x]$ gives an example of a Unique Factorization Domain that is not a Principal Ideal Domain.
(2) Similarly, $\mathbb{Q}[x]$, $\mathbb{Q}[x, y]$, etc. are Unique Factorization Domains.

We saw earlier that if R is a Unique Factorization Domain with field of fractions F and $p(x) \in R[x]$, then we can factor out the greatest common divisor d of the coefficients of $p(x)$ to obtain $p(x) = dp'(x)$, where $p'(x)$ is irreducible in both $R[x]$ and $F[x]$. Suppose now that R is an *arbitrary* integral domain with field of fractions F. In R the notion of greatest common divisor may not make sense, however one might still ask if, say, a *monic* polynomial which is irreducible in $R[x]$ is still irreducible in $F[x]$ (i.e., whether the last statement in Corollary 6 is true).

Note first that if a monic polynomial $p(x)$ is reducible, it must have a factorization $p(x) = a(x)b(x)$ in $R[x]$ with both $a(x)$ and $b(x)$ *monic, nonconstant* polynomials (recall that the leading term of $p(x)$ is the product of the leading terms of the factors, so the leading coefficients of both $a(x)$ and $b(x)$ are units — we can thus arrange these to be 1). In other words, a nonconstant *monic* polynomial $p(x)$ is irreducible if and only if it cannot be factored as a product of two *monic* polynomials of smaller degree.

We now see that it is not true that if R is an arbitrary integral domain and $p(x)$ is a monic irreducible polynomial in $R[x]$, then $p(x)$ is irreducible in $F[x]$. For example, let $R = \mathbb{Z}[2i] = \{a + 2bi \mid a, b \in \mathbb{Z}\}$ (a subring of the complex numbers) and let $p(x) = x^2 + 1$. Then the fraction field of R is $F = \{a + bi \mid a, b \in \mathbb{Q}\}$. The polynomial $p(x)$ factors uniquely into a product of two linear factors in $F[x]$: $x^2 + 1 = (x - i)(x + i)$ so in particular, $p(x)$ *is reducible in* $F[x]$. Neither of these factors lies in $R[x]$ (because $i \notin R$) so $p(x)$ *is irreducible in* $R[x]$. In particular, by Corollary 6, $\mathbb{Z}[2i]$ *is not a Unique Factorization Domain*.

EXERCISES

1. Let R be an integral domain with quotient field F and let $p(x)$ be a monic polynomial in $R[x]$. Assume that $p(x) = a(x)b(x)$ where $a(x)$ and $b(x)$ are monic polynomials in $F[x]$ of smaller degree than $p(x)$. Prove that if $a(x) \notin R[x]$ then R is not a Unique Factorization Domain. Deduce that $\mathbb{Z}[2\sqrt{2}]$ is not a U.F.D.

2. Prove that if $f(x)$ and $g(x)$ are polynomials with rational coefficients whose product $f(x)g(x)$ has integer coefficients, then the product of any coefficient of $g(x)$ with any coefficient of $f(x)$ is an integer.

3. Let F be a field. Prove that the set R of polynomials in $F[x]$ whose coefficient of x is equal to 0 is a subring of $F[x]$ and that R is not a U.F.D. [Show that $x^6 = (x^2)^3 = (x^3)^2$ gives two distinct factorizations of x^6 into irreducibles.]

4. Let $R = \mathbb{Z} + x\mathbb{Q}[x] \subset \mathbb{Q}[x]$ be the set of polynomials in x with rational coefficients whose constant term is an integer.
   (a) Prove that R is an integral domain and its units are ± 1.
   (b) Show that the irreducibles in R are $\pm p$ where p is a prime in $\mathbb{Z}$ and the polynomials $f(x)$ that are irreducible in $\mathbb{Q}[x]$ and have constant term ± 1. Prove that these irreducibles are prime in R.
   (c) Show that x cannot be written as the product of irreducibles in R (in particular, x is not irreducible) and conclude that R is not a U.F.D.
   (d) Show that x is not a prime in R and describe the quotient ring $R/(x)$.

5. Let $R = \mathbb{Z} + x\mathbb{Q}[x] \subset \mathbb{Q}[x]$ be the ring considered in the previous exercise.
   (a) Suppose that $f(x), g(x) \in \mathbb{Q}[x]$ are two nonzero polynomials with rational coefficients and that x^r is the largest power of x dividing both $f(x)$ and $g(x)$ in $\mathbb{Q}[x]$, (i.e., r is the degree of the lowest order term appearing in either $f(x)$ or $g(x)$). Let f_r and

g_r be the coefficients of x^r in $f(x)$ and $g(x)$, respectively (one of which is nonzero by definition of r). Then $\mathbb{Z}f_r + \mathbb{Z}g_r = \mathbb{Z}d_r$ for some nonzero $d_r \in \mathbb{Q}$ (cf. Exercise 14 in Section 2.4). Prove that there is a polynomial $d(x) \in \mathbb{Q}[x]$ that is a g.c.d. of $f(x)$ and $g(x)$ in $\mathbb{Q}[x]$ and whose term of minimal degree is $d_r x^r$.

(b) Prove that $f(x) = d(x)q_1(x)$ and $g(x) = d(x)q_2(x)$ where $q_1(x)$ and $q_2(x)$ are elements of the subring R of $\mathbb{Q}[x]$.

(c) Prove that $d(x) = a(x)f(x) + b(x)g(x)$ for polynomials $a(x), b(x)$ in R. [The existence of $a(x), b(x)$ in the Euclidean Domain $\mathbb{Q}[x]$ is immediate. Use Exercise 11 in Section 2 to show that $a(x)$ and $b(x)$ can be chosen to lie in R.]

(d) Conclude from (a) and (b) that $Rf(x) + Rg(x) = Rd(x)$ in $\mathbb{Q}[x]$ and use this to prove that R is a Bezout Domain (cf. Exercise 7 in Section 8.2).

(e) Show that (d), the results of the previous exercise, and Exercise 11 of Section 8.3 imply that R must contain ideals that are not principal (hence not finitely generated). Prove that in fact $I = x\mathbb{Q}[x]$ is an ideal of R that is not finitely generated.

9.4 IRREDUCIBILITY CRITERIA

If R is a Unique Factorization Domain, then by Corollary 8 a polynomial ring in any number of variables with coefficients in R is also a Unique Factorization Domain. It is of interest then to determine the irreducible elements in such a polynomial ring, particularly in the ring $R[x]$. In the one-variable case, a nonconstant monic polynomial is irreducible in $R[x]$ if it cannot be factored as the product of two other polynomials of smaller degrees. Determining whether a polynomial has factors is frequently difficult to check, particularly for polynomials of large degree in several variables. The purpose of irreducibility criteria is to give an easier mechanism for determining when some types of polynomials are irreducible.

For the most part we restrict attention to polynomials in one variable where the coefficient ring is a Unique Factorization Domain. By Gauss' Lemma it suffices to consider factorizations in $F[x]$ where F is the field of fractions of R (although we shall occasionally consider questions of irreducibility when the coefficient ring is just an integral domain). The next proposition considers when there is a factor of degree one (a *linear* factor).

Proposition 9. Let F be a field and let $p(x) \in F[x]$. Then $p(x)$ has a factor of degree one if and only if $p(x)$ has a root in F, i.e., there is an $\alpha \in F$ with $p(\alpha) = 0$.

Proof: If $p(x)$ has a factor of degree one, then since F is a field, we may assume the factor is monic, i.e., is of the form $(x - \alpha)$ for some $\alpha \in F$. But then $p(\alpha) = 0$. Conversely, suppose $p(\alpha) = 0$. By the Division Algorithm in $F[x]$ we may write

$$p(x) = q(x)(x - \alpha) + r$$

where r is a constant. Since $p(\alpha) = 0$, r must be 0, hence $p(x)$ has $(x - \alpha)$ as a factor.

Proposition 9 gives a criterion for irreducibility for polynomials of small degree:

Proposition 10. A polynomial of degree two or three over a field F is reducible if and only if it has a root in F.

Proof: This follows immediately from the previous proposition, since a polynomial of degree two or three is reducible if and only if it has at least one linear factor.

The next result limits the possibilities for roots of polynomials with integer coefficients (it is stated for $\mathbb{Z}[x]$ for convenience although it clearly generalizes to $R[x]$, where R is any Unique Factorization Domain).

Proposition 11. Let $p(x) = a_n x^n + a_{n-1}x^{n-1} + \cdots + a_0$ be a polynomial of degree n with integer coefficients. If $r/s \in \mathbb{Q}$ is in lowest terms (i.e., r and s are relatively prime integers) and r/s is a root of $p(x)$, then r divides the constant term and s divides the leading coefficient of $p(x)$: $r \mid a_0$ and $s \mid a_n$. In particular, if $p(x)$ is a *monic* polynomial with integer coefficients and $p(d) \neq 0$ for all integers d dividing the constant term of $p(x)$, then $p(x)$ has no roots in $\mathbb{Q}$.

Proof: By hypothesis, $p(r/s) = 0 = a_n(r/s)^n + a_{n-1}(r/s)^{n-1} + \cdots + a_0$. Multiplying through by s^n gives

$$0 = a_n r^n + a_{n-1}r^{n-1}s + \cdots + a_0 s^n.$$

Thus $a_n r^n = s(-a_{n-1}r^{n-1} - \cdots - a_0 s^{n-1})$, so s divides $a_n r^n$. By assumption, s is relatively prime to r and it follows that $s \mid a_n$. Similarly, solving the equation for $a_0 s^n$ shows that $r \mid a_0$. The last assertion of the proposition follows from the previous ones.

Examples

(1) The polynomial $x^3 - 3x - 1$ is irreducible in $\mathbb{Z}[x]$. To prove this, by Gauss' Lemma and Proposition 10 it suffices to show it has no rational roots. By Proposition 11 the only candidates for rational roots are integers which divide the constant term 1, namely ± 1. Substituting both 1 and -1 into the polynomial shows that these are not roots.

(2) For p any prime the polynomials $x^2 - p$ and $x^3 - p$ are irreducible in $\mathbb{Q}[x]$. This is because they have degrees ≤ 3 so it suffices to show they have no rational roots. By Proposition 11 the only candidates for roots are ± 1 and $\pm p$, but none of these give 0 when they are substituted into the polynomial.

(3) The polynomial $x^2 + 1$ is reducible in $\mathbb{Z}/2\mathbb{Z}[x]$ since it has 1 as a root, and it factors as $(x + 1)^2$.

(4) The polynomial $x^2 + x + 1$ is irreducible in $\mathbb{Z}/2\mathbb{Z}[x]$ since it does not have a root in $\mathbb{Z}/2\mathbb{Z}$: $0^2 + 0 + 1 = 1$ and $1^2 + 1 + 1 = 1$.

(5) Similarly, the polynomial $x^3 + x + 1$ is irreducible in $\mathbb{Z}/2\mathbb{Z}[x]$.

This technique is limited to polynomials of low degree because it relies on the presence of a factor of degree one. A polynomial of degree 4, for example, may be the product of two irreducible quadratics, hence be reducible but have no linear factor. One fairly general technique for checking irreducibility uses Proposition 2 above and consists of reducing the coefficients modulo some ideal.

Proposition 12. Let I be a proper ideal in the integral domain R and let $p(x)$ be a nonconstant monic polynomial in $R[x]$. If the image of $p(x)$ in $(R/I)[x]$ cannot be factored in $(R/I)[x]$ into two polynomials of smaller degree, then $p(x)$ is irreducible in $R[x]$.

Proof: Suppose $p(x)$ cannot be factored in $(R/I)[x]$ but that $p(x)$ is reducible in $R[x]$. As noted at the end of the preceding section this means there are monic, nonconstant polynomials $a(x)$ and $b(x)$ in $R[x]$ such that $p(x) = a(x)b(x)$. By Proposition 2, reducing the coefficients modulo I gives a factorization in $(R/I)[x]$ with nonconstant factors, a contradiction.

This proposition indicates that if it is possible to find a proper ideal I such that the *reduced* polynomial cannot be factored, then the polynomial is itself irreducible. Unfortunately, there are examples of polynomials even in $\mathbb{Z}[x]$ which are irreducible but whose reductions modulo every ideal are reducible (so their irreducibility is not detectable by this technique). For example, the polynomial $x^4 + 1$ is irreducible in $\mathbb{Z}[x]$ but is reducible modulo every prime (we shall verify this in Chapter 14) and the polynomial $x^4 - 72x^2 + 4$ is irreducible in $\mathbb{Z}[x]$ but is reducible modulo every integer.

Examples

(1) Consider the polynomial $p(x) = x^2 + x + 1$ in $\mathbb{Z}[x]$. Reducing modulo 2, we see from Example 4 above that $p(x)$ is irreducible in $\mathbb{Z}[x]$. Similarly, $x^3 + x + 1$ is irreducible in $\mathbb{Z}[x]$ because it is irreducible in $\mathbb{Z}/2\mathbb{Z}[x]$.

(2) The polynomial $x^2 + 1$ is irreducible in $\mathbb{Z}[x]$ since it is irreducible in $\mathbb{Z}/3\mathbb{Z}[x]$ (no root in $\mathbb{Z}/3\mathbb{Z}$), but is reducible mod 2. This shows that the converse to Proposition 12 does not hold.

(3) The idea of reducing modulo an ideal to determine irreducibility can be used also in several variables, but some care must be exercised. For example, the polynomial $x^2 + xy + 1$ in $\mathbb{Z}[x, y]$ is irreducible since modulo the ideal (y) it is $x^2 + 1$ in $\mathbb{Z}[x]$, which is irreducible and of the same degree. In this sort of argument it is necessary to be careful about "collapsing." For example, the polynomial $xy + x + y + 1$ (which is $(x + 1)(y + 1)$) is reducible, but appears irreducible modulo both (x) and (y). The reason for this is that nonunit polynomials in $\mathbb{Z}[x, y]$ can reduce to units in the quotient. To take account of this it is necessary to determine which elements in the original ring become units in the quotient. The elements in $\mathbb{Z}[x, y]$ which are units modulo (y), for example, are the polynomials in $\mathbb{Z}[x, y]$ with constant term ± 1 and all nonconstant terms divisible by y. The fact that $x^2 + xy + 1$ and its reduction mod (y) have the same degree therefore eliminates the possibility of a factor which is a unit modulo (y), but not a unit in $\mathbb{Z}[x, y]$ and gives the irreducibility of this polynomial.

A special case of reducing modulo an ideal to test for irreducibility which is frequently useful is known as *Eisenstein's Criterion* (although originally proved earlier by Schönemann, so more properly known as the *Eisenstein-Schönemann Criterion*):

Proposition 13. (*Eisenstein's Criterion*) Let P be a prime ideal of the integral domain R and let $f(x) = x^n + a_{n-1}x^{n-1} + \cdots + a_1 x + a_0$ be a polynomial in $R[x]$ (here $n \geq 1$). Suppose $a_{n-1}, \ldots, a_1, a_0$ are all elements of P and suppose a_0 is not an element of P^2. Then $f(x)$ is irreducible in $R[x]$.

Proof: Suppose $f(x)$ were reducible, say $f(x) = a(x)b(x)$ in $R[x]$, where $a(x)$ and $b(x)$ are nonconstant polynomials. Reducing this equation modulo P and using the assumptions on the coefficients of $f(x)$ we obtain the equation $x^n = \overline{a(x)b(x)}$ in $(R/P)[x]$, where the bar denotes the polynomials with coefficients reduced mod P. Since P is a prime ideal, R/P is an integral domain, and it follows that both $\overline{a(x)}$ and $\overline{b(x)}$ have 0 constant term, i.e., the constant terms of both $a(x)$ and $b(x)$ are elements of P. But then the constant term a_0 of $f(x)$ as the product of these two would be an element of P^2, a contradiction.

Eisenstein's Criterion is most frequently applied to $\mathbb{Z}[x]$ so we state the result explicitly for this case:

Corollary 14. *(Eisenstein's Criterion for $\mathbb{Z}[x]$)* Let p be a prime in $\mathbb{Z}$ and let $f(x) = x^n + a_{n-1}x^{n-1} + \cdots + a_1 x + a_0 \in \mathbb{Z}[x]$, $n \geq 1$. Suppose p divides a_i for all $i \in \{0, 1, \ldots, n-1\}$ but that p^2 does not divide a_0. Then $f(x)$ is irreducible in both $\mathbb{Z}[x]$ and $\mathbb{Q}[x]$.

Proof: This is simply a restatement of Proposition 13 in the case of the prime ideal (p) in $\mathbb{Z}$ together with Corollary 6.

Examples

(1) The polynomial $x^4 + 10x + 5$ in $\mathbb{Z}[x]$ is irreducible by Eisenstein's Criterion applied for the prime 5.

(2) If a is any integer which is divisible by some prime p but not divisible by p^2, then $x^n - a$ is irreducible in $\mathbb{Z}[x]$ by Eisenstein's Criterion. In particular, $x^n - p$ is irreducible for all positive integers n and so for $n \geq 2$ the nth roots of p are not rational numbers (i.e., this polynomial has no root in $\mathbb{Q}$).

(3) Consider the polynomial $f(x) = x^4 + 1$ mentioned previously. Eisenstein's Criterion does not apply directly to $f(x)$. The polynomial $g(x) = f(x+1)$ is $(x+1)^4 + 1$, i.e., $x^4 + 4x^3 + 6x^2 + 4x + 2$, and Eisenstein's Criterion for the prime 2 shows that this polynomial is irreducible. It follows then that $f(x)$ must also be irreducible, since any factorization for $f(x)$ would provide a factorization for $g(x)$ (just replace x by $x + 1$ in each of the factors). This example shows that Eisenstein's Criterion can sometimes be used to verify the irreducibility of a polynomial to which it does not immediately apply.

(4) As another example of this, let p be a prime and consider the polynomial

$$\Phi_p(x) = \frac{x^p - 1}{x - 1} = x^{p-1} + x^{p-2} + \cdots + x + 1,$$

an example of a *cyclotomic polynomial* which we shall consider more thoroughly in Part IV. Again, Eisenstein's Criterion does not immediately apply, but it does apply for the prime p to the polynomial

$$\Phi_p(x + 1) = \frac{(x + 1)^p - 1}{x} = x^{p-1} + px^{p-2} + \cdots + \frac{p(p-1)}{2}x + p \in \mathbb{Z}[x]$$

since all the coefficients except the first are divisible by p by the Binomial Theorem. As before, this shows $\Phi_p(x)$ is irreducible in $\mathbb{Z}[x]$.

(5) As an example of the use of the more general Eisenstein's Criterion in Proposition 13 we mimic Example 2 above. Let $R = \mathbb{Q}[x]$ and let n be any positive integer. Consider the polynomial $X^n - x$ in the ring $R[X]$. The ideal (x) is prime in the coefficient ring R since $R/(x) = \mathbb{Q}[x]/(x)$ is the integral domain $\mathbb{Q}$. Eisenstein's Criterion for the ideal (x) of R applies directly to show that $X^n - x$ is irreducible in $R[X]$. Note that this construction works with $\mathbb{Q}$ replaced by any field or, indeed, by any integral domain.

There are now efficient algorithms for factoring polynomials over certain fields. For polynomials with integer coefficients these algorithms have been implemented in a number of computer packages. An efficient algorithm for factoring polynomials over $\mathbb{F}_p$, called the Berlekamp Algorithm, is described in detail in the exercises at the end of Section 14.3.

EXERCISES

1. Determine whether the following polynomials are irreducible in the rings indicated. For those that are reducible, determine their factorization into irreducibles. The notation $\mathbb{F}_p$ denotes the finite field $\mathbb{Z}/p\mathbb{Z}$, p a prime.
   (a) $x^2 + x + 1$ in $\mathbb{F}_2[x]$.
   (b) $x^3 + x + 1$ in $\mathbb{F}_3[x]$.
   (c) $x^4 + 1$ in $\mathbb{F}_5[x]$.
   (d) $x^4 + 10x^2 + 1$ in $\mathbb{Z}[x]$.

2. Prove that the following polynomials are irreducible in $\mathbb{Z}[x]$:
   (a) $x^4 - 4x^3 + 6$
   (b) $x^6 + 30x^5 - 15x^3 + 6x - 120$
   (c) $x^4 + 4x^3 + 6x^2 + 2x + 1$ [Substitute $x - 1$ for x.]
   (d) $\dfrac{(x+2)^p - 2^p}{x}$, where p is an odd prime.

3. Show that the polynomial $(x-1)(x-2) \cdots (x-n) - 1$ is irreducible over $\mathbb{Z}$ for all $n \geq 1$. [If the polynomial factors consider the values of the factors at $x = 1, 2, \ldots, n$.]

4. Show that the polynomial $(x - 1)(x - 2) \cdots (x - n) + 1$ is irreducible over $\mathbb{Z}$ for all $n \geq 1$, $n \neq 4$.

5. Find all the monic irreducible polynomials of degree ≤ 3 in $\mathbb{F}_2[x]$, and the same in $\mathbb{F}_3[x]$.

6. Construct fields of each of the following orders: (a) 9, (b) 49, (c) 8, (d) 81 (you may exhibit these as $F[x]/(f(x))$ for some F and f). [Use Exercises 2 and 3 in Section 2.]

7. Prove that $\mathbb{R}[x]/(x^2 + 1)$ is a field which is isomorphic to the complex numbers.

8. Prove that $K_1 = \mathbb{F}_{11}[x]/(x^2 + 1)$ and $K_2 = \mathbb{F}_{11}[y]/(y^2 + 2y + 2)$ are both fields with 121 elements. Prove that the map which sends the element $p(\bar{x})$ of K_1 to the element $p(\bar{y} + 1)$ of K_2 (where p is any polynomial with coefficients in $\mathbb{F}_{11}$) is well defined and gives a ring (hence field) isomorphism from K_1 to K_2.

9. Prove that the polynomial $x^2 - \sqrt{2}$ is irreducible over $\mathbb{Z}[\sqrt{2}]$ (you may use the fact that $\mathbb{Z}[\sqrt{2}]$ is a U.F.D. — cf. Exercise 9 of Section 8.1).

10. Prove that the polynomial $p(x) = x^4 - 4x^2 + 8x + 2$ is irreducible over the quadratic field $F = \mathbb{Q}(\sqrt{-2}) = \{a + b\sqrt{-2} \mid a, b \in \mathbb{Q}\}$. [First use the method of Proposition 11 for the Unique Factorization Domain $\mathbb{Z}[\sqrt{-2}]$ (cf. Exercise 8, Section 8.1) to show that if $\alpha \in \mathbb{Z}[\sqrt{-2}]$ is a root of $p(x)$ then α is a divisor of 2 in $\mathbb{Z}[\sqrt{-2}]$. Conclude that α must be ± 1, $\pm\sqrt{-2}$ or ± 2, and hence show $p(x)$ has no linear factor over F. Show similarly that $p(x)$ is not the product of two quadratics with coefficients in F.]

11. Prove that $x^2 + y^2 - 1$ is irreducible in $\mathbb{Q}[x, y]$.

12. Prove that $x^{n-1} + x^{n-2} + \cdots + x + 1$ is irreducible over $\mathbb{Z}$ if and only if n is a prime.

13. Prove that $x^3 + nx + 2$ is irreducible over $\mathbb{Z}$ for all integers $n \neq 1, -3, -5$.

14. Factor each of the two polynomials: $x^8 - 1$ and $x^6 - 1$ into irreducibles over each of the following rings: **(a)** $\mathbb{Z}$, **(b)** $\mathbb{Z}/2\mathbb{Z}$, **(c)** $\mathbb{Z}/3\mathbb{Z}$.

15. Prove that if F is a field then the polynomial $X^n - x$ which has coefficients in the ring $F[[x]]$ of formal power series (cf. Exercise 3 of Section 7.2) is irreducible over $F[[x]]$. [Recall that $F[[x]]$ is a Euclidean Domain — cf. Exercise 5, Section 7.2 and Example 4, Section 8.1.]

16. Let F be a field and let $f(x)$ be a polynomial of degree n in $F[x]$. The polynomial $g(x) = x^n f(1/x)$ is called the *reverse* of $f(x)$.
    (a) Describe the coefficients of g in terms of the coefficients of f.
    (b) If $f(0) \neq 0$ prove that f is irreducible if and only if g is irreducible.

17. Prove the following variant of Eisenstein's Criterion: let P be a prime ideal in the Unique Factorization Domain R and let $f(x) = a_n x^n + a_{n-1} x^{n-1} + \cdots + a_1 x + a_0$ be a polynomial in $R[x]$, $n \geq 1$. Suppose $a_n \notin P$, $a_{n-1}, \ldots, a_0 \in P$ and $a_0 \notin P^2$. Prove that $f(x)$ is irreducible in $F[x]$, where F is the quotient field of R.

18. Show that $6x^5 + 14x^3 - 21x + 35$ and $18x^5 - 30x^2 + 120x + 360$ are irreducible in $\mathbb{Q}[x]$.

19. Let F be a field and let $f(x) = a_n x^n + a_{n-1} x^{n-1} + \cdots + a_0 \in F[x]$. The *derivative*, $D_x(f(x))$, of $f(x)$ is defined by

$$D_x(f(x)) = na_n x^{n-1} + (n-1)a_{n-1} x^{n-2} + \cdots + a_1$$

where, as usual, $na = a + a + \cdots + a$ (n times). Note that $D_x(f(x))$ is again a polynomial with coefficients in F.

The polynomial $f(x)$ is said to have a *multiple root* if there is some field E containing F and some $\alpha \in E$ such that $(x - \alpha)^2$ divides $f(x)$ in $E[x]$. For example, the polynomial $f(x) = (x - 1)^2(x - 2) \in \mathbb{Q}[x]$ has $\alpha = 1$ as a multiple root and the polynomial $f(x) = x^4 + 2x^2 + 1 = (x^2 + 1)^2 \in \mathbb{R}[x]$ has $\alpha = \pm i \in \mathbb{C}$ as multiple roots. We shall prove in Section 13.5 that a nonconstant polynomial $f(x)$ has a multiple root if and only if $f(x)$ is not relatively prime to its derivative (which can be detected by the Euclidean Algorithm in $F[x]$). Use this criterion to determine whether the following polynomials have multiple roots:
    (a) $x^3 - 3x - 2 \in \mathbb{Q}[x]$
    (b) $x^3 + 3x + 2 \in \mathbb{Q}[x]$
    (c) $x^6 - 4x^4 + 6x^3 + 4x^2 - 12x + 9 \in \mathbb{Q}[x]$
    (d) Show for any prime p and any $a \in \mathbb{F}_p$ that the polynomial $x^p - a$ has a multiple root.

20. Show that the polynomial $f(x) = x$ in $\mathbb{Z}/6\mathbb{Z}[x]$ factors as $(3x + 4)(4x + 3)$, hence is not an irreducible polynomial.
    (a) Show that the reduction of $f(x)$ modulo both of the nontrivial ideals (2) and (3) of $\mathbb{Z}/6\mathbb{Z}$ is an irreducible polynomial, showing that the condition that R be an integral domain in Proposition 12 is necessary.
    (b) Show that in any factorization $f(x) = g(x)h(x)$ in $\mathbb{Z}/6\mathbb{Z}[x]$ the reduction of $g(x)$ modulo (2) is either 1 or x and the reduction of $h(x)$ modulo (2) is then either x or 1, and similarly for the reductions modulo (3). Determine all the factorizations of $f(x)$ in $\mathbb{Z}/6\mathbb{Z}[x]$. [Use the Chinese Remainder Theorem.]
    (c) Show that the ideal $(3, x)$ is a principal ideal in $\mathbb{Z}/6\mathbb{Z}[x]$.
    (d) Show that over the ring $\mathbb{Z}/30\mathbb{Z}[x]$ the polynomial $f(x) = x$ has the factorization

$f(x) = (10x+21)(15x+16)(6x+25)$. Prove that the product of any of these factors is again of the same degree. Prove that the reduction of $f(x)$ modulo any prime in $\mathbb{Z}/30\mathbb{Z}$ is an irreducible polynomial. Determine all the factorizations of $f(x)$ in $\mathbb{Z}/30\mathbb{Z}[x]$. [Consider the reductions modulo (2), (3) and (5) and use the Chinese Remainder Theorem.]

(e) Generalize part (d) to $\mathbb{Z}/n\mathbb{Z}[x]$ where n is the product of k distinct primes.

9.5 POLYNOMIAL RINGS OVER FIELDS II

Let F be a field. We prove here some additional results for the one-variable polynomial ring $F[x]$. The first is a restatement of results obtained earlier.

Proposition 15. The maximal ideals in $F[x]$ are the ideals $(f(x))$ generated by irreducible polynomials $f(x)$. In particular, $F[x]/(f(x))$ is a field if and only if $f(x)$ is irreducible.

Proof: This follows from Proposition 7 of Section 8.2 applied to the Principal Ideal Domain $F[x]$.

Proposition 16. Let $g(x)$ be a nonconstant monic element of $F[x]$ and let

$$g(x) = f_1(x)^{n_1} f_2(x)^{n_2} \cdots f_k(x)^{n_k}$$

be its factorization into irreducibles, where the $f_i(x)$ are distinct. Then we have the following isomorphism of rings:

$$F[x]/(g(x)) \cong F[x]/(f_1(x)^{n_1}) \times F[x]/(f_2(x)^{n_2}) \times \cdots \times F[x]/(f_k(x)^{n_k}).$$

Proof: This follows from the Chinese Remainder Theorem (Theorem 7.17), since the ideals $(f_i(x)^{n_i})$ and $(f_j(x)^{n_j})$ are comaximal if $f_i(x)$ and $f_j(x)$ are distinct (they are relatively prime in the Euclidean Domain $F[x]$, hence the ideal generated by them is $F[x]$).

The next result concerns the number of roots of a polynomial over a field F. By Proposition 9, a root α corresponds to a linear factor $(x - \alpha)$ of $f(x)$. If $f(x)$ is divisible by $(x - \alpha)^m$ but not by $(x - \alpha)^{m+1}$, then α is said to be a root of *multiplicity m*.

Proposition 17. If the polynomial $f(x)$ has roots $\alpha_1, \alpha_2, \ldots, \alpha_k$ in F (not necessarily distinct), then $f(x)$ has $(x - \alpha_1) \cdots (x - \alpha_k)$ as a factor. In particular, a polynomial of degree n in one variable over a field F has at most n roots in F, even counted with multiplicity.

Proof: The first statement follows easily by induction from Proposition 9. Since linear factors are irreducible, the second statement follows since $F[x]$ is a Unique Factorization Domain.

This last result has the following interesting consequence.

Proposition 18. A finite subgroup of the multiplicative group of a field is cyclic. In particular, if F is a finite field, then the multiplicative group $F^\times$ of nonzero elements of F is a cyclic group.

Proof: We give a proof of this result using the Fundamental Theorem of Finitely Generated Abelian Groups (Theorem 3 in Section 5.2). A more number-theoretic proof is outlined in the exercises, or Proposition 5 in Section 6.1 may be used in place of the Fundamental Theorem. By the Fundamental Theorem, the finite subgroup can be written as the direct product of cyclic groups

$$\mathbb{Z}/n_1\mathbb{Z} \times \mathbb{Z}/n_2\mathbb{Z} \times \cdots \times \mathbb{Z}/n_k\mathbb{Z}$$

where $n_k \mid n_{k-1} \mid \cdots \mid n_2 \mid n_1$. In general, if G is a cyclic group and $d \mid |G|$ then G contains precisely d elements of order dividing d. Since n_k divides the order of each of the cyclic groups in the direct product, it follows that each direct factor contains n_k elements of order dividing n_k. If k were greater than 1, there would therefore be a total of more than n_k such elements. But then there would be more than n_k roots of the polynomial $x^{n_k} - 1$ in the field F, contradicting Proposition 17. Hence $k = 1$ and the group is cyclic.

Corollary 19. Let p be a prime. The multiplicative group $(\mathbb{Z}/p\mathbb{Z})^\times$ of nonzero residue classes mod p is cyclic.

Proof: This is the multiplicative group of the finite field $\mathbb{Z}/p\mathbb{Z}$.

Corollary 20. Let $n \geq 2$ be an integer with factorization $n = p_1^{\alpha_1} p_2^{\alpha_2} \cdots p_r^{\alpha_r}$ in $\mathbb{Z}$, where $p_1, \ldots, p_r$ are distinct primes. We have the following isomorphisms of (multiplicative) groups:
  (1) $(\mathbb{Z}/n\mathbb{Z})^\times \cong (\mathbb{Z}/p_1^{\alpha_1}\mathbb{Z})^\times \times (\mathbb{Z}/p_2^{\alpha_2}\mathbb{Z})^\times \times \cdots \times (\mathbb{Z}/p_r^{\alpha_r}\mathbb{Z})^\times$
  (2) $(\mathbb{Z}/2^\alpha\mathbb{Z})^\times$ is the direct product of a cyclic group of order 2 and a cyclic group of order $2^{\alpha-2}$, for all $\alpha \geq 2$
  (3) $(\mathbb{Z}/p^\alpha\mathbb{Z})^\times$ is a cyclic group of order $p^{\alpha-1}(p - 1)$, for all odd primes p.

Remark: These isomorphisms describe the group-theoretic structure of the automorphism group of the cyclic group, Z_n, of order n since $\mathrm{Aut}(Z_n) \cong (\mathbb{Z}/n\mathbb{Z})^\times$ (cf. Proposition 16 in Section 4.4). In particular, for p a prime the automorphism group of the cyclic group of order p is cyclic of order $p - 1$.

Proof: This is mainly a matter of collecting previous results. The isomorphism in (1) follows from the Chinese Remainder Theorem (see Corollary 18, Section 7.6). The isomorphism in (2) follows directly from Exercises 22 and 23 of Section 2.3.

For p an odd prime, $(\mathbb{Z}/p^\alpha\mathbb{Z})^\times$ is an abelian group of order $p^{\alpha-1}(p - 1)$. By Exercise 21 of Section 2.3 the Sylow p-subgroup of this group is cyclic. The map

$$\mathbb{Z}/p^\alpha\mathbb{Z} \to \mathbb{Z}/p\mathbb{Z} \qquad \text{defined by} \qquad a + (p^\alpha) \mapsto a + (p)$$

is a ring homomorphism (reduction mod p) which gives a surjective group homomorphism from $(\mathbb{Z}/p^\alpha\mathbb{Z})^\times$ onto $(\mathbb{Z}/p\mathbb{Z})^\times$. The latter group is cyclic of order $p - 1$

(Corollary 19). The kernel of this map is of order $p^{\alpha-1}$, hence for all primes $q \neq p$, the Sylow q-subgroup of $(\mathbb{Z}/p^{\alpha}\mathbb{Z})^{\times}$ maps isomorphically into the cyclic group $(\mathbb{Z}/p\mathbb{Z})^{\times}$. All Sylow subgroups of $(\mathbb{Z}/p^{\alpha}\mathbb{Z})^{\times}$ are therefore cyclic, so (3) holds, completing the proof.

EXERCISES

1. Let F be a field and let $f(x)$ be a nonconstant polynomial in $F[x]$. Describe the nilradical of $F[x]/(f(x))$ in terms of the factorization of $f(x)$ (cf. Exercise 29, Section 7.3).

2. For each of the fields constructed in Exercise 6 of Section 4 exhibit a generator for the (cyclic) multiplicative group of nonzero elements.

3. Let p be an odd prime in $\mathbb{Z}$ and let n be a positive integer. Prove that $x^n - p$ is irreducible over $\mathbb{Z}[i]$. [Use Proposition 18 in Chapter 8 and Eisenstein's Criterion.]

4. Prove that $x^3 + 12x^2 + 18x + 6$ is irreducible over $\mathbb{Z}[i]$. [Use Proposition 8.18 and Eisenstein's Criterion.]

5. Let φ denote Euler's φ-function. Prove the identity $\sum_{d|n} \varphi(d) = n$, where the sum is extended over all the divisors d of n. [First observe that the identity is valid when $n = p^m$ is the power of a prime p since the sum telescopes. Write $n = p^m n'$ where p does not divide n'. Prove that $\sum_{d|n} \varphi(d) = \sum_{d''|p^m} \varphi(d'') \sum_{d'|n'} \varphi(d')$ by multiplying out the right hand side and using the multiplicativity $\varphi(ab) = \varphi(a)\varphi(b)$ when a and b are relatively prime. Use induction to complete the proof. This problem may be done alternatively by letting Z be the cyclic group of order n and showing that since Z contains a unique subgroup of order d for each d dividing n, the number of elements of Z of order d is $\varphi(d)$. Then $|Z|$ is the sum of $\varphi(d)$ as d runs over all divisors of n.]

6. Let G be a finite subgroup of order n of the multiplicative group $F^{\times}$ of nonzero elements of the field F. Let φ denote Euler's φ-function and let $\psi(d)$ denote the number of elements of G of order d. Prove that $\psi(d) = \varphi(d)$ for every divisor d of n. In particular conclude that $\psi(n) \geq 1$, so that G is a cyclic group. [Observe that for any integer $N \geq 1$ the polynomial $x^N - 1$ has at most N roots in F. Conclude that for any integer N we have $\sum_{d|N} \psi(d) \leq N$. Since $\sum_{d|N} \varphi(d) = N$ by the previous exercise, show by induction that $\psi(d) \leq \varphi(d)$ for every divisor d of n. Since $\sum_{d|n} \psi(d) = n = \sum_{d|n} \varphi(d)$ show that this implies $\psi(d) = \varphi(d)$ for every divisor d of n.]

7. Prove that the additive and multiplicative groups of a field are never isomorphic. [Consider three cases: when $|F|$ is finite, when $-1 \neq 1$ in F, and when $-1 = 1$ in F.]

9.6 POLYNOMIALS IN SEVERAL VARIABLES OVER A FIELD AND GRÖBNER BASES

In this section we consider polynomials in many variables, present some basic computational tools, and indicate some applications. The results of this section are not required in Chapters 10 through 14. Additional applications will be given in Chapter 15.

We proved in Section 2 that a polynomial ring $F[x]$ in a variable x over a field F is a Euclidean Domain, and Corollary 8 showed that the polynomial ring $F[x_1, \ldots, x_n]$ is a U.F.D. However it follows from Corollary 8 in Section 8.2 that the latter ring is not a P.I.D. unless $n = 1$. Our first result below shows that ideals in such polynomial rings, although not necessarily principal, are always finitely generated. General rings with this property are given a special name:

Definition. A commutative ring R with 1 is called *Noetherian* if every ideal of R is finitely generated.

Noetherian rings will be studied in greater detail in Chapters 15 and 16. In this section we develop some of the basic theory and resulting algorithms for working with (finitely generated) ideals in $F[x_1, \ldots, x_n]$.

As we saw in Section 1, a polynomial ring in n variables can be considered as a polynomial ring in one variable with coefficients in a polynomial ring in $n-1$ variables. By following this inductive approach—as we did in Theorem 7 and Corollary 8—we can deduce that $F[x_1, x_2, \ldots, x_n]$ is Noetherian from the following more general result.

Theorem 21. *(Hilbert's Basis Theorem)* If R is a Noetherian ring then so is the polynomial ring $R[x]$.

Proof: Let I be an ideal in $R[x]$ and let L be the set of all leading coefficients of the elements in I. We first show that L is an ideal of R, as follows. Since I contains the zero polynomial, $0 \in L$. Let $f = ax^d + \cdots$ and $g = bx^e + \cdots$ be polynomials in I of degrees d, e and leading coefficients $a, b \in R$. Then for any $r \in R$ either $ra - b$ is zero or it is the leading coefficient of the polynomial $rx^e f - x^d g$. Since the latter polynomial is in I we have $ra - b \in L$, which shows L is an ideal of R. Since R is assumed Noetherian, the ideal L in R is finitely generated, say by $a_1, a_2, \ldots, a_n \in R$. For each $i = 1, \ldots, n$ let f_i be an element of I whose leading coefficient is a_i. Let e_i denote the degree of f_i, and let N be the maximum of $e_1, e_2, \ldots, e_n$.

For each $d \in \{0, 1, \ldots, N-1\}$, let L_d be the set of all leading coefficients of polynomials in I of degree d together with 0. A similar argument as that for L shows each L_d is also an ideal of R, again finitely generated since R is Noetherian. For each nonzero ideal L_d let $b_{d,1}, b_{d,2}, \ldots, b_{d,n_d} \in R$ be a set of generators for L_d, and let $f_{d,i}$ be a polynomial in I of degree d with leading coefficient $b_{d,i}$.

We show that the polynomials $f_1, \ldots, f_n$ together with all the polynomials $f_{d,i}$ for all the nonzero ideals L_d are a set of generators for I, i.e., that

$$I = (\{f_1, \ldots, f_n\} \cup \{f_{d,i} \mid 0 \le d < N, \ 1 \le i \le n_d\}).$$

By construction, the ideal I' on the right above is contained in I since all the generators were chosen in I. If $I' \ne I$, there exists a nonzero polynomial $f \in I$ of minimum degree with $f \notin I'$. Let $d = \deg f$ and let a be the leading coefficient of f.

Suppose first that $d \ge N$. Since $a \in L$ we may write a as an R-linear combination of the generators of L: $a = r_1 a_1 + \cdots + r_n a_n$. Then $g = r_1 x^{d-e_1} f_1 + \cdots + r_n x^{d-e_n} f_n$ is an element of I' with the same degree d and the same leading coefficient a as f. Then $f - g \in I$ is a polynomial in I of smaller degree than f. By the minimality of f, we must have $f - g = 0$, so $f = g \in I'$, a contradiction.

Suppose next that $d < N$. In this case $a \in L_d$ for some $d < N$, and so we may write $a = r_1 b_{d,1} + \cdots + r_{n_d} b_{n_d}$ for some $r_i \in R$. Then $g = r_1 f_{d,1} + \cdots + r_{n_d} f_{n_d}$ is a polynomial in I' with the same degree d and the same leading coefficient a as f, and we have a contradiction as before.

It follows that $I = I'$ is finitely generated, and since I was arbitrary, this completes the proof that $R[x]$ is Noetherian.

316

Since a field is clearly Noetherian, Hilbert's Basis Theorem and induction immediately give:

Corollary 22. Every ideal in the polynomial ring $F[x_1, x_2, \ldots, x_n]$ with coefficients from a field F is finitely generated.

If I is an ideal in $F[x_1, \ldots, x_n]$ generated by a (possibly infinite) set $\mathcal{S}$ of polynomials, Corollary 22 shows that I is finitely generated, and in fact I is generated by a finite number of the polynomials from the set $\mathcal{S}$ (cf. Exercise 1).

As the proof of Hilbert's Basis Theorem shows, the collection of leading coefficients of the polynomials in an ideal I in $R[x]$ forms an extremely useful ideal in R that can be used to understand I. This suggests studying "leading terms" in $F[x_1, x_2, \ldots, x_n]$ more generally (and somewhat more intrinsically). To do this we need to specify a total ordering on the monomials, since without some sort of ordering we cannot in general tell which is the "leading" term of a polynomial. We implicitly chose such an ordering in the inductive proof of Corollary 22—we first viewed a polynomial f as a polynomial in x_1 with coefficients in $R = F[x_2, \ldots, x_n]$, say, then viewed its "leading coefficient" in $F[x_2, \ldots, x_n]$ as a polynomial in x_2 with coefficients in $F[x_3, \ldots, x_n]$, etc. This is an example of a *lexicographic* monomial ordering on the polynomial ring $F[x_1, \ldots, x_n]$ which is defined by first declaring an ordering of the variables, for example $x_1 > x_2 > \cdots > x_n$ and then declaring that the monomial term $Ax_1^{a_1} x_2^{a_2} \cdots x_n^{a_n}$ with exponents $(a_1, a_2, \ldots, a_n)$ has higher order than the monomial term $Bx_1^{b_1} x_2^{b_2} \cdots x_n^{b_n}$ with exponents $(b_1, b_2, \ldots, b_n)$ if the first component where the n-tuples differ has $a_i > b_i$. This is analogous to the ordering used in a dictionary (hence the name), where the letter "a" comes before "b" which in turn comes before "c", etc., and then "aardvark" comes before "abacus" (although the 'word' $a^2 = aa$ comes before a in the lexicographical order). Note that the ordering is only defined up to multiplication by units (elements of $F^\times$) and that multiplying two monomials by the same nonzero monomial does not change their ordering. This can be formalized in general.

Definition. A *monomial ordering* is a well ordering "$\geq$" on the set of monomials that satisfies $mm_1 \geq mm_2$ whenever $m_1 \geq m_2$ for monomials m, m_1, m_2. Equivalently, a monomial ordering may be specified by defining a well ordering on the n-tuples $\alpha = (a_1, \ldots, a_n) \in \mathbb{Z}^n$ of multidegrees of monomials $Ax_1^{a_1} \cdots x_n^{a_n}$ that satisfies $\alpha + \gamma \geq \beta + \gamma$ if $\alpha \geq \beta$.

It is easy to show for any monomial ordering that $m \geq 1$ for every monomial m (cf. Exercise 2). It is not difficult to show, using Hilbert's Basis Theorem, that any total ordering on monomials which for every monomial m satisfies $m \geq 1$ and $mm_1 \geq mm_2$ whenever $m_1 \geq m_2$, is necessarily a well ordering (hence a monomial ordering)—this equivalent set of axioms for a monomial ordering may be easier to verify. For simplicity we shall limit the examples to the particularly easy and intuitive lexicographic ordering, but it is important to note that there are useful computational advantages to using other monomial orderings in practice. Some additional commonly used monomial orderings are introduced in the exercises.

As mentioned, once we have a monomial ordering we can define the leading term of a polynomial:

Definition. Fix a monomial ordering on the polynomial ring $F[x_1, x_2, \ldots, x_n]$.
   (1) The *leading term* of a nonzero polynomial f in $F[x_1, x_2, \ldots, x_n]$, denoted $LT(f)$, is the monomial term of maximal order in f and the leading term of $f = 0$ is 0. Define the *multidegree of f*, denoted $\partial(f)$, to be the multidegree of the leading term of f.
   (2) If I is an ideal in $F[x_1, x_2, \ldots, x_n]$, the *ideal of leading terms*, denoted $LT(I)$, is the ideal generated by the leading terms of all the elements in the ideal, i.e., $LT(I) = (LT(f) \mid f \in I)$.

The leading term and the multidegree of a polynomial clearly depend on the choice of the ordering. For example $LT(2xy + y^3) = 2xy$ with multidegree $(1, 1)$ if $x > y$, but $LT(2xy + y^3) = y^3$ with multidegree $(0, 3)$ if $y > x$. In particular, the leading term of a polynomial need not be the term of largest total degree. Similarly, the ideal of leading terms $LT(I)$ of an ideal I in general depends on the ordering used. Note also that the multidegree of a polynomial satisfies $\partial(fg) = \partial f + \partial g$ when f and g are nonzero, and that in this case $LT(fg) = LT(f) + LT(g)$ (cf. Exercise 2).

The ideal $LT(I)$ is by definition generated by monomials. Such ideals are called *monomial ideals* and are typically much easier to work with than generic ideals. For example, a polynomial is contained in a monomial ideal if and only if each of its monomial terms is a multiple of one of the generators for the ideal (cf. Exercise 10).

It was important in the proof of Hilbert's Basis Theorem to have *all* of the leading terms of the ideal I. If $I = (f_1, \ldots, f_m)$, then $LT(I)$ contains the leading terms $LT(f_1), \ldots, LT(f_m)$ of the generators for I by definition. Since $LT(I)$ is an ideal, it contains the ideal generated by these leading terms:

$$(LT(f_1), \ldots, LT(f_m)) \subseteq LT(I).$$

The first of the following examples shows that the ideal $LT(I)$ of leading terms can in general be strictly larger than the ideal generated just by the leading terms of some generators for I.

Examples
   (1) Choose the lexicographic ordering $x > y$ on $F[x, y]$. The leading terms of the polynomials $f_1 = x^3 y - xy^2 + 1$ and $f_2 = x^2 y^2 - y^3 - 1$ are $LT(f_1) = x^3 y$ (so the multidegree of f_1 is $\partial(f_1) = (3, 1)$) and $LT(f_2) = x^2 y^2$ (so $\partial(f_2) = (2, 2)$). If $I = (f_1, f_2)$ is the ideal generated by f_1 and f_2 then the leading term ideal $LT(I)$ contains $LT(f_1) = x^3 y$ and $LT(f_2) = x^2 y^2$, so $(x^3 y, x^2 y^2) \subseteq LT(I)$. Since

$$y f_1 - x f_2 = y(x^3 y - xy^2 + 1) - x(x^2 y^2 - y^3 - 1) = x + y$$

we see that $g = x + y$ is an element of I and so the ideal $LT(I)$ also contains the leading term $LT(g) = x$. This shows that $LT(I)$ is strictly larger than $(LT(f_1), LT(f_2))$, since every element in $(LT(f_1), LT(f_2)) = (x^3 y, x^2 y^2)$ has total degree at least 4. We shall see later that in this case $LT(I) = (x, y^4)$.

(2) With respect to the lexicographic ordering $y > x$, the leading terms of f_1 and f_2 in the previous example are $LT(f_1) = -xy^2$ (which one could write as $-y^2x$ to emphasize the chosen ordering) and $LT(f_2) = -y^3$. We shall see later that in this ordering $LT(I) = (x^4, y)$, which is a different ideal than the ideal $LT(I)$ obtained in the previous example using the ordering $x > y$, and is again strictly larger than $(LT(f_1), LT(f_2))$.

(3) Choose any ordering on $F[x, y]$ and let $f = f(x, y)$ be any nonzero polynomial. The leading term of every element of the principal ideal $I = (f)$ is then a multiple of the leading term of f, so in this case $LT(I) = (LT(f))$.

In the case of one variable, leading terms are used in the Division Algorithm to reduce one polynomial g modulo another polynomial f to get a unique remainder r, and this remainder is 0 if and only if g is contained in the ideal (f). Since $F[x_1, x_2, \ldots, x_n]$ is not a Euclidean Domain if $n \geq 2$ (since it is not a P.I.D.), the situation is more complicated for polynomials in more than one variable. In the first example above, neither f_1 nor f_2 divides g in $F[x, y]$ (by degree considerations, for example), so attempting to first divide g by one of f_1 or f_2 and then by the other to try to reduce g modulo the ideal I would produce a (nonzero) "remainder" of g itself. In particular, this would suggest that $g = yf_1 - xf_2$ is not an element of the ideal I even though it is. The reason the polynomial g of degree 1 can be a linear combination of the two polynomials f_1 and f_2 of degree 4 is that the leading terms in yf_1 and xf_2 cancel in the difference, and this is reflected in the fact that $LT(f_1)$ and $LT(f_2)$ are not sufficient to generate $LT(I)$. A set of generators for an ideal I in $F[x_1, \ldots, x_n]$ whose leading terms generate the leading terms of *all* the elements in I is given a special name.

Definition. A *Gröbner basis* for an ideal I in the polynomial ring $F[x_1, \ldots, x_n]$ is a finite set of generators $\{g_1, \ldots, g_m\}$ for I whose leading terms generate the ideal of all leading terms in I, i.e.,

$$I = (g_1, \ldots, g_m) \quad \text{and} \quad LT(I) = (LT(g_1), \ldots, LT(g_m)).$$

Remark: Note that a Gröbner "basis" is in fact a set of *generators* for I (that depends on the choice of ordering), i.e., every element in I is a linear combination of the generators, and not a basis in the sense of vector spaces (where the linear combination would be *unique*, cf. Sections 10.3 and 11.1). Although potentially misleading, the terminology "Gröbner basis" has been so widely adopted that it would be hazardous to introduce a different nomenclature.

One of the most important properties of a Gröbner basis (proved in Theorem 23 following) is that every polynomial g can be written *uniquely* as the sum of an element in I and a remainder r obtained by a general polynomial division. In particular, we shall see that g is an element of I if and only if this remainder r is 0. While there is a similar decomposition in general, we shall see that if we do not use a Gröbner basis the uniqueness is lost (and we cannot detect membership in I by checking whether the remainder is 0) because there are leading terms not accounted for by the leading terms of the generators.

We first use the leading terms of polynomials defined by a monomial ordering on $F[x_1, \ldots, x_n]$ to extend the one variable Division Algorithm to a noncanonical polynomial division in several variables. Recall that for polynomials in one variable, the usual Division Algorithm determines the quotient $q(x)$ and remainder $r(x)$ in the equation $f(x) = q(x)g(x) + r(x)$ by successively testing whether the leading term of the dividend $f(x)$ is divisible by the leading term of $g(x)$: if $LT(f) = a(x)LT(g)$, the monomial term $a(x)$ is added to the quotient and the process is iterated with $f(x)$ replaced by the dividend $f(x) - a(x)g(x)$, which is of smaller degree since the leading terms cancel (by the choice of $a(x)$). The process terminates when the leading term of the divisor $g(x)$ no longer divides the leading term of the dividend, leaving the remainder $r(x)$. We can extend this to division by a finite number of polynomials in several variables simply by allowing successive divisions, resulting in a remainder and several quotients, as follows.

General Polynomial Division

Fix a monomial ordering on $F[x_1, \ldots, x_n]$, and suppose $g_1, \ldots, g_m$ is a set of nonzero polynomials in $F[x_1, \ldots, x_n]$. If f is any polynomial in $F[x_1, \ldots, x_n]$, start with a set of quotients $q_1, \ldots, q_m$ and a remainder r initially all equal to 0 and successively test whether the leading term of the dividend f is divisible by the leading terms of the divisors $g_1, \ldots, g_m$, in that order. Then

i. If $LT(f)$ is divisible by $LT(g_i)$, say, $LT(f) = a_i LT(g_i)$, add a_i to the quotient q_i, replace f by the dividend $f - a_i g_i$ (a polynomial with lower order leading term), and reiterate the entire process.

ii. If the leading term of the dividend f is not divisible by any of the leading terms $LT(g_1), \ldots, LT(g_m)$, add the leading term of f to the remainder r, replace f by the dividend $f - LT(f)$ (i.e., remove the leading term of f), and reiterate the entire process.

The process terminates (cf. Exercise 3) when the dividend is 0 and results in a set of quotients $q_1, \ldots, q_m$ and a remainder r with

$$f = q_1 g_1 + \cdots + q_m g_m + r.$$

Each $q_i g_i$ has multidegree less than or equal to the multidegree of f and the remainder r has the property that no nonzero term in r is divisible by any of the leading terms $LT(g_1), \ldots, LT(g_m)$ (since only terms with this property are added to r in (ii)).

Examples

Fix the lexicographic ordering $x > y$ on $F[x, y]$.

(1) Suppose $f = x^3 y^3 + 3x^2 y^4$ and $g = xy^4$. The leading term of f is $x^3 y^3$, which is not divisible by (the leading term of) g, so $x^3 y^3$ is added to the remainder r (so now $r = x^3 y^3$) and f is replaced by $f - LT(f) = 3x^2 y^4$ and we start over. Since $3x^2 y^4$ is divisible by $LT(g) = xy^4$, with quotient $a = 3x$, we add $3x$ to the quotient q (so $q = 3x$), and replace $3x^2 y^4$ by $3x^2 y^4 - aLT(g) = 0$, at which point the process terminates. The result is the quotient $q = 3x$ and remainder $r = x^3 y^3$ and

$$x^3 y^3 + 3x^2 y^4 = f = qg + r = (3x)(xy^4) + x^3 y^3.$$

Note that if we had terminated at the first step because the leading term of f is not divisible by the leading term of g (which terminates the Division Algorithm for polynomials in one variable), then we would have been left with a 'remainder' of f itself, even though 'more' of f is divisible by g. This is the reason for step 2 in the division process (which is not necessary for polynomials in one variable).

(2) Let $f = x^2 + x - y^2 + y$, and suppose $g_1 = xy + 1$ and $g_2 = x + y$. In the first iteration, the leading term x^2 of f is not divisible by the leading term of g_1, but is divisible by the leading term of g_2, so the quotient q_2 is x and the dividend f is replaced by the dividend $f - xg_2 = -xy + x - y^2 + y$. In the second iteration, the leading term of $-xy + x - y^2 + y$ is divisible by $LT(g_1)$, with quotient -1, so $q_1 = -1$ and the dividend is replaced by $(-xy + x - y^2 + y) - (-1)g_1 = x - y^2 + y + 1$. In the third iteration, the leading term of $x - y^2 + y + 1$ is not divisible by the leading term of g_1, but is divisible by the leading term of g_2, with quotient 1, so 1 is added to q_2 (which is now $q_2 = x + 1$) and the dividend becomes $(x - y^2 + y + 1) - (1)(g_2) = -y^2 + 1$. The leading term is now $-y^2$, which is not divisible by either $LT(g_1) = xy$ or $LT(g_2) = x$, so $-y^2$ is added to the remainder r (which is now $-y^2$) and the dividend becomes simply 1. Finally, 1 is not divisible by either $LT(g_1)$ or $LT(g_2)$, so is added to the remainder (so r is now $-y^2 + 1$), and the process terminates. The result is

$$q_1 = -1, \qquad q_2 = x + 1, \qquad r = -y^2 + 1 \quad \text{and}$$

$$f = x^2 + x - y^2 + y = (-1)(xy + 1) + (x + 1)(x + y) + (-y^2 + 1)$$
$$= q_1 g_1 + q_2 g_2 + r.$$

(3) Let $f = x^2 + x - y^2 + y$ as in the previous example and interchange the divisors g_1 and g_2: $g_1 = x + y$ and $g_2 = xy + 1$. In this case an easy computation gives

$$q_1 = x - y + 1, \qquad q_2 = 0, \qquad r = 0 \quad \text{and}$$
$$f = x^2 + x - y^2 + y = (x - y + 1)(x + y) = q_1 g_1 + q_2 g_2 + r,$$

showing that the quotients q_i and the remainder r are in general not unique and depend on the order of the divisors $g_1, \ldots, g_m$.

The computation in Example 3 shows that the polynomial $f = x^2 + x - y^2 + y$ is an element of the ideal $I = (x + y, xy + 1)$ since the remainder obtained in this case was 0 (in fact f is just a multiple of the first generator). In Example 2, however, the same polynomial resulted in a nonzero remainder $-y^2 + 1$ when divided by $xy + 1$ and $x + y$, and it was not at all clear from that computation that f was an element of I.

The next theorem shows that if we use a Gröbner basis for the ideal I then these difficulties do not arise: we obtain a *unique* remainder, which in turn can be used to determine whether a polynomial f is an element of the ideal I.

Theorem 23. Fix a monomial ordering on $R = F[x_1, \ldots, x_n]$ and suppose $\{g_1, \ldots, g_m\}$ is a Gröbner basis for the nonzero ideal I in R. Then

(1) Every polynomial $f \in R$ can be written uniquely in the form

$$f = f_I + r$$

where $f_I \in I$ and no nonzero monomial term of the 'remainder' r is divisible by any of the leading terms $LT(g_1), \ldots, LT(g_m)$.

(2) Both f_I and r can be computed by general polynomial division by $g_1, \ldots, g_m$ and are independent of the order in which these polynomials are used in the division.

(3) The remainder r provides a unique representative for the coset of f in the quotient ring $F[x_1, \ldots, x_n]/I$. In particular, $f \in I$ if and only if $r = 0$.

Proof: Letting $f_I = \sum_{i=1}^{m} q_i g_i \in I$ in the general polynomial division of f by $g_1, \ldots, g_m$ immediately gives a decomposition $f = f_I + r$ for any generators $g_1, \ldots, g_m$. Suppose now that $\{g_1, \ldots, g_m\}$ is a Gröbner basis, and $f = f_I + r = f'_I + r'$. Then $r - r' = f'_I - f_I \in I$, so its leading term $LT(r - r')$ is an element of $LT(I)$, which is the ideal $(LT(g_1), \ldots, LT(g_m))$ since $\{g_1, \ldots, g_m\}$ is a Gröbner basis for I. Every element in this ideal is a sum of multiples of the monomial terms $LT(g_1), \ldots, LT(g_m)$, so is a sum of terms each of which is divisible by one of the $LT(g_i)$. But both r and r', hence also $r - r'$, are sums of monomial terms none of which is divisible by $LT(g_1), \ldots, LT(g_m)$, which is a contradiction unless $r - r' = 0$. It follows that $r = r'$ is unique, hence so is $f_I = f - r$, which proves (1).

We have already seen that f_I and r can be computed algorithmically by polynomial division, and the uniqueness in (1) implies that r is independent of the order in which the polynomials $g_1, \ldots, g_m$ are used in the division. Similarly $f_I = \sum_{i=1}^{m} q_i g_i$ is uniquely determined (even though the individual quotients q_i are not in general unique), which gives (2).

The first statement in (3) is immediate from the uniqueness in (1). If $r = 0$, then $f = f_I \in I$. Conversely, if $f \in I$, then $f = f + 0$ together with the uniqueness of r implies that $r = 0$, and the final statement of the theorem follows.

As previously mentioned, the importance of Theorem 23, and one of the principal uses of Gröbner bases, is the uniqueness of the representative r, which allows effective computation in the quotient ring $F[x_1, \ldots, x_n]/I$.

We next prove that a set of polynomials in an ideal whose leading terms generate all the leading terms of an ideal is in fact a set of generators for the ideal itself (and so is a Gröbner basis—in some works this is taken as the definition of a Gröbner basis), and this shows in particular that a Gröbner basis always exists.

Proposition 24. Fix a monomial ordering on $R = F[x_1, \ldots, x_n]$ and let I be a nonzero ideal in R.

(1) If $g_1, \ldots, g_m$ are any elements of I such that $LT(I) = (LT(g_1), \ldots, LT(g_m))$, then $\{g_1, \ldots, g_m\}$ is a Gröbner basis for I.

(2) The ideal I has a Gröbner basis.

Proof: Suppose $g_1, \ldots, g_m \in I$ with $LT(I) = (LT(g_1), \ldots, LT(g_m))$. We need to see that $g_1, \ldots, g_m$ generate the ideal I. If $f \in I$, use general polynomial division to write $f = \sum_{i=1}^{m} q_i g_i + r$ where no nonzero term in the remainder r is divisible by any $LT(g_i)$. Since $f \in I$, also $r \in I$, which means $LT(r)$ is in $LT(I)$. But then $LT(r)$ would be divisible by one of $LT(g_1), \ldots, LT(g_m)$, which is a contradiction unless $r = 0$. Hence $f = \sum_{i=1}^{m} q_i g_i$ and $g_1, \ldots, g_m$ generate I, so are a Gröbner basis for I, which proves (1).

For (2), note that the ideal $LT(I)$ of leading terms of any ideal I is a monomial ideal generated by all the leading terms of the polynomials in I. By Exercise 1 a finite number of those leading terms suffice to generate $LT(I)$, say $LT(I) = (LT(h_1), \ldots, LT(h_k))$ for some $h_1, \ldots, h_k \in I$. By (1), the polynomials $h_1, \ldots, h_k$ are a Gröbner basis of I, completing the proof.

Proposition 24 proves that Gröbner bases always exist. We next prove a criterion that determines whether a given set of generators of an ideal I is a Gröbner basis, which we then use to provide an algorithm to find a Gröbner basis. The basic idea is very simple: additional elements in $LT(I)$ can arise by taking linear combinations of generators that cancel leading terms, as we saw in taking $yf_1 - xf_2$ in the first example in this section. We shall see that obtaining new leading terms from generators in this simple manner is the only obstruction to a set of generators being a Gröbner basis.

In general, if f_1, f_2 are two polynomials in $F[x_1, \ldots, x_n]$ and M is the monic least common multiple of the monomial terms $LT(f_1)$ and $LT(f_2)$ then we can cancel the leading terms by taking the difference

$$S(f_1, f_2) = \frac{M}{LT(f_1)} f_1 - \frac{M}{LT(f_2)} f_2. \tag{9.1}$$

The next lemma shows that these elementary linear combinations account for all cancellation in leading terms of polynomials of the same multidegree.

Lemma 25. Suppose $f_1, \ldots, f_m \in F[x_1, \ldots, x_n]$ are polynomials with the same multidegree α and that the linear combination $h = a_1 f_1 + \cdots + a_m f_m$ with constants $a_i \in F$ has strictly smaller multidegree. Then

$$h = \sum_{i=2}^{m} b_i S(f_{i-1}, f_i), \quad \text{for some constants } b_i \in F.$$

Proof: Write $f_i = c_i f_i'$ where $c_i \in F$ and f_i' is a monic polynomial of multidegree α. We have

$$h = \sum a_i c_i f_i' = a_1 c_1 (f_1' - f_2') + (a_1 c_1 + a_2 c_2)(f_2' - f_3') + \cdots$$
$$+ (a_1 c_1 + \cdots + a_{m-1} c_{m-1})(f_{m-1}' - f_m') + (a_1 c_1 + \cdots + a_m c_m) f_m'.$$

Note that $f_{i-1}' - f_i' = S(f_{i-1}, f_i)$. Then since h and each $f_{i-1}' - f_i'$ has multidegree strictly smaller than α, we have $a_1 c_1 + \cdots + a_m c_m = 0$, so the last term on the right hand side is 0 and the lemma follows.

The next proposition shows that a set of generators $g_1, \ldots, g_m$ is a Gröbner basis if there are no new leading terms among the remainders of the differences $S(g_i, g_j)$ not already accounted for by the g_i. This result provides the principal ingredient in an algorithm to construct a Gröbner basis.

For a fixed monomial ordering on $R = F[x_1, \ldots, x_n]$ and ordered set of polynomials $G = \{g_1, \ldots, g_m\}$ in R, write $f \equiv r \mod G$ if r is the remainder obtained by general polynomial division of $f \in R$ by $g_1, \ldots, g_m$ (in that order).

Proposition 26. *(Buchberger's Criterion)* Let $R = F[x_1, \ldots, x_n]$ and fix a monomial ordering on R. If $I = (g_1, \ldots, g_m)$ is a nonzero ideal in R, then $G = \{g_1, \ldots, g_m\}$ is a Gröbner basis for I if and only if $S(g_i, g_j) \equiv 0 \bmod G$ for $1 \leq i < j \leq m$.

Proof: If $\{g_1, \ldots, g_m\}$ is a Gröbner basis for I, then $S(g_i, g_j) \equiv 0 \bmod G$ by Theorem 23 since each $S(g_i, g_j)$ is an element of I.

Suppose now that $S(g_i, g_j) \equiv 0 \bmod G$ for $1 \leq i < j \leq m$ and take any element $f \in I$. To see that G is a Gröbner basis we need to see that $(LT(g_1), \ldots, LT(g_m))$ contains $LT(f)$. Since $f \in I$, we can write $f = \sum_{i=1}^{m} h_i g_i$ for some polynomials $h_1, \ldots, h_m$. Such a representation is not unique. Among all such representations choose one for which the largest multidegree of any summand (i.e., $\max_{i=1,\ldots,m} \partial(h_i g_i)$) is minimal, say α. It is clear that the multidegree of f is no worse than the largest multidegree of all the summands $h_i g_i$, so $\partial(f) \leq \alpha$. Write

$$f = \sum_{i=1}^{m} h_i g_i = \sum_{\partial(h_i g_i) = \alpha} h_i g_i + \sum_{\partial(h_i g_i) < \alpha} h_i g_i$$

$$= \sum_{\partial(h_i g_i) = \alpha} LT(h_i) g_i + \sum_{\partial(h_i g_i) = \alpha} (h_i - LT(h_i)) g_i + \sum_{\partial(h_i g_i) < \alpha} h_i g_i. \tag{9.2}$$

Suppose that $\partial(f) < \alpha$. Then since the multidegree of the second two sums is also strictly smaller than α it follows that the multidegree of the first sum is strictly smaller than α. If $a_i \in F$ denotes the constant coefficient of the monomial term $LT(h_i)$ then $LT(h_i) = a_i h_i'$ where h_i' is a monomial. We can apply Lemma 25 to $\sum a_i(h_i' g_i)$ to write the first sum above as $\sum b_i S(h_{i-1}' g_{i-1}, h_i' g_i)$ with $\partial(h_{i-1}' g_{i-1}) = \partial(h_i' g_i) = \alpha$. Let $\beta_{i-1,i}$ be the multidegree of the monic least common multiple of $LT(g_{i-1})$ and $LT(g_i)$. Then an easy computation shows that $S(h_{i-1}' g_{i-1}, h_i' g_i)$ is just $S(g_{i-1}, g_i)$ multiplied by the monomial of multidegree $\alpha - \beta_{i-1,i}$. The polynomial $S(g_{i-1}, g_i)$ has multidegree less than $\beta_{i-1,i}$ and, by assumption, $S(g_{i-1}, g_i) \equiv 0 \bmod G$. This means that after general polynomial division of $S(g_{i-1}, g_i)$ by $g_1, \ldots, g_m$, each $S(g_{i-1}, g_i)$ can be written as a sum $\sum q_j g_j$ with $\partial(q_j g_j) < \beta_{i-1,i}$. It follows that each $S(h_{i-1}' g_{i-1}, h_i' g_i)$ is a sum $\sum q_j' g_j$ with $\partial(q_j' g_j) < \alpha$. But then all the sums on the right hand side of equation (2) can be written as a sum of terms of the form $p_i g_i$ with polynomials p_i satisfying $\partial(p_i g_i) < \alpha$. This contradicts the minimality of α and shows that in fact $\partial(f) = \alpha$, i.e., the leading term of f has multidegree α.

If we now take the terms in equation (2) of multidegree α we see that

$$LT(f) = \sum_{\partial(h_i g_i) = \alpha} LT(h_i) LT(g_i),$$

so indeed $LT(f) \in (LT(g_1), \ldots, LT(g_m))$. It follows that $G = \{g_1, \ldots, g_m\}$ is a Gröbner basis.

Buchberger's Algorithm

Buchberger's Criterion can be used to provide an algorithm to find a Gröbner basis for an ideal I, as follows. If $I = (g_1, \ldots, g_m)$ and each $S(g_i, g_j)$ leaves a remainder of 0 when divided by $G = \{g_1, \ldots, g_m\}$ using general polynomial division then G

is a Gröbner basis. Otherwise $S(g_i, g_j)$ has a nonzero remainder r. Increase G by appending the polynomial $g_{m+1} = r$: $G' = \{g_1, \ldots, g_m, g_{m+1}\}$ and begin again (note that this is again a set of generators for I since $g_{m+1} \in I$). It is not hard to check that this procedure terminates after a finite number of steps in a generating set G that satisfies Buchberger's Criterion, hence is a Gröbner basis for I (cf. Exercise 16). Note that once an $S(g_i, g_j)$ yields a remainder of 0 after division by the polynomials in G it also yields a remainder of 0 when additional polynomials are appended to G.

If $\{g_1, \ldots, g_m\}$ is a Gröbner basis for the ideal I and $LT(g_j)$ is divisible by $LT(g_i)$ for some $j \neq i$, then $LT(g_j)$ is not needed as a generator for $LT(I)$. By Proposition 24 we may therefore delete g_j and still retain a Gröbner basis for I. We may also assume without loss that the leading term of each g_i is monic. A Gröbner basis $\{g_1, \ldots, g_m\}$ for I where each $LT(g_i)$ is monic and where $LT(g_j)$ is not divisible by $LT(g_i)$ for $i \neq j$ is called a *minimal Gröbner basis*. While a minimal Gröbner basis is not unique, the number of elements and their leading terms are unique (cf. Exercise 15).

Examples

(1) Choose the lexicographic ordering $x > y$ on $F[x, y]$ and consider the ideal I generated by $f_1 = x^3y - xy^2 + 1$ and $f_2 = x^2y^2 - y^3 - 1$ as in Example 1 at the beginning of this section. To test whether $G = \{f_1, f_2\}$ is a Gröbner basis we compute $S(f_1, f_2) = yf_1 - xf_2 = x + y$, which is its own remainder when divided by $\{f_1, f_2\}$, so G is not a Gröbner basis for I. Set $f_3 = x + y$, and increase the generating set: $G' = \{f_1, f_2, f_3\}$. Now $S(f_1, f_2) \equiv 0 \mod G'$, and a brief computation yields

$$S(f_1, f_3) = f_1 - x^2yf_3 = -x^2y^2 - xy^2 + 1 \equiv 0 \mod G'$$

$$S(f_2, f_3) = f_2 - xy^2f_3 = -xy^3 - y^3 - 1 \equiv y^4 - y^3 - 1 \mod G'.$$

Let $f_4 = y^4 - y^3 - 1$ and increase the generating set to $G'' = \{f_1, f_2, f_3, f_4\}$. The previous 0 remainder is still 0, and now $S(f_2, f_3) \equiv 0 \mod G''$ by the choice of f_4. Some additional computation yields

$$S(f_1, f_4) \equiv S(f_2, f_4) \equiv S(f_3, f_4) \equiv 0 \mod G''$$

and so $\{x^3y - xy^2 + 1, x^2y^2 - y^3 - 1, x + y, y^4 - y^3 - 1\}$ is a Gröbner basis for I. In particular, $LT(I)$ is generated by the leading terms of these four polynomials, so $LT(I) = (x^3y, x^2y^2, x, y^4) = (x, y^4)$, as previously mentioned. Then $x + y$ and $y^4 - y^3 - 1$ in I have leading terms generating $LT(I)$, so by Proposition 24, $\{x + y, y^4 - y^3 - 1\}$ gives a minimal Gröbner basis for I:

$$I = (x + y, y^4 - y^3 - 1).$$

This description of I is much simpler than $I = (x^3y - xy^2 + 1, x^2y^2 - y^3 - 1)$.

(2) Choose the lexicographic ordering $y > x$ on $F[x, y]$ and consider the ideal I in the previous example. In this case, $S(f_1, f_2)$ produces a remainder of $f_3 = -x - y$; then $S(f_1, f_3)$ produces a remainder of $f_4 = -x^4 - x^3 + 1$, and then all remainders are 0 with respect to the Gröbner basis $\{x^3y - xy^2 + 1, x^2y^2 - y^3 - 1, -x - y, -x^4 - x^3 + 1\}$. Here $LT(I) = (-xy^2, -y^3, -y, -x^4) = (y, x^4)$, as previously mentioned, and $\{x + y, x^4 + x^3 - 1\}$ gives a minimal Gröbner basis for I with respect to this ordering:

$$I = (x + y, x^4 + x^3 - 1),$$

a different simpler description of I.

In Example 1 above it is easy to check that $\{x+y^4-y^3+y-1,\ y^4-y^3-1\}$ is again a minimal Gröbner basis for I (this is just $\{f_3+f_4,\ f_4\}$), so even with a fixed monomial ordering on $F[x_1,\ldots,x_n]$ a minimal Gröbner basis for an ideal I is not unique. We can obtain an important uniqueness property by strengthening the condition on divisibility by the leading terms of the basis.

Definition. Fix a monomial ordering on $R = F[x_1,\ldots,x_n]$. A Gröbner basis $\{g_1,\ldots,g_m\}$ for the nonzero ideal I in R is called a *reduced Gröbner basis* if
    (a) each g_i has monic leading term, i.e., $LT(g_i)$ is monic, $i=1,\ldots,m$, and
    (b) no term in g_j is divisible by $LT(g_i)$ for $j \neq i$.

Note that a reduced Gröbner basis is, in particular, a minimal Gröbner basis. If $G = \{g_1,\ldots,g_m\}$ is a minimal Gröbner basis for I, then the leading term $LT(g_j)$ is not divisible by $LT(g_i)$ for any $i \neq j$. As a result, if we use polynomial division to divide g_j by the other polynomials in G we obtain a remainder g'_j in the ideal I with the same leading term as g_j (the remainder g'_j does not depend on the order of the polynomials used in the division by (2) of Theorem 23). By Proposition 24, replacing g_j by g'_j in G again gives a minimal Gröbner basis for I, and in this basis no term of g'_j is divisible by $LT(g_i)$ for any $i \neq j$. Replacing each element in G by its remainder after division by the other elements in G therefore results in a reduced Gröbner basis for I. The importance of reduced Gröbner bases is that they are unique (for a given monomial ordering), as the next result shows.

Theorem 27. Fix a monomial ordering on $R = F[x_1,\ldots,x_n]$. Then there is a unique reduced Gröbner basis for every nonzero ideal I in R.

Proof: By Exercise 15, two reduced bases have the same number of elements and the same leading terms since reduced bases are also minimal bases. If $G = \{g_1,\ldots,g_m\}$ and $G' = \{g'_1,\ldots,g'_m\}$ are two reduced bases for the same nonzero ideal I, then after a possible rearrangement we may assume $LT(g_i) = LT(g'_i) = h_i$ for $i = 1,\ldots,m$. For any fixed i, consider the polynomial $f_i = g_i - g'_i$. If f_i is nonzero, then since $f_i \in I$, its leading term must be divisible by some h_j. By definition of a reduced basis, h_j for $j \neq i$ does not divide any of the terms in either g_i or g'_i, hence does not divide $LT(f_i)$. But h_i also does not divide $LT(f_i)$ since all the terms in f_i have strictly smaller multidegree. This forces $f_i = 0$, i.e., $g_i = g'_i$ for every i, so $G = G'$.

One application of the uniqueness of the reduced Gröbner basis is a computational method to determine when two ideals in a polynomial ring are equal.

Corollary 28. Let I and J be two ideals in $F[x_1,\ldots,x_n]$. Then $I = J$ if and only if I and J have the same reduced Gröbner basis with respect to any fixed monomial ordering on $F[x_1,\ldots,x_n]$.

Examples

    (1) Consider the ideal $I = (h_1, h_2, h_3)$ with $h_1 = x^2+xy^5+y^4$, $h_2 = xy^6-xy^3+y^5-y^2$, and $h_3 = xy^5-xy^2$ in $F[x,y]$. Using the lexicographic ordering $x > y$ we find

$S(h_1, h_2) \equiv S(h_1, h_3) \equiv 0 \bmod \{h_1, h_2, h_3\}$ and $S(h_2, h_3) \equiv y^5 - y^2 \bmod \{h_1, h_2, h_3\}$. Setting $h_4 = y^5 - y^2$ we find $S(h_i, h_j) \equiv 0 \bmod \{h_1, h_2, h_3, h_4\}$ for $1 \le i < j \le 4$, so

$$x^2 + xy^5 + y^4, \quad xy^6 - xy^3 + y^5 - y^2, \quad xy^5 - xy^2, \quad y^5 - y^2$$

is a Gröbner basis for I. The leading terms of this basis are x^2, xy^6, xy^5, y^5. Since y^5 divides both xy^6 and xy^5, we may remove the second and third generators to obtain a minimal Gröbner basis $\{x^2 + xy^5 + y^4, y^5 - y^2\}$ for I. The second term in the first generator is divisible by the leading term y^5 of the second generator, so this is not a reduced Gröbner basis. Replacing $x^2 + xy^5 + y^4$ by its remainder $x^2 + xy^2 + y^4$ after division by the other polynomials in the basis (which in this case is only the polynomial $y^5 - y^2$), we are left with the reduced Gröbner basis $\{x^2 + xy^2 + y^4, y^5 - y^2\}$ for I.

(2) Consider the ideal $J = (h_1, h_2, h_3)$ with $h_1 = xy^3 + y^3 + 1$, $h_2 = x^3 y - x^3 + 1$, and $h_3 = x + y$ in $F[x, y]$. Using the lexicographic monomial ordering $x > y$ we find $S(h_1, h_2) \equiv 0 \bmod \{h_1, h_2, h_3\}$ and $S(h_1, h_3) \equiv y^4 - y^3 - 1 \bmod \{h_1, h_2, h_3\}$. Setting $h_4 = y^4 - y^3 - 1$ we find $S(h_i, h_j) \equiv 0 \bmod \{h_1, h_2, h_3, h_4\}$ for $1 \le i < j \le 4$, so

$$xy^3 + y^3 + 1, \quad x^3 y - x^3 + 1, \quad x + y, \quad y^4 - y^3 - 1$$

is a Gröbner basis for J. The leading terms of this basis are $xy^3, x^3 y, x$, and y^4, so $\{x + y, y^4 - y^3 - 1\}$ is a minimal Gröbner basis for J. In this case none of the terms in $y^4 - y^3 - 1$ are divisible by the leading term of $x + y$ and none of the terms in $x + y$ are divisible by the leading term in $y^4 - y^3 - 1$, so $\{x + y, y^4 - y^3 - 1\}$ is the reduced Gröbner basis for J. This is the basis for the ideal I in Example 1 following Proposition 26, so these two ideals are equal:

$$(x^3 y - xy^2 + 1, x^2 y^2 - y^3 - 1) = (xy^3 + y^3 + 1, x^3 y - x^3 + 1, x + y)$$

(and both are equal to the ideal $(x + y, y^4 - y^3 - 1)$).

Gröbner Bases and Solving Algebraic Equations: Elimination

The theory of Gröbner bases is very useful in explicitly solving systems of algebraic equations, and is the basis by which computer algebra programs attempt to solve systems of equations. Suppose $S = \{f_1, \ldots, f_m\}$ is a collection of polynomials in n variables $x_1, \ldots, x_n$ and we are trying to find the solutions of the system of equations $f_1 = 0$, $f_2 = 0, \ldots, f_m = 0$ (i.e., the common set of zeros of the polynomials in S). If $(a_1, \ldots, a_n)$ is any solution to this system, then every element f of the ideal I generated by S also satisfies $f(a_1, \ldots, a_n) = 0$. Furthermore, it is an easy exercise to see that if $S' = \{g_1, \ldots, g_s\}$ is *any* set of generators for the ideal I then the set of solutions to the system $g_1 = 0, \ldots, g_s = 0$ is the *same* as the original solution set.

In the situation where $f_1, \ldots, f_m$ are *linear* polynomials, a solution to the system of equations can be obtained by successively eliminating the variables $x_1, x_2, \ldots$ by elementary means—using linear combinations of the original equations to eliminate the variable x_1, then using these equations to eliminate x_2, etc., producing a system of equations that can be easily solved (this is "Gauss-Jordan elimination" in linear algebra, cf. the exercises in Section 11.2).

The situation for polynomial equations that are nonlinear is naturally more complicated, but the basic principle is the same. If there is a nonzero polynomial in the ideal I involving only one of the variables, say $p(x_n)$, then the last coordinate a_n is a solution of $p(x_n) = 0$. If now there is a polynomial in I involving only x_{n-1} and x_n, say $q(x_{n-1}, x_n)$, then the coordinate a_{n-1} would be a solution of $q(x_{n-1}, a_n) = 0$, etc. If we can successively find polynomials in I that eliminate the variables $x_1, x_2, \ldots$ then we will be able to determine all the solutions $(a_1, \ldots, a_n)$ to our original system of equations explicitly.

Finding equations that follow from the system of equations in S, i.e., finding elements of the ideal I that do not involve some of the variables, is referred to as *elimination theory*. The polynomials in I that do not involve the variables $x_1, \ldots, x_i$, i.e., $I \cap F[x_{i+1}, \ldots, x_n]$, is easily seen to be an ideal in $F[x_{i+1}, \ldots, x_n]$ and is given a name.

Definition. If I is an ideal in $F[x_1, \ldots, x_n]$ then $I_i = I \cap F[x_{i+1}, \ldots, x_n]$ is called the ith *elimination ideal* of I with respect to the ordering $x_1 > \cdots > x_n$.

The success of using elimination to solve a system of equations depends on being able to determine the elimination ideals (and, ultimately, on whether these elimination ideals are nonzero).

The following fundamental proposition shows that if the lexicographic monomial ordering $x_1 > \cdots > x_n$ is used to compute a Gröbner basis for I then the elements in the resulting basis not involving the variables $x_1, ..., x_i$ not only determine the ith elimination ideal, but in fact give a Gröbner basis for the ith elimination ideal of I.

Proposition 29. (*Elimination*) Suppose $G = \{g_1, \ldots, g_m\}$ is a Gröbner basis for the nonzero ideal I in $F[x_1, \ldots, x_n]$ with respect to the lexicographic monomial ordering $x_1 > \cdots > x_n$. Then $G \cap F[x_{i+1}, \ldots, x_n]$ is a Gröbner basis of the ith elimination ideal $I_i = I \cap F[x_{i+1}, \ldots, x_n]$ of I. In particular, $I \cap F[x_{i+1}, \ldots, x_n] = 0$ if and only if $G \cap F[x_{i+1}, \ldots, x_n] = \emptyset$.

Proof: Denote $G_i = G \cap F[x_{i+1}, \ldots, x_n]$. Then $G_i \subseteq I_i$, so by Proposition 24, to see that G_i is a Gröbner basis of I_i it suffices to see that $LT(G_i)$, the leading terms of the elements in G_i, generate $LT(I_i)$ as an ideal in $F[x_{i+1}, \ldots, x_n]$. Certainly $(LT(G_i)) \subseteq LT(I_i)$ as ideals in $F[x_{i+1}, \ldots, x_n]$. To show the reverse containment, let f be any element in I_i. Then $f \in I$ and since G is a Gröbner basis for I we have

$$LT(f) = a_1(x_1, \ldots, x_n)LT(g_1) + \cdots + a_m(x_1, \ldots, x_n)LT(g_m)$$

for some polynomials $a_1, \ldots, a_m \in F[x_1, \ldots, x_n]$. Writing each polynomial a_i as a sum of monomial terms we see that $LT(f)$ is a sum of monomial terms of the form $ax_1^{s_1} \ldots x_n^{s_n} LT(g_i)$. Since $LT(f)$ involves only the variables $x_{i+1}, \ldots, x_n$, the sum of all such terms containing any of the variables $x_1, \ldots, x_i$ must be 0, so $LT(f)$ is also the sum of those monomial terms only involving $x_{i+1}, \ldots, x_n$. It follows that $LT(f)$ can be written as a $F[x_{i+1}, \ldots, x_n]$-linear combination of some monomial terms $LT(g_t)$ where $LT(g_t)$ does not involve the variables $x_1, \ldots, x_i$. But by the choice of the ordering, if $LT(g_t)$ does not involve $x_1, \ldots, x_i$, then neither do any of the other terms in g_t, i.e., $g_t \in G_i$. Hence $LT(f)$ can be written as a $F[x_{i+1}, \ldots, x_n]$-linear combination of elements $LT(G_i)$, completing the proof.

Note also that Gröbner bases can be used to eliminate any variables simply by using an appropriate monomial ordering.

Examples

(1) The ellipse $2x^2 + 2xy + y^2 - 2x - 2y = 0$ intersects the circle $x^2 + y^2 = 1$ in two points. To find them we compute a Gröbner basis for the ideal $I = (2x^2 + 2xy + y^2 - 2x - 2y, x^2 + y^2 - 1) \subset \mathbb{R}[x, y]$ using the lexicographic monomial order $x > y$ to eliminate x, obtaining $g_1 = 2x + y^2 + 5y^3 - 2$ and $g_2 = 5y^4 - 4y^3$. Hence $5y^4 = 4y^3$ and $y = 0$ or $y = 4/5$. Substituting these values into $g_1 = 0$ and solving for x we find the two intersection points are $(1, 0)$ and $(-3/5, 4/5)$.

Instead using the lexicographic monomial order $y > x$ to eliminate y results in the Gröbner basis $\{y^2 + x^2 - 1, 2yx - 2y + x^2 - 2x + 1, 5x^3 - 7x^2 - x + 3\}$. Then $5x^3 - 7x^2 - x + 3 = (x - 1)^2(5x + 3)$ shows that x is 1 or $-3/5$ and we obtain the same solutions as before, although with more effort.

(2) In the previous example the solutions could also have been found by elementary means. Consider now the solutions in $\mathbb{C}$ to the system of two equations

$$x^3 - 2xy + y^3 = 0 \quad \text{and} \quad x^5 - 2x^2y^2 + y^5 = 0.$$

Computing a Gröbner basis for the ideal generated by $f_1 = x^3 - 2xy + y^3$ and $f_2 = x^5 - 2x^2y^2 + y^5$ with respect to the lexicographic monomial order $x > y$ we obtain the basis

$$g_1 = x^3 - 2xy + y^3$$
$$g_2 = 200xy^2 + 193y^9 + 158y^8 - 45y^7 - 456y^6 + 50y^5 - 100y^4$$
$$g_3 = y^{10} - y^8 - 2y^7 + 2y^6.$$

Any solution to our original equations would satisfy $g_1 = g_2 = g_3 = 0$. Since $g_3 = y^6(y - 1)^2(y^2 + 2y + 2)$, we have $y = 0$, $y = 1$ or $y = -1 \pm i$. Since $g_1(x, 0) = x^3$ and $g_2(x, 0) = 0$, we see that $(0, 0)$ is the only solution with $y = 0$. Since $g_1(x, 1) = x^3 - 2x + 1$ and $g_2(x, 1) = 200(x - 1)$ have only $x = 1$ as a common zero, the only solution with $y = 1$ is $(1, 1)$. Finally,

$$g_1(x, -1 \pm i) = x^3 + (2 \mp 2i)x + (2 \pm 2i)$$
$$g_2(x, -1 \pm i) = -400i(x + 1 \pm i),$$

and a quick check shows the common zero $x = -1 \mp i$ when $y = -1 \pm i$, respectively. Hence, there are precisely four solutions to the original pair of equations, namely

$$(x, y) = (0, 0), \quad (1, 1), \quad (-1 + i, -1 - i), \quad \text{or} \quad (-1 - i, -1 + i).$$

(3) Consider the solutions in $\mathbb{C}$ to the system of equations

$$x + y + z = 1$$
$$x^2 + y^2 + z^2 = 2$$
$$x^3 + y^3 + z^3 = 3.$$

The reduced Gröbner basis with respect to the lexicographic ordering $x > y > z$ is

$$\{x + y + z - 1, \ y^2 + yz - y + z^2 - z - (1/2), \ z^3 - z^2 - (1/2)z - (1/6)\}$$

and so z is a root of the polynomial $t^3 - t^2 - (1/2)t - (1/6)$ (by symmetry, also x and y are roots of this same polynomial). For each of the three roots of this polynomial, there are two values of y and one corresponding value of x making the first two polynomials in the Gröbner basis equal to 0. The resulting six solutions are quickly checked to be the three distinct roots of the polynomial $t^3 - t^2 - (1/2)t - (1/6)$ (which is irreducible over $\mathbb{Q}$) in some order.

As the previous examples show, the study of solutions to systems of polynomial equations $f_1 = 0$, $f_2 = 0, \ldots, f_m = 0$ is intimately related to the study of the ideal $I = (f_1, f_2, \ldots, f_m)$ the polynomials generate in $F[x_1, \ldots, x_n]$. This fundamental connection is the starting point for the important and active branch of mathematics called "algebraic geometry", introduced in Chapter 15, where additional applications of Gröbner bases are given.

We close this section by showing how to compute the basic set-theoretic operations of sums, products and intersections of ideals in polynomial rings. Suppose $I = (f_1, \ldots, f_s)$ and $J = (h_1, \ldots, h_t)$ are two ideals in $F[x_1, \ldots, x_n]$. Then $I + J = (f_1, \ldots, f_s, h_1, \ldots, h_t)$ and $IJ = (f_1 h_1, \ldots, f_i h_j, \ldots, f_s h_t)$. The following proposition shows how to compute the intersection of any two ideals.

Proposition 30. If I and J are any two ideals in $F[x_1, \ldots, x_n]$ then $tI + (1 - t)J$ is an ideal in $F[t, x_1, \ldots, x_n]$ and $I \cap J = (tI + (1 - t)J) \cap F[x_1, \ldots, x_n]$. In particular, $I \cap J$ is the first elimination ideal of $tI + (1 - t)J$ with respect to the ordering $t > x_1 > \cdots > x_n$.

Proof: First, tI and $(1-t)J$ are clearly ideals in $F[x_1, \ldots, x_n, t]$, so also their sum $tI + (1-t)J$ is an ideal in $F[x_1, \ldots, x_n, t]$. If $f \in I \cap J$, then $f = tf + (1-t)f$ shows $I \cap J \subseteq (tI + (1 - t)J) \cap F[x_1, \ldots, x_n]$. Conversely, suppose $f = tf_1 + (1 - t)f_2$ is an element of $F[x_1, \ldots, x_n]$, where $f_1 \in I$ and $f_2 \in J$. Then $t(f_1 - f_2) = f - f_2 \in F[x_1, \ldots, x_n]$ shows that $f_1 - f_2 = 0$ and $f = f_2$, so $f = f_1 = f_2 \in I \cap J$. Since $I \cap J = (tI + (1 - t)J) \cap F[x_1, \ldots, x_n]$, $I \cap J$ is the first elimination ideal of $tI + (1 - t)J$ with respect to the ordering $t > x_1 > \cdots > x_n$.

We have $tI + (1-t)J = (tf_1, \ldots, tf_s, (1-t)h_1, \ldots, (1-t)h_t)$ if $I = (f_1, \ldots, f_s)$ and $J = (h_1, \ldots, h_t)$. By Proposition 29, the elements not involving t in a Gröbner basis for this ideal in $F[t, x_1, \ldots, x_n]$, computed for the lexicographic monomial ordering $t > x_1 > \cdots > x_n$, give a Gröbner basis for the ideal $I \cap J$ in $F[x_1, \ldots, x_n]$.

Example

Let $I = (x, y)^2 = (x^2, xy, y^2)$ and let $J = (x)$. For the lexicographic monomial ordering $t > x > y$ the reduced Gröbner basis for $tI + (1-t)J$ in $F[t, x, y]$ is $\{tx - x, ty^2, x^2, xy\}$ and so $I \cap J = (x^2, xy)$.

EXERCISES

1. Suppose I is an ideal in $F[x_1, \ldots, x_n]$ generated by a (possibly infinite) set S of polynomials. Prove that a finite subset of the polynomials in S suffice to generate I. [Use Theorem 21 to write $I = (f_1, \ldots, f_m)$ and then write each $f_i \in I$ using polynomials in S.]

2. Let $\geq$ be any monomial ordering.
   (a) Prove that $LT(fg) = LT(f)LT(g)$ and $\partial(fg) = \partial(f) + \partial(g)$ for any nonzero polynomials f and g.
   (b) Prove that $\partial(f + g) \leq \max(\partial(f), \partial(g))$ with equality if $\partial(f) \neq \partial(g)$.

(c) Prove that $m \geq 1$ for every monomial m.

(d) Prove that if m_1 divides m_2 then $m_2 \geq m_1$. Deduce that the leading term of a polynomial does not divide any of its lower order terms.

3. Prove that if $\geq$ is any total or partial ordering on a nonempty set then the following are equivalent:

(i) Every nonempty subset contains a minimum element.

(ii) There is no infinite strictly decreasing sequence $a_1 > a_2 > a_3 > \cdots$ (this is called the *descending chain condition* or *D.C.C.*).

Deduce that General Polynomial Division always terminates in finitely many steps.

4. Let $\geq$ be a monomial ordering, and for monomials m_1, m_2 define $m_1 \geq_g m_2$ if either $\deg m_1 > \deg m_2$, or $\deg m_1 = \deg m_2$ and $m_1 \geq m_2$.

(a) Prove that $\geq_g$ is also a monomial ordering. (The relation $\geq_g$ is called the *grading* of $\geq$. An ordering in which the most important criterion for comparison is degree is sometimes called a *graded* or a *degree* ordering, so this exercise gives a method for constructing graded orderings.)

(b) The grading of the lexicographic ordering $x_1 > \cdots > x_n$ is called the *grlex* monomial ordering. Show that $x_2^4 > x_1^2 x_2 > x_1 x_2^2 > x_2^2 > x_1$ with respect to the grlex ordering and $x_1^2 x_2 > x_1 x_2^2 > x_1 > x_2^4 > x_2^2$ with respect to the lexicographic ordering.

5. The *grevlex* monomial ordering is defined by first choosing an ordering of the variables $\{x_1, x_2, \ldots, x_n\}$, then defining $m_1 \geq m_2$ for monomials m_1, m_2 if either $\deg m_1 > \deg m_2$ or $\deg m_1 = \deg m_2$ and the first exponent of $x_n, x_{n-1}, \ldots, x_1$ (in that order) where m_1 and m_2 differ is *smaller* in m_1.

(a) Prove that grevlex is a monomial ordering that satisfies $x_1 > x_2 > \cdots > x_n$.

(b) Prove that the grevlex ordering on $F[x_1, x_2]$ with respect to $\{x_1, x_2\}$ is the graded lexicographic ordering with $x_1 > x_2$, but that the grevlex ordering on $F[x_1, x_2, x_3]$ is not the grading of any lexicographic ordering.

(c) Show that $x_1 x_2^2 x_3 > x_1^2 x_3^2 > x_2^2 x_3^2 > x_2 x_3^2 > x_1 x_2 > x_2^2 > x_1 x_3 > x_3^2 > x_1 > x_2$ for the grevlex monomial ordering with respect to $\{x_1, x_2, x_3\}$.

6. Show that $x^3 y > x^3 z^2 > x^3 z > x^2 y^2 z > x^2 y > x z^2 > y^2 z^2 > y^2 z$ with respect to the lexicographic monomial ordering $x > y > z$. Show that for the corresponding grlex monomial ordering $x^3 z^2 > x^2 y^2 z > x^3 y > x^3 z > y^2 z^2 > x^2 y > x z^2 > y^2 z$, and that $x^2 y^2 z > x^3 z^2 > x^3 y > x^3 z > y^2 z^2 > x^2 y > y^2 z > x z^2$ for the grevlex monomial ordering with respect to $\{x, y, z\}$.

7. Order the monomials $x^2 z, x^2 y^2 z, x y^2 z, x^3 y, x^3 z^2, x^2, x^2 y z^2, x^2 z^2$ for the lexicographic monomial ordering $x > y > z$, for the corresponding grlex monomial order, and for the grevlex monomial ordering with respect to $\{x, y, z\}$.

8. Show there are $n!$ distinct lexicographic monomial orderings on $F[x_1, \ldots, x_n]$. Show similarly that there are $n!$ distinct grlex and grevlex monomial orderings.

9. It can be shown that any monomial ordering on $F[x_1, \ldots, x_n]$ may be obtained as follows. For $k \leq n$ let $v_1, v_2, \ldots, v_k$ be nonzero vectors in Euclidean n-space, $\mathbb{R}^n$, that are pairwise orthogonal: $v_i \cdot v_j = 0$ for all $i \neq j$, where $\cdot$ is the usual dot product, and suppose also that all the coordinates of v_1 are nonnegative. Define an order, $\geq$, on monomials by $m_1 > m_2$ if and only if for some $t \leq k$ we have $v_i \cdot \partial(m_1) = v_i \cdot \partial(m_2)$ for all $i \in \{1, 2, \ldots, t-1\}$ and $v_t \cdot \partial(m_1) > v_t \cdot \partial(m_2)$.

(a) Let $k = n$ and let $v_i = (0, \ldots, 0, 1, 0, \ldots, 0)$ with 1 in the ith position. Show that $\geq$ defines the lexicographic order with $x_1 > x_2 > \cdots > x_n$.

(b) Let $k = n$ and define $v_1 = (1, 1, \ldots, 1)$ and $v_i = (1, 1, \ldots, 1, -n+i-1, 0, \ldots, 0)$, where there are $i - 2$ trailing zeros, $2 \leq i \leq n$. Show that $\geq$ defines the grlex order with respect to $\{x_1, \ldots, x_n\}$.

10. Suppose I is a monomial ideal generated by monomials $m_1, \ldots, m_k$. Prove that the polynomial $f \in F[x_1, \ldots, x_n]$ is in I if and only if every monomial term f_i of f is a multiple of one of the m_j. [For polynomials $a_1, \ldots, a_k \in F[x_1, \ldots, x_n]$ expand the polynomial $a_1 m_1 + \cdots + a_k m_k$ and note that every monomial term is a multiple of at least one of the m_j.] Show that $x^2 yz + 3xy^2$ is an element of the ideal $I = (xyz, y^2) \subset F[x, y, z]$ but is not an element of the ideal $I' = (xz^2, y^2)$.

11. Fix a monomial ordering on $R = F[x_1, \ldots, x_n]$ and suppose $\{g_1, \ldots, g_m\}$ is a Gröbner basis for the ideal I in R. Prove that $h \in LT(I)$ if and only if h is a sum of monomial terms each divisible by some $LT(g_i)$, $1 \leq i \leq m$. [Use the previous exercise.]

12. Suppose I is a monomial ideal with monomial generators $g_1, \ldots, g_m$. Use the previous exercise to prove directly that $\{g_1, \ldots, g_m\}$ is a Gröbner basis for I.

13. Suppose I is a monomial ideal with monomial generators $g_1, \ldots, g_m$. Use Buchberger's Criterion to prove that $\{g_1, \ldots, g_m\}$ is a Gröbner basis for I.

14. Suppose I is a monomial ideal in $R = F[x_1, \ldots, x_n]$ and suppose $\{m_1, \ldots, m_k\}$ is a minimal set of monomials generating I, i.e., each m_i is a monomial and no proper subset of $\{m_1, \ldots, m_k\}$ generates I. Prove that the m_i, $1 \leq i \leq k$ are unique. [Use Exercise 10.]

15. Fix a monomial ordering on $R = F[x_1, \ldots, x_n]$.
    (a) Prove that $\{g_1, \ldots, g_m\}$ is a minimal Gröbner basis for the ideal I in R if and only if $\{LT(g_1), \ldots, LT(g_m)\}$ is a minimal generating set for $LT(I)$.
    (b) Prove that the leading terms of a minimal Gröbner basis for I are uniquely determined and the number of elements in any two minimal Gröbner bases for I is the same. [Use (a) and the previous exercise.]

16. Fix a monomial ordering on $F[x_1, \ldots, x_n]$ and suppose $G = \{g_1, \ldots, g_m\}$ is a set of generators for the nonzero ideal I. Show that if $S(g_i, g_j) \not\equiv 0 \bmod G$ then the ideal $(LT(g_1), \ldots, LT(g_m), LT(r))$ is strictly larger than the ideal $(LT(g_1), \ldots, LT(g_m))$ where $S(g_i, g_j) \equiv r \bmod G$. Deduce that the algorithm for computing a Gröbner basis described following Proposition 26 terminates after a finite number of steps. [Use Exercise 1.]

17. Fix the lexicographic ordering $x > y$ on $F[x, y]$. Use Buchberger's Criterion to show that $\{x^2 y - y^2, x^3 - xy\}$ is a Gröbner basis for the ideal $I = (x^2 y - y^2, x^3 - xy)$.

18. Show $\{x - y^3, y^5 - y^6\}$ is the reduced Gröbner basis for the ideal $I = (x - y^3, -x^2 + xy^2)$ with respect to the lexicographic ordering defined by $x > y$ in $F[x, y]$.

19. Fix the lexicographic ordering $x > y$ on $F[x, y]$.
    (a) Show that $\{x^3 - y, x^2 y - y^2, xy^2 - y^2, y^3 - y^2\}$ is the reduced Gröbner basis for the ideal $I = (-x^3 + y, x^2 y - y^2)$.
    (b) Determine whether the polynomial $f = x^6 - x^5 y$ is an element of the ideal I.

20. Fix the lexicographic ordering $x > y > z$ on $F[x, y, z]$. Show that $\{x^2 + xy + z, xyz + z^2, xz^2, z^3\}$ is the reduced Gröbner basis for the ideal $I = (x^2 + xy + z, xyz + z^2)$ and in particular conclude that the leading term ideal $LT(I)$ requires four generators.

21. Fix the lexicographic ordering $x > y$ on $F[x, y]$. Use Buchberger's Criterion to show that $\{x^2 y - y^2, x^3 - xy\}$ is a Gröbner basis for the ideal $I = (x^2 y - y^2, x^3 - xy)$.

22. Let $I = (x^2 - y, x^2 y - z)$ in $F[x, y, z]$.
    (a) Show that $\{x^2 - y, y^2 - z\}$ is the reduced Gröbner basis for I with respect to the lexicographic ordering defined by $x > y > z$.
    (b) Show that $\{x^2 - y, z - y^2\}$ is the reduced Gröbner basis for I with respect to the

lexicographic ordering defined by $z > x > y$ (note these are essentially the same polynomials as in (a)).

(c) Show that $\{y - x^2, z - x^4\}$ is the reduced Gröbner basis for I with respect to the lexicographic ordering defined by $z > y > x$.

23. Show that the ideals $I = (x^2y + xy^2 - 2y, x^2 + xy - x + y^2 - 2y, xy^2 - x - y + y^3)$ and $J = (x - y^2, xy - y, x^2 - y)$ in $F[x, y]$ are equal.

24. Use reduced Gröbner bases to show that the ideal $I = (x^3 - yz, yz + y)$ and the ideal $J = (x^3z + x^3, x^3 + y)$ in $F[x, y, z]$ are equal.

25. Show that the reduced Gröbner basis using the lexicographic ordering $x > y$ for the ideal $I = (x^2 + xy^2, x^2 - y^3, y^3 - y^2)$ is $\{x^2 - y^2, y^3 - y^2, xy^2 + y^2\}$.

26. Show that the reduced Gröbner basis for the ideal $I = (xy + y^2, x^2y + xy^2 + x^2)$ is $\{x^2, xy + y^2, y^3\}$ with respect to the lexicographic ordering $x > y$ and is $\{y^2 + yx, x^2\}$ with respect to the lexicographic ordering $y > x$.

There are generally substantial differences in computational complexity when using different monomial orders. The grevlex monomial ordering often provides the most efficient computation and produces simpler polynomials.

27. Show that $\{x^3 - y^3, x^2 + xy^2 + y^4, x^2y + xy^3 + y^2\}$ is a reduced Gröbner basis for the ideal I in the example following Corollary 28 with respect to the grlex monomial ordering. (Note that while this gives three generators for I rather than two for the lexicographic ordering as in the example, the degrees are smaller.)

28. Let $I = (x^4 - y^4 + z^3 - 1, x^3 + y^2 + z^2 - 1)$. Show that there are five elements in a reduced Gröbner basis for I with respect to the lexicographic ordering with $x > y > z$ (the maximum degree among the five generators is 12 and the maximum number of monomial terms among the five generators is 35), that there are two elements for the lexicographic ordering $y > z > x$ (maximum degree is 6 and maximum number of terms is 8), and that $\{x^3 + y^2 + z^2 - 1, xy^2 + xz^2 - x + y^4 - z^3 + 1\}$ is the reduced Gröbner basis for the grevlex monomial ordering.

29. Solve the system of equations $x^2 - yz = 3$, $y^2 - xz = 4$, $z^2 - xy = 5$ over $\mathbb{C}$.

30. Find a Gröbner basis for the ideal $I = (x^2 + xy + y^2 - 1, x^2 + 4y^2 - 4)$ for the lexicographic ordering $x > y$ and use it to find the four points of intersection of the ellipse $x^2 + xy + y^2 = 1$ with the ellipse $x^2 + 4y^2 = 4$ in $\mathbb{R}^2$.

31. Use Gröbner bases to find all six solutions to the system of equations $2x^3 + 2x^2y^2 + 3y^3 = 0$ and $3x^5 + 2x^3y^3 + 2y^5 = 0$ over $\mathbb{C}$.

32. Use Gröbner bases to show that $(x, z) \cap (y^2, x - yz) = (xy, x - yz)$ in $F[x, y, z]$.

33. Use Gröbner bases to compute the intersection of the ideals $(x^3y - xy^2 + 1, x^2y^2 - y^3 - 1)$ and $(x^2 - y^2, x^3 + y^3)$ in $F[x, y]$.

The following four exercises deal with the *ideal quotient* of two ideals I and J in a ring R.

Definition. The *ideal quotient* $(I : J)$ of two ideals I, J in a ring R is the ideal

$$(I : J) = \{r \in R \mid rJ \subseteq I\}.$$

34. (a) Suppose R is an integral domain, $0 \neq f \in R$ and I is an ideal in R. Show that if $\{g_1, \ldots, g_s\}$ are generators for the ideal $I \cap (f)$, then $\{g_1/f, \ldots, g_s/f\}$ are generators for the ideal quotient $(I : (f))$.

(b) If I is an ideal in the commutative ring R and $f_1, \ldots, f_s \in R$, show that the ideal quotient $(I : (f_1, \ldots f_s))$ is the ideal $\cap_{i=1}^{s}(I : (f_i))$.

35. If $I = (x^2y + z^3, x + y^3 - z, 2y^4z - yz^2 - z^3)$ and $J = (x^2y^5, x^3z^4, y^3z^7)$ in $\mathbb{Q}[x, y, z]$ show $(I : J)$ is the ideal $(z^2, y + z, x - z)$. [Use the previous exercise and Proposition 30.]

36. Suppose that K is an ideal in R, that I is an ideal containing K, and J is any ideal. If $\bar{I}$ and $\bar{J}$ denote the images of I and J in the quotient ring R/K, show that $\overline{(I : J)} = (\bar{I} : \bar{J})$ where $\overline{(I : J)}$ is the image in R/K of the ideal quotient $(I : J)$.

37. Let K be the ideal $(y^5 - z^4)$ in $R = \mathbb{Q}[y, z]$. For each of the following pairs of ideals I and J, use the previous two exercises together with Proposition 30 to verify the ideal quotients $(\bar{I} : \bar{J})$ in the ring R/K:

    i. $I = (y^3, y^5 - z^4)$, $J = (z)$, $(\bar{I} : \bar{J}) = (\bar{y}^3, \bar{z}^3)$.
    ii. $I = (y^3, z, y^5 - z^4)$, $J = (y)$, $(\bar{I} : \bar{J}) = (\bar{y}^2, \bar{z})$.
    iii. $I = (y, y^3, z, y^5 - z^4)$, $J = (1)$, $(\bar{I} : \bar{J}) = (\bar{y}, \bar{z})$.

Exercises 38 to 44 develop some additional elementary properties of monomial ideals in $F[x_1, \ldots, x_n]$. It follows from Hilbert's Basis Theorem that ideals are finitely generated, however one need not assume this in these exercises—the arguments are the same for finitely or infinitely generated ideals. These exercises may be used to give an independent proof of Hilbert's Basis Theorem (Exercise 44). In these exercises, M and N are monomial ideals with monomial generators $\{m_i \mid i \in I\}$ and $\{n_j \mid j \in J\}$ for some index sets I and J respectively.

38. Prove that the sum and product of two monomial ideals is a monomial ideal by showing that $M + N = (m_i, n_j \mid i \in I, j \in J)$, and $MN = (m_i n_j \mid i \in I, j \in J)$.

39. Show that if $\{M_s \mid s \in S\}$ is any nonempty collection of monomial ideals that is totally ordered under inclusion then $\cup_{s \in S} M_s$ is a monomial ideal. (In particular, the union of any increasing sequence of monomial ideals is a monomial ideal, cf. Exercise 19, Section 7.3.)

40. Prove that the intersection of two monomial ideals is a monomial ideal by showing that $M \cap N = (e_{i,j} \mid i \in I, j \in J)$, where $e_{i,j}$ is the least common multiple of m_i and n_j. [Use Exercise 10.]

41. Prove that for any monomial n, the ideal quotient $(M : (n))$ is $(m_i/d_i \mid i \in I)$, where d_i is the greatest common divisor of m_i and n (cf. Exercise 34). Show that if N is finitely generated, then the ideal quotient $(M : N)$ of two monomial ideals is a monomial ideal.

42. **(a)** Show that M is a monomial prime ideal if and only if $M = (S)$ for some subset S of $\{x_1, x_2, \ldots, x_n\}$. (In particular, there are only finitely many monomial prime ideals, and each is finitely generated.)
    (b) Show that $(x_1, \ldots, x_n)$ is the only monomial maximal ideal.

43. *(Dickson's Lemma*—a special case of Hilbert's Basis Theorem) Prove that every monomial ideal in $F[x_1, \ldots, x_n]$ is finitely generated as follows.
Let $\mathcal{S} = \{N \mid N$ is a monomial ideal that is not finitely generated$\}$, and assume by way of contradiction $\mathcal{S} \neq \emptyset$.
    (a) Show that $\mathcal{S}$ contains a maximal element M. [Use Exercise 39 and Zorn's Lemma.]
    (b) Show that there are monomials x, y not in M with $xy \in M$. [Use Exercise 42(a).]
    (c) For x as in (b), show that M contains a finitely generated monomial ideal M_0 such that $M_0 + (x) = M + (x)$ and $M = M_0 + (x)(M : (x))$, where $(M : (x))$ is the (monomial) ideal quotient (cf. Exercise 41), and $(x)(M : (x))$ is the product of these two ideals. Deduce that M is finitely generated, a contradiction which proves $\mathcal{S} = \emptyset$. [Use the maximality of M and previous exercises.]

44. If I is a nonzero ideal in $F[x_1, \ldots, x_n]$, use Dickson's Lemma to prove that $LT(I)$ is finitely generated. Conclude that I has a Gröbner basis and deduce Hilbert's Basis Theorem. [cf. Proposition 24.]

45. (*n-colorings of graphs*) A finite *graph* $\mathcal{G}$ of size N is a set of vertices $i \in \{1, 2, \ldots, N\}$ and a collection of edges (i, j) connecting vertex i with vertex j. An *n-coloring* of $\mathcal{G}$ is an assignment of one of n colors to each vertex in such a way that vertices connected by an edge have distinct colors. Let F be any field containing at least n elements. If we introduce a variable x_i for each vertex i and represent the n colors by choosing a set S of n distinct elements from F, then an n-coloring of $\mathcal{G}$ is equivalent to assigning a value $x_i = \alpha_i$ for each $i = 1, 2, \ldots, N$ where $\alpha_i \in S$ and $\alpha_i \neq \alpha_j$ if (i, j) is an edge in $\mathcal{G}$. If $f(x) = \prod_{\alpha \in S}(x - \alpha)$ is the polynomial in $F[x]$ of degree n whose roots are the elements in S, then $x_i = \alpha_i$ for some $\alpha_i \in S$ is equivalent to the statement that x_i is a solution to the equation $f(x_i) = 0$. The statement $\alpha_i \neq \alpha_j$ is then the statement that $f(x_i) = f(x_j)$ but $x_i \neq x_j$, so x_i and x_j satisfy the equation $g(x_i, x_j) = 0$, where $g(x_i, x_j)$ is the polynomial $(f(x_i) - f(x_j))/(x_i - x_j)$ in $F[x_i, x_j]$. It follows that finding an *n-coloring* of $\mathcal{G}$ is equivalent to solving the system of equations

$$\begin{cases} f(x_i) = 0, & \text{for } i = 1, 2, \ldots, N, \\ g(x_i, x_j) = 0, & \text{for all edges } (i, j) \text{ in } \mathcal{G} \end{cases}$$

(note also we may use any polynomial g satisfying $\alpha_i \neq \alpha_j$ if $g(\alpha_i, \alpha_j) = 0$). It follows by "Hilbert's Nullstellensatz" (cf. Corollary 33 in Section 15.3) that this system of equations has a solution, hence $\mathcal{G}$ has an n-coloring, unless the ideal I in $F[x_1, x_2, \ldots, x_N]$ generated by the polynomials $f(x_i)$ for $i = 1, 2, \ldots, N$, together with the polynomials $g(x_i, x_j)$ for all the edges (i, j) in the graph $\mathcal{G}$, is not a proper ideal. This in turn is equivalent to the statement that the reduced Gröbner basis for I (with respect to any monomial ordering) is simply $\{1\}$. Further, when an n-coloring does exist, solving this system of equations as in the examples following Proposition 29 provides an explicit coloring for $\mathcal{G}$.

There are many possible choices of field F and set S. For example, use any field F containing a set S of distinct n^{th} roots of unity, in which case $f(x) = x^n - 1$ and we may take $g(x_i, x_j) = (x_i^n - x_j^n)/(x_i - x_j) = x_i^{n-1} + x_i^{n-2}x_j + \cdots + x_i x_j^{n-2} + x_j^{n-1}$, or use any subset S of $F = \mathbb{F}_p$ with a prime $p \geq n$ (in the special case $n = p$, then, by Fermat's Little Theorem, we have $f(x) = x^p - x$ and $g(x_i, x_j) = (x_i - x_j)^{p-1} - 1$).

(a) Consider a possible 3-coloring of the graph $\mathcal{G}$ with eight vertices and 14 edges $(1, 3)$, $(1, 4)$, $(1, 5)$, $(2, 4)$, $(2, 7)$, $(2, 8)$, $(3, 4)$, $(3, 6)$, $(3, 8)$, $(4, 5)$, $(5, 6)$, $(6, 7)$, $(6, 8)$, $(7, 8)$. Take $F = \mathbb{F}_3$ with 'colors' $0, 1, 2 \in \mathbb{F}_3$ and suppose vertex 1 is colored by 0. In this case $f(x) = x(x - 1)(x - 2) = x^3 - x \in \mathbb{F}_3[x]$ and $g(x_i, x_j) = x_i^2 + x_i x_j + x_j^2 - 1$. If I is the ideal generated by $x_1, x_i^3 - x_i, 2 \leq i \leq 8$ and $g(x_i, x_j)$ for the edges (i, j) in $\mathcal{G}$, show that the reduced Gröbner basis for I with respect to the lexicographic monomial ordering $x_1 > x_2 > \cdots > x_8$ is $\{x_1, x_2, x_3 + x_8, x_4 + 2x_8, x_5 + x_8, x_6, x_7 + x_8, x_8^2 + 2\}$. Deduce that $\mathcal{G}$ has two distinct 3-colorings, determined by the coloring of vertex 8 (which must be colored by a nonzero element in $\mathbb{F}_3$), and exhibit the colorings of $\mathcal{G}$.

Show that if the edge $(3, 7)$ is added to $\mathcal{G}$ then the graph cannot be 3-colored.

(b) Take $F = \mathbb{F}_5$ with four 'colors' $1, 2, 3, 4 \in \mathbb{F}_5$, so $f(x) = x^4 - 1$ and we may use $g(x_i, x_j) = x_i^3 + x_i^2 x_j + x_i x_j^2 + x_j^3$. Show that the graph $\mathcal{G}$ with five vertices having 9 edges $(1, 3)$, $(1, 4)$, $(1, 5)$, $(2, 3)$, $(2, 4)$, $(2, 5)$, $(3, 4)$, $(3, 5)$, $(4, 5)$ (the "complete graph on five vertices" with one edge removed) can be 4-colored but cannot be 3-colored.

(c) Use Gröbner bases to show that the graph $\mathcal{G}$ with nine vertices and 22 edges $(1, 4)$, $(1, 6)$, $(1, 7)$, $(1, 8)$, $(2, 3)$, $(2, 4)$, $(2, 6)$, $(2, 7)$, $(3, 5)$, $(3, 7)$, $(3, 9)$, $(4, 5)$, $(4, 6)$, $(4, 7)$, $(4, 9)$, $(5, 6)$, $(5, 7)$, $(5, 8)$, $(5, 9)$, $(6, 7)$, $(6, 9)$, $(7, 8)$ has precisely four 4-colorings up to a permutation of the colors (so a total of 96 total 4-colorings). Show that if the edge $(1, 5)$ is added then $\mathcal{G}$ cannot be 4-colored.

Part III

MODULES AND VECTOR SPACES

In Part III we study the mathematical objects called modules. The use of modules was pioneered by one of the most prominent mathematicians of the first part of this century, Emmy Noether, who led the way in demonstrating the power and elegance of this structure. We shall see that vector spaces are just special types of modules which arise when the underlying ring is a field. If R is a ring, the definition of an R-module M is closely analogous to the definition of a group action where R plays the role of the group and M the role of the set. The additional axioms for a module require that M itself have more structure (namely that M be an abelian group). Modules are the "representation objects" for rings, i.e., they are, by definition, algebraic objects on which rings act. As the theory develops it will become apparent how the structure of the ring R (in particular, the structure and wealth of its ideals) is reflected by the structure of its modules and vice versa in the same way that the structure of the collection of normal subgroups of a group was reflected by its permutation representations.

CHAPTER 10

Introduction to Module Theory

10.1 BASIC DEFINITIONS AND EXAMPLES

We start with the definition of a module.

Definition. Let R be a ring (not necessarily commutative nor with 1). A *left R-module* or a *left module over R* is a set M together with
   (1) a binary operation $+$ on M under which M is an abelian group, and
   (2) an action of R on M (that is, a map $R \times M \to M$) denoted by rm, for all $r \in R$ and for all $m \in M$ which satisfies
      (a) $(r + s)m = rm + sm$,   for all $r, s \in R, m \in M$,
      (b) $(rs)m = r(sm)$,   for all $r, s \in R, m \in M$, and
      (c) $r(m + n) = rm + rn$,   for all $r \in R, m, n \in M$.
   If the ring R has a 1 we impose the additional axiom:
      (d) $1m = m$,   for all $m \in M$.

The descriptor "left" in the above definition indicates that the ring elements appear on the left; "right" R-modules can be defined analogously. If the ring R is *commutative* and M is a left R-module we can make M into a right R-module by defining $mr = rm$ for $m \in M$ and $r \in R$. If R is not commutative, axiom 2(b) in general will not hold with this definition (so not every left R-module is also a right R-module). Unless explicitly mentioned otherwise the term "module" will always mean "left module." Modules satisfying axiom 2(d) are called *unital* modules and in this book all our modules will be unital (this is to avoid "pathologies" such as having $rm = 0$ for all $r \in R$ and $m \in M$).

When R is a field F the axioms for an R-module are precisely the same as those for a vector space over F, so that

modules over a field F and vector spaces over F are the same.

Before giving other examples of R-modules we record the obvious definition of submodules.

Definition. Let R be a ring and let M be an R-module. An *R-submodule* of M is a subgroup N of M which is closed under the action of ring elements, i.e., $rn \in N$, for all $r \in R, n \in N$.

Submodules of M are therefore just subsets of M which are themselves modules under the restricted operations. In particular, if $R = F$ is a field, submodules are the same as subspaces. Every R-module M has the two submodules M and 0 (the latter is called the *trivial submodule*).

Examples

(1) Let R be any ring. Then $M = R$ is a left R-module, where the action of a ring element on a module element is just the usual multiplication in the ring R (similarly, R is a right module over itself). In particular, every field can be considered as a (1-dimensional) vector space over itself. When R is considered as a left module over itself in this fashion, the submodules of R are precisely the left ideals of R (and if R is considered as a right R-module over itself, its submodules are the right ideals). Thus if R is not commutative it has a left and right module structure over itself and these structures may be different (e.g., the submodules may be different) — Exercise 21 at the end of this section gives a specific example of this.

(2) Let $R = F$ be a field. As noted above, every vector space over F is an F-module and vice versa. Let $n \in \mathbb{Z}^+$ and let

$$F^n = \{(a_1, a_2, \ldots, a_n) \mid a_i \in F, \text{ for all } i\}$$

(called *affine n-space over F*). Make F^n into a vector space by defining addition and scalar multiplication componentwise:

$$(a_1, a_2, \ldots, a_n) + (b_1, b_2, \ldots, b_n) = (a_1 + b_1, a_2 + b_2, \ldots, a_n + b_n)$$
$$\alpha(a_1, \ldots, a_n) = (\alpha a_1, \ldots, \alpha a_n), \qquad \alpha \in F.$$

As in the case of Euclidean n-space (i.e., when $F = \mathbb{R}$), affine n-space is a vector space of dimension n over F (we shall discuss the notion of dimension more thoroughly in the next chapter).

(3) Let R be a ring with 1 and let $n \in \mathbb{Z}^+$. Following Example 2 define

$$R^n = \{(a_1, a_2, \ldots, a_n) \mid a_i \in R, \text{ for all } i\}.$$

Make R^n into an R-module by componentwise addition and multiplication by elements of R in the same manner as when R was a field. The module R^n is called *the free module of rank n over R*. (We shall see shortly that free modules have the same "universal property" in the context of R-modules that free groups were seen to have in Section 6.3. We shall also soon discuss direct products of R-modules.) An obvious submodule of R^n is given by the ith component, namely the set of n-tuples with arbitrary ring elements in the ith component and zeros in the jth component for all $j \neq i$.

(4) The same abelian group may have the structure of an R-module for a number of different rings R and each of these module structures may carry useful information. Specifically, if M is an R-module and S is a subring of R with $1_S = 1_R$, then M is automatically an S-module as well. For instance the field $\mathbb{R}$ is an $\mathbb{R}$-module, a $\mathbb{Q}$-module and a $\mathbb{Z}$-module.

(5) If M is an R-module and for some (2-sided) ideal I of R, $am = 0$, for all $a \in I$ and all $m \in M$, we say M is *annihilated by I*. In this situation we can make M into an (R/I)-module by defining an action of the quotient ring R/I on M as follows: for each $m \in M$ and coset $r + I$ in R/I let

$$(r + I)m = rm.$$

Since $am = 0$ for all $a \in I$ and all $m \in M$ this is well defined and one easily checks that it makes M into an (R/I)-module. In particular, when I is a maximal ideal in the commutative ring R and $IM = 0$, then M is a vector space over the field R/I (cf. the following example).

The next example is of sufficient importance as to be singled out. It will form the basis for our proof of the Fundamental Theorem of Finitely Generated Abelian Groups in Chapter 12.

Example: ($\mathbb{Z}$-modules)

Let R $= \mathbb{Z}$, let A be *any* abelian group (finite or infinite) and write the operation of A as $+$. Make A into a $\mathbb{Z}$-module as follows: for any $n \in \mathbb{Z}$ and $a \in A$ define

$$na = \begin{cases} a + a + \cdots + a \ \ (n \text{ times}) & \text{if } n > 0 \\ 0 & \text{if } n = 0 \\ -a - a - \cdots - a \ \ (-n \text{ times}) & \text{if } n < 0 \end{cases}$$

(here 0 is the identity of the additive group A). This definition of an action of the integers on A makes A into a $\mathbb{Z}$-module, and the module axioms show that this is the only possible action of $\mathbb{Z}$ on A making it a (unital) $\mathbb{Z}$-module. Thus every abelian group is a $\mathbb{Z}$-module. Conversely, if M is any $\mathbb{Z}$-module, a fortiori M is an abelian group, so

$\mathbb{Z}$-*modules are the same as abelian groups.*

Furthermore, it is immediate from the definition that

$\mathbb{Z}$-*submodules are the same as subgroups.*

Note that for the cyclic group $\langle a \rangle$ written multiplicatively the additive notation na becomes a^n, that is, we have all along been using the fact that $\langle a \rangle$ is a right $\mathbb{Z}$-module (checking that this "exponential" notation satisfies the usual laws of exponents is equivalent to checking the $\mathbb{Z}$-module axioms — this was given as an exercise at the end of Section 1.1). Note that since $\mathbb{Z}$ is commutative these definitions of left and right actions by ring elements give the same module structure.

If A is an abelian group containing an element x of finite order n then $nx = 0$. Thus, in contrast to vector spaces, a $\mathbb{Z}$-module may have nonzero elements x such that $nx = 0$ for some nonzero ring element n. In particular, if A has order m, then by Lagrange's Theorem (Corollary 9, Section 3.2) $mx = 0$, for *all* $x \in A$. Note that then A is a module over $\mathbb{Z}/m\mathbb{Z}$.

In particular, if p is a prime and A is an abelian group (written additively) such that $px = 0$, for all $x \in A$, then (as noted in Example 5) A is a $\mathbb{Z}/p\mathbb{Z}$-module, i.e., can be considered as a vector space over the field $\mathbb{F}_p = \mathbb{Z}/p\mathbb{Z}$. For instance, the Klein 4-group is a (2-dimensional) vector space over $\mathbb{F}_2$. These groups are the *elementary abelian p-groups* discussed in Section 4.4 (see, in particular, Proposition 17(3)).

The next example is also of fundamental importance and will form the basis for our study of canonical forms of matrices in Sections 12.2 and 12.3.

Example: ($F[x]$-modules)

Let F be a field, let x be an indeterminate and let R be the polynomial ring $F[x]$. Let V be a vector space over F and let T be a linear transformation from V to V (we shall review the theory of linear transformations in the next chapter — for the purposes of this example one only needs to know the definition of a linear transformation). We have already seen that V is an F-module; the linear map T will enable us to make V into an $F[x]$-module.

First, for the nonnegative integer n, define

$$T^0 = I,$$

$$\vdots$$

$$T^n = T \circ T \circ \cdots \circ T \qquad (n \text{ times})$$

where I is the identity map from V to V and $\circ$ denotes function composition (which makes sense because the domain and codomain of T are the same). Also, for any two linear transformations A, B from V to V and elements $\alpha, \beta \in F$, let $\alpha A + \beta B$ be defined by

$$(\alpha A + \beta B)(v) = \alpha(A(v)) + \beta(B(v))$$

(i.e., addition and scalar multiplication of linear transformations are defined pointwise). Then $\alpha A + \beta B$ is easily seen to be a linear transformation from V to V, so that linear combinations of linear transformations are again linear transformations.

We now define the action of any polynomial in x on V. Let $p(x)$ be the polynomial

$$p(x) = a_n x^n + a_{n-1} x^{n-1} + \cdots + a_1 x + a_0,$$

where $a_0, \ldots, a_n \in F$. For each $v \in V$ define an action of the ring element $p(x)$ on the module element v by

$$p(x)v = (a_n T^n + a_{n-1} T^{n-1} + \cdots + a_1 T + a_0)(v)$$

$$= a_n T^n(v) + a_{n-1} T^{n-1}(v) + \cdots + a_1 T(v) + a_0 v$$

(i.e., $p(x)$ acts by substituting the linear transformation T for x in $p(x)$ and applying the resulting linear transformation to v). Put another way, x acts on V as the linear transformation T and we extend this to an action of all of $F[x]$ on V in a natural way. It is easy to check that this definition of an action of $F[x]$ on V satisfies all the module axioms and makes V into an $F[x]$-module.

The field F is naturally a subring of $F[x]$ (the constant polynomials) and the action of these field elements is by definition the same as their action when viewed as constant polynomials. In other words, the definition of the $F[x]$ action on V is consistent with the given action of the field F on the vector space V, i.e., the definition *extends* the action of F to an action of the larger ring $F[x]$.

The way $F[x]$ acts on V depends on the choice of T so that there are in general many different $F[x]$-module structures on the same vector space V. For instance, if $T = 0$, and $p(x)$, v are as above, then $p(x)v = a_0 v$, that is, the polynomial $p(x)$ acts on v simply by multiplying by the constant term of $p(x)$, so that the $F[x]$-module structure is just the F-module structure. If, on the other hand, T is the identity transformation (so $T^n(v) = v$, for all n and v), then $p(x)v = a_n v + a_{n-1} v + \cdots + a_0 v = (a_n + \cdots + a_0)v$, so that now $p(x)$ multiplies v by the sum of the coefficients of $p(x)$.

To give another specific example, let V be affine n-space F^n and let T be the "shift operator"

$$T(x_1, x_2, \ldots, x_n) = (x_2, x_3, \ldots, x_n, 0).$$

Let e_i be the usual i^{th} basis vector $(0, 0, \ldots, 0, 1, 0, \ldots, 0)$ where the 1 is in position i. Then

$$T^k(e_i) = \begin{cases} e_{i-k} & \text{if } i > k \\ 0 & \text{if } i \leq k \end{cases}$$

so for example, if $m < n$,

$$(a_m x^m + a_{m-1} x^{m-1} + \cdots + a_0) e_n = (0, \ldots, 0, a_m, a_{m-1}, \ldots, a_0).$$

From this we can determine the action of any polynomial on any vector.

The construction of an $F[x]$-module from a vector space V over F and a linear transformation T from V to V in fact describes *all* $F[x]$-modules; namely, an $F[x]$-module is a vector space together with a linear transformation which specifies the action of x. This is because if V is any $F[x]$-module, then V is an F-module and the action of the ring element x on V is a linear transformation from V to V. The axioms for a module ensure that the actions of F and x on V uniquely determine the action of any element of $F[x]$ on V. Thus there is a bijection between the collection of $F[x]$-modules and the collection of pairs V, T

$$\left\{ V \text{ an } F[x]\text{-module} \right\} \quad \longleftrightarrow \quad \left\{ \begin{array}{c} V \text{ a vector space over } F \\ \text{and} \\ T : V \to V \text{ a linear transformation} \end{array} \right\}$$

given by

the element x acts on V as the linear transformation T.

Now we consider $F[x]$-submodules of V where, as above, V is any $F[x]$-module and T is the linear transformation from V to V given by the action of x. An $F[x]$-submodule W of V must first be an F-submodule, i.e., W must be a vector subspace of V. Secondly, W must be sent to itself under the action of the ring element x, i.e., we must have $T(w) \in W$, for all $w \in W$. Any vector subspace U of V such that $T(U) \subseteq U$ is called T-*stable* or T-*invariant*. If U is any T-stable subspace of V it follows that $T^n(U) \subseteq U$, for all $n \in \mathbb{Z}^+$ (for example, $T(U) \subseteq U$ implies $T^2(U) = T(T(U)) \subseteq T(U) \subseteq U$). Moreover any linear combination of powers of T then sends U into U so that U is also stable by the action of *any* polynomial in T. Thus U is an $F[x]$-submodule of V. This shows that

the $F[x]$-*submodules of V are precisely the T-stable subspaces of V.*

In terms of the bijection above,

$$\left\{ W \text{ an } F[x]\text{-submodule} \right\} \quad \longleftrightarrow \quad \left\{ \begin{array}{c} W \text{ a subspace of } V \\ \text{and} \\ W \text{ is } T\text{-stable} \end{array} \right\}$$

which gives a complete dictionary between $F[x]$-modules V and vector spaces V together with a given linear transformation T from V to V.

For instance, if T is the shift operator defined on affine n-space above and k is any integer in the range $0 \leq k \leq n$, then the subspace

$$U_k = \{ (x_1, x_2, \ldots, x_k, 0, \ldots, 0) \mid x_i \in F \}$$

is clearly T-stable so is an $F[x]$-submodule of V.

We emphasize that an abelian group M may have many different R-module structures, even if the ring R does not vary (in the same way that a given group G may act in many ways as a permutation group on some fixed set Ω). We shall see that the structure of an R-module is reflected by the ideal structure of R. When R is a field (the subject of the next chapter) all R-modules will be seen to be products of copies of R (as in Example 3 above).

We shall see in Chapter 12 that the relatively simple ideal structure of the ring $F[x]$ (recall that $F[x]$ is a Principal Ideal Domain) forces the $F[x]$-module structure of V to be correspondingly uncomplicated, and this in turn provides a great deal of information about the linear transformation T (in particular, gives some nice matrix representations for T: its *rational canonical form* and its *Jordan canonical form*). Moreover, the same arguments which *classify* finitely generated $F[x]$-modules apply to any Principal Ideal Domain R, and when these are invoked for $R = \mathbb{Z}$, we obtain the Fundamental Theorem of Finitely Generated Abelian Groups. These results generalize the theorem that every finite dimensional vector space has a basis.

In Part VI of the book we shall study modules over certain noncommutative rings (group rings) and see that this theory in some sense generalizes both the study of $F[x]$-modules in Chapter 12 and the notion of a permutation representation of a finite group.

We establish a submodule criterion analogous to that for subgroups of a group in Section 2.1.

Proposition 1. *(The Submodule Criterion)* Let R be a ring and let M be an R-module. A subset N of M is a submodule of M if and only if

(1) $N \neq \emptyset$, and
(2) $x + ry \in N$ for all $r \in R$ and for all $x, y \in N$.

Proof: If N is a submodule, then $0 \in N$ so $N \neq \emptyset$. Also N is closed under addition and is sent to itself under the action of elements of R. Conversely, suppose (1) and (2) hold. Let $r = -1$ and apply the subgroup criterion (in additive form) to see that N is a subgroup of M. In particular, $0 \in N$. Now let $x = 0$ and apply hypothesis (2) to see that N is sent to itself under the action of R. This establishes the proposition.

We end this section with an important definition and some examples.

Definition. Let R be a commutative ring with identity. An R-*algebra* is a ring A with identity together with a ring homomorphism $f : R \to A$ mapping 1_R to 1_A such that the subring $f(R)$ of A is contained in the center of A.

If A is an R-algebra then it is easy to check that A has a natural left and right (unital) R-module structure defined by $r \cdot a = a \cdot r = f(r)a$ where $f(r)a$ is just the multiplication in the ring A (and this is the same as $af(r)$ since by assumption $f(r)$ lies in the center of A). In general it is possible for an R-algebra A to have other left (or right) R-module structures, but unless otherwise stated, this natural module structure on an algebra will be assumed.

Definition. If A and B are two R-algebras, an *R-algebra homomorphism* (or isomorphism) is a ring homomorphism (isomorphism, respectively) $\varphi : A \to B$ mapping 1_A to 1_B such that $\varphi(r \cdot a) = r \cdot \varphi(a)$ for all $r \in R$ and $a \in A$.

Examples

Let R be a commutative ring with 1.

(1) Any ring with identity is a $\mathbb{Z}$-algebra.

(2) For any ring A with identity, if R is a subring of the center of A containing the identity of A then A is an R-algebra. In particular, a commutative ring A containing 1 is an R-algebra for any subring R of A containing 1. For example, the polynomial ring $R[x]$ is an R-algebra, the polynomial ring over R in any number of variables is an R-algebra, and the group ring RG for a finite group G is an R-algebra (cf. Section 7.2).

(3) If A is an R-algebra then the R-module structure of A depends only on the subring $f(R)$ contained in the center of A as in the previous example. If we replace R by its image $f(R)$ we see that "up to a ring homomorphism" every algebra A arises from a subring of the center of A that contains 1_A.

(4) A special case of the previous example occurs when $R = F$ is a *field*. In this case F is *isomorphic* to its image under f, so we can identify F itself as a subring of A. Hence, saying that A is an algebra over a field F is the same as saying that the ring A contains the field F in its center and the identity of A and of F are the same (this last condition is necessary, cf. Exercise 23).

Suppose that A is an R-algebra. Then A is a ring with identity that is a (unital) left R-module satisfying $r \cdot (ab) = (r \cdot a)b = a(r \cdot b)$ for all $r \in R$ and $a, b \in A$ (these are all equal to the product $f(r)ab$ in the ring A—recall that $f(R)$ is contained in the center of A). Conversely, these conditions on a ring A define an R-algebra, and are sometimes used as the definition of an R-algebra (cf. Exercise 22).

EXERCISES

In these exercises R is a ring with 1 and M is a left R-module.

1. Prove that $0m = 0$ and $(-1)m = -m$ for all $m \in M$.

2. Prove that $R^\times$ and M satisfy the two axioms in Section 1.7 for a *group action* of the multiplicative group $R^\times$ on the set M.

3. Assume that $rm = 0$ for some $r \in R$ and some $m \in M$ with $m \neq 0$. Prove that r does not have a left inverse (i.e., there is no $s \in R$ such that $sr = 1$).

4. Let M be the module R^n described in Example 3 and let $I_1, I_2, \ldots, I_n$ be left ideals of R. Prove that the following are submodules of M:
   (a) $\{(x_1, x_2, \ldots, x_n) \mid x_i \in I_i\}$
   (b) $\{(x_1, x_2, \ldots, x_n) \mid x_i \in R \text{ and } x_1 + x_2 + \cdots + x_n = 0\}$.

5. For any left ideal I of R define
$$IM = \{\sum_{\text{finite}} a_i m_i \mid a_i \in I, \ m_i \in M\}$$
to be the collection of all finite sums of elements of the form am where $a \in I$ and $m \in M$. Prove that IM is a submodule of M.

6. Show that the intersection of any nonempty collection of submodules of an R-module is a submodule.

7. Let $N_1 \subseteq N_2 \subseteq \cdots$ be an ascending chain of submodules of M. Prove that $\cup_{i=1}^{\infty} N_i$ is a submodule of M.

8. An element m of the R-module M is called a *torsion element* if $rm = 0$ for some nonzero element $r \in R$. The set of torsion elements is denoted

$$\text{Tor}(M) = \{m \in M \mid rm = 0 \text{ for some nonzero } r \in R\}.$$

   (a) Prove that if R is an integral domain then $\text{Tor}(M)$ is a submodule of M (called the *torsion* submodule of M).
   (b) Give an example of a ring R and an R-module M such that $\text{Tor}(M)$ is not a submodule. [Consider the torsion elements in the R-module R.]
   (c) If R has zero divisors show that every nonzero R-module has nonzero torsion elements.

9. If N is a submodule of M, the *annihilator of N in R* is defined to be $\{r \in R \mid rn = 0 \text{ for all } n \in N\}$. Prove that the annihilator of N in R is a 2-sided ideal of R.

10. If I is a right ideal of R, the *annihilator of I in M* is defined to be $\{m \in M \mid am = 0 \text{ for all } a \in I\}$. Prove that the annihilator of I in M is a submodule of M.

11. Let M be the abelian group (i.e., $\mathbb{Z}$-module) $\mathbb{Z}/24\mathbb{Z} \times \mathbb{Z}/15\mathbb{Z} \times \mathbb{Z}/50\mathbb{Z}$.
   (a) Find the annihilator of M in $\mathbb{Z}$ (i.e., a generator for this principal ideal).
   (b) Let $I = 2\mathbb{Z}$. Describe the annihilator of I in M as a direct product of cyclic groups.

12. In the notation of the preceding exercises prove the following facts about annihilators.
   (a) Let N be a submodule of M and let I be its annihilator in R. Prove that the annihilator of I in M contains N. Give an example where the annihilator of I in M does not equal N.
   (b) Let I be a right ideal of R and let N be its annihilator in M. Prove that the annihilator of N in R contains I. Give an example where the annihilator of N in R does not equal I.

13. Let I be an ideal of R. Let M' be the subset of elements a of M that are annihilated by some power, I^k, of the ideal I, where the power may depend on a. Prove that M' is a submodule of M. [Use Exercise 7.]

14. Let z be an element of the center of R, i.e., $zr = rz$ for all $r \in R$. Prove that zM is a submodule of M, where $zM = \{zm \mid m \in M\}$. Show that if R is the ring of 2×2 matrices over a field and e is the matrix with a 1 in position 1,1 and zeros elsewhere then eR is *not* a left R-submodule (where $M = R$ is considered as a left R-module as in Example 1) — in this case the matrix e is not in the center of R.

15. If M is a finite abelian group then M is naturally a $\mathbb{Z}$-module. Can this action be extended to make M into a $\mathbb{Q}$-module?

16. Prove that the submodules U_k described in the example of $F[x]$-modules are all of the $F[x]$-submodules for the shift operator.

17. Let T be the shift operator on the vector space V and let $e_1, \ldots, e_n$ be the usual basis vectors described in the example of $F[x]$-modules. If $m \geq n$ find $(a_m x^m + a_{m-1} x^{m-1} + \cdots + a_0) e_n$.

18. Let $F = \mathbb{R}$, let $V = \mathbb{R}^2$ and let T be the linear transformation from V to V which is rotation clockwise about the origin by $\pi/2$ radians. Show that V and 0 are the only $F[x]$-submodules for this T.

19. Let $F = \mathbb{R}$, let $V = \mathbb{R}^2$ and let T be the linear transformation from V to V which is projection onto the y-axis. Show that V, 0, the x-axis and the y-axis are the only $F[x]$-submodules for this T.

20. Let $F = \mathbb{R}$, let $V = \mathbb{R}^2$ and let T be the linear transformation from V to V which is rotation clockwise about the origin by π radians. Show that *every* subspace of V is an

$F[x]$-submodule for this T.

21. Let $n \in \mathbb{Z}^+, n > 1$ and let R be the ring of $n \times n$ matrices with entries from a field F. Let M be the set of $n \times n$ matrices with arbitrary elements of F in the first column and zeros elsewhere. Show that M is a submodule of R when R is considered as a left module over itself, but M is not a submodule of R when R is considered as a right R-module.

22. Suppose that A is a ring with identity 1_A that is a (unital) left R-module satisfying $r \cdot (ab) = (r \cdot a)b = a(r \cdot b)$ for all $r \in R$ and $a, b \in A$. Prove that the map $f : R \to A$ defined by $f(r) = r \cdot 1_A$ is a ring homomorphism mapping 1_R to 1_A and that $f(R)$ is contained in the center of A. Conclude that A is an R-algebra and that the R-module structure on A induced by its algebra structure is precisely the original R-module structure.

23. Let A be the direct product ring $\mathbb{C} \times \mathbb{C}$ (cf. Section 7.6). Let τ_1 denote the identity map on $\mathbb{C}$ and let τ_2 denote complex conjugation. For any pair $p, q \in \{1, 2\}$ (not necessarily distinct) define

$$f_{p,q} : \mathbb{C} \to \mathbb{C} \times \mathbb{C} \qquad \text{by} \qquad f_{p,q}(z) = (\tau_p(z), \tau_q(z)).$$

So, for example, $f_{2,1} : z \mapsto (\bar{z}, z)$, where $\bar{z}$ is the complex conjugate of z, i.e., $\tau_2(z)$.

   (a) Prove that each $f_{p,q}$ is an injective ring homomorphism, and that they all agree on the subfield $\mathbb{R}$ of $\mathbb{C}$. Deduce that A has four distinct $\mathbb{C}$-algebra structures. Explicitly give the action $z \cdot (u, v)$ of a complex number z on an ordered pair in A in each case.
   (b) Prove that if $f_{p,q} \neq f_{p',q'}$ then the identity map on A is not a $\mathbb{C}$-algebra homomorphism from A considered as a $\mathbb{C}$-algebra via $f_{p,q}$ to A considered a $\mathbb{C}$-algebra via $f_{p',q'}$ (although the identity is an $\mathbb{R}$-algebra isomorphism).
   (c) Prove that for any pair p, q there is some ring isomorphism from A to itself such that A is isomorphic as a $\mathbb{C}$-algebra via $f_{p,q}$ to A considered as $\mathbb{C}$-algebra via $f_{1,1}$ (the "natural" $\mathbb{C}$-algebra structure on A).

Remark: In the preceding exercise $A = \mathbb{C} \times \mathbb{C}$ is not a $\mathbb{C}$-algebra over either of the direct factor component copies of $\mathbb{C}$ (for example the subring $\mathbb{C} \times 0 \cong \mathbb{C}$) since it is not a unital module over these copies of $\mathbb{C}$ (the 1 of these subrings is not the same as the 1 of A).

10.2 QUOTIENT MODULES AND MODULE HOMOMORPHISMS

This section contains the basic theory of quotient modules and module homomorphisms.

Definition. Let R be a ring and let M and N be R-modules.
   (1) A map $\varphi : M \to N$ is an *R-module homomorphism* if it respects the R-module structures of M and N, i.e.,
       (a) $\varphi(x + y) = \varphi(x) + \varphi(y)$, for all $x, y \in M$ and
       (b) $\varphi(rx) = r\varphi(x)$, for all $r \in R, x \in M$.
   (2) An R-module homomorphism is an *isomorphism (of R-modules)* if it is both injective and surjective. The modules M and N are said to be *isomorphic*, denoted $M \cong N$, if there is some R-module isomorphism $\varphi : M \to N$.
   (3) If $\varphi : M \to N$ is an R-module homomorphism, let $\ker \varphi = \{m \in M \mid \varphi(m) = 0\}$ (the *kernel* of φ) and let $\varphi(M) = \{n \in N \mid n = \varphi(m)$ for some $m \in M\}$ (the *image* of φ, as usual).
   (4) Let M and N be R-modules and define $\operatorname{Hom}_R(M, N)$ to be the set of all R-module homomorphisms from M into N.

Any R-module homomorphism is also a homomorphism of the additive groups, but not every group homomorphism need be a module homomorphism (because condition (b) may not be satisfied). The unqualified term "isomorphism" when applied to R-modules will always mean R-module isomorphism. When the symbol $\cong$ is used without qualification it will denote an isomorphism of the respective structures (which will be evident from the context).

It is an easy exercise using the submodule criterion (Proposition 1) to show that kernels and images of R-module homomorphisms are submodules.

Examples

(1) If R is a ring and $M = R$ is a module over itself, then R-module homomorphisms (even from R to itself) need not be ring homomorphisms and ring homomorphisms need not be R-module homomorphisms. For example, when $R = \mathbb{Z}$ the $\mathbb{Z}$-module homomorphism $x \mapsto 2x$ is not a ring homomorphism (1 does not map to 1). When $R = F[x]$ the ring homomorphism $\varphi : f(x) \mapsto f(x^2)$ is not an $F[x]$-module homomorphism (if it were, we would have $x^2 = \varphi(x) = \varphi(x \cdot 1) = x\varphi(1) = x$).

(2) Let R be a ring, let $n \in \mathbb{Z}^+$ and let $M = R^n$. One easily checks that for each $i \in \{1, \ldots, n\}$ the projection map

$$\pi_i : R^n \to R \quad \text{by} \quad \pi_i(x_1, \ldots, x_n) = x_i$$

is a surjective R-module homomorphism with kernel equal to the submodule of n-tuples which have a zero in position i.

(3) If R is a field, R-module homomorphisms are called *linear transformations*. These will be studied extensively in Chapter 11.

(4) For the ring $R = \mathbb{Z}$ the action of ring elements (integers) on any $\mathbb{Z}$-module amounts to just adding and subtracting within the (additive) abelian group structure of the module so that in this case condition (b) of a homomorphism is implied by condition (a). For example, $\varphi(2x) = \varphi(x + x) = \varphi(x) + \varphi(x) = 2\varphi(x)$, etc. It follows that

$\mathbb{Z}$-module homomorphisms are the same as abelian group homomorphisms.

(5) Let R be a ring, let I be a 2-sided ideal of R and suppose M and N are R-modules annihilated by I (i.e., $am = 0$ and $an = 0$ for all $a \in I$, $n \in N$ and $m \in M$). Any R-module homomorphism from N to M is then automatically a homomorphism of (R/I)-modules (see Example 5 of Section 1). In particular, if A is an additive abelian group such that for some prime p, $px = 0$ for all $x \in A$, then any group homomorphism from A to itself is a $\mathbb{Z}/p\mathbb{Z}$-module homomorphism, i.e., is a linear transformation over the field $\mathbb{F}_p$. In particular, the group of all (group) automorphisms of A is the group of invertible linear transformations from A to itself: $GL(A)$.

Proposition 2. Let M, N and L be R-modules.
(1) A map $\varphi : M \to N$ is an R-module homomorphism if and only if
$\varphi(rx + y) = r\varphi(x) + \varphi(y)$ for all $x, y \in M$ and all $r \in R$.
(2) Let φ, ψ be elements of $\text{Hom}_R(M, N)$. Define $\varphi + \psi$ by

$$(\varphi + \psi)(m) = \varphi(m) + \psi(m) \qquad \text{for all } m \in M.$$

Then $\varphi + \psi \in \text{Hom}_R(M, N)$ and with this operation $\text{Hom}_R(M, N)$ is an abelian group. If R is a commutative ring then for $r \in R$ define $r\varphi$ by

$$(r\varphi)(m) = r(\varphi(m)) \qquad \text{for all } m \in M.$$

Then $r\varphi \in \text{Hom}_R(M, N)$ and with this action of the commutative ring R the abelian group $\text{Hom}_R(M, N)$ is an R-module.

(3) If $\varphi \in \text{Hom}_R(L, M)$ and $\psi \in \text{Hom}_R(M, N)$ then $\psi \circ \varphi \subset \text{Hom}_R(L, N)$.

(4) With addition as above and multiplication defined as function composition, $\text{Hom}_R(M, M)$ is a ring with 1. When R is commutative $\text{Hom}_R(M, M)$ is an R-algebra.

Proof: (1) Certainly $\varphi(rx+y) = r\varphi(x)+\varphi(y)$ if φ is an R-module homomorphism. Conversely, if $\varphi(rx + y) = r\varphi(x) + \varphi(y)$, take $r = 1$ to see that φ is additive and take $y = 0$ to see that φ commutes with the action of R on M (i.e., is *homogeneous*).

(2) It is straightforward to check that all the abelian group and R-module axioms hold with these definitions — the details are left as an exercise. We note that the commutativity of R is used to show that $r\varphi$ satisfies the second axiom of an R-module homomorphism, namely,

$$
\begin{aligned}
(r_1\varphi)(r_2m) &= r_1\varphi(r_2m) && \text{(by definition of } r_1\varphi) \\
&= r_1r_2(\varphi(m)) && \text{(since } \varphi \text{ is a homomorphism)} \\
&= r_2r_1\varphi(m) && \text{(since } R \text{ is commutative)} \\
&= r_2(r_1\varphi)(m) && \text{(by definition of } r_1\varphi).
\end{aligned}
$$

Verification of the axioms relies ultimately on the hypothesis that N is an R-module. The domain M could in fact be any set — it does not have to be an R-module nor an abelian group.

(3) Let φ and ψ be as given and let $r \in R$, $x, y \in L$. Then

$$
\begin{aligned}
(\psi \circ \varphi)(rx + y) &= \psi(\varphi(rx + y)) \\
&= \psi(r\varphi(x) + \varphi(y)) && \text{(by (1) applied to } \varphi) \\
&= r\psi(\varphi(x)) + \psi(\varphi(y)) && \text{(by (1) applied to } \psi) \\
&= r(\psi \circ \varphi)(x) + (\psi \circ \varphi)(y)
\end{aligned}
$$

so, by (1), $\psi \circ \varphi$ is an R-module homomorphism.

(4) Note that since the domain and codomain of the elements of $\text{Hom}_R(M, M)$ are the same, function composition is defined. By (3), it is a binary operation on $\text{Hom}_R(M, M)$. As usual, function composition is associative. The remaining ring axioms are straightforward to check — the details are left as an exercise. The identity function, I, (as usual, $I(x) = x$, for all $x \in M$) is seen to be the multiplicative identity of $\text{Hom}_R(M, M)$. If R is commutative, then (2) shows that the ring $\text{Hom}_R(M, M)$ is a left R-module and defining $\varphi r = r\varphi$ for all $\varphi \in \text{Hom}_R(M, M)$ and $r \in R$ makes $\text{Hom}_R(M, M)$ into an R-algebra.

Definition. The ring $\text{Hom}_R(M, M)$ is called the *endomorphism ring of M* and will often be denoted by $\text{End}_R(M)$, or just $\text{End}(M)$ when the ring R is clear from the context. Elements of $\text{End}(M)$ are called *endomorphisms*.

When R is commutative there is a natural map from R into $\text{End}(M)$ given by $r \mapsto rI$, where the latter endomorphism of M is just multiplication by r on M (cf. Exercise 7). The image of R is contained in the center of $\text{End}(M)$ so if R has an identity, $\text{End}(M)$ is an R-algebra. The ring homomorphism (cf. Exercise 7) from R to $\text{End}_R(M)$ may not be injective since for some r we may have $rm = 0$ for all $m \in M$ (e.g., $R = \mathbb{Z}$, $M = \mathbb{Z}/2\mathbb{Z}$, and $r = 2$). When R is a field, however, this map is injective (in general, no unit is in the kernel of this map) and the copy of R in $\text{End}_R(M)$ is called the (subring of) *scalar transformations*.

Next we prove that every submodule N of an R-module M is "normal" in the sense that we can *always* form the quotient module M/N, and the natural projection $\pi : M \to M/N$ is an R-module homomorphism with kernel N. The proof of this fact and, more generally, the subsequent proofs of the isomorphism theorems for modules follow easily from the corresponding facts for groups. The reason for this is because a module is first of all an *abelian* group and so *every* submodule is automatically a normal subgroup and any module homomorphism is, in particular, a homomorphism of abelian groups, all of which we have already considered in Chapter 3. What remains to be proved in order to extend results on abelian groups to corresponding results on modules is to check that the action of R is compatible with these group quotients and homomorphisms. For example, the map π above was shown to be a group homomorphism in Chapter 3 but the abelian group M/N must be shown to be an R-module (i.e., to have an action by R) and property (b) in the definition of a module homomorphism must be checked for π.

Proposition 3. Let R be a ring, let M be an R-module and let N be a submodule of M. The (additive, abelian) quotient group M/N can be made into an R-module by defining an action of elements of R by

$$r(x + N) = (rx) + N, \qquad \text{for all } r \in R, \ x + N \in M/N.$$

The natural projection map $\pi : M \to M/N$ defined by $\pi(x) = x + N$ is an R-module homomorphism with kernel N.

Proof: Since M is an abelian group under $+$ the quotient group M/N is defined and is an abelian group. To see that the action of the ring element r on the coset $x + N$ is well defined, suppose $x + N = y + N$, i.e., $x - y \in N$. Since N is a (left) R-submodule, $r(x - y) \in N$. Thus $rx - ry \in N$ and $rx + N = ry + N$, as desired. Now since the operations in M/N are "compatible" with those of M, the axioms for an R-module are easily checked in the same way as was done for quotient groups. For example, axiom 2(b) holds as follows: for all $r_1, r_2 \in R$ and $x + N \in M/N$, by definition of the action of ring elements on elements of M/N

$$(r_1 r_2)(x + N) = (r_1 r_2 x) + N$$
$$= r_1(r_2 x + N)$$
$$= r_1(r_2(x + N)).$$

The other axioms are similarly checked — the details are left as an exercise. Finally, the natural projection map π described above is, in particular, the natural projection of the abelian group M onto the abelian group M/N hence is a group homomorphism with kernel N. The kernel of any module homomorphism is the same as its kernel when viewed as a homomorphism of the abelian group structures. It remains only to show π is a module homomorphism, i.e., $\pi(rm) = r\pi(m)$. But

$$
\begin{aligned}
\pi(rm) &= rm + N \\
&= r(m + N) \qquad \text{(by definition of the action of R on M/N)} \\
&= r\pi(m).
\end{aligned}
$$

This completes the proof.

All the isomorphism theorems stated for groups also hold for R-modules. The proofs are similar to that of Proposition 3 above in that they begin by invoking the corresponding theorem for groups and then prove that the group homomorphisms are also R-module homomorphisms. To state the Second Isomorphism Theorem we need the following.

Definition. Let A, B be submodules of the R-module M. The *sum* of A and B is the set

$$
A + B = \{a + b \mid a \in A, b \in B\}.
$$

One can easily check that the sum of two submodules A and B is a submodule and is the smallest submodule which contains both A and B.

Theorem 4. (Isomorphism Theorems)
  (1) *(The First Isomorphism Theorem for Modules)* Let M, N be R-modules and let $\varphi : M \to N$ be an R-module homomorphism. Then $\ker \varphi$ is a submodule of M and $M/\ker \varphi \cong \varphi(M)$.
  (2) *(The Second Isomorphism Theorem)* Let A, B be submodules of the R-module M. Then $(A + B)/B \cong A/(A \cap B)$.
  (3) *(The Third Isomorphism Theorem)* Let M be an R-module, and let A and B be submodules of M with $A \subseteq B$. Then $(M/A)/(B/A) \cong M/B$.
  (4) *(The Fourth or Lattice Isomorphism Theorem)* Let N be a submodule of the R-module M. There is a bijection between the submodules of M which contain N and the submodules of M/N. The correspondence is given by $A \leftrightarrow A/N$, for all $A \supseteq N$. This correspondence commutes with the processes of taking sums and intersections (i.e., is a lattice isomorphism between the lattice of submodules of M/N and the lattice of submodules of M which contain N).

Proof: Exercise.

EXERCISES

In these exercises R is a ring with 1 and M is a left R-module.

1. Use the submodule criterion to show that kernels and images of R-module homomorphisms are submodules.

2. Show that the relation "is R-module isomorphic to" is an equivalence relation on any set of R-modules.

3. Give an explicit example of a map from one R-module to another which is a group homomorphism but not an R-module homomorphism.

4. Let A be any $\mathbb{Z}$-module, let a be any element of A and let n be a positive integer. Prove that the map $\varphi_a : \mathbb{Z}/n\mathbb{Z} \to A$ given by $\varphi_a(\bar{k}) = ka$ is a well defined $\mathbb{Z}$-module homomorphism if and only if $na = 0$. Prove that $\text{Hom}_{\mathbb{Z}}(\mathbb{Z}/n\mathbb{Z}, A) \cong A_n$, where $A_n = \{a \in A \mid na = 0\}$ (so A_n is the annihilator in A of the ideal (n) of $\mathbb{Z}$ — cf. Exercise 10, Section 1).

5. Exhibit all $\mathbb{Z}$-module homomorphisms from $\mathbb{Z}/30\mathbb{Z}$ to $\mathbb{Z}/21\mathbb{Z}$.

6. Prove that $\text{Hom}_{\mathbb{Z}}(\mathbb{Z}/n\mathbb{Z}, \mathbb{Z}/m\mathbb{Z}) \cong \mathbb{Z}/(n, m)\mathbb{Z}$.

7. Let z be a fixed element of the center of R. Prove that the map $m \mapsto zm$ is an R-module homomorphism from M to itself. Show that for a commutative ring R the map from R to $\text{End}_R(M)$ given by $r \mapsto rI$ is a ring homomorphism (where I is the identity endomorphism).

8. Let $\varphi : M \to N$ be an R-module homomorphism. Prove that $\varphi(\text{Tor}(M)) \subseteq \text{Tor}(N)$ (cf. Exercise 8 in Section 1).

9. Let R be a commutative ring. Prove that $\text{Hom}_R(R, M)$ and M are isomorphic as left R-modules. [Show that each element of $\text{Hom}_R(R, M)$ is determined by its value on the identity of R.]

10. Let R be a commutative ring. Prove that $\text{Hom}_R(R, R)$ and R are isomorphic as rings.

11. Let $A_1, A_2, \ldots, A_n$ be R-modules and let B_i be a submodule of A_i for each $i = 1, 2, \ldots, n$. Prove that

$$(A_1 \times \cdots \times A_n)/(B_1 \times \cdots \times B_n) \cong (A_1/B_1) \times \cdots \times (A_n/B_n).$$

[Recall Exercise 14 in Section 5.1.]

12. Let I be a left ideal of R and let n be a positive integer. Prove

$$R^n/IR^n \cong R/IR \times \cdots \times R/IR \quad (n \text{ times})$$

where IR^n is defined as in Exercise 5 of Section 1. [Use the preceding exercise.]

13. Let I be a nilpotent ideal in a commutative ring R (cf. Exercise 37, Section 7.3), let M and N be R-modules and let $\varphi : M \to N$ be an R-module homomorphism. Show that if the induced map $\bar{\varphi} : M/IM \to N/IN$ is surjective, then φ is surjective.

14. Let $R = \mathbb{Z}[x]$ be the ring of polynomials in x and let $A = \mathbb{Z}[t_1, t_2, \ldots]$ be the ring of polynomials in the independent indeterminates $t_1, t_2, \ldots$. Define an action of R on A as follows: 1) let $1 \in R$ act on A as the identity, 2) for $n \geq 1$ let $x^n \circ 1 = t_n$, let $x^n \circ t_i = t_{n+i}$ for $i = 1, 2, \ldots$, and let x^n act as 0 on monomials in A of (total) degree at least two, and 3) extend $\mathbb{Z}$-linearly, i.e., so that the module axioms 2(a) and 2(c) are satisfied.
   - (a) Show that $x^{p+q} \circ t_i = x^p \circ (x^q \circ t_i) = t_{p+q+i}$ and use this to show that under this action the ring A is a (unital) R-module.
   - (b) Show that the map $\varphi : R \to A$ defined by $\varphi(r) = r \circ 1_A$ is an R-module homomorphism of the ring R into the ring A mapping 1_R to 1_A, but is not a ring homomorphism from R to A.

10.3 GENERATION OF MODULES, DIRECT SUMS, AND FREE MODULES

Let R be a ring with 1. As in the preceding sections the term "module" will mean "left module." We first extend the notion of the sum of two submodules to sums of any finite number of submodules and define the submodule generated by a subset.

Definition. Let M be an R-module and let $N_1, \ldots, N_n$ be submodules of M.

(1) The *sum* of $N_1, \ldots, N_n$ is the set of all finite sums of elements from the sets N_i: $\{a_1 + a_2 + \cdots + a_n \mid a_i \in N_i \text{ for all } i\}$. Denote this sum by $N_1 + \cdots + N_n$.

(2) For any subset A of M let

$$RA = \{r_1 a_1 + r_2 a_2 + \cdots + r_m a_m \mid r_1, \ldots, r_m \in R, \ a_1, \ldots, a_m \in A, \ m \in \mathbb{Z}^+\}$$

(where by convention $RA = \{0\}$ if $A = \emptyset$). If A is the finite set $\{a_1, a_2, \ldots, a_n\}$ we shall write $Ra_1 + Ra_2 + \cdots + Ra_n$ for RA. Call RA the *submodule of M generated by A*. If N is a submodule of M (possibly $N = M$) and $N = RA$, for some subset A of M, we call A a *set of generators* or *generating set* for N, and we say *N is generated by A*.

(3) A submodule N of M (possibly $N = M$) is *finitely generated* if there is some finite subset A of M such that $N = RA$, that is, if N is generated by some finite subset.

(4) A submodule N of M (possibly $N = M$) is *cyclic* if there exists an element $a \in M$ such that $N = Ra$, that is, if N is generated by one element:

$$N = Ra = \{ra \mid r \in R\}.$$

Note that these definitions do not require that the ring R contain a 1, however this condition ensures that A is contained in RA. It is easy to see using the Submodule Criterion that for any subset A of M, RA is indeed a submodule of M and is the smallest submodule of M which contains A (i.e., any submodule of M which contains A also contains RA). In particular, for submodules $N_1, \ldots, N_n$ of M, $N_1 + \cdots + N_n$ is just the submodule generated by the set $N_1 \cup \cdots \cup N_n$ and is the smallest submodule of M containing N_i, for all i. If $N_1, \ldots, N_n$ are generated by sets $A_1, \ldots, A_n$ respectively, then $N_1 + \cdots + N_n$ is generated by $A_1 \cup \cdots \cup A_n$. Note that cyclic modules are, a fortiori, finitely generated.

A submodule N of an R-module M may have many different generating sets (for instance the set N itself always generates N). If N is finitely generated, then there is a smallest nonnegative integer d such that N is generated by d elements (and no fewer). Any generating set consisting of d elements will be called a *minimal set of generators for N* (it is not unique in general). If N is not finitely generated, it need not have a minimal generating set.

The process of generating submodules of an R-module M by taking subsets A of M and forming all finite "R-linear combinations" of elements of A will be our primary way of producing submodules (this notion is perhaps familiar from vector space theory where it is referred to as taking the *span* of A). The obstruction which made the analogous process so difficult for groups in general was the noncommutativity of group operations. For abelian groups, G, however, it was much simpler to control the subgroup $\langle A \rangle$ generated by A, for a subset A of G (see Section 2.4 for the complete discussion of this). The situation for R-modules is similar to that of abelian groups (even if R is a noncommutative ring) because we can always collect "like terms" in elements of A, i.e., terms such as $r_1a_1 + r_2a_2 + s_1a_1$ can always be simplified to $(r_1 + s_1)a_1 + r_2a_2$. This again reflects the underlying abelian group structure of modules.

Examples

(1) Let $R = \mathbb{Z}$ and let M be any R-module, that is, any abelian group. If $a \in M$, then $\mathbb{Z}a$ is just the cyclic subgroup of M generated by a: $\langle a \rangle$ (compare Definition 4 above with the definition of a cyclic group). More generally, M is generated as a $\mathbb{Z}$-module by a set A if and only if M is generated as a group by A (that is, the action of ring elements in this instance produces no elements that cannot already be obtained from A by addition and subtraction). The definition of finitely generated for $\mathbb{Z}$-modules is identical to that for abelian groups found in Chapter 5.

(2) Let R be a ring with 1 and let M be the (left) R-module R itself. Note that R is a finitely generated, in fact cyclic, R-module because $R = R1$ (i.e., we can take $A = \{1\}$). Recall that the submodules of R are precisely the left ideals of R, so saying I is a cyclic R-submodule of the left R-module R is the same as saying I is a principal ideal of R (usually the term "principal ideal" is used in the context of commutative rings). Also, saying I is a finitely generated R-submodule of R is the same as saying I is a finitely generated ideal. When R is a commutative ring we often write AR or aR for the submodule (ideal) generated by A or a respectively, as we have been doing for $\mathbb{Z}$ when we wrote $n\mathbb{Z}$. In this situation $AR = RA$ and $aR = Ra$ (elementwise as well). Thus a Principal Ideal Domain is a (commutative) integral domain R with identity in which every R-submodule of R is cyclic.

Submodules of a finitely generated module need not be finitely generated: take M to be the cyclic R-module R itself where R is the polynomial ring in infinitely many variables $x_1, x_2, x_3, \dots$ with coefficients in some field F. The submodule (i.e., 2-sided ideal) generated by $\{x_1, x_2, \dots\}$ cannot be generated by any finite set (note that one must show that *no* finite subset of this ideal will generate it).

(3) Let R be a ring with 1 and let M be the free module of rank n over R, as described in the first section. For each $i \in \{1, 2, \dots, n\}$ let $e_i = (0, 0, \dots, 0, 1, 0, \dots, 0)$, where the 1 appears in position i. Since

$$(s_1, s_2, \dots, s_n) = \sum_{i=1}^{n} s_i e_i$$

it is clear that M is generated by $\{e_1, \dots, e_n\}$. If R is commutative then this is a *minimal* generating set (cf. Exercises 2 and 27).

(4) Let F be a field, let x be an indeterminate, let V be a vector space over F and let T be a linear transformation from V to V. Make V into an $F[x]$-module via T. Then V is a *cyclic* $F[x]$-module (with generator v) if and only if $V = \{p(x)v \mid p(x) \in F[x]\}$, that is, if and only if every element of V can be written as an F-linear combination of elements of the set $\{T^n(v) \mid n \geq 0\}$. This in turn is equivalent to saying $\{v, T(v), T^2(v), \dots\}$ span V as a vector space over F.

For instance if T is the identity linear transformation from V to V or the zero linear transformation, then for every $v \in V$ and every $p(x) \in F[x]$ we have $p(x)v = \alpha v$ for some $\alpha \in F$. Thus if V has dimension > 1, V cannot be a cyclic $F[x]$-module.

For another example suppose V is affine n-space and T is the "shift operator" described in Section 1. Let e_i be the ith basis vector (as usual) numbered so that T is defined by $T^k(e_n) = e_{n-k}$ for $1 \le k < n$. Thus V is spanned by the elements $e_n, T(e_n), \dots, T^{n-1}(e_n)$, that is, V is a cyclic $F[x]$-module with generator e_n. For $n > 1$, V is not, however, a cyclic F-module (i.e., is not a 1-dimensional vector space over F).

Definition. Let $M_1, \dots, M_k$ be a collection of R-modules. The collection of k-tuples $(m_1, m_2, \dots, m_k)$ where $m_i \in M_i$ with addition and action of R defined componentwise is called the *direct product* of $M_1, \dots, M_k$, denoted $M_1 \times \cdots \times M_k$.

It is evident that the direct product of a collection of R-modules is again an R-module. The direct product of $M_1, \dots, M_k$ is also referred to as the *(external) direct sum* of $M_1, \dots, M_k$ and denoted $M_1 \oplus \cdots \oplus M_k$. The direct product and direct sum of an infinite number of modules (which are different in general) are defined in Exercise 20.

The next proposition indicates when a module is isomorphic to the direct product of some of its submodules and is the analogue for modules of Theorem 9 in Section 5.4 (which determines when a group is the direct product of two of its subgroups).

Proposition 5. Let $N_1, N_2, \dots, N_k$ be submodules of the R-module M. Then the following are equivalent:
(1) The map $\pi : N_1 \times N_2 \times \cdots \times N_k \to N_1 + N_2 + \cdots + N_k$ defined by
$$\pi(a_1, a_2, \dots, a_k) = a_1 + a_2 + \cdots + a_k$$
is an isomorphism (of R-modules): $N_1 + N_2 + \cdots + N_k \cong N_1 \times N_2 \times \cdots \times N_k$.
(2) $N_j \cap (N_1 + N_2 + \cdots + N_{j-1} + N_{j+1} + \cdots + N_k) = 0$ for all $j \in \{1, 2, \dots, k\}$.
(3) Every $x \in N_1 + \cdots + N_k$ can be written *uniquely* in the form $a_1 + a_2 + \cdots + a_k$ with $a_i \in N_i$.

Proof: To prove (1) implies (2), suppose for some j that (2) fails to hold and let $a_j \in (N_1 + \cdots + N_{j-1} + N_{j+1} + \cdots + N_k) \cap N_j$, with $a_j \ne 0$. Then
$$a_j = a_1 + \cdots + a_{j-1} + a_{j+1} + \cdots + a_k$$
for some $a_i \in N_i$, and $(a_1, \dots, a_{j-1}, -a_j, a_{j+1}, \dots, a_k)$ would be a nonzero element of $\ker \pi$, a contradiction.

Assume now that (2) holds. If for some module elements $a_i, b_i \in N_i$ we have
$$a_1 + a_2 + \cdots + a_k = b_1 + b_2 + \cdots + b_k$$
then for each j we have
$$a_j - b_j = (b_1 - a_1) + \cdots + (b_{j-1} - a_{j-1}) + (b_{j+1} - a_{j+1}) + \cdots + (b_k - a_k).$$
The left hand side is in N_j and the right side belongs to $N_1 + \cdots + N_{j-1} + N_{j+1} + \cdots + N_k$. Thus
$$a_j - b_j \in N_j \cap (N_1 + \cdots + N_{j-1} + N_{j+1} + \cdots + N_k) = 0.$$
This shows $a_j = b_j$ for all j, and so (2) implies (3).

Finally, to see that (3) implies (1) observe first that the map π is clearly a surjective R-module homomorphism. Then (3) simply implies π is injective, hence is an isomorphism, completing the proof.

If an R-module $M = N_1 + N_2 + \cdots + N_k$ is the sum of submodules $N_1, N_2, \ldots, N_k$ of M satisfying the equivalent conditions of the proposition above, then M is said to be the *(internal) direct sum* of $N_1, N_2, \ldots, N_k$, written

$$M = N_1 \oplus N_2 \oplus \cdots \oplus N_k.$$

By the proposition, this is equivalent to the assertion that every element m of M can be written *uniquely* as a sum of elements $m = n_1 + n_2 + \cdots + n_k$ with $n_i \in N_i$. (Note that part (1) of the proposition is the statement that the internal direct sum of $N_1, N_2, \ldots, N_k$ is isomorphic to their external direct sum, which is the reason we identify them and use the same notation for both.)

Definition. An R-module F is said to be *free* on the subset A of F if for every nonzero element x of F, there exist unique nonzero elements $r_1, r_2, \ldots, r_n$ of R and unique $a_1, a_2, \ldots, a_n$ in A such that $x = r_1 a_1 + r_2 a_2 + \cdots + r_n a_n$, for some $n \in \mathbb{Z}^+$. In this situation we say A is a *basis* or *set of free generators* for F. If R is a commutative ring the cardinality of A is called the *rank* of F (cf. Exercise 27).

One should be careful to note the difference between the uniqueness property of direct sums (Proposition 5(3)) and the uniqueness property of free modules. Namely, in the direct sum of two modules, say $N_1 \oplus N_2$, each element can be written uniquely as $n_1 + n_2$; here the uniqueness refers to the *module elements* n_1 and n_2. In the case of free modules, the uniqueness is on the *ring elements as well as the module elements*. For example, if $R = \mathbb{Z}$ and $N_1 = N_2 = \mathbb{Z}/2\mathbb{Z}$, then each element of $N_1 \oplus N_2$ has a unique representation in the form $n_1 + n_2$ where each $n_i \in N_i$, however n_1 (for instance) can be expressed as n_1 or $3n_1$ or $5n_1 \ldots$ etc., so each element does not have a unique representation in the form $r_1 a_1 + r_2 a_2$, where $r_1, r_2 \in R$, $a_1 \in N_1$ and $a_2 \in N_2$. Thus $\mathbb{Z}/2\mathbb{Z} \oplus \mathbb{Z}/2\mathbb{Z}$ is not a free $\mathbb{Z}$-module on the set $\{(1, 0), (0, 1)\}$. Similarly, it is not free on any set.

Theorem 6. For any set A there is a free R-module $F(A)$ on the set A and $F(A)$ satisfies the following *universal property:* if M is any R-module and $\varphi : A \to M$ is any map of sets, then there is a unique R-module homomorphism $\Phi : F(A) \to M$ such that $\Phi(a) = \varphi(a)$, for all $a \in A$, that is, the following diagram commutes.

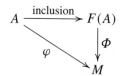

When A is the finite set $\{a_1, a_2, \ldots, a_n\}$, $F(A) = Ra_1 \oplus Ra_2 \oplus \cdots \oplus Ra_n \cong R^n$. (Compare: Section 6.3, free groups.)

Proof: Let $F(A) = \{0\}$ if $A = \emptyset$. If A is nonempty let $F(A)$ be the collection of all set functions $f : A \to R$ such that $f(a) = 0$ for all but finitely many $a \in A$. Make

354

$F(A)$ into an R-module by pointwise addition of functions and pointwise multiplication of a ring element times a function, i.e.,

$$(f + g)(a) = f(a) + g(a) \quad \text{and}$$
$$(rf)(a) = r(f(a)), \quad \text{for all } a \in A, \ r \in R \text{ and } f, g \in F(A).$$

It is an easy matter to check that all the R-module axioms hold (the details are omitted). Identify A as a subset of $F(A)$ by $a \mapsto f_a$, where f_a is the function which is 1 at a and zero elsewhere. We can, in this way, think of $F(A)$ as all finite R-linear combinations of elements of A by identifying each function f with the sum $r_1 a_1 + r_2 a_2 + \cdots + r_n a_n$, where f takes on the value r_i at a_i and is zero at all other elements of A. Moreover, each element of $F(A)$ has a unique expression as such a formal sum. To establish the universal property of $F(A)$ suppose $\varphi : A \to M$ is a map of the set A into the R-module M. Define $\Phi : F(A) \to M$ by

$$\Phi : \sum_{i=1}^{n} r_i a_i \mapsto \sum_{i=1}^{n} r_i \varphi(a_i).$$

By the uniqueness of the expression for the elements of $F(A)$ as linear combinations of the a_i we see easily that Φ is a well defined R-module homomorphism (the details are left as an exercise). By definition, the restriction of Φ to A equals φ. Finally, since $F(A)$ is generated by A, once we know the values of an R-module homomorphism on A its values on every element of $F(A)$ are uniquely determined, so Φ is the unique extension of φ to all of $F(A)$.

When A is the finite set $\{a_1, a_2, \ldots, a_n\}$ Proposition 5(3) shows that $F(A) = Ra_1 \oplus Ra_2 \oplus \cdots \oplus Ra_n$. Since $R \cong Ra_i$ for all i (under the map $r \mapsto ra_i$) Proposition 5(1) shows that the direct sum is isomorphic to R^n.

Corollary 7.
 (1) If F_1 and F_2 are free modules on the same set A, there is a unique isomorphism between F_1 and F_2 which is the identity map on A.
 (2) If F is any free R-module with basis A, then $F \cong F(A)$. In particular, F enjoys the same universal property with respect to A as $F(A)$ does in Theorem 6.

Proof: Exercise.

If F is a free R-module with basis A, we shall often (particularly in the case of vector spaces) define R-module homomorphisms from F into other R-modules simply by specifying their values on the elements of A and then saying "*extend by linearity.*" Corollary 7(2) ensures that this is permissible.

When $R = \mathbb{Z}$, the free module on a set A is called the *free abelian group on A*. If $|A| = n$, $F(A)$ is called the free abelian group of *rank n* and is isomorphic to $\mathbb{Z} \oplus \cdots \oplus \mathbb{Z}$ (n times). These definitions agree with the ones given in Chapter 5.

EXERCISES

In these exercises R is a ring with 1 and M is a left R-module.

1. Prove that if A and B are sets of the same cardinality, then the free modules $F(A)$ and $F(B)$ are isomorphic.

2. Assume R is commutative. Prove that $R^n \cong R^m$ if and only if $n = m$, i.e., two free R-modules of finite rank are isomorphic if and only if they have the same rank. [Apply Exercise 12 of Section 2 with I a maximal ideal of R. You may assume that if F is a field, then $F^n \cong F^m$ if and only if $n = m$, i.e., two finite dimensional vector spaces over F are isomorphic if and only if they have the same dimension — this will be proved later in Section 11.1.]

3. Show that the $F[x]$-modules in Exercises 18 and 19 of Section 1 are both cyclic.

4. An R-module M is called a *torsion* module if for each $m \in M$ there is a nonzero element $r \in R$ such that $rm = 0$, where r may depend on m (i.e., $M = \mathrm{Tor}(M)$ in the notation of Exercise 8 of Section 1). Prove that every finite abelian group is a torsion $\mathbb{Z}$-module. Give an example of an infinite abelian group that is a torsion $\mathbb{Z}$-module.

5. Let R be an integral domain. Prove that every finitely generated torsion R-module has a nonzero annihilator i.e., there is a nonzero element $r \in R$ such that $rm = 0$ for all $m \in M$ — here r does not depend on m (the annihilator of a module was defined in Exercise 9 of Section 1). Give an example of a torsion R-module whose annihilator is the zero ideal.

6. Prove that if M is a finitely generated R-module that is generated by n elements then every quotient of M may be generated by n (or fewer) elements. Deduce that quotients of cyclic modules are cyclic.

7. Let N be a submodule of M. Prove that if both M/N and N are finitely generated then so is M.

8. Let S be the collection of sequences $(a_1, a_2, a_3, \dots)$ of integers $a_1, a_2, a_3, \dots$ where all but finitely many of the a_i are 0 (called the *direct sum* of infinitely many copies of $\mathbb{Z}$). Recall that S is a ring under componentwise addition and multiplication and S does not have a multiplicative identity — cf. Exercise 20, Section 7.1. Prove that S is not finitely generated as a module over itself.

9. An R-module M is called *irreducible* if $M \neq 0$ and if 0 and M are the only submodules of M. Show that M is irreducible if and only if $M \neq 0$ and M is a cyclic module with any nonzero element as generator. Determine all the irreducible $\mathbb{Z}$-modules.

10. Assume R is commutative. Show that an R-module M is irreducible if and only if M is isomorphic (as an R-module) to R/I where I is a maximal ideal of R. [By the previous exercise, if M is irreducible there is a natural map $R \to M$ defined by $r \mapsto rm$, where m is any fixed nonzero element of M.]

11. Show that if M_1 and M_2 are irreducible R-modules, then any nonzero R-module homomorphism from M_1 to M_2 is an isomorphism. Deduce that if M is irreducible then $\mathrm{End}_R(M)$ is a division ring (this result is called *Schur's Lemma*). [Consider the kernel and the image.]

12. Let R be a commutative ring and let A, B and M be R-modules. Prove the following isomorphisms of R-modules:
   (a) $\mathrm{Hom}_R(A \times B, M) \cong \mathrm{Hom}_R(A, M) \times \mathrm{Hom}_R(B, M)$
   (b) $\mathrm{Hom}_R(M, A \times B) \cong \mathrm{Hom}_R(M, A) \times \mathrm{Hom}_R(M, B)$.

13. Let R be a commutative ring and let F be a free R-module of finite rank. Prove the following isomorphism of R-modules: $\mathrm{Hom}_R(F, R) \cong F$.

14. Let R be a commutative ring and let F be the free R-module of rank n. Prove that $\text{Hom}_R(F, M) \cong M \times \cdots \times M$ (n times). [Use Exercise 9 in Section 2 and Exercise 12.]

15. An element $e \in R$ is called a *central idempotent* if $e^2 = e$ and $er = re$ for all $r \in R$. If e is a central idempotent in R, prove that $M = eM \oplus (1-e)M$. [Recall Exercise 14 in Section 1.]

The next two exercises establish the Chinese Remainder Theorem for modules (cf. Section 7.6).

16. For any ideal I of R let IM be the submodule defined in Exercise 5 of Section 1. Let $A_1, \ldots, A_k$ be any ideals in the ring R. Prove that the map

$$M \to M/A_1 M \times \cdots \times M/A_k M \quad \text{defined by} \quad m \mapsto (m + A_1 M, \ldots, m + A_k M)$$

is an R-module homomorphism with kernel $A_1 M \cap A_2 M \cap \cdots \cap A_k M$.

17. In the notation of the preceding exercise, assume further that the ideals $A_1, \ldots, A_k$ are pairwise comaximal (i.e., $A_i + A_j = R$ for all $i \neq j$). Prove that

$$M/(A_1 \cdots A_k)M \cong M/A_1 M \times \cdots \times M/A_k M.$$

[See the proof of the Chinese Remainder Theorem for rings in Section 7.6.]

18. Let R be a Principal Ideal Domain and let M be an R-module that is annihilated by the nonzero, proper ideal (a). Let $a = p_1^{\alpha_1} p_2^{\alpha_2} \cdots p_k^{\alpha_k}$ be the unique factorization of a into distinct prime powers in R. Let M_i be the annihilator of $p_i^{\alpha_i}$ in M, i.e., M_i is the set $\{m \in M \mid p_i^{\alpha_i} m = 0\}$ — called the p_i-*primary component* of M. Prove that

$$M = M_1 \oplus M_2 \oplus \cdots \oplus M_k.$$

19. Show that if M is a finite abelian group of order $a = p_1^{\alpha_1} p_2^{\alpha_2} \cdots p_k^{\alpha_k}$ then, considered as a $\mathbb{Z}$-module, M is annihilated by (a), the p_i-primary component of M is the unique Sylow p_i-subgroup of M and M is isomorphic to the direct product of its Sylow subgroups.

20. Let I be a nonempty index set and for each $i \in I$ let M_i be an R-module. The *direct product* of the modules M_i is defined to be their direct product as abelian groups (cf. Exercise 15 in Section 5.1) with the action of R componentwise multiplication. The *direct sum* of the modules M_i is defined to be the restricted direct product of the abelian groups M_i (cf. Exercise 17 in Section 5.1) with the action of R componentwise multiplication. In other words, the direct sum of the M_i's is the subset of the direct product, $\prod_{i \in I} M_i$, which consists of all elements $\prod_{i \in I} m_i$ such that only finitely many of the components m_i are nonzero; the action of R on the direct product or direct sum is given by $r \prod_{i \in I} m_i = \prod_{i \in I} rm_i$ (cf. Appendix I for the definition of Cartesian products of infinitely many sets). The direct sum will be denoted by $\oplus_{i \in I} M_i$.
  (a) Prove that the direct product of the M_i's is an R-module and the direct sum of the M_i's is a submodule of their direct product.
  (b) Show that if $R = \mathbb{Z}$, $I = \mathbb{Z}^+$ and M_i is the cyclic group of order i for each i, then the direct sum of the M_i's is not isomorphic to their direct product. [Look at torsion.]

21. Let I be a nonempty index set and for each $i \in I$ let N_i be a submodule of M. Prove that the following are equivalent:
  (i) the submodule of M generated by all the N_i's is isomorphic to the direct sum of the N_i's
  (ii) if $\{i_1, i_2, \ldots, i_k\}$ is any finite subset of I then $N_{i_1} \cap (N_{i_2} + \cdots + N_{i_k}) = 0$
  (iii) if $\{i_1, i_2, \ldots, i_k\}$ is any finite subset of I then $N_1 + \cdots + N_k = N_1 \oplus \cdots \oplus N_k$
  (iv) for every element x of the submodule of M generated by the N_i's there are unique elements $a_i \in N_i$ for all $i \in I$ such that all but a finite number of the a_i are zero and x is the (finite) sum of the a_i.

22. Let R be a Principal Ideal Domain, let M be a torsion R-module (cf. Exercise 4) and let p be a prime in R (do not assume M is finitely generated, hence it need not have a nonzero annihilator — cf. Exercise 5). The *p-primary component* of M is the set of all elements of M that are annihilated by some positive power of p.

   (a) Prove that the p-primary component is a submodule. [See Exercise 13 in Section 1.]

   (b) Prove that this definition of p-primary component agrees with the one given in Exercise 18 when M has a nonzero annihilator.

   (c) Prove that M is the (possibly infinite) direct sum of its p-primary components, as p runs over all primes of R.

23. Show that any direct sum of free R-modules is free.

24. *(An arbitrary direct product of free modules need not be free)* For each positive integer i let M_i be the free $\mathbb{Z}$-module $\mathbb{Z}$, and let M be the direct product $\prod_{i \in \mathbb{Z}^+} M_i$ (cf. Exercise 20). Each element of M can be written uniquely in the form $(a_1, a_2, a_3, \dots)$ with $a_i \in \mathbb{Z}$ for all i. Let N be the submodule of M consisting of all such tuples with only finitely many nonzero a_i. Assume M is a free $\mathbb{Z}$-module with basis $\mathcal{B}$.

   (a) Show that N is countable.

   (b) Show that there is some countable subset $\mathcal{B}_1$ of $\mathcal{B}$ such that N is contained in the submodule, N_1, generated by $\mathcal{B}_1$. Show also that N_1 is countable.

   (c) Let $\overline{M} = M/N_1$. Show that $\overline{M}$ is a free $\mathbb{Z}$-module. Deduce that if $\overline{x}$ is any nonzero element of $\overline{M}$ then there are only finitely many distinct positive integers k such that $\overline{x} = k\overline{m}$ for some $m \in M$ (depending on k).

   (d) Let $\mathcal{S} = \{(b_1, b_2, b_3, \dots) \mid b_i = \pm i! \text{ for all } i\}$. Prove that $\mathcal{S}$ is uncountable. Deduce that there is some $s \in \mathcal{S}$ with $s \notin N_1$.

   (e) Show that the assumption M is free leads to a contradiction: By (d) we may choose $s \in \mathcal{S}$ with $s \notin N_1$. Show that for each positive integer k there is some $m \in M$ with $\overline{s} = k\overline{m}$, contrary to (c). [Use the fact that $N \subseteq N_1$.]

25. In the construction of direct limits, Exercise 8 of Section 7.6, show that if all A_i are R-modules and the maps ρ_{ij} are R-module homomorphisms, then the direct limit $A = \varinjlim A_i$ may be given the structure of an R-module in a natural way such that the maps $\rho_i : A_i \to A$ are all R-module homomorphisms. Verify the corresponding universal property (part (e)) for R-module homomorphisms $\varphi_i : A_i \to C$ commuting with the ρ_{ij}.

26. Carry out the analysis of the preceding exercise corresponding to inverse limits to show that an inverse limit of R-modules is an R-module satisfying the appropriate universal property (cf. Exercise 10 of Section 7.6).

27. *(Free modules over noncommutative rings need not have a unique rank)* Let M be the $\mathbb{Z}$-module $\mathbb{Z} \times \mathbb{Z} \times \cdots$ of Exercise 24 and let R be its endomorphism ring, $R = \mathrm{End}_{\mathbb{Z}}(M)$ (cf. Exercises 29 and 30 in Section 7.1). Define $\varphi_1, \varphi_2 \in R$ by

$$\varphi_1(a_1, a_2, a_3, \dots) = (a_1, a_3, a_5, \dots)$$
$$\varphi_2(a_1, a_2, a_3, \dots) = (a_2, a_4, a_6, \dots)$$

   (a) Prove that $\{\varphi_1, \varphi_2\}$ is a free basis of the left R-module R. [Define the maps ψ_1 and ψ_2 by $\psi_1(a_1, a_2, \dots) = (a_1, 0, a_2, 0, \dots)$ and $\psi_2(a_1, a_2, \dots) = (0, a_1, 0, a_2, \dots)$. Verify that $\varphi_i \psi_i = 1$, $\varphi_1 \psi_2 = 0 = \varphi_2 \psi_1$ and $\psi_1 \varphi_1 + \psi_2 \varphi_2 = 1$. Use these relations to prove that φ_1, φ_2 are independent and generate R as a left R-module.]

   (b) Use (a) to prove that $R \cong R^2$ and deduce that $R \cong R^n$ for all $n \in \mathbb{Z}^+$.

10.4 TENSOR PRODUCTS OF MODULES

In this section we study the tensor product of two modules M and N over a ring (not necessarily commutative) containing 1. Formation of the tensor product is a general construction that, loosely speaking, enables one to form another module in which one can take "products" mn of elements $m \in M$ and $n \in N$. The general construction involves various left- and right- module actions, and it is instructive, by way of motivation, to first consider an important special case: the question of "extending scalars" or "changing the base."

Suppose that the ring R is a subring of the ring S. Throughout this section, we always assume that $1_R = 1_S$ (this ensures that S is a unital R-module).

If N is a left S-module, then N can also be naturally considered as a left R-module since the elements of R (being elements of S) act on N by assumption. The S-module axioms for N include the relations

$$(s_1 + s_2)n = s_1 n + s_2 n \quad \text{and} \quad s(n_1 + n_2) = sn_1 + sn_2 \qquad (10.1)$$

for all $s, s_1, s_2 \in S$ and all $n, n_1, n_2 \in N$, and the relation

$$(s_1 s_2)n = s_1(s_2 n) \quad \text{for all } s_1, s_2 \in S, \text{ and all } n \in N. \qquad (10.2)$$

A particular case of the latter relation is

$$(sr)n = s(rn) \quad \text{for all } s \in S, r \in R \text{ and } n \in N. \qquad (10.2')$$

More generally, if $f : R \to S$ is a ring homomorphism from R into S with $f(1_R) = 1_S$ (for example the injection map if R is a subring of S as above) then it is easy to see that N can be considered as an R-module with $rn = f(r)n$ for $r \in R$ and $n \in N$. In this situation S can be considered as an *extension* of the ring R and the resulting R-module is said to be obtained from N by *restriction of scalars* from S to R.

Suppose now that R is a subring of S and we try to reverse this, namely we start with an R-module N and attempt to define an S-module structure on N that extends the action of R on N to an action of S on N (hence "extending the scalars" from R to S). In general this is impossible, even in the simplest situation: the ring R itself is an R-module but is usually not an S-module for the larger ring S. For example, $\mathbb{Z}$ is a $\mathbb{Z}$-module but it cannot be made into a $\mathbb{Q}$-module (if it could, then $\frac{1}{2} \circ 1 = z$ would be an element of $\mathbb{Z}$ with $z + z = 1$, which is impossible). Although $\mathbb{Z}$ itself cannot be made into a $\mathbb{Q}$-module it is *contained* in a $\mathbb{Q}$-module, namely $\mathbb{Q}$ itself. Put another way, there is an injection (also called an *embedding*) of the $\mathbb{Z}$-module $\mathbb{Z}$ into the $\mathbb{Q}$-module $\mathbb{Q}$ (and similarly the ring R can always be embedded as an R-submodule of the S-module S). This raises the question of whether an arbitrary R-module N can be embedded as an R-submodule of some S-module, or more generally, the question of what R-module homomorphisms exist from N to S-modules. For example, suppose N is a nontrivial *finite* abelian group, say $N = \mathbb{Z}/2\mathbb{Z}$, and consider possible $\mathbb{Z}$-module homomorphisms (i.e., abelian group homomorphisms) of N into some $\mathbb{Q}$-module. A $\mathbb{Q}$-module is just a vector space over $\mathbb{Q}$ and every nonzero element in a vector space over $\mathbb{Q}$ has infinite (additive) order. Since every element of N has finite order, every element of N must map to 0 under such a homomorphism. In other words there are *no* nonzero $\mathbb{Z}$-module homomorphisms from this N to *any* $\mathbb{Q}$-module, much less embeddings of N identifying

N as a submodule of a $\mathbb{Q}$-module. The two $\mathbb{Z}$-modules $\mathbb{Z}$ and $\mathbb{Z}/2\mathbb{Z}$ exhibit extremely different behaviors when we try to "extend scalars" from $\mathbb{Z}$ to $\mathbb{Q}$: the first module maps injectively into some $\mathbb{Q}$-module, the second always maps to 0 in a $\mathbb{Q}$-module.

We now construct for a general R-module N an S-module that is the "best possible" target in which to try to embed N. We shall also see that this module determines *all* of the possible R-module homomorphisms of N into S-modules, in particular determining when N is contained in some S-module (cf. Corollary 9). In the case of $R = \mathbb{Z}$ and $S = \mathbb{Q}$ this construction will give us $\mathbb{Q}$ when applied to the module $N = \mathbb{Z}$, and will give us 0 when applied to the module $N = \mathbb{Z}/2\mathbb{Z}$ (Examples 2 and 3 following Corollary 9).

If the R-module N were already an S-module then of course there is no difficulty in "extending" the scalars from R to S, so we begin the construction by returning to the basic module axioms in order to examine whether we can define "products" of the form sn, for $s \in S$ and $n \in N$. These axioms start with an abelian group N together with a map from $S \times N$ to N, where the image of the pair (s, n) is denoted by sn. It is therefore natural to consider the free $\mathbb{Z}$-module (i.e., , the free abelian group) on the set $S \times N$, i.e., the collection of all finite commuting sums of elements of the form (s_i, n_i) where $s_i \in S$ and $n_i \in N$. This is an abelian group where there are no relations between any distinct pairs (s, n) and (s', n'), i.e., no relations between the "formal products" sn, and in this abelian group the original module N has been thoroughly distinguished from the new "coefficients" from S. To satisfy the relations necessary for an S-module structure imposed in equation (1) and the compatibility relation with the action of R on N in (2'), we must take the quotient of this abelian group by the subgroup H generated by all elements of the form

$$
\begin{gathered}
(s_1 + s_2, n) - (s_1, n) - (s_2, n), \\
(s, n_1 + n_2) - (s, n_1) - (s, n_2), \text{ and} \\
(sr, n) - (s, rn),
\end{gathered}
\tag{10.3}
$$

for $s, s_1, s_2 \in S$, $n, n_1, n_2 \in N$ and $r \in R$, where rn in the last element refers to the R-module structure already defined on N.

The resulting quotient group is denoted by $S \otimes_R N$ (or just $S \otimes N$ if R is clear from the context) and is called the *tensor product of S and N over R*. If $s \otimes n$ denotes the coset containing (s, n) in $S \otimes_R N$ then by definition of the quotient we have forced the relations

$$
\begin{gathered}
(s_1 + s_2) \otimes n = s_1 \otimes n + s_2 \otimes n, \\
s \otimes (n_1 + n_2) = s \otimes n_1 + s \otimes n_2, \text{ and} \\
sr \otimes n = s \otimes rn.
\end{gathered}
\tag{10.4}
$$

The elements of $S \otimes_R N$ are called *tensors* and can be written (non-uniquely in general) as finite sums of "simple tensors" of the form $s \otimes n$ with $s \in S, n \in N$.

We now show that the tensor product $S \otimes_R N$ is naturally a left S-module under the action defined by

$$
s \left(\sum_{\text{finite}} s_i \otimes n_i \right) = \sum_{\text{finite}} (s s_i) \otimes n_i.
\tag{10.5}
$$

We first check this is well defined, i.e., independent of the representation of the element of $S \otimes_R N$ as a sum of simple tensors. Note first that if s' is any element of S then

$$(s'(s_1 + s_2), n) - (s's_1, n) - (s's_2, n) \quad \big(= (s's_1 + s's_2, n) - (s's_1, n) - (s's_2, n) \big),$$

$$(s's, n_1 + n_2) - (s's, n_1) - (s's, n_2), \text{ and}$$

$$(s'(sr), n) - (s's, rn) \quad \big(= ((s's)r, n) - (s's, rn) \big)$$

each belongs to the set of generators in (3), so in particular each lies in the subgroup H. This shows that multiplying the first entries of the generators in (3) on the left by s' gives another element of H (in fact another generator). Since any element of H is a sum of elements as in (3), it follows that for any element $\sum(s_i, n_i)$ in H also $\sum(s's_i, n_i)$ lies in H. Suppose now that $\sum s_i \otimes n_i = \sum s'_i \otimes n'_i$ are two representations for the same element in $S \otimes_R N$. Then $\sum(s_i, n_i) - \sum(s'_i, n'_i)$ is an element of H, and by what we have just seen, for any $s \in S$ also $\sum(ss_i, n_i) - \sum(ss'_i, n'_i)$ is an element of H. But this means that $\sum ss_i \otimes n_i = \sum ss'_i \otimes n'_i$ in $S \otimes_R N$, so the expression in (5) is indeed well defined.

It is now straightforward using the relations in (4) to check that the action defined in (5) makes $S \otimes_R N$ into a left S-module. For example, on the simple tensor $s_i \otimes n_i$,

$$
\begin{aligned}
(s + s')(s_i \otimes n_i) &= ((s + s')s_i) \otimes n_i && \text{by definition (5)} \\
&= (ss_i + s's_i) \otimes n_i \\
&= ss_i \otimes n_i + s's_i \otimes n_i && \text{by the first relation in (4)} \\
&= s(s_i \otimes n_i) + s'(s_i \otimes n_i) && \text{by definition (5)} .
\end{aligned}
$$

The module $S \otimes_R N$ is called *the (left) S-module obtained by extension of scalars from the (left) R-module N*.

There is a natural map $\iota : N \to S \otimes_R N$ defined by $n \mapsto 1 \otimes n$ (i.e., first map $n \in N$ to the element $(1, n)$ in the free abelian group and then pass to the quotient group). Since $1 \otimes rn = r \otimes n = r(1 \otimes n)$ by (4) and (5), it is easy to check that ι is an R-module homomorphism from N to $S \otimes_R N$. Since we have passed to a quotient group, however, ι is not injective in general. Hence, while there is a natural R-module homomorphism from the original left R-module N to the left S-module $S \otimes_R N$, in general $S \otimes_R N$ need not contain (an isomorphic copy of) N. On the other hand, the relations in equation (3) were the *minimal* relations that we had to impose in order to obtain an S-module, so it is reasonable to expect that the tensor product $S \otimes_R N$ is the "best possible" S-module to serve as target for an R-module homomorphism from N. The next theorem makes this more precise by showing that any other R-module homomorphism from N factors through this one, and is referred to as the *universal property* for the tensor product $S \otimes_R N$. The analogous result for the general tensor product is given in Theorem 10.

Theorem 8. Let R be a subring of S, let N be a left R-module and let $\iota : N \to S \otimes_R N$ be the R-module homomorphism defined by $\iota(n) = 1 \otimes n$. Suppose that L is any left S-module (hence also an R-module) and that $\varphi : N \to L$ is an R-module homomorphism from N to L. Then there is a unique S-module homomorphism $\Phi : S \otimes_R N \to L$ such that φ factors through Φ, i.e., $\varphi = \Phi \circ \iota$ and the diagram

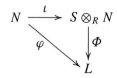

commutes. Conversely, if $\Phi : S \otimes_R N \to L$ is an S-module homomorphism then $\varphi = \Phi \circ \iota$ is an R-module homomorphism from N to L.

Proof: Suppose $\varphi : N \to L$ is an R-module homomorphism to the S-module L. By the universal property of free modules (Theorem 6 in Section 3) there is a $\mathbb{Z}$-module homomorphism from the free $\mathbb{Z}$-module F on the set $S \times N$ to L that sends each generator (s, n) to $s\varphi(n)$. Since φ is an R-module homomorphism, the generators of the subgroup H in equation (3) all map to zero in L. Hence this $\mathbb{Z}$-module homomorphism factors through H, i.e., there is a well defined $\mathbb{Z}$-module homomorphism Φ from $F/H = S \otimes_R N$ to L satisfying $\Phi(s \otimes n) = s\varphi(n)$. Moreover, on simple tensors we have

$$s'\Phi(s \otimes n) = s'(s\varphi(n)) = (s's)\varphi(n) = \Phi((s's) \otimes n) = \Phi(s'(s \otimes n)).$$

for any $s' \in S$. Since Φ is additive it follows that Φ is an S-module homomorphism, which proves the existence statement of the theorem. The module $S \otimes_R N$ is generated as an S-module by elements of the form $1 \otimes n$, so any S-module homomorphism is uniquely determined by its values on these elements. Since $\Phi(1 \otimes n) = \varphi(n)$, it follows that the S-module homomorphism Φ is uniquely determined by φ, which proves the uniqueness statement of the theorem. The converse statement is immediate.

The universal property of $S \otimes_R N$ in Theorem 8 shows that R-module homomorphisms of N into S-modules arise from S-module homomorphisms from $S \otimes_R N$. In particular this determines when it is possible to map N injectively into some S-module:

Corollary 9. Let $\iota : N \to S \otimes_R N$ be the R-module homomorphism in Theorem 8. Then $N/\ker \iota$ is the unique largest quotient of N that can be embedded in any S-module. In particular, N can be embedded as an R-submodule of some left S-module if and only if ι is injective (in which case N is isomorphic to the R-submodule $\iota(N)$ of the S-module $S \otimes_R N$).

Proof: The quotient $N/\ker \iota$ is mapped injectively (by ι) into the S-module $S \otimes_R N$. Suppose now that φ is an R-module homomorphism injecting the quotient $N/\ker \varphi$ of N into an S-module L. Then, by Theorem 8, $\ker \iota$ is mapped to 0 by φ, i.e., $\ker \iota \subseteq \ker \varphi$. Hence $N/\ker \varphi$ is a quotient of $N/\ker \iota$ (namely, the quotient by the submodule $\ker \varphi/\ker \iota$). It follows that $N/\ker \iota$ is the unique largest quotient of N that can be embedded in any S-module. The last statement in the corollary follows immediately.

Examples

(1) For any ring R and any left R-module N we have $R \otimes_R N \cong N$ (so "extending scalars from R to R" does not change the module). This follows by taking φ to be the identity map from N to itself (and $S = R$) in Theorem 8: ι is then an isomorphism with inverse isomorphism given by Φ. In particular, if A is any abelian group (i.e., a $\mathbb{Z}$-module), then $\mathbb{Z} \otimes_{\mathbb{Z}} A = A$.

(2) Let $R = \mathbb{Z}$, $S = \mathbb{Q}$ and let A be a finite abelian group of order n. In this case the $\mathbb{Q}$-module $\mathbb{Q} \otimes_{\mathbb{Z}} A$ obtained by extension of scalars from the $\mathbb{Z}$-module A is 0. To see this, observe first that in any tensor product $1 \otimes 0 = 1 \otimes (0 + 0) = 1 \otimes 0 + 1 \otimes 0$, by the second relation in (4), so
$$1 \otimes 0 = 0.$$

Now, for any simple tensor $q \otimes a$ we can write the rational number q as $(q/n)n$. Then since $na = 0$ in A by Lagrange's Theorem, we have
$$q \otimes a = (\frac{q}{n} \cdot n) \otimes a = \frac{q}{n} \otimes (na) = (q/n) \otimes 0 = (q/n)(1 \otimes 0) = 0.$$

It follows that $\mathbb{Q} \otimes_{\mathbb{Z}} A = 0$. In particular, the map $\iota : A \to S \otimes_R A$ is the zero map. By Theorem 8, we see again that any homomorphism of a finite abelian group into a rational vector space is the zero map. In particular, if A is nontrivial, then the original $\mathbb{Z}$-module A is not contained in the $\mathbb{Q}$-module obtained by extension of scalars.

(3) *Extension of scalars for free modules:* If $N \cong R^n$ is a free module of rank n over R then $S \otimes_R N \cong S^n$ is a free module of rank n over S. We shall prove this shortly (Corollary 18) when we discuss tensor products of direct sums. For example, $\mathbb{Q} \otimes_{\mathbb{Z}} \mathbb{Z}^n \cong \mathbb{Q}^n$. In this case the module obtained by extension of scalars contains (an isomorphic copy of) the original R-module N. For example, $\mathbb{Q} \otimes_{\mathbb{Z}} \mathbb{Z}^n \cong \mathbb{Q}^n$ and $\mathbb{Z}^n$ is a subgroup of the abelian group $\mathbb{Q}^n$.

(4) *Extension of scalars for vector spaces:* As a special case of the previous example, let F be a subfield of the field K and let V be an n-dimensional vector space over F (i.e., $V \cong F^n$). Then $K \otimes_F V \cong K^n$ is a vector space over the larger field K of the same dimension, and the original vector space V is contained in $K \otimes_F V$ as an F-vector subspace.

(5) *Induced modules for finite groups:* Let R be a commutative ring with 1, let G be a finite group and let H be a subgroup of G. As in Section 7.2 we may form the group ring RG and its subring RH. For any RH-module N define the *induced module* $RG \otimes_{RH} N$. In this way we obtain an RG-module for each RH-module N. We shall study properties of induced modules and some of their important applications to group theory in Chapters 17 and 19.

The general tensor product construction follows along the same lines as the extension of scalars above, but before describing it we make two observations from this special case. The first is that the construction of $S \otimes_R N$ as an *abelian group* involved only the elements in equation (3), which in turn only required S to be a *right R-module* and N to be a *left R-module*. In a similar way we shall construct an *abelian group* $M \otimes_R N$ for *any right R-module M* and *any left R-module N*. The second observation is that the S-*module* structure on $S \otimes_R N$ defined by equation (5) required only a *left* S-module structure on S together with a "compatibility relation"
$$s'(sr) = (s's)r \qquad \text{for } s, s' \in S, r \in R,$$

between this left S-module structure and the right R-module structure on S (this was needed in order to deduce that (5) was well defined). We first consider the general construction of $M \otimes_R N$ as an abelian group, after which we shall return to the question of when this abelian group can be given a module structure.

Suppose then that N is a left R-module and that M is a right R-module. The quotient of the free $\mathbb{Z}$-module on the set $M \times N$ by the subgroup generated by all elements of the form

$$(m_1 + m_2, n) - (m_1, n) - (m_2, n),$$
$$(m, n_1 + n_2) - (m, n_1) - (m, n_2), \text{ and} \tag{10.6}$$
$$(mr, n) - (m, rn),$$

for $m, m_1, m_2 \in M, n, n_1, n_2 \in N$ and $r \in R$ is an abelian group, denoted by $M \otimes_R N$, or simply $M \otimes N$ if the ring R is clear from the context, and is called the *tensor product of M and N over R*. The elements of $M \otimes_R N$ are called *tensors*, and the coset, $m \otimes n$, of (m, n) in $M \otimes_R N$ is called a simple tensor. We have the relations

$$(m_1 + m_2) \otimes n = m_1 \otimes n + m_2 \otimes n,$$
$$m \otimes (n_1 + n_2) = m \otimes n_1 + m \otimes n_2, \text{ and} \tag{10.7}$$
$$mr \otimes n = m \otimes rn.$$

Every tensor can be written (non-uniquely in general) as a finite sum of simple tensors.

Remark: We emphasize that care must be taken when working with tensors, since each $m \otimes n$ represents a *coset* in some quotient group, and so we may have $m \otimes n = m' \otimes n'$ where $m \neq m'$ or $n \neq n'$. More generally, an element of $M \otimes N$ may be expressible in many different ways as a sum of simple tensors. In particular, care must be taken when defining maps from $M \otimes_R N$ to another group or module, since a map from $M \otimes N$ which is described on the generators $m \otimes n$ in terms of m and n is not well defined unless it is shown to be independent of the particular choice of $m \otimes n$ as a coset representative.

Another point where care must be exercised is in reference to the element $m \otimes n$ when the modules M and N or the ring R are not clear from the context. The first two examples of extension of scalars give an instance where M is a submodule of a larger module M', and for some $m \in M$ and $n \in N$ we have $m \otimes n = 0$ in $M' \otimes_R N$ but $m \otimes n$ is *nonzero* in $M \otimes_R N$. This is possible because the symbol "$m \otimes n$" represents different cosets, hence possibly different elements, in the two tensor products. In particular, these two examples show that $M \otimes_R N$ need not be a subgroup of $M' \otimes_R N$ even when M is a submodule of M' (cf. also Exercise 2).

Mapping $M \times N$ to the free $\mathbb{Z}$-module on $M \times N$ and then passing to the quotient defines a map $\iota : M \times N \to M \otimes_R N$ with $\iota(m, n) = m \otimes n$. This map is in general not a group homomorphism, but it is additive in both m and n separately and satisfies $\iota(mr, n) = mr \otimes n = m \otimes rn = \iota(m, rn)$. Such maps are given a name:

Definition. Let M be a right R-module, let N be a left R-module and let L be an abelian group (written additively). A map $\varphi : M \times N \to L$ is called R-*balanced* or *middle linear with respect to R* if

$$\varphi(m_1 + m_2, n) = \varphi(m_1, n) + \varphi(m_2, n)$$
$$\varphi(m, n_1 + n_2) = \varphi(m, n_1) + \varphi(m, n_2)$$
$$\varphi(m, rn) = \varphi(mr, n)$$

for all $m, m_1, m_2 \in M$, $n, n_1, n_2 \in N$, and $r \in R$.

With this terminology, it follows immediately from the relations in (7) that the map $\iota : M \times N \to M \otimes_R N$ is R-balanced. The next theorem proves the extremely useful *universal property of the tensor product* with respect to balanced maps.

Theorem 10. Suppose R is a ring with 1, M is a right R-module, and N is a left R-module. Let $M \otimes_R N$ be the tensor product of M and N over R and let $\iota : M \times N \to M \otimes_R N$ be the R-balanced map defined above.

   (1) If $\Phi : M \otimes_R N \to L$ is any group homomorphism from $M \otimes_R N$ to an abelian group L then the composite map $\varphi = \Phi \circ \iota$ is an R-balanced map from $M \times N$ to L.
   (2) Conversely, suppose L is an abelian group and $\varphi : M \times N \to L$ is any R-balanced map. Then there is a unique group homomorphism $\Phi : M \otimes_R N \to L$ such that φ factors through ι, i.e., $\varphi = \Phi \circ \iota$ as in (1).

Equivalently, the correspondence $\varphi \leftrightarrow \Phi$ in the commutative diagram

$$
\begin{array}{ccc}
M \times N & \xrightarrow{\ \iota\ } & M \otimes_R N \\
& {\scriptstyle \varphi} \searrow & \big\downarrow {\scriptstyle \Phi} \\
& & L
\end{array}
$$

establishes a bijection

$$\left\{ \begin{array}{c} R\text{-balanced maps} \\ \varphi : M \times N \to L \end{array} \right\} \longleftrightarrow \left\{ \begin{array}{c} \text{group homomorphisms} \\ \Phi : M \otimes_R N \to L \end{array} \right\}.$$

Proof: The proof of (1) is immediate from the properties of ι above. For (2), the map φ defines a unique $\mathbb{Z}$-module homomorphism $\tilde{\varphi}$ from the free group on $M \times N$ to L (Theorem 6 in Section 3) such that $\tilde{\varphi}(m, n) = \varphi(m, n) \in L$. Since φ is R-balanced, $\tilde{\varphi}$ maps each of the elements in equation (6) to 0; for example

$$\tilde{\varphi}\left((mr, n) - (m, rn)\right) = \varphi(mr, n) - \varphi(m, rn) = 0.$$

It follows that the kernel of $\tilde{\varphi}$ contains the subgroup generated by these elements, hence $\tilde{\varphi}$ induces a homomorphism Φ on the quotient group $M \otimes_R N$ to L. By definition we then have

$$\Phi(m \otimes n) = \tilde{\varphi}(m, n) = \varphi(m, n),$$

i.e., $\varphi = \Phi \circ \iota$. The homomorphism Φ is uniquely determined by this equation since the elements $m \otimes n$ generate $M \otimes_R N$ as an abelian group. This completes the proof.

Theorem 10 is extremely useful in defining homomorphisms on $M \otimes_R N$ since it replaces the often tedious check that maps defined on simple tensors $m \otimes n$ are well defined with a check that a related map defined on ordered pairs (m, n) is balanced.

The first consequence of the universal property in Theorem 10 is a characterization of the tensor product $M \otimes_R N$ as an abelian group:

Corollary 11. Suppose D is an abelian group and $\iota' : M \times N \to D$ is an R-balanced map such that

   (i) the image of ι' generates D as an abelian group, and
   (ii) every R-balanced map defined on $M \times N$ factors through ι' as in Theorem 10.
Then there is an isomorphism $f : M \otimes_R N \cong D$ of abelian groups with $\iota' = f \circ \iota$.

Proof: Since $\iota' : M \times N \to D$ is a balanced map, the universal property in (2) of Theorem 10 implies there is a (unique) homomorphism $f : M \otimes_R N \to D$ with $\iota' = f \circ \iota$. In particular $\iota'(m, n) = f(m \otimes n)$ for every $m \in M$, $n \in N$. By the first assumption on ι', these elements generate D as an abelian group, so f is a surjective map. Now, the balanced map $\iota : M \times N \to M \otimes_R N$ together with the second assumption on ι' implies there is a (unique) homomorphism $g : D \to M \otimes_R N$ with $\iota = g \circ \iota'$. Then $m \otimes n = (g \circ f)(m \otimes n)$. Since the simple tensors $m \otimes n$ generate $M \otimes_R N$, it follows that $g \circ f$ is the identity map on $M \otimes_R N$ and so f is injective, hence an isomorphism. This establishes the corollary.

We now return to the question of giving the abelian group $M \otimes_R N$ a *module* structure. As we observed in the special case of extending scalars from R to S for the R-module N, the S-module structure on $S \otimes_R N$ required only a left S-module structure on S together with the compatibility relation $s'(sr) = (s's)r$ for $s, s' \in S$ and $r \in R$. In this special case this relation was simply a consequence of the associative law in the ring S. To obtain an S-module structure on $M \otimes_R N$ more generally we impose a similar structure on M:

Definition. Let R and S be any rings with 1. An abelian group M is called an (S, R)-*bimodule* if M is a left S-module, a right R-module, and $s(mr) = (sm)r$ for all $s \in S$, $r \in R$ and $m \in M$.

Examples

   (1) Any ring S is an (S, R)-bimodule for any subring R with $1_R = 1_S$ by the associativity of the multiplication in S. More generally, if $f : R \to S$ is any ring homomorphism with $f(1_R) = 1_S$ then S can be considered as a right R-module with the action $s \cdot r = sf(r)$, and with respect to this action S becomes an (S, R)-bimodule.
   (2) Let I be an ideal (two-sided) in the ring R. Then the quotient ring R/I is an $(R/I, R)$-bimodule. This is easy to see directly and is also a special case of the previous example (with respect to the canonical projection homomorphism $R \to R/I$).
   (3) Suppose that R is a commutative ring. Then a left (respectively, right) R-module M can always be given the structure of a right (respectively, left) R-module by defining $mr = rm$ (respectively, $rm = mr$), for all $m \in M$ and $r \in R$, and this makes M into

an (R, R)-bimodule. Hence every module (right or left) over a commutative ring R has at least one natural (R, R)-bimodule structure.

(4) Suppose that M is a left S-module and R is a subring contained in the *center* of S (for example, if S is commutative). Then in particular R is commutative so M can be given a right R-module structure as in the previous example. Then for any $s \in S, r \in R$ and $m \in M$ by definition of the right action of R we have

$$(sm)r = r(sm) = (rs)m = (sr)m = s(rm) = s(mr)$$

(note that we have used the fact that r commutes with s in the middle equality). Hence M is an (S, R)-bimodule with respect to this definition of the right action of R.

Since the situation in Example 3 occurs so frequently, we give this bimodule structure a name:

Definition. Suppose M is a left (or right) R-module over the commutative ring R. Then the (R, R)-bimodule structure on M defined by letting the left and right R-actions coincide, i.e., $mr = rm$ for all $m \in M$ and $r \in R$, will be called the *standard R-module* structure on M.

Suppose now that N is a left R-module and M is an (S, R)-bimodule. Then just as in the example of extension of scalars the (S, R)-bimodule structure on M implies that

$$s\left(\underset{\text{finite}}{\sum} m_i \otimes n_i \right) = \underset{\text{finite}}{\sum} (sm_i) \otimes n_i \tag{10.8}$$

gives a well defined action of S under which $M \otimes_R N$ *is a left S-module*. Note that Theorem 10 may be used to give an alternate proof that (8) is well defined, replacing the direct calculations on the relations defining the tensor product with the easier check that a map is R-balanced, as follows. It is very easy to see that for each fixed $s \in S$ the map $(m, n) \mapsto sm \otimes n$ is an R-balanced map from $M \times N$ to $M \otimes_R N$. By Theorem 10 there is a well defined group homomorphism λ_s from $M \otimes_R N$ to itself such that $\lambda_s(m \otimes n) = sm \otimes n$. Since the right side of (8) is then $\lambda_s(\sum m_i \otimes n_i)$, the fact that λ_s is well defined shows that this expression is indeed independent of the representation of the tensor $\sum m_i \otimes n_i$ as a sum of simple tensors. Because λ_s is additive, equation (8) holds.

By a completely parallel argument, if M is a right R-module and N is an (R, S)-bimodule then the tensor product $M \otimes_R N$ has the structure of a *right S-module*, where $(\sum m_i \otimes n_i) s = \sum m_i \otimes (n_i s)$.

Before giving some more examples of tensor products it is worthwhile to highlight one frequently encountered special case of the previous discussion, namely the case when M and N are two left modules over a *commutative* ring R and $S = R$ (in some works on tensor products this is the only case considered). Then the standard R-module structure on M defined previously gives M the structure of an (R, R)-bimodule, so in this case the tensor product $M \otimes_R N$ always has the structure of a left R-module.

The corresponding map $\iota : M \times N \to M \otimes_R N$ maps $M \times N$ into an R-module and is additive in each factor. Since $r(m \otimes n) = rm \otimes n = mr \otimes n = m \otimes rn$ it also satisfies

$$r\iota(m, n) = \iota(rm, n) = \iota(m, rn).$$

Such maps are given a name:

Definition. Let R be a commutative ring with 1 and let M, N, and L be left R-modules. The map $\varphi : M \times N \to L$ is called R-*bilinear* if it is R-linear in each factor, i.e., if

$$\varphi(r_1 m_1 + r_2 m_2, n) = r_1 \varphi(m_1, n) + r_2 \varphi(m_2, n), \quad \text{and}$$

$$\varphi(m, r_1 n_1 + r_2 n_2) = r_1 \varphi(m, n_1) + r_2 \varphi(m, n_2)$$

for all $m, m_1, m_2 \in M$, $n, n_1, n_2 \in N$ and $r_1, r_2 \in R$.

With this terminology Theorem 10 gives

Corollary 12. Suppose R is a commutative ring. Let M and N be two left R-modules and let $M \otimes_R N$ be the tensor product of M and N over R, where M is given the standard R-module structure. Then $M \otimes_R N$ is a left R-module with

$$r(m \otimes n) = (rm) \otimes n = (mr) \otimes n = m \otimes (rn),$$

and the map $\iota : M \times N \to M \otimes_R N$ with $\iota(m, n) = m \otimes n$ is an R-bilinear map. If L is any left R-module then there is a bijection

$$\left\{ \begin{array}{c} R\text{-bilinear maps} \\ \varphi : M \times N \to L \end{array} \right\} \longleftrightarrow \left\{ \begin{array}{c} R\text{-module homomorphisms} \\ \varPhi : M \otimes_R N \to L \end{array} \right\}$$

where the correspondence between φ and $\varPhi$ is given by the commutative diagram

$$\begin{array}{ccc} M \times N & \xrightarrow{\ \iota\ } & M \otimes_R N \\ & {\scriptstyle \varphi} \searrow & \big\downarrow {\scriptstyle \varPhi} \\ & & L \end{array}$$

Proof: We have shown $M \otimes_R N$ is an R-module and that ι is bilinear. It remains only to check that in the bijective correspondence in Theorem 10 the bilinear maps correspond with the R-module homomorphisms. If $\varphi : M \times N \to L$ is bilinear then it is an R-balanced map, so the corresponding $\varPhi : M \otimes_R N$ is a group homomorphism. Moreover, on simple tensors $\varPhi((rm) \otimes n) = \varphi(rm, n) = r\varphi(m, n) = r\varPhi(m \otimes n)$, where the middle equality holds because φ is R-linear in the first variable. Since $\varPhi$ is additive this extends to sums of simple tensors to show $\varPhi$ is an R-module homomorphism. Conversely, if $\varPhi$ is an R-module homomorphism it is an exercise to see that the corresponding balanced map φ is bilinear.

Examples

(1) In any tensor product $M \otimes_R N$ we have $m \otimes 0 = m \otimes (0 + 0) = (m \otimes 0) + (m \otimes 0)$, so $m \otimes 0 = 0$. Likewise $0 \otimes n = 0$.

(2) We have $\mathbb{Z}/2\mathbb{Z} \otimes_{\mathbb{Z}} \mathbb{Z}/3\mathbb{Z} = 0$, since $3a = a$ for $a \in \mathbb{Z}/2\mathbb{Z}$ so that

$$a \otimes b = 3a \otimes b = a \otimes 3b = a \otimes 0 = 0$$

and every simple tensor is reduced to 0. In particular $1 \otimes 1 = 0$. It follows that there are no nonzero balanced (or bilinear) maps from $\mathbb{Z}/2\mathbb{Z} \times \mathbb{Z}/3\mathbb{Z}$ to any abelian group.

On the other hand, consider the tensor product $\mathbb{Z}/2\mathbb{Z} \otimes_{\mathbb{Z}} \mathbb{Z}/2\mathbb{Z}$, which is generated as an abelian group by the elements $0 \otimes 0 = 1 \otimes 0 = 0 \otimes 1 = 0$ and $1 \otimes 1$. In this case $1 \otimes 1 \neq 0$ since, for example, the map $\mathbb{Z}/2\mathbb{Z} \times \mathbb{Z}/2\mathbb{Z} \to \mathbb{Z}/2\mathbb{Z}$ defined by $(a, b) \mapsto ab$ is clearly nonzero and linear in both a and b. Since $2(1 \otimes 1) = 2 \otimes 1 = 0 \otimes 1 = 0$, the element $1 \otimes 1$ is of order 2. Hence $\mathbb{Z}/2\mathbb{Z} \otimes_{\mathbb{Z}} \mathbb{Z}/2\mathbb{Z} \cong \mathbb{Z}/2\mathbb{Z}$.

(3) In general,

$$\mathbb{Z}/m\mathbb{Z} \otimes_{\mathbb{Z}} \mathbb{Z}/n\mathbb{Z} \cong \mathbb{Z}/d\mathbb{Z},$$

where d is the g.c.d. of the integers m and n. To see this, observe first that

$$a \otimes b = a \otimes (b \cdot 1) = (ab) \otimes 1 = ab(1 \otimes 1),$$

from which it follows that $\mathbb{Z}/m\mathbb{Z} \otimes_{\mathbb{Z}} \mathbb{Z}/n\mathbb{Z}$ is a cyclic group with $1 \otimes 1$ as generator. Since $m(1 \otimes 1) = m \otimes 1 = 0 \otimes 1 = 0$ and similarly $n(1 \otimes 1) = 1 \otimes n = 0$, we have $d(1 \otimes 1) = 0$, so the cyclic group has order dividing d. The map $\varphi : \mathbb{Z}/m\mathbb{Z} \times \mathbb{Z}/n\mathbb{Z} \to \mathbb{Z}/d\mathbb{Z}$ defined by $\varphi(a \bmod m, b \bmod n) = ab \bmod d$ is well defined since d divides both m and n. It is clearly $\mathbb{Z}$-bilinear. The induced map $\Phi : \mathbb{Z}/m\mathbb{Z} \otimes_{\mathbb{Z}} \mathbb{Z}/n\mathbb{Z} \to \mathbb{Z}/d\mathbb{Z}$ from Corollary 12 maps $1 \otimes 1$ to the element $1 \in \mathbb{Z}/d\mathbb{Z}$, which is an element of order d. In particular $\mathbb{Z}/m\mathbb{Z} \otimes_{\mathbb{Z}} \mathbb{Z}/n\mathbb{Z}$ has order at least d. Hence $1 \otimes 1$ is an element of order d and Φ gives an isomorphism $\mathbb{Z}/m\mathbb{Z} \otimes_{\mathbb{Z}} \mathbb{Z}/n\mathbb{Z} \cong \mathbb{Z}/d\mathbb{Z}$.

(4) In $\mathbb{Q}/\mathbb{Z} \otimes_{\mathbb{Z}} \mathbb{Q}/\mathbb{Z}$ a simple tensor has the form $(a/b \bmod \mathbb{Z}) \otimes (c/d \bmod \mathbb{Z})$ for some rational numbers a/b and c/d. Then

$$\begin{aligned}
(\frac{a}{b} \bmod \mathbb{Z}) \otimes (\frac{c}{d} \bmod \mathbb{Z}) &= d(\frac{a}{bd} \bmod \mathbb{Z}) \otimes (\frac{c}{d} \bmod \mathbb{Z}) \\
&= (\frac{a}{bd} \bmod \mathbb{Z}) \otimes d(\frac{c}{d} \bmod \mathbb{Z}) = (\frac{a}{bd} \bmod \mathbb{Z}) \otimes 0 = 0
\end{aligned}$$

and so

$$\mathbb{Q}/\mathbb{Z} \otimes_{\mathbb{Z}} \mathbb{Q}/\mathbb{Z} = 0.$$

In a similar way, $A \otimes_{\mathbb{Z}} B = 0$ for any *divisible* abelian group A and *torsion* abelian group B (an abelian group in which every element has finite order). For example

$$\mathbb{Q} \otimes_{\mathbb{Z}} \mathbb{Q}/\mathbb{Z} = 0.$$

(5) The structure of a tensor product can vary considerably depending on the ring over which the tensors are taken. For example $\mathbb{Q} \otimes_{\mathbb{Q}} \mathbb{Q}$ and $\mathbb{Q} \otimes_{\mathbb{Z}} \mathbb{Q}$ are isomorphic as left $\mathbb{Q}$-modules (both are one dimensional vector spaces over $\mathbb{Q}$) — cf. the exercises. On the other hand we shall see at the end of this section that $\mathbb{C} \otimes_{\mathbb{C}} \mathbb{C}$ and $\mathbb{C} \otimes_{\mathbb{R}} \mathbb{C}$ are not isomorphic $\mathbb{C}$-modules (the former is a 1-dimensional vector space over $\mathbb{C}$ and the latter is 2-dimensional over $\mathbb{C}$).

(6) *General extension of scalars or change of base:* Let $f : R \to S$ be a ring homomorphism with $f(1_R) = 1_S$. Then $s \cdot r = sf(r)$ gives S the structure of a right R-module with respect to which S is an (S, R)-bimodule. Then for any left R-module N, the resulting tensor product $S \otimes_R N$ is a left S-module obtained by *changing the base* from R to S. This gives a slight generalization of the notion of extension of scalars (where R was a subring of S).

(7) Let $f : R \to S$ be a ring homomorphism as in the preceding example. Then we have $S \otimes_R R \cong S$ as left S-modules, as follows. The map $\varphi : S \times R \to S$ defined by $(s, r) \mapsto sr$ (where $sr = sf(r)$ by definition of the right R-action on S), is an R-balanced map, as is easily checked. For example,

$$\varphi(s_1 + s_2, r) = (s_1 + s_2)r = s_1 r + s_2 r = \varphi(s_1, r) + \varphi(s_2, r)$$

and
$$\varphi(sr, r') = (sr)r' = s(rr') = \varphi(s, rr').$$

By Theorem 10 we have an associated group homomorphism $\Phi : S \otimes_R R \to S$ with $\Phi(s \otimes r) = sr$. Since $\Phi(s'(s \otimes r)) = \Phi(s's \otimes r) = s'sr = s'\Phi(s \otimes r)$, it follows that Φ is also an S-module homomorphism. The map $\Phi' : S \to S \otimes_R R$ with $s \mapsto s \otimes 1$ is an S-module homomorphism that is inverse to Φ because $\Phi \circ \Phi'(s) = \Phi(s \otimes 1) = s$ gives $\Phi\Phi' = 1$, and
$$\Phi' \circ \Phi(s \otimes r) = \Phi'(sr) = sr \otimes 1 = s \otimes r$$
shows that $\Phi'\Phi$ is the identity on simple tensors, hence $\Phi'\Phi = 1$.

(8) Let R be a ring (not necessarily commutative), let I be a two sided ideal in R, and let N be a left R-module. Then as previously mentioned, R/I is an $(R/I, R)$-bimodule, so the tensor product $R/I \otimes_R N$ is a left R/I-module. This is an example of "extension of scalars" with respect to the natural projection homomorphism $R \to R/I$.

Define
$$IN = \left\{ \sum_{\text{finite}} a_i n_i \mid a_i \in I, n_i \in N \right\},$$

which is easily seen to be a left R-submodule of N (cf. Exercise 5, Section 1). Then
$$(R/I) \otimes_R N \cong N/IN,$$

as left R-modules, as follows. The tensor product is generated as an abelian group by the simple tensors $(r \bmod I) \otimes n = r(1 \otimes n)$ for $r \in R$ and $n \in N$ (viewing the R/I-module tensor product as an R-module on which I acts trivially). Hence the elements $1 \otimes n$ generate $(R/I) \otimes_R N$ as an R/I-module. The map $N \to (R/I) \otimes_R N$ defined by $n \mapsto 1 \otimes n$ is a left R-module homomorphism and, by the previous observation, is surjective. Under this map $a_i n_i$ with $a_i \in I$ and $n_i \in N$ maps to $1 \otimes a_i n_i = a_i \otimes n_i = 0$, and so IN is contained in the kernel. This induces a surjective R-module homomorphism $f : N/IN \to (R/I) \otimes_R N$ with $f(n \bmod I) = 1 \otimes n$. We show f is an isomorphism by exhibiting its inverse. The map $(R/I) \times N \to N/IN$ defined by mapping $(r \bmod I, n)$ to $(rn \bmod IN)$ is well defined and easily checked to be R-balanced. It follows by Theorem 10 that there is an associated group homomorphism $g : (R/I) \otimes N \to N/IN$ with $g((r \bmod I) \otimes n) = rn \bmod IN$. As usual, $fg = 1$ and $gf = 1$, so f is a bijection and $(R/I) \otimes_R N \cong N/IN$, as claimed.

As an example, let $R = \mathbb{Z}$ with ideal $I = m\mathbb{Z}$ and let N be the $\mathbb{Z}$-module $\mathbb{Z}/n\mathbb{Z}$. Then $IN = m(\mathbb{Z}/n\mathbb{Z}) = (m\mathbb{Z} + n\mathbb{Z})/n\mathbb{Z} = d\mathbb{Z}/n\mathbb{Z}$ where d is the g.c.d. of m and n. Then $N/IN \cong \mathbb{Z}/d\mathbb{Z}$ and we recover the isomorphism $\mathbb{Z}/m\mathbb{Z} \otimes_{\mathbb{Z}} \mathbb{Z}/n\mathbb{Z} \cong \mathbb{Z}/d\mathbb{Z}$ of Example 3 above.

We now establish some of the basic properties of tensor products. Note the frequent application of Theorem 10 to establish the existence of homomorphisms.

Theorem 13. *(The "Tensor Product" of Two Homomorphisms)* Let M, M' be right R-modules, let N, N' be left R-modules, and suppose $\varphi : M \to M'$ and $\psi : N \to N'$ are R-module homomorphisms.

(1) There is a unique group homomorphism, denoted by $\varphi \otimes \psi$, mapping $M \otimes_R N$ into $M' \otimes_R N'$ such that $(\varphi \otimes \psi)(m \otimes n) = \varphi(m) \otimes \psi(n)$ for all $m \in M$ and $n \in N$.

(2) If M, M' are also (S, R)-bimodules for some ring S and φ is also an S-module homomorphism, then $\varphi \otimes \psi$ is a homomorphism of left S-modules. In particular, if R is commutative then $\varphi \otimes \psi$ is always an R-module homomorphism for the standard R-module structures.

(3) If $\lambda : M' \to M''$ and $\mu : N' \to N''$ are R-module homomorphisms then
$$(\lambda \otimes \mu) \circ (\varphi \otimes \psi) = (\lambda \circ \varphi) \otimes (\mu \circ \psi).$$

Proof: The map $(m, n) \mapsto \varphi(m) \otimes \psi(n)$ from $M \times N$ to $M' \otimes_R N'$ is clearly R-balanced, so (1) follows immediately from Theorem 10.

In (2) the definition of the (left) action of S on M together with the assumption that φ is an S-module homomorphism imply that on simple tensors
$$(\varphi \otimes \psi)(s(m \otimes n)) = (\varphi \otimes \psi)(sm \otimes n) = \varphi(sm) \otimes \psi(n) = s\varphi(m) \otimes \psi(n).$$

Since $\varphi \otimes \psi$ is additive, this extends to sums of simple tensors to show that $\varphi \otimes \psi$ is an S-module homomorphism. This gives (2).

The uniqueness condition in Theorem 10 implies (3), which completes the proof.

The next result shows that we may write $M \otimes N \otimes L$, or more generally, an n-fold tensor product $M_1 \otimes M_2 \otimes \cdots \otimes M_n$, unambiguously whenever it is defined.

Theorem 14. *(Associativity of the Tensor Product)* Suppose M is a right R-module, N is an (R, T)-bimodule, and L is a left T-module. Then there is a unique isomorphism
$$(M \otimes_R N) \otimes_T L \cong M \otimes_R (N \otimes_T L)$$
of abelian groups such that $(m \otimes n) \otimes l \mapsto m \otimes (n \otimes l)$. If M is an (S, R)-bimodule, then this is an isomorphism of S-modules.

Proof: Note first that the (R, T)-bimodule structure on N makes $M \otimes_R N$ into a right T-module and $N \otimes_T L$ into a left R-module, so both sides of the isomorphism are well defined. For each fixed $l \in L$, the mapping $(m, n) \mapsto m \otimes (n \otimes l)$ is R-balanced, so by Theorem 10 there is a homomorphism $M \otimes_R N \to M \otimes_R (N \otimes_T L)$ with $m \otimes n \mapsto m \otimes (n \otimes l)$. This shows that the map from $(M \otimes_R N) \times L$ to $M \otimes_R (N \otimes_T L)$ given by $(m \otimes n, l) \mapsto m \otimes (n \otimes l)$ is well defined. Since it is easily seen to be T-balanced, another application of Theorem 10 implies that it induces a homomorphism $(M \otimes_R N) \otimes_T L \to M \otimes_R (N \otimes_T L)$ such that $(m \otimes n) \otimes l \mapsto m \otimes (n \otimes l)$. In a similar way we can construct a homomorphism in the opposite direction that is inverse to this one. This proves the group isomorphism.

Assume in addition M is an (S, R)-bimodule. Then for $s \in S$ and $t \in T$ we have
$$s((m \otimes n)t) = s(m \otimes nt) = sm \otimes nt = (sm \otimes n)t = (s(m \otimes n))t$$
so that $M \otimes_R N$ is an (S, T)-bimodule. Hence $(M \otimes_R N) \otimes_T L$ is a left S-module. Since $N \otimes_T L$ is a left R-module, also $M \otimes_R (N \otimes_T L)$ is a left S-module. The group isomorphism just established is easily seen to be a homomorphism of left S-modules by the same arguments used in previous proofs: it is additive and is S-linear on simple tensors since $s((m \otimes n) \otimes l) = s(m \otimes n) \otimes l = (sm \otimes n) \otimes l$ maps to the element $sm \otimes (n \otimes l) = s(m \otimes (n \otimes l))$. The proof is complete.

Corollary 15. Suppose R is commutative and M, N, and L are left R-modules. Then
$$(M \otimes N) \otimes L \cong M \otimes (N \otimes L)$$
as R-modules for the standard R-module structures on M, N and L.

There is a natural extension of the notion of a bilinear map:

Definition. Let R be a commutative ring with 1 and let $M_1, M_2, \ldots, M_n$ and L be R-modules with the standard R-module structures. A map $\varphi : M_1 \times \cdots \times M_n \to L$ is called *n-multilinear over R* (or simply *multilinear* if n and R are clear from the context) if it is an R-module homomorphism in each component when the other component entries are kept constant, i.e., for each i

$$\varphi(m_1, \ldots, m_{i-1}, rm_i + r'm_i', m_{i+1}, \ldots, m_n)$$
$$= r\varphi(m_1, \ldots, m_i, \ldots, m_n) + r'\varphi(m_1, \ldots, m_i', \ldots, m_n)$$

for all $m_i, m_i' \in M_i$ and $r, r' \in R$. When $n = 2$ (respectively, 3) one says φ is *bilinear* (respectively, *trilinear*) rather than 2-multilinear (or 3-multilinear).

One may construct the n-fold tensor product $M_1 \otimes M_2 \otimes \cdots \otimes M_n$ from first principles and prove its analogous universal property with respect to multilinear maps from $M_1 \times \cdots \times M_n$ to L. By the previous theorem and corollary, however, an n-fold tensor product may be obtained unambiguously by iterating the tensor product of pairs of modules since any bracketing of $M_1 \otimes \cdots \otimes M_n$ into tensor products of pairs gives an isomorphic R-module. The universal property of the tensor product of a pair of modules in Theorem 10 and Corollary 12 then implies that multilinear maps factor uniquely through the R-module $M_1 \otimes \cdots \otimes M_n$, i.e., this tensor product is the universal object with respect to multilinear functions:

Corollary 16. Let R be a commutative ring and let $M_1, \ldots, M_n, L$ be R-modules. Let $M_1 \otimes M_2 \otimes \cdots \otimes M_n$ denote any bracketing of the tensor product of these modules and let
$$\iota : M_1 \times \cdots \times M_n \to M_1 \otimes \cdots \otimes M_n$$
be the map defined by $\iota(m_1, \ldots, m_n) = m_1 \otimes \cdots \otimes m_n$. Then
> **(1)** for every R-module homomorphism $\Phi : M_1 \otimes \cdots \otimes M_n \to L$ the map $\varphi = \Phi \circ \iota$ is n-multilinear from $M_1 \times \cdots \times M_n$ to L, and
> **(2)** if $\varphi : M_1 \times \cdots \times M_n \to L$ is an n-multilinear map then there is a unique R-module homomorphism $\Phi : M_1 \otimes \cdots \otimes M_n \to L$ such that $\varphi = \Phi \circ \iota$.

Hence there is a bijection
$$\left\{ \begin{matrix} n\text{-multilinear maps} \\ \varphi : M_1 \times \cdots \times M_n \to L \end{matrix} \right\} \longleftrightarrow \left\{ \begin{matrix} R\text{-module homomorphisms} \\ \Phi : M_1 \otimes \cdots \otimes M_n \to L \end{matrix} \right\}$$
with respect to which the following diagram commutes:

$$
\begin{array}{ccc}
M_1 \times \cdots \times M_n & \xrightarrow{\;\iota\;} & M_1 \otimes \cdots \otimes M_n \\
& {\scriptstyle \varphi} \searrow & \downarrow {\scriptstyle \Phi} \\
& & L
\end{array}
$$

We have already seen examples where $M_1 \otimes_R N$ is not contained in $M \otimes_R N$ even when M_1 is an R-submodule of M. The next result shows in particular that (an isomorphic copy of) $M_1 \otimes_R N$ *is* contained in $M \otimes_R N$ if M_1 is an R-module *direct summand* of M.

Theorem 17. *(Tensor Products of Direct Sums)* Let M, M' be right R-modules and let N, N' be left R-modules. Then there are unique group isomorphisms

$$(M \oplus M') \otimes_R N \cong (M \otimes_R N) \oplus (M' \otimes_R N)$$

$$M \otimes_R (N \oplus N') \cong (M \otimes_R N) \oplus (M \otimes_R N')$$

such that $(m, m') \otimes n \mapsto (m \otimes n, m' \otimes n)$ and $m \otimes (n, n') \mapsto (m \otimes n, m \otimes n')$ respectively. If M, M' are also (S, R)-bimodules, then these are isomorphisms of left S-modules. In particular, if R is commutative, these are isomorphisms of R-modules.

Proof: The map $(M \oplus M') \times N \to (M \otimes_R N) \oplus (M' \otimes_R N)$ defined by $((m, m'), n) \mapsto (m \otimes n, m' \otimes n)$ is well defined since m and m' in $M \oplus M'$ are uniquely defined in the direct sum. The map is clearly R-balanced, so induces a homomorphism f from $(M \oplus M') \otimes N$ to $(M \otimes_R N) \oplus (M' \otimes_R N)$ with

$$f((m, m') \otimes n) = (m \otimes n, m' \otimes n).$$

In the other direction, the R-balanced maps $M \times N \to (M \oplus M') \otimes_R N$ and $M' \times N \to (M \oplus M') \otimes_R N$ given by $(m, n) \mapsto (m, 0) \otimes n$ and $(m', n) \mapsto (0, m') \otimes n$, respectively, define homomorphisms from $M \otimes_R N$ and $M' \otimes_R N$ to $(M \oplus M') \otimes_R N$. These in turn give a homomorphism g from the direct sum $(M \otimes_R N) \oplus (M' \otimes_R N)$ to $(M \oplus M') \otimes_R N$ with

$$g((m \otimes n_1, m' \otimes n_2)) = (m, 0) \otimes n_1 + (0, m') \otimes n_2.$$

An easy check shows that f and g are inverse homomorphisms and are S-module isomorphisms when M and M' are (S, R)-bimodules. This completes the proof.

The previous theorem clearly extends by induction to any finite direct sum of R-modules. The corresponding result is also true for arbitrary direct sums. For example

$$M \otimes (\oplus_{i \in I} N_i) \cong \oplus_{i \in I} (M \otimes N_i),$$

where I is any index set (cf. the exercises). This result is referred to by saying that *tensor products commute with direct sums*.

Corollary 18. *(Extension of Scalars for Free Modules)* The module obtained from the free R-module $N \cong R^n$ by extension of scalars from R to S is the free S-module S^n, i.e.,

$$S \otimes_R R^n \cong S^n$$

as left S-modules.

Proof: This follows immediately from Theorem 17 and the isomorphism $S \otimes_R R \cong S$ proved in Example 7 previously.

Corollary 19. Let R be a commutative ring and let $M \cong R^s$ and $N \cong R^t$ be free R-modules with bases $m_1, \ldots, m_s$ and $n_1, \ldots, n_t$, respectively. Then $M \otimes_R N$ is a free R-module of rank st, with basis $m_i \otimes n_j$, $1 \le i \le s$ and $1 \le j \le t$, i.e.,

$$R^s \otimes_R R^t \cong R^{st}.$$

Remark: More generally, the tensor product of two free modules of arbitrary rank over a commutative ring is free (cf. the exercises).

Proof: This follows easily from Theorem 17 and the first example following Corollary 9.

Proposition 20. Suppose R is a commutative ring and M, N are left R-modules, considered with the standard R-module structures. Then there is a unique R-module isomorphism

$$M \otimes_R N \cong N \otimes_R M$$

mapping $m \otimes n$ to $n \otimes m$.

Proof: The map $M \times N \to N \otimes M$ defined by $(m, n) \mapsto n \otimes m$ is R-balanced. Hence it induces a unique homomorphism f from $M \otimes N$ to $N \otimes M$ with $f(m \otimes n) = n \otimes m$. Similarly, we have a unique homomorphism g from $N \otimes M$ to $M \otimes N$ with $g(n \otimes m) = m \otimes n$ giving the inverse of f, and both maps are easily seen to be R-module isomorphisms.

Remark: When $M = N$ it is not in general true that $a \otimes b = b \otimes a$ for $a, b \in M$. We shall study "symmetric tensors" in Section 11.5.

We end this section by showing that the tensor product of R-algebras is again an R-algebra.

Proposition 21. Let R be a commutative ring and let A and B be R-algebras. Then the multiplication $(a \otimes b)(a' \otimes b') = aa' \otimes bb'$ is well defined and makes $A \otimes_R B$ into an R-algebra.

Proof: Note first that the definition of an R-algebra shows that

$$r(a \otimes b) = ra \otimes b = ar \otimes b = a \otimes rb = a \otimes br = (a \otimes b)r$$

for every $r \in R, a \in A$ and $b \in B$. To show that $A \otimes B$ is an R-algebra the main task is, as usual, showing that the specified multiplication is well defined. One way to proceed is to use two applications of Corollary 16, as follows. The map $\varphi : A \times B \times A \times B \to A \otimes B$ defined by $f(a, b, a', b') = aa' \otimes bb'$ is multilinear over R. For example,

$$f(a, r_1 b_1 + r_2 b_2, a', b') = aa' \otimes (r_1 b_1 + r_2 b_2) b'$$
$$= aa' \otimes r_1 b_1 b' + aa' \otimes r_2 b_2 b'$$
$$= r_1 f(a, b_1, a', b') + r_2 f(a, b_2, a', b').$$

By Corollary 16, there is a corresponding R-module homomorphism Φ from $A \otimes B \otimes A \otimes B$ to $A \otimes B$ with $\Phi(a \otimes b \otimes a' \otimes b') = aa' \otimes bb'$. Viewing $A \otimes B \otimes A \otimes B$ as $(A \otimes B) \otimes (A \otimes B)$, we can apply Corollary 16 once more to obtain a well defined R-bilinear mapping φ' from $(A \otimes B) \times (A \otimes B)$ to $A \otimes B$ with $\varphi'(a \otimes b, a' \otimes b') = aa' \otimes bb'$. This shows that the multiplication is indeed well defined (and also that it satisfies the distributive laws). It is now a simple matter (left to the exercises) to check that with this multiplication $A \otimes B$ is an R-algebra.

Example

The tensor product $\mathbb{C} \otimes_{\mathbb{R}} \mathbb{C}$ is free of rank 4 as a module over $\mathbb{R}$ with basis given by $e_1 = 1 \otimes 1$, $e_2 = 1 \otimes i$, $e_3 = i \otimes 1$, and $e_4 = i \otimes i$ (by Corollary 19). By Proposition 21, this tensor product is also a (commutative) ring with $e_1 = 1$, and, for example,

$$e_4^2 = (i \otimes i)(i \otimes i) = i^2 \otimes i^2 = (-1) \otimes (-1) = (-1)(-1) \otimes 1 = 1.$$

Then $(e_4 - 1)(e_4 + 1) = 0$, so $\mathbb{C} \otimes_{\mathbb{R}} \mathbb{C}$ is not an integral domain.

The ring $\mathbb{C} \otimes_{\mathbb{R}} \mathbb{C}$ is an $\mathbb{R}$-algebra and the left and right $\mathbb{R}$-actions are the same: $xr = rx$ for every $r \in \mathbb{R}$ and $x \in \mathbb{C} \otimes_{\mathbb{R}} \mathbb{C}$. The ring $\mathbb{C} \otimes_{\mathbb{R}} \mathbb{C}$ has a structure of a left $\mathbb{C}$-module because the first $\mathbb{C}$ is a $(\mathbb{C}, \mathbb{R})$-bimodule. It also has a right $\mathbb{C}$-module structure because the second $\mathbb{C}$ is an $(\mathbb{R}, \mathbb{C})$-bimodule. For example,

$$i \cdot e_1 = i \cdot (1 \otimes 1) = (i \cdot 1) \otimes 1 = i \otimes 1 = e_3$$

and

$$e_1 \cdot i = (1 \otimes 1) \cdot i = 1 \otimes (1 \cdot i) = 1 \otimes i = e_2.$$

This example also shows that even when the rings involved are commutative there may be natural left and right module structures (over some ring) that are not the same.

EXERCISES

Let R be a ring with 1.

1. Let $f : R \to S$ be a ring homomorphism from the ring R to the ring S with $f(1_R) = 1_S$. Verify the details that $sr = sf(r)$ defines a right R-action on S under which S is an (S, R)-bimodule.

2. Show that the element "$2 \otimes 1$" is 0 in $\mathbb{Z} \otimes_{\mathbb{Z}} \mathbb{Z}/2\mathbb{Z}$ but is nonzero in $2\mathbb{Z} \otimes_{\mathbb{Z}} \mathbb{Z}/2\mathbb{Z}$.

3. Show that $\mathbb{C} \otimes_{\mathbb{R}} \mathbb{C}$ and $\mathbb{C} \otimes_{\mathbb{C}} \mathbb{C}$ are both left $\mathbb{R}$-modules but are not isomorphic as $\mathbb{R}$-modules.

4. Show that $\mathbb{Q} \otimes_{\mathbb{Z}} \mathbb{Q}$ and $\mathbb{Q} \otimes_{\mathbb{Q}} \mathbb{Q}$ are isomorphic left $\mathbb{Q}$-modules. [Show they are both 1-dimensional vector spaces over $\mathbb{Q}$.]

5. Let A be a finite abelian group of order n and let p^k be the largest power of the prime p dividing n. Prove that $\mathbb{Z}/p^k\mathbb{Z} \otimes_{\mathbb{Z}} A$ is isomorphic to the Sylow p-subgroup of A.

6. If R is any integral domain with quotient field Q, prove that $(Q/R) \otimes_R (Q/R) = 0$.

7. If R is any integral domain with quotient field Q and N is a left R-module, prove that every element of the tensor product $Q \otimes_R N$ can be written as a simple tensor of the form $(1/d) \otimes n$ for some nonzero $d \in R$ and some $n \in N$.

8. Suppose R is an integral domain with quotient field Q and let N be any R-module. Let $U = R^{\times}$ be the set of nonzero elements in R and define $U^{-1}N$ to be the set of equivalence classes of ordered pairs of elements (u, n) with $u \in U$ and $n \in N$ under the equivalence relation $(u, n) \sim (u', n)$ if and only if $u'n = un'$ in N.

(a) Prove that $U^{-1}N$ is an abelian group under the addition defined by $\overline{(u_1, n_1)} + \overline{(u_2, n_2)} = \overline{(u_1 u_2, u_2 n_1 + u_1 n_2)}$. Prove that $r\overline{(u, n)} = \overline{(u, rn)}$ defines an action of R on $U^{-1}N$ making it into an R-module. [This is an example of *localization* considered in general in Section 4 of Chapter 15, cf. also Section 5 in Chapter 7.]

(b) Show that the map from $Q \times N$ to $U^{-1}N$ defined by sending $(a/b, n)$ to $\overline{(b, an)}$ for $a \in R$, $b \in U$, $n \in N$, is an R-balanced map, so induces a homomorphism f from $Q \otimes_R N$ to $U^{-1}N$. Show that the map g from $U^{-1}N$ to $Q \otimes_R N$ defined by $g(\overline{(u, n)}) = (1/u) \otimes n$ is well defined and is an inverse homomorphism to f. Conclude that $Q \otimes_R N \cong U^{-1}N$ as R-modules.

(c) Conclude from (b) that $(1/d) \otimes n$ is 0 in $Q \otimes_R N$ if and only if $rn = 0$ for some nonzero $r \in R$.

(d) If A is an abelian group, show that $\mathbb{Q} \otimes_{\mathbb{Z}} A = 0$ if and only if A is a torsion abelian group (i.e., every element of A has finite order).

9. Suppose R is an integral domain with quotient field Q and let N be any R-module. Let $Q \otimes_R N$ be the module obtained from N by extension of scalars from R to Q. Prove that the kernel of the R-module homomorphism $\iota : N \to Q \otimes_R N$ is the torsion submodule of N (cf. Exercise 8 in Section 1). [Use the previous exercise.]

10. Suppose R is commutative and $N \cong R^n$ is a free R-module of rank n with R-module basis $e_1, \ldots, e_n$.

(a) For any nonzero R-module M show that every element of $M \otimes N$ can be written uniquely in the form $\sum_{i=1}^n m_i \otimes e_i$ where $m_i \in M$. Deduce that if $\sum_{i=1}^n m_i \otimes e_i = 0$ in $M \otimes N$ then $m_i = 0$ for $i = 1, \ldots, n$.

(b) Show that if $\sum m_i \otimes n_i = 0$ in $M \otimes N$ where the n_i are merely assumed to be R-linearly independent then it is not necessarily true that all the m_i are 0. [Consider $R = \mathbb{Z}$, $n = 1$, $M = \mathbb{Z}/2\mathbb{Z}$, and the element $1 \otimes 2$.]

11. Let $\{e_1, e_2\}$ be a basis of $V = \mathbb{R}^2$. Show that the element $e_1 \otimes e_2 + e_2 \otimes e_1$ in $V \otimes_{\mathbb{R}} V$ cannot be written as a simple tensor $v \otimes w$ for any $v, w \in \mathbb{R}^2$.

12. Let V be a vector space over the field F and let v, v' be nonzero elements of V. Prove that $v \otimes v' = v' \otimes v$ in $V \otimes_F V$ if and only if $v = av'$ for some $a \in F$.

13. Prove that the usual dot product of vectors defined by letting $(a_1, \ldots, a_n) \cdot (b_1, \ldots, b_n)$ be $a_1 b_1 + \cdots + a_n b_n$ is a bilinear map from $\mathbb{R}^n \times \mathbb{R}^n$ to $\mathbb{R}$.

14. Let I be an arbitrary nonempty index set and for each $i \in I$ let N_i be a left R-module. Let M be a right R-module. Prove the group isomorphism: $M \otimes (\oplus_{i \in I} N_i) \cong \oplus_{i \in I} (M \otimes N_i)$, where the direct sum of an arbitrary collection of modules is defined in Exercise 20, Section 3. [Use the same argument as for the direct sum of two modules, taking care to note where the direct *sum* hypothesis is needed — cf. the next exercise.]

15. Show that tensor products do not commute with direct products in general. [Consider the extension of scalars from $\mathbb{Z}$ to $\mathbb{Q}$ of the direct product of the modules $M_i = \mathbb{Z}/2^i \mathbb{Z}$, $i = 1, 2, \ldots$]

16. Suppose R is commutative and let I and J be ideals of R, so R/I and R/J are naturally R-modules.

(a) Prove that every element of $R/I \otimes_R R/J$ can be written as a simple tensor of the form $(1 \bmod I) \otimes (r \bmod J)$.

(b) Prove that there is an R-module isomorphism $R/I \otimes_R R/J \cong R/(I + J)$ mapping $(r \bmod I) \otimes (r' \bmod J)$ to $rr' \bmod (I + J)$.

17. Let $I = (2, x)$ be the ideal generated by 2 and x in the ring $R = \mathbb{Z}[x]$. The ring $\mathbb{Z}/2\mathbb{Z} = R/I$ is naturally an R-module annihilated by both 2 and x.

(a) Show that the map $\varphi : I \times I \to \mathbb{Z}/2\mathbb{Z}$ defined by

$$\varphi(a_0 + a_1 x + \cdots + a_n x^n, b_0 + b_1 x + \cdots + b_m x^m) = \frac{a_0}{2} b_1 \bmod 2$$

is R-bilinear.

(b) Show that there is an R-module homomorphism from $I \otimes_R I \to \mathbb{Z}/2\mathbb{Z}$ mapping

$p(x) \otimes q(x)$ to $\dfrac{p(0)}{2} q'(0)$ where q' denotes the usual polynomial derivative of q.

(c) Show that $2 \otimes x \neq x \otimes 2$ in $I \otimes_R I$.

18. Suppose I is a principal ideal in the integral domain R. Prove that the R-module $I \otimes_R I$ has no nonzero torsion elements (i.e., $rm = 0$ with $0 \neq r \in R$ and $m \in I \otimes_R I$ implies that $m = 0$).

19. Let $I = (2, x)$ be the ideal generated by 2 and x in the ring $R = \mathbb{Z}[x]$ as in Exercise 17. Show that the nonzero element $2 \otimes x - x \otimes 2$ in $I \otimes_R I$ is a torsion element. Show in fact that $2 \otimes x - x \otimes 2$ is annihilated by both 2 and x and that the submodule of $I \otimes_R I$ generated by $2 \otimes x - x \otimes 2$ is isomorphic to R/I.

20. Let $I = (2, x)$ be the ideal generated by 2 and x in the ring $R = \mathbb{Z}[x]$. Show that the element $2 \otimes 2 + x \otimes x$ in $I \otimes_R I$ is not a simple tensor, i.e., cannot be written as $a \otimes b$ for some $a, b \in I$.

21. Suppose R is commutative and let I and J be ideals of R.
   (a) Show there is a surjective R-module homomorphism from $I \otimes_R J$ to the product ideal IJ mapping $i \otimes j$ to the element ij.
   (b) Give an example to show that the map in (a) need not be injective (cf. Exercise 17).

22. Suppose that M is a left and a right R-module such that $rm = mr$ for all $r \in R$ and $m \in M$. Show that the elements $r_1 r_2$ and $r_2 r_1$ act the same on M for every $r_1, r_2 \in R$. (This explains why the assumption that R is commutative in the definition of an R-algebra is a fairly natural one.)

23. Verify the details that the multiplication in Proposition 19 makes $A \otimes_R B$ into an R-algebra.

24. Prove that the extension of scalars from $\mathbb{Z}$ to the Gaussian integers $\mathbb{Z}[i]$ of the ring $\mathbb{R}$ is isomorphic to $\mathbb{C}$ as a ring: $\mathbb{Z}[i] \otimes_{\mathbb{Z}} \mathbb{R} \cong \mathbb{C}$ as rings.

25. Let R be a subring of the commutative ring S and let x be an indeterminate over S. Prove that $S[x]$ and $S \otimes_R R[x]$ are isomorphic as S-algebras.

26. Let S be a commutative ring containing R (with $1_S = 1_R$) and let $x_1, \ldots, x_n$ be independent indeterminates over the ring S. Show that for every ideal I in the polynomial ring $R[x_1, \ldots, x_n]$ that $S \otimes_R (R[x_1, \ldots, x_n]/I) \cong S[x_1, \ldots, x_n]/I S[x_1, \ldots, x_n]$ as S-algebras.

The next exercise shows the ring $\mathbb{C} \otimes_{\mathbb{R}} \mathbb{C}$ introduced at the end of this section is isomorphic to $\mathbb{C} \times \mathbb{C}$. One may also prove this via Exercise 26 and Proposition 16 in Section 9.5, since $\mathbb{C} \cong \mathbb{R}[x]/(x^2 + 1)$. The ring $\mathbb{C} \times \mathbb{C}$ is also discussed in Exercise 23 of Section 1.

27. **(a)** Write down a formula for the multiplication of two elements $a \cdot 1 + b \cdot e_2 + c \cdot e_3 + d \cdot e_4$ and $a' \cdot 1 + b' \cdot e_2 + c' \cdot e_3 + d' \cdot e_4$ in the example $A = \mathbb{C} \otimes_{\mathbb{R}} \mathbb{C}$ following Proposition 21 (where $1 = 1 \otimes 1$ is the identity of A).
   (b) Let $\epsilon_1 = \frac{1}{2}(1 \otimes 1 + i \otimes i)$ and $\epsilon_2 = \frac{1}{2}(1 \otimes 1 - i \otimes i)$. Show that $\epsilon_1 \epsilon_2 = 0, \epsilon_1 + \epsilon_2 = 1$, and $\epsilon_j^2 = \epsilon_j$ for $j = 1, 2$ (ϵ_1 and ϵ_2 are called *orthogonal idempotents* in A). Deduce that A is isomorphic as a ring to the direct product of two principal ideals: $A \cong A\epsilon_1 \times A\epsilon_2$ (cf. Exercise 1, Section 7.6).
   (c) Prove that the map $\varphi : \mathbb{C} \times \mathbb{C} \to \mathbb{C} \times \mathbb{C}$ by $\varphi(z_1, z_2) = (z_1 z_2, z_1 \overline{z_2})$, where $\overline{z_2}$ denotes the complex conjugate of z_2, is an $\mathbb{R}$-bilinear map.

(d) Let Φ be the $\mathbb{R}$-module homomorphism from A to $\mathbb{C} \times \mathbb{C}$ obtained from φ in (c). Show that $\Phi(\epsilon_1) = (0, 1)$ and $\Phi(\epsilon_2) = (1, 0)$. Show also that Φ is $\mathbb{C}$-linear, where the action of $\mathbb{C}$ is on the left tensor factor in A and on both factors in $\mathbb{C} \times \mathbb{C}$. Deduce that Φ is surjective. Show that Φ is a $\mathbb{C}$-algebra isomorphism.

10.5 EXACT SEQUENCES—PROJECTIVE, INJECTIVE, AND FLAT MODULES

One of the fundamental results for studying the structure of an algebraic object B (e.g., a group, a ring, or a module) is the First Isomorphism Theorem, which relates the subobjects of B (the normal subgroups, the ideals, or the submodules, respectively) with the possible homomorphic images of B. We have already seen many examples applying this theorem to understand the structure of B from an understanding of its "smaller" constituents—for example in analyzing the structure of the dihedral group D_8 by determining its center and the resulting quotient by the center.

In most of these examples we began *first* with a given B and then determined some of its basic properties by constructing a homomorphism φ (often given implicitly by the specification of $\ker \varphi \subseteq B$) and examining both $\ker \varphi$ and the resulting quotient $B / \ker \varphi$. We now consider in some greater detail the reverse situation, namely whether we may *first* specify the "smaller constituents." More precisely, we consider whether, given two modules A and C, there exists a module B containing (an isomorphic copy of) A such that the resulting quotient module B/A is isomorphic to C—in which case B is said to be an *extension of C by A*. It is then natural to ask how many such B exist for a given A and C, and the extent to which properties of B are determined by the corresponding properties of A and C. There are, of course, analogous problems in the contexts of groups and rings. This is the *extension problem* first discussed (for groups) in Section 3.4; in this section we shall be primarily concerned with left modules over a ring R, making note where necessary of the modifications required for some other structures, notably noncommutative groups. As in the previous section, throughout this section all rings contain a 1.

We first introduce a very convenient notation. To say that A is isomorphic to a submodule of B, is to say that there is an injective homomorphism $\psi : A \rightarrow B$ (so then $A \cong \psi(A) \subseteq B$). To say that C is isomorphic to the resulting quotient is to say that there is a surjective homomorphism $\varphi : B \rightarrow C$ with $\ker \varphi = \psi(A)$. In particular this gives us a pair of homomorphisms:

$$A \xrightarrow{\psi} B \xrightarrow{\varphi} C$$

with image $\psi = \ker \varphi$. A pair of homomorphisms with this property is given a name:

Definition.

(1) The pair of homomorphisms $X \xrightarrow{\alpha} Y \xrightarrow{\beta} Z$ is said to be *exact* (at Y) if image $\alpha = \ker \beta$.

(2) A sequence $\cdots \rightarrow X_{n-1} \rightarrow X_n \rightarrow X_{n+1} \rightarrow \cdots$ of homomorphisms is said to be an *exact sequence* if it is exact at every X_n between a pair of homomorphisms.

With this terminology, the pair of homomorphisms $A \xrightarrow{\psi} B \xrightarrow{\varphi} C$ above is exact at B. We can also use this terminology to express the fact that for these maps ψ is injective and φ is surjective:

Proposition 22. Let A, B and C be R-modules over some ring R. Then

(1) The sequence $0 \to A \xrightarrow{\psi} B$ is exact (at A) if and only if ψ is injective.

(2) The sequence $B \xrightarrow{\varphi} C \to 0$ is exact (at C) if and only if φ is surjective.

Proof: The (uniquely defined) homomorphism $0 \to A$ has image 0 in A. This will be the kernel of ψ if and only if ψ is injective. Similarly, the kernel of the (uniquely defined) zero homomorphism $C \to 0$ is all of C, which is the image of φ if and only if φ is surjective.

Corollary 23. The sequence $0 \to A \xrightarrow{\psi} B \xrightarrow{\varphi} C \to 0$ is exact if and only if ψ is injective, φ is surjective, and image $\psi = \ker \varphi$, i.e., B is an extension of C by A.

Definition. The exact sequence $0 \to A \xrightarrow{\psi} B \xrightarrow{\varphi} C \to 0$ is called a *short exact sequence*.

In terms of this notation, the extension problem can be stated succinctly as follows: given modules A and C, determine all the short exact sequences

$$0 \longrightarrow A \xrightarrow{\psi} B \xrightarrow{\varphi} C \longrightarrow 0. \tag{10.9}$$

We shall see below that the exact sequence notation is also extremely convenient for analyzing the extent to which properties of A and C determine the corresponding properties of B. If A, B and C are groups written multiplicatively, the sequence (9) will be written

$$1 \longrightarrow A \xrightarrow{\psi} B \xrightarrow{\varphi} C \longrightarrow 1 \tag{10.9'}$$

where 1 denotes the trivial group. Both Proposition 22 and Corollary 23 are valid with the obvious notational changes.

Note that any exact sequence can be written as a succession of short exact sequences since to say $X \xrightarrow{\alpha} Y \xrightarrow{\beta} Z$ is exact at Y is the same as saying that the sequence $0 \to \alpha(X) \to Y \to Y/\ker \beta \to 0$ is a short exact sequence.

Examples

(1) Given modules A and C we can always form their direct sum $B = A \oplus C$ and the sequence

$$0 \to A \xrightarrow{\iota} A \oplus C \xrightarrow{\pi} C \to 0$$

where $\iota(a) = (a, 0)$ and $\pi(a, c) = c$ is a short exact sequence. In particular, it follows that there always exists at least one extension of C by A.

(2) As a special case of the previous example, consider the two $\mathbb{Z}$-modules $A = \mathbb{Z}$ and $C = \mathbb{Z}/n\mathbb{Z}$:

$$0 \longrightarrow \mathbb{Z} \xrightarrow{\iota} \mathbb{Z} \oplus (\mathbb{Z}/n\mathbb{Z}) \xrightarrow{\varphi} \mathbb{Z}/n\mathbb{Z} \longrightarrow 0,$$

giving one extension of $\mathbb{Z}/n\mathbb{Z}$ by $\mathbb{Z}$.

Another extension of $\mathbb{Z}/n\mathbb{Z}$ by $\mathbb{Z}$ is given by the short exact sequence

$$0 \to \mathbb{Z} \xrightarrow{n} \mathbb{Z} \xrightarrow{\pi} \mathbb{Z}/n\mathbb{Z} \to 0$$

where n denotes the map $x \mapsto nx$ given by multiplication by n, and π denotes the natural projection. Note that the modules in the middle of the previous two exact sequences are not isomorphic even though the respective "A" and "C" terms are isomorphic. Thus there are (at least) two "essentially different" or "inequivalent" ways of extending $\mathbb{Z}/n\mathbb{Z}$ by $\mathbb{Z}$.

(3) If $\varphi : B \to C$ is any homomorphism we may form an exact sequence:

$$0 \longrightarrow \ker\varphi \xrightarrow{\iota} B \xrightarrow{\varphi} \text{image } \varphi \longrightarrow 0$$

where ι is the inclusion map. In particular, if φ is surjective, the sequence $\varphi : B \to C$ may be extended to a short exact sequence with $A = \ker\varphi$.

(4) One particularly important instance of the preceding example is when M is an R-module and S is a set of generators for M. Let $F(S)$ be the free R-module on S. Then

$$0 \longrightarrow K \xrightarrow{\iota} F(S) \xrightarrow{\varphi} M \longrightarrow 0$$

is the short exact sequence where φ is the unique R-module homomorphism which is the identity on S (cf. Theorem 6) and $K = \ker\varphi$.

More generally, when M is any group (possibly non-abelian) the above short exact sequence (with 1's at the ends, if M is written multiplicatively) describes a *presentation* of M, where K is the normal subgroup of $F(S)$ generated by the *relations* defining M (cf. Section 6.3).

(5) Two "inequivalent" extensions G of the Klein 4-group by the cyclic group Z_2 of order two are

$$1 \longrightarrow Z_2 \xrightarrow{\iota} D_8 \xrightarrow{\varphi} Z_2 \times Z_2 \longrightarrow 1, \text{ and}$$

$$1 \longrightarrow Z_2 \xrightarrow{\iota} Q_8 \xrightarrow{\varphi} Z_2 \times Z_2 \longrightarrow 1,$$

where in each case ι maps Z_2 injectively into the center of G (recall that both D_8 and Q_8 have centers of order two), and φ is the natural projection of G onto $G/Z(G)$.

Two other inequivalent extensions G of the Klein 4-group by Z_2 occur when G is either of the abelian groups $Z_2 \times Z_2 \times Z_2$ or $Z_2 \times Z_4$ for appropriate maps ι and φ.

Examples 2 and 5 above show that, for a fixed A and C, in general there may be several extensions of C by A. To distinguish different extensions we define the notion of a homomorphism (and isomorphism) between two exact sequences. Recall first that a diagram involving various homomorphisms is said to *commute* if any compositions of homomorphisms with the same starting and ending points are equal, i.e., the composite map defined by following a path of homomorphisms in the diagram depends only on the starting and ending points and not on the choice of the path taken.

Definition. Let $0 \to A \to B \to C \to 0$ and $0 \to A' \to B' \to C' \to 0$ be two short exact sequences of modules.

(1) A *homomorphism of short exact sequences* is a triple α, β, γ of module homomorphisms such that the following diagram commutes:

$$
\begin{array}{ccccccccc}
0 & \longrightarrow & A & \longrightarrow & B & \longrightarrow & C & \longrightarrow & 0 \\
& & \downarrow{\scriptstyle\alpha} & & \downarrow{\scriptstyle\beta} & & \downarrow{\scriptstyle\gamma} & & \\
0 & \longrightarrow & A' & \longrightarrow & B' & \longrightarrow & C' & \longrightarrow & 0
\end{array}
$$

The homomorphism is an *isomorphism of short exact sequences* if α, β, γ are all isomorphisms, in which case the extensions B and B' are said to be *isomorphic extensions*.

(2) The two exact sequences are called *equivalent* if $A = A'$, $C = C'$, and there is an isomorphism between them as in (1) that is the identity maps on A and C (i.e., α and γ are the identity). In this case the corresponding extensions B and B' are said to be *equivalent* extensions.

If B and B' are isomorphic extensions then in particular B and B' are isomorphic as R-modules, but more is true: there is an R-module isomorphism between B and B' that restricts to an isomorphism from A to A' and induces an isomorphism on the quotients C and C'. For a given A and C the condition that two extensions B and B' of C by A are equivalent is stronger still: there must exist an R-module isomorphism between B and B' that restricts to the *identity* map on A and induces the *identity* map on C. The notion of isomorphic extensions measures how many different extensions of C by A there are, allowing for C and A to be changed by an isomorphism. The notion of equivalent extensions measures how many different extensions of C by A there are when A and C are rigidly fixed.

Homomorphisms and isomorphisms between short exact sequences of multiplicative groups $(9')$ are defined similarly.

It is an easy exercise to see that the composition of homomorphisms of short exact sequences is also a homomorphism. Likewise, if the triple α, β, γ is an isomorphism (or equivalence) then α^{-1}, β^{-1}, γ^{-1} is an isomorphism (equivalence, respectively) in the reverse direction. It follows that "isomorphism" (or equivalence) is an equivalence relation on any set of short exact sequences.

Examples

(1) Let m and n be integers greater than 1. Assume n divides m and let $k = m/n$. Define a map from the exact sequence of $\mathbb{Z}$-modules in Example 2 of the preceding set of examples:

$$
\begin{array}{ccccccccc}
0 & \longrightarrow & \mathbb{Z} & \xrightarrow{\;n\;} & \mathbb{Z} & \xrightarrow{\;\pi\;} & \mathbb{Z}/n\mathbb{Z} & \longrightarrow & 0 \\
& & \downarrow{\scriptstyle\alpha} & & \downarrow{\scriptstyle\beta} & & \downarrow{\scriptstyle\gamma} & & \\
0 & \longrightarrow & \mathbb{Z}/k\mathbb{Z} & \xrightarrow{\;\iota\;} & \mathbb{Z}/m\mathbb{Z} & \xrightarrow{\;\pi'\;} & \mathbb{Z}/n\mathbb{Z} & \longrightarrow & 0
\end{array}
$$

where α and β are the natural projections, γ is the identity map, ι maps $a \bmod k$ to $na \bmod m$, and π' is the natural projection of $\mathbb{Z}/m\mathbb{Z}$ onto its quotient $(\mathbb{Z}/m\mathbb{Z})/(n\mathbb{Z}/m\mathbb{Z})$

(which is isomorphic to $\mathbb{Z}/n\mathbb{Z}$). One easily checks that this is a homomorphism of short exact sequences.

(2) If again $0 \to \mathbb{Z} \xrightarrow{n} \mathbb{Z} \xrightarrow{\pi} \mathbb{Z}/n\mathbb{Z} \to 0$ is the short exact sequence of $\mathbb{Z}$-modules defined previously, map each module to itself by $x \mapsto -x$. This triple of homomorphisms gives an isomorphism of the exact sequence with itself. This isomorphism is not an equivalence of sequences since it is not the identity on the first $\mathbb{Z}$.

(3) The short exact sequences in Examples 1 and 2 following Corollary 23 are not isomorphic—the extension modules are not isomorphic $\mathbb{Z}$-modules (abelian groups). Likewise the two extensions, D_8 and Q_8, in Example 5 of the same set are not isomorphic (hence not equivalent), even though the two end terms "A" and "C" are the same for both sequences.

(4) Consider the maps

$$
\begin{array}{ccccccccc}
0 & \longrightarrow & \mathbb{Z}/2\mathbb{Z} & \xrightarrow{\psi} & \mathbb{Z}/2\mathbb{Z} \oplus \mathbb{Z}/2\mathbb{Z} & \xrightarrow{\varphi} & \mathbb{Z}/2\mathbb{Z} & \longrightarrow & 0 \\
& & \downarrow{\scriptstyle \text{id}} & & \downarrow{\scriptstyle \beta} & & \downarrow{\scriptstyle \text{id}} & & \\
0 & \longrightarrow & \mathbb{Z}/2\mathbb{Z} & \xrightarrow{\psi'} & \mathbb{Z}/2\mathbb{Z} \oplus \mathbb{Z}/2\mathbb{Z} & \xrightarrow{\varphi'} & \mathbb{Z}/2\mathbb{Z} & \longrightarrow & 0
\end{array}
$$

where ψ maps $\mathbb{Z}/2\mathbb{Z}$ injectively into the first component of the direct sum and φ projects the direct sum onto its second component. Also ψ' embeds $\mathbb{Z}/2\mathbb{Z}$ into the *second* component of the direct sum and φ' projects the direct sum onto its *first* component. If β maps the direct sum $\mathbb{Z}/2\mathbb{Z} \oplus \mathbb{Z}/2\mathbb{Z}$ to itself by interchanging the two factors, then this diagram is seen to commute, hence giving an equivalence of the two exact sequences that is not the identity isomorphism.

(5) We exhibit two isomorphic but inequivalent $\mathbb{Z}$-module extensions. For $i = 1, 2$ define

$$
0 \longrightarrow \mathbb{Z}/2\mathbb{Z} \xrightarrow{\psi} \mathbb{Z}/4\mathbb{Z} \oplus \mathbb{Z}/2\mathbb{Z} \xrightarrow{\varphi_i} \mathbb{Z}/2\mathbb{Z} \oplus \mathbb{Z}/2\mathbb{Z} \longrightarrow 0
$$

where $\psi : 1 \mapsto (2, 0)$ in both sequences, φ_1 is defined by $\varphi_1(a \bmod 4, b \bmod 2) = (a \bmod 2, b \bmod 2)$, and $\varphi_2(a \bmod 4, b \bmod 2) = (b \bmod 2, a \bmod 2)$. It is easy to see that the resulting two sequences are both short exact sequences.

An evident isomorphism between these two exact sequences is provided by the triple of maps id, id, γ, where $\gamma : \mathbb{Z}/2\mathbb{Z} \oplus \mathbb{Z}/2\mathbb{Z} \to \mathbb{Z}/2\mathbb{Z} \oplus \mathbb{Z}/2\mathbb{Z}$ is the map $\gamma((c, d)) = (d, c)$ that interchanges the two direct factors.

We now check that these two isomorphic sequences are *not equivalent*, as follows. Since $\varphi_1(0, 1) = (0, 1)$, any equivalence, id, β, id, from the first sequence to the second must map $(0, 1) \in \mathbb{Z}/4\mathbb{Z} \oplus \mathbb{Z}/2\mathbb{Z}$ to either $(1, 0)$ or $(3, 0)$ in $\mathbb{Z}/4\mathbb{Z} \oplus \mathbb{Z}/2\mathbb{Z}$, since these are the two possible elements mapping to $(0, 1)$ by φ_2. This is impossible, however, since the isomorphism β cannot send an element of order 2 to an element of order 4.

Put another way, equivalences involving the same extension module B are automorphisms of B that restrict to the identity on both $\psi(A)$ and $B/\psi(A)$. Any such automorphism of $B = \mathbb{Z}/4\mathbb{Z} \oplus \mathbb{Z}/2\mathbb{Z}$ must fix the coset $(0, 1) + \psi(A)$ since this is the unique nonidentity coset containing elements of order 2. Thus maps which send this coset to different elements in C give inequivalent extensions. In particular, there is yet a third inequivalent extension involving the same modules $A = \mathbb{Z}/2\mathbb{Z}$, $B = \mathbb{Z}/4\mathbb{Z} \oplus \mathbb{Z}/2\mathbb{Z}$ and $C = \mathbb{Z}/2\mathbb{Z} \oplus \mathbb{Z}/2\mathbb{Z}$, that maps the coset $(0, 1) + \psi(A)$ to the element $(1, 1) \in \mathbb{Z}/2\mathbb{Z} \oplus \mathbb{Z}/2\mathbb{Z}$.

By similar reasoning there are three inequivalent but isomorphic group extensions of $Z_2 \times Z_2$ by Z_2 with $B \cong D_8$ (cf. the exercises).

The homomorphisms α, β, γ in a homomorphism of short exact sequences are not independent. The next result gives some relations among these three homomorphisms.

Proposition 24. *(The Short Five Lemma)* Let α, β, γ be a homomorphism of short exact sequences

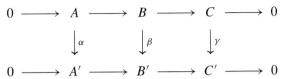

(1) If α and γ are injective then so is β.
(2) If α and γ are surjective then so is β.
(3) If α and γ are isomorphisms then so is β (and then the two sequences are isomorphic).

Remark: These results hold also for short exact sequences of (possibly non-abelian) groups (as the proof demonstrates).

Proof: We shall prove (1), leaving the proof of (2) as an exercise (and (3) follows immediately from (1) and (2)). Suppose then that α and γ are injective and suppose $b \in B$ with $\beta(b) = 0$. Let $\psi : A \to B$ and $\varphi : B \to C$ denote the homomorphisms in the first short exact sequence. Since $\beta(b) = 0$, it follows in particular that the image of $\beta(b)$ in the quotient C' is also 0. By the commutativity of the diagram this implies that $\gamma(\varphi(b)) = 0$, and since γ is assumed injective, we obtain $\varphi(b) = 0$, i.e., b is in the kernel of φ. By the exactness of the first sequence, this means that b is in the image of ψ, i.e., $b = \psi(a)$ for some $a \in A$. Then, again by the commutativity of the diagram, the image of $\alpha(a)$ in B' is the same as $\beta(\psi(a)) = \beta(b) = 0$. But α and the map from A' to B' are injective by assumption, and it follows that $a = 0$. Finally, $b = \psi(a) = \psi(0) = 0$ and we see that β is indeed injective.

We have already seen that there is always at least one extension of a module C by A, namely the direct sum $B = A \oplus C$. In this case the module B contains a submodule C' isomorphic to C (namely $C' = 0 \oplus C$) as well as the submodule A, and this submodule complement to A "splits" B into a direct sum. In the case of groups the existence of a subgroup complement C' to a normal subgroup in B implies that B is a semidirect product (cf. Section 5 in Chapter 5). The fact that B is a direct sum in the context of modules is a reflection of the fact that the underlying group structure in this case is *abelian*; for abelian groups semidirect products are direct products. In either case the corresponding short exact sequence is said to "split":

Definition.
(1) Let R be a ring and let $0 \to A \xrightarrow{\psi} B \xrightarrow{\varphi} C \to 0$ be a short exact sequence of R-modules. The sequence is said to be *split* if there is an R-module complement to $\psi(A)$ in B. In this case, up to isomorphism, $B = A \oplus C$ (more precisely, $B = \psi(A) \oplus C'$ for some submodule C', and C' is mapped isomorphically onto C by φ: $\varphi(C') \cong C$).

(2) If $1 \to A \overset{\psi}{\to} B \overset{\varphi}{\to} C \to 1$ is a short exact sequence of groups, then the sequence is said to be *split* if there is a subgroup complement to $\psi(A)$ in B. In this case, up to isomorphism, $B = A \rtimes C$ (more precisely, $B = \psi(A) \rtimes C'$ for some subgroup C', and C' is mapped isomorphically onto C by φ: $\varphi(C') \cong C$).

In either case the extension B is said to be a *split extension* of C by A.

The question of whether an extension splits is the question of the existence of a complement to $\psi(A)$ in B isomorphic (by φ) to C, so the notion of a split extension may equivalently be phrased in the language of homomorphisms:

Proposition 25. The short exact sequence $0 \to A \overset{\psi}{\to} B \overset{\varphi}{\to} C \to 0$ of R-modules is split if and only if there is an R-module homomorphism $\mu : C \to B$ such that $\varphi \circ \mu$ is the identity map on C. Similarly, the short exact sequence $1 \to A \overset{\psi}{\to} B \overset{\varphi}{\to} C \to 1$ of groups is split if and only if there is a group homomorphism $\mu : C \to B$ such that $\varphi \circ \mu$ is the identity map on C.

Proof: This follows directly from the definitions: if μ is given define $C' = \mu(C) \subseteq B$ and if C' is given define $\mu = \varphi^{-1} : C \cong C' \subseteq B$.

Definition. With notation as in Proposition 25, any set map $\mu : C \to B$ such that $\varphi \circ \mu = $ id is called a *section* of φ. If μ is a *homomorphism* as in Proposition 25 then μ is called a *splitting homomorphism* for the sequence.

Note that a section of φ is nothing more than a choice of coset representatives in B for the quotient $B/\ker \varphi \cong C$. A section is a (splitting) homomorphism if this set of coset representatives forms a *submodule* (respectively, *subgroup*) in B, in which case this submodule (respectively, subgroup) gives a complement to $\psi(A)$ in B.

Examples

(1) The split short exact sequence $0 \to A \overset{\iota}{\to} A \oplus C \overset{\pi}{\to} C \to 0$ has the evident splitting homomorphism $\mu(c) = (0, c)$.

(2) The extension $0 \to \mathbb{Z} \overset{\iota}{\to} \mathbb{Z} \oplus (\mathbb{Z}/n\mathbb{Z}) \overset{\varphi}{\to} \mathbb{Z}/n\mathbb{Z} \to 0$, of $\mathbb{Z}/n\mathbb{Z}$ by $\mathbb{Z}$ is split (with splitting homomorphism μ mapping $\mathbb{Z}/n\mathbb{Z}$ isomorphically onto the second factor of the direct sum). On the other hand, the exact sequence of $\mathbb{Z}$-modules $0 \to \mathbb{Z} \overset{n}{\to} \mathbb{Z} \overset{\pi}{\to} \mathbb{Z}/n\mathbb{Z} \to 0$ is not split since there is no nonzero homomorphism of $\mathbb{Z}/n\mathbb{Z}$ into $\mathbb{Z}$.

(3) Neither D_8 nor Q_8 is a split extension of $Z_2 \times Z_2$ by Z_2 because in neither group is there a subgroup complement to the center (Section 2.5 gives the subgroup structures of these groups).

(4) The group D_8 *is* a split extension of Z_2 by Z_4, i.e., there is a split short exact sequence

$$1 \longrightarrow Z_4 \overset{\iota}{\longrightarrow} D_8 \overset{\pi}{\longrightarrow} Z_2 \longrightarrow 1,$$

namely,

$$1 \longrightarrow \langle r \rangle \overset{\iota}{\longrightarrow} D_8 \overset{\pi}{\longrightarrow} \langle \bar{s} \rangle \longrightarrow 1,$$

using our usual set of generators for D_8. Here ι is the inclusion map and $\pi : r^a s^b \mapsto \bar{s}^b$ is the projection onto the quotient $D_8/\langle r \rangle \cong Z_2$. The splitting homomorphism μ

maps $\langle \bar{s} \rangle$ isomorphically onto the complement $\langle s \rangle$ for $\langle r \rangle$ in D_8. Equivalently, D_8 is the semidirect product of the normal subgroup $\langle r \rangle$ (isomorphic to Z_4) with $\langle s \rangle$ (isomorphic to Z_2).

On the other hand, while Q_8 is also an extension of Z_2 by Z_4 (for example, $\langle i \rangle \cong Z_4$ has quotient isomorphic to Z_2), Q_8 is *not* a split extension of Z_2 by Z_4: no cyclic subgroup of Q_8 of order 4 has a complement in Q_8.

Section 5.5 contains many more examples of split extensions of groups.

Proposition 25 shows that an extension B of C by A is a split extension if and only if there is a splitting homomorphism μ of the projection map $\varphi : B \to C$ from B to the quotient C. The next proposition shows in particular that for modules this is equivalent to the existence of a splitting homomorphism for ψ at the other end of the sequence.

Proposition 26. Let $0 \to A \xrightarrow{\psi} B \xrightarrow{\varphi} C \to 0$ be a short exact sequence of modules (respectively, $1 \to A \xrightarrow{\psi} B \xrightarrow{\varphi} C \to 1$ a short exact sequence of groups). Then $B = \psi(A) \oplus C'$ for some submodule C' of B with $\varphi(C') \cong C$ (respectively, $B = \psi(A) \times C'$ for some subgroup C' of B with $\varphi(C') \cong C$) if and only if there is a homomorphism $\lambda : B \to A$ such that $\lambda \circ \psi$ is the identity map on A.

Proof: This is similar to the proof of Proposition 25. If λ is given, define $C' =$ ker $\lambda \subseteq B$ and if C' is given define $\lambda : B = \psi(A) \oplus C' \to A$ by $\lambda((\psi(a), c') = a$. Note that in this case $C' =$ ker λ is *normal* in B, so that C' is a *normal* complement to $\psi(A)$ in B, which in turn implies that B is the *direct sum* of $\psi(A)$ and C' (cf. Theorem 9 of Section 5.4).

Proposition 26 shows that for general group extensions, the existence of a splitting homomorphism λ on the *left* end of the sequence is stronger than the condition that the extension splits: in this case the extension group is a *direct* product, and not just a *semidirect* product. The fact that these two notions are equivalent in the context of modules is again a reflection of the abelian nature of the underlying groups, where semidirect products are always direct products.

Projective Modules and $Hom_R(D, \_)$

Let R be a ring with 1 and suppose the R-module M is an extension of N by L, with

$$0 \longrightarrow L \xrightarrow{\psi} M \xrightarrow{\varphi} N \longrightarrow 0$$

the corresponding short exact sequence of R-modules. It is natural to ask whether properties for L and N imply related properties for the extension M. The first situation we shall consider is whether an R-module homomorphism from some fixed R-module D to either L or N implies there is also an R-module homomorphism from D to M.

The question of obtaining a homomorphism from D to M given a homomorphism from D to L is easily disposed of: if $f \in Hom_R(D, L)$ is an R-module homomorphism from D to L then the composite $f' = \psi \circ f$ is an R-module homomorphism from D to

M. The relation between these maps can be indicated pictorially by the commutative diagram

$$\begin{array}{ccc} & D & \\ f \downarrow & \searrow f' & \\ L & \xrightarrow{\psi} & M \end{array}$$

Put another way, composition with ψ induces a map

$$\psi' : \operatorname{Hom}_R(D, L) \longrightarrow \operatorname{Hom}_R(D, M)$$
$$f \longmapsto f' = \psi \circ f.$$

Recall that, by Proposition 2, $\operatorname{Hom}_R(D, L)$ and $\operatorname{Hom}_R(D, M)$ are abelian groups.

Proposition 27. Let D, L and M be R-modules and let $\psi : L \to M$ be an R-module homomorphism. Then the map

$$\psi' : \operatorname{Hom}_R(D, L) \longrightarrow \operatorname{Hom}_R(D, M)$$
$$f \longmapsto f' = \psi \circ f$$

is a homomorphism of abelian groups. If ψ is injective, then ψ' is also injective, i.e.,

$$\text{if} \quad 0 \longrightarrow L \xrightarrow{\psi} M \quad \text{is exact,}$$

$$\text{then} \quad 0 \longrightarrow \operatorname{Hom}_R(D, L) \xrightarrow{\psi'} \operatorname{Hom}_R(D, M) \quad \text{is also exact.}$$

Proof: The fact that ψ' is a homomorphism is immediate. If ψ is injective, then distinct homomorphisms f and g from D into L give distinct homomorphisms $\psi \circ f$ and $\psi \circ g$ from D into M, which is to say that ψ' is also injective.

While obtaining homomorphisms into M from homomorphisms into the submodule L is straightforward, the situation for homomorphisms into the quotient N is much less evident. More precisely, given an R-module homomorphism $f : D \to N$ the question is whether there exists an R-module homomorphism $F : D \to M$ that *extends* or *lifts* f to M, i.e., that makes the following diagram commute:

$$\begin{array}{ccc} & & D \\ F \nearrow & & \downarrow f \\ M & \xrightarrow{\varphi} & N \end{array}$$

As before, composition with the homomorphism φ induces a homomorphism of abelian groups

$$\varphi' : \operatorname{Hom}_R(D, M) \longrightarrow \operatorname{Hom}_R(D, N)$$
$$F \longmapsto F' = \varphi \circ F.$$

In terms of φ', the homomorphism f to N lifts to a homomorphism to M if and only if f is in the image of φ' (namely, f is the image of the lift F).

In general it may not be possible to lift a homomorphism f from D to N to a homomorphism from D to M. For example, consider the nonsplit exact sequence $0 \to \mathbb{Z} \xrightarrow{2} \mathbb{Z} \xrightarrow{\pi} \mathbb{Z}/2\mathbb{Z} \to 0$ from the previous set of examples. Let $D = \mathbb{Z}/2\mathbb{Z}$ and let f be the identity map from D into N. Any homomorphism F of D into $M = \mathbb{Z}$ must map D to 0 (since $\mathbb{Z}$ has no elements of order 2), hence $\pi \circ F$ maps D to 0 in N, and in particular, $\pi \circ F \neq f$. Phrased in terms of the map φ', this shows that

$$\text{if} \quad M \xrightarrow{\varphi} N \longrightarrow 0 \quad \text{is exact,}$$

$$\text{then} \quad \text{Hom}_R(D, M) \xrightarrow{\varphi'} \text{Hom}_R(D, N) \longrightarrow 0 \quad \text{is } \textit{not necessarily} \text{ exact.}$$

These results relating the homomorphisms into L and N to the homomorphisms into M can be neatly summarized as part of the following theorem.

Theorem 28. Let D, L, M, and N be R-modules. If

$$0 \longrightarrow L \xrightarrow{\psi} M \xrightarrow{\varphi} N \longrightarrow 0 \quad \text{is exact,}$$

then the associated sequence

$$0 \to \text{Hom}_R(D, L) \xrightarrow{\psi'} \text{Hom}_R(D, M) \xrightarrow{\varphi'} \text{Hom}_R(D, N) \quad \text{is exact.} \qquad (10.10)$$

A homomorphism $f : D \to N$ lifts to a homomorphism $F : D \to M$ if and only if $f \in \text{Hom}_R(D, N)$ is in the image of φ'. In general $\varphi' : \text{Hom}_R(D, M) \to \text{Hom}_R(D, N)$ need not be surjective; the map φ' is surjective if and only if every homomorphism from D to N lifts to a homomorphism from D to M, in which case the sequence (10) can be extended to a short exact sequence.

The sequence (10) is exact for *all* R-modules D if and only if the sequence

$$0 \to L \xrightarrow{\psi} M \xrightarrow{\varphi} N \quad \text{is exact.}$$

Proof: The only item in the first statement that has not already been proved is the exactness of (10) at $\text{Hom}_R(D, M)$, i.e., $\ker \varphi' = \text{image } \psi'$. Suppose $F : D \to M$ is an element of $\text{Hom}_R(D, M)$ lying in the kernel of φ', i.e., with $\varphi \circ F = 0$ as homomorphisms from D to N. If $d \in D$ is any element of D, this implies that $\varphi(F(d)) = 0$ and $F(d) \in \ker \varphi$. By the exactness of the sequence defining the extension M we have $\ker \varphi = \text{image } \psi$, so there is some element $l \in L$ with $F(d) = \psi(l)$. Since ψ is injective, the element l is unique, so this gives a well defined map $F' : D \to L$ given by $F'(d) = l$. It is an easy check to verify that F' is a homomorphism, i.e., $F' \in \text{Hom}_R(D, L)$. Since $\psi \circ F'(d) = \psi(l) = F(d)$, we have $F = \psi'(F')$ which shows that F is in the image of ψ', proving that $\ker \varphi' \subseteq \text{image } \psi'$. Conversely, if F is in the image of ψ' then $F = \psi'(F')$ for some $F' \in \text{Hom}_R(D, L)$ and so $\varphi(F(d)) = \varphi(\psi(F'(d)))$ for any $d \in D$. Since $\ker \varphi = \text{image } \psi$ we have $\varphi \circ \psi = 0$, and it follows that $\varphi(F(d)) = 0$ for any $d \in D$, i.e., $\varphi'(F) = 0$. Hence F is in the kernel of φ', proving the reverse containment: image $\psi' \subseteq \ker \varphi'$.

For the last statement in the theorem, note first that the surjectivity of φ was not required for the proof that (10) is exact, so the "if" portion of the statement has already been proved. For the converse, suppose that the sequence (10) is exact for all R-modules D. In general, $\mathrm{Hom}_R(R, X) \cong X$ for any left R-module X, the isomorphism being given by mapping a homomorphism to its value on the element $1 \in R$ (cf. Exercise 10(b)). Taking $D = R$ in (10), the exactness of the sequence $0 \to L \overset{\psi}{\to} M \overset{\varphi}{\to} N$ follows easily.

By Theorem 28, the sequence

$$0 \longrightarrow \mathrm{Hom}_R(D, L) \overset{\psi'}{\longrightarrow} \mathrm{Hom}_R(D, M) \overset{\varphi'}{\longrightarrow} \mathrm{Hom}_R(D, N) \longrightarrow 0 \qquad (10.11)$$

is in general *not* a short exact sequence since the homomorphism φ' need not be surjective. The question of whether this sequence is exact precisely measures the extent to which the homomorphisms from D into M are uniquely determined by pairs of homomorphisms from D into L and D into N. More precisely, this sequence is exact if and only if there is a bijection $F \leftrightarrow (g, f)$ between homomorphisms $F : D \to M$ and pairs of homomorphisms $g : D \to L$ and $f : D \to N$ given by $F|_{\psi(L)} = \psi'(g)$ and $f = \varphi'(F)$.

One situation in which the sequence (11) is exact occurs when the original sequence $0 \to L \to M \to N \to 0$ is a *split* exact sequence, i.e., when $M = L \oplus N$. In this case the sequence (11) is also a split exact sequence, as the first part of the following proposition shows.

Proposition 29. Let D, L and N be R-modules. Then
    (1) $\mathrm{Hom}_R(D, L \oplus N) \cong \mathrm{Hom}_R(D, L) \oplus \mathrm{Hom}_R(D, N)$, and
    (2) $\mathrm{Hom}_R(L \oplus N, D) \cong \mathrm{Hom}_R(L, D) \oplus \mathrm{Hom}_R(N, D)$.

Proof: Let $\pi_1 : L \oplus N \to L$ be the natural projection from $L \oplus N$ to L and similarly let π_2 be the natural projection to N. If $f \in \mathrm{Hom}_R(D, L \oplus N)$ then the compositions $\pi_1 \circ f$ and $\pi_2 \circ f$ give elements in $\mathrm{Hom}_R(D, L)$ and $\mathrm{Hom}_R(D, N)$, respectively. This defines a map from $\mathrm{Hom}_R(D, L \oplus N)$ to $\mathrm{Hom}_R(D, L) \oplus \mathrm{Hom}_R(D, N)$ which is easily seen to be a homomorphism. Conversely, given $f_1 \in \mathrm{Hom}_R(D, L)$ and $f_2 \in \mathrm{Hom}_R(D, N)$, define the map $f \in \mathrm{Hom}_R(D, L \oplus N)$ by $f(d) = (f_1(d), f_2(d))$. This defines a map from $\mathrm{Hom}_R(D, L) \oplus \mathrm{Hom}_R(D, N)$ to $\mathrm{Hom}_R(D, L \oplus N)$ that is easily checked to be a homomorphism inverse to the map above, proving the isomorphism in (1). The proof of (2) is similar and is left as an exercise.

The results in Proposition 29 extend immediately by induction to any finite direct sum of R-modules. These results are referred to by saying that Hom *commutes with finite direct sums in either variable* (compare to Theorem 17 for a corresponding result for tensor products). For infinite direct sums the situation is more complicated. Part (1) remains true if $L \oplus N$ is replaced by an arbitrary direct sum and the direct sum on the right hand side is replaced by a direct product (Exercise 13 shows that the direct product is necessary). Part (2) remains true if the direct sums on both sides are replaced by direct products.

This proposition shows that if the sequence

$$0 \longrightarrow L \overset{\psi}{\longrightarrow} M \overset{\varphi}{\longrightarrow} N \longrightarrow 0$$

is a split short exact sequence of R-modules, then

$$0 \longrightarrow \text{Hom}_R(D, L) \xrightarrow{\psi'} \text{Hom}_R(D, M) \xrightarrow{\varphi'} \text{Hom}_R(D, N) \longrightarrow 0$$

is also a split short exact sequence of abelian groups for every R-module D. Exercise 14 shows that a converse holds: if $0 \to \text{Hom}_R(D, L) \xrightarrow{\psi'} \text{Hom}_R(D, M) \xrightarrow{\varphi'} \text{Hom}_R(D, N) \to 0$ is exact for every R-module D then $0 \to L \xrightarrow{\psi} M \xrightarrow{\varphi} N \to 0$ is a split short exact sequence (which then implies that if the original Hom sequence is exact for every D, then in fact it is split exact for every D).

Proposition 29 identifies a situation in which the sequence (11) is exact in terms of the modules L, M, and N. The next result adopts a slightly different perspective, characterizing instead the modules D having the property that the sequence (10) in Theorem 28 can *always* be extended to a short exact sequence:

Proposition 30. Let P be an R-module. Then the following are equivalent:
(1) For any R-modules L, M, and N, if

$$0 \longrightarrow L \xrightarrow{\psi} M \xrightarrow{\varphi} N \longrightarrow 0$$

is a short exact sequence, then

$$0 \longrightarrow \text{Hom}_R(P, L) \xrightarrow{\psi'} \text{Hom}_R(P, M) \xrightarrow{\varphi'} \text{Hom}_R(P, N) \longrightarrow 0$$

is also a short exact sequence.
(2) For any R-modules M and N, if $M \xrightarrow{\varphi} N \to 0$ is exact, then every R-module homomorphism from P into N lifts to an R-module homomorphism into M, i.e., given $f \in \text{Hom}_R(P, N)$ there is a lift $F \in \text{Hom}_R(P, M)$ making the following diagram commute:

(3) If P is a quotient of the R-module M then P is isomorphic to a direct summand of M, i.e., every short exact sequence $0 \to L \to M \to P \to 0$ splits.
(4) P is a direct summand of a free R-module.

Proof: The equivalence of (1) and (2) is a restatement of a result in Theorem 28. Suppose now that (2) is satisfied, and let $0 \to L \xrightarrow{\psi} M \xrightarrow{\varphi} P \to 0$ be exact. By (2), the identity map from P to P lifts to a homomorphism μ making the following diagram commute:

Then $\varphi \circ \mu = 1$, so μ is a splitting homomorphism for the sequence, which proves (3). Every module P is the quotient of a free module (for example, the free module on the set of elements in P), so there is always an exact sequence $0 \to \ker \varphi \to \mathcal{F} \xrightarrow{\varphi} P \to 0$ where $\mathcal{F}$ is a free R-module (cf. Example 4 following Corollary 23). If (3) is satisfied, then this sequence splits, so $\mathcal{F}$ is isomorphic to the direct sum of $\ker \varphi$ and P, which proves (4).

Finally, to prove (4) implies (2), suppose that P is a direct summand of a free R-module on some set S, say $\mathcal{F}(S) = P \oplus K$, and that we are given a homomorphism f from P to N as in (2). Let π denote the natural projection from $\mathcal{F}(S)$ to P, so that $f \circ \pi$ is a homomorphism from $\mathcal{F}(S)$ to N. For any $s \in S$ define $n_s = f \circ \pi(s) \in N$ and let $m_s \in M$ be any element of M with $\varphi(m_s) = n_s$ (which exists because φ is surjective). By the universal property for free modules (Theorem 6 of Section 3), there is a unique R-module homomorphism F' from $\mathcal{F}(S)$ to M with $F'(s) = m_s$. The diagram is the following:

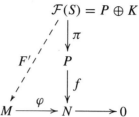

By definition of the homomorphism F' we have $\varphi \circ F'(s) = \varphi(m_s) = n_s = f \circ \pi(s)$, from which it follows that $\varphi \circ F' = f \circ \pi$ on $\mathcal{F}(S)$, i.e., the diagram above is commutative. Now define a map $F : P \to M$ by $F(d) = F'((d, 0))$. Since F is the composite of the injection $P \to \mathcal{F}(S)$ with the homomorphism F', it follows that F is an R-module homomorphism. Then

$$\varphi \circ F(d) = \varphi \circ F'((d, 0)) = f \circ \pi((d, 0)) = f(d)$$

i.e., $\varphi \circ F = f$, so the diagram

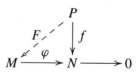

commutes, which proves that (4) implies (2) and completes the proof.

Definition. An R-module P is called *projective* if it satisfies any of the equivalent conditions of Proposition 30.

The third statement in Proposition 30 can be rephrased as saying that any module M that projects onto P has (an isomorphic copy of) P as a direct summand, which explains the terminology.

The following result is immediate from Proposition 30 (and its proof):

Corollary 31. *Free modules are projective. A finitely generated module is projective if and only if it is a direct summand of a finitely generated free module. Every module is a quotient of a projective module.*

If D is fixed, then given any R-module X we have an associated abelian group $\text{Hom}_R(D, X)$. Further, an R-module homomorphism $\alpha : X \to Y$ induces an abelian group homomorphism $\alpha' : \text{Hom}_R(D, X) \to \text{Hom}_R(D, Y)$, defined by $\alpha'(f) = \alpha \circ f$. Put another way, the map $\text{Hom}_R(D, \_)$ is a *covariant functor* from the category of R-modules to the category of abelian groups (cf. Appendix II). Theorem 28 shows that applying this functor to the terms in the exact sequence

$$0 \longrightarrow L \overset{\psi}{\longrightarrow} M \overset{\varphi}{\longrightarrow} N \longrightarrow 0$$

produces an exact sequence

$$0 \to \text{Hom}_R(D, L) \overset{\psi'}{\to} \text{Hom}_R(D, M) \overset{\varphi'}{\to} \text{Hom}_R(D, N).$$

This is referred to by saying that $\text{Hom}_R(D, \_)$ is a *left exact* functor. By Proposition 30, the functor $\text{Hom}_R(D, \_)$ is *exact*, i.e., always takes short exact sequences to short exact sequences, if and only if D is projective. We summarize this as

Corollary 32. If D is an R-module, then the functor $\text{Hom}_R(D, \_)$ from the category of R-modules to the category of abelian groups is left exact. It is exact if and only if D is a projective R-module.

Note that if $\text{Hom}_R(D, \_)$ takes short exact sequences to short exact sequences, then it takes exact sequences of any length to exact sequences since any exact sequence can be broken up into a succession of short exact sequences.

As we have seen, the functor $\text{Hom}_R(D, \_)$ is in general not exact on the right. Measuring the extent to which functors such as $\text{Hom}_R(D, \_)$ fail to be exact leads to the notions of "homological algebra," considered in Chapter 17.

Examples

(1) We shall see in Section 11.1 that if $R = F$ is a field then every F-module is projective (although we only prove this for finitely generated modules).

(2) By Corollary 31, $\mathbb{Z}$ is a projective $\mathbb{Z}$-module. This can be seen directly as follows: suppose f is a map from $\mathbb{Z}$ to N and $M \overset{\varphi}{\to} N \to 0$ is exact. The homomorphism f is uniquely determined by the value $n = f(1)$. Then f can be lifted to a homomorphism $F : \mathbb{Z} \to M$ by first defining $F(1) = m$, where m is any element in M mapped to n by φ, and then extending F to all of $\mathbb{Z}$ by additivity.

By the first statement in Proposition 30, since $\mathbb{Z}$ is projective, if

$$0 \longrightarrow L \overset{\psi}{\longrightarrow} M \overset{\varphi}{\longrightarrow} N \longrightarrow 0$$

is an exact sequence of $\mathbb{Z}$-modules, then

$$0 \longrightarrow \text{Hom}_{\mathbb{Z}}(\mathbb{Z}, L) \overset{\psi'}{\longrightarrow} \text{Hom}_{\mathbb{Z}}(\mathbb{Z}, M) \overset{\varphi'}{\longrightarrow} \text{Hom}_{\mathbb{Z}}(\mathbb{Z}, N) \longrightarrow 0$$

is also an exact sequence. This can also be seen directly using the isomorphism $\text{Hom}_{\mathbb{Z}}(\mathbb{Z}, M) \cong M$ of abelian groups, which shows that the two exact sequences above are essentially the same.

(3) Free $\mathbb{Z}$-modules have no nonzero elements of finite order so no nonzero finite abelian group can be isomorphic to a submodule of a free module. By Corollary 31 it follows that no nonzero finite abelian group is a projective $\mathbb{Z}$-module.

(4) As a particular case of the preceding example, we see that for $n \geq 2$ the $\mathbb{Z}$-module $\mathbb{Z}/n\mathbb{Z}$ is not projective. By Theorem 28 it must be possible to find a short exact sequence which after applying the functor $\text{Hom}_{\mathbb{Z}}(\mathbb{Z}/n\mathbb{Z}, \_\_)$ is no longer exact on the right. One such sequence is the exact sequence of Example 2 following Corollary 23:

$$0 \longrightarrow \mathbb{Z} \xrightarrow{n} \mathbb{Z} \xrightarrow{\pi} \mathbb{Z}/n\mathbb{Z} \longrightarrow 0,$$

for $n \geq 2$. Note first that $\text{Hom}_{\mathbb{Z}}(\mathbb{Z}/n\mathbb{Z}, \mathbb{Z}) = 0$ since there are no nonzero $\mathbb{Z}$-module homomorphisms from $\mathbb{Z}/n\mathbb{Z}$ to $\mathbb{Z}$. It is also easy to see that $\text{Hom}_{\mathbb{Z}}(\mathbb{Z}/n\mathbb{Z}, \mathbb{Z}/n\mathbb{Z}) \cong \mathbb{Z}/n\mathbb{Z}$, as follows. Every homomorphism f is uniquely determined by $f(1) = a \in \mathbb{Z}/n\mathbb{Z}$, and given any $a \in \mathbb{Z}/n\mathbb{Z}$ there is a unique homomorphism f_a with $f_a(1) = a$; the map $f_a \mapsto a$ is easily checked to be an isomorphism from $\text{Hom}_{\mathbb{Z}}(\mathbb{Z}/n\mathbb{Z}, \mathbb{Z}/n\mathbb{Z})$ to $\mathbb{Z}/n\mathbb{Z}$.

Applying $\text{Hom}_{\mathbb{Z}}(\mathbb{Z}/n\mathbb{Z}, \_\_)$ to the short exact sequence above thus gives the sequence

$$0 \longrightarrow 0 \xrightarrow{n'} 0 \xrightarrow{\pi'} \mathbb{Z}/n\mathbb{Z} \longrightarrow 0$$

which is not exact at its only nonzero term.

(5) Since $\mathbb{Q}/\mathbb{Z}$ is a torsion $\mathbb{Z}$-module it is not a submodule of a free $\mathbb{Z}$-module, hence is not projective. Note also that the exact sequence $0 \to \mathbb{Z} \to \mathbb{Q} \xrightarrow{\pi} \mathbb{Q}/\mathbb{Z} \to 0$ does not split since $\mathbb{Q}$ contains no submodule isomorphic to $\mathbb{Q}/\mathbb{Z}$.

(6) The $\mathbb{Z}$-module $\mathbb{Q}$ is not projective (cf. the exercises).

(7) We shall see in Chapter 12 that a finitely generated $\mathbb{Z}$-module is projective if and only if it is free.

(8) Let R be the commutative ring $\mathbb{Z}/2\mathbb{Z} \times \mathbb{Z}/2\mathbb{Z}$ under componentwise addition and multiplication. If P_1 and P_2 are the principal ideals generated by $(1, 0)$ and $(0, 1)$ respectively then $R = P_1 \oplus P_2$, hence both P_1 and P_2 are projective R-modules by Proposition 30. Neither P_1 nor P_2 is free, since any free module has order a multiple of four.

(9) The direct sum of two projective modules is again projective (cf. Exercise 3).

(10) We shall see in Part VI that if F is any field and $n \in \mathbb{Z}^+$ then the ring $R = M_n(F)$ of all $n \times n$ matrices with entries from F has the property that every R-module is projective. We shall also see that if G is a finite group of order n and $n \neq 0$ in the field F then the group ring FG also has the property that every module is projective.

Injective Modules and $\text{Hom}_R(\_\_, D)$

If $0 \longrightarrow L \xrightarrow{\psi} M \xrightarrow{\varphi} N \longrightarrow 0$ is a short exact sequence of R-modules then, instead of considering maps *from* an R-module D into L or N and the extent to which these determine maps from D into M, we can consider the "dual" question of maps from L or N *to* D. In this case, it is easy to dispose of the situation of a map from N to D: an R-module map from N to D immediately gives a map from M to D simply by composing with φ. It is easy to check that this defines an injective homomorphism of abelian groups

$$\varphi' : \text{Hom}_R(N, D) \longrightarrow \text{Hom}_R(M, D)$$

$$f \longmapsto f' = f \circ \varphi,$$

or, put another way,

$$\text{if} \quad M \xrightarrow{\varphi} N \to 0 \quad \text{is exact,}$$

$$\text{then} \quad 0 \to \operatorname{Hom}_R(N, D) \xrightarrow{\varphi'} \operatorname{Hom}_R(M, D) \quad \text{is exact.}$$

(Note that the associated maps on the homomorphism groups are in the reverse direction from the original maps.)

On the other hand, given an R-module homomorphism f from L to D it may not be possible to extend f to a map F from M to D, i.e., given f it may not be possible to find a map F making the following diagram commute:

$$
\begin{array}{ccc}
L & \xrightarrow{\psi} & M \\
\ {\scriptstyle f}\downarrow & \swarrow {\scriptstyle F} & \\
D & &
\end{array}
$$

For example, consider the exact sequence $0 \longrightarrow \mathbb{Z} \xrightarrow{\psi} \mathbb{Z} \xrightarrow{\varphi} \mathbb{Z}/2\mathbb{Z} \longrightarrow 0$ of $\mathbb{Z}$-modules, where ψ is multiplication by 2 and φ is the natural projection. Take $D = \mathbb{Z}/2\mathbb{Z}$ and let $f : \mathbb{Z} \to \mathbb{Z}/2\mathbb{Z}$ be reduction modulo 2 on the first $\mathbb{Z}$ in the sequence. There is only one nonzero homomorphism F from the second $\mathbb{Z}$ in the sequence to $\mathbb{Z}/2\mathbb{Z}$ (namely, reduction modulo 2), but this F does not lift the map f since $F \circ \psi(\mathbb{Z}) = F(2\mathbb{Z}) = 0$, so $F \circ \psi \neq f$.

Composition with ψ induces an abelian group homomorphism ψ' from $\operatorname{Hom}_R(M,D)$ to $\operatorname{Hom}_R(L, D)$, and in terms of the map ψ', the homomorphism $f \in \operatorname{Hom}_R(L, D)$ can be lifted to a homomorphism from M to D if and only if f is in the image of ψ'. The example above shows that

$$\text{if} \quad 0 \longrightarrow L \xrightarrow{\psi} M \quad \text{is exact,}$$

$$\text{then} \quad \operatorname{Hom}_R(M, D) \xrightarrow{\psi'} \operatorname{Hom}_R(L, D) \longrightarrow 0 \quad \text{is \textit{not necessarily} exact.}$$

We can summarize these results in the following dual version of Theorem 28:

Theorem 33. Let D, L, M, and N be R-modules. If

$$0 \longrightarrow L \xrightarrow{\psi} M \xrightarrow{\varphi} N \longrightarrow 0 \quad \text{is exact,}$$

then the associated sequence

$$0 \to \operatorname{Hom}_R(N, D) \xrightarrow{\varphi'} \operatorname{Hom}_R(M, D) \xrightarrow{\psi'} \operatorname{Hom}_R(L, D) \quad \text{is exact.} \qquad (10.12)$$

A homomorphism $f : L \to D$ lifts to a homomorphism $F : M \to D$ if and only if $f \in \operatorname{Hom}_R(L, D)$ is in the image of ψ'. In general $\psi' : \operatorname{Hom}_R(M, D) \to \operatorname{Hom}_R(L, D)$ need not be surjective; the map ψ' is surjective if and only if every homomorphism from L to D lifts to a homomorphism from M to D, in which case the sequence (12) can be extended to a short exact sequence.

The sequence (12) is exact for *all* R-modules D if and only if the sequence

$$L \xrightarrow{\psi} M \xrightarrow{\varphi} N \to 0 \quad \text{is exact.}$$

Proof: The only item remaining to be proved in the first statement is the exactness of (12) at $\text{Hom}_R(M, D)$. The proof of this statement is very similar to the proof of the corresponding result in Theorem 28 and is left as an exercise. Note also that the injectivity of ψ is not required, which proves the "if" portion of the final statement of the theorem.

Suppose now that the sequence (12) is exact for all R-modules D. We first show that $\varphi : M \to N$ is a surjection. Take $D = N/\varphi(M)$. If $\pi_1 : N \to N/\varphi(M)$ is the natural projection homomorphism, then $\pi_1 \circ \varphi(M) = 0$ by definition of π_1. Since $\pi_1 \circ \varphi = \varphi'(\pi_1)$, this means that the element $\pi_1 \in \text{Hom}_R(N, N/\varphi(M))$ is mapped to 0 by φ'. Since φ' is assumed to be injective for all modules D, this means π_1 is the zero map, i.e., $N = \varphi(M)$ and so φ is a surjection. We next show that $\varphi \circ \psi = 0$, which will imply that image $\psi \subseteq \ker \varphi$. For this we take $D = N$ and observe that the identity map id_N on N is contained in $\text{Hom}_R(N, N)$, hence $\varphi'(id_N) \in \text{Hom}_R(M, N)$. Then the exactness of (12) for $D = N$ implies that $\varphi'(id_N) \in \ker \psi'$, so $\psi'(\varphi'(id_N)) = 0$. Then $id_N \circ \psi \circ \varphi = 0$, i.e., $\psi \circ \varphi = 0$, as claimed. Finally, we show that $\ker \varphi \subseteq$ image ψ. Let $D = M/\psi(L)$ and let $\pi_2 : M \to M/\psi(L)$ be the natural projection. Then $\psi'(\pi_2) = 0$ since $\pi_2(\psi(L)) = 0$ by definition of π_2. The exactness of (12) for this D then implies that π_2 is in the image of φ', say $\pi_2 = \varphi'(f)$ for some homomorphism $f \in \text{Hom}_R(N, M/\psi(L))$, i.e., $\pi_2 = f \circ \varphi$. If $m \in \ker \varphi$ then $\pi_2(m) = f(\varphi(m)) = 0$, which means that $m \in \psi(L)$ since π_2 is just the projection from M into the quotient $M/\psi(L)$. Hence $\ker \varphi \subseteq$ image ψ, completing the proof.

By Theorem 33, the sequence

$$0 \longrightarrow \text{Hom}_R(N, D) \xrightarrow{\varphi'} \text{Hom}_R(M, D) \xrightarrow{\psi'} \text{Hom}_R(L, D) \longrightarrow 0$$

is in general *not* a short exact sequence since ψ' need not be surjective, and the question of whether this sequence is exact precisely measures the extent to which homomorphisms from M to D are uniquely determined by pairs of homomorphisms from L and N to D.

The second statement in Proposition 29 shows that this sequence is exact when the original exact sequence $0 \to L \to M \to N \to 0$ is a *split* exact sequence. In fact in this case the sequence $0 \to \text{Hom}_R(N, D) \xrightarrow{\varphi'} \text{Hom}_R(M, D) \xrightarrow{\psi'} \text{Hom}_R(L, D) \to 0$ is also a split exact sequence of abelian groups for every R-module D. Exercise 14 shows that a converse holds: if $0 \to \text{Hom}_R(N, D) \xrightarrow{\varphi'} \text{Hom}_R(M, D) \xrightarrow{\psi'} \text{Hom}_R(L, D) \to 0$ is exact for every R-module D then $0 \to L \xrightarrow{\psi} M \xrightarrow{\varphi} N \to 0$ is a split short exact sequence (which then implies that if the Hom sequence is exact for every D, then in fact it is split exact for every D).

There is also a dual version of the first three parts of Proposition 30, which describes the R-modules D having the property that the sequence (12) in Theorem 33 can *always* be extended to a short exact sequence:

Proposition 34. Let Q be an R-module. Then the following are equivalent:
    (1) For any R-modules L, M, and N, if

$$0 \longrightarrow L \xrightarrow{\psi} M \xrightarrow{\varphi} N \longrightarrow 0$$

is a short exact sequence, then

$$0 \longrightarrow \operatorname{Hom}_R(N, Q) \xrightarrow{\varphi'} \operatorname{Hom}_R(M, Q) \xrightarrow{\psi'} \operatorname{Hom}_R(L, Q) \longrightarrow 0$$

is also a short exact sequence.

(2) For any R-modules L and M, if $0 \to L \xrightarrow{\psi} M$ is exact, then every R-module homomorphism from L into Q lifts to an R-module homomorphism of M into Q, i.e., given $f \in \operatorname{Hom}_R(L, Q)$ there is a lift $F \in \operatorname{Hom}_R(M, Q)$ making the following diagram commute:

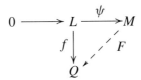

(3) If Q is a submodule of the R-module M then Q is a direct summand of M, i.e., every short exact sequence $0 \to Q \to M \to N \to 0$ splits.

Proof: The equivalence of (1) and (2) is part of Theorem 33. Suppose now that (2) is satisfied and let $0 \to Q \xrightarrow{\psi} M \xrightarrow{\varphi} N \to 0$ be exact. Taking $L = Q$ and f the identity map from Q to itself, it follows by (2) that there is a homomorphism $F : M \to Q$ with $F \circ \psi = 1$, so F is a splitting homomorphism for the sequence, which proves (3). The proof that (3) implies (2) is outlined in the exercises.

Definition. An R-module Q is called *injective* if it satisfies any of the equivalent conditions of Proposition 34.

The third statement in Proposition 34 can be rephrased as saying that any module M into which Q injects has (an isomorphic copy of) Q as a direct summand, which explains the terminology.

If D is fixed, then given any R-module X we have an associated abelian group $\operatorname{Hom}_R(X, D)$. Further, an R-module homomorphism $\alpha : X \to Y$ induces an abelian group homomorphism $\alpha' : \operatorname{Hom}_R(Y, D) \to \operatorname{Hom}_R(X, D)$, defined by $\alpha'(f) = f \circ \alpha$, that "reverses" the direction of the arrow. Put another way, the map $\operatorname{Hom}_R(D, \_\_)$ is a *contravariant functor* from the category of R-modules to the category of abelian groups (cf. Appendix II). Theorem 33 shows that applying this functor to the terms in the exact sequence

$$0 \longrightarrow L \xrightarrow{\psi} M \xrightarrow{\varphi} N \longrightarrow 0$$

produces an exact sequence

$$0 \to \operatorname{Hom}_R(N, D) \xrightarrow{\varphi'} \operatorname{Hom}_R(M, D) \xrightarrow{\psi'} \operatorname{Hom}_R(L, D).$$

This is referred to by saying that $\operatorname{Hom}_R(\_\_, D)$ is a *left exact* (contravariant) functor. Note that the functor $\operatorname{Hom}_R(\_\_, D)$ and the functor $\operatorname{Hom}_R(D, \_\_)$ considered earlier are both left exact; the former reverses the directions of the maps in the original short exact sequence, the latter maintains the directions of the maps.

By Proposition 34, the functor $\operatorname{Hom}_R(\_\,, D)$ is *exact*, i.e., always takes short exact sequences to short exact sequences (and hence exact sequences of any length to exact sequences), if and only if D is injective. We summarize this in the following proposition, which is dual to the covariant result of Corollary 32.

Corollary 35. If D is an R-module, then the functor $\operatorname{Hom}_R(\_\,, D)$ from the category of R-modules to the category of abelian groups is left exact. It is exact if and only if D is an injective R-module.

We have seen that an R-module is projective if and only if it is a direct summand of a free R-module. Providing such a simple characterization of injective R-modules is not so easy. The next result gives a criterion for Q to be an injective R-module (a result due to Baer, who introduced the notion of injective modules around 1940), and using it we can give a characterization of injective modules when $R = \mathbb{Z}$ (or, more generally, when R is a P.I.D.). Recall that a $\mathbb{Z}$-module A (i.e., an abelian group, written additively) is said to be *divisible* if $A = nA$ for all nonzero integers n. For example, both $\mathbb{Q}$ and $\mathbb{Q}/\mathbb{Z}$ are divisible (cf. Exercises 19 and 20 in Section 2.4 and Exercise 15 in Section 3.1).

Proposition 36. Let Q be an R-module.

(1) *(Baer's Criterion)* The module Q is injective if and only if for every left ideal I of R any R-module homomorphism $g : I \to Q$ can be extended to an R-module homomorphism $G : R \to Q$.

(2) If R is a P.I.D. then Q is injective if and only if $rQ = Q$ for every nonzero $r \in R$. In particular, a $\mathbb{Z}$-module is injective if and only if it is divisible. When R is a P.I.D., quotient modules of injective R-modules are again injective.

Proof: If Q is injective and $g : I \to Q$ is an R-module homomorphism from the nonzero ideal I of R into Q, then g can be extended to an R-module homomorphism from R into Q by Proposition 34(2) applied to the exact sequence $0 \to I \to R$, which proves the "only if" portion of (1). Suppose conversely that every homomorphism $g : I \to Q$ can be lifted to a homomorphism $G : R \to Q$. To show that Q is injective we must show that if $0 \to L \to M$ is exact and $f : L \to Q$ is an R-module homomorphism then there is a lift $F : M \to Q$ extending f. If $\mathcal{S}$ is the collection (f', L') of lifts $f' : L' \to Q$ of f to a submodule L' of M containing L, then the ordering $(f', L') \le (f'', L'')$ if $L' \subseteq L''$ and $f'' = f'$ on L' partially orders $\mathcal{S}$. Since $\mathcal{S} \ne \emptyset$, by Zorn's Lemma there is a maximal element (F, M') in $\mathcal{S}$. The map $F : M' \to Q$ is a lift of f and it suffices to show that $M' = M$. Suppose that there is some element $m \in M$ not contained in M' and let $I = \{r \in R \mid rm \in M'\}$. It is easy to check that I is a left ideal in R, and the map $g : I \to Q$ defined by $g(x) = F(xm)$ is an R-module homomorphism from I to Q. By hypothesis, there is a lift $G : R \to Q$ of g. Consider the submodule $M' + Rm$ of M, and define the map $F' : M' + Rm \to Q$ by $F'(m' + rm) = F(m') + G(r)$. If $m_1 + r_1 m = m_2 + r_2 m$ then $(r_1 - r_2)m = m_2 - m_1$

shows that $r_1 - r_2 \in I$, so that
$$G(r_1 - r_2) = g(r_1 - r_2) = F((r_1 - r_2)m) = F(m_2 - m_1),$$
and so $F(m_1) + G(r_1) = F(m_2) + G(r_2)$. Hence F' is well defined and it is then immediate that F' is an R-module homomorphism extending f to $M' + Rm$. This contradicts the maximality of M', so that $M' = M$, which completes the proof of (1).

To prove (2), suppose R is a P.I.D. Any nonzero ideal I of R is of the form $I = (r)$ for some nonzero element r of R. An R-module homomorphism $f : I \to Q$ is completely determined by the image $f(r) = q$ in Q. This homomorphism can be extended to a homomorphism $F : R \to Q$ if and only if there is an element q' in Q with $F(1) = q'$ satisfying $q = f(r) = F(r) = rq'$. It follows that Baer's criterion for Q is satisfied if and only if $rQ = Q$, which proves the first two statements in (2). The final statement follows since a quotient of a module Q with $rQ = Q$ for all $r \neq 0$ in R has the same property.

Examples

(1) Since $\mathbb{Z}$ is not divisible, $\mathbb{Z}$ is not an injective $\mathbb{Z}$-module. This also follows from the fact that the exact sequence $0 \longrightarrow \mathbb{Z} \xrightarrow{2} \mathbb{Z} \longrightarrow \mathbb{Z}/2\mathbb{Z} \longrightarrow 0$ corresponding to multiplication by 2 does not split.

(2) The rational numbers $\mathbb{Q}$ is an injective $\mathbb{Z}$-module.

(3) The quotient $\mathbb{Q}/\mathbb{Z}$ of the injective $\mathbb{Z}$-module $\mathbb{Q}$ is an injective $\mathbb{Z}$-module.

(4) It is immediate that a direct sum of divisible $\mathbb{Z}$-modules is again divisible, hence a direct sum of injective $\mathbb{Z}$-modules is again injective. For example, $\mathbb{Q} \oplus \mathbb{Q}/\mathbb{Z}$ is an injective $\mathbb{Z}$-module. (See also Exercise 4).

(5) We shall see in Chapter 12 that no nonzero finitely generated $\mathbb{Z}$-module is injective.

(6) Suppose that the ring R is an integral domain. An R-module A is said to be a *divisible* R-module if $rA = A$ for every nonzero $r \in R$. The proof of Proposition 36 shows that in this case an injective R-module is divisible.

(7) We shall see in Section 11.1 that if $R = F$ is a field then every F-module is injective.

(8) We shall see in Part VI that if F is any field and $n \in \mathbb{Z}^+$ then the ring $R = M_n(F)$ of all $n \times n$ matrices with entries from F has the property that every R-module is injective (and also projective). We shall also see that if G is a finite group of order n and $n \neq 0$ in the field F then the group ring FG also has the property that every module is injective (and also projective).

Corollary 37. Every $\mathbb{Z}$-module is a submodule of an injective $\mathbb{Z}$-module.

Proof: Let M be a $\mathbb{Z}$-module and let A be any set of $\mathbb{Z}$-module generators of M. Let $\mathcal{F} = F(A)$ be the free $\mathbb{Z}$-module on the set A. Then by Theorem 6 there is a surjective $\mathbb{Z}$-module homomorphism from $\mathcal{F}$ to M and if $\mathcal{K}$ denotes the kernel of this homomorphism then $\mathcal{K}$ is a $\mathbb{Z}$-submodule of $\mathcal{F}$ and we can identify $M = \mathcal{F}/\mathcal{K}$. Let $\mathcal{Q}$ be the free $\mathbb{Q}$-module on the set A. Then $\mathcal{Q}$ is a direct sum of a number of copies of $\mathbb{Q}$, so is a divisible, hence (by Proposition 36) injective, $\mathbb{Z}$-module containing $\mathcal{F}$. Then $\mathcal{K}$ is also a $\mathbb{Z}$-submodule of $\mathcal{Q}$, so the quotient $\mathcal{Q}/\mathcal{K}$ is injective, again by Proposition 36. Since $M = \mathcal{F}/\mathcal{K} \subseteq \mathcal{Q}/\mathcal{K}$, it follows that M is contained in an injective $\mathbb{Z}$-module.

Corollary 37 can be used to prove the following more general version valid for arbitrary R-modules. This theorem is the injective analogue of the results in Theorem 6 and Corollary 31 showing that every R-module is a quotient of a projective R-module.

Theorem 38. Let R be a ring with 1 and let M be an R-module. Then M is contained in an injective R-module.

Proof: A proof is outlined in Exercises 15 and 16.

It is possible to prove a sharper result than Theorem 38, namely that there is a *minimal* injective R-module H containing M in the sense that any injective map of M into an injective R-module Q factors through H. More precisely, if $M \subseteq Q$ for an injective R-module Q then there is an injection $\iota : H \hookrightarrow Q$ that restricts to the identity map on M; using ι to identify H as a subset of Q we have $M \subseteq H \subseteq Q$. (cf. Theorem 57.13 in *Representation Theory of Finite Groups and Associative Algebras* by C. Curtis and I. Reiner, John Wiley & Sons, 1966). This module H is called the *injective hull* or *injective envelope* of M. The universal property of the injective hull of M with respect to inclusions of M into injective R-modules should be compared to the universal property with respect to homomorphisms of M of the free module $F(A)$ on a set of generators A for M in Theorem 6. For example, the injective hull of $\mathbb{Z}$ is $\mathbb{Q}$, and the injective hull of any field is itself (cf. the exercises).

Flat Modules and $D \otimes_R \underline{\ \ }$

We now consider the behavior of extensions $0 \longrightarrow L \xrightarrow{\psi} M \xrightarrow{\varphi} N \longrightarrow 0$ of R-modules with respect to tensor products.

Suppose that D is a *right* R-module. For any homomorphism $f : X \to Y$ of left R-modules we obtain a homomorphism $1 \otimes f : D \otimes_R X \to D \otimes_R Y$ of abelian groups (Theorem 13). If in addition D is an (S, R)-bimodule (for example, when $S = R$ is commutative and D is given the standard (R, R)-bimodule structure as in Section 4), then $1 \otimes f$ is a homomorphism of left S-modules. Put another way,

$$D \otimes_R \underline{\ \ } : X \longrightarrow D \otimes_R X$$

is a *covariant functor* from the category of left R-modules to the category of abelian groups (respectively, to the category of left S-modules when D is an (S, R)-bimodule), cf. Appendix II. In a similar way, if D is a left R-module then $\underline{\ \ } \otimes_R D$ is a covariant functor from the category of right R-modules to the category of abelian groups (respectively, to the category of right S-modules when D is an (R, S)-bimodule). Note that, unlike Hom, the tensor product is covariant in both variables, and we shall therefore concentrate on $D \otimes_R \underline{\ \ }$, leaving as an exercise the minor alterations necessary for $\underline{\ \ } \otimes_R D$.

We have already seen examples where the map $1 \otimes \psi : D \otimes_R L \to D \otimes_R M$ induced by an injective map $\psi : L \hookrightarrow M$ is no longer injective (for example the injection $\mathbb{Z} \hookrightarrow \mathbb{Q}$ of $\mathbb{Z}$-modules induces the zero map from $\mathbb{Z}/2\mathbb{Z} \otimes_{\mathbb{Z}} \mathbb{Z} = \mathbb{Z}/2\mathbb{Z}$ to $\mathbb{Z}/2\mathbb{Z} \otimes_{\mathbb{Z}} \mathbb{Q} = 0$). On the other hand, suppose that $\varphi : M \to N$ is a surjective R-module homomorphism. The tensor product $D \otimes_R N$ is generated as an abelian group by the simple tensors $d \otimes n$ for $d \in D$ and $n \in N$. The surjectivity of φ implies that $n = \varphi(m)$ for some $m \in M$, and then $1 \otimes \varphi(d \otimes m) = d \otimes \varphi(m) = d \otimes n$ shows that $1 \otimes \varphi$ is a surjective homomorphism of abelian groups from $D \otimes_R M$ to $D \otimes_R N$. This proves most of the following theorem.

Theorem 39. Suppose that D is a right R-module and that L, M and N are left R-modules. If

$$0 \longrightarrow L \overset{\psi}{\longrightarrow} M \overset{\varphi}{\longrightarrow} N \longrightarrow 0 \quad \text{is exact,}$$

then the associated sequence of abelian groups

$$D \otimes_R L \overset{1 \otimes \psi}{\longrightarrow} D \otimes_R M \overset{1 \otimes \varphi}{\longrightarrow} D \otimes_R N \longrightarrow 0 \quad \text{is exact.} \qquad (10.13)$$

If D is an (S, R)-bimodule then (13) is an exact sequence of left S-modules. In particular, if $S = R$ is a commutative ring, then (13) is an exact sequence of R-modules with respect to the standard R-module structures. The map $1 \otimes \psi$ is not in general injective, i.e., the sequence (13) cannot in general be extended to a short exact sequence.

The sequence (13) is exact for *all* right R-modules D if and only if

$$L \overset{\psi}{\to} M \overset{\varphi}{\to} N \to 0 \quad \text{is exact.}$$

Proof: For the first statement it remains to prove the exactness of (13) at $D \otimes_R M$. Since $\varphi \circ \psi = 0$, we have

$$(1 \otimes \varphi) \left(\sum d_i \otimes \psi(l_i) \right) = \sum d_i \otimes (\varphi \circ \psi(l_i)) = 0$$

and it follows that image$(1 \otimes \psi) \subseteq \ker(1 \otimes \varphi)$. In particular, there is a natural projection $\pi : (D \otimes_R M) / \text{image}(1 \otimes \psi) \to (D \otimes_R M) / \ker(1 \otimes \varphi) = D \otimes_R N$. The composite of the two projection homomorphisms

$$D \otimes_R M \to (D \otimes_R M) / \text{image}(1 \otimes \psi) \overset{\pi}{\to} D \otimes_R N$$

is the quotient of $D \otimes_R M$ by $\ker(1 \otimes \varphi)$, so is just the map $1 \otimes \varphi$. We shall show that π is an isomorphism, which will show that the kernel of $1 \otimes \varphi$ is just the kernel of the first projection above, i.e., image$(1 \otimes \psi)$, giving the exactness of (13) at $D \otimes_R M$. To see that π is an isomorphism we define an inverse map. First define $\pi' : D \times N \to (D \otimes_R M) / \text{image}(1 \otimes \psi)$ by $\pi'((d, n)) = d \otimes m$ for any $m \in M$ with $\varphi(m) = n$. Note that this is well defined: any other element $m' \in M$ mapping to n differs from m by an element in $\ker \varphi = \text{image } \psi$, i.e., $m' = m + \psi(l)$ for some $l \in L$, and $d \otimes \psi(l) \in \text{image}(1 \otimes \psi)$. It is easy to check that π' is a balanced map, so induces a homomorphism $\tilde{\pi} : D \times N \to (D \otimes_R M) / \text{image}(1 \otimes \psi)$ with $\tilde{\pi}(d \otimes n) = d \otimes m$. Then $\tilde{\pi} \circ \pi(d \otimes m) = \tilde{\pi}(d \otimes \varphi(m)) = d \otimes m$ shows that $\tilde{\pi} \circ \pi = 1$. Similarly, $\pi \circ \tilde{\pi} = 1$, so that π and $\tilde{\pi}$ are inverse isomorphisms, completing the proof that (13) is exact. Note also that the injectivity of ψ was not required for the proof.

Finally, suppose (13) is exact for every right R-module D. In general, $R \otimes_R X \cong X$ for any left R-module X (Example 1 following Corollary 9). Taking $D = R$ the exactness of the sequence $L \overset{\psi}{\to} M \overset{\varphi}{\to} N \to 0$ follows.

By Theorem 39, the sequence

$$0 \longrightarrow D \otimes_R L \overset{1 \otimes \psi}{\longrightarrow} D \otimes_R M \overset{1 \otimes \varphi}{\longrightarrow} D \otimes_R N \longrightarrow 0$$

is not in general exact since $1 \otimes \psi$ need not be injective. If $0 \to L \xrightarrow{\psi} M \xrightarrow{\varphi} N \to 0$ is a *split* short exact sequence, however, then since tensor products commute with direct sums by Theorem 17, it follows that

$$0 \longrightarrow D \otimes_R L \xrightarrow{1 \otimes \psi} D \otimes_R M \xrightarrow{1 \otimes \varphi} D \otimes_R N \longrightarrow 0$$

is also a split short exact sequence.

The following result relating to modules D having the property that (13) can always be extended to a short exact sequence is immediate from Theorem 39:

Proposition 40. Let A be a right R-module. Then the following are equivalent:
  (1) For any left R-modules L, M, and N, if

$$0 \longrightarrow L \xrightarrow{\psi} M \xrightarrow{\varphi} N \longrightarrow 0$$

is a short exact sequence, then

$$0 \longrightarrow A \otimes_R L \xrightarrow{1 \otimes \psi} A \otimes_R M \xrightarrow{1 \otimes \varphi} A \otimes_R N \longrightarrow 0$$

is also a short exact sequence.

  (2) For any left R-modules L and M, if $0 \to L \xrightarrow{\psi} M$ is an exact sequence of left R-modules (i.e., $\psi : L \to M$ is injective) then $0 \to A \otimes_R L \xrightarrow{1 \otimes \psi} A \otimes_R M$ is an exact sequence of abelian groups (i.e., $1 \otimes \psi : A \otimes_R L \to A \otimes_R M$ is injective).

Definition. A right R-module A is called *flat* if it satisfies either of the two equivalent conditions of Proposition 40.

For a fixed right R-module D, the first part of Theorem 39 is referred to by saying that the functor $D \otimes_R \underline{}$ is *right exact*.

Corollary 41. If D is a right R-module, then the functor $D \otimes_R \underline{}$ from the category of left R-modules to the category of abelian groups is right exact. If D is an (S, R)-bimodule (for example when $S = R$ is commutative and D is given the standard R-module structure), then $D \otimes_R \underline{}$ is a right exact functor from the category of left R-modules to the category of left S-modules. The functor is exact if and only if D is a flat R-module.

We have already seen some flat modules:

Corollary 42. Free modules are flat; more generally, projective modules are flat.

Proof: To show that the free R-module F is flat it suffices to show that for any injective map $\psi : L \to M$ of R-modules L and M the induced map $1 \otimes \psi : F \otimes_R L \to F \otimes_R M$ is also injective. Suppose first that $F \cong R^n$ is a finitely generated free R-module. In this case $F \otimes_R L = R^n \otimes_R L \cong L^n$ since $R \otimes_R L \cong L$ and tensor products commute with direct sums. Similarly $F \otimes_R M \cong M^n$ and under these isomorphisms the map $1 \otimes \psi : F \otimes_R L \to F \otimes_R M$ is just the natural map of L^n to M^n induced by the inclusion ψ in each component. In particular, $1 \otimes \psi$ is injective and it follows that any finitely generated free module is flat. Suppose now that F is an arbitrary free module and that the element $\sum f_i \otimes l_i \in F \otimes_R L$ is mapped to 0 by $1 \otimes \psi$. This means that the element $\sum (f_i, \psi(l_i))$ can be written as a sum of generators as in equation (6) in the previous section in the free group on $F \times M$. Since this sum of elements is finite, all of the first coordinates of the resulting equation lie in some finitely generated free submodule F' of F. Then this equation implies that $\sum f_i \otimes l_i \in F' \otimes_R L$ is mapped to 0 in $F' \otimes_R M$. Since F' is a finitely generated free module, the injectivity we proved above shows that $\sum f_i \otimes l_i$ is 0 in $F' \otimes_R L$ and so also in $F \otimes_R L$. It follows that $1 \otimes \psi$ is injective and hence that F is flat.

Suppose now that P is a projective module. Then P is a direct summand of a free module F (Proposition 30), say $F = P \oplus P'$. If $\psi : L \to M$ is injective then $1 \otimes \psi : F \otimes_R L \to F \otimes_R M$ is also injective by what we have already shown. Since $F = P \oplus P'$ and tensor products commute with direct sums, this shows that

$$1 \otimes \psi : (P \otimes_R L) \oplus (P' \otimes_R L) \to (P \otimes_R M) \oplus (P' \otimes_R M)$$

is injective. Hence $1 \otimes \psi : P \otimes_R L \to P \otimes_R M$ is injective, proving that P is flat.

Examples

(1) Since $\mathbb{Z}$ is a projective $\mathbb{Z}$-module it is flat. The example before Theorem 39 shows that $\mathbb{Z}/2\mathbb{Z}$ is not a flat $\mathbb{Z}$-module.

(2) The $\mathbb{Z}$-module $\mathbb{Q}$ is a flat $\mathbb{Z}$-module, as follows. Suppose $\psi : L \to M$ is an injective map of $\mathbb{Z}$-modules. Every element of $\mathbb{Q} \otimes_{\mathbb{Z}} L$ can be written in the form $(1/d) \otimes l$ for some nonzero integer d and some $l \in L$ (Exercise 7 in Section 4). If $(1/d) \otimes l$ is in the kernel of $1 \otimes \psi$ then $(1/d) \otimes \psi(l)$ is 0 in $\mathbb{Q} \otimes_{\mathbb{Z}} M$. By Exercise 8 in Section 4 this means $c\psi(l) = 0$ in M for some nonzero integer c. Then $\psi(c \cdot l) = 0$, and the injectivity of ψ implies $c \cdot l = 0$ in L. But this implies that $(1/d) \otimes l = (1/cd) \otimes (c \cdot l) = 0$ in L, which shows that $1 \otimes \psi$ is injective.

(3) The $\mathbb{Z}$-module $\mathbb{Q}/\mathbb{Z}$ is injective (by Proposition 36), but is not flat: the injective map $\psi(z) = 2z$ from $\mathbb{Z}$ to $\mathbb{Z}$ does not remain injective after tensoring with $\mathbb{Q}/\mathbb{Z}$ ($1 \otimes \psi : \mathbb{Q}/\mathbb{Z} \otimes_{\mathbb{Z}} \mathbb{Z} \to \mathbb{Q}/\mathbb{Z} \otimes \mathbb{Z}$ has the nonzero element $(\frac{1}{2} + \mathbb{Z}) \otimes 1$ in its kernel — identifying $\mathbb{Q}/\mathbb{Z} = \mathbb{Q}/\mathbb{Z} \otimes_{\mathbb{Z}} \mathbb{Z}$ this is the statement that multiplication by 2 has the element $1/2$ in its kernel).

(4) The direct sum of flat modules is flat (Exercise 5). In particular, $\mathbb{Q} \oplus \mathbb{Z}$ is flat. This module is neither projective nor injective (since $\mathbb{Q}$ is not projective by Exercise 8 and $\mathbb{Z}$ is not injective by Proposition 36 (cf. Exercises 3 and 4).

We close this section with an important relation between Hom and tensor products:

Theorem 43. *(Adjoint Associativity)* Let R and S be rings, let A be a right R-module, let B be an (R, S)-bimodule and let C be a right S-module. Then there is an isomorphism of abelian groups:

$$\text{Hom}_S(A \otimes_R B, C) \cong \text{Hom}_R(A, \text{Hom}_S(B, C))$$

(the homomorphism groups are right module homomorphisms—note that $\text{Hom}_S(B, C)$ has the structure of a right R-module, cf. the exercises). If $R = S$ is commutative this is an isomorphism of R-modules with the standard R-module structures.

Proof: Suppose $\varphi : A \otimes_R B \to C$ is a homomorphism. For any fixed $a \in A$ define the map $\Phi(a)$ from B to C by $\Phi(a)(b) = \varphi(a \otimes b)$. It is easy to check that $\Phi(a)$ is a homomorphism of right S-modules and that the map Φ from A to $\text{Hom}_S(B, C)$ given by mapping a to $\Phi(a)$ is a homomorphism of right R-modules. Then $f(\varphi) = \Phi$ defines a group homomorphism from $\text{Hom}_S(A \otimes_R B, C)$ to $\text{Hom}_R(A, \text{Hom}_S(B, C))$. Conversely, suppose $\Phi : A \to \text{Hom}_S(B, C)$ is a homomorphism. The map from $A \times B$ to C defined by mapping (a, b) to $\Phi(a)(c)$ is an R-balanced map, so induces a homomorphism φ from $A \otimes_R B$ to C. Then $g(\Phi) = \varphi$ defines a group homomorphism inverse to f and gives the isomorphism in the theorem.

As a first application of Theorem 43 we give an alternate proof of the first result in Theorem 39 that the tensor product is right exact in the case where $S = R$ is a commutative ring. If $0 \longrightarrow L \longrightarrow M \longrightarrow N \longrightarrow 0$ is exact, then by Theorem 33 the sequence

$$0 \longrightarrow \text{Hom}_R(N, E) \longrightarrow \text{Hom}_R(M, E) \longrightarrow \text{Hom}_R(L, E)$$

is exact for every R-module E. Then by Theorem 28, the sequence

$$0 \to \text{Hom}_R(D, \text{Hom}_R(N, E)) \to \text{Hom}_R(D, \text{Hom}_R(M, E)) \to \text{Hom}_R(D, \text{Hom}_R(L, E))$$

is exact for all D and all E. By adjoint associativity, this means the sequence

$$0 \longrightarrow \text{Hom}_R(D \otimes_R N, E) \longrightarrow \text{Hom}_R(D \otimes_R M, E) \longrightarrow \text{Hom}_R(D \otimes_R L, E)$$

is exact for any D and all E. Then, by the second part of Theorem 33, it follows that the sequence

$$D \otimes_R L \longrightarrow D \otimes_R M \longrightarrow D \otimes_R N \longrightarrow 0$$

is exact for all D, which is the right exactness of the tensor product.

As a second application of Theorem 43 we prove that the tensor product of two projective modules over a commutative ring R is again projective (see also Exercise 9 for a more direct proof).

Corollary 44. *If R is commutative then the tensor product of two projective R-modules is projective.*

Proof: Let P_1 and P_2 be projective modules. Then by Corollary 32, $\text{Hom}_R(P_2, \_)$ is an exact functor from the category of R-modules to the category of R-modules. Then the composition $\text{Hom}_R(P_1, \text{Hom}_R(P_2, \_))$ is an exact functor by the same corollary. By Theorem 43 this means that $\text{Hom}_R(P_1 \otimes_R P_2, \_)$ is an exact functor on R-modules. It follows again from Corollary 32 that $P_1 \otimes_R P_2$ is projective.

Summary

Each of the functors $\text{Hom}_R(A, \_)$, $\text{Hom}_R(\_, A)$, and $A \otimes_R \_$, map left R-modules to abelian groups; the functor $\_ \otimes_R A$ maps right R-modules to abelian groups. When R is commutative all four functors map R-modules to R-modules.

(1) Let A be a left R-module. The functor $\text{Hom}_R(A, \_)$ is covariant and left exact; the module A is projective if and only if $\text{Hom}_R(A, \_)$ is exact (i.e., is also right exact).

(2) Let A be a left R-module. The functor $\text{Hom}_R(\_\_, A)$ is contravariant and left exact; the module A is injective if and only if $\text{Hom}_R(\_\_, A)$ is exact.

(3) Let A be a right R-module. The functor $A \otimes_R \_\_$ is covariant and right exact; the module A is flat if and only if $A \otimes_R \_\_$ is exact (i.e., is also left exact).

(4) Let A be a left R-module. The functor $\_\_ \otimes_R A$ is covariant and right exact; the module A is flat if and only if $\_\_ \otimes_R A$ is exact.

(5) Projective modules are flat. The $\mathbb{Z}$-module $\mathbb{Q}/\mathbb{Z}$ is injective but not flat. The $\mathbb{Z}$-module $\mathbb{Z} \oplus \mathbb{Q}$ is flat but neither projective nor injective.

EXERCISES

Let R be a ring with 1.

1. Suppose that

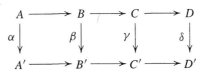

is a commutative diagram of groups and that the rows are exact. Prove that
   (a) if φ and α are surjective, and β is injective then γ is injective. [If $c \in \ker \gamma$, show there is a $b \in B$ with $\varphi(b) = c$. Show that $\varphi'(\beta(b)) = 0$ and deduce that $\beta(b) = \psi'(a')$ for some $a' \in A'$. Show there is an $a \in A$ with $\alpha(a) = a'$ and that $\beta(\psi(a)) = \beta(b)$. Conclude that $b = \psi(a)$ and hence $c = \varphi(b) = 0$.]
   (b) if ψ', α, and γ are injective, then β is injective,
   (c) if φ, α, and γ are surjective, then β is surjective,
   (d) if β is injective, α and φ are surjective, then γ is injective,
   (e) if β is surjective, γ and ψ' are injective, then α is surjective.

2. Suppose that

$$A \longrightarrow B \longrightarrow C \longrightarrow D$$
$$\alpha \downarrow \quad \beta \downarrow \quad \gamma \downarrow \quad \delta \downarrow$$
$$A' \longrightarrow B' \longrightarrow C' \longrightarrow D'$$

is a commutative diagram of groups, and that the rows are exact. Prove that
   (a) if α is surjective, and β, δ are injective, then γ is injective.
   (b) if δ is injective, and α, γ are surjective, then β is surjective.

3. Let P_1 and P_2 be R-modules. Prove that $P_1 \oplus P_2$ is a projective R-module if and only if both P_1 and P_2 are projective.

4. Let Q_1 and Q_2 be R-modules. Prove that $Q_1 \oplus Q_2$ is an injective R-module if and only if both Q_1 and Q_2 are injective.

5. Let A_1 and A_2 be R-modules. Prove that $A_1 \oplus A_2$ is a flat R-module if and only if both A_1 and A_2 are flat. More generally, prove that an arbitrary direct sum $\sum A_i$ of R-modules is flat if and only if each A_i is flat. [Use the fact that tensor product commutes with arbitrary direct sums.]

6. Prove that the following are equivalent for a ring R:
   (i) Every R-module is projective.
   (ii) Every R-module is injective.

7. Let A be a nonzero finite abelian group.
   (a) Prove that A is not a projective $\mathbb{Z}$-module.
   (b) Prove that A is not an injective $\mathbb{Z}$-module.

8. Let Q be a nonzero divisible $\mathbb{Z}$-module. Prove that Q is not a projective $\mathbb{Z}$-module. Deduce that the rational numbers $\mathbb{Q}$ is not a projective $\mathbb{Z}$-module. [Show first that if F is any free module then $\cap_{n=1}^{\infty} nF = 0$ (use a basis of F to prove this). Now suppose to the contrary that Q is projective and derive a contradiction from Proposition 30(4).]

9. Assume R is commutative with 1.
   (a) Prove that the tensor product of two free R-modules is free. [Use the fact that tensor products commute with direct sums.]
   (b) Use (a) to prove that the tensor product of two projective R-modules is projective.

10. Let R and S be rings with 1 and let M and N be left R-modules. Assume also that M is an (R, S)-bimodule.
   (a) For $s \in S$ and for $\varphi \in \operatorname{Hom}_R(M, N)$ define $(s\varphi) : M \to N$ by $(s\varphi)(m) = \varphi(ms)$. Prove that $s\varphi$ is a homomorphism of left R-modules, and that this action of S on $\operatorname{Hom}_R(M, N)$ makes it into a *left* S-module.
   (b) Let $S = R$ and let $M = R$ (considered as an (R, R)-bimodule by left and right ring multiplication on itself). For each $n \in N$ define $\varphi_n : R \to N$ by $\varphi_n(r) = rn$, i.e., φ_n is the unique R-module homomorphism mapping 1_R to n. Show that $\varphi_n \in \operatorname{Hom}_R(R, N)$. Use part (a) to show that the map $n \mapsto \varphi_n$ is an isomorphism of left R-modules: $N \cong \operatorname{Hom}_R(R, N)$.
   (c) Deduce that if N is a free (respectively, projective, injective, flat) left R-module, then $\operatorname{Hom}_R(R, N)$ is also a free (respectively, projective, injective, flat) left R-module.

11. Let R and S be rings with 1 and let M and N be left R-modules. Assume also that N is an (R, S)-bimodule.
   (a) For $s \in S$ and for $\varphi \in \operatorname{Hom}_R(M, N)$ define $(\varphi s) : M \to N$ by $(\varphi s)(m) = \varphi(m)s$. Prove that φs is a homomorphism of left R-modules, and that this action of S on $\operatorname{Hom}_R(M, N)$ makes it into a *right* S-module. Deduce that $\operatorname{Hom}_R(M, R)$ is a right R-module, for any R-module M—called the *dual module* to M.
   (b) Let $N = R$ be considered as an (R, R)-bimodule as usual. Under the action defined in part (a) show that the map $r \mapsto \varphi_r$ is an isomorphism of right R-modules: $\operatorname{Hom}_R(R, R) \cong R$, where φ_r is the homomorphism that maps 1_R to r. Deduce that if M is a finitely generated free left R-module, then $\operatorname{Hom}_R(M, R)$ is a free right R-module of the same rank. (cf. also Exercise 13.)
   (c) Show that if M is a finitely generated projective R-module then its dual module $\operatorname{Hom}_R(M, R)$ is also projective.

12. Let A be an R-module, let I be any nonempty index set and for each $i \in I$ let B_i be an R-module. Prove the following isomorphisms of abelian groups; when R is commutative prove also that these are R-module isomorphisms. (Arbitrary direct sums and direct products of modules are introduced in Exercise 20 of Section 3.)
   (a) $\operatorname{Hom}_R(\bigoplus_{i \in I} B_i, A) \cong \prod_{i \in I} \operatorname{Hom}_R(B_i, A)$
   (b) $\operatorname{Hom}_R(A, \prod_{i \in I} B_i) \cong \prod_{i \in I} \operatorname{Hom}_R(A, B_i)$.

13. **(a)** Show that the dual of the free $\mathbb{Z}$-module with countable basis is not free. [Use the preceding exercise and Exercise 24, Section 3.] (See also Exercise 5 in Section 11.3.)
   (b) Show that the dual of the free $\mathbb{Z}$-module with countable basis is also not projective. [You may use the fact that any submodule of a free $\mathbb{Z}$-module is free.]

14. Let $0 \longrightarrow L \overset{\psi}{\longrightarrow} M \overset{\varphi}{\longrightarrow} N \longrightarrow 0$ be a sequence of R-modules.

(a) Prove that the associated sequence

$$0 \longrightarrow \operatorname{Hom}_R(D, L) \xrightarrow{\psi'} \operatorname{Hom}_R(D, M) \xrightarrow{\varphi'} \operatorname{Hom}_R(D, N) \longrightarrow 0$$

is a short exact sequence of abelian groups for all R-modules D if and only if the original sequence is a split short exact sequence. [To show the sequence splits, take $D = N$ and show the lift of the identity map in $\operatorname{Hom}_R(N, N)$ to $\operatorname{Hom}_R(N, M)$ is a splitting homomorphism for φ.]

(b) Prove that the associated sequence

$$0 \longrightarrow \operatorname{Hom}_R(N, D) \xrightarrow{\varphi'} \operatorname{Hom}_R(M, D) \xrightarrow{\psi'} \operatorname{Hom}_R(L, D) \longrightarrow 0$$

is a short exact sequence of abelian groups for all R-modules D if and only if the original sequence is a split short exact sequence.

15. Let M be a left $\mathbb{Z}$-module and let R be a ring with 1.
  (a) Show that $\operatorname{Hom}_{\mathbb{Z}}(R, M)$ is a left R-module under the action $(r\varphi)(r') = \varphi(r'r)$ (see Exercise 10).
  (b) Suppose that $0 \to A \xrightarrow{\psi} B$ is an exact sequence of R-modules. Prove that if every $\mathbb{Z}$-module homomorphism f from A to M lifts to a $\mathbb{Z}$-module homomorphism F from B to M with $f = F \circ \psi$, then every R-module homomorphism f' from A to $\operatorname{Hom}_{\mathbb{Z}}(R, M)$ lifts to an R-module homomorphism F' from B to $\operatorname{Hom}_{\mathbb{Z}}(R, M)$ with $f' = F' \circ \psi$. [Given f', show that $f(a) = f'(a)(1_R)$ defines a $\mathbb{Z}$-module homomorphism of A to M. If F is the associated lift of f to B, show that $F'(b)(r) = F(rb)$ defines an R-module homomorphism from B to $\operatorname{Hom}_{\mathbb{Z}}(R, M)$ that lifts f'.]
  (c) Prove that if Q is an injective $\mathbb{Z}$-module then $\operatorname{Hom}_{\mathbb{Z}}(R, Q)$ is an injective R-module.

16. This exercise proves Theorem 38 that every left R-module M is contained in an injective left R-module.
  (a) Show that M is contained in an injective $\mathbb{Z}$-module Q. [M is a $\mathbb{Z}$-module—use Corollary 37.]
  (b) Show that $\operatorname{Hom}_R(R, M) \subseteq \operatorname{Hom}_{\mathbb{Z}}(R, M) \subseteq \operatorname{Hom}_{\mathbb{Z}}(R, Q)$.
  (c) Use the R-module isomorphism $M \cong \operatorname{Hom}_R(R, M)$ (Exercise 10) and the previous exercise to conclude that M is contained in an injective R-module.

17. This exercise completes the proof of Proposition 34. Suppose that Q is an R-module with the property that every short exact sequence $0 \to Q \to M_1 \to N \to 0$ splits and suppose that the sequence $0 \to L \xrightarrow{\psi} M$ is exact. Prove that every R-module homomorphism f from L to Q can be lifted to an R-module homomorphism F from M to Q with $f = F \circ \psi$. [By the previous exercise, Q is contained in an injective R-module. Use the splitting property together with Exercise 4 (noting that Exercise 4 can be proved using (2) in Proposition 34 as the definition of an injective module).]

18. Prove that the injective hull of the $\mathbb{Z}$-module $\mathbb{Z}$ is $\mathbb{Q}$. [Let H be the injective hull of $\mathbb{Z}$ and argue that $\mathbb{Q}$ contains an isomorphic copy of H. Use the divisibility of H to show $1/n \in H$ for all nonzero integers n, and deduce that $H = \mathbb{Q}$.]

19. If F is a field, prove that the injective hull of F is F.

20. Prove that the polynomial ring $R[x]$ in the indeterminate x over the commutative ring R is a flat R-module.

21. Let R and S be rings with 1 and suppose M is a right R-module, and N is an (R, S)-bimodule. If M is flat over R and N is flat as an S-module prove that $M \otimes_R N$ is flat as a right S-module.

22. Suppose that R is a commutative ring and that M and N are flat R-modules. Prove that $M \otimes_R N$ is a flat R-module. [Use the previous exercise.]

23. Prove that the (right) module $M \otimes_R S$ obtained by changing the base from the ring R to the ring S (by some homomorphism $f : R \to S$ with $f(1_R) = 1_S$, cf. Example 6 following Corollary 12 in Section 4) of the flat (right) R-module M is a flat S-module.

24. Prove that A is a flat R-module if and only if for any left R-modules L and M where L is *finitely generated*, then $\psi : L \to M$ injective implies that also $1 \otimes \psi : A \otimes_R L \to A \otimes_R M$ is injective. [Use the techniques in the proof of Corollary 42.]

25. *(A Flatness Criterion)* Parts (a)-(c) of this exercise prove that A is a flat R-module if and only if for every finitely generated ideal I of R, the map from $A \otimes_R I \to A \otimes_R R \cong A$ induced by the inclusion $I \subseteq R$ is again injective (or, equivalently, $A \otimes_R I \cong AI \subseteq A$).
   (a) Prove that if A is flat then $A \otimes_R I \to A \otimes_R R$ is injective.
   (b) If $A \otimes_R I \to A \otimes_R R$ is injective for every finitely generated ideal I, prove that $A \otimes_R I \to A \otimes_R R$ is injective for every ideal I. Show that if K is any submodule of a finitely generated free module F then $A \otimes_R K \to A \otimes_R F$ is injective. Show that the same is true for any free module F. [Cf. the proof of Corollary 42.]
   (c) Under the assumption in (b), suppose L and M are R-modules and $L \overset{\psi}{\to} M$ is injective. Prove that $A \otimes_R L \overset{1 \otimes \psi}{\to} A \otimes_R M$ is injective and conclude that A is flat. [Write M as a quotient of the free module F, giving a short exact sequence

$$0 \longrightarrow K \longrightarrow F \overset{f}{\longrightarrow} M \longrightarrow 0.$$

Show that if $J = f^{-1}(\psi(L))$ and $\iota : J \to F$ is the natural injection, then the diagram

$$
\begin{array}{ccccccccc}
0 & \longrightarrow & K & \longrightarrow & J & \longrightarrow & L & \longrightarrow & 0 \\
& & \downarrow{\scriptstyle id} & & \downarrow{\scriptstyle \iota} & & \downarrow{\scriptstyle \psi} & & \\
0 & \longrightarrow & K & \longrightarrow & F & \longrightarrow & M & \longrightarrow & 0
\end{array}
$$

is commutative with exact rows. Show that the induced diagram

$$
\begin{array}{ccccccc}
A \otimes_R K & \longrightarrow & A \otimes_R J & \longrightarrow & A \otimes_R L & \longrightarrow & 0 \\
\downarrow{\scriptstyle id} & & \downarrow{\scriptstyle 1 \otimes \iota} & & \downarrow{\scriptstyle 1 \otimes \psi} & & \\
A \otimes_R K & \longrightarrow & A \otimes_R F & \longrightarrow & A \otimes_R M & \longrightarrow & 0
\end{array}
$$

is commutative with exact rows. Use (b) to show that $1 \otimes \iota$ is injective, then use Exercise 1 to conclude that $1 \otimes \psi$ is injective.]
   (d) *(A Flatness Criterion for quotients)* Suppose $A = F/K$ where F is flat (e.g., if F is free) and K is an R-submodule of F. Prove that A is flat if and only if $FI \cap K = KI$ for every finitely generated ideal I of R. [Use (a) to prove $F \otimes_R I \cong FI$ and observe the image of $K \otimes_R I$ is KI; tensor the exact sequence $0 \to K \to F \to A \to 0$ with I to prove that $A \otimes_R I \cong FI/KI$, and apply the flatness criterion.]

26. Suppose R is a P.I.D. This exercise proves that A is a flat R-module if and only if A is torsion free R-module (i.e., if $a \in A$ is nonzero and $r \in R$, then $ra = 0$ implies $r = 0$).
   (a) Suppose that A is flat and for fixed $r \in R$ consider the map $\psi_r : R \to R$ defined by multiplication by r: $\psi_r(x) = rx$. If r is nonzero show that ψ_r is an injection. Conclude from the flatness of A that the map from A to A defined by mapping a to ra is injective and that A is torsion free.
   (b) Suppose that A is torsion free. If I is a nonzero ideal of R, then $I = rR$ for some nonzero $r \in R$. Show that the map ψ_r in (a) induces an isomorphism $R \cong I$ of

R-modules and that the composite $R \xrightarrow{\psi} I \xrightarrow{\iota} R$ of ψ_r with the inclusion $\iota : I \subseteq R$ is multiplication by r. Prove that the composite $A \otimes_R R \xrightarrow{1 \otimes \psi_r} A \otimes_R I \xrightarrow{1 \otimes \iota} A \otimes_R R$ corresponds to the map $a \mapsto ra$ under the identification $A \otimes_R R = A$ and that this composite is injective since A is torsion free. Show that $1 \otimes \psi_r$ is an isomorphism and deduce that $1 \otimes \iota$ is injective. Use the previous exercise to conclude that A is flat.

27. Let M, A and B be R-modules.
   (a) Suppose $f : A \to M$ and $g : B \to M$ are R-module homomorphisms. Prove that $X = \{(a, b) \mid a \in A,\ b \in B \text{ with } f(a) = g(b)\}$ is an R-submodule of the direct sum $A \oplus B$ (called the *pullback* or *fiber product* of f and g) and that there is a commutative diagram

$$\begin{array}{ccc} X & \xrightarrow{\pi_2} & B \\ {\scriptstyle \pi_1}\downarrow & & \downarrow{\scriptstyle g} \\ A & \xrightarrow{f} & M \end{array}$$

   where π_1 and π_2 are the natural projections onto the first and second components.
   (b) Suppose $f' : M \to A$ and $g' : M \to B$ are R-module homomorphisms. Prove that the quotient Y of $A \oplus B$ by $\{(f'(m), -g'(m)) \mid m \in M\}$ is an R-module (called the *pushout* or *fiber sum* of f' and g') and that there is a commutative diagram

$$\begin{array}{ccc} M & \xrightarrow{g'} & B \\ {\scriptstyle f'}\downarrow & & \downarrow{\scriptstyle \pi_2'} \\ A & \xrightarrow{\pi_1'} & Y \end{array}$$

   where π_1' and π_2' are the natural maps to the quotient induced by the maps into the first and second components.

28. (a) *(Schanuel's Lemma)* If $0 \to K \to P \xrightarrow{\varphi} M \to 0$ and $0 \to K' \to P' \xrightarrow{\varphi'} M \to 0$ are exact sequences of R-modules where P and P' are projective, prove $P \oplus K' \cong P' \oplus K$ as R-modules. [Show that there is an exact sequence $0 \to \ker \pi \to X \xrightarrow{\pi} P \to 0$ with $\ker \pi \cong K'$, where X is the fiber product of φ and φ' as in the previous exercise. Deduce that $X \cong P \oplus K'$. Show similarly that $X \cong P' \oplus K$.]
   (b) If $0 \to M \to Q \xrightarrow{\psi} L \to 0$ and $0 \to M \to Q' \xrightarrow{\psi'} L' \to 0$ are exact sequences of R-modules where Q and Q' are injective, prove $Q \oplus L' \cong Q' \oplus L$ as R-modules.

The R-modules M and N are said to be *projectively equivalent* if $M \oplus P \cong N \oplus P'$ for some projective modules P, P'. Similarly, M and N are *injectively equivalent* if $M \oplus Q \cong N \oplus Q'$ for some injective modules Q, Q'. The previous exercise shows K and K' are projectively equivalent and L and L' are injectively equivalent.

CHAPTER 11

Vector Spaces

In this chapter we review the basic theory of finite dimensional vector spaces over an arbitrary field F (some infinite dimensional vector space theory is covered in the exercises). Since the proofs are identical to the corresponding arguments for real vector spaces our treatment is very terse. For the most part we include only those results which are used in other parts of the text so basic topics such as Gauss–Jordan elimination, row echelon forms, methods for finding bases of subspaces, elementary properties of matrices, etc., are not covered or are discussed in the exercises. The reader should therefore consider this chapter as a refresher in linear algebra and as a prelude to field theory and Galois theory. Characteristic polynomials and eigenvalues will be reviewed and treated in a larger context in the next chapter.

11.1 DEFINITIONS AND BASIC THEORY

The terminology for vector spaces is slightly different from that of modules, that is, when the ring R is a field there are different names for many of the properties of R-modules which we defined in the last chapter. The following is a dictionary of these new terms (many of which may already be familiar). The definition of each corresponding vector space property is the same (verbatim) as the module-theoretic definition with the only added assumption being that the ring R is a field (so these definitions are not repeated here).

Terminology for R any Ring	Terminology for R a Field
M is an R-module	M is a vector space over R
m is an element of M	m is a vector in M
α is a ring element	α is a scalar
N is a submodule of M	N is a subspace of M
M/N is a quotient module	M/N is a quotient space
M is a free module of rank n	M is a vector space of dimension n
M is a finitely generated module	M is a finite dimensional vector space
M is a nonzero cyclic module	M is a 1-dimensional vector space
$\varphi : M \to N$ is an R-module homomorphism	$\varphi : M \to N$ is a linear transformation
M and N are isomorphic as R-modules	M and N are isomorphic vector spaces
the subset A of M generates M	the subset A of M spans M
$M = RA$	each element of M is a linear combination of elements of A i.e., $M = \mathrm{Span}(A)$

For the remainder of this chapter F is a field and V is a vector space over F.

One of the first results we shall prove about vector spaces is that they are free F-modules, that is, they have bases. Although our arguments treat only the case of finite dimensional spaces, the corresponding result for arbitrary vector spaces is proved in the exercises as an application of Zorn's Lemma. The reader may first wish to review the section in the previous chapter on free modules, especially their properties pertaining to homomorphisms.

Definition.
   (1) A subset S of V is called a set of *linearly independent* vectors if an equation $\alpha_1 v_1 + \alpha_2 v_2 + \cdots + \alpha_n v_n = 0$ with $\alpha_1, \alpha_2, \ldots, \alpha_n \in F$ and $v_1, v_2, \ldots, v_n \in S$ implies $\alpha_1 = \alpha_2 = \cdots = \alpha_n = 0$.
   (2) A *basis* of a vector space V is an ordered set of linearly independent vectors which span V. In particular two bases will be considered different even if one is simply a rearrangement of the other. This is sometimes referred to as an *ordered basis*.

Examples
   (1) The space $V = F[x]$ of polynomials in the variable x with coefficients from the field F is in particular a vector space over F. The elements $1, x, x^2, \ldots$ are linearly independent by definition (i.e., a polynomial is 0 if and only if all its coefficients are 0). Since these elements also span V by definition, they are a basis for V.
   (2) The collection of solutions of a linear, homogeneous, constant coefficient differential equation (for example, $y'' - 3y' + 2y = 0$) over $\mathbb{C}$ form a vector space over $\mathbb{C}$ since differentiation is a linear operator. Elements of this vector space are linearly independent if they are linearly independent as functions. For example, e^t and e^{2t} are easily seen to be solutions of the equation $y'' - 3y' + 2y = 0$ (differentiation with respect to t). They are linearly independent functions since $ae^t + be^{2t} = 0$ implies $a + b = 0$ (let $t = 0$) and $ae + be^2 = 0$ (let $t = 1$) and the only solution to these two equations is $a = b = 0$. It is a theorem in differential equations that these elements span the set of solutions of this equation, hence are a basis for this space.

Proposition 1. Assume the set $\mathcal{A} = \{v_1, v_2, \ldots, v_n\}$ spans the vector space V but no proper subset of $\mathcal{A}$ spans V. Then $\mathcal{A}$ is a basis of V. In particular, any finitely generated (i.e., finitely spanned) vector space over F is a free F-module.

Proof: It is only necessary to prove that $v_1, v_2, \ldots, v_n$ are linearly independent. Suppose $\alpha_1 v_1 + \alpha_2 v_2 + \cdots + \alpha_n v_n = 0$ where not all of the α_i are 0. By reordering, we may assume that $\alpha_1 \neq 0$ and then

$$v_1 = -\frac{1}{\alpha_1}(\alpha_2 v_2 + \cdots + \alpha_n v_n).$$

It follows that $\{v_2, v_3, \ldots, v_n\}$ also spans V since any linear combination of $v_1, v_2, \ldots, v_n$ can be written as a linear combination of $v_2, v_3, \ldots, v_n$ using the equation above. This is a contradiction.

Example

Let F be a field and consider $F[x]/(f(x))$ where $f(x) = x^n + a_{n-1}x^{n-1} + \cdots + a_1 x + a_0$. The ideal $(f(x))$ is a subspace of the vector space $F[x]$ and the quotient $F[x]/(f(x))$ is also a vector space over F. By the Euclidean Algorithm, every polynomial $a(x) \in F[x]$ can be written uniquely in the form $a(x) = q(x)f(x) + r(x)$ where $r(x) \in F[x]$ and $0 \le \deg r(x) \le n - 1$. Since $q(x)f(x) \in (f(x))$, it follows that every element of the quotient is represented by a polynomial $r(x)$ of degree $\le n - 1$. Two distinct such polynomials cannot be the same in the quotient since this would say their difference (which is a nonzero polynomial of degree at most $n - 1$) would be divisible by $f(x)$ (which is of degree n). It follows that the elements $\bar{1}, \bar{x}, \overline{x^2}, \ldots, \overline{x^{n-1}}$ (the bar denotes the image of these elements in the quotient, as usual) span $F[x]/(f(x))$ as a vector space over F and that no proper subset of these elements also spans, hence these elements give a basis for $F[x]/(f(x))$.

Corollary 2. Assume the finite set $\mathcal{A}$ spans the vector space V. Then $\mathcal{A}$ contains a basis of V.

Proof: Any subset $\mathcal{B}$ of $\mathcal{A}$ spanning V such that no proper subset of $\mathcal{B}$ also spans V (there clearly exist such subsets) is a basis for V by Proposition 1.

Theorem 3. (*A Replacement Theorem*) Assume $\mathcal{A} = \{a_1, a_2, \ldots, a_n\}$ is a basis for V containing n elements and $\{b_1, b_2, \ldots, b_m\}$ is a set of linearly independent vectors in V. Then there is an ordering $a_1, a_2, \ldots, a_n$ such that for each $k \in \{1, 2, \ldots, m\}$ the set $\{b_1, b_2, \ldots, b_k, a_{k+1}, a_{k+2}, \ldots, a_n\}$ is a basis of V. In other words, the elements $b_1, b_2, \ldots, b_m$ can be used to successively replace the elements of the basis $\mathcal{A}$, still retaining a basis. In particular, $n \ge m$.

Proof: Proceed by induction on k. If $k = 0$ there is nothing to prove, since $\mathcal{A}$ is given as a basis for V. Suppose now that $\{b_1, b_2, \ldots, b_k, a_{k+1}, a_{k+2}, \ldots, a_n\}$ is a basis for V. Then in particular this is a spanning set, so b_{k+1} is a linear combination:

$$b_{k+1} = \beta_1 b_1 + \cdots + \beta_k b_k + \alpha_{k+1} a_{k+1} + \cdots + \alpha_n a_n. \tag{11.1}$$

Not all of the α_i can be 0, since this would imply b_{k+1} is a linear combination of $b_1, b_2, \ldots, b_k$, contrary to the linear independence of these elements. By reordering if necessary, we may assume $\alpha_{k+1} \ne 0$. Then solving this last equation for a_{k+1} as a linear combination of b_{k+1} and $b_1, b_2, \ldots, b_k, a_{k+2}, \ldots, a_n$ shows

$$\text{Span}\{b_1, b_2, \ldots, b_k, b_{k+1}, a_{k+2}, \ldots, a_n\} = \text{Span}\{b_1, b_2, \ldots, b_k, a_{k+1}, a_{k+2}, \ldots, a_n\}$$

and so this is a spanning set for V. It remains to show $b_1, \ldots, b_k, b_{k+1}, a_{k+2}, \ldots, a_n$ are linearly independent. If

$$\beta_1 b_1 + \cdots + \beta_k b_k + \beta_{k+1} b_{k+1} + \alpha_{k+2} a_{k+2} + \cdots + \alpha_n a_n = 0 \tag{11.2}$$

then substituting for b_{k+1} from the expression for b_{k+1} in equation (1), we obtain a linear combination of $\{b_1, b_2, \ldots, b_k, a_{k+1}, a_{k+2}, \ldots, a_n\}$ equal to 0, where the coefficient of a_{k+1} is β_{k+1}. Since this last set is a basis by induction, all the coefficients in this linear combination, in particular β_{k+1}, must be 0. But then equation (2) is

$$\beta_1 b_1 + \cdots + \beta_k b_k + \alpha_{k+2} a_{k+2} + \cdots + \alpha_n a_n = 0.$$

Again by the induction hypothesis all the other coefficients must be 0 as well. Thus $\{b_1, b_2, \ldots, b_k, b_{k+1}, a_{k+2}, \ldots, a_n\}$ is a basis for V, and the induction is complete.

Corollary 4.
(1) Suppose V has a finite basis with n elements. Any set of linearly independent vectors has $\leq n$ elements. Any spanning set has $\geq n$ elements.
(2) If V has some finite basis then any two bases of V have the same cardinality.

Proof: (1) This is a restatement of the last result of Theorem 3 and Corollary 2. (2) This is immediate from (1) since a basis is both a spanning set and a linearly independent set.

Definition. If V is a finitely generated F-module (i.e., has a finite basis) the cardinality of any basis is called the *dimension* of V and is denoted by $\dim_F V$, or just $\dim V$ when F is clear from the context, and V is said to be *finite dimensional* over F. If V is not finitely generated, V is said to be infinite dimensional (written $\dim V = \infty$).

Examples
(1) The dimension of the space of solutions to the differential equation $y'' - 3y' + 2y = 0$ over $\mathbb{C}$ is 2 (with basis e^t, e^{2t}, for example). In general, it is a theorem in differential equations that the space of solutions of an nth order linear, homogeneous, constant coefficient differential equation of degree n over $\mathbb{C}$ form a vector space over $\mathbb{C}$ of dimension n.
(2) The dimension over F of the quotient $F[x]/(f(x))$ by the nonzero polynomial $f(x)$ considered above is $n = \deg f(x)$. The space $F[x]$ and its subspace $(f(x))$ are infinite dimensional vector spaces over F.

Corollary 5. *(Building–Up Lemma)* If A is a set of linearly independent vectors in the finite dimensional space V then there exists a basis of V containing A.

Proof: This is also immediate from Theorem 3, since we can use the elements of A to successively replace the elements of any given basis for V (which exists by the assumption that V is finite dimensional).

Theorem 6. If V is an n dimensional vector space over F, then $V \cong F^n$. In particular, any two finite dimensional vector spaces over F of the same dimension are isomorphic.

Proof: Let $v_1, v_2, \ldots, v_n$ be a basis for V. Define the map

$$\varphi : F^n \to V \qquad \text{by} \qquad \varphi(\alpha_1, \alpha_2, \ldots, \alpha_n) = \alpha_1 v_1 + \alpha_2 v_2 + \cdots + \alpha_n v_n.$$

The map φ is clearly F-linear, is surjective since the v_i span V, and is injective since the v_i are linearly independent, hence is an isomorphism.

Examples

(1) Let $\mathbb{F}$ be a finite field with q elements and let W be a k-dimensional vector space over $\mathbb{F}$. We show that the number of distinct bases of W is

$$(q^k - 1)(q^k - q)(q^k - q^2) \dots (q^k - q^{k-1}).$$

Every basis of W can be built up as follows. Any nonzero vector w_1 can be the first element of a basis. Since W is isomorphic to $\mathbb{F}^k$, $|W| = q^k$, so there are $q^k - 1$ choices for w_1. Any vector not in the 1-dimensional space spanned by w_1 is linearly independent from w_1 and so may be chosen for the second basis element, w_2. A 1-dimensional space is isomorphic to $\mathbb{F}$ and so has q elements. Thus there are $q^k - q$ choices for w_2. Proceeding in this way one sees that at the ith stage any vector not in the $(i - 1)$-dimensional space spanned by $w_1, w_2, \dots, w_{i-1}$ will be linearly independent from $w_1, w_2, \dots, w_{i-1}$ and so may be chosen for the ith basis vector w_i. An $(i - 1)$-dimensional space is isomorphic to $\mathbb{F}^{i-1}$ and so has q^{i-1} elements. Thus there are $q^k - q^{i-1}$ choices for w_i. The process terminates when w_k is chosen, for then we have k linear independent vectors in a k-dimensional space, hence a basis.

(2) Let $\mathbb{F}$ be a finite field with q elements and let V be an n-dimensional vector space over $\mathbb{F}$. For each $k \in \{1, 2, \dots, n\}$ we show that the number of subspaces of V of dimension k is

$$\frac{(q^n - 1)(q^n - q) \dots (q^n - q^{k-1})}{(q^k - 1)(q^k - q) \dots (q^k - q^{k-1})}.$$

Any k-dimensional space is spanned by k independent vectors. By arguing as in the preceding example the numerator of the above expression is the number of ways of picking k independent vectors from an n-dimensional space. Two sets of k independent vectors span the same space W if and only if they are both bases of the k-dimensional space W. In order to obtain the formula for the number of distinct subspaces of dimension k we must divide by the number of repetitions, i.e., the number of bases of a fixed k-dimensional space. This factor which appears in the denominator is precisely the number computed in Example 1.

Next, we prove an important relation between the dimension of a subspace, the dimension of its associated quotient space and the dimension of the whole space:

Theorem 7. Let V be a vector space over F and let W be a subspace of V. Then V/W is a vector space with $\dim V = \dim W + \dim V/W$ (where if one side is infinite then both are).

Proof: Suppose W has dimension m and V has dimension n over F and let $w_1, w_2, \dots, w_m$ be a basis for W. By Corollary 5, these linearly independent elements of V can be extended to a basis $w_1, w_2, \dots, w_m, v_{m+1}, \dots, v_n$ of V. The natural surjective projection map of V into V/W maps each w_i to 0. No linear combination of the v_i is mapped to 0, since this would imply this linear combination is an element of W, contrary to the choice of the v_i. Hence, the image V/W of this projection map is isomorphic to the subspace of V spanned by the v_i, hence $\dim V/W = n - m$, which is the theorem when the dimensions are finite. If either side is infinite it is an easy exercise to produce an infinite number of linearly independent vectors showing the other side is also infinite.

Corollary 8. Let $\varphi : V \to U$ be a linear transformation of vector spaces over F. Then ker φ is a subspace of V, $\varphi(V)$ is a subspace of U and dim $V = $ dim ker $\varphi +$ dim $\varphi(V)$.

Proof: This follows immediately from Theorem 7. Note that the proof of Theorem 7 is in fact the special case of Corollary 8 where U is the quotient V/W and φ is the natural projection homomorphism.

Corollary 9. Let $\varphi : V \to W$ be a linear transformation of vector spaces of the same finite dimension. Then the following are equivalent:
- **(1)** φ is an isomorphism
- **(2)** φ is injective, i.e., ker $\varphi = 0$
- **(3)** φ is surjective, i.e., $\varphi(V) = W$
- **(4)** φ sends a basis of V to a basis of W.

Proof: The equivalence of these conditions follows from Corollary 8 by counting dimensions.

Definition. If $\varphi : V \to U$ is a linear transformation of vector spaces over F, ker φ is sometimes called the *null space* of φ and the dimension of ker φ is called the *nullity* of φ. The dimension of $\varphi(V)$ is called the *rank* of φ. If ker $\varphi = 0$, the transformation is said to be *nonsingular*.

Example

Let F be a finite field with q elements and let V be an n-dimensional vector space over F. Recall that the *general linear group* $GL(V)$ is the group of all nonsingular linear transformations from V to V (the group operation being composition). We show that the order of this group is

$$|GL(V)| = (q^n - 1)(q^n - q)(q^n - q^2)\ldots(q^n - q^{n-1}).$$

To see this, fix a basis $v_1, \ldots, v_n$ of V. A linear transformation is nonsingular if and only if it sends this basis to another basis of V. Moreover, if $w_1 \ldots, w_n$ is any basis of V, by Theorem 6 in Section 10.3 there is a unique linear transformation which sends v_i to w_i, $1 \le i \le n$. Thus the number of nonsingular linear transformations from V to itself equals the number of distinct bases of V. This number, which was computed in Example 1 above (with $k = n$), is the order of $GL(V)$.

EXERCISES

1. Let $V = \mathbb{R}^n$ and let $(a_1, a_2, \ldots, a_n)$ be a fixed vector in V. Prove that the collection of elements $(x_1, x_2, \ldots, x_n)$ of V with $a_1x_1 + a_2x_2 + \ldots + a_nx_n = 0$ is a subspace of V. Determine the dimension of this subspace and find a basis.

2. Let V be the collection of polynomials with coefficients in $\mathbb{Q}$ in the variable x of degree at most 5. Prove that V is a vector space over $\mathbb{Q}$ of dimension 6, with $1, x, x^2, \ldots, x^5$ as basis. Prove that $1, 1+x, 1+x+x^2, \ldots, 1+x+x^2+x^3+x^4+x^5$ is also a basis for V.

3. Let φ be the linear transformation $\varphi : \mathbb{R}^4 \to \mathbb{R}^1$ such that

$$\varphi((1, 0, 0, 0)) = 1 \qquad \varphi((1, -1, 0, 0)) = 0$$
$$\varphi((1, -1, 1, 0)) = 1 \qquad \varphi((1, -1, 1, -1)) = 0.$$

Determine $\varphi((a, b, c, d))$.

4. Prove that the space of real-valued functions on the closed interval $[a, b]$ is an infinite dimensional vector space over $\mathbb{R}$, where $a < b$.

5. Prove that the space of continuous real-valued functions on the closed interval $[a, b]$ is an infinite dimensional vector space over $\mathbb{R}$, where $a < b$.

6. Let V be a vector space of finite dimension. If φ is any linear transformation from V to V prove there is an integer m such that the intersection of the image of φ^m and the kernel of φ^m is $\{0\}$.

7. Let φ be a linear transformation from a vector space V of dimension n to itself that satisfies $\varphi^2 = 0$. Prove that the image of φ is contained in the kernel of φ and hence that the rank of φ is at most $n/2$.

8. Let V be a vector space over F and let φ be a linear transformation of the vector space V to itself. A nonzero element $v \in V$ satisfying $\varphi(v) = \lambda v$ for some $\lambda \in F$ is called an *eigenvector* of φ with *eigenvalue* λ. Prove that for any fixed $\lambda \in F$ the collection of eigenvectors of φ with eigenvalue λ together with 0 forms a subspace of V.

9. Let V be a vector space over F and let φ be a linear transformation of the vector space V to itself. Suppose for $i = 1, 2, \ldots, k$ that $v_i \in V$ is an eigenvector for φ with eigenvalue $\lambda_i \in F$ (cf. the preceding exercise) and that all the eigenvalues λ_i are distinct. Prove that $v_1, v_2, \ldots, v_k$ are linearly independent. [Use induction on k: write a linear dependence relation among the v_i and apply φ to get another linear dependence relation among the v_i involving the eigenvalues — now subtract a suitable multiple of the first linear relation to get a linear dependence relation on fewer elements.] Conclude that any linear transformation on an n-dimensional vector space has at most n distinct eigenvalues.

In the following exercises let V be a vector space of arbitrary dimension over a field F.

10. Prove that any vector space V has a basis (by convention the null set is the basis for the zero space). [Let S be the set of subsets of V consisting of linearly independent vectors, partially ordered under inclusion; apply Zorn's Lemma to S and show a maximal element of S is a basis.]

11. Refine your argument in the preceding exercise to prove that any set of linearly independent vectors of V is contained in a basis of V.

12. If F is a field with a finite or countable number of elements and V is an infinite dimensional vector space over F with basis $\mathcal{B}$, prove that the cardinality of V equals the cardinality of $\mathcal{B}$. Deduce in this case that any two bases of V have the same cardinality.

13. Prove that as vector spaces over $\mathbb{Q}$, $\mathbb{R}^n \cong \mathbb{R}$, for all $n \in \mathbb{Z}^+$ (note that, in particular, this means $\mathbb{R}^n$ and $\mathbb{R}$ are isomorphic as additive abelian groups).

14. Let $\mathcal{A}$ be a basis for the infinite dimensional space V. Prove that V is isomorphic to the *direct sum* of copies of the field F indexed by the set $\mathcal{A}$. Prove that the *direct product* of copies of F indexed by $\mathcal{A}$ is a vector space over F and it has strictly larger dimension than the dimension of V (see the exercises in Section 10.3 for the definitions of direct sum and direct product of infinitely many modules).

11.2 THE MATRIX OF A LINEAR TRANSFORMATION

Throughout this section let V, W be vector spaces over the same field F, let $\mathcal{B} = \{v_1, v_2, \ldots, v_n\}$ be an (ordered) basis of V, let $\mathcal{E} = \{w_1, w_2, \ldots, w_m\}$ be an (ordered) basis of W and let $\varphi \in \text{Hom}(V, W)$ be a linear transformation from V to W. For each $j \in \{1, 2, \ldots, n\}$ write the image of v_j under φ in terms of the basis $\mathcal{E}$:

$$\varphi(v_j) = \sum_{i=1}^{m} \alpha_{ij} w_i. \tag{11.3}$$

Let $M_{\mathcal{B}}^{\mathcal{E}}(\varphi) = (a_{ij})$ be the $m \times n$ matrix whose i, j entry is α_{ij} (that is, use the coefficients of the w_i's in the above computation of $\varphi(v_j)$ for the jth column of this matrix). The matrix $M_{\mathcal{B}}^{\mathcal{E}}(\varphi)$ is called the *matrix of φ with respect to the bases $\mathcal{B}$, $\mathcal{E}$*. The domain basis is the lower and the codomain basis the upper letters appearing after the "M." Given this matrix, we can recover the linear transformation φ as follows: to compute $\varphi(v)$ for $v \in V$, write v in terms of the basis $\mathcal{B}$:

$$v = \sum_{i=1}^{n} \alpha_i v_i, \qquad \alpha_i \in F,$$

and then calculate the product of the $m \times n$ and $n \times 1$ matrices

$$M_{\mathcal{B}}^{\mathcal{E}}(\varphi) \times \begin{pmatrix} \alpha_1 \\ \alpha_2 \\ \vdots \\ \alpha_n \end{pmatrix} = \begin{pmatrix} \beta_1 \\ \beta_2 \\ \vdots \\ \beta_m \end{pmatrix}.$$

The image of v under φ is given by

$$\varphi(v) = \sum_{i=1}^{m} \beta_i w_i \ ,$$

i.e., the column vector of coordinates of $\varphi(v)$ with respect to the basis $\mathcal{E}$ are obtained by multiplying the matrix $M_{\mathcal{B}}^{\mathcal{E}}(\varphi)$ by the column vector of coordinates of v with respect to the basis $\mathcal{B}$ (sometimes denoted $[\varphi(v)]_{\mathcal{E}} = M_{\mathcal{B}}^{\mathcal{E}}(\varphi)[v]_{\mathcal{B}}$).

Definition. The $m \times n$ matrix $A = (a_{ij})$ associated to the linear transformation φ above is said to *represent* the linear transformation φ with respect to the bases $\mathcal{B}$, $\mathcal{E}$. Similarly, φ is the linear transformation represented by A with respect to the bases $\mathcal{B}$, $\mathcal{E}$.

Examples

(1) Let $V = \mathbb{R}^3$ with the standard basis $\mathcal{B} = \{(1, 0, 0), (0, 1, 0), (0, 0, 1)\}$ and let $W = \mathbb{R}^2$ with the standard basis $\mathcal{E} = \{(1, 0), (0, 1)\}$. Let φ be the linear transformation $\varphi(x, y, z) = (x + 2y, x + y + z)$. Since $\varphi(1, 0, 0) = (1, 1)$, $\varphi(0, 1, 0) = (2, 1)$, $\varphi(0, 0, 1) = (0, 1)$, the matrix $A = M_{\mathcal{B}}^{\mathcal{E}}(\varphi)$ is the matrix $\begin{pmatrix} 1 & 2 & 0 \\ 1 & 1 & 1 \end{pmatrix}$.

(2) Let $V = W$ be the 2-dimensional space of solutions of the differential equation $y'' - 3y' + 2y = 0$ over $\mathbb{C}$ and let $\mathcal{B} = \mathcal{E}$ be the basis $v_1 = e^t$, $v_2 = e^{2t}$. Since the coefficients of this equation are constants it is easy to check that if y is a solution then its derivative y' is also a solution. It follows that the map $\varphi = d/dt = $ differentiation (with respect to t) is a linear transformation from V to itself. Since $\varphi(v_1) = d(e^t)/dt = e^t = v_1$ and $\varphi(v_2) = d(e^{2t})/dt = 2e^{2t} = 2v_2$ we see that the corresponding matrix with respect to these bases is the diagonal matrix $\begin{pmatrix} 1 & 0 \\ 0 & 2 \end{pmatrix}$.

(3) Let $V = W = \mathbb{Q}^3 = \{(x, y, z) \mid x, y, z \in \mathbb{Q}\}$ be the usual 3-dimensional vector space of ordered 3-tuples with entries from the field $F = \mathbb{Q}$ of rational numbers and suppose φ is the linear transformation

$$\varphi(x, y, z) = (9x + 4y + 5z, -4x - 3z, -6x - 4y - 2z), \qquad x, y, z \in \mathbb{Q}$$

from V to itself. Take the standard basis $e_1 = (1, 0, 0)$, $e_2 = (0, 1, 0)$, $e_3 = (0, 0, 1)$ for V and for $W = V$. Since $\varphi(1, 0, 0) = (9, -4, -6)$, $\varphi(0, 1, 0) = (4, 0, -4)$, $\varphi(0, 0, 1) = (5, -3, -2)$, the matrix A representing this linear transformation with respect to these bases is

$$A = \begin{pmatrix} 9 & 4 & 5 \\ -4 & 0 & -3 \\ -6 & -4 & -2 \end{pmatrix}.$$

Theorem 10. Let V be a vector space over F of dimension n and let W be a vector space over F of dimension m, with bases $\mathcal{B}$, $\mathcal{E}$ respectively. Then the map $\text{Hom}_F(V, W) \to M_{m \times n}(F)$ from the space of linear transformations from V to W to the space of $m \times n$ matrices with coefficients in F defined by $\varphi \mapsto M_{\mathcal{B}}^{\mathcal{E}}(\varphi)$ is a vector space isomorphism. In particular, there is a bijective correspondence between linear transformations and their associated matrices with respect to a fixed choice of bases.

Proof: The columns of the matrix $M_{\mathcal{B}}^{\mathcal{E}}(\varphi)$ are determined by the action of φ on the basis $\mathcal{B}$ as in equation (3). This shows in particular that the map $\varphi \mapsto M_{\mathcal{B}}^{\mathcal{E}}(\varphi)$ is an F-linear map since φ is F-linear. This map is *surjective* since given a matrix M, the map φ *defined* by equation (3) on a basis and then extended by linearity is a linear transformation with matrix M. The map is *injective* since two linear transformations agreeing on a basis are the same.

Note that different choices of bases give rise to different isomorphisms, so in the same sense that there is no natural choice of basis for a vector space, there is no natural isomorphism between $\text{Hom}_F(V, W)$ and $M_{m \times n}(F)$.

Corollary 11. The dimension of $\text{Hom}_F(V, W)$ is $(\dim V)(\dim W)$.

Proof: The dimension of $M_{m \times n}(F)$ is mn.

Definition. An $m \times n$ matrix A is called *nonsingular* if $Ax = 0$ with $x \in F^n$ implies $x = 0$.

416

The connection of the term *nonsingular* applied to matrices and to linear transformations is the following: let $A = M_B^{\mathcal{E}}(\varphi)$ be the matrix associated to the linear transformation φ (with some choice of bases $\mathcal{B}$, $\mathcal{E}$). Then independently of the choice of bases, the $m \times n$ matrix A is nonsingular if and only if the linear transformation φ is a nonsingular linear transformation from the n-dimensional space V to the m-dimensional space W (cf. the exercises).

Assume now that U, V and W are all finite dimensional vector spaces over F with ordered bases $\mathcal{D}$, $\mathcal{B}$ and $\mathcal{E}$ respectively, where $\mathcal{B}$ and $\mathcal{E}$ are as before and suppose $\mathcal{D} = \{u_1, u_2, \ldots, u_k\}$. Assume $\psi : U \to V$ and $\varphi : V \to W$ are linear transformations. Their composite, $\varphi \circ \psi$, is a linear transformation from U to W, so we can compute its matrix with respect to the appropriate bases; namely, $M_{\mathcal{D}}^{\mathcal{E}}(\varphi \circ \psi)$ is found by computing

$$\varphi \circ \psi(u_j) = \sum_{i=1}^{m} \gamma_{ij} w_i$$

and putting the coefficients γ_{ij} down the jth column of $M_{\mathcal{D}}^{\mathcal{E}}(\varphi \circ \psi)$. Next, compute the matrices of ψ and φ separately:

$$\psi(u_j) = \sum_{p=1}^{n} \alpha_{pj} v_p \qquad \text{and} \qquad \varphi(v_p) = \sum_{i=1}^{m} \beta_{ip} w_i$$

so that $M_{\mathcal{D}}^{\mathcal{B}}(\psi) = (\alpha_{pj})$ and $M_{\mathcal{B}}^{\mathcal{E}}(\varphi) = (\beta_{ip})$.

Using these coefficients we can find an expression for the γ's in terms of the α's and β's as follows:

$$\varphi \circ \psi(u_j) = \varphi\left(\sum_{p=1}^{n} \alpha_{pj} v_p \right)$$

$$= \sum_{p=1}^{n} \alpha_{pj} \varphi(v_p)$$

$$= \sum_{p=1}^{n} \alpha_{pj} \sum_{i=1}^{m} \beta_{ip} w_i$$

$$= \sum_{p=1}^{n} \sum_{i=1}^{m} \alpha_{pj} \beta_{ip} w_i.$$

By interchanging the order of summation in the above double sum we see that γ_{ij}, which is the coefficient of w_i in the above expression, is

$$\gamma_{ij} = \sum_{p=1}^{n} \alpha_{pj} \beta_{ip}.$$

Computing the product of the matrices for φ and ψ (in that order) we obtain

$$(\beta_{ij})(\alpha_{ij}) = (\delta_{ij}), \quad \text{where} \quad \delta_{ij} = \sum_{p=1}^{m} \beta_{ip} \alpha_{pj}.$$

By comparing the two sums above and using the commutativity of field multiplication, we see that for all i and j, $\gamma_{ij} = \delta_{ij}$. This computation proves the following result:

Theorem 12. With notations as above, $M_{\mathcal{D}}^{\mathcal{E}}(\varphi \circ \psi) = M_{\mathcal{B}}^{\mathcal{E}}(\varphi) M_{\mathcal{D}}^{\mathcal{B}}(\psi)$, i.e., with respect to a compatible choice of bases, the product of the matrices representing the linear transformations φ and ψ is the matrix representing the composite linear transformation $\varphi \circ \psi$.

Corollary 13. Matrix multiplication is associative and distributive (whenever the dimensions are such as to make products defined). An $n \times n$ matrix A is nonsingular if and only if it is invertible.

Proof: Let A, B and C be matrices such that the products $(AB)C$ and $A(BC)$ are defined, and let S, T and R denote the associated linear transformations. By Theorem 12, the linear transformation corresponding to AB is the composite $S \circ T$ so the linear transformation corresponding to $(AB)C$ is the composite $(S \circ T) \circ R$. Similarly, the linear transformation corresponding to $A(BC)$ is the composite $S \circ (T \circ R)$. Since function composition is associative, these two linear transformations are the same, and so $(AB)C = A(BC)$ by Theorem 10. The distributivity is proved similarly. Note also that it is possible to prove these results by straightforward (albeit tedious) calculations with matrices.

If A is invertible, then $Ax = 0$ implies $x = A^{-1}Ax = A^{-1}0 = 0$, so A is nonsingular. Conversely, if A is nonsingular, fix bases $\mathcal{B}$, $\mathcal{E}$ for V and let φ be the linear transformation of V to itself represented by A with respect to these bases. By Corollary 9, φ is an isomorphism of V to itself, hence has an inverse, φ^{-1}. Let B be the matrix representing φ^{-1} with respect to the bases $\mathcal{E}$, $\mathcal{B}$ (note the order). Then $AB = M_{\mathcal{B}}^{\mathcal{E}}(\varphi) M_{\mathcal{E}}^{\mathcal{B}}(\varphi^{-1}) = M_{\mathcal{E}}^{\mathcal{E}}(\varphi \circ \varphi^{-1}) = M_{\mathcal{E}}^{\mathcal{E}}(1) = I$. Similarly, $BA = I$ so B is the inverse of A.

Corollary 14.
 (1) If $\mathcal{B}$ is a basis of the n-dimensional space V, the map $\varphi \mapsto M_{\mathcal{B}}^{\mathcal{B}}(\varphi)$ is a ring and a vector space isomorphism of $\mathrm{Hom}_F(V, V)$ onto the space $M_n(F)$ of $n \times n$ matrices with coefficients in F.
 (2) $GL(V) \cong GL_n(F)$ where $\dim V = n$. In particular, if F is a finite field the order of the finite group $GL_n(F)$ (which equals $|GL(V)|$) is given by the formula at the end of Section 1.

Proof: (1) We have already seen in Theorem 10 that this map is an isomorphism of vector spaces over F. Corollary 13 shows that $M_n(F)$ is a ring under matrix multiplication, and then Theorem 12 shows that multiplication is preserved under this map, hence it is also a ring isomorphism.
 (2) This is immediate from (1) since a ring isomorphism sends units to units.

Definition. If A is any $m \times n$ matrix with entries from F, the *row rank* (respectively, *column rank*) of A is the maximal number of linearly independent rows (respectively,

Chap. 11 Vector Spaces

columns) of A (where the rows or columns of A are considered as vectors in affine n-space, m-space, respectively).

The relation between the rank of a matrix and the rank of the associated linear transformation is the following: the rank of φ as a linear transformation equals the column rank of the matrix $M_{\mathcal{B}}^{\mathcal{E}}(\varphi)$ (cf. the exercises). We shall also see that the row rank and the column rank of any matrix are the same.

We now consider the relation of two matrices associated to the same linear transformation of a vector space to itself but with respect to two different choices of bases (cf. the exercises for the general statement regarding a linear transformation from a vector space V to another vector space W).

Definition. Two $n \times n$ matrices A and B are said to be *similar* if there is an invertible (i.e., nonsingular) $n \times n$ matrix P such that $P^{-1}AP = B$. Two linear transformations φ and ψ from a vector space V to itself are said to be *similar* if there is a nonsingular linear transformation ξ from V to V such that $\xi^{-1}\varphi\xi = \psi$.

Suppose $\mathcal{B}$ and $\mathcal{E}$ are two bases of the same vector space V and let $\varphi \in \mathrm{Hom}_F(V, V)$. Let I be the identity map from V to V and let $P = M_{\mathcal{E}}^{\mathcal{B}}(I)$ be its associated matrix (in other words, write the elements of the basis $\mathcal{E}$ in terms of the basis $\mathcal{B}$ — note the order — and use the resulting coordinates for the columns of the matrix P). Note that if $\mathcal{B} \neq \mathcal{E}$ then P is *not* the identity matrix. Then $P^{-1}M_{\mathcal{B}}^{\mathcal{B}}(\varphi)P = M_{\mathcal{E}}^{\mathcal{E}}(\varphi)$. If $[v]_{\mathcal{B}}$ is the $n \times 1$ matrix of coordinates for $v \in V$ with respect to the basis $\mathcal{B}$, and similarly $[v]_{\mathcal{E}}$ is the $n \times 1$ matrix of coordinates for $v \in V$ with respect to the basis $\mathcal{E}$, then $[v]_{\mathcal{B}} = P[v]_{\mathcal{E}}$. The matrix P is called the *transition* or *change of basis* matrix from $\mathcal{B}$ to $\mathcal{E}$ and this similarity action on $M_{\mathcal{B}}^{\mathcal{B}}(\varphi)$ is called a *change of basis*. This shows that the matrices associated to the same linear transformation with respect to two different bases are similar.

Conversely, suppose A and B are $n \times n$ matrices similar by a nonsingular matrix P. Let $\mathcal{B}$ be a basis for the n-dimensional vector space V. Define the linear transformation φ of V (with basis $\mathcal{B}$) to V (again with basis $\mathcal{B}$) by equation (3) using the given matrix A, i.e.,

$$\varphi(v_j) = \sum_{i=1}^{n} \alpha_{ij} v_i.$$

Then $A = M_{\mathcal{B}}^{\mathcal{B}}(\varphi)$ by definition of φ. Define a new basis $\mathcal{E}$ of V by using the ith column of P for the coordinates of w_i in terms of the basis $\mathcal{B}$ (so $P = M_{\mathcal{E}}^{\mathcal{B}}(I)$ by definition). Then $B = P^{-1}AP = P^{-1}M_{\mathcal{B}}^{\mathcal{B}}(\varphi)P = M_{\mathcal{E}}^{\mathcal{E}}(\varphi)$ is the matrix associated to φ with respect to the basis $\mathcal{E}$. This shows that any two similar $n \times n$ matrices arise in this fashion as the matrices representing the same linear transformation with respect to two different choices of bases.

Note that change of basis for a linear transformation from V to itself is the same as conjugation by some element of the group $GL(V)$ of nonsingular linear transformations of V to V. In particular, the relation "similarity" is an equivalence relation whose equivalence classes are the orbits of $GL(V)$ acting by conjugation on $\mathrm{Hom}_F(V, V)$. If

$\varphi \in GL(V)$ (i.e., φ is an invertible linear transformation), then the similarity class of φ is none other than the conjugacy class of φ in the group $GL(V)$.

Example

Let $V = \mathbb{Q}^3$ and let φ be the linear transformation

$$\varphi(x, y, z) = (9x + 4y + 5z, -4x - 3z, -6x - 4y - 2z), \qquad x, y, z \in \mathbb{Q}$$

from V to itself we considered in an earlier example. With respect to the standard basis, $\mathcal{B}$, $b_1 = (1, 0, 0)$, $b_2 = (0, 1, 0)$, $b_3 = (0, 0, 1)$ we saw that the matrix A representing this linear transformation is

$$A = M_{\mathcal{B}}^{\mathcal{B}}(\varphi) = \begin{pmatrix} 9 & 4 & 5 \\ -4 & 0 & -3 \\ -6 & -4 & -2 \end{pmatrix}.$$

Take now the basis, $\mathcal{E}$, $e_1 = (2, -1, -2)$, $e_2 = (1, 0, -1)$, $e_3 = (3, -2, -2)$ for V (we shall see that this is in fact a basis momentarily). Since

$$\varphi(e_1) = \varphi(2, -1, -2) = (4, -2, -4) = 2 \cdot e_1 + 0 \cdot e_2 + 0 \cdot e_3$$
$$\varphi(e_2) = \varphi(1, 0, -1) = (4, -1, -4) = 1 \cdot e_1 + 2 \cdot e_2 + 0 \cdot e_3$$
$$\varphi(e_3) = \varphi(3, -2, -2) = (9, -6, -6) = 0 \cdot e_1 + 0 \cdot e_2 + 3 \cdot e_3,$$

the matrix representing φ with respect to this basis is the matrix

$$B = M_{\mathcal{E}}^{\mathcal{E}}(\varphi) = \begin{pmatrix} 2 & 1 & 0 \\ 0 & 2 & 0 \\ 0 & 0 & 3 \end{pmatrix}.$$

Writing the elements of the basis $\mathcal{E}$ in terms of the basis $\mathcal{B}$ we have

$$e_1 = 2b_1 - b_2 - 2b_3$$
$$e_2 = b_1 - b_3$$
$$e_3 = 3b_1 - 2b_2 - 2b_3$$

so the matrix $P = M_{\mathcal{E}}^{\mathcal{B}}(I) = \begin{pmatrix} 2 & 1 & 3 \\ -1 & 0 & -2 \\ -2 & -1 & -2 \end{pmatrix}$ with inverse $P^{-1} = \begin{pmatrix} -2 & -1 & -2 \\ 2 & 2 & 1 \\ 1 & 0 & 1 \end{pmatrix}$

conjugates A into B, i.e., $P^{-1}AP = B$, as can easily be checked. (Note incidentally that since P is invertible this proves that $\mathcal{E}$ is indeed a basis for V.)

We observe in passing that the matrix B representing this linear transformation φ is much simpler than the matrix A representing φ. The study of the simplest possible matrix representing a given linear transformation (and which basis to choose to realize it) is the study of *canonical forms* considered in the next chapter.

Linear Transformations on Tensor Products of Vector Spaces

For convenience we reiterate Corollaries 18 and 19 of Section 10.4 for the special case of vector spaces.

Proposition 15. Let F be a subfield of the field K. If W is an m-dimensional vector space over F with basis $w_1, \ldots, w_m$, then $K \otimes_F W$ is an m-dimensional vector space over K with basis $1 \otimes w_1, \ldots, 1 \otimes w_m$.

Proposition 16. Let V and W be finite dimensional vector spaces over the field F with bases $v_1, \ldots, v_n$ and $w_1, \ldots, w_m$ respectively. Then $V \otimes_F W$ is a vector space over F of dimension nm with basis $v_i \otimes w_j$, $1 \le i \le n$ and $1 \le j \le m$.

Remark: If v and w are nonzero elements of V and W, respectively, then it follows from the proposition that $v \otimes w$ is a nonzero element of $V \otimes_F W$, because we may always build bases of V and W whose first basis vectors are v, w, respectively. In a tensor product $M \otimes_R N$ of two R-modules where R is not a field it is in general substantially more difficult to determine when the tensor product $m \otimes n$ of two nonzero elements is zero.

Now let V, W, X, Y be finite dimensional vector spaces over F and let

$$\varphi : V \to X \qquad \text{and} \qquad \psi : W \to Y$$

be linear transformations. We compute a matrix of the linear transformation

$$\varphi \otimes \psi : V \otimes W \to X \otimes Y.$$

Let $\mathcal{B}_1 = \{v_1, \ldots, v_n\}$ and $\mathcal{B}_2 = \{w_1, \ldots, w_m\}$ be (ordered) bases of V and W respectively, and let $\mathcal{E}_1 = \{x_1, \ldots, x_r\}$ and $\mathcal{E}_2 = \{y_1, \ldots, y_s\}$ be (ordered) bases of X and Y respectively. Let $\mathcal{B} = \{v_i \otimes w_j\}$ and $\mathcal{E} = \{x_i \otimes y_j\}$ be the bases of $V \otimes W$ and $X \otimes Y$ given by Proposition 16; we shall order these shortly. Suppose

$$\varphi(v_i) = \sum_{p=1}^{r} \alpha_{pi} x_p \qquad \text{and} \qquad \psi(w_j) = \sum_{q=1}^{s} \beta_{qj} y_q.$$

Then

$$
\begin{aligned}
(\varphi \otimes \psi)(v_i \otimes w_j) &= (\varphi(v_i)) \otimes (\psi(w_j)) \\
&= \Big(\sum_{p=1}^{r} \alpha_{pi} x_p \Big) \otimes \Big(\sum_{q=1}^{s} \beta_{qj} y_q \Big) \\
&= \sum_{p=1}^{r} \sum_{q=1}^{s} \alpha_{pi} \beta_{qj} (x_p \otimes y_q).
\end{aligned}
\tag{11.8}
$$

In view of the order of summation in (11.8) we order the basis $\mathcal{E}$ into r ordered sets, with the pth list being $x_p \otimes y_1, x_p \otimes y_2, \ldots, x_p \otimes y_s$, and similarly order the basis $\mathcal{B}$. Then equation (8) determines the column entries for the corresponding matrix of $\varphi \otimes \psi$. The resulting matrix $M_{\mathcal{B}}^{\mathcal{E}}(\varphi \otimes \psi)$ is an $r \times n$ block matrix whose p, q block is the $s \times m$ matrix $\alpha_{p,q} M_{\mathcal{B}_2}^{\mathcal{E}_2}(\psi)$. In other words, the matrix for $\varphi \otimes \psi$ is obtained by taking the matrix for φ and multiplying each entry by the matrix for ψ. Such matrices have a name:

Definition. Let $A = (\alpha_{ij})$ and B be $r \times n$ and $s \times m$ matrices, respectively, with coefficients from any commutative ring. The *Kronecker product* or *tensor product* of A and B, denoted by $A \otimes B$, is the $rs \times nm$ matrix consisting of an $r \times n$ block matrix whose i, j block is the $s \times m$ matrix $\alpha_{ij} B$.

With this terminology we have

Proposition 17. Let $\varphi : V \to X$ and $\psi : W \to Y$ be linear transformations of finite dimensional vector spaces. Then the Kronecker product of matrices representing φ and ψ is a matrix representation of $\varphi \otimes \psi$.

Example

Let $V = X = \mathbb{R}^3$, both with basis v_1, v_2, v_3, and $W = Y = \mathbb{R}^2$, both with basis w_1, w_2. Suppose $\varphi : \mathbb{R}^3 \to \mathbb{R}^3$ is the linear transformation given by $\varphi(av_1 + bv_2 + cv_3) = cv_1 + 2av_2 - 3bv_3$ and $\psi : \mathbb{R}^2 \to \mathbb{R}^2$ is the linear transformation given by $\psi(aw_1 + bw_2) = (a + 3b)w_1 + (4b - 2a)w_2$. With respect to the chosen bases, the matrices for φ and ψ are

$$\begin{pmatrix} 0 & 0 & 1 \\ 2 & 0 & 0 \\ 0 & -3 & 0 \end{pmatrix} \quad \text{and} \quad \begin{pmatrix} 1 & 3 \\ -2 & 4 \end{pmatrix},$$

respectively. Then with respect to the ordered basis

$$\mathcal{B} = \{v_1 \otimes w_1, \ v_1 \otimes w_2, \ v_2 \otimes w_1, \ v_2 \otimes w_2, \ v_3 \otimes w_1, \ v_3 \otimes w_2\}$$

we have

$$M_{\mathcal{B}}^{\mathcal{B}}(\varphi \otimes \psi) = \left(\begin{array}{cc:cc:cc} 0 & 0 & 0 & 0 & 1 & 3 \\ 0 & 0 & 0 & 0 & -2 & 4 \\ \hdashline 2 & 6 & 0 & 0 & 0 & 0 \\ -4 & 8 & 0 & 0 & 0 & 0 \\ \hdashline 0 & 0 & -3 & -9 & 0 & 0 \\ 0 & 0 & 6 & -12 & 0 & 0 \end{array} \right),$$

obtained (as indicated by the dashed lines) by multiplying the 2×2 matrix for ψ successively by the entries in the matrix for φ.

EXERCISES

1. Let V be the collection of polynomials with coefficients in $\mathbb{Q}$ in the variable x of degree at most 5. Determine the transition matrix from the basis $1, x, x^2, \ldots, x^5$ for V to the basis $1, 1 + x, 1 + x + x^2, \ldots, 1 + x + x^2 + x^3 + x^4 + x^5$ for V.

2. Let V be the vector space of the preceding exercise. Let $\varphi = d/dx$ be the linear transformation of V to itself given by usual differentiation of a polynomial with respect to x. Determine the matrix of φ with respect to the two bases for V in the previous exercise.

3. Let V be the collection of polynomials with coefficients in F in the variable x of degree at most n. Determine the transition matrix from the basis $1, x, x^2, \ldots, x^n$ for V to the elements

$$1, x - \lambda, \ldots, (x - \lambda)^{n-1}, (x - \lambda)^n$$

where λ is a fixed element of F. Conclude that these elements are a basis for V.

4. Let φ be the linear transformation of $\mathbb{R}^2$ to itself given by rotation counterclockwise around the origin through an angle θ. Show that the matrix of φ with respect to the standard basis for $\mathbb{R}^2$ is $\begin{pmatrix} \cos\theta & -\sin\theta \\ \sin\theta & \cos\theta \end{pmatrix}$.

5. Show that the $m \times n$ matrix A is nonsingular if and only if the linear transformation φ is a nonsingular linear transformation from the n-dimensional space V to the m-dimensional space W, where $A = M_{\mathcal{B}}^{\mathcal{E}}(\varphi)$, regardless of the choice of bases $\mathcal{B}$ and $\mathcal{E}$.

6. Prove if $\varphi \in \text{Hom}_F(F^n, F^m)$, and $\mathcal{B}, \mathcal{E}$ are the natural bases of F^n, F^m respectively, then the range of φ equals the span of the set of columns of $M_{\mathcal{B}}^{\mathcal{E}}(\varphi)$. Deduce that the rank of φ (as a linear transformation) equals the column rank of $M_{\mathcal{B}}^{\mathcal{E}}(\varphi)$.

7. Prove that any two similar matrices have the same row rank and the same column rank.

8. Let V be an n-dimensional vector space over F and let φ be a linear transformation of the vector space V to itself.
   (a) Prove that if V has a basis consisting of eigenvectors for φ (cf. Exercise 8 of Section 1) then the matrix representing φ with respect to this basis (for both domain and range) is diagonal with the eigenvalues as diagonal entries.
   (b) If A is the $n \times n$ matrix representing φ with respect to a given basis for V (for both domain and range) prove that A is similar to a diagonal matrix if and only if V has a basis of eigenvectors for φ.

9. If W is a subspace of the vector space V stable under the linear transformation φ (i.e., $\varphi(W) \subseteq W$), show that φ induces linear transformations $\varphi|_W$ on W and $\bar{\varphi}$ on the quotient vector space V/W. If $\varphi|_W$ and $\bar{\varphi}$ are nonsingular prove φ is nonsingular. Prove the converse holds if V has finite dimension and give a counterexample with V infinite dimensional.

10. Let V be an n-dimensional vector space and let φ be a linear transformation of V to itself. Suppose W is a subspace of V of dimension m that is stable under φ.
    (a) Prove that there is a basis for V with respect to which the matrix for φ is of the form

    $$\begin{pmatrix} A & B \\ 0 & C \end{pmatrix}$$

    where A is an $m \times m$ matrix, B is an $m \times (n-m)$ matrix and C is an $(n-m) \times (n-m)$ matrix (such a matrix is called *block upper triangular*).
    (b) Prove that if there is a subspace W' invariant under φ so that $V = W \oplus W'$ decomposes as a direct sum then the bases for W and W' give a basis for V with respect to which the matrix for φ is *block diagonal*:

    $$\begin{pmatrix} A & 0 \\ 0 & C \end{pmatrix}$$

    where A is an $m \times m$ matrix and C is an $(n-m) \times (n-m)$ matrix.
    (c) Prove conversely that if there is a basis for V with respect to which φ is block diagonal as in (b) then there are φ-invariant subspaces W and W' of dimensions m and $n-m$, respectively, with $V = W \oplus W'$.

11. Let φ be a linear transformation from the finite dimensional vector space V to itself such that $\varphi^2 = \varphi$.
    (a) Prove that image $\varphi \cap \ker \varphi = 0$.
    (b) Prove that $V = $ image $\varphi \oplus \ker \varphi$.
    (c) Prove that there is a basis of V such that the matrix of φ with respect to this basis is a diagonal matrix whose entries are all 0 or 1.

    A linear transformation φ satisfying $\varphi^2 = \varphi$ is called an *idempotent* linear transformation. This exercise proves that idempotent linear transformations are simply projections onto some subspace.

12. Let $V = \mathbb{R}^2$, $v_1 = (1,0)$, $v_2 = (0,1)$, so that v_1, v_2 are a basis for V. Let φ be the linear transformation of V to itself whose matrix with respect to this basis is $\begin{pmatrix} 2 & 1 \\ 0 & 2 \end{pmatrix}$. Prove that if W is the subspace generated by v_1 then W is stable under the action of φ. Prove that there is no subspace W' invariant under φ so that $V = W \oplus W'$.

13. Let V be a vector space of dimension n and let W be a vector space of dimension m over a field F. Suppose A is the $m \times n$ matrix representing a linear transformation φ from V to W with respect to the bases $\mathcal{B}_1$ for V and $\mathcal{E}_1$ for W. Suppose similarly that B is the $m \times n$ matrix representing φ with respect to the bases $\mathcal{B}_2$ for V and $\mathcal{E}_2$ for W. Let $P = M_{\mathcal{B}_2}^{\mathcal{B}_1}(I)$ where I denotes the identity map from V to V, and let $Q = M_{\mathcal{E}_2}^{\mathcal{E}_1}(I)$ where I denotes the identity map from W to W. Prove that $Q^{-1} = M_{\mathcal{E}_1}^{\mathcal{E}_2}(I)$ and that $Q^{-1}AP = B$, giving the general relation between matrices representing the same linear transformation but with respect to different choices of bases.

The following exercises recall the *Gauss–Jordan* elimination process. This is one of the fastest computational methods for the solution of a number of problems involving vector spaces — solving systems of linear equations, determining inverses of matrices, computing determinants, determining the span of a set of vectors, determining linear independence of a set of vectors etc.

Consider the system of m linear equations

$$
\begin{aligned}
a_{11}x_1 + a_{12}x_2 + \ldots + a_{1n}x_n &= c_1 \\
a_{21}x_1 + a_{22}x_2 + \ldots + a_{2n}x_n &= c_2 \\
&\vdots \\
a_{m1}x_1 + a_{m2}x_2 + \ldots + a_{mn}x_n &= c_m
\end{aligned}
\tag{11.4}
$$

in the n unknowns $x_1, x_2, \ldots, x_n$ where $a_{ij}, c_i, i = 1, 2, \ldots, m, j = 1, 2, \ldots, n$ are elements of the field F. Associated to this system is the *coefficient matrix*:

$$
A = \begin{pmatrix}
a_{11} & a_{12} & \cdots & a_{1n} \\
a_{21} & a_{22} & \cdots & a_{2n} \\
\vdots & \vdots & & \vdots \\
a_{m1} & a_{m2} & \cdots & a_{mn}
\end{pmatrix}
$$

and the *augmented matrix*:

$$
(A \mid C) = \left(
\begin{array}{cccc|c}
a_{11} & a_{12} & \cdots & a_{1n} & c_1 \\
a_{21} & a_{22} & \cdots & a_{2n} & c_2 \\
\vdots & \vdots & & \vdots & \vdots \\
a_{m1} & a_{m2} & \cdots & a_{mn} & c_m
\end{array}
\right)
$$

(the term *augmented* refers to the presence of the column matrix $C = (c_i)$ in addition to the coefficient matrix $A = (a_{ij})$). The set of solutions in F of this system of equations is not altered if we perform any of the following three operations:

(1) interchange any two equations
(2) add a multiple of one equation to another
(3) multiply any equation by a nonzero element from F,

which correspond to the following three *elementary row operations* on the augmented matrix:

(1) interchange any two rows
(2) add a multiple of one row to another
(3) multiply any row by a unit in F, i.e., by any nonzero element in F.

If a matrix A can be transformed into a matrix C by a series of elementary row operations then A is said to be *row reduced* to C.

424

14. Prove that if A can be row reduced to C then C can be row reduced to A. Prove that the relation "$A \sim C$ if and only if A can be row reduced to C" is an equivalence relation. [Observe that the elementary row operations are reversible.]

Matrices lying in the same equivalence class under this equivalence relation are said to be *row equivalent*.

15. Prove that the row rank of two row equivalent matrices is the same. [It suffices to prove this for two matrices differing by an elementary row operation.]

An $m \times n$ matrix is said to be in *reduced row echelon form* if

(a) the first nonzero entry a_{ij_i} in row i is 1 and all other entries in the corresponding j_i^{th} column are zero, and

(b) $j_1 < j_2 < \ldots < j_r$ where r is the number of nonzero rows, i.e., the number of initial zeros in each row is strictly increasing (hence the term *echelon*).

An augmented matrix $(A \mid C)$ is said to be in reduced row echelon form if its coefficient matrix A is in reduced row echelon form. For example, the following two matrices are in reduced row echelon form:

$$\left(\begin{array}{cccccc|c} 1 & 0 & 5 & 7 & 0 & 3 & 0 \\ 0 & 1 & -1 & 1 & 0 & -4 & -1 \\ 0 & 0 & 0 & 0 & 1 & 6 & 1 \\ 0 & 0 & 0 & 0 & 0 & 0 & 0 \end{array} \right) \qquad \left(\begin{array}{cccc|c} 0 & 1 & -1 & 0 & 0 \\ 0 & 0 & 0 & 1 & 2 \\ 0 & 0 & 0 & 0 & -3 \end{array} \right)$$

(with $j_1 = 1$, $j_2 = 2$, $j_3 = 5$ for the first matrix and $j_1 = 2$, $j_2 = 4$ for the second matrix). The first nonzero entry in any given row of the coefficient matrix of a reduced row echelon augmented matrix (in position (i, j_i) by definition) is sometimes referred to as a *pivotal* element (so the pivotal elements in the first matrix are in positions $(1,1)$, $(2,2)$ and $(3,5)$ and the pivotal elements in the second matrix are in positions $(1,2)$ and $(2,4)$). The columns containing pivotal elements will be called *pivotal* columns and the columns of the coefficient matrix not containing pivotal elements will be called *nonpivotal*.

16. Prove by induction that any augmented matrix can be put in reduced row echelon form by a series of elementary row operations.

17. Let A and C be two matrices in reduced row echelon form. Prove that if A and C are row equivalent then $A = C$.

18. Prove that the row rank of a matrix in reduced row echelon form is the number of nonzero rows.

19. Prove that the reduced row echelon forms of the matrices

$$\left(\begin{array}{cccccc|c} 1 & 1 & 4 & 8 & 0 & -1 & -1 \\ 1 & 2 & 3 & 9 & 0 & -5 & -2 \\ 0 & -2 & 2 & -2 & 1 & 14 & 3 \\ 1 & 4 & 1 & 11 & 0 & -13 & -4 \end{array} \right) \qquad \left(\begin{array}{cccc|c} 0 & -3 & 3 & 1 & 5 \\ 0 & 1 & -1 & 0 & 0 \\ 0 & 2 & -2 & 0 & -3 \end{array} \right)$$

are the two matrices preceding Exercise 16.

The point of the reduced row echelon form is that the corresponding system of linear equations is in a particularly simple form, from which the solutions to the system $AX = C$ in (4) can be determined immediately:

20. (*Solving Systems of Linear Equations*) Let $(A' \mid C')$ be the reduced row echelon form of the augmented matrix $(A \mid C)$. The number of zero rows of A' is clearly at least as great as the number of zero rows of $(A' \mid C')$.

(a) Prove that if the number of zero rows of A' is strictly larger than the number of zero rows of $(A' \mid C')$ then there are no solutions to $AX = C$.

By (a) we may assume that A' and $(A' \mid C')$ have the same number, r, of nonzero rows (so $n \geq r$).

(b) Prove that if $r = n$ then there is precisely one solution to the system of equations $AX = C$.

(c) Prove that if $r < n$ then there are infinitely many solutions to the system of equations $AX = C$. Prove in fact that the values of the $n - r$ variables corresponding to the nonpivotal columns of $(A' \mid C')$ can be chosen arbitrarily and that the remaining r variables corresponding to the pivotal columns of $(A' \mid C')$ are then determined uniquely.

21. Determine the solutions of the following systems of equations:

(a)
$$\begin{aligned} -3x + 3y + z &= 5 \\ x - y &= 0 \\ 2x - 2y &= -3 \end{aligned}$$

(b)
$$\begin{aligned} x - 2y + z &= 5 \\ x - 4y + 6z &= 10 \\ 4x - 11y + 11z &= 12 \end{aligned}$$

(c)
$$\begin{aligned} x - 2y + z &= 5 \\ y - 2z &= 17 \\ 2x - 3y &= 27 \end{aligned}$$

(d)
$$\begin{aligned} x + y - 3z + 2u &= 2 \\ 3x - 2y + 5z + u &= 1 \\ 6x + y - 4z + 3u &= 7 \\ 2x + 2y - 6z &= 4 \end{aligned}$$

(e)
$$\begin{aligned} x + y + 4z + 8u - w &= -1 \\ x + 2y + 3z + 9u - 5w &= -2 \\ -2y + 2z - 2u + v + 14w &= 3 \\ x + 4y + z + 11u - 13w &= -4 \end{aligned}$$

22. Suppose A and B are two row equivalent $m \times n$ matrices.

(a) Prove that the set

$$X = \begin{pmatrix} x_1 \\ x_2 \\ \vdots \\ x_n \end{pmatrix}$$

of solutions to the homogeneous linear equations $AX = 0$ as in equation (4) above are the same as the set of solutions to the homogeneous linear equations $BX = 0$. [It suffices to prove this for two matrices differing by an elementary row operation.]

(b) Prove that any linear dependence relation satisfied by the columns of A viewed as vectors in F^m is also satisfied by the columns of B.

(c) Conclude from (b) that the number of linearly independent columns of A is the same as the number of linearly independent columns of B.

23. Let A' be a matrix in reduced row echelon form.

(a) Prove that the nonzero rows of A' are linearly independent. Prove that the pivotal columns of A' are linearly independent and that the nonpivotal columns of A' are linearly dependent on the pivotal columns. (Note the role the pivotal elements play.)

(b) Prove that the number of linearly independent columns of a matrix in reduced row echelon form is the same as the number of linearly independent rows, i.e., the row rank and the column rank of such a matrix are the same.

24. Use the previous two exercises and Exercise 15 above to prove in general that the row rank and the column rank of a matrix are the same.

25. (*Computing Inverses of Matrices*) Let A be an $n \times n$ matrix.

(a) Show that A has an inverse matrix B with columns $B_1, B_2, \ldots, B_n$ if and only if the systems of equations:

$$AB_1 = \begin{pmatrix} 1 \\ 0 \\ \vdots \\ 0 \\ 0 \end{pmatrix}, \quad AB_2 = \begin{pmatrix} 0 \\ 1 \\ \vdots \\ 0 \\ 0 \end{pmatrix}, \quad \ldots, \quad AB_n = \begin{pmatrix} 0 \\ 0 \\ \vdots \\ 0 \\ 1 \end{pmatrix}$$

have solutions.

(b) Prove that A has an inverse if and only if A is row equivalent to the $n \times n$ identity matrix.

(c) Prove that A has an inverse B if and only if the augmented matrix $(A \mid I)$ can be row reduced to the augmented matrix $(I \mid B)$ where I is the $n \times n$ identity matrix.

26. Determine the inverses of the following matrices using row reduction:

$$A = \begin{pmatrix} -7 & -1 & -4 \\ 7 & 1 & 3 \\ 1 & 0 & 0 \end{pmatrix} \qquad B = \begin{pmatrix} 1 & 1 & 0 & 2 \\ 0 & 2 & 1 & -1 \\ 0 & 2 & 0 & 0 \\ -1 & 1 & 1 & 0 \end{pmatrix}.$$

27. (*Computing Spans, Linear Independence and Linear Dependencies in Vector Spaces*) Let V be an m-dimensional vector space with basis $e_1, e_2, \ldots, e_m$ and let $v_1, v_2, \ldots, v_n$ be vectors in V. Let A be the $m \times n$ matrix whose columns are the coordinates of the vectors v_i (with respect to the basis $e_1, e_2, \ldots, e_m$) and let A' be the reduced row echelon form of A.

(a) Let B be any matrix row equivalent to A. Let $w_1, w_2, \ldots, w_n$ be the vectors whose coordinates (with respect to the basis $e_1, e_2, \ldots, e_m$) are the columns of B. Prove that any linear relation

$$x_1 v_1 + x_2 v_2 + \ldots + x_n v_n = 0 \tag{11.5}$$

satisfied by $v_1, v_2, \ldots, v_n$ is also satisfied when v_i is replaced by w_i, $i = 1, 2, \ldots, n$.

(b) Prove that the vectors whose coordinates are given by the pivotal columns of A' are linearly independent and that the vectors whose coordinates are given by the nonpivotal columns of A' are linearly dependent on these.

(c) (*Determining Linear Independence of Vectors*) Prove that the vectors $v_1, v_2, \ldots, v_n$ are linearly independent if and only if A' has n nonzero rows (i.e., has rank n).

(d) (*Determining Linear Dependencies of Vectors*) By (c), the vectors $v_1, v_2, \ldots, v_n$ are linearly dependent if and only if A' has nonpivotal columns. The solutions to (5) defining linear dependence relations among $v_1, v_2, \ldots, v_n$ are given by the linear equations defined by A'. Show that each of the variables $x_1, x_2, \ldots, x_n$ in (5) corresponding to the nonpivotal columns of A' can be prescribed arbitrarily and the values of the remaining variables are then uniquely determined to give a linear dependence relation among $v_1, v_2, \ldots, v_n$ as in (5).

(e) (*Determining the Span of a Set of Vectors*) Prove that the subspace W spanned by $v_1, v_2, \ldots, v_n$ has dimension r where r is the number of nonzero rows of A' and that a basis for W is given by the original vectors v_{j_i} ($i = 1, 2, \ldots, r$) corresponding to the pivotal columns of A'.

28. Let $V = \mathbb{R}^5$ with the standard basis and consider the vectors

$$v_1 = (1, 1, 3, -2, 3), \quad v_2 = (0, 1, 0, -1, 0), \quad v_3 = (2, 3, 6, -5, 6)$$

$$v_4 = (0, 3, 1, -3, 1), \quad v_5 = (2, -1, -1, -1, -1).$$

(a) Show that the reduced row echelon form of the matrix

$$A = \begin{pmatrix} 1 & 0 & 2 & 0 & 2 \\ 1 & 1 & 3 & 3 & -1 \\ 3 & 0 & 6 & 1 & -1 \\ -2 & -1 & -5 & -3 & -1 \\ 3 & 0 & 6 & 1 & -1 \end{pmatrix}$$

whose columns are the coordinates of v_1, v_2, v_3, v_4, v_5 is the matrix

$$A' = \begin{pmatrix} 1 & 0 & 2 & 0 & 2 \\ 0 & 1 & 1 & 0 & 18 \\ 0 & 0 & 0 & 1 & -7 \\ 0 & 0 & 0 & 0 & 0 \\ 0 & 0 & 0 & 0 & 0 \end{pmatrix}$$

where the 1$^\text{st}$, 2$^\text{nd}$ and 4$^\text{th}$ columns are pivotal and the remaining two are nonpivotal.

(b) Conclude that these vectors are linearly dependent, that the subspace W spanned by v_1, v_2, v_3, v_4, v_5 is 3-dimensional and that the vectors

$$v_1 = (1, 1, 3, -2, 3), \quad v_2 = (0, 1, 0, -1, 0) \quad \text{and} \quad v_4 = (0, 3, 1, -3, 1)$$

are a basis for W.

(c) Conclude from (a) that the coefficients x_1, x_2, x_3, x_4, x_5 of any linear relation

$$x_1 v_1 + x_2 v_2 + x_3 v_3 + x_4 v_4 + x_5 v_5 = 0$$

satisfied by v_1, v_2, v_3, v_4, v_5 are given by the equations

$$\begin{aligned} x_1 \quad + 2x_3 \quad + \ 2x_5 &= 0 \\ x_2 + \ x_3 \quad + 18x_5 &= 0 \\ x_4 - \ 7x_5 &= 0. \end{aligned}$$

Deduce that the 3$^\text{rd}$ and 5$^\text{th}$ variables, namely x_3 and x_5, corresponding to the nonpivotal columns of A', can be prescribed arbitrarily and the remaining variables are then uniquely determined as:

$$x_1 = -2x_3 - 2x_5$$

$$x_2 = -x_3 - 18x_5$$

$$x_4 = 7x_5$$

to give all the linear dependence relations satisfied by v_1, v_2, v_3, v_4, v_5. In particular show that

$$-2v_1 - v_2 + v_3 = 0$$

and

$$-2v_1 - 18v_2 + 7v_4 + v_5 = 0$$

corresponding to ($x_3 = 1$, $x_5 = 0$) and ($x_3 = 0$, $x_5 = 1$), respectively.

29. For each exercise below, determine whether the given vectors in $\mathbb{R}^4$ are linearly independent. If they are linearly dependent, determine an explicit linear dependence among them.
  (a) $(1, -4, 3, 0)$, $(0, -1, 4, -3)$, $(1, -1, 1, -1)$, $(2, 2, -1, -3)$.
  (b) $(1, -2, 4, 1)$, $(2, -3, 9, -1)$, $(1, 0, 6, -5)$, $(2, -5, 7, 5)$.
  (c) $(1, -2, 0, 1)$, $(2, -2, 0, 0)$, $(-1, 3, 0, -2)$, $(-2, 1, 0, 1)$.
  (d) $(0, 1, 1, 0)$, $(1, 0, 1, 1)$, $(2, 2, 2, 0)$, $(0, -1, 1, 1)$.

30. For each exercise below, determine the subspace spanned in $\mathbb{R}^4$ by the given vectors and give a basis for this subspace.
  (a) $(1, -2, 5, 3)$, $(2, 3, 1, -4)$, $(3, 8, -3, -5)$.
  (b) $(2, -5, 3, 0)$, $(0, -2, 5, -3)$, $(1, -1, 1, -1)$, $(-3, 2, -1, 2)$.
  (c) $(1, -2, 0, 1)$, $(2, -2, 0, 0)$, $(-1, 3, 0, -2)$, $(-2, 1, 0, 1)$.
  (d) $(1, 1, 0, -1)$, $(1, 2, 3, 0)$, $(2, 3, 3, -1)$, $(1, 2, 2, -2)$, $(2, 3, 2, -3)$, $(1, 3, 4, -3)$.

31. (*Computing the Image and Kernel of a Linear Transformation*) Let V be an n-dimensional vector space with basis $e_1, e_2, \ldots, e_n$ and let W be an m-dimensional vector space with basis $f_1, f_2, \ldots, f_m$. Let φ be a linear transformation from V to W and let A be the corresponding $m \times n$ matrix with respect to these bases: $A = (a_{ij})$ where

$$\varphi(e_j) = \sum_{i=1}^{m} a_{ij} f_i, \qquad j = 1, 2, \ldots, n,$$

i.e., the columns of A are the coordinates of the vectors $\varphi(e_1)$, $\varphi(e_2)$, $\ldots$, $\varphi(e_n)$ with respect to the basis $f_1, f_2, \ldots, f_m$ of W. Let A' be the reduced row echelon form of A.
  (a) (*Determining the Image of a Linear Transformation*) Prove that the image $\varphi(V)$ of V under φ has dimension r where r is the number of nonzero rows of A' and that a basis for $\varphi(V)$ is given by the vectors $\varphi(e_{j_i})$ ($i = 1, 2, \ldots, r$), i.e., the columns of A corresponding to the pivotal columns of A' give the coordinates of a basis for the image of φ.
  (b) (*Determining the Kernel of a Linear Transformation*) The elements in the kernel of φ are the vectors in V whose coordinates $(x_1, x_2, \ldots, x_n)$ with respect to the basis $e_1, e_2, \ldots, e_n$ satisfy the equation

$$A \begin{pmatrix} x_1 \\ x_2 \\ \vdots \\ x_n \end{pmatrix} = 0,$$

and the solutions $x_1, x_2, \ldots, x_n$ to this system of linear equations are determined by the matrix A'.
  (i) Prove that φ is injective if and only if A' has n nonzero rows (i.e., has rank n).
  (ii) By (i), the kernel of φ is nontrivial if and only if A' has nonpivotal columns. Show that each of the variables $x_1, x_2, \ldots, x_n$ above corresponding to the nonpivotal columns of A' can be prescribed arbitrarily and the values of the remaining variables are then

uniquely determined to give an element $x_1 e_1 + x_2 e_2 + \ldots + x_n e_n$ in the kernel of φ. In particular, show that the coordinates of a basis for the kernel are obtained by successively setting one nonpivotal variable equal to 1 and all other nonpivotal variables to 0 and solving for the remaining pivotal variables. Conclude that the kernel of φ has dimension $n - r$ where r is the rank of A.

32. Let $V = \mathbb{R}^5$ and $W = \mathbb{R}^4$ with the standard bases. Let φ be the linear transformation $\varphi : V \to W$ defined by

$$\varphi(x, y, z, u, v) = (x + 2y + 3z + 4u + 4v, -2x - 4y + 2v, x + 2y + u - 2v, x + 2y - v).$$

(a) Prove that the matrix A corresponding to φ and these bases is

$$A = \begin{pmatrix} 1 & 2 & 3 & 4 & 4 \\ -2 & -4 & 0 & 0 & 2 \\ 1 & 2 & 0 & 1 & -2 \\ 1 & 2 & 0 & 0 & -1 \end{pmatrix}$$

and that the reduced row echelon matrix A' row equivalent to A is

$$A' = \begin{pmatrix} 1 & 2 & 0 & 0 & -1 \\ 0 & 0 & 1 & 0 & 3 \\ 0 & 0 & 0 & 1 & -1 \\ 0 & 0 & 0 & 0 & 0 \end{pmatrix}$$

where the 1st, 3rd and 4th columns are pivotal and the remaining two are nonpivotal.

(b) Conclude that the image of φ is 3-dimensional and that the image of the 1st, 3rd and 4th basis elements of V, namely, $(1, -2, 1, 1)$, $(3, 0, 0, 0)$ and $(4, 0, 1, 0)$ give a basis for the image $\varphi(V)$ of V.

(c) Conclude from (a) that the elements in the kernel of φ are the vectors (x, y, z, u, v) satisfying the equations

$$
\begin{aligned}
x + 2y \quad &- \quad v = 0 \\
z \quad &+ 3v = 0 \\
u &- \quad v = 0.
\end{aligned}
$$

Deduce that the 2nd and 5th variables, namely y and v, corresponding to the nonpivotal columns of A' can be prescribed arbitrarily and the remaining variables are then uniquely determined as

$$
\begin{aligned}
x &= -2y + v \\
z &= -3v \\
u &= v.
\end{aligned}
$$

Show that $(-2, 1, 0, 0, 0)$ and $(1, 0, -3, 1, 1)$ give a basis for the 2-dimensional kernel of φ, corresponding to $(y = 1, v = 0)$ and $(y = 0, v = 1)$, respectively.

33. Let φ be the linear transformation from $\mathbb{R}^4$ to itself defined by the matrix

$$A = \begin{pmatrix} 1 & -1 & 0 & 3 \\ -1 & 2 & 1 & -1 \\ -1 & 1 & 0 & -3 \\ 1 & -2 & -1 & 1 \end{pmatrix}$$

with respect to the standard basis for $\mathbb{R}^4$. Determine a basis for the image and for the kernel of φ.

430

34. Let φ be the linear transformation $\varphi : \mathbb{R}^4 \to \mathbb{R}^2$ such that

$$\varphi((1, 0, 0, 0)) = (1, -1) \qquad \varphi((1, -1, 0, 0)) = (0, 0)$$
$$\varphi((1, -1, 1, 0)) = (1, -1) \qquad \varphi((1, -1, 1, -1)) = (0, 0).$$

Determine a basis for the image and for the kernel of φ.

35. Let V be the set of all 2×2 matrices with real entries and let $\varphi : V \to \mathbb{R}$ be the map defined by sending a matrix $A \in V$ to the sum of the diagonal entries of A (the *trace* of A).

   (a) Show that

$$\begin{pmatrix} 1 & 0 \\ 0 & 0 \end{pmatrix}, \quad \begin{pmatrix} 0 & 1 \\ 0 & 0 \end{pmatrix}, \quad \begin{pmatrix} 0 & 0 \\ 1 & 0 \end{pmatrix}, \quad \begin{pmatrix} 0 & 0 \\ 0 & 1 \end{pmatrix}$$

   is a basis for V.

   (b) Prove that φ is a linear transformation and determine the matrix of φ with respect to the basis in (a) for V. Determine the dimension of and a basis for the kernel of φ.

36. Let V be the 6-dimensional vector space over $\mathbb{Q}$ consisting of the polynomials in the variable x of degree at most 5. Let φ be the map of V to itself defined by $\varphi(f) = x^2 f'' - 6xf' + 12f$, where f'' denotes the usual second derivative (with respect to x) of the polynomial $f \in V$ and f' similarly denotes the usual first derivative.

   (a) Prove that φ is a linear transformation of V to itself.

   (b) Determine a basis for the image and for the kernel of φ.

37. Let V be the 7-dimensional vector space over the field F consisting of the polynomials in the variable x of degree at most 6. Let φ be the linear transformation of V to itself defined by $\varphi(f) = f'$, where f' denotes the usual derivative (with respect to x) of the polynomial $f \in V$. For each of the fields below, determine a basis for the image and for the kernel of φ:

   (a) $F = \mathbb{R}$

   (b) $F = \mathbb{F}_2$, the finite field of 2 elements (note that, for example, $(x^2)' = 2x = 0$ over this field)

   (c) $F = \mathbb{F}_3$

   (d) $F = \mathbb{F}_5$.

38. Let A and B be square matrices. Prove that the trace of their Kronecker product is the product of their traces: $\text{tr}\,(A \otimes B) = \text{tr}\,(A)\,\text{tr}\,(B)$. (Recall that the trace of a square matrix is the sum of its diagonal entries.)

39. Let F be a subfield of K and let $\psi : V \to W$ be a linear transformation of finite dimensional vector spaces over F.

   (a) Prove that $1 \otimes \psi$ is a K–linear transformation from the vector spaces $K \otimes_F V$ to $K \otimes_F W$ over K. (Here 1 denotes the identity map from K to itself.)

   (b) Let $\mathcal{B} = \{v_1, \ldots, v_n\}$ and $\mathcal{E} = \{w_1, \ldots, w_m\}$ be bases of V and W respectively. Prove that the matrix of $1 \otimes \psi$ with respect to the bases $\{1 \otimes v_1, \ldots, 1 \otimes v_n\}$ and $\{1 \otimes w_1, \ldots, 1 \otimes w_m\}$ is the same as the matrix of ψ with respect to $\mathcal{B}$ and $\mathcal{E}$.

11.3 DUAL VECTOR SPACES

Definition.

(1) For V any vector space over F let $V^* = \text{Hom}_F(V, F)$ be the space of linear transformations from V to F, called the *dual space* of V. Elements of V^* are called *linear functionals*.

(2) If $\mathcal{B} = \{v_1, v_2, \ldots, v_n\}$ is a basis of the finite dimensional space V, define $v_i^* \in V^*$ for each $i \in \{1, 2, \ldots, n\}$ by its action on the basis $\mathcal{B}$:

$$v_i^*(v_j) = \begin{cases} 1, & \text{if } i = j \\ 0, & \text{if } i \neq j \end{cases} \qquad 1 \leq j \leq n. \qquad (11.6)$$

Proposition 18. With notations as above, $\{v_1^*, v_2^*, \ldots, v_n^*\}$ is a basis of V^*. In particular, if V is finite dimensional then V^* has the same dimension as V.

Proof: Observe that since V is finite dimensional, $\dim V^* = \dim \operatorname{Hom}_F(V, F) = \dim V = n$ (Corollary 11), so since there are n of the v_i^*'s it suffices to prove that they are linearly independent. If

$$\alpha_1 v_1^* + \alpha_2 v_2^* + \cdots + \alpha_n v_n^* = 0 \quad \text{in } \operatorname{Hom}_F(V, F),$$

then applying this element to v_i and using equation (6) above we obtain $\alpha_i = 0$. Since i is arbitrary these elements are linearly independent.

Definition. The basis $\{v_1^*, v_2^*, \ldots, v_n^*\}$ of V^* is called the *dual basis* to $\{v_1, v_2, \ldots, v_n\}$.

The exercises later show that if V is infinite dimensional it is always true that $\dim V < \dim V^*$. For spaces of arbitrary dimension the space V^* is the "algebraic" dual space to V. If V has some additional structure, for example a continuous structure (i.e., a topology), then one may define other types of dual spaces (e.g., the continuous dual of V, defined by requiring the linear functionals to be *continuous* maps). One has to be careful when reading other works (particularly analysis books) to ascertain what qualifiers are implicit in the use of the terms "dual space" and "linear functional."

Example

Let $[a, b]$ be a closed interval in $\mathbb{R}$ and let V be the real vector space of all continuous functions $f : [a, b] \to \mathbb{R}$. If $a < b$, V is infinite dimensional. For each $g \in V$ the function $\varphi_g : V \to \mathbb{R}$ defined by $\varphi_g(f) = \int_a^b f(t)g(t)dt$ is a linear functional on V.

Definition. The dual of V^*, namely V^{**}, is called the *double dual* or *second dual* of V.

Note that for a finite dimensional space V, $\dim V = \dim V^*$ and also $\dim V^* = \dim V^{**}$, hence V and V^{**} are isomorphic vector spaces. For infinite dimensional spaces $\dim V < \dim V^{**}$ (cf. the exercises) so V and V^{**} cannot be isomorphic. In the case of finite dimensional spaces there is a *natural*, i.e., basis independent or coordinate free way of exhibiting the isomorphism between a vector space and its second dual. The basic idea, in a more general setting, is as follows: if X is any set and S is any set of functions of X into the field F, we normally think of choosing or fixing an $f \in S$ and computing $f(x)$ as x ranges over all of X. Alternatively, we could think of fixing a point x in X and computing $f(x)$ as f ranges over all of S. The latter process, called *evaluation at x* shows that for each $x \in X$ there is a function $E_x : S \to F$ defined by

$E_x(f) = f(x)$ (i.e., evaluate f at x). This gives a map $x \mapsto E_x$ of X into the set of F-valued functions on S. If S "separates points" in the sense that for distinct points x and y of X there is some $f \in S$ such that $f(x) \neq f(y)$, then the map $x \mapsto E_x$ is injective. The proof of the next lemma applies this "role reversal" process to the situation where $X = V$ and $S = V^*$, proves E_x is a linear F-valued function on S, that is, E_x belongs to the dual space of V^*, and proves the map $x \mapsto E_x$ is a linear transformation from V into V^{**}. Note that throughout this process there is no mention of the word "basis" (although it is convenient to know the dimension of V^{**} — a fact we established by picking bases). In particular, the proof does not start with the familiar phrase "pick a basis of V . . . ."

Theorem 19. There is a natural injective linear transformation from V to V^{**}. If V is finite dimensional then this linear transformation is an isomorphism.

Proof: Let $v \in V$. Define the map (*evaluation at v*)

$$E_v : V^* \to F \qquad \text{by} \qquad E_v(f) = f(v).$$

Then $E_v(f + \alpha g) = (f + \alpha g)(v) = f(v) + \alpha g(v) = E_v(f) + \alpha E_v(g)$, so that E_v is a linear transformation from V^* to F. Hence E_v is an element of $\mathrm{Hom}_F(V^*, F) = V^{**}$. This defines a natural map

$$\varphi : V \to V^{**} \qquad \text{by} \qquad \varphi(v) = E_v.$$

The map φ is a *linear* map, as follows: for $v, w \in V$ and $\alpha \in F$,

$$E_{v+\alpha w}(f) = f(v + \alpha w) = f(v) + \alpha f(w) = E_v(f) + \alpha E_w(f)$$

for every $f \in V^*$, and so

$$\varphi(v + \alpha w) = E_{v+\alpha w} = E_v + \alpha E_w = \varphi(v) + \alpha \varphi(w).$$

To see that φ is injective let v be any nonzero vector in V. By the Building Up Lemma there is a basis $\mathcal{B}$ containing v. Let f be the linear transformation from V to F defined by sending v to 1 and every element of $\mathcal{B} - \{v\}$ to zero. Then $f \in V^*$ and $E_v(f) = f(v) = 1$. Thus $\varphi(v) = E_v$ is not zero in V^{**}. This proves $\ker \varphi = 0$, i.e., φ is injective.

If V has finite dimension n then by Proposition 18, V^* and hence also V^{**} has dimension n. In this case φ is an injective linear transformation from V to a finite dimensional vector space of the same dimension, hence is an isomorphism.

Let V, W be finite dimensional vector spaces over F with bases $\mathcal{B}, \mathcal{E}$, respectively and let $\mathcal{B}^*, \mathcal{E}^*$ be the dual bases. Fix some $\varphi \in \mathrm{Hom}_F(V, W)$. Then for each $f \in W^*$, the composite $f \circ \varphi$ is a linear transformation from V to F, that is $f \circ \varphi \in V^*$. Thus the map $f \mapsto f \circ \varphi$ defines a function from W^* to V^*. We denote this induced function on dual spaces by φ^*.

Theorem 20. With notations as above, φ^* is a linear transformation from W^* to V^* and $M_{\mathcal{E}^*}^{\mathcal{B}^*}(\varphi^*)$ is the transpose of the matrix $M_{\mathcal{B}}^{\mathcal{E}}(\varphi)$ (recall that the transpose of the matrix (a_{ij}) is the matrix (a_{ji})).

Proof: The map φ^* is linear because $(f + \alpha g) \circ \varphi = (f \circ \varphi) + \alpha(g \circ \varphi)$. The equations which define φ are (from its matrix)

$$\varphi(v_j) = \sum_{i=1}^{m} \alpha_{ij} w_i \qquad 1 \le j \le n.$$

To compute the matrix for φ^*, observe that by the definitions of φ^* and w_k^*

$$\varphi^*(w_k^*)(v_j) = (w_k^* \circ \varphi)(v_j) = w_k^*\left(\sum_{i=1}^{m} \alpha_{ij} w_i \right) = \alpha_{kj}.$$

Also

$$\left(\sum_{i=1}^{n} \alpha_{ki} v_i^*\right)(v_j) = \alpha_{kj}$$

for all j. This shows that the two linear functionals below agree on a basis of V, hence they are the same element of V^*:

$$\varphi^*(w_k^*) = \sum_{i=1}^{n} \alpha_{ki} v_i^*.$$

This determines the matrix for φ^* with respect to the bases $\mathcal{E}^*$ and $\mathcal{B}^*$ as the transpose of the matrix for φ.

Corollary 21. For any matrix A, the row rank of A equals the column rank of A.

Proof: Let $\varphi : V \to W$ be a linear transformation whose matrix with respect to some fixed bases of V and W is A. By Theorem 20 the matrix of $\varphi^* : W^* \to V^*$ with respect to the dual bases is the transpose of A. The column rank of A is the rank of φ and the row rank of A (= the column rank of the transpose of A) is the rank of φ^* (cf. Exercise 6 of Section 2). It therefore suffices to show that φ and φ^* have the same rank. Now

$$f \in \ker \varphi^* \Leftrightarrow \varphi^*(f) = 0 \Leftrightarrow f \circ \varphi(v) = 0, \quad \text{for all } v \in V$$
$$\Leftrightarrow \varphi(V) \subseteq \ker f \Leftrightarrow f \in \mathrm{Ann}(\varphi(V)),$$

where $\mathrm{Ann}(S)$ is the annihilator of S described in Exercise 3 below. Thus $\mathrm{Ann}(\varphi(V)) = \ker \varphi^*$. By Exercise 3, $\dim \mathrm{Ann}(\varphi(V)) = \dim W - \dim \varphi(V)$. By Corollary 8, $\dim \ker \varphi^* = \dim W^* - \dim \varphi^*(W^*)$. Since W and W^* have the same dimension, $\dim \varphi(V) = \dim \varphi^*(W^*)$ as needed.

EXERCISES

1. Let V be a vector space over F of dimension $n < \infty$. Prove that the map $\varphi \mapsto \varphi^*$ in Theorem 20 is a vector space isomorphism of $\text{End}(V)$ with $\text{End}(V^*)$, but is not a ring homomorphism when $n > 1$. Exhibit an F-algebra isomorphism from $\text{End}(V)$ to $\text{End}(V^*)$.

2. Let V be the collection of polynomials with coefficients in $\mathbb{Q}$ in the variable x of degree at most 5 with $1, x, x^2, \ldots, x^5$ as basis. Prove that the following are elements of the dual space of V and express them as linear combinations of the dual basis:
   (a) $E : V \to \mathbb{Q}$ defined by $E(p(x)) = p(3)$ (i.e., evaluation at $x = 3$).
   (b) $\varphi : V \to \mathbb{Q}$ defined by $\varphi(p(x)) = \int_0^1 p(t)dt$.
   (c) $\varphi : V \to \mathbb{Q}$ defined by $\varphi(p(x)) = \int_0^1 t^2 p(t)dt$.
   (d) $\varphi : V \to \mathbb{Q}$ defined by $\varphi(p(x)) = p'(5)$ where $p'(x)$ denotes the usual derivative of the polynomial $p(x)$ with respect to x.

3. Let S be any subset of V^* for some finite dimensional space V. Define $\text{Ann}(S) = \{v \in V \mid f(v) = 0 \text{ for all } f \in S\}$. ($\text{Ann}(S)$ is called the *annihilator of S in V*).
   (a) Prove that $\text{Ann}(S)$ is a subspace of V.
   (b) Let W_1 and W_2 be subspaces of V^*. Prove that $\text{Ann}(W_1 + W_2) = \text{Ann}(W_1) \cap \text{Ann}(W_2)$ and $\text{Ann}(W_1 \cap W_2) = \text{Ann}(W_1) + \text{Ann}(W_2)$.
   (c) Let W_1 and W_2 be subspaces of V^*. Prove that $W_1 = W_2$ if and only if $\text{Ann}(W_1) = \text{Ann}(W_2)$.
   (d) Prove that the annihilator of S is the same as the annihilator of the subspace of V^* spanned by S.
   (e) Assume V is finite dimensional with basis $v_1, \ldots, v_n$. Prove that if $S = \{v_1^*, \ldots, v_k^*\}$ for some $k \leq n$, then $\text{Ann}(S)$ is the subspace spanned by $\{v_{k+1}, \ldots, v_n\}$.
   (f) Assume V is finite dimensional. Prove that if W^* is any subspace of V^* then $\dim \text{Ann}(W^*) = \dim V - \dim W^*$.

4. If V is infinite dimensional with basis $\mathcal{A}$, prove that $\mathcal{A}^* = \{v^* \mid v \in \mathcal{A}\}$ does *not* span V^*.

5. If V is infinite dimensional with basis $\mathcal{A}$, prove that V^* is isomorphic to the direct product of copies of F indexed by $\mathcal{A}$. Deduce that $\dim V^* > \dim V$. [Use Exercise 14, Section 1.]

11.4 DETERMINANTS

Although we shall be using the theory primarily for vector spaces over a field, the theory of determinants can be developed with no extra effort over arbitrary commutative rings with 1. Thus in this section R is any commutative ring with 1 and $V_1, V_2, \ldots, V_n$, V and W are R-modules. For convenience we repeat the definition of multilinear functions from Section 10.4.

Definition.
   (1) A map $\varphi : V_1 \times V_2 \times \cdots \times V_n \to W$ is called *multilinear* if for each fixed i and fixed elements $v_j \in V_j$, $j \neq i$, the map

$$V_i \to W \qquad \text{defined by} \qquad x \mapsto \varphi(v_1, \ldots, v_{i-1}, x, v_{i+1}, \ldots, v_n)$$

   is an R-module homomorphism. If $V_i = V$, $i = 1, 2, \ldots, n$, then φ is called an *n-multilinear function on V*, and if in addition $W = R$, φ is called an *n-multilinear form on V*.

(2) An n-multilinear function φ on V is called *alternating* if $\varphi(v_1, v_2, \ldots, v_n) = 0$ whenever $v_i = v_{i+1}$ for some $i \in \{1, 2, \ldots, n-1\}$ (i.e., φ is zero whenever two consecutive arguments are equal). The function φ is called *symmetric* if interchanging v_i and v_j for any i and j in $(v_1, v_2, \ldots, v_n)$ does not alter the value of φ on this n-tuple.

When $n = 2$ (respectively, 3) one says φ is *bilinear* (respectively, *trilinear*) rather than 2-multilinear (respectively, 3-multilinear). Also, when n is clear from the context we shall simply say φ is multilinear.

Example

For any fixed $m \geq 0$ the usual dot product on $V = \mathbb{R}^m$ is a bilinear form (here the ring R is the field of real numbers).

Proposition 22. Let φ be an n-multilinear alternating function on V. Then
- **(1)** $\varphi(v_1, \ldots, v_{i-1}, v_{i+1}, v_i, v_{i+2}, \ldots, v_n) = -\varphi(v_1, v_2, \ldots, v_n)$ for any $i \in \{1, 2, \ldots, n-1\}$, i.e., the value of φ on an n-tuple is negated if two adjacent components are interchanged.
- **(2)** For each $\sigma \in S_n$, $\varphi(v_{\sigma(1)}, v_{\sigma(2)}, \ldots, v_{\sigma(n)}) = \epsilon(\sigma)\varphi(v_1, v_2, \ldots, v_n)$, where $\epsilon(\sigma)$ is the sign of the permutation σ (cf. Section 3.5).
- **(3)** If $v_i = v_j$ for any pair of distinct $i, j \in \{1, 2, \ldots, n\}$ then $\varphi(v_1, v_2, \ldots, v_n) = 0$.
- **(4)** If v_i is replaced by $v_i + \alpha v_j$ in $(v_1, \ldots, v_n)$ for any $j \neq i$ and any $\alpha \in R$, the value of φ on this n-tuple is not changed.

Proof: (1) Let $\psi(x, y)$ be the function φ with variable entries x and y in positions i and $i + 1$ respectively and fixed entries v_j in position j, for all other j. Thus (1) is the same as showing $\psi(y, x) = -\psi(x, y)$. Since φ is alternating $\psi(x + y, x + y) = 0$. Expanding $x + y$ in each variable in turn gives $\psi(x + y, x + y) = \psi(x, x) + \psi(x, y) + \psi(y, x) + \psi(y, y)$. Again, by the alternating property of φ, the first and last terms on the right hand side of the latter equation are zero. Thus $0 = \psi(x, y) + \psi(y, x)$, which gives (1).

(2) Every permutation can be written as a product of transpositions (cf. Section 3.5). Furthermore, every transposition may be written as a product of transpositions which interchange two successive integers (cf. Exercise 3 of Section 3.5). Thus every permutation σ can be written as $\tau_1 \cdots \tau_m$, where τ_k is a transposition interchanging two successive integers, for all k. It follows from m applications of (1) that

$$\varphi(v_{\sigma(1)}, v_{\sigma(2)}, \ldots, v_{\sigma(n)}) = \epsilon(\tau_m) \cdots \epsilon(\tau_1)\varphi(v_1, v_2, \ldots, v_n).$$

Finally, since ϵ is a homomorphism into the abelian group ± 1 (so the order of the factors ± 1 does not matter), $\epsilon(\tau_1) \cdots \epsilon(\tau_m) = \epsilon(\tau_1 \cdots \tau_m) = \epsilon(\sigma)$. This proves (2).

(3) Choose σ to be any permutation which fixes i and moves j to $i + 1$. Thus $(v_{\sigma(1)}, v_{\sigma(2)}, \ldots, v_{\sigma(n)})$ has two equal adjacent components so φ is zero on this n-tuple. By (2), $\varphi(v_{\sigma(1)}, v_{\sigma(2)}, \ldots, v_{\sigma(n)}) = \pm\varphi(v_1, v_2, \ldots, v_n)$. This implies (3).

(4) This follows immediately from (3) on expanding by linearity in the ith position.

Proposition 23. Assume φ is an n-multilinear alternating function on V and that for some $v_1, v_2, \ldots, v_n$ and $w_1, w_2, \ldots, w_n \in V$ and some $\alpha_{ij} \in R$ we have

$$w_1 = \alpha_{11}v_1 + \alpha_{21}v_2 + \cdots + \alpha_{n1}v_n$$
$$w_2 = \alpha_{12}v_1 + \alpha_{22}v_2 + \cdots + \alpha_{n2}v_n$$
$$\vdots$$
$$w_n = \alpha_{1n}v_1 + \alpha_{2n}v_2 + \cdots + \alpha_{nn}v_n$$

(we have purposely written the indices of the α_{ij} in "column format"). Then

$$\varphi(w_1, w_2, \ldots, w_n) = \sum_{\sigma \in S_n} \epsilon(\sigma)\alpha_{\sigma(1)\,1}\alpha_{\sigma(2)\,2} \cdots \alpha_{\sigma(n)\,n}\varphi(v_1, v_2, \ldots, v_n).$$

Proof: If we expand $\varphi(w_1, w_2, \ldots, w_n)$ by multilinearity we obtain a sum of n^n terms of the form $\alpha_{i_1\,1}\alpha_{i_2\,2} \ldots \alpha_{i_n\,n}\varphi(v_{i_1}, v_{i_2}, \ldots, v_{i_n})$, where the indices $i_1, i_2, \ldots, i_n$ each run over $1, 2, \ldots, n$. By Proposition 22(3), φ is zero on the terms where two or more of the i_j's are equal. Thus in this expansion we need only consider the terms where $i_1, \ldots, i_n$ are distinct. Such sequences are in bijective correspondence with permutations in S_n, so each nonzero term may be written as $\alpha_{\sigma(1)\,1}\alpha_{\sigma(2)\,2} \cdots \alpha_{\sigma(n)\,n}\varphi(v_{\sigma(1)}, v_{\sigma(2)}, \ldots, v_{\sigma(n)})$, for some $\sigma \in S_n$. Applying (2) of the previous proposition to each of these terms in the expansion of $\varphi(w_1, w_2, \ldots, w_n)$ gives the expression in the proposition.

Definition. An $n \times n$ *determinant function* on R is any function

$$\det : M_{n \times n}(R) \to R$$

that satisfies the following two axioms:
  (1) det is an n-multilinear alternating form on $R^n (= V)$, where the n-tuples are the n columns of the matrices in $M_{n \times n}(R)$
  (2) $\det(I) = 1$, where I is the $n \times n$ identity matrix.

On occasion we shall write $\det(A_1, A_2, \ldots, A_n)$ for $\det A$, where $A_1, A_2, \ldots, A_n$ are the columns of A.

Theorem 24. There is a unique $n \times n$ determinant function on R and it can be computed for any $n \times n$ matrix (α_{ij}) by the formula:

$$\det(\alpha_{ij}) = \sum_{\sigma \in S_n} \epsilon(\sigma)\alpha_{\sigma(1)\,1}\alpha_{\sigma(2)\,2} \cdots \alpha_{\sigma(n)\,n}.$$

Proof: Let $A_1, A_2, \ldots, A_n$ be the column vectors in a general $n \times n$ matrix (α_{ij}). We leave it as an exercise to check that the formula given in the statement of the theorem does satisfy the axioms of a determinant function — this gives existence of a determinant function. To prove uniqueness let e_i be the column n-tuple with 1 in position i and zeros in all other positions. Then

$$A_1 = \alpha_{11}e_1 + \alpha_{21}e_2 + \cdots + \alpha_{n1}e_n$$
$$A_2 = \alpha_{12}e_1 + \alpha_{22}e_2 + \cdots + \alpha_{n2}e_n$$
$$\vdots$$
$$A_n = \alpha_{1n}e_1 + \alpha_{2n}e_2 + \cdots + \alpha_{nn}e_n.$$

By Proposition 23, $\det A = \sum_{\sigma \in S_n} \epsilon(\sigma)\alpha_{\sigma(1)\,1}\alpha_{\sigma(2)\,2}\cdots\alpha_{\sigma(n)\,n} \det(e_1, e_2, \ldots, e_n)$. Since by axiom (2) of a determinant function $\det(e_1, e_2, \ldots, e_n) = 1$, the value of $\det A$ is as claimed.

Corollary 25. The determinant is an n-multilinear function of the rows of $M_{n \times n}(R)$ and for any $n \times n$ matrix A, $\det A = \det(A^t)$, where A^t is the transpose of A.

Proof: The first statement is an immediate consequence of the second, so it suffices to prove that a matrix and its transpose have the same determinant. For $A = (\alpha_{ij})$ one calculates that

$$\det A^t = \sum_{\sigma \in S_n} \epsilon(\sigma)\alpha_{1\,\sigma(1)}\alpha_{2\,\sigma(2)} \cdots \alpha_{n\,\sigma(n)}.$$

Each number from 1 to n appears exactly once among $\sigma(1), \ldots, \sigma(n)$ so we may rearrange the product $\alpha_{1\,\sigma(1)}\alpha_{2\,\sigma(2)} \cdots \alpha_{n\,\sigma(n)}$ as $\alpha_{\sigma^{-1}(1)\,1}\alpha_{\sigma^{-1}(2)\,2} \cdots \alpha_{\sigma^{-1}(n)\,n}$. Also, the homomorphism ϵ takes values in $\{\pm 1\}$ so $\epsilon(\sigma) = \epsilon(\sigma^{-1})$. Thus the sum for $\det A^t$ may be rewritten as

$$\sum_{\sigma \in S_n} \epsilon(\sigma^{-1})\alpha_{\sigma^{-1}(1)\,1}\alpha_{\sigma^{-1}(2)\,2} \cdots \alpha_{\sigma^{-1}(n)\,n}.$$

The latter sum is over all permutations, so the index σ^{-1} may be replaced by σ. The resulting expression is the sum for $\det A$. This completes the proof.

Theorem 26. *(Cramer's Rule)* If $A_1, A_2, \ldots, A_n$ are the columns of an $n \times n$ matrix A and $B = \beta_1 A_1 + \beta_2 A_2 + \cdots + \beta_n A_n$, for some $\beta_1, \ldots, \beta_n \in R$, then

$$\beta_i \det A = \det(A_1, \ldots, A_{i-1}, B, A_{i+1}, \ldots, A_n).$$

Proof: This follows immediately from Proposition 22(3) on replacing the given expression for B in the ith position and expanding by multilinearity in that position.

Corollary 27. If R is an integral domain, then $\det A = 0$ for $A \in M_n(R)$ if and only if the columns of A are R-linearly dependent as elements of the free R-module of rank n. Also, $\det A = 0$ if and only if the rows of A are R-linearly dependent.

Proof: Since $\det A = \det A^t$ the first sentence implies the second.
Assume first that the columns of A are linearly dependent and

$$0 = \beta_1 A_1 + \beta_2 A_2 + \cdots + \beta_n A_n$$

is a dependence relation on the columns of A with, say, $\beta_i \neq 0$. By Cramer's Rule, $\beta_i \det A = 0$. Since R is an integral domain and $\beta_i \neq 0$, $\det A = 0$.

Conversely, assume the columns of A are independent. Consider the integral domain R as embedded in its quotient field F so that $M_{n \times n}(R)$ may be considered as a subring of $M_{n \times n}(F)$ (and note that the determinant function on the subring is the restriction of the determinant function from $M_{n \times n}(F)$). The columns of A in this way become elements of F^n. Any nonzero F-linear combination of the columns of A which is zero in F^n gives, by multiplying the coefficients by a common denominator, a nonzero R-linear dependence relation. The columns of A must therefore be independent vectors in F^n. Since A has n columns, these form a basis of F^n. Thus there are elements β_{ij} of F such that for each i, the ith basis vector e_i in F^n may be expressed as

$$e_i = \beta_{1i} A_1 + \beta_{2i} A_2 + \cdots + \beta_{ni} A_n.$$

The $n \times n$ identity matrix is the one whose columns are $e_1, e_2, \ldots, e_n$. By Proposition 23 (with $\varphi = \det$), the determinant of the identity matrix is some F-multiple of $\det A$. Since the determinant of the identity matrix is 1, $\det A$ cannot be zero. This completes the proof.

Theorem 28. For matrices $A, B \in M_{n \times n}(R)$, $\det AB = (\det A)(\det B)$.

Proof: Let $B = (\beta_{ij})$ and let $A_1, A_2, \ldots, A_n$ be the columns of A. Then $C = AB$ is the $n \times n$ matrix whose j^{th} column is $C_j = \beta_{1j} A_1 + \beta_{2j} A_2 + \cdots + \beta_{nj} A_n$. By Proposition 23 applied to the multilinear function $\det$ we obtain

$$\det C = \det(C_1, \ldots, C_n) = \left[\sum_{\sigma \in S_n} \epsilon(\sigma) \beta_{\sigma(1)\,1} \beta_{\sigma(2)\,2} \cdots \beta_{\sigma(n)\,n} \right] \det(A_1, \ldots, A_n).$$

The sum inside the brackets is the formula for $\det B$, hence $\det C = (\det B)(\det A)$, as required ($R$ is commutative).

Definition. Let $A = (\alpha_{ij})$ be an $n \times n$ matrix. For each i, j, let A_{ij} be the $n-1 \times n-1$ matrix obtained from A by deleting its ith row and j^{th} column (an $n-1 \times n-1$ *minor* of A). Then $(-1)^{i+j} \det(A_{ij})$ is called the ij *cofactor of A*.

Theorem 29. *(The Cofactor Expansion Formula along the ith row)* If $A = (\alpha_{ij})$ is an $n \times n$ matrix, then for each fixed $i \in \{1, 2, \ldots, n\}$ the determinant of A can be computed from the formula

$$\det A = (-1)^{i+1} \alpha_{i1} \det A_{i1} + (-1)^{i+2} \alpha_{i2} \det A_{i2} + \cdots + (-1)^{i+n} \alpha_{in} \det A_{in}.$$

Proof: For each A let $D(A)$ be the element of R obtained from the cofactor expansion formula described above. We prove that D satisfies the axioms of a determinant function, hence is *the* determinant function. Proceed by induction on n. If $n = 1$, $D((\alpha)) = \alpha$, for all 1×1 matrices (α) and the result holds. Assume therefore that $n \geq 2$. To show that D is an alternating multilinear function of the columns, fix an index k and consider the kth column as varying and all other columns as fixed. If $j \neq k$,

α_{ij} does not depend on k and $D(A_{ij})$ is linear in the kth column by induction. Also, as the kth column varies linearly so does α_{ik}, whereas $D(A_{ik})$ remains unchanged (the kth column has been deleted from A_{ik}). Thus each term in the formula for D varies linearly in the kth column. This proves D is multilinear in the columns.

To prove D is alternating assume columns k and $k+1$ of A are equal. If $j \neq k$ or $k+1$, the two equal columns of A become two equal columns in the matrix A_{ij}. By induction $D(A_{ij}) = 0$. The formula for D therefore has at most two nonzero terms: when $j = k$ and when $j = k+1$. The minor matrices A_{ik} and $A_{i\,k+1}$ are identical and $\alpha_{ik} = \alpha_{i\,k+1}$. Then the two remaining terms in the expansion for D, $(-1)^{i+k}\alpha_{ik}D(A_{ik})$ and $(-1)^{i+k+1}\alpha_{i\,k+1}D(A_{i\,k+1})$ are equal and appear with opposite signs, hence they cancel. Thus $D(A) = 0$ if A has two adjacent columns which are equal, i.e., D is alternating.

Finally, it follows easily from the formula and induction that $D(I) = 1$, where I is the identity matrix. This completes the induction.

Theorem 30. *(Cofactor Formula for the Inverse of a Matrix)* Let $A = (\alpha_{ij})$ be an $n \times n$ matrix and let B be the transpose of its matrix of cofactors, i.e., $B = (\beta_{ij})$, where $\beta_{ij} = (-1)^{i+j} \det A_{ji}$, $1 \leq i, j \leq n$. Then $AB = BA = (\det A)I$. Moreover, $\det A$ is a unit in R if and only if A is a unit in $M_{n \times n}(R)$; in this case the matrix $\dfrac{1}{\det A} B$ is the inverse of A.

Proof: The i, j entry of AB is $\alpha_{i1}\beta_{1j} + \alpha_{i2}\beta_{2j} + \cdots + \alpha_{in}\beta_{nj}$. By definition of the entries of B this equals

$$\alpha_{i1}(-1)^{j+1}D(A_{j1}) + \alpha_{i2}(-1)^{j+2}D(A_{j2}) + \cdots + \alpha_{in}(-1)^{j+n}D(A_{jn}). \qquad (11.7)$$

If $i = j$, this is the cofactor expansion for $\det A$ along the ith row. The diagonal entries of AB are thus all equal to $\det A$. If $i \neq j$, let $\overline{A}$ be the matrix A with the jth row replaced by the ith row, so $\det \overline{A} = 0$. By inspection $\overline{A}_{jk} = A_{jk}$ and $\alpha_{ik} = \overline{\alpha}_{jk}$ for every $k \in \{1, 2, \ldots, n\}$. By making these substitutions in equation (7) for each $k = 1, 2, \ldots, n$ one sees that the i, j entry in AB equals $\overline{\alpha}_{j1}(-1)^{1+j}D(\overline{A}_{j1}) + \cdots + \overline{\alpha}_{jn}(-1)^{n+j}D(\overline{A}_{jn})$. This expression is the cofactor expansion for $\det \overline{A}$ along the jth row. Since, as noted above, $\det \overline{A} = 0$, this proves that all off diagonal terms of AB are zero, which proves that $AB = (\det A)I$.

It follows directly from the definition of B that the pair (A^t, B^t) satisfies the same hypotheses as the pair (A, B). By what has already been shown it follows that $(BA)^t = A^t B^t = (\det A^t)I$. Since $\det A^t = \det A$ and the transpose of a diagonal matrix is itself, we obtain $BA = (\det A)I$ as well.

If $d = \det A$ is a unit in R, then $d^{-1}B$ is a matrix with entries in R whose product with A (on either side) is the identity, i.e., A is a unit in $M_{n \times n}(R)$. Conversely, assume that A is a unit in R with (2-sided) inverse matrix C. Since $\det C \in R$ and

$$1 = \det I = \det AC = (\det A)(\det C) = (\det C)(\det A),$$

it follows that $\det A$ has a 2-sided inverse in R, as needed. This completes all parts of the proof.

EXERCISES

1. Formulate and prove the cofactor expansion formula along the jth column of a square matrix A.

2. Let F be a field and let $A_1, A_2, \ldots, A_n$ be (column) vectors in F^n. Form the matrix A whose i^{th} column is A_i. Prove that these vectors form a basis of F^n if and only if $\det A \neq 0$.

3. Let R be any commutative ring with 1, let V be an R-module and let $x_1, x_2, \ldots, x_n \in V$. Assume that for some $A \in M_{n \times n}(R)$,

$$A \begin{pmatrix} x_1 \\ \vdots \\ x_n \end{pmatrix} = 0.$$

Prove that $(\det A)x_i = 0$, for all $i \in \{1, 2, \ldots, n\}$.

4. (*Computing Determinants of Matrices*) This exercise outlines the use of Gauss–Jordan elimination (cf. the exercises in Section 2) to compute determinants. This is the most efficient general procedure for computing large determinants. Let A be an $n \times n$ matrix.
   (a) Prove that the elementary row operations have the following effect on determinants:
      (i) interchanging two rows changes the sign of the determinant
      (ii) adding a multiple of one row to another does not alter the determinant
      (iii) multiplying any row by a nonzero element u from F multiplies the determinant by u.
   (b) Prove that $\det A$ is nonzero if and only if A is row equivalent to the $n \times n$ identity matrix. Suppose A can be row reduced to the identity matrix using a total of s row interchanges as in (i) and by multiplying rows by the nonzero elements $u_1, u_2, \ldots, u_t$ as in (iii). Prove that $\det A = (-1)^s (u_1 u_2 \ldots u_t)^{-1}$.

5. Compute the determinants of the following matrices using row reduction:

$$A = \begin{pmatrix} 5 & 4 & -6 \\ -2 & 0 & 2 \\ 3 & 4 & -2 \end{pmatrix} \qquad B = \begin{pmatrix} 1 & 2 & -4 & 4 \\ 2 & -1 & 4 & -8 \\ 1 & 0 & 1 & -2 \\ 0 & 1 & -2 & 3 \end{pmatrix}.$$

6. (*Minkowski's Criterion*) Suppose A is an $n \times n$ matrix with real entries such that the diagonal elements are all positive, the off-diagonal elements are all negative and the row sums are all positive. Prove that $\det A \neq 0$. [Consider the corresponding system of equations $AX = 0$ and suppose there is a nontrivial solution $(x_1, \ldots, x_n)$. If x_i has the largest absolute value show that the ith equation leads to a contradiction.]

11.5 TENSOR ALGEBRAS, SYMMETRIC AND EXTERIOR ALGEBRAS

In this section R is any commutative ring with 1, and we assume the left and right actions of R on each R-module are the same. We shall primarily be interested in the special case when $R = F$ is a field, but the basic constructions hold in general.

Suppose M is an R-module. When tensor products were first introduced in Section 10.4 we spoke heuristically of forming "products" $m_1 m_2$ of elements of M, and we constructed a new module $M \otimes M$ generated by such "products" $m_1 \otimes m_2$. The "value" of this product is not in M, so this does not give a ring structure on M itself. If, however, we iterate this by taking the "products" $m_1 m_2 m_3$ and $m_1 m_2 m_3 m_4$, and all finite sums of such products, we can construct a ring containing M that is "universal" with respect to rings containing M (and, more generally, with respect to homomorphic images of M), as we now show.

For each integer $k \geq 1$, define

$$\mathcal{T}^k(M) = M \otimes_R M \otimes_R \cdots \otimes_R M \quad (k \text{ factors}),$$

and set $\mathcal{T}^0(M) = R$. The elements of $\mathcal{T}^k(M)$ are called k-*tensors*. Define

$$\mathcal{T}(M) = R \oplus \mathcal{T}^1(M) \oplus \mathcal{T}^2(M) \oplus \mathcal{T}^3(M) \cdots = \bigoplus_{k=0}^{\infty} \mathcal{T}^k(M).$$

Every element of $\mathcal{T}(M)$ is a finite linear combination of k-tensors for various $k \geq 0$. We identify M with $\mathcal{T}^1(M)$, so that M is an R-submodule of $\mathcal{T}(M)$.

Theorem 31. If M is any R-module over the commutative ring R then
  (1) $\mathcal{T}(M)$ is an R-algebra containing M with multiplication defined by mapping

$$(m_1 \otimes \cdots \otimes m_i)(m_1' \otimes \cdots \otimes m_j') = m_1 \otimes \cdots \otimes m_i \otimes m_1' \otimes \cdots \otimes m_j'$$

  and extended to sums via the distributive laws. With respect to this multiplication $\mathcal{T}^i(M)\mathcal{T}^j(M) \subseteq \mathcal{T}^{i+j}(M)$.
  (2) *(Universal Property)* If A is any R-algebra and $\varphi : M \to A$ is an R-module homomorphism, then there is a unique R-algebra homomorphism $\Phi : \mathcal{T}(M) \to A$ such that $\Phi|_M = \varphi$.

Proof: The map

$$\underbrace{M \times M \times \cdots \times M}_{i \text{ factors}} \times \underbrace{M \times M \times \cdots \times M}_{j \text{ factors}} \to \mathcal{T}^{i+j}(M)$$

defined by

$$(m_1, \ldots, m_i, m_1', \ldots, m_j') \mapsto m_1 \otimes \ldots \otimes m_i \otimes m_1' \otimes \ldots \otimes m_j'$$

is R-multilinear, so induces a bilinear map $\mathcal{T}^i(M) \times \mathcal{T}^j(M)$ to $\mathcal{T}^{i+j}(M)$ which is easily checked to give a well defined multiplication satisfying (1) (cf. the proof of Proposition 21 in Section 10.4). To prove (2), assume that $\varphi : M \to A$ is an R-module homomorphism. Then

$$(m_1, m_2, \ldots, m_k) \mapsto \varphi(m_1)\varphi(m_2) \ldots \varphi(m_k)$$

defines an R-multilinear map from $M \times \cdots \times M$ (k times) to A. This in turn induces a unique R-module homomorphism Φ from $\mathcal{T}^k(M)$ to A (Corollary 16 of Section 10.4) mapping $m_1 \otimes \ldots \otimes m_k$ to the element on the right hand side above. It is easy to check from the definition of the multiplication in (1) that the resulting uniquely defined map $\Phi : \mathcal{T}(M) \to A$ is an R-algebra homomorphism.

442

Definition. The ring $\mathcal{T}(M)$ is called the *tensor algebra* of M.

Proposition 32. Let V be a finite dimensional vector space over the field F with basis $\mathcal{B} = \{v_1, \ldots, v_n\}$. Then the k-tensors

$$v_{i_1} \otimes v_{i_2} \otimes \cdots \otimes v_{i_k} \qquad \text{with } v_{i_j} \in \mathcal{B}$$

are a vector space basis of $\mathcal{T}^k(V)$ over F (with the understanding that the basis vector is the element $1 \in F$ when $k = 0$). In particular, $\dim_F(\mathcal{T}^k(V)) = n^k$.

Proof: This follows immediately from Proposition 16 of Section 2.

Theorem 31 and Proposition 32 show that the space $\mathcal{T}(V)$ may be regarded as the *noncommutative polynomial algebra* over F in the (noncommuting) variables $v_1, \ldots, v_n$. The analogous result also holds for finitely generated free modules over any commutative ring (using Corollary 19 in Section 10.4).

Examples

(1) Let $R = \mathbb{Z}$ and let $M = \mathbb{Q}/\mathbb{Z}$. Then $(\mathbb{Q}/\mathbb{Z}) \otimes_{\mathbb{Z}} (\mathbb{Q}/\mathbb{Z}) = 0$ (Example 4 following Corollary 12 in Section 10.4). Thus $\mathcal{T}(\mathbb{Q}/\mathbb{Z}) = \mathbb{Z} \oplus (\mathbb{Q}/\mathbb{Z})$, where addition is componentwise and the multiplication is given by $(r, \overline{p})(s, \overline{q}) = (rs, \overline{rq + sp})$. The ring $R/(x)$ of Exercise 4(d) in Section 9.3 is isomorphic to $\mathcal{T}(\mathbb{Q}/\mathbb{Z})$.

(2) Let $R = \mathbb{Z}$ and let $M = \mathbb{Z}/n\mathbb{Z}$. Then $(\mathbb{Z}/n\mathbb{Z}) \otimes_{\mathbb{Z}} (\mathbb{Z}/n\mathbb{Z}) \cong \mathbb{Z}/n\mathbb{Z}$ (Example 3 following Corollary 12 in Section 10.4). Thus $\mathcal{T}^i(M) \cong M$ for all $i > 0$ and so $\mathcal{T}(\mathbb{Z}/n\mathbb{Z}) \cong \mathbb{Z} \oplus (\mathbb{Z}/n\mathbb{Z}) \oplus (\mathbb{Z}/n\mathbb{Z}) \cdots$. It follows easily that $\mathcal{T}(\mathbb{Z}/n\mathbb{Z}) \cong \mathbb{Z}[x]/(nx)$.

Since $\mathcal{T}^i(M)\mathcal{T}^j(M) \subseteq \mathcal{T}^{i+j}(M)$, the tensor algebra $\mathcal{T}(M)$ has a natural "grading" or "degree" structure reminiscent of a polynomial ring.

Definition.

(1) A ring S is called a *graded ring* if it is the direct sum of additive subgroups: $S = S_0 \oplus S_1 \oplus S_2 \oplus \cdots$ such that $S_i S_j \subseteq S_{i+j}$ for all $i, j \geq 0$. The elements of S_k are said to be *homogeneous of degree* k, and S_k is called the *homogeneous component of S of degree* k.

(2) An ideal I of the graded ring S is called a *graded ideal* if $I = \oplus_{k=0}^{\infty}(I \cap S_k)$.

(3) A ring homomorphism $\varphi : S \to T$ between two graded rings is called a *homomorphism of graded rings* if it respects the grading structures on S and T, i.e., if $\varphi(S_k) \subseteq T_k$ for $k = 0, 1, 2, \ldots$.

Note that $S_0 S_0 \subseteq S_0$, which implies that S_0 is a subring of the graded ring S and then S is an S_0-module. If S_0 is in the center of S and it contains an identity of S, then S is an S_0-algebra. Note also that the ideal I is graded if whenever a sum $i_{k_1} + \cdots + i_{k_n}$ of homogeneous elements with distinct degrees $k_1, \ldots, k_n$ is in I then each of the individual summands $i_{k_1}, \ldots, i_{k_n}$ is itself in I.

Example

The polynomial ring $S = R[x_1, x_2, \ldots, x_n]$ in n variables over the commutative ring R is an example of a graded ring. Here $S_0 = R$ and the homogeneous component of degree k is the subgroup of all R-linear combinations of monomials of degree k.

The ideal I generated by $x_1, \ldots, x_n$ is a graded ideal: every polynomial with zero constant term may be written uniquely as a sum of homogeneous polynomials of degree $k > 1$, and each of these has zero constant term hence lies in I. More generally, an ideal is a graded ideal if and only if it can be generated by homogeneous polynomials (cf. Exercise 17 in Section 9.1).

Not every ideal of a graded ring need be a graded ideal. For example in the graded ring $\mathbb{Z}[x]$ the principal ideal J generated by $1 + x$ is not graded: $1 + x \in J$ and $1 \notin J$ so $1 + x$ cannot be written as a sum of homogeneous polynomials each of which belongs to J.

The next result shows that quotients of graded rings by graded ideals are again graded rings.

Proposition 33. Let S be a graded ring, let I be a graded ideal in S and let $I_k = I \cap S_k$ for all $k \geq 0$. Then S/I is naturally a graded ring whose homogeneous component of degree k is isomorphic to S_k/I_k.

Proof: The map

$$S = \oplus_{k=0}^{\infty} S_k \longrightarrow \oplus_{k=0}^{\infty} (S_k/I_k)$$

$$(\ldots, s_k, \ldots) \longmapsto (\ldots, s_k \bmod I_k, \ldots)$$

is surjective with kernel $I = \oplus_{k=0}^{\infty} I_k$ and defines an isomorphism of graded rings. The details are left for the exercises.

Symmetric Algebras

The first application of Proposition 33 is in the construction of a commutative quotient ring of $\mathcal{T}(M)$ through which R-module homomorphisms from M to any *commutative* R-algebra must factor. This gives an "abelianized" version of Theorem 31. The construction is analogous to forming the commutator quotient G/G' of a group (cf. Section 5.4).

Definition. The *symmetric algebra* of an R-module M is the R-algebra obtained by taking the quotient of the tensor algebra $\mathcal{T}(M)$ by the ideal $\mathcal{C}(M)$ generated by all elements of the form $m_1 \otimes m_2 - m_2 \otimes m_1$, for all $m_1, m_2 \in M$. The symmetric algebra $\mathcal{T}(M)/\mathcal{C}(M)$ is denoted by $\mathcal{S}(M)$.

The tensor algebra $\mathcal{T}(M)$ is generated as a ring by $R = \mathcal{T}^0(M)$ and $M = \mathcal{T}^1(M)$, and these elements commute in the quotient ring $\mathcal{S}(M)$ by definition. It follows that the symmetric algebra $\mathcal{S}(M)$ is a commutative ring. The ideal $\mathcal{C}(M)$ is generated by homogeneous tensors of degree 2 and it follows easily that $\mathcal{C}(M)$ is a graded ideal. Then by Proposition 33 the symmetric algebra is a graded ring whose homogeneous component of degree k is $\mathcal{S}^k(M) = \mathcal{T}^k(M)/\mathcal{C}^k(M)$. Since $\mathcal{C}(M)$ consists of k-tensors with $k \geq 2$, we have $\mathcal{C}(M) \cap M = 0$ and so the image of $M = \mathcal{T}^1(M)$ in $\mathcal{S}(M)$ is isomorphic to M. Identifying M with its image we see that $\mathcal{S}^1(M) = M$ and the symmetric algebra contains M. In a similar way $\mathcal{S}^0(M) = R$, so the symmetric algebra is also an R-algebra. The R-module $\mathcal{S}^k(M)$ is called the kth *symmetric power* of M.

The first part of the next theorem shows that the elements of the kth symmetric power of M can be considered as finite sums of simple tensors $m_1 \otimes \cdots \otimes m_k$ where tensors with the order of the factors permuted are identified. Recall also from Section 4 that a k-multilinear map $\varphi : M \times \cdots \times M \rightarrow N$ is said to be *symmetric* if $\varphi(m_1, \ldots, m_k) = \varphi(m_{\sigma(1)}, \ldots, m_{\sigma(k)})$ for all permutations σ of $1, 2, \ldots, k$. (The definition is the same for modules over any commutative ring R as for vector spaces.)

Theorem 34. Let M be an R-module over the commutative ring R and let $\mathcal{S}(M)$ be its symmetric algebra.

  (1) The kth symmetric power, $\mathcal{S}^k(M)$, of M is equal to $M \otimes \cdots \otimes M$ (k factors) modulo the submodule generated by all elements of the form

$$(m_1 \otimes m_2 \otimes \cdots \otimes m_k) - (m_{\sigma(1)} \otimes m_{\sigma(2)} \otimes \cdots \otimes m_{\sigma(k)})$$

  for all $m_i \in M$ and all permutations σ in the symmetric group S_k.
  (2) *(Universal Property for Symmetric Multilinear Maps)* If $\varphi : M \times \cdots \times M \rightarrow N$ is a symmetric k-multilinear map over R then there is a unique R-module homomorphism $\Phi : \mathcal{S}^k(M) \rightarrow N$ such that $\varphi = \Phi \circ \iota$, where

$$\iota : M \times \cdots \times M \rightarrow \mathcal{S}^k(M)$$

  is the map defined by

$$\iota(m_1, \ldots, m_k) = m_1 \otimes \cdots \otimes m_n \bmod \mathcal{C}(M).$$

  (3) *(Universal Property for maps to commutative R-algebras)* If A is any commutative R-algebra and $\varphi : M \rightarrow A$ is an R-module homomorphism, then there is a unique R-algebra homomorphism $\Phi : \mathcal{S}(M) \rightarrow A$ such that $\Phi|_M = \varphi$.

Proof: The k-tensors $\mathcal{C}^k(M)$ in the ideal $\mathcal{C}(M)$ are finite sums of elements of the form

$$m_1 \otimes \ldots \otimes m_{i-1} \otimes (m_i \otimes m_{i+1} - m_{i+1} \otimes m_i) \otimes m_{i+2} \otimes \ldots \otimes m_k$$

with $m_1, \ldots, m_k \in M$ (where $k \geq 2$ and $1 \leq i < k$). This product gives a difference of two k-tensors which are equal except that two entries (in positions i and $i + 1$) have been transposed, i.e., gives the element in (1) of the theorem corresponding to the transposition $(i\ i+1)$ in the symmetric group S_k. Conversely, since any permutation σ in S_k can be written as a product of such transpositions it is easy to see that every element in (1) can be written as a sum of elements of the form above. This gives (1).

The proofs of (2) and (3) are very similar to the proofs of the corresponding "asymmetric" results (Corollary 16 of Section 10.4 and Theorem 31) noting that $\mathcal{C}^k(M)$ is contained in the kernel of any symmetric map from $\mathcal{T}^k(M)$ to N by part (1).

Corollary 35. Let V be an n-dimensional vector space over the field F. Then $\mathcal{S}(V)$ is isomorphic as a graded F-algebra to the ring of polynomials in n variables over F (i.e., the isomorphism is also a vector space isomorphism from $\mathcal{S}^k(V)$ onto the space of all homogeneous polynomials of degree k). In particular, $\dim_F(\mathcal{S}^k(V)) = \binom{k+n-1}{n-1}$.

Proof: Let $\mathcal{B} = \{v_1, \ldots, v_n\}$ be a basis of V. By Proposition 32 there is a bijection between a basis of $\mathcal{T}^k(V)$ and the set $\mathcal{B}^k$ of ordered k-tuples of elements from $\mathcal{B}$. Define two k-tuples in $\mathcal{B}^k$ to be equivalent if there is some permutation of the entries of one that gives the other — this is easily seen to be an equivalence relation on $\mathcal{B}^k$. Let $S(\mathcal{B}^k)$ denote the corresponding set of equivalence classes. Any symmetric k-multilinear function from V^k to a vector space over F will be constant on all of the basis tensors whose corresponding k-tuples lie in the same equivalence class; conversely, any function from $S(\mathcal{B}^k)$ can be uniquely extended to a symmetric k-multilinear function on V^k. It follows that the vector space over F with basis $S(\mathcal{B}^k)$ satisfies the universal property of $\mathcal{S}^k(V)$ in Theorem 34(2), hence is isomorphic to $\mathcal{S}^k(V)$. Each equivalence class has a unique representative of the form $(v_1^{a_1}, v_2^{a_2}, \ldots, v_n^{a_n})$, where v_i^a denotes the sequence $v_i, v_i, \ldots, v_i$ taken a times, each $a_i \geq 0$, and $a_1 + \cdots + a_n = k$. Thus there is a bijection between the basis $S^k(\mathcal{B})$ and the set $x_1^{a_1} \cdots x_n^{a_n}$ of monic monomials of degree k in the polynomial ring $F[x_1, \ldots, x_n]$. This bijection extends to an isomorphism of graded F-algebras, proving the first part of the corollary. The computation of the dimension of $\mathcal{S}^k(V)$ (i.e., the number of monic monomials of degree k) is left as an exercise.

Exterior Algebras

Recall from Section 4 that a multilinear map $\varphi : M \times \cdots \times M \to N$ is called *alternating* if $\varphi(m_1, \ldots, m_k) = 0$ whenever $m_i = m_{i+1}$ for some i. (The definition is the same for any R-module as for vector spaces.) We saw that the determinant map was alternating, and was uniquely determined by some additional constraints. We can apply Proposition 33 to construct an algebra through which alternating multilinear maps must factor in a manner similar to the construction of the symmetric algebra (through which symmetric multilinear maps factor).

Definition. The *exterior algebra* of an R-module M is the R-algebra obtained by taking the quotient of the tensor algebra $\mathcal{T}(M)$ by the ideal $\mathcal{A}(M)$ generated by all elements of the form $m \otimes m$, for $m \in M$. The exterior algebra $\mathcal{T}(M)/\mathcal{A}(M)$ is denoted by $\bigwedge(M)$ and the image of $m_1 \otimes m_2 \otimes \cdots \otimes m_k$ in $\bigwedge(M)$ is denoted by $m_1 \wedge m_2 \wedge \cdots \wedge m_k$.

As with the symmetric algebra, the ideal $\mathcal{A}(M)$ is generated by homogeneous elements hence is a graded ideal. By Proposition 33 the exterior algebra is graded, with k^{th} homogeneous component $\bigwedge^k(M) = \mathcal{T}^k(M)/\mathcal{A}^k(M)$. We can again identify R with $\bigwedge^0(M)$ and M with $\bigwedge^1(M)$ and so consider M as an R-submodule of the R-algebra $\bigwedge(M)$. The R-module $\bigwedge^k(M)$ is called the kth *exterior power* of M.

The multiplication

$$(m_1 \wedge \cdots \wedge m_i) \wedge (m_1' \wedge \cdots \wedge m_j') = m_1 \wedge \cdots \wedge m_i \wedge m_1' \wedge \cdots \wedge m_j'$$

in the exterior algebra is called the *wedge* (or *exterior*) *product*. By definition of the quotient, this multiplication is alternating in the sense that the product $m_1 \wedge \cdots \wedge m_k$ is 0 in $\bigwedge(M)$ if $m_i = m_{i+1}$ for any $1 \le i < k$. Then

$$0 = (m + m') \wedge (m + m')$$
$$= (m \wedge m) + (m \wedge m') + (m' \wedge m) + (m' \wedge m')$$
$$= (m \wedge m') + (m' \wedge m)$$

shows that the multiplication is also anticommutative on simple tensors:

$$m \wedge m' = -m' \wedge m \qquad \text{for all } m, m' \in M.$$

This anticommutativity does not extend to arbitrary products, however, i.e., we need not have $ab = -ba$ for all $a, b \in \bigwedge(M)$ (cf. Exercise 4).

Theorem 36. Let M be an R-module over the commutative ring R and let $\bigwedge(M)$ be its exterior algebra.

(1) The kth exterior power, $\bigwedge^k(M)$, of M is equal to $M \otimes \cdots \otimes M$ (k factors) modulo the submodule generated by all elements of the form

$$m_1 \otimes m_2 \otimes \cdots \otimes m_k \quad \text{where } m_i = m_j \text{ for some } i \ne j.$$

In particular,

$$m_1 \wedge m_2 \wedge \cdots \wedge m_k = 0 \quad \text{if } m_i = m_j \text{ for some } i \ne j.$$

(2) *(Universal Property for Alternating Multilinear Maps)* If $\varphi : M \times \cdots \times M \to N$ is an alternating k-multilinear map then there is a unique R-module homomorphism $\Phi : \bigwedge^k(M) \to N$ such that $\varphi = \Phi \circ \iota$, where

$$\iota : M \times \cdots \times M \to \bigwedge^k(M)$$

is the map defined by

$$\iota(m_1, \ldots, m_k) = m_1 \wedge \cdots \wedge m_k.$$

Remark: The exterior algebra also satisfies a universal property similar to (3) of Theorem 34, namely with respect to R-module homomorphisms from M to R-algebras A satisfying $a^2 = 0$ for all $a \in A$ (cf. Exercise 6).

Proof: The k-tensors $\mathcal{A}^k(M)$ in the ideal $\mathcal{A}(M)$ are finite sums of elements of the form

$$m_1 \otimes \ldots \otimes m_{i-1} \otimes (m \otimes m) \otimes m_{i+2} \otimes \ldots \otimes m_k$$

with $m_1, \ldots, m_k, m \in M$ (where $k \ge 2$ and $1 \le i < k$), which is a k-tensor with two equal entries (in positions i and $i+1$), so is of the form in (1). For the reverse inclusion, note that since

$$m' \otimes m = -m \otimes m' + \left[(m + m') \otimes (m + m') - m \otimes m - m' \otimes m' \right]$$
$$\equiv -m \otimes m' \mod \mathcal{A}(M),$$

interchanging any two consecutive entries and multiplying by -1 in a simple k-tensor gives an equivalent tensor modulo $\mathcal{A}^k(M)$. Using such a sequence of interchanges and sign changes we can arrange for the equal entries m_i and m_j of a simple tensor as in (1) to be adjacent, which gives an element of $\mathcal{A}^k(M)$. It follows that the generators in (1) are contained in $\mathcal{A}^k(M)$, which proves the first part of the theorem.

As in Theorem 34, the proof of (2) follows easily from the corresponding result for the tensor algebra in Theorem 31 since $\mathcal{A}^k(M)$ is contained in the kernel of any alternating map from $\mathcal{T}^k(M)$ to N.

Examples

(1) Suppose V is a one-dimensional vector space over F with basis element v. Then $\bigwedge^k(V)$ consists of finite sums of elements of the form $\alpha_1 v \wedge \alpha_2 v \wedge \cdots \wedge \alpha_k v$, i.e., $\alpha_1 \alpha_2 \cdots \alpha_k (v \wedge v \wedge \cdots \wedge v)$ for $\alpha_1, \ldots, \alpha_k \in F$. Since $v \wedge v = 0$, it follows that $\bigwedge^0(V) = F$, $\bigwedge^1(V) = V$, and $\bigwedge^i(V) = 0$ for $i \geq 2$, so as a graded F-algebra we have

$$\bigwedge(V) = F \oplus V \oplus 0 \oplus 0 \oplus \ldots.$$

(2) Suppose now that V is a two-dimensional vector space over F with basis v, v'. Here $\bigwedge^k(V)$ consists of finite sums of elements of the form $(\alpha_1 v + \alpha_1' v') \wedge \cdots \wedge (\alpha_k v + \alpha_k' v')$. Such an element is a sum of elements that are simple wedge products involving only v and v'. For example, an element in $\bigwedge^2(V)$ is a sum of elements of the form

$$(av + bv') \wedge (cv + dv') = ac(v \wedge v) + ad(v \wedge v') + bc(v' \wedge v)$$
$$+ bd(v' \wedge v')$$
$$= (ad - bc)v \wedge v'.$$

It follows that $\bigwedge^i(V) = 0$ for $i \geq 3$ since then at least one of v, v' appears twice in such simple products.

We can see directly from $\bigwedge^2(V) = \mathcal{T}^2(V)/\mathcal{A}^2(V)$ that $v \wedge v' \neq 0$, as follows. The vector space $\mathcal{T}^2(V)$ is 4-dimensional with $v \otimes v$, $v \otimes v'$, $v' \otimes v$, $v' \otimes v'$ as basis (Proposition 16). The elements $v \otimes v$, $v \otimes v' + v' \otimes v$, $v' \otimes v'$ and $v \otimes v'$ are therefore also a basis for $\mathcal{T}^2(V)$. The subspace $\mathcal{A}^2(V)$ consists of all the 2-tensors in the ideal generated by the tensors

$$(av + bv') \otimes (av + bv') = a^2(v \otimes v) + ab(v \otimes v' + v' \otimes v) + b^2(v' \otimes v'),$$

from which it is clear that $\mathcal{A}^2(V)$ is contained in the 3-dimensional subspace having $v \otimes v$, $v \otimes v' + v' \otimes v$, and $v' \otimes v'$ as basis. In particular, the basis element $v \otimes v'$ of $\mathcal{T}^2(V)$ is not contained in $\mathcal{A}^2(V)$, i.e., $v \wedge v' \neq 0$ in $\bigwedge^2(V)$.

It follows that $\bigwedge^0(V) = F$, $\bigwedge^1(V) = V$, $\bigwedge^2(V) = F(v \wedge v')$, and $\bigwedge^i(V) = 0$ for $i \geq 3$, so as a graded F-algebra we have

$$\bigwedge(V) = F \oplus V \oplus F(v \wedge v') \oplus 0 \oplus \ldots.$$

As the previous examples illustrate, unlike the tensor and symmetric algebras, for finite dimensional vector spaces the exterior algebra is finite dimensional:

Corollary 37. Let V be a finite dimensional vector space over the field F with basis $\mathcal{B} = \{v_1, \ldots, v_n\}$. Then the vectors

$$v_{i_1} \wedge v_{i_2} \wedge \cdots \wedge v_{i_k} \qquad \text{for } 1 \le i_1 < i_2 < \cdots < i_k \le n$$

are a basis of $\bigwedge^k(V)$, and $\bigwedge^k(V) = 0$ when $k > n$ (when $k = 0$ the basis vector is the element $1 \in F$). In particular, $\dim_F(\bigwedge^k(V)) = \binom{n}{k}$.

Proof: As the proof of Theorem 36 shows, modulo $\mathcal{A}^k(M)$, the order of the terms in any simple k-tensor can be rearranged up to introducing a sign change. It follows that the k-tensors in the corollary (which have been arranged with increasing subscripts on the v_i and with no repeated entries) are generators for $\bigwedge^k(V)$. To show these vectors are linearly independent it suffices to exhibit an alternating k-multilinear function from V^k to F which is 1 on a given $v_{i_1} \wedge v_{i_2} \wedge \cdots \wedge v_{i_k}$ and zero on all other generators. Such a function f is defined on the basis of $\mathcal{T}^k(V)$ in Proposition 32 by $f(v_{j_1} \otimes v_{j_2} \otimes \cdots \otimes v_{j_k}) = \epsilon(\sigma)$ if σ is the unique permutation of $(j_1, j_2, \ldots, j_k)$ into $(i_1, i_2, \ldots, i_k)$, and f is zero on every basis tensor whose k-tuple of indices cannot be permuted to $(i_1, i_2, \ldots, i_k)$ (where $\epsilon(\sigma)$ is the sign of σ). Note that f is zero on any basis tensor with repeated entries. The value $\epsilon(\sigma)$ ensures that when f is extended to all elements of $\mathcal{T}^k(V)$ it gives an alternating map, i.e., f factors through $\mathcal{A}^k(V)$. Hence f is the desired function. The computation of the dimension of $\bigwedge^k(V)$ (i.e., of the number of increasing sequences of k-tuples of indices) is left to the exercises.

The results in Corollary 37 are true for any *free* R-module of rank n. In particular if $M \cong R^n$ with R-module basis $m_1, \ldots, m_n$ then

$$\bigwedge{}^n(M) = R(m_1 \wedge \cdots \wedge m_n)$$

is a free (rank 1) R-module with generator $m_1 \wedge \cdots \wedge m_n$ and

$$\bigwedge{}^{n+1}(M) = \bigwedge{}^{n+2}(M) = \cdots = 0.$$

Example

Let R be the polynomial ring $\mathbb{Z}[x, y]$ in the variables x and y. If $M = R$, then $\bigwedge^2(M) = 0$ so, for example, there are no nontrivial alternating bilinear maps on $R \times R$ by the universal property of $\bigwedge^2(R)$ with respect to such maps (Theorem 36).

Suppose now that $M = I$ is the ideal (x, y) generated by x and y in R. Then $I \bigwedge I \ne 0$. Perhaps the easiest way to see this is to construct a nontrivial alternating bilinear map on $I \times I$. The map

$$\varphi(ax + by, cx + dy) = (ad - bc) \bmod (x, y)$$

is a well defined alternating R-bilinear map from $I \times I$ to $\mathbb{Z} = R/I$ (cf. Exercise 7). Since $\varphi(x, y) = 1$, it follows that $x \wedge y \in \bigwedge^2(I)$ is nonzero. Unlike the situation of free modules as in the examples following Theorem 36 (where arguments involving *bases* could be used), in this case it is not at all a trivial matter to give a direct verification that $x \wedge y \ne 0$ in $\bigwedge^2(I)$.

Remark: The ideal I is an example of a rank 1 (but *not* free) R-module (the rank of a module over an integral domain is defined in Section 12.1), and this example shows that the results of Corollary 37 are not true in general if the R-module is not free over R.

Homomorphisms of Tensor Algebras

If $\varphi : M \to N$ is any R-module homomorphism, then there is an induced map on the kth tensor power:

$$\mathcal{T}^k(\varphi) : m_1 \otimes m_2 \otimes \cdots \otimes m_k \longmapsto \varphi(m_1) \otimes \varphi(m_2) \otimes \cdots \otimes \varphi(m_k).$$

It follows directly that this map sends generators of each of the homogeneous components of the ideals $\mathcal{C}(M)$ and $\mathcal{A}(M)$ to themselves. Thus φ induces R-module homomorphisms on the quotients:

$$\mathcal{S}^k(\varphi) : \mathcal{S}^k(M) \longrightarrow \mathcal{S}^k(N) \qquad \text{and} \qquad \textstyle\bigwedge^k(\varphi) : \bigwedge^k(M) \longrightarrow \bigwedge^k(N).$$

Moreover, each of these three maps is a ring homomorphism (hence they are graded R-algebra homomorphisms).

Of particular interest is the case when $M = V$ is an n-dimensional vector space over the field F and $\varphi : V \to V$ is an endomorphism. In this case by Corollary 37, $\bigwedge^n(\varphi)$ maps the 1-dimensional space $\bigwedge^n(V)$ to itself. Let $v_1, \ldots, v_n$ be a basis of V, so that $v_1 \wedge \cdots \wedge v_n$ is a basis of $\bigwedge^n(V)$. Then

$$\textstyle\bigwedge^n(\varphi)(v_1 \wedge \cdots \wedge v_n) = \varphi(v_1) \wedge \cdots \wedge \varphi(v_n) = D(\varphi)v_1 \wedge \cdots \wedge v_n$$

for some scalar $D(\varphi) \in F$.

For any $n \times n$ matrix A over F we can define the associated endomorphism φ (with respect to the given basis $v_1, \ldots, v_n$), which gives a map $D : M_{n \times n}(F) \to F$ where $D(A) = D(\varphi)$. It is easy to check that this map D satisfies the three axioms for a determinant function in Section 4. Then the uniqueness statement of Theorem 24 gives:

Proposition 38. If φ is an endomorphism on a n-dimensional vector space V, then $\bigwedge^n(\varphi)(w) = \det(\varphi)w$ for all $w \in \bigwedge^n(V)$.

Note that Proposition 38 characterizes the determinant of the endomorphism φ as a certain naturally induced *linear* map on $\bigwedge^n(V)$. The fact that the determinant arises naturally when considering alternating multilinear maps also explains the source of the map φ in the example above.

As with the tensor product, the maps $\mathcal{S}^k(\varphi)$ and $\bigwedge^k(\varphi)$ induced from an injective map from M to N need not remain injective (so $\bigwedge^2(M)$ need not be a submodule of $\bigwedge^2(N)$ when M is a submodule of N, for example).

Example

The inclusion $\varphi : I \hookrightarrow R$ of the ideal (x, y) into the ring $R = \mathbb{Z}[x, y]$, both considered as R-modules, induces a map

$$\textstyle\bigwedge^2(\varphi) : \bigwedge^2(I) \to \bigwedge^2(R).$$

Since $\bigwedge^2(R) = 0$ and $\bigwedge^2(I) \neq 0$, the map cannot be injective.

One can show that if M is an R-module *direct summand* of N, then $\mathcal{T}(M)$ (respectively, $\mathcal{S}(M)$ and $\bigwedge(M)$) is an R-subalgebra of $\mathcal{T}(N)$ (respectively, $\mathcal{S}(N)$ and $\bigwedge(N)$) (cf. the exercises). When $R = F$ is a field then *every* subspace M of N is a direct summand of N and so the corresponding algebra for M is a subalgebra of the algebra for N.

Symmetric and Alternating Tensors

The symmetric and exterior algebras can in some instances also be defined in terms of *symmetric* and *alternating* tensors (defined below), which identify these algebras as *sub*algebras of the tensor algebra rather than as quotient algebras.

For any R-module M there is a natural left group action of the symmetric group S_k on $M \times M \times \cdots \times M$ (k factors) given by permuting the factors:

$$\sigma(m_1, m_2, \ldots, m_k) = (m_{\sigma^{-1}(1)}, m_{\sigma^{-1}(2)}, \ldots, m_{\sigma^{-1}(k)}) \qquad \text{for each } \sigma \in S_k$$

(the reason for σ^{-1} is to make this a *left* group action, cf. Exercise 8 of Section 5.1). This map is clearly R-multilinear, so there is a well defined R-linear left group action of S_k on $\mathcal{T}^k(M)$ which is defined on simple tensors by

$$\sigma(m_1 \otimes m_2 \otimes \cdots \otimes m_k) = m_{\sigma^{-1}(1)} \otimes m_{\sigma^{-1}(2)} \otimes \cdots \otimes m_{\sigma^{-1}(k)} \qquad \text{for each } \sigma \in S_k.$$

Definition.
  (1) An element $z \in \mathcal{T}^k(M)$ is called a *symmetric k-tensor* if $\sigma z = z$ for all σ in the symmetric group S_k.
  (2) An element $z \in \mathcal{T}^k(M)$ is called an *alternating k-tensor* if $\sigma z = \epsilon(\sigma)z$ for all σ in the symmetric group S_k, where $\epsilon(\sigma)$ is the sign, ± 1, of the permutation σ.

It is immediate from the definition that the collection of symmetric (respectively, alternating) k-tensors is an R-submodule of the module of all k-tensors.

Example

The elements $m \otimes m$ and $m_1 \otimes m_2 + m_2 \otimes m_1$ are symmetric 2-tensors. The element $m_1 \otimes m_2 - m_2 \otimes m_1$ is an alternating 2-tensor.

It is also clear from the definition that both $\mathcal{C}^k(M)$ and $\mathcal{A}^k(M)$ are stable under the action of S_k, hence there is an induced action on the quotients $\mathcal{S}^k(M)$ and $\bigwedge^k(M)$.

Proposition 39. Let σ be an element in the symmetric group S_k and let $\epsilon(\sigma)$ be the sign of the permutation σ. Then
  (1) for every $w \in \mathcal{S}^k(M)$ we have $\sigma w = w$, and
  (2) for every $w \in \bigwedge^k(M)$ we have $\sigma w = \epsilon(\sigma)w$.

Proof: The first statement is immediate from (1) in Theorem 34. We showed in the course of the proof of Theorem 36 that

$$m_1 \wedge \cdots \wedge m_i \wedge m_{i+1} \wedge \cdots \wedge m_k = -m_1 \wedge \cdots \wedge m_{i+1} \wedge m_i \wedge \cdots \wedge m_k,$$

which shows that the formula in (2) is valid on simple products for the transposition $\sigma = (i\ i{+}1)$. Since these transpositions generate S_k and ϵ is a group homomorphism it follows that (2) is valid for any $\sigma \in S_k$ on simple products w. Since both sides are R-linear in w, it follows that (2) holds for all $w \in \bigwedge^k(M)$.

By Proposition 39, the symmetric group S_k acts trivially on both the *sub*module of symmetric k-tensors and the *quotient* module $S^k(M)$, the kth symmetric power of M. Similarly, S_k acts the same way on the submodule of alternating k-tensors as on $\bigwedge^k(M)$, the kth exterior power of M. We now show that when $k!$ is a unit in R that these respective submodules and quotient modules are isomorphic (where $k!$ is the sum of the 1 of R with itself $k!$ times).

For any k-tensor $z \in T^k(M)$ define

$$Sym(z) = \sum_{\sigma \in S_k} \sigma z$$

$$Alt(z) = \sum_{\sigma \in S_k} \epsilon(\sigma)\, \sigma z.$$

For any k-tensor z, the k-tensor $Sym(z)$ is symmetric and the k-tensor $Alt(z)$ is alternating. For example, for any $\tau \in S_k$

$$
\begin{aligned}
\tau\, Alt(z) &= \sum_{\sigma \in S_k} \epsilon(\sigma)\, \tau\sigma\, z \\
&= \sum_{\sigma' \in S_k} \epsilon(\tau^{-1}\sigma')\, \sigma' z \quad \text{(letting } \sigma' = \tau\sigma) \\
&= \epsilon(\tau^{-1}) \sum_{\sigma' \in S_k} \epsilon(\sigma')\, \sigma' z = \epsilon(\tau)Alt(z).
\end{aligned}
$$

The tensor $Sym(z)$ is sometimes called the *symmetrization* of z and $Alt(z)$ the *skew-symmetrization* of z.

If z is already a symmetric (respectively, alternating) tensor then $Sym(z)$ (respectively, $Alt(z)$) is just $k!z$. It follows that Sym (respectively, Alt) is an R-module endomorphism of $T^k(M)$ whose image lies in the submodule of symmetric (respectively, alternating) tensors. In general these maps are not surjective, but if $k!$ is a unit in R then

$$\frac{1}{k!}Sym(z) = z \quad \text{for any symmetric tensor } z, \text{ and}$$

$$\frac{1}{k!}Alt(z) = z \quad \text{for any alternating tensor } z$$

so that in this case the maps $(1/k!)Sym$ and $(1/k!)Alt$ give surjective R-module homomorphisms from $T^k(M)$ to the submodule of symmetric (respectively, alternating) tensors.

452

Proposition 40. Suppose $k!$ is a unit in the ring R and M is an R-module. Then

(1) The map $(1/k!)Sym$ induces an R-module isomorphism between the kth symmetric power of M and the R-submodule of symmetric k-tensors:

$$\frac{1}{k!}Sym : \mathcal{S}^k(M) \cong \{\text{symmetric } k\text{-tensors}\}.$$

(2) The map $(1/k!)Alt$ induces an R-module isomorphism between the kth exterior power of M and the R-submodule of alternating k-tensors:

$$\frac{1}{k!}Alt : \bigwedge^k(M) \cong \{\text{alternating } k\text{-tensors}\}.$$

Proof: We have seen that the respective maps are surjective R-homomorphisms from $\mathcal{T}^k(M)$ so to prove the proposition it suffices to check that their kernels are $\mathcal{C}^k(M)$ and $\mathcal{A}^k(M)$, respectively. We show the first and leave the second to the exercises. It is clear that Sym is 0 on any difference of two k-tensors which differ only in the order of their factors, so $\mathcal{C}^k(M)$ is contained in the kernel of $(1/k!)Sym$ by (1) of Theorem 34. For the reverse inclusion, observe that

$$z - \frac{1}{k!}Sym(z) = \frac{1}{k!}\sum_{\sigma \in S_k}(z - \sigma z)$$

for any k-tensor z. If z is in the kernel of Sym then the left hand side of this equality is just z; and since $z - \sigma z \in \mathcal{C}^k(M)$ for every $\sigma \in S_k$ (again by (1) of Theorem 34), it follows that $z \in \mathcal{C}^k(M)$, completing the proof.

The maps $(1/k!)Sym$ and $(1/k!)Alt$ are *projections* (cf. Exercise 11 in Section 2) onto the submodules of symmetric and antisymmetric tensors, respectively. Equivalently, if $k!$ is a unit in R, we have R-module direct sums

$$\mathcal{T}^k(M) = \ker(\pi) \oplus \operatorname{image}(\pi)$$

for $\pi = (1/k!)Sym$ or $\pi = (1/k!)Alt$. In the former case the kernel consists of $\mathcal{C}^k(M)$ and the image is the collection of symmetric tensors (in which case $\mathcal{C}^k(M)$ is said to form an R-module *complement* to the symmetric tensors). In the latter case the kernel is $\mathcal{A}^k(M)$ and the image consists of the alternating tensors.

The R-linear left group action of S_k on $\mathcal{T}^k(M)$ makes $\mathcal{T}^k(M)$ into a module over the group ring RS_k (analogous to the formation of $F[x]$-modules described in Section 10.1). In terms of this module structure these projections give RS_k-submodule complements to the RS_k-submodules $\mathcal{C}^k(M)$ and $\mathcal{A}^k(M)$. The "averaging" technique used to construct these maps can be used to prove a very general result (Maschke's Theorem in Section 18.1) related to actions of finite groups on vector spaces (which is the subject of the "representation theory" of finite groups in Part VI).

If $k!$ is not invertible in R then in general we do not have such S_k-invariant direct sum decompositions so it is not in general possible to identify, for example, the kth exterior power of M with the alternating k-tensors of M.

Note also that when $k!$ is invertible it is possible to *define* the kth exterior power of M as the collection of alternating k-tensors (this equivalent approach is sometimes found in the literature when the theory is developed over fields such as $\mathbb{R}$ and $\mathbb{C}$). In this case the multiplication of two alternating tensors z and w is defined by first taking the product $zw = z \otimes w$ in $\mathcal{T}(M)$ and then projecting the resulting tensor into the submodule of alternating tensors. Note that the simple product of two alternating tensors need not be alternating (for example, the square of an alternating tensor is a symmetric tensor).

Example

Let V be a vector space over a field F in which $k! \neq 0$. There are many *vector space complements* to $\mathcal{A}^k(V)$ in $\mathcal{T}^k(V)$ (just extend a basis for the subspace $\mathcal{A}^k(V)$ to a basis for $\mathcal{T}^k(V)$, for example). These complements depend on choices of bases for $\mathcal{T}^k(V)$ and so are indistinguishable from each other from vector space considerations alone. The additional structure on $\mathcal{T}^k(V)$ given by the action of S_k singles out a unique complement to $\mathcal{A}^k(V)$, namely the subspace of alternating tensors in Proposition 40.

Suppose that $k! \neq 0$ in F for all $k \geq 2$ (i.e., the field F has "characteristic 0," cf. Exercise 26 in Section 7.3), for example, $F = \mathbb{Q}$. Then the full exterior algebra $\bigwedge(V) = \oplus_{k \geq 0} \bigwedge^k(V)$ can be identified with the collection of tensors whose homogeneous components are alternating (with respect to the appropriate symmetric groups S_k).

Multiplication in $\bigwedge(V)$ in terms of alternating tensors is rather cumbersome, however. For example let v_1, v_2, v_3 be distinct basis vectors in V. The product of the two alternating tensors $z = v_1$ and $w = v_2 \otimes v_3 - v_3 \otimes v_2$ is obtained by first computing

$$z \otimes w = v_1 \otimes v_2 \otimes v_3 - v_1 \otimes v_3 \otimes v_2$$

in the full tensor algebra. This 3-tensor is not alternating — for example,

$$(1\,2)(z \otimes w) = v_2 \otimes v_1 \otimes v_3 - v_3 \otimes v_1 \otimes v_2 \neq -z \otimes w$$

and also $(1\,2\,3)(z \otimes w) = v_3 \otimes v_1 \otimes v_2 - v_2 \otimes v_1 \otimes v_3 \neq z \otimes w$. The multiplication requires that we project this tensor into the subspace of alternating tensors. This projection is given by $(1/3!)Alt(z \otimes w)$ and an easy computation shows that

$$\frac{1}{6}Alt(z \otimes w) = \frac{1}{3}[v_1 \otimes v_2 \otimes v_3 + v_2 \otimes v_3 \otimes v_1 + v_3 \otimes v_1 \otimes v_2$$
$$-v_1 \otimes v_3 \otimes v_2 - v_2 \otimes v_1 \otimes v_3 - v_3 \otimes v_2 \otimes v_1],$$

so the right hand side is the product of z and w in terms of alternating tensors. The same product in terms of the quotient algebra $\bigwedge(V)$ is simply

$$v_1 \wedge (2v_2 \wedge v_3) = 2v_1 \wedge v_2 \wedge v_3.$$

EXERCISES

In these exercises R is a commutative ring with 1 and M is an R-module; F is a field and V is a finite dimensional vector space over F.

1. Prove that if M is a cyclic R-module then $\mathcal{T}(M) = \mathcal{S}(M)$, i.e., the tensor algebra $\mathcal{T}(M)$ is commutative.

2. Fill in the details for the proof of Proposition 33 that $S/I = \oplus_{k=0}^{\infty} S_k/I_k$. [Show first that $S_i I_j \subseteq I_{i+j}$. Use this to show that the multiplication $(S_i/I_i)(S_j/I_j) \subseteq S_{i+j}/I_{i+j}$ is well defined, and then check the ring axioms and verify the statements made in the proof of Proposition 33.]

3. Show that the image of the map Sym_2 for the $\mathbb{Z}$-module $\mathbb{Z}$ consists of the 2-tensors $a(1 \otimes 1)$ where a is an even integer. Conclude in particular that the symmetric tensor $1 \otimes 1$ in $\mathbb{Z} \otimes_\mathbb{Z} \mathbb{Z}$ is not contained in the image of the map Sym.

4. Prove that $m \wedge n_1 \wedge n_2 \wedge \cdots \wedge n_k = (-1)^k (n_1 \wedge n_2 \wedge \cdots \wedge n_k \wedge m)$. In particular, $x \wedge (y \wedge z) = (y \wedge z) \wedge x$ for all $x, y, z \in M$.

5. Prove that if M is a free R-module of rank n then $\bigwedge^i(M)$ is a free R-module of rank $\binom{n}{i}$ for $i = 0, 1, 2, \ldots$.

6. If A is any R-algebra in which $a^2 = 0$ for all $a \in A$ and $\varphi : M \to A$ is an R-module homomorphism, prove there is a unique R-algebra homomorphism $\Phi : \bigwedge(M) \to A$ such that $\Phi|_M = \varphi$.

7. Let $R = \mathbb{Z}[x, y]$ and $I = (x, y)$.
   (a) Prove that if $ax + by = a'x + b'y$ in R then $a' = a + yf$ and $b' = b - xf$ for some polynomial $f(x, y) \in R$.
   (b) Prove that the map $\varphi(ax+by, cx+dy) = ad - bc \bmod (x, y)$ in the example following Corollary 37 is a well defined alternating R-bilinear map from $I \times I$ to $\mathbb{Z} = R/I$.

8. Let R be an integral domain and let F be its field of fractions.
   (a) Considering F as an R-module, prove that $\bigwedge^2 F = 0$.
   (b) Let I be any R-submodule of F (for example, any ideal in R). Prove that $\bigwedge^i I$ is a torsion R-module for $i \geq 2$ (i.e., for every $x \in \bigwedge^i I$ there is some nonzero $r \in R$ with $rx = 0$).
   (c) Give an example of an integral domain R and an R-module I in F with $\bigwedge^i I \neq 0$ for every $i \geq 0$ (cf. the example following Corollary 37).

9. Let $R = \mathbb{Z}[G]$ be the group ring of the group $G = \{1, \sigma\}$ of order 2. Let $M = \mathbb{Z}e_1 + \mathbb{Z}e_2$ be a free $\mathbb{Z}$-module of rank 2 with basis e_1 and e_2. Define $\sigma(e_1) = e_1 + 2e_2$ and $\sigma(e_2) = -e_2$. Prove that this makes M into an R-module and that the R-module $\bigwedge^2 M$ is a group of order 2 with $e_1 \wedge e_2$ as generator.

10. Prove that $z - (1/k!) Alt(z) = (1/k!) \sum_{\sigma \in S_k} (z - \epsilon(\sigma)\sigma z)$ for any k-tensor z and use this to prove that the kernel of the R-module homomorphism $(1/k!) Alt$ in Proposition 40 is $\mathcal{A}^k(M)$.

11. Prove that the image of Alt_k is the unique largest subspace of $\mathcal{T}^k(V)$ on which each permutation σ in the symmetric group S_k acts as multiplication by the scalar $\epsilon(\sigma)$.

12. (a) Prove that if $f(x, y)$ is an alternating bilinear map on V (i.e., $f(x, x) = 0$ for all $x \in V$) then $f(x, y) = -f(y, x)$ for all $x, y \in V$.
    (b) Suppose that $-1 \neq 1$ in F. Prove that $f(x, y)$ is an alternating bilinear map on V (i.e., $f(x, x) = 0$ for all $x \in V$) if and only if $f(x, y) = -f(y, x)$ for all $x, y \in V$.
    (c) Suppose that $-1 = 1$ in F. Prove that every alternating bilinear form $f(x, y)$ on V is symmetric (i.e., $f(x, y) = f(y, x)$ for all $x, y \in V$). Prove that there is a symmetric bilinear map on V that is not alternating. [One approach: show that $C^2(V) \subset \mathcal{A}^2(V)$ and $C^2(V) \neq \mathcal{A}^2(V)$ by counting dimensions. Alternatively, construct an explicit symmetric map that is not alternating.]

13. Let F be any field in which $-1 \neq 1$ and let V be a vector space over F. Prove that $V \otimes_F V = \mathcal{S}^2(V) \oplus \bigwedge^2(V)$ i.e., that every 2-tensor may be written uniquely as a sum of a symmetric and an alternating tensor.

14. Prove that if M is an R-module *direct factor* of the R-module N then $\mathcal{T}(M)$ (respectively, $\mathcal{S}(M)$ and $\bigwedge(M)$) is an R-subalgebra of $\mathcal{T}(N)$ (respectively, $\mathcal{S}(N)$ and $\bigwedge(N)$).

CHAPTER 12

Modules over
Principal Ideal Domains

The main purpose of this chapter is to prove a structure theorem for finitely generated modules over particularly nice rings, namely Principal Ideal Domains. This theorem is an example of the ideal structure of the ring (which is particularly simple for P.I.D.s) being reflected in the structure of its modules. If we apply this result in the case where the P.I.D. is the ring of integers $\mathbb{Z}$ then we obtain a proof of the Fundamental Theorem of Finitely Generated Abelian Groups (which we examined in Chapter 5 without proof). If instead we apply this structure theorem in the case where the P.I.D. is the ring $F[x]$ of polynomials in x with coefficients in a field F we shall obtain the basic results on the so-called rational and Jordan canonical forms for a matrix. Before proceeding to the proof we briefly discuss these two important applications.

We have already discussed in Chapter 5 the result that any finitely generated abelian group is isomorphic to the direct sum of cyclic abelian groups, either $\mathbb{Z}$ or $\mathbb{Z}/n\mathbb{Z}$ for some positive integer $n \neq 0$. Recall also that an abelian group is the same thing as a $\mathbb{Z}$-module. Since the ideals of $\mathbb{Z}$ are precisely the trivial ideal (0) and the principal ideals $(n) = n\mathbb{Z}$ generated by positive integers n, we see that the Fundamental Theorem of Finitely Generated Abelian Groups in the language of modules says that any finitely generated $\mathbb{Z}$-module is the direct sum of modules of the form $\mathbb{Z}/I$ where I is an ideal of $\mathbb{Z}$ (these are the cyclic $\mathbb{Z}$-modules), together with a uniqueness statement when the direct sum is written in a particular form. Note the correspondence between the ideal structure of $\mathbb{Z}$ and the structure of its (finitely generated) modules, the finitely generated abelian groups.

The Fundamental Theorem of Finitely Generated Modules over a P.I.D. states that the same result holds when the Principal Ideal Domain $\mathbb{Z}$ is replaced by *any* P.I.D. In particular, we have seen in Chapter 10 that a module over the ring $F[x]$ of polynomials in x with coefficients in the field F is the same thing as a vector space V together with a fixed linear transformation T of V (where the element x acts on V by the linear transformation T). The Fundamental Theorem in this case will say that such a vector space is the direct sum of modules of the form $F[x]/I$ where I is an ideal of $F[x]$, hence is either the trivial ideal (0) or a principal ideal $(f(x))$ generated by some nonzero polynomial $f(x)$ (these are the cyclic $F[x]$-modules), again with a uniqueness statement when the direct sum is written in a particular form. If this is translated back into the language of vector spaces and linear transformations we can obtain information on the linear transformation T.

For example, suppose V is a vector space of dimension n over F and we choose a basis for V. Then giving a linear transformation T of V to itself is the same thing as giving an $n \times n$ matrix A with coefficients in F (and choosing a different basis for V gives a different matrix B for T which is similar to A i.e., is of the form $P^{-1}AP$ for some invertible matrix P which defines the change of basis). We shall see that the Fundamental Theorem in this situation implies (under the assumption that the field F contains all the "eigenvalues" for the given linear transformation T) that there is a basis for V so that the associated matrix for T is *as close to being a diagonal matrix as possible* and so has a particularly simple form. This is the *Jordan canonical form*. The *rational canonical form* is another simple form for the matrix for T (that does not require the eigenvalues for T to be elements of F). In this way we shall be able to give canonical forms for arbitrary $n \times n$ matrices over fields F, that is, find matrices which are similar to a given $n \times n$ matrix and which are particularly simple (almost diagonal, for example).

Example

Let $V = \mathbb{Q}^3 = \{(x, y, z) \mid x, y, z \in \mathbb{Q}\}$ be the usual 3-dimensional vector space of ordered 3-tuples with entries from the field $F = \mathbb{Q}$ of rational numbers and suppose T is the linear transformation

$$T(x, y, z) = (9x + 4y + 5z, -4x - 3z, -6x - 4y - 2z), \qquad x, y, z \in \mathbb{Q}.$$

If we take the standard basis $e_1 = (1, 0, 0)$, $e_2 = (0, 1, 0)$, $e_3 = (0, 0, 1)$ for V then the matrix A representing this linear transformation is

$$A = \begin{pmatrix} 9 & 4 & 5 \\ -4 & 0 & -3 \\ -6 & -4 & -2 \end{pmatrix}.$$

We shall see that the Jordan canonical form for this matrix A is the much simpler matrix

$$B = \begin{pmatrix} 2 & 1 & 0 \\ 0 & 2 & 0 \\ 0 & 0 & 3 \end{pmatrix}$$

obtained by taking instead the basis $f_1 = (2, -1, -2)$, $f_2 = (1, 0, -1)$, $f_3 = (3, -2, -2)$ for V, since in this case

$$T(f_1) = T(2, -1, -2) = (4, -2, -4) = 2 \cdot f_1 + 0 \cdot f_2 + 0 \cdot f_3$$
$$T(f_2) = T(1, 0, -1) = (4, -1, -4) = 1 \cdot f_1 + 2 \cdot f_2 + 0 \cdot f_3$$
$$T(f_3) = T(3, -2, -2) = (9, -6, -6) = 0 \cdot f_1 + 0 \cdot f_2 + 3 \cdot f_3,$$

so the columns of the matrix representing T with respect to this basis are $(2, 0, 0)$, $(1, 2, 0)$ and $(0, 0, 3)$, i.e., T has matrix B with respect to this basis. In particular A is similar to the simpler matrix B.

In fact this linear transformation T *cannot* be diagonalized (i.e., there is no choice of basis for V for which the corresponding matrix is a diagonal matrix) so that the matrix B is as close to a diagonal matrix for T as is possible.

The first section below gives some general definitions and states and proves the Fundamental Theorem over an arbitrary P.I.D., after which we return to the application to canonical forms (the application to abelian groups appears in Chapter 5). These applications can be read independently of the general proof. An alternate and computationally useful proof valid for Euclidean Domains (so in particular for the rings $\mathbb{Z}$ and $F[x]$) along the lines of row and column operations is outlined in the exercises.

12.1 THE BASIC THEORY

We first describe some general finiteness conditions. Let R be a ring and let M be a left R-module.

Definition.

(1) The left R-module M is said to be a *Noetherian R-module* or to satisfy the *ascending chain condition on submodules* (or *A.C.C. on submodules*) if there are no infinite increasing chains of submodules, i.e., whenever

$$M_1 \subseteq M_2 \subseteq M_3 \subseteq \cdots$$

is an increasing chain of submodules of M, then there is a positive integer m such that for all $k \geq m$, $M_k = M_m$ (so the chain becomes stationary at stage m: $M_m = M_{m+1} = M_{m+2} = \dots$).

(2) The ring R is said to be *Noetherian* if it is Noetherian as a left module over itself, i.e., if there are no infinite increasing chains of left ideals in R.

One can formulate analogous notions of A.C.C. on right and on two-sided ideals in a (possibly noncommutative) ring R. For noncommutative rings these properties need not be related.

Theorem 1. Let R be a ring and let M be a left R-module. Then the following are equivalent:

(1) M is a Noetherian R-module.

(2) Every nonempty set of submodules of M contains a maximal element under inclusion.

(3) Every submodule of M is finitely generated.

Proof: [(1) implies (2)] Assume M is Noetherian and let Σ be any nonempty collection of submodules of M. Choose any $M_1 \in \Sigma$. If M_1 is a maximal element of Σ, (2) holds, so assume M_1 is not maximal. Then there is some $M_2 \in \Sigma$ such that $M_1 \subset M_2$. If M_2 is maximal in Σ, (2) holds, so we may assume there is an $M_3 \in \Sigma$ properly containing M_2. Proceeding in this way one sees that if (2) fails we can produce by the Axiom of Choice an infinite strictly increasing chain of elements of Σ, contrary to (1).

[(2) implies (3)] Assume (2) holds and let N be any submodule of M. Let Σ be the collection of all finitely generated submodules of N. Since $\{0\} \in \Sigma$, this collection is nonempty. By (2) Σ contains a maximal element N'. If $N' \neq N$, let $x \in N - N'$. Since $N' \in \Sigma$, the submodule N' is finitely generated by assumption, hence also the submodule generated by N' and x is finitely generated. This contradicts the maximality of N', so $N = N'$ is finitely generated.

[(3) implies (1)] Assume (3) holds and let $M_1 \subseteq M_2 \subseteq M_3 \ldots$ be a chain of submodules of M. Let

$$N = \bigcup_{i=1}^{\infty} M_i$$

and note that N is a submodule. By (3) N is finitely generated by, say, $a_1, a_2, \ldots, a_n$. Since $a_i \in N$ for all i, each a_i lies in one of the submodules in the chain, say M_{j_i}. Let $m = \max\{j_1, j_2, \ldots, j_n\}$. Then $a_i \in M_m$ for all i so the module they generate is contained in M_m, i.e., $N \subseteq M_m$. This implies $M_m = N = M_k$ for all $k \geq m$, which proves (1).

Corollary 2. If R is a P.I.D. then every nonempty set of ideals of R has a maximal element and R is a Noetherian ring.

Proof: The P.I.D. R satisfies condition (3) in the theorem with $M = R$.

Recall that even if M itself is a finitely generated R-module, submodules of M need not be finitely generated, so the condition that M be a Noetherian R-module is in general stronger than the condition that M be a finitely generated R-module.

We require a result on "linear dependence" before turning to the main results of this chapter.

Proposition 3. Let R be an integral domain and let M be a free R-module of rank $n < \infty$. Then any $n + 1$ elements of M are R-linearly dependent, i.e., for any $y_1, y_2, \ldots, y_{n+1} \in M$ there are elements $r_1, r_2, \ldots, r_{n+1} \in R$, not all zero, such that

$$r_1 y_1 + r_2 y_2 + \ldots + r_{n+1} y_{n+1} = 0.$$

Proof: The quickest way of proving this is to embed R in its quotient field F (since R is an integral domain) and observe that since $M \cong R \oplus R \oplus \cdots \oplus R$ (n times) we obtain $M \subseteq F \oplus F \oplus \cdots \oplus F$. The latter is an n-dimensional vector space over F so any $n + 1$ elements of M are F-linearly dependent. By clearing the denominators of the scalars (by multiplying through by the product of all the denominators, for example), we obtain an R-linear dependence relation among the $n + 1$ elements of M.

Alternatively, let $e_1, \ldots, e_n$ be a basis of the free R-module M and let $y_1, \ldots, y_{n+1}$ be any $n + 1$ elements of M. For $1 \leq i \leq n + 1$ write $y_i = a_{1i} e_1 + a_{2i} e_2 + \ldots + a_{ni} e_i$ in terms of the basis $e_1, e_2, \ldots, e_n$. Let A be the $(n + 1) \times (n + 1)$ matrix whose i, j entry is a_{ij}, $1 \leq i \leq n$, $1 \leq j \leq n + 1$ and whose last row is zero, so certainly $\det A = 0$. Since R is an integral domain, Corollary 27 of Section 11.4 shows that the columns of A are R-linearly dependent. Any dependence relation on the columns of A gives a dependence relation on the y_i's, completing the proof.

If R is any integral domain and M is any R-module recall that

$$\mathrm{Tor}(M) = \{x \in M \mid rx = 0 \text{ for some nonzero } r \in R\}$$

is a submodule of M (called *the* torsion submodule of M) and if N is any submodule of $\text{Tor}(M)$, N is called *a* torsion submodule of M (so the torsion submodule of M is the union of all torsion submodules of M, i.e., is the maximal torsion submodule of M). If $\text{Tor}(M) = 0$, the module M is said to be *torsion free*.

For any submodule N of M, the *annihilator* of N is the ideal of R defined by

$$\text{Ann}(N) = \{r \in R \mid rn = 0 \text{ for all } n \in N\}.$$

Note that if N is not a torsion submodule of M then $\text{Ann}(N) = (0)$. It is easy to see that if N, L are submodules of M with $N \subseteq L$, then $\text{Ann}(L) \subseteq \text{Ann}(N)$. If R is a P.I.D. and $N \subseteq L \subseteq M$ with $\text{Ann}(N) = (a)$ and $\text{Ann}(L) = (b)$, then $a \mid b$. In particular, the annihilator of any element x of M divides the annihilator of M (this is implied by Lagrange's Theorem when $R = \mathbb{Z}$).

Definition. For any integral domain R the *rank* of an R-module M is the maximum number of R-linearly independent elements of M.

The preceding proposition states that for a free R-module M over an integral domain the rank of a submodule is bounded by the rank of M. This notion of rank agrees with previous uses of the same term. If the ring $R = F$ is a field, then the rank of an R-module M is the dimension of M as a vector space over F and any maximal set of F-linearly independent elements is a basis for M. For a general integral domain, however, an R-module M of rank n need not have a "basis," i.e., need not be a *free* R-module even if M is torsion free, so some care is necessary with the notion of rank, particularly with respect to the torsion elements of M. Exercises 1 to 6 and 20 give an alternate characterization of the rank and provide some examples of (torsion free) R-modules (of rank 1) that are not free.

The next important result shows that if N is a submodule of a free module of finite rank over a P.I.D. then N is again a free module of finite rank and furthermore it is possible to choose generators for the two modules which are related in a simple way.

Theorem 4. Let R be a Principal Ideal Domain, let M be a free R-module of finite rank n and let N be a submodule of M. Then
  (1) N is free of rank m, $m \le n$ and
  (2) there exists a basis $y_1, y_2, \ldots, y_n$ of M so that $a_1 y_1, a_2 y_2, \ldots, a_m y_m$ is a basis of N where $a_1, a_2, \ldots, a_m$ are nonzero elements of R with the divisibility relations

$$a_1 \mid a_2 \mid \cdots \mid a_m.$$

Proof: The theorem is trivial for $N = \{0\}$, so assume $N \ne \{0\}$. For each R-module homomorphism φ of M into R, the image $\varphi(N)$ of N is a submodule of R, i.e., an ideal in R. Since R is a P.I.D. this ideal must be principal, say $\varphi(N) = (a_\varphi)$, for some $a_\varphi \in R$. Let

$$\Sigma = \{(a_\varphi) \mid \varphi \in \text{Hom}_R(M, R)\}$$

be the collection of the principal ideals in R obtained in this way from the R-module homomorphisms of M into R. The collection Σ is certainly nonempty since taking φ

to be the trivial homomorphism shows that $(0) \in \Sigma$. By Corollary 2, Σ has at least one maximal element i.e., there is at least one homomorphism ν of M to R so that the principal ideal $\nu(N) = (a_\nu)$ is not properly contained in any other element of Σ. Let $a_1 = a_\nu$ for this maximal element and let $y \in N$ be an element mapping to the generator a_1 under the homomorphism ν: $\nu(y) = a_1$.

We now show the element a_1 is nonzero. Let $x_1, x_2, \ldots, x_n$ be any basis of the free module M and let $\pi_i \in \mathrm{Hom}_R(M, R)$ be the natural projection homomorphism onto the ith coordinate with respect to this basis. Since $N \neq \{0\}$, there exists an i such that $\pi_i(N) \neq 0$, which in particular shows that Σ contains more than just the trivial ideal (0). Since (a_1) is a maximal element of Σ it follows that $a_1 \neq 0$.

We next show that this element a_1 divides $\varphi(y)$ for every $\varphi \in \mathrm{Hom}_R(M, R)$. To see this let d be a generator for the principal ideal generated by a_1 and $\varphi(y)$. Then d is a divisor of both a_1 and $\varphi(y)$ in R and $d = r_1 a_1 + r_2 \varphi(y)$ for some $r_1, r_2 \in R$. Consider the homomorphism $\psi = r_1 \nu + r_2 \varphi$ from M to R. Then $\psi(y) = (r_1 \nu + r_2 \varphi)(y) = r_1 a_1 + r_2 \varphi(y) = d$ so that $d \in \psi(N)$, hence also $(d) \subseteq \psi(N)$. But d is a divisor of a_1 so we also have $(a_1) \subseteq (d)$. Then $(a_1) \subseteq (d) \subseteq \psi(N)$ and by the maximality of (a_1) we must have equality: $(a_1) = (d) = \psi(N)$. In particular $(a_1) = (d)$ shows that $a_1 \mid \varphi(y)$ since d divides $\varphi(y)$.

If we apply this to the projection homomorphisms π_i we see that a_1 divides $\pi_i(y)$ for all i. Write $\pi_i(y) = a_1 b_i$ for some $b_i \in R$, $1 \leq i \leq n$ and define

$$y_1 = \sum_{i=1}^{n} b_i x_i.$$

Note that $a_1 y_1 = y$. Since $a_1 = \nu(y) = \nu(a_1 y_1) = a_1 \nu(y_1)$ and a_1 is a nonzero element of the integral domain R this shows

$$\nu(y_1) = 1.$$

We now verify that this element y_1 can be taken as one element in a basis for M and that $a_1 y_1$ can be taken as one element in a basis for N, namely that we have
(a) $M = R y_1 \oplus \ker \nu$, and
(b) $N = R a_1 y_1 \oplus (N \cap \ker \nu)$.

To see (a) let x be an arbitrary element in M and write $x = \nu(x) y_1 + (x - \nu(x) y_1)$. Since

$$\nu(x - \nu(x) y_1) = \nu(x) - \nu(x) \nu(y_1)$$
$$= \nu(x) - \nu(x) \cdot 1$$
$$= 0$$

we see that $x - \nu(x) y_1$ is an element in the kernel of ν. This shows that x can be written as the sum of an element in $R y_1$ and an element in the kernel of ν, so $M = R y_1 + \ker \nu$. To see that the sum is direct, suppose $r y_1$ is also an element in the kernel of ν. Then $0 = \nu(r y_1) = r \nu(y_1) = r$ shows that this element is indeed 0.

For (b) observe that $\nu(x')$ is divisible by a_1 for every $x' \in N$ by the definition of a_1 as a generator for $\nu(N)$. If we write $\nu(x') = b a_1$ where $b \in R$ then the decomposition we used in (a) above is $x' = \nu(x') y_1 + (x' - \nu(x') y_1) = b a_1 y_1 + (x' - b a_1 y_1)$ where the second summand is in the kernel of ν and is an element of N. This shows that

$N = Ra_1 y_1 + (N \cap \ker \nu)$. The fact that the sum in (b) is direct is a special case of the directness of the sum in (a).

We now prove part (1) of the theorem by induction on the rank, m, of N. If $m = 0$, then N is a torsion module, hence $N = 0$ since a free module is torsion free, so (1) holds trivially. Assume then that $m > 0$. Since the sum in (b) above is direct we see easily that $N \cap \ker \nu$ has rank $m - 1$ (cf. Exercise 3). By induction $N \cap \ker \nu$ is then a free R-module of rank $m - 1$. Again by the directness of the sum in (b) we see that adjoining $a_1 y_1$ to any basis of $N \cap \ker \nu$ gives a basis of N, so N is also free (of rank m), which proves (1).

Finally, we prove (2) by induction on n, the rank of M. Applying (1) to the submodule $\ker \nu$ shows that this submodule is free and because the sum in (a) is direct it is free of rank $n - 1$. By the induction assumption applied to the module $\ker \nu$ (which plays the role of M) and its submodule $\ker \nu \cap N$ (which plays the role of N), we see that there is a basis $y_2, y_3, \ldots, y_n$ of $\ker \nu$ such that $a_2 y_2, a_3 y_3, \ldots, a_m y_m$ is a basis of $N \cap \ker \nu$ for some elements $a_2, a_3, \ldots, a_m$ of R with $a_2 \mid a_3 \mid \cdots \mid a_m$. Since the sums (a) and (b) are direct, $y_1, y_2, \ldots, y_n$ is a basis of M and $a_1 y_1, a_2 y_2, \ldots, a_m y_m$ is a basis of N. To complete the induction it remains to show that a_1 divides a_2. Define a homomorphism φ from M to R by defining $\varphi(y_1) = \varphi(y_2) = 1$ and $\varphi(y_i) = 0$, for all $i > 2$, on the basis for M. Then for this homomorphism φ we have $a_1 = \varphi(a_1 y_1)$ so $a_1 \in \varphi(N)$ hence also $(a_1) \subseteq \varphi(N)$. By the maximality of (a_1) in Σ it follows that $(a_1) = \varphi(N)$. Since $a_2 = \varphi(a_2 y_2) \in \varphi(N)$ we then have $a_2 \in (a_1)$ i.e., $a_1 \mid a_2$. This completes the proof of the theorem.

Recall that the left R-module C is a *cyclic* R-module (for any ring R, not necessarily commutative nor with 1) if there is an element $x \in C$ such that $C = Rx$. We can then define an R-module homomorphism

$$\pi : R \rightarrow C$$

by $\pi(r) = rx$, which will be surjective by the assumption $C = Rx$. The First Isomorphism Theorem gives an isomorphism of (left) R-modules

$$R / \ker \pi \cong C.$$

If R is a P.I.D., $\ker \pi$ is a principal ideal, (a), so we see that the cyclic R-modules C are of the form $R/(a)$ where $(a) = \text{Ann}(C)$.

The cyclic modules are the simplest modules (since they require only one generator). The existence portion of the Fundamental Theorem states that any finitely generated module over a P.I.D. is isomorphic to the direct sum of finitely many cyclic modules.

Theorem 5. *(Fundamental Theorem, Existence: Invariant Factor Form)* Let R be a P.I.D. and let M be a finitely generated R-module.

  (1) Then M is isomorphic to the direct sum of finitely many cyclic modules. More precisely,

$$M \cong R^r \oplus R/(a_1) \oplus R/(a_2) \oplus \cdots \oplus R/(a_m)$$

  for some integer $r \geq 0$ and nonzero elements $a_1, a_2, \ldots, a_m$ of R which are not units in R and which satisfy the divisibility relations

$$a_1 \mid a_2 \mid \cdots \mid a_m.$$

(2) M is torsion free if and only if M is free.

(3) In the decomposition in (1),

$$\text{Tor}(M) \cong R/(a_1) \oplus R/(a_2) \oplus \cdots \oplus R/(a_m).$$

In particular M is a torsion module if and only if $r = 0$ and in this case the annihilator of M is the ideal (a_m).

Proof: The module M can be generated by a finite set of elements by assumption so let $x_1, x_2, \ldots, x_n$ be a set of generators of M of minimal cardinality. Let R^n be the free R-module of rank n with basis $b_1, b_2, \ldots, b_n$ and define the homomorphism $\pi : R^n \to M$ by defining $\pi(b_i) = x_i$ for all i, which is automatically surjective since $x_1, \ldots, x_n$ generate M. By the First Isomorphism Theorem for modules we have $R^n/\ker \pi \cong M$. Now, by Theorem 4 applied to R^n and the submodule $\ker \pi$ we can choose another basis $y_1, y_2, \ldots, y_n$ of R^n so that $a_1 y_1, a_2 y_2, \ldots, a_m y_m$ is a basis of $\ker \pi$ for some elements $a_1, a_2, \ldots, a_m$ of R with $a_1 \mid a_2 \mid \cdots \mid a_m$. This implies

$$M \cong R^n/\ker \pi = (R y_1 \oplus R y_2 \oplus \cdots \oplus R y_n)/(R a_1 y_1 \oplus R a_2 y_2 \oplus \cdots \oplus R a_m y_m).$$

To identify the quotient on the right hand side we use the natural surjective R-module homomorphism

$$R y_1 \oplus R y_2 \oplus \cdots \oplus R y_n \to R/(a_1) \oplus R/(a_2) \oplus \cdots \oplus R/(a_m) \oplus R^{n-m}$$

that maps $(\alpha_1 y_1, \ldots, \alpha_n y_n)$ to $(\alpha_1 \bmod (a_1), \ldots, \alpha_m \bmod (a_m), \alpha_{m+1}, \ldots, \alpha_n)$. The kernel of this map is clearly the set of elements where a_i divides α_i, $i = 1, 2, \ldots, m$, i.e., $R a_1 y_1 \oplus R a_2 y_2 \oplus \cdots \oplus R a_m y_m$ (cf. Exercise 7). Hence we obtain

$$M \cong R/(a_1) \oplus R/(a_2) \oplus \cdots \oplus R/(a_m) \oplus R^{n-m}.$$

If a is a unit in R then $R/(a) = 0$, so in this direct sum we may remove any of the initial a_i which are units. This gives the decomposition in (1) (with $r = n - m$).

Since $R/(a)$ is a torsion R-module for any nonzero element a of R, (1) immediately implies M is a torsion free module if and only if $M \cong R^r$, which is (2). Part (3) is immediate from the definitions since the annihilator of $R/(a)$ is evidently the ideal (a).

We shall shortly prove the uniqueness of the decomposition in Theorem 5, namely that if we have

$$M \cong R^{r'} \oplus R/(b_1) \oplus R/(b_2) \oplus \cdots \oplus R/(b_{m'})$$

for some integer $r' \geq 0$ and nonzero elements $b_1, b_2, \ldots, b_{m'}$ of R which are not units with

$$b_1 \mid b_2 \mid \cdots \mid b_{m'},$$

then $r = r'$, $m = m'$ and $(a_i) = (b_i)$ (so $a_i = b_i$ up to units) for all i. It is precisely the divisibility condition $a_1 \mid a_2 \mid \cdots \mid a_m$ which gives this uniqueness.

Definition. The integer r in Theorem 5 is called the *free rank* or the *Betti number* of M and the elements $a_1, a_2, \ldots, a_m \in R$ (defined up to multiplication by units in R) are called the *invariant factors* of M.

Note that until we have proved that the invariant factors of M are unique we should properly refer to *a* set of invariant factors for M (and similarly for the free rank), by which we mean any elements giving a decomposition for M as in (1) of the theorem above.

Using the Chinese Remainder Theorem it is possible to decompose the cyclic modules in Theorem 5 further so that M is the direct sum of cyclic modules whose annihilators are as simple as possible (namely (0) or generated by powers of primes in R). This gives an alternate decomposition which we shall also see is unique and which we now describe.

Suppose a is a nonzero element of the Principal Ideal Domain R. Then since R is also a Unique Factorization Domain we can write

$$a = u p_1^{\alpha_1} p_2^{\alpha_2} \cdots p_s^{\alpha_s}$$

where the p_i are distinct primes in R and u is a unit. This factorization is unique up to units, so the ideals $(p_i^{\alpha_i})$, $i = 1, \ldots, s$ are uniquely defined. For $i \neq j$ we have $(p_i^{\alpha_i}) + (p_j^{\alpha_j}) = R$ since the sum of these two ideals is generated by a greatest common divisor, which is 1 for distinct primes p_i, p_j. Put another way, the ideals $(p_i^{\alpha_i})$, $i = 1, \ldots, s$, are comaximal in pairs. The intersection of all these ideals is the ideal (a) since a is the least common multiple of $p_1^{\alpha_1}, p_2^{\alpha_2}, \ldots, p_s^{\alpha_s}$. Then the Chinese Remainder Theorem (Theorem 7.17) shows that

$$R\big/(a) \cong R\big/(p_1^{\alpha_1}) \oplus R\big/(p_2^{\alpha_2}) \oplus \cdots \oplus R\big/(p_s^{\alpha_s})$$

as rings and also as R-modules.

Applying this to the modules in Theorem 5 allows us to write each of the direct summands $R\big/(a_i)$ for the invariant factor a_i of M as a direct sum of cyclic modules whose annihilators are the prime power divisors of a_i. This proves:

Theorem 6. *(Fundamental Theorem, Existence: Elementary Divisor Form)* Let R be a P.I.D. and let M be a finitely generated R-module. Then M is the direct sum of a finite number of cyclic modules whose annihilators are either (0) or generated by powers of primes in R, i.e.,

$$M \cong R^r \oplus R\big/(p_1^{\alpha_1}) \oplus R\big/(p_2^{\alpha_2}) \oplus \cdots \oplus R\big/(p_t^{\alpha_t})$$

where $r \geq 0$ is an integer and $p_1^{\alpha_1}, \ldots, p_t^{\alpha_t}$ are positive powers of (not necessarily distinct) primes in R.

We proved Theorem 6 by using the prime power factors of the invariant factors for M. In fact we shall see that the decomposition of M into a direct sum of cyclic modules whose annihilators are (0) or prime powers as in Theorem 6 is unique, i.e., the integer r and the ideals $(p_1^{\alpha_1}), \ldots, (p_t^{\alpha_t})$ are uniquely defined for M. These prime powers are given a name:

Definition. Let R be a P.I.D. and let M be a finitely generated R-module as in Theorem 6. The prime powers $p_1^{\alpha_1}, \ldots, p_t^{\alpha_t}$ (defined up to multiplication by units in R) are called the *elementary divisors* of M.

Suppose M is a finitely generated torsion module over the Principal Ideal Domain R. If for the *distinct* primes $p_1, p_2, \ldots, p_n$ occurring in the decomposition in Theorem 6 we group together all the cyclic factors corresponding to the same prime p_i we see in particular that M can be written as a direct sum

$$M = N_1 \oplus N_2 \oplus \cdots \oplus N_n$$

where N_i consists of all the elements of M which are annihilated by some power of the prime p_i. This result holds also for modules over R which may not be finitely generated:

Theorem 7. *(The Primary Decomposition Theorem)* Let R be a P.I.D. and let M be a nonzero torsion R-module (not necessarily finitely generated) with nonzero annihilator a. Suppose the factorization of a into distinct prime powers in R is

$$a = u p_1^{\alpha_1} p_2^{\alpha_2} \cdots p_n^{\alpha_n}$$

and let $N_i = \{x \in M \mid p_i^{\alpha_i} x = 0\}$, $1 \le i \le n$. Then N_i is a submodule of M with annihilator $p_i^{\alpha_i}$ and is the submodule of M of all elements annihilated by some power of p_i. We have

$$M = N_1 \oplus N_2 \oplus \cdots \oplus N_n.$$

If M is finitely generated then each N_i is the direct sum of finitely many cyclic modules whose annihilators are divisors of $p_i^{\alpha_i}$.

Proof: We have already proved these results in the case where M is finitely generated over R. In the general case it is clear that N_i is a submodule of M with annihilator dividing $p_i^{\alpha_i}$. Since R is a P.I.D. the ideals $(p_i^{\alpha_i})$ and $(p_j^{\alpha_j})$ are comaximal for $i \neq j$, so the direct sum decomposition of M can be proved easily by modifying the argument in the proof of the Chinese Remainder Theorem to apply it to modules. Using this direct sum decomposition it is easy to see that the annihilator of N_i is precisely $p_i^{\alpha_i}$.

Definition. The submodule N_i in the previous theorem is called the p_i-*primary component* of M.

Notice that with this terminology the elementary divisors of a finitely generated module M are just the invariant factors of the primary components of $\text{Tor}(M)$.

We now prove the uniqueness statements regarding the decompositions in the Fundamental Theorem.

Note that if M is any module over a commutative ring R and a is an element of R then $aM = \{am \mid m \in M\}$ is a submodule of M. Recall also that in a Principal Ideal Domain R the nonzero prime ideals are maximal, hence the quotient of R by a nonzero prime ideal is a field.

Lemma 8. Let R be a P.I.D. and let p be a prime in R. Let F denote the field $R/(p)$.

  (1) Let $M = R^r$. Then $M/pM \cong F^r$.

  (2) Let $M = R/(a)$ where a is a nonzero element of R. Then

$$M/pM \cong \begin{cases} F & \text{if } p \text{ divides } a \text{ in } R \\ 0 & \text{if } p \text{ does not divide } a \text{ in } R. \end{cases}$$

  (3) Let $M = R/(a_1) \oplus R/(a_2) \oplus \cdots \oplus R/(a_k)$ where each a_i is divisible by p. Then $M/pM \cong F^k$.

Proof: (1) There is a natural map from R^r to $(R/(p))^r$ defined by mapping $(\alpha_1, \ldots, \alpha_r)$ to $(\alpha_1 \bmod (p), \ldots, \alpha_r \bmod (p))$. This is clearly a surjective R-module homomorphism with kernel consisting of the r-tuples all of whose coordinates are divisible by p, i.e., pR^r, so $R^r/pR^r \cong (R/(p))^r$, which is (1).

   (2) This follows from the Isomorphism Theorems: note first that $p(R/(a))$ is the image of the ideal (p) in the quotient $R/(a)$, hence is $(p)+(a)/(a)$. The ideal $(p)+(a)$ is generated by a greatest common divisor of p and a, hence is (p) if p divides a and is $R = (1)$ otherwise. Hence $pM = (p)/(a)$ if p divides a and is $R/(a) = M$ otherwise. If p divides a then $M/pM = (R/(a))/((p)/(a)) \cong R/(p)$, and if p does not divide a then $M/pM = M/M = 0$, which proves (2).

   (3) This follows from (2) as in the proof of part (1) of Theorem 5.

Theorem 9. *(Fundamental Theorem, Uniqueness)* Let R be a P.I.D.

  (1) Two finitely generated R-modules M_1 and M_2 are isomorphic if and only if they have the same free rank and the same list of invariant factors.

  (2) Two finitely generated R-modules M_1 and M_2 are isomorphic if and only if they have the same free rank and the same list of elementary divisors.

Proof: If M_1 and M_2 have the same free rank and list of invariant factors or the same free rank and list of elementary divisors then they are clearly isomorphic.

   Suppose that M_1 and M_2 are isomorphic. Any isomorphism between M_1 and M_2 maps the torsion in M_1 to the torsion in M_2 so we must have $\text{Tor}(M_1) \cong \text{Tor}(M_2)$. Then $R^{r_1} \cong M_1/\text{Tor}(M_1) \cong M_2/\text{Tor}(M_2) \cong R^{r_2}$ where r_1 is the free rank of M_1 and r_2 is the free rank of M_2. Let p be any nonzero prime in R. Then from $R^{r_1} \cong R^{r_2}$ we obtain $R^{r_1}/pR^{r_1} \cong R^{r_2}/pR^{r_2}$. By (1) of the previous lemma, this implies $F^{r_1} \cong F^{r_2}$ where F is the field R/pR. Hence we have an isomorphism of an r_1-dimensional vector space over F with an r_2-dimensional vector space over F, so that $r_1 = r_2$ and M_1 and M_2 have the same free rank.

   We are reduced to showing that M_1 and M_2 have the same lists of invariant factors and elementary divisors. To do this we need only work with the isomorphic torsion modules $\text{Tor}(M_1)$ and $\text{Tor}(M_2)$, i.e., we may as well assume that both M_1 and M_2 are torsion R-modules.

   We first show they have the same elementary divisors. It suffices to show that for any fixed prime p the elementary divisors which are a power of p are the same for both M_1 and M_2. If $M_1 \cong M_2$ then the p-primary submodule of M_1 (= the direct sum of the cyclic factors whose elementary divisors are powers of p) is isomorphic to the p-primary submodule of M_2, since these are the submodules of elements which are annihilated by some power of p. We are therefore reduced to the case of proving that if two modules M_1 and M_2 which have annihilator a power of p are isomorphic then they have the same elementary divisors.

We proceed by induction on the power of p in the annihilator of M_1 (which is the same as the annihilator of M_2 since M_1 and M_2 are isomorphic). If this power is 0, then both M_1 and M_2 are 0 and we are done. Otherwise M_1 (and M_2) have nontrivial elementary divisors. Suppose the elementary divisors of M_1 are given by

$$\text{elementary divisors of } M_1: \underbrace{p, p, \dots, p}_{m \text{ times}}, p^{\alpha_1}, p^{\alpha_2}, \dots, p^{\alpha_s},$$

where $2 \le \alpha_1 \le \alpha_2 \le \cdots \le \alpha_s$, i.e., M_1 is the direct sum of cyclic modules with generators $x_1, x_2, \dots, x_m, x_{m+1}, \dots, x_{m+s}$, say, whose annihilators are $(p), (p), \dots, (p)$, $(p^{\alpha_1}), \dots, (p^{\alpha_s})$, respectively. Then the submodule pM_1 has elementary divisors

$$\text{elementary divisors of } pM_1: p^{\alpha_1-1}, p^{\alpha_2-1}, \dots, p^{\alpha_s-1}$$

since pM_1 is the direct sum of the cyclic modules with generators $px_1, px_2, \dots, px_m$, $px_{m+1}, \dots, px_{m+s}$ whose annihilators are $(1), (1), \dots, (1), (p^{\alpha_1-1}), \dots, (p^{\alpha_s-1})$, respectively. Similarly, if the elementary divisors of M_2 are given by

$$\text{elementary divisors of } M_2: \underbrace{p, p, \dots, p}_{n \text{ times}}, p^{\beta_1}, p^{\beta_2}, \dots, p^{\beta_t},$$

where $2 \le \beta_1 \le \beta_2 \le \cdots \le \beta_t$, then pM_2 has elementary divisors

$$\text{elementary divisors of } pM_2: p^{\beta_1-1}, p^{\beta_2-1}, \dots, p^{\beta_t-1}.$$

Since $M_1 \cong M_2$, also $pM_1 \cong pM_2$ and the power of p in the annihilator of pM_1 is one less than the power of p in the annihilator of M_1. By induction, the elementary divisors for pM_1 are the same as the elementary divisors for pM_2, i.e., $s = t$ and $\alpha_i - 1 = \beta_i - 1$ for $i = 1, 2, \dots, s$, hence $\alpha_i = \beta_i$ for $i = 1, 2, \dots, s$. Finally, since also $M_1/pM_1 \cong M_2/pM_2$ we see from (3) of the lemma above that $F^{m+s} \cong F^{n+t}$, which shows that $m + s = n + t$ hence $m = n$ since we have already seen $s = t$. This proves that the set of elementary divisors for M_1 is the same as the set of elementary divisors for M_2.

We now show that M_1 and M_2 must have the same invariant factors. Suppose $a_1 \mid a_2 \mid \cdots \mid a_m$ are invariant factors for M_1. We obtain a set of elementary divisors for M_1 by taking the prime power factors of these elements. Note that then the divisibility relations on the invariant factors imply that a_m is the product of the largest of the prime powers among these elementary divisors, a_{m-1} is the product of the largest prime powers among these elementary divisors once the factors for a_m have been removed, and so on. If $b_1 \mid b_2 \mid \cdots \mid b_n$ are invariant factors for M_2 then we similarly obtain a set of elementary divisors for M_2 by taking the prime power factors of these elements. But we showed above that the elementary divisors for M_1 and M_2 are the same, and it follows that the same is true of the invariant factors.

Corollary 10. Let R be a P.I.D. and let M be a finitely generated R-module.

(1) The elementary divisors of M are the prime power factors of the invariant factors of M.

(2) The largest invariant factor of M is the product of the largest of the distinct prime powers among the elementary divisors of M, the next largest invariant factor is the product of the largest of the distinct prime powers among the remaining elementary divisors of M, and so on.

Proof: The procedure in (1) gives *a* set of elementary divisors and since the elementary divisors for M are unique by the theorem, it follows that the procedure in (1) gives *the* set of elementary divisors. Similarly for (2).

Corollary 11. *(The Fundamental Theorem of Finitely Generated Abelian Groups)* See Theorem 5.3 and Theorem 5.5.

Proof: Take $R = \mathbb{Z}$ in Theorems 5, 6 and 9 (note however that the invariant factors are listed in reverse order in Chapter 5 for computational convenience).

The procedure for passing between elementary divisors and invariant factors in Corollary 10 is described in some detail in Chapter 5 in the case of finitely generated abelian groups.

Note also that if a finitely generated module M is written as a direct sum of cyclic modules of the form $R/(a)$ then the ideals (a) which occur are not in general unique unless some additional conditions are imposed (such as the divisibility condition for the invariant factors or the condition that a be the power of a prime in the case of the elementary divisors). To decide whether two modules are isomorphic it is necessary to first write them in such a standard (or *canonical*) form.

EXERCISES

1. Let M be a module over the integral domain R.
   (a) Suppose x is a nonzero torsion element in M. Show that x and 0 are "linearly dependent." Conclude that the rank of $\text{Tor}(M)$ is 0, so that in particular any torsion R-module has rank 0.
   (b) Show that the rank of M is the same as the rank of the (torsion free) quotient $M/\text{Tor}M$.

2. Let M be a module over the integral domain R.
   (a) Suppose that M has rank n and that $x_1, x_2, \ldots, x_n$ is any maximal set of linearly independent elements of M. Let $N = R x_1 + \ldots + R x_n$ be the submodule generated by $x_1, x_2, \ldots, x_n$. Prove that N is isomorphic to R^n and that the quotient M/N is a torsion R-module (equivalently, the elements $x_1, \ldots, x_n$ are linearly independent and for any $y \in M$ there is a nonzero element $r \in R$ such that ry can be written as a linear combination $r_1 x_1 + \ldots + r_n x_n$ of the x_i).
   (b) Prove conversely that if M contains a submodule N that is free of rank n (i.e., $N \cong R^n$) such that the quotient M/N is a torsion R-module then M has rank n. [Let $y_1, y_2, \ldots, y_{n+1}$ be any $n + 1$ elements of M. Use the fact that M/N is torsion to write $r_i y_i$ as a linear combination of a basis for N for some nonzero elements $r_1, \ldots, r_{n+1}$ of R. Use an argument as in the proof of Proposition 3 to see that the $r_i y_i$, and hence also the y_i, are linearly dependent.]

3. Let R be an integral domain and let A and B be R-modules of ranks m and n, respectively. Prove that the rank of $A \oplus B$ is $m + n$. [Use the previous exercise.]

4. Let R be an integral domain, let M be an R-module and let N be a submodule of M. Suppose M has rank n, N has rank r and the quotient M/N has rank s. Prove that $n = r + s$. [Let $x_1, x_2, \ldots, x_s$ be elements of M whose images in M/N are a maximal set of independent elements and let $x_{s+1}, x_{s+2}, \ldots, x_{s+r}$ be a maximal set of independent elements in N. Prove that $x_1, x_2, \ldots, x_{s+r}$ are linearly independent in M and that for any element $y \in M$ there is a nonzero element $r \in R$ such that ry is a linear combination of these elements. Then use Exercise 2.]

5. Let $R = \mathbb{Z}[x]$ and let $M = (2, x)$ be the ideal generated by 2 and x, considered as a submodule of R. Show that $\{2, x\}$ is not a basis of M. [Find a nontrivial R-linear dependence between these two elements.] Show that the rank of M is 1 but that M is not free of rank 1 (cf. Exercise 2).

6. Show that if R is an integral domain and M is any nonprincipal ideal of R then M is torsion free of rank 1 but is not a free R-module.

7. Let R be any ring, let $A_1, A_2, \ldots, A_m$ be R-modules and let B_i be a submodule of A_i, $1 \le i \le m$. Prove that

$$(A_1 \oplus A_2 \oplus \cdots \oplus A_m)/(B_1 \oplus B_2 \oplus \cdots \oplus B_m) \cong (A_1/B_1) \oplus (A_2/B_2) \oplus \cdots \oplus (A_m/B_m).$$

8. Let R be a P.I.D., let B be a torsion R-module and let p be a prime in R. Prove that if $pb = 0$ for some nonzero $b \in B$, then $\text{Ann}(B) \subseteq (p)$.

9. Give an example of an integral domain R and a nonzero torsion R-module M such that $\text{Ann}(M) = 0$. Prove that if N is a finitely generated torsion R-module then $\text{Ann}(N) \ne 0$.

10. For p a prime in the P.I.D. R and N an R-module prove that the p-primary component of N is a submodule of N and prove that N is the direct sum of its p-primary components (there need not be finitely many of them).

11. Let R be a P.I.D., let a be a nonzero element of R and let $M = R/(a)$. For any prime p of R prove that

$$p^{k-1}M/p^k M \cong \begin{cases} R/(p) & \text{if } k \le n \\ 0 & \text{if } k > n, \end{cases}$$

where n is the power of p dividing a in R.

12. Let R be a P.I.D. and let p be a prime in R.
    (a) Let M be a finitely generated torsion R-module. Use the previous exercise to prove that $p^{k-1}M/p^k M \cong F^{n_k}$ where F is the field $R/(p)$ and n_k is the number of elementary divisors of M which are powers p^α with $\alpha \ge k$.
    (b) Suppose M_1 and M_2 are isomorphic finitely generated torsion R-modules. Use (a) to prove that, for every $k \ge 0$, M_1 and M_2 have the same number of elementary divisors p^α with $\alpha \ge k$. Prove that this implies M_1 and M_2 have the same set of elementary divisors.

13. If M is a finitely generated module over the P.I.D. R, describe the structure of $M/\text{Tor}(M)$.

14. Let R be a P.I.D. and let M be a torsion R-module. Prove that M is irreducible (cf. Exercises 9 to 11 of Section 10.3) if and only if $M = Rm$ for any nonzero element $m \in M$ where the annihilator of m is a nonzero prime ideal (p).

15. Prove that if R is a Noetherian ring then R^n is a Noetherian R-module. [Fix a basis of R^n. If M is a submodule of R^n show that the collection of first coordinates of elements of M is a submodule of R hence is finitely generated. Let $m_1, m_2, \ldots, m_k$ be elements of M

whose first coordinates generate this submodule of R. Show that any element of M can be written as an R-linear combination of $m_1, m_2, \ldots, m_k$ plus an element of M whose first coordinate is 0. Prove that $M \cap R^{n-1}$ is a submodule of R^{n-1} where R^{n-1} is the set of elements of R^n with first coordinate 0 and then use induction on n.

The following set of exercises outlines a proof of Theorem 5 in the special case where R is a Euclidean Domain using a matrix argument involving row and column operations. This applies in particular to the cases $R = \mathbb{Z}$ and $R = F[x]$ of interest in the applications and is computationally useful.

Let R be a Euclidean Domain and let M be an R-module.

16. Prove that M is finitely generated if and only if there is a surjective R-homomorphism $\varphi : R^n \to M$ for some integer n (this is true for any ring R).

Suppose $\varphi : R^n \to M$ is a surjective R-module homomorphism. By Exercise 15, $\ker \varphi$ is finitely generated. If $x_1, x_2, \ldots, x_n$ is a basis for R^n and $y_1, \ldots, y_m$ are generators for $\ker \varphi$ we have

$$y_i = a_{i1}x_1 + a_{i2}x_2 + \cdots + a_{in}x_n \quad i = 1, 2, \ldots, m$$

with coefficients $a_{ij} \in R$. It follows that the homomorphism φ (hence the module structure of M) is determined by the choice of generators for R^n and the matrix $A = (a_{ij})$. Such a matrix A will be called a *relations matrix*.

17. (a) Show that interchanging x_i and x_j in the basis for R^n interchanges the ith column with the jth column in the corresponding relations matrix.

(b) Show that, for any $a \in R$, replacing the element x_j by $x_j - ax_i$ in the basis for R^n gives another basis for R^n and that the corresponding relations matrix for this basis is the same as the original relations matrix except that a times the jth column has been added to the ith column. [Note that $\cdots + a_ix_i + \cdots + a_jx_j + \cdots = \cdots + (a_i + aa_j)x_i + \cdots + a_j(x_j - ax_i) + \ldots .$]

18. (a) Show that interchanging the generators y_i and y_j interchanges the ith row with the jth row in the relations matrix.

(b) Show that, for any $a \in R$, replacing the element y_j by $y_j - ay_i$ gives another set of generators for $\ker \varphi$ and that the corresponding relations matrix for this choice of generators is the same as the original relations matrix except that $-a$ times the ith row has been added to the jth row.

19. By the previous two exercises we may perform elementary row and column operations on a given relations matrix by choosing different generators for R^n and $\ker \varphi$. If all relation matrices are the zero matrix then $\ker \varphi = 0$ and $M \cong R^n$. Otherwise let a_1 be the (nonzero) g.c.d. (recall R is a Euclidean Domain) of all the entries in a fixed initial relations matrix for M.

(a) Prove that by elementary row and column operations we may assume a_1 occurs in a relations matrix of the form

$$\begin{pmatrix} a_1 & a_{12} & \cdots & a_{1n} \\ a_{21} & a_{22} & \cdots & a_{2n} \\ \vdots & \vdots & \ddots & \vdots \\ a_{m1} & a_{m2} & \cdots & a_{mn} \end{pmatrix}$$

where a_1 divides a_{ij}, $i = 1, 2, \ldots, m$, $j = 1, 2, \ldots, n$.

470

(b) Prove that there is a relations matrix of the form

$$\begin{pmatrix} a_1 & 0 & \cdots & 0 \\ 0 & a_{22} & \cdots & a_{2n} \\ \vdots & \vdots & \ddots & \vdots \\ 0 & a_{m2} & \cdots & a_{mn} \end{pmatrix}$$

where a_1 divides all the entries.

(c) Let a_2 be a g.c.d. of all the entries except the element a_1 in the relations matrix in (b). Prove that there is a relations matrix of the form

$$\begin{pmatrix} a_1 & 0 & 0 & \cdots & 0 \\ 0 & a_2 & 0 & \cdots & 0 \\ 0 & 0 & a_{33} & \cdots & a_{3n} \\ \vdots & \vdots & \vdots & \ddots & \vdots \\ 0 & 0 & a_{m3} & \cdots & a_{mn} \end{pmatrix}$$

where a_1 divides a_2 and a_2 divides all the other entries of the matrix.

(d) Prove that there is a relations matrix of the form $\begin{pmatrix} D & 0 \\ 0 & 0 \end{pmatrix}$ where D is a diagonal matrix with nonzero entries $a_1, a_2, \ldots, a_k, k \le n$, satisfying

$$a_1 \mid a_2 \mid \cdots \mid a_k.$$

Conclude that

$$M \cong R/(a_1) \oplus R/(a_2) \oplus \cdots \oplus R/(a_k) \oplus R^{n-k}.$$

If n is not the minimal number of generators required for M then some of the initial elements $a_1, a_2, \ldots$ above will be units, so the corresponding direct summands above will be 0. If we remove these irrelevant factors we have produced the invariant factors of the module M. Further, the image of the new generators for R^n corresponding to the direct summands above will then be a set of R-generators for the cyclic submodules of M in its invariant factor decomposition (note that the image in M of the generators corresponding to factors with a_i a unit will be 0). The *column* operations performed in the relations matrix reduction correspond to changing the basis used for R^n as described in Exercise 17:

(a) Interchanging the ith column with the jth column corresponds to interchanging the ith and j^{th} elements in the basis for R^n.

(b) For any $a \in R$, adding a times the jth column to the ith column corresponds to subtracting a times the ith basis element from the jth basis element.

Keeping track of the column operations performed and changing the initial choice of generators for M in the same way therefore gives a set of R-generators for the cyclic submodules of M in its invariant factor decomposition.

This process is quite fast computationally once an initial set of generators for M and initial relations matrix are determined. The element a_1 is determined using the Euclidean Algorithm as the g.c.d. of the elements in the initial relations matrix. Using the row and column operations we can obtain the appropriate linear combination of the entries to produce this g.c.d. in the (1,1)-position of a new relations matrix. One then subtracts the appropriate multiple of the first column and first row to obtain a matrix as in Exercise 19(b), then iterates this process. Some examples of this procedure in a special case are given at the end of the following section.

20. Let R be an integral domain with quotient field F and let M be any R-module. Prove that the rank of M equals the dimension of the vector space $F \otimes_R M$ over F.

21. Prove that a finitely generated module over a P.I.D. is projective if and only if it is free.

22. Let R be a P.I.D. that is not a field. Prove that no finitely generated R-module is injective. [Use Exercise 4, Section 10.5 to consider torsion and free modules separately.]

12.2 THE RATIONAL CANONICAL FORM

We now apply our results on finitely generated modules in the special case where the P.I.D. is the ring $F[x]$ of polynomials in x with coefficients in a field F.

Let V be a finite dimensional vector space over F of dimension n and let T be a fixed linear transformation of V (i.e., from V to itself). As we saw in Chapter 10 we can consider V as an $F[x]$-module where the element x acts on V as the linear transformation T (and so any polynomial in x acts on V as the same polynomial in T). Since V has finite dimension over F by assumption, it is by definition finitely generated as an F-module, hence certainly finitely generated as an $F[x]$-module, so the classification theorems of the preceding section apply.

Any nonzero free $F[x]$-module (being isomorphic to a direct sum of copies of $F[x]$) is an infinite dimensional vector space over F, so if V has finite dimension over F then it must in fact be a torsion $F[x]$-module (i.e., its free rank is 0). It follows from the Fundamental Theorem that then V is isomorphic as an $F[x]$-module to the direct sum of cyclic, torsion $F[x]$-modules. We shall see that this decomposition of V will allow us to choose a basis for V with respect to which the matrix representation for the linear transformation T is in a specific simple form. When we use the invariant factor decomposition of V we obtain the *rational canonical form* for the matrix for T, which we analyze in this section. When we use the elementary divisor decomposition (and when F contains all the eigenvalues of T) we obtain the *Jordan canonical form*, considered in the following section and mentioned earlier as the matrix representing T which is as close to being a diagonal matrix as possible. The uniqueness portion of the Fundamental Theorem ensures that the rational and Jordan canonical forms are unique (which is why they are referred to as *canonical*).

One important use of these canonical forms is to classify the distinct linear transformations of V. In particular they allow us to determine when two matrices represent the same linear transformation, i.e., when two given $n \times n$ matrices are similar.

Note that this will be another instance where the structure of the space being acted upon (the invariant factor decomposition of V for example) is used to obtain significant information on the algebraic objects (in this case the linear transformations) which are acting. This will be considered in the case of *groups* acting on vector spaces in Chapter 18 (and goes under the name of Representation Theory of Groups).

Before describing the rational canonical form in detail we first introduce some linear algebra.

Definition.

(1) An element λ of F is called an *eigenvalue* of the linear transformation T if there is a nonzero vector $v \in V$ such that $T(v) = \lambda v$. In this situation v is called an *eigenvector* of T with corresponding eigenvalue λ.

(2) If A is an $n \times n$ matrix with coefficients in F, an element λ is called an *eigenvalue* of A with corresponding eigenvector v if v is a nonzero $n \times 1$ column vector such that $Av = \lambda v$.

(3) If λ is an eigenvalue of the linear transformation T, the set $\{v \in V \mid T(v) = \lambda v\}$ is called the *eigenspace* of T corresponding to the eigenvalue λ. Similarly, if λ is an eigenvalue of the $n \times n$ matrix A, the set of $n \times 1$ matrices v with $Av = \lambda v$ is called the *eigenspace* of A corresponding to the eigenvalue λ.

Note that if we fix a basis $\mathcal{B}$ of V then any linear transformation T of V has an associated $n \times n$ matrix A. Conversely, if A is any $n \times n$ matrix then the map T defined by $T(v) = Av$ for $v \in V$, where the v on the right is the $n \times 1$ vector consisting of the coordinates of v with respect to the fixed basis $\mathcal{B}$ of V, is a linear transformation of V. Then v is an eigenvector of T with corresponding eigenvalue λ if and only if the coordinate vector of v with respect to $\mathcal{B}$ is an eigenvector of A with eigenvalue λ. In other words, the eigenvalues for the linear transformation T are the same as the eigenvalues for the matrix A of T with respect to any fixed basis for V.

Definition. The determinant of a linear transformation from V to V is the determinant of any matrix representing the linear transformation (note that this does not depend on the choice of the basis used).

Proposition 12. The following are equivalent:
  (1) λ is an eigenvalue of T
  (2) $\lambda I - T$ is a singular linear transformation of V
  (3) $\det(\lambda I - T) = 0$.

Proof: Since λ is an eigenvalue of T with corresponding eigenvector v if and only if v is a nonzero vector in the kernel of $\lambda I - T$, it follows that (1) and (2) are equivalent. (2) and (3) are equivalent by our results on determinants.

Definition. Let x be an indeterminate over F. The polynomial $\det(xI - T)$ is called the *characteristic polynomial* of T and will be denoted $c_T(x)$. If A is an $n \times n$ matrix with coefficients in F, $\det(xI - A)$ is called the *characteristic polynomial* of A and will be denoted $c_A(x)$.

It is easy to see by expanding the determinant that the characteristic polynomial of either T or A is a monic polynomial of degree $n = \dim V$. Proposition 12 says that the set of eigenvalues of T (or A) is precisely the set of roots of the characteristic polynomial of T (of A, respectively). In particular, T has at most n distinct eigenvalues.

We have seen that V considered as a module over $F[x]$ via the linear transformation T is a torsion $F[x]$-module. Let $m(x) \in F[x]$ be the unique monic polynomial generating the annihilator of V in $F[x]$. Equivalently, $m(x)$ is the unique monic polynomial of minimal degree annihilating V (i.e., such that $m(T)$ is the 0 linear transformation), and if $f(x) \in F[x]$ is any polynomial annihilating V, $m(x)$ divides $f(x)$. Since the ring of all $n \times n$ matrices over F is isomorphic to the collection of all linear transformations of V to itself (an isomorphism is obtained by choosing a basis for V), it follows that for any $n \times n$ matrix A over F there is similarly a unique monic polynomial of minimal degree with $m(A)$ the zero matrix.

Definition. The unique monic polynomial which generates the ideal $\text{Ann}(V)$ in $F[x]$ is called the *minimal polynomial* of T and will be denoted $m_T(x)$. The unique monic polynomial of smallest degree which when evaluated at the matrix A is the zero matrix is called the *minimal polynomial* of A and will be denoted $m_A(x)$.

It is easy to see (cf. Exercise 5) that the degrees of these minimal polynomials are at most n^2 where n is the dimension of V. We shall shortly prove that the minimal polynomial for T is a divisor of the characteristic polynomial for T (this is the *Cayley–Hamilton Theorem*), and similarly for A, so in fact the degrees of these polynomials are at most n.

We now describe the *rational canonical form* of the linear transformation T (respectively, of the $n \times n$ matrix A). By Theorem 5 we have an isomorphism

$$V \cong F[x]/(a_1(x)) \oplus F[x]/(a_2(x)) \oplus \cdots \oplus F[x]/(a_m(x)) \qquad (12.1)$$

of $F[x]$-modules where $a_1(x), a_2(x), \ldots, a_m(x)$ are polynomials in $F[x]$ of degree at least one with the divisibility conditions

$$a_1(x) \mid a_2(x) \mid \cdots \mid a_m(x).$$

These invariant factors $a_i(x)$ are only determined up to a unit in $F[x]$ but since the units of $F[x]$ are precisely the nonzero elements of F (i.e., the nonzero constant polynomials), we may make these polynomials *unique* by stipulating that they be *monic*.

Since the annihilator of V is the ideal $(a_m(x))$ (part (3) of Theorem 5), we immediately obtain:

Proposition 13. The minimal polynomial $m_T(x)$ is the largest invariant factor of V. All the invariant factors of V divide $m_T(x)$.

We shall see below how to calculate not only the minimal polynomial for T but also the other invariant factors.

We now choose a basis for each of the direct summands for V in the decomposition (1) above for which the matrix for T is quite simple. Recall that the linear transformation T acting on the left side of (1) is the element x acting by multiplication on each of the factors on the right side of the isomorphism in (1).

We have seen in the example following Proposition 1 of Chapter 11 that the elements $1, \bar{x}, \bar{x}^2, \ldots, \bar{x}^{k-1}$ give a basis for the vector space $F[x]/(a(x))$ where $a(x) = x^k + b_{k-1}x^{k-1} + \cdots + b_1 x + b_0$ is any monic polynomial in $F[x]$ and $\bar{x} = x \bmod (a(x))$. With respect to this basis the linear transformation of multiplication by x acts in a simple manner:

$$
x : \quad
\begin{aligned}
1 &\mapsto \bar{x} \\
\bar{x} &\mapsto \bar{x}^2 \\
\bar{x}^2 &\mapsto \bar{x}^3 \\
&\vdots \\
\bar{x}^{k-2} &\mapsto \bar{x}^{k-1} \\
\bar{x}^{k-1} &\mapsto \bar{x}^k = -b_0 - b_1 \bar{x} - \cdots - b_{k-1} \bar{x}^{k-1}
\end{aligned}
$$

where the last equality is because $\bar{x}^k + b_{k-1}\bar{x}^{k-1} + \cdots + b_1\bar{x} + b_0 = 0$ since $a(\bar{x}) = 0$ in $F[x]/(a(x))$. With respect to this basis, the matrix for multiplication by x is therefore

$$\begin{pmatrix} 0 & 0 & \cdots & \cdots & \cdots & -b_0 \\ 1 & 0 & \cdots & \cdots & \cdots & -b_1 \\ 0 & 1 & \cdots & \cdots & \cdots & -b_2 \\ 0 & 0 & \ddots & & & \vdots \\ \vdots & \vdots & & \ddots & & \vdots \\ 0 & 0 & \cdots & \cdots & 1 & -b_{k-1} \end{pmatrix}.$$

Such matrices are given a name:

Definition. Let $a(x) = x^k + b_{k-1}x^{k-1} + \cdots + b_1x + b_0$ be any monic polynomial in $F[x]$. The *companion matrix* of $a(x)$ is the $k \times k$ matrix with 1's down the first subdiagonal, $-b_0, -b_1, \ldots, -b_{k-1}$ down the last column and zeros elsewhere. The companion matrix of $a(x)$ will be denoted by $\mathcal{C}_{a(x)}$.

We apply this to each of the cyclic modules on the right side of (1) above and let $\mathcal{B}_i$ be the elements of V corresponding to the basis chosen above for the cyclic factor $F[x]/(a_i(x))$ under the isomorphism in (1). Then by definition the linear transformation T acts on $\mathcal{B}_i$ by the companion matrix for $a_i(x)$ since we have seen that this is how multiplication by x acts. The union $\mathcal{B}$ of the $\mathcal{B}_i$'s gives a basis for V since the sum on the right of (1) is direct and with respect to this basis the linear transformation T has as matrix the *direct sum* of the companion matrices for the invariant factors, i.e.,

$$\begin{pmatrix} \mathcal{C}_{a_1(x)} & & & \\ & \mathcal{C}_{a_2(x)} & & \\ & & \ddots & \\ & & & \mathcal{C}_{a_m(x)} \end{pmatrix}. \tag{12.2}$$

Notice that this matrix is uniquely determined from the invariant factors of the $F[x]$-module V and, by Theorem 9, the list of invariant factors uniquely determines the module V up to isomorphism as an $F[x]$-module.

Definition.
   (1) A matrix is said to be in *rational canonical form* if it is the direct sum of companion matrices for monic polynomials $a_1(x), \ldots, a_m(x)$ of degree at least one with $a_1(x) \mid a_2(x) \mid \cdots \mid a_m(x)$. The polynomials $a_i(x)$ are called the *invariant factors* of the matrix. Such a matrix is also said to be a *block diagonal* matrix with blocks the companion matrices for the $a_i(x)$.
   (2) A *rational canonical form* for a linear transformation T is a matrix representing T which is in rational canonical form.

We have seen that any linear transformation T has a rational canonical form. We now see that this rational canonical form is unique (hence is called *the* rational canonical form for T). To see this note that the process we used to determine the matrix of T

from the direct sum decomposition is reversible. Suppose $b_1(x), b_2(x), \ldots, b_t(x)$ are monic polynomials in $F[x]$ of degree at least one such that $b_i(x) \mid b_{i+1}(x)$ for all i and suppose for some basis $\mathcal{E}$ of V, that the matrix of T with respect to the basis $\mathcal{E}$ is the direct sum of the companion matrices of the $b_i(x)$. Then V must be a direct sum of T-stable subspaces D_i, one for each $b_i(x)$ in such a way that the matrix of T on each D_i is the companion matrix of $b_i(x)$. Let $\mathcal{E}_i$ be the corresponding (ordered) basis of D_i (so $\mathcal{E}$ is the union of the $\mathcal{E}_i$) and let e_i be the first basis element in $\mathcal{E}_i$. Then it is easy to see that D_i is a cyclic $F[x]$-module with generator e_i and that the annihilator of D_i is $b_i(x)$. Thus the torsion $F[x]$-module V decomposes into a direct sum of cyclic $F[x]$-modules in two ways, both of which satisfy the conditions of Theorem 5, i.e., both of which give lists of invariant factors. Since the invariant factors are unique by Theorem 9, $a_i(x)$ and $b_i(x)$ must differ by a unit factor in $F[x]$ and since the polynomials are monic by assumption, we must have $a_i(x) = b_i(x)$ for all i. This proves the following result:

Theorem 14. *(Rational Canonical Form for Linear Transformations)* Let V be a finite dimensional vector space over the field F and let T be a linear transformation of V.
  (1) There is a basis for V with respect to which the matrix for T is in rational canonical form, i.e., is a block diagonal matrix whose diagonal blocks are the companion matrices for monic polynomials $a_1(x), a_2(x), \ldots, a_m(x)$ of degree at least one with $a_1(x) \mid a_2(x) \mid \cdots \mid a_m(x)$.
  (2) The rational canonical form for T is unique.

The use of the word *rational* is to indicate that this canonical form is calculated entirely within the field F and exists for any linear transformation T. This is not the case for the Jordan canonical form (considered later), which only exists if the field F contains the eigenvalues for T (cf. also the remarks following Corollary 18).

The following result translates the notion of similar linear transformations (i.e., the same linear transformation up to a change of basis) into the language of modules and relates this notion to rational canonical forms.

Theorem 15. Let S and T be linear transformations of V. Then the following are equivalent:
  (1) S and T are similar linear transformations
  (2) the $F[x]$-modules obtained from V via S and via T are isomorphic $F[x]$-modules
  (3) S and T have the same rational canonical form.

Proof: [(1) implies (2)] Assume there is a nonsingular linear transformation U such that $S = UTU^{-1}$. The vector space isomorphism $U : V \to V$ is also an $F[x]$-module homomorphism, where x acts on the first V via T and on the second via S, since for example $U(xv) = U(Tv) = UT(v) = SU(v) = x(Uv)$. Hence this is an $F[x]$-module isomorphism of the two modules in (2).

[(2) implies (3)] Assume (2) holds and denote by V_1 the vector space V made into an $F[x]$-module via S and denote by V_2 the space V made into an $F[x]$-module via T. Since $V_1 \cong V_2$ as $F[x]$-modules they have the same list of invariant factors. Thus S and T have a common rational canonical form.

[(3) implies (1)] Assume (3) holds. Since S and T have the same matrix representation with respect to some choice of (possibly different) bases of V by assumption, they are, up to a change of basis, the same linear transformation of V, hence are similar.

Let A be any $n \times n$ matrix with entries from F. Let V be an n-dimensional vector space over F. Recall we can then *define* a linear transformation T on V by choosing a basis for V and setting $T(v) = Av$ where v on the right hand side means the $n \times 1$ column vector of coordinates of v with respect to our chosen basis (this is just the usual identification of linear transformations with matrices). Then (of course) the matrix for this T with respect to this basis is the given matrix A. Put another way, any $n \times n$ matrix A with entries from the field F arises as the matrix for some linear transformation T of an n-dimensional vector space.

This dictionary between linear transformations of vector spaces and matrices allows us to state our previous two results in the language of matrices:

Theorem 16. *(Rational Canonical Form for Matrices)* Let A be an $n \times n$ matrix over the field F.
  (1) The matrix A is similar to a matrix in rational canonical form, i.e., there is an invertible $n \times n$ matrix P over F such that $P^{-1}AP$ is a block diagonal matrix whose diagonal blocks are the companion matrices for monic polynomials $a_1(x), a_2(x), \ldots, a_m(x)$ of degree at least one with $a_1(x) \mid a_2(x) \mid \cdots \mid a_m(x)$.
  (2) The rational canonical form for A is unique.

Definition. The *invariant factors* of an $n \times n$ matrix over a field F are the invariant factors of its rational canonical form.

Theorem 17. Let A and B be $n \times n$ matrices over the field F. Then A and B are similar if and only if A and B have the same rational canonical form.

If A is a matrix with entries from a field F and F is a subfield of a larger field K then we may also consider A as a matrix over K. The next result shows that the rational canonical form for A and questions of similarity do not depend on which field contains the entries of A.

Corollary 18. Let A and B be two $n \times n$ matrices over a field F and suppose F is a subfield of the field K.
  (1) The rational canonical form of A is the same whether it is computed over K or over F. The minimal and characteristic polynomials and the invariant factors of A are the same whether A is considered as a matrix over F or as a matrix over K.
  (2) The matrices A and B are similar over K if and only if they are similar over F, i.e., there exists an invertible $n \times n$ matrix P with entries from K such that $B = P^{-1}AP$ if and only if there exists an (in general different) invertible $n \times n$ matrix Q with entries from F such that $B = Q^{-1}AQ$.

Proof: (1) Let M be the rational canonical form of A when computed over the smaller field F. Since M satisfies the conditions in the definition of the rational canonical form over K, the uniqueness of the rational canonical form implies that M is also the rational canonical form of A over K. Hence the invariant factors of A are the same whether A is viewed over F or over K. In particular, since the minimal polynomial is the largest invariant factor of A it also does not depend on the field over which A is viewed. It is clear from the determinant definition of the characteristic polynomial of A that this polynomial depends only on the entries of A (we shall see shortly that the characteristic polynomial is the product of all the invariant factors for A, which will give an alternate proof of this result).

(2) If A and B are similar over the smaller field F they are clearly similar over K. Conversely, if A and B are similar over K, they have the same rational canonical form over K. By (1) they have the same rational canonical form over F, hence are similar over F by Theorem 17.

This corollary asserts in particular that the rational canonical form for an $n \times n$ matrix A is an $n \times n$ matrix with entries in the smallest field containing the entries of A. Further, this canonical form is the same matrix even if we allow conjugation of A by nonsingular matrices whose entries come from larger fields. This explains the terminology of *rational* canonical form.

The next proposition gives the connection between the characteristic polynomial of a matrix (or of a linear transformation) and its invariant factors and is quite useful for determining these invariant factors (particularly for matrices of small size).

Lemma 19. Let $a(x) \in F[x]$ be any monic polynomial.
  (1) The characteristic polynomial of the companion matrix of $a(x)$ is $a(x)$.
  (2) If M is the block diagonal matrix

$$M = \begin{pmatrix} A_1 & 0 & \cdots & 0 \\ 0 & A_2 & \cdots & 0 \\ \vdots & \vdots & \ddots & \vdots \\ 0 & 0 & \cdots & A_k \end{pmatrix},$$

given by the direct sum of matrices $A_1, A_2, \ldots, A_k$ then the characteristic polynomial of M is the product of the characteristic polynomials of $A_1, A_2, \ldots, A_k$.

Proof: These are both straightforward exercises.

Proposition 20. Let A be an $n \times n$ matrix over the field F.
  (1) The characteristic polynomial of A is the product of all the invariant factors of A.
  (2) *(The Cayley–Hamilton Theorem)* The minimal polynomial of A divides the characteristic polynomial of A.
  (3) The characteristic polynomial of A divides some power of the minimal polynomial of A. In particular these polynomials have the same roots, not counting multiplicities.
The same statements are true if the matrix A is replaced by a linear transformation T of an n-dimensional vector space over F.

Proof: Let B be the rational canonical form of A. By the previous lemma the block diagonal form of B shows that the characteristic polynomial of B is the product of the characteristic polynomials of the companion matrices of the invariant factors of A. By the first part of the lemma above, the characteristic polynomial of the companion matrix $\mathcal{C}_{a(x)}$ for $a(x)$ is just $a(x)$, which implies that the characteristic polynomial for B is the product of the invariant factors of A. Since A and B are similar, they have the same characteristic polynomial, which proves (1). Assertion (2) is immediate from (1) since the minimal polynomial for A is the largest invariant factor of A. The fact that all the invariant factors divide the largest one immediately implies (3). The final assertion is clear from the dictionary between linear transformations of vector spaces and matrices.

Note that part (2) of the proposition is the assertion that the matrix A satisfies its own characteristic polynomial, i.e., $c_A(A) = 0$ as matrices, which is the usual formulation for the Cayley–Hamilton Theorem. Note also that it implies the minimal polynomial for A has degree at most n, a result mentioned before.

The relations in Proposition 20 are frequently quite useful in the determination of the invariant factors for a matrix A, particularly for matrices of small degree (cf. Exercises 3 and 4 and the examples). The following result (which relies on Exercises 16 to 19 in the previous section and whose proof we outline in the exercises) computes the invariant factors in general.

Let A be an $n \times n$ matrix over the field F. Then $xI - A$ is an $n \times n$ matrix with entries in $F[x]$. The three operations
(a) interchanging two rows or columns
(b) adding a multiple (in $F[x]$) of one row or column to another
(c) multiplying any row or column by a unit in $F[x]$, i.e., by a nonzero element in F,
are called *elementary row and column operations.*

Theorem 21. Let A be an $n \times n$ matrix over the field F. Using the three elementary row and column operations above, the $n \times n$ matrix $xI - A$ with entries from $F[x]$ can be put into the diagonal form (called the *Smith Normal Form* for A)

$$\begin{pmatrix} 1 \\ & \ddots \\ & & 1 \\ & & & a_1(x) \\ & & & & a_2(x) \\ & & & & & \ddots \\ & & & & & & a_m(x) \end{pmatrix}$$

with monic nonzero elements $a_1(x), a_2(x), \ldots, a_m(x)$ of $F[x]$ with degrees at least one and satisfying $a_1(x) \mid a_2(x) \mid \cdots \mid a_m(x)$. The elements $a_1(x), \ldots, a_m(x)$ are the invariant factors of A.

Proof: cf. the exercises.

Invariant Factor Decomposition Algorithm: Converting to Rational Canonical Form

As mentioned in the exercises near the end of the previous section, keeping track of the operations necessary to diagonalize $xI - A$ will explicitly give a matrix P such that $P^{-1}AP$ is in rational canonical form. Equivalently, if V is a given $F[x]$-module with vector space basis $[e_1, e_2, \ldots, e_n]$, then P defines the change of basis giving the Invariant Factor Decomposition of V into a direct sum of cyclic $F[x]$-modules. In particular, if A is the matrix of the linear transformation T of the $F[x]$-module V defined by x (i.e., $T(e_j) = xe_j = \sum_{i=1}^{n} a_{ij}e_i$ where $A = (a_{ij})$), then the matrix P defines the change of basis for V with respect to which the matrix for T is in rational canonical form.

We first describe the algorithm in the general context of determining the Invariant Factor Decomposition of a given $F[x]$-module V with vector space basis $[e_1, e_2, \ldots, e_n]$ (the proof is outlined in the exercises). We then describe the algorithm to convert a given $n \times n$ matrix A to rational canonical form (in which reference to an underlying vector space and associated linear transformation are suppressed).

Explicit numerical examples of this algorithm are given in Examples 2 and 3 following.

Invariant Factor Decomposition Algorithm

Let V be an $F[x]$-module with vector space basis $[e_1, e_2, \ldots, e_n]$ (so in particular these elements are generators for V as an $F[x]$-module). Let T be the linear transformation of V to itself defined by x and let A be the $n \times n$ matrix associated to T and this choice of basis for V, i.e.,

$$T(e_j) = xe_j = \sum_{i=1}^{n} a_{ij}e_i \qquad \text{where} \qquad A = (a_{ij}).$$

(1) Use the following three elementary row and column operations to diagonalize the matrix $xI - A$ over $F[x]$, keeping track of the *row* operations used:
   (a) interchange two rows or columns (which will be denoted by $R_i \leftrightarrow R_j$ for the interchange of the ith and j^{th} rows and similarly by $C_i \leftrightarrow C_j$ for columns),
   (b) add a multiple (in $F[x]$) of one row or column to another (which will be denoted by $R_i + p(x)R_j \mapsto R_i$ if $p(x)$ times the jth row is added to the ith row, and similarly by $C_i + p(x)C_j \mapsto C_i$ for columns),
   (c) multiply any row or column by a unit in $F[x]$, i.e., by a nonzero element in F (which will be denoted by uR_i if the ith row is multiplied by $u \in F^{\times}$, and similarly by uC_i for columns).

(2) Beginning with the $F[x]$-module generators $[e_1, e_2, \ldots, e_n]$, for each row operation used in (1), change the set of generators by the following rules:
   (a) If the ith row is interchanged with the jth row then interchange the ith and j^{th} generators.
   (b) If $p(x)$ times the jth row is added to the ith row then subtract $p(x)$ times the ith generator from the jth generator (note the indices).

(c) If the ith row is multiplied by the unit $u \in F$ then divide the ith generator by u.

(3) When $xI - A$ has been diagonalized to the form in Theorem 21 the generators $[e_1, e_2, \ldots, e_n]$ for V will be in the form of $F[x]$-linear combinations of $e_1, e_2, \ldots, e_n$. Use $xe_j = T(e_j) = \sum_{i=1}^{n} a_{ij} e_i$ to write these elements as F-linear combinations of $e_1, e_2, \ldots, e_n$. When $xI - A$ has been diagonalized, the first $n - m$ of these linear combinations are 0 (providing a useful numerical check on the computations) and the remaining m linear combinations are nonzero, i.e., the generators for V are in the form $[0, \ldots, 0, f_1, \ldots, f_m]$ corresponding precisely to the diagonal elements in Theorem 21. The elements $f_1, \ldots, f_m$ are a set of $F[x]$-module generators for the cyclic factors in the invariant factor decomposition of V (with annihilators $(a_1(x)), \ldots, (a_m(x))$, respectively):

$$V = F[x] f_1 \oplus F[x] f_2 \oplus \ldots \oplus F[x] f_m,$$

$$F[x] f_i \cong F[x]/(a_i(x)) \qquad i = 1, 2, \ldots, m,$$

giving the Invariant Factor Decomposition of the $F[x]$-module V.

(4) The corresponding *vector space* basis for each cyclic factor of V is then given by the elements $f_i, T f_i, T^2 f_i, \ldots, T^{\deg a_i(x)-1} f_i$.

(5) Write the kth element of the vector space basis computed in (4) in terms of the original vector space basis $[e_1, e_2, \ldots, e_n]$ and use the coordinates for the kth column of an $n \times n$ matrix P. Then $P^{-1}AP$ is in rational canonical form (with diagonal blocks the companion matrices for the $a_i(x)$). This is the matrix for the linear transformation T with respect to the vector space basis in (4).

We now describe the algorithm to convert a given $n \times n$ matrix A to rational canonical form, i.e., to determine an $n \times n$ matrix P so that $P^{-1}AP$ is in rational canonical form. This is nothing more than the algorithm above applied to the vector space $V = F^n$ of $n \times 1$ column vectors with standard basis $[e_1, e_2, \ldots, e_n]$ (where e_i is the column vector with 1 in the ith position and 0's elsewhere) and T is the linear transformation defined by A and this choice of basis. Explicit reference to this underlying vector space and associated linear transformation are suppressed, so the algorithm is purely matrix theoretic.

Converting an $n \times n$ Matrix to Rational Canonical Form

Let A be an $n \times n$ matrix with entries in the field F.

(1) Use the following three elementary row and column operations to diagonalize the matrix $xI - A$ over $F[x]$, keeping track of the *row* operations used:

   (a) interchange two rows or columns (which will be denoted by $R_i \leftrightarrow R_j$ for the interchange of the ith and j^{th} rows and similarly by $C_i \leftrightarrow C_j$ for columns),

   (b) add a multiple (in $F[x]$) of one row or column to another (which will be denoted by $R_i + p(x)R_j \mapsto R_i$ if $p(x)$ times the jth row is added to the ith row, and similarly by $C_i + p(x)C_j \mapsto C_i$ for columns),

   (c) multiply any row or column by a unit in $F[x]$, i.e., by a nonzero element in F (which will be denoted by uR_i if the ith row is multiplied by $u \in F^\times$, and similarly by uC_i for columns).

Define $d_1, \ldots, d_m$ to be the degrees of the monic nonconstant polynomials $a_1(x), \ldots, a_m(x)$ appearing on the diagonal, respectively.

(2) Beginning with the $n \times n$ identity matrix P', for each row operation used in (1), change the matrix P' by the following rules:

   (a) If $R_i \leftrightarrow R_j$ then interchange the ith and j^{th} columns of P' (i.e., $C_i \leftrightarrow C_j$ for P').

   (b) If $R_i + p(x)R_j \mapsto R_i$ then subtract the product of the matrix $p(A)$ times the ith column of P' from the jth column of P' (i.e., $C_j - p(A)C_i \mapsto C_j$ for P' — note the indices).

   (c) If $u R_i$ then divide the elements of the ith column of P' by u (i.e., $u^{-1}C_i$ for P').

(3) When $xI - A$ has been diagonalized to the form in Theorem 21 the first $n - m$ columns of the matrix P' are 0 (providing a useful numerical check on the computations) and the remaining m columns of P' are nonzero. For each $i = 1, 2, \ldots, m$, multiply the ith nonzero column of P' successively by $A^0 = I, A^1, A^2, \ldots, A^{d_i-1}$, where d_i is the integer in (1) above and use the resulting column vectors (in this order) as the next d_i columns of an $n \times n$ matrix P. Then $P^{-1}AP$ is in rational canonical form (whose diagonal blocks are the companion matrices for the polynomials $a_1(x), \ldots, a_m(x)$ in (1)).

In the theory of canonical forms for linear transformations (or matrices) the characteristic polynomial plays the role of the order of a finite abelian group and the minimal polynomial plays the role of the exponent (after all, they are the same invariants, one for modules over the Principal Ideal Domain $\mathbb{Z}$ and the other for modules over the Principal Ideal Domain $F[x]$) so we can solve problems directly analogous to those we considered for finite abelian groups in Chapter 5. In particular, this includes the following:

(A) determine the rational canonical form of a given matrix (analogous to decomposing a finite abelian group as a direct product of cyclic groups)

(B) determine whether two given matrices are similar (analogous to determining whether two given finite abelian groups are isomorphic)

(C) determine all similarity classes of matrices over F with a given characteristic polynomial (analogous to determining all abelian groups of a given order)

(D) determine all similarity classes of $n \times n$ matrices over F with a given minimal polynomial (analogous to determining all abelian groups of rank at most n of a given exponent).

Examples

(1) We find the rational canonical forms of the following matrices over $\mathbb{Q}$ and determine if they are similar:

$$A = \begin{pmatrix} 2 & -2 & 14 \\ 0 & 3 & -7 \\ 0 & 0 & 2 \end{pmatrix} \quad B = \begin{pmatrix} 0 & -4 & 85 \\ 1 & 4 & -30 \\ 0 & 0 & 3 \end{pmatrix} \quad C = \begin{pmatrix} 2 & 2 & 1 \\ 0 & 2 & -1 \\ 0 & 0 & 3 \end{pmatrix}.$$

A direct computation shows that all three of these matrices have the same characteristic polynomial: $c_A(x) = c_B(x) = c_C(x) = (x-2)^2(x-3)$. Since the minimal and char-

acteristic polynomials have the same roots, the only possibilities for the minimal polynomials are $(x-2)(x-3)$ or $(x-2)^2(x-3)$. We quickly find that $(A-2I)(A-3I) = 0$,
$(B - 2I)(B - 3I) \neq 0$ (the 1,1-entry is nonzero) and $(C - 2I)(C - 3I) \neq 0$ (the
1,2-entry is nonzero). It follows that

$$m_A(x) = (x - 2)(x - 3), \quad m_B(x) = m_C(x) = (x - 2)^2(x - 3).$$

It follows immediately that there are no additional invariant factors for B and C.
Since the invariant factors for A divide the minimal polynomial and have product
the characteristic polynomial, we see that A has for invariant factors the polynomials
$x - 2$, $(x - 2)(x - 3) = x^2 - 5x + 6$. (For 2×2 and 3×3 matrices the determination
of the characteristic and minimal polynomials determines all the invariant factors, cf.
Exercises 3 and 4.) We conclude that B and C are similar and neither is similar to A.
The rational canonical forms are (note $(x - 2)^2(x - 3) = x^3 - 7x^2 + 16x - 12$)

$$\begin{pmatrix} 2 & 0 & 0 \\ 0 & 0 & -6 \\ 0 & 1 & 5 \end{pmatrix} \quad \begin{pmatrix} 0 & 0 & 12 \\ 1 & 0 & -16 \\ 0 & 1 & 7 \end{pmatrix} \quad \begin{pmatrix} 0 & 0 & 12 \\ 1 & 0 & -16 \\ 0 & 1 & 7 \end{pmatrix}.$$

(2) In the example above the rational canonical forms were obtained simply by determining
the characteristic and minimal polynomials for the matrices. As mentioned, this is
sufficient for 2×2 and 3×3 matrices since this information is sufficient to determine
all of the invariant factors. For larger matrices, however, this is in general not sufficient
(cf. the next example) and more work is required to determine the invariant factors. In
this example we again compute the rational canonical form for the matrix A in Example
1 following the two algorithms outlined above. While this is computationally more
difficult for this small matrix (as will be apparent), it has the advantage even in this
case that it also explicitly computes a matrix P with $P^{-1}AP$ in rational canonical
form.

I. (*Invariant Factor Decomposition*) We use row and column operations (in $\mathbb{Q}[x]$) to
reduce the matrix

$$xI - A = \begin{pmatrix} x - 2 & 2 & -14 \\ 0 & x - 3 & 7 \\ 0 & 0 & x - 2 \end{pmatrix}$$

to diagonal form. As in the invariant factor decomposition algorithm, we shall use the
notation $R_i \leftrightarrow R_j$ to denote the interchange of the ith and j^{th} rows, $R_i + aR_j \mapsto R_i$
if a times the jth row is added to the ith row, simply uR_i if the ith row is multiplied
by u (and similarly for columns, using C instead of R). Note also that the first two
operations we perform below are rather *ad hoc* and were chosen simply to have integers
everywhere in the computation:

$$\begin{pmatrix} x-2 & 2 & -14 \\ 0 & x-3 & 7 \\ 0 & 0 & x-2 \end{pmatrix} \xrightarrow[\mapsto R_1]{R_1+R_2} \begin{pmatrix} x-2 & x-1 & -7 \\ 0 & x-3 & 7 \\ 0 & 0 & x-2 \end{pmatrix} \longrightarrow$$

$$\xrightarrow[\mapsto C_1]{C_1-C_2} \begin{pmatrix} -1 & x-1 & -7 \\ -x+3 & x-3 & 7 \\ 0 & 0 & x-2 \end{pmatrix} \xrightarrow{-R_1} \begin{pmatrix} 1 & -x+1 & 7 \\ -x+3 & x-3 & 7 \\ 0 & 0 & x-2 \end{pmatrix} \longrightarrow$$

$$\xrightarrow[\substack{R_2+(x-3)R_1 \\ \mapsto R_2}]{} \begin{pmatrix} 1 & -x+1 & 7 \\ 0 & -x^2+5x-6 & 7(x-2) \\ 0 & 0 & x-2 \end{pmatrix} \xrightarrow[\substack{C_2+(x-1)C_1 \\ \mapsto C_2}]{} \begin{pmatrix} 1 & 0 & 7 \\ 0 & -x^2+5x-6 & 7(x-2) \\ 0 & 0 & x-2 \end{pmatrix} \longrightarrow$$

$$\xrightarrow[\substack{C_3-7C_1 \\ \mapsto C_3}]{} \begin{pmatrix} 1 & 0 & 0 \\ 0 & -x^2+5x-6 & 7(x-2) \\ 0 & 0 & x-2 \end{pmatrix} \xrightarrow{-C_2} \begin{pmatrix} 1 & 0 & 0 \\ 0 & x^2-5x+6 & 7(x-2) \\ 0 & 0 & x-2 \end{pmatrix} \longrightarrow$$

$$\xrightarrow[\substack{R_2-7R_3 \\ \mapsto R_2}]{} \begin{pmatrix} 1 & 0 & 0 \\ 0 & x^2-5x+6 & 0 \\ 0 & 0 & x-2 \end{pmatrix} \xrightarrow[\substack{R_2\leftrightarrow R_3 \\ C_2\leftrightarrow C_3}]{} \begin{pmatrix} 1 & 0 & 0 \\ 0 & x-2 & 0 \\ 0 & 0 & x^2-5x+6 \end{pmatrix}.$$

This determines the invariant factors $x - 2, x^2 - 5x + 6$ for this matrix, which we determined in Example 1 above. Let now V be a 3-dimensional vector space over $\mathbb{Q}$ with basis e_1, e_2, e_3 and let T be the corresponding linear transformation (which defines the action of x on V), i.e.,

$$xe_1 = T(e_1) = 2e_1$$
$$xe_2 = T(e_2) = -2e_1 + 3e_2$$
$$xe_3 = T(e_3) = 14e_1 - 7e_2 + 2e_3.$$

The row operations used in the reduction above were

$$R_1 + R_2 \mapsto R_1, \quad -R_1, \quad R_2 + (x - 3)R_1 \mapsto R_2, \quad R_2 - 7R_3 \mapsto R_2, \quad R_2 \leftrightarrow R_3.$$

Starting with the basis $[e_1, e_2, e_3]$ for V and changing it according to the rules given in the text, we obtain

$$[e_1, e_2, e_3] \longrightarrow [e_1, e_2-e_1, e_3] \longrightarrow [-e_1, e_2-e_1, e_3]$$
$$\longrightarrow [-e_1-(x-3)(e_2-e_1), e_2-e_1, e_3]$$
$$\longrightarrow [-e_1-(x-3)(e_2-e_1), e_2-e_1, e_3+7(e_2-e_1)]$$
$$\longrightarrow [-e_1-(x-3)(e_2-e_1), e_3+7(e_2-e_1), e_2-e_1].$$

Using the formulas above for the action of x, we see that these last elements are the elements $[0, -7e_1 + 7e_2 + e_3, -e_1 + e_2]$ of V corresponding to the elements $1, x - 2$ and $x^2 - 5x + 6$ in the diagonalized form of $xI - A$, respectively. The elements $f_1 = -7e_1 + 7e_2 + e_3$ and $f_2 = -e_1 + e_2$ are therefore $\mathbb{Q}[x]$-module generators for the two cyclic factors of V in its invariant factor decomposition as a $\mathbb{Q}[x]$-module. The corresponding $\mathbb{Q}$-vector space bases for these two factors are then f_1 and $f_2, xf_2 = Tf_2$, i.e., $-7e_1+7e_2+e_3$ and $-e_1+e_2, T(-e_1+e_2) = -4e_1+3e_2$. Then the matrix

$$P = \begin{pmatrix} -7 & -1 & -4 \\ 7 & 1 & 3 \\ 1 & 0 & 0 \end{pmatrix}$$

conjugates A into its rational canonical form:

$$P^{-1}AP = \begin{pmatrix} 2 & 0 & 0 \\ 0 & 0 & -6 \\ 0 & 1 & 5 \end{pmatrix},$$

as one easily checks.

II. (*Converting A Directly to Rational Canonical Form*) We use the row operations involved in the diagonalization of $xI - A$ to determine the matrix P' of the algorithm above:

$$\begin{pmatrix} 1 & 0 & 0 \\ 0 & 1 & 0 \\ 0 & 0 & 1 \end{pmatrix} \underset{\substack{C_2 - C_1 \\ \mapsto C_2}}{\longrightarrow} \begin{pmatrix} 1 & -1 & 0 \\ 0 & 1 & 0 \\ 0 & 0 & 1 \end{pmatrix} \underset{-C_1}{\longrightarrow} \begin{pmatrix} -1 & -1 & 0 \\ 0 & 1 & 0 \\ 0 & 0 & 1 \end{pmatrix} \longrightarrow$$

$$\underset{\substack{C_1 - (A - 3I)C_2 \\ \mapsto C_1}}{\longrightarrow} \begin{pmatrix} 0 & -1 & 0 \\ 0 & 1 & 0 \\ 0 & 0 & 1 \end{pmatrix} \underset{\substack{C_3 + 7C_2 \\ \mapsto C_3}}{\longrightarrow} \begin{pmatrix} 0 & -1 & -7 \\ 0 & 1 & 7 \\ 0 & 0 & 1 \end{pmatrix} \underset{C_2 \leftrightarrow C_3}{\longrightarrow} \begin{pmatrix} 0 & -7 & -1 \\ 0 & 7 & 1 \\ 0 & 1 & 0 \end{pmatrix} = P'.$$

Here we have $d_1 = 1$ and $d_2 = 2$, corresponding to the second and third nonzero columns of P', respectively. The columns of P are therefore given by

$$\begin{pmatrix} -7 \\ 7 \\ 1 \end{pmatrix} \quad \text{and} \quad \begin{pmatrix} -1 \\ 1 \\ 0 \end{pmatrix}, \quad A\begin{pmatrix} -1 \\ 1 \\ 0 \end{pmatrix} = \begin{pmatrix} -4 \\ 3 \\ 0 \end{pmatrix},$$

respectively, which again gives the matrix P above.

(3) For the 3×3 matrix A it was not necessary to perform the lengthy calculations above merely to determine the rational canonical form (equivalently, the invariant factors), as we saw in Example 1. For $n \times n$ matrices with $n \geq 4$, however, the computation of the characteristic and minimal polynomials is in general not sufficient for the determination of all the invariant factors, so the more extensive calculations of the previous example may become necessary. For example, consider the matrix

$$D = \begin{pmatrix} 1 & 2 & -4 & 4 \\ 2 & -1 & 4 & -8 \\ 1 & 0 & 1 & -2 \\ 0 & 1 & -2 & 3 \end{pmatrix}.$$

A short computation shows that the characteristic polynomial of D is $(x - 1)^4$. The possible minimal polynomials are then $x - 1, (x - 1)^2, (x - 1)^3$ and $(x - 1)^4$. Clearly $D - I \neq 0$ and another short computation shows that $(D - I)^2 = 0$, so the minimal polynomial for D is $(x - 1)^2$. There are then two possible sets of invariant factors:

$$x - 1, x - 1, (x - 1)^2 \quad \text{and} \quad (x - 1)^2, (x - 1)^2.$$

To determine the invariant factors for D we apply the procedure of the previous example to the 4×4 matrix

$$xI - D = \begin{pmatrix} x-1 & -2 & 4 & -4 \\ -2 & x+1 & -4 & 8 \\ -1 & 0 & x-1 & 2 \\ 0 & -1 & 2 & x-3 \end{pmatrix}.$$

The diagonal matrix obtained from this matrix by elementary row and column operations is the matrix

$$\begin{pmatrix} 1 & 0 & 0 & 0 \\ 0 & 1 & 0 & 0 \\ 0 & 0 & (x-1)^2 & 0 \\ 0 & 0 & 0 & (x-1)^2 \end{pmatrix},$$

which shows that the invariant factors for D are $(x - 1)^2, (x - 1)^2$ (one series of elementary row and column operations which diagonalize $xI - D$ are $R_1 \leftrightarrow R_3, -R_1$,

$R_2 + 2R_1 \mapsto R_2$, $R_3 - (x-1)R_1 \mapsto R_3$, $C_3 + (x-1)C_1 \mapsto C_3$, $C_4 + 2C_1 \mapsto C_4$, $R_2 \leftrightarrow R_4$, $-R_2$, $R_3 + 2R_2 \mapsto R_3$, $R_4 - (x+1)R_2 \mapsto R_4$, $C_3 + 2C_2 \mapsto C_3$, $C_4 + (x-3)C_2 \mapsto C_4$).

I. (*Invariant Factor Decomposition*) If e_1, e_2, e_3, e_4 is a basis for V in this case, then using the row operations in this diagonalization as in the previous example we see that the generators of V corresponding to the factors above are $(x-1)e_1 - 2e_2 - e_3 = 0$, $-2e_1 + (x+1)e_2 - e_4 = 0$, e_1, e_2. Hence a vector space basis for the two direct factors in the invariant decomposition of V in this case is given by e_1, Te_1 and e_2, Te_2 where T is the linear transformation defined by D, i.e., $e_1, e_1 + 2e_2 + e_3$ and $e_2, 2e_1 - e_2 + e_4$. The corresponding matrix P relating these bases is

$$P = \begin{pmatrix} 1 & 1 & 0 & 2 \\ 0 & 2 & 1 & -1 \\ 0 & 1 & 0 & 0 \\ 0 & 0 & 0 & 1 \end{pmatrix}$$

so that $P^{-1}DP$ is in rational canonical form:

$$P^{-1}DP = \begin{pmatrix} 0 & -1 & 0 & 0 \\ 1 & 2 & 0 & 0 \\ 0 & 0 & 0 & -1 \\ 0 & 0 & 1 & 2 \end{pmatrix}$$

as can easily be checked.

II. (*Converting D Directly to Rational Canonical Form*) As in Example 2 we determine the matrix P' of the algorithm from the row operations used in the diagonalization of $xI - D$:

$$\begin{pmatrix} 1 & 0 & 0 & 0 \\ 0 & 1 & 0 & 0 \\ 0 & 0 & 1 & 0 \\ 0 & 0 & 0 & 1 \end{pmatrix} \xrightarrow{C_1 \leftrightarrow C_3} \begin{pmatrix} 0 & 0 & 1 & 0 \\ 0 & 1 & 0 & 0 \\ 1 & 0 & 0 & 0 \\ 0 & 0 & 0 & 1 \end{pmatrix} \xrightarrow{-C_1} \begin{pmatrix} 0 & 0 & 1 & 0 \\ 0 & 1 & 0 & 0 \\ -1 & 0 & 0 & 0 \\ 0 & 0 & 0 & 1 \end{pmatrix} \rightarrow$$

$$\xrightarrow[\substack{C_1 - 2C_2 \\ \mapsto C_1}]{} \begin{pmatrix} 0 & 0 & 1 & 0 \\ -2 & 1 & 0 & 0 \\ -1 & 0 & 0 & 0 \\ 0 & 0 & 0 & 1 \end{pmatrix} \xrightarrow[\substack{C_1 + (D-I)C_3 \\ \mapsto C_1}]{} \begin{pmatrix} 0 & 0 & 1 & 0 \\ 0 & 1 & 0 & 0 \\ 0 & 0 & 0 & 0 \\ 0 & 0 & 0 & 1 \end{pmatrix} \xrightarrow{C_2 \leftrightarrow C_4} \begin{pmatrix} 0 & 0 & 1 & 0 \\ 0 & 0 & 0 & 1 \\ 0 & 0 & 0 & 0 \\ 0 & 1 & 0 & 0 \end{pmatrix} \rightarrow$$

$$\xrightarrow{-C_2} \begin{pmatrix} 0 & 0 & 1 & 0 \\ 0 & 0 & 0 & 1 \\ 0 & 0 & 0 & 0 \\ 0 & -1 & 0 & 0 \end{pmatrix} \xrightarrow[\substack{C_2 - 2C_3 \\ \mapsto C_2}]{} \begin{pmatrix} 0 & -2 & 1 & 0 \\ 0 & 0 & 0 & 1 \\ 0 & 0 & 0 & 0 \\ 0 & -1 & 0 & 0 \end{pmatrix} \xrightarrow[\substack{C_2 + (D+I)C_4 \\ \mapsto C_2}]{} \begin{pmatrix} 0 & 0 & 1 & 0 \\ 0 & 0 & 0 & 1 \\ 0 & 0 & 0 & 0 \\ 0 & 0 & 0 & 0 \end{pmatrix} = P'.$$

Here we have $d_1 = 2$ and $d_2 = 2$, corresponding to the third and fourth nonzero columns of P'. The columns of P are therefore given by

$$\begin{pmatrix} 1 \\ 0 \\ 0 \\ 0 \end{pmatrix}, \quad D\begin{pmatrix} 1 \\ 0 \\ 0 \\ 0 \end{pmatrix} = \begin{pmatrix} 1 \\ 2 \\ 1 \\ 0 \end{pmatrix} \quad \text{and} \quad \begin{pmatrix} 0 \\ 1 \\ 0 \\ 0 \end{pmatrix}, \quad D\begin{pmatrix} 0 \\ 1 \\ 0 \\ 0 \end{pmatrix} = \begin{pmatrix} 2 \\ -1 \\ 0 \\ 1 \end{pmatrix},$$

respectively, which again gives the matrix P above.

(4) In this example we determine all similarity classes of matrices A with entries from $\mathbb{Q}$ with characteristic polynomial $(x^4 - 1)(x^2 - 1)$. First note that any matrix with a degree

486

6 characteristic polynomial must be a 6×6 matrix. The polynomial $(x^4 - 1)(x^2 - 1)$ factors into irreducibles in $\mathbb{Q}[x]$ as $(x - 1)^2(x + 1)^2(x^2 + 1)$. Since the minimal polynomial $m_A(x)$ for A has the same roots as $c_A(x)$ it follows that $(x-1)(x+1)(x^2+1)$ divides $m_A(x)$. Suppose $a_1(x), \ldots, a_m(x)$ are the invariant factors of some A, so $a_m(x) = m_A(x)$, $a_i(x) \mid a_{i+1}(x)$ (in particular, all the invariant factors divide $m_A(x)$) and $a_1(x)a_2(x) \cdots a_m(x) = (x^4 - 1)(x^2 - 1)$. One easily sees that the only permissible lists under these constraints are

(a) $(x - 1)(x + 1), \quad (x - 1)(x + 1)(x^2 + 1)$
(b) $x - 1, \quad (x - 1)(x + 1)^2(x^2 + 1)$
(c) $x + 1, \quad (x - 1)^2(x + 1)(x^2 + 1)$
(d) $(x - 1)^2(x + 1)^2(x^2 + 1)$.

One can now easily write out the corresponding direct sums of companion matrices to obtain representatives of the 4 similarity classes. We shall see in the next section that there are still only 4 similarity classes even in $M_6(\mathbb{C})$.

(5) In this example we find all similarity classes of 3×3 matrices A with entries from $\mathbb{Q}$ satisfying $A^6 = I$. For each such A, its minimal polynomial divides $x^6 - 1$ and in $\mathbb{Q}[x]$ the complete factorization of this polynomial is

$$x^6 - 1 = (x - 1)(x + 1)(x^2 - x + 1)(x^2 + x + 1).$$

Conversely, if B is any 3×3 matrix whose minimal polynomial divides $x^6 - 1$, then $B^6 = I$. The only restriction on the minimal polynomial for B is that its degree is at most 3 (by the Cayley–Hamilton Theorem). The only possibilities for the minimal polynomial of such a matrix A are therefore

(a) $x - 1$
(b) $x + 1$
(c) $x^2 - x + 1$
(d) $x^2 + x + 1$
(e) $(x - 1)(x + 1)$
(f) $(x - 1)(x^2 - x + 1)$
(g) $(x - 1)(x^2 + x + 1)$
(h) $(x + 1)(x^2 - x + 1)$
(i) $(x + 1)(x^2 + x + 1)$.

Under the constraints of the rational canonical form these give rise to the following permissible lists of invariant factors:

(i) $x - 1, \quad x - 1, \quad x - 1$
(ii) $x + 1, \quad x + 1, \quad x + 1$
(iii) $x - 1, \quad (x - 1)(x + 1)$
(iv) $x + 1, \quad (x - 1)(x + 1)$
(v) $(x - 1)(x^2 - x + 1)$
(vi) $(x - 1)(x^2 + x + 1)$
(vii) $(x + 1)(x^2 - x + 1)$
(viii) $(x + 1)(x^2 + x + 1)$.

Note that it is impossible to have a suitable set of invariant factors if the minimal polynomial is $x^2 + x + 1$ or $x^2 - x + 1$. One can now write out the corresponding rational canonical forms; for example, (i) is I, (ii) is $-I$, and (iii) is

$$\begin{pmatrix} 1 & 0 & 0 \\ 0 & 0 & 1 \\ 0 & 1 & 0 \end{pmatrix}.$$

Note also that another way of phrasing this result is that any 3×3 matrix with entries from $\mathbb{Q}$ whose order (multiplicatively, of course) divides 6 is similar to one of these 8 matrices, so this example determines all elements of orders 1,2,3 and 6 in the group $GL_3(\mathbb{Q})$ (up to similarity).

EXERCISES

1. Prove that similar linear transformations of V (or $n \times n$ matrices) have the same characteristic and the same minimal polynomial.

2. Let M be as in Lemma 19. Prove that the minimal polynomial of M is the least common multiple of the minimal polynomials of $A_1, \ldots, A_k$.

3. Prove that two 2×2 matrices over F which are not scalar matrices are similar if and only if they have the same characteristic polynomial.

4. Prove that two 3×3 matrices are similar if and only if they have the same characteristic and same minimal polynomials. Give an explicit counterexample to this assertion for 4×4 matrices.

5. Prove directly from the fact that the collection of *all* linear transformations of an n dimensional vector space V over F to itself form a vector space over F of dimension n^2 that the minimal polynomial of a linear transformation T has degree at most n^2.

6. Prove that the constant term in the characteristic polynomial of the $n \times n$ matrix A is $(-1)^n \det A$ and that the coefficient of x^{n-1} is the negative of the sum of the diagonal entries of A (the sum of the diagonal entries of A is called the *trace* of A). Prove that $\det A$ is the product of the eigenvalues of A and that the trace of A is the sum of the eigenvalues of A.

7. Determine the eigenvalues of the matrix

$$\begin{pmatrix} 0 & 1 & 0 & 0 \\ 0 & 0 & 1 & 0 \\ 0 & 0 & 0 & 1 \\ 1 & 0 & 0 & 0 \end{pmatrix}.$$

8. Verify that the characteristic polynomial of the companion matrix

$$\begin{pmatrix} 0 & 0 & 0 & \cdots & 0 & -a_0 \\ 1 & 0 & 0 & \cdots & 0 & -a_1 \\ 0 & 1 & 0 & \cdots & 0 & -a_2 \\ \vdots & \vdots & \vdots & & \vdots & \vdots \\ 0 & 0 & 0 & \cdots & 1 & -a_{n-1} \end{pmatrix}$$

is

$$x^n + a_{n-1}x^{n-1} + \cdots + a_1 x + a_0.$$

9. Find the rational canonical forms of
$$\begin{pmatrix} 0 & -1 & -1 \\ 0 & 0 & 0 \\ -1 & 0 & 0 \end{pmatrix}, \quad \begin{pmatrix} c & 0 & -1 \\ 0 & c & 1 \\ -1 & 1 & c \end{pmatrix} \quad \text{and} \quad \begin{pmatrix} 422 & 465 & 15 & -30 \\ -420 & -463 & -15 & 30 \\ 840 & 930 & 32 & -60 \\ -140 & -155 & -5 & 12 \end{pmatrix}.$$

10. Find all similarity classes of 6×6 matrices over $\mathbb{Q}$ with minimal polynomial $(x+2)^2(x-1)$ (it suffices to give all lists of invariant factors and write out some of their corresponding matrices).

11. Find all similarity classes of 6×6 matrices over $\mathbb{C}$ with characteristic polynomial $(x^4 - 1)(x^2 - 1)$.

12. Find all similarity classes of 3×3 matrices A over $\mathbb{F}_2$ satisfying $A^6 = I$ (compare with the answer we computed over $\mathbb{Q}$). Do the same for 4×4 matrices B satisfying $B^{20} = I$.

13. Prove that the number of similarity classes of 3×3 matrices over $\mathbb{Q}$ with a given characteristic polynomial in $\mathbb{Q}[x]$ is the same as the number of similarity classes over any extension field of $\mathbb{Q}$. Give an example to show that this is not true in general for 4×4 matrices.

14. Determine all possible rational canonical forms for a linear transformation with characteristic polynomial $x^2(x^2 + 1)^2$.

15. Determine up to similarity all 2×2 rational matrices (i.e., $\in M_2(\mathbb{Q})$) of precise order 4 (multiplicatively, of course). Do the same if the matrix has entries from $\mathbb{C}$.

16. Show that $x^5 - 1 = (x - 1)(x^2 - 4x + 1)(x^2 + 5x + 1)$ in $\mathbb{F}_{19}[x]$. Use this to determine up to similarity all 2×2 matrices with entries from $\mathbb{F}_{19}$ of (multiplicative) order 5.

17. Determine representatives for the conjugacy classes for $GL_3(\mathbb{F}_2)$. [Compare your answer with Theorem 15 and Proposition 14 of Chapter 6.]

18. Let V be a finite dimensional vector space over $\mathbb{Q}$ and suppose T is a nonsingular linear transformation of V such that $T^{-1} = T^2 + T$. Prove that the dimension of V is divisible by 3. If the dimension of V is precisely 3 prove that all such transformations T are similar.

19. Let V be the infinite dimensional real vector space
$$\mathbb{R}^{\infty} = \{(a_0, a_1, a_2, \ldots) \mid a_0, a_1, a_2, \cdots \in \mathbb{R}\}.$$
Define the map $T : V \to V$ by $T(a_0, a_1, a_2, \ldots) = (0, a_0, a_1, a_2, \ldots)$. Prove that T has no eigenvectors.

20. Let ℓ be a prime and let $\Phi_\ell(x) = \frac{x^\ell - 1}{x - 1} = x^{\ell-1} + x^{\ell-2} + \ldots + x + 1 \in \mathbb{Z}[x]$ be the ℓth cyclotomic polynomial, which is irreducible over $\mathbb{Q}$ (Example 4 following Corollary 9.14). This exercise determines the smallest degree of a factor of $\Phi_\ell(x)$ modulo p for any prime p and so in particular determines when $\Phi_\ell(x)$ is irreducible modulo p. (This actually determines the complete factorization of $\Phi_\ell(x)$ modulo p — cf. Exercise 8 of Section 13.6.)
   (a) Show that if $p = \ell$ then $\Phi_\ell(x)$ is divisible by $x - 1$ in $\mathbb{F}_\ell[x]$.
   (b) Suppose $p \neq \ell$ and let f denote the order of p in $\mathbb{F}_\ell^{\times}$, i.e., f is the smallest power of p with $p^f \equiv 1 \bmod \ell$. Show that $m = f$ is the first value of m for which the group $GL_m(\mathbb{F}_p)$ contains an element A of order ℓ. [Use the formula for the order of this group at the end of Section 11.1.]
   (c) Show that $\Phi_\ell(x)$ is not divisible by any polynomial of degree smaller than f in $\mathbb{F}_p[x]$ [consider the companion matrix for such a divisor and use (b)]. Let $m_A(x) \in \mathbb{F}_p[x]$ denote the minimal polynomial for the matrix A in (b) and conclude that $m_A(x)$ is irreducible of degree f and divides $\Phi_\ell(x)$ in $\mathbb{F}_p[x]$.

(d) In particular, prove that $\Phi_\ell(x)$ is irreducible modulo p if and only if $l-1$ is the smallest power of p which is congruent to 1 modulo ℓ, i.e., p is a primitive root modulo ℓ.

21. Prove that the first two elementary row and column operations described before Theorem 21 do not change the determinant of the matrix and the third elementary operation multiplies the determinant by a unit. Conclude from Theorem 21 that the characteristic polynomial of A differs by a unit from the product of the invariant factors of A. Since both these polynomials are monic by definition, conclude that they are equal (this gives an alternate proof of Proposition 20).

The following exercises outline the proof of Theorem 21. They carry out explicitly the construction described in Exercises 16 to 19 of the previous section for the Euclidean Domain $F[x]$. Let V be an n-dimensional vector space with basis $v_1, v_2, \ldots, v_n$ and let T be the linear transformation of V defined by the matrix A and this choice of basis, i.e., T is the linear transformation with

$$T(v_j) = \sum_{i=1}^{n} a_{ij} v_i, \qquad j = 1, 2, \ldots, n$$

where $A = (a_{ij})$. Let $F[x]^n$ be the free module of rank n over $F[x]$ and let $\xi_1, \xi_2, \ldots, \xi_n$ denote a basis. Then we have a natural surjective $F[x]$-module homomorphism

$$\varphi : F[x]^n \to V$$

defined by mapping ξ_i to v_i, $i = 1, 2, \ldots, n$. As indicated in the exercises of the previous section the invariant factors for the $F[x]$-module V can be determined once we have determined a set of generators and the corresponding relations matrix for ker φ. Since by definition x acts on V by the linear transformation T, we have

$$x(v_j) = \sum_{i=1}^{n} a_{ij} v_i, \qquad j = 1, 2, \ldots, n.$$

22. Show that the elements

$$v_j = -a_{1j}\xi_1 - \cdots - a_{j-1\,j}\xi_{j-1} + (x - a_{jj})\xi_j - a_{j+1\,j}\xi_{j+1} - \cdots - a_{nj}\xi_n$$

for $j = 1, 2, \ldots, n$ are elements of the kernel of φ.

23. (a) Show that $x\xi_j = v_j + f_j$ where $f_j \in F\xi_1 + \cdots + F\xi_n$ is an element in the F-vector space spanned by $\xi_1, \ldots, \xi_n$.

(b) Show that

$$F[x]\xi_1 + \cdots + F[x]\xi_n = (F[x]v_1 + \cdots + F[x]v_n) + (F\xi_1 + \cdots + F\xi_n).$$

24. Show that $v_1, v_2, \ldots, v_n$ generate the kernel of φ. [Use the previous result to show that any element of ker φ is the sum of an element in the module generated by $v_1, v_2, \ldots, v_n$ and an element of the form $b_1\xi_1 + \cdots + b_n\xi_n$ where the b_i are elements of F. Then show that such an element is in ker φ if and only if all the b_i are 0 since $v_1, \ldots, v_n$ are a basis for V over F.]

25. Show that the generators $v_1, v_2, \ldots, v_n$ of ker φ have corresponding relations matrix

$$\begin{pmatrix} x - a_{11} & -a_{21} & \cdots & -a_{n1} \\ -a_{12} & x - a_{22} & \cdots & -a_{n2} \\ \vdots & \vdots & \ddots & \vdots \\ -a_{1n} & -a_{2n} & \cdots & x - a_{nn} \end{pmatrix} = xI - A^t,$$

where A^t is the transpose of A. Conclude that Theorem 21 and the algorithm for determining the invariant factors of A follows by Exercises 16 to 19 in the previous section (note that the row and column operations necessary to diagonalize this relations matrix are the column and row operations necessary to diagonalize the matrix in Theorem 21, which explains why the invariant factor algorithm keeps track of the *row* operations used).

12.3 THE JORDAN CANONICAL FORM

We continue with the notation in the previous section: F is a field, $F[x]$ is the ring of polynomials in x with coefficients in F, V is a finite dimensional vector space over F of dimension n, T is a fixed linear transformation of V by which we make V into an $F[x]$-module, and A is an $n \times n$ matrix with coefficients in F. Recall that once a basis for V has been fixed any linear transformation T defines a matrix A and conversely any matrix A defines a linear transformation T.

In the previous section we used the invariant factor form of the Fundamental Theorem for finitely generated modules over the Principal Ideal Domain $F[x]$ to obtain the rational canonical form for such a linear transformation T and the rational canonical form for such an $n \times n$ matrix A. In this section we use the elementary divisor form of the Fundamental Theorem to obtain the *Jordan canonical form*. We shall see that matrices in this canonical form are as close to being diagonal matrices as possible, so the matrices are simpler than in the rational canonical form (but we lose some of the "rationality" results).

The elementary divisors of a module are the prime power divisors of its invariant factors (this was Corollary 10). For the $F[x]$-module V the invariant factors were monic polynomials $a_1(x), a_2(x), \ldots, a_m(x)$ of degree at least one (with $a_1(x) \mid a_2(x) \mid \cdots \mid a_m(x)$), so the associated elementary divisors are the powers of the irreducible polynomial factors of these polynomials. These polynomials are only defined up to multiplication by a unit and, as in the case of the invariant factors, we can specify them uniquely by requiring that they be monic.

To obtain the simplest possible elementary divisors we shall assume that the polynomials $a_1(x), a_2(x), \ldots, a_m(x)$ factor completely into linear factors, i.e., that the elementary divisors of V are powers $(x - \lambda)^k$ of linear polynomials. Since the product of the elementary divisors is the characteristic polynomial, this is equivalent to the assumption that the field F contains all the eigenvalues of the linear transformation T (equivalently, of the matrix A representing the linear transformation T).

Under this assumption on F, it follows immediately from Theorem 6 that V is the direct sum of finitely many cyclic $F[x]$-modules of the form $F[x]/(x - \lambda)^k$ where $\lambda \in F$ is one of the eigenvalues of T, corresponding to the elementary divisors of V.

We now choose a vector space basis for each of the direct summands corresponding to the elementary divisors of V for which the corresponding matrix for T is particularly simple. Recall that by definition of the $F[x]$-module structure the linear transformation T acting on V is the element x acting by multiplication on each of the direct summands $F[x]/(x - \lambda)^k$.

Consider the elements

$$(\bar{x} - \lambda)^{k-1},\ (\bar{x} - \lambda)^{k-2}, \ldots, \bar{x} - \lambda,\ 1,$$

in the quotient $F[x]/(x - \lambda)^k$. Expanding each of these polynomials in $\bar{x}$ we see that the matrix relating these elements to the F-basis $\bar{x}^{k-1}, \bar{x}^{k-2}, \ldots, \bar{x}, 1$ of $F[x]/(x - \lambda)^k$ is upper triangular with 1's along the diagonal. Since this is an invertible matrix (having determinant 1), it follows that the elements above are an F-basis for $F[x]/(x - \lambda)^k$. With respect to this basis the linear transformation of multiplication by x acts in a particularly simple manner (note that $x = \lambda + (x - \lambda)$ and that $(\bar{x} - \lambda)^k = 0$ in the quotient):

$$
x : \quad
\begin{aligned}
(\bar{x} - \lambda)^{k-1} &\mapsto \lambda \cdot (\bar{x} - \lambda)^{k-1} + (\bar{x} - \lambda)^k = \lambda \cdot (\bar{x} - \lambda)^{k-1} \\
(\bar{x} - \lambda)^{k-2} &\mapsto \lambda \cdot (\bar{x} - \lambda)^{k-2} + (\bar{x} - \lambda)^{k-1} \\
&\quad\vdots \\
\bar{x} - \lambda &\mapsto \lambda \cdot (\bar{x} - \lambda) + (\bar{x} - \lambda)^2 \\
1 &\mapsto \lambda \cdot 1 + (\bar{x} - \lambda).
\end{aligned}
$$

With respect to this basis, the matrix for multiplication by x is therefore

$$
\begin{pmatrix}
\lambda & 1 & & & \\
 & \lambda & \ddots & & \\
 & & \ddots & 1 & \\
 & & & \lambda & 1 \\
 & & & & \lambda
\end{pmatrix}
$$

where the blank entries are all zero. Such matrices are given a name:

Definition. The $k \times k$ matrix with λ along the main diagonal and 1 along the first superdiagonal depicted above is called the $k \times k$ *elementary Jordan matrix with eigenvalue* λ or the *Jordan block of size k with eigenvalue* λ.

Applying this to each of the cyclic factors of V in its elementary divisor decomposition we obtain a vector space basis for V with respect to which the linear transformation T has as matrix the direct sum of the Jordan blocks corresponding to the elementary divisors of V, i.e., is block diagonal with Jordan blocks along the diagonal:

$$
\begin{pmatrix}
J_1 & & & \\
 & J_2 & & \\
 & & \ddots & \\
 & & & J_t
\end{pmatrix}.
$$

Notice that this matrix is uniquely determined up to permutation of the blocks along the diagonal by the elementary divisors of the $F[x]$-module V and conversely, by Theorem 9, the list of elementary divisors uniquely determines the module V up to $F[x]$-module isomorphism.

Definition.
  (1) A matrix is said to be in *Jordan canonical form* if it is a block diagonal matrix with Jordan blocks along the diagonal.
  (2) A *Jordan canonical form* for a linear transformation T is a matrix representing T which is in Jordan canonical form.

We have proved that any linear transformation T has a Jordan canonical form. As in the case of the rational canonical form, it follows from the uniqueness of the elementary divisors that the Jordan canonical form is unique up to a permutation of the Jordan blocks along the diagonal (hence is called *the* Jordan canonical form for T). We summarize this in the following theorem.

Theorem 22. *(Jordan Canonical Form for Linear Transformations)* Let V be a finite dimensional vector space over the field F and let T be a linear transformation of V. Assume F contains all the eigenvalues of T.
  (1) There is a basis for V with respect to which the matrix for T is in Jordan canonical form, i.e., is a block diagonal matrix whose diagonal blocks are the Jordan blocks for the elementary divisors of V.
  (2) The Jordan canonical form for T is unique up to a permutation of the Jordan blocks along the diagonal.

As for the rational canonical form, the following theorem gives the corresponding statement for $n \times n$ matrices over F.

Theorem 23. *(Jordan Canonical Form for Matrices)* Let A be an $n \times n$ matrix over the field F and assume F contains all the eigenvalues of A.
  (1) The matrix A is similar to a matrix in Jordan canonical form, i.e., there is an invertible $n \times n$ matrix P over F such that $P^{-1}AP$ is a block diagonal matrix whose diagonal blocks are the Jordan blocks for the elementary divisors of A.
  (2) The Jordan canonical form for A is unique up to a permutation of the Jordan blocks along the diagonal.

The Jordan canonical form differs from a diagonal matrix only by the possible presence of some 1's along the first superdiagonal (and then only if there are Jordan blocks of size greater than one), hence is close to being a diagonal matrix. The following result shows in particular that the Jordan canonical form for a matrix A is as close to being a diagonal matrix as possible.

Corollary 24.
  (1) If a matrix A is similar to a diagonal matrix D, then D is the Jordan canonical form of A.
  (2) Two diagonal matrices are similar if and only if their diagonal entries are the same up to a permutation.

Proof: The first assertion is immediate from the uniqueness of Jordan canonical forms because a diagonal matrix is itself in Jordan form (with Jordan blocks of size 1). The uniqueness of the Jordan canonical form gives (2).

The next corollary gives a criterion to determine when a matrix A can be diagonalized.

Corollary 25. If A is an $n \times n$ matrix with entries from F and F contains all the eigenvalues of A, then A is similar to a diagonal matrix over F if and only if the minimal polynomial of A has no repeated roots.

Proof: Suppose A is similar to a diagonal matrix. The minimal polynomial of a diagonal matrix has no repeated roots (its roots are precisely the distinct elements along the diagonal). Since similar matrices have the same minimal polynomial it follows that the minimal polynomial for A has no repeated roots.

Conversely, suppose the minimal polynomial for A has no repeated roots and let B be the Jordan canonical form of A. The matrix B is a block diagonal matrix with elementary Jordan matrices down the diagonal. By the exercises at the end of the preceding section the minimal polynomial for B is the least common multiple of the minimal polynomials of the Jordan blocks. It is easy to see directly that a Jordan block of size k with eigenvalue λ has minimal polynomial $(x - \lambda)^k$ (note that this is immediate from the fact that each elementary Jordan matrix gives the action on a *cyclic* $F[x]$-submodule whose annihilator is $(x - \lambda)^k$). Since A and B have the same minimal polynomial, the least common multiple of the $(x - \lambda)^k$ cannot have any repeated roots. It follows that k must be 1, i.e., that each Jordan block must be of size one and B is a diagonal matrix.

Changing From One Canonical Form to Another

We continue to assume that the field F contains all the eigenvalues of T (or A) so both the rational and Jordan canonical forms exist over F. The process of passing from one form to the other is exactly the same algorithm described in Section 5.2 for finite abelian groups (where the elementary divisors were determined from the list of invariant factors and vice versa).

In brief summary, recall that the elementary divisors are the prime power divisors of the invariant factors. They are obtained from the invariant factors by writing each invariant factor as a product of distinct linear factors to powers; the resulting set of powers of linear polynomials is the set of elementary divisors. For example, if the invariant factors of T are

$$(x - 1)(x - 3)^3, \quad (x - 1)(x - 2)(x - 3)^3, \quad (x - 1)(x - 2)^2(x - 3)^3$$

then the elementary divisors are

$$(x-1), \quad (x-3)^3, \quad (x-1), \quad (x-2), \quad (x-3)^3, \quad (x-1), \quad (x-2)^2, \quad (x-3)^3.$$

The largest invariant factor is the product of the largest of the distinct prime powers among the elementary divisors, the next largest invariant factor is the product of the largest of the distinct prime powers among the remaining elementary divisors, and so on. Given a list of elementary divisors we can find the list of invariant factors by first arranging the elementary divisors into n separate lists, one for each eigenvalue. In each of these n lists arrange the polynomials in increasing (i.e., nondecreasing) degree. Next arrange for all n lists to have the same length by appending an appropriate number of the constant polynomial 1. Now form the ith invariant factor by taking the product of the ith polynomial in each of these lists. For example, if the elementary divisors of T are

$$(x-1)^3, \ (x+4), \ (x+4)^2, \ (x-5)^2, \ (x-1)^5, \ (x-1)^3, \ (x-5)^3, \ (x-1)^4, \ (x+4)^3$$

then the intermediate lists are

$$
\begin{array}{llll}
(1) & (x-1)^3, & (x-1)^3, & (x-1)^4, & (x-1)^5 \\
(2) & 1, & x+4, & (x+4)^2, & (x+4)^3 \\
(3) & 1, & 1, & (x-5)^2, & (x-5)^3
\end{array}
$$

so the list of invariant factors is

$$(x-1)^3, \quad (x-1)^3(x+4), \quad (x-1)^4(x+4)^2(x-5)^2, \quad (x-1)^5(x+4)^3(x-5)^3.$$

Elementary Divisor Decomposition Algorithm: Converting to Jordan Canonical Forms

Theorem 21 indicates a computational procedure to determine the invariant factors of any given matrix A. Factorization of these invariant factors produces the elementary divisors of A, hence determines the Jordan canonical form for A as above.

The Invariant Factor Decomposition Algorithm following Theorem 21 starts with a basis $e_1, \ldots, e_n$ for V and produces a set $f_1, \ldots, f_m$ of elements of V which are $F[x]$-module generators for the cyclic factors in the invariant factor decomposition of V (with annihilators $(a_1(x)), \ldots, (a_m(x))$, respectively). Since the elementary divisor decomposition is obtained from the invariant factor decomposition by applying the Chinese Remainder Theorem to the cyclic modules $F[x]/(a_i(x))$, this gives a set of $F[x]$-module generators for the cyclic factors in the elementary divisor decomposition of V. These elements then give rise to an explicit vector space basis for V with respect to which the linear transformation corresponding to A is in Jordan canonical form (equivalently, an explicit matrix P such that $P^{-1}AP$ is in Jordan canonical form). As for the Invariant Factor Decomposition Algorithm we state the result first in the general context of decomposing a vector space and then describe the algorithm to convert a given $n \times n$ matrix A to Jordan canonical form.

Explicit numerical examples of this algorithm are given later in Examples 2 and 3.

Elementary Divisor Decomposition Algorithm

(1) to (3): The first three steps in the algorithm are those from the Invariant Factor Decomposition Algorithm following Theorem 21.

(4) For each invariant factor $a(x)$ computed for A write

$$a(x) = (x-\lambda_1)^{\alpha_1}(x-\lambda_2)^{\alpha_2} \ldots (x-\lambda_s)^{\alpha_s}$$

where $\lambda_1, \ldots, \lambda_s \in F$ are distinct. Let $f \in V$ be the $F[x]$-module generator for the cyclic factor corresponding to the invariant factor $a(x)$ computed in (3). Then the elements

$$\frac{a(x)}{(x-\lambda_1)^{\alpha_1}}f, \quad \frac{a(x)}{(x-\lambda_2)^{\alpha_2}}f, \quad \ldots, \quad \frac{a(x)}{(x-\lambda_s)^{\alpha_s}}f$$

(note that the $\dfrac{a(x)}{(x - \lambda_i)^{\alpha_i}} \in F[x]$ are polynomials) are $F[x]$-module generators for the cyclic factors of V corresponding to the elementary divisors

$$(x - \lambda_1)^{\alpha_1}, \quad (x - \lambda_2)^{\alpha_2}, \quad \ldots, \quad (x - \lambda_s)^{\alpha_s},$$

respectively.

(5) If $g_i = \dfrac{a(x)}{(x - \lambda_i)^{\alpha_i}} f$ is the $F[x]$-module generator for the cyclic factor of V corresponding to the elementary divisor $(x - \lambda_i)^{\alpha_i}$ then the corresponding *vector space* basis for this cyclic factor of V is given by the elements

$$(T - \lambda_i)^{\alpha_i - 1} g_i, \quad (T - \lambda_i)^{\alpha_i - 2} g_i, \quad \ldots, \quad (T - \lambda_i) g_i, \quad g_i.$$

(6) Write the kth element of the vector space basis computed in (5) in terms of the original vector space basis $[e_1, e_2, \ldots, e_n]$ for V and use the coordinates for the kth column of an $n \times n$ matrix P. Then $P^{-1} A P$ is in Jordan canonical form (with Jordan blocks appearing in the order used in (5) for the cyclic factors of V).

Converting an $n \times n$ Matrix to Jordan Canonical Form

(1) to **(2)**: The first two steps are those from the algorithm for Converting an $n \times n$ matrix to Rational Canonical Form following Theorem 21.

(3) When $xI - A$ has been diagonalized to the form in Theorem 21 the first $n - m$ columns of the matrix P' are 0 (providing a useful numerical check on the computations) and the remaining m columns of P' are nonzero. For each successive $i = 1, 2, \ldots, m$:

(a) Factor the ith nonconstant diagonal element (which is of degree d_i):

$$a(x) = (x - \lambda_1)^{\alpha_1} (x - \lambda_2)^{\alpha_2} \ldots (x - \lambda_s)^{\alpha_s}$$

where $\lambda_1, \ldots, \lambda_s \in F$ are distinct (here $a(x) = a_i(x)$ is the ith nonconstant diagonal element and s depends on i).

(b) Multiply the ith nonzero column of P' successively by the d_i matrices:

$$(A - \lambda_1 I)^{\alpha_1 - 1} (A - \lambda_2 I)^{\alpha_2} \ldots (A - \lambda_s I)^{\alpha_s}$$
$$(A - \lambda_1 I)^{\alpha_1 - 2} (A - \lambda_2 I)^{\alpha_2} \ldots (A - \lambda_s I)^{\alpha_s}$$
$$\vdots$$
$$(A - \lambda_1 I)^{0} \quad (A - \lambda_2 I)^{\alpha_2} \ldots (A - \lambda_s I)^{\alpha_s}$$

$$(A - \lambda_1 I)^{\alpha_1} \quad (A - \lambda_2 I)^{\alpha_2 - 1} \ldots (A - \lambda_s I)^{\alpha_s}$$
$$(A - \lambda_1 I)^{\alpha_1} \quad (A - \lambda_2 I)^{\alpha_2 - 2} \ldots (A - \lambda_s I)^{\alpha_s}$$
$$\vdots$$
$$(A - \lambda_1 I)^{\alpha_1} \quad (A - \lambda_2 I)^{0} \quad \ldots (A - \lambda_s I)^{\alpha_s}$$
$$\vdots$$

$$(A - \lambda_1 I)^{\alpha_1} (A - \lambda_2 I)^{\alpha_2} \ldots (A - \lambda_s I)^{\alpha_s - 1}$$
$$(A - \lambda_1 I)^{\alpha_1} (A - \lambda_2 I)^{\alpha_2} \ldots (A - \lambda_s I)^{\alpha_s - 2}$$
$$\vdots$$
$$(A - \lambda_1 I)^{\alpha_1} (A - \lambda_2 I)^{\alpha_2} \ldots (A - \lambda_s I)^0.$$

(c) Use the column vectors resulting from (b) (in that order) as the next d_i columns of an $n \times n$ matrix P.

Then $P^{-1}AP$ is in Jordan canonical form (whose Jordan blocks correspond to the ordering of the factors in (a)).

Examples

We can use Jordan canonical forms to carry out the same analysis of matrices that we did as examples of the use of rational canonical forms. In some instances, when the field is enlarged, the number of similarity classes increases (the number of similarity classes can never decrease when we extend the field by Corollary 18(2)).

(1) Let A, B and C be the matrices in Example 1 of the previous section and let $F = \mathbb{Q}$. Note that $\mathbb{Q}$ contains all the eigenvalues for these matrices. Since we have already determined the invariant factors of these matrices we can immediately obtain their elementary divisors. The elementary divisors of A are $x - 2$, $x - 2$ and $x - 3$ and the elementary divisors of B and C are $(x - 2)^2$ and $x - 3$ so the respective Jordan canonical forms are:

$$\begin{pmatrix} 2 & 0 & 0 \\ 0 & 2 & 0 \\ 0 & 0 & 3 \end{pmatrix} \qquad \begin{pmatrix} 2 & 1 & 0 \\ 0 & 2 & 0 \\ 0 & 0 & 3 \end{pmatrix} \qquad \begin{pmatrix} 2 & 1 & 0 \\ 0 & 2 & 0 \\ 0 & 0 & 3 \end{pmatrix}.$$

Notice that A is similar to a diagonal matrix but, by Corollary 25, B and C are not.

(2) For the matrix A, we determined in Example 2 of the previous section that $f_1 = -7e_1 + 7e_2 + e_3$ and $f_2 = -e_1 + e_2$ were $\mathbb{Q}[x]$-module generators for the two cyclic factors of V in its invariant factor decomposition, corresponding to the invariant factors $x - 2$ and $(x - 2)(x - 3)$, respectively. Using the first algorithm described above, the elements f_1, $(x - 3)f_2$ and $(x - 2)f_2$ are therefore $\mathbb{Q}[x]$-module generators for the three cyclic factors of V in its elementary divisor decomposition, corresponding to the elementary divisors $x - 2$, $x - 2$, and $x - 3$. An easy computation shows that these are the elements $-7e_1 + 7e_2 + e_3$, $-e_1$ and $-2e_1 + e_2$, respectively. Then the matrix

$$P = \begin{pmatrix} -7 & -1 & -2 \\ 7 & 0 & 1 \\ 1 & 0 & 0 \end{pmatrix}$$

conjugates A into its Jordan canonical form:

$$P^{-1}AP = \begin{pmatrix} 2 & 0 & 0 \\ 0 & 2 & 0 \\ 0 & 0 & 3 \end{pmatrix},$$

as one easily checks.

The columns of this matrix can also be obtained following the second algorithm above, using the nonzero columns of the matrix P' computed in Example 2 of the previous section:

$$(A - 2I)^0 \begin{pmatrix} -7 \\ 7 \\ 1 \end{pmatrix} = \begin{pmatrix} -7 \\ 7 \\ 1 \end{pmatrix}$$

and

$$(A - 2I)^0(A - 3I)^1 \begin{pmatrix} -1 \\ 1 \\ 0 \end{pmatrix} = \begin{pmatrix} -1 \\ 0 \\ 0 \end{pmatrix}, \quad (A - 2I)^1(A - 3I)^0 \begin{pmatrix} -1 \\ 1 \\ 0 \end{pmatrix} = \begin{pmatrix} -2 \\ 1 \\ 0 \end{pmatrix},$$

respectively, which again gives the matrix P.

(3) For the 4×4 matrix D of Example 3 of the previous section, the invariant factors were $(x - 1)^2$, $(x - 1)^2$, with corresponding $\mathbb{Q}[x]$-module generators $f_1 = e_1$ and $f_2 = e_2$, respectively. These are also the elementary divisors for this matrix. The corresponding vector space bases for these two factors are given by $(T - 1)f_1, f_1$ and $(T - 1)f_2, f_2$, respectively. An easy computation shows these are the elements $2e_2 + e_3, e_1$ and $2e_1 - e_2 + e_4, e_2$, respectively. Then the matrix

$$P = \begin{pmatrix} 0 & 1 & 2 & 0 \\ 2 & 0 & -2 & 1 \\ 1 & 0 & 0 & 0 \\ 0 & 0 & 1 & 0 \end{pmatrix}$$

conjugates D into its Jordan canonical form:

$$P^{-1}DP = \begin{pmatrix} 1 & 1 & 0 & 0 \\ 0 & 1 & 0 & 0 \\ 0 & 0 & 1 & 1 \\ 0 & 0 & 0 & 1 \end{pmatrix}$$

as can easily be checked.

The columns of this matrix can also be obtained following the second algorithm above, using the nonzero columns of the matrix P' computed in Example 3 of the previous section:

$$(D - I)^1 \begin{pmatrix} 1 \\ 0 \\ 0 \\ 0 \end{pmatrix} = \begin{pmatrix} 0 \\ 2 \\ 1 \\ 0 \end{pmatrix}, \quad (D - I)^0 \begin{pmatrix} 1 \\ 0 \\ 0 \\ 0 \end{pmatrix} = \begin{pmatrix} 1 \\ 0 \\ 0 \\ 0 \end{pmatrix},$$

and

$$(D - I)^1 \begin{pmatrix} 0 \\ 1 \\ 0 \\ 0 \end{pmatrix} = \begin{pmatrix} 2 \\ -2 \\ 0 \\ 1 \end{pmatrix}, \quad (D - I)^0 \begin{pmatrix} 0 \\ 1 \\ 0 \\ 0 \end{pmatrix} = \begin{pmatrix} 0 \\ 1 \\ 0 \\ 0 \end{pmatrix},$$

respectively, which again gives the matrix P.

(4) The set of similarity classes of 6×6 matrices with entries from $\mathbb{C}$ with characteristic polynomial $(x^4 - 1)(x^2 - 1)$ consists of the 4 classes represented by the rational canonical forms in the preceding set of examples (there are no additional lists of invariant factors over $\mathbb{C}$). Their Jordan canonical forms cannot all be written over $\mathbb{Q}$, however. For instance, if the invariant factors are

$$(x - 1)(x + 1) \quad \text{and} \quad (x - 1)(x + 1)(x^2 + 1)$$

then the elementary divisors are

$$x - 1, \quad x + 1, \quad x - 1, \quad x + 1, \quad x - i, \quad x + i,$$

where i is a square root of -1 in $\mathbb{C}$, so the Jordan form for this matrix is a diagonal matrix with diagonal entries $1, 1, -1, -1, i, -i$.

(5) In contrast, the set of similarity classes of 3×3 matrices, A, over $\mathbb{C}$ satisfying $A^6 = I$ is considerably larger than that over $\mathbb{Q}$. If A is any such matrix, $m_A(x) \mid x^6 - 1$ so since the latter polynomial has no repeated roots in $\mathbb{C}$, the minimal polynomial of A has no repeated roots. By Corollary 25 the Jordan canonical form of A is a diagonal matrix. Since this diagonal matrix has the same minimal polynomial, its 6th power is also the identity, and so each diagonal entry is a 6th root of unity. For each list $\zeta_1, \zeta_2, \zeta_3$ of 6th roots of unity we obtain a Jordan canonical form, and two such forms are the same (i.e., give rise to similar matrices) if and only if the lists are permuted versions of each other. One finds that there are, up to similarity, 56 classes of such A's.

EXERCISES

1. Suppose the vector space V is the direct sum of cyclic $F[x]$-modules whose annihilators are $(x + 1)^2$, $(x - 1)(x^2 + 1)^2$, $(x^4 - 1)$ and $(x + 1)(x^2 - 1)$. Determine the invariant factors and elementary divisors for V.

2. Prove that if $\lambda_1, \ldots, \lambda_n$ are the eigenvalues of the $n \times n$ matrix A then $\lambda_1^k, \ldots, \lambda_n^k$ are the eigenvalues of A^k for any $k \geq 0$.

3. Use the method of Example 2 above to determine explicit matrices P_1 and P_2 with $P_1^{-1} B P_1$ and $P_2^{-1} C P_2$ in Jordan canonical form. Use this to explicitly construct a matrix Q which conjugates B into C (proving directly that these matrices are similar).

4. Prove that the Jordan canonical form for the matrix

$$\begin{pmatrix} 9 & 4 & 5 \\ -4 & 0 & -3 \\ -6 & -4 & -2 \end{pmatrix}$$

is that stated at the beginning of this chapter. Explicitly determine a matrix P which conjugates this matrix to its Jordan canonical form. Explain why this matrix cannot be diagonalized.

5. Compute the Jordan canonical form for the matrix

$$\begin{pmatrix} 1 & 0 & 0 \\ 0 & 0 & -2 \\ 0 & 1 & 3 \end{pmatrix}.$$

6. Determine which of the following matrices are similar:

$$\begin{pmatrix} -1 & 4 & -4 \\ 2 & -1 & 3 \\ 0 & -4 & 3 \end{pmatrix} \quad \begin{pmatrix} -3 & -4 & 0 \\ 2 & 3 & 0 \\ 8 & 8 & 1 \end{pmatrix} \quad \begin{pmatrix} -3 & 2 & -4 \\ 2 & 1 & 0 \\ 3 & -1 & 3 \end{pmatrix} \quad \begin{pmatrix} -1 & 4 & -4 \\ 0 & -3 & 2 \\ 0 & -4 & 3 \end{pmatrix}.$$

7. Determine the Jordan canonical forms for the following matrices:

$$\begin{pmatrix} 5 & 4 & 1 \\ -1 & 0 & 0 \\ -3 & -4 & 1 \end{pmatrix} \quad \begin{pmatrix} 3 & 4 & 2 \\ -2 & -3 & -1 \\ -4 & -4 & -3 \end{pmatrix}.$$

8. Prove that the matrices

$$A = \begin{pmatrix} 5 & 6 & 0 \\ -3 & -4 & 0 \\ -2 & 0 & 1 \end{pmatrix} \qquad B = \begin{pmatrix} 3 & -1 & 2 \\ -10 & 6 & -14 \\ -6 & 3 & -7 \end{pmatrix}$$

are similar. Prove that both A and B can be diagonalized and determine explicit matrices P_1 and P_2 with $P_1^{-1}AP_1$ and $P_2^{-1}BP_2$ in diagonal form.

9. Prove that the matrices

$$A = \begin{pmatrix} -8 & -10 & -1 \\ 7 & 9 & 1 \\ 3 & 2 & 0 \end{pmatrix} \qquad B = \begin{pmatrix} -3 & 2 & -4 \\ 4 & -1 & 4 \\ 4 & -2 & 5 \end{pmatrix}$$

both have $(x-1)^2(x+1)$ as characteristic polynomial but that one can be diagonalized and the other cannot. Determine the Jordan canonical form for both matrices.

10. Find all Jordan canonical forms of 2×2, 3×3 and 4×4 matrices over $\mathbb{C}$.

11. Verify that the characteristic polynomial of

$$A = \begin{pmatrix} 1 & 0 & 0 & 0 \\ 0 & 1 & 0 & 0 \\ -2 & -2 & 0 & 1 \\ -2 & 0 & -1 & -2 \end{pmatrix}$$

is a product of linear factors over $\mathbb{Q}$. Determine the rational and Jordan canonical forms for A over $\mathbb{Q}$.

12. Determine the Jordan canonical form for the matrix

$$\begin{pmatrix} 1 & 2 & 0 & 0 \\ 0 & 1 & 2 & 0 \\ 0 & 0 & 1 & 2 \\ 0 & 0 & 0 & 1 \end{pmatrix}.$$

13. Determine the Jordan canonical form for the matrix

$$\begin{pmatrix} 3 & 0 & -2 & -3 \\ 4 & -8 & 14 & -15 \\ 2 & -4 & 7 & -7 \\ 0 & 2 & -4 & 3 \end{pmatrix}.$$

14. Prove that the matrices

$$A = \begin{pmatrix} 2 & 0 & 0 & 0 \\ -4 & -1 & -4 & 0 \\ 2 & 1 & 3 & 0 \\ -2 & 4 & 9 & 1 \end{pmatrix} \qquad B = \begin{pmatrix} 5 & 0 & -4 & -7 \\ 3 & -8 & 15 & -13 \\ 2 & -4 & 7 & -7 \\ 1 & 2 & -5 & 1 \end{pmatrix}$$

are similar.

15. Prove that the matrices

$$A = \begin{pmatrix} 0 & 1 & 1 & 1 \\ 1 & 0 & 1 & 1 \\ 1 & 1 & 0 & 1 \\ 1 & 1 & 1 & 0 \end{pmatrix} \qquad B = \begin{pmatrix} 5 & 2 & -8 & -8 \\ -6 & -3 & 8 & 8 \\ -3 & -1 & 3 & 4 \\ 3 & 1 & -4 & -5 \end{pmatrix}$$

both have characteristic polynomial $(x-3)(x+1)^3$. Determine whether they are similar and determine the Jordan canonical form for each matrix.

500

16. Determine the Jordan canonical form for the matrix

$$\begin{pmatrix} 1 & 1 & 1 & 1 \\ 0 & 1 & 0 & -1 \\ 0 & 0 & 1 & 1 \\ 0 & 0 & 0 & 1 \end{pmatrix}$$

and determine a matrix P which conjugates this matrix into its Jordan canonical form.

17. Prove that any matrix A is similar to its transpose A^t.

18. Determine all possible Jordan canonical forms for a linear transformation with characteristic polynomial $(x - 2)^3 (x - 3)^2$.

19. Prove that all $n \times n$ matrices with characteristic polynomial $f(x)$ are similar if and only if $f(x)$ has no repeated factors in its unique factorization in $F[x]$.

20. Show that the following matrices are similar in $M_p(\mathbb{F}_p)$ ($p \times p$ matrices with entries from $\mathbb{F}_p$):

$$\begin{pmatrix} 0 & 0 & 0 & \cdots & 0 & 1 \\ 1 & 0 & 0 & \cdots & 0 & 0 \\ 0 & 1 & 0 & \cdots & 0 & 0 \\ \vdots & \vdots & \vdots & & \vdots & \vdots \\ 0 & 0 & 0 & \cdots & 1 & 0 \end{pmatrix} \quad \text{and} \quad \begin{pmatrix} 1 & 1 & 0 & \cdots & 0 & 0 \\ 0 & 1 & 1 & \cdots & 0 & 0 \\ 0 & 0 & 1 & \cdots & 0 & 0 \\ \vdots & \vdots & \vdots & & \vdots & \vdots \\ 0 & 0 & 0 & \cdots & 1 & 1 \\ 0 & 0 & 0 & \cdots & 0 & 1 \end{pmatrix}.$$

21. Show that if $A^2 = A$ then A is similar to a diagonal matrix which has only 0's and 1's along the diagonal.

22. Prove that an $n \times n$ matrix A with entries from $\mathbb{C}$ satisfying $A^3 = A$ can be diagonalized. Is the same statement true over *any* field F?

23. Suppose A is a 2×2 matrix with entries from $\mathbb{Q}$ for which $A^3 = I$ but $A \neq I$. Write A in rational canonical form and in Jordan canonical form viewed as a matrix over $\mathbb{C}$.

24. Prove there are no 3×3 matrices A over $\mathbb{Q}$ with $A^8 = I$ but $A^4 \neq I$.

25. Determine the Jordan canonical form for the $n \times n$ matrix over $\mathbb{Q}$ whose entries are all equal to 1.

26. Determine the Jordan canonical form for the $n \times n$ matrix over $\mathbb{F}_p$ whose entries are all equal to 1 (the answer depends on whether or not p divides n).

27. Determine the Jordan canonical form for the $n \times n$ matrix over $\mathbb{Q}$ whose entries are all equal to 1 except that the entries along the main diagonal are all equal to 0.

28. Determine the Jordan canonical form for the $n \times n$ matrix over $\mathbb{F}_p$ whose entries are all equal to 1 except that the entries along the main diagonal are all equal to 0.

The direct sum of the cyclic submodules of V corresponding to all the elementary divisors of V which are powers of the same $x - \lambda$ is called the *generalized eigenspace of T* corresponding to the eigenvalue λ. Note that this is the p-primary component of V for the prime $p = x - \lambda$ of $F[x]$ and consists of the elements of V which are annihilated by some power of the linear transformation $T - \lambda$. The matrix for T on the generalized eigenspace for λ is the block diagonal matrix of all Jordan blocks for T with the same eigenvalue λ.

29. Suppose V_i is the generalized eigenspace of T corresponding to eigenvalue λ_i. For any $k \geq 0$, prove that the nullity of $T - \lambda_i$ on the subspace $(T - \lambda_i)^k V_i$ is the same as the nullity of $T - \lambda_i$ on $(T - \lambda_i)^k V$ and equals the number of Jordan blocks of T having eigenvalue λ_i and size greater than k (so for $k = 0$ this gives the number of Jordan blocks).

30. Let λ be an eigenvalue of the linear transformation T on the finite dimensional vector space V over the field F. Let $r_k = \dim_F (T - \lambda)^k V$ be the rank of the linear transformation $(T - \lambda)^k$ on V. For any $k \geq 1$, prove that $r_{k-1} - 2r_k + r_{k+1}$ is the number of Jordan blocks of T corresponding to λ of size k [use Exercise 12 in Section 1]. (This gives an efficient method for determining the Jordan canonical form for T by computing the ranks of the matrices $(A - \lambda I)^k$ for a matrix A representing T, cf. Exercise 31(a) in Section 11.2.)

31. Let N be an $n \times n$ matrix with coefficients in the field F. The matrix N is said to be *nilpotent* if some power of N is the zero matrix, i.e., $N^k = 0$ for some k. Prove that any nilpotent matrix is similar to a block diagonal matrix whose blocks are matrices with 1's along the first superdiagonal and 0's elsewhere.

32. Prove that if N is an $n \times n$ nilpotent matrix then in fact $N^n = 0$.

33. Let A be a strictly upper triangular $n \times n$ matrix (all entries on and below the main diagonal are zero). Prove that A is nilpotent.

34. Prove that the trace of a nilpotent $n \times n$ matrix is 0 (recall the trace of a matrix is the sum of the diagonal elements).

35. For $0 \leq i \leq n$, let d_i be the g.c.d. of the determinants of all the $i \times i$ minors of $xI - A$, for A as in Theorem 21 (take the 0×0 minor to be 1). Prove that the ith element along the diagonal of the Smith Normal Form for A is d_i / d_{i-1}. This gives the invariant factors for A. [Show these g.c.d.s do not change under elementary row and column operations.]

36. Let $V = \mathbb{C}^n$ be the usual n-dimensional vector space of n-tuples $(\alpha_1, \alpha_2, \ldots, \alpha_n)$ of complex numbers. Let T be the linear transformation defined by setting $T(\alpha_1, \alpha_2, \ldots, \alpha_n)$ equal to $(0, \alpha_1, \alpha_2, \ldots, \alpha_{n-1})$. Determine the Jordan canonical form for T.

37. Let J be a Jordan block of size n with eigenvalue λ over $\mathbb{C}$.
  (a) Prove that the Jordan canonical form for the matrix J^2 is the Jordan block of size n with eigenvalue λ^2 if $\lambda \neq 0$.
  (b) If $\lambda = 0$ prove that the Jordan canonical form for J^2 has two blocks (with eigenvalues 0) of size $\dfrac{n}{2}, \dfrac{n}{2}$ if n is even and of size $\dfrac{n-1}{2}, \dfrac{n+1}{2}$ if n is odd.

38. Determine necessary and sufficient conditions for a matrix $A \in M_n(\mathbb{C})$ to have a square root, i.e., for there to exist another matrix $B \in M_n(\mathbb{C})$ such that $A = B^2$. [Suppose B is in Jordan canonical form and consider the Jordan canonical form for B^2 using the previous exercise.]

39. Let J be a Jordan block of size n with eigenvalue λ over a field F of characteristic 2. Determine the Jordan canonical form for the matrix J^2. Determine necessary and sufficient conditions for a matrix $A \in M_n(F)$ to have a square root, i.e., for there to exist another matrix $B \in M_n(F)$ such that $A = B^2$.

The remaining exercises explore functions (power series) of a matrix and introduce some applications of the Jordan canonical form to the theory of differential equations.

Throughout these exercises the matrices are assumed to be $n \times n$ matrices with entries from the field K, where K is either the real or complex numbers. Let

$$G(x) = \sum_{k=0}^{\infty} \alpha_k x^k$$

be a power series with coefficients from K. Let $G_N(x) = \sum_{k=0}^{N} \alpha_k x^k$ be the Nth partial sum of $G(x)$ and for each $A \in M_n(K)$ let $G_N(A)$ be the element of $M_n(K)$ obtained (as usual) by substituting A in this polynomial. For each fixed i, j we obtain a sequence of real or complex numbers c_{ij}^N, $N = 0, 1, 2, \ldots$ by taking c_{ij}^N to be the i, j entry of the matrix $G_N(A)$. The series

$$G(A) = \sum_{k=0}^{\infty} \alpha_k A^k$$

is said to *converge* to the matrix C in $M_n(K)$ if for each i, $j \in \{1, 2, \ldots, n\}$ the sequence c_{ij}^N, $N = 0, 1, 2, \ldots$ converges to the i, j entry of C (in which case we write $G(A) = C$). Say $G(A)$ *converges* if there is some $C \in M_n(K)$ such that $G(A) = C$. If A is a 1×1 matrix, this is the usual notion of convergence of a series in K.

For $A = (a_{ij}) \in M_n(K)$ define

$$\| A \| = \sum_{i,j=1}^{n} |a_{ij}|$$

i.e., $\| A \|$ is the sum of the absolute values of all the entries of A.

40. Prove that for all $A, B \in M_n(K)$ and all $\alpha \in K$

    (a) $\| A + B \| \le \| A \| + \| B \|$

    (b) $\| AB \| \le \| A \| \cdot \| B \|$

    (c) $\| \alpha A \| = |\alpha| \cdot \| A \|$.

41. Let R be the radius of convergence of the real or complex power series $G(x)$ (where $R = \infty$ if $G(x)$ converges for all $x \in K$).

    (a) Prove that if $\| A \| < R$ then $G(A)$ converges.

    (b) Deduce that for *all* matrices A the following power series converge:

$$\sin(A) = A - \frac{A^3}{3!} + \frac{A^5}{5!} + \cdots + (-1)^k \frac{A^{2k+1}}{(2k+1)!} + \cdots$$

$$\cos(A) = I - \frac{A^2}{2!} + \frac{A^4}{4!} + \cdots + (-1)^k \frac{A^{2k}}{(2k)!} + \cdots$$

$$\exp(A) = I + A + \frac{A^2}{2!} + \frac{A^3}{3!} + \cdots + \frac{A^k}{k!} + \cdots$$

    where I is the $n \times n$ identity matrix.

In view of applications to the theory of differential equations we introduce a variable t at this point, so that for $A \in M_n(K)$ the matrix At is obtained from A by multiplying each entry by t (which is the same as multiplying A by the "scalar" matrix tI). We obtain a function from a subset of K into $M_n(K)$ defined by $t \mapsto G(At)$ at all points t where the series $G(At)$ converges. In particular, $\sin(At)$, $\cos(At)$ and $\exp(At)$ converge for all $t \in K$.

42. Let P be a nonsingular $n \times n$ matrix.

    (a) Prove that $PG(At)P^{-1} = G(PAtP^{-1}) = G(PAP^{-1}t)$. (This implies that, up to a change of basis, it suffices to compute $G(At)$ for matrices A in canonical form). [Take limits of partial sums to get the first equality. The second equality is immediate because the matrix tI commutes with every matrix.]

    (b) Prove that if A is the direct sum of matrices $A_1, A_2, \ldots, A_m$, then $G(At)$ is the direct sum of the matrices $G(A_1 t), G(A_2 t), \ldots, G(A_m t)$.

    (c) Show that if Z is the diagonal matrix with entries $z_1, z_2, \ldots, z_n$ then $G(Zt)$ is the diagonal matrix with entries $G(z_1 t), G(z_2 t), \ldots, G(z_n t)$.

The matrix $\exp(A)$ defined in Exercise 41(b) is called the *exponential* of A and is often denoted by e^A. The next three exercises lead to a formula for the matrix $\exp(Jt)$, where J is an elementary Jordan matrix.

43. Prove that if A and B are *commuting* matrices then $\exp(A + B) = \exp(A)\exp(B)$. [Treat A and B as commuting indeterminates and deduce this by comparing the power series on the left hand side with the product of the two power series on the right hand side.]

44. Use the preceding exercise to show that if M is any matrix and λ is any element of K then
$$\exp(\lambda I t + M) = e^{\lambda t}\exp(M).$$

45. Let N be the $r \times r$ matrix with 1's on the first superdiagonal and zeros elsewhere. Compute the exponential of the following nilpotent $r \times r$ matrix:

$$\text{if } Nt = \begin{pmatrix} 0 & t & & & \\ & 0 & t & & \\ & & & \ddots & \\ & & & & t \\ & & & & 0 \end{pmatrix} \quad \text{then } \exp(Nt) = \begin{pmatrix} 1 & t & \frac{t^2}{2!} & \cdots & \cdots & \frac{t^{r-1}}{(r-1)!} \\ & 1 & t & \frac{t^2}{2!} & & \vdots \\ & & \ddots & \ddots & \ddots & \vdots \\ & & & \ddots & t & \frac{t^2}{2!} \\ & & & & 1 & t \\ & & & & & 1 \end{pmatrix}.$$

Deduce that if J is the $r \times r$ elementary Jordan matrix with eigenvalue λ then

$$\exp(Jt) = \begin{pmatrix} e^{\lambda t} & te^{\lambda t} & \frac{t^2}{2!}e^{\lambda t} & \cdots & \cdots & \frac{t^{r-1}}{(r-1)!}e^{\lambda t} \\ & e^{\lambda t} & te^{\lambda t} & \frac{t^2}{2!}e^{\lambda t} & & \vdots \\ & & \ddots & \ddots & \ddots & \vdots \\ & & & \ddots & te^{\lambda t} & \frac{t^2}{2!}e^{\lambda t} \\ & & & & e^{\lambda t} & te^{\lambda t} \\ & & & & & e^{\lambda t} \end{pmatrix}.$$

[To do the first part use the observation that since Nt is a nilpotent matrix, $\exp(Nt)$ is a *polynomial* in Nt, i.e., all but a finite number of the terms in the power series are zero. To compute the exponential of Jt write Jt as $\lambda I t + Nt$ and use Exercise 44 with $M = Nt$.]

Let $A \in M_n(K)$ and let P be a change of basis matrix such that $P^{-1}AP$ is in Jordan canonical form. Suppose $P^{-1}AP$ is the sum of elementary Jordan matrices $J_1, \ldots, J_m$. The preceding exercises (with $t = 1$) show that $\exp(A)$ can easily be found by writing $E = \exp(P^{-1}AP)$ as the direct sum of the matrices $\exp(J_1), \ldots, \exp(J_m)$ and then changing the basis back again to obtain $\exp(A) = PEP^{-1}$.

46. For the 4×4 matrices D and P given in Example 3 of this section:

$$D = \begin{pmatrix} 1 & 2 & -4 & 4 \\ 2 & -1 & 4 & -8 \\ 1 & 0 & 1 & -2 \\ 0 & 1 & -2 & 3 \end{pmatrix} \qquad P = \begin{pmatrix} 0 & 1 & 2 & 0 \\ 2 & 0 & -2 & 1 \\ 1 & 0 & 0 & 0 \\ 0 & 0 & 1 & 0 \end{pmatrix}$$

show that

$$E = \begin{pmatrix} e & e & 0 & 0 \\ 0 & e & 0 & 0 \\ 0 & 0 & e & e \\ 0 & 0 & 0 & e \end{pmatrix} \qquad \text{and} \qquad \exp(D) = \begin{pmatrix} e & 2e & -4e & 4e \\ 2e & -e & 4e & -8e \\ e & 0 & e & -2e \\ 0 & e & -2e & 3e \end{pmatrix}.$$

47. Compute the exponential of each of the following matrices:
  (a) the matrix A in Example 2 of this section
  (b) the matrix in Exercise 4 (where you computed the Jordan canonical form and a change of basis matrix)
  (c) the matrix in Exercise 16.

48. Show that $\exp(0) = I$ (here 0 is the zero matrix and I is the identity matrix). Deduce that $\exp(A)$ is nonsingular with inverse $\exp(-A)$ for all matrices $A \in M_n(K)$.

49. Prove that $\det(\exp(A)) = e^{\text{tr}(A)}$, where $\text{tr}(A)$ is the trace of A (the sum of the diagonal entries of A).

50. Fix any $A \in M_n(K)$. Prove that the map

$$K \rightarrow GL_n(K) \qquad \text{defined by} \qquad t \mapsto \exp(At)$$

is a group homomorphism (here K is the additive group of the field). (Note how this generalizes the familiar exponential map from K to $K^\times$, which is the $n = 1$ case. The subgroup $\{\exp(At) \mid t \in K\}$ is called a *1-parameter subgroup* of $GL_n(K)$. These subgroups and the exponential map play an important role in the theory of *Lie groups* — $GL_n(K)$ being a particular example of a Lie group.).

Let $G(x)$ be a power series having an infinite radius of convergence and fix a matrix $A \in M_n(K)$. The entries of the matrix $G(At)$ are K-valued functions of the variable t that are defined for all t. Let $c_{ij}(t)$ be the function of t in the i, j entry of $G(At)$. The *derivative* of $G(At)$ with respect to t, denoted by $\dfrac{d}{dt}G(At)$, is the matrix whose i, j entry is $\dfrac{d}{dt}c_{ij}(t)$ obtained by differentiating each of the entries of $G(At)$. In other words, if we identify $M_n(K)$ with K^{n^2} by considering each $n \times n$ matrix as an n^2-tuple, then $t \mapsto G(At)$ is a map from K to K^{n^2} (i.e., is a vector valued function of t) whose derivative is just the usual (componentwise) derivative of this vector valued function.

51. Establish the following properties of derivatives:
  (a) If $G(x) = \sum\limits_{k=0}^{\infty} \alpha_k x^k$ then $\dfrac{d}{dt}G(At) = A\sum\limits_{k=1}^{\infty} k\alpha_k(At)^{k-1}$.
  (b) If v is an $n \times 1$ matrix with (constant) entries from K then

$$\frac{d}{dt}(G(At)v) = \left(\frac{d}{dt}G(At)\right)v.$$

52. Deduce from part (a) of the preceding exercise that

$$\frac{d}{dt}\exp(At) = A\exp(At).$$

Now let $y_1(t), \ldots, y_n(t)$ be differentiable functions of the real variable t that are related by the following linear system of first order differential equations with constant coefficients $a_{ij} \in K$:

$$
\begin{aligned}
y_1' &= a_{11}y_1 + a_{12}y_2 + \ldots + a_{1n}y_n \\
y_2' &= a_{21}y_1 + a_{22}y_2 + \ldots + a_{2n}y_n \\
&\;\;\vdots \\
y_n' &= a_{n1}y_1 + a_{n2}y_2 + \ldots + a_{nn}y_n
\end{aligned}
\qquad (*)
$$

(here the primes denote derivatives with respect to t). Let A be the matrix whose i, j entry is a_{ij}, so that $(*)$ may be written as

$$\begin{pmatrix} y_1' \\ y_2' \\ \vdots \\ y_n' \end{pmatrix} = A \begin{pmatrix} y_1 \\ y_2 \\ \vdots \\ y_n \end{pmatrix}$$

or, more succinctly, as $y' = Ay$, where y is the column vector of functions $y_1(t), \ldots, y_n(t)$.

An $n \times n$ matrix whose entries are functions of t and whose columns are independent solutions to the system $(*)$ is called a *fundamental matrix* of $(*)$. By the theory of differential equations, the set of vectors y that are solutions to the system $(*)$ form an n-dimensional vector space over K and so the columns of a fundamental matrix are a *basis for the vector space of all solutions to $(*)$.*

53. Prove that $\exp(At)$ is a fundamental matrix of $(*)$. Show also that if C is the $n \times 1$ constant vector whose entries are $y_1(0), \ldots, y_n(0)$ then $y(t) = \exp(At)C$ is the particular solution to the system $(*)$ satisfying the initial condition $y(0) = C$. (Note how this generalizes the 1-dimensional result that the single differential equation $y' = ay$ has e^{at} as a basis for the 1-dimensional space of solutions and the unique solution to this differential equation satisfying the initial condition $y(0) = c$ is $y = ce^{at}$.) [Use the preceding exercises.]

54. Prove that if M is a fundamental matrix of $(*)$ and if Q is a nonsingular matrix in $M_n(K)$, then MQ is also a fundamental matrix of $(*)$. [The columns of MQ are linear combinations of the columns of M.]

Now apply the preceding two exercises to solve some specific systems of differential equations as follows: given the matrix A in a system $(*)$, calculate a change of basis matrix P such that $B = P^{-1}AP$ is in Jordan canonical form. Then $\exp(At) = P \exp(Bt)P^{-1}$ is a fundamental matrix for $(*)$. By the preceding exercise, $P \exp(Bt)$ is also a fundamental matrix for $(*)$ and $\exp(Bt)$ can be calculated by the method described in the discussion following Exercise 45 (in particular, one does not have to find the inverse of the matrix P to obtain a fundamental matrix for $(*)$). Thus, for example, if $A = D$ and P are the matrices given in Exercise 46, then we saw that the Jordan canonical form for A is the matrix $B = P^{-1}AP$ consisting of two 2×2 Jordan blocks with eigenvalues 1. A fundamental matrix for the system $y' = Ay$ is therefore

$$P \exp(B) = \begin{pmatrix} 0 & 1 & 2 & 0 \\ 2 & 0 & -2 & 1 \\ 1 & 0 & 0 & 0 \\ 0 & 0 & 1 & 0 \end{pmatrix} \begin{pmatrix} e^t & te^t & 0 & 0 \\ 0 & e^t & 0 & 0 \\ 0 & 0 & e^t & te^t \\ 0 & 0 & 0 & e^t \end{pmatrix} = \begin{pmatrix} 0 & e^t & 2e^t & 2te^t \\ 2e^t & 2te^t & -2e^t & e^t(1-2t) \\ e^t & te^t & 0 & 0 \\ 0 & 0 & e^t & te^t \end{pmatrix}.$$

Writing this out more explicitly, this shows that the general solution to the system of differential equations

$$y_1' = y_1 + 2y_2 - 4y_3 + 4y_4$$
$$y_2' = 2y_1 - y_2 + 4y_3 - 8y_4$$
$$y_3' = y_1 + y_3 - 2y_4$$
$$y_4' = y_2 - 2y_3 + 3y_4$$

is given by

$$\begin{pmatrix} y_1 \\ y_2 \\ y_3 \\ y_4 \end{pmatrix} = \alpha_1 \begin{pmatrix} 0 \\ 2e^t \\ e^t \\ 0 \end{pmatrix} + \alpha_2 \begin{pmatrix} e^t \\ 2te^t \\ te^t \\ 0 \end{pmatrix} + \alpha_3 \begin{pmatrix} 2e^t \\ -2e^t \\ 0 \\ e^t \end{pmatrix} + \alpha_4 \begin{pmatrix} 2te^t \\ e^t(1-2t) \\ 0 \\ te^t \end{pmatrix}$$

where $\alpha_1, \ldots, \alpha_4$ are arbitrary elements of the field K (this describes the 4-dimensional vector space of solutions).

55. In each of Parts (a) to (c) find a fundamental matrix for the system $(*)$, where the coefficient matrix A of $(*)$ is specified.

   (a) A is the matrix in Part (a) of Exercise 47.
   (b) A is the matrix in Part (b) of Exercise 47.
   (c) A is the matrix in Part (c) of Exercise 47.

56. Consider the system $(*)$ whose coefficient matrix A is the matrix D listed in Exercise 46 and whose fundamental matrix was computed just before the preceding exercise. Find the particular solution to $(*)$ that satisfies the initial condition $y_i(0) = 1$ for $i = 1, 2, 3, 4$.

Next we explore a special case of $(*)$. Given the linear n^{th} *order* differential equation with constant coefficients

$$y^{(n)} + a_{n-1}y^{(n-1)} + \cdots + a_1 y' + a_0 y = 0 \qquad (**)$$

(where $y^{(k)}$ is the kth derivative of y and $y^{(0)} = y$) one can form a *system* of linear *first order* differential equations by letting $y_i = y^{(i-1)}$ for $1 \le i \le n$ (the coefficient matrix of this system is described in the next exercise). A basis for the n-dimensional vector space of solutions to the nth order equation $(**)$ may then obtained from a fundamental matrix for the linear system. Specifically, in each of the $n \times 1$ columns of functions in a fundamental matrix for the system, the $1, 1$ entry is a solution to $(**)$ and so the n functions in the first row of the fundamental matrix for the system form a basis for the solutions to $(**)$.

57. Prove that the matrix, A, of coefficients of the system of n first order equations obtained from $(**)$ is the transpose of the companion matrix of the polynomial $x^n + a_{n-1}x^{n-1} + \cdots + a_1 x + a_0$.

58. Use the above methods to find a basis for the vector space of solutions to the following differential equations

   (a) $y''' - 3y' + 2y = 0$
   (b) $y'''' + 4y''' + 6y'' + 4y' + y = 0.$

A system of differential equations

$$y_1' = F_1(y_1, y_2, \ldots, y_n)$$
$$y_2' = F_2(y_1, y_2, \ldots, y_n)$$
$$\vdots$$
$$y_n' = F_n(y_1, y_2, \ldots, y_n)$$

where $F_1, F_2, \ldots, F_n$ are functions of n variables, is called an *autonomous* system and it will be written more succinctly as $y' = F(y)$, where $F = (F_1, \ldots, F_n)$. (The expression autonomous means "independent of time" and it indicates that the variable t — which may be thought of as a time variable — does not appear explicitly on the right hand side.) The system $(*)$ is the special type of autonomous system in which each F_i is a linear function. In many instances it is desirable to analyze the behavior of solutions to an autonomous system of differential equations without explicitly finding these solutions (indeed, it is unlikely that it will be possible to find explicit solutions for a given nonlinear system). This investigation falls under the rubric "qualitative analysis" of autonomous differential equations and the rudiments of this study are often treated in basic calculus courses for 1×1 systems. The first step in a qualitative analysis of an $n \times n$ autonomous system is to find the *steady states*, namely the constant solutions (these are called steady states since they do not change with t). Note that a constant function $y = c$, where c is the $n \times 1$ constant vector with entries $c_1, \ldots, c_n$, is a solution to $y' = F(y)$ if and only if

$$c_i' = 0 = F_i(c_1, \ldots, c_n) \quad \text{for } i = 1, 2, \ldots, n,$$

so the steady states are found by computing the zeros of F (in the case of a nonlinear system this may require numerical methods). Next, given the initial value of some solution, one wishes to analyze the behavior of this solution as $t \to \infty$. This is called the *asymptotic behavior* of the solution. Again, it may not be possible to find the solution explicitly, although by the general theory of differential equations a solution to the initial value problem is unique provided the functions F_i are differentiable. A steady state $y = c$ is called *globally asymptotically stable* if every solution tends to c as $t \to \infty$, i.e., for any solution $y(t)$ we have $\lim_{t \to \infty} y_i(t) = c_i$ for all $i = 1, 2, \ldots, n$.

In the case of the linear autonomous system ($*$) the solutions form a vector space, so the only constant solution is the zero solution. The next exercise gives a *sufficient* condition for zero to be globally asymptotically stable and it gives one example of how the behavior of a linear system may be analyzed in terms of the eigenvalues of its coefficient matrix. Nonlinear systems can be approximated by linear systems in some neighborhood of a steady state by considering $y' = Ty$, where $T = \left(\dfrac{\partial F_i}{\partial y_j} \right)$ is the $n \times n$ Jacobian matrix of F evaluated at the steady state point. In this way the analysis of linear systems plays an important role in the local analysis of general autonomous systems.

59. Prove that the solution of ($*$) given by $y_i(t) = 0$ for all $i \in \{1, \ldots, n\}$ (i.e., the zero solution) is globally asymptotically stable if all the eigenvalues of A have negative real parts. [For those unfamiliar with the behavior of the complex exponential function, assume all eigenvalues are real (hence are negative real numbers). Use the explicit nature of the solutions to show that they all tend to zero as $t \to \infty$.]

Part IV

FIELD THEORY AND GALOIS THEORY

The previous sections have developed the theory of some of the basic algebraic structures of groups, rings and fields. The next two chapters consider properties of fields, particularly fields which arise from trying to solve equations (such as the simple equation $x^2 + 1 = 0$), and fields which naturally arise in trying to perform "arithmetic" (adding, subtracting, multiplying and dividing). The elegant and beautiful Galois Theory relates the structure of *fields* to certain related *groups* and is one of the basic algebraic tools. Applications include solutions of classical compass and straightedge construction questions, finite fields and Abel's famous theorem on the insolvability (by radicals) of the general quintic polynomial.

CHAPTER 13

Field Theory

13.1 BASIC THEORY OF FIELD EXTENSIONS

Recall that a field F is a commutative ring with identity in which every nonzero element has an inverse. Equivalently, the set $F^\times = F - \{0\}$ of nonzero elements of F is an abelian group under multiplication.

One of the first invariants associated with any field F is its *characteristic*, defined as follows: If 1_F denotes the identity of F, then F contains the elements 1_F, $1_F + 1_F$, $1_F + 1_F + 1_F, \ldots$ of the additive subgroup of F generated by 1_F, which may not all be distinct. For n a positive integer, let $n \cdot 1_F = 1_F + \cdots + 1_F$ (n times). Then two possibilities arise: either all the elements $n \cdot 1_F$ are distinct, or else $n \cdot 1_F = 0$ for some positive integer n.

Definition. The *characteristic* of a field F, denoted $\mathrm{ch}(F)$, is defined to be the smallest positive integer p such that $p \cdot 1_F = 0$ if such a p exists and is defined to be 0 otherwise.

It is easy to see that

$$n \cdot 1_F + m \cdot 1_F = (m + n) \cdot 1_F \qquad \text{and that}$$
$$(n \cdot 1_F)(m \cdot 1_F) = mn \cdot 1_F \tag{13.1}$$

for positive integers m and n. It follows that the characteristic of a field is either 0 or a *prime* p (hence the choice of p in the definition above), since if $n = ab$ is composite with $n \cdot 1_F = 0$, then $ab \cdot 1_F = (a \cdot 1_F)(b \cdot 1_F) = 0$ and since F is a field, one of $a \cdot 1_F$ or $b \cdot 1_F$ is 0, so the smallest such integer is necessarily a prime. It also follows that if $n \cdot 1_F = 0$, then n is divisible by p.

Proposition 1. The characteristic of a field F, $\mathrm{ch}(F)$, is either 0 or a prime p. If $\mathrm{ch}(F) = p$ then for any $\alpha \in F$,

$$p \cdot \alpha = \underbrace{\alpha + \alpha + \cdots + \alpha}_{p \text{ times}} = 0.$$

Proof: Only the second statement has not been proved, and this follows immediately from the evident equality $p \cdot \alpha = p \cdot (1_F \alpha) = (p \cdot 1_F)(\alpha)$ in F.

510

Remark: This notion of a characteristic makes sense also for any integral domain and its characteristic will be the same as for its field of fractions.

Examples

(1) The fields $\mathbb{Q}$ and $\mathbb{R}$ both have characteristic 0: $\mathrm{ch}(\mathbb{Q}) = \mathrm{ch}(\mathbb{R}) = 0$. The integral domain $\mathbb{Z}$ also has characteristic 0.
(2) The (finite) field $\mathbb{F}_p = \mathbb{Z}/p\mathbb{Z}$ has characteristic p for any prime p.
(3) The integral domain $\mathbb{F}_p[x]$ of polynomials in the variable x with coefficients in the field $\mathbb{F}_p$ has characteristic p, as does its field of fractions $\mathbb{F}_p(x)$ (the field of rational functions in x with coefficients in $\mathbb{F}_p$).

If we define $(-n) \cdot 1_F = -(n \cdot 1_F)$ for positive n and $0 \cdot 1_F = 0$, then we have a natural ring homomorphism (by equation (1))

$$\varphi : \mathbb{Z} \longrightarrow F$$
$$n \longmapsto n \cdot 1_F$$

and we can interpret the characteristic of F by noting that $\ker(\varphi) = \mathrm{ch}(F)\mathbb{Z}$. Taking the quotient by the kernel gives us an *injection* of either $\mathbb{Z}$ or $\mathbb{Z}/p\mathbb{Z}$ into F (depending on whether $\mathrm{ch}(F) = 0$ or $\mathrm{ch}(F) = p$). Since F is a field, we see that F contains a subfield isomorphic either to $\mathbb{Q}$ (the field of fractions of $\mathbb{Z}$) or to $\mathbb{F}_p = \mathbb{Z}/p\mathbb{Z}$ (the field of fractions of $\mathbb{Z}/p\mathbb{Z}$) depending on the characteristic of F, and in either case is the smallest subfield of F containing 1_F (the field *generated* by 1_F in F).

Definition. The *prime subfield* of a field F is the subfield of F generated by the multiplicative identity 1_F of F. It is (isomorphic to) either $\mathbb{Q}$ (if $\mathrm{ch}(F) = 0$) or $\mathbb{F}_p$ (if $\mathrm{ch}(F) = p$).

Remark: We shall usually denote the identity 1_F of a field F simply by 1. Then in a field of characteristic p, one has $p \cdot 1 = 0$, frequently written simply $p = 0$ (for example, $2 = 0$ in a field of characteristic 2). It should be kept in mind, however, that this is a shorthand statement — the element "p" is really $p \cdot 1_F$ and is not a distinct element in F. This notation is useful in light of the second statement in Proposition 1.

Examples

(1) The prime subfield of both $\mathbb{Q}$ and $\mathbb{R}$ is $\mathbb{Q}$.
(2) The prime subfield of the field $\mathbb{F}_p(x)$ is isomorphic to $\mathbb{F}_p$, given by the constant polynomials.

Definition. If K is a field containing the subfield F, then K is said to be an *extension field* (or simply an *extension*) of F, denoted K/F or by the diagram

$$
\begin{array}{c}
K \\
| \\
F
\end{array}
$$

In particular, every field F is an extension of its prime subfield. The field F is sometimes called the *base field* of the extension.

The notation K/F for a field extension is a shorthand for "K over F" and is not the quotient of K by F.

If K/F is any extension of fields, then the multiplication defined in K makes K into a *vector space* over F. In particular every field F can be considered as a vector space over its prime field.

Definition. The *degree* (or *relative degree* or *index*) of a field extension K/F, denoted $[K : F]$, is the dimension of K as a vector space over F (i.e., $[K : F] = \dim_F K$). The extension is said to be *finite* if $[K : F]$ is finite and is said to be *infinite* otherwise.

An important class of field extensions are those obtained by trying to solve equations over a given field F. For example, if $F = \mathbb{R}$ is the field of real numbers, then the simple equation $x^2 + 1 = 0$ does not have a solution in F. The question arises whether there is some larger field containing $\mathbb{R}$ in which this equation does have a solution, and it was this question that led Gauss to introduce the *complex numbers* $\mathbb{C} = \mathbb{R} + \mathbb{R}i$, where i is defined so that $i^2 + 1 = 0$. One then defines addition and multiplication in $\mathbb{C}$ by the usual rules familiar from elementary algebra and checks that in fact $\mathbb{C}$ so defined is a *field*, i.e., it is possible to find an inverse for every nonzero element of $\mathbb{C}$.

Given any field F and any polynomial $p(x) \in F[x]$ one can ask a similar question: does there exist an extension K of F containing a solution of the equation $p(x) = 0$ (i.e., containing a *root* of $p(x)$)? Note that we may assume here that the polynomial $p(x)$ is irreducible in $F[x]$ since a root of any factor of $p(x)$ is certainly a root of $p(x)$ itself. The answer is yes and follows almost immediately from our work on the polynomial ring $F[x]$. We first recall the following useful result on homomorphisms of fields (Corollary 10 of Chapter 7) which follows from the fact that the only ideals of a field F are 0 and F.

Proposition 2. Let $\varphi : F \to F'$ be a homomorphism of fields. Then φ is either identically 0 or is injective, so that the image of φ is either 0 or isomorphic to F.

Theorem 3. Let F be a field and let $p(x) \in F[x]$ be an irreducible polynomial. Then there exists a field K containing an isomorphic copy of F in which $p(x)$ has a root. Identifying F with this isomorphic copy shows that there exists an extension of F in which $p(x)$ has a root.

Proof: Consider the quotient

$$K = F[x]/(p(x))$$

of the polynomial ring $F[x]$ by the ideal generated by $p(x)$. Since by assumption $p(x)$ is an irreducible polynomial in the P.I.D. $F[x]$, the ideal $(p(x))$ is a *maximal* ideal. Hence K is actually a *field* (this is Proposition 12 of Chapter 7). The canonical projection π of $F[x]$ to the quotient $F[x]/(p(x))$ restricted to $F \subset F[x]$ gives a homomorphism $\varphi = \pi|_F : F \to K$ which is not identically 0 since it maps the identity 1 of F to the identity 1 of K. Hence by the proposition above, $\varphi(F) \cong F$ is an isomorphic copy

512

of F contained in K. We identify F with its isomorphic image in K and view F as a *subfield* of K. If $\bar{x} = \pi(x)$ denotes the image of x in the quotient K, then

$$
\begin{aligned}
p(\bar{x}) &= \overline{p(x)} && \text{(since } \pi \text{ is a homomorphism)} \\
&= p(x) \pmod{p(x)} && \text{in } F[x]/(p(x)) \\
&= 0 && \text{in } F[x]/(p(x))
\end{aligned}
$$

so that K does indeed contain a root of the polynomial $p(x)$. Then K is an extension of F in which the polynomial $p(x)$ has a root.

We shall use this result later to construct extensions of F containing *all* the roots of $p(x)$ (this is the notion of a *splitting field* and one of the central objects of interest in Galois theory).

To understand the field $K = F[x]/(p(x))$ constructed above more fully, it is useful to have a simple representation for the elements of this field. Since F is a subfield of K, we might in particular ask for a basis for K as a vector space over F.

Theorem 4. Let $p(x) \in F[x]$ be an irreducible polynomial of degree n over the field F and let K be the field $F[x]/(p(x))$. Let $\theta = x \bmod (p(x)) \in K$. Then the elements

$$1, \theta, \theta^2, \ldots, \theta^{n-1}$$

are a basis for K as a vector space over F, so the degree of the extension is n, i.e., $[K : F] = n$. Hence

$$K = \{a_0 + a_1\theta + a_2\theta^2 + \cdots + a_{n-1}\theta^{n-1} \mid a_0, a_1, \ldots, a_{n-1} \in F\}$$

consists of all polynomials of degree $< n$ in θ.

Proof: Let $a(x) \in F[x]$ be any polynomial with coefficients in F. Since $F[x]$ is a Euclidean Domain (this is Theorem 3 of Chapter 9), we may divide $a(x)$ by $p(x)$:

$$a(x) = q(x)p(x) + r(x) \qquad q(x), r(x) \in F[x] \text{ with } \deg r(x) < n.$$

Since $q(x)p(x)$ lies in the ideal $(p(x))$, it follows that $a(x) \equiv r(x) \bmod (p(x))$, which shows that every residue class in $F[x]/(p(x))$ is represented by a polynomial of degree less than n. Hence the images $1, \theta, \theta^2, \ldots, \theta^{n-1}$ of $1, x, x^2, \ldots, x^{n-1}$ in the quotient *span* the quotient as a vector space over F. It remains to see that these elements are linearly independent, so form a *basis* for the quotient over F.

If the elements $1, \theta, \theta^2, \ldots, \theta^{n-1}$ were not linearly independent in K, then there would be a linear combination

$$b_0 + b_1\theta + b_2\theta^2 + \cdots + b_{n-1}\theta^{n-1} = 0$$

in K, with $b_0, b_1, \ldots, b_{n-1} \in F$, not all 0. This is equivalent to

$$b_0 + b_1 x + b_2 x^2 + \cdots + b_{n-1}x^{n-1} \equiv 0 \bmod (p(x))$$

i.e.,

$$p(x) \text{ divides } b_0 + b_1 x + b_2 x^2 + \cdots + b_{n-1}x^{n-1}$$

in $F[x]$. But this is impossible, since $p(x)$ is of degree n and the degree of the nonzero polynomial on the right is $< n$. This proves that $1, \theta, \theta^2, \ldots, \theta^{n-1}$ are a basis for K over F, so that $[K : F] = n$ by definition. The last statement of the theorem is clear.

This theorem provides an easy description of the elements of the field $F[x]/(p(x))$ as polynomials of degree $< n$ in θ where θ is an element (in K) with $p(\theta) = 0$. It remains only to see how to add and multiply elements written in this form. The addition in the quotient $F[x]/(p(x))$ is just usual addition of polynomials. The multiplication of polynomials $a(x)$ and $b(x)$ in the quotient $F[x]/(p(x))$ is performed by finding the product $a(x)b(x)$ in $F[x]$, then finding the representative of degree $< n$ for the coset $a(x)b(x) + (p(x))$ (as in the proof above) by dividing $a(x)b(x)$ by $p(x)$ and finding the remainder.

This can also be done easily in terms of θ as follows: We may suppose $p(x)$ is monic (since its roots and the ideal it generates do not change by multiplying by a constant), say $p(x) = x^n + p_{n-1}x^{n-1} + \cdots + p_1 x + p_0$. Then in K, since $p(\theta) = 0$, we have

$$\theta^n = -(p_{n-1}\theta^{n-1} + \cdots + p_1\theta + p_0)$$

i.e., θ^n is a linear combination of lower powers of θ. Multiplying both sides by θ and replacing the θ^n on the right hand side by these lower powers again, we see that also θ^{n+1} is a polynomial of degree $< n$ in θ. Similarly, any positive power of θ can be written as a polynomial of degree $< n$ in θ, hence *any* polynomial in θ can be written as a polynomial of degree $< n$ in θ. Multiplication in K is now easily performed: one simply writes the product of two polynomials of degree $< n$ in θ as another polynomial of degree $< n$ in θ.

We summarize this as:

Corollary 5. Let K be as in Theorem 4, and let $a(\theta), b(\theta) \in K$ be two polynomials of degree $< n$ in θ. Then addition in K is defined simply by usual polynomial addition and multiplication in K is defined by

$$a(\theta)b(\theta) = r(\theta)$$

where $r(x)$ is the remainder (of degree $< n$) obtained after dividing the polynomial $a(x)b(x)$ by $p(x)$ in $F[x]$.

By the results proved above, this definition of addition and multiplication on the polynomials of degree $< n$ in θ make K into a *field*, so that one can also *divide* by nonzero elements as well, which is not so immediately obvious from the definitions of the operations.

It is also important in Theorem 4 that the polynomial $p(x)$ be *irreducible* over F. In general the addition and multiplication in Corollary 5 (which can be defined in the same way for any polynomial $p(x)$) do *not* make the polynomials of degree $< n$ in θ into a field if $p(x)$ is not irreducible. In fact, this set is not even an integral domain in general (its structure is given by Proposition 16 of Chapter 9). To describe the *field* containing a root θ of a general polynomial $f(x)$ over F, $f(x)$ is factored into irreducibles in $F[x]$ and the results above are applied to an irreducible factor $p(x)$ of $f(x)$ having θ as a root. We shall consider this more in the following sections.

514

Examples

(1) If we apply this construction to the special case $F = \mathbb{R}$ and $p(x) = x^2 + 1$ then we obtain the field

$$\mathbb{R}[x]/(x^2 + 1)$$

which is an extension of degree 2 of $\mathbb{R}$ in which $x^2 + 1$ has a root. The elements of this field are of the form $a + b\theta$ for $a, b \in \mathbb{R}$. Addition is defined by

$$(a + b\theta) + (c + d\theta) = (a + c) + (b + d)\theta. \tag{13.2a}$$

To multiply we use the fact that $\theta^2 + 1 = 0$, i.e., $\theta^2 = -1$ in K. (Alternatively, note that -1 is also the remainder when x^2 is divided by $x^2 + 1$ in $\mathbb{R}[x]$.) Then

$$\begin{aligned}
(a + b\theta)(c + d\theta) &= ac + (ad + bc)\theta + bd\theta^2 \\
&= ac + (ad + bc)\theta + bd(-1) \\
&= (ac - bd) + (ad + bc)\theta. \tag{13.2b}
\end{aligned}$$

These are, up to changing θ to i, the formulas for adding and multiplying in $\mathbb{C}$. Put another way, the map

$$\varphi : \mathbb{R}[x]/(x^2 + 1) \longrightarrow \mathbb{C}$$
$$a + bx \mapsto a + bi$$

is a homomorphism. Since it is bijective (as a map of vector spaces over the reals, for example), it is an isomorphism. Notice that instead of taking the existence of $\mathbb{C}$ for granted (along with the fairly tedious verification that it is in fact a field), we could have *defined* $\mathbb{C}$ by this isomorphism. Then the fact that it is a field is a consequence of Theorem 4.

(2) Take now $F = \mathbb{Q}$ to be the field of rational numbers and again take $p(x) = x^2 + 1$ (still irreducible over $\mathbb{Q}$, of course). Then the same construction, with the same addition and multiplication formulas as (2a) and (2b) above, except that now a and b are elements of $\mathbb{Q}$, defines a field extension $\mathbb{Q}(i)$ of $\mathbb{Q}$ of degree 2 containing a root i of $x^2 + 1$.

(3) Take $F = \mathbb{Q}$ and $p(x) = x^2 - 2$, irreducible over $\mathbb{Q}$ by Eisenstein's Criterion, for example. Then we obtain a field extension of $\mathbb{Q}$ of degree 2 containing a square root θ of 2, denoted $\mathbb{Q}(\theta)$. If we denote θ by $\sqrt{2}$, the elements of this field are of the form

$$a + b\sqrt{2}, \qquad a, b \in \mathbb{Q}$$

with addition defined by

$$(a + b\sqrt{2}) + (c + d\sqrt{2}) = (a + c) + (b + d)\sqrt{2}$$

and multiplication defined by

$$(a + b\sqrt{2})(c + d\sqrt{2}) = (ac + 2bd) + (ad + bc)\sqrt{2}.$$

(4) Let $F = \mathbb{Q}$ and $p(x) = x^3 - 2$, irreducible again by Eisenstein. Denoting a root of $p(x)$ by θ, we obtain the field

$$\mathbb{Q}[x]/(x^3 - 2) \cong \{a + b\theta + c\theta^2 \mid a, b, c \in \mathbb{Q}\}$$

with $\theta^3 = 2$, an extension of degree 3. To find the inverse of, say, $1 + \theta$ in this field, we can proceed as follows: By the Euclidean Algorithm in $\mathbb{Q}[x]$ there are polynomials $a(x)$ and $b(x)$ with

$$a(x)(1 + x) + b(x)(x^3 - 2) = 1$$

(since $p(x) = x^3 - 2$ is irreducible, it is relatively prime to every polynomial of smaller degree). In the quotient field this equation implies that $a(\theta)$ is the inverse of $1 + \theta$. In this case, a simple computation shows that we can take $a(x) = \frac{1}{3}(x^2 - x + 1)$ (and $b(x) = -\frac{1}{3}$), so that

$$(1 + \theta)^{-1} = \frac{\theta^2 - \theta + 1}{3}.$$

(5) In general, if $\theta \in K$ is a root of the irreducible polynomial

$$p(x) = p_n x^n + p_{n-1} x^{n-1} + \cdots + p_1 x + p_0$$

we can compute $\theta^{-1} \in K$ from

$$\theta(p_n \theta^{n-1} + p_{n-1} \theta^{n-2} + \cdots + p_1) = -p_0$$

namely

$$\theta^{-1} = \frac{-1}{p_0}(p_n \theta^{n-1} + p_{n-1} \theta^{n-2} + \cdots + p_1) \in K$$

(note that $p_0 \neq 0$ since $p(x)$ is irreducible).

Remark: Determining inverses in extensions of this type may be familiar from elementary algebra in the case of $\mathbb{C}$ or Example 3 under the name "rationalizing denominators." The last two examples indicates a procedure which is much more general than the ad hoc procedures of elementary algebra.

(6) Take $F = \mathbb{F}_2$, the finite field with two elements, and $p(x) = x^2 + x + 1$, which we have previously checked is irreducible over $\mathbb{F}_2$. Here we obtain a degree 2 extension of $\mathbb{F}_2$

$$\mathbb{F}_2[x]/(x^2 + x + 1) \cong \{a + b\theta \mid a, b \in \mathbb{F}_2\}$$

where $\theta^2 = -\theta - 1 = \theta + 1$. Multiplication in this field $\mathbb{F}_2(\theta)$ (which contains four elements) is defined by

$$\begin{aligned}
(a + b\theta)(c + d\theta) &= ac + (ad + bc)\theta + bd\theta^2 \\
&= ac + (ad + bc)\theta + bd(\theta + 1) \\
&= (ac + bd) + (ad + bc + bd)\theta.
\end{aligned}$$

(7) Let $F = k(t)$ be the field of rational functions in the variable t over a field k (for example, $k = \mathbb{Q}$ or $k = \mathbb{F}_p$). Let $p(x) = x^2 - t \in F[x]$. Then $p(x)$ is irreducible (it is Eisenstein at the prime (t) in $k[t]$). If we denote a root by θ, the corresponding degree 2 field extension $F(\theta)$ consists of the elements

$$\{a(t) + b(t)\theta \mid a(t), b(t) \in F\}$$

where the coefficients $a(t)$ and $b(t)$ are rational functions in t with coefficients in k and where $\theta^2 = t$.

Suppose F is a subfield of a field K and $\alpha \in K$ is an element of K. Then the collection of subfields of K containing both F and α is nonempty (K is such a field, for example). Since the intersection of subfields is again a subfield, it follows that there is a unique minimal subfield of K containing both F and α (the intersection of all subfields with this property). Similar remarks apply if α is replaced by a collection $\alpha, \beta, \ldots$ of elements of K.

Definition. Let K be an extension of the field F and let $\alpha, \beta, \cdots \in K$ be a collection of elements of K. Then the smallest subfield of K containing both F and the elements $\alpha, \beta, \ldots$, denoted $F(\alpha, \beta, \ldots)$ is called the field *generated by* $\alpha, \beta, \ldots$ *over* F.

Definition. If the field K is generated by a single element α over F, $K = F(\alpha)$, then K is said to be a *simple* extension of F and the element α is called a *primitive element* for the extension.

We shall later characterize which extensions of a field F are simple. In particular we shall prove that every finite extension of a field of characteristic 0 is a simple extension.

The connection between the simple extension $F(\alpha)$ generated by α over F where α is a root of some irreducible polynomial $p(x)$ and the field constructed in Theorem 3 is provided by the following:

Theorem 6. Let F be a field and let $p(x) \in F[x]$ be an irreducible polynomial. Suppose K is an extension field of F containing a root α of $p(x)$: $p(\alpha) = 0$. Let $F(\alpha)$ denote the subfield of K generated over F by α. Then

$$F(\alpha) \cong F[x]/(p(x)).$$

Remark: This theorem says that *any* field over F in which $p(x)$ contains a root contains a subfield isomorphic to the extension of F constructed in Theorem 3 and that this field is (up to isomorphism) the smallest extension of F containing such a root. The difference between this result and Theorem 3 is that Theorem 6 *assumes* the existence of a root α of $p(x)$ in some field K and the major point of Theorem 3 is *proving* that there exists such an extension field K.

Proof: There is a natural homomorphism

$$\varphi : F[x] \longrightarrow F(\alpha) \subseteq K$$
$$a(x) \longmapsto a(\alpha)$$

obtained by mapping F to F by the identity map and sending x to α and then extending so that the map is a ring homomorphism (i.e., the polynomial $a(x)$ in x maps to the polynomial $a(\alpha)$ in α). Since $p(\alpha) = 0$ by assumption, the element $p(x)$ is in the kernel of φ, so we obtain an induced homomorphism (also denoted φ):

$$\varphi : F[x]/(p(x)) \longrightarrow F(\alpha).$$

But since $p(x)$ is irreducible, the quotient on the left is a *field*, and φ is not the 0 map (it is the identity on F, for example), hence φ is an isomorphism of the field on the left with its image. Since this image is then a subfield of $F(\alpha)$ containing F and containing α, by the definition of $F(\alpha)$ the map must be surjective, proving the theorem.

Combined with Corollary 5, this determines the field $F(\alpha)$ when α is a root of an irreducible polynomial $p(x)$:

Corollary 7. Suppose in Theorem 6 that $p(x)$ is of degree n. Then

$$F(\alpha) = \{a_0 + a_1\alpha + a_2\alpha^2 + \cdots + a_{n-1}\alpha^{n-1} \mid a_0, a_1, \ldots, a_{n-1} \in F\} \subseteq K.$$

Describing fields generated by more than one element is more complicated and we shall return to this question in the following section.

Examples

(1) In Example 3 above, we have determined the field $\mathbb{Q}(\sqrt{2})$ generated over $\mathbb{Q}$ by the element $\sqrt{2} \in \mathbb{R}$, having suggestively denoted the abstract solution θ of the equation $x^2 - 2 = 0$ by the symbol $\sqrt{2}$, which has an independent meaning in the field $\mathbb{R}$ (namely the *positive* square root of 2 in $\mathbb{R}$).

(2) The equation $x^2 - 2 = 0$ has another solution in $\mathbb{R}$, namely $-\sqrt{2}$, the *negative* square root of 2 in $\mathbb{R}$. The field generated over $\mathbb{Q}$ by this solution consists of the elements $\{a + b(-\sqrt{2}) \mid a, b \in \mathbb{Q}\}$, and is again isomorphic to the field in Example 3 above (hence also isomorphic to the field just considered, the isomorphism given explicitly by $a + b\sqrt{2} \mapsto a - b\sqrt{2}$). As a subset of $\mathbb{R}$ this is the same set of elements as in Example 1.

(3) Similarly, if we use the symbol $\sqrt[3]{2}$ to denote the (positive) cube root of 2 in $\mathbb{R}$, then the field generated by $\sqrt[3]{2}$ over $\mathbb{Q}$ in $\mathbb{R}$ consists of the elements

$$\{a + b\sqrt[3]{2} + c(\sqrt[3]{2})^2 \mid a, b, c \in \mathbb{Q}\}$$

and is isomorphic to the field constructed in Example 4 above.

(4) The equation $x^3 - 2 = 0$ has no further solutions in $\mathbb{R}$, but there are two additional solutions in $\mathbb{C}$ given by $\sqrt[3]{2}(\dfrac{-1 + i\sqrt{3}}{2})$ and $\sqrt[3]{2}(\dfrac{-1 - i\sqrt{3}}{2})$ ($\sqrt{3}$ denoting the positive real square root of 3) as can easily be checked. The fields generated by either of these two elements over $\mathbb{Q}$ are subfields of $\mathbb{C}$ (but not of $\mathbb{R}$) and are both isomorphic to the field constructed in the previous example (and to Example 4 earlier).

As Theorem 6 indicates, the roots of an irreducible polynomial $p(x)$ are *algebraically indistinguishable* in the sense that the fields obtained by adjoining any root of an irreducible polynomial are isomorphic. In the last two examples above, the fields obtained by adjoining one of the three possible (complex) roots of $x^3 - 2 = 0$ to $\mathbb{Q}$ were all algebraically isomorphic. The fields were distinguished not by their algebraic properties, but by whether their elements were *real*, which involves *continuous* operations.

The fact that different roots of the same irreducible polynomial have the same algebraic properties can be extended slightly, as follows:

Let $\varphi : F \xrightarrow{\sim} F'$ be an isomorphism of fields. The map φ induces a ring isomorphism (also denoted φ)

$$\varphi : F[x] \xrightarrow{\sim} F'[x]$$

defined by applying φ to the coefficients of a polynomial in $F[x]$. Let $p(x) \in F[x]$ be an irreducible polynomial and let $p'(x) \in F'[x]$ be the polynomial obtained by applying the map φ to the coefficients of $p(x)$, i.e., the image of $p(x)$ under φ. The isomorphism φ maps the maximal ideal $(p(x))$ to the ideal $(p'(x))$, so this ideal is also

518

maximal, which shows that $p'(x)$ is also irreducible in $F'[x]$. The following theorem shows that the fields obtained by adjoining a root of $p(x)$ to F and a root of $p'(x)$ to F' have the same algebraic structure (i.e., are isomorphic):

Theorem 8. Let $\varphi : F \overset{\sim}{\to} F'$ be an isomorphism of fields. Let $p(x) \in F[x]$ be an irreducible polynomial and let $p'(x) \in F'[x]$ be the irreducible polynomial obtained by applying the map φ to the coefficients of $p(x)$. Let α be a root of $p(x)$ (in some extension of F) and let β be a root of $p'(x)$ (in some extension of F'). Then there is an isomorphism

$$\sigma : F(\alpha) \overset{\sim}{\longrightarrow} F'(\beta)$$
$$\alpha \longmapsto \beta$$

mapping α to β and extending φ, i.e., such that σ restricted to F is the isomorphism φ.

Proof: As noted above, the isomorphism φ induces a natural isomorphism from $F[x]$ to $F'[x]$ which maps the maximal ideal $(p(x))$ to the maximal ideal $(p'(x))$. Taking the quotients by these ideals, we obtain an isomorphism of fields

$$F[x]/(p(x)) \overset{\sim}{\longrightarrow} F'[x]/(p'(x)).$$

By Theorem 6 the field on the left is isomorphic to $F(\alpha)$ and by the same theorem the field on the right is isomorphic to $F'(\beta)$. Composing these isomorphisms, we obtain the isomorphism σ. It is clear that the restriction of this isomorphism to F is φ, completing the proof.

This extension theorem will be of considerable use when we consider Galois Theory later. It can be represented pictorially by the diagram

$$\sigma : \quad F(\alpha) \quad \overset{\sim}{\longrightarrow} \quad F'(\beta)$$
$$| \qquad\qquad |$$
$$\varphi : \qquad F \quad \overset{\sim}{\longrightarrow} \quad F'$$

EXERCISES

1. Show that $p(x) = x^3 + 9x + 6$ is irreducible in $\mathbb{Q}[x]$. Let θ be a root of $p(x)$. Find the inverse of $1 + \theta$ in $\mathbb{Q}(\theta)$.

2. Show that $x^3 - 2x - 2$ is irreducible over $\mathbb{Q}$ and let θ be a root. Compute $(1+\theta)(1+\theta+\theta^2)$ and $\dfrac{1+\theta}{1+\theta+\theta^2}$ in $\mathbb{Q}(\theta)$.

3. Show that $x^3 + x + 1$ is irreducible over $\mathbb{F}_2$ and let θ be a root. Compute the powers of θ in $\mathbb{F}_2(\theta)$.

4. Prove directly that the map $a + b\sqrt{2} \mapsto a - b\sqrt{2}$ is an isomorphism of $\mathbb{Q}(\sqrt{2})$ with itself.

5. Suppose α is a rational root of a monic polynomial in $\mathbb{Z}[x]$. Prove that α is an integer.

6. Show that if α is a root of $a_n x^n + a_{n-1}x^{n-1} + \cdots + a_1 x + a_0$ then $a_n \alpha$ is a root of the monic polynomial $x^n + a_{n-1}x^{n-1} + a_n a_{n-2}x^{n-2} + \cdots + a_n^{n-2}a_1 x + a_n^{n-1}a_0$.

7. Prove that $x^3 - nx + 2$ is irreducible for $n \neq -1, 3, 5$.

8. Prove that $x^5 - ax - 1 \in \mathbb{Z}[x]$ is irreducible unless $a = 0, 2$ or -1. The first two correspond to linear factors, the third corresponds to the factorization $(x^2 - x + 1)(x^3 + x^2 - 1)$.

13.2 ALGEBRAIC EXTENSIONS

Let F be a field and let K be an extension of F.

Definition. The element $\alpha \in K$ is said to be *algebraic* over F if α is a root of some nonzero polynomial $f(x) \in F[x]$. If α is not algebraic over F (i.e., is not the root of any nonzero polynomial with coefficients in F) then α is said to be *transcendental* over F. The extension K/F is said to be *algebraic* if every element of K is algebraic over F.

Note that if α is algebraic over a field F then it is algebraic over any extension field L of F (if $f(x)$ having α as a root has coefficients in F then it also has coefficients in L).

Proposition 9. Let α be algebraic over F. Then there is a unique monic irreducible polynomial $m_{\alpha,F}(x) \in F[x]$ which has α as a root. A polynomial $f(x) \in F[x]$ has α as a root if and only if $m_{\alpha,F}(x)$ divides $f(x)$ in $F[x]$.

Proof: Let $g(x) \in F[x]$ be a polynomial of minimal degree having α as a root. Multiplying $g(x)$ by a constant, we may assume $g(x)$ is monic. Suppose $g(x)$ were reducible in $F[x]$, say $g(x) = a(x)b(x)$ with $a(x), b(x) \in F[x]$ both of degree smaller than the degree of $g(x)$. Then $g(\alpha) = a(\alpha)b(\alpha)$ in K, and since K is a field, either $a(\alpha) = 0$ or $b(\alpha) = 0$, contradicting the minimality of the degree of $g(x)$. It follows that $g(x)$ is a monic irreducible polynomial having α as a root. Suppose now that $f(x) \in F[x]$ is any polynomial having α as a root. By the Euclidean Algorithm in $F[x]$ there are polynomials $q(x), r(x) \in F[x]$ such that

$$f(x) = q(x)g(x) + r(x) \quad \text{with} \quad \deg r(x) < \deg g(x).$$

Then $f(\alpha) = q(\alpha)g(\alpha) + r(\alpha)$ in K and since α is a root of both $f(x)$ and $g(x)$, we obtain $r(\alpha) = 0$, which contradicts the minimality of $g(x)$ unless $r(x) = 0$. Hence $g(x)$ divides any polynomial $f(x)$ in $F[x]$ having α as a root and, in particular, would divide any other monic irreducible polynomial in $F[x]$ having α as a root. This proves that $m_{\alpha,F}(x) = g(x)$ is unique and completes the proof of the proposition.

Corollary 10. If L/F is an extension of fields and α is algebraic over both F and L, then $m_{\alpha,L}(x)$ divides $m_{\alpha,F}(x)$ in $L[x]$.

Proof: This is immediate from the second statement in Proposition 9 applied to L, since $m_{\alpha,F}(x)$ is a polynomial in $L[x]$ having α as a root.

Definition. The polynomial $m_{\alpha,F}(x)$ (or just $m_\alpha(x)$ if the field F is understood) in Proposition 9 is called the *minimal polynomial* for α over F. The *degree* of $m_\alpha(x)$ is called the *degree* of α.

Note that by the proposition, a monic polynomial over F with α as a root is the minimal polynomial for α over F if and only if it is irreducible over F. Exercise 20 gives one method for computing the minimal polynomial for α over F, and the theory of Gröbner bases can be used to compute the minimal polynomial for other elements in $F(\alpha)$ (cf. Proposition 10 and Exercise 48 in Section 15.1).

Proposition 11. Let α be algebraic over the field F and let $F(\alpha)$ be the field generated by α over F. Then

$$F(\alpha) \cong F[x]/(m_\alpha(x))$$

so that in particular

$$[F(\alpha) : F] = \deg\, m_\alpha(x) = \deg\, \alpha,$$

i.e., the degree of α over F is the degree of the extension it generates over F.

Proof: This follows immediately from Theorem 6.

Examples

(1) The minimal polynomial for $\sqrt{2}$ over $\mathbb{Q}$ is $x^2 - 2$ and $\sqrt{2}$ is of degree 2 over $\mathbb{Q}$: $[\mathbb{Q}(\sqrt{2}) : \mathbb{Q}] = 2$.

(2) The minimal polynomial for $\sqrt[3]{2}$ over $\mathbb{Q}$ is $x^3 - 2$ and $\sqrt[3]{2}$ is of degree 3 over $\mathbb{Q}$: $[\mathbb{Q}(\sqrt[3]{2}) : \mathbb{Q}] = 3$.

(3) Similarly, for any $n > 1$, the polynomial $x^n - 2$ is irreducible over $\mathbb{Q}$ since it is Eisenstein. Denoting a root of this polynomial by $\sqrt[n]{2}$ (where as usual we reserve this symbol to denote the *positive* nth root of 2 if we want to view this root as an element of $\mathbb{R}$, and where the symbol denotes any one of the algebraically indistinguishable abstract solutions in general), we have $[\mathbb{Q}(\sqrt[n]{2}) : \mathbb{Q}] = n$.

(4) The minimal polynomial and the degree of an element α depend on the base field. For example, over $\mathbb{R}$, the element $\sqrt[n]{2}$ is of degree *one*, with minimal polynomial $m_{\sqrt[n]{2}, \mathbb{R}}(x) = x - \sqrt[n]{2}$.

(5) Consider the polynomial $p(x) = x^3 - 3x - 1$ over $\mathbb{Q}$, which is irreducible over $\mathbb{Q}$ since it is a cubic which has no rational root (cf. Proposition 11 of Chapter 9). Hence $[\mathbb{Q}(\alpha) : \mathbb{Q}] = 3$ for any root α of $p(x)$. For future reference we note that a quick sketch of the graph of this function over the real numbers shows that the graph crosses the x-axis precisely once in the interval $[0,2]$, i.e., there is precisely one real number $\alpha, 0 < \alpha < 2$ satisfying $\alpha^3 - 3\alpha - 1 = 0$.

Proposition 12. The element α is algebraic over F if and only if the simple extension $F(\alpha)/F$ is finite. More precisely, if α is an element of an extension of degree n over F then α satisfies a polynomial of degree at most n over F and if α satisfies a polynomial of degree n over F then the degree of $F(\alpha)$ over F is at most n.

Proof: If α is algebraic over F, then the degree of the extension $F(\alpha)/F$ is the degree of the minimal polynomial for α over F. Hence the extension is finite, of degree $\leq n$ if α satisfies a polynomial of degree n. Conversely, suppose α is an element of an extension of degree n over F (for example, if $[F(\alpha) : F] = n$). Then the $n + 1$ elements

$$1, \alpha, \alpha^2, \ldots, \alpha^n$$

of $F(\alpha)$ are linearly dependent over F, say

$$b_0 + b_1\alpha + b_2\alpha^2 + \cdots + b_n\alpha^n = 0$$

with $b_0, b_1, b_2, \ldots, b_n \in F$ not all 0. Hence α is the root of a nonzero polynomial with coefficients in F (of degree $\leq n$), which proves α is algebraic over F and also proves the second statement of the proposition.

Corollary 13. If the extension K/F is finite, then it is algebraic.

Proof: If $\alpha \in K$, then the subfield $F(\alpha)$ is in particular a subspace of the vector space K over F. Hence $[F(\alpha) : F] \leq [K : F]$ and so α is algebraic over F by the proposition.

Remark: We shall prove below a sort of converse to this result (Theorem 17), but note that there are infinite algebraic extensions (we shall have an example later), so the literal converse of this corollary is not true.

Example: (Quadratic Extensions over Fields of Characteristic $\neq 2$)

Let F be a field of characteristic $\neq 2$ (for example, any field of characteristic 0, such as $\mathbb{Q}$) and let K be an extension of F of degree 2, $[K : F] = 2$. Let α be any element of K not contained in F. By the proposition above, α satisfies an equation of degree at most 2 over F. This equation cannot be of degree 1, since α is not an element of F by assumption. It follows that the minimal polynomial of α is a monic quadratic

$$m_\alpha(x) = x^2 + bx + c \qquad b, c \in F.$$

Since $F \subset F(\alpha) \subseteq K$ and $F(\alpha)$ is already a vector space over F of dimension 2, we have $K = F(\alpha)$.

The roots of this quadratic equation can be determined by the quadratic formula, which is valid over any field of characteristic $\neq 2$ (the formula is obtained as in elementary algebra by completing the square):

$$\alpha = \frac{-b \pm \sqrt{b^2 - 4c}}{2}$$

(the reason for requiring the characteristic of F not be 2 is that we must divide by 2). Here $b^2 - 4c$ is not a square in F since α is not an element of F and the symbol $\sqrt{b^2 - 4c}$ denotes a root of the equation $x^2 - (b^2 - 4c) = 0$ in K (see the end of the next paragraph). Note that here there is no natural choice of one of the roots analogous to choosing the *positive* square root of 2 in $\mathbb{R}$ — the roots are algebraically indistinguishable.

Now $F(\alpha) = F(\sqrt{b^2 - 4c})$ as follows: by the formula above, α is an element of the field on the right, hence $F(\alpha) \subseteq F(\sqrt{b^2 - 4c})$. Conversely, $\sqrt{b^2 - 4c} = \mp(b+2\alpha)$ shows that $\sqrt{b^2 - 4c}$ is an element of $F(\alpha)$, which gives the reverse inclusion $F(\sqrt{b^2 - 4c}) \subseteq F(\alpha)$ (and incidentally shows that the equation $x^2 - (b^2 - 4c) = 0$ does have a solution in K).

It follows that any extension K of F of degree 2 is of the form $F(\sqrt{D})$ where D is an element of F which is not a square in F, and conversely, every such extension is an extension of degree 2 of F. For this reason, extensions of degree 2 of a field F are called *quadratic* extensions of F.

Suppose that F is a subfield of a field K which in turn is a subfield of a field L. Then there are three associated extension degrees — the dimension of K and L as vector spaces over F, and the dimension of L as a vector space over K.

Theorem 14. Let $F \subseteq K \subseteq L$ be fields. Then

$$[L : F] = [L : K][K : F],$$

i.e. extension degrees are multiplicative, where if one side of the equation is infinite, the other side is also infinite. Pictorially,

Proof: Suppose first that $[L : K] = m$ and $[K : F] = n$ are finite. Let $\alpha_1, \alpha_2, \ldots, \alpha_m$ be a basis for L over K and let $\beta_1, \beta_2, \ldots, \beta_n$ be a basis for K over F. Then every element of L can be written as a linear combination

$$a_1\alpha_1 + a_2\alpha_2 + \cdots + a_m\alpha_m$$

where $a_1, \ldots, a_m$ are elements of K, hence are F-linear combinations of $\beta_1, \ldots, \beta_n$:

$$a_i = b_{i1}\beta_1 + b_{i2}\beta_2 + \cdots + b_{in}\beta_n \qquad i = 1, 2, \ldots, m \qquad (13.3)$$

where the b_{ij} are elements of F. Substituting these expressions in for the coefficients a_i above, we see that every element of L can be written as a linear combination

$$\sum_{\substack{i=1,2,\ldots,m \\ j=1,2,\ldots,n}} b_{ij}\alpha_i\beta_j$$

of the mn elements $\alpha_i\beta_j$ with coefficients in F. Hence these elements *span* L as a vector space over F.

Suppose now that we had a linear relation in L

$$\sum_{\substack{i=1,2,\ldots,m \\ j=1,2,\ldots,n}} b_{ij}\alpha_i\beta_j = 0$$

with coefficients b_{ij} in F. Then defining the elements $a_i \in K$ by equation (3) above, this linear relation could be written

$$a_1\alpha_1 + a_2\alpha_2 + \cdots + a_m\alpha_m = 0.$$

Since the α_i are a basis for L over K, it follows that all the coefficients $a_i, i = 1, 2, \ldots, m$ must be 0, i.e., that

$$b_{i1}\beta_1 + b_{i2}\beta_2 + \cdots + b_{in}\beta_n = 0 \qquad i = 1, 2, \ldots, m$$

in K. Since now the $\beta_j, j = 1, 2, \ldots, n$ form a basis for K over F, this implies $b_{ij} = 0$ for all i and j. Hence the elements $\alpha_i\beta_j$ are linearly independent over F, so form a basis for L over F and $[L : F] = mn = [L : K][K : F]$, as claimed.

If $[K : F]$ is infinite, then there are infinitely many elements of K, hence of L, which are linearly independent over F, so that $[L : F]$ is also infinite. Similarly, if $[L : K]$ is infinite, there are infinitely many elements of L linearly independent over K, so certainly linearly independent over F, so again $[L : F]$ is infinite. Finally, if $[L : K]$ and $[K : F]$ are both finite, then the proof above shows $[L : F]$ is finite, so that $[L : F]$ infinite implies at least one of $[L : K]$ and $[K : F]$ is infinite, completing the proof.

Remark: Note the similarity of this result with the result on group orders proved in Part I. As with diagrams involving groups we shall frequently indicate the relative degrees of extensions in field diagrams.

The multiplicativity of extension degrees is extremely useful in computations. A particular application is the following:

Corollary 15. Suppose L/F is a finite extension and let K be any subfield of L containing F, $F \subseteq K \subseteq L$. Then $[K : F]$ divides $[L : F]$.

Proof: This is immediate.

Examples

(1) The element $\sqrt{2}$ is not contained in the field $\mathbb{Q}(\alpha)$ where α is the real root of $x^3 - 3x - 1$ between 0 and 2, since we have already determined that $[\mathbb{Q}(\sqrt{2}) : \mathbb{Q}] = 2$ and $[\mathbb{Q}(\alpha) : \mathbb{Q}] = 3$ and 2 does not divide 3. Note that it is not so easy to prove directly that $\sqrt{2}$ cannot be written as a rational linear combination of $1, \alpha, \alpha^2$.

(2) Let as usual $\sqrt[6]{2}$ denote the positive real 6th root of 2. Then $[\mathbb{Q}(\sqrt[6]{2}) : \mathbb{Q}] = 6$. Since $(\sqrt[6]{2})^3 = \sqrt{2}$ we have $\mathbb{Q}(\sqrt{2}) \subset \mathbb{Q}(\sqrt[6]{2})$ and by the multiplicativity of extension degrees, $[\mathbb{Q}(\sqrt[6]{2}) : \mathbb{Q}(\sqrt{2})] = 3$. This gives us the field diagram

$$\overbrace{\mathbb{Q} \quad \underbrace{\subset \quad \mathbb{Q}(\sqrt{2})}_{2} \quad \underbrace{\subset \quad \mathbb{Q}(\sqrt[6]{2})}_{3}}^{6}$$

In particular, this shows that the minimal polynomial for $\sqrt[6]{2}$ over $\mathbb{Q}(\sqrt{2})$ is of degree 3. It is therefore the polynomial $x^3 - \sqrt{2}$. Note that showing directly that this polynomial is irreducible over $\mathbb{Q}(\sqrt{2})$ is not completely trivial.

By Theorem 14 a finite extension of a finite extension is finite. The next results use this to show that an extension generated by a finite number of algebraic elements is finite (extending Proposition 12).

Definition. An extension K/F is *finitely generated* if there are elements $\alpha_1, \alpha_2, \ldots, \alpha_k$ in K such that $K = F(\alpha_1, \alpha_2, \ldots, \alpha_k)$.

Recall that the field generated over F by a collection of elements in a field K is the smallest subfield of K containing these elements and F. The next lemma will show that for finitely generated extensions this field can be obtained recursively by a series of simple extensions.

Lemma 16. $F(\alpha, \beta) = (F(\alpha))(\beta)$, i.e., the field generated over F by α and β is the field generated by β over the field $F(\alpha)$ generated by α.

Proof: This follows by the minimality of the fields in question. The field $F(\alpha, \beta)$ contains F and α, hence contains the field $F(\alpha)$, and since it also contains β, we have the inclusion $(F(\alpha))(\beta) \subseteq F(\alpha, \beta)$ by the minimality of the field $(F(\alpha))(\beta)$. Since the field $(F(\alpha))(\beta)$ contains F, α and β, by the minimality of $F(\alpha, \beta)$ we have the reverse inclusion $F(\alpha, \beta) \subseteq (F(\alpha))(\beta)$, which proves the lemma.

By the lemma we have

$$K = F(\alpha_1, \alpha_2, \ldots, \alpha_k) = (F(\alpha_1, \alpha_2, \ldots, \alpha_{k-1}))(\alpha_k)$$

and so by iterating, we see that K is obtained by taking the field F_1 generated over F by α_1, then the field F_2 generated *over* F_1 (this is important) by α_2, and so on, with $F_k = K$. This gives a sequence of fields:

$$F = F_0 \subseteq F_1 \subseteq F_2 \subseteq \cdots \subseteq F_k = K$$

where

$$F_{i+1} = F_i(\alpha_{i+1}) \qquad i = 0, 1, \ldots, k - 1.$$

Suppose now that the elements $\alpha_1, \alpha_2, \ldots, \alpha_k$ are algebraic over F of degrees $n_1, n_2, \ldots, n_k$ (so a priori are algebraic over any extension of F). Then the extensions in this sequence are simple extensions of the type considered in Proposition 11. The relative extension degree $[F_{i+1} : F_i]$ is equal to the degree of the minimal polynomial of α_{i+1} over F_i, which is at most n_{i+1} (and equals n_{i+1} if and only if the minimal polynomial of α_{i+1} over F remains irreducible over F_i). By the multiplicativity of extension degrees, we see that

$$[K : F] = [F_k : F_{k-1}][F_{k-1} : F_{k-2}] \cdots [F_1 : F_0]$$

is also finite, and $\leq n_1 n_2 \cdots n_k$.

This also gives a description of the elements of $F(\alpha_1, \alpha_2, \ldots, \alpha_k)$. For simplicity, consider the case of the field $F(\alpha, \beta)$ where α and β are algebraic over F. Then the elements of this field are of the form

$$b_0 + b_1 \beta + b_2 \beta^2 + \cdots + b_{d-1} \beta^{d-1}$$

where $d = [F(\alpha)(\beta) : F(\alpha)]$ is the degree of β over $F(\alpha)$ (which may be strictly smaller than the degree of β over F), and where the coefficients $b_0, b_1, \ldots, b_{d-1}$ are elements of $F(\alpha)$. The coefficients $b_i \in F(\alpha)$, $i = 0, \ldots, d - 1$, are of the form

$$a_{0i} + a_{1i}\alpha + a_{2i}\alpha^2 + \cdots + a_{n-1i}\alpha^{n-1}$$

where $n = [F(\alpha) : F]$ is the degree of α over F and the a_{ij} are elements of F. Hence the elements of $F(\alpha, \beta)$ are of the form

$$\sum_{\substack{i=0,1,\ldots,n-1 \\ j=0,1,\ldots,d-1}} a_{ij}\alpha^i \beta^j \qquad a_{ij} \in F.$$

Since $[F(\alpha, \beta) : F] = [F(\alpha, \beta) : F(\alpha)][F(\alpha) : F] = dn$, the elements $\alpha^i \beta^j$ are in fact an F basis for $F(\alpha, \beta)$.

In practice the field $F(\alpha)$ generated by the algebraic α is obtained by adjoining the element α to F and then "closing" the resulting set with respect to addition and multiplication, which amounts to adjoining the powers α^2, α^3, ... of α and taking linear combinations (with coefficients from F) of these elements. The process terminates when a power of α is a linear combination of lower powers of α which amounts to knowing the minimal polynomial for α. The previous discussion shows a similar process gives the field $F(\alpha, \beta)$ generated by two elements, and by recursion, the field generated by any finite number of algebraic elements. This shows in particular that "closing" with respect to addition and multiplication also closes with respect to division for algebraic elements (cf. Example 5 following Corollary 5 above). If the elements are not algebraic, one must also "close" with respect to inverses. The difficulty in this procedure is determining the degrees of the *relative* extensions — for example the degree d for $F(\alpha, \beta)$ over $F(\alpha)$ above, for which one has only an a priori upper bound (the degree of β over F).

This is the analogue of "closing" a set of elements in a group G to determine the subgroup they generate.

Examples

(1) The extension $\mathbb{Q}(\sqrt[6]{2}, \sqrt{2})$ is simply the extension $\mathbb{Q}(\sqrt[6]{2})$ since $\sqrt{2}$ is already an element of this field. Put another way, the degree d of $\sqrt{2}$ over $\mathbb{Q}(\sqrt[6]{2})$ is 1, which is strictly smaller than the degree of $\sqrt{2}$ over $\mathbb{Q}$. We shall later have less obvious examples where this occurs.

(2) Consider the field $\mathbb{Q}(\sqrt{2}, \sqrt{3})$ generated over $\mathbb{Q}$ by $\sqrt{2}$ and $\sqrt{3}$. Since $\sqrt{3}$ is of degree 2 over $\mathbb{Q}$ the degree of the extension $\mathbb{Q}(\sqrt{2}, \sqrt{3})/\mathbb{Q}(\sqrt{2})$ is at most 2 and is precisely 2 if and only if $x^2 - 3$ is irreducible over $\mathbb{Q}(\sqrt{2})$. Since this polynomial is of degree 2, it is reducible only if it has a root, i.e., if and only if $\sqrt{3} \in \mathbb{Q}(\sqrt{2})$. Suppose $\sqrt{3} = a + b\sqrt{2}$ with $a, b \in \mathbb{Q}$. Squaring this we obtain $3 = (a^2 + 2b^2) + 2ab\sqrt{2}$. If $ab \neq 0$, then we can solve this equation for $\sqrt{2}$ in terms of a and b which implies that $\sqrt{2}$ is rational, which it is not. If $b = 0$, then we would have that $\sqrt{3} = a$ is rational, a contradiction. Finally, if $a = 0$, we have $\sqrt{3} = b\sqrt{2}$ and multiplying both sides by $\sqrt{2}$ we see that $\sqrt{6}$ would be rational, again a contradiction. This shows $\sqrt{3} \notin \mathbb{Q}(\sqrt{2})$, proving

$$[\mathbb{Q}(\sqrt{2}, \sqrt{3}) : \mathbb{Q}] = 4.$$

Elements in this field (by "closing " $1, \sqrt{2}, \sqrt{3}$) include $1, \sqrt{2}, \sqrt{3}, \sqrt{6}$ and by the computations above, these form a basis for this field:

$$\mathbb{Q}(\sqrt{2}, \sqrt{3}) = \{a + b\sqrt{2} + c\sqrt{3} + d\sqrt{6} \mid a, b, c, d \in \mathbb{Q}\}.$$

We can now characterize the finite extensions of a field F:

Theorem 17. The extension K/F is finite if and only if K is generated by a finite number of algebraic elements over F. More precisely, a field generated over F by a finite number of algebraic elements of degrees $n_1, n_2, \ldots, n_k$ is algebraic of degree $\leq n_1 n_2 \cdots n_k$.

Proof: If K/F is finite of degree n, let $\alpha_1, \alpha_2, \ldots, \alpha_n$ be a basis for K as a vector space over F. By Corollary 15, $[F(\alpha_i) : F]$ divides $[K : F] = n$ for $i = 1, 2, \ldots, n$, so that Proposition 12 implies each α_i is algebraic over F. Since K is obviously generated over F by $\alpha_1, \alpha_2, \ldots, \alpha_n$, we see that K is generated by a finite number of algebraic elements over F. The converse was proved above. The second statement of the theorem is immediate from Corollary 13 and the computation above.

The first example above shows that the inequality for the degree of the extension given in the theorem may be strict. We remark that information helpful in the determination of this degree can often be obtained by determining subfields and then applying Corollary 15.

Corollary 18. Suppose α and β are algebraic over F. Then $\alpha \pm \beta$, $\alpha\beta$, α/β (for $\beta \neq 0$), (in particular α^{-1} for $\alpha \neq 0$) are all algebraic.

Proof: All of these elements lie in the extension $F(\alpha, \beta)$, which is finite over F by the theorem, hence they are algebraic by Corollary 13.

Corollary 19. Let L/F be an arbitrary extension. Then the collection of elements of L that are algebraic over F form a subfield K of L.

Proof: This is immediate from the previous corollary.

Examples

(1) Consider the extension $\mathbb{C}/\mathbb{Q}$ and let $\overline{\mathbb{Q}}$ denote the subfield of all elements in $\mathbb{C}$ that are algebraic over $\mathbb{Q}$. In particular, the elements $\sqrt[n]{2}$ (the positive n^{th} roots of 2 in $\mathbb{R}$) are all elements of $\overline{\mathbb{Q}}$, so that $[\overline{\mathbb{Q}} : \mathbb{Q}] \geq n$ for all integers $n > 1$. Hence $\overline{\mathbb{Q}}$ is an *infinite* algebraic extension of $\mathbb{Q}$, called the field of *algebraic numbers*.

(2) Consider the field $\overline{\mathbb{Q}} \cap \mathbb{R}$, the subfield of $\mathbb{R}$ consisting of elements algebraic over $\mathbb{Q}$. The field $\mathbb{Q}$ is *countable*. The number of polynomials in $\mathbb{Q}[x]$ of any given degree n is therefore also countable (since such a polynomial is determined by specifying $n + 1$ coefficients from $\mathbb{Q}$). Since these polynomials have at most n roots in $\mathbb{R}$, the number of algebraic elements of $\mathbb{R}$ of degree n is countable. Finally, the collection of all algebraic elements in $\mathbb{R}$ is the countable union (indexed by n) of countable sets, hence is countable. Since $\mathbb{R}$ is uncountable, it follows that there exist (in fact many) elements of $\mathbb{R}$ which are not algebraic, i.e., are transcendental, over $\mathbb{Q}$. In particular the subfield $\overline{\mathbb{Q}} \cap \mathbb{R}$ of algebraic elements of $\mathbb{R}$ is a *proper* subfield of $\mathbb{R}$, so also $\overline{\mathbb{Q}}$ is a proper subfield of $\mathbb{C}$.

It is extremely difficult in general to prove that a given real number is not algebraic. For example, it is known (these are theorems) that $\pi = 3.14159\ldots$ and $e = 2.71828\ldots$ are transcendental elements of $\mathbb{R}$. Even the proofs that these elements are not *rational* are not too easy.

Theorem 20. If K is algebraic over F and L is algebraic over K, then L is algebraic over F.

Proof: Let α be any element of L. Then α is algebraic over K, so α satisfies some polynomial equation

$$a_n\alpha^n + a_{n-1}\alpha^{n-1} + \cdots + a_1\alpha + a_0 = 0$$

where the coefficients $a_0, a_1, \ldots, a_n$ are in K. Consider the field $F(\alpha, a_0, a_1, \ldots, a_n)$ generated over F by α and the coefficients of this polynomial. Since K/F is algebraic, the elements $a_0, a_1, \ldots, a_n$ are algebraic over F, so the extension $F(a_0, a_1, \ldots, a_n)/F$ is finite by Theorem 17. By the equation above, we see that α generates an extension of this field of degree at most n, since its minimal polynomial over this field is a divisor of the polynomial above. Therefore

$$[F(\alpha, a_0, a_1, \ldots, a_n) : F] = [F(\alpha, a_0, \ldots, a_n) : F(a_0, \ldots, a_n)][F(a_0, \ldots, a_n) : F]$$

is also finite and $F(\alpha, a_0, a_1, \ldots, a_n)/F$ is an algebraic extension. In particular the element α is algebraic over F, which proves that L is algebraic over F.

The subfield $F(\alpha_1, \alpha_2, \ldots, \alpha_k)$ generated by a finite set of elements $\alpha_1, \alpha_2, \ldots, \alpha_k$ of a field K contains each of the fields $F(\alpha_i)$, $i = 1, 2, \ldots, k$. By the definitions, it is also the smallest subfield of K containing these fields.

Definition. Let K_1 and K_2 be two subfields of a field K. Then the *composite field* of K_1 and K_2, denoted $K_1 K_2$, is the smallest subfield of K containing both K_1 and K_2. Similarly, the composite of any collection of subfields of K is the smallest subfield containing all the subfields.

Note that the composite $K_1 K_2$ can also be described as the intersection of all the subfields of K containing both K_1 and K_2 and similarly for the composite of more than two fields, analogous to the subgroup generated by a subset of a group (cf. Section 2.4).

Example

The composite of the two fields $\mathbb{Q}(\sqrt{2})$ and $\mathbb{Q}(\sqrt[3]{2})$ is the field $\mathbb{Q}(\sqrt[6]{2})$. This is because this field contains both of these subfields ($(\sqrt[6]{2})^3 = \sqrt{2}$ and $(\sqrt[6]{2})^2 = \sqrt[3]{2}$) and conversely, any field containing both $\sqrt{2}$ and $\sqrt[3]{2}$ contains their quotient, which is $\sqrt[6]{2}$.

Suppose now that K_1 and K_2 are finite extensions of F in K. Let $\alpha_1, \alpha_2, \ldots, \alpha_n$ be an F-basis for K_1 and let $\beta_1, \beta_2, \ldots, \beta_m$ be an F-basis for K_2 (so that $[K_1 : F] = n$ and $[K_2 : F] = m$). Then it is clear that these give generators for the composite $K_1 K_2$ over F:

$$K_1 K_2 = F(\alpha_1, \alpha_2, \ldots, \alpha_n, \beta_1, \beta_2, \ldots, \beta_m).$$

Since $\alpha_1, \alpha_2, \ldots, \alpha_n$ is an F-basis for K_1 any power α_i^k of one of the α's is a *linear* combination with coefficients in F of the α's and a similar statement holds for the β's. It follows that the collection of linear combinations

$$\sum_{\substack{i=1,2,\ldots,n \\ j=1,2,\ldots,m}} a_{ij} \alpha_i \beta_j$$

with coefficients in F is *closed* under multiplication and addition since in a product of two such elements any higher powers of the α's and β's can be replaced by linear expressions. Hence, the elements $\alpha_i \beta_j$ for $i = 1, 2, \ldots, n$ and $j = 1, 2, \ldots, m$ *span* the composite extension $K_1 K_2$ over F. In particular, $[K_1 K_2 : F] \leq mn$. We summarize this as:

Proposition 21. Let K_1 and K_2 be two finite extensions of a field F contained in K. Then

$$[K_1 K_2 : F] \le [K_1 : F][K_2 : F]$$

with equality if and only if an F-basis for one of the fields remains linearly independent over the other field. If $\alpha_1, \alpha_2, \ldots, \alpha_n$ and $\beta_1, \beta_2, \ldots, \beta_m$ are bases for K_1 and K_2 over F, respectively, then the elements $\alpha_i \beta_j$ for $i = 1, 2, \ldots, n$ and $j = 1, 2, \ldots, m$ span $K_1 K_2$ over F.

Proof: From $K_1 K_2 = F(\alpha_1, \alpha_2, \ldots, \alpha_n, \beta_1, \beta_2, \ldots, \beta_m) = K_1(\beta_1, \beta_2, \ldots, \beta_m)$, we see as above that $\beta_1, \beta_2, \ldots, \beta_m$ span $K_1 K_2$ over K_1. Hence $[K_1 K_2 : K_1] \le m = [K_2 : F]$ with equality if and only if these elements are linearly independent over K_1. Since $[K_1 K_2 : F] = [K_1 K_2 : K_1][K_1 : F]$ this proves the proposition.

By the proposition (and its proof), we have the following diagram:

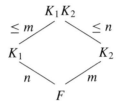

We shall have examples shortly where the inequality in the proposition is strict. One useful situation where one can be certain of equality is the following:

Corollary 22. Suppose that $[K_1 : F] = n$, $[K_2 : F] = m$ in Proposition 21, where n and m are relatively prime: $(n, m) = 1$. Then $[K_1 K_2 : F] = [K_1 : F][K_2 : F] = nm$.

Proof: In general the extension degree $[K_1 K_2 : F]$ is divisible by both n and m since K_1 and K_2 are subfields of $K_1 K_2$, hence is divisible by their least common multiple. In this case, since $(n, m) = 1$, this means $[K_1 K_2 : F]$ is divisible by nm, which together with the inequality $[K_1 K_2 : F] \le nm$ of the proposition proves the corollary.

Example

The composite of the two fields $\mathbb{Q}(\sqrt{2})$ and $\mathbb{Q}(\sqrt[3]{2})$ is of degree 6 over $\mathbb{Q}$, which we determined earlier by actually computing the composite $\mathbb{Q}(\sqrt[6]{2})$.

EXERCISES

1. Let $\mathbb{F}$ be a finite field of characteristic p. Prove that $|\mathbb{F}| = p^n$ for some positive integer n.

2. Let $g(x) = x^2 + x - 1$ and let $h(x) = x^3 - x + 1$. Obtain fields of 4, 8, 9 and 27 elements by adjoining a root of $f(x)$ to the field F where $f(x) = g(x)$ or $h(x)$ and $F = \mathbb{F}_2$ or $\mathbb{F}_3$. Write down the multiplication tables for the fields with 4 and 9 elements and show that the nonzero elements form a cyclic group.

3. Determine the minimal polynomial over $\mathbb{Q}$ for the element $1 + i$.

4. Determine the degree over $\mathbb{Q}$ of $2 + \sqrt{3}$ and of $1 + \sqrt[3]{2} + \sqrt[3]{4}$.

5. Let $F = \mathbb{Q}(i)$. Prove that $x^3 - 2$ and $x^3 - 3$ are irreducible over F.

6. Prove directly from the definitions that the field $F(\alpha_1, \alpha_2, \ldots, \alpha_n)$ is the composite of the fields $F(\alpha_1), F(\alpha_2), \ldots, F(\alpha_n)$.

7. Prove that $\mathbb{Q}(\sqrt{2} + \sqrt{3}) = \mathbb{Q}(\sqrt{2}, \sqrt{3})$ [one inclusion is obvious, for the other consider $(\sqrt{2} + \sqrt{3})^2$, etc.]. Conclude that $[\mathbb{Q}(\sqrt{2} + \sqrt{3}) : \mathbb{Q}] = 4$. Find an irreducible polynomial satisfied by $\sqrt{2} + \sqrt{3}$.

8. Let F be a field of characteristic $\neq 2$. Let D_1 and D_2 be elements of F, neither of which is a square in F. Prove that $F(\sqrt{D_1}, \sqrt{D_2})$ is of degree 4 over F if $D_1 D_2$ is not a square in F and is of degree 2 over F otherwise. When $F(\sqrt{D_1}, \sqrt{D_2})$ is of degree 4 over F the field is called a *biquadratic extension of F*.

9. Let F be a field of characteristic $\neq 2$. Let a, b be elements of the field F with b not a square in F. Prove that a necessary and sufficient condition for $\sqrt{a + \sqrt{b}} = \sqrt{m} + \sqrt{n}$ for some m and n in F is that $a^2 - b$ is a square in F. Use this to determine when the field $\mathbb{Q}(\sqrt{a + \sqrt{b}})$ $(a, b \in \mathbb{Q})$ is biquadratic over $\mathbb{Q}$.

10. Determine the degree of the extension $\mathbb{Q}(\sqrt{3 + 2\sqrt{2}})$ over $\mathbb{Q}$.

11. (a) Let $\sqrt{3 + 4i}$ denote the square root of the complex number $3 + 4i$ that lies in the first quadrant and let $\sqrt{3 - 4i}$ denote the square root of $3 - 4i$ that lies in the fourth quadrant. Prove that $[\mathbb{Q}(\sqrt{3 + 4i} + \sqrt{3 - 4i}) : \mathbb{Q}] = 1$.

    (b) Determine the degree of the extension $\mathbb{Q}(\sqrt{1 + \sqrt{-3}} + \sqrt{1 - \sqrt{-3}})$ over $\mathbb{Q}$.

12. Suppose the degree of the extension K/F is a prime p. Show that any subfield E of K containing F is either K or F.

13. Suppose $F = \mathbb{Q}(\alpha_1, \alpha_2, \ldots, \alpha_n)$ where $\alpha_i^2 \in \mathbb{Q}$ for $i = 1, 2, \ldots, n$. Prove that $\sqrt[3]{2} \notin F$.

14. Prove that if $[F(\alpha) : F]$ is odd then $F(\alpha) = F(\alpha^2)$.

15. A field F is said to be *formally real* if -1 is not expressible as a sum of squares in F. Let F be a formally real field, let $f(x) \in F[x]$ be an irreducible polynomial of odd degree and let α be a root of $f(x)$. Prove that $F(\alpha)$ is also formally real. [Pick α a counterexample of minimal degree. Show that $-1 + f(x)g(x) = (p_1(x))^2 + \cdots + (p_m(x))^2$ for some $p_i(x), g(x) \in F[x]$ where $g(x)$ has odd degree $< \deg f$. Show that some root β of g has odd degree over F and $F(\beta)$ is not formally real, violating the minimality of α.]

16. Let K/F be an algebraic extension and let R be a *ring* contained in K and containing F. Show that R is a subfield of K containing F.

17. Let $f(x)$ be an irreducible polynomial of degree n over a field F. Let $g(x)$ be any polynomial in $F[x]$. Prove that every irreducible factor of the composite polynomial $f(g(x))$ has degree divisible by n.

18. Let k be a field and let $k(x)$ be the field of rational functions in x with coefficients from k. Let $t \in k(x)$ be the rational function $\dfrac{P(x)}{Q(x)}$ with relatively prime polynomials $P(x), Q(x) \in k[x]$, with $Q(x) \neq 0$. Then $k(x)$ is an extension of $k(t)$ and to compute its degree it is necessary to compute the minimal polynomial with coefficients in $k(t)$ satisfied by x.

    (a) Show that the polynomial $P(X) - tQ(X)$ in the variable X and coefficients in $k(t)$ is irreducible over $k(t)$ and has x as a root. [By Gauss' Lemma this polynomial is irreducible in $(k(t))[X]$ if and only if it is irreducible in $(k[t])[X]$. Then note that $(k[t])[X] = (k[X])[t]$.]

(b) Show that the degree of $P(X) - tQ(X)$ as a polynomial in X with coefficients in $k(t)$ is the maximum of the degrees of $P(x)$ and $Q(x)$.

(c) Show that $[k(x) : k(t)] = [k(x) : k(\frac{P(x)}{Q(x)})] = \max \, (\deg P(x), \deg Q(x))$.

19. Let K be an extension of F of degree n.

    (a) For any $\alpha \in K$ prove that α acting by left multiplication on K is an F-linear transformation of K.

    (b) Prove that K is isomorphic to a subfield of the ring of $n \times n$ matrices over F, so the ring of $n \times n$ matrices over F contains an isomorphic copy of *every* extension of F of degree $\leq n$.

20. Show that if the matrix of the linear transformation "multiplication by α" considered in the previous exercise is A then α is a root of the characteristic polynomial for A. This gives an effective procedure for determining an equation of degree n satisfied by an element α in an extension of F of degree n. Use this procedure to obtain the monic polynomial of degree 3 satisfied by $\sqrt[3]{2}$ and by $1 + \sqrt[3]{2} + \sqrt[3]{4}$.

21. Let $K = \mathbb{Q}(\sqrt{D})$ for some squarefree integer D. Let $\alpha = a + b\sqrt{D}$ be an element of K. Use the basis $1, \sqrt{D}$ for K as a vector space over $\mathbb{Q}$ and show that the matrix of the linear transformation "multiplication by α" on K considered in the previous exercises has the matrix $\begin{pmatrix} a & bD \\ b & a \end{pmatrix}$. Prove directly that the map $a + b\sqrt{D} \mapsto \begin{pmatrix} a & bD \\ b & a \end{pmatrix}$ is an isomorphism of the field K with a subfield of the ring of 2×2 matrices with coefficients in $\mathbb{Q}$.

22. Let K_1 and K_2 be two finite extensions of a field F contained in the field K. Prove that the F-algebra $K_1 \otimes_F K_2$ is a field if and only if $[K_1 K_2 : F] = [K_1 : F][K_2 : F]$.

13.3 CLASSICAL STRAIGHTEDGE AND COMPASS CONSTRUCTIONS

As a simple application of the results we have obtained on algebraic extensions, and in particular on the multiplicativity of extension degrees, we can answer (in the negative) the following geometric problems posed by the Greeks:

  I. *(Doubling the Cube)* Is it possible using only straightedge and compass to construct a cube with precisely twice the volume of a given cube?

 II. *(Trisecting an Angle)* Is it possible using only straightedge and compass to trisect any given angle θ?

III. *(Squaring the Circle)* Is it possible using only straightedge and compass to construct a square whose area is precisely the area of a given circle?

To answer these questions we must translate the construction of lengths by compass and straightedge into algebraic terms. Let 1 denote a fixed given unit distance. Then any distance is determined by its length $a \in \mathbb{R}$, which allows us to view geometric distances as elements of the real numbers $\mathbb{R}$. Using the given unit distance 1 to define the scale on the axes, we can then construct the usual Cartesian plane $\mathbb{R}^2$ and view all of our constructions as occurring in $\mathbb{R}^2$. A point $(x, y) \in \mathbb{R}^2$ is then constructible starting with the given distance 1 if and only if its coordinates x and y are constructible elements of $\mathbb{R}$. The problems above then amount to determining whether particular lengths in $\mathbb{R}$ can be obtained by compass and straightedge constructions from a fixed unit distance. The collection of such real numbers together with their negatives will be called the *constructible* elements of $\mathbb{R}$, and we shall not distinguish between the lengths that are constructible and the real numbers that are constructible.

Each straightedge and compass construction consists of a series of operations of the following four types: (1) connecting two given points by a straight line, (2) finding a point of intersection of two straight lines, (3) drawing a circle with given radius and center, and (4) finding the point(s) of intersection of a straight line and a circle or the intersection of two circles.

It is an elementary fact from geometry that if two lengths a and b are given one may construct using straightedge and compass the lengths $a \pm b$, ab and a/b (the first two are clear and the latter two are given by the construction of parallel lines (Figure 1)).

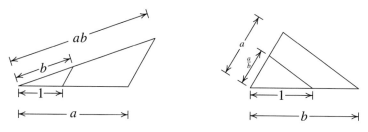

Fig. 1

It is also an elementary geometry construction to construct $\sqrt{a}$ if a is given: construct the circle with diameter $1 + a$ and erect the perpendicular to the diameter as indicated in Figure 2. Then $\sqrt{a}$ is the length of this perpendicular.

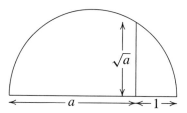

Fig. 2

It follows that straightedge and compass constructions give all the algebraic operations of addition, subtraction, multiplication and division (by nonzero elements) in the reals so the collection of constructible elements is a *subfield* of $\mathbb{R}$. One can also take square roots of constructible elements. We shall now see that these are essentially the only operations possible.

From the given length 1 it is possible to construct by these operations all the rational numbers $\mathbb{Q}$. Hence we may construct all of the points $(x, y) \in \mathbb{R}^2$ whose coordinates are rational. We may construct additional elements of $\mathbb{R}$ by taking square roots, so the collection of elements constructible from 1 of $\mathbb{R}$ form a field strictly larger than $\mathbb{Q}$.

The usual formula ("two point form") for the straight line connecting two points with coordinates in some field F gives an equation for the line of the form $ax+by-c = 0$ with $a, b, c \in F$. Solving two such equations simultaneously to determine the point of intersection of two such lines gives solutions also in F. It follows that if the coordinates of two points lie in the field F then straightedge constructions alone will not produce additional points whose coordinates are not also in F.

A compass construction (type (3) or (4) above) defines points obtained by the intersection of a circle with either a straight line or another circle. A circle with center (h, k) and radius r has equation

$$(x - h)^2 + (y - k)^2 = r^2$$

so when we consider the effect of compass constructions on elements of a field F we are considering simultaneous solutions of such an equation with a linear equation $ax + by - c = 0$ where $a, b, c, h, k, r \in F$, or the simultaneous solutions of two quadratic equations.

In the case of a linear equation and the equation for the circle, solving for y, say, in the linear equation and substituting gives a *quadratic* equation for x (and y is given linearly in terms of x). Hence the coordinates of the point of intersection are at worst in a *quadratic extension* of F.

In the case of the intersection of two circles, say

$$(x - h)^2 + (y - k)^2 = r^2$$
$$\text{and} \quad (x - h')^2 + (y - k')^2 = r'^2,$$

subtraction of the second equation from the first shows that we have the same intersection by considering the two equations

$$(x - h)^2 + (y - k)^2 = r^2$$
$$\text{and} \quad 2(h' - h)x + 2(k' - k)y = r^2 - h^2 - k^2 - r'^2 + h'^2 + k'^2$$

which is the intersection of a circle and a straight line (the straight line connecting the two points of intersection, in fact) of the type just considered.

It follows that if a collection of constructible elements is given, then one can construct all the elements in the subfield F of $\mathbb{R}$ generated by these elements and that any straightedge and compass operation on elements of F produces elements in at worst a *quadratic* extension of F. Since quadratic extensions have degree 2 and extension degrees are multiplicative, it follows that if $\alpha \in \mathbb{R}$ is obtained from elements in a field F by a (finite) series of straightedge and compass operations then α is an element of an extension K of F of degree a power of 2: $[K : F] = 2^m$ for some m. Since $[F(\alpha) : F]$ divides this extension degree, it must also be a power of 2.

Proposition 23. If the element $\alpha \in \mathbb{R}$ is obtained from a field $F \subset \mathbb{R}$ by a series of compass and straightedge constructions then $[F(\alpha) : F] = 2^k$ for some integer $k \geq 0$.

Theorem 24. None of the classical Greek problems: (I) Doubling the Cube, (II) Trisecting an Angle, and (III) Squaring the Circle, is possible.

Proof: (I) Doubling the cube amounts to constructing $\sqrt[3]{2}$ in the reals starting with the unit 1. Since $[\mathbb{Q}(\sqrt[3]{2}) : \mathbb{Q}] = 3$ is not a power of 2, this is impossible.

(II) If an angle θ can be constructed, then determining the point at distance 1 from the origin and angle θ from the positive x axis in $\mathbb{R}^2$ shows that $\cos \theta$ (the x-coordinate of this point) can be constructed (so then $\sin\theta$ can also be constructed). Conversely if $\cos\theta$, then $\sin\theta$, can be constructed, the point with those coordinates gives the angle θ.

The problem of trisecting the angle θ is then equivalent to the problem: given $\cos\theta$ construct $\cos\theta/3$.

To see that this is not always possible (it is certainly occasionally possible, for example for $\theta = 180°$), consider $\theta = 60°$. Then $\cos\theta = \frac{1}{2}$. By the triple angle formula for cosines:

$$\cos\theta = 4\cos^3\theta/3 - 3\cos\theta/3,$$

substituting $\theta = 60°$, we see that $\beta = \cos 20°$ satisfies the equation

$$4\beta^3 - 3\beta - 1/2 = 0$$

or $8(\beta)^3 - 6\beta - 1 = 0$. This can be written $(2\beta)^3 - 3(2\beta) - 1 = 0$. Let $\alpha = 2\beta$. Then α is a real number between 0 and 2 satisfying the equation

$$\alpha^3 - 3\alpha - 1 = 0.$$

But we considered this equation in the last section and determined $[\mathbb{Q}(\alpha) : \mathbb{Q}] = 3$, and as before we see that α is not constructible.

(III) Squaring the circle is equivalent to determining whether the real number $\pi = 3.14159\ldots$ is constructible. As mentioned previously, it is a difficult problem even to prove that this number is not rational. It is in fact transcendental (which we shall assume without proof), so that $[\mathbb{Q}(\pi) : \mathbb{Q}]$ is not even finite, much less a power of 2, showing the impossibility of squaring the circle by straightedge and compass.

Remark: The proof above shows that $\cos 20°$ and $\sin 20°$ cannot be constructed. The question arises as to which integer angles (measured in degrees) are constructible? The angles $1°$ and $2°$ are not constructible, since otherwise the addition formulae for sines and cosines would give the constructibility for $20°$. On the other hand, elementary geometric constructions (of the regular 5-gon for an angle of $72°$ and the equilateral triangle for an angle of $60°$) together with the addition formulae and the half-angle formulae show that $\cos 3°$ and $\sin 3°$ are constructible. It follows from this that the trigonometric functions of an integer degree angle are constructible precisely when the angle is a multiple of $3°$. Explicitly,

$$\cos 3° = \frac{1}{8}(\sqrt{3}+1)\sqrt{5+\sqrt{5}} + \frac{1}{16}(\sqrt{6}-\sqrt{2})(\sqrt{5}-1)$$

$$\sin 3° = \frac{1}{16}(\sqrt{6}+\sqrt{2})(\sqrt{5}-1) - \frac{1}{8}(\sqrt{3}-1)\sqrt{5+\sqrt{5}},$$

showing that these are obtained from $\mathbb{Q}$ by successive extractions of square roots and field operations.

After discussing the cyclotomic fields in Section 14.5 we shall consider another classical geometric question: "which regular n-gons can be constructed by straightedge and compass?" (cf. Proposition 14.29).

We have been careful here to consider constructions using a *straightedge* rather than a *ruler*, the distinction being that a ruler has marks on it. If one uses a ruler, it is possible to construct many additional algebraic elements. For example, suppose θ is a given angle and the unit distance 1 is marked on the ruler. Draw a circle of radius 1 with central angle θ as shown in Figure 3 and then slide the ruler until the distance between points A and B on the circle is 1. Then some elementary geometry shows that (cf. the exercises) the angle α indicated is $\theta/3$, i.e., this construction (due to Archimedes) trisects θ. In particular, the second classical problem in Theorem 24 (Trisecting an Angle) can be solved with *ruler* and compass.

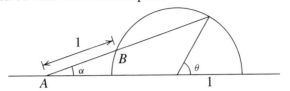

Fig. 3

The first of the classical problems in Theorem 24 (Duplication of the Cube), which amounts to the construction of $\sqrt[3]{2}$, can also be solved with ruler and compass. The following gives a construction for $k^{1/3}$ for any given positive real k which is less than 1. This construction was shown to us by J.H. Conway.

Drawing a circle of radius 1 and using the point $A = (k, 0)$ as center, construct the point $B = (0, \sqrt{1-k^2})$. Dividing this distance by 3, construct the point $(0, -\frac{1}{3}\sqrt{1-k^2})$ and draw the line connecting this point with A. Slide the ruler with marked unit length 1 so that it passes through the point B and so that the distance from the intersection point C to the intersection point D with the x-axis is of length 1, as indicated in Figure 4.

Then the distance between A and D is $2k^{1/3}$ and the distance between B and C is $2k^{2/3}$ (cf. the exercises).

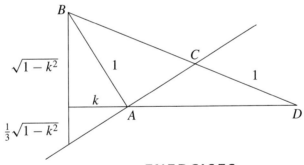

Fig. 4

EXERCISES

1. Prove that it is impossible to construct the regular 9-gon.

2. Prove that Archimedes' construction actually trisects the angle θ. [Note the isosceles triangles in Figure 5 to prove that $\beta = \gamma = 2\alpha$.]

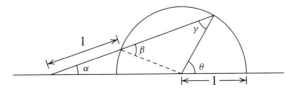

Fig. 5

3. Prove that Conway's construction indicated in the text actually constructs $2k^{1/3}$ and $2k^{2/3}$. [One method: let (x, y) be the coordinates of the point C, a the distance from B to C and b the distance from A to D; use similar triangles to prove (a) $\dfrac{y}{1} = \dfrac{\sqrt{1-k^2}}{1+a}$, (b) $\dfrac{x}{a} = \dfrac{b+k}{1+a}$, (c) $\dfrac{y}{x-k} = \dfrac{\sqrt{1-k^2}}{3k}$, and also show that (d) $(1-k^2)+(b+k)^2 = (1+a)^2$; solve these equations for a and b.]

4. The construction of the regular 7-gon amounts to the constructibility of $\cos(2\pi/7)$. We shall see later (Section 14.5 and Exercise 2 of Section 14.7) that $\alpha = 2\cos(2\pi/7)$ satisfies the equation $x^3 + x^2 - 2x - 1 = 0$. Use this to prove that the regular 7-gon is not constructible by straightedge and compass.

5. Use the fact that $\alpha = 2\cos(2\pi/5)$ satisfies the equation $x^2 + x - 1 = 0$ to conclude that the regular 5-gon is constructible by straightedge and compass.

13.4 SPLITTING FIELDS AND ALGEBRAIC CLOSURES

Let F be a field.

If $f(x)$ is any polynomial in $F[x]$ then we have seen in Section 2 that there exists a field K which can (by identifying F with an isomorphic copy of F) be considered an extension of F in which $f(x)$ has a root α. This is equivalent to the statement that $f(x)$ has a linear factor $x - \alpha$ in $K[x]$ (this is Proposition 9 of Chapter 9).

Definition. The extension field K of F is called a *splitting field* for the polynomial $f(x) \in F[x]$ if $f(x)$ factors completely into linear factors (or *splits completely*) in $K[x]$ and $f(x)$ does not factor completely into linear factors over any proper subfield of K containing F.

If $f(x)$ is of degree n, then $f(x)$ has at most n roots in F (Proposition 17 of Chapter 9) and has precisely n roots (counting multiplicities) in F if and only if $f(x)$ splits completely in $F[x]$.

Theorem 25. For any field F, if $f(x) \in F[x]$ then there exists an extension K of F which is a splitting field for $f(x)$.

Proof: We first show that there is an extension E of F over which $f(x)$ splits completely into linear factors by induction on the degree n of $f(x)$. If $n = 1$, then take $E = F$. Suppose now that $n > 1$. If the irreducible factors of $f(x)$ over F are all of degree 1, then F is the splitting field for $f(x)$ and we may take $E = F$. Otherwise, at least one of the irreducible factors, say $p(x)$ of $f(x)$ in $F[x]$ is of degree at least 2. By Theorem 3 there is an extension E_1 of F containing a root α of $p(x)$. Over E_1 the polynomial $f(x)$ has the linear factor $x - \alpha$. The degree of the remaining factor $f_1(x)$ of $f(x)$ is $n - 1$, so by induction there is an extension E of E_1 containing all the roots of $f_1(x)$. Since $\alpha \in E$, E is an extension of F containing all the roots of $f(x)$. Now let K be the intersection of all the subfields of E containing F which also contain all the roots of $f(x)$. Then K is a field which is a splitting field for $f(x)$.

We shall see shortly that any two splitting fields for $f(x)$ are isomorphic (which extends Theorem 8), so (by abuse) we frequently refer to *the* splitting field of a polynomial.

Definition. If K is an algebraic extension of F which is the splitting field over F for a collection of polynomials $f(x) \in F[x]$ then K is called a *normal* extension of F.

We shall generally use the term "splitting field" rather than "normal extension" (cf. also Section 14.9).

Examples

(1) The splitting field for $x^2 - 2$ over $\mathbb{Q}$ is just $\mathbb{Q}(\sqrt{2})$, since the two roots are $\pm\sqrt{2}$ and $-\sqrt{2} \in \mathbb{Q}(\sqrt{2})$.

(2) The splitting field for $(x^2 - 2)(x^2 - 3)$ is the field $\mathbb{Q}(\sqrt{2}, \sqrt{3})$ generated over $\mathbb{Q}$ by $\sqrt{2}$ and $\sqrt{3}$ since the roots of the polynomial are $\pm\sqrt{2}, \pm\sqrt{3}$. We have already seen that this is an extension of degree 4 over $\mathbb{Q}$ and we have the following diagram of known subfields:

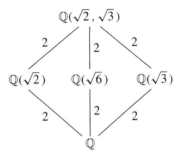

(3) The splitting field of $x^3 - 2$ over $\mathbb{Q}$ is not just $\mathbb{Q}(\sqrt[3]{2})$ since as previously noted the three roots of this polynomial in $\mathbb{C}$ are

$$\sqrt[3]{2}, \quad \sqrt[3]{2}\left(\frac{-1+i\sqrt{3}}{2}\right), \quad \sqrt[3]{2}\left(\frac{-1-i\sqrt{3}}{2}\right)$$

and the latter two roots are not elements of $\mathbb{Q}(\sqrt[3]{2})$, since the elements of this field are of the form $a + b\sqrt[3]{2} + c\sqrt[3]{4}$ with rational a, b, c and all such numbers are real.

The splitting field K of this polynomial is obtained by adjoining all three of these roots to $\mathbb{Q}$. Note that since K contains the first two roots above, then it contains their quotient $\dfrac{-1+\sqrt{-3}}{2}$ hence K contains the element $\sqrt{-3}$. On the other hand, any field containing $\sqrt[3]{2}$ and $\sqrt{-3}$ contains all three of the roots above. It follows that

$$K = \mathbb{Q}(\sqrt[3]{2}, \sqrt{-3})$$

is the splitting field of $x^3 - 2$ over $\mathbb{Q}$. Since $\sqrt{-3}$ satisfies the equation $x^2 + 3 = 0$, the degree of this extension over $\mathbb{Q}(\sqrt[3]{2})$ is at most 2, hence must be 2 since we observed above that $\mathbb{Q}(\sqrt[3]{2})$ is not the splitting field. It follows that

$$[\mathbb{Q}(\sqrt[3]{2}, \sqrt{-3}) : \mathbb{Q}] = 6.$$

Note that we could have proceeded slightly differently at the end by noting that $\mathbb{Q}(\sqrt{-3})$ is a subfield of K, so that the degree $[\mathbb{Q}(\sqrt{-3}) : \mathbb{Q}] = 2$ divides $[K : \mathbb{Q}]$.

Since this extension degree is also divisible by 3 (because $\mathbb{Q}(\sqrt[3]{2}) \subset K$), the degree is divisible by 6, hence must be 6.

This gives us the diagram of known subfields:

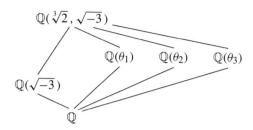

where

$$\theta_1 = \sqrt[3]{2}, \quad \theta_2 = \sqrt[3]{2}\left(\frac{-1 + i\sqrt{3}}{2}\right), \quad \theta_3 = \sqrt[3]{2}\left(\frac{-1 - i\sqrt{3}}{2}\right).$$

(4) One must be careful in computing splitting fields. The splitting field for the polynomial $x^4 + 4$ over $\mathbb{Q}$ is smaller than one might at first suspect. In fact this polynomial factors over $\mathbb{Q}$:

$$x^4 + 4 = x^4 + 4x^2 + 4 - 4x^2 = (x^2 + 2)^2 - 4x^2$$
$$= (x^2 + 2x + 2)(x^2 - 2x + 2)$$

where these two factors are irreducible (Eisenstein again). Solving for the roots of the two factors by the quadratic formula, we find the four roots

$$\pm 1 \pm i$$

so that the splitting field of this polynomial is just the field $\mathbb{Q}(i)$, an extension of degree 2 of $\mathbb{Q}$.

In general, if $f(x) \in F[x]$ is a polynomial of degree n, then adjoining one root of $f(x)$ to F generates an extension F_1 of degree at most n (and equal to n if and only if $f(x)$ is irreducible). Over F_1 the polynomial $f(x)$ now has at least one linear factor, so that any other root of $f(x)$ satisfies an equation of degree at most $n - 1$ over F_1. Adjoining such a root to F_1 we therefore obtain an extension of degree at most $n - 1$ of F_1, etc. Using the multiplicativity of extension degrees, this proves

Proposition 26. A splitting field of a polynomial of degree n over F is of degree at most $n!$ over F.

As the examples above show, the degree of a splitting field may be smaller than $n!$. It will be proved later using Galois Theory that a "general" polynomial of degree n (in a well defined sense) over $\mathbb{Q}$ has a splitting field of degree $n!$, so this may be viewed as the "generic" situation (although most of the interesting examples we shall consider have splitting fields of smaller degree).

Example: (Splitting Field of $x^n - 1$: Cyclotomic Fields)

Consider the splitting field of the polynomial $x^n - 1$ over $\mathbb{Q}$. The roots of this polynomial are called the nth *roots of unity*.

Recall that every nonzero complex number $a + bi \in \mathbb{C}$ can be written uniquely in the form

$$re^{i\theta} = r(\cos\theta + i\sin\theta) \qquad r > 0, \quad 0 \le \theta < 2\pi$$

which is simply representing the point $a + bi$ in the complex plane in terms of polar coordinates: r is the distance of (a, b) from the origin and θ is the angle made with the real positive axis.

Over $\mathbb{C}$ there are n distinct solutions of the equation $x^n = 1$, namely the elements

$$e^{2\pi k i/n} = \cos\left(\frac{2\pi k}{n}\right) + i\sin\left(\frac{2\pi k}{n}\right)$$

for $k = 0, 1, \ldots, n - 1$. These points are given geometrically by n equally spaced points starting with the point $(1,0)$ (corresponding to $k = 0$) on a circle of radius 1 in the complex plane (see Figure 6). The fact that these are all n^{th} roots of unity is immediate, since

$$(e^{2\pi k i/n})^n = e^{(2\pi k i/n)n} = e^{2\pi k i} = 1.$$

It follows that $\mathbb{C}$ contains a splitting field for $x^n - 1$ and we shall frequently view the splitting field for $x^n - 1$ over $\mathbb{Q}$ as the field generated over $\mathbb{Q}$ in $\mathbb{C}$ by the numbers above.

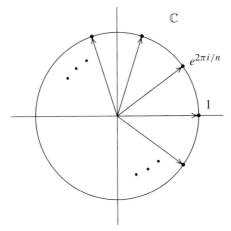

Fig. 6

In *any* abstract splitting field $K/\mathbb{Q}$ for $x^n - 1$ the collection of nth roots of unity form a *group* under multiplication since if $\alpha^n = 1$ and $\beta^n = 1$ then $(\alpha\beta)^n = 1$, so this subset of $K^\times$ is closed under multiplication. It follows that this is a *cyclic* group (Proposition 18 of Chapter 9); we shall see that there are n distinct roots in K so it has order n.

Definition. A generator of the cyclic group of all n^{th} roots of unity is called a *primitive n^{th} root of unity.*

Let ζ_n denote a primitive n^{th} root of unity. The other *primitive* n^{th} roots of unity are then the elements ζ_n^a where $1 \le a < n$ is an integer relatively prime to n, since these are the other generators for a cyclic group of order n. In particular there are precisely $\varphi(n)$ primitive n^{th} roots of unity, where $\varphi(n)$ denotes the Euler φ-function.

Over $\mathbb{C}$ we can see all of this directly by letting

$$\zeta_n = e^{2\pi i/n}$$

(the first n^{th} root of unity counterclockwise from 1). Then all the other roots of unity are powers of ζ_n:

$$e^{2\pi k i/n} = \zeta_n^k$$

so that ζ_n is one possible generator for the multiplicative group of nth roots of unity. When we view the roots of unity in $\mathbb{C}$ we shall usually use ζ_n to denote this choice of a primitive n^{th} root of unity. The primitive roots of unity in $\mathbb{C}$ for some small values of n are

$$\zeta_1 = 1$$
$$\zeta_2 = -1$$
$$\zeta_3 = \frac{-1 + i\sqrt{3}}{2}$$
$$\zeta_4 = i$$
$$\zeta_5 = \frac{\sqrt{5} - 1}{4} + i\left(\frac{\sqrt{10 + 2\sqrt{5}}}{4}\right)$$
$$\zeta_6 = \frac{1 + i\sqrt{3}}{2}$$
$$\zeta_8 = \frac{\sqrt{2}}{2} + i\frac{\sqrt{2}}{2}$$

(these formulas follow from the elementary geometry of n-gons and in any case can be verified directly by raising them to the appropriate power).

The splitting field of $x^n - 1$ over $\mathbb{Q}$ is the field $\mathbb{Q}(\zeta_n)$ and this field is given a name:

Definition. The field $\mathbb{Q}(\zeta_n)$ is called the *cyclotomic field of nth roots of unity*.

Determining the degree of this extension requires some analysis of the minimal polynomial of ζ_n over $\mathbb{Q}$ and will be postponed until later (Section 6). One important special case which we have in fact already considered is when $n = p$ is a *prime*. In this case, we have the factorization

$$x^p - 1 = (x - 1)(x^{p-1} + x^{p-2} + \cdots + x + 1)$$

and since $\zeta_p \neq 1$ it follows that ζ_p is a root of the polynomial

$$\Phi_p(x) = \frac{x^p - 1}{x - 1} = x^{p-1} + x^{p-2} + \cdots + x + 1$$

which we showed was irreducible in Section 9.4. It follows that $\Phi_p(x)$ is the minimal polynomial of ζ_p over $\mathbb{Q}$, so that

$$[\mathbb{Q}(\zeta_p) : \mathbb{Q}] = p - 1.$$

We shall see later that in general $[\mathbb{Q}(\zeta_n) : \mathbb{Q}] = \varphi(n)$, where $\varphi(n)$ is the Euler phi-function of n (so that $\varphi(p) = p - 1$).

Example: (Splitting Field of $x^p - 2$, p a prime)

Let p be a prime and consider the splitting field of $x^p - 2$. If α is a root of this equation, i.e., $\alpha^p = 2$, then $(\zeta\alpha)^p = 2$ where ζ is any p^{th} root of unity. Hence the solutions of this equation are

$$\zeta \sqrt[p]{2}, \qquad \zeta \text{ a } p^{\text{th}} \text{ root of unity}$$

where as usual the symbol $\sqrt[p]{2}$ denotes the positive real p^{th} root of 2 if we wish to view these elements as complex numbers, and denotes any one solution of $x^p = 2$ if we view these roots abstractly. Since the ratio of the two solutions $\zeta_p \sqrt[p]{2}$ and $\sqrt[p]{2}$ for ζ_p a primitive p^{th} root of unity is just ζ_p, the splitting field of $x^p - 2$ over $\mathbb{Q}$ contains $\mathbb{Q}(\sqrt[p]{2}, \zeta_p)$. On the other hand, all the roots above lie in this field, so that the splitting field is precisely

$$\mathbb{Q}(\sqrt[p]{2}, \zeta_p).$$

This field contains the cyclotomic field of pth roots of unity and is generated over it by $\sqrt[p]{2}$, hence is an extension of degree at most p. It follows that the degree of this extension over $\mathbb{Q}$ is $\leq p(p-1)$. Since both $\mathbb{Q}(\sqrt[p]{2})$ and $\mathbb{Q}(\zeta_p)$ are subfields, the degree of the extension over $\mathbb{Q}$ is divisible by p and by $p-1$. Since these two numbers are relatively prime it follows that the extension degree is divisible by $p(p-1)$ so that we must have

$$[\mathbb{Q}(\sqrt[p]{2}, \zeta_p) : \mathbb{Q}] = p(p-1)$$

(this is Corollary 22). Note in particular that we have proved $x^p - 2$ remains irreducible over $\mathbb{Q}(\zeta_p)$, which is not at all obvious. We have the following diagram of known subfields:

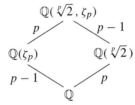

The special case $p = 3$ was Example 3 above, where we simply indicated the 3rd roots of unity explicitly.

We now return to the problem of proving it makes no difference how the splitting field of a polynomial $f(x)$ over a field F is constructed. As in Theorem 8 it is convenient to state the result for an arbitrary isomorphism $\varphi : F \xrightarrow{\sim} F'$ between two fields.

Theorem 27. Let $\varphi : F \xrightarrow{\sim} F'$ be an isomorphism of fields. Let $f(x) \in F[x]$ be a polynomial and let $f'(x) \in F'[x]$ be the polynomial obtained by applying φ to the coefficients of $f(x)$. Let E be a splitting field for $f(x)$ over F and let E' be a splitting field for $f'(x)$ over F'. Then the isomorphism φ extends to an isomorphism $\sigma : E \xrightarrow{\sim} E'$, i.e., σ restricted to F is the isomorphism φ:

$$\begin{array}{ccc} \sigma : & E & \xrightarrow{\sim} & E' \\ & | & & | \\ \varphi : & F & \xrightarrow{\sim} & F' \end{array}$$

Proof: We shall proceed by induction on the degree n of $f(x)$. As in the discussion before Theorem 8, recall that an isomorphism φ from one field F to another field

F' induces a natural isomorphism between the polynomial rings $F[x]$ and $F'[x]$. In particular, if $f(x)$ and $f'(x)$ correspond to one another under this isomorphism then the irreducible factors of $f(x)$ in $F[x]$ correspond to the irreducible factors of $f'(x)$ in $F'[x]$.

If $f(x)$ has all its roots in F then $f(x)$ splits completely in $F[x]$ and $f'(x)$ splits completely in $F'[x]$ (with its linear factors being the images of the linear factors for $f(x)$). Hence $E = F$ and $E' = F'$, and in this case we may take $\sigma = \varphi$. This shows the result is true for $n = 1$ and in the case where all the irreducible factors of $f(x)$ have degree 1.

Assume now by induction that the theorem has been proved for any field F, isomorphism φ, and polynomial $f(x) \in F[x]$ of degree $< n$. Let $p(x)$ be an irreducible factor of $f(x)$ in $F[x]$ of degree at least 2 and let $p'(x)$ be the corresponding irreducible factor of $f'(x)$ in $F'[x]$. If $\alpha \in E$ is a root of $p(x)$ and $\beta \in E'$ is a root of $p'(x)$, then by Theorem 8 we can extend φ to an isomorphism $\sigma' : F(\alpha) \overset{\sim}{\to} F'(\beta)$:

$$\sigma' : \quad F(\alpha) \quad \overset{\sim}{\longrightarrow} \quad F'(\beta)$$
$$\mid \qquad\qquad\qquad \mid$$
$$\varphi : \quad\quad F \quad \overset{\sim}{\longrightarrow} \quad F'.$$

Let $F_1 = F(\alpha)$, $F_1' = F'(\beta)$, so that we have the isomorphism $\sigma' : F_1 \overset{\sim}{\to} F_1'$. We have $f(x) = (x - \alpha)f_1(x)$ over F_1 where $f_1(x)$ has degree $n - 1$ and $f'(x) = (x - \beta)f_1'(x)$. The field E is a splitting field for $f_1(x)$ over F_1: all the roots of $f_1(x)$ are in E and if they were contained in any smaller extension L containing F_1, then, since F_1 contains α, L would also contain all the roots of $f(x)$, which would contradict the minimality of E as the splitting field of $f(x)$ over F. Similarly E' is a splitting field for $f_1'(x)$ over F_1'. Since the degrees of $f_1(x)$ and $f_1'(x)$ are less than n, by induction there exists a map $\sigma : E \overset{\sim}{\to} E'$ extending the isomorphism $\sigma' : F_1 \overset{\sim}{\to} F_1'$. This gives the extended diagram:

$$\sigma : \quad E \quad \overset{\sim}{\longrightarrow} \quad E'$$
$$\mid \qquad\qquad \mid$$
$$\sigma' : \quad F_1 \quad \overset{\sim}{\longrightarrow} \quad F_1'$$
$$\mid \qquad\qquad \mid$$
$$\varphi : \quad F \quad \overset{\sim}{\longrightarrow} \quad F'.$$

Then as the diagram indicates, σ restricted to F_1 is the isomorphism σ', so in particular σ restricted to F is σ' restricted to F, which is φ, showing that σ is an extension of φ, completing the proof.

Corollary 28. *(Uniqueness of Splitting Fields)* Any two splitting fields for a polynomial $f(x) \in F[x]$ over a field F are isomorphic.

Proof: Take φ to be the identity mapping from F to itself and E and E' to be two splitting fields for $f(x)(= f'(x))$.

As we mentioned before, this result justifies the terminology of *the* splitting field for $f(x)$ over F, since any two are isomorphic. Splitting fields play a natural role in the study of algebraic elements (if you are adjoining one root of a polynomial, why not adjoin *all* the roots?) and so take a particularly important role in Galois Theory.

We end this section with a discussion of field extensions of F which contain all the roots of *all* polynomials over F.

Definition. The field $\overline{F}$ is called an *algebraic closure* of F if $\overline{F}$ is algebraic over F and if every polynomial $f(x) \in F[x]$ splits completely over $\overline{F}$ (so that $\overline{F}$ can be said to contain all the elements algebraic over F).

Definition. A field K is said to be *algebraically closed* if every polynomial with coefficients in K has a root in K.

It is not obvious that algebraically closed fields exist nor that there exists an algebraic closure of a given field F (we shall prove this shortly).

Note that if K is algebraically closed, then in fact every $f(x) \in K[x]$ has *all* its roots in K, since by definition $f(x)$ has a root $\alpha \in K$, hence has a factor $x - \alpha$ in $K[x]$. The remaining factor of $f(x)$ then is a polynomial in $K[x]$, hence has a root, so has a linear factor etc., so that $f(x)$ must split completely. Hence if K is algebraically closed, then K itself is an algebraic closure of K and the converse is obvious, so that $K = \overline{K}$ if and only if K is algebraically closed.

The next result shows that the process of "taking the algebraic closure" actually stops after one step — taking the algebraic closure of an algebraic closure does not give a larger field: the field is already algebraically closed (notationally: $\overline{\overline{F}} = \overline{F}$).

Proposition 29. Let $\overline{F}$ be an algebraic closure of F. Then $\overline{F}$ is algebraically closed.

Proof: Let $f(x)$ be a polynomial in $\overline{F}[x]$ and let α be a root of $f(x)$. Then α generates an algebraic extension $\overline{F}(\alpha)$ of $\overline{F}$, and $\overline{F}$ is algebraic over F. By Theorem 20, $\overline{F}(\alpha)$ is algebraic over F so in particular its element α is algebraic over F. But then $\alpha \in \overline{F}$, showing $\overline{F}$ is algebraically closed.

Given a field F we have already shown how to construct (finite) extensions of F containing all the roots of any given polynomial $f(x) \in F[x]$. Intuitively, an algebraic closure of F is given by the field "generated" by all of these fields. The difficulty with this is "generated" *where*?, since they are not all subfields of a given field. For a *finite* collection of polynomials $f_1(x), \ldots, f_k(x)$, we can identify their splitting fields as subfields of the splitting field of the product polynomial $f_1(x) \cdots f_k(x)$, but the same idea used for an *infinite* number of polynomials requires numerous "bookkeeping" identifications and an application of Zorn's Lemma.

We shall instead construct an algebraic closure of F by first constructing an algebraically closed field containing F. The proof uses a clever idea of Artin which very neatly solves the "bookkeeping" problem of constructing a field containing the appropriate roots of polynomials (which also ultimately relies on Zorn's Lemma) by introducing a separate variable for every polynomial.

Proposition 30. For any field F there exists an algebraically closed field K containing F.

Proof: For every nonconstant monic polynomial $f = f(x)$ with coefficients in F, let x_f denote an indeterminate and consider the polynomial ring $F[\ldots, x_f, \ldots]$ generated over F by the variables x_f. In this polynomial ring consider the ideal I generated by the polynomials $f(x_f)$. If this ideal is not proper, then 1 is an element of the ideal, hence we have a relation

$$g_1 f_1(x_{f_1}) + g_2 f_2(x_{f_2}) + \cdots + g_n f_n(x_{f_n}) = 1$$

where the g_i, $i = 1, 2, \ldots, n$, are polynomials in the x_f. For $i = 1, 2, \ldots, n$ let $x_{f_i} = x_i$ and let $x_{n+1}, \ldots, x_m$ be the remaining variables occurring in the polynomials g_j, $j = 1, 2, \ldots, n$. Then the relation above reads

$$g_1(x_1, x_2, \ldots, x_m) f_1(x_1) + \cdots + g_n(x_1, x_2, \ldots, x_m) f_n(x_n) = 1.$$

Let F' be a finite extension of F containing a root α_i of $f_i(x)$ for $i = 1, 2, \ldots, n$. Letting $x_i = \alpha_i$, $i = 1, 2, \ldots, n$ and setting $x_{n+1} = \cdots = x_m = 0$, say, in the polynomial equation above would imply that $0 = 1$ in F', clearly impossible.

Since the ideal I is a proper ideal, it is contained in a maximal ideal $\mathcal{M}$ (this is where Zorn's Lemma is used). Then the quotient

$$K_1 = F[\ldots, x_f, \ldots]/\mathcal{M}$$

is a field containing (an isomorphic copy of) F. Each of the polynomials f has a root in K_1 by construction, namely the image of x_f, since $f(x_f) \in I \subseteq \mathcal{M}$. We have constructed a field K_1 in which every polynomial with coefficients from F has a root. Performing the same construction with K_1 instead of F gives a field K_2 containing K_1 in which all polynomials with coefficients from K_1 have a root. Continuing in this fashion we obtain a sequence of fields

$$F = K_0 \subseteq K_1 \subseteq K_2 \subseteq \cdots \subseteq K_j \subseteq K_{j+1} \subseteq \cdots$$

where every polynomial with coefficients in K_j has a root in K_{j+1}, $j = 0, 1, \ldots$. Let

$$K = \bigcup_{j \geq 0} K_j$$

be the union of these fields. Then K is clearly a field containing F. Since K is the union of the fields K_j, the coefficients of any polynomial $h(x)$ in $K[x]$ all lie in some field K_N for N sufficiently large. But then $h(x)$ has a root in K_{N+1}, so has a root in K. It follows that K is algebraically closed, completing the proof.

We now use the algebraically closed field containing F to construct an algebraic closure of F:

Proposition 31. Let K be an algebraically closed field and let F be a subfield of K. Then the collection of elements $\overline{F}$ of K that are algebraic over F is an algebraic closure of F. An algebraic closure of F is unique up to isomorphism.

Proof: By definition, $\overline{F}$ is an algebraic extension of F. Every polynomial $f(x) \in F[x]$ splits completely over K into linear factors $x - \alpha$ (the same is true for every polynomial even in $K[x]$). But each α is a root of $f(x)$, so is algebraic over F, hence is an element of $\overline{F}$. It follows that all the linear factors $x - \alpha$ have coefficients in $\overline{F}$, i.e., $f(x)$ splits completely in $\overline{F}[x]$ and $\overline{F}$ is an algebraic closure of F.

The uniqueness (up to isomorphism) of the algebraic closure is natural in light of the uniqueness (up to isomorphism) of splitting fields, and is proved along the same lines together with an application of Zorn's Lemma and will be omitted.

We shall prove later using Galois theory the following result (purely analytic proofs using complex analysis also exist).

Theorem. *(Fundamental Theorem of Algebra)* The field $\mathbb{C}$ is algebraically closed.

By Proposition 31, we immediately obtain:

Corollary 32. The field $\mathbb{C}$ contains an algebraic closure for any of its subfields. In particular, $\overline{\mathbb{Q}}$, the collection of complex numbers algebraic over $\mathbb{Q}$, is an algebraic closure of $\mathbb{Q}$.

The point of these considerations is that all the computations involving elements algebraic over a field F may be viewed as taking place in one (large) field, namely $\overline{F}$. Similarly, we can speak sensibly of the composite of any collection of algebraic extensions by viewing them all as subfields of an algebraic closure. In the case of $\mathbb{Q}$ or finite extensions of $\mathbb{Q}$ we may consider all of our computations as occurring in $\mathbb{C}$.

EXERCISES

1. Determine the splitting field and its degree over $\mathbb{Q}$ for $x^4 - 2$.
2. Determine the splitting field and its degree over $\mathbb{Q}$ for $x^4 + 2$.
3. Determine the splitting field and its degree over $\mathbb{Q}$ for $x^4 + x^2 + 1$.
4. Determine the splitting field and its degree over $\mathbb{Q}$ for $x^6 - 4$.
5. Let K be a finite extension of F. Prove that K is a splitting field over F if and only if every irreducible polynomial in $F[x]$ that has a root in K splits completely in $K[x]$. [Use Theorems 8 and 27.]
6. Let K_1 and K_2 be finite extensions of F contained in the field K, and assume both are splitting fields over F.
   (a) Prove that their composite $K_1 K_2$ is a splitting field over F.
   (b) Prove that $K_1 \cap K_2$ is a splitting field over F. [Use the preceding exercise.]

13.5 SEPARABLE AND INSEPARABLE EXTENSIONS

Let F be a field and let $f(x) \in F[x]$ be a polynomial with leading coefficient a_n. Over a splitting field for $f(x)$ we have the factorization

$$f(x) = a_n(x - \alpha_1)^{n_1}(x - \alpha_2)^{n_2} \cdots (x - \alpha_k)^{n_k}$$

where $\alpha_1, \alpha_2, \ldots, \alpha_k$ are distinct elements of the splitting field and $n_i \geq 1$ for all i. Recall that α_i is called a *multiple* root if $n_i > 1$ and is called a *simple* root if $n_i = 1$. The integer n_i is called the *multiplicity* of the root α_i.

Definition. A polynomial over F is called *separable* if it has no multiple roots (i.e., all its roots are distinct). A polynomial which is not separable is called *inseparable*.

Note that if a polynomial $f(x)$ has distinct roots in one splitting field then $f(x)$ has distinct roots in any splitting field (since this is equivalent to $f(x)$ factoring into distinct linear factors, and there is an isomorphism over F between any two splitting fields of $f(x)$ that is bijective on its roots), so that we need not specify the field containing all the roots of $f(x)$.

Examples

(1) The polynomial $x^2 - 2$ is separable over $\mathbb{Q}$ since its two roots $\pm\sqrt{2}$ are distinct. The polynomial $(x^2 - 2)^n$ for any $n \geq 2$ is inseparable since it has the multiple roots $\pm\sqrt{2}$, each with multiplicity n.

(2) The polynomial $x^2 - t$ $(= x^2 + t)$ over the field $F = \mathbb{F}_2(t)$ of rational functions in t with coefficients from $\mathbb{F}_2$ is irreducible as we've seen before, but is not separable. If $\sqrt{t}$ denotes a root in some extension field (note that $\sqrt{t} \notin F$), then

$$(x - \sqrt{t})^2 = x^2 - 2x\sqrt{t} + t = x^2 + t = x^2 - t$$

since F is a field of characteristic 2. Hence this irreducible polynomial has only one root (with multiplicity 2), so is not separable over F.

There is a simple criterion to check whether a polynomial has multiple roots.

Definition. The *derivative* of the polynomial

$$f(x) = a_n x^n + a_{n-1} x^{n-1} + \cdots + a_1 x + a_0 \in F[x]$$

is defined to be the polynomial

$$D_x f(x) = n a_n x^{n-1} + (n-1)a_{n-1} x^{n-2} + \cdots + 2a_2 x + a_1 \in F[x].$$

This formula is nothing but the usual formula for the derivative of a polynomial familiar from calculus. It is purely algebraic and so can be applied to a polynomial over an arbitrary field F, where the analytic notion of derivative (involving limits — a *continuous* operation) may not exist.

The usual (calculus) formulas for derivatives hold for derivatives in this situation as well, for example the formulas for the derivative of a sum and of a product:

$$D_x(f(x) + g(x)) = D_x f(x) + D_x g(x)$$
$$D_x(f(x)g(x)) = f(x)D_x g(x) + (D_x f(x))g(x).$$

These formulas can be proved directly from the definition for polynomials and do not require any limiting operations and are left as an exercise.

The next proposition shows that the separability of $f(x)$ can be determined by the Euclidean Algorithm in the field where the coefficients of $f(x)$ lie, without passing to a splitting field and factoring $f(x)$.

546

Proposition 33. A polynomial $f(x)$ has a multiple root α if and only if α is also a root of $D_x f(x)$, i.e., $f(x)$ and $D_x f(x)$ are both divisible by the minimal polynomial for α. In particular, $f(x)$ is separable if and only if it is relatively prime to its derivative: $(f(x), D_x f(x)) = 1$.

Proof: Suppose first that α is a multiple root of $f(x)$. Then over a splitting field,

$$f(x) = (x - \alpha)^n g(x)$$

for some integer $n \geq 2$ and some polynomial $g(x)$. Taking derivatives we obtain

$$D_x f(x) = n(x - \alpha)^{n-1} g(x) + (x - \alpha)^n D_x g(x)$$

which shows ($n \geq 2$) that $D_x f(x)$ has α as a root.

Conversely, suppose that α is a root of both $f(x)$ and $D_x f(x)$. Then write

$$f(x) = (x - \alpha) h(x)$$

for some polynomial $h(x)$ and take the derivative:

$$D_x f(x) = h(x) + (x - \alpha) D_x h(x).$$

Since $D_x f(\alpha) = 0$ by assumption, substituting α into the last equation shows that $h(\alpha) = 0$. Hence $h(x) = (x - \alpha) h_1(x)$ for some polynomial $h_1(x)$, and

$$f(x) = (x - \alpha)^2 h_1(x)$$

showing that α is a multiple root of $f(x)$.

The equivalence with divisibility by the minimal polynomial for α follows from Proposition 9. The last statement is then clear (let α denote any root of a common factor of $f(x)$ and $D_x f(x)$).

Examples

(1) The polynomial $x^{p^n} - x$ over $\mathbb{F}_p$ has derivative $p^n x^{p^n-1} - 1 = -1$ since the field has characteristic p. Since in this case the derivative has no roots at all, it follows that the polynomial has no multiple roots, hence is separable.

(2) The polynomial $x^n - 1$ has derivative nx^{n-1}. Over any field of characteristic not dividing n (including characteristic 0) this polynomial has only the root 0 (of multiplicity $n - 1$), which is not a root of $x^n - 1$. Hence $x^n - 1$ is separable and there are n distinct n^{th} roots of unity. We saw this directly over $\mathbb{Q}$ by exhibiting n distinct solutions over $\mathbb{C}$.

(3) If F is of characteristic p and p divides n, then there are fewer than n distinct n^{th} roots of unity over F: in this case the derivative is identically 0 since $n = 0$ in F. In fact *every* root of $x^n - 1$ is multiple in this case.

Corollary 34. Every irreducible polynomial over a field of characteristic 0 (for example, $\mathbb{Q}$) is separable. A polynomial over such a field is separable if and only if it is the product of distinct irreducible polynomials.

Proof: Suppose F is a field of characteristic 0 and $p(x) \in F[x]$ is irreducible of degree n. Then the derivative $D_x p(x)$ is a polynomial of degree $n - 1$. Up to constant factors the only factors of $p(x)$ in $F[x]$ are 1 and $p(x)$, so $D_x p(x)$ must be relatively prime to $p(x)$. This shows that any irreducible polynomial over a field of characteristic 0 is separable. The second statement of the corollary is then clear since distinct irreducibles never have zeros in common (by Proposition 9).

The point in the proof of the corollary that can fail in characteristic p is the statement that the derivative $D_x p(x)$ is of degree $n - 1$. In characteristic p the derivative of any power x^{pm} of x^p is identically 0:

$$D_x(x^{pm}) = pmx^{pm-1} = 0$$

so it is possible for the degree of the derivative to decrease by more than one. If the derivative $D_x p(x)$ of the *irreducible* polynomial $p(x)$ is nonzero, however, then just as before we conclude that $p(x)$ must be separable.

It is clear from the definition of the derivative that if $p(x)$ is a polynomial whose derivative is 0, then every exponent of x in $p(x)$ must be a multiple of p where p is the characteristic of F:

$$p(x) = a_m x^{mp} + a_{m-1} x^{(m-1)p} + \cdots + a_1 x^p + a_0.$$

Letting

$$p_1(x) = a_m x^m + a_{m-1} x^{m-1} + \cdots + a_1 x + a_0$$

we see that $p(x)$ is a polynomial in x^p, namely $p(x) = p_1(x^p)$.

We now prove a simple but important result about raising to the pth power in a field of characteristic p.

Proposition 35. Let F be a field of characteristic p. Then for any $a, b \in F$,

$$(a + b)^p = a^p + b^p, \quad \text{and} \quad (ab)^p = a^p b^p.$$

Put another way, the pth-power map defined by $\varphi(a) = a^p$ is an injective field homomorphism from F to F.

Proof: The Binomial Theorem for expanding $(a + b)^n$ for any positive integer n holds (by the standard induction proof) over any commutative ring:

$$(a + b)^n = a^n + \binom{n}{1} a^{n-1} b + \cdots + \binom{n}{i} a^{n-i} b^i + \cdots + b^n .$$

It should be observed that the binomial coefficients

$$\binom{n}{i} = \frac{n!}{i!(n - i)!}$$

are integers (recall that $m\alpha$ for $m \in \mathbb{Z}$ is defined for α an element of any ring) and here are elements of the prime field.

If p is a prime, then the binomial coefficients $\binom{p}{i}$ for $i = 1, 2, \ldots, p - 1$ are all divisible by p since for these values of i the numbers $i!$ and $(p - i)!$ only involve factors smaller than p, hence are relatively prime to p and so cannot cancel the factor of p in the numerator of the expression $\dfrac{p!}{i!(p - i)!}$. It follows that over a field of characteristic p all the intermediate terms in the expansion of $(a + b)^p$ are 0, which gives the first equation of the proposition. The second equation is trivial, as is the fact that φ is injective.

548

Definition. The map in Proposition 35 is called the *Frobenius endomorphism* of F.

Corollary 36. Suppose that $\mathbb{F}$ is a finite field of characteristic p. Then every element of $\mathbb{F}$ is a pth power in $\mathbb{F}$ (notationally, $\mathbb{F} = \mathbb{F}^p$).

Proof: The injectivity of the Frobenius endomorphism of $\mathbb{F}$ implies that it is also surjective when $\mathbb{F}$ is finite, which is the statement of the corollary.

We now prove the analogue of Corollary 34 for finite fields.

Let $\mathbb{F}$ be a finite field and suppose that $p(x) \in \mathbb{F}[x]$ is an irreducible polynomial with coefficients in $\mathbb{F}$. If $p(x)$ were inseparable then we have seen that $p(x) = q(x^p)$ for some polynomial $q(x) \in \mathbb{F}[x]$. Let

$$q(x) = a_m x^m + a_{m-1} x^{m-1} + \cdots + a_1 x + a_0.$$

By Corollary 36, each a_i, $i = 1, 2, \ldots, m$ is a pth power in $\mathbb{F}$, say $a_i = b_i^p$. Then by Proposition 35 we have

$$
\begin{aligned}
p(x) = q(x^p) &= a_m (x^p)^m + a_{m-1}(x^p)^{m-1} + \cdots + a_1 x^p + a_0 \\
&= b_m^p (x^p)^m + b_{m-1}^p (x^p)^{m-1} + \cdots + b_1^p x^p + b_0^p \\
&= (b_m x^m)^p + (b_{m-1} x^{m-1})^p + \cdots + (b_1 x)^p + (b_0)^p \\
&= (b_m x^m + b_{m-1} x^{m-1} + \cdots + b_1 x + b_0)^p
\end{aligned}
$$

which shows that $p(x)$ is the pth power of a polynomial in $\mathbb{F}[x]$, a contradiction to the irreducibility of $p(x)$. This proves:

Proposition 37. Every irreducible polynomial over a finite field $\mathbb{F}$ is separable. A polynomial in $\mathbb{F}[x]$ is separable if and only if it is the product of distinct irreducible polynomials in $\mathbb{F}[x]$.

The important part of the proof of this result is the fact that every element in the characteristic p field $\mathbb{F}$ was a pth power in $\mathbb{F}$. This suggests the following definition:

Definition. A field K of characteristic p is called *perfect* if every element of K is a pth power in K, i.e., $K = K^p$. Any field of characteristic 0 is also called perfect.

With this definition, we see that we have proved that every irreducible polynomial over a perfect field is separable. It is not hard to see that if K is not perfect then there are inseparable irreducible polynomials.

Example: (Existence and Uniqueness of Finite Fields)

Let $n > 0$ be any positive integer and consider the splitting field of the polynomial $x^{p^n} - x$ over $\mathbb{F}_p$. We have already seen that this polynomial is separable, hence has precisely p^n roots. Let α and β be any two roots of this polynomial, so that $\alpha^{p^n} = \alpha$ and $\beta^{p^n} = \beta$. Then $(\alpha\beta)^{p^n} = \alpha\beta$, $(\alpha^{-1})^{p^n} = \alpha^{-1}$ and by Proposition 35 also

$$(\alpha + \beta)^{p^n} = \alpha^{p^n} + \beta^{p^n} = \alpha + \beta.$$

Hence the set $\mathbb{F}$ consisting of the p^n distinct roots of $x^{p^n} - x$ over $\mathbb{F}_p$ is *closed* under addition, multiplication and inverses in its splitting field. It follows that $\mathbb{F}$ is a subfield, hence in fact must be the splitting field. Since the number of elements is p^n, we have $[\mathbb{F} : \mathbb{F}_p] = n$, which shows that there exist finite fields of degree n over $\mathbb{F}_p$ for any $n > 0$.

Let now $\mathbb{F}$ be any finite field of characteristic p. If $\mathbb{F}$ is of dimension n over its prime subfield $\mathbb{F}_p$, then $\mathbb{F}$ has precisely p^n elements. Since the multiplicative group $\mathbb{F}^\times$ is (in fact cyclic) of order $p^n - 1$, we have $\alpha^{p^n - 1} = 1$ for every $\alpha \neq 0$ in $\mathbb{F}$, so that $\alpha^{p^n} = \alpha$ for every $\alpha \in \mathbb{F}$. But this means α is a root of $x^{p^n} - x$, hence $\mathbb{F}$ is contained in a splitting field for this polynomial. Since we have seen that the splitting field has order p^n this shows that $\mathbb{F}$ is a splitting field for $x^{p^n} - x$. Since splitting fields are unique up to isomorphism, this proves that *finite fields of any order p^n exist and are unique up to isomorphism.* We shall denote the finite field of order p^n by $\mathbb{F}_{p^n}$.

We shall consider finite fields more later.

We now investigate further the structure of inseparable irreducible polynomials over fields of characteristic p. We have seen above that if $p(x)$ is an irreducible polynomial which is not separable, then its derivative $D_x p(x)$ is identically 0, so that $p(x) = p_1(x^p)$ for some polynomial $p_1(x)$. The polynomial $p_1(x)$ may or may not itself be separable. If not, then it too is a polynomial in x^p, $p_1(x) = p_2(x^p)$, so that $p(x)$ is a polynomial in x^{p^2}: $p(x) = p_2(x^{p^2})$. Continuing in this fashion we see that there is a uniquely defined power p^k of p such that $p(x) = p_k(x^{p^k})$ where $p_k(x)$ has nonzero derivative. It is clear that $p_k(x)$ is irreducible since any factorization of $p_k(x)$ would, after replacing x by x^{p^k}, immediately imply a factorization of the irreducible $p(x)$. It follows that $p_k(x)$ is separable. We summarize this as:

Proposition 38. Let $p(x)$ be an irreducible polynomial over a field F of characteristic p. Then there is a unique integer $k \geq 0$ and a unique irreducible separable polynomial $p_{sep}(x) \in F[x]$ such that
$$p(x) = p_{sep}(x^{p^k}).$$

Definition. Let $p(x)$ be an irreducible polynomial over a field of characteristic p. The degree of $p_{sep}(x)$ in the last proposition is called the *separable degree* of $p(x)$, denoted $\deg_s p(x)$. The integer p^k in the proposition is called the *inseparable degree* of $p(x)$, denoted $\deg_i p(x)$.

From the definitions and the proposition we see that $p(x)$ is separable if and only if its inseparability degree is 1 if and only if its degree is equal to its separable degree. Also, computing degrees in the relation $p(x) = p_{sep}(x^{p^k})$ we see that
$$\deg p(x) = \deg_s p(x) \deg_i p(x).$$

Examples

(1) In general, $p(x) = x^p - t$ over $\mathbb{F}_p(t)$ is irreducible (it's Eisenstein, cf. Example 5 in Section 9.4), with derivative 0, hence is inseparable. Here $p_{sep}(x) = x - t$ and the separable degree of $p(x)$ is 1 ($p(x)$ is said to be "purely inseparable"). There is a single root of multiplicity p: $x^p - t = (x - \sqrt[p]{t})^p$.

(2) The polynomial $p(x) = x^{p^m} - t$ over $F = \mathbb{F}_p(t)$ is irreducible with the same separable polynomial part as in (1), but with inseparability degree p^m.

(3) The polynomial $(x^{p^2} - t)(x^p - t)$ over $F = \mathbb{F}_p(t)$ has (two) inseparable irreducible factors so is inseparable. This polynomial cannot be written in the form $f_{sep}(x^{p^k})$ where $f_{sep}(x)$ is separable, which is the reason we restricted to *irreducible* polynomials above. This example also shows that there is no analogous factorization to define the separable and inseparable degrees of a general polynomial.

The notion of separability carries over to the fields generated by the roots of these polynomials.

Definition. The field K is said to be *separable* (or *separably algebraic*) over F if every element of K is the root of a separable polynomial over F (equivalently, the minimal polynomial over F of every element of K is separable). A field which is not separable is *inseparable*.

We have seen that the issue of separability is straightforward for finite extensions of perfect fields since for these fields the minimal polynomial of an algebraic element is irreducible hence separable.

Corollary 39. Every finite extension of a perfect field is separable. In particular, every finite extension of either $\mathbb{Q}$ or a finite field is separable.

We shall consider separable and inseparable extensions more after developing some Galois Theory, in particular defining the separable and inseparable *degree* of the extension K/F.

EXERCISES

1. Prove that the derivative D_x of a polynomial satisfies $D_x(f(x) + g(x)) = D_x(f(x)) + D_x(g(x))$ and $D_x(f(x)g(x)) = D_x(f(x))g(x) + D_x(g(x))f(x)$ for any two polynomials $f(x)$ and $g(x)$.

2. Find all irreducible polynomials of degrees 1, 2 and 4 over $\mathbb{F}_2$ and prove that their product is $x^{16} - x$.

3. Prove that d divides n if and only if $x^d - 1$ divides $x^n - 1$. [Note that if $n = qd + r$ then $x^n - 1 = (x^{qd+r} - x^r) + (x^r - 1)$.]

4. Let $a > 1$ be an integer. Prove for any positive integers n, d that d divides n if and only if $a^d - 1$ divides $a^n - 1$ (cf. the previous exercise). Conclude in particular that $\mathbb{F}_{p^d} \subseteq \mathbb{F}_{p^n}$ if and only if d divides n.

5. For any prime p and any nonzero $a \in \mathbb{F}_p$ prove that $x^p - x + a$ is irreducible and separable over $\mathbb{F}_p$. [For the irreducibility: One approach — prove first that if α is a root then $\alpha + 1$ is also a root. Another approach — suppose it's reducible and compute derivatives.]

6. Prove that $x^{p^n-1} - 1 = \prod_{\alpha \in \mathbb{F}_{p^n}^\times} (x - \alpha)$. Conclude that $\prod_{\alpha \in \mathbb{F}_{p^n}^\times} \alpha = (-1)^{p^n}$ so the product of the nonzero elements of a finite field is $+1$ if $p = 2$ and -1 if p is odd. For p odd and $n = 1$ derive *Wilson's Theorem*: $(p - 1)! \equiv -1 \pmod{p}$.

7. Suppose K is a field of characteristic p which is not a perfect field: $K \neq K^p$. Prove there exist irreducible inseparable polynomials over K. Conclude that there exist inseparable finite extensions of K.

8. Prove that $f(x)^p = f(x^p)$ for any polynomial $f(x) \in \mathbb{F}_p[x]$.

9. Show that the binomial coefficient $\binom{pn}{pi}$ is the coefficient of x^{pi} in the expansion of $(1+x)^{pn}$. Working over $\mathbb{F}_p$ show that this is the coefficient of $(x^p)^i$ in $(1 + x^p)^n$ and hence prove that $\binom{pn}{pi} \equiv \binom{n}{i} \pmod{p}$.

10. Let $f(x_1, x_2, \ldots, x_n) \in \mathbb{Z}[x_1, x_2, \ldots, x_n]$ be a polynomial in the variables $x_1, x_2, \ldots, x_n$ with integer coefficients. For any prime p prove that the polynomial
$$f(x_1, x_2, \ldots, x_n)^p - f(x_1^p, x_2^p, \ldots, x_n^p) \in \mathbb{Z}[x_1, x_2, \ldots, x_n]$$
has all its coefficients divisible by p.

11. Suppose $K[x]$ is a polynomial ring over the field K and F is a subfield of K. If F is a perfect field and $f(x) \in F[x]$ has no repeated irreducible factors in $F[x]$, prove that $f(x)$ has no repeated irreducible factors in $K[x]$.

13.6 CYCLOTOMIC POLYNOMIALS AND EXTENSIONS

The purpose of this section is to prove that the cyclotomic extension
$$\mathbb{Q}(\zeta_n)/\mathbb{Q}$$
generated by the nth roots of unity over $\mathbb{Q}$ introduced in Section 4 is of degree $\varphi(n)$ where φ denotes Euler's phi-function (= the number of integers a, $1 \leq a < n$ relatively prime to n = the order of the group $(\mathbb{Z}/n\mathbb{Z})^\times$).

Definition. Let μ_n denote the *group of nth roots of unity over* $\mathbb{Q}$.

Then as we have already observed, $\mathbb{Z}/n\mathbb{Z} \cong \mu_n$ as groups (under multiplication on the right, addition on the left), given explicitly by the map $a \mapsto (\zeta_n)^a$ for a fixed primitive n^{th} root of unity. The primitive n^{th} roots of unity are given by the residue classes prime to n so there are precisely $\varphi(n)$ primitive n^{th} roots of unity.

If d is a divisor of n and ζ is a dth root of unity, then ζ is also an nth root of unity since $\zeta^n = (\zeta^d)^{n/d} = 1$. Hence
$$\mu_d \subseteq \mu_n \qquad \text{for all } d \mid n.$$
Conversely, the order of any element of the group μ_n is a divisor of n so that if ζ is an nth root of unity which is also a dth root of unity for some smaller d then $d \mid n$.

Definition. Define the nth *cyclotomic polynomial* $\Phi_n(x)$ to be the polynomial whose roots are the primitive n^{th} roots of unity:
$$\Phi_n(x) = \prod_{\zeta \text{ primitive } \in \mu_n} (x - \zeta) = \prod_{\substack{1 \leq a < n \\ (a,n)=1}} (x - \zeta_n^a)$$
(which is of degree $\varphi(n)$).

552

The roots of the polynomial $x^n - 1$ are precisely the nth roots of unity so we have the factorization

$$x^n - 1 = \prod_{\substack{\zeta^n = 1 \\ \text{i.e. } \zeta \in \mu_n}} (x - \zeta).$$

If we group together the factors $(x - \zeta)$ where ζ is an element of order d in μ_n (i.e., ζ is a primitive d^{th} root of unity) we obtain

$$x^n - 1 = \prod_{d|n} \prod_{\substack{\zeta \in \mu_d \\ \zeta \text{ primitive}}} (x - \zeta).$$

The inner product is $\Phi_d(x)$ by definition so we have the factorization

$$x^n - 1 = \prod_{d|n} \Phi_d(x). \tag{13.4}$$

Note incidentally that comparing degrees gives the identity

$$n = \sum_{d|n} \varphi(d).$$

This factorization allows us to compute $\Phi_n(x)$ for any n recursively: clearly $\Phi_1(x) = x - 1$ and $\Phi_2(x) = x + 1$. Then

$$x^3 - 1 = \Phi_1(x)\Phi_3(x) = (x - 1)\Phi_3(x)$$

which gives

$$\Phi_3(x) = x^2 + x + 1.$$

Similarly

$$x^4 - 1 = \Phi_1(x)\Phi_2(x)\Phi_4(x) = (x - 1)(x + 1)\Phi_4(x)$$

gives

$$\Phi_4(x) = x^2 + 1$$

(in these cases these could also be obtained directly from the explicit roots of unity). Continuing in this fashion we can compute $\Phi_n(x)$ for any n. Note also that for p a prime we recover our polynomial

$$\Phi_p(x) = x^{p-1} + x^{p-2} + \cdots + x + 1.$$

For some small values of n the polynomials are

$$\Phi_5(x) = x^4 + x^3 + x^2 + x + 1$$
$$\Phi_6(x) = x^2 - x + 1$$
$$\Phi_7(x) = x^6 + x^5 + x^4 + x^3 + x^2 + x + 1$$
$$\Phi_8(x) = x^4 + 1$$
$$\Phi_9(x) = x^6 + x^3 + 1$$
$$\Phi_{10}(x) = x^4 - x^3 + x^2 - x + 1$$
$$\Phi_{11}(x) = x^{10} + x^9 + \cdots + x + 1$$
$$\Phi_{12}(x) = x^4 - x^2 + 1.$$

For all the values computed above, $\Phi_n(x)$ was a (monic) polynomial with integer coefficients. This is always the case:

Lemma 40. The cyclotomic polynomial $\Phi_n(x)$ is a monic polynomial in $\mathbb{Z}[x]$ of degree $\varphi(n)$.

Proof: It is clear that $\Phi_n(x)$ is monic and has degree $\varphi(n)$. We must show the coefficients lie in $\mathbb{Z}$. We use induction on n. The result is true for $n = 1$ (and $n \leq 12$). Assume by induction that $\Phi_d(x) \in \mathbb{Z}[x]$ for all $1 \leq d < n$. Then $x^n - 1 = f(x)\Phi_n(x)$ where $f(x) = \prod_{\substack{d \mid n \\ d < n}} \Phi_d(x)$ is monic and has coefficients in $\mathbb{Z}$. Since $f(x)$ clearly divides $x^n - 1$ in $F[x]$ where $F = \mathbb{Q}(\zeta_n)$ is the field of nth roots of unity and both $f(x)$ and $x^n - 1$ have coefficients in $\mathbb{Q}$, $f(x)$ divides $x^n - 1$ in $\mathbb{Q}[x]$ by the Division Algorithm (cf. the remark at the end of Section 9.2). By Gauss' Lemma, $f(x)$ divides $x^n - 1$ in $\mathbb{Z}[x]$, hence $\Phi_n(x) \in \mathbb{Z}[x]$.

We remark in passing that while all the coefficients of $\Phi_n(x)$ in the examples computed above were $0, \pm 1$, it is known that there are cyclotomic polynomials with arbitrarily large coefficients.

Theorem 41. The cyclotomic polynomial $\Phi_n(x)$ is an irreducible monic polynomial in $\mathbb{Z}[x]$ of degree $\varphi(n)$.

Proof: We must show that $\Phi_n(x)$ is irreducible. If not then we have a factorization

$$\Phi_n(x) = f(x)g(x) \qquad \text{with } f(x), g(x) \text{ monic in } \mathbb{Z}[x]$$

where we take $f(x)$ to be an *irreducible* factor of $\Phi_n(x)$. Let ζ be a primitive n^{th} root of 1 which is a root of $f(x)$ (so then $f(x)$ is the minimal polynomial for ζ over $\mathbb{Q}$) and let p denote *any* prime not dividing n. Then ζ^p is again a primitive n^{th} root of 1, hence is a root of either $f(x)$ or $g(x)$.

Suppose $g(\zeta^p) = 0$. Then ζ is a root of $g(x^p)$ and since $f(x)$ is the minimal polynomial for ζ, $f(x)$ must divide $g(x^p)$ in $\mathbb{Z}[x]$, say

$$g(x^p) = f(x)h(x), \qquad h(x) \in \mathbb{Z}[x].$$

If we reduce this equation mod p, we obtain

$$\bar{g}(x^p) = \bar{f}(x)\bar{h}(x) \qquad \text{in } \mathbb{F}_p[x].$$

By the remarks of the last section,

$$\bar{g}(x^p) = (\bar{g}(x))^p$$

so we have the equation

$$(\bar{g}(x))^p = \bar{f}(x)\bar{h}(x)$$

in the U.F.D. $\mathbb{F}_p[x]$. It follows that $\bar{f}(x)$ and $\bar{g}(x)$ have a factor in common in $\mathbb{F}_p[x]$.

Now, from $\Phi_n(x) = f(x)g(x)$ we see by reducing mod p that $\overline{\Phi}_n(x) = \bar{f}(x)\bar{g}(x)$, and so by the above it follows that $\overline{\Phi}_n(x) \in \mathbb{F}_p[x]$ has a multiple root. But then also $x^n - 1$ would have a multiple root over $\mathbb{F}_p$ since it has $\overline{\Phi}_n(x)$ as a factor. This is a contradiction since we have seen in the last section that there are n distinct roots of $x^n - 1$ over any field of characteristic not dividing n.

Hence ζ^p must be a root of $f(x)$. Since this applies to every root ζ of $f(x)$, it follows that ζ^a is a root of $f(x)$ for every integer a relatively prime to n: write $a = p_1 p_2 \cdots p_k$ as a product of (not necessarily distinct) primes not dividing n so that ζ^{p_1} is a root of $f(x)$, so also $(\zeta^{p_1})^{p_2}$ is a root of $f(x)$, etc. But this means that *every* primitive n^{th} root of unity is a root of $f(x)$, i.e., $f(x) = \Phi_n(x)$, showing $\Phi_n(x)$ is irreducible.

Corollary 42. The degree over $\mathbb{Q}$ of the cyclotomic field of nth roots of unity is $\varphi(n)$:

$$[\mathbb{Q}(\zeta_n) : \mathbb{Q}] = \varphi(n).$$

Proof: By the theorem, $\Phi_n(x)$ is the minimal polynomial for any primitive n^{th} root of unity ζ_n.

Example

The cyclotomic field $\mathbb{Q}(\zeta_8)$ of the 8th roots of unity is of degree $\varphi(8) = 4$ over $\mathbb{Q}$. This field contains the 4th roots of unity, i.e., $\mathbb{Q}(i) \subset \mathbb{Q}(\zeta_8)$ as well as the element $\zeta_8 + \zeta_8^7 = \sqrt{2}$ (recall the explicit roots of unity in Section 4). It follows that

$$\mathbb{Q}(\zeta_8) = \mathbb{Q}(i, \sqrt{2}).$$

One interesting number-theoretic application of the cyclotomic polynomials outlined in the exercises is the proof that for any n there are infinitely many primes which are congruent to 1 modulo n. The complete factorization in $\mathbb{F}_p[x]$ of $\Phi_\ell(x)$ for a prime ℓ (which is irreducible in $\mathbb{Z}[x]$) is described in Exercise 8 below.

We shall return to the example of cyclotomic fields after we have developed some Galois Theory.

EXERCISES

1. Suppose m and n are relatively prime positive integers. Let ζ_m be a primitive m^{th} root of unity and let ζ_n be a primitive n^{th} root of unity. Prove that $\zeta_m \zeta_n$ is a primitive mn^{th} root of unity.

2. Let ζ_n be a primitive n^{th} root of unity and let d be a divisor of n. Prove that ζ_n^d is a primitive $(n/d)^{\text{th}}$ root of unity.

3. Prove that if a field contains the nth roots of unity for n odd then it also contains the $2n^{\text{th}}$ roots of unity.

4. Prove that if $n = p^k m$ where p is a prime and m is relatively prime to p then there are precisely m distinct n^{th} roots of unity over a field of characteristic p.

5. Prove there are only a finite number of roots of unity in any finite extension K of $\mathbb{Q}$.

6. Prove that for n odd, $n > 1$, $\Phi_{2n}(x) = \Phi_n(-x)$.

7. Use the Möbius Inversion formula indicated in Section 14.3 to prove

$$\Phi_m(x) = \prod_{d \mid n} (x^d - 1)^{\mu(m/d)}.$$

8. Let ℓ be a prime and let $\Phi_\ell(x) = \frac{x^\ell - 1}{x - 1} = x^{\ell-1} + x^{\ell-2} + \ldots + x + 1 \in \mathbb{Z}[x]$ be the ℓth cyclotomic polynomial, which is irreducible over $\mathbb{Z}$ by Theorem 41. This exercise determines the factorization of $\Phi_\ell(x)$ modulo p for any prime p. Let ζ denote any fixed primitive ℓth root of unity.

   (a) Show that if $p = l$ then $\Phi_\ell(x) = (x - 1)^{\ell-1} \in \mathbb{F}_\ell[x]$.

   (b) Suppose $p \neq \ell$ and let f denote the order of $p \bmod \ell$, i.e., f is the smallest power of p with $p^f \equiv 1 \bmod \ell$. Use the fact that $\mathbb{F}_{p^n}^\times$ is a cyclic group to show that $n = f$ is the smallest power p^n of p with $\zeta \in \mathbb{F}_{p^n}$. Conclude that the minimal polynomial of ζ over $\mathbb{F}_p$ has degree f.

   (c) Show that $\mathbb{F}_p(\zeta) = \mathbb{F}_p(\zeta^a)$ for any integer a not divisible by ℓ. [One inclusion is obvious. For the other, note that $\zeta = (\zeta^a)^b$ where b is the multiplicative inverse of $a \bmod \ell$.] Conclude using (b) that, in $\mathbb{F}_p[x]$, $\Phi_\ell(x)$ is the product of $\frac{\ell-1}{f}$ distinct irreducible polynomials of degree f.

   (d) In particular, prove that, viewed in $\mathbb{F}_p[x]$, $\Phi_7(x) = x^6 + x^5 + \ldots + x + 1$ is $(x - 1)^6$ for $p = 7$, a product of distinct linear factors for $p \equiv 1 \bmod 7$, a product of 3 irreducible quadratics for $p \equiv 6 \bmod 7$, a product of 2 irreducible cubics for $p \equiv 2, 4 \bmod 7$, and is irreducible for $p \equiv 3, 5 \bmod 7$.

9. Suppose A is an $n \times n$ matrix over $\mathbb{C}$ for which $A^k = I$ for some integer $k \geq 1$. Show that A can be diagonalized. Show that the matrix $A = \begin{pmatrix} 1 & \alpha \\ 0 & 1 \end{pmatrix}$ where α is an element of a field of characteristic p satisfies $A^p = I$ and cannot be diagonalized if $\alpha \neq 0$.

10. Let φ denote the Frobenius map $x \mapsto x^p$ on the finite field $\mathbb{F}_{p^n}$. Prove that φ gives an isomorphism of $\mathbb{F}_{p^n}$ to itself (such an isomorphism is called an *automorphism*). Prove that φ^n is the identity map and that no lower power of φ is the identity.

11. Let φ denote the Frobenius map $x \mapsto x^p$ on the finite field $\mathbb{F}_{p^n}$ as in the previous exercise. Determine the rational canonical form over $\mathbb{F}_p$ for φ considered as an $\mathbb{F}_p$-linear transformation of the n-dimensional $\mathbb{F}_p$-vector space $\mathbb{F}_{p^n}$.

12. Let φ denote the Frobenius map $x \mapsto x^p$ on the finite field $\mathbb{F}_{p^n}$ as in the previous exercise. Determine the Jordan canonical form (over a field containing all the eigenvalues) for φ considered as an $\mathbb{F}_p$-linear transformation of the n-dimensional $\mathbb{F}_p$-vector space $\mathbb{F}_{p^n}$.

13. (*Wedderburn's Theorem on Finite Division Rings*) This exercise outlines a proof (following Witt) of Wedderburn's Theorem that a finite division ring D is a field (i.e., is commutative).

   (a) Let Z denote the center of D (i.e., the elements of D which commute with every element of D). Prove that Z is a field containing $\mathbb{F}_p$ for some prime p. If $Z = \mathbb{F}_q$ prove that D has order q^n for some integer n [D is a vector space over Z].

   (b) The nonzero elements $D^\times$ of D form a multiplicative group. For any $x \in D^\times$ show that the elements of D which commute with x form a division ring which contains Z. Show that this division ring is of order q^m for some integer m and that $m < n$ if x is not an element of Z.

   (c) Show that the class equation (Theorem 4.7) for the group $D^\times$ is

$$q^n - 1 = (q - 1) + \sum_{i=1}^r \frac{q^n - 1}{|C_{D^\times}(x_i)|}$$

   where $x_1, x_2, \ldots, x_r$ are representatives of the distinct conjugacy classes in $D^\times$ not contained in the center of $D^\times$. Conclude from (b) that for each i, $|C_{D^\times}(x_i)| = q^{m_i} - 1$ for some $m_i < n$.

(d) Prove that since $\dfrac{q^n - 1}{q^{m_i} - 1}$ is an integer (namely, the index $|D^\times : C_{D^\times}(x_i)|$) then m_i divides n (cf. Exercise 4 of Section 5). Conclude that $\Phi_n(x)$ divides $(x^n - 1)/(x^{m_i} - 1)$ and hence that the integer $\Phi_n(q)$ divides $(q^n - 1)/(q^{m_i} - 1)$ for $i = 1, 2, \ldots, r$.

(e) Prove that (c) and (d) imply that $\Phi_n(q) = \prod_{\zeta \text{ primitive}} (q - \zeta)$ divides $q - 1$. Prove that $|q - \zeta| > q - 1$ (complex absolute value) for any root of unity $\zeta \ne 1$ [note that 1 is the closest point on the unit circle in $\mathbb{C}$ to the point q on the real line]. Conclude that $n = 1$, i.e., that $D = Z$ is a field.

The following exercises provide a proof that for any positive integer m there are infinitely many primes p with $p \equiv 1 \pmod{m}$. This is a special case of *Dirichlet's Theorem on Primes in Arithmetic Progressions* which states more generally that there are infinitely many primes p with $p \equiv a \pmod{m}$ for any a relatively prime to m.

14. Given any monic polynomial $P(x) \in \mathbb{Z}[x]$ of degree at least one show that there are infinitely many distinct prime divisors of the integers

$$P(1),\ P(2),\ P(3),\ \ldots,\ P(n),\ \ldots.$$

[Suppose $p_1, p_2, \ldots, p_k$ are the only primes dividing the values $P(n)$, $n = 1, 2, \ldots$. Let N be an integer with $P(N) = a \ne 0$. Show that $Q(x) = a^{-1}P(N + a\, p_1 p_2 \ldots p_k\, x)$ is an element of $\mathbb{Z}[x]$ and that $Q(n) \equiv 1 \pmod{p_1 p_2 \ldots p_k}$ for $n = 1, 2, \ldots$. Conclude that there is some integer M such that $Q(M)$ has a prime factor different from $p_1, p_2, \ldots, p_k$ and hence that $P(N + a p_1 p_2 \cdots p_k M)$ has a prime factor different from $p_1, p_2, \ldots, p_k$.]

15. Let p be an odd prime not dividing m and let $\Phi_m(x)$ be the mth cyclotomic polynomial. Suppose $a \in \mathbb{Z}$ satisfies $\Phi_m(a) \equiv 0 \pmod{p}$. Prove that a is relatively prime to p and that the order of a in $(\mathbb{Z}/p\mathbb{Z})^\times$ is precisely m. [Since

$$x^m - 1 = \prod_{d \mid m} \Phi_d(x) = \Phi_m(x) \prod_{\substack{d \mid m \\ d < m}} \Phi_d(x)$$

we see first that $a^m - 1 \equiv 0 \pmod{p}$ i.e., $a^m \equiv 1 \pmod{p}$. If the order of $a \bmod p$ were less than m, then $a^d \equiv 1 \pmod{p}$ for some d dividing m, so then $\Phi_d(a) \equiv 0 \pmod{p}$ for some $d < m$. But then $x^m - 1$ would have a as a multiple root mod p, a contradiction.]

16. Let $a \in \mathbb{Z}$. Show that if p is an odd prime dividing $\Phi_m(a)$ then either p divides m or $p \equiv 1 \pmod{m}$.

17. Prove there are infinitely many primes p with $p \equiv 1 \pmod{m}$.

CHAPTER 14

Galois Theory

14.1 BASIC DEFINITIONS

In the previous chapter we proved the existence of a finite extension of a field F which contains all the roots of a given polynomial $f(x)$ whose coefficients are in F. The main idea of Galois Theory (named for Évariste Galois, 1811–1832) is to consider the relation of the group of permutations of the roots of $f(x)$ to the algebraic structure of its splitting field. The connection is given by the Fundamental Theorem of the next section. It can be viewed as another (extremely elegant) application of the important idea in mathematics that one (in our case algebraic) object *acting* on another provides structural information about both.

In this section we introduce the terminology and basic properties of the objects of interest. Let K be a field.

Definition.
   (1) An isomorphism σ of K with itself is called an *automorphism* of K. The collection of automorphisms of K is denoted $\mathrm{Aut}(K)$. If $\alpha \in K$ we shall write $\sigma\alpha$ for $\sigma(\alpha)$.
   (2) An automorphism $\sigma \in \mathrm{Aut}(K)$ is said to *fix* an element $\alpha \in K$ if $\sigma\alpha = \alpha$. If F is a subset of K (for example, a subfield), then an automorphism σ is said to *fix* F if it fixes all the elements of F, i.e., $\sigma a = a$ for all $a \in F$.

Note that any field has at least one automorphism, the identity map, denoted by 1 and sometimes called the *trivial* automorphism.

The prime field of K is generated by $1 \in K$ and since any automorphism σ takes 1 to 1 (and 0 to 0), i.e., $\sigma(1) = 1$, it follows that $\sigma a = a$ for all a in the prime field. Hence any automorphism of a field K fixes its prime subfield. In particular we see that $\mathbb{Q}$ and $\mathbb{F}_p$ have only the trivial automorphism: $\mathrm{Aut}(\mathbb{Q}) = \{1\}$ and $\mathrm{Aut}(\mathbb{F}_p) = \{1\}$.

Definition. Let K/F be an extension of fields. Let $\mathrm{Aut}(K/F)$ be the collection of automorphisms of K which fix F.

Note that if F is the prime subfield of K then $\mathrm{Aut}(K) = \mathrm{Aut}(K/F)$ since every automorphism of K automatically fixes F.

558

If σ and τ are automorphisms of K then the composite $\sigma\tau$ (and also the composite $\tau\sigma$, which may not be the same) is defined and is again an automorphism of K.

Proposition 1. $\text{Aut}(K)$ is a group under composition and $\text{Aut}(K/F)$ is a subgroup.

Proof: It is clear that $\text{Aut}(K)$ is a group. If σ and τ are automorphisms of K which fix F then also $\sigma\tau$ and σ^{-1} are the identity on F, which shows that $\text{Aut}(K/F)$ is a subgroup.

The following proposition is extremely useful for determining the automorphisms of algebraic extensions.

Proposition 2. Let K/F be a field extension and let $\alpha \in K$ be algebraic over F. Then for any $\sigma \in \text{Aut}(K/F)$, $\sigma\alpha$ is a root of the minimal polynomial for α over F i.e., $\text{Aut}(K/F)$ permutes the roots of irreducible polynomials. Equivalently, any polynomial with coefficients in F having α as a root also has $\sigma\alpha$ as a root.

Proof: Suppose α satisfies the equation

$$\alpha^n + a_{n-1}\alpha^{n-1} + \cdots + a_1\alpha + a_0 = 0$$

where $a_0, a_1, \ldots, a_{n-1}$ are elements of F. Applying the automorphism σ we obtain (using the fact that σ is an additive homomorphism)

$$\sigma(\alpha^n) + \sigma(a_{n-1}\alpha^{n-1}) + \cdots + \sigma(a_1\alpha) + \sigma(a_0) = \sigma(0) = 0.$$

Using the fact that σ is also a multiplicative homomorphism this becomes

$$(\sigma(\alpha))^n + \sigma(a_{n-1})(\sigma(\alpha))^{n-1} + \cdots + \sigma(a_1)(\sigma(\alpha)) + \sigma(a_0) = 0.$$

By assumption, σ fixes all the elements of F, so $\sigma(a_i) = a_i$, $i = 0, 1, \ldots, n-1$. Hence

$$(\sigma\alpha)^n + a_{n-1}(\sigma\alpha)^{n-1} + \cdots + a_1(\sigma\alpha) + a_0 = 0.$$

But this says precisely that $\sigma\alpha$ is a root of the same polynomial over F as α. This proves the proposition.

Examples

(1) Let $K = \mathbb{Q}(\sqrt{2})$. If $\tau \in \text{Aut}(\mathbb{Q}(\sqrt{2})) = \text{Aut}(\mathbb{Q}(\sqrt{2})/\mathbb{Q})$, then $\tau(\sqrt{2}) = \pm\sqrt{2}$ since these are the two roots of the minimal polynomial for $\sqrt{2}$. Since τ fixes $\mathbb{Q}$, this determines τ completely:

$$\tau(a + b\sqrt{2}) = a \pm b\sqrt{2}.$$

The map $\sqrt{2} \mapsto \sqrt{2}$ is just the identity automorphism 1 of $\mathbb{Q}(\sqrt{2})$. The map $\sigma : \sqrt{2} \mapsto -\sqrt{2}$ is the isomorphism considered in Example 2 following Corollary 13.7. Hence $\text{Aut}(\mathbb{Q}(\sqrt{2})) = \text{Aut}(\mathbb{Q}(\sqrt{2})/\mathbb{Q}) = \{1, \sigma\}$ is a cyclic group of order 2 generated by σ.

(2) Let $K = \mathbb{Q}(\sqrt[3]{2})$. As before, if $\tau \in \text{Aut}(K/\mathbb{Q})$, then τ is completely determined by its action on $\sqrt[3]{2}$ since

$$\tau(a + b\sqrt[3]{2} + c(\sqrt[3]{2})^2) = a + b\tau\sqrt[3]{2} + c(\tau\sqrt[3]{2})^2.$$

Since $\tau\sqrt[3]{2}$ must be a root of $x^3 - 2$ and the other two roots of this equation are not elements of K (recall the splitting field of this polynomial is degree 6 over $\mathbb{Q}$), the only possibility is $\tau\sqrt[3]{2} = \sqrt[3]{2}$ i.e., $\tau = 1$. Hence $\text{Aut}(\mathbb{Q}(\sqrt[3]{2})/\mathbb{Q}) = 1$ is the trivial group.

In general, if K is generated over F by some collection of elements, then any automorphism $\sigma \in \text{Aut}(K/F)$ is completely determined by what it does to the generators. If K/F is finite then K is finitely generated over F by algebraic elements so by the proposition the number of automorphisms of K fixing F is finite, i.e., $\text{Aut}(K/F)$ is a finite group. In particular, the automorphisms of a finite extension can be considered as permutations of the roots of a finite number of equations (not every permutation gives rise to an automorphism, however, as Example 2 above illustrates). It was the investigation of permutations of the roots of equations that led Galois to the theory we are describing.

We have associated to each field extension K/F (equivalently, with a subfield F of K) a *group*, $\text{Aut}(K/F)$, the group of automorphisms of K which fix F. One can also reverse this process and associate to each group of automorphisms a field extension.

Proposition 3. Let $H \leq \text{Aut}(K)$ be a subgroup of the group of automorphisms of K. Then the collection F of elements of K fixed by all the elements of H is a subfield of K.

Proof: Let $h \in H$ and let $a, b \in F$. Then by definition $h(a) = a$, $h(b) = b$ so that $h(a \pm b) = h(a) \pm h(b) = a \pm b$, $h(ab) = h(a)h(b) = ab$ and $h(a^{-1}) = h(a)^{-1} = a^{-1}$, so that F is closed, hence a subfield of K.

Note that it is not important in this proposition that H actually be a *subgroup* of $\text{Aut}(K)$ — the collection of elements of K fixed by all the elements of a *subset* of $\text{Aut}(K)$ is also a subfield of K.

Definition. If H is a subgroup of the group of automorphisms of K, the subfield of K fixed by all the elements of H is called the *fixed field* of H.

Proposition 4. The association of groups to fields and fields to groups defined above is inclusion reversing, namely
(1) if $F_1 \subseteq F_2 \subseteq K$ are two subfields of K then $\text{Aut}(K/F_2) \leq \text{Aut}(K/F_1)$, and
(2) if $H_1 \leq H_2 \leq \text{Aut}(K)$ are two subgroups of automorphisms with associated fixed fields F_1 and F_2, respectively, then $F_2 \subseteq F_1$.

Proof: Any automorphism of K that fixes F_2 also fixes its subfield F_1, which gives (1). The second assertion is proved similarly.

Examples

(1) Suppose $K = \mathbb{Q}(\sqrt{2})$ as in Example 1 above. Then the fixed field of $\text{Aut}(\mathbb{Q}(\sqrt{2})) = \text{Aut}(\mathbb{Q}(\sqrt{2})/\mathbb{Q}) = \{1, \sigma\}$ will be the set of elements of $\mathbb{Q}(\sqrt{2})$ with
$$\sigma(a + b\sqrt{2}) = a + b\sqrt{2}$$
since everything is fixed by the identity automorphism. This is the equation
$$a - b\sqrt{2} = a + b\sqrt{2}.$$
which is equivalent to $b = 0$, so the fixed field of $\text{Aut}(\mathbb{Q}(\sqrt{2})/\mathbb{Q})$ is just $\mathbb{Q}$.

(2) Suppose now that $K = \mathbb{Q}(\sqrt[3]{2})$ as in Example 2 above. In this case $\text{Aut}(K) = 1$, so that every element of K is fixed, i.e., the fixed field of $\text{Aut}(\mathbb{Q}(\sqrt[3]{2})/\mathbb{Q})$ is $\mathbb{Q}(\sqrt[3]{2})$.

Given a subfield F of K, the associated group is the collection of automorphisms of K which fix F. Given a group of automorphisms of K, the associated extension is defined by taking F to be the fixed field of the automorphisms. In the first example above, starting with the subfield $\mathbb{Q}$ of $\mathbb{Q}(\sqrt{2})$ one obtains the group $\{1, \sigma\}$ and starting with the group $\{1, \sigma\}$ one obtains the subfield $\mathbb{Q}$, so there is a "duality" between the two. In the second example, however, starting with the subfield $\mathbb{Q}$ of $\mathbb{Q}(\sqrt[3]{2})$ one obtains only the trivial group and starting with the trivial group one obtains the full field $\mathbb{Q}(\sqrt[3]{2})$.

An examination of the two examples suggests that for the second example there are "not enough" automorphisms to force the fixed field to be $\mathbb{Q}$ rather than the full $\mathbb{Q}(\sqrt[3]{2})$. This in turn seems to be due to the fact that the other roots of $x^3 - 2$, which are the only possible images of $\sqrt[3]{2}$ under an automorphism, are not elements of $\mathbb{Q}(\sqrt[3]{2})$. (Although even if they were we would need to check that the additional maps we could define were *automorphisms*.) We now make precise the notion of fields with "enough" automorphisms (leading to the definition of a *Galois* extension). As one might suspect even from these two examples (and we prove in the next section) these are related to splitting fields.

We first investigate the size of the automorphism group in the case of splitting fields.

Let F be a field and let E be the splitting field over F of $f(x) \in F[x]$. The main tool is Theorem 13.27 on the existence of extensions of isomorphisms, which states that any isomorphism $\varphi : F \xrightarrow{\sim} F'$ of F with F' can be extended to an isomorphism $\sigma : E \xrightarrow{\sim} E'$ between E and the splitting field E' for $f'(x) = \varphi(f(x)) \in F'[x]$.

We now show by induction on $[E : F]$ that the number of such extensions is at most $[E : F]$, with equality if $f(x)$ is separable over F. If $[E : F] = 1$ then $E = F$, $E' = F'$, $\sigma = \varphi$ and the number of extensions is 1. If $[E : F] > 1$ then $f(x)$ has at least one irreducible factor $p(x)$ of degree > 1 with corresponding irreducible factor $p'(x)$ of $f'(x)$. Let α be a fixed root of $p(x)$. If σ is any extension of φ to E, then σ restricted to the subfield $F(\alpha)$ of E is an isomorphism τ of $F(\alpha)$ with some subfield of E'. The isomorphism τ is completely determined by its action on α, i.e., by $\tau\alpha$, since α generates $F(\alpha)$ over F. Just as in Proposition 2, we see that $\tau\alpha$ must be some root β of $p'(x)$. Then we have a diagram

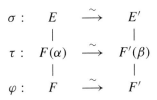

Conversely, for any β a root of $p'(x)$ there are extensions τ and σ giving such a diagram (this is Theorem 13.8 and Theorem 13.27). Hence to count the number of extensions σ we need only count the possible number of these diagrams.

The number of extensions of φ to an isomorphism τ is equal to the number of distinct roots β of $p'(x)$. Since the degree of $p(x)$ and $p'(x)$ are both equal to $[F(\alpha) : F]$, we see that the number of extensions of φ to a τ is at most $[F(\alpha) : F]$, with equality if the roots of $p(x)$ are distinct.

Since E is also the splitting field of $f(x)$ over $F(\alpha)$, E' is the splitting field of $f'(x)$

over $F'(\beta)$, and $[E : F(\alpha)] < [E : F]$, we may apply our induction hypothesis to these field extensions. By induction, the number of extensions of τ to σ is $\leq [E : F(\alpha)]$, with equality if $f(x)$ has distinct roots.

From $[E : F] = [E : F(\alpha)][F(\alpha) : F]$ it follows that the number of extensions of φ to σ is $\leq [E : F]$. We have equality if $p(x)$ and $f(x)$ have distinct roots, which is equivalent to $f(x)$ having distinct roots since $p(x)$ is a factor of $f(x)$, completing the proof by induction.

In the particular case when $F = F'$ and φ is the identity map we have $f(x) = f'(x)$ and $E = E'$ so the isomorphisms of E to E' restricting to φ on F are the automorphisms of E fixing F. We state this as follows:

Proposition 5. Let E be the splitting field over F of the polynomial $f(x) \in F[x]$. Then

$$|\mathrm{Aut}(E/F)| \leq [E : F]$$

with equality if $f(x)$ is separable over F.

Remark: While we were primarily interested in counting the automorphisms of E which fix F (which is the situation of $F = F'$, $\varphi = 1$ above), it would still have been necessary to consider the situation of more general φ (and different fields F') because of the induction step in the proof (which involves the fields $F(\alpha)$ and $F(\beta)$ for two roots of the same polynomial $p(x)$).

One can modify the proof above to show more generally that $|\mathrm{Aut}(K/F)| \leq [K : F]$ for *any* finite extension K/F (we shall prove this in the next section from a slightly different point of view). This gives us a notion of field extensions with "enough" automorphisms.

Definition. Let K/F be a finite extension. Then K is said to be *Galois* over F and K/F is a *Galois* extension if $|\mathrm{Aut}(K/F)| = [K : F]$. If K/F is Galois the group of automorphisms $\mathrm{Aut}(K/F)$ is called the *Galois group of K/F*, denoted $\mathrm{Gal}(K/F)$.

Remark: The Galois group of an extension K/F is sometimes defined to be the group of automorphisms $\mathrm{Aut}(K/F)$ for all K/F. We have chosen the definition above so that the notation $\mathrm{Gal}(K/F)$ will emphasize that the extension K/F has the maximal number of automorphisms.

Corollary 6. If K is the splitting field over F of a separable polynomial $f(x)$ then K/F is Galois.

We shall see in the next section that the converse is also true, which will completely characterize Galois extensions.

Note also that Corollary 6 implies that the splitting field of *any* polynomial over $\mathbb{Q}$ is Galois, since the splitting field of $f(x)$ is clearly the same as the splitting field of the product of the irreducible factors of $f(x)$ (i.e., the polynomial obtained by removing multiple factors), which is separable (Corollary 13.34).

Definition. If $f(x)$ is a separable polynomial over F, then the *Galois group of $f(x)$ over F* is the Galois group of the splitting field of $f(x)$ over F.

Examples

(1) The extension $\mathbb{Q}(\sqrt{2})/\mathbb{Q}$ is Galois with Galois group $\mathrm{Gal}(\mathbb{Q}(\sqrt{2})/\mathbb{Q}) = \{1, \sigma\} \cong \mathbb{Z}/2\mathbb{Z}$ where σ is the automorphism

$$\sigma : \mathbb{Q}(\sqrt{2}) \xrightarrow{\sim} \mathbb{Q}(\sqrt{2})$$
$$a + b\sqrt{2} \longmapsto a - b\sqrt{2}.$$

(2) More generally, any quadratic extension K of any field F of characteristic different from 2 is Galois. This follows from the discussion of quadratic extensions following Corollary 13.13, which shows that any extension K of degree 2 of F (where the characteristic of F is not 2) is of the form $F(\sqrt{D})$ for some D hence is the splitting field of $x^2 - D$ (since if $\sqrt{D} \in K$ then also $-\sqrt{D} \in K$).

(3) The extension $\mathbb{Q}(\sqrt[3]{2})/\mathbb{Q}$ is not Galois since its group of automorphisms is only of order 1.

(4) The extension $\mathbb{Q}(\sqrt{2}, \sqrt{3})$ is Galois over $\mathbb{Q}$ since it is the splitting field of the polynomial $(x^2 - 2)(x^2 - 3)$. Any automorphism σ is completely determined by its action on the generators $\sqrt{2}$ and $\sqrt{3}$, which must be mapped to $\pm\sqrt{2}$ and $\pm\sqrt{3}$, respectively. Hence the only possibilities for automorphisms are the maps

$$\begin{cases} \sqrt{2} \mapsto \sqrt{2} \\ \sqrt{3} \mapsto \sqrt{3} \end{cases} \begin{cases} \sqrt{2} \mapsto -\sqrt{2} \\ \sqrt{3} \mapsto \sqrt{3} \end{cases} \begin{cases} \sqrt{2} \mapsto \sqrt{2} \\ \sqrt{3} \mapsto -\sqrt{3} \end{cases} \begin{cases} \sqrt{2} \mapsto -\sqrt{2} \\ \sqrt{3} \mapsto -\sqrt{3} \end{cases}.$$

Since the Galois group is of order 4, *all* these elements are in fact automorphisms of $\mathbb{Q}(\sqrt{2}, \sqrt{3})$ over $\mathbb{Q}$.

Define the automorphisms σ and τ by

$$\sigma : \begin{cases} \sqrt{2} \mapsto -\sqrt{2} \\ \sqrt{3} \mapsto \sqrt{3} \end{cases} \qquad \tau : \begin{cases} \sqrt{2} \mapsto \sqrt{2} \\ \sqrt{3} \mapsto -\sqrt{3} \end{cases}$$

or, more explicitly, by

$$\sigma : a + b\sqrt{2} + c\sqrt{3} + d\sqrt{6} \mapsto a - b\sqrt{2} + c\sqrt{3} - d\sqrt{6}$$
$$\tau : a + b\sqrt{2} + c\sqrt{3} + d\sqrt{6} \mapsto a + b\sqrt{2} - c\sqrt{3} - d\sqrt{6}$$

(since, for example,

$$\sigma(\sqrt{6}) = \sigma(\sqrt{2}\sqrt{3}) = \sigma(\sqrt{2})\sigma(\sqrt{3}) = (-\sqrt{2})(\sqrt{3}) = -\sqrt{6} \quad).$$

Then $\sigma^2(\sqrt{2}) = \sigma(\sigma\sqrt{2}) = \sigma(-\sqrt{2}) = \sqrt{2}$ and clearly $\sigma^2(\sqrt{3}) = \sqrt{3}$. Hence $\sigma^2 = 1$ is the identity automorphism. Similarly, $\tau^2 = 1$. The automorphism $\sigma\tau$ can be easily computed:

$$\sigma\tau(\sqrt{2}) = \sigma(\tau(\sqrt{2})) = \sigma(\sqrt{2}) = -\sqrt{2}$$

and

$$\sigma\tau(\sqrt{3}) = \sigma(\tau(\sqrt{3})) = \sigma(-\sqrt{3}) = -\sqrt{3}$$

so that $\sigma\tau$ is the remaining nontrivial automorphism in the Galois group. Since this automorphism also evidently has order 2 in the Galois group, we have

$$\mathrm{Gal}(\mathbb{Q}(\sqrt{2}, \sqrt{3})/\mathbb{Q}) = \{1, \sigma, \tau, \sigma\tau\}$$

i.e., the Galois group is isomorphic to the Klein 4-group.

Associated to each subgroup of $\mathrm{Gal}(\mathbb{Q}(\sqrt{2},\sqrt{3})/\mathbb{Q})$ is the corresponding fixed subfield of $\mathbb{Q}(\sqrt{2},\sqrt{3})$. For example, the subfield corresponding to $\{1,\sigma\tau\}$ is the set of elements fixed by the map

$$\sigma\tau : a + b\sqrt{2} + c\sqrt{3} + d\sqrt{6} \mapsto a - b\sqrt{2} - c\sqrt{3} + d\sqrt{6}$$

which is the set of elements $a + d\sqrt{6}$, i.e., the field $\mathbb{Q}(\sqrt{6})$. One can similarly determine the fixed fields for the other subgroups of the Galois group:

subgroup	fixed field
$\{1\}$	$\mathbb{Q}(\sqrt{2},\sqrt{3})$
$\{1,\sigma\}$	$\mathbb{Q}(\sqrt{3})$
$\{1,\sigma\tau\}$	$\mathbb{Q}(\sqrt{6})$
$\{1,\tau\}$	$\mathbb{Q}(\sqrt{2})$
$\{1,\sigma,\tau,\sigma\tau\}$	$\mathbb{Q}$

(5) The splitting field of $x^3 - 2$ over $\mathbb{Q}$ is Galois of degree 6. The roots of this equation are $\sqrt[3]{2}$, $\rho\sqrt[3]{2}$, $\rho^2\sqrt[3]{2}$ where $\rho = \zeta_3 = \dfrac{-1 + \sqrt{-3}}{2}$ is a primitive cube root of unity. Hence the splitting field can be written $\mathbb{Q}(\sqrt[3]{2}, \rho\sqrt[3]{2})$. Any automorphism maps each of these two elements to one of the roots of $x^3 - 2$, giving 9 possibilities, but since the Galois group has order 6 not every such map is an automorphism of the field.

To determine the Galois group we use a more convenient set of generators, namely $\sqrt[3]{2}$ and ρ. Then any automorphism σ maps $\sqrt[3]{2}$ to one of $\sqrt[3]{2}$, $\rho\sqrt[3]{2}$, $\rho^2\sqrt[3]{2}$ and maps ρ to ρ or $\rho^2 = \dfrac{-1 - \sqrt{-3}}{2}$ since these are the roots of the cyclotomic polynomial $\Phi_3(x) = x^2 + x + 1$. Since σ is completely determined by its action on these two elements this gives only 6 possibilities and so each of these possibilities is actually an automorphism. To give these automorphisms explicitly, let σ and τ be the automorphisms defined by

$$\sigma : \begin{cases} \sqrt[3]{2} \mapsto \rho\sqrt[3]{2} \\ \rho \mapsto \rho \end{cases} \qquad \tau : \begin{cases} \sqrt[3]{2} \mapsto \sqrt[3]{2} \\ \rho \mapsto \rho^2 = -1 - \rho. \end{cases}$$

As before, these can be given explicitly on the elements of $\mathbb{Q}(\sqrt[3]{2}, \rho)$, which are linear combinations of the basis $\{1, \sqrt[3]{2}, (\sqrt[3]{2})^2, \rho, \rho\sqrt[3]{2}, \rho(\sqrt[3]{2})^2\}$. For example

$$\sigma(\rho\sqrt[3]{2}) = (\rho)(\rho\sqrt[3]{2}) = \rho^2\sqrt[3]{2} = (-1 - \rho)\sqrt[3]{2}$$
$$= -\sqrt[3]{2} - \rho\sqrt[3]{2}$$

and we may similarly determine the action of σ on the other basis elements. This gives

$$\sigma : \quad a + b\sqrt[3]{2} + c\sqrt[3]{4} + d\rho + e\rho\sqrt[3]{2} + f\rho\sqrt[3]{4} \quad \longmapsto$$
$$a - e\sqrt[3]{2} + (f - c)\sqrt[3]{4} + d\rho + (b - e)\rho\sqrt[3]{2} - c\rho\sqrt[3]{4}.$$

$$(14.1)$$

The other elements of the Galois group are

$$1 : \begin{cases} \sqrt[3]{2} \mapsto \sqrt[3]{2} \\ \rho \mapsto \rho \end{cases} \qquad \sigma^2 : \begin{cases} \sqrt[3]{2} \mapsto \rho^2\sqrt[3]{2} \\ \rho \mapsto \rho \end{cases}$$

564

$$\tau\sigma : \begin{cases} \sqrt[3]{2} \mapsto \rho^2 \sqrt[3]{2} \\ \rho \mapsto \rho^2 \end{cases} \qquad \tau\sigma^2 : \begin{cases} \sqrt[3]{2} \mapsto \rho \sqrt[3]{2} \\ \rho \mapsto \rho^2 \end{cases}$$

Computing $\sigma\tau$ we have

$$\sigma\tau : \begin{cases} \sqrt[3]{2} \overset{\tau}{\mapsto} \sqrt[3]{2} \overset{\sigma}{\mapsto} \rho\sqrt[3]{2} \\ \rho \overset{\tau}{\mapsto} \rho^2 \overset{\sigma}{\mapsto} \rho^2 \end{cases}$$

i.e.,

$$\sigma\tau : \begin{cases} \sqrt[3]{2} \mapsto \rho \sqrt[3]{2} \\ \rho \mapsto \rho^2 \end{cases}$$

so that $\sigma\tau = \tau\sigma^2$. Similarly one computes that $\sigma^3 = \tau^2 = 1$. Hence

$$\mathrm{Gal}(\mathbb{Q}(\sqrt[3]{2}, \zeta_3)/\mathbb{Q}) = \langle \sigma, \tau \rangle \cong S_3$$

is the symmetric group on 3 letters. Alternatively (and less computationally), since $G = \mathrm{Gal}(\mathbb{Q}(\sqrt[3]{2}, \zeta_3)/\mathbb{Q})$ acts as permutations of the 3 roots of $x^3 - 2$, G is a subgroup of S_3, hence must be S_3 since it is of order 6. The computations above explicitly identify the automorphisms in G and give an explicit isomorphism of G with S_3.

As in the previous example we can determine the fixed fields for any of the subgroups of the Galois group. For example, consider the fixed field of the subgroup $\{1, \sigma, \sigma^2\}$ generated by σ. These are just the elements fixed by σ (given explicitly in equation (1)) since if an element is fixed by σ then it is also fixed by σ^2. (In general, the fixed field of some subgroup is the field fixed by a set of generators for the subgroup.) The elements fixed by σ are those with

$$a = a \quad b = -e \quad c = f - c \quad d = d \quad e = b - e \quad f = -c$$

which is equivalent to $b = c = f = e = 0$. Hence the fixed field of $\{1, \sigma, \sigma^2\}$ is the field $\mathbb{Q}(\rho)$.

Remark: This example shows that some care must be exercised in determining Galois groups from the actions on generators. As mentioned, not every map taking $\sqrt[3]{2}$ and $\rho\sqrt[3]{2}$ to roots of $x^3 - 2$ gives rise to an automorphism of the field (for example, the map

$$\sqrt[3]{2} \mapsto \rho\sqrt[3]{2}$$
$$\rho\sqrt[3]{2} \mapsto \rho\sqrt[3]{2}$$

clearly cannot be an automorphism since it is evidently not an injection). The point is that there may be (sometimes very subtle) algebraic relations among the generators and these relations must be respected by an automorphism. For example, the quotient of the generators here is ρ, which is mapped to 1 and not to a root of the minimal polynomial for ρ. Put another way, the quotient of these generators satisfies a quadratic equation and this map does not respect that property.

For another (less trivial) example, compare with the discussion of the splitting field of $x^8 - 2$ in Section 2.

(6) As in Example 3, the field $\mathbb{Q}(\sqrt[4]{2})$ is not Galois over $\mathbb{Q}$ since any automorphism is determined by where it sends $\sqrt[4]{2}$ and of the four possibilities $\{\pm\sqrt[4]{2}, \pm i\sqrt[4]{2}\}$, only two are elements of the field (the two real roots).

Note that we have

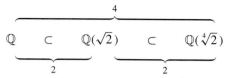

where $\mathbb{Q}(\sqrt{2})/\mathbb{Q}$ and $\mathbb{Q}(\sqrt[4]{2})/\mathbb{Q}(\sqrt{2})$ are both Galois extensions by Example 2 since both are quadratic extensions. This shows that a Galois extension of a Galois extension is not necessarily Galois.

(7) The extension of finite fields $\mathbb{F}_{p^n}/\mathbb{F}_p$ constructed after Proposition 13.37 is Galois by Corollary 6 since $\mathbb{F}_{p^n}$ is the splitting field over $\mathbb{F}_p$ of the separable polynomial $x^{p^n} - x$. It follows that the group of automorphisms for this extension is of order n. The injective homomorphism

$$\sigma : \mathbb{F}_{p^n} \to \mathbb{F}_{p^n}$$
$$\alpha \mapsto \alpha^p$$

of Proposition 13.35 is surjective in this case since $\mathbb{F}_{p^n}$ is finite, hence is an isomorphism. This gives an automorphism of $\mathbb{F}_{p^n}$, called the *Frobenius* automorphism, which we shall denote by σ_p. Iterating σ_p we have $\sigma_p^2(\alpha) = \sigma_p(\sigma_p(\alpha)) = (\alpha^p)^p = \alpha^{p^2}$. Similarly we have

$$\sigma_p^i(\alpha) = \alpha^{p^i} \qquad i = 0, 1, 2, \dots$$

Since $\alpha^{p^n} = \alpha$, we see that $\sigma_p^n = 1$ is the identity automorphism. No lower power of σ_p can be the identity, since this would imply $\alpha^{p^i} = \alpha$ for all $\alpha \in \mathbb{F}_{p^n}$ for some $i < n$, which is impossible since there are only p^i roots of this equation. It follows that σ_p is of order n in the Galois group, which means that $\mathrm{Gal}(\mathbb{F}_{p^n}/\mathbb{F}_p)$ is *cyclic* of order n, with the Frobenius automorphism σ_p as generator.

(8) The inseparable extension $\mathbb{F}_2(x)$ over $\mathbb{F}_2(t)$ where $x^2 - t = 0$ considered in Section 13.5 is not Galois. Any automorphism of this degree 2 extension is determined by its action on x, which must be sent to a root of the equation $x^2 - t$. We have already seen that there is only one root of this equation (with multiplicity 2) since we are in a field of characteristic 2. Hence the extension has only the trivial automorphism. Note that $\mathbb{F}_2(x)$ is the splitting field for $x^2 - t$ over $\mathbb{F}_2(t)$, so this example shows the separability condition in Corollary 6 is necessary.

EXERCISES

1. (a) Show that if the field K is generated over F by the elements $\alpha_1, \dots, \alpha_n$ then an automorphism σ of K fixing F is uniquely determined by $\sigma(\alpha_1), \dots, \sigma(\alpha_n)$. In particular show that an automorphism fixes K if and only if it fixes a set of generators for K.

(b) Let $G \leq \mathrm{Gal}(K/F)$ be a subgroup of the Galois group of the extension K/F and suppose $\sigma_1, \dots, \sigma_k$ are generators for G. Show that the subfield E/F is fixed by G if and only if it is fixed by the generators $\sigma_1, \dots, \sigma_k$.

2. Let τ be the map $\tau : \mathbb{C} \to \mathbb{C}$ defined by $\tau(a + bi) = a - bi$ *(complex conjugation)*. Prove that τ is an automorphism of $\mathbb{C}$.

3. Determine the fixed field of complex conjugation on $\mathbb{C}$.

4. Prove that $\mathbb{Q}(\sqrt{2})$ and $\mathbb{Q}(\sqrt{3})$ are not isomorphic.

5. Determine the automorphisms of the extension $\mathbb{Q}(\sqrt[4]{2})/\mathbb{Q}(\sqrt{2})$ explicitly.

6. Let k be a field.
   (a) Show that the mapping $\varphi : k[t] \to k[t]$ defined by $\varphi(f(t)) = f(at + b)$ for fixed $a, b \in k$, $a \neq 0$ is an automorphism of $k[t]$ which is the identity on k.
   (b) Conversely, let φ be an automorphism of $k[t]$ which is the identity on k. Prove that there exist $a, b \in k$ with $a \neq 0$ such that $\varphi(f(t)) = f(at + b)$ as in (a).

7. This exercise determines $\mathrm{Aut}(\mathbb{R}/\mathbb{Q})$.
   (a) Prove that any $\sigma \in \mathrm{Aut}(\mathbb{R}/\mathbb{Q})$ takes squares to squares and takes positive reals to positive reals. Conclude that $a < b$ implies $\sigma a < \sigma b$ for every $a, b \in \mathbb{R}$.
   (b) Prove that $-\dfrac{1}{m} < a - b < \dfrac{1}{m}$ implies $-\dfrac{1}{m} < \sigma a - \sigma b < \dfrac{1}{m}$ for every positive integer m. Conclude that σ is a continuous map on $\mathbb{R}$.
   (c) Prove that any continuous map on $\mathbb{R}$ which is the identity on $\mathbb{Q}$ is the identity map, hence $\mathrm{Aut}(\mathbb{R}/\mathbb{Q}) = 1$.

8. Prove that the automorphisms of the rational function field $k(t)$ which fix k are precisely the *fractional linear transformations* determined by $t \mapsto \dfrac{at + b}{ct + d}$ for $a, b, c, d \in k$, $ad - bc \neq 0$ (so $f(t) \in k(t)$ maps to $f(\dfrac{at + b}{ct + d})$) (cf. Exercise 18 of Section 13.2).

9. Determine the fixed field of the automorphism $t \mapsto t + 1$ of $k(t)$.

10. Let K be an extension of the field F. Let $\varphi : K \to K'$ be an isomorphism of K with a field K' which maps F to the subfield F' of K'. Prove that the map $\sigma \mapsto \varphi \sigma \varphi^{-1}$ defines a group isomorphism $\mathrm{Aut}(K/F) \xrightarrow{\sim} \mathrm{Aut}(K'/F')$.

14.2 THE FUNDAMENTAL THEOREM OF GALOIS THEORY

In the Galois extension $\mathrm{Gal}(\mathbb{Q}(\sqrt{2}, \sqrt{3})/\mathbb{Q})$ considered in the previous section, there was a strong similarity between the diagram of subgroups of the Galois group:

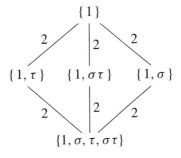

and the diagram of corresponding fixed fields

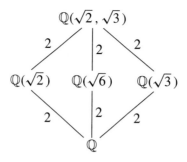

(we have inverted the lattice of subgroups because of the inclusion-reversing nature of the correspondence).

Note that this is also the diagram of *all* known subfields of the extension and that in this case each of the subfields is also a Galois extension of $\mathbb{Q}$.

In a similar way there is a strong similarity between the diagram

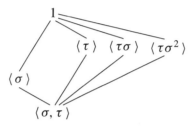

of subgroups of the Galois group and the diagram of known subfields for the splitting field of $x^3 - 2$:

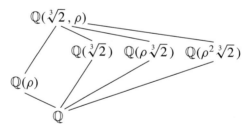

where the subfields in the second diagram are precisely the fixed fields of the subgroups in the first diagram.

Note in this pair of diagrams only the subgroup $\langle \sigma \rangle$ generated by σ is normal in S_3 and that the subfield $\mathbb{Q}(\rho)$ is the only subfield Galois over $\mathbb{Q}$.

The Fundamental Theorem of Galois Theory states that the relations observed in the two examples above are not coincidental and hold for any Galois extension. Before proving this we first develop some preliminary results on *group characters*, of which field automorphisms give particular examples.

Definition. A *character*[1] χ of a group G with values in a field L is a homomorphism from G to the multiplicative group of L:

$$\chi : G \to L^{\times}$$

i.e., $\chi(g_1 g_2) = \chi(g_1)\chi(g_2)$ for all $g_1, g_2 \in G$ and $\chi(g)$ is a nonzero element of L for all $g \in G$.

Definition. The characters $\chi_1, \chi_2, \ldots, \chi_n$ of G are said to be *linearly independent* over L if they are linearly independent as functions on G, i.e., if there is no nontrivial relation

$$a_1\chi_1 + a_2\chi_2 + \cdots + a_n\chi_n = 0 \qquad (a_1, \ldots, a_n \in L \text{ not all } 0) \qquad (14.2)$$

as a function on G (that is, $a_1\chi_1(g) + a_2\chi_2(g) + \cdots + a_n\chi_n(g) = 0$ for all $g \in G$).

Theorem 7. *(Linear Independence of Characters)* If $\chi_1, \chi_2, \ldots, \chi_n$ are distinct characters of G with values in L then they are linearly independent over L.

Proof: Suppose the characters were linearly dependent. Among all the linear dependence relations (2) above, choose one with the minimal number m of nonzero coefficients a_i. We may suppose (by renumbering, if necessary) that the m nonzero coefficients are $a_1, a_2, \ldots, a_m$:

$$a_1\chi_1 + a_2\chi_2 + \cdots + a_m\chi_m = 0.$$

Then for any $g \in G$ we have

$$a_1\chi_1(g) + a_2\chi_2(g) + \cdots + a_m\chi_m(g) = 0. \qquad (14.3)$$

Let g_0 be an element with $\chi_1(g_0) \neq \chi_m(g_0)$ (which exists, since $\chi_1 \neq \chi_m$). Since (3) holds for every element of G, in particular we have

$$a_1\chi_1(g_0 g) + a_2\chi_2(g_0 g) + \cdots + a_m\chi_m(g_0 g) = 0$$

i.e.,

$$a_1\chi_1(g_0)\chi_1(g) + a_2\chi_2(g_0)\chi_2(g) + \cdots + a_m\chi_m(g_0)\chi_m(g) = 0. \qquad (14.4)$$

Multiplying equation (3) by $\chi_m(g_0)$ and subtracting from equation (4) we obtain

$$[\chi_m(g_0) - \chi_1(g_0)]a_1\chi_1(g) + [\chi_m(g_0) - \chi_2(g_0)]a_2\chi_2(g) + \cdots$$
$$+ [\chi_m(g_0) - \chi_{m-1}(g_0)]a_{m-1}\chi_{m-1}(g) = 0,$$

which holds for all $g \in G$. But the first coefficient is nonzero and this is a relation with fewer nonzero coefficients, a contradiction.

Consider now an injective homomorphism σ of a field K into a field L, called an *embedding* of K into L. Then in particular σ is a homomorphism of the multiplicative group $G = K^{\times}$ into the multiplicative group $L^{\times}$, so σ may be viewed as a character of $K^{\times}$ with values in L. Note also that this character contains all of the useful information about the values of σ viewed simply as a *function* on K, since the only point of K not considered in $K^{\times}$ is 0, and we know σ maps 0 to 0.

---

[1]This is the definition of a *linear* character. More general characters will be studied in Chapter 18.

Corollary 8. If $\sigma_1, \sigma_2, \ldots, \sigma_n$ are distinct embeddings of a field K into a field L, then they are linearly independent as functions on K. In particular distinct automorphisms of a field K are linearly independent as functions on K.

We now use Corollary 8 to prove the fundamental relation between the orders of subgroups of the automorphism group of a field K and the degrees of the extensions over their fixed fields.

Theorem 9. Let $G = \{\sigma_1 = 1, \sigma_2, \ldots, \sigma_n\}$ be a subgroup of automorphisms of a field K and let F be the fixed field. Then

$$[K : F] = n = |G|.$$

Proof: Suppose first that $n > [K : F]$ and let $\omega_1, \omega_2, \ldots, \omega_m$ be a basis for K over F ($m = [K : F]$). Then the system

$$\sigma_1(\omega_1)x_1 + \sigma_2(\omega_1)x_2 + \cdots + \sigma_n(\omega_1)x_n = 0$$
$$\vdots$$
$$\sigma_1(\omega_m)x_1 + \sigma_2(\omega_m)x_2 + \cdots + \sigma_n(\omega_m)x_n = 0$$

of m equations in n unknowns $x_1, x_2, \ldots, x_n$ has a nontrivial solution $\beta_1, \beta_2, \ldots, \beta_n$ in K since by assumption there are more unknowns than equations.

Let $a_1, a_2, \ldots, a_m$ be m arbitrary elements of F. The field F is by definition fixed by $\sigma_1, \ldots, \sigma_n$ so each of these elements is fixed by every σ_i, i.e., $\sigma_i(a_j) = a_j$, $i = 1, 2, \ldots, n, j = 1, 2, \ldots, m$. Multiplying the first equation above by a_1, the second by $a_2, \ldots$, the last by a_m then gives the system of equations

$$\sigma_1(a_1\omega_1)\beta_1 + \sigma_2(a_1\omega_1)\beta_2 + \cdots + \sigma_n(a_1\omega_1)\beta_n = 0$$
$$\vdots$$
$$\sigma_1(a_m\omega_m)\beta_1 + \sigma_2(a_m\omega_m)\beta_2 + \cdots + \sigma_n(a_m\omega_m)\beta_n = 0.$$

Adding these equations we see that there are elements $\beta_1, \ldots, \beta_n$ in K, not all 0, satisfying

$$\sigma_1(a_1\omega_1 + a_2\omega_2 + \cdots + a_m\omega_m)\beta_1 + \cdots + \sigma_n(a_1\omega_1 + a_2\omega_2 + \cdots + a_m\omega_m)\beta_n = 0$$

for all choices of $a_1, \ldots, a_m$ in F. Since $\omega_1, \ldots, \omega_m$ is an F-basis for K, every $\alpha \in K$ is of the form $a_1\omega_1 + a_2\omega_2 + \cdots + a_m\omega_m$, so the previous equation means

$$\sigma_1(\alpha)\beta_1 + \cdots + \sigma_n(\alpha)\beta_n = 0$$

for all $\alpha \in K$. But this means the distinct automorphisms $\sigma_1, \ldots, \sigma_n$ are linearly dependent over K, contradicting Corollary 8.

We have proved $n \le [K : F]$. Note that we have so far not used the fact that $\sigma_1, \sigma_2, \ldots, \sigma_n$ are the elements of a *group*.

Suppose now that $n < [K : F]$. Then there are more than n F-linearly independent elements of K, say $\alpha_1, \ldots, \alpha_{n+1}$. The system

$$\sigma_1(\alpha_1)x_1 + \sigma_1(\alpha_2)x_2 + \cdots + \sigma_1(\alpha_{n+1})x_{n+1} = 0$$
$$\vdots \tag{14.5}$$
$$\sigma_n(\alpha_1)x_1 + \sigma_n(\alpha_2)x_2 + \cdots + \sigma_n(\alpha_{n+1})x_{n+1} = 0$$

of n equations in $n + 1$ unknowns $x_1, \ldots, x_{n+1}$ has a solution $\beta_1, \ldots, \beta_{n+1}$ in K where not all the $\beta_i, i = 1, 2, \ldots, n + 1$ are 0. If all the elements of the solution $\beta_1, \ldots, \beta_{n+1}$ were elements of F then the first equation (recall $\sigma_1 = 1$ is the identity automorphism) would contradict the linear independence over F of $\alpha_1, \alpha_2, \ldots, \alpha_{n+1}$. Hence at least one $\beta_i, i = 1, 2, \ldots, n + 1$, is not an element of F.

Among all the nontrivial solutions $(\beta_1, \ldots, \beta_{n+1})$ of the system (5) choose one with the minimal number r of nonzero β_i. By renumbering if necessary we may assume $\beta_1, \ldots, \beta_r$ are nonzero. Dividing the equations by β_r we may also assume $\beta_r = 1$. We have already seen that at least one of $\beta_1, \ldots, \beta_{r-1}, 1$ is not an element of F (which shows in particular that $r > 1$), say $\beta_1 \notin F$. Then our system of equations reads

$$\sigma_1(\alpha_1)\beta_1 + \cdots + \sigma_1(\alpha_{r-1})\beta_{r-1} + \sigma_1(\alpha_r) = 0$$
$$\vdots \tag{14.6}$$
$$\sigma_n(\alpha_1)\beta_1 + \cdots + \sigma_n(\alpha_{r-1})\beta_{r-1} + \sigma_n(\alpha_r) = 0$$

or more briefly

$$\sigma_i(\alpha_1)\beta_1 + \cdots + \sigma_i(\alpha_{r-1})\beta_{r-1} + \sigma_i(\alpha_r) = 0 \qquad i = 1, 2, \ldots, n. \tag{14.7}$$

Since $\beta_1 \notin F$, there is an automorphism σ_{k_0} ($k_0 \in \{1, 2, \ldots, n\}$) with $\sigma_{k_0}\beta_1 \neq \beta_1$. If we apply the automorphism σ_{k_0} to the equations in (6), we obtain the system of equations

$$\sigma_{k_0}\sigma_j(\alpha_1)\sigma_{k_0}(\beta_1) + \cdots + \sigma_{k_0}\sigma_j(\alpha_{r-1})\sigma_{k_0}(\beta_{r-1}) + \sigma_{k_0}\sigma_j(\alpha_r) = 0 \tag{14.8}$$

for $j = 1, 2, \ldots, n$. But the elements

$$\sigma_{k_0}\sigma_1, \quad \sigma_{k_0}\sigma_2, \quad \ldots, \quad \sigma_{k_0}\sigma_n$$

are the same as the elements

$$\sigma_1, \quad \sigma_2, \quad \ldots, \quad \sigma_n$$

in some order since these elements form a *group*. In other words, if we define the index i by $\sigma_{k_0}\sigma_j = \sigma_i$ then i and j both run over the set $\{1, 2, \ldots, n\}$. Hence the equations in (8) can be written

$$\sigma_i(\alpha_1)\sigma_{k_0}(\beta_1) + \cdots + \sigma_i(\alpha_{r-1})\sigma_{k_0}(\beta_{r-1}) + \sigma_i(\alpha_r) = 0. \tag{14.8'}$$

If we now subtract the equations in (8') from those in (7) we obtain the system

$$\sigma_i(\alpha_1)[\beta_1 - \sigma_{k_0}(\beta_1)] + \cdots + \sigma_i(\alpha_{r-1})[\beta_{r-1} - \sigma_{k_0}(\beta_{r-1})] = 0$$

for $i = 1, 2, \ldots, n$. But this is a solution to the system of equations (5) with

$$x_1 = \beta_1 - \sigma_{k_0}(\beta_1) \neq 0$$

(by the choice of k_0), hence is nontrivial and has fewer than r nonzero x_i. This is a contradiction and completes the proof.

Our first use of this result is to prove that the inequality of Proposition 5 holds for any finite extension K/F.

Corollary 10. Let K/F be any finite extension. Then

$$|\mathrm{Aut}(K/F)| \leq [K : F]$$

with equality if and only if F is the fixed field of $\mathrm{Aut}(K/F)$. Put another way, K/F is Galois if and only if F is the fixed field of $\mathrm{Aut}(K/F)$.

Proof: Let F_1 be the fixed field of $\mathrm{Aut}(K/F)$, so that

$$F \subseteq F_1 \subseteq K.$$

By Theorem 9, $[K : F_1] = |\mathrm{Aut}(K/F)|$. Hence $[K : F] = |\mathrm{Aut}(K/F)||F_1 : F]$, which proves the corollary.

Corollary 11. Let G be a finite subgroup of automorphisms of a field K and let F be the fixed field. Then every automorphism of K fixing F is contained in G, i.e., $\mathrm{Aut}(K/F) = G$, so that K/F is Galois, with Galois group G.

Proof: By definition F is fixed by all the elements of G so we have $G \leq \mathrm{Aut}(K/F)$ (and the question is whether there are any automorphisms of K fixing F not in G i.e., whether this containment is proper). Hence $|G| \leq |\mathrm{Aut}(K/F)|$. By the theorem we have $|G| = [K : F]$ and by the previous corollary $|\mathrm{Aut}(K/F)| \leq [K : F]$. This gives

$$[K : F] = |G| \leq |\mathrm{Aut}(K/F)| \leq [K : F]$$

and it follows that we must have equalities throughout, proving the corollary.

Corollary 12. If $G_1 \neq G_2$ are distinct finite subgroups of automorphisms of a field K then their fixed fields are also distinct.

Proof: Suppose F_1 is the fixed field of G_1 and F_2 is the fixed field of G_2. If $F_1 = F_2$ then by definition F_1 is fixed by G_2. By the previous corollary any automorphism fixing F_1 is contained in G_1, hence $G_2 \leq G_1$. Similarly $G_1 \leq G_2$ and so $G_1 = G_2$.

By the corollaries above we see that taking the fixed fields for distinct finite subgroups of $\mathrm{Aut}(K)$ gives distinct subfields of K over which K is Galois. Further, the degrees of the extensions are given by the orders of the subgroups. We saw this explicitly for the fields $K = \mathbb{Q}(\sqrt{2}, \sqrt{3})$ and $K = \mathbb{Q}(\sqrt[3]{2}, \rho)$ above. A portion of the Fundamental Theorem states that these are *all* the subfields of K.

The next result provides the converse of Proposition 5 and characterizes Galois extensions.

Theorem 13. The extension K/F is Galois if and only if K is the splitting field of some separable polynomial over F. Furthermore, if this is the case then every irreducible polynomial with coefficients in F which has a root in K is separable and has all its roots in K (so in particular K/F is a separable extension).

Proof: Proposition 5 proves that the splitting field of a separable polynomial is Galois.

We now show that if K/F is Galois then every irreducible polynomial $p(x)$ in $F[x]$ having a root in K splits completely in K. Set $G = \text{Gal}(K/F)$. Let $\alpha \in K$ be a root of $p(x)$ and consider the elements

$$\alpha, \sigma_2(\alpha), \ldots, \sigma_n(\alpha) \in K \tag{14.9}$$

where $\{1, \sigma_2, \ldots, \sigma_n\}$ are the elements of $\text{Gal}(K/F)$. Let

$$\alpha, \alpha_2, \alpha_3, \ldots, \alpha_r$$

denote the *distinct* elements in (9). If $\tau \in G$ then since G is a group the elements $\{\tau, \tau\sigma_2, \ldots, \tau\sigma_n\}$ are the same as the elements $\{1, \sigma_2, \ldots, \sigma_n\}$ in some order. It follows that applying $\tau \in G$ to the elements in (9) simply permutes them, so in particular applying τ to $\alpha, \alpha_2, \alpha_3, \ldots, \alpha_r$ also permutes these elements. The polynomial

$$f(x) = (x - \alpha)(x - \alpha_2) \cdots (x - \alpha_r)$$

therefore has coefficients which are fixed by all the elements of G since the elements of G simply permute the factors. Hence the coefficients lie in the fixed field of G, which by Corollary 10 is the field F. Hence $f(x) \in F[x]$.

Since $p(x)$ is irreducible and has α as a root, $p(x)$ is the minimal polynomial for α over F, hence divides any polynomial with coefficients in F having α as a root (this is Proposition 13.9). It follows that $p(x)$ divides $f(x)$ in $F[x]$ and since $f(x)$ obviously divides $p(x)$ in $K[x]$ by Proposition 2, we have

$$p(x) = f(x).$$

In particular, this shows that $p(x)$ is separable and that all its roots lie in K (in fact they are among the elements $\alpha, \sigma_2\alpha, \ldots, \sigma_n\alpha$), proving the last statement of the theorem.

To complete the proof, suppose K/F is Galois and let $\omega_1, \omega_2, \ldots, \omega_n$ be a basis for K/F. Let $p_i(x)$ be the minimal polynomial for ω_i over F, $i = 1, 2, \ldots, n$. Then by what we have just proved, $p_i(x)$ is separable and has all its roots in K. Let $g(x)$ be the polynomial obtained by removing any multiple factors in the product $p_1(x) \cdots p_n(x)$ (the "squarefree part"). Then the splitting field of the two polynomials is the same and this field is K (all the roots lie in K, so K contains the splitting field, but $\omega_1, \omega_2, \ldots, \omega_n$ are among the roots, so the splitting field contains K). Hence K is the splitting field of the separable polynomial $g(x)$.

Definition. Let K/F be a Galois extension. If $\alpha \in K$ the elements $\sigma\alpha$ for σ in $\text{Gal}(K/F)$ are called the *conjugates* (or *Galois conjugates*) of α over F. If E is a subfield of K containing F, the field $\sigma(E)$ is called the *conjugate field* of E over F.

The proof of the theorem shows that in a Galois extension K/F the other roots of the minimal polynomial over F of any element $\alpha \in K$ are precisely the distinct conjugates of α under the Galois group of K/F.

The second statement in this theorem also shows that K is not Galois over F if we can find even one irreducible polynomial over F having a root in K but not having *all* its roots in K. This justifies in a very strong sense the intuition from earlier examples that Galois extensions are extensions with "enough" distinct roots of irreducible polynomials (namely, if it contains one root then it contains all the roots).

Finally, notice that we now have 4 characterizations of Galois extensions K/F:

(1) splitting fields of separable polynomials over F

(2) fields where F is precisely the set of elements fixed by $\text{Aut}(K/F)$ (in general, the fixed field may be larger than F)

(3) fields with $[K : F] = |\text{Aut}(K/F)|$ (the original definition)

(4) finite, normal and separable extensions.

Theorem 14. *(Fundamental Theorem of Galois Theory)* Let K/F be a Galois extension and set $G = \text{Gal}(K/F)$. Then there is a bijection

$$
\left\{ \begin{matrix} \text{subfields } E \\ \text{of } K \\ \text{containing } F \end{matrix} \quad \begin{matrix} K \\ | \\ E \\ | \\ F \end{matrix} \right\} \quad \longleftrightarrow \quad \left\{ \begin{matrix} \text{subgroups } H \\ \text{of } G \end{matrix} \quad \begin{matrix} 1 \\ | \\ H \\ | \\ G \end{matrix} \right\}
$$

given by the correspondences

$$
E \quad \longrightarrow \quad \left\{ \begin{matrix} \text{the elements of } G \\ \text{fixing } E \end{matrix} \right\}
$$

$$
\left\{ \begin{matrix} \text{the fixed field} \\ \text{of } H \end{matrix} \right\} \quad \longleftarrow \quad H
$$

which are inverse to each other. Under this correspondence,

(1) (inclusion reversing) If E_1, E_2 correspond to H_1, H_2, respectively, then $E_1 \subseteq E_2$ if and only if $H_2 \le H_1$

(2) $[K : E] = |H|$ and $[E : F] = |G : H|$, the index of H in G:

$$
\begin{matrix} K \\ | \quad \} \quad |H| \\ E \\ | \quad \} \quad |G : H| \\ F \end{matrix}
$$

(3) K/E is always Galois, with Galois group $\text{Gal}(K/E) = H$:

$$
\begin{matrix} K \\ | \quad H \\ E \end{matrix}
$$

(4) E is Galois over F if and only if H is a normal subgroup in G. If this is the case, then the Galois group is isomorphic to the quotient group

$$
\text{Gal}(E/F) \cong G/H.
$$

More generally, even if H is not necessarily normal in G, the isomorphisms of E (into a fixed algebraic closure of F containing K) which fix F are in one to one correspondence with the cosets $\{\sigma H\}$ of H in G.

(5) If E_1, E_2 correspond to H_1, H_2, respectively, then the intersection $E_1 \cap E_2$ corresponds to the group $\langle H_1, H_2 \rangle$ generated by H_1 and H_2 and the composite field $E_1 E_2$ corresponds to the intersection $H_1 \cap H_2$. Hence the lattice of subfields

574

of K containing F and the lattice of subgroups of G are "dual" (the lattice diagram for one is the lattice diagram for the other turned upside down).

Proof: Given any subgroup H of G we obtain a unique fixed field $E = K_H$ by Corollary 12. This shows that the correspondence above is injective from right to left.

If K is the splitting field of the separable polynomial $f(x) \in F[x]$ then we may also view $f(x)$ as an element of $E[x]$ for any subfield E of K containing F. Then K is also the splitting field of $f(x)$ over E, so the extension K/E is Galois. By Corollary 10, E is the fixed field of $\mathrm{Aut}(K/E) \le G$, showing that *every* subfield of K containing F arises as the fixed field for some subgroup of G. Hence the correspondence above is surjective from right to left, hence a bijection. The correspondences are inverse to each other since the automorphisms fixing E are precisely $\mathrm{Aut}(K/E)$ by Corollary 10.

We have already seen that the Galois correspondence is inclusion reversing in Proposition 4, which gives (1).

If $E = K_H$ is the fixed field of H, then Theorem 9 gives $[K : E] = |H|$ and $[K : F] = |G|$. Taking the quotient gives $[E : F] = |G : H|$, which proves (2).

Corollary 11 gives (3) immediately.

Suppose $E = K_H$ is the fixed field of the subgroup H. Every $\sigma \in G = \mathrm{Gal}(K/F)$ when restricted to E is an embedding $\sigma|_E$ of E with the subfield $\sigma(E)$ of K. Conversely, let $\tau : E \xrightarrow{\sim} \tau(E) \subseteq \overline{F}$ be any embedding of E (into a fixed algebraic closure $\overline{F}$ of F containing K) which fixes F. Then $\tau(E)$ is in fact contained in K: if $\alpha \in E$ has minimal polynomial $m_\alpha(x)$ over F then $\tau(\alpha)$ is another root of $m_\alpha(x)$ and K contains all these roots by Theorem 13. As above K is the splitting field of $f(x)$ over E and so also the splitting field of $\tau f(x)$ (which is the same as $f(x)$ since $f(x)$ has coefficients in F) over $\tau(E)$. Theorem 13.27 on extending isomorphisms then shows that we can extend τ to an isomorphism σ:

$$
\begin{array}{ccc}
\sigma : & K \xrightarrow{\ \sim\ } & K \\
& | & | \\
\tau : & E \xrightarrow{\ \sim\ } & \tau(E).
\end{array}
$$

Since σ fixes F (because τ does), it follows that *every* embedding τ of E fixing F is the restriction to E of some automorphism σ of K fixing F, in other words, every embedding of E is of the form $\sigma|_E$ for some $\sigma \in G$.

Two automorphisms $\sigma, \sigma' \in G$ restrict to the *same* embedding of E if and only if $\sigma^{-1}\sigma'$ is the identity map on E. But then $\sigma^{-1}\sigma' \in H$ (i.e., $\sigma' \in \sigma H$) since by (3) the automorphisms of K which fix E are precisely the elements in H. Hence the distinct embeddings of E are in bijection with the cosets σH of H in G. In particular this gives

$$|\mathrm{Emb}(E/F)| = [G : H] = [E : F]$$

where $\mathrm{Emb}(E/F)$ denotes the set of embeddings of E (into a fixed algebraic closure of F) which fix F. Note that $\mathrm{Emb}(E/F)$ contains the automorphisms $\mathrm{Aut}(E/F)$.

The extension E/F will be Galois if and only if $|\mathrm{Aut}(E/F)| = [E : F]$. By the equality above, this will be the case if and only if each of the *embeddings* of E is actually an *automorphism* of E, i.e., if and only if $\sigma(E) = E$ for every $\sigma \in G$.

If $\sigma \in G$, then the subgroup of G fixing the field $\sigma(E)$ is the group $\sigma H \sigma^{-1}$, i.e.,

$$\sigma(E) = K_{\sigma H \sigma^{-1}}.$$

To see this observe that if $\sigma\alpha \in \sigma(E)$ then

$$(\sigma h \sigma^{-1})(\sigma\alpha) = \sigma(h\alpha) = \sigma\alpha \qquad \text{for all } h \in H,$$

since h fixes $\alpha \in E$, which shows that $\sigma H \sigma^{-1}$ fixes $\sigma(E)$. The group fixing $\sigma(E)$ has order equal to the degree of K over $\sigma(E)$. But this is the same as the degree of K over E since the fields are isomorphic, hence the same as the order of H. Hence $\sigma H \sigma^{-1}$ is precisely the group fixing $\sigma(E)$ since we have shown containment and their orders are the same.

Because of the bijective nature of the Galois correspondence already proved we know that two subfields of K containing F are equal if and only if their fixing subgroups are equal in G. Hence $\sigma(E) = E$ for all $\sigma \in G$ if and only if $\sigma H \sigma^{-1} = H$ for all $\sigma \in G$, in other words E is Galois over F if and only if H is a normal subgroup of G.

We have already identified the embeddings of E over F as the set of cosets of H in G and when H is normal in G seen that the embeddings are automorphisms. It follows that in this case the *group* of cosets G/H is identified with the *group* of automorphisms of the Galois extension E/F by the definition of the group operation (composition of automorphisms). Hence $G/H \cong \text{Gal}(E/F)$ when H is normal in G, which completes the proof of (4).

Suppose H_1 is the subgroup of elements of G fixing the subfield E_1 and H_2 is the subgroup of elements of G fixing the subfield E_2. Any element in $H_1 \cap H_2$ fixes both E_1 and E_2, hence fixes every element in the composite $E_1 E_2$, since the elements in this field are algebraic combinations of the elements of E_1 and E_2. Conversely, if an automorphism σ fixes the composite $E_1 E_2$ then in particular σ fixes E_1, i.e., $\sigma \in H_1$, and σ fixes E_2, i.e., $\sigma \in H_2$, hence $\sigma \in H_1 \cap H_2$. This proves that the composite $E_1 E_2$ corresponds to the intersection $H_1 \cap H_2$. Similarly, the intersection $E_1 \cap E_2$ corresponds to the group $\langle H_1, H_2 \rangle$ generated by H_1 and H_2, completing the proof of the theorem.

Example: $(\mathbb{Q}(\sqrt{2}, \sqrt{3})$ and $\mathbb{Q}(\sqrt[3]{2}, \rho))$

We have already seen examples of this theorem at the beginning of this section. We now see that the diagrams of subfields for the two fields $\mathbb{Q}(\sqrt{2}, \sqrt{3})$ and $\mathbb{Q}(\sqrt[3]{2}, \rho)$ given before indicate *all* the subfields for these two fields.

Since every subgroup of the Klein 4-group is normal, all the subfields of $\mathbb{Q}(\sqrt{2}, \sqrt{3})$ are Galois extensions of $\mathbb{Q}$.

Similarly, since the only nontrivial normal subgroup of S_3 is the subgroup of order 3, we see that only the subfield $\mathbb{Q}(\rho)$ of $K = \mathbb{Q}(\sqrt[3]{2}, \rho)$ is Galois over $\mathbb{Q}$, with Galois group isomorphic to $S_3/\langle \sigma \rangle$, i.e., the cyclic group of order 2. For example, the nontrivial automorphism of $\mathbb{Q}(\rho)$ is induced by restricting any element (τ, for instance) in the nontrivial coset of $\langle \sigma \rangle$ to $\mathbb{Q}(\rho)$. This is clear from the explicit descriptions of these automorphisms given before — each of the elements $\tau, \tau\sigma, \tau\sigma^2$ in this coset map ρ to ρ^2. The restrictions of the elements of $\text{Gal}(K/\mathbb{Q})$ to the (non-Galois) cubic subfields do not give automorphisms of these fields in general, rather giving isomorphisms of these fields with each other, in accordance with (4) of the theorem.

Example: $(\mathbb{Q}(\sqrt{2} + \sqrt{3}))$

Consider the field $\mathbb{Q}(\sqrt{2} + \sqrt{3})$. This is clearly a subfield of the Galois extension $\mathbb{Q}(\sqrt{2}, \sqrt{3})$. The other roots of the minimal polynomial for $\sqrt{2} + \sqrt{3}$ over $\mathbb{Q}$ are therefore the distinct conjugates of $\sqrt{2} + \sqrt{3}$ under the Galois group. The conjugates are

$$\pm\sqrt{2} \pm \sqrt{3}$$

which are easily seen to be distinct. The minimal polynomial is therefore

$$[x - (\sqrt{2} + \sqrt{3})][x - (\sqrt{2} - \sqrt{3})][x - (-\sqrt{2} + \sqrt{3})][x - (-\sqrt{2} - \sqrt{3})]$$

which is quickly computed to be the polynomial $x^4 - 10x^2 + 1$. It follows that this polynomial is irreducible and that

$$\mathbb{Q}(\sqrt{2}, \sqrt{3}) = \mathbb{Q}(\sqrt{2} + \sqrt{3}),$$

either by degree considerations or by noting that only the automorphism 1 of $\{1, \sigma, \tau, \sigma\tau\}$ fixes $\sqrt{2} + \sqrt{3}$ so the fixing group for this field is the same as for $\mathbb{Q}(\sqrt{2}, \sqrt{3})$.

Example: (Splitting Field of $x^8 - 2$)

The splitting field of $x^8 - 2$ over $\mathbb{Q}$ is generated by $\theta = \sqrt[8]{2}$ (any fixed 8th root of 2, say the real one) and a primitive 8th root of unity $\zeta = \zeta_8$. Recall from Section 13.6 that

$$\mathbb{Q}(\zeta_8) = \mathbb{Q}(i, \sqrt{2}).$$

Since $\theta^4 = \sqrt{2}$ we see that the splitting field is generated by θ and i. The subfield $\mathbb{Q}(\theta)$ is of degree 8 over $\mathbb{Q}$ (since $x^8 - 2$ is irreducible, being Eisenstein), and all the elements of this field are real. Hence $i \notin \mathbb{Q}(\theta)$ and since i generates at most a quadratic extension of this field, the splitting field

$$\mathbb{Q}(\sqrt[8]{2}, \zeta_8) = \mathbb{Q}(\sqrt[8]{2}, i)$$

is of degree 16 over $\mathbb{Q}$.

The Galois group is determined by the action on the generators θ and i which gives the possibilities

$$\begin{cases} \theta \mapsto \zeta^a\theta & a = 0, 1, 2, \ldots, 7 \\ i \mapsto \pm i \end{cases}$$

Since we have already seen that the degree of the extension is 16 and there are only 16 possible such maps, it follows that in fact each of the maps above is an automorphism of $\mathbb{Q}(\sqrt[8]{2}, i)$ over $\mathbb{Q}$.

Define the two automorphisms

$$\sigma : \begin{cases} \theta \mapsto \zeta\theta \\ i \mapsto i \end{cases} \qquad \tau : \begin{cases} \theta \mapsto \theta \\ i \mapsto -i \end{cases}$$

(τ is the map induced by complex conjugation). Since

$$\zeta = \zeta_8 = \frac{\sqrt{2}}{2} + i\frac{\sqrt{2}}{2} = \frac{1}{2}(1 + i)\sqrt{2}$$
$$= \frac{1}{2}(1 + i)\theta^4$$

we can easily compute what happens to ζ from the explicit expressions for the powers of ζ in the following Figure 1.

Using these explicit values we find

$$\sigma : \begin{cases} \theta \mapsto \zeta\theta \\ i \mapsto i \\ \zeta \mapsto -\zeta = \zeta^5 \end{cases} \qquad \tau : \begin{cases} \theta \mapsto \theta \\ i \mapsto -i \\ \zeta \mapsto \zeta^7 \end{cases}$$

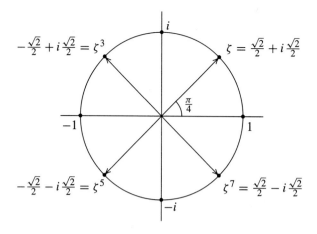

$$-\frac{\sqrt{2}}{2} + i\frac{\sqrt{2}}{2} = \zeta^3 \qquad \zeta = \frac{\sqrt{2}}{2} + i\frac{\sqrt{2}}{2}$$

$$-\frac{\sqrt{2}}{2} - i\frac{\sqrt{2}}{2} = \zeta^5 \qquad \zeta^7 = \frac{\sqrt{2}}{2} - i\frac{\sqrt{2}}{2}$$

Fig. 1

Note that the reason we are interested in also keeping track of the action on the element ζ is that it will be needed in computing the composites of automorphisms, for example in computing

$$\sigma^2(\theta) = \sigma(\zeta\theta) = \sigma(\zeta)\sigma(\theta) = (-\zeta)(\zeta\theta) = -\zeta^2\theta$$
$$= -i\theta.$$

We can similarly compute the following automorphisms:

$$\sigma : \begin{cases} \theta \mapsto \zeta\theta \\ i \mapsto i \\ \zeta \mapsto \zeta^5 \end{cases} \qquad \tau\sigma : \begin{cases} \theta \mapsto \zeta^7\theta \\ i \mapsto -i \\ \zeta \mapsto \zeta^3 \end{cases}$$

$$\sigma^2 : \begin{cases} \theta \mapsto \zeta^6\theta \\ i \mapsto i \\ \zeta \mapsto \zeta \end{cases} \qquad \tau\sigma^2 : \begin{cases} \theta \mapsto \zeta^2\theta \\ i \mapsto -i \\ \zeta \mapsto \zeta^7 \end{cases}$$

$$\sigma^3 : \begin{cases} \theta \mapsto \zeta^7\theta \\ i \mapsto i \\ \zeta \mapsto -\zeta \end{cases} \qquad \tau\sigma^3 : \begin{cases} \theta \mapsto \zeta\theta \\ i \mapsto -i \\ \zeta \mapsto \zeta^3 \end{cases}$$

$$\sigma^4 : \begin{cases} \theta \mapsto -\theta \\ i \mapsto i \\ \zeta \mapsto \zeta \end{cases} \qquad \tau\sigma^4 : \begin{cases} \theta \mapsto -\theta \\ i \mapsto -i \\ \zeta \mapsto \zeta^7 \end{cases}$$

$$\sigma^5 : \begin{cases} \theta \mapsto \zeta^5\theta \\ i \mapsto i \\ \zeta \mapsto -\zeta \end{cases} \qquad \tau\sigma^5 : \begin{cases} \theta \mapsto \zeta^3\theta \\ i \mapsto -i \\ \zeta \mapsto \zeta^3 \end{cases}$$

$$\sigma^6 : \begin{cases} \theta \mapsto \zeta^2\theta \\ i \mapsto i \\ \zeta \mapsto \zeta \end{cases} \qquad \tau\sigma^6 : \begin{cases} \theta \mapsto \zeta^6\theta \\ i \mapsto -i \\ \zeta \mapsto \zeta^7 \end{cases}$$

$$\sigma^7 : \begin{cases} \theta \mapsto \zeta^3\theta \\ i \mapsto i \\ \zeta \mapsto -\zeta \end{cases} \qquad \tau\sigma^7 : \begin{cases} \theta \mapsto \zeta^5\theta \\ i \mapsto -i \\ \zeta \mapsto \zeta^3. \end{cases}$$

Since this exhausts the possibilities, these elements (together with 1 and τ) are the Galois group. We see in particular that σ and τ generate the Galois group. To determine the relations satisfied by these elements, we observe first that clearly $\tau^2 = 1$ and $(\sigma^4)^2 = 1$, so that

$$\sigma^8 = \tau^2 = 1.$$

Also, we compute

$$\sigma\tau : \begin{cases} \theta \mapsto \zeta\theta \\ i \mapsto -i \\ \zeta \mapsto \zeta^3 \end{cases}$$

so that

$$\sigma\tau = \tau\sigma^3.$$

It is not too difficult to show that these relations define the group completely, i.e.,

$$\mathrm{Gal}(\mathbb{Q}(\sqrt[8]{2}, i)/\mathbb{Q}) = \langle \, \sigma, \tau \mid \sigma^8 = \tau^2 = 1, \sigma\tau = \tau\sigma^3 \, \rangle.$$

Such a group is called a *quasidihedral group* (recall that the dihedral group of order 16 would have the relation $\sigma\tau = \tau\sigma^7$ instead of $\sigma\tau = \tau\sigma^3$) and is a subgroup of S_8 since the Galois group is a subgroup of the permutations of the 8 roots of $x^8 - 2$.

This example again illustrates that one must take care in determining Galois groups from the actions on generators. We first computed the degree of the Galois extension above to determine the number of elements in the Galois group. Had we proceeded directly from the original generators $\theta = \sqrt[8]{2}$ and $\zeta = \zeta_8$ we might have (incorrectly) concluded that there were a total of 32 elements in the Galois group, since the first generator is mapped to any of 8 possible roots of $x^8 - 2$ and the second generator is mapped to any of 4 possible roots of its minimal polynomial $\Phi_4(x) = x^4 + 1$. The problem, as previously indicated, is that these choices are not independent. Here the reason is provided by the algebraic relation

$$\theta^4 = \sqrt{2} = \zeta + \zeta^7$$

which shows that one cannot specify the images of θ and ζ independently — their images must again satisfy this algebraic relation. This relation is perhaps sufficiently subtle to serve as a caution against rashly concluding maps are automorphisms. We note that in general it is necessary to provide justification that maps are automorphisms. This can be accomplished for example by using the extension theorems or by using degree considerations as we did here.

Determining the lattice of subgroups of this group G is a straightforward problem.

The lattice is the following:

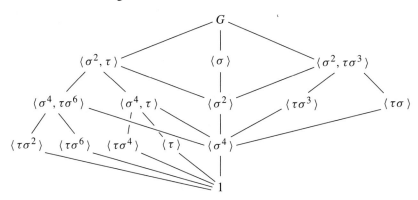

Determining the subfields corresponding to these subgroups (which by the Fundamental Theorem gives *all* the subfields of $\mathbb{Q}(\sqrt[8]{2}, i)$) is quite simple for a number of the subgroups above using (2) of the Fundamental Theorem, which states that the degree of the extension over $\mathbb{Q}$ is equal to the *index* of the fixing subgroup. It then suffices to find a subfield of the right degree which is fixed by the subgroup in question. Remember also that if a subfield is fixed by the *generators* of a subgroup, then it is fixed by the subgroup. For example, from the explicit description for the automorphism σ we see that $\mathbb{Q}(i)$ is fixed by the group generated by σ. Since this is a subgroup of index 2 and $\mathbb{Q}(i)$ is of degree 2 over $\mathbb{Q}$, it must be the full fixed field. Most of the fixed fields for the subgroups above can be determined in as simple a manner.

For the subgroups of order 4 on the right (namely, generated by $\tau\sigma^3$ and by $\tau\sigma$), it is perhaps not so easy to see how to determine the corresponding fixed field. For the subgroup H generated by $\tau\sigma^3$ we may proceed as follows: the element $\theta^2 = \sqrt[4]{2}$ is clearly fixed by σ^4. By the diagram above, σ^4 is a normal subgroup of H of index 2, with representatives $1, \tau\sigma^3$ for the cosets. Consider the element

$$\alpha = (1 + \tau\sigma^3)\theta^2 = \theta^2 + \tau\sigma^3\theta^2.$$

Then α is fixed by σ^4 (we are in a commutative group H of order 4, so σ^4 commutes with 1 and $\tau\sigma^3$ and we already know θ^2 is fixed by σ^4). But (and this is the point), α is also fixed by $\tau\sigma^3$:

$$\tau\sigma^3\alpha = \tau\sigma^3(1 + \tau\sigma^3)\theta^2 = [\tau\sigma^3 + (\tau\sigma^3)^2]\theta^2$$
$$= (\tau\sigma^3 + \sigma^4)\theta^2$$

and the last expression is just α since $\sigma^4\theta^2 = \theta^2$. Hence α is an element of the fixed field for H. Explicitly

$$\alpha = \sqrt[4]{2} + i\sqrt[4]{2} = (1 + i)\sqrt[4]{2}.$$

A quick check shows that α is not fixed by the automorphism σ^2, so by the diagram of subgroups above, it follows that the fixing subgroup for the field $\mathbb{Q}(\alpha)$ is no larger than H, hence is precisely H, which gives us our fixed field. This also gives the fixed field for $\langle \tau\sigma \rangle$ by recalling that in general if E is the fixed field of H then the fixed field of $\tau H\tau^{-1}$ is the field $\tau(E)$. For $H = \langle \tau\sigma^3 \rangle$, $\tau H\tau^{-1} = \langle \tau\sigma \rangle$, with fixed field given by $\tau(\alpha) = (1 - i)\sqrt[4]{2}$.

In general one tries to determine elements which are fixed by a given subgroup H of the Galois group (cf. the exercises, which indicate where the element above arose) and attempts to generate a sufficiently large field to give the full fixed field. In our case we were able to accomplish this with a single generator. We shall see later that every finite extension of $\mathbb{Q}$ is a simple extension, so there will be a single generator of this type, but in general it may be difficult to produce it directly.

The element α is a root of the polynomial

$$x^4 + 8$$

which must therefore be irreducible since we have already determined that a root of this polynomial generates an extension of degree 4 over $\mathbb{Q}$.

In a similar way it is possible to complete the diagram of subfields of $\mathbb{Q}(\sqrt[8]{2}, i)$, which we have inverted to emphasize its relation with the subgroup diagram above ($\theta = \sqrt[8]{2}$):

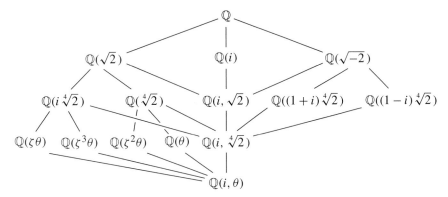

Note that the group $\langle \sigma^4 \rangle$ is normal in G (in fact it is the center of G) with quotient $G/\langle \sigma^4 \rangle \cong D_8$, so the corresponding fixed field $\mathbb{Q}(i, \sqrt[4]{2})$ is Galois over $\mathbb{Q}$ with D_8 as Galois group. Being Galois it is a splitting field, evidently the splitting field for $x^4 - 2$. The lattice of subfields for this field is then immediate from the lattice above.

We end this example with the following amusing aspect of this Galois extension. It is an easy exercise to verify that

$$\langle \sigma^2, \tau \rangle \cong D_8 \quad \langle \sigma \rangle \cong \mathbb{Z}/8\mathbb{Z} \quad \langle \sigma^2, \tau\sigma^3 \rangle \cong Q_8$$

where D_8 is the dihedral group of order 8 and Q_8 is the quaternion group of order 8. It follows that the field $\mathbb{Q}(\sqrt[8]{2}, i)$ is Galois of degree 8 over its three quadratic subfields

$$\mathbb{Q}(\sqrt{2}) \qquad \mathbb{Q}(i) \qquad \mathbb{Q}(\sqrt{-2})$$

with dihedral, cyclic and quaternion Galois groups, respectively, so that three of the 5 possible groups of order 8 (and both non-abelian ones) appear as Galois groups in this extension.

We shall consider additional examples and applications in the following sections.

EXERCISES

1. Determine the minimal polynomial over $\mathbb{Q}$ for the element $\sqrt{2} + \sqrt{5}$.

2. Determine the minimal polynomial over $\mathbb{Q}$ for the element $1 + \sqrt[3]{2} + \sqrt[3]{4}$.

3. Determine the Galois group of $(x^2 - 2)(x^2 - 3)(x^2 - 5)$. Determine *all* the subfields of the splitting field of this polynomial.

4. Let p be a prime. Determine the elements of the Galois group of $x^p - 2$.

5. Prove that the Galois group of $x^p - 2$ for p a prime is isomorphic to the group of matrices $\begin{pmatrix} a & b \\ 0 & 1 \end{pmatrix}$ where $a, b \in \mathbb{F}_p$, $a \neq 0$.

6. Let $K = \mathbb{Q}(\sqrt[8]{2}, i)$ and let $F_1 = \mathbb{Q}(i)$, $F_2 = \mathbb{Q}(\sqrt{2})$, $F_3 = \mathbb{Q}(\sqrt{-2})$. Prove that $\mathrm{Gal}(K/F_1) \cong Z_8$, $\mathrm{Gal}(K/F_2) \cong D_8$, $\mathrm{Gal}(K/F_3) \cong Q_8$.

7. Determine all the subfields of the splitting field of $x^8 - 2$ which are Galois over $\mathbb{Q}$.

8. Suppose K is a Galois extension of F of degree p^n for some prime p and some $n \geq 1$. Show there are Galois extensions of F contained in K of degrees p and p^{n-1}.

9. Give an example of fields F_1, F_2, F_3 with $\mathbb{Q} \subset F_1 \subset F_2 \subset F_3$, $[F_3 : \mathbb{Q}] = 8$ and each field is Galois over all its subfields with the exception that F_2 is not Galois over $\mathbb{Q}$.

10. Determine the Galois group of the splitting field over $\mathbb{Q}$ of $x^8 - 3$.

11. Suppose $f(x) \in \mathbb{Z}[x]$ is an irreducible quartic whose splitting field has Galois group S_4 over $\mathbb{Q}$ (there are many such quartics, cf. Section 6). Let θ be a root of $f(x)$ and set $K = \mathbb{Q}(\theta)$. Prove that K is an extension of $\mathbb{Q}$ of degree 4 which has no proper subfields. Are there any Galois extensions of $\mathbb{Q}$ of degree 4 with no proper subfields?

12. Determine the Galois group of the splitting field over $\mathbb{Q}$ of $x^4 - 14x^2 + 9$.

13. Prove that if the Galois group of the splitting field of a cubic over $\mathbb{Q}$ is the cyclic group of order 3 then all the roots of the cubic are real.

14. Show that $\mathbb{Q}(\sqrt{2 + \sqrt{2}})$ is a cyclic quartic field, i.e., is a Galois extension of degree 4 with cyclic Galois group.

15. (*Biquadratic Extensions*) Let F be a field of characteristic $\neq 2$.
   (a) If $K = F(\sqrt{D_1}, \sqrt{D_2})$ where $D_1, D_2 \in F$ have the property that none of D_1, D_2 or $D_1 D_2$ is a square in F, prove that K/F is a Galois extension with $\mathrm{Gal}(K/F)$ isomorphic to the Klein 4-group.
   (b) Conversely, suppose K/F is a Galois extension with $\mathrm{Gal}(K/F)$ isomorphic to the Klein 4-group. Prove that $K = F(\sqrt{D_1}, \sqrt{D_2})$ where $D_1, D_2 \in F$ have the property that none of D_1, D_2 or $D_1 D_2$ is a square in F.

16. (a) Prove that $x^4 - 2x^2 - 2$ is irreducible over $\mathbb{Q}$.
   (b) Show the roots of this quartic are
$$\alpha_1 = \sqrt{1 + \sqrt{3}} \qquad \alpha_3 = -\sqrt{1 + \sqrt{3}}$$
$$\alpha_2 = \sqrt{1 - \sqrt{3}} \qquad \alpha_4 = -\sqrt{1 - \sqrt{3}}.$$
   (c) Let $K_1 = \mathbb{Q}(\alpha_1)$ and $K_2 = \mathbb{Q}(\alpha_2)$. Show that $K_1 \neq K_2$, and $K_1 \cap K_2 = \mathbb{Q}(\sqrt{3}) = F$.
   (d) Prove that K_1, K_2 and $K_1 K_2$ are Galois over F with $\mathrm{Gal}(K_1 K_2/F)$ the Klein 4-group. Write out the elements of $\mathrm{Gal}(K_1 K_2/F)$ explicitly. Determine all the subgroups of the Galois group and give their corresponding fixed subfields of $K_1 K_2$ containing F.
   (e) Prove that the splitting field of $x^4 - 2x^2 - 2$ over $\mathbb{Q}$ is of degree 8 with dihedral Galois group.

The following two exercises indicate one method for constructing elements in subfields of a given field and are quite useful in many computations.

17. Let K/F be any finite extension and let $\alpha \in K$. Let L be a Galois extension of F containing K and let $H \leq \mathrm{Gal}(L/F)$ be the subgroup corresponding to K. Define the *norm* of α from

K to F to be

$$N_{K/F}(\alpha) = \prod_{\sigma} \sigma(\alpha),$$

where the product is taken over all the embeddings of K into an algebraic closure of F (so over a set of coset representatives for H in $\mathrm{Gal}(L/F)$ by the Fundamental Theorem of Galois Theory). This is a product of Galois conjugates of α. In particular, if K/F is Galois this is $\prod_{\sigma \in \mathrm{Gal}(K/F)} \sigma(\alpha)$.

(a) Prove that $N_{K/F}(\alpha) \in F$.

(b) Prove that $N_{K/F}(\alpha\beta) = N_{K/F}(\alpha)N_{K/F}(\beta)$, so that the norm is a multiplicative map from K to F.

(c) Let $K = F(\sqrt{D})$ be a quadratic extension of F. Show that $N_{K/F}(a + b\sqrt{D}) = a^2 - Db^2$.

(d) Let $m_\alpha(x) = x^d + a_{d-1}x^{d-1} + \cdots + a_1 x + a_0 \in F[x]$ be the minimal polynomial for $\alpha \in K$ over F. Let $n = [K : F]$. Prove that d divides n, that there are d distinct Galois conjugates of α which are all repeated n/d times in the product above and conclude that $N_{K/F}(\alpha) = (-1)^n a_0^{n/d}$.

18. With notation as in the previous problem, define the *trace* of α from K to F to be

$$\mathrm{Tr}_{K/F}(\alpha) = \sum_{\sigma} \sigma(\alpha),$$

a sum of Galois conjugates of α.

(a) Prove that $\mathrm{Tr}_{K/F}(\alpha) \in F$.

(b) Prove that $\mathrm{Tr}_{K/F}(\alpha + \beta) = \mathrm{Tr}_{K/F}(\alpha) + \mathrm{Tr}_{K/F}(\beta)$, so that the trace is an additive map from K to F.

(c) Let $K = F(\sqrt{D})$ be a quadratic extension of F. Show that $\mathrm{Tr}_{K/F}(a + b\sqrt{D}) = 2a$.

(d) Let $m_\alpha(x)$ be as in the previous problem. Prove that $\mathrm{Tr}_{K/F}(\alpha) = -\dfrac{n}{d} a_{d-1}$.

19. With notation as in the previous problems show that $N_{K/F}(a\alpha) = a^n N_{K/F}(\alpha)$ and $\mathrm{Tr}_{K/F}(a\alpha) = a\mathrm{Tr}_{K/F}(\alpha)$ for all a in the base field F. In particular show that $N_{K/F}(a) = a^n$ and $\mathrm{Tr}_{K/F}(a) = na$ for all $a \in F$.

20. With notation as in the previous problems show more generally that $\prod_{\sigma}(x - \sigma(\alpha)) = (m_\alpha(x))^{n/d}$.

21. Use the linear independence of characters to show that for any Galois extension K of F there is an element $\alpha \in K$ with $\mathrm{Tr}_{K/F}(\alpha) \neq 0$.

22. Suppose K/F is a Galois extension and let σ be an element of the Galois group.

(a) Suppose $\alpha \in K$ is of the form $\alpha = \dfrac{\beta}{\sigma\beta}$ for some nonzero $\beta \in K$. Prove that $N_{K/F}(\alpha) = 1$.

(b) Suppose $\alpha \in K$ is of the form $\alpha = \beta - \sigma\beta$ for some $\beta \in K$. Prove that $\mathrm{Tr}_{K/F}(\alpha) = 0$.

The next exercise and Exercise 26 following establish the multiplicative and additive forms of Hilbert's Theorem 90. These are instances of the vanishing of a first cohomology group, as will be discussed in Section 17.3.

23. (*Hilbert's Theorem 90*) Let K be a Galois extension of F with cyclic Galois group of order n generated by σ. Suppose $\alpha \in K$ has $N_{K/F}(\alpha) = 1$. Prove that α is of the form $\alpha = \dfrac{\beta}{\sigma\beta}$ for some nonzero $\beta \in K$. [By the linear independence of characters show there exists some $\theta \in K$ such that

$$\beta = \theta + \alpha\sigma(\theta) + (\alpha\,\sigma\alpha)\sigma^2(\theta) + \cdots + (\alpha\,\sigma\alpha\ldots\sigma^{n-2}\alpha)\sigma^{n-1}(\theta)$$

is nonzero. Compute $\dfrac{\beta}{\sigma\beta}$ using the fact that α has norm 1 to F.]

24. Prove that the rational solutions $a, b \in \mathbb{Q}$ of Pythagoras' equation $a^2 + b^2 = 1$ are of the form $a = \dfrac{s^2 - t^2}{s^2 + t^2}$ and $b = \dfrac{2st}{s^2 + t^2}$ for some $s, t \in \mathbb{Q}$. Deduce that any right triangle with integer sides has sides of lengths $((m^2 - n^2)d, 2mnd, (m^2 + n^2)d)$ for some integers m, n, d. [Note that $a^2 + b^2 = 1$ is equivalent to $N_{\mathbb{Q}(i)/\mathbb{Q}}(a + ib) = 1$, then use Hilbert's Theorem 90 above with $\beta = s + it$.]

25. Generalize the previous problem to determine all the rational solutions of the equation $a^2 + Db^2 = 1$ for $D \in \mathbb{Z}$, $D > 0$, D not a perfect square in $\mathbb{Z}$.

26. (*Additive Hilbert's Theorem 90*) Let K be a Galois extension of F with cyclic Galois group of order n generated by σ. Suppose $\alpha \in K$ has $\mathrm{Tr}_{K/F}(\alpha) = 0$. Prove that α is of the form $\alpha = \beta - \sigma\beta$ for some $\beta \in K$. [Let $\theta \in K$ be an element with $\mathrm{Tr}_{K/F}(\theta) \neq 0$ by a previous exercise, let

$$\beta = \frac{1}{\mathrm{Tr}_{K/F}(\theta)} [\alpha\sigma(\theta) + (\alpha + \sigma\alpha)\sigma^2(\theta) + \cdots + (\alpha + \sigma\alpha + \cdots + \sigma^{n-2}\alpha)\sigma^{n-1}(\theta)]$$

and compute $\beta - \sigma\beta$.]

27. Let $\alpha = \sqrt{(2 + \sqrt{2})(3 + \sqrt{3})}$ (positive real square roots for concreteness) and consider the extension $E = \mathbb{Q}(\alpha)$.

   (a) Show that $a = (2 + \sqrt{2})(3 + \sqrt{3})$ is not a square in $F = \mathbb{Q}(\sqrt{2}, \sqrt{3})$. [If $a = c^2$, $c \in F$, then $a\varphi(a) = (2 + \sqrt{2})^2(6) = (c\,\varphi c)^2$ for the automorphism $\varphi \in \mathrm{Gal}(F/\mathbb{Q})$ fixing $\mathbb{Q}(\sqrt{2})$. Since $c\,\varphi c = N_{F/\mathbb{Q}(\sqrt{2})}(c) \in \mathbb{Q}(\sqrt{2})$ conclude that this implies $\sqrt{6} \in \mathbb{Q}(\sqrt{2})$, a contradiction.]

   (b) Conclude from (a) that $[E : \mathbb{Q}] = 8$. Prove that the roots of the minimal polynomial over $\mathbb{Q}$ for α are the 8 elements $\pm\sqrt{(2 \pm \sqrt{2})(3 \pm \sqrt{3})}$.

   (c) Let $\beta = \sqrt{(2 - \sqrt{2})(3 + \sqrt{3})}$. Show that $\alpha\beta = \sqrt{2}(3 + \sqrt{3}) \in F$ so that $\beta \in E$. Show similarly that the other roots are also elements of E so that E is a Galois extension of $\mathbb{Q}$. Show that the elements of the Galois group are precisely the maps determined by mapping α to one of the eight elements in (b).

   (d) Let $\sigma \in \mathrm{Gal}(E/\mathbb{Q})$ be the automorphism which maps α to β. Show that since $\sigma(\alpha^2) = \beta^2$ that $\sigma(\sqrt{2}) = -\sqrt{2}$ and $\sigma(\sqrt{3}) = \sqrt{3}$. From $\alpha\beta = \sqrt{2}(3 + \sqrt{3})$ conclude that $\sigma(\alpha\beta) = -\alpha\beta$ and hence $\sigma(\beta) = -\alpha$. Show that σ is an element of order 4 in $\mathrm{Gal}(E/\mathbb{Q})$.

   (e) Show similarly that the map τ defined by $\tau(\alpha) = \sqrt{(2 + \sqrt{2})(3 - \sqrt{3})}$ is an element of order 4 in $\mathrm{Gal}(E/\mathbb{Q})$. Prove that σ and τ generate the Galois group, $\sigma^4 = \tau^4 = 1$, $\sigma^2 = \tau^2$ and that $\sigma\tau = \tau\sigma^3$.

   (f) Conclude that $\mathrm{Gal}(E/\mathbb{Q}) \cong Q_8$, the quaternion group of order 8.

28. Let $f(x) \in F[x]$ be an irreducible polynomial of degree n over the field F, let L be the splitting field of $f(x)$ over F and let α be a root of $f(x)$ in L. If K is any Galois extension of F contained in L, show that the polynomial $f(x)$ splits into a product of m irreducible polynomials each of degree d over K, where $m = [F(\alpha) \cap K : F]$ and $d = [K(\alpha) : K]$ (cf. also the generalization in Exercise 4 of Section 4). [If H is the subgroup of the Galois group of L over F corresponding to K then the factors of $f(x)$ over K correspond to the orbits of H on the roots of $f(x)$. Then use Exercise 9 of Section 4.1.]

29. Let k be a field and let $k(t)$ be the field of rational functions in the variable t. Define the maps σ and τ of $k(t)$ to itself by $\sigma f(t) = f(\frac{1}{1-t})$ and $\tau f(t) = f(\frac{1}{t})$ for $f(t) \in k(t)$.

    (a) Prove that σ and τ are automorphisms of $k(t)$ (cf. Exercise 8 of Section 1) and that the group $G = \langle \sigma, \tau \rangle$ they generate is isomorphic to S_3.

    (b) Prove that the element $s = \dfrac{(t^2 - t + 1)^3}{t^2(t-1)^2}$ is fixed by all the elements of G.

    (c) Prove that $k(s)$ is precisely the fixed field of G in $k(t)$ [compute the degree of the extension].

30. Prove that the fixed field of the subgroup of automorphisms generated by τ in the previous problem is $k(t + \frac{1}{t})$. Prove that the fixed field of the subgroup generated by the automorphism $\tau\sigma^2$ (which maps t to $1 - t$) is $k(t(1-t))$. Determine the fixed field of the subgroup generated by $\tau\sigma$ and the fixed field of the subgroup generated by σ.

31. Let K be a finite extension of F of degree n. Let α be an element of K.

    (a) Prove that α acting by left multiplication on K is an F-linear transformation T_α of K.

    (b) Prove that the minimal polynomial for α over F is the same as the minimal polynomial for the linear transformation T_α.

    (c) Prove that the trace $\mathrm{Tr}_{K/F}(\alpha)$ is the trace of the $n \times n$ matrix defined by T_α (which justifies these two uses of the same word "trace"). Prove that the norm $\mathrm{N}_{K/F}(\alpha)$ is the determinant of T_α.

14.3 FINITE FIELDS

A finite field $\mathbb{F}$ has characteristic p for some prime p so is a finite dimensional vector space over $\mathbb{F}_p$. If the dimension is n, i.e., $[\mathbb{F} : \mathbb{F}_p] = n$, then $\mathbb{F}$ has precisely p^n elements. We have already seen (following Proposition 13.37) that $\mathbb{F}$ is then isomorphic to the splitting field of the polynomial $x^{p^n} - x$, hence is unique up to isomorphism. We denote the finite field of order p^n by $\mathbb{F}_{p^n}$.

The field $\mathbb{F}_{p^n}$ is Galois over $\mathbb{F}_p$, with cyclic Galois group of order n generated by the Frobenius automorphism

$$\mathrm{Gal}(\mathbb{F}_{p^n}/\mathbb{F}_p) = \langle \sigma_p \rangle \cong \mathbb{Z}/n\mathbb{Z}$$

where

$$\sigma_p : \mathbb{F}_{p^n} \to \mathbb{F}_{p^n}$$
$$\alpha \mapsto \alpha^p$$

(Example 7 following Corollary 6). By the Fundamental Theorem, every subfield of $\mathbb{F}_{p^n}$ corresponds to a subgroup of $\mathbb{Z}/n\mathbb{Z}$. Hence for every divisor d of n there is precisely one subfield of $\mathbb{F}_{p^n}$ of degree d over $\mathbb{F}_p$, namely the fixed field of the subgroup generated by σ_p^d of order n/d, and there are no other subfields. This field is isomorphic to $\mathbb{F}_{p^d}$, the unique finite field of order p^d.

Since the Galois group is abelian, every subgroup is normal, so each of the subfields $\mathbb{F}_{p^d}$ (d a divisor of n) is Galois over $\mathbb{F}_p$ (which is also clear from the fact that these are themselves splitting fields). Further, the Galois group $\mathrm{Gal}(\mathbb{F}_{p^d}/\mathbb{F}_p)$ is generated by the image of σ_p in the quotient group $\mathrm{Gal}(\mathbb{F}_{p^n}/\mathbb{F}_p)/\langle \sigma_p^d \rangle$. If we denote this element again by σ_p, we recover the Frobenius automorphism for the extension $\mathbb{F}_{p^d}/\mathbb{F}_p$. (Note, however, that σ_p has order n in $\mathrm{Gal}(\mathbb{F}_{p^n}/\mathbb{F}_p)$ and order d in $\mathrm{Gal}(\mathbb{F}_{p^d}/\mathbb{F}_p)$.)

We summarize this in the following proposition.

Proposition 15. Any finite field is isomorphic to $\mathbb{F}_{p^n}$ for some prime p and some integer $n \geq 1$. The field $\mathbb{F}_{p^n}$ is the splitting field over $\mathbb{F}_p$ of the polynomial $x^{p^n} - x$, with cyclic Galois group of order n generated by the Frobenius automorphism σ_p. The subfields of $\mathbb{F}_{p^n}$ are all Galois over $\mathbb{F}_p$ and are in one to one correspondence with the divisors d of n. They are the fields $\mathbb{F}_{p^d}$, the fixed fields of $\sigma_p{}^d$.

The corresponding statements for the finite extensions of any finite field are easy consequences of Proposition 15 and are outlined in the exercises.

As an elementary application we have the following result on the polynomial $x^4 + 1$ in $\mathbb{Z}[x]$.

Corollary 16. The irreducible polynomial $x^4 + 1 \in \mathbb{Z}[x]$ is reducible modulo every prime p.

Proof: Consider the polynomial $x^4 + 1$ over $\mathbb{F}_p[x]$ for the prime p. If $p = 2$ we have $x^4 + 1 = (x + 1)^4$ and the polynomial is reducible. Assume now that p is odd. Then $p^2 - 1$ is divisible by 8 since p is congruent mod 8 to 1, 3, 5 or 7 and all of these square to 1 mod 8. Hence $x^{p^2-1} - 1$ is divisible by $x^8 - 1$. Then we have the divisibilities

$$x^4 + 1 \mid x^8 - 1 \mid x^{p^2-1} - 1 \mid x^{p^2} - x$$

which shows that all the roots of $x^4 + 1$ are roots of $x^{p^2} - x$. (Equivalently, these roots are fixed by the square of the Frobenius automorphism σ_p^2.) Since the roots of $x^{p^2} - x$ are the field $\mathbb{F}_{p^2}$, it follows that the extension generated by any root of $x^4 + 1$ is at most of degree 2 over $\mathbb{F}_p$, which means that $x^4 + 1$ cannot be irreducible over $\mathbb{F}_p$.

The multiplicative group $\mathbb{F}_{p^n}{}^\times$ is obviously a finite subgroup of the multiplicative group of a field. By Proposition 9.18, this is a *cyclic* group. If θ is any generator, then clearly $\mathbb{F}_{p^n} = \mathbb{F}_p(\theta)$. This proves the following result.

Proposition 17. The finite field $\mathbb{F}_{p^n}$ is simple. In particular, there exists an irreducible polynomial of degree n over $\mathbb{F}_p$ for every $n \geq 1$.

We have described the finite fields $\mathbb{F}_{p^n}$ above as the splitting fields of the polynomials $x^{p^n} - x$. By the previous proposition, this field can also be described as a quotient of $\mathbb{F}_p[x]$, namely by the minimal polynomial for θ. Since θ is necessarily a root of $x^{p^n} - x$, we see that the minimal polynomial for θ is a divisor of $x^{p^n} - x$ of degree n.

Conversely, let $p(x)$ be any irreducible polynomial of degree d, say, dividing $x^{p^n} - x$. If α is a root of $p(x)$, then the extension $\mathbb{F}_p(\alpha)$ is a subfield of $\mathbb{F}_{p^n}$ of degree d. Hence d is a divisor of n and the extension is Galois by Proposition 15 (in fact, the extension $\mathbb{F}_{p^d}$) so in particular all the roots of $p(x)$ are contained in $\mathbb{F}_p(\alpha)$.

586

The elements of $\mathbb{F}_{p^n}$ are precisely the roots of $x^{p^n} - x$. If we group together the factors $x - \alpha$ of this polynomial according to the degree d of their minimal polynomials over $\mathbb{F}_p$, we obtain

Proposition 18. The polynomial $x^{p^n} - x$ is precisely the product of all the distinct irreducible polynomials in $\mathbb{F}_p[x]$ of degree d where d runs through all divisors of n.

This proposition can be used to produce irreducible polynomials over $\mathbb{F}_p$ recursively. For example, the irreducible quadratics over $\mathbb{F}_2$ are the divisors of

$$\frac{x^4 - x}{x(x - 1)}$$

which gives the single polynomial $x^2 + x + 1$. Similarly, the irreducible cubics over this field are the divisors of

$$\frac{x^8 - x}{x(x - 1)} = x^6 + x^5 + x^4 + x^3 + x^2 + x + 1$$

which factors into the two cubics $x^3 + x + 1$ and $x^3 + x^2 + 1$. The irreducible quartics are given by dividing $x^{16} - x$ by $x(x - 1)$ and the irreducible quadratic $x^2 + x + 1$ above and then factoring into irreducible quartics:

$$\frac{x^{16} - x}{x(x - 1)(x^2 + x + 1)} = (x^4 + x^3 + x^2 + x + 1)(x^4 + x^3 + 1)(x^4 + x + 1).$$

This gives a method for determining the product of all the irreducible polynomials over $\mathbb{F}_p$ of a given degree. There exist efficient algorithms for factorization of polynomials mod p which will give the individual irreducible polynomials (cf. the exercises) in practice. The importance of having irreducible polynomials at hand is that they give a representation of the finite fields $\mathbb{F}_{p^n}$ (as quotients $\mathbb{F}_p[x]/(f(x))$ for $f(x)$ irreducible of degree n) conducive to explicit computations.

Note also that since the finite field $\mathbb{F}_{p^n}$ is unique up to isomorphism, the quotients of $\mathbb{F}_p[x]$ by any of the irreducible polynomials of degree n are all isomorphic. If $f_1(x)$ and $f_2(x)$ are irreducible of degree n, then $f_2(x)$ splits completely in the field $\mathbb{F}_{p^n} \cong \mathbb{F}_p[x]/(f_1(x))$. If we denote a root of $f_2(x)$ by $\alpha(x)$ (to emphasize that it is a polynomial of degree $< n$ in x in $\mathbb{F}_p[x]/(f_1(x))$), then the isomorphism is given by

$$\mathbb{F}_p[x]/(f_2(x)) \cong \mathbb{F}_p[x]/(f_1(x))$$
$$x \mapsto \alpha(x)$$

(we have mapped a root of $f_2(x)$ in the first field to a root of $f_2(x)$ in the second field). For example, if $f_1(x) = x^4 + x^3 + 1$, $f_2(x) = x^4 + x + 1$ are two of the irreducible quartics over $\mathbb{F}_2$ determined above, then a simple computation verifies that

$$\alpha(x) = x^3 + x^2$$

is a root of $f_2(x)$ in $\mathbb{F}_{16} = \mathbb{F}_2[x]/(x^4 + x^3 + 1)$. Then we have

$$\mathbb{F}_2[x]/(x^4 + x + 1) \cong \mathbb{F}_2[x]/(x^4 + x^3 + 1) \quad (\cong \mathbb{F}_{16})$$
$$x \mapsto x^3 + x^2.$$

If we assume a result from elementary number theory we can give a formula for the number of irreducible polynomials of degree n. Define the *Möbius* μ-function by

$$\mu(n) = \begin{cases} 1 & \text{for } n = 1 \\ 0 & \text{if } n \text{ has a square factor} \\ (-1)^r & \text{if } n \text{ has } r \text{ distinct prime factors.} \end{cases}$$

If now $f(n)$ is a function defined for all nonnegative integers n and $F(n)$ is defined by

$$F(n) = \sum_{d \mid n} f(d) \qquad n = 1, 2, \ldots$$

then the *Möbius inversion formula* states that one can recover the function $f(n)$ from $F(n)$:

$$f(n) = \sum_{d \mid n} \mu(d) F\left(\frac{n}{d}\right) \qquad n = 1, 2, \ldots.$$

This is an elementary result from number theory which we take for granted. Define

$$\psi(n) = \text{the number of irreducible polynomials of degree } n \text{ in } \mathbb{F}_p[x].$$

Counting degrees in Proposition 18 we have

$$p^n = \sum_{d \mid n} d \psi(d).$$

Applying the Möbius inversion formula (for $f(n) = n\psi(n)$) we obtain

$$n\psi(n) = \sum_{d \mid n} \mu(d) p^{n/d}$$

which gives us a formula for the number of irreducible polynomials of degree n over $\mathbb{F}_p$:

$$\psi(n) = \frac{1}{n} \sum_{d \mid n} \mu(d) p^{n/d}.$$

For example, in the case $p = 2, n = 4$ we have

$$\psi(4) = \frac{1}{4}[\mu(1)2^4 + \mu(2)2^2 + \mu(4)2^1] = \frac{1}{4}(16 - 4 + 0) = 3$$

as we determined directly above.

We have seen above that

$$\mathbb{F}_{p^m} \subseteq \mathbb{F}_{p^n} \text{ if and only if } m \text{ divides } n.$$

In particular, given any two finite fields $\mathbb{F}_{p^{n_1}}$ and $\mathbb{F}_{p^{n_2}}$ there is a third finite field containing (an isomorphic copy of) them, namely $\mathbb{F}_{p^{n_1 n_2}}$. This gives us a partial ordering on these fields and allows us to think of their union. Since these give *all* the finite extensions of $\mathbb{F}_p$, we see that the union of $\mathbb{F}_{p^n}$ for all n is an algebraic closure of $\mathbb{F}_p$, unique up to isomorphism:

$$\overline{\mathbb{F}_p} = \bigcup_{n \geq 1} \mathbb{F}_{p^n}.$$

This provides a simple description of the algebraic closure of $\mathbb{F}_p$.

EXERCISES

1. Factor $x^8 - x$ into irreducibles in $\mathbb{Z}[x]$ and in $\mathbb{F}_2[x]$.

2. Write out the multiplication table for $\mathbb{F}_4$ and $\mathbb{F}_8$.

3. Prove that an algebraically closed field must be infinite.

4. Construct the finite field of 16 elements and find a generator for the multiplicative group. How many generators are there?

5. Exhibit an explicit isomorphism between the splitting fields of $x^3 - x + 1$ and $x^3 - x - 1$ over $\mathbb{F}_3$.

6. Suppose $K = \mathbb{Q}(\theta) = \mathbb{Q}(\sqrt{D_1}, \sqrt{D_2})$ with $D_1, D_2 \in \mathbb{Z}$, is a biquadratic extension and that $\theta = a + b\sqrt{D_1} + c\sqrt{D_2} + d\sqrt{D_1 D_2}$ where $a, b, c, d \in \mathbb{Z}$ are integers. Prove that the minimal polynomial $m_\theta(x)$ for θ over $\mathbb{Q}$ is irreducible of degree 4 over $\mathbb{Q}$ but is reducible modulo every prime p. In particular show that the polynomial $x^4 - 10x^2 + 1$ is irreducible in $\mathbb{Z}[x]$ but is reducible modulo every prime. [Use the fact that there are no biquadratic extensions over finite fields.]

7. Prove that one of 2, 3 or 6 is a square in $\mathbb{F}_p$ for every prime p. Conclude that the polynomial

$$x^6 - 11x^4 + 36x^2 - 36 = (x^2 - 2)(x^2 - 3)(x^2 - 6)$$

has a root modulo p for every prime p but has no root in $\mathbb{Z}$.

8. Determine the splitting field of the polynomial $x^p - x - a$ over $\mathbb{F}_p$ where $a \neq 0, a \in \mathbb{F}_p$. Show explicitly that the Galois group is cyclic. [Show $\alpha \mapsto \alpha + 1$ is an automorphism.] Such an extension is called an *Artin–Schreier extension* (cf. Exercise 9 of Section 7).

9. Let $q = p^m$ be a power of the prime p and let $\mathbb{F}_q = \mathbb{F}_{p^m}$ be the finite field with q elements. Let $\sigma_q = \sigma_p^m$ be the mth power of the Frobenius automorphism σ_p, called the q-Frobenius automorphism.
   (a) Prove that σ_q fixes $\mathbb{F}_q$.
   (b) Prove that every finite extension of $\mathbb{F}_q$ of degree n is the splitting field of $x^{q^n} - x$ over $\mathbb{F}_q$, hence is unique.
   (c) Prove that every finite extension of $\mathbb{F}_q$ of degree n is cyclic with σ_q as generator.
   (d) Prove that the subfields of the unique extension of $\mathbb{F}_q$ of degree n are in bijective correspondence with the divisors d of n.

10. Prove that n divides $\varphi(p^n - 1)$. [Observe that $\varphi(p^n - 1)$ is the order of the group of automorphisms of a cyclic group of order $p^n - 1$.]

11. Prove that $x^{p^n} - x + 1$ is irreducible over $\mathbb{F}_p$ only when $n = 1$ or $n = p = 2$. [Note that if α is a root, then so is $\alpha + a$ for any $a \in \mathbb{F}_{p^n}$. Show that this implies $\mathbb{F}_p(\alpha)$ contains $\mathbb{F}_{p^n}$ and that $[\mathbb{F}_p(\alpha) : \mathbb{F}_{p^n}] = p$.]

(*Berlekamp's Factorization Algorithm*) The following exercises outline the Berlekamp factorization algorithm for factoring polynomials in $\mathbb{F}_p[x]$. The efficiency of this algorithm is based on the efficiency of computing greatest common divisors in $\mathbb{F}_p[x]$ by the Euclidean Algorithm and on the efficiency of row-reduction matrix algorithms for solving systems of linear equations.

Let $f(x) \in \mathbb{F}_p[x]$ be a monic polynomial of degree n and let $f(x) = p_1(x)p_2(x)\dots p_k(x)$ where $p_1(x), p_2(x), \dots, p_k(x)$ are powers of distinct monic irreducibles in $\mathbb{F}_p[x]$.

12. Show that in order to write $f(x)$ as a product of irreducible polynomials in $\mathbb{F}_p[x]$ it suffices to determine the factors $p_1(x), \dots, p_k(x)$. [If $p(x) = q(x)^N \in \mathbb{F}_p[x]$ with $q(x)$ monic

and irreducible, show that $q(x)$ can be determined from $p(x)$ by checking for p^{th} powers and by computing greatest common divisors with derivatives.]

13. Let $g(x) \in \mathbb{F}_p[x]$ be any polynomial of degree $< n$. Denote by $R(h(x))$ the remainder of $h(x)$ after division by $f(x)$. Prove the following are equivalent:
    (a) $R(g(x^p)) = g(x)$.
    (b) $f(x)$ divides $[g(x)-0][g(x)-1]\ldots[g(x)-(p-1)]$. [Use the fact that $g(x^p) = g(x)^p$ together with the factorization of $x^p - x$ in $\mathbb{F}_p[x]$.]
    (c) $p_i(x)$ divides the product in (b) for $i = 1, 2, \ldots, k$.
    (d) For each i, $i = 1, 2, \ldots, k$ there is an $s_i \in \mathbb{F}_p$ such that $p_i(x)$ divides $g(x) - s_i$, i.e., $g(x) \equiv s_i \pmod{p_i(x)}$.

14. Prove that the polynomials $g(x)$ of degree $< n$ satisfying the equivalent conditions of the previous exercise form a vector space V over $\mathbb{F}_p$ of dimension k. [Use the Chinese Remainder Theorem applied to the p^k possible choices for the s_i in 13(d)].

15. Let $g(x) = b_0 + b_1 x + \cdots + b_{n-1}x^{n-1} \in V$. For $j = 0, 1, \ldots, n-1$ let

$$R(x^{pj}) = a_{0,j} + a_{1,j}x + \cdots + a_{n-1,j}x^{n-1}$$

and let A be the the $n \times n$ matrix

$$A = \begin{pmatrix} a_{0,0} & a_{0,1} & \cdots & a_{0,n-1} \\ a_{1,0} & a_{1,1} & \cdots & a_{1,n-1} \\ \vdots & \vdots & \ddots & \vdots \\ a_{n-1,0} & a_{n-1,1} & \cdots & a_{n-1,n-1} \end{pmatrix}. \tag{*}$$

Show that condition (a) of Exercise 13 for $g(x) \in V$ is equivalent to

$$(A - I)B = 0 \tag{**}$$

where B is the column matrix with entries $b_0, b_1, \ldots, b_{n-1}$. Conclude that the rank of the matrix $A - I$ is $n - k$. Note that this already suffices to determine if $f(x)$ is irreducible, without actually determining the factors.

16. Let $g_1(x), g_2(x), \ldots, g_k(x)$ be a basis of solutions to (**) (so a basis for V), where we may take $g_1(x) = 1$. Beginning with $w(x) = f(x)$, compute the greatest common divisor $(w(x), g_i(x)-s)$ for $i = 2, 3, \ldots, k$ and $s \in \mathbb{F}_p$ for every factor of $f(x)$ already computed. Note by Exercise 13(d) that every factor $p_i(x)$ of $f(x)$ divides such a g.c.d. The process terminates when k relatively prime factors have been determined.

  Prove that this procedure actually gives all the factors $p_1(x), p_2(x), \ldots, p_k(x)$, i.e., one can separate the individual factors $p_1(x), p_2(x), \ldots, p_k(x)$ by this procedure, as follows:

  If this were not the case, then for two of the factors, say $p_1(x)$ and $p_2(x)$, for each $i = 1, 2, \ldots, k$ there would exist $s_i \in \mathbb{F}_p$ such that $g_i(x) - s_i$ is divisible by both $p_1(x)$ and $p_2(x)$. By the Chinese Remainder Theorem, choose a $g(x) \in V$ satisfying $g(x) \equiv 0 \pmod{p_1(x)}$ and $g(x) \equiv 1 \pmod{p_2(x)}$. Write $g(x) = \sum_{i=1}^{k} c_i g_i(x)$ in terms of the basis for V and let $s = \sum_{i=1}^{k} c_i s_i(x) \in \mathbb{F}_p$. Show that $s \equiv 0 \pmod{p_1(x)}$ so that $s = 0$ and $s \equiv 1 \pmod{p_2(x)}$ so that $s = 1$, a contradiction.

17. This exercise follows Berlekamp's Factorization Algorithm outlined in the previous exercises to determine the factorization of $f(x) = x^5 + x^2 + 4x + 6$ in $\mathbb{F}_7[x]$.
    (a) Show that $x^7 \equiv x^2 + 3x^3 + 6x^4 \pmod{f(x)}$. Similarly compute x^{14}, x^{21}, and x^{28} modulo $f(x)$ (note that x^{14} can most easily be computed by squaring the result for

590

x^7 and then reducing, etc.) to show that in this case the matrix A in Exercise 15 is

$$\begin{pmatrix} 1 & 0 & 5 & 1 & 4 \\ 0 & 0 & 1 & 1 & 2 \\ 0 & 1 & 3 & 3 & 3 \\ 0 & 3 & 4 & 2 & 2 \\ 0 & 6 & 3 & 1 & 1 \end{pmatrix}.$$

(b) Show that the reduced row echelon form for $A - I$ is the matrix

$$\begin{pmatrix} 0 & 1 & 0 & 0 & 6 \\ 0 & 0 & 1 & 0 & 6 \\ 0 & 0 & 0 & 1 & 2 \\ 0 & 0 & 0 & 0 & 0 \\ 0 & 0 & 0 & 0 & 0 \end{pmatrix}.$$

Conclude that $k = 2$ (so $f(x)$ is the product of precisely two factors which are powers of irreducible polynomials) and that $g_1(x) = 1$ and $g_2(x) = x^4 + 5x^3 + x^2 + x$ give a basis for the solutions to $(\ast\ast)$ in Exercise 15.

(c) Following the procedure in Exercise 16, show that $(f(x), g_2(x) - 1) = x^2 + 3x + 5 = p_1(x)$, with $f(x)/p_1(x) = x^3 + 4x^2 + 4x + 4 = p_2(x)$, giving the powers of the irreducible polynomials dividing $f(x)$ in $\mathbb{F}_7[x]$. Show that neither factor is a 7th power in $\mathbb{F}_7[x]$ and that each is relatively prime to its derivative to conclude that both factors are irreducible polynomials, giving the complete factorization of $f(x)$ into irreducible polynomials:

$$f(x) = (x^2 + 3x + 5)(x^3 + 4x^2 + 4x + 4) \in \mathbb{F}_7[x].$$

14.4 COMPOSITE EXTENSIONS AND SIMPLE EXTENSIONS

We now consider the effect of taking composites with Galois extensions. The first result states that "sliding up" a Galois extension gives a Galois extension.

Proposition 19. Suppose K/F is a Galois extension and F'/F is any extension. Then KF'/F' is a Galois extension, with Galois group

$$\text{Gal}(KF'/F') \cong \text{Gal}(K/K \cap F')$$

isomorphic to a subgroup of $\text{Gal}(K/F)$. Pictorially,

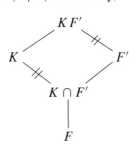

Proof: If K/F is Galois, then K is the splitting field of some separable polynomial $f(x)$ in $F[x]$. Then KF'/F' is the splitting field of $f(x)$ viewed as a polynomial in

$F'[x]$, hence this extension is Galois. Since K/F is Galois, every embedding of K fixing F is an automorphism of K, so the map

$$\varphi : \text{Gal}(KF'/F') \to \text{Gal}(K/F)$$

$$\sigma \mapsto \sigma|_K$$

defined by restricting an automorphism σ to the subfield K is well defined. It is clearly a homomorphism, with kernel

$$\ker \varphi = \{\sigma \in \text{Gal}(KF'/F') \mid \sigma|_K = 1\}.$$

Since an element in $\text{Gal}(KF'/F')$ is trivial on F', the elements in the kernel are trivial both on K and on F', hence on their composite, so the kernel consists only of the identity automorphism. Hence φ is injective.

Let H denote the image of φ in $\text{Gal}(K/F)$ and let K_H denote the corresponding fixed subfield of K containing F. Since every element in H fixes F', K_H contains $K \cap F'$. On the other hand, the composite $K_H F'$ is fixed by $\text{Gal}(KF'/F')$ (any $\sigma \in \text{Gal}(KF'/F')$ fixes F' and acts on $K_H \subseteq K$ via its restriction $\sigma|_K \in H$, which fixes K_H by definition). By the Fundamental Theorem it follows that $K_H F' = F'$, so that $K_H \subseteq F'$, which gives the reverse inclusion $K_H \subseteq K \cap F'$. Hence $K_H = K \cap F'$, so again by the Fundamental Theorem, $H = \text{Gal}(K/K \cap F')$, completing the proof.

Corollary 20. Suppose K/F is a Galois extension and F'/F is any finite extension. Then

$$[KF' : F] = \frac{[K : F][F' : F]}{[K \cap F' : F]}.$$

Proof: This follows by the proposition from the equality $[KF' : F'] = [K : K \cap F']$ given by the orders of the Galois groups in the proposition.

The example $F = \mathbb{Q}$, $K = \mathbb{Q}(\sqrt[3]{2})$, $F' = \mathbb{Q}(\rho\sqrt[3]{2})$, ρ a primitive 3rd root of unity, shows that the formula of Corollary 20 does not hold in general if neither of the two extensions is Galois.

Proposition 21. Let K_1 and K_2 be Galois extensions of a field F. Then
  (1) The intersection $K_1 \cap K_2$ is Galois over F.
  (2) The composite $K_1 K_2$ is Galois over F. The Galois group is isomorphic to the subgroup

$$H = \{(\sigma, \tau) \mid \sigma|_{K_1 \cap K_2} = \tau|_{K_1 \cap K_2}\}$$

of the direct product $\text{Gal}(K_1/F) \times \text{Gal}(K_2/F)$ consisting of elements whose restrictions to the intersection $K_1 \cap K_2$ are equal.

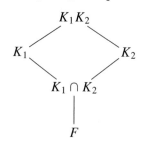

Proof: (1) Suppose $p(x)$ is an irreducible polynomial in $F[x]$ with a root α in $K_1 \cap K_2$. Since $\alpha \in K_1$ and K_1/F is Galois, all the roots of $p(x)$ lie in K_1. Similarly all the roots lie in K_2, hence all the roots of $p(x)$ lie in $K_1 \cap K_2$. It follows easily that $K_1 \cap K_2$ is Galois as in Theorem 13.

(2) If K_1 is the splitting field of the separable polynomial $f_1(x)$ and K_2 is the splitting field of the separable polynomial $f_2(x)$ then the composite is the splitting field for the squarefree part of the polynomial $f_1(x)f_2(x)$, hence is Galois over F.

The map

$$\varphi : \text{Gal}(K_1 K_2/F) \to \text{Gal}(K_1/F) \times \text{Gal}(K_2/F)$$
$$\sigma \mapsto (\sigma|_{K_1}, \sigma|_{K_2})$$

is clearly a homomorphism. The kernel consists of the elements σ which are trivial on both K_1 and K_2, hence trivial on the composite, so the map is injective. The image lies in the subgroup H, since

$$(\sigma|_{K_1})|_{K_1 \cap K_2} = \sigma|_{K_1 \cap K_2} = (\sigma|_{K_2})|_{K_1 \cap K_2}.$$

The order of H can be computed by observing that for every $\sigma \in \text{Gal}(K_1/F)$ there are $|\text{Gal}(K_2/K_1 \cap K_2)|$ elements $\tau \in \text{Gal}(K_2/F)$ whose restrictions to $K_1 \cap K_2$ are $\sigma|_{K_1 \cap K_2}$. Hence

$$|H| = |\text{Gal}(K_1/F)| \cdot |\text{Gal}(K_2/K_1 \cap K_2)|$$
$$= |\text{Gal}(K_1/F)| \frac{|\text{Gal}(K_2/F)|}{|\text{Gal}(K_1 \cap K_2/F)|}.$$

By Corollary 20 and the diagram above we see that the orders of H and $\text{Gal}(K_1 K_2/F)$ are then both equal to

$$[K_1 K_2 : F] = \frac{[K_1 : F][K_2 : F]}{[K_1 \cap K_2 : F]}.$$

Hence the image of φ is precisely H, completing the proof.

Corollary 22. Let K_1 and K_2 be Galois extensions of a field F with $K_1 \cap K_2 = F$. Then

$$\text{Gal}(K_1 K_2/F) \cong \text{Gal}(K_1/F) \times \text{Gal}(K_2/F).$$

Conversely, if K is Galois over F and $G = \text{Gal}(K/F) = G_1 \times G_2$ is the direct product of two subgroups G_1 and G_2, then K is the composite of two Galois extensions K_1 and K_2 of F with $K_1 \cap K_2 = F$.

Proof: The first part follows immediately from the proposition. For the second, let K_1 be the fixed field of $G_1 \subset G$ and let K_2 be the fixed field of $G_2 \subset G$. Then $K_1 \cap K_2$ is the field corresponding to the subgroup $G_1 G_2$, which is all of G in this case, so $K_1 \cap K_2 = F$. The composite $K_1 K_2$ is the field corresponding to the subgroup $G_1 \cap G_2$, which is the identity here, so $K_1 K_2 = K$, completing the proof.

Corollary 23. Let E/F be any finite separable extension. Then E is contained in an extension K which is Galois over F and is minimal in the sense that in a fixed algebraic closure of K any other Galois extension of F containing E contains K.

Proof: There exists a Galois extension of F containing E, for example the composite of the splitting fields of the minimal polynomials for a basis for E over F (which are all separable since E is separable over F). Then the intersection of all the Galois extensions of F containing E is the field K.

Definition. The Galois extension K of F containing E in the previous corollary is called the *Galois closure* of E over F.

It is often simpler to work in a Galois extension (for example in computing degrees as in Corollary 20). The existence of a Galois closure for a separable extension is frequently useful for reducing computations to consideration of Galois extensions.

Recall that an extension K of F is called *simple* if $K = F(\theta)$ for some element θ, in which case θ is called a *primitive element* for K.

Proposition 24. Let K/F be a finite extension. Then $K = F(\theta)$ if and only if there exist only finitely many subfields of K containing F.

Proof: Suppose first that $K = F(\theta)$ is simple. Let E be a subfield of K containing F: $F \subseteq E \subseteq K$. Let $f(x) \in F[x]$ be the minimal polynomial for θ over F and let $g(x) \in E[x]$ be the minimal polynomial for θ over E. Then $g(x)$ divides $f(x)$ in $E[x]$. Let E' be the field generated over F by the coefficients of $g(x)$. Then $E' \subseteq E$ and clearly the minimal polynomial for θ over E' is still $g(x)$. But then

$$[K : E] = \deg g(x) = [K : E']$$

implies that $E = E'$. It follows that the subfields of K containing F are the subfields generated by the coefficients of the monic factors of $f(x)$, hence there are finitely many such subfields.

Suppose conversely that there are finitely many subfields of K containing F. If F is a finite field, then we have already seen that K is a simple extension (Proposition 17). Hence we may suppose F is infinite. It clearly suffices to show that $F(\alpha, \beta)$ is generated by a single element since K is finitely generated over F. Consider the subfields

$$F(\alpha + c\beta), \qquad c \in F.$$

Then since there are infinitely many choices for $c \in F$ and only finitely many such subfields, there exist c, c' in F, $c \neq c'$, with

$$F(\alpha + c\beta) = F(\alpha + c'\beta).$$

Then $\alpha + c\beta$ and $\alpha + c'\beta$ both lie in $F(\alpha + c\beta)$, and taking their difference shows that $(c - c')\beta \in F(\alpha + c\beta)$ Hence $\beta \in F(\alpha + c\beta)$ and then also $\alpha \in F(\alpha + c\beta)$. Therefore $F(\alpha, \beta) \subseteq F(\alpha + c\beta)$ and since the reverse inclusion is obvious, we have

$$F(\alpha, \beta) = F(\alpha + c\beta),$$

completing the proof.

Theorem 25. *(The Primitive Element Theorem)* If K/F is finite and separable, then K/F is simple. In particular, any finite extension of fields of characteristic 0 is simple.

Proof: Let L be the Galois closure of K over F. Then any subfield of K containing F corresponds to a subgroup of the Galois group $\text{Gal}(L/F)$ by the Fundamental Theorem. Since there are only finitely many such subgroups, the previous proposition shows that K/F is simple. The last statement follows since any finite extension of fields in characteristic 0 is separable.

As the proof of the proposition indicates, a primitive element for an extension can be obtained as a simple linear combination of the generators for the extension. In the case of Galois extensions it is only necessary to determine a linear combination which is not fixed by any nontrivial element of the Galois group since then by the Fundamental Theorem this linear combination could not lie in any proper subfield.

Examples

(1) The element $\sqrt{2} + \sqrt{3}$ generates the field $\mathbb{Q}(\sqrt{2}, \sqrt{3})$ as we have already seen (it is not fixed by any of the four Galois automorphisms of this field).

(2) The field $\overline{\mathbb{F}_p}(x, y)$ of rational functions in the variables x and y over the algebraic closure $\overline{\mathbb{F}_p}$ of $\mathbb{F}_p$ is not a simple extension of the subfield $F = \overline{\mathbb{F}_p}(x^p, y^p)$. It is easy to see that

$$[\overline{\mathbb{F}_p}(x, y) : \overline{\mathbb{F}_p}(x^p, y^p)] = p^2$$

and that the subfields

$$F(x + cy), \qquad c \in \overline{\mathbb{F}_p}$$

are all of degree p over $\overline{\mathbb{F}_p}(x^p, y^p)$ (note that $(x + cy)^p = x^p + c^p y^p \in \overline{\mathbb{F}_p}(x^p, y^p)$). If any two of these subfields were equal, then just as in the proof of Proposition 24 we would have

$$\overline{\mathbb{F}_p}(x, y) = F(x + cy)$$

which is impossible by degree considerations. Hence there are infinitely many such subfields and the extension cannot be simple.

EXERCISES

1. Determine the Galois closure of the field $\mathbb{Q}(\sqrt{1 + \sqrt{2}})$ over $\mathbb{Q}$.

2. Find a primitive generator for $\mathbb{Q}(\sqrt{2}, \sqrt{3}, \sqrt{5})$ over $\mathbb{Q}$.

3. Let F be a field contained in the ring of $n \times n$ matrices over $\mathbb{Q}$. Prove that $[F : \mathbb{Q}] \leq n$. (Note that, by Exercise 19 of Section 13.2, the ring of $n \times n$ matrices over $\mathbb{Q}$ does contain fields of degree n over $\mathbb{Q}$.)

4. For *any* Galois extension K of F, show the irreducible polynomial $f(x) \in F[x]$ factors in $K[x]$ as in Exercise 28 of Section 2 (whether or not K is contained in the Galois closure L of $f(x)$). [Show first that the factorization of $f(x)$ over K is the same as its factorization over $L \cap K$. Then show the factors of $f(x)$ over $L \cap K$ correspond to the orbits of $H = \text{Gal}(L/L \cap K)$ on the roots of $f(x)$ and use Exercise 9 of Section 4.1.]

5. Let p be a prime and let F be a field. Let K be a Galois extension of F whose Galois group is a p-group (i.e., the degree $[K : F]$ is a power of p). Such an extension is called a *p-extension* (note that p-extensions are Galois by definition).

   (a) Let L be a p-extension of K. Prove that the Galois closure of L over F is a p-extension of F.

   (b) Give an example to show that (a) need not hold if $[K : F]$ is a power of p but K/F is not Galois.

6. Prove that $\mathbb{F}_p(x, y)/\mathbb{F}_p(x^p, y^p)$ is not a simple extension by explicitly exhibiting an infinite number of intermediate subfields.

7. Let $F \subseteq K \subseteq L$ and let $\theta \in L$ with $p(x) = m_{\theta, F}(x)$. Prove that $K \otimes_F F(\theta) \cong K[x]/(p(x))$ as K-algebras.

8. Let K_1 and K_2 be two algebraic extensions of a field F contained in the field L of characteristic zero. Prove that the F-algebra $K_1 \otimes_F K_2$ has no nonzero nilpotent elements. [Use the preceding exercise.]

9. Suppose K/F is Galois with Galois group G and θ is a primitive element for K, i.e., $K = F(\theta)$. For any subgroup H of G, let $f(x) = \prod_{\sigma \in H}(x - \sigma(\theta))$. Show $f(x) \in E[x]$ where E is the fixed field of H in K, and that $f(x)$ is the minimal polynomial for θ over E. Prove that the coefficients of $f(x)$ generate E over F (these coefficients are the 'elementary symmetric functions' of the conjugates $\sigma(\theta)$ of θ for $\sigma \in H$, cf. Section 6).

14.5 CYCLOTOMIC EXTENSIONS AND ABELIAN EXTENSIONS OVER $\mathbb{Q}$

We have already determined that the cyclotomic field $\mathbb{Q}(\zeta_n)$ of nth roots of unity is a Galois extension of $\mathbb{Q}$ of degree $\varphi(n)$ where φ denotes the Euler φ-function. Any automorphism of this field is uniquely determined by its action on the primitive n^{th} root of unity ζ_n. This element must be mapped to another primitive n^{th} root of unity (recall these are the roots of the irreducible cyclotomic polynomial $\Phi_n(x)$). Hence $\sigma(\zeta_n) = \zeta_n^a$ for some integer a, $1 \leq a < n$, relatively prime to n. Since there are precisely $\varphi(n)$ such integers a it follows that in fact each of these maps is indeed an automorphism of $\mathbb{Q}(\zeta_n)$. Note also that we can define σ_a for any integer a relatively prime to n by the same formula and that σ_a depends only on the residue class of a modulo n.

Theorem 26. The Galois group of the cyclotomic field $\mathbb{Q}(\zeta_n)$ of nth roots of unity is isomorphic to the multiplicative group $(\mathbb{Z}/n\mathbb{Z})^\times$. The isomorphism is given explicitly by the map

$$(\mathbb{Z}/n\mathbb{Z})^\times \xrightarrow{\sim} \text{Gal}(\mathbb{Q}(\zeta_n)/\mathbb{Q})$$
$$a \ (\text{mod } n) \longmapsto \sigma_a$$

where σ_a is the automorphism defined by

$$\sigma_a(\zeta_n) = \zeta_n^a.$$

Proof: The discussion above shows that σ_a is an automorphism for any $a \ (\text{mod } n)$, so the map above is well defined. It is a homomorphism since

$$(\sigma_a \sigma_b)(\zeta_n) = \sigma_a(\zeta_n^b) = (\zeta_n^b)^a = \zeta_n^{ab} = \sigma_{ab}(\zeta_n)$$

which shows that $\sigma_a\sigma_b = \sigma_{ab}$. The map is bijective by the discussion above since we know that every Galois automorphism is of the form σ_a for a uniquely defined a (mod n). Hence the map is an isomorphism.

Examples

(1) The field $\mathbb{Q}(\zeta_5)$ is Galois over $\mathbb{Q}$ with Galois group $(\mathbb{Z}/5\mathbb{Z})^\times \cong \mathbb{Z}/4\mathbb{Z}$. This is our first example of a Galois extension of $\mathbb{Q}$ of degree 4 with a *cyclic* Galois group. The elements of the Galois group are $\{\sigma_1 = 1, \sigma_2, \sigma_3, \sigma_4\}$ in the notation above. A generator for this cyclic group is $\sigma_2 : \zeta_5 \mapsto \zeta_5^2$ (since 2 has order 4 in $(\mathbb{Z}/5\mathbb{Z})^\times$).

There is precisely one nontrivial subfield, a quadratic extension of $\mathbb{Q}$, the fixed field of the subgroup $\{1, \sigma_4 = \sigma_{-1}\}$. An element in this subfield is given by

$$\alpha = \zeta_5 + \sigma_{-1}\zeta_5 = \zeta_5 + \zeta_5^{-1}$$

since this element is clearly fixed by σ_{-1}. The element ζ_5 satisfies

$$\zeta_5^4 + \zeta_5^3 + \zeta_5^2 + \zeta_5 + 1 = 0.$$

Notice then that

$$\begin{aligned}
\alpha^2 + \alpha - 1 &= (\zeta_5^2 + 2 + \zeta_5^{-2}) + (\zeta_5 + \zeta_5^{-1}) - 1 \\
&= \zeta_5^2 + 2 + \zeta_5^3 + \zeta_5 + \zeta_5^4 - 1 = 0.
\end{aligned}$$

Solving explicitly for α we see that the quadratic extension of $\mathbb{Q}$ generated by α is $\mathbb{Q}(\sqrt{5})$:

$$\mathbb{Q}(\zeta_5 + \zeta_5^{-1}) = \mathbb{Q}(\sqrt{5}).$$

It can be shown in general (this is not completely trivial) that for p an odd prime the field $\mathbb{Q}(\zeta_p)$ contains the quadratic field $\mathbb{Q}(\sqrt{\pm p})$, where the $+$ sign is correct if $p \equiv 1 \bmod 4$ and the $-$ sign is correct if $p \equiv 3 \bmod 4$ (cf. Exercise 11 in Section 7).

(2) $\mathbb{Q}(\zeta_{13})$. For p an odd prime we can construct a primitive element for any of the subfields of $\mathbb{Q}(\zeta_p)$ as in the previous example. A basis for $\mathbb{Q}(\zeta_p)$ over $\mathbb{Q}$ is given by

$$1, \zeta_p, \zeta_p^2, \dots, \zeta_p^{p-2}.$$

Since

$$\zeta_p^{p-1} + \zeta_p^{p-2} + \cdots + \zeta_p + 1 = 0$$

we see that also the elements

$$\zeta_p, \zeta_p^2, \dots, \zeta_p^{p-2}, \zeta_p^{p-1}$$

form a basis. The reason for choosing this basis is that any σ in the Galois group $\text{Gal}(\mathbb{Q}(\zeta_p)/\mathbb{Q})$ simply *permutes* these basis elements since these are precisely the primitive p^{th} roots of unity. Note that it is at this point that we need p to be a prime — in general the primitive n^{th} roots of unity do not give a basis for the cyclotomic field of nth roots of unity over $\mathbb{Q}$ (for example, the primitive 4th roots of unity, $\pm i$, are not linearly independent).

Let H be any subgroup of the Galois group of $\mathbb{Q}(\zeta_p)$ over $\mathbb{Q}$ and let

$$\alpha_H = \sum_{\sigma \in H} \sigma \zeta_p, \tag{14.10}$$

the sum of the conjugates of ζ_p by the elements in H. For any $\tau \in H$, the elements $\tau\sigma$ run over the elements of H as σ runs over the elements of H. It follows that $\tau\alpha = \alpha$, so that α lies in the fixed field for H. If now τ is *not* an element of H, then $\tau\alpha$ is the sum of basis elements (recall that any automorphism permutes the basis elements here), one of which is $\tau(\zeta_p)$. If we had $\tau\alpha = \alpha$ then since these elements are a basis, we must have $\tau(\zeta_p) = \sigma(\zeta_p)$ for one of the terms $\sigma\zeta_p$ in (10). But this implies $\tau\sigma^{-1} = 1$ since this automorphism is the identity on ζ_p. Then $\tau = \sigma \in H$, a contradiction. This shows that α is not fixed by any automorphism not contained in H, so that $\mathbb{Q}(\alpha)$ is precisely the fixed field of H.

For a specific example, consider the subfields of $\mathbb{Q}(\zeta_{13})$, which correspond to the subgroups of $(\mathbb{Z}/13\mathbb{Z})^\times \cong \mathbb{Z}/12\mathbb{Z}$. A generator for this cyclic group is the automorphism $\sigma = \sigma_2$ which maps ζ_{13} to ζ_{13}^2. The nontrivial subgroups correspond to the nontrivial divisors of 12, hence are of orders 2, 3, 4, and 6 with generators $\sigma^6, \sigma^4, \sigma^3$ and σ^2, respectively. The corresponding fixed fields will be of degrees 6, 4, 3 and 2 over $\mathbb{Q}$, respectively. Generators are given by ($\zeta = \zeta_{13}$)

$$\zeta + \sigma^6\zeta = \zeta + \zeta^{2^6} = \zeta + \zeta^{-1}$$
$$\zeta + \sigma^4\zeta + \sigma^8\zeta = \zeta + \zeta^{2^4} + \zeta^{2^8} = \zeta + \zeta^3 + \zeta^9$$
$$\zeta + \sigma^3\zeta + \sigma^6\zeta + \sigma^9\zeta = \zeta + \zeta^8 + \zeta^{12} + \zeta^5$$
$$\zeta + \sigma^2\zeta + \sigma^4\zeta + \sigma^6\zeta + \sigma^8\zeta + \sigma^{10}\zeta = \zeta + \zeta^4 + \zeta^3 + \zeta^{12} + \zeta^9 + \zeta^{10}.$$

The lattice of subfields for this extension is the following:

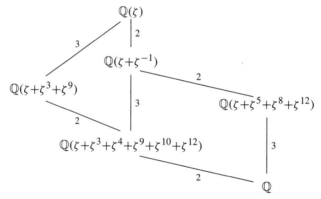

The elements constructed in equation (10) and their conjugates are called the *periods* of ζ and are useful in the study of the arithmetic of the cyclotomic fields. The study of their combinatorial properties is referred to as *cyclotomy*.

Suppose that $n = p_1^{a_1} p_2^{a_2} \cdots p_k^{a_k}$ is the decomposition of n into distinct prime powers. Since $\zeta_n^{p_2^{a_2}\cdots p_k^{a_k}}$ is a primitive $p_1^{a_1}$-th root of unity, the field $K_1 = \mathbb{Q}(\zeta_{p_1^{a_1}})$ is a subfield of $\mathbb{Q}(\zeta_n)$. Similarly, each of the fields $K_i = \mathbb{Q}(\zeta_{p_i^{a_i}})$, $i = 1, 2, \ldots, k$ is a subfield of $\mathbb{Q}(\zeta_n)$. The composite of the fields contains the product $\zeta_{p_1^{a_1}}\zeta_{p_2^{a_2}} \cdots \zeta_{p_k^{a_k}}$, which is a primitive n^{th} root of unity, hence the composite field is $\mathbb{Q}(\zeta_n)$. Since the extension degrees $[K_i : \mathbb{Q}]$ equal $\varphi(p_i^{a_i})$, $i = 1, 2, \ldots, k$ and $\varphi(n) = \varphi(p_1^{a_1})\varphi(p_2^{a_2}) \cdots \varphi(p_k^{a_k})$, the degree of the composite of the fields K_i is precisely the product of the degrees of the K_i. It follows from Proposition 21 (and a simple induction from the two fields considered in the proposition to the k fields here) that the intersection of all these fields is precisely $\mathbb{Q}$. Then Corollary 22 shows that the Galois group for $\mathbb{Q}(\zeta_n)$ is the direct product of the Galois groups over $\mathbb{Q}$ for the subfields K_i. We summarize this as the following corollary.

Corollary 27. Let $n = p_1^{a_1} p_2^{a_2} \cdots p_k^{a_k}$ be the decomposition of the positive integer n into distinct prime powers. Then the cyclotomic fields $\mathbb{Q}(\zeta_{p_i^{a_i}})$, $i = 1, 2, \ldots, k$ intersect only in the field $\mathbb{Q}$ and their composite is the cyclotomic field $\mathbb{Q}(\zeta_n)$. We have

$$\mathrm{Gal}(\mathbb{Q}(\zeta_n)/\mathbb{Q}) \cong \mathrm{Gal}(\mathbb{Q}(\zeta_{p_1^{a_1}})/\mathbb{Q}) \times \mathrm{Gal}(\mathbb{Q}(\zeta_{p_2^{a_2}})/\mathbb{Q}) \times \cdots \times \mathrm{Gal}(\mathbb{Q}(\zeta_{p_k^{a_k}})/\mathbb{Q})$$

which under the isomorphism in Theorem 26 is the Chinese Remainder Theorem:

$$(\mathbb{Z}/n\mathbb{Z})^\times \cong (\mathbb{Z}/p_1^{a_1}\mathbb{Z})^\times \times (\mathbb{Z}/p_2^{a_2}\mathbb{Z})^\times \times \cdots \times (\mathbb{Z}/p_k^{a_k}\mathbb{Z})^\times.$$

Proof: The only statement which has not been proved is the identification of the isomorphism of Galois groups with the statement of the Chinese Remainder Theorem on the group $(\mathbb{Z}/n\mathbb{Z})^\times$, which is quite simple and is left for the exercises.

By Theorem 26 the Galois group of $\mathbb{Q}(\zeta_n)/\mathbb{Q}$ is in particular an abelian group.

Definition. The extension K/F is called an *abelian* extension if K/F is Galois and $\mathrm{Gal}(K/F)$ is an abelian group.

Since all the subgroups and quotient groups of abelian groups are abelian, we see by the Fundamental Theorem of Galois Theory that every subfield containing F of an abelian extension of F is again an abelian extension of F. By the results on composites of extensions in the last section, we also see that the composite of abelian extensions is again an abelian extension (since the Galois group of the composite is isomorphic to a subgroup of the direct product of the Galois groups, hence is abelian).

It is an open problem to determine which groups arise as the Galois groups of Galois extensions of $\mathbb{Q}$. Using the results above we can see that every *abelian* group appears as the Galois group of some extension of $\mathbb{Q}$, in fact as the Galois group of some subfield of a cyclotomic field.

Let $n = p_1 p_2 \cdots p_k$ be the product of distinct primes. Then by the Chinese Remainder Theorem

$$\begin{aligned}(\mathbb{Z}/n\mathbb{Z})^\times &\cong (\mathbb{Z}/p_1\mathbb{Z})^\times \times (\mathbb{Z}/p_2\mathbb{Z})^\times \times \cdots \times (\mathbb{Z}/p_k\mathbb{Z})^\times \\ &\cong Z_{p_1-1} \times Z_{p_2-1} \times \cdots \times Z_{p_k-1}.\end{aligned} \qquad (14.11)$$

Now, suppose G is any finite abelian group. By the Fundamental Theorem for Abelian Groups,

$$G \cong Z_{n_1} \times Z_{n_2} \times \cdots \times Z_{n_k}$$

for some integers $n_1, n_2, \ldots, n_k$. We take as known that given any integer m there are infinitely many primes p with $p \equiv 1 \bmod m$ (see the exercises following Section 13.6

for one proof using cyclotomic polynomials). Given this result, choose distinct primes $p_1, p_2, \ldots, p_k$ such that

$$p_1 \equiv 1 \bmod n_1$$
$$p_2 \equiv 1 \bmod n_2$$
$$\vdots$$
$$p_k \equiv 1 \bmod n_k$$

and let $n = p_1 p_2 \cdots p_k$ as above.

By construction, n_i divides $p_i - 1$ for $i = 1, 2, \ldots, k$, so the group Z_{p_i-1} has a subgroup H_i of order $\dfrac{p_i - 1}{n_i}$ for $i = 1, 2, \ldots, k$, and the quotient by this subgroup is cyclic of order n_i. Hence the quotient of $(\mathbb{Z}/n\mathbb{Z})^\times$ in equation (11) by $H_1 \times H_2 \times \cdots \times H_k$ is isomorphic to the group G.

By Theorem 26 and the Fundamental Theorem of Galois Theory, we see that there is a subfield of $\mathbb{Q}(\zeta_{p_1 p_2 \cdots p_k})$ which is Galois over $\mathbb{Q}$ with G as Galois group. We summarize this in the following corollary.

Corollary 28. Let G be any finite abelian group. Then there is a subfield K of a cyclotomic field with $\mathrm{Gal}(K/\mathbb{Q}) \cong G$.

There is a converse to this result (whose proof is beyond our scope), the celebrated Kronecker–Weber Theorem:

Theorem *(Kronecker–Weber)* Let K be a finite abelian extension of $\mathbb{Q}$. Then K is contained in a cyclotomic extension of $\mathbb{Q}$.

The abelian extensions of $\mathbb{Q}$ are the "easiest" Galois extensions (at least in so far as the structure of their Galois groups is concerned) and the previous result shows they can be classified by the cyclotomic extensions of $\mathbb{Q}$. For other finite extensions of $\mathbb{Q}$ as base field, it is more difficult to describe the abelian extensions. The study of the abelian extensions of an arbitrary finite extension F of $\mathbb{Q}$ is referred to as *class field theory*. There is a classification of the abelian extensions of F by invariants associated to F which greatly generalizes the results on cyclotomic fields over $\mathbb{Q}$. In general, however, the construction of abelian extensions is not nearly as explicit as in the case of the cyclotomic fields. One case where such a description is possible is for the abelian extensions of an imaginary quadratic field ($\mathbb{Q}(\sqrt{-D})$ for D positive), where the abelian extensions can be constructed by adjoining values of certain elliptic functions (this is the analogue of adjoining the roots of unity, which are the values of the exponential function e^x for certain x). The study of the arithmetic of such abelian extensions and the search for similar results for non-abelian extensions are rich and fascinating areas of current mathematical research.

We end our discussion of the cyclotomic fields with the problem of the constructibility of the regular n-gon by straightedge and compass.

Recall (cf. Section 13.3) that an element α is constructible over $\mathbb{Q}$ if and only if the field $\mathbb{Q}(\alpha)$ is contained in a field K obtained by a series of quadratic extensions:

$$\mathbb{Q} = K_0 \subset K_1 \subset \cdots \subset K_i \subset K_{i+1} \subset \cdots \subset K_m = K \qquad (14.12)$$

with

$$[K_{i+1} : K_i] = 2, \qquad i = 0, 1, \ldots, m - 1.$$

The construction of the regular n-gon in $\mathbb{R}^2$ is evidently equivalent to the construction of the nth roots of unity, since the nth roots of unity form the vertices of a regular n-gon on the unit circle in $\mathbb{C}$ with one vertex at the point 1.

The construction of ζ_n is equivalent to the constructibility of the first coordinate x in $\mathbb{R}^2$ of ζ_n, namely the real part of ζ_n. Since the complex conjugate of ζ_n is just ζ_n^{-1}, the real part of ζ_n is $x = \dfrac{1}{2}(\zeta_n + \zeta_n^{-1})$. Note that ζ_n satisfies the quadratic equation $\zeta_n^2 - 2x\zeta_n + 1 = 0$ over $\mathbb{Q}(x)$. Since $\mathbb{Q}(x)$ consists only of real numbers, it follows that $[\mathbb{Q}(\zeta_n) : \mathbb{Q}(x)] = 2$, so that $\mathbb{Q}(x)$ is an extension of degree $\varphi(n)/2$ of $\mathbb{Q}$.

It follows that if the regular n-gon can be constructed by straightedge and compass then $\varphi(n)$ must be a power of 2. Conversely, if $\varphi(n) = 2^m$ is a power of 2, then the Galois group $\mathrm{Gal}(\mathbb{Q}(\zeta_n)/\mathbb{Q})$ is an abelian group whose order is a power of 2, so the same is true for the Galois group $\mathrm{Gal}(\mathbb{Q}(x)/\mathbb{Q})$. It is easy to see by the Fundamental Theorem for Abelian Groups that an abelian group G of order 2^m has a chain of subgroups

$$G = G_m > G_{m-1} > \cdots > G_{i+1} > G_i > \cdots > G_0 = 1$$

with

$$[G_{i+1} : G_i] = 2, \qquad i = 0, 1, 2, \ldots, m - 1.$$

Applying this to the group $G = \mathrm{Gal}(\mathbb{Q}(x)/\mathbb{Q})$ and taking the fixed fields for the subgroups G_i, $i = 0, 1, \ldots, m - 1$, we obtain (by the Fundamental Theorem of Galois Theory) a sequence of quadratic extensions as in (12) above.

We conclude that the regular n-gon can be constructed by straightedge and compass if and only if $\varphi(n)$ is a power of 2. Decomposing n into prime powers to compute $\varphi(n)$ we see that this means $n = 2^k p_1 \cdots p_r$ is the product of a power of 2 and distinct odd primes p_i where $p_i - 1$ is a power of 2. It is an elementary exercise to see that a prime p with $p - 1$ a power of 2 must be of the form

$$p = 2^{2^s} + 1$$

for some integer s. Such primes are called *Fermat primes*. The first few are

$$3 = 2^1 + 1$$
$$5 = 2^2 + 1$$
$$17 = 2^4 + 1$$
$$257 = 2^8 + 1$$
$$65537 = 2^{16} + 1$$

(but $2^{32} + 1$ is not a prime, being divisible by 641). It is not known if there are infinitely many Fermat primes. We summarize this in the following proposition.

Proposition 29. The regular n-gon can be constructed by straightedge and compass if and only if $n = 2^k p_1 \cdots p_r$ is the product of a power of 2 and distinct Fermat primes.

The proof above actually indicates a procedure for constructing the regular n-gon as a succession of square roots. For example, the construction of the regular 17-gon (solved by Gauss in 1796 at age 19) requires the construction of the subfields of degrees 2, 4, 8 and 16 in $\mathbb{Q}(\zeta_{17})$. These subfields can be constructed by forming the *periods* of ζ_{17} as in the example of the 13th roots of unity above. In this case, the fact that $\mathbb{Q}(\zeta_{17})$ is obtained by a series of quadratic extensions reflects itself in the fact that the periods can be "halved" successively (i.e., if $H_1 < H_2$ are subgroups with $[H_2 : H_1] = 2$ then the periods for H_1 satisfy a quadratic equation whose coefficients involve the periods for H_2). For example, the periods for the subgroup of index 2 (generated by σ_2) in the Galois group are ($\zeta = \zeta_{17}$)

$$\eta_1 = \zeta + \zeta^2 + \zeta^4 + \zeta^8 + \zeta^9 + \zeta^{13} + \zeta^{15} + \zeta^{16}$$
$$\eta_2 = \zeta^3 + \zeta^5 + \zeta^6 + \zeta^7 + \zeta^{10} + \zeta^{11} + \zeta^{12} + \zeta^{14}$$

which "halve" the period for the full Galois group and which satisfy

$$\eta_1 + \eta_2 = -1$$

(from the minimal polynomial satisfied by ζ_{17}) and

$$\eta_1 \eta_2 = -4$$

(which requires computation — we know that it must be rational by Galois Theory, since this product is fixed by all the elements of the Galois group). Hence these two periods are the roots of the quadratic equation

$$x^2 + x - 4 = 0$$

which we can solve explicitly. In a similar way, the periods for the subgroup of index 4 (generated by σ_4) naturally halve these periods, so are quadratic over these, etc. In this way one can determine ζ_{17} explicitly in terms of iterated square roots. For example, one finds that $8(\zeta + \zeta^{-1}) = 16\cos(\frac{2\pi}{17})$ (which is enough to construct the regular 17-gon) is given explicitly by

$$-1 + \sqrt{17} + \sqrt{2(17 - \sqrt{17})} + 2\sqrt{17 + 3\sqrt{17} - \sqrt{2(17 - \sqrt{17})} - 2\sqrt{2(17 + \sqrt{17})}}.$$

A relatively simple construction of the regular 17-gon (shown to us by J.H. Conway) is indicated in the exercises.

While we have seen that it is not possible to solve for ζ_n using only successive square roots in general, by definition it is possible to obtain ζ_n by successive extraction of higher roots (namely, taking an nth root of 1). This is not the case for solutions of general equations of degree n, where one cannot generally determine solutions by radicals, as we shall see in the next sections.

602

EXERCISES

1. Determine the minimal polynomials satisfied by the primitive generators given in the text for the subfields of $\mathbb{Q}(\zeta_{13})$.

2. Determine the subfields of $\mathbb{Q}(\zeta_8)$ generated by the periods of ζ_8 and in particular show that not every subfield has such a period as primitive element.

3. Determine the quadratic equation satisfied by the period $\alpha = \zeta_5 + \zeta_5^{-1}$ of the 5th root of unity ζ_5. Determine the quadratic equation satisfied by ζ_5 over $\mathbb{Q}(\alpha)$ and use this to explicitly solve for the 5th root of unity.

4. Let $\sigma_a \in \text{Gal}(\mathbb{Q}(\zeta_n)/\mathbb{Q})$ denote the automorphism of the cyclotomic field of nth roots of unity which maps ζ_n to ζ_n^a where a is relatively prime to n and ζ_n is a primitive n^{th} root of unity. Show that $\sigma_a(\zeta) = \zeta^a$ for *every* n^{th} root of unity.

5. Let p be a prime and let $\epsilon_1, \epsilon_2, \dots, \epsilon_{p-1}$ denote the primitive p^{th} roots of unity. Set $p_n = \epsilon_1^n + \epsilon_2^n + \cdots + \epsilon_{p-1}^n$, the sum of the nth powers of the ϵ_i. Prove that $p_n = -1$ if p does not divide n and that $p_n = p - 1$ if p does divide n. [One approach: $p_1 = -1$ from $\Phi_p(x)$; show that p_n is a Galois conjugate of p_1 for p not dividing n, hence is also -1.]

6. Let ζ_n denote a primitive n^{th} root of unity and let $K = \mathbb{Q}(\zeta_n)$ be the associated cyclotomic field. Let a denote the trace of ζ_n from K to $\mathbb{Q}$ (cf. Exercise 18 of Section 2). Prove that $a = 1$ if $n = 1$, $a = 0$ if n is divisible by the square of a prime, and $a = (-1)^r$ if n is the product of r distinct primes.

7. Show that complex conjugation restricts to the automorphism $\sigma_{-1} \in \text{Gal}(\mathbb{Q}(\zeta_n)/\mathbb{Q})$ of the cyclotomic field of nth roots of unity. Show that the field $K^+ = \mathbb{Q}(\zeta_n + \zeta_n^{-1})$ is the subfield of real elements in $K = \mathbb{Q}(\zeta_n)$, called the *maximal real subfield of K*.

8. Let $K_n = \mathbb{Q}(\zeta_{2^{n+2}})$ be the cyclotomic field of 2^{n+2}-th roots of unity, $n \geq 0$. Set $\alpha_n = \zeta_{2^{n+2}} + \zeta_{2^{n+2}}^{-1}$ and $K_n^+ = \mathbb{Q}(\alpha_n)$, the maximal real subfield of K_n.

   (a) Show that for all $n \geq 0$, $[K_n : \mathbb{Q}] = 2^{n+1}$, $[K_n : K_n^+] = 2$, $[K_n^+ : \mathbb{Q}] = 2^n$, and $[K_{n+1}^+ : K_n^+] = 2$.

   (b) Determine the quadratic equation satisfied by $\zeta_{2^{n+2}}$ over K_n^+ in terms of α_n.

   (c) Show that for $n \geq 0$, $\alpha_{n+1}^2 = 2 + \alpha_n$ and hence show that

   $$\alpha_n = \sqrt{2 + \sqrt{2 + \sqrt{\cdots + \sqrt{2}}}} \qquad (n \text{ times}),$$

   giving an explicit formula for the (constructible) 2^{n+2}-th roots of unity.

9. Notation as in the previous exercise.

   (a) Prove that K_n^+ is a cyclic extension of $\mathbb{Q}$ of degree 2^n. [Use an explicit isomorphism $(\mathbb{Z}/2^{n+2}\mathbb{Z})^\times \cong \mathbb{Z}/2\mathbb{Z} \times \mathbb{Z}/2^n\mathbb{Z}$ as abelian groups (i.e., $(\mathbb{Z}/2^{n+2}\mathbb{Z})^\times$ is isomorphic to a cyclic group of order 2 and a cyclic group of order 2^n — cf. Exercises 22 and 23 of Section 2.3]

   (b) Prove that K_n is a biquadratic extension of K_{n-1}^+ and that two of the three intermediate subfields are K_n^+ and K_{n-1}. Prove that the remaining field intermediate between K_{n-1}^+ and K_n is a cyclic extension of $\mathbb{Q}$ of degree 2^n.

10. Prove that $\mathbb{Q}(\sqrt[3]{2})$ is not a subfield of any cyclotomic field over $\mathbb{Q}$.

11. Prove that the primitive n^{th} roots of unity form a basis over $\mathbb{Q}$ for the cyclotomic field of nth roots of unity if and only if n is squarefree (i.e., n is not divisible by the square of any prime).

12. Let σ_p denote the Frobenius automorphism $x \mapsto x^p$ of the finite field $\mathbb{F}_q$ of $q = p^n$ elements. Viewing $\mathbb{F}_q$ as a vector space V of dimension n over $\mathbb{F}_p$ we can consider σ_p as a linear transformation of V to V. Determine the characteristic polynomial of σ_p and prove that the linear transformation σ_p is diagonalizable over $\mathbb{F}_p$ if and only if n divides $p - 1$, and is diagonalizable over the algebraic closure of $\mathbb{F}_p$ if and only if $(n, p) = 1$.

13. Let $n = p_1^{a_1} p_2^{a_2} \ldots p_k^{a_k}$ be the prime factorization of n and let ζ_n be a primitive n^{th} root of unity. For each $i = 1, 2, \ldots, k$ define d_i by $n = p_i^{a_i} d_i$ and let $\zeta_{p_i^{a_i}} = \zeta_n^{d_i}$, so that $\zeta_{p_i^{a_i}}$ is a particular primitive $p_i^{a_i}$-th root of unity. Let $\sigma_a \in \mathrm{Gal}(\mathbb{Q}(\zeta_n)/\mathbb{Q})$ be the automorphism mapping ζ_n to ζ_n^a for a relatively prime to n.
   (a) Prove that for $i = 1, 2, \ldots, k$, σ_a maps $\zeta_{p_i^{a_i}}$ to $\zeta_{p_i^{a_i}}^a$ and gives an automorphism of $\mathbb{Q}(\zeta_{p_i^{a_i}})/\mathbb{Q}$ which depends only on $a \pmod{p_i^{a_i}}$, which we may denote $\sigma_{a \pmod{p_i^{a_i}}}$.
   (b) Prove that the map $\sigma_a \mapsto (\sigma_{a \pmod{p_1^{a_1}}}, \ldots, \sigma_{a \pmod{p_k^{a_k}}})$ is the isomorphism of Corollary 27 corresponding to the Chinese Remainder Theorem for $(\mathbb{Z}/n\mathbb{Z})^\times$.

The following Exercises 14 to 18 determine the periods associated to a primitive 17th root of unity and provide a proof for the simple geometric construction indicated in Exercise 17 for the regular 17-gon. Let $\zeta = \zeta_{17} = \cos \dfrac{2\pi}{17} + i \sin \dfrac{2\pi}{17}$ be a fixed primitive 17th root of unity in $\mathbb{C}$.

14. Define the *periods* of ζ as follows:

$$\eta_1 = \zeta + \zeta^2 + \zeta^4 + \zeta^8 + \zeta^9 + \zeta^{13} + \zeta^{15} + \zeta^{16}$$
$$\eta_2 = \zeta^3 + \zeta^5 + \zeta^6 + \zeta^7 + \zeta^{10} + \zeta^{11} + \zeta^{12} + \zeta^{14}$$
$$\eta_1' = \zeta + \zeta^4 + \zeta^{13} + \zeta^{16}$$
$$\eta_2' = \zeta^2 + \zeta^8 + \zeta^9 + \zeta^{15}$$

$$\eta_3' = \zeta^6 + \zeta^7 + \zeta^{10} + \zeta^{11}$$
$$\eta_4' = \zeta^3 + \zeta^5 + \zeta^{12} + \zeta^{14}$$
$$\eta_1'' = \zeta + \zeta^{16}$$
$$\eta_2'' = \zeta^4 + \zeta^{13}.$$

   (a) Show that all of these periods are real numbers and that $\eta_1'' = 2 \cos \dfrac{2\pi}{17}$. Show that as real numbers these periods are approximately

$$\begin{array}{llll}
\eta_1 \sim 1.562 & \eta_1' \sim 2.049 & \eta_3' \sim -2.906 & \eta_1'' \sim 1.865 \\
\eta_2 \sim -2.562 & \eta_2' \sim -0.488 & \eta_4' \sim 0.344 & \eta_2'' \sim 0.185.
\end{array}$$

   (b) Prove that η_1 and η_2 are roots of the equation $x^2 + x - 4 = 0$.
   (c) Prove that η_1' and η_2' are roots of the equation $x^2 - \eta_1 x - 1 = 0$ and that η_3' and η_4' are roots of the equation $x^2 - \eta_2 x - 1 = 0$.
   (d) Prove that η_1'' and η_2'' are roots of the equation $x^2 - \eta_1' x + \eta_4' = 0$.

15. Prove that if $\tan 2\theta = a$ $(0 < 2\theta < \dfrac{\pi}{2})$ then $\tan \theta$ satisfies the equation $x^2 - \dfrac{2}{a} x - 1 = 0$.

16. Let C be the circle in $\mathbb{R}^2$ having the points (h, k) and $(0, 1)$ as a diameter. Prove that this circle intersects the x-axis if and only if $h^2 - 4k \geq 0$ and in this case the two intercepts are the roots of the equation $x^2 - hx + k = 0$.

17. (*Construction of the Regular 17-gon*) Draw a circle of radius 2 centered at the origin $(0, 0)$.
   (a) Join the point $(4, 0)$ to the point $(0, 1)$ and construct the line ℓ_1 bisecting the angle

604

between this line and the y-axis. Construct the line ℓ_2 perpendicular to ℓ_1 in Figure 2.

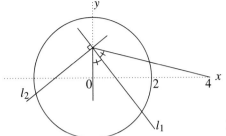

Fig. 2

(b) Using the intersection of ℓ_1 and the x-axis as center and radius equal to the distance to $(0, 1)$, construct the circle C_1 and let $A = (s, 0)$ be the right-hand point of intersection of C_1 with the x-axis. Similarly, let $B = (t, 0)$ denote the right-hand point of intersection of the x-axis and the circle C_2 whose center is the intersection of ℓ_2 and the x-axis and whose radius is equal to the distance to $(0, 1)$ as in Figure 3.

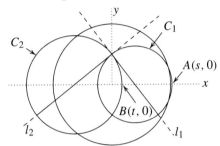

Fig. 3

(c) Construct a perpendicular to the x-axis at the point A and mark off the distance t from $(0, 0)$ to B to construct the point (s, t). Construct the circle with (s, t) and $(0, 1)$ as a diameter and let P denote the right-hand point of intersection of this circle with the x-axis. The perpendicular to the x-axis at P intersects the circle of radius 2 at the second vertex of a regular 17-gon whose first vertex is at $(2,0)$, hence constructs the regular 17-gon by straightedge and compass as in Figure 4.

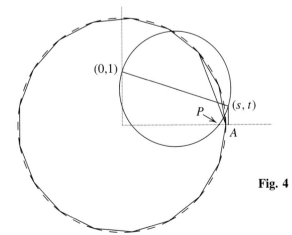

Fig. 4

18. Notation as in the previous exercises.

    (a) Prove that ℓ_1 intersects the x-axis in the point $(\eta_1/2, 0)$ and that ℓ_2 intersects the x-axis in the point $(\eta_2/2, 0)$.

    (b) Prove that C_1 is the circle having the points $(\eta_1, -1)$ and $(0, 1)$ as diameter. Prove that $s = \eta_1'$. Similarly prove that C_2 is the circle having the points $(\eta_2, -1)$ and $(0, 1)$ as diameter and that $t = \eta_4'$.

    (c) Prove that P has coordinates $(\eta_1'', 0)$ and hence that the construction in the previous problem constructs the regular 17-gon by straightedge and compass.

14.6 GALOIS GROUPS OF POLYNOMIALS

Recall that the Galois group of a separable polynomial $f(x) \in F[x]$ is defined to be the Galois group of the splitting field of $f(x)$ over F.

If K is a Galois extension of F then K is the splitting field for some separable polynomial $f(x)$ over F. Any automorphism $\sigma \in \mathrm{Gal}(K/F)$ maps a root of an irreducible factor of $f(x)$ to another root of the irreducible factor and σ is uniquely determined by its action on these roots (since they generate K over F). If we fix a labelling of the roots $\alpha_1, \ldots, \alpha_n$ of $f(x)$ we see that any $\sigma \in \mathrm{Gal}(K/F)$ defines a unique permutation of $\alpha_1, \ldots, \alpha_n$, hence defines a unique permutation of the subscripts $\{1, 2, \ldots, n\}$ (which depends on the fixed labelling of the roots). This gives an injection

$$\mathrm{Gal}(K/F) \hookrightarrow S_n$$

of the Galois group into the symmetric group on n letters which is clearly a homomorphism (both group operations are composition). We may therefore think of Galois groups as subgroups of symmetric groups. Since the degree of the splitting field is the same as the order of the Galois group by the Fundamental Theorem, this explains from the group-theoretic side why the splitting field for a polynomial of degree n over F is of degree at most $n!$ over F (Proposition 13.26).

In general, if the factorization of $f(x)$ into irreducibles is $f(x) = f_1(x) \cdots f_k(x)$ where $f_i(x)$ has degree n_i, $i = 1, 2, \ldots, k$, then since the Galois group permutes the roots of the irreducible factors among themselves we have $\mathrm{Gal}(K/F) \leq S_{n_1} \times \cdots \times S_{n_k}$.

If $f(x)$ is irreducible, then given any two roots of $f(x)$ there is an automorphism in the Galois group G of $f(x)$ which maps the first root to the second (this follows from our extension Theorem 13.27). Such a group is said to be *transitive* on the roots, i.e., you can get from any given root to any other root by applying some element of G. The fact that the Galois group must be transitive on blocks of roots (namely, the roots of the irreducible factors) can often be helpful in reducing the number of possibilities for the structure of G (cf. the discussion of Galois groups of polynomials of degree 4 below).

Examples

    (1) Consider the biquadratic extension $\mathbb{Q}(\sqrt{2}, \sqrt{3})$ over $\mathbb{Q}$, which is the splitting field of $(x^2 - 2)(x^2 - 3)$. Label the roots as $\alpha_1 = \sqrt{2}$, $\alpha_2 = -\sqrt{2}$, $\alpha_3 = \sqrt{3}$ and $\alpha_4 = -\sqrt{3}$. The elements of the Galois group are $\{1, \sigma, \tau, \sigma\tau\}$ where σ maps $\sqrt{2}$ to $-\sqrt{2}$ and fixes $\sqrt{3}$ and τ fixes $\sqrt{2}$ and maps $\sqrt{3}$ to $-\sqrt{3}$. As permutations of the roots for this labelling we see that σ interchanges the first two and fixes the second two and τ fixes the first two and interchanges the second two, i.e.,

$$\sigma = (12) \qquad \text{and} \qquad \tau = (34)$$

as elements of S_4. Similarly, or by taking the product of these two elements, we see that

$$\sigma\tau = (12)(34) \in S_4.$$

Hence

$$\text{Gal}(\mathbb{Q}(\sqrt{2}, \sqrt{3})/\mathbb{Q}) \cong \{1, (12), (34), (12)(34)\} \subset S_4$$

identifying this Galois group with the Klein-4 subgroup of S_4. Note that if we had changed the labelling of the roots above we would have obtained a different (isomorphic) representation of the Galois group as a subgroup of S_4 (for example, interchanging the second and third roots would have given the subgroup $\{1, (13), (24), (13)(24)\}$).

(2) The Galois group of $x^3 - 2$ acts as permutations on the three roots $\sqrt[3]{2}$, $\rho \sqrt[3]{2}$ and $\rho^2 \sqrt[3]{2}$ where ρ is a primitive 3rd root of unity. With this ordering, the generators σ and τ we have defined earlier give the permutations

$$\sigma = (123) \qquad \tau = (23)$$

which gives

$$\{1, \sigma, \sigma^2, \tau, \tau\sigma, \tau\sigma^2\} = \{1, (123), (132), (23), (13), (12)\} = S_3,$$

in this case the full symmetric group on 3 letters.

Recall that every finite group is isomorphic to a subgroup of some symmetric group S_n. It is an open problem to determine whether every finite group appears as the Galois group for some polynomial over $\mathbb{Q}$. We have seen in the last section that every abelian group is a Galois group over $\mathbb{Q}$ (for some subfield of a cyclotomic field). We shall explicitly determine the Galois groups for polynomials of small degree (≤ 4) below which will in particular show that every subgroup of S_4 arises as a Galois group.

We first introduce some definitions and show that the "general" polynomial of degree n has S_n as Galois group (so the second example above should be viewed as "typical").

Definition. Let $x_1, x_2, \ldots, x_n$ be indeterminates. The *elementary symmetric functions* $s_1, s_2, \ldots, s_n$ are defined by

$$s_1 = x_1 + x_2 + \cdots + x_n$$
$$s_2 = x_1 x_2 + x_1 x_3 + \cdots + x_2 x_3 + x_2 x_4 + \cdots + x_{n-1} x_n$$
$$\vdots$$
$$s_n = x_1 x_2 \cdots x_n$$

i.e., the ith symmetric function s_i of $x_1, x_2, \ldots, x_n$ is the sum of all products of the x_j's taken i at a time.

Definition. The *general polynomial of degree n* is the polynomial

$$(x - x_1)(x - x_2) \cdots (x - x_n)$$

whose roots are the indeterminates $x_1, x_2, \ldots, x_n$.

It is easy to see by induction that the coefficients of the general polynomial of degree n are given by the elementary symmetric functions in the roots:

$$(x - x_1)(x - x_2) \cdots (x - x_n) = x^n - s_1 x^{n-1} + s_2 x^{n-2} + \cdots + (-1)^n s_n. \quad (14.13)$$

For any field F, the extension $F(x_1, x_2, \ldots, x_n)$ is then a Galois extension of the field $F(s_1, s_2, \ldots, s_n)$ since it is the splitting field of the general polynomial of degree n.

If $\sigma \in S_n$ is any permutation of $\{1, 2, \ldots, n\}$, then σ acts on the rational functions in $F(x_1, x_2, \ldots, x_n)$ by permuting the subscripts of the variables $x_1, x_2, \ldots, x_n$. It is clear that this gives an automorphism of $F(x_1, x_2, \ldots, x_n)$. Identifying $\sigma \in S_n$ with this automorphism of $F(x_1, x_2, \ldots, x_n)$ identifies S_n as a subgroup of $\text{Aut}(F(x_1, x_2, \ldots, x_n))$.

The elementary symmetric functions $s_1, s_2, \ldots, s_n$ are fixed under any permutation of their subscripts (this is the reason they are called *symmetric*), which shows that the subfield $F(s_1, s_2, \ldots, s_n)$ is contained in the fixed field of S_n. By the Fundamental Theorem of Galois Theory, the fixed field of S_n has index precisely $n!$ in $F(x_1, x_2, \ldots, x_n)$. Since $F(x_1, x_2, \ldots, x_n)$ is the splitting field over $F(s_1, s_2, \ldots, s_n)$ of the polynomial of degree n in (13), we have

$$[F(x_1, x_2, \ldots, x_n) : F(s_1, s_2, \ldots, s_n)] \le n! \,. \quad (14.14)$$

It follows that we actually have equality and that $F(s_1, s_2, \ldots, s_n)$ is precisely the fixed field of S_n. This proves the following result.

Proposition 30. The fixed field of the symmetric group S_n acting on the field of rational functions in n variables $F(x_1, x_2, \ldots, x_n)$ is the field of rational functions in the elementary symmetric functions $F(s_1, s_2, \ldots, s_n)$.

Definition. A rational function $f(x_1, x_2, \ldots, x_n)$ is called *symmetric* if it is not changed by any permutation of the variables $x_1, x_2, \ldots, x_n$.

Corollary 31. (*Fundamental Theorem on Symmetric Functions*) Any symmetric function in the variables $x_1, x_2, \ldots, x_n$ is a rational function in the elementary symmetric functions $s_1, s_2, \ldots, s_n$.

Proof: A symmetric function lies in the fixed field of S_n above, hence is a rational function in $s_1, \ldots, s_n$.

This corollary explains why these are called the *elementary* symmetric functions.

Remark: If $f(x_1, \ldots, x_n)$ is a *polynomial* in $x_1, x_2, \ldots, x_n$ which is symmetric then it can be seen that f is actually a polynomial in $s_1, s_2, \ldots, s_n$, which strengthens the statement of the corollary. It is in fact true that a symmetric polynomial whose coefficients lie in R, where R is any commutative ring with identity, is a polynomial in the elementary symmetric functions with coefficients in R. A proof of this fact is implicit in the algorithm outlined in the exercises for writing a symmetric polynomial as a polynomial in the elementary symmetric functions.

Examples

   (1) The expression $(x_1 - x_2)^2$ is symmetric in x_1, x_2. We have

$$(x_1 - x_2)^2 = (x_1 + x_2)^2 - 4x_1x_2 = s_1^2 - 4s_2,$$

    a polynomial in the elementary symmetric functions.

   (2) The polynomial $x_1^2 + x_2^2 + x_3^2$ is symmetric in x_1, x_2, x_3, and in this case we have

$$x_1^2 + x_2^2 + x_3^2 = (x_1 + x_2 + x_3)^2 - 2(x_1x_2 + x_1x_3 + x_2x_3)$$
$$= s_1^2 - 2s_2.$$

   (3) The polynomial $x_1^2x_2^2 + x_1^2x_3^2 + x_2^2x_3^2$ is symmetric. Since

$$(x_1x_2 + x_1x_3 + x_2x_3)^2 = x_1^2x_2^2 + x_1^2x_3^2 + x_2^2x_3^2 + 2(x_1^2x_2x_3 + x_2^2x_1x_3 + x_3^2x_1x_2)$$
$$= x_1^2x_2^2 + x_1^2x_3^2 + x_2^2x_3^2 + 2x_1x_2x_3(x_1 + x_2 + x_3)$$

    we have

$$x_1^2x_2^2 + x_1^2x_3^2 + x_2^2x_3^2 = s_2^2 - 2s_1s_3.$$

Suppose now we *start* with the general polynomial

$$x^n - s_1x^{n-1} + s_2x^{n-2} + \cdots + (-1)^n s_n$$

over the field $F(s_1, s_2, \ldots, s_n)$ where we view the $s_i, i = 1, 2, \ldots, n$ as indeterminates. If we define the roots of this polynomial to be $x_1, x_2, \ldots, x_n$ then the s_i are precisely the elementary symmetric functions in the roots $x_1, \ldots, x_n$. Moreover, these roots are indeterminates as well in the sense that there are no polynomial relations over F between them. For suppose $p(t_1, \ldots, t_n)$ is a nonzero polynomial in n variables with coefficients in F such that $p(x_1, \ldots, x_n) = 0$. Then the product, $\widetilde{p}$, over all σ in S_n of $p(t_{\sigma(1)}, \ldots, t_{\sigma(n)})$ is a nonzero symmetric polynomial with $\widetilde{p}(x_1, \ldots, x_n) = 0$. This gives a nonzero polynomial relation over F among $s_1, \ldots, s_n$, a contradiction. Conversely, if the roots of a polynomial $f(x)$ are independent indeterminates over F, then so are the coefficients of $f(x)$ — cf. the beginning of Section 9. Thus defining the general polynomial over F as having indeterminate roots or indeterminate coefficients is equivalent. From this point of view our result can be stated in the following form.

Theorem 32. The general polynomial

$$x^n - s_1x^{n-1} + s_2x^{n-2} + \cdots + (-1)^n s_n$$

over the field $F(s_1, s_2, \ldots, s_n)$ is separable with Galois group S_n.

    This result says that if there are no relations among the coefficients of a polynomial of degree n (which is what we mean when we say the s_i are indeterminates above) then the Galois group of this polynomial over the field generated by its coefficients is the full symmetric group S_n. Loosely speaking, this means that the "generic" polynomial of degree n will have S_n as Galois group. Note, however, that over finite fields every polynomial has a *cyclic* Galois group (all extensions of finite fields are cyclic), so that "generic" polynomials in this sense do not exist. Over $\mathbb{Q}$ one can make precise the notion of "generic" polynomial and then it is true that most polynomials have the full symmetric group as Galois group.

For $n \geq 5$ there is only one normal subgroup of S_n, namely the subgroup A_n of index 2. Hence in general there is only one normal subfield of $F(x_1, x_2, \ldots, x_n)$ containing $F(s_1, s_2, \ldots, s_n)$ and it is an extension of degree 2.

Definition. Define the *discriminant* D of $x_1, x_2, \ldots, x_n$ by the formula

$$D = \prod_{i<j} (x_i - x_j)^2.$$

Define the discriminant of a polynomial to be the discriminant of the roots of the polynomial.

The discriminant D is a symmetric function in $x_1, \ldots, x_n$, hence is an element of $K = F(s_1, s_2, \ldots, s_n)$.

When we first defined the alternating group A_n we saw that a permutation $\sigma \in S_n$ is an element of the subgroup A_n if and only if σ fixes the product

$$\sqrt{D} = \prod_{i<j} (x_i - x_j) \in \mathbb{Z}[x_1, x_2, \ldots, x_n].$$

It follows (by the Fundamental Theorem) that if F has characteristic different from 2 then $\sqrt{D}$ generates the fixed field of A_n and generates a quadratic extension of K. This proves the following proposition.

Proposition 33. If $\mathrm{ch}(F) \neq 2$ then the permutation $\sigma \in S_n$ is an element of A_n if and only if it fixes the square root of the discriminant D.

We now consider the Galois groups of separable polynomials of small degree (≤ 4) over a field F which we assume is of characteristic different from 2 and 3. Note that over $\mathbb{Q}$ or over a finite field (or, more generally, over any perfect field) the splitting field of an arbitrary polynomial $f(x)$ is the same as the splitting field for the product of the irreducible factors of $f(x)$ taken precisely once, which is a separable polynomial.

If the roots of the polynomial $f(x) = x^n + a_{n-1}x^{n-1} + \cdots + a_1 x + a_0$ are $\alpha_1, \alpha_2, \ldots, \alpha_n$, then the discriminant of $f(x)$ is[2]

$$D = \prod_{i<j} (\alpha_i - \alpha_j)^2.$$

Note that $D = 0$ if and only if $f(x)$ is not separable, i.e., if the roots $\alpha_1, \ldots, \alpha_n$ are not distinct. Recall that over a perfect field (e.g., $\mathbb{Q}$ or a finite field) this implies $f(x)$ is reducible since every irreducible polynomial over a perfect field is separable.

The discriminant D is symmetric in the roots of $f(x)$, hence is fixed by all the automorphisms of the Galois group of $f(x)$. By the Fundamental Theorem it follows that

---

[2]If $f(x) = a_n x^n + \cdots + a_0$ is not monic then its discriminant is defined to be a_n^{2n-2} times the D defined above.

$D \in F$. The discriminant can in general be written as a polynomial in the coefficients of $f(x)$ (by Corollary 31) which are fairly complicated for larger degrees (we shall give formulas for $n \le 4$ below). Finally, note that since

$$\sqrt{D} = \prod_{i<j}(\alpha_i - \alpha_j)$$

we have the useful fact that $\sqrt{D}$ is always contained in the splitting field for $f(x)$.

If the roots of $f(x)$ are distinct, fix some ordering of the roots and view the Galois group of $f(x)$ as a subgroup of S_n as above.

Proposition 34. The Galois group of $f(x) \in F[x]$ is a subgroup of A_n if and only if the discriminant $D \in F$ is the square of an element of F.

Proof: This is a restatement of Proposition 33 in this case. The Galois group is contained in A_n if and only if every element of the Galois group fixes

$$\sqrt{D} = \prod_{i<j}(\alpha_i - \alpha_j)$$

i.e., if and only if $\sqrt{D} \in F$.

This property, together with the fact that $D = 0$ determines the presence of multiple roots, is the reason D is called the *discriminant*.

Polynomials of Degree 2

Consider the polynomial $x^2 + ax + b$ with roots α, β. The discriminant D for this polynomial is $(\alpha - \beta)^2$, which can be written as a polynomial in the elementary symmetric functions of the roots. We did this in Example 1 above:

$$D = s_1^2 - 4s_2 = (-a)^2 - 4(b) = a^2 - 4b,$$

the usual discriminant for this quadratic.

The polynomial is separable if and only if $a^2 - 4b \ne 0$. The Galois group is a subgroup of S_2, the cyclic group of order 2 and is trivial (i.e., A_2 in this case) if and only if $a^2 - 4b$ is a rational square, which completely determines the possible Galois groups.

Note that this restates results we obtained previously by explicitly solving for the roots: if the polynomial is reducible (namely D is a square in F), then the Galois group is trivial (the splitting field is just F), while if the polynomial is irreducible the Galois group is isomorphic to $\mathbb{Z}/2\mathbb{Z}$ since the splitting field is the quadratic extension $F(\sqrt{D})$.

Polynomials of degree 3

Suppose the cubic polynomial is

$$f(x) = x^3 + ax^2 + bx + c. \tag{14.15}$$

If we make the substitution $x = y - a/3$ the polynomial becomes

$$g(y) = y^3 + py + q \tag{14.16}$$

where
$$p = \frac{1}{3}(3b - a^2) \qquad q = \frac{1}{27}(2a^3 - 9ab + 27c). \qquad (14.17)$$

The splitting fields for these two polynomials are the same since their roots differ by the constant $a/3 \in F$ and since the formula for the discriminant involves the *differences* of roots, we see that these two polynomials also have the *same* discriminant.

Let the roots of the polynomial in (16) be α, β, and γ. We first compute the discriminant of this polynomial in terms of p and q. Note that

$$g(y) = (y - \alpha)(y - \beta)(y - \gamma)$$

so that if we differentiate we have

$$D_y g(y) = (y - \alpha)(y - \beta) + (y - \alpha)(y - \gamma) + (y - \beta)(y - \gamma).$$

Then

$$D_y g(\alpha) = (\alpha - \beta)(\alpha - \gamma)$$
$$D_y g(\beta) = (\beta - \alpha)(\beta - \gamma)$$
$$D_y g(\gamma) = (\gamma - \alpha)(\gamma - \beta).$$

Taking the product we see that

$$D = [(\alpha - \beta)(\alpha - \gamma)(\beta - \gamma)]^2 = -D_y g(\alpha) D_y g(\beta) D_y g(\gamma).$$

Since $D_y g(y) = 3y^2 + p$, we have

$$-D = (3\alpha^2 + p)(3\beta^2 + p)(3\gamma^2 + p)$$
$$= 27\alpha^2\beta^2\gamma^2 + 9p(\alpha^2\beta^2 + \alpha^2\gamma^2 + \beta^2\gamma^2) + 3p^2(\alpha^2 + \beta^2 + \gamma^2) + p^3.$$

The corresponding expressions in the elementary symmetric functions of the roots were determined in Examples 2 and 3 above. Note that here $s_1 = 0$, $s_2 = p$ and $s_3 = -q$. We obtain

$$-D = 27(-q)^2 + 9p(p^2) + 3p^2(-2p) + p^3$$

so that

$$D = -4p^3 - 27q^2. \qquad (14.18)$$

This is the same as the discriminant of $f(x)$ in (15). Expressing D in terms of a, b, c using (17) we obtain

$$D = a^2b^2 - 4b^3 - 4a^3c - 27c^2 + 18abc \qquad (14.18')$$

(Galois Group of a Cubic)

a. If the cubic polynomial $f(x)$ is reducible, then it splits either into three linear factors or into a linear factor and an irreducible quadratic. In the first case the Galois group is trivial and in the second case the Galois group is of order 2.

b. If the cubic polynomial $f(x)$ is irreducible then a root of $f(x)$ generates an extension of degree 3 over F, so the degree of the splitting field over F is divisible by 3. Since the Galois group is a subgroup of S_3, there are only two possibilities, namely

612

A_3 or S_3. The Galois group is A_3 (i.e., cyclic of order 3) if and only if the discriminant D in (18) is a square.

Explicitly, if D is the square of an element of F, then the splitting field of the irreducible cubic $f(x)$ is obtained by adjoining any single root of $f(x)$ to F. The resulting field is Galois over F of degree 3 with a cyclic group of order 3 as Galois group. If D is not the square of an element of F then the splitting field of $f(x)$ is of degree 6 over F, hence is the field $F(\theta, \sqrt{D})$ for any one of the roots θ of $f(x)$. This extension is Galois over F with Galois group S_3 (generators are given by σ, which takes θ to one of the other roots of $f(x)$ and fixes $\sqrt{D}$, and τ, which takes $\sqrt{D}$ to $-\sqrt{D}$ and fixes θ).

We see that in both cases the splitting field for the irreducible cubic $f(x)$ is obtained by adjoining $\sqrt{D}$ and a root of $f(x)$ to F.

We shall give explicit formulas for the roots of (16) (*Cardano's Formulas*) in the next section after introducing the notion of a *Lagrange Resolvent*.

Polynomials of Degree 4

Let the quartic polynomial be

$$f(x) = x^4 + ax^3 + bx^2 + cx + d$$

which under the substitution $x = y - a/4$ becomes the quartic

$$g(y) = y^4 + py^2 + qy + r$$

with

$$p = \frac{1}{8}(-3a^2 + 8b)$$

$$q = \frac{1}{8}(a^3 - 4ab + 8c)$$

$$r = \frac{1}{256}(-3a^4 + 16a^2b - 64ac + 256d).$$

Let the roots of $g(y)$ be α_1, α_2, α_3, and α_4 and let G denote the Galois group for the splitting field of $g(y)$ (or of $f(x)$).

Suppose first that $g(y)$ is reducible. If $g(y)$ splits into a linear and a cubic, then G is the Galois group of the cubic, which we determined above. Suppose then that $g(y)$ splits into two irreducible quadratics. Then the splitting field is the extension $F(\sqrt{D_1}, \sqrt{D_2})$ where D_1 and D_2 are the discriminants of the two quadratics. If D_1 and D_2 do not differ by a square factor then this extension is a biquadratic extension and G is isomorphic to the Klein 4-subgroup of S_4. If D_1 is a square times D_2 then this extension is a quadratic extension and G is isomorphic to $\mathbb{Z}/2\mathbb{Z}$.

We are reduced to the situation where $g(y)$ is irreducible. In this case recall that the Galois group is transitive on the roots, i.e., it is possible to get from a given root to any other root by applying some automorphism of the Galois group. Examining the possibilities we see that the only transitive subgroups of S_4, hence the only possibilities for our Galois group G, are the groups

S_4, A_4

$D_8 = \{1, (1324), (12)(34), (1423), (13)(24), (14)(23), (12), (34)\}$ and its conjugates

$V = \{1, (12)(34), (13)(24), (14)(23)\}$

$C = \{1, (1234), (13)(24), (1432)\}$ and its conjugates.

(D_8 is the dihedral group, a Sylow 2-subgroup of S_4, with 3 (isomorphic) conjugate subgroups in S_4, V is the Klein 4-subgroup of S_4, normal in S_4, and C is a cyclic group, with 3 (isomorphic) conjugates in S_4).

Consider the elements

$$\theta_1 = (\alpha_1 + \alpha_2)(\alpha_3 + \alpha_4)$$
$$\theta_2 = (\alpha_1 + \alpha_3)(\alpha_2 + \alpha_4)$$
$$\theta_3 = (\alpha_1 + \alpha_4)(\alpha_2 + \alpha_3)$$

in the splitting field for $g(y)$. These elements are permuted amongst themselves by the permutations in S_4. The stabilizer of θ_1 in S_4 is the dihedral group D_8. The stabilizers in S_4 of θ_2 and θ_3 are the conjugate dihedral subgroups of order 8. The subgroup of S_4 which stabilizes all three of these elements is the intersection of these subgroups, namely the Klein 4-group V.

Since S_4 merely permutes $\theta_1, \theta_2, \theta_3$ it follows that the elementary symmetric functions in the θ's are fixed by all the elements of S_4, hence are in F. An elementary computation in symmetric functions shows that these elementary symmetric functions are $2p$, $p^2 - 4r$, and $-q^2$, which shows that $\theta_1, \theta_2, \theta_3$ are the roots of

$$h(x) = x^3 - 2px^2 + (p^2 - 4r)x + q^2$$

called the *resolvent cubic* for the quartic $g(y)$. Since

$$\theta_1 - \theta_2 = \alpha_1\alpha_3 + \alpha_2\alpha_4 - \alpha_1\alpha_2 - \alpha_3\alpha_4$$
$$= -(\alpha_1 - \alpha_4)(\alpha_2 - \alpha_3)$$

and similarly

$$\theta_1 - \theta_3 = -(\alpha_1 - \alpha_3)(\alpha_2 - \alpha_4)$$
$$\theta_2 - \theta_3 = -(\alpha_1 - \alpha_2)(\alpha_3 - \alpha_4)$$

we see that the discriminant of the resolvent cubic is the *same* as the discriminant of the quartic $g(y)$, hence also as the discriminant of the quartic $f(x)$. Using our formula for the discriminant of the cubic, we can easily compute the discriminant in terms of p, q, r:

$$D = 16p^4r - 4p^3q^2 - 128p^2r^2 + 144pq^2r - 27q^4 + 256r^3$$

from which one can give the formula for D in terms of a, b, c, d:

$$D = -128b^2d^2 - 4a^3c^3 + 16b^4d - 4b^3c^2 - 27a^4d^2 + 18abc^3$$
$$+ 144a^2bd^2 - 192acd^2 + a^2b^2c^2 - 4a^2b^3d - 6a^2c^2d$$
$$+ 144bc^2d + 256d^3 - 27c^4 - 80ab^2cd + 18a^3bcd.$$

The splitting field for the resolvent cubic is a subfield of the splitting field of the quartic, so the Galois group of the resolvent cubic is a quotient of G. Hence knowing the action of the Galois group on the roots of the resolvent cubic $h(x)$ gives information about the Galois group of $g(y)$, as follows:

(Galois group of a quartic)

a. Suppose first that the resolvent cubic is irreducible. If D is not a square, then G is not contained in A_4 and the Galois group of the resolvent cubic is S_3, which implies that the degree of the splitting field for $g(y)$ is divisible by 6. The only possibility is then $G = S_4$.

b. If the resolvent cubic is irreducible and D is a square, then G is a subgroup of A_4 and 3 divides the order of G (the Galois group of the resolvent cubic is A_3). The only possibility is $G = A_4$.

c1. We are left with the case where the resolvent cubic is reducible. The first possibility is that $h(x)$ has 3 roots in F (i.e., splits completely). Since each of the elements $\theta_1, \theta_2, \theta_3$ is in F, every element of G fixes all three of these elements, which means $G \subseteq V$. The only possibility is $G = V$.

c2. If $h(x)$ splits into a linear and a quadratic, then precisely one of $\theta_1, \theta_2, \theta_3$ is in F, say θ_1. Then G stabilizes θ_1 but not θ_2 and θ_3, so we have $G \subseteq D_8$ and $G \not\subseteq V$. This leaves two possibilities: $G = D_8$ or $G = C$. One way to distinguish between these is to observe that $F(\sqrt{D})$ is the fixed field of the elements of G in A_4. For the two cases being considered, we have $D_8 \cap A_4 = V$, $C \cap A_4 = \{1, (13)(24)\}$. The first group is transitive on the roots of $g(y)$, the second is not. It follows that the first case occurs if and only if $g(y)$ is irreducible over $F(\sqrt{D})$. We may therefore determine G completely by factoring $g(y)$ in $F(\sqrt{D})$, and so completely determine the Galois group in all cases. (cf. the exercises following and in the next section, where it is shown that over $\mathbb{Q}$ the Galois group cannot be cyclic of degree 4 if D is not the sum of two squares — so in particular if $D < 0$.)

We shall give explicit formulas for the roots of a quartic polynomial at the end of the next section.

The Fundamental Theorem of Algebra

We end this section with two proofs of the Fundamental Theorem of Algebra. We need two facts regarding the field $\mathbb{C}$:

(a) Every polynomial with real coefficients of odd degree has a root in the reals. Equivalently, there are no nontrivial finite extensions of $\mathbb{R}$ of odd degree.

(b) Quadratic polynomials with coefficients in $\mathbb{C}$ have roots in $\mathbb{C}$. Equivalently, there are no quadratic extensions of $\mathbb{C}$.

The first result follows from the Intermediate Value Theorem in calculus, since the graph of a monic polynomial $f(x) \in \mathbb{R}[x]$ of odd degree is negative for large negative values of x and positive for large positive values of x, hence crosses the axis somewhere. The equivalence with the second statement follows since a finite extension of $\mathbb{R}$ is a simple extension and the minimal polynomial of a primitive element would have odd degree, hence would be both irreducible over $\mathbb{R}$ and have a root in $\mathbb{R}$, hence must be of degree 1.

The second result follows by a direct computation. By the quadratic formula it suffices to show that every complex number $\alpha = a + bi$, $a, b \in \mathbb{R}$, has a square root in $\mathbb{C}$. Write $\alpha = re^{i\theta}$ for some $r \geq 0$ and some $\theta \in [0, 2\pi)$. Then $\sqrt{r}e^{i\theta/2}$ is a square root of α. (Explicitly, let $c \in \mathbb{R}$ be a square root of the real number $\dfrac{a + \sqrt{a^2 + b^2}}{2}$ and let $d \in \mathbb{R}$ be a square root of the real number $\dfrac{-a + \sqrt{a^2 + b^2}}{2}$ where the signs of the two square roots are chosen so that cd has the same sign as b. Then multiplying out we see that $(c + di)^2 = a + bi$.)

Theorem 35. *(Fundamental Theorem of Algebra)* Every polynomial $f(x) \in \mathbb{C}[x]$ of degree n has precisely n roots in $\mathbb{C}$ (counted with multiplicity). Equivalently, $\mathbb{C}$ is algebraically closed.

Proof: I. It suffices to prove that every polynomial $f(x) \in \mathbb{C}[x]$ has a root in $\mathbb{C}$. Let τ denote the automorphism complex conjugation. If $f(x)$ has no root in $\mathbb{C}$ then neither does the conjugate polynomial $\bar{f}(x) = \tau f(x)$ obtained by applying τ to the coefficients of $f(x)$ (since its roots are the conjugates of the roots of $f(x)$). The product $f(x)\bar{f}(x)$ has coefficients which are invariant under complex conjugation, hence has real coefficients. It suffices then to prove that a polynomial with real coefficients has a root in $\mathbb{C}$.

Suppose that $f(x)$ is a polynomial of degree n with real coefficients and write $n = 2^k m$ where m is odd. We prove that $f(x)$ has a root in $\mathbb{C}$ by induction on k. For $k = 0$, $f(x)$ has odd degree and by (a) above $f(x)$ has a root in $\mathbb{R}$ so we are done. Suppose now that $k \geq 1$. Let $\alpha_1, \alpha_2, \ldots, \alpha_n$ be the roots of $f(x)$ and set $K = \mathbb{R}(\alpha_1, \alpha_2, \ldots, \alpha_n, i)$. Then K is a Galois extension of $\mathbb{R}$ containing $\mathbb{C}$ and the roots of $f(x)$. For any $t \in \mathbb{R}$ consider the polynomial

$$L_t = \prod_{1 \leq i < j \leq n} [x - (\alpha_i + \alpha_j + t\alpha_i\alpha_j)].$$

Any automorphism of $K/\mathbb{R}$ permutes the terms in this product so the coefficients of L_t are invariant under all the elements of $\mathrm{Gal}(K/\mathbb{R})$. Hence L_t is a polynomial with real coefficients. The degree of L_t is

$$\frac{n(n-1)}{2} = 2^{k-1}m(2^k m - 1) = 2^{k-1}m'$$

where m' is odd (since $k \geq 1$). The power of 2 in this degree is therefore less than k, so by induction the polynomial L_t has a root in $\mathbb{C}$. Hence for each $t \in \mathbb{R}$ one of the elements $\alpha_i + \alpha_j + t\alpha_i\alpha_j$ for some i, j ($1 \leq i < j \leq n$) is an element of $\mathbb{C}$. Since there are infinitely many choices for t and only finitely many values of i and j we see that for some i and j (say, $i = 1$ and $j = 2$) there are distinct real numbers s and t with

$$\alpha_1 + \alpha_2 + s\alpha_1\alpha_2 \in \mathbb{C} \qquad \alpha_1 + \alpha_2 + t\alpha_1\alpha_2 \in \mathbb{C}.$$

616

Since $s \neq t$ it follows that $a = \alpha_1 + \alpha_2 \in \mathbb{C}$ and $b = \alpha_1\alpha_2 \in \mathbb{C}$. But then α_1 and α_2 are the roots of the quadratic $x^2 - ax + b$ with coefficients in $\mathbb{C}$, hence are elements of $\mathbb{C}$ by (b) above, completing the proof.

II. The second proof again uses (a) and (b) above, but replaces the computations with the polynomials L_t above with a simple group-theoretic argument involving the nilpotency of a Sylow 2-subgroup of the Galois group:

Let $f(x)$ be a polynomial of degree n with real coefficients and let K be the splitting field of $f(x)$ over $\mathbb{R}$. Then $K(i)$ is a Galois extension of $\mathbb{R}$. Let G denote its Galois group and let P_2 denote a Sylow 2-subgroup of G. The fixed field of P_2 is an extension of $\mathbb{R}$ of odd degree, hence by (a) is trivial.

It follows that $\mathrm{Gal}(K(i)/\mathbb{C})$ is a 2-group. Since 2-groups have subgroups of all orders (recall this is true of a finite p-group for any prime p, cf. Theorem 6.1), if this group is nontrivial, there would exist a quadratic extension of $\mathbb{C}$, impossible by (b), completing the proof.

The Fundamental Theorem of Algebra was first rigorously proved by Gauss in 1816 (his doctoral dissertation in 1798 provides a proof using geometric considerations requiring some topological justification). The first proof above is essentially due to Laplace in 1795 (hence the reason for naming the polynomials L_t). The reason Laplace's proof was deemed unacceptable was that he assumed the existence of a splitting field for polynomials (i.e., that the roots existed *somewhere* in *some* field), which had not been established at that time. The elegant second proof is a simplification due to Artin.

EXERCISES

1. Show that a cubic with a multiple root has a linear factor. Is the same true for quartics?

2. Determine the Galois groups of the following polynomials:
   (a) $x^3 - x^2 - 4$
   (b) $x^3 - 2x + 4$
   (c) $x^3 - x + 1$
   (d) $x^3 + x^2 - 2x - 1$.

3. Prove for any $a, b \in \mathbb{F}_{p^n}$ that if $x^3 + ax + b$ is irreducible then $-4a^3 - 27b^2$ is a square in $\mathbb{F}_{p^n}$.

4. Determine the Galois group of $x^4 - 25$.

5. Determine the Galois group of $x^4 + 4$.

6. Determine the Galois group of $x^4 + 3x^3 - 3x - 2$.

7. Determine the Galois group of $x^4 + 2x^2 + x + 3$.

8. Determine the Galois group of $x^4 + 8x + 12$.

9. Determine the Galois group of $x^4 + 4x - 1$ (cf. Exercise 19).

10. Determine the Galois group of $x^5 + x - 1$.

11. Let F be an extension of $\mathbb{Q}$ of degree 4 that is not Galois over $\mathbb{Q}$. Prove that the Galois closure of F has Galois group either S_4, A_4 or the dihedral group D_8 of order 8. Prove that the Galois group is dihedral if and only if F contains a quadratic extension of $\mathbb{Q}$.

12. Prove that an extension F of $\mathbb{Q}$ of degree 4 can be generated by the root of an irreducible biquadratic $x^4 + ax^2 + b$ over $\mathbb{Q}$ if and only if F contains a quadratic extension of $\mathbb{Q}$.

13. **(a)** Let $\pm\alpha, \pm\beta$ denote the roots of the polynomial $f(x) = x^4 + ax^2 + b \in \mathbb{Z}[x]$. Prove that $f(x)$ is irreducible if and only if $\alpha^2, \alpha \pm \beta$ are not elements of $\mathbb{Q}$.[3]

    (b) Suppose $f(x)$ is irreducible and let G be the Galois group of $f(x)$. Prove that

        (i) $G \cong V$, the Klein 4-group, if and only if b is a square in $\mathbb{Q}$ if and only if $\alpha\beta \in \mathbb{Q}$ is rational.

        (ii) $G \cong C$, the cyclic group of order 4, if and only if $b(a^2 - 4b)$ is a square in $\mathbb{Q}$ if and only if $\mathbb{Q}(\alpha\beta) = \mathbb{Q}(\alpha^2)$.

        (iii) $G \cong D_8$, the dihedral group of order 8, if and only if b and $b(a^2 - 4b)$ are not squares in $\mathbb{Q}$ if and only if $\alpha\beta \notin \mathbb{Q}(\alpha^2)$.

14. Prove the polynomial $x^4 - px^2 + q \in \mathbb{Q}[x]$ is irreducible for any distinct odd primes p and q and has as Galois group the dihedral group of order 8.[4]

15. Prove the polynomial $x^4 + px + p \in \mathbb{Q}[x]$ is irreducible for every prime p and for $p \neq 3, 5$ has Galois group S_4. Prove the Galois group for $p = 3$ is dihedral of order 8 and for $p = 5$ is cyclic of order 4.[5]

16. Determine the Galois group over $\mathbb{Q}$ of the polynomial $x^4 + 8x^2 + 8x + 4$. Determine which of the subfields of this field are Galois over $\mathbb{Q}$ and for those which are Galois determine a polynomial $f(x) \in \mathbb{Q}[x]$ for which they are the splitting field over $\mathbb{Q}$.

17. Find the Galois group of $x^4 - 7$ over $\mathbb{Q}$ explicitly as a permutation group on the roots.

18. Let θ be a root of $x^3 - 3x + 1$. Prove that the splitting field of this polynomial is $\mathbb{Q}(\theta)$ and that the Galois group is cyclic of order 3. In particular the other roots of this polynomial can be written in the form $a + b\theta + c\theta^2$ for some $a, b, c \in \mathbb{Q}$. Determine the other roots explicitly in terms of θ.

19. Let $f(x)$ be an irreducible polynomial of degree 4 in $\mathbb{Q}[x]$ with discriminant D. Let K denote the splitting field of $f(x)$, viewed as a subfield of the complex numbers $\mathbb{C}$.

    (a) Prove that $\mathbb{Q}(\sqrt{D}) \subset K$.

    (b) Let τ denote complex conjugation and let τ_K denote the restriction of complex conjugation to K. Prove that τ_K is an element of $\mathrm{Gal}(K/\mathbb{Q})$ of order 1 or 2 depending on whether every element of K is real or not.

    (c) Prove that if $D < 0$ then K cannot be cyclic of degree 4 over $\mathbb{Q}$ (i.e., $\mathrm{Gal}(K/\mathbb{Q})$ cannot be a cyclic group of order 4).

    (d) Prove generally that $\mathbb{Q}(\sqrt{D})$ for squarefree $D < 0$ is not a subfield of a cyclic quartic field (cf. also Exercise 19 of Section 7).

20. Determine the Galois group of $(x^3 - 2)(x^3 - 3)$ over $\mathbb{Q}$. Determine all the subfields which contain $\mathbb{Q}(\rho)$ where ρ is a primitive 3rd root of unity.

21. Let $G \leq S_n$ be a subgroup of the symmetric group and suppose $\sigma_1, \ldots, \sigma_k$ are generators for G. If the function $f(x_1, x_2, \ldots, x_n)$ is fixed by the generators σ_i show it is fixed by G.

22. *(Newton's Formulas)* Let $f(x)$ be a monic polynomial of degree n with roots $\alpha_1, \ldots, \alpha_n$. Let s_i be the elementary symmetric function of degree i in the roots and define $s_i = 0$ for $i > n$. Let $p_i = \alpha_1^i + \cdots + \alpha_n^i$, $i \geq 0$, be the sum of the ith powers of the roots of $f(x)$.

---

[3] cf. the note *An Elementary Test for the Galois Group of a Quartic Polynomial*, Luise-Charlotte Kappe and Bette Warren, Amer. Math. Monthly, 96(1989), pp. 133–137.

[4] Ibid.

[5] Ibid.

Prove *Newton's Formulas*:

$$p_1 - s_1 = 0$$
$$p_2 - s_1 p_1 + 2s_2 = 0$$
$$p_3 - s_1 p_2 + s_2 p_1 - 3s_3 = 0$$

$$\vdots$$

$$p_i - s_1 p_{i-1} + s_2 p_{i-2} - \cdots + (-1)^{i-1} s_{i-1} p_1 + (-1)^i i s_i = 0$$

23. (a) If $x + y + z = 1$, $x^2 + y^2 + z^2 = 2$ and $x^3 + y^3 + z^3 = 3$, determine $x^4 + y^4 + z^4$.

   (b) Prove generally that x, y, z are not rational but that $x^n + y^n + z^n$ is rational for every positive integer n.

24. Prove that an $n \times n$ matrix A over a field of characteristic 0 is nilpotent if and only if the trace of A^k is 0 for all $k \geq 0$.

25. Prove that two $n \times n$ matrices A and B over a field of characteristic 0 have the same characteristic polynomial if and only if the trace of A^k equals the trace of B^k for all $k \geq 0$.

26. Use the fact that the trace of AB is the same as the trace of BA for any two $n \times n$ matrices A and B to show that AB and BA have the same characteristic polynomial over a field of characteristic 0 (the same result is true over a field of arbitrary characteristic).

27. Let $f(x)$ be a monic polynomial of degree n with roots $\alpha_1, \alpha_2, \ldots, \alpha_n$.

   (a) Show that the discriminant D of $f(x)$ is the square of the Vandermonde determinant

$$\begin{vmatrix} 1 & \alpha_1 & \alpha_1^2 & \cdots & \alpha_1^{n-1} \\ 1 & \alpha_2 & \alpha_2^2 & \cdots & \alpha_2^{n-1} \\ \vdots & \vdots & \vdots & \ddots & \vdots \\ 1 & \alpha_n & \alpha_n^2 & \cdots & \alpha_n^{n-1} \end{vmatrix} = \prod_{i > j} (\alpha_i - \alpha_j).$$

   (b) Taking the Vandermonde matrix above, multiplying on the left by its transpose and taking the determinant show that one obtains

$$D = \begin{vmatrix} p_0 & p_1 & p_2 & \cdots & p_{n-1} \\ p_1 & p_2 & p_3 & \cdots & p_n \\ \vdots & \vdots & \vdots & \ddots & \vdots \\ p_{n-1} & p_n & p_{n+1} & \cdots & p_{2n-2} \end{vmatrix}$$

   where $p_i = \alpha_1^i + \cdots + \alpha_n^i$ is the sum of the ith powers of the roots of $f(x)$, which can be computed in terms of the coefficients of $f(x)$ using Newton's formulas above. This gives an efficient procedure for calculating the discriminant of a polynomial.

28. Let α be a root of the irreducible polynomial $f(x) \in F[x]$ and let $K = F(\alpha)$. Let D be the discriminant of $f(x)$. Prove that $D = (-1)^{n(n-1)/2} N_{K/F}(f'(\alpha))$, where $f'(x) = D_x f(x)$ is the derivative of $f(x)$.

The following exercises describe the *resultant* of two polynomials and in particular provide another efficient method for calculating the discriminant of a polynomial.

29. Let F be a field and let $f(x) = a_n x^n + a_{n-1} x^{n-1} + \cdots + a_1 x + a_0$ and $g(x) = b_m x^m + b_{m-1} x^{m-1} + \cdots + b_1 x + b_0$ be two polynomials in $F[x]$.

   (a) Prove that a necessary and sufficient condition for $f(x)$ and $g(x)$ to have a common root (or, equivalently, a common divisor in $F[x]$) is the existence of a polynomial

$a(x) \in F[x]$ of degree at most $m - 1$ and a polynomial $b(x) \in F[x]$ of degree at most $n - 1$ with $a(x)f(x) = b(x)g(x)$.

(b) Writing $a(x)$ and $b(x)$ explicitly as polynomials show that equating coefficients in the equation $a(x)f(x) = b(x)g(x)$ gives a system of $n + m$ linear equations for the coefficients of $a(x)$ and $b(x)$. Prove that this system has a nontrivial solution (hence $f(x)$ and $g(x)$ have a common zero) if and only if the determinant

$$
R(f, g) =
\begin{vmatrix}
a_n & a_{n-1} & \cdots & a_0 & & & & \\
 & a_n & a_{n-1} & \cdots & a_0 & & & \\
 & & a_n & a_{n-1} & \cdots & a_0 & & \\
 & & & \ddots & & & & \\
 & & & & a_n & a_{n-1} & \cdots & a_0 \\
b_m & b_{m-1} & \cdots & b_0 & & & & \\
 & b_m & b_{m-1} & \cdots & b_0 & & & \\
 & & b_m & b_{m-1} & \cdots & b_0 & & \\
 & & & \ddots & & & & \\
 & & & & b_m & b_{m-1} & \cdots & b_0
\end{vmatrix}
$$

is zero. Here $R(f, g)$, called the *resultant* of the two polynomials, is the determinant of an $(n+m) \times (n+m)$ matrix R with m rows involving the coefficients of $f(x)$ and n rows involving the coefficients of $g(x)$.

30. (a) With notations as in the previous problem, show that we have the matrix equation

$$
R
\begin{pmatrix}
x^{n+m-1} \\
x^{n+m-2} \\
\vdots \\
x \\
1
\end{pmatrix}
=
\begin{pmatrix}
x^{m-1}f(x) \\
x^{m-2}f(x) \\
\vdots \\
f(x) \\
x^{n-1}g(x) \\
x^{n-2}g(x) \\
\vdots \\
g(x)
\end{pmatrix}.
$$

(b) Let R' denote the matrix of cofactors of R as in Theorem 30 of Section 11.4, so $R'R = R(f, g)I$, where I is the identity matrix. Multiply both sides of the matrix equation above by R' and equate the bottom entry of the resulting column matrices to prove that there are polynomials $r(x), s(x) \in F[x]$ such that $R(f, g)$ is equal to $r(x)f(x) + s(x)g(x)$, i.e., the resultant of two polynomials is a linear combination (in $F[x]$) of the polynomials.

31. Consider $f(x)$ and $g(x)$ as general polynomials and suppose the roots of $f(x)$ are $x_1, \ldots, x_n$ and the roots of $g(x)$ are $y_1, \ldots, y_m$. The coefficients of $f(x)$ are powers of a_n times the elementary symmetric functions in $x_1, x_2, \ldots, x_n$ and the coefficients of $g(x)$ are powers of b_m times the elementary symmetric functions in $y_1, y_2, \ldots, y_m$.
   (a) By expanding the determinant show that $R(f, g)$ is homogeneous of degree m in the coefficients a_i and homogeneous of degree n in the coefficients b_j.
   (b) Show that $R(f, g)$ is $a_n^m b_m^n$ times a symmetric function in $x_1, \ldots, x_n$ and $y_1, \ldots, y_m$.
   (c) Since $R(f, g)$ is 0 if $f(x)$ and $g(x)$ have a common root, say $x_i = y_j$, show that $R(f, g)$ is divisible by $x_i - y_j$ for $i = 1, 2, \ldots, n$, $j = 1, 2, \ldots, m$. Conclude by

degree considerations that

$$R = a_n^m b_m^n \prod_{i=1}^{n} \prod_{j=1}^{m} (x_i - y_j).$$

(d) Show that the product in (c) can be also be written

$$R(f, g) = a_n^m \prod_{i=1}^{n} g(x_i) = (-1)^{nm} b_m^n \prod_{j=1}^{m} f(y_j).$$

This gives an interesting *reciprocity* between the product of g evaluated at the roots of f and the product of f evaluated at the roots of g.

32. Consider now the special case where $g(x) = f'(x)$ is the derivative of the polynomial $f(x) = x^n + a_{n-1}x^{n-1} + \cdots + a_1 x + a_0$ and suppose the roots of $f(x)$ are $\alpha_1, \alpha_2, \ldots, \alpha_n$. Using the formula

$$R(f, f') = \prod_{i=1}^{n} f'(\alpha_i)$$

of the previous exercise, prove that

$$D = (-1)^{n(n-1)/2} R(f, f')$$

where D is the discriminant of $f(x)$.

33. (a) Prove that the discriminant of the cyclotomic polynomial $\Phi_p(x)$ of the pth roots of unity for an odd prime p is $(-1)^{(p-1)/2} p^{p-2}$ [One approach: use Exercise 5 of the previous section together with the determinant form for the discriminant in terms of the power sums p_i.]
    (b) Prove that $\mathbb{Q}(\sqrt{(-1)^{(p-1)/2} p}) \subset \mathbb{Q}(\zeta_p)$ for p an odd prime. (Cf. also Exercise 11 of Section 7.)

34. Use the previous exercise to prove that every quadratic extension of $\mathbb{Q}$ is contained in a cyclotomic extension (a special case of the Kronecker–Weber Theorem).

35. Prove that the discriminant D of the polynomial $x^n + px + q$ is given by the formula $(-1)^{n(n-1)/2} n^n q^{n-1} + (-1)^{(n-1)(n-2)/2} (n-1)^{n-1} p^n$.

36. Prove that the discriminant of $x^n + nx^{n-1} + n(n-1)x^{n-2} + \cdots + n(n-1) \ldots (3)(2)x + n!$ is $(-1)^{n(n-1)/2} (n!)^n$.

The following exercises 37 to 43 outline two procedures for writing a symmetric function in terms of the elementary symmetric functions. Let $f(x_1, \ldots, x_n)$ be a polynomial which is symmetric in $x_1, \ldots, x_n$. Recall that the degree (sometimes called the *weight*) of the monomial $Ax_1^{a_1} x_2^{a_2} \ldots x_n^{a_n}$ ($a_i \geq 0$) is $a_1 + a_2 + \cdots + a_n$ and that a polynomial is *homogeneous (of degree m)* if every monomial has the same degree (m).

37. (a) Show that every polynomial $f(x_1, \ldots, x_n)$ can be written as a sum of homogeneous polynomials. Show that if $f(x_1, \ldots, x_n)$ is symmetric then each of these homogeneous polynomials is also symmetric.
    (b) Show that the monomial $Bs_1^{a_1} s_2^{a_2} \ldots s_n^{a_n}$ in the elementary symmetric functions is a homogeneous polynomial in $x_1, x_2, \ldots, x_n$ of degree $a_1 + 2a_2 + \cdots + na_n$.

In writing $f(x_1, \ldots, x_n)$ as a polynomial in the symmetric functions it therefore suffices to assume that $f(x_1, \ldots, x_n)$ is homogeneous.

Recall the *lexicographic monomial order* with $x_1 > x_2 > \cdots > x_n$ defined in Section 9.6, where the nonzero monomial term with exponents $(a_1, a_2, \ldots, a_n)$ comes before the nonzero monomial term with exponents $(b_1, b_2, \ldots, b_n)$ if the initial components of the two n-tuples of exponents are equal and the first component where they differ has $a_i > b_i$. If $f(x_1, \ldots, x_n)$ contains the monomial $Ax_1^{a_1} x_2^{a_2} \ldots x_n^{a_n}$ then since $f(x_1, \ldots, x_n)$ is symmetric it also contains all the permuted monomials. Among these choose the lexicographically largest monomial, which therefore satisfies $a_1 \geq a_2 \geq \cdots \geq a_n \geq 0$.

38. (a) Show that the monomial $As_1^{a_1-a_2} s_2^{a_2-a_3} \cdots s_n^{a_n}$ in the elementary symmetric functions has the same lexicographic initial term.

(b) Show that subtracting $As_1^{a_1-a_2} s_2^{a_2-a_3} \cdots s_n^{a_n}$ from $f(x)$ yields either 0 or a symmetric polynomial of the same degree whose terms are lexicographically smaller than the terms in $f(x_1, \ldots, x_n)$.

(c) Show that the iteration of this procedure (lexicographic ordering, choosing the lexicographically largest term, subtracting the associated monomial in the elementary symmetric functions) terminates, expressing $f(x_1, \ldots, x_n)$ as a polynomial in the elementary symmetric functions.

39. Use the algorithm described in Exercise 38 to prove that a polynomial $f(x_1, \ldots, x_n)$ that is symmetric in $x_1, \ldots, x_n$ can be expressed *uniquely* as a polynomial in the elementary symmetric functions.

40. Use the procedure in Exercise 38 to express each of the following symmetric functions as a polynomial in the elementary symmetric functions:

(a) $(x_1 - x_2)^2$
(b) $x_1^2 + x_2^2 + x_3^2$
(c) $x_1^2 x_2^2 + x_1^2 x_3^2 + x_2^2 x_3^2$.

41. Use the procedure in Exercise 38 to express $\sum_{i \neq j} x_i^2 x_j$ as a polynomial in the elementary symmetric functions.

We now know that a symmetric polynomial $f(x_1, \ldots, x_n)$ can be written uniquely as a polynomial in the elementary symmetric functions. Using this existence and uniqueness we can describe an alternate and computationally useful method for determining the coefficients of the elementary symmetric functions in this polynomial. As in Exercise 37 we may assume that $f(x_1, \ldots, x_n)$ is homogeneous of degree M. Let N be the maximum degree of any of the variables $x_1, \ldots, x_n$ in $f(x_1, \ldots, x_n)$.

(a) Determine all of the possible monomials $A_i s_1^{a_1} s_2^{a_2} \cdots s_n^{a_n}$ appearing in $f(x_1, \ldots, x_n)$ from the constraints

$$a_1 + 2a_2 + \cdots + na_n = M$$
$$a_1 + a_2 + \cdots + a_n \leq N.$$

(b) Since $f(x_1, \ldots, x_n) = \sum A_i s_1^{a_1} s_2^{a_2} \cdots s_n^{a_n}$ is a polynomial *identity*, it is valid for any substitution of values for $x_1, \ldots, x_n$. Each substitution into this equation gives a linear relation on the coefficients A_i and so a sufficient number of substitutions will determine the A_i.

42. Show that the function $(x_1 + x_2 - x_3 - x_4)(x_1 + x_3 - x_2 - x_4)(x_1 + x_4 - x_2 - x_3)$ is symmetric in x_1, x_2, x_3, x_4 and use the preceding procedure to prove it can be expressed as a polynomial in the elementary symmetric functions as $s_1^3 - 4s_1 s_2 + 8s_3$.

43. Express each of the following in terms of the elementary symmetric functions:

(a) $\sum_{i \neq j} x_i^2 x_j$ (b) $\sum_{i,j,k \text{ distinct}} x_i^2 x_j x_k$ (c) $\sum_{i,j,k \text{ distinct}} x_i^2 x_j^2 x_k^2$.

622

44. Let $\alpha_1, \alpha_2, \alpha_3, \alpha_4$ be the roots of a quartic polynomial $f(x)$ over $\mathbb{Q}$. Show that the quantities $\alpha_1\alpha_2 + \alpha_3\alpha_4$, $\alpha_1\alpha_3 + \alpha_2\alpha_4$, and $\alpha_1\alpha_4 + \alpha_2\alpha_3$ are permuted by the Galois group of $f(x)$. Conclude that these elements are the roots of a cubic polynomial with coefficients in $\mathbb{Q}$ (also sometimes referred to as the *resolvent cubic* of $f(x)$).

45. If $f(x) = x^3 + px + q \in \mathbb{Z}[x]$ is irreducible, prove that its discriminant $D = -4p^3 - 27q^2$ is an integer not equal to $0, \pm 1$.

46. Prove that every finite group occurs as the Galois group of a field extension of the form $F(x_1, x_2, \ldots, x_n)/E$.

47. Let F be a field of characteristic 0 in which every cubic polynomial has a root. Let $f(x)$ be an irreducible quartic polynomial over F whose discriminant is a square in F. Determine the Galois group of $f(x)$.

48. This exercise determines the splitting field K for the polynomial $f(x) = x^6 - 2x^3 - 2$ over $\mathbb{Q}$ (cf. also Exercise 2 of Section 8).
  (a) Prove that $f(x)$ is irreducible over $\mathbb{Q}$ with roots the three cube roots of $1 \pm \sqrt{3}$.
  (b) Prove that K contains the field $\mathbb{Q}(\sqrt{-3})$ of 3rd roots of unity and contains $\mathbb{Q}(\sqrt{3})$, hence contains the biquadratic field $F = \mathbb{Q}(i, \sqrt{3})$. Take the product of two of the roots in (a) to prove that K contains $\sqrt[3]{2}$ and conclude that K is an extension of the field $L = \mathbb{Q}(\sqrt[3]{2}, i, \sqrt{3})$.
  (c) Prove that $[L : \mathbb{Q}] = 12$ and that K is obtained from L by adjoining the cube root of an element in L, so that $[K : \mathbb{Q}] = 12$ or 36.
  (d) Prove that if $[K : \mathbb{Q}] = 12$ then $K = \mathbb{Q}(\sqrt[3]{2}, i, \sqrt{3})$ and that $\mathrm{Gal}(K/\mathbb{Q})$ is isomorphic to the direct product of the cyclic group of order 2 and S_3. Prove that if $[K : \mathbb{Q}] = 12$ then there is a unique real cubic subfield in K, namely $\mathbb{Q}(\sqrt[3]{2})$.
  (e) Take the quotient of the two real roots in (a) to show that $\sqrt[3]{2 + \sqrt{3}}$ and $\sqrt[3]{2 - \sqrt{3}}$ (real roots) are both elements of K. Show that $\alpha = \sqrt[3]{2 + \sqrt{3}} + \sqrt[3]{2 - \sqrt{3}}$ is a real root of the irreducible cubic equation $x^3 - 3x - 4$ whose discriminant is $-2^2 3^4$. Conclude that the Galois closure of $\mathbb{Q}(\alpha)$ contains $\mathbb{Q}(i)$ so in particular $\mathbb{Q}(\alpha) \neq \mathbb{Q}(\sqrt[3]{2})$.
  (f) Conclude from (e) that $G = \mathrm{Gal}(K/\mathbb{Q})$ is of order 36. Determine all the elements of G explicitly and in particular show that G is isomorphic to $S_3 \times S_3$.

49. Prove that the Galois group over $\mathbb{Q}$ of $x^6 - 4x^3 + 1$ is isomorphic to the dihedral group of order 12. [Observe that the two real roots are inverses of each other.]

50. (*Criterion for the Galois Group of an Irreducible Cubic over an Arbitrary Field*) Suppose K is a field and $f(x) = x^3 + ax^2 + bx + c \in K[x]$ is irreducible, so the Galois group of $f(x)$ over K is either S_3 or A_3.
  (a) Show that the Galois group of $f(x)$ is A_3 if and only if the resultant quadratic polynomial $g(x) = x^2 + (ab - 3c)x + (b^3 + a^3c - 6abc + 9c^2)$ has a root in K. [If α, β, γ are the roots of $f(x)$ show that the Galois group is A_3 if and only if the element $\theta = \alpha\beta^2 + \beta\gamma^2 + \gamma\alpha^2$ is an element of K and that θ is a root of $g(x)$.] Show that the discriminant of $g(x)$ is the same as the discriminant of $f(x)$.
  (b) (ch$(K) \neq 2$) If K has characteristic different from 2 show either from (a) or directly from the definition of the discriminant that the Galois group of $f(x)$ is A_3 if and only if the discriminant of $f(x)$ is a square in K.
  (c) (ch$(K) = 2$) If K has characteristic 2 show that the discriminant of $f(x)$ is always a square. Show that $f(x)$ can be taken to be of the form $x^3 + px + q$ and that the Galois group of $f(x)$ is A_3 if and only if the quadratic $x^2 + qx + (p^3 + q^2)$ has a root in K (equivalently, if $(p^3 + q^2)/q^2 \in K$ is in the image of the *Artin–Schreier map* $x \mapsto x^2 - x$ mapping K to K).

(d) If $K = \mathbb{F}_2(t)$ where t is transcendental over $\mathbb{F}_2$. Prove that the polynomials $x^3 + t^2 x + t^3$, $x^3 + (t^2 + t + 1)x + (t^2 + t + 1)$, and $x^3 + (t^2 + t + 1)x + (t^3 + t^2 + t)$ have A_3 as Galois group while $x^3 + t^2 x + t$ and $x^3 = x + t$ have S_3 as Galois group.

51. This exercise proves *Sturm's Theorem* determining the number of real roots of a polynomial $f(x) \in \mathbb{R}[x]$ in an interval $[a, b]$. The multiple roots of $f(x)$ are zeros of the g.c.d. of $f(x)$ and its derivative $f'(x)$, and it follows that to determine the real roots of $f(x)$ in $[a, b]$ we may assume that the roots of $f(x)$ are *simple*.

Apply the Euclidean algorithm to $f_0(x) = f(x)$ and its derivative $f_1(x) = f'(x)$ using the *negative* of the remainder at each stage to find a sequence of polynomials $f(x), f'(x), f_2(x), \ldots, f_n(x)$ with

$$f_{i-1}(x) = q_i(x) f_i(x) - f_{i+1}(x) \qquad i = 0, 1, \ldots, n - 1$$

where $f_n(x) \in \mathbb{R}$ is a nonzero constant.

(a) Prove that consecutive polynomials $f_i(x)$, $f_{i+1}(x)$ for $i = 0, 1, \ldots, n - 1$ have no common zeros. [Show that otherwise $f_{i+2}(c) = f_{i+3}(c) = \cdots = 0$, and derive a contradiction.]

(b) If $f_i(c) = 0$ for some $i = 0, 1, \ldots, n - 1$, prove that one of the two values $f_{i-1}(c)$, $f_{i+1}(c)$ is strictly negative and the other is strictly positive.

For any real number α, let $V(\alpha)$ denote the number of sign changes in the *Sturm sequence* of real numbers

$$f(\alpha), f'(\alpha), f_2(\alpha), \ldots, f_n(\alpha),$$

ignoring any 0's that appear (for example $-1, -2, 0, +3, -4$ has signs $--+-$ disregarding the 0, so there are 2 sign changes, the first from -2 to $+3$, the second from $+3$ to -4).

(c) Suppose $\alpha < \beta$ and that all the elements in the Sturm sequences for α and for β are nonzero. Prove that unless $f_i(c) = 0$ for some $\alpha < c < \beta$ and some $i = 0, 1, \ldots, n - 1$, then the signs of all the elements in these two Sturm sequences are the same, so in particular $V(\alpha) = V(\beta)$.

(d) If $f_j(c) = 0$ prove that there is a sufficiently small interval (α, β) containing c so that $f_j(x)$ has no zero other than c for $\alpha < x < \beta$.

(e) If $j \geq 1$ in (d), prove that the number of sign changes in $f_{j-1}(\alpha), f_j(\alpha), f_{j+1}(\alpha)$ and in $f_{j-1}(\beta), f_j(\beta), f_{j+1}(\beta)$ are the same. [Observe that $f_{j-1}(c)$ and $f_{j+1}(c)$ have opposite signs by (b) and $f_{j-1}(x)$ and $f_{j+1}(x)$ do not change sign in (α, β).]

(f) If $j = 0$ in (d) show that the number of sign changes in $f(\alpha), f'(\alpha)$ is one more than the number of sign changes in $f(\beta), f'(\beta)$. [If $f'(c) > 0$ then $f(x)$ is increasing at c, so that $f(\alpha) < 0$, $f(\beta) > 0$, and $f'(x)$ does not change sign in (α, β), so the signs change from $-+$ to $++$. Similarly if $f'(c) < 0$.]

(g) Prove *Sturm's Theorem*: if $f(x)$ is a polynomial with real coefficients all of whose real roots are simple then the number of real zeros of $f(x)$ in an interval $[a, b]$ where $f(a)$ and $f(b)$ are both nonzero is given by $V(a) - V(b)$. [Use (c), (e) and (f) to see that as α runs from a to b the number $V(\alpha)$ of sign changes is constant unless α passes through a zero of $f(x)$, in which case it decreases by precisely 1.]

(h) Suppose $f(x) = x^5 + px + q \in \mathbb{R}[x]$ has simple roots. Show that the sequence of polynomials above is given by $f(x)$, $5x^4 + p$, $(-4p/5)x + q$, and $-D/(256p^4)$ where $D = 256p^5 + 3125q^4$ is the discriminant of $f(x)$. Conclude for $p > 0$ that $f(x)$ has precisely one real root and for $p < 0$ that $f(x)$ has precisely 1 or 3 real roots depending on whether $D > 0$ or $D < 0$, respectively. [E.g., if $p < 0$ and $D < 0$ then at $-\infty$ the signs are $-+-+$ with 3 sign changes and at $+\infty$ the signs are $++++$ with no sign changes.]

14.7 SOLVABLE AND RADICAL EXTENSIONS: INSOLVABILITY OF THE QUINTIC

We now investigate the question of solving for the roots of a polynomial by *radicals*, that is, in terms of the algebraic operations of addition, subtraction, multiplication, division and the extraction of nth roots. The quadratic formula for the roots of a polynomial of degree 2 is familiar from elementary algebra and we shall derive below similar formulas for the roots of cubic and quartic polynomials. For polynomials of degree ≥ 5, however, we shall see that such formulas are not possible — this is Abel's Theorem on the insolvability of the general quintic. The reason for this is quite simple: we shall see that a polynomial is solvable by radicals if and only if its Galois group is a solvable group (which explains the terminology) and for $n \geq 5$ the group S_n is not solvable.

We first discuss *simple radical extensions*, namely extensions obtained by adjoining to a field F the nth root of an element a in F. Since all the roots of the polynomial $x^n - a$ for $a \in F$ differ by factors of the nth roots of unity, adjoining one such root will give a Galois extension if and only if this field contains the nth roots of unity. Simple radical extensions are best behaved when the base field F already contains the appropriate roots of unity. The symbol $\sqrt[n]{a}$ for $a \in F$ will be used to denote any root of the polynomial $x^n - a \in F[x]$.

Definition. The extension K/F is said to be *cyclic* if it is Galois with a cyclic Galois group.

Proposition 36. Let F be a field of characteristic not dividing n which contains the nth roots of unity. Then the extension $F(\sqrt[n]{a})$ for $a \in F$ is cyclic over F of degree dividing n.

Proof: The extension $K = F(\sqrt[n]{a})$ is Galois over F if F contains the nth roots of unity since it is the splitting field for $x^n - a$. For any $\sigma \in \mathrm{Gal}(K/F)$, $\sigma(\sqrt[n]{a})$ is another root of this polynomial, hence $\sigma(\sqrt[n]{a}) = \zeta_\sigma \sqrt[n]{a}$ for some n^{th} root of unity ζ_σ. This gives a map

$$\mathrm{Gal}(K/F) \to \mu_n$$

$$\sigma \mapsto \zeta_\sigma$$

where μ_n denotes the group of nth roots of unity. Since F contains μ_n, every n^{th} root of unity is fixed by every element of $\mathrm{Gal}(K/F)$. Hence

$$\sigma\tau(\sqrt[n]{a}) = \sigma(\zeta_\tau \sqrt[n]{a})$$
$$= \zeta_\tau \sigma(\sqrt[n]{a})$$
$$= \zeta_\tau \zeta_\sigma \sqrt[n]{a} = \zeta_\sigma \zeta_\tau \sqrt[n]{a}$$

which shows that $\zeta_{\sigma\tau} = \zeta_\sigma \zeta_\tau$, so the map above is a homomorphism. The kernel consists precisely of the automorphisms which fix $\sqrt[n]{a}$, namely the identity. This gives an injection of $\mathrm{Gal}(K/F)$ into the cyclic group μ_n of order n, which proves the proposition.

Let now K be any cyclic extension of degree n over a field F of characteristic not dividing n which contains the nth roots of unity. Let σ be a generator for the cyclic group $\mathrm{Gal}(K/F)$.

Definition. For $\alpha \in K$ and any n^{th} root of unity ζ, define the *Lagrange resolvent* $(\alpha, \zeta) \in K$ by

$$(\alpha, \zeta) = \alpha + \zeta\sigma(\alpha) + \zeta^2\sigma^2(\alpha) + \cdots + \zeta^{n-1}\sigma^{n-1}(\alpha).$$

If we apply the automorphism σ to (α, ζ) we obtain

$$\sigma(\alpha, \zeta) = \sigma\alpha + \zeta\sigma^2(\alpha) + \zeta^2\sigma^3(\alpha) + \cdots + \zeta^{n-1}\sigma^n(\alpha)$$

since ζ is an element of the base field F so is fixed by σ. We have $\zeta^n = 1$ in μ_n and $\sigma^n = 1$ in $\text{Gal}(K/F)$ so this can be written

$$\begin{aligned}
\sigma(\alpha, \zeta) &= \sigma\alpha + \zeta\sigma^2(\alpha) + \zeta^2\sigma^3(\alpha) + \cdots + \zeta^{-1}\alpha \\
&= \zeta^{-1}(\alpha + \zeta\sigma(\alpha) + \zeta^2\sigma^2(\alpha) + \cdots + \zeta^{n-1}\sigma^{n-1}(\alpha)) \\
&= \zeta^{-1}(\alpha, \zeta).
\end{aligned} \tag{14.19}$$

It follows that

$$\sigma(\alpha, \zeta)^n = (\zeta^{-1})^n(\alpha, \zeta)^n = (\alpha, \zeta)^n$$

so that $(\alpha, \zeta)^n$ is fixed by $\text{Gal}(K/F)$, hence is an element of F for any $\alpha \in K$.

Let ζ be a primitive n^{th} root of unity. By the linear independence of the automorphisms $1, \sigma, \ldots, \sigma^{n-1}$ (Theorem 7), there is an element $\alpha \in K$ with $(\alpha, \zeta) \neq 0$. Iterating (19) we have

$$\sigma^i(\alpha, \zeta) = \zeta^{-i}(\alpha, \zeta), \qquad i = 0, 1, \ldots,$$

and it follows that σ^i does not fix (α, ζ) for any $i < n$. Hence this element cannot lie in any proper subfield of K, so $K = F((\alpha, \zeta))$. Since we proved $(\alpha, \zeta)^n = a \in F$ above, we have $F(\sqrt[n]{a}) = F((\alpha, \zeta)) = K$. This proves the following converse of Proposition 36.

Proposition 37. Any cyclic extension of degree n over a field F of characteristic not dividing n which contains the nth roots of unity is of the form $F(\sqrt[n]{a})$ for some $a \in F$.

Remark: The two propositions above form a part of what is referred to as *Kummer theory*. A group G is said to have *exponent* n if $g^n = 1$ for every $g \in G$. Let F be a field of characteristic not dividing n which contains the nth roots of unity. If we take elements $a_1, \ldots, a_k \in F^\times$ then as in Proposition 36 we can see that the extension

$$F(\sqrt[n]{a_1}, \sqrt[n]{a_2}, \ldots, \sqrt[n]{a_k}) \tag{14.20}$$

is an abelian extension of F whose Galois group is of exponent n. Conversely, any abelian extension of exponent n is of this form.

Denote by $(F^\times)^n$ the subgroup of the multiplicative group $F^\times$ consisting of the nth powers of nonzero elements of F. The quotient group $F^\times/(F^\times)^n$ is an abelian group of exponent n. The Galois group of the extension in (20) is isomorphic to the group generated in $F^\times/(F^\times)^n$ by the elements $a_1, \ldots, a_k$ and two extensions as in (20) are equal if and only if their associated groups in $F^\times/(F^\times)^n$ are equal.

Hence the (finitely generated) subgroups of $F^\times/(F^\times)^n$ classify the abelian extensions of exponent n over fields containing the nth roots of unity (and characteristic not dividing n). Such extensions are called *Kummer extensions* (cf. Section 17.3).

These results generalize the case $k = 1$ above and can be proved in a similar way.

For simplicity we now consider the situation of a base field F of characteristic 0. As in the previous propositions the results are valid over fields whose characteristics do not divide any of the orders of the roots that will be taken.

Definition.

(1) An element α which is algebraic over F can be *expressed by radicals* or *solved for in terms of radicals* if α is an element of a field K which can be obtained by a succession of simple radical extensions

$$F = K_0 \subset K_1 \subset \cdots \subset K_i \subset K_{i+1} \subset \cdots \subset K_s = K \qquad (14.21)$$

where $K_{i+1} = K_i(\sqrt[n_i]{a_i})$ for some $a_i \in K_i$, $i = 0, 1, \ldots, s - 1$. Here $\sqrt[n_i]{a_i}$ denotes some root of the polynomial $x^{n_i} - a_i$. Such a field K will be called a *root extension* of F.

(2) A polynomial $f(x) \in F[x]$ can be *solved by radicals* if all its roots can be solved for in terms of radicals.

This gives a precise meaning to the intuitive notion that α is obtained by successive algebraic operations (addition, subtraction, multiplication and division) and successive root extractions. For example, the element

$$-1 + \sqrt{17} + \sqrt{2(17 - \sqrt{17})} + 2\sqrt{17 + 3\sqrt{17} - \sqrt{2(17 - \sqrt{17})} - 2\sqrt{2(17 + \sqrt{17})}}$$

encountered at the end of Section 5 (used to construct the regular 17-gon) is expressed by radicals and is contained in the field K_4, where

$$K_0 = \mathbb{Q}$$
$$K_1 = K_0(\sqrt{a_0}) \qquad a_0 = 17$$
$$K_2 = K_1(\sqrt{a_1}) \qquad a_1 = 2(17 - \sqrt{17})$$
$$K_3 = K_2(\sqrt{a_2}) \qquad a_2 = 2(17 + \sqrt{17})$$
$$K_4 = K_3(\sqrt{a_3}) \qquad a_3 = 17 + 3\sqrt{17} - \sqrt{2(17 - \sqrt{17})} - 2\sqrt{2(17 + \sqrt{17})}.$$

Each of these extensions is a radical extension. The fact that no roots other than square roots are required reflects the fact that the regular 17-gon is constructible by straightedge and compass.

In considering radical extensions one may always adjoin roots of unity, since by definition the roots of unity are radicals. This is useful because then cyclic extensions become radical extensions and conversely. In particular we have:

Lemma 38. If α is contained in a root extension K as in (21) above, then α is contained in a root extension which is Galois over F and where each extension K_{i+1}/K_i is cyclic.

Proof: Let L be the Galois closure of K over F. For any $\sigma \in \mathrm{Gal}(L/F)$ we have the chain of subfields

$$F = \sigma K_0 \subset \sigma K_1 \subset \cdots \subset \sigma K_i \subset \sigma K_{i+1} \subset \cdots \subset \sigma K_s = \sigma K$$

where $\sigma K_{i+1}/\sigma K_i$ is again a simple radical extension (since it is generated by the element $\sigma(\sqrt[n_i]{a_i})$, which is a root of the equation $x^{n_i} - \sigma(a_i)$ over $\sigma(K_i)$). It is easy to see that the composite of two root extensions is again a root extension (if K' is another root extension with subfields K_i', first take the composite of K_1' with the fields $K_0, K_1, \ldots, K_s$, then the composite of these fields with K_2', etc. so that each individual extension in this process is a simple radical extension). It follows that the composite of all the conjugate fields $\sigma(K)$ for $\sigma \in \mathrm{Gal}(L/F)$ is again a root extension. Since this field is precisely L, we see that α is contained in a Galois root extension.

We now adjoin to F the n_i-th roots of unity for all the roots $\sqrt[n_i]{a_i}$ of the simple radical extensions in the Galois root extension K/F, obtaining the field F', say, and then form the composite of F' with the root extension:

$$F \subseteq F' = F'K_0 \subseteq F'K_1 \subseteq \cdots \subseteq F'K_i \subseteq F'K_{i+1} \subseteq \cdots \subseteq F'K_s = F'K.$$

The field $F'K$ is a Galois extension of F since it is the composite of two Galois extensions. The extension from F to $F' = F'K_0$ can be given as a chain of subfields with each individual extension cyclic (this is true for any abelian extension). Each extension $F'K_{i+1}/F'K_i$ is a simple radical extension and since we now have the appropriate roots of unity in the base fields, each of these individual extensions from F' to $F'K$ is a cyclic extension by Proposition 36. Hence $F'K/F$ is a root extension which is Galois over F with cyclic intermediate extensions, completing the proof.

Recall from Section 3.4 (cf. also Section 6.1) that a finite group G is *solvable* if there exists a chain of subgroups

$$1 = G_s \leq G_{s-1} \leq \cdots \leq G_{i+1} \leq G_i \leq \cdots \leq G_0 = G \tag{14.22}$$

with G_i/G_{i+1} cyclic, $i = 0, 1, \ldots, s - 1$. We have proved that subgroups and quotient groups of solvable groups are solvable and that if $H \leq G$ and G/H are both solvable, then G is solvable.

We now prove Galois' fundamental connection between solving for the roots of polynomials in terms of radicals and the Galois group of the polynomial. We continue to work over a field F of characteristic 0, but it is easy to see that the proof is valid over any field of characteristic not dividing the order of the Galois group or the orders of the radicals involved.

Theorem 39. The polynomial $f(x)$ can be solved by radicals if and only if its Galois group is a solvable group.

Proof: Suppose first that $f(x)$ can be solved by radicals. Then each root of $f(x)$ is contained in an extension as in the lemma. The composite L of such extensions is again of the same type by Proposition 21. Let G_i be the subgroups corresponding to the subfields $K_i, i = 0, 1, \ldots, s - 1$. Since

$$\text{Gal}(K_{i+1}/K_i) = G_i/G_{i+1} \qquad i = 0, 1, \ldots, s - 1$$

it follows that the Galois group $G = \text{Gal}(L/F)$ is a solvable group. The field L contains the splitting field of $f(x)$ so the Galois group of $f(x)$ is a quotient group of the solvable group G, hence is solvable.

Suppose now that the Galois group G of $f(x)$ is a solvable group and let K be the splitting field for $f(x)$. Taking the fixed fields of the subgroups in a chain (22) for G gives a chain

$$F = K_0 \subset K_1 \subset \cdots \subset K_i \subset K_{i+1} \subset \cdots \subset K_s = K$$

where $K_{i+1}/K_i, i = 0, 1, \ldots, s - 1$ is a cyclic extension of degree n_i. Let F' be the cyclotomic field over F of all roots of unity of order $n_i, i = 0, 1, \ldots, s - 1$ and form the composite fields $K_i' = F'K_i$. We obtain a sequence of extensions

$$F \subseteq F' = F'K_0 \subseteq F'K_1 \subseteq \cdots \subseteq F'K_i \subseteq F'K_{i+1} \subseteq \cdots \subseteq F'K_s = F'K.$$

The extension $F'K_{i+1}/F'K_i$ is cyclic of degree dividing $n_i, i = 0, 1, \ldots, s - 1$ (by Proposition 19). Since we now have the appropriate roots of unity in the base fields, each of these cyclic extensions is a simple radical extension by Proposition 37. Each of the roots of $f(x)$ is therefore contained in the root extension $F'K$ so that $f(x)$ can be solved by radicals.

Corollary 40. The general equation of degree n cannot be solved by radicals for $n \geq 5$.

Proof: For $n \geq 5$ the group S_n is not solvable as we showed in Chapter 4. The corollary follows immediately from Theorems 32 and 39.

This corollary shows that there is no formula involving radicals analogous to the quadratic formula for polynomials of degree 2 for the roots of a polynomial of degree 5. To give an example of a *specific* polynomial over $\mathbb{Q}$ of degree 5 whose roots cannot be expressed in terms of radicals we must demonstrate a polynomial of degree 5 with rational coefficients having S_5 (or A_5, which is also not solvable) as Galois group (cf. also Exercise 21, which gives a criterion for the solvability of a quintic).

Example

Consider the polynomial $f(x) = x^5 - 6x + 3 \in \mathbb{Q}[x]$. This polynomial is irreducible since it is Eisenstein at 3. The splitting field K for this polynomial therefore has degree divisible by 5, since adjoining one root of $f(x)$ to $\mathbb{Q}$ generates an extension of degree 5. The Galois group G is therefore a subgroup of S_5 of order divisible by 5 so contains an element of order 5. The only elements in S_5 of order 5 are 5-cycles, so G contains a 5-cycle.

Since $f(-2) = -17$, $f(0) = 3$, $f(1) = -2$, and $f(2) = 23$ we see that $f(x)$ has a real root in each of the intervals $(-2, 0)$, $(0, 1)$ and $(1, 2)$. By the Mean Value Theorem, if there were 4 real roots then the derivative $f'(x) = 5x^4 - 6$ would have at least 3 real zeros, which it does not. Hence these are the only real roots. (This also follows easily by Descartes' rule of signs.) By the Fundamental Theorem of Algebra $f(x)$ has 5 roots in $\mathbb{C}$. Hence $f(x)$ has two complex roots which are not real. Let τ denote the automorphism of complex conjugation in $\mathbb{C}$. Since the coefficients of $f(x)$ are real, the two complex roots must be interchanged by τ (since they are not fixed, not being real). Hence the restriction of complex conjugation to K fixes three of the roots of $f(x)$ and interchanges the other two. As an element of G, $\tau|_K$ is therefore a transposition.

It is now a simple exercise to show that any 5-cycle together with any transposition generate all of S_5. It follows that $G = S_5$, so the roots of $x^5 - 6x + 3$ cannot be expressed by radicals.

As indicated in this example, a great deal of information regarding the Galois group can be obtained by understanding the *cycle types* of the automorphisms in G considered as a subgroup of S_n. In practice this is the most efficient way of determining the Galois groups of polynomials of degrees ≥ 5 (becoming more difficult the larger the degree, of course, if only because the possible subgroups of S_n are vastly more numerous). We describe this procedure in the next section.

By Theorem 39, any polynomial of degree $n \leq 4$ can be solved by radicals, since S_n is a solvable group for these n. For $n = 2$ this is just the familiar quadratic formula. For $n = 3$ the formula is known as *Cardano's Formula* (named for Geronimo Cardano (1501–1576)) and the formula for $n = 4$ can be reduced to this one. The formulas are valid over any field F of characteristic $\neq 2, 3$, which are the characteristics dividing the orders of the radicals necessary and the orders of the possible Galois groups (which are subgroups of S_3 and S_4). For simplicity we shall derive the formulas over $\mathbb{Q}$.

Solution of Cubic Equations by Radicals: Cardano's Formulas

From the proof of Theorem 39 and the fact that a composition series for S_3 as in equation (22) is given by $1 \leq A_3 \leq S_3$ we should expect that the solution of the cubic

$$f(x) = x^3 + ax^2 + bx + c$$

(or equivalently, under the substitution $x = y - a/3$,

$$g(y) = y^3 + py + q,$$

where

$$p = \frac{1}{3}(3b - a^2) \qquad q = \frac{1}{27}(2a^3 - 9ab + 27c))$$

to involve adjoining the 3rd roots of unity and the formation of Lagrange resolvents involving these roots of unity.

Let ρ denote a primitive 3rd root of unity, so that $\rho^2 + \rho + 1 = 0$. Let the roots of $g(y)$ be α, β, and γ, so that

$$\alpha + \beta + \gamma = 0$$

(one of the reasons for changing from $f(x)$ to $g(x)$). Over the field $\mathbb{Q}(\sqrt{D})$ where D is the discriminant (computed in the last section) the Galois group of $g(y)$ is A_3, i.e., a cyclic group of order 3. If we adjoin ρ then this extension is a radical extension of degree 3, with generator given by a Lagrange Resolvent, as in the proof of Proposition 37. Consider therefore the elements

$$(\alpha, 1) = \alpha + \beta + \gamma = 0$$
$$\theta_1 = (\alpha, \rho) = \alpha + \rho\beta + \rho^2\gamma$$
$$\theta_2 = (\alpha, \rho^2) = \alpha + \rho^2\beta + \rho\gamma.$$

Note that the sum of these resolvents is

$$\theta_1 + \theta_2 = 3\alpha \tag{14.23}$$

since $1 + \rho + \rho^2 = 0$. Similarly

$$\rho^2\theta_1 + \rho\theta_2 = 3\beta$$
$$\rho\theta_1 + \rho^2\theta_2 = 3\gamma. \tag{14.23'}$$

We also showed in general before Proposition 37 that the cube of these resolvents must lie in $\mathbb{Q}(\sqrt{D}, \rho)$. Expanding θ_1^3 we obtain

$$\alpha^3 + \beta^3 + \gamma^3 + 3\rho(\alpha^2\beta + \beta^2\gamma + \alpha\gamma^2)$$
$$+ 3\rho^2(\alpha\beta^2 + \beta\gamma^2 + \alpha^2\gamma) + 6\alpha\beta\gamma. \tag{14.24}$$

We have

$$\sqrt{D} = (\alpha - \beta)(\alpha - \gamma)(\beta - \gamma)$$
$$= (\alpha^2\beta + \beta^2\gamma + \alpha\gamma^2) - (\alpha\beta^2 + \beta\gamma^2 + \alpha^2\gamma).$$

Using this equation we see that (24) can be written

$$\alpha^3 + \beta^3 + \gamma^3 + 3\rho[\tfrac{1}{2}(S + \sqrt{D})] + 3\rho^2[\tfrac{1}{2}(S - \sqrt{D})] + 6\alpha\beta\gamma \tag{14.24'}$$

where for simplicity we have denoted by S the expression

$$(\alpha^2\beta + \beta^2\gamma + \alpha\gamma^2) + (\alpha\beta^2 + \beta\gamma^2 + \alpha^2\gamma).$$

Since S is symmetric in the roots, each of the expressions in (24') is a symmetric polynomial in α, β and γ, hence is a polynomial in the elementary symmetric functions $s_1 = 0$, $s_2 = p$, and $s_3 = -q$. After a short calculation one finds

$$\alpha^3 + \beta^3 + \gamma^3 = -3q \qquad S = 3q$$

so that from (24') we find ($\rho + \rho^2 = -1$ and $\rho - \rho^2 = \sqrt{-3}$)

$$\theta_1^3 = -3q + \frac{3}{2}\rho(3q + \sqrt{D}) + \frac{3}{2}\rho^2(3q - \sqrt{D}) - 6q$$
$$= \frac{-27}{2}q + \frac{3}{2}\sqrt{-3D}. \tag{14.25}$$

Similarly, we find

$$\theta_2^3 = \frac{-27}{2}q - \frac{3}{2}\sqrt{-3D}. \tag{14.25'}$$

Equations (25) and (23) essentially give the solutions of our cubic. One small point remains, however, namely the issue of extracting the cube roots of the expressions in (25) to obtain θ_1 and θ_2. There are 3 possible cube roots, which might suggest a total of 9 expressions in (23). This is not the case since θ_1 and θ_2 are not independent (adjoining one of them already gives the Galois extension containing all of the roots). A computation like the one above (but easier) shows that

$$\theta_1\theta_2 = -3p \tag{14.26}$$

showing that the choice of cube root for θ_1 determines θ_2. Using $D = -4p^3 - 27q^2$, we obtain Cardano's explicit formulas, as follows.

Let

$$A = \sqrt[3]{\frac{-27}{2}q + \frac{3}{2}\sqrt{-3D}}$$

$$B = \sqrt[3]{\frac{-27}{2}q - \frac{3}{2}\sqrt{-3D}}$$

where the cube roots are chosen so that $AB = -3p$. Then the roots of the equation

$$y^3 + py + q = 0$$

are

$$\alpha = \frac{A + B}{3} \qquad \beta = \frac{\rho^2 A + \rho B}{3} \qquad \gamma = \frac{\rho A + \rho^2 B}{3} \tag{14.27}$$

where $\rho = -\frac{1}{2} + \frac{1}{2}\sqrt{-3}$.

Examples

(1) Consider the cubic equation $x^3 - x + 1 = 0$. The discriminant of this cubic is

$$D = -4(-1)^3 - 27(1)^2 = -23$$

which is not the square of a rational number, so the Galois group for this polynomial is S_3. Substituting into the formulas above we have

$$A = \sqrt[3]{\frac{-27}{2} + \frac{3}{2}\sqrt{69}}$$

$$B = \sqrt[3]{\frac{-27}{2} - \frac{3}{2}\sqrt{69}}$$

where we choose A to be the real cube root and then from $AB = 3$ we see that B is also real. The roots of the cubic are given by (27) and we see that there is one real root and two (conjugate) complex roots (which we could have determined without solving for the roots, of course).

(2) Consider the equation $x^3 + x^2 - 2x - 1 = 0$. Letting $x = s - 1/3$ the equation becomes $s^3 - \frac{7}{3}s - \frac{7}{27} = 0$. Multiplying through by 27 to clear denominators and letting $y = 3s$ we see that y satisfies the cubic equation

$$y^3 - 21y - 7 = 0.$$

The discriminant D for this cubic is

$$D = -4(-21)^3 - 27(-7)^2 = 3^6 7^2$$

which shows that the Galois group for this (Eisenstein at 7) cubic is A_3. Substituting into the formulas above we have

$$A = 3 \sqrt[3]{\frac{7}{2} + \frac{21}{2}\sqrt{-3}}$$

$$B = 3 \sqrt[3]{\frac{7}{2} - \frac{21}{2}\sqrt{-3}}$$

and the roots of our cubics can be expressed in terms of A and B using the formulas above. This cubic arises from trying to express a primitive 7th root of unity ζ_7 in terms of radicals similar to the explicit formulas for the other roots of unity of small order (cf. the exercises).

In this case we have $g(-5) = -27$, $g(-1) = 13$, $g(0) = -7$ and $g(5) = 13$, so that this cubic has 3 *real* roots. The expressions above for these roots are sums of the conjugates of *complex* numbers. We shall see later that this is necessary, namely that it is impossible to solve for these real roots using only radicals involving real numbers.

A cubic with rational coefficients has either one real root and two complex conjugate imaginary roots or has three real roots. These two cases can be distinguished by the sign of the discriminant:

Suppose in the first case that the roots are a and $b \pm ic$ where a, b, and c are real and $c \neq 0$. Then

$$\sqrt{D} = [a - (b + ic)][a - (b - ic)][(b + ic) - (b - ic)]$$
$$= 2ic[(a - b)^2 + c^2]$$

is purely imaginary, so that the discriminant D is negative. Then in the formulas for A and B above we may choose both to be real. The first root in (27) is then real and the second two are complex conjugates.

If all three roots are real, then clearly $\sqrt{D}$ is real, so $D \geq 0$ is a nonnegative real number. If $D = 0$ then the cubic has repeated roots. For $D > 0$ (sometimes called the *Casus irreducibilis*), the formulas for the roots involve radicals of nonreal numbers, as in Example 2. We now show that for irreducible cubics this is necessary. The exercises outline the proof of the following generalization: if all the roots of the irreducible polynomial $f(x) \in \mathbb{Q}[x]$ are real and if one of these roots can be expressed by *real* radicals, then the degree of $f(x)$ is a power of 2, the Galois group of $f(x)$ is a 2-group, and the roots of $f(x)$ can be constructed by straightedge and compass.

Suppose that the irreducible cubic $f(x)$ has three real roots and that it were possible to express one of these roots by radicals involving only real numbers. Then the splitting field for the cubic would be contained in a root extension

$$\mathbb{Q} = K_0 \subset K_1 = \mathbb{Q}(\sqrt{D}) \subset \cdots \subset K_i \subset K_{i+1} \subset \cdots \subset K_s = K$$

where each field K_i, $i = 0, 1, \ldots, s$, is contained in the real numbers $\mathbb{R}$ and $s \geq 2$ since the quadratic extension $\mathbb{Q}(\sqrt{D})$ cannot contain the root of an irreducible cubic. We have begun this root extension with $\mathbb{Q}(\sqrt{D})$ because over this field the Galois group of the polynomial is cyclic of degree 3.

Note that for any field F the extension $F(\sqrt[mn]{a})$ of F can be obtained by two smaller simple radical extensions: let

$$F_1 = F(\sqrt[n]{a})$$

and let $b = \sqrt[n]{a} \in F_1$, so that

$$F(\sqrt[mn]{a}) = F_1(\sqrt[m]{b}).$$

We may therefore always assume our radical extensions are of the form $F(\sqrt[p]{a})$ where p is a prime.

Suppose now that F is a subfield of the real numbers $\mathbb{R}$ and let a be an element of F. Let p be a prime and let $\alpha = \sqrt[p]{a}$ denote a real p^{th} root of a. Then $[F(\sqrt[p]{a}) : F]$ must be either 1 or p, as follows. The conjugates of α over F all differ from α by a pth root of unity. It follows that the constant term of the minimal polynomial of α over F is $\alpha^d \zeta$ where $d = [F(\sqrt[p]{a}) : F]$ is the degree of the minimal polynomial and ζ is some p^{th} root of unity. Since α is real and $\alpha^d \zeta \in F$ is real, it follows that $\zeta = \pm 1$, so that $\alpha^d \in F$. Then, if $d \neq p$, $\alpha^d \in F$ and $\alpha^p = a \in F$ implies $\alpha \in F$, so $d = 1$.

Hence we may assume for the radical extensions above that $[K_{i+1} : K_i]$ is a prime p_i and $K_{i+1} = K_i(\sqrt[p_i]{a_i})$ for some $a_i \in K_i$, $i = 0, 1, \ldots, s-1$. In other words, the original tower of real radical extensions can be refined to a tower where each of the successive radical extensions has prime degree.

If any field containing $\sqrt{D}$ contains one of the roots of $f(x)$ then it contains the splitting field for $f(x)$, hence contains all the roots of the cubic. We suppose s is chosen so that K_{s-1} does not contain any of the roots of the cubic.

Consider the extension K_s/K_{s-1}. The field K_s contains all the roots of the cubic $f(x)$ and the field K_{s-1} contains none of these roots. It follows that $f(x)$ is irreducible over K_{s-1}, so $[K_s : K_{s-1}]$ is divisible by 3. Since we have reduced to the case where this extension degree is a prime, it follows that the extension degree is precisely 3 and that the extension K_s/K_{s-1} is Galois (being the splitting field of $f(x)$ over K_{s-1}). Since also $K_s = K_{s-1}(\sqrt[3]{a})$ for some $a \in K_{s-1}$, the Galois extension K_s must also contain the other cube roots of a. This implies that K_s contains ρ, a primitive 3rd root of unity. This contradicts the assumption that K_s is a subfield of $\mathbb{R}$ and shows that it is impossible to express the roots of this cubic in terms of real radicals only.

Solution of Quartic Equations by Radicals

Consider now the case of a quartic polynomial

$$f(x) = x^4 + ax^3 + bx^2 + cx + d$$

which under the substitution $x = y - a/4$ becomes the quartic

$$g(y) = y^4 + py^2 + qy + r$$

with

$$p = \frac{1}{8}(-3a^2 + 8b)$$

$$q = \frac{1}{8}(a^3 - 4ab + 8c)$$

$$r = \frac{1}{256}(-3a^4 + 16a^2 b - 64ac + 256d).$$

Let the roots of $g(y)$ be α_1, α_2, α_3, and α_4. The resolvent cubic introduced in the previous section for this quartic is

$$h(x) = x^3 - 2px^2 + (p^2 - 4r)x + q^2$$

and has roots

$$\theta_1 = (\alpha_1 + \alpha_2)(\alpha_3 + \alpha_4)$$
$$\theta_2 = (\alpha_1 + \alpha_3)(\alpha_2 + \alpha_4)$$
$$\theta_3 = (\alpha_1 + \alpha_4)(\alpha_2 + \alpha_3).$$

The Galois group of the splitting field for $f(x)$ (or $g(y)$) over the splitting field of the resolvent cubic $h(x)$ is the Klein 4-group. Such extensions are biquadratic, which means that it is possible to solve for the roots α_1, α_2, α_3, and α_4 in terms of square roots of expressions involving the roots θ_1, θ_2, and θ_3 of the resolvent cubic. In this case we evidently have

$$(\alpha_1 + \alpha_2)(\alpha_3 + \alpha_4) = \theta_1 \qquad (\alpha_1 + \alpha_2) + (\alpha_3 + \alpha_4) = 0$$

which gives

$$\alpha_1 + \alpha_2 = \sqrt{-\theta_1} \qquad \alpha_3 + \alpha_4 = -\sqrt{-\theta_1}.$$

Similarly,

$$\alpha_1 + \alpha_3 = \sqrt{-\theta_2} \qquad \alpha_2 + \alpha_4 = -\sqrt{-\theta_2}$$
$$\alpha_1 + \alpha_4 = \sqrt{-\theta_3} \qquad \alpha_2 + \alpha_3 = -\sqrt{-\theta_3}.$$

An easy computation shows that $\sqrt{-\theta_1}\sqrt{-\theta_2}\sqrt{-\theta_3} = -q$, so that the choice of two of the square roots determines the third. Since $\alpha_1 + \alpha_2 + \alpha_3 + \alpha_4 = 0$, if we add the left-hand equations above we obtain $2\alpha_1$, and similarly we may solve for the other roots of $g(y)$. We find

$$2\alpha_1 = \sqrt{-\theta_1} + \sqrt{-\theta_2} + \sqrt{-\theta_3}$$
$$2\alpha_2 = \sqrt{-\theta_1} - \sqrt{-\theta_2} - \sqrt{-\theta_3}$$
$$2\alpha_3 = -\sqrt{-\theta_1} + \sqrt{-\theta_2} - \sqrt{-\theta_3}$$
$$2\alpha_4 = -\sqrt{-\theta_1} - \sqrt{-\theta_2} + \sqrt{-\theta_3}$$

which reduces the solution of the quartic equation to the solution of the associated resolvent cubic.

EXERCISES

1. Use Cardano's Formulas to solve the equation $x^3 + x^2 - 2 = 0$. In particular show that the equation has the real root

$$\frac{1}{3}(\sqrt[3]{26 + 15\sqrt{3}} + \sqrt[3]{26 - 15\sqrt{3}} - 1).$$

Show directly that the roots of this cubic are $1, -1 \pm i$. Explain this by proving that

$$\sqrt[3]{26 + 15\sqrt{3}} = 2 + \sqrt{3} \qquad \sqrt[3]{26 - 15\sqrt{3}} = 2 - \sqrt{3}$$

so that

$$\sqrt[3]{26 + 15\sqrt{3}} + \sqrt[3]{26 - 15\sqrt{3}} = 4.$$

2. Let ζ_7 be a primitive 7th root of unity and let $\alpha = \zeta + \zeta^{-1}$.
   (a) Show that ζ_7 is a root of the quadratic $z^2 - \alpha z + 1$ over $\mathbb{Q}(\alpha)$.
   (b) Show using the minimal polynomial for ζ_7 that α is a root of the cubic $x^3 + x^2 - 2x - 1$.
   (c) Use (a) and (b) together with the explicit solution of the cubic in (b) in the text to express ζ_7 in terms of radicals similar to the expressions given earlier for the other roots of unity of small order. (The complicated nature of the expression explains why we did not include ζ_7 earlier in our list of explicit roots of unity.)

3. Let F be a field of characteristic $\neq 2$. State and prove a necessary and sufficient condition on $\alpha, \beta \in F$ so that $F(\sqrt{\alpha}) = F(\sqrt{\beta})$. Use this to determine whether $\mathbb{Q}(\sqrt{1 - \sqrt{2}}) = \mathbb{Q}(i, \sqrt{2})$.

4. Let $K = \mathbb{Q}(\sqrt[n]{a})$, where $a \in \mathbb{Q}$, $a > 0$ and suppose $[K : \mathbb{Q}] = n$ (i.e., $x^n - a$ is irreducible). Let E be any subfield of K and let $[E : \mathbb{Q}] = d$. Prove that $E = \mathbb{Q}(\sqrt[d]{a})$. [Consider $N_{K/E}(\sqrt[n]{a}) \in E$.]

5. Let K be as in the previous exercise. Prove that if n is odd then K has no nontrivial subfields which are Galois over $\mathbb{Q}$ and if n is even then the only nontrivial subfield of K which is Galois over $\mathbb{Q}$ is $\mathbb{Q}(\sqrt{a})$.

6. Let L be the Galois closure of K in the previous two exercises (i.e., the splitting field of $x^n - a$). Prove that $[L : \mathbb{Q}] = n\varphi(n)$ or $\frac{1}{2}n\varphi(n)$. [Note that $\mathbb{Q}(\zeta_n) \cap K$ is a Galois extension of $\mathbb{Q}$.]

7. (*Kummer Generators for Cyclic Extensions*) Let F be a field of characteristic not dividing n containing the nth roots of unity and let K be a cyclic extension of degree d dividing n. Then $K = F(\sqrt[n]{a})$ for some nonzero $a \in F$. Let σ be a generator for the cyclic group $\text{Gal}(K/F)$.
   (a) Show that $\sigma(\sqrt[n]{a}) = \zeta \sqrt[n]{a}$ for some primitive d^{th} root of unity ζ.
   (b) Suppose $K = F(\sqrt[n]{a}) = F(\sqrt[n]{b})$. Use (a) to show that $\dfrac{\sigma(\sqrt[n]{a})}{\sqrt[n]{a}} = \left(\dfrac{\sigma(\sqrt[n]{b})}{\sqrt[n]{b}}\right)^i$ for some integer i relatively prime to d. Conclude that σ fixes the element $\dfrac{\sqrt[n]{a}}{(\sqrt[n]{b})^i}$ so this is an element of F.
   (c) Prove that $K = F(\sqrt[n]{a}) = F(\sqrt[n]{b})$ if and only if $a = b^i c_1^n$ and $b = a^j c_2^n$ for some $c_1, c_2 \in F$, i.e., if and only if a and b generate the same subgroup of $F^\times$ modulo n^{th} powers.

8. Let p, q and r be primes in $\mathbb{Z}$ with $q \neq r$. Let $\sqrt[p]{q}$ denote any root of $x^p - q$ and let $\sqrt[p]{r}$ denote any root of $x^p - r$. Prove that $\mathbb{Q}(\sqrt[p]{q}) \neq \mathbb{Q}(\sqrt[p]{r})$.

9. (*Artin–Schreier Extensions*) Let F be a field of characteristic p and let K be a cyclic extension of F of degree p. Prove that $K = F(\alpha)$ where α is a root of the polynomial $x^p - x - a$ for some $a \in F$. [Note that $\text{Tr}_{K/F}(-1) = 0$ since F is of characteristic p so that $-1 = \alpha - \sigma\alpha$ for some $\alpha \in K$ where σ is a generator of $\text{Gal}(K/F)$ by Exercise 26 of Section 2. Show that $a = \alpha^p - \alpha$ is an element of F.] Note that since F contains the pth roots of unity (namely, 1) that this completes the description of all cyclic extensions of prime degree p over fields containing the pth roots of unity in all characteristics.

10. Let $K = \mathbb{Q}(\zeta_p)$ be the cyclotomic field of pth roots of unity for the prime p and let

$G = \mathrm{Gal}(K/\mathbb{Q})$. Let ζ denote any p^{th} root of unity. Prove that $\sum_{\sigma \in G} \sigma(\zeta)$ (the trace from K to $\mathbb{Q}$ of ζ) is -1 or $p-1$ depending on whether ζ is or is not a primitive p^{th} root of unity.

11. (*The Classical Gauss Sum*) Let $K = \mathbb{Q}(\zeta_p)$ be the cyclotomic field of pth roots of unity for the odd prime p, viewed as a subfield of $\mathbb{C}$, and let $G = \mathrm{Gal}(K/\mathbb{Q})$. Let H denote the subgroup of index 2 in the cyclic group G. Define $\eta_0 = \sum_{\tau \in H} \tau(\zeta_p)$, $\eta_1 = \sum_{\tau \in \sigma H} \tau(\zeta_p)$, where σ is a generator of $\mathrm{Gal}(K/\mathbb{Q})$ (the two *periods* of ζ_p with respect to H, i.e., the sum of the conjugates of ζ_p with respect to the two cosets of H in G, cf. Section 5).

(a) Prove that $\sigma(\eta_0) = \eta_1$, $\sigma(\eta_1) = \eta_0$ and that

$$\eta_0 = \sum_{a=\text{square}} \zeta_p^a \quad , \quad \eta_1 = \sum_{b \neq \text{square}} \zeta_p^b,$$

where the sums are over the squares and nonsquares (respectively) in $(\mathbb{Z}/p\mathbb{Z})^\times$. [Observe that H is the subgroup of squares in $(\mathbb{Z}/p\mathbb{Z})^\times$.]

(b) Prove that $\eta_0 + \eta_1 = (\zeta_p, 1) = -1$ and $\eta_0 - \eta_1 = (\zeta_p, -1)$ where $(\zeta_p, 1)$ and $(\zeta_p, -1)$ are two of the Lagrange resolvents of ζ_p.

(c) Let $g = \sum_{i=0}^{p-1} \zeta_p^{i^2}$ (the classical *Gauss sum*). Prove that

$$g = (\zeta_p, -1) = \sum_{i=0}^{p-2} (-1)^i \, \sigma^i(\zeta_p).$$

(d) Prove that $\tau g = g$ if $\tau \in H$ and $\tau g = -g$ if $\tau \notin H$. Conclude in particular that $[\mathbb{Q}(g) : \mathbb{Q}] = 2$. Recall that complex conjugation is the automorphism σ_{-1} on K (cf. Exercise 7 of Section 5). Conclude that $\bar{g} = g$ if -1 is a square mod p (i.e., if $p \equiv 1 \bmod 4$) and $\bar{g} = -g$ if -1 is not a square mod p (i.e., if $p \equiv 3 \bmod 4$) where $\bar{g}$ denotes the complex conjugate of g.

(e) Prove that $g\bar{g} = p$. [The complex conjugate of a root of unity is its reciprocal. Then $\bar{g} = \sum_{j=0}^{p-2} (-1)^j \, (\sigma^j(\zeta_p))^{-1}$ gives

$$g\bar{g} = \sum_{i,j=0}^{p-2} (-1)^i (-1)^j \frac{\sigma^i(\zeta_p)}{\sigma^j(\zeta_p)} = \sum_{i,j=0}^{p-2} (-1)^{i-j} \sigma^j \left[\frac{\sigma^{i-j}(\zeta_p)}{\zeta_p} \right]$$

$$= \sum_{k=0}^{p-2} (-1)^k \sum_{j=0}^{p-2} \sigma^j \left[\frac{\sigma^k(\zeta_p)}{\zeta_p} \right]$$

where $k = i - j$. If $k = 0$ the element $\dfrac{\sigma^k(\zeta_p)}{\zeta_p}$ is 1, and if $k \neq 0$ then this is a primitive p^{th} root of unity. Use the previous exercise to conclude that the inner sum is $p - 1$ when $k = 0$ and is -1 otherwise.]

(f) Conclude that $g^2 = (-1)^{(p-1)/2} p$ and that $\mathbb{Q}(\sqrt{(-1)^{(p-1)/2} p})$ is the unique quadratic subfield of $\mathbb{Q}(\zeta_p)$. (Cf. also Exercise 33 of Section 6.)

12. Let L be the Galois closure of the finite extension $\mathbb{Q}(\alpha)$ of $\mathbb{Q}$. For any prime p dividing the order of $\mathrm{Gal}(L/\mathbb{Q})$ prove there is a subfield F of L with $[L : F] = p$ and $L = F(\alpha)$.

13. Let F be a subfield of the real numbers $\mathbb{R}$. Let a be an element of F and let $K = F(\sqrt[n]{a})$ where $\sqrt[n]{a}$ denotes a real n^{th} root of a. Prove that if L is any Galois extension of F contained in K then $[L : F] \leq 2$.

14. This exercise shows that in general it is necessary to use complex numbers when expressing real roots in terms of radicals and generalizes the *Casus irreducibilis* of cubic equations.

Let $f(x) \in \mathbb{Q}[x]$ be an irreducible polynomial all of whose roots are real. Suppose further that one of the roots, α, of $f(x)$ can be expressed in terms of *real* radicals (i.e., there is a root extension of real fields $\mathbb{Q} = K_0 \subset K_1 \subset \ldots \subset K_m \subset \mathbb{R}$ with $K_{i+1} = K_i(\sqrt[n_i]{a_i})$, $i = 1, 2, \ldots, m-1$, for some integers n_i and some $a_i \in K_i$ and $\alpha \in K_m$). Prove that the Galois group of $f(x)$ is a 2-group. Conclude in particular that the degree of $f(x)$ is a power of 2 and that the real roots of such a polynomial can be expressed entirely in terms of real radicals if and only if these roots can be constructed by straightedge and compass. [The argument is similar to the case of cubics. Let $L \in \mathbb{R}$ be the Galois closure of $\mathbb{Q}(\alpha)$ and suppose the order of $\mathrm{Gal}(L/\mathbb{Q})$ is divisible by some odd prime p. Let F be a subfield of L with $[L : F] = p$ and $L = F(\alpha)$ (by Exercise 12) and consider the composite fields $K_i' = FK_i$, $i = 0, 1, \ldots, m$. These are again real radical extensions and by the argument in the text for the *Casus irreducibilis*, we may assume each $[K_{i+1}' : K_i']$ is a prime. Since $\alpha \notin F = FK_0$, there is an integer s with $\alpha \notin K_{s-1}'$, $\alpha \in K_s'$. Since the extensions are of prime degree, we have $K_s' = K_{s-1}'(\alpha)$. Since $L = F(\alpha)$ is Galois of degree p, K_s' is a Galois extension of K_{s-1}' of degree p, contradicting the previous exercise.]

15. *('Cardano's Formulas' for a Cubic in Characteristic 2)* Suppose $f(x) = x^3 + px + q$ is an irreducible cubic over a field of characteristic 2. Let ρ be a primitive 3rd root of unity and let θ, θ' be the roots of the quadratic $x^2 + qx + (p^3 + q^2)$ (cf. Exercise 50 of Section 6). Let θ_1 and θ_2 be cube roots of $\rho q + \theta$ and $\rho q + \theta'$, respectively, where the cube roots are chosen so that $\theta_1\theta_2 = p$. Prove that the roots of $f(x)$ are given by $\alpha = \theta_1 + \theta_2$, $\beta = \rho\alpha + \theta_1$, and $\gamma = \rho\alpha + \theta_2 = \alpha + \beta$.

16. Let a be a nonzero rational number.
  (a) Determine when the extension $\mathbb{Q}(\sqrt{ai})$ $(i^2 = -1)$ is of degree 4 over $\mathbb{Q}$.
  (b) When $K = \mathbb{Q}(\sqrt{ai})$ is of degree 4 over $\mathbb{Q}$ show that K is Galois over $\mathbb{Q}$ with the Klein 4-group as Galois group. In this case determine the quadratic extensions of $\mathbb{Q}$ contained in K.

17. Let $D \in \mathbb{Z}$ be a squarefree integer and let $a \in \mathbb{Q}$ be a nonzero rational number. Show that $\mathbb{Q}(\sqrt{a\sqrt{D}})$ cannot be a cyclic extension of degree 4 over $\mathbb{Q}$.

18. Let $D \in \mathbb{Z}$ be a squarefree integer and let $a \in \mathbb{Q}$ be a nonzero rational number. Prove that if $\mathbb{Q}(\sqrt{a\sqrt{D}})$ is Galois over $\mathbb{Q}$ then $D = -1$.

19. Let $D \in \mathbb{Z}$ be a squarefree integer and let $K = \mathbb{Q}(\sqrt{D})$.
  (a) Prove that if $D = s^2 + t^2$ is the sum of two rational squares then there exists an extension $L/\mathbb{Q}$ containing K which is Galois over $\mathbb{Q}$ with a cyclic Galois group of order 4. [Consider the extension $\mathbb{Q}(\sqrt{D + s\sqrt{D}})$.] (Note also that D is the sum of two rational squares if and only if D is also the sum of two integer squares, so one may assume s and t are integral without loss.)
  (b) Prove conversely that if K can be embedded in a cyclic extension L of degree 4 as in (a) then D is the sum of two squares. [One approach: (i) observe first that L is quadratic over K, so $L = K(\sqrt{a + b\sqrt{D}})$ for some $a, b \in \mathbb{Q}$, (ii) show that L contains the quadratic subfield $\mathbb{Q}(\sqrt{a^2 - b^2 D})$, which must be $\mathbb{Q}(\sqrt{D})$ if $L/\mathbb{Q}$ is cyclic, and use Exercise 7.]
  (c) Conclude in particular that $\mathbb{Q}(\sqrt{3})$ is not a subfield of any cyclic extension of degree 4 over $\mathbb{Q}$. Similarly conclude that the fields $\mathbb{Q}(\sqrt{D})$ for squarefree integers $D < 0$ are never contained in cyclic extensions of degree 4 over $\mathbb{Q}$ (this gives an alternate proof for Exercise 19, Section 6).

20. Let p be a prime. Show that any solvable subgroup of S_p of order divisible by p is contained in the normalizer of a Sylow p-subgroup of S_p (a Frobenius group of order $p(p-1)$). Conclude that an irreducible polynomial $f(x) \in \mathbb{Q}[x]$ of degree p is solvable by radicals if and only if its Galois group is contained in the Frobenius group of order $p(p-1)$. [Let $G \le S_p$ be a solvable subgroup of order divisible by p. Then G contains a p-cycle, hence is transitive on $\{1, 2, \dots, p\}$. Let $H < G$ be the stabilizer in G of the element 1, so H has index p in G. Show that H contains no nontrivial normal subgroups of G (note that the conjugates of H are the stabilizers of the other points). Let $G^{(n-1)}$ be the last nontrivial subgroup in the derived series for G. Show that $H \cap G^{(n-1)} = 1$ and conclude that $|G^{(n-1)}| = p$, so that the Sylow p-subgroup of G (which is also a Sylow p-subgroup in S_p) is normal in G.]

21. (*Criterion for the Solvability of a Quintic*) By the previous exercise, an irreducible polynomial $f(x)$ in $\mathbb{Q}[x]$ of degree 5 can be solved by radicals if and only if its Galois group (considered as a subgroup of S_5) is contained in the Frobenius group of order 20. It is known that this is the case if and only if an associated polynomial $g(x)$ of degree 6 has a rational root (cf. Dummit, *Solving Solvable Quintics*, Math. Comp., 57(1991), pp. 387–401). If the quintic is in the general form (where a translation is performed so that the coefficient of x^4 is zero)

$$f(x) = x^5 + px^3 + qx^2 + rx + s \qquad p, q, r, s \in \mathbb{Q}$$

then the associated polynomial of degree 6 is

$$
\begin{aligned}
g(x) = {} & x^6 + 8rx^5 + (2pq^2 - 6p^2r + 40r^2 - 50qs)\, x^4 \\
& + (-2q^4 + 21pq^2r - 40p^2r^2 + 160r^3 - 15p^2qs - 400qrs + 125ps^2)\, x^3 \\
& + (p^2q^4 - 8q^4r + 9p^4r^2 - 136p^2r^3 + 625q^2s^2 + 400r^4 - 6p^3q^2r \\
& \quad + 76pq^2r^2 - 50pq^3s - 1400qr^2s + 500prs^2 + 90p^2qrs)\, x^2 \\
& + (-108p^5s^2 + 32p^4r^3 - 256p^2r^4 - 3125s^4 + 512r^5 - 2pq^6 + 3q^4r^2 \\
& \quad - 58q^5s + 2750q^2rs^2 - 31p^3q^3s - 500pr^2s^2 + 19p^2q^4r \\
& \quad - 51p^3q^2r^2 + 76pq^2r^3 - 2400qr^3s - 325p^2q^2s^2 + 525p^3rs^2 \\
& \quad + 625pqs^3 + 117p^4qrs + 105pq^3rs + 260p^2qr^2s)\, x \\
& + (q^8 + 256r^6 + 17q^4r^3 - 27p^7s^2 - 4p^6r^3 + 48p^4r^4 - 192p^2r^5 \\
& \quad + 3125p^2s^4 - 9375rs^4 - 1600qr^4s - 99p^5rs^2 - 125pq^4s^2 \\
& \quad - 124q^5rs + 3250q^2r^2s^2 - 2000pr^3s^2 - 13pq^6r + p^5q^2r^2 \\
& \quad + 65p^2q^4r^2 - 128p^3q^2r^3 - 16pq^2r^4 - 4p^5q^3s - 12p^2q^5s \\
& \quad - 150p^4q^2s^2 + 1200p^3r^2s^2 + 18p^6qrs + 12p^3q^3rs + 196p^4qr^2s \\
& \quad + 590pq^3r^2s - 160p^2qr^3s - 725p^2q^2rs^2 - 1250pqrs^3).
\end{aligned}
$$

In the particular case where $f(x) = x^5 + Ax + B$ this polynomial is simply

$$g(x) = x^6 + 8Ax^5 + 40A^2x^4 + 160A^3x^3 + 400A^4x^2 + (512A^5 - 3125B^4)x - 9375AB^4 + 256A^6.$$

(a) Use this criterion to prove that the Galois group over $\mathbb{Q}$ of the polynomial $x^5 - 5x + 12$ is the dihedral group of order 10. [Show the associated sixth degree polynomial is

$$x^6 - 40x^5 + 1000x^4 - 20000x^3 + 250000x^2 - 66400000x + 976000000$$

and has $x = 40$ as a rational root. Cf. also Exercise 35 in Section 6.]

(b) Use this criterion to prove that $x^5 - x - 1$ is not solvable by radicals.

14.8 COMPUTATION OF GALOIS GROUPS OVER ℚ

In the determination of the Galois groups of polynomials of degrees ≤ 4 in Section 6 and in the determination of the Galois group of the polynomial $x^5 - 6x + 3$ in the previous section we observed that it was possible to obtain useful information regarding the Galois group from the *cycle types* of the automorphisms as elements in S_n. This is very useful in computing Galois groups of polynomials over ℚ and we now briefly describe the theoretical justification.

Let $f(x)$ be a polynomial with rational coefficients. In determining the Galois group of $f(x)$ we may assume that $f(x)$ is separable and has integer coefficients. Then the discriminant D of $f(x)$ is an integer and is nonzero.

For any prime p, consider the reduction $\overline{f}(x) \in \mathbb{F}_p[x]$ of $f(x)$ modulo p. If p divides D then the reduced polynomial $\overline{f}(x)$ has discriminant $\overline{D} = 0$ in $\mathbb{F}_p$, so is not separable.

If p does not divide D, then $\overline{f}(x)$ is a separable polynomial over $\mathbb{F}_p$ and we can factor $\overline{f}(x)$ into distinct irreducibles

$$\overline{f}(x) = \overline{f}_1(x)\overline{f}_2(x) \cdots \overline{f}_k(x) \qquad \text{in } \mathbb{F}_p[x].$$

Let n_i be the degree of $\overline{f}_i(x)$, $i = 1, 2, \ldots, k$.

The importance of this reduction is provided by the following theorem from algebraic number theory which is an elementary consequence of the study of the arithmetic in finite extensions of ℚ (and which we take for granted).

Theorem. For any prime p not dividing the discriminant D of $f(x) \in \mathbb{Z}[x]$, the Galois group over $\mathbb{F}_p$ of the reduction $\overline{f}(x) = f(x) \pmod{p}$ is permutation group isomorphic to a subgroup of the Galois group over ℚ of $f(x)$.

The meaning of the statement "permutation group isomorphic" in the theorem is that not only is the Galois group of the reduction $\overline{f}(x)$ mod p of $f(x)$ isomorphic to a subgroup of the Galois group of $f(x)$ but that there is an ordering of the roots of $\overline{f}(x)$ and of $f(x)$ (depending on p) so that under this isomorphism the action of the corresponding automorphisms as permutations of these roots is the same. In particular there are automorphisms in the Galois group of $f(x)$ with the same cycle types as the automorphisms of $\overline{f}(x)$.

The Galois group of $\overline{f}(x)$ is a *cyclic* group since every finite extension of $\mathbb{F}_p$ is a cyclic extension. Let σ be a generator for this Galois group over $\mathbb{F}_p$ (for example, the Frobenius automorphism). The roots of $\overline{f}_1(x)$ are permuted amongst themselves by the Galois group, and given any two of these roots there is a Galois automorphism taking the first root to the second (recall that the group is said to be *transitive* on the roots when this is the case). Similarly, the Galois group permutes the roots of each of the factors $\overline{f}_i(x)$, $i = 1, 2, \ldots, k$ transitively. Since these factors are relatively prime we also see that no root of one factor is mapped to a root of any other factor by any element of the Galois group.

View σ as an element in S_n by labelling the n roots of $\overline{f}(x)$ and consider the cycle decomposition of σ, which is a product of k distinct permutations since σ permutes the roots of each of the factors $\overline{f}_i(x)$ amongst themselves. By the observations we just made, the action of σ on the roots of $\overline{f}_1(x)$ must be a cycle of length n_i since otherwise the powers of σ could not be transitive on the roots of $\overline{f}_1(x)$. Similarly the action of σ on the roots of $\overline{f}_i(x)$ gives a cycle of length n_i, $i = 1, 2, \ldots, k$.

We see that the automorphism σ generating the Galois group of $\overline{f}(x)$ has cycle decomposition $(n_1, n_2, \ldots, n_k)$ where $n_1, n_2, \ldots, n_k$ are the degrees of the irreducible factors of $f(x)$ reduced modulo p, which gives us the following result.

Corollary 41. For any prime p not dividing the discriminant of $f(x) \in \mathbb{Z}[x]$, the Galois group of $f(x)$ over $\mathbb{Q}$ contains an element with cycle decomposition $(n_1, n_2, \ldots, n_k)$ where $n_1, n_2, \ldots, n_k$ are the degrees of the irreducible factors of $f(x)$ reduced modulo p.

Example

Consider the polynomial $x^5 - x - 1$. The discriminant of this polynomial is $2869 = 19 \cdot 151$ so we reduce at primes $\neq 19, 151$. Reducing mod 2 the polynomial $x^5 - x - 1$ factors as $(x^2 + x + 1)(x^3 + x^2 + 1)$ (mod 2) so the Galois group has a (2,3)-cycle. Cubing this element we see the Galois group contains a transposition.

Reducing mod 3 the polynomial is irreducible, as follows: $x^5 - x - 1$ has no roots mod 3 so if it were reducible mod 3 then it would have an irreducible quadratic factor, hence would have a factor in common with $x^9 - x$ (which is the product of all irreducible polynomials of degrees 1 and 2 over $\mathbb{F}_3$), hence a factor in common with either $x^4 - 1$ or $x^4 + 1$, hence a factor in common with either $x^5 - x$ or $x^5 + x$, hence a factor in common with either -1 or $2x + 1$ which it obviously does not. This shows both that $x^5 - x - 1$ is irreducible in $\mathbb{Z}[x]$ and that there is a 5-cycle in its Galois group.

Since S_5 is generated by any 5-cycle and any transposition, it follows that the Galois group of $x^5 - x - 1$ is S_5 (so in particular this polynomial cannot be solved by radicals, (cf. Exercise 21 of Section 7).

The arguments in the example above indicate how to construct polynomials with S_n as Galois group. We use the fact that a transitive subgroup of S_n containing a transposition and an $n - 1$-cycle is S_n. Let f_1 be an irreducible polynomial of degree n over $\mathbb{F}_2$. Let $f_2 \in \mathbb{F}_3[x]$ be the product of an irreducible polynomial of degree 2 with irreducible polynomials of odd degree (for example, an irreducible polynomial of degree $n - 3$ and x if n is even and an irreducible polynomial of degree $n - 2$ if n is odd). Let $f_3 \in \mathbb{F}_5[x]$ be the product of x with an irreducible polynomial of degree $n - 1$. Finally, let $f(x) \in \mathbb{Z}[x]$ be any polynomial with

$$f(x) \equiv f_1(x) \pmod{2}$$
$$\equiv f_2(x) \pmod{3}$$
$$\equiv f_3(x) \pmod{5}.$$

The reduction of $f(x)$ mod 2 shows that $f(x)$ is irreducible in $\mathbb{Z}[x]$, hence the Galois group is transitive on the n roots of $f(x)$. Raising the element given by the factorization of $f(x)$ mod 3 to a suitable odd power shows the Galois group contains a transposition. The factorization mod 5 shows the Galois group contains an $n - 1$-cycle, hence the Galois group is S_n.

Proposition 42. For each $n \in \mathbb{Z}^+$ there exist infinitely many polynomials $f(x) \in \mathbb{Z}[x]$ with S_n as Galois group over $\mathbb{Q}$.

There are extremely efficient algorithms for factoring polynomials $f(x) \in \mathbb{Z}[x]$ modulo p (cf. Exercises 12 to 17 of Section 3), so the corollary above is an effective procedure for determining some of the cycle types of the elements of the Galois group. In using Corollary 41 some care should be taken not to assume that a *particular* cycle is an element of the Galois group. For example, one factorization might imply the existence of a (2,2) cycle, say (12)(34) and another factorization imply the existence of a transposition. One cannot conclude that the transposition is necessarily (12), however (nor (34), nor (13), etc.). The choice of (12)(34) to represent the first cycle fixes a particular ordering on the roots and this may not be the ordering with respect to which the transposition appears as (12).

Corollary 41 is particularly efficient in determining when the Galois group is large (e.g., S_n), since a transitive group containing sufficiently many cycle types must be S_n (for example, a transitive subgroup of S_n containing a transposition and an $n - 1$-cycle is S_n, as used above). The most difficult Galois groups to determine in this way are the *small* Galois groups (e.g., a cyclic group of order n), since factorization after factorization will produce only elements of orders dividing n and one is not sure whether there will be some p yet to come producing a cycle type inconsistent with the assumption of a cyclic Galois group. If one could "compute forever" one could at least be sure of the precise distribution of cycle types among the elements of the Galois group in the following sense: suppose the Galois group $G \subseteq S_n$ has order N and that there are n_T elements of G with cycle type T (e.g., (2,2)-cycles, transpositions, etc.) so that the "density" of cycle type T in G is $d_T = n_T / N$. Then it is possible to define a density on the set of prime numbers (so that it makes sense to speak of "1/2" the primes, etc.) and we have the following result (which relies on the Tchebotarov Density Theorem in algebraic number theory).

Theorem. The density of primes p for which $f(x)$ splits into type T modulo p is precisely d_T.

This says that if we knew the factorization of $f(x)$ modulo every prime we could at least determine the number of elements of G with a given cycle type. Unfortunately, even this would not be sufficient to determine G (up to isomorphism): it is known that there are nonisomorphic groups containing the same number of elements of all cycle types (there are two nonisomorphic groups of order 96 in S_8 both having cycle type distributions: 1 1-cycle, 6 (2,2)-cycles, 13 (2,2,2,2)-cycles, 32 (3,3)-cycles, 12 (4,4)-cycles, 32 (2,6)-cycles). There are infinitely many such examples (the regular representation of the elementary abelian group of order p^3 and for the nonabelian group of order p^3 of exponent p give two nonisomorphic groups in S_{p^3} whose nonidentity elements are all the product of p^2 p-cycles for any prime p).

In practice one uses the factorizations of $f(x)$ modulo small primes to get an idea of the probable Galois group (based on the previous result). One then tries to prove this is indeed the Galois group — often a difficult problem. For polynomials of small degree, definitive algorithms exist, based in part on the computation of *resolvent* polynomials.

642

These are analogues of the cubic resolvent used in the previous sections to determine the Galois group of quartic polynomials. These resolvent polynomials have rational coefficients and have as roots certain combinations of the roots of $f(x)$ (similar to the combinations $(\alpha_1 + \alpha_2)(\alpha_3 + \alpha_4)$ for the cubic resolvent). One then determines the factorization of these resolvent polynomials to obtain information on the Galois group of $f(x)$ — for example the existence of a linear factor implies the Galois group lies in the stabilizer in S_n of the combination of the roots of $f(x)$ chosen (for example, the dihedral group of order 8 for our resolvent cubic). It should be observed, however, that the degree of the resolvent polynomials constructed, unlike the situation of the resolvent cubic for quartic polynomials, are in general much larger than the degree of $f(x)$. The effectiveness of this computational technique also depends heavily on the explicit knowledge of the possible transitive subgroups of S_n. For $n = 2, 3, \ldots, 8$ the number of isomorphism classes of transitive subgroups of S_n is 1, 2, 5, 5, 16, 7, 50, respectively. There is a great deal of interest in the computation of Galois groups, motivated in part by the problem of determining which groups occur as Galois groups over $\mathbb{Q}$.

We illustrate these techniques with some easier examples (from *The Computation of Galois Groups*, L. Soicher, Master's Thesis, Concordia University, Montreal, 1981).

Examples

(1) There are 5 isomorphism classes of transitive subgroups of S_5 given by the groups Z_5, D_{10}, F_{20}, the so-called Frobenius group of order 20 (the Galois group of $x^5 - 2$ with generators $(1\,2\,3\,4\,5)$ and $(2\,3\,5\,4)$ in S_5), A_5 and S_5. The cycle type distributions for these groups are as follows:

cycle type :	1	2	(2, 2)	3	(2, 3)	4	5
Z_5	1						4
D_{10}	1		5				4
F_{20}	1		5			10	4
A_5	1		15	20			24
S_5	1	10	15	20	20	30	24.

Given this information, the irreducibility of $x^5 - x - 1$ (giving the transitivity on the 5 roots) and the cycle type (2,3) immediately shows that the Galois group of $x^5 - x - 1$ is S_5.

Consider now the polynomial $x^5 + 15x + 12$. The discriminant is $2^{10}3^45^5$ so the Galois group is not contained in A_5. There are two possibilities: S_5 or F_{20}. One can easily determine which is more likely by factoring the polynomial modulo a number of small primes and comparing the distribution of cycle types with those in the table above. This does not *prove* the probable Galois group is actually correct. To decide which of S_5 and F_{20} is correct one can compute the resolvent polynomial $R(x)$ of degree 15 whose roots are the distinct permutations under S_5 of $(\alpha_1 + \alpha_2 - \alpha_3 - \alpha_4)^2$ for 4 of the roots $\alpha_1, \alpha_2, \alpha_3, \alpha_4$ of $f(x)$. By definition, S_5 is transitive on the roots of $R(x)$ and it is not difficult to check using the explicit generators for F_{20} given above that F_{20} is not transitive on these 15 values. It follows that $R(x)$ will be a reducible polynomial over $\mathbb{Q}$ if and only if the Galois group of the quintic is F_{20}. One finds that for $x^5 + 15x + 12$ the resolvent polynomial $R(x)$ factors into a polynomial of degree 5 and a polynomial of degree 10, hence the Galois group for this quintic is F_{20}. One can also use Exercise 21 of the previous section (cf. Exercise 6), which is also based on the computation of a related resolvent polynomial.

(2) Consider the polynomial $x^7 - 14x^5 + 56x^3 - 56x + 22$. The discriminant is computed to be $2^6 7^{10}$ so the Galois group is contained in A_7.

Factoring the polynomial for the 42 primes not equal to 7 between 3 and 193 gives a cycle type distribution of 1 1-cycle (2.38 %), 30 (3,3)-cycles (71.43 %), 11 7-cycles (26.19 %). There are 7 isomorphism classes of transitive subgroups of S_7, 4 of them contained in A_7. Of these, one contains no (3,3)-cycles, which leaves the three possibilities A_7, $GL_3(\mathbb{F}_2)$, or F_{21}, the Frobenius group of order 21 (which has generators $(1\,2\,3\,4\,5\,6\,7)$ and $(2\,3\,5)(4\,7\,6)$ in S_7). The cycle type distributions for these three are as follows:

cycle type:	1	2	(2, 2)	3	(2, 2, 3)	(3, 3)	(2, 4)	5	7
F_{21}	1					14			6
$GL_3(\mathbb{F}_2)$	1		21			56	42		48
A_7	1	21	105	70	210	280	630	504	720

It follows that there is a strong probability that the Galois group of this polynomial is the Frobenius group of order 21. This is actually the case (the verification requires computation of a resolvent of degree 35 and factoring it over $\mathbb{Z}$ — there are three factors, of degrees 7,7, and 21).

EXERCISES

1. Let p be a prime. Prove that the polynomial $x^4 + 1$ splits mod p either into two irreducible quadratics or into 4 linear factors using Corollary 41 together with the knowledge that the Galois group of this polynomial is the Klein 4-group.

2. (Cf. Exercise 48 of Section 6).
   (a) Let K be the splitting field of $x^6 - 2x^3 - 2$. Prove that if $[K : \mathbb{Q}] = 12$ then $K = \mathbb{Q}(\sqrt[3]{2}, i, \sqrt{3})$ and K is generated over the biquadratic field $F = \mathbb{Q}(i, \sqrt{3})$ by $\alpha = \sqrt[3]{1 + \sqrt{3}}$ and by $\beta = \sqrt[3]{1 - \sqrt{3}}$. Show that if this is the case then the elements of order 3 in $\mathrm{Gal}(K/\mathbb{Q})$ lie in $\mathrm{Gal}(K/F)$. Conclude that any element of $\mathrm{Gal}(K/\mathbb{Q})$ of order 3 maps α to another cube root of $1 + \sqrt{3}$ and maps β to another cube root of $1 - \sqrt{3}$ and if it is the identity on α or β then it is the identity on all of K.
   (b) Show that the factorization of $f(x)$ into irreducibles over $\mathbb{F}_{13}$ is the polynomial $(x - 7)(x - 8)(x - 11)(x^3 + 3)$ and use Corollary 41 to show that $[K : \mathbb{Q}] = 36$.
   (c) Knowing that $G = \mathrm{Gal}(K/\mathbb{Q})$ is of order 36 determine all the elements of G explicitly and in particular show that G is isomorphic to $S_3 \times S_3$.

3. Prove that the Galois group of $x^5 + 20x + 16$ is A_5.

4. Prove that the Galois group of $x^5 + x^4 - 4x^3 - 3x^2 + 3x + 1$ is cyclic of order 5. [Show this is the minimal polynomial of $\zeta_{11} + \zeta_{11}^{-1}$.]

5. Prove that the Galois group of $x^5 + 11x + 44$ is the dihedral group D_{10} (cf. Exercise 21 of Section 7).

6. Prove that the Galois group of $x^5 + 15x + 12$ is F_{20}, the Frobenius group of order 20 (cf. Exercise 21 of Section 7).

7. Prove that the Galois group of $x^6 + 24x - 20$ is A_6.

8. Prove that the Galois group of $x^7 + 7x^4 + 14x + 3$ is A_7.

644

9. Determine a polynomial of degree 7 whose Galois group is cyclic of order 7.

10. Determine the probable Galois group of $x^7 - 7x + 3$.

14.9 TRANSCENDENTAL EXTENSIONS, INSEPARABLE EXTENSIONS, INFINITE GALOIS GROUPS

This section collects some results on arbitrary extensions E/F. These results supplement those of the preceding sections and complete the basic picture of how an arbitrary (possibly infinite) extension decomposes. Since this section is primarily intended as a survey, none of the proofs are included; whenever these proofs can be easily supplied by the reader we indicate this either in the text or (with hints) in the exercises.

Throughout this section E/F is an extension of fields. Recall that an element of E which is not algebraic over F is called *transcendental* over F. Keep in mind that extensions involving transcendentals are always of infinite degree. We generally reserve the expression "t is an 'indeterminate' over F", when we are thinking of evaluating t. Field theoretically, however, the terms transcendental and indeterminate are synonymous (so that the subfield $\mathbb{Q}(\pi)$ of $\mathbb{R}$ and the field $\mathbb{Q}(t)$ are isomorphic).

Definition.
  (1) A subset $\{a_1, a_2, \ldots, a_n\}$ of E is called *algebraically independent* over F if there is no nonzero polynomial $f(x_1, x_2, \ldots, x_n) \in F[x_1, x_2, \ldots, x_n]$ such that $f(a_1, a_2, \ldots, a_n) = 0$. An arbitrary subset S of E is called *algebraically independent* over F if every finite subset of S is algebraically independent. The elements of S are called *independent transcendentals* over F.
  (2) A *transcendence base* for E/F is a maximal subset (with respect to inclusion) of E which is algebraically independent over F.

Note that if E/F is algebraic, the empty set is the only algebraically independent subset of E. In particular, elements of an algebraically independent set are necessarily transcendental. Moreover, one easily checks that $S \subseteq E$ is an algebraically independent set over F if and only if each $s \in S$ is transcendental over $F(S - \{s\})$. It is also an easy exercise to see that S is a transcendence base for E/F if and only if S is a set of algebraically independent transcendentals over F and E is algebraic over $F(S)$.

Theorem. The extension E/F has a transcendence base and any two transcendence bases of E/F have the same cardinality.

Proof: The first statement is a standard Zorn's Lemma argument. The proof of the second uses the same "Replacement Lemma" idea as was used to prove that any two bases of a vector space have the same cardinality.

Definition. The cardinality of a transcendence base for E/F is called the *transcendence degree* of E/F.

Algebraic extensions are precisely the extensions of transcendence degree 0.

One special case of this theorem is when E is *finitely generated* over F, that is, $E = F(\alpha_1, \alpha_2, \ldots, \alpha_n)$, for some (not necessarily algebraically independent) elements $\alpha_1, \ldots, \alpha_n$ of E. It is clear that we may renumber $\alpha_1, \ldots, \alpha_n$ so that $\alpha_1, \ldots, \alpha_m$ are independent transcendentals and $\alpha_{m+1}, \ldots, \alpha_n$ are algebraic over $F(\alpha_1, \ldots, \alpha_m)$ (so E is a finite extension of the latter field). In this case E is called a *function field in m variables* over F. Such fields play a fundamental role in algebraic geometry as fields of functions on m-dimensional surfaces. For instance, when $F = \mathbb{C}$ and $m = 1$, these fields arise in analysis as fields of meromorphic functions on compact Riemann surfaces.

Note that if S_1 and S_2 are transcendence bases for E/F it is not necessarily the case that $F(S_1) = F(S_2)$. For example, if t is transcendental over $\mathbb{Q}$, $\{t\}$ and $\{t^2\}$ are both transcendence bases for $\mathbb{Q}(t)/\mathbb{Q}$ but (as we shall see shortly) $\mathbb{Q}(t^2)$ is a proper subfield of $\mathbb{Q}(t)$.

We now see that if $x_1, x_2, \ldots, x_n$ are indeterminates over F and

$$f(x) = (x - x_1)(x - x_2) \cdots (x - x_n) \tag{14.28}$$

is the general polynomial of degree n, then the set of n elementary symmetric functions $s_1, s_2, \ldots, s_n$ in the x_i's are also independent transcendentals over F. This is because $x_1, \ldots, x_n$ is a transcendence base for $E = F(x_1, \ldots, x_n)$ over F (so the transcendence degree is n) and E is algebraic over $F(s_1, \ldots, s_n)$ (of degree $n!$). The theorem forces $s_1, \ldots, s_n$ to be a transcendence base for this extension as well (in particular, they are independent transcendentals). The general polynomial of degree n over F may therefore equivalently be defined by taking $a_1, \ldots, a_n$ to be any independent transcendentals (or indeterminates) and letting

$$f(x) = x^n + a_1 x^{n-1} + \cdots + a_n \tag{14.29}$$

where the roots of f are denoted by $x_1, \ldots, x_n$ (and $s_i = (-1)^i a_i$).

Definition. An extension E/F is called *purely transcendental* if it has a transcendence base S such that $E = F(S)$.

In the preceding discussion, both $F(x_1, \ldots, x_n)$ and $F(s_1, \ldots, s_n)$ are purely transcendental over F. As an exercise (following) one can show that $\mathbb{Q}(t, \sqrt{t^3 - t})$ is not a purely transcendental extension of $\mathbb{Q}$ even though it contains no elements that are algebraic over $\mathbb{Q}$ other than those in $\mathbb{Q}$ itself (i.e., the process of decomposing a general extension into a purely transcendental extension followed by an algebraic extension cannot generally be reversed so that the algebraic piece occurs first).

If E is a purely transcendental extension of F of transcendence degree $n = 1$ or 2 and L is an intermediate field, $F \subseteq L \subseteq E$ with the same transcendence degree, then L is again a purely transcendental extension of F (Lüroth ($n = 1$), Castelnuovo ($n = 2$)). This result is not true if the transcendence degree is ≥ 3, however, although examples where L fails to be purely transcendental are difficult to construct. For extensions of transcendence degree 1 the intermediate fields are described by the following theorem.

Theorem. Let t be transcendental over F.

(1) (Lüroth) If $F \subseteq K \subseteq F(t)$, then $K = F(r)$, for some $r \in F(t)$. In particular, every nontrivial extension of F contained in $F(t)$ is purely transcendental over F.

(2) If $P = P(t)$, $Q = Q(t)$ are nonzero relatively prime polynomials in $F[t]$ which are not both constant,

$$[F(t) : F(P/Q)] = \max(\deg P, \deg Q).$$

Proof: The proof of (2) is outlined in Exercise 18 of Section 13.2.

By part (2) of this theorem we see that $F(P/Q) = F(t)$ if and only if P, Q are nonzero relatively prime polynomials of degree ≤ 1 (not both constant). Thus $F(r) = F(t)$ if and only if $r = \dfrac{at + b}{ct + d}$, where $a, b, c, d \in F$ and $ad - bc \neq 0$ (called a *fractional linear transformation of t*). For any $r \in F(t) - F$ the map $t \mapsto r$ extends to an embedding of $F(t)$ into itself which is the identity on F. This embedding is surjective (i.e., is an automorphism of $F(t)$) precisely for the fractional linear transformations. Furthermore, the map

$$GL_2(F) \to \mathrm{Aut}(F(t)/F) \quad \text{defined by} \quad A = \begin{pmatrix} a & c \\ b & d \end{pmatrix} \mapsto \sigma_A,$$

where σ_A denotes the automorphism of $F(t)$ defined by mapping t to $(at + b)/(ct + d)$, is a surjective homomorphism with kernel consisting of the scalar matrices. Thus

$$\mathrm{Aut}(F(t)/F) \cong PGL_2(F)$$

where $PGL_2(F) = GL_2(F)/\{\lambda I \mid \lambda \in F^\times\}$ gives the group of automorphisms of this transcendental extension (cf. Exercise 8 of Section 1).

When $\mathbb{F}$ is a finite field of order q, $\mathrm{Aut}(\mathbb{F}(t)/\mathbb{F}) \cong PGL_2(\mathbb{F})$ is a finite group of order $q(q - 1)(q + 1)$. By Corollary 11 if K is the fixed field of $\mathrm{Aut}(\mathbb{F}(t)/\mathbb{F})$, then $\mathbb{F}(t)$ is Galois over K with Galois group equal to $\mathrm{Aut}(\mathbb{F}(t)/\mathbb{F})$. In particular, the fixed field of $\mathrm{Aut}(\mathbb{F}(t)/\mathbb{F})$ is not $\mathbb{F}$ in this case.

This also provides further examples of the Galois correspondence which can be written out completely for small values of q. For instance, if $q = |\mathbb{F}| = 2$, $PGL_2(\mathbb{F})$ is nonabelian of order 6, hence is isomorphic to S_3, and has the following lattice of subgroups:

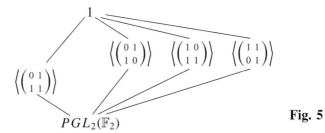

Fig. 5

The field $\mathbb{F}(t)$ is of degree 6 over the fixed field K of $\mathrm{Aut}(\mathbb{F}(t)/\mathbb{F})$ and the lattice of subfields $K \subseteq L \subseteq \mathbb{F}(t)$ is dual to the lattice of subgroups of S_3. The fixed field of a cyclic subgroup $\langle \sigma \rangle$ is easily found (via the preceding theorem) by finding a rational function r in t which is fixed by σ such that $[\mathbb{F}(t) : \mathbb{F}(r)] = |\sigma|$. For example, if $\sigma : t \mapsto 1/(1 + t)$, then σ has order 3. The rational function

$$r = t + \sigma(t) + \sigma^2(t) = \frac{t^3 + t + 1}{t(t + 1)}$$

is fixed by σ and $[\mathbb{F}(t) : \mathbb{F}(r)] = 3$ (by part (2) of the theorem). Since $\mathbb{F}(r)$ is contained in the fixed field of $\langle \sigma \rangle$ and the degree of $\mathbb{F}(t)$ over the fixed field is 3, $\mathbb{F}(r)$ is the fixed field of $\langle \sigma \rangle$. In this way one can explicitly describe the lattice of all subfields of $\mathbb{F}(t)$ containing K shown in Figure 6.

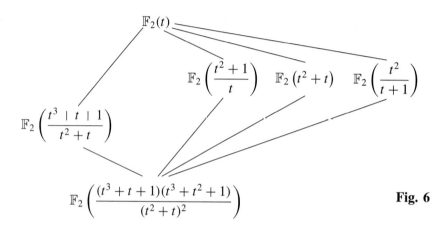

Fig. 6

Purely transcendental extensions of $\mathbb{Q}$ play an important role in the problem of realizing finite groups as Galois groups over $\mathbb{Q}$. We describe a deep result of Hilbert which is fundamental to this area of research. If $a_1, a_2, \ldots, a_n$ are independent indeterminates over a field F, we may evaluate (or *specialize*) $a_1, \ldots, a_n$ at any elements of F, i.e., substitute values in F for the "variables" $a_1, a_2, \ldots, a_n$. If E is a Galois extension of $F(a_1, \ldots, a_n)$, then E is obtained as a splitting field of a polynomial whose coefficients lie in $F[a_1, \ldots, a_n]$. Any specialization of $a_1, \ldots, a_n$ into F maps this polynomial into one whose coefficients lie in F. The specialization of E is the splitting field of the resulting specialized polynomial.

Theorem. (Hilbert) Let $x_1, x_2, \ldots, x_n$ be independent transcendentals over $\mathbb{Q}$, let $E = \mathbb{Q}(x_1, \ldots, x_n)$ and let G be a finite group of automorphisms of E with fixed field K. If K is a purely transcendental extension of $\mathbb{Q}$ with transcendence basis $a_1, a_2, \ldots, a_n$, then there are infinitely many specializations of $a_1, \ldots, a_n$ in $\mathbb{Q}$ such that E specializes to a Galois extension of $\mathbb{Q}$ with Galois group isomorphic to G.

Hilbert's Theorem gives a sufficient condition for the specialized extension not to collapse. In general, the Galois group of the specialized extension is a subgroup of G (cf. Proposition 19) and may be a proper subgroup of G. It is also known that the fixed field K need not always be a purely transcendental extension of $\mathbb{Q}$. An example of this occurs when G is the cyclic group of order 47.

This theorem can be used to give another proof of Proposition 42:

Corollary. S_n is a Galois group over $\mathbb{Q}$, for all n.

Proof of the Corollary: We have already proved that the fixed field of S_n acting in the obvious fashion on $\mathbb{Q}(x_1, \ldots, x_n)$ is purely transcendental over $\mathbb{Q}$ (with the elementary symmetric functions as a transcendence base), so Hilbert's Theorem immediately implies the corollary.

The hypothesis that K be purely transcendental over $\mathbb{Q}$ is crucial to the proof of Hilbert's Theorem. Every finite group is isomorphic to a subgroup of S_n and so acts on $\mathbb{Q}(x_1, \ldots, x_n)$ for some n. It is not known, however, even for the subgroup A_n of S_n whether its fixed field under the obvious action is a purely transcendental extension of $\mathbb{Q}$ (although it is known by other means that A_n is a Galois group over $\mathbb{Q}$ for all n). Thus there are a number of important open problems in this area of research.

One should also notice that Hilbert's Theorem does not work when the base field $\mathbb{Q}$ is replaced by an arbitrary field F (suppose F were algebraically closed, for instance). In particular, as noted earlier, the general polynomial $f(x)$ in Section 6 has Galois group S_n over $F(a_1, \ldots, a_n)$ for any F, but when F is a finite field, the specialized extension obtained from its splitting field is always cyclic.

We next expand on the theory of inseparable extensions described in Section 13.5. Let p be a prime and let F be a field of characteristic p.

Definition. An algebraic extension E/F is called *purely inseparable* if for each $\alpha \in E$ the minimal polynomial of α over F has only one distinct root.

It is easy to see that the following are equivalent:

(1) E/F is purely inseparable
(2) if $\alpha \in E$ is separable over F, then $\alpha \in F$
(3) if $\alpha \in E$, then $\alpha^{p^n} \in F$ for some n (depending on α), and $m_{\alpha, F}(x) = x^{p^n} - \alpha^{p^n}$.

The following easy proposition describes composites of separable and purely inseparable extensions.

Proposition. If E_1 and E_2 are subfields of E which are both separable (or both purely inseparable) extensions of F, then their composite $E_1 E_2$ is separable (purely inseparable, respectively) over F.

Proof: Exercise.

One immediate consequence of this is the following result.

Proposition. Let E/F be an algebraic extension. Then there is a unique field E_{sep} with $F \subseteq E_{sep} \subseteq E$ such that E_{sep} is separable over F and E is purely inseparable over E_{sep}. The field E_{sep} is the set of elements of E which are separable over F.

The degree of E_{sep}/F is called the *separable degree* of E/F and the degree of E/E_{sep} is called the *inseparable degree* of E/F (often denoted as $[E : F]_s$ and $[E : F]_i$ respectively). The product of these two degrees is the (ordinary) degree. The propositions immediately give the following corollary.

Corollary. Separable degrees (respectively inseparable degrees) are multiplicative.

When E is generated over F by the root of an irreducible polynomial $p(x) \in F[x]$ the separable and inseparable degrees of the extension E/F are the same as the separable and inseparable degrees of the polynomial $p(x)$ defined in Section 13.5.

The proposition asserts that any algebraic extension may be decomposed into a separable extension followed by a purely inseparable one. Exercise 3 at the end of this section outlines an example illustrating that this decomposition cannot generally be reversed, namely an extension which is not a separable extension of a purely inseparable extension. We shall shortly state conditions on an extension under which the decomposition into separable and purely inseparable subextensions may be reversed.

We now know that an arbitrary extension E/F can be decomposed into a purely transcendental extension $F(S)$ of F followed by a separable extension E_1 of $F(S)$ followed by a purely inseparable extension E/E_1. In certain instances the inseparability in the algebraic extension at the "top" may be removed by a judicious choice of transcendence base:

Proposition. If E is a finitely generated extension of a perfect field F, then there is a transcendence base T of E/F such that E is a separable (algebraic) extension of $F(T)$.

A transcendence base T as described in the proposition is called a *separating transcendence base*. Exercise 4 at the end of this section illustrates this with a nontrivial example.

Recall that an extension E/F is *normal* if it is the splitting field of some (possibly infinite) set of polynomials in $F[x]$ (in particular, normal extensions are algebraic but not necessarily finite or separable). We previously used the synonymous term splitting field and the term normal is reintroduced here in the context of arbitrary algebraic extensions since it is used frequently in the literature, often in the context of embeddings of a field into an algebraic closure. Although the following set of equivalences can be gleaned from the preceding sections, the reader should write out a complete proof, checking that the arguments work for both infinite and inseparable extensions:

Proposition. Let E/F be an arbitrary algebraic extension and let Ω be an algebraic closure of E. The following are equivalent:
   (1) E/F is a normal extension (i.e., is the splitting field over F of some set of polynomials in $F[x]$)

(2) whenever $\sigma : E \to \Omega$ is an embedding such that $\sigma|_F$ is the identity, $\sigma(E) = E$

(3) whenever an irreducible polynomial $f(x) \in F[x]$ has one root in E, it has all its roots in E.

In general, any embedding of a normal extension E/F into an algebraic closure of E which extends the identity embedding of F is an automorphism of E, i.e., is an element of $\mathrm{Aut}(E/F)$. Moreover, the number of such automorphisms equals the separable degree of E/F, provided the latter is finite:

if E/F is a normal extension and $[E : F]_s$ is finite, $|\mathrm{Aut}(E/F)| = [E : F]_s$.

If $[E : F]_s$ is infinite we shall see shortly that $|\mathrm{Aut}(E/F)|$ is also infinite but need not be of the same cardinality.

If E/F is a normal extension whose separable degree is finite, let E_0 be the fixed field of $\mathrm{Aut}(E/F)$. By Corollary 11, E/E_0 is a (separable) Galois extension whose degree equals $|\mathrm{Aut}(E/F)|$. It follows that E_0/F must be purely inseparable (of degree equal to $[E : F]_i$), i.e., the separable and purely inseparable pieces of the extension may be reversed for normal extensions. More precisely, we easily obtain the following proposition.

Proposition. If E/F is normal with $[E : F]_s < \infty$, then $E = E_{sep}E_{pi}$, where E_{pi} is a purely inseparable extension of F (E_{pi} consists of all purely inseparable elements of E over F) and $E_{sep} \cap E_{pi} = F$.

Finally, we mention how Galois Theory generalizes to infinite extensions.

Definition. An extension E/F is called *Galois* if it is algebraic, normal and separable. In this case $\mathrm{Aut}(E/F)$ is called the *Galois group* of the extension and is denoted by $\mathrm{Gal}(E/F)$.

For infinite extensions there need not be a bijection between the set of all subgroups of the Galois group and the set of all subfields of E containing F, as the following example illustrates.

Let E be the subfield of $\mathbb{R}$ obtained by adjoining to $\mathbb{Q}$ all square roots of positive rational numbers. One easily sees that E may also be described as the splitting field of the set of polynomials $x^2 - p$, where p runs over all primes in $\mathbb{Z}^+$. Note that E is a (countably) infinite Galois extension of $\mathbb{Q}$. Since every automorphism σ of E is determined by its action on the square roots of the primes and σ either fixes or negates each of these, σ^2 is the identity automorphism. It follows that $\mathrm{Aut}(E)$ is an infinite elementary abelian 2-group. Thus $\mathrm{Aut}(E)$ is an infinite dimensional vector space over $\mathbb{F}_2$. By an exercise in the section on dual spaces (Section 11.3) the number of nonzero homomorphisms of $\mathrm{Aut}(E)$ into $\mathbb{F}_2$ is uncountable, whence their kernels (which are subspaces of co-dimension 1) are uncountable in number (and distinct). Thus $\mathrm{Aut}(E)$ has *uncountably* many subgroups of index 2, whereas $\mathbb{Q}$ has only a *countable* number of quadratic extensions.

The basic problem is that many (most) subgroups of $\mathrm{Gal}(E/F)$ do not correspond (in a bijective fashion) to subfields of E containing F. In order to pick out the "right" set of subgroups of $\mathrm{Gal}(E/F)$ we must introduce a topology on this group (called the Krull topology). The axioms for the collection of (topologically) closed subsets of a topological space are precisely the bookkeeping devices which single out the relevant subgroups (these are listed in Section 15.2). Galois theory for finite extensions force certain subgroups of finite index to be closed sets and these in turn determine the topology on the entire group (as we might expect since every extension of F inside E is a composite of finite extensions). Moreover, the Galois group of E/F is the inverse limit of the collection of finite groups $\mathrm{Gal}(K/F)$, where K runs over all finite Galois extensions of F contained in E (cf. Exercise 10, Section 7.6).

Theorem. (Krull) Let E/F be a Galois extension with Galois group G. Topologize G by taking as a base for the closed sets the subgroups of G which are the fixing subgroups of the finite extensions of F in E, together with all left and right cosets of these subgroups. Then with this ("Krull") topology the closed subgroups of G correspond bijectively with the subfields of E containing F and the corresponding lattices are dual. Closed normal subgroups of G correspond to normal extensions of F in E.

One important area of current research is to describe (as a topological group) the Galois group of certain field extensions such as $\overline{F}/F$, where $\overline{F}$ is the algebraic closure of F. Little is known about the latter group when $F = \mathbb{Q}$ (in particular, its normal subgroups of finite index, i.e., which finite groups occur as Galois groups over $\mathbb{Q}$, are not known). If E is the algebraic closure of the finite field $\mathbb{F}_p$, the Galois group of this extension is the topologically cyclic group $\widehat{\mathbb{Z}}$ with the Frobenius automorphism as a topological generator. The group $\widehat{\mathbb{Z}}$ is an uncountable group (in particular, is not isomorphic to $\mathbb{Z}$) with the property that every closed subgroup of finite index is normal with cyclic quotient. Note that $\widehat{\mathbb{Z}}$ must also have nontrivial infinite closed subgroups (unlike $\mathbb{Z}$) since E contains proper subfields which are infinite over $\mathbb{F}_p$ (such as the composite of all extensions of $\mathbb{F}_p$ of q-power degree, for any prime q — this Galois extension of $\mathbb{F}_p$ has Galois group $\mathbb{Z}_q$, the q-adic integers, as described in Exercise 11 of Section 7.6).

EXERCISES

1. Prove that every purely inseparable extension is normal.

2. Let p be a prime and let $K = \mathbb{F}_p(x, y)$ with x and y independent transcendentals over $\mathbb{F}_p$. Let $F = \mathbb{F}_p(x^p - x, y^p - x)$.
   (a) Prove that $[K : F] = p^2$ and the separable degree and inseparable degree of K/F are both equal to p.
   (b) Prove that there is a subfield E of K containing F which is purely inseparable over F of degree p (so then K is a separable extension of E of degree p). [Let $s = x^p - x \in F$ and $t = y^p - x \in F$ and consider $s - t$.]

3. Let p be an odd prime, let s and t be independent transcendentals over $\mathbb{F}_p$, and let F be the field $\mathbb{F}_p(s, t)$. Let β be a root of $x^2 - sx + t = 0$ and let α be a root of $x^p - \beta = 0$ (in some algebraic closure of F). Set $E = F(\beta)$ and $K = F(\alpha)$.
   (a) Prove that E is a Galois extension of F of degree 2 and that K is a purely inseparable extension of E of degree p.

(b) Prove that K is not a normal extension of F. [If it were, conjugate β over F to show that K would contain a pth root of s and then also a pth root of t, so $[K : F] \geq p^2$, a contradiction.]

(c) Prove that there is no field K_0 such that $F \subseteq K_0 \subseteq K$ with K_0/F purely inseparable and K/K_0 separable. [If there were such a field, use Exercise 1 and the fact that the composite of two normal extensions is again normal to show that K would be a normal extension of F.]

4. Under the notation of the previous exercise prove that α, s is a separating transcendence base for K over $\mathbb{F}_p$.

5. Let p be a prime, let t be transcendental over $\mathbb{F}_p$ and let K be obtained by adjoining to $\mathbb{F}_p(t)$ all p-power roots of t. Prove that K has transcendence degree 1 over $\mathbb{F}_p$ and has no separating transcendence base.

6. Show that if t is transcendental over $\mathbb{Q}$ then $\mathbb{Q}(t, \sqrt{t^3 - t})$ is not a purely transcendental extension of $\mathbb{Q}$. (This is an example of what is called an *elliptic* function field.)

7. Let k be the field with 4 elements, t a transcendental over k, $F = k(t^4 + t)$ and $K = k(t)$.
(a) Show that $[K : F] = 4$.
(b) Show that K is separable over F.
(c) Show that K is Galois over F.
(d) Describe the lattice of subgroups of the Galois group and the corresponding lattice of subfields of K, giving each subfield in the form $k(r)$, for some rational function r.

8. Let p be an odd prime, k an algebraically closed field of characteristic p and let t be transcendental over k. Suppose F is a degree 2 field extension of $k(t)$. Show that F can be written in the form $k(t, y)$, for some $y \in F$ with $y^2 \in k(t)$ and y transcendental over k. If $y^2 = 4t^3 - t - 1$, find $[F : k(y)]$ and describe $k(t) \cap k(y)$ as $k(r)$, for some $r \in k(t)$.

9. Let t be transcendental over $\mathbb{F}_3$, let $K = \mathbb{F}_3(t)$, let $G = \text{Aut}(K/\mathbb{F}_3)$ and let F be the fixed field of G.
(a) Prove $G \cong S_4$ and deduce that there is a unique field E with $F \subseteq E \subseteq K$ and $[E : F] = 2$. [Recall that $G \cong PGL_2(\mathbb{F}_3)$; show that $GL_2(\mathbb{F}_3)$ permutes the 4 lines in a 2-dimensional vector space over $\mathbb{F}_3$ and the kernel of this permutation representation is the scalar matrices.]
(b) Complete the description of the lattice of subfields of K containing E:

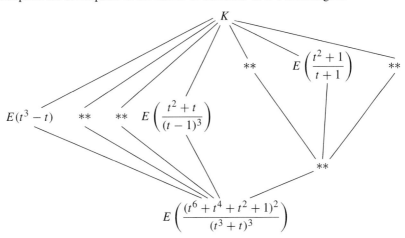

Give each subfield in the form $E(r)$ for some rational function r. (The lattice of subgroups of A_4 appears in Section 3.5).

10. Prove that a purely transcendental proper extension of a field is never algebraically closed.

11. Let S be a set of independent transcendentals over a field F and let Ω be an algebraic closure of $F(S)$. Prove that any permutation on S extends to an element of $\operatorname{Aut}(F(S)/F)$. Prove that any such automorphism of $F(S)$ extends to an automorphism of Ω. Deduce that $\mathbb{C}$ has infinitely many automorphisms.

12. Let K be a subfield of $\mathbb{C}$ maximal with respect to the property "$\sqrt{2} \notin K$."
    (a) Show such a field K exists.
    (b) Show that $\mathbb{C}$ is algebraic over K.
    (c) Prove every finite extension of K in $\mathbb{C}$ is Galois with Galois group a cyclic 2-group.
    (d) Deduce that $[\mathbb{C} : K]$ is countable (and not finite).

13. Let K be the fixed field in $\mathbb{C}$ of an automorphism of $\mathbb{C}$. Prove that every finite extension of K in $\mathbb{C}$ is cyclic.

14. Let K_n be the splitting field of $(x^2 - p_1)(x^2 - p_2) \cdots (x^2 - p_n)$ over $\mathbb{Q}$, where $p_1, \ldots, p_n$ are the first n primes. Prove that the Galois group of $K_n/\mathbb{Q}$ is an elementary abelian 2-group of order 2^n.

15. Let $K_0 = \mathbb{Q}$ and for $n \geq 0$ define the field K_{n+1} as the extension of K_n obtained by adjoining to K_n all roots of all cubic polynomials over K_n. Let K be the union of the subfields K_n, $n \geq 0$. Prove that K is a Galois extension of $\mathbb{Q}$. Prove that every cubic polynomial over K splits completely over K. Prove that there are nontrivial algebraic extensions of K.

16. Let F be the composite of all the splitting fields of irreducible cubics over $\mathbb{Q}$. Prove that F contains all quadratic extensions of $\mathbb{Q}$. [One approach: start by considering $x^3 + 3ax + 2a$ for positive integers a.]

17. Let $K_0 = \mathbb{Q}$ and for $n \geq 0$ define the field K_{n+1} as the extension of K_n obtained by adjoining to K_n all radicals of elements in K_n. Let K be the union of the subfields K_n, $n \geq 0$. Prove that K is a Galois extension of $\mathbb{Q}$. Prove that there are no nontrivial solvable Galois extensions of K. Prove that there are nontrivial Galois extensions of K.

18. Let $F_0 = \mathbb{Q}$ and for $n \geq 0$ define the field F_{n+1} as the extension of F_n obtained by adjoining to F_n all real radicals of elements in F_n. Let F be the union of the subfields F_n, $n \geq 0$. Let K^+ be the fixed field of complex conjugation restricted to the field K in the previous exercise (the maximal real subfield of K). Prove that $F \neq K^+$.

19. This exercise proves that if K/F is a Galois extension of fields, then $\operatorname{Gal}(K/F)$ is isomorphic to $\varprojlim \operatorname{Gal}(L/F)$, where the inverse limit is taken over all the finite Galois extensions L of F contained in K.
    (a) Show that K is the union of the fields L.
    (b) Prove that $\varphi : \operatorname{Gal}(K/F) \to \varprojlim \operatorname{Gal}(L/F)$ defined by mapping σ in $\operatorname{Gal}(K/F)$ to $(\ldots, \sigma|_L, \ldots)$, where $\sigma|_L$ is the restriction of σ to L, is a homomorphism.
    (c) Show that φ is injective.
    (d) If $(\ldots, \sigma_L, \ldots) \in \varprojlim \operatorname{Gal}(L/F)$, define $\sigma \in \operatorname{Gal}(K/F)$ by $\sigma(\alpha) = \sigma_L(\alpha)$ if $\alpha \in L$. Prove that σ is a well defined automorphism and deduce that φ is surjective.

20. Let t be transcendental over the field F, let $L_1 = F(t^2)$ and let $L_2 = F((t+1)^2)$.
    (a) If F has characteristic 0, show $L_1 \cap L_2 = F$. (So in this case $[F(t) : L_1 \cap L_2]$ is infinite even though $[F(t) : L_1] = [F(t) : L_2] = 2$ are both finite.)
    (b) If F has characteristic $p \neq 2$, show that $F(t)$ is Galois over $L_1 \cap L_2$ with the dihedral group D_{2p} as Galois group. Find an explicit $r \in F(t)$ with $L_1 \cap L_2 = F(r)$.

Part V

INTRODUCTION TO COMMUTATIVE RINGS, ALGEBRAIC GEOMETRY, AND HOMOLOGICAL ALGEBRA

In this part of the book we continue the study of rings and modules, concentrating first on commutative rings. The topic of commutative algebra, which is of interest in its own right, is also a basic foundation for other areas of algebra. To indicate some of the importance of the algebraic topics introduced, we parallel the development of the ring theory in Chapter 15 with an introduction to affine algebraic geometry. Each section first presents the basic algebraic theory and then follows with an application of those ideas to geometry together with an indication of computational methods using the theory of Gröbner bases from Chapter 9. The purpose here is twofold: the first is to present an application of algebraic techniques in the important branch of mathematics called Algebraic Geometry, and the second is to indicate some of the motivations for the algebraic concepts introduced from their origins in geometric questions.

This connection of geometry and algebra shows a rich interplay between these two areas of mathematics and demonstrates again how results and structures in one circle of mathematical ideas provide insights into another.

In Chapter 16 we continue with some of the fundamental structures involving commutative rings, culminating with Dedekind Domains and a structure theorem for modules over such rings which is a generalization of the structure theorem for modules over P.I.D.s in Chapter 12.

In Chapter 17 we describe some of the basic techniques of "homological algebra," which continues with some of the questions raised by the failure of exactness of some of the sequences considered in Chapter 10. The cohomology of groups in this chapter is intended to serve both as a more in-depth application of homological algebra to see its uses in practice, and as a relatively self contained exposition of this important topic.

CHAPTER 15

Commutative Rings
and Algebraic Geometry

Throughout this chapter R will denote a commutative ring with $1 \neq 0$.

15.1 NOETHERIAN RINGS AND AFFINE ALGEBRAIC SETS

In this section we study Noetherian rings in greater detail. These are a natural generalization of Principal Ideal Domains and were introduced briefly in Chapter 12. Note
that when R is considered as a left module over itself, its R-submodules are precisely
its ideals, so the definition in Section 1 of Chapter 12 may be phrased in the following
form:

Definition. A commutative ring R is said to be *Noetherian* or to satisfy the *ascending
chain condition on ideals* (or *A.C.C. on ideals*) if there is no infinite increasing chain
of ideals in R, i.e., whenever $I_1 \subseteq I_2 \subseteq I_3 \subseteq \cdots$ is an increasing chain of ideals of R,
then there is a positive integer m such that $I_k = I_m$ for all $k \geq m$.

Proposition 1. If I is an ideal of the Noetherian ring R, then the quotient R/I is a
Noetherian ring. Any homomorphic image of a Noetherian ring is Noetherian.

Proof: If R is a ring and I is an ideal in R, then any infinite ascending chain of
ideals in the quotient R/I would correspond by the Lattice Isomorphism Theorem to
an infinite ascending chain of ideals in R. This gives the first statement, and the second
follows by the first Isomorphism Theorem.

Theorem 2. The following are equivalent:
   (1) R is a Noetherian ring.
   (2) Every nonempty set of ideals of R contains a maximal element under inclusion.
   (3) Every ideal of R is finitely generated.

Proof: The proof is identical to that of Theorem 1 in Section 12.1 in the special
case where the R-module M is R itself (and submodules are ideals).

656

Examples

Every Principal Ideal Domain is Noetherian since it satisfies condition (3) of Theorem 2. In particular, $\mathbb{Z}$, the polynomial ring $k[x]$ where k is a field, and the Gaussian integers $\mathbb{Z}[i]$, are Noetherian rings. The ring $\mathbb{Z}[x_1, x_2, \dots]$ is not Noetherian since the ideal $(x_1, x_2, \dots)$ cannot be generated by any finite set (any finite set of generators involves only finitely many of the x_i). Exercise 33(d) in Section 7.4 shows that the ring of continuous real valued functions on $[0, 1]$ is not Noetherian.

A Noetherian ring may have arbitrarily long ascending chains of ideals and may have infinitely long descending chains of ideals. For example, $\mathbb{Z}$ has the infinite descending chain

$$(2) \supset (4) \supset (8) \supset \cdots$$

i.e., a Noetherian ring need not satisfy the *descending chain condition on ideals (D.C.C.)*. We shall see, however, that a commutative ring satisfying D.C.C. on ideals necessarily also satisfies A.C.C., i.e., is Noetherian; such rings are called *Artinian* and are studied in Chapter 16.

The following theorem and its corollary, which we record here for completeness, were proved in Section 9.6 (Theorem 21 and Corollary 22, respectively).

Theorem 3. *(Hilbert's Basis Theorem)* If R is a Noetherian ring then so is the polynomial ring $R[x]$.

Note that Hilbert's Basis Theorem shows how larger Noetherian rings may be built from existing ones in a manner analogous to Theorem 7 of Section 9.3 (which proved that if R is a U.F.D., then so is $R[x]$).

Corollary 4. The polynomial ring $k[x_1, x_2, \dots, x_n]$ with coefficients from a field k is a Noetherian ring.

Let k be a field. Recall that a ring R is a *k-algebra* if k is contained in the center of R and the identity of k is the identity of R.

Definition.
   (1) The ring R is a *finitely generated k-algebra* if R is generated as a ring by k together with some finite set $r_1, r_2, \dots, r_n$ of elements of R.
   (2) Let R and S be k-algebras. A map $\psi : R \to S$ is a *k-algebra homomorphism* if ψ is a ring homomorphism that is the identity on k.

If R is a k-algebra then R is both a ring and a vector space over k, and it is important to distinguish the sense in which elements of R are generators for R. For example, the polynomial ring $k[x_1, \dots, x_n]$ in a finite number of variables over k is a finitely generated k-algebra since $x_1, \dots, x_n$ are ring generators, but for $n > 0$ this ring is an *infinite* dimensional vector space over k.

Corollary 5. The ring R is a finitely generated k-algebra if and only if there is some surjective k-algebra homomorphism

$$\varphi : k[x_1, x_2, \ldots, x_n] \to R$$

from the polynomial ring in a finite number of variables onto R that is the identity map on k. Any finitely generated k-algebra is therefore Noetherian.

Proof: If R is generated as a k-algebra by $r_1, \ldots, r_n$, then we may define the map $\varphi : k[x_1, \ldots, x_n] \to R$ by $\varphi(x_i) = r_i$ for all i and $\varphi(a) = a$ for all $a \in k$. Then φ extends uniquely to a surjective ring homomorphism. Conversely, given a surjective homomorphism φ, the images of $x_1, \ldots, x_n$ under φ then generate R as a k-algebra, proving that R is finitely generated. Since $k[x_1, \ldots, x_n]$ is Noetherian by the previous corollary, any finitely generated k-algebra is therefore the quotient of a Noetherian ring, hence also Noetherian by Proposition 1.

Example

Suppose the k-algebra R is finite dimensional as a vector space over k, for example when $R = k[x]/(f(x))$, where f is any nonzero polynomial in $k[x]$. Then in particular R is a finitely generated k-algebra since a vector space basis also generates R as a ring. In this case since ideals are also k-subspaces any ascending or descending chain of ideals has at most $\dim_k R + 1$ distinct terms, hence R satisfies both A.C.C. and D.C.C. on ideals.

The basic idea behind "algebraic geometry" is to equate geometric questions with algebraic questions involving ideals in rings such as $k[x_1, \ldots, x_n]$. The Noetherian nature of these rings reduces many questions to consideration of finitely many algebraic equations (and this was in turn one of the main original motivations for Hilbert's Basis Theorem). We first consider the principal geometric object, the notion of an "algebraic set" of points.

Affine Algebraic Sets

Recall that the set $\mathbb{A}^n$ of n-tuples of elements of the field k is called *affine n-space over k* (cf. Section 10.1). If $x_1, x_2, \ldots, x_n$ are independent variables over k, then the polynomials f in $k[x_1, x_2, \ldots, x_n]$ can be viewed as k-valued functions $f : \mathbb{A}^n \to k$ on $\mathbb{A}^n$ by evaluating f at the points in $\mathbb{A}^n$:

$$f : (a_1, a_2, \ldots, a_n) \mapsto f(a_1, a_2, \ldots, a_n) \in k.$$

This gives a ring of k-valued functions on $\mathbb{A}^n$, denoted by $k[\mathbb{A}^n]$ and called the *coordinate ring of $\mathbb{A}^n$*. For instance, when $k = \mathbb{R}$ and $n = 2$, the coordinate ring of Euclidean 2-space $\mathbb{R}^2$ is denoted by $\mathbb{R}[\mathbb{A}^2]$ and is the ring of polynomials in two variables, say x and y, acting as real valued functions on $\mathbb{R}^2$ (the usual "coordinate functions").

Each subset S of functions in the coordinate ring $k[\mathbb{A}^n]$ determines a subset $\mathcal{Z}(S)$ of affine space, namely the set of points where all functions in S are simultaneously zero:

$$\mathcal{Z}(S) = \{(a_1, a_2, \ldots, a_n) \in \mathbb{A}^n \mid f(a_1, a_2, \ldots, a_n) = 0 \text{ for all } f \in S\},$$

where $\mathcal{Z}(\emptyset) = \mathbb{A}^n$.

Definition. A subset V of $\mathbb{A}^n$ is called an *affine algebraic set* (or just an algebraic set) if V is the set of common zeros of some set S of polynomials, i.e., if $V = \mathcal{Z}(S)$ for some $S \subseteq k[\mathbb{A}^n]$. In this case $V = \mathcal{Z}(S)$ is called the *locus of S* in $\mathbb{A}^n$.

If $S = \{f\}$ or $\{f_1, \ldots, f_m\}$ we shall simply write $\mathcal{Z}(f)$ or $\mathcal{Z}(f_1, \ldots, f_m)$ for $\mathcal{Z}(S)$ and call it the locus of f or $f_1, \ldots, f_m$, respectively. Note that the locus of a single polynomial of the form $f - g$ is the same as the solutions in affine n-space of the equation $f = g$, so affine algebraic sets are the solution sets to systems of polynomial equations, and as a result occur frequently in mathematics.

Examples

(1) If $n = 1$ then the locus of a single polynomial $f \in k[x]$ is the set of roots of f in k. The algebraic sets in $\mathbb{A}^1$ are $\emptyset$, any finite set, and k (cf. the exercises).
(2) The one point subsets of $\mathbb{A}^n$ for any n are affine algebraic since $\{(a_1, a_2, \ldots, a_n)\}$ is $\mathcal{Z}(x_1 - a_1, \ x_2 - a_2, \ \ldots, \ x_n - a_n)$. More generally, any finite subset of $\mathbb{A}^n$ is an affine algebraic set.
(3) One may define lines, planes, etc. in $\mathbb{A}^n$ — these are *linear algebraic sets*, the loci of sets of linear (degree 1) polynomials of $k[x_1, \ldots, x_n]$. For example, a line in $\mathbb{A}^2$ is defined by an equation $ax + by = c$ (which is the locus of the polynomial $f(x, y) = ax + by - c \in k[x, y]$). A line in $\mathbb{A}^3$ is the locus of two linear polynomials of $k[x, y, z]$ that are not multiples of each other. In particular, the coordinate axes, coordinate planes, etc. in $\mathbb{A}^n$ are all affine algebraic sets. For instance, the x-axis in $\mathbb{A}^3$ is the zero set $\mathcal{Z}(y, z)$ and the x,y plane is the zero set $\mathcal{Z}(z)$.
(4) In general the algebraic set $\mathcal{Z}(f)$ of a nonconstant polynomial f is called a *hypersurface* in $\mathbb{A}^n$. Conic sections are familiar algebraic sets in the Euclidean plane $\mathbb{R}^2$. For example, the locus of $y - x^2$ is the parabola $y = x^2$, the locus of $x^2 + y^2 - 1$ is the unit circle, and $\mathcal{Z}(xy - 1)$ is the hyperbola $y = 1/x$. The x- and y-axes are the algebraic sets $\mathcal{Z}(y)$ and $\mathcal{Z}(x)$ respectively. Likewise, quadric surfaces such as the ellipsoid defined by the equation $x^2 + \dfrac{y^2}{4} + \dfrac{z^2}{9} = 1$ are affine algebraic sets in $\mathbb{R}^3$.

We leave as exercises the straightforward verification of the following properties of affine algebraic sets. Let S and T be subsets of $k[\mathbb{A}^n]$.

(1) If $S \subseteq T$ then $\mathcal{Z}(T) \subseteq \mathcal{Z}(S)$ (i.e., $\mathcal{Z}$ is inclusion reversing or *contravariant*).
(2) $\mathcal{Z}(S) = \mathcal{Z}(I)$, where $I = (S)$ is the ideal in $k[\mathbb{A}^n]$ generated by the subset S.
(3) The intersection of two affine algebraic sets is again an affine algebraic set, in fact $\mathcal{Z}(S) \cap \mathcal{Z}(T) = \mathcal{Z}(S \cup T)$. More generally an arbitrary intersection of affine algebraic sets is an algebraic set: if $\{S_j\}$ is any collection of subsets of $k[\mathbb{A}^n]$, then

$$\cap \mathcal{Z}(S_j) = \mathcal{Z}(\cup S_j).$$

(4) The union of two affine algebraic sets is again an affine algebraic set, in fact $\mathcal{Z}(I) \cup \mathcal{Z}(J) = \mathcal{Z}(IJ)$, where I and J are ideals and IJ is their product.
(5) $\mathcal{Z}(0) = \mathbb{A}^n$ and $\mathcal{Z}(1) = \emptyset$ (here 0 and 1 denote constant functions).

By (2), every affine algebraic set is the algebraic set corresponding to an *ideal* of the coordinate ring. Thus we may consider

$$\mathcal{Z} : \{\text{ ideals of } k[\mathbb{A}^n]\} \rightarrow \{\text{ affine algebraic sets in } \mathbb{A}^n\}.$$

Since every ideal I in the Noetherian ring $k[x_1, x_2, \ldots, x_n]$ is finitely generated, say $I = (f_1, f_2, \ldots, f_q)$, it follows from (3) that $\mathcal{Z}(I) = \mathcal{Z}(f_1) \cap \mathcal{Z}(f_2) \cap \cdots \cap \mathcal{Z}(f_q)$, i.e., *each affine algebraic set is the intersection of a finite number of hypersurfaces in* $\mathbb{A}^n$. Note that this "geometric" property in affine n-space is a consequence of an "algebraic" property of the corresponding coordinate ring (namely, Hilbert's Basis Theorem).

If V is an algebraic set in affine n-space, then there may be many ideals I such that $V = \mathcal{Z}(I)$. For example, in affine 2-space over $\mathbb{R}$ the y-axis is the locus of the ideal (x) of $\mathbb{R}[x, y]$, and also the locus of (x^2), (x^3), etc. More generally, the zeros of any polynomial are the same as the zeros of all its positive powers, and it follows that $\mathcal{Z}(I) = \mathcal{Z}(I^k)$ for all $k \geq 1$. We shall study the relationship between ideals that determine the same affine algebraic set in the next section when we discuss radicals of ideals.

While the ideal whose locus determines a particular algebraic set V is not unique, there is a unique largest ideal that determines V, given by the set of *all* polynomials that vanish on V. In general, for any subset A of $\mathbb{A}^n$ define

$$\mathcal{I}(A) = \{ f \in k[x_1, \ldots, x_n] \mid f(a_1, a_2, \ldots, a_n) = 0 \text{ for all } (a_1, a_2, \ldots, a_n) \in A \}.$$

It is immediate that $\mathcal{I}(A)$ is an *ideal*, and is the unique largest ideal of functions that are identically zero on A. This defines a correspondence

$$\mathcal{I} : \{\text{subsets in } \mathbb{A}^n\} \to \{\text{ideals of } k[\mathbb{A}^n]\}.$$

Examples

(1) In the Euclidean plane, $\mathcal{I}$(the x-axis) is the ideal generated by y in the coordinate ring $\mathbb{R}[x, y]$.

(2) Over any field k, the ideal of functions vanishing at $(a_1, a_2, \ldots, a_n) \in \mathbb{A}^n$ is a maximal ideal since it is the kernel of the surjective ring homomorphism from $k[x_1, \ldots, x_n]$ to the field k given by evaluation at $(a_1, a_2, \ldots, a_n)$. It follows that

$$\mathcal{I}((a_1, a_2, \ldots, a_n)) = (x_1 - a_1, \ x_2 - a_2, \ \ldots, \ x_n - a_n).$$

(3) Let $V = \mathcal{Z}(x^3 - y^2)$ in $\mathbb{A}^2$. If $(a, b) \in \mathbb{A}^2$ is an element of V then $a^3 = b^2$. If $a \neq 0$, then also $b \neq 0$ and we can write $a = (b/a)^2$, $b = (b/a)^3$. It follows that V is the set $\{(a^2, a^3) \mid a \in k\}$. For any polynomial $f(x, y) \in k[x, y]$ we can write $f(x, y) = f_0(x) + f_1(x)y + (x^3 - y^2)g(x, y)$. For $f(x, y) \in \mathcal{I}(V)$, i.e., $f(a^2, a^3) = 0$ for all $a \in k$, it follows that $f_0(a^2) + f_1(a^2)a^3 = 0$ for all $a \in k$. If $f_0(x) = a_r x^r + \cdots + a_0$ and $f_1(x) = b_s x^s + \cdots + b_0$ then

$$f_0(x^2) + x^3 f_1(x^2) = (a_r x^{2r} + \cdots + a_0) + (b_s x^{2s+3} + \cdots + b_0 x^3)$$

and this polynomial is 0 for every $a \in k$. If k is infinite, this polynomial has infinitely many zeros, which can happen only if all of the coefficients are zero. The coefficients of the terms of even degree are the coefficients of $f_0(x)$ and the coefficients of the terms of odd degree are the coefficients of $f_1(x)$, so it follows that $f_0(x)$ and $f_1(x)$ are both 0. It follows that $f(x, y) = (x^3 - y^2)g(x, y)$, and so

$$\mathcal{I}(V) = (x^3 - y^2) \subset k[x, y].$$

If k is finite, however, there may be elements in $\mathcal{I}(V)$ not lying in the ideal $(x^3 - y^2)$. For example, if $k = \mathbb{F}_2$, then V is simply the set $\{(0, 0), (1, 1)\}$ and so $\mathcal{I}(V)$ contains the polynomial $x(x - 1)$ (cf. Exercise 15).

The following properties of the map $\mathcal{I}$ are very easy exercises. Let A and B be subsets of $\mathbb{A}^n$.

(6) If $A \subseteq B$ then $\mathcal{I}(B) \subseteq \mathcal{I}(A)$ (i.e., $\mathcal{I}$ is also *contravariant*).
(7) $\mathcal{I}(A \cup B) = \mathcal{I}(A) \cap \mathcal{I}(B)$.
(8) $\mathcal{I}(\emptyset) = k[x_1, \ldots, x_n]$ and, if k is infinite, $\mathcal{I}(\mathbb{A}^n) = 0$.

Moreover, there are easily verified relations between the maps $\mathcal{Z}$ and $\mathcal{I}$:

(9) If A is any subset of $\mathbb{A}^n$ then $A \subseteq \mathcal{Z}(\mathcal{I}(A))$, and if I is any ideal then $I \subseteq \mathcal{I}(\mathcal{Z}(I))$.
(10) If $V = \mathcal{Z}(I)$ is an affine algebraic set then $V = \mathcal{Z}(\mathcal{I}(V))$, and if $I = \mathcal{I}(A)$ then $\mathcal{I}(\mathcal{Z}(I)) = I$, i.e., $\mathcal{Z}(\mathcal{I}(\mathcal{Z}(I))) = \mathcal{Z}(I)$ and $\mathcal{I}(\mathcal{Z}(\mathcal{I}(A))) = \mathcal{I}(A)$.

The last relation shows that the maps $\mathcal{Z}$ and $\mathcal{I}$ act as inverses of each other provided one restricts to the collection of affine algebraic sets $V = \mathcal{Z}(I)$ in $\mathbb{A}^n$ and to the set of ideals in $k[\mathbb{A}^n]$ of the form $\mathcal{I}(V)$. In the case where the field k is algebraically closed we shall (in the following two sections) characterize those ideals I that are of the form $\mathcal{I}(V)$ for some affine algebraic set V in terms of purely ring-theoretic properties of the ideal I (this is the famous "Zeros Theorem" of Hilbert, cf. Theorem 32).

Definition. If $V \subseteq \mathbb{A}^n$ is an affine algebraic set the quotient ring $k[\mathbb{A}^n]/\mathcal{I}(V)$ is called the *coordinate ring of* V, and is denoted by $k[V]$.

Note that for $V = \mathbb{A}^n$ and k infinite we have $\mathcal{I}(V) = 0$, so this definition extends the previous terminology. The polynomials in $k[\mathbb{A}^n]$ define k-valued functions on V simply by restricting these functions on $\mathbb{A}^n$ to the subset V. Two such polynomial functions f and g define the *same* function on V if and only if $f - g$ is identically 0 on V, which is to say that $f - g \in \mathcal{I}(V)$. Hence the cosets $\overline{f} = f + \mathcal{I}(V)$ giving the elements of the quotient $k[V]$ are precisely the restrictions to V of ordinary polynomial functions f from $\mathbb{A}^n$ to k (which helps to explain the notation $k[V]$). If x_i denotes the ith coordinate function on $\mathbb{A}^n$ (projecting an n-tuple onto its ith component), then the restriction $\overline{x_i}$ of x_i to V (which also just gives the ith component of the elements in V viewed as a subset of $\mathbb{A}^n$) is an element of $k[V]$, and $k[V]$ is finitely generated as a k-algebra by $\overline{x_1}, \ldots, \overline{x_n}$ (although this need not be a minimal generating set).

Example

If $V = \mathcal{Z}(xy - 1)$ is the hyperbola $y = 1/x$ in $\mathbb{R}^2$, then $\mathbb{R}[V] = \mathbb{R}[x, y]/(xy - 1)$. The polynomials $f(x, y) = x$ (the x-coordinate function) and $g(x, y) = x + (xy - 1)$, which are different functions on $\mathbb{R}^2$, define the same function on the subset V. On the point $(1/2, 2) \in V$, for example, both give the value $1/2$. In the quotient ring $\mathbb{R}[V]$ we have $\overline{x}\,\overline{y} = 1$, so $\mathbb{R}[V] \cong \mathbb{R}[x, 1/x]$. For any function $\overline{f} \in \mathbb{R}[V]$ and any $(a, b) \in V$ we have $\overline{f}(a, b) = f(a, 1/a)$ for any polynomial $f \in k[x, y]$ mapping to $\overline{f}$ in the quotient.

Suppose now that $V \subseteq \mathbb{A}^n$ and $W \subseteq \mathbb{A}^m$ are two affine algebraic sets. Since V and W are defined by the vanishing of polynomials, the most natural algebraic maps between V and W are those defined by polynomials:

Definition. A map $\varphi : V \to W$ is called a *morphism* (or *polynomial map* or *regular map*) of algebraic sets if there are polynomials $\varphi_1, \ldots, \varphi_m \in k[x_1, x_2, \ldots, x_n]$ such that

$$\varphi((a_1, \ldots, a_n)) = (\varphi_1(a_1, \ldots, a_n), \ldots, \varphi_m(a_1, \ldots, a_n))$$

for all $(a_1, \ldots, a_n) \in V$. The map $\varphi : V \to W$ is an *isomorphism* of algebraic sets if there is a morphism $\psi : W \to V$ with $\varphi \circ \psi = 1_W$ and $\psi \circ \varphi = 1_V$.

Note that in general $\varphi_1, \varphi_2, \ldots, \varphi_m$ are not uniquely defined. For example, both $f = x$ and $g = x + (xy - 1)$ in the example above define the same morphism from $V = \mathcal{Z}(xy - 1)$ to $W = \mathbb{A}^1$.

Suppose F is a polynomial in $k[x_1, \ldots, x_m]$. Then $F \circ \varphi = F(\varphi_1, \varphi_2, \ldots, \varphi_m)$ is a polynomial in $k[x_1, \ldots, x_n]$ since $\varphi_1, \varphi_2, \ldots, \varphi_m$ are polynomials in $x_1, \ldots, x_n$. If $F \in \mathcal{I}(W)$, then $F \circ \varphi((a_1, a_2, \ldots, a_n)) = 0$ for every $(a_1, a_2, \ldots, a_n) \in V$ since $\varphi((a_1, a_2, \ldots, a_n)) \in W$. Thus $F \circ \varphi \in \mathcal{I}(V)$. It follows that φ induces a well defined map from the quotient ring $k[x_1, \ldots, x_m]/\mathcal{I}(W)$ to the quotient ring $k[x_1, \ldots, x_n]/\mathcal{I}(V)$:

$$\widetilde{\varphi} : k[W] \to k[V]$$
$$f \mapsto f \circ \varphi$$

where $f \circ \varphi$ is given by $F \circ \varphi + \mathcal{I}(V)$ for any polynomial $F = F(x_1, \ldots, x_m)$ with $f = F + \mathcal{I}(W)$. It is easy to check that $\tilde{\varphi}$ is a k-algebra homomorphism (for example, $\widetilde{\varphi}(f + g) = (f + g) \circ \varphi = f \circ \varphi + g \circ \varphi = \widetilde{\varphi}(f) + \widetilde{\varphi}(g)$ shows that $\widetilde{\varphi}$ is additive). Note also the contravariant nature of $\widetilde{\varphi}$: the morphism from V to W induces a k-algebra homomorphism from $k[W]$ to $k[V]$.

Suppose conversely that Φ is any k-algebra homomorphism from the coordinate ring $k[W] = k[x_1, \ldots, x_m]/\mathcal{I}(W)$ to $k[V] = k[x_1, \ldots, x_n]/\mathcal{I}(V)$. Let F_i be a representative in $k[x_1, \ldots, x_n]$ for the image under Φ of $\bar{x}_i \in k[W]$ (i.e., $\Phi(x_i \bmod \mathcal{I}(W))$ is $F_i \bmod \mathcal{I}(V)$). Then $\varphi = (F_1, \ldots, F_m)$ defines a polynomial map from $\mathbb{A}^n$ to $\mathbb{A}^m$, and in fact φ is a morphism from V to W. To see this it suffices to check that φ maps a point of V to a point of W since by definition φ is already defined by polynomials. If $g \in \mathcal{I}(W) \subset k[x_1, \ldots, x_m]$, then in $k[W]$ we have

$$g(x_1 + \mathcal{I}(W), \ldots, x_m + \mathcal{I}(W)) = g(x_1, \ldots, x_m) + \mathcal{I}(W) = \mathcal{I}(W) = 0 \in k[W],$$

and so

$$\Phi(g(x_1 + \mathcal{I}(W), \ldots, x_m + \mathcal{I}(W))) = 0 \in k[V].$$

Since Φ is a k-algebra homomorphism, it follows that

$$g(\Phi(x_1 + \mathcal{I}(W)), \ldots, \Phi(x_m + \mathcal{I}(W)) = 0 \in k[V].$$

By definition, $\Phi(x_i + \mathcal{I}(W)) = F_i \bmod \mathcal{I}(V)$, so

$$g(F_1 \bmod \mathcal{I}(V), \ldots, F_m \bmod \mathcal{I}(V)) = 0 \in k[V],$$

i.e.,

$$g(F_1, \ldots, F_m) \in \mathcal{I}(V).$$

It follows that $g(F_1(a_1, \ldots, a_n), \ldots, F_m(a_1, \ldots, a_n)) = 0$ for every $(a_1, \ldots, a_n)$ in V. This shows that if $(a_1, \ldots, a_n) \in V$, then every polynomial in $\mathcal{I}(W)$ vanishes on $\varphi(a_1, \ldots, a_n)$. By property (10) of the maps $\mathcal{Z}$ and $\mathcal{I}$ above, this means that $\varphi(a_1, \ldots, a_n) \in \mathcal{Z}(\mathcal{I}(W)) = W$, which proves that φ maps a point in V to a point in W. It follows that $\varphi = (F_1, \ldots, F_m)$ is a morphism from V to W. Since the F_i are well defined modulo $\mathcal{I}(V)$, this morphism from V to W does not depend on the choice of the F_i. Furthermore, the morphism φ induces the original k-algebra homomorphism Φ from $k[W]$ to $k[V]$, i.e., $\widetilde{\varphi} = \Phi$, since both homomorphisms take the value $F_i + \mathcal{I}(V)$ on $x_i + \mathcal{I}(W) \in k[W]$. This proves the first two statements in the following theorem.

Theorem 6. Let $V \subseteq \mathbb{A}^n$ and $W \subseteq \mathbb{A}^m$ be affine algebraic sets. Then there is a bijective correspondence

$$\left\{ \begin{array}{c} \text{morphisms from } V \text{ to } W \\ \text{as algebraic sets} \end{array} \right\} \longleftrightarrow \left\{ \begin{array}{c} k\text{-algebra homomorphisms} \\ \text{from } k[W] \text{ to } k[V] \end{array} \right\}.$$

More precisely,
  (1) Every morphism $\varphi : V \to W$ induces an associated k-algebra homomorphism $\widetilde{\varphi} : k[W] \to k[V]$ defined by $\widetilde{\varphi}(f) = f \circ \varphi$.
  (2) Every k-algebra homomorphism $\Phi : k[W] \to k[V]$ is induced by a unique morphism $\varphi : V \to W$, i.e., $\Phi = \widetilde{\varphi}$.
  (3) If $\varphi : V \to W$ and $\psi : W \to U$ are morphisms of affine algebraic sets, then $\widetilde{\psi \circ \varphi} = \widetilde{\varphi} \circ \widetilde{\psi} : k[U] \to k[V]$.
  (4) The morphism $\varphi : V \to W$ is an isomorphism if and only if $\widetilde{\varphi} : k[W] \to k[V]$ is a k-algebra isomorphism.

Proof: The proof of (3) is left as an exercise and (4) is then immediate.

Example

For any infinite field k let $V = \mathbb{A}^1$ and let $W = \mathcal{Z}(x^3 - y^2) = \{(a^2, a^3) \mid a \in k\}$. The map $\varphi : V \to W$ defined by $\varphi(a) = (a^2, a^3)$ is a morphism from V to W. Note that φ is a bijection. The coordinate rings are $k[V] = k[x]$ and $k[W] = k[x, y]/(x^3 - y^2)$ (by the computations in a previous example — it is at this point we need k to be infinite) and the associated k-algebra homomorphism of coordinate rings is determined by

$$\widetilde{\varphi} : k[W] \longrightarrow k[V]$$
$$x \mapsto x^2$$
$$y \mapsto x^3.$$

The image of $\widetilde{\varphi}$ is the subalgebra $k[x^2, x^3] = k + x^2 k[x]$ of $k[x]$, so in particular $\widetilde{\varphi}$ is not surjective. Hence $\widetilde{\varphi}$ is not an isomorphism of coordinate rings, and it follows that φ is not an isomorphism of algebraic sets, even though the morphism φ is a bijective map. The inverse map is given by $\psi(0, 0) = 0$ and $\psi(a, b) = b/a$ for $b \neq 0$, and this cannot be achieved by a polynomial map.

The bijection in Theorem 6 gives a translation from maps between two geometrically defined algebraic sets V and W into algebraic maps between their coordinate rings. It also allows us to define a morphism intrinsically in terms of V and W without explicit reference to the ambient affine spaces containing them:

Corollary 7. Suppose $\varphi : V \to W$ is a map of affine algebraic sets. Then φ is a morphism if and only if for every $f \in k[W]$ the composite map $f \circ \varphi$ is an element of $k[V]$ (as a k-valued function on V). When φ is a morphism, $\varphi(v) = w$ with $v \in V$ and $w \in W$ if and only if $\widetilde{\varphi}^{-1}(\mathcal{I}(\{v\})) = \mathcal{I}(\{w\})$.

Proof: We first prove that if φ is any map from V to W such that $\widetilde{\varphi}$ is a k-algebra homomorphism then $\varphi(v) = w$ if and only if $\widetilde{\varphi}^{-1}(\mathcal{I}(\{v\})) = \mathcal{I}(\{w\})$, which will in particular establish the second statement. Note that $\varphi(v) = w$ if and only if every polynomial f vanishing at w vanishes at $\varphi(v)$ (by property (10) above: $\{w\} = \mathcal{Z}(\mathcal{I}(\{w\}))$). Since f vanishes at $\varphi(v)$ if and only if $\widetilde{\varphi}(f)$ vanishes at v, this is equivalent to the statement that $\widetilde{\varphi}(f) \in \mathcal{I}(\{v\})$ for every $f \in \mathcal{I}(\{w\})$, i.e., $\widetilde{\varphi}(\mathcal{I}(\{w\})) \subseteq \mathcal{I}(\{v\})$, or $\mathcal{I}(\{w\}) \subseteq \widetilde{\varphi}^{-1}(\mathcal{I}(\{v\}))$. Since both $\mathcal{I}(\{w\})$ and $\mathcal{I}(\{v\})$ are maximal ideals, this is equivalent to $\widetilde{\varphi}^{-1}(\mathcal{I}(\{v\})) = \mathcal{I}(\{w\})$.

We now prove the first statement. If φ is a morphism, then $f \circ \varphi \in k[V]$ for every $f \in k[W]$. For the converse, observe first that composition with any map $\varphi : V \to W$ defines a k-algebra homomorphism $\widetilde{\varphi}$ from the k-algebra of k-valued functions on W to the k-algebra of k-valued functions on V (this is immediate from the pointwise definition of the addition and multiplication of functions). If $f \circ \varphi \in k[V]$ for every $f \in k[W]$, then $\widetilde{\varphi}$ is a k-algebra homomorphism from $k[W]$ to $k[V]$, so by the proposition, $\widetilde{\varphi} = \widetilde{\Phi}$ for a unique morphism $\Phi : V \to W$. Also, since $\widetilde{\varphi}$ is a k-algebra homomorphism from $k[W]$ to $k[V]$ it follows by what we have already shown that $\varphi(v) = w$ if and only if $\widetilde{\varphi}^{-1}(\mathcal{I}(\{v\})) = \mathcal{I}(\{w\})$. Because $\widetilde{\varphi} = \widetilde{\Phi}$, this is equivalent to $\widetilde{\Phi}^{-1}(\mathcal{I}(\{v\})) = \mathcal{I}(\{w\})$, and so $\Phi(v) = w$. Hence φ and Φ define the same map on V and so φ is a morphism, completing the proof.

Corollary 7 and the last part of Theorem 6 show that the isomorphism type of the coordinate ring of V (as a k-algebra) does not depend on the embedding of V in a particular affine n-space.

Computations in Affine Algebraic Sets and k-algebras

The theory of Gröbner bases developed in Section 9.6 is very useful in computations involving affine algebraic sets, for example in computing in the coordinate rings $k[\mathbb{A}^n]/\mathcal{I}(V)$. When $n > 1$ it can be difficult to describe the elements in this quotient ring explicitly. By Theorem 23 in Section 9.6, each polynomial f in $k[\mathbb{A}^n]$ has a unique remainder after general polynomial division by the elements in a Gröbner basis for $\mathcal{I}(V)$, and this remainder therefore serves as a unique representative for the coset $\bar{f}$ of f in the quotient $k[\mathbb{A}^n]/\mathcal{I}(V)$.

Examples

(1) In the example $W = \mathcal{Z}(x^3 - y^2)$ above, we showed $I = \mathcal{I}(W) = (x^3 - y^2)$ for any infinite field k and so $k[W] = k[x, y]/(x^3 - y^2)$. Here $x^3 - y^2$ gives a Gröbner basis for I with respect to the lexicographic monomial ordering with $y > x$, so every polynomial $f = f(x, y)$ can be written uniquely in the form $f(x, y) = f_0(x) + f_1(x)y + f_I$ with $f_0(x), f_1(x) \in k[x]$ and $f_I \in I$. Then $f_0(x) + f_1(x)y$ gives a unique representative for $\bar{f}$ in $k[W]$. With respect to the lexicographic monomial ordering with $x > y$,

$x^3 - y^2$ is again a Gröbner basis for I, but now the remainder representing $\bar{f}$ in $k[W]$ is of the form $h_0(y) + h_1(y)x + h_2(y)x^2$.

(2) Let $V = \mathcal{Z}(xz + y^2 + z^2, xy - xz + yz - 2z^2) \subset \mathbb{C}^3$ and $W = \mathcal{Z}(u^3 - uv^2 + v^3) \subset \mathbb{C}^2$. We shall show later that $I = \mathcal{I}(V) = (xz + y^2 + z^2, xy - xz + yz - 2z^2) \subset \mathbb{C}[x, y, z]$ and $J = \mathcal{I}(W) = (u^3 - uv^2 + v^3) \subset \mathbb{C}[u, v]$. In this case $u^3 - uv^2 + v^3$ gives a Gröbner basis for J for the lexicographic monomial ordering with $u > v$ similar to the previous example. The situation for I is more complicated. With respect to the lexicographic monomial ordering with $x > y > z$ the reduced Gröbner basis for I is given by

$$g_1 = xy + y^2 + yz - z^2, \qquad g_2 = xz + y^2 + z^2, \qquad g_3 = y^3 - y^2z + z^3.$$

Unique representatives for $\mathbb{C}[V] = \mathbb{C}[x, y, z]/(x^2 + xz + y^2, 2x^2 - xy + xz - yz)$ are given by the remainders after general polynomial division by $\{g_1, g_2, g_3\}$.

We saw already in Section 9.6 that Gröbner bases and elimination theory can be used in the explicit computation of affine algebraic sets $\mathcal{Z}(S)$, or, equivalently, in explicitly solving systems of algebraic equations. The same theory can be used to determine explicitly a set of generators for the image and kernel of a k-algebra homomorphism

$$\Phi : k[y_1, \ldots, y_m]/J \longrightarrow k[x_1, \ldots, x_n]/I$$

where I and J are ideals. In the particular case when $I = \mathcal{I}(V)$ and $J = \mathcal{I}(W)$ are the ideals associated to affine algebraic sets $V \subseteq \mathbb{A}^n$ and $W \subseteq \mathbb{A}^m$ then by Theorem 6, the k-algebra homomorphism Φ corresponds to a morphism from V to W, and we shall apply the results here to affine algebraic sets in Section 3.

For $1 \le i \le m$, let $\varphi_i \in k[x_1, \ldots, x_n]$ be any polynomial representing the coset $\Phi(\bar{y}_i)$, where as usual we use a bar to denote the coset of an element in a quotient. The polynomials $\varphi_1, \ldots, \varphi_n$ are unique up to elements of I. Then the image of a coset $f(y_1, \ldots, y_m) + J$ under Φ is the coset $f(\varphi_1, \ldots, \varphi_m) + I$. Given any $\varphi_1, \ldots, \varphi_n$, the map sending y_i to φ_i induces a k-algebra homomorphism Φ if and only if $f(y_1, \ldots, y_m) \in I$ for every $f \in J$, a condition which can be checked on a set of generators for J.

Proposition 8. With notation as above, let $R = k[y_1, \ldots, y_m, x_1, \ldots, x_n]$ and let $\mathcal{A}$ be the ideal generated by $y_1 - \varphi_1, \ldots, y_m - \varphi_m$ together with generators for I. Let G be the reduced Gröbner basis of $\mathcal{A}$ with respect to the lexicographic monomial ordering $x_1 > \cdots > x_n > y_1 > \cdots > y_m$. Then

(a) The kernel of Φ is $\mathcal{A} \cap k[y_1, \ldots, y_m]$ modulo J. The elements of G in $k[y_1, \ldots, y_m]$ (taken modulo J) generate $\ker \Phi$.

(b) If $f \in k[x_1, \ldots, x_n]$, then $\bar{f}$ is in the image of Φ if and only if the remainder after general polynomial division of f by the elements in G is an element $h \in k[y_1, \ldots, y_m]$, in which case $\Phi(\bar{h}) = \bar{f}$.

Proof: If we show $\ker \Phi = \mathcal{A} \cap k[y_1, \ldots, y_m]$ modulo J then (a) follows by Proposition 30 in Section 9.6. Suppose first that $f \in \mathcal{A} \cap k[y_1, \ldots, y_m]$. If $f_1, \ldots, f_s$ are generators for I in $k[x_1, \ldots, x_n]$, then

$$f(y_1, \ldots, y_m) = \sum_{i=1}^{n} a_i(y_i - \varphi_i) + \sum_{j=1}^{s} b_i f_i$$

as polynomials in R, where $a_1, \ldots, a_n, b_1, \ldots, b_s \in R$. Substituting $y_i = \varphi_i$ we see that $f(\varphi_1, \ldots, \varphi_m)$ is an element of I. Since $\Phi(\bar{f}) = f(\varphi_1, \ldots, \varphi_m)$ modulo I, it follows that f represents a coset in the kernel of Φ. Conversely, suppose $f \in k[y_1, \ldots, y_m]$ represents an element in $\ker \Phi$. Then $f(\varphi_1, \ldots, \varphi_m) \in I$ (in $k[x_1, \ldots, x_n]$) and so also $f(\varphi_1, \ldots, \varphi_m) \in \mathcal{A}$ (in R). Since $y_i - \varphi_i \in \mathcal{A}$,

$$f(y_1, \ldots, y_m) \equiv f(\varphi_1, \ldots, \varphi_m) \equiv 0 \bmod \mathcal{A}$$

so $f \in \mathcal{A} \cap k[y_1, \ldots, y_m]$.

For (b), suppose first that $f \in k[x_1, \ldots, x_n]$ represents an element in the image of Φ, i.e., $\bar{f} = \Phi(\bar{h})$ for some polynomial $h \in k[y_1, \ldots, y_m]$. Then

$$f(x_1, \ldots, x_n) - h(\varphi_1, \ldots, \varphi_m) \in I$$

as polynomials in $k[x_1, \ldots, x_n]$, and so $f(x_1, \ldots, x_n) - h(\varphi_1, \ldots, \varphi_m) \in \mathcal{A}$ as polynomials in R. As before, since each $y_i - \varphi_i \in \mathcal{A}$ it follows that

$$f(x_1, \ldots, x_n) - h(y_1, \ldots, y_m) \in \mathcal{A}.$$

Then $f(x_1, \ldots, x_n)$ and $h(y_1, \ldots, y_m)$ leave the same remainder after general polynomial division by the elements in G. Since $x_1 > \cdots > x_n > y_1 > \cdots > y_m$, the remainder of $h(y_1, \ldots, y_m)$ is again a polynomial h_0 only involving $y_1, \ldots, y_m$. Note also that $h - h_0 \in \mathcal{A} \cap k[y_1, \ldots, y_m]$ so $\bar{h}$ and $\bar{h}_0$ differ by an element in $\ker \Phi$ by (a), so $\Phi(\bar{h}_0) = \Phi(\bar{h}) = \bar{f}$. For the converse, if f leaves the remainder $h \in k[y_1, \ldots, y_m]$ after general polynomial division by the elements in G then $f(x_1, \ldots, x_n) - h(y_1, \ldots, y_m) \in \mathcal{A}$, i.e.,

$$f(x_1, \ldots, x_n) - h(y_1, \ldots, y_m) = \sum_{i=1}^{n} a_i(y_i - \varphi_i) + \sum_{j=1}^{s} b_i f_i$$

as polynomials in R, where $a_1, \ldots, a_n, b_1, \ldots, b_s \in R$. Substituting $y_i = \varphi_i$ we obtain

$$f(x_1, \ldots, x_n) - h(\varphi_1, \ldots, \varphi_m) \in I$$

as polynomials in $x_1, \ldots x_n$, and so $\bar{f} = \Phi(\bar{h})$.

It follows in particular from Proposition 8 that Φ will be a surjective homomorphism if and only if for each $i = 1, 2, \ldots, n$, dividing x_i by the elements in the Gröbner basis G leaves a remainder h_i in $k[y_1, \ldots, y_m]$. In particular, $x_n - h_n$ leaves a remainder of 0. But this means the leading term of some element g_n in G divides the leading term of $x_n - h_n$ and since $x_1 > \cdots > x_n > y_1 > \cdots > y_m$ by the choice of the ordering, the leading term of $x_n - h_n$ is just x_n. It follows that $LT(g_n) = x_n$ and so $g_n = x_n - h_{n,0} \in G$ for some $h_{n,0} \in k[y_1, \ldots, y_m]$ (in fact $h_{n,0}$ is the remainder of h_n after division by the elements in G). Next, since $x_{n-1} - h_{n-1}$ leaves a remainder of 0, there is an element g_{n-1} in G whose leading term is x_{n-1}. Since G is a reduced Gröbner basis and $g_n \in G$, the leading term of g_n, i.e., x_n, does not divide any of the terms in g_{n-1} and it follows that $g_{n-1} = x_{n-1} - h_{n-1,0} \in G$ for some $h_{n-1,0} \in k[y_1, \ldots, y_m]$. Proceeding in a similar fashion we obtain the following corollary, showing that whether Φ is surjective can be seen immediately from the elements in the reduced Gröbner basis.

Corollary 9. The map Φ is surjective if and only if for each i, $1 \le i \le n$, the reduced Gröbner basis G contains a polynomial $x_i - h_i$ where $h_i \in k[y_1, \ldots, y_m]$.

Examples

(1) Let $\Phi : \mathbb{Q}[u, v] \to \mathbb{Q}[x]$ be defined by $\Phi(u) = x^2 + x$ and $\Phi(v) = x^3$. The reduced Gröbner basis G for the ideal $\mathcal{A} = (u - x^2 - x, v - x^3)$ with respect to the lexicographic monomial ordering $x > u > v$ is

$$g_1 = x^2 + x - u, \qquad g_3 = vx - x - u^2 + u + 2v,$$
$$g_2 = ux + x - u - v, \qquad g_4 = u^3 - 3uv - v^2 - v.$$

The kernel of Φ is the ideal generated by $G \cap \mathbb{Q}[u, v] = \{g_4\}$. By Corollary 9, we see that Φ is not surjective. The remainder after general polynomial division of x^4 by $\{g_1, g_s, g_3, g_4\}$ is $x + u^2 - u - 2v \notin \mathbb{Q}[u, v]$, so x^4 is not in the image of Φ. The remainder of $x^5 + x$ is $-u^2 + uv + u + 2v \in \mathbb{Q}[u, v]$ so $x^5 + x = \Phi(-u^2 + uv + u + 2v)$ is in the image of Φ, as a quick check will confirm.

(2) Let $V = \mathcal{Z}(I) \subset \mathbb{C}^3$ and $W = \mathcal{Z}(J) \subset \mathbb{C}^2$ where $I = (xz + y^2 + z^2, xy - xz + yz - 2z^2)$ and $J = (u^3 - uv^2 + v^3)$ as in Example 2 following Corollary 7. Then the map $\varphi : V \to W$ defined by $\varphi((a, b, c)) = (c, b)$ is a morphism from V to W. To see this, we must check that $(c, b) \in W$ if $(a, b, c) \in V$. Equivalently, by Theorem 6, we must check that the map

$$\widetilde{\varphi} : \mathbb{C}[u, v]/(u^3 - uv^2 + v^3) \longrightarrow \mathbb{C}[x, y, z]/(xz + y^2 + z^2, xy - xz + yz - 2z^2)$$

induced by mapping u to z and v to y is a $\mathbb{C}$-algebra homomorphism. This in turn is equivalent to verifying that $f = z^3 - zy^2 + y^3$ is an element of the ideal I. In this case f is actually an element in the reduced Gröbner basis for I:

$$xy + y^2 + yz - z^2, \qquad xz + y^2 + z^2, \qquad y^3 - y^2z + z^3,$$

so certainly $f \in I$. (Note that dividing f by the original two generators for I leaves the nonzero remainder f itself, from which it is much less clear that $f \in I$, so it is important to use a Gröbner basis when working in coordinate rings.)

(3) In the previous example, let $\mathcal{A} = (u - z, v - y, xz + y^2 + z^2, xy - xz + yz - 2z^2) \subset \mathbb{C}[u, v, x, y, z]$ as in Proposition 8. With respect to the lexicographic monomial ordering $x > y > z > u > v$ the reduced Gröbner basis G for $\mathcal{A}$ is

$$xu + u^2 + v^2, \quad xv - u^2 + uv + v^2, \quad y - v, \quad z - u, \quad u^3 - uv^2 + v^3.$$

By Proposition 8, we see that $\ker \widetilde{\varphi}$ is generated by $u^3 - uv^2 + v^3 \equiv 0 \bmod J$, so $\widetilde{\varphi}$ is injective. Since there is no element of the form $x - h(u, v)$ in G, $\widetilde{\varphi}$ is not surjective (in fact x is not in the image).

As a final example, we use the determination of the kernel of k-algebra homomorphisms to compute minimal polynomials of elements in simple algebraic field extensions.

Proposition 10. Suppose α is a root of the irreducible polynomial $p(x) \in k[x]$ and $\beta \in k(\alpha)$, say $\beta = f(\alpha)$ for the polynomial $f \in k[x]$. Let G be the reduced Gröbner basis for the ideal $(p, y - f)$ in $k[x, y]$ for the lexicographic monomial ordering $x > y$. Then the minimal polynomial of β over k is the monic polynomial in $G \cap k[y]$.

Proof: The kernel of the k-algebra homomorphism $k[y] \to k[x]/(p) \cong k(\alpha)$ defined by mapping y first to f and then to β is the principal ideal generated by the minimal polynomial of β in $k[y]$, and the result follows by Proposition 8.

Example

Take $k = \mathbb{Q}$, and let $\beta = 1 + \sqrt[3]{2} + 3\sqrt[3]{4} \in \mathbb{Q}(\sqrt[3]{2})$. Then the ideal $(x^3 - 2, \, y - (1 + x + 3x^2))$ in $\mathbb{Q}[x, y]$ has reduced Gröbner basis $\{53x - 3y^2 + 7y + 32, \, y^3 - 3y^2 - 15y - 93\}$ for the lexicographic monomial ordering $x > y$, so the minimal polynomial for β is $y^3 - 3y^2 - 15y - 93$.

EXERCISES

Let R be a commutative ring with $1 \neq 0$ and let k be a field.

1. Prove the converse to Hilbert's Basis Theorem: if the polynomial ring $R[x]$ is Noetherian, then R is Noetherian.

2. Show that each of the following rings are not Noetherian by exhibiting an explicit infinite increasing chain of ideals:
   (a) the ring of continuous real valued functions on $[0, 1]$,
   (b) the ring of all functions from any infinite set X to $\mathbb{Z}/2\mathbb{Z}$.

3. Prove that the field $k(x)$ of rational functions over k in the variable x is not a finitely generated k-algebra. (Recall that $k(x)$ is the field of fractions of the polynomial ring $k[x]$. Note that $k(x)$ *is* a finitely generated *field extension* over k.)

4. Prove that if R is Noetherian, then so is the ring $R[[x]]$ of formal power series in the variable x with coefficients from R (cf. Exercise 3, Section 7.2). [Mimic the proof of Hilbert's Basis Theorem.]

5. (*Fitting's Lemma*) Suppose M is a Noetherian R-module and $\varphi : M \to M$ is an R-module endomorphism of M. Prove that $\ker(\varphi^n) \cap \mathrm{image}(\varphi^n) = 0$ for n sufficiently large. Show that if φ is surjective, then φ is an isomorphism. [Observe that $\ker(\varphi) \subseteq \ker(\varphi^2) \subseteq \ldots$]

6. Suppose that $0 \longrightarrow M' \longrightarrow M \longrightarrow M'' \longrightarrow 0$ is an exact sequence of R-modules. Prove that M is a Noetherian R-module if and only if M' and M'' are Noetherian R-modules.

7. Prove that submodules, quotient modules, and finite direct sums of Noetherian R-modules are again Noetherian R-modules.

8. If R is a Noetherian ring, prove that M is a Noetherian R-module if and only if M is a finitely generated R-module. (Thus any submodule of a finitely generated module over a Noetherian ring is also finitely generated.)

9. For k a field show that any subring of the polynomial ring $k[x]$ containing k is Noetherian. Give an example to show such subrings need not be U.F.D.s. [If $k \subset R \subseteq k[x]$ and $y \in R - k$ show that $k[x]$ is a finitely generated $k[y]$-module; then use the previous two exercises. For the second, consider $k[x^2, x^3]$.]

10. Prove that the subring $k[x, x^2y, x^3y^2, \ldots, x^i y^{i-1}, \ldots]$ of the polynomial ring $k[x, y]$ is *not* a Noetherian ring, hence not a finitely generated k-algebra. (Thus subrings of Noetherian rings need not be Noetherian and subalgebras of finitely generated k-algebras need not be finitely generated.)

11. Suppose R is a commutative ring in which all the prime ideals are finitely generated. This exercise proves that R is Noetherian.

(a) Prove that if the collection of ideals of R that are not finitely generated is nonempty, then it contains a maximal element I, and that R/I is a Noetherian ring.

(b) Prove that there are finitely generated ideals J_1 and J_2 containing I with $J_1 J_2 \subseteq I$ and that $J_1 J_2$ is finitely generated. [Observe that I is not a prime ideal.]

(c) Prove that $I/J_1 J_2$ is a finitely generated R/I-submodule of $J_1/J_1 J_2$. [Use Exercise 8.]

(d) Show that (c) implies the contradiction that I would be finitely generated over R and deduce that R is Noetherian.

12. Suppose R is a Noetherian ring and S is a finitely generated R-algebra. If $T \subseteq S$ is an R-algebra such that S is a finitely generated T-*module*, prove that T is a finitely generated R-algebra. [If $s_1, \ldots, s_n$ generate S as an R-algebra, and $s_1', \ldots, s_m'$ generate S as a T-module, show that the elements s_i and $s_j' s_k'$ can be written as finite T-linear combinations of the s_i'. If T_0 is the R-subalgebra generated by the coefficients of these linear combinations, show S (hence T_0) is finitely generated (by the s_i') as a T_0-module, and conclude that T is finitely generated as an R-algebra.]

13. Verify properties (1) to (10) of the maps $\mathcal{Z}$ and $\mathcal{I}$.

14. Show that the affine algebraic sets in $\mathbb{A}^1$ over any field k are $\emptyset$, k, and finite subsets of k.

15. If $k = \mathbb{F}_2$ and $V = \{(0,0), (1,1)\} \subset \mathbb{A}^2$, show that $\mathcal{I}(V)$ is the product ideal $\mathfrak{m}_1 \mathfrak{m}_2$ where $\mathfrak{m}_1 = (x, y)$ and $\mathfrak{m}_2 = (x-1, y-1)$.

16. Suppose that V is a finite algebraic set in $\mathbb{A}^n$. If V has m points, prove that $k[V]$ is isomorphic as a k-algebra to k^m. [Use the Chinese Remainder Theorem.]

17. If k is a finite field show that every subset of $\mathbb{A}^n$ is an affine algebraic set.

18. If $k = \mathbb{F}_q$ is the finite field with q elements show that $\mathcal{I}(\mathbb{A}^1) = (x^q - x) \subset k[x]$.

19. For each nonconstant $f \in k[x]$ describe $\mathcal{Z}(f) \subseteq \mathbb{A}^1$ in terms of the unique factorization of f in $k[x]$, and then use this to describe $\mathcal{I}(\mathcal{Z}(f))$. Deduce that $\mathcal{I}(\mathcal{Z}(f)) = (f)$ if and only if f is the product of distinct linear factors in $k[x]$.

20. If f and g are irreducible polynomials in $k[x, y]$ that are not associates (do not divide each other), show that $\mathcal{Z}((f, g))$ is either $\emptyset$ or a finite set in $\mathbb{A}^2$. [If $(f, g) \neq (1)$, show (f, g) contains a nonzero polynomial in $k[x]$ (and similarly a nonzero polynomial in $k[y]$) by letting $R = k[x]$, $F = k(x)$, and applying Gauss's Lemma to show f and g are relatively prime in $F[y]$.]

21. Identify each 2×2 matrix $\begin{pmatrix} a & b \\ c & d \end{pmatrix}$ with entries from k with the point (a, b, c, d) in $\mathbb{A}^4$. Show that the group $SL_2(k)$ of matrices of determinant 1 is an algebraic set in $\mathbb{A}^4$.

22. Prove that $SL_n(k)$ is an affine algebraic set in $\mathbb{A}^{n^2}$. [Generalize the preceding exercise.]

23. Let V be any line in $\mathbb{R}^2$ (the zero set of any nonzero linear polynomial $ax + by - c$). Prove that $\mathbb{R}[V]$ is isomorphic as an $\mathbb{R}$-algebra to the polynomial ring $\mathbb{R}[x]$, and give the corresponding isomorphism from $\mathbb{A}^1$ to V.

24. Let $V = \mathcal{Z}(xy - z) \subseteq \mathbb{A}^3$. Prove that V is isomorphic to $\mathbb{A}^2$ and provide an explicit isomorphism φ and associated k-algebra isomorphism $\widetilde{\varphi}$ from $k[V]$ to $k[\mathbb{A}^2]$, along with their inverses. Is $V = \mathcal{Z}(xy - z^2)$ isomorphic to $\mathbb{A}^2$?

25. Suppose $V \subseteq \mathbb{A}^n$ is an affine algebraic set and $f \in k[V]$. The *graph* of f is the collection of points $\{(a_1, \ldots, a_n, f(a_1, \ldots, a_n))\}$ in $\mathbb{A}^{n+1}$. Prove that the graph of f is an affine algebraic set isomorphic to V. [The morphism in one direction maps $(a_1, \ldots, a_n)$ to $(a_1, \ldots, a_n, f(a_1, \ldots, a_n))$.]

26. Let $V = \mathcal{Z}(xz - y^2, yz - x^3, z^2 - x^2y) \subseteq \mathbb{A}^3$.

    (a) Prove that the map $\varphi : \mathbb{A}^1 \to V$ defined by $\varphi(t) = (t^3, t^4, t^5)$ is a surjective morphism. [For the surjectivity, if $(x, y, z) \neq (0, 0, 0)$, let $t = y/x$.]

    (b) Describe the corresponding k-algebra homomorphism $\widetilde{\varphi} : k[V] \to k[\mathbb{A}^1]$ explicitly.

    (c) Prove that φ is not an isomorphism.

27. Suppose $\varphi : V \to W$ is a morphism of affine algebraic sets. If W' is an affine algebraic subset of W prove that the preimage $V' = \varphi^{-1}(W')$ of W' in V is an affine algebraic subset of V. If $W' = \mathcal{Z}(I)$ show that $V' = \mathcal{Z}(\widetilde{\varphi}(I))$ for the corresponding morphism $\widetilde{\varphi} : k[W] \to k[V]$.

28. Prove that if V and W are affine algebraic sets, then so is $V \times W$ and $k[V \times W] \cong k[V] \otimes_k k[W]$.

The following seven exercises introduce the notion of the *associated primes* of an R-module M. Cf. also Exercises 30–40 in Section 4 and Exercises 25–30 in Section 5.

Definition. A prime ideal P of R is said to be *associated* to the R-module M (sometimes called an *assassin* for M) if P is the annihilator of some element m of M, i.e., if M contains a submodule Rm isomorphic to R/P. The collection of associated primes for M is denoted $\mathrm{Ass}_R(M)$.

    When $M = I$ is an ideal in R, it is customary to abuse the terminology and refer instead to the elements of $\mathrm{Ass}_R(R/I)$ (rather than the less interesting collection $\mathrm{Ass}_R(I)$) as the *primes associated to I*. (Cf. Exercises 28–29 in Section 5.)

29. If $R = \mathbb{Z}$ and $M = \mathbb{Z}/n\mathbb{Z}$, show that $\mathrm{Ass}_R(M)$ consists of the prime ideals (p) for the prime divisors p of n.

30. If M is the union of some collection of submodules M_i, prove that $\mathrm{Ass}_R(M)$ is the union of the collection $\mathrm{Ass}_R(M_i)$.

31. Suppose that $\mathrm{Ann}(m) = P$, i.e., that $Rm \cong R/P$. Prove that if $0 \neq m' \in Rm$ then $\mathrm{Ann}(m') = P$. Deduce that $\mathrm{Ass}_R(R/P) = \{P\}$. [Observe that R/P is an integral domain.]

32. Suppose that M is an R-module and that P is a maximal element in the collection of ideals of the form $\mathrm{Ann}(m)$, for $m \in M$. Prove that P is a prime ideal. [If $P = \mathrm{Ann}(m)$ and $ab \in P$, show that $bm \neq 0$ implies $\mathrm{Ann}(m) \subseteq \mathrm{Ann}(bm)$ and use the maximality of P to deduce that $a \in \mathrm{Ann}(bm) = P$.]

33. Suppose R is a Noetherian ring and $M \neq 0$ is an R-module. Prove that $\mathrm{Ass}_R(M) \neq \emptyset$. [Use Exercise 32.]

34. If L is a submodule of M with quotient $N \cong M/L$, prove that there are containments $\mathrm{Ass}_R(L) \subseteq \mathrm{Ass}_R(M) \subseteq \mathrm{Ass}_R(L) \cup \mathrm{Ass}_R(N)$, and show that both containments can be proper. [If $Rm \cong R/P$, show that $Rm \cap L = 0$ implies $P \in \mathrm{Ass}_R(N)$ and if $Rm \cap L \neq 0$ then $P \in \mathrm{Ass}_R(L)$ (by Exercise 31). For the second statement, consider $n\mathbb{Z} \subset \mathbb{Z}$.]

35. Suppose M is an R-module and let $\mathcal{S}$ be a subset of the prime ideals in $\mathrm{Ass}_R(M)$. Prove there is a submodule N of M with $\mathrm{Ass}_R(N) = \mathcal{S}$ and $\mathrm{Ass}_R(M/N) = \mathrm{Ass}_R(M) - \mathcal{S}$. [Consider the collection of submodules N' of M with $\mathrm{Ass}_R(N') \subseteq \mathcal{S}$. Use Exercise 30 and Zorn's Lemma to show that there is a maximal submodule N subject to $\mathrm{Ass}_R(N) \subseteq \mathcal{S}$. If $P \in \mathrm{Ass}_R(M/N)$, there is a submodule $M'/N \cong R/P$. Use the previous exercise to show that $\mathrm{Ass}_R(M') \subseteq \mathrm{Ass}_R(N) \cup \{P\}$ and then use maximality of N to show $P \in \mathrm{Ass}_R(M) - \mathcal{S}$, so that $\mathrm{Ass}_R(M/N) \subseteq \mathrm{Ass}_R(M) - \mathcal{S}$ and $\mathrm{Ass}_R(N) \subseteq \mathcal{S}$. Use the previous exercise again to conclude that equality holds in each.]

Suppose M is a finitely generated module over the commutative ring R with generators $m_1, \ldots, m_n$. The *Fitting ideal* $\mathcal{F}_R(M)$ (of level 0) of M (also called a *determinant ideal*) is the ideal in R generated by the determinants of all $n \times n$ matrices $A = (r_{ij})$ where $r_{ij} \in R$ and $r_{i1}m_1 + \cdots + r_{in}m_n = 0$ in M, i.e., the rows of A consist of the coefficients in R of relations among the generators m_i (A is called an $n \times n$ "relations matrix" for M). The following five exercises outline some of the properties of the Fitting ideal.

36. **(a)** Show that the Fitting ideal of M is also the ideal in R generated by all the $n \times n$ minors of all $p \times n$ matrices $A = (r_{ij})$ for $p \geq 1$ whose rows consist of the coefficients in R of relations among the generators m_i.
  (b) Let A be a fixed $p \times n$ matrix as in (a) and let A' be a $p \times n$ matrix obtained from A by any elementary row or column operation. Show that the ideal in R generated by all the $n \times n$ minors of A is the same as the ideal in R generated by all the $n \times n$ minors of A'.

37. Suppose $m_1, \ldots, m_n$ and $m_1', \ldots, m_{n'}'$ are two sets of R-module generators for M. Let $\mathcal{F}$ denote the Fitting ideal for M computed using the generators $m_1, \ldots, m_n$ and let $\mathcal{F}'$ denote the Fitting ideal for M computed using the generators $m_1, \ldots, m_n, m_1', \ldots, m_{n'}'$.
  (a) Show that $m_s' = a_{s'1}m_1 + \cdots + a_{s'n}m_n$ for some $a_{s'1}, \ldots, a_{s'n} \in R$, and deduce that $(-a_{s'1}, \ldots, -a_{s'n}, 0, \ldots, 0, 1, 0, \ldots 0)$ is a relation among $m_1, \ldots, m_n, m_1', \ldots, m_{n'}'$.
  (b) If $A = (r_{ij})$ is an $n \times n$ matrix whose rows are the coefficients of relations among $m_1, \ldots, m_n$ show that $\det A = \det A'$ where A' is an $(n + n') \times (n + n')$ matrix whose rows are the coefficients of relations among $m_1, \ldots, m_n, m_1', \ldots, m_{n'}'$. Deduce that $\mathcal{F} \subseteq \mathcal{F}'$. [Use (a) to find a block upper triangular A' having A in the upper left block and the $n' \times n'$ identity matrix in the lower right block.]
  (c) Prove that $\mathcal{F}' \subseteq \mathcal{F}$ and conclude that $\mathcal{F}' = \mathcal{F}$. [Use the previous exercise.]
  (d) Deduce from (c) that the Fitting ideal $\mathcal{F}_R(M)$ of M is an invariant of M that does not depend on the choice of generators for M used to compute it.

38. All modules in this exercise are assumed finitely generated.
  (a) If M can be generated by n elements prove that $\text{Ann}(M)^n \subseteq \mathcal{F}_R(M) \subseteq \text{Ann}(M)$, where $\text{Ann}(M)$ is the annihilator of M in R. [If A is an $n \times n$ relations matrix for M, then $AX = 0$, where X is the column matrix whose entries are $m_1, \ldots, m_n$. Multiply by the adjoint of A to deduce that $\det A$ annihilates M.]
  (b) If $M = M_1 \times M_2$ is the direct product of the R-modules M_1 and M_2 prove that $\mathcal{F}_R(M) = \mathcal{F}_R(M_1)\mathcal{F}_R(M_2)$.
  (c) If $M = (R/I_1) \times \cdots \times (R/I_n)$ is the direct product of cyclic R-modules for ideals I_i in R prove that $\mathcal{F}_R(M) = I_1 I_2 \ldots I_n$.
  (d) If $R = \mathbb{Z}$ and M is a finitely generated abelian group show that $\mathcal{F}_{\mathbb{Z}}(M) = 0$ if M is infinite and $\mathcal{F}_{\mathbb{Z}}(M) = |M|\mathbb{Z}$ if M is finite.
  (e) If I is an ideal in R prove that the image of $\mathcal{F}_R(M)$ in the quotient R/I is $\mathcal{F}_{R/I}(M/IM)$.
  (f) Prove that $\mathcal{F}_R(M/IM) \subseteq (\mathcal{F}_R(M), I) \subseteq R$.
  (g) If $\varphi : M \to M'$ is a surjective R-module homomorphism prove $\mathcal{F}_R(M) \subseteq \mathcal{F}_R(M')$.
  (h) If $0 \to L \to M \to N \to 0$ is a short exact sequence of R-modules, prove that $\mathcal{F}_R(L)\mathcal{F}_R(N) \subseteq \mathcal{F}_R(M)$.
  (i) Suppose R is the polynomial ring $k[x, y, z]$ over the field k. Let $M = R/(x, y^2, yz, z^2)$ and let L be the submodule $(x, y, z)/(x, y^2, yz, z^2)$ of M. Prove that $\mathcal{F}_R(M)$ is (x, y^2, yz, z^2) and $\mathcal{F}_R(L)$ is $(x, y, z)^2$. (This shows that in general the Fitting ideal of a submodule L of M need not contain the Fitting ideal for M.)

39. Suppose M is an R-module and that $\varphi : R^n \to M$ is a surjective R-module homomorphism (i.e., M can be generated by n elements). Let $L = \ker \varphi$. Prove that the image of the

R-module homomorphism from $\bigwedge^n(L) \to \bigwedge^n(R^n) \cong R$ induced by the inclusion of L in R^n is the Fitting ideal $\mathcal{F}_R(M)$.

40. Suppose R and S are commutative rings, $\varphi : R \to S$ is a ring homomorphism, M is a finitely generated R-module, and $M' = S \otimes_R M$ is the S-module obtained by extending scalars from R to S. Prove that the Fitting ideal $\mathcal{F}_S(M')$ for M' over S is the extension to S of the Fitting ideal $\mathcal{F}_R(M)$ for M over R.

The following two exercises indicate how the remainder in Theorem 23 of Chapter 9 can be used to effect computations in quotients of polynomial rings.

41. Suppose $\{g_1, \ldots, g_m\}$ is a Gröbner basis for the ideal I in $k[x_1, \ldots, x_n]$. Prove that the monomials m not divisible by any $LT(g_i)$, $1 \le i \le m$, give a k-vector space basis for the quotient $k[x_1, \ldots, x_n]/I$.

42. Let $I = (x^3y - xy^2 + 1, x^2y^2 - y^3 - 1)$ as in Example 1 following Proposition 9.26.
   (a) Use the previous exercise to show that $\{1, y, y^2, y^3\}$ is a basis for the k-vector space $k[x, y]/I$.
   (b) Compute the 4×4 multiplication table for the basis vectors in (a).

43. Suppose $K[x_1, \ldots, x_n]$ is a polynomial ring in n variables over a field K and k is a subfield of K. If I is an ideal in $k[x_1, \ldots, x_n]$, let I' be the ideal generated by I in $K[x_1, \ldots, x_n]$.
   (a) If G is a Gröbner basis for the ideal I in $k[x_1, \ldots, x_n]$ with respect to some monomial ordering, show that G is also a Gröbner basis for the ideal I' in $K[x_1, \ldots, x_n]$ with respect to the same monomial ordering. [Use Buchberger's Criterion.]
   (b) Prove that the dimension of the quotient $k[x_1, \ldots, x_n]/I$ as a vector space over k is the same as the dimension of the quotient $K[x_1, \ldots, x_n]/I'$ as a vector space over K. [One method: use (a) and Exercise 41.]
   (c) Prove that $I = k[x_1, \ldots, x_n]$ if and only if $I' = K[x_1, \ldots, x_n]$.

44. Let $V = \mathcal{Z}(x^3 - x^2z - y^2z)$ and $W = \mathcal{Z}(x^2 + y^2 - z^2)$ in $\mathbb{C}^3$. Then $\mathcal{I}(V) = (x^3 - x^2z - y^2z)$ and $\mathcal{I}(W) = (x^2 + y^2 - z^2)$ in $\mathbb{C}[x, y, z]$ (cf. Exercise 23 in Section 3). Show that $\varphi((a, b, c)) = (a^2c - b^2c, 2abc, -a^3)$ defines a morphism from V to W.

45. Let $V = \mathcal{Z}(x^3 + y^3 + 7z^3) \subset \mathbb{C}^3$. Then $\mathcal{I}(V) = (x^3 + y^3 + 7z^3)$ in $\mathbb{C}[x, y, z]$ (cf. Exercise 24 in Section 3).
   (a) Show that
$$\widetilde{\varphi}(x) = x(y^3 - 7z^3), \qquad \widetilde{\varphi}(y) = y(7z^3 - x^3), \qquad \widetilde{\varphi}(z) = z(x^3 - y^3)$$
   defines a $\mathbb{C}$-algebra homomorphism from $k[V]$ to itself.
   (b) Let $\varphi : V \to V$ be the morphism corresponding to $\widetilde{\varphi}$. Observe that $(-2, 1, 1) \in V$ and compute $\varphi((-2, 1, 1)) \in V$.
   (c) Prove there are infinitely many points (a, b, c) on V with $a, b, c \in \mathbb{Z}$ and the greatest common divisor of a, b, and c is 1.

46. Let $V = \mathcal{Z}(xz + y^2 + z^2, xy - xz + yz - 2z^2) \subset \mathbb{C}^3$ and $W = \mathcal{Z}(u^3 - uv^2 + v^3) \subset \mathbb{C}^2$ as in Example 2 following Corollary 9. Show that the map $\varphi((a, b)) = (-2a^2 + ab, ab - b^2, a^2 - ab)$ defines a morphism from W to V. Show the corresponding $\mathbb{C}$-algebra homomorphism from $k[V]$ to $k[W]$ has a kernel generated by $x^2 - 3y^2 + yz$.

47. Define $\Phi : \mathbb{Q}[u, v, w] \to \mathbb{Q}[x, y]$ by $\Phi(u) = x^2 + y$, $\Phi(v) = x + y^2$, and $\Phi(w) = x - y$. Show that neither x nor y is in the image of Φ. Show that $f = 2x^3 - 4xy - 2y^3 - 4y$ is in the image of Φ and find a polynomial in $\mathbb{Q}[u, v, w]$ mapping to f. Show that $\ker \Phi$ is the ideal generated by
$$u^2 - 2uv - 2uw^2 + 4uw + v^2 - 2vw^2 - 4vw + w^4 + 3w^2.$$

48. Suppose α is a root of the irreducible polynomial $p(x) \in k[x]$ and $\beta = f(\alpha)/g(\alpha)$ with polynomials $f(x), g(x) \in k[x]$ where $g(\alpha) \neq 0$.

    (a) Show $ag + bp = 1$ for some polynomials $a, b \in k[x]$ and show $\beta = h(\alpha)$ where $h = af$.

    (b) Show that the ideals $(p, y - h)$ and $(p, gy - f)$ are equal in $k[x, y]$.

    (c) Conclude that the minimal polynomial for β is the monic polynomial in $G \cap k[y]$ where G is the reduced Gröbner basis for the ideal $(p, gy - f)$ in $k[x, y]$ for the lexicographic monomial ordering $x > y$.

    (d) Find the minimal polynomial over $\mathbb{Q}$ of $(3 - \sqrt[3]{2} + \sqrt[3]{4})/(1 + 3\sqrt[3]{2} - 3\sqrt[3]{4})$.

15.2 RADICALS AND AFFINE VARIETIES

Since the zeros of a polynomial f are the same as the zeros of the powers $f^2, f^3, \ldots$ in general there are many different ideals in the ring $k[x_1, x_2, \ldots, x_n]$ whose zero locus define the same algebraic set V in affine n-space. This leads to the notion of the radical of an ideal, which can be defined in any commutative ring:

Definition. Let I be an ideal in a commutative ring R.

    (1) The *radical* of I, denoted by $\operatorname{rad} I$, is the collection of elements in R some power of which lie in I, i.e.,

$$\operatorname{rad} I = \{a \in R \mid a^k \in I \text{ for some } k \geq 1\}.$$

    (2) The radical of the zero ideal is called the *nilradical* of R.

    (3) An ideal I is called a *radical* ideal if $I = \operatorname{rad} I$.

Note that $a \in R$ is in the nilradical of R if and only if some power of a is 0, so the nilradical of R is the set of all nilpotent elements of R.

Proposition 11. Let I be an ideal in the commutative ring R. Then $\operatorname{rad} I$ is an ideal containing I, and $(\operatorname{rad} I)/I$ is the nilradical of R/I. In particular, R/I has no nilpotent elements if and only if $I = \operatorname{rad} I$ is a radical ideal.

Proof: It is clear that $I \subseteq \operatorname{rad} I$. By definition, the nilradical of R/I consists of the elements in the quotient some power of which is 0. Under the Lattice Isomorphism Theorem for rings this collection of elements corresponds to the elements of R some power of which lie in I, i.e., $\operatorname{rad} I$. It is therefore sufficient to prove that the nilradical N of any commutative ring R is an ideal. Since $0 \in N$, $N \neq \emptyset$. If $a \in N$ and $r \in R$, then since $a^n = 0$ for some $n \geq 1$, the commutativity of R implies that $(ra)^n = r^n a^n = 0$, so $ra \in N$. It remains to see that if $a, b \in N$ then $a + b \in N$. Suppose $a^n = 0$ and $b^m = 0$. Since the Binomial Theorem holds in the commutative ring R (cf. Exercise 25 in Section 7.3),

$$(a + b)^{n+m} = \sum_{i=0}^{n+m} r_i a^i b^{n+m-i}$$

for some ring elements r_i (the binomial coefficients in R). For each term in this sum either $i \geq n$ (in which case $a^i = 0$) or $n + m - i \geq m$, (in which case $b^{n+m-i} = 0$). Hence $(a + b)^{n+m} = 0$, which shows that $a + b$ is nilpotent, i.e., $a + b \in N$.

Proposition 12. The radical of a proper ideal I is the intersection of all prime ideals containing I. In particular, the nilradical is the intersection of all the prime ideals in R.

Proof: Passing to R/I, Proposition 11 shows that it suffices to prove this result for $I = 0$, and in this case the statement is that the nilradical N of R is the intersection of all the prime ideals in R. Let N' denote the intersection of all the prime ideals in R.

Let a be any nilpotent element in R and let P be any prime ideal. Since $a^k = 0$ for some k, there is a smallest positive power n such that $a^n \in P$. Then the product $a^{n-1}a \in P$, and since P is prime, either $a^{n-1} \in P$ or $a \in P$. The former contradicts the minimality of n, and so $a \in P$. Since P was arbitrary, $a \in N'$, which shows that $N \subseteq N'$.

We prove the reverse containment $N' \subseteq N$ by showing that if $a \notin N$, then $a \notin N'$. If a is an element of R not contained in N, let $\mathcal{S}$ be the family of all proper ideals not containing any positive power of a. The collection $\mathcal{S}$ is not empty since $0 \in \mathcal{S}$. Also, if a^k is not contained in any ideal in the chain $I_1 \subseteq I_2 \subseteq \cdots$, then a^k is also not contained in the union of these ideals, which shows that chains in $\mathcal{S}$ have upper bounds. By Zorn's Lemma, $\mathcal{S}$ has a maximal element, P. The ideal P must in fact be a prime ideal, as follows. Suppose for some x and y not contained in P, the product xy is an element of P. By the maximality of P, $a^n \in (x) + P$ and $a^m \in (y) + P$ for some positive integers n and m. Then $a^{n+m} \in (xy) + P = P$ contradicting the fact that P is an element of $\mathcal{S}$. This shows that P is indeed a prime ideal not containing a, and hence $a \notin N'$, completing the proof.

Note that in Noetherian rings, Theorem 2 can be used to circumvent the appeal to Zorn's Lemma in the preceding proof.

Corollary 13. Prime (and hence also maximal) ideals are radical.

Proof: If P is a prime ideal, then P is clearly the intersection of all the prime ideals containing P, so $P = \operatorname{rad} P$ by the proposition.

Examples

(1) In the ring of integers $\mathbb{Z}$, the ideal (a) is a radical ideal if and only if a is square-free or zero. More generally, if $a = p_1^{a_1} p_2^{a_2} \cdots p_r^{a_r}$ with $a_i \geq 1$ for all i, is the prime factorization of the positive integer a, then $\operatorname{rad}(a) = (p_1 p_2 \cdots p_r)$. For instance, $\operatorname{rad}(180) = (30)$. Note that $(p_1), (p_2), \ldots, (p_r)$ are precisely the prime ideals containing the ideal (a) and that their intersection is the ideal $(p_1 p_2 \cdots p_r)$. More generally, in any U.F.D. R, $\operatorname{rad}(a) = (p_1 p_2 \cdots p_r)$ if $a = p_1^{a_1} p_2^{a_2} \cdots p_r^{a_r}$ is the unique factorization of a into distinct irreducibles.

(2) The ideal $(x^3 - y^2)$ in $k[x, y]$ is a prime ideal (Exercise 14, Section 9.1), hence is radical.

(3) If $l_1, \ldots, l_m$ are linear polynomials in $k[x_1, x_2, \ldots, x_n]$ then $I = (l_1, \ldots, l_m)$ is either $k[x_1, x_2, \ldots, x_n]$ or a prime ideal, hence I is a radical ideal.

Proposition 14. If R is a Noetherian ring then for any ideal I some positive power of $\operatorname{rad} I$ is contained in I. In particular, the nilradical, N, of a Noetherian ring is a nilpotent ideal: $N^k = 0$ for some $k \geq 1$.

Proof: For any ideal I, the ideal rad I is finitely generated since R is Noetherian. If $a_1, \ldots, a_m$ are generators of rad I, then by definition of the radical, for each i we have $a_i^{k_i} \in I$ for some positive integer k_i. Let k be the maximum of all the k_i. Then the ideal $(\text{rad } I)^{km}$ is generated by elements of the form $a_1^{d_1} a_2^{d_2} \cdots a_m^{d_m}$ where $d_1 + \cdots + d_m = km$, and each of these elements has at least one factor $a_i^{d_i}$ with $d_i \geq k$. Then $a_i^{d_i} \in I$, hence each generator of $(\text{rad } I)^{km}$ lies in I, and so $(\text{rad } I)^{km} \subseteq I$.

The Zariski Topology

We saw in the preceding section that if we restrict to the set of ideals I of $k[\mathbb{A}^n]$ arising as the ideals associated with some algebraic set V, i.e., with $I = \mathcal{I}(V)$, then the maps $\mathcal{Z}$ (from such ideals to algebraic sets) and $\mathcal{I}$ (from algebraic sets to ideals) are inverses of each other: $\mathcal{Z}(\mathcal{I}(V)) = V$ and $\mathcal{I}(\mathcal{Z}(I)) = I$. The elements of the ring $k[\mathbb{A}^n]/\mathcal{I}(V)$ give k-valued functions on V and, since k has no nilpotent elements, powers of nonzero functions are also nonzero functions. Put another way, the ring $k[\mathbb{A}^n]/\mathcal{I}(V)$ has no nilpotent elements, so by Proposition 11, the ideal $\mathcal{I}(V)$ is always a radical ideal.

For arbitrary fields k, it is in general not true that every radical ideal is the ideal of some algebraic set, i.e., of the form $\mathcal{I}(V)$ for some algebraic set V. For example, the ideal $(x^2 + 1)$ in $\mathbb{R}[x]$ is maximal, hence is a radical ideal (by Corollary 13), but is not the ideal of any algebraic set — if it were, then $x^2 + 1$ would have to vanish on that set, but $x^2 + 1$ has no zeros in $\mathbb{R}$. A similar construction works for any field k that is not algebraically closed — there exists an irreducible polynomial $p(x)$ of degree at least 2 in $k[x]$, which then generates the maximal (hence radical) ideal $(p(x))$ in $k[x]$ that has no zeros in k. It is perhaps surprising that the presence of polynomials in one variable that have no zeros is the *only* obstruction to a radical ideal (in *any* number of variables) not being of the form $\mathcal{I}(V)$. This is shown by the next theorem, which provides a fundamental connection between "geometry" and "algebra" and shows that over an *algebraically closed* field (such as $\mathbb{C}$) every radical ideal is of the form $\mathcal{I}(V)$. Over these fields the "geometrically defined" ideals $I = \mathcal{I}(V)$ are therefore the same as the radical ideals, which is a "purely algebraic" property of the ideal I (namely that $I = \text{rad } I$).

Theorem. *(Hilbert's Nullstellensatz)* Let E be an algebraically closed field. Then $\mathcal{I}(\mathcal{Z}(I)) = \text{rad } I$ for every ideal I of $E[x_1, x_2, \ldots, x_n]$. Moreover, the maps $\mathcal{Z}$ and $\mathcal{I}$ in the correspondence

$$\{\text{affine algebraic sets}\} \quad \underset{\mathcal{Z}}{\overset{\mathcal{I}}{\rightleftarrows}} \quad \{\text{radical ideals}\}$$

are bijections that are inverses of each other.

Proof: This will be proved in the next section (cf. Theorem 32).

Example

The maps $\mathcal{I}$ and $\mathcal{Z}$ in the Nullstellensatz are defined over any field k, and as mentioned are not bijections if k is not algebraically closed. For any field k, however, the map $\mathcal{Z}$ is always surjective and the map $\mathcal{I}$ is always injective (cf. Exercise 9).

One particular consequence of the Nullstellensatz is that for any *proper* ideal I we have $\mathcal{Z}(I) \neq \emptyset$ since rad $I \neq k[\mathbb{A}^n]$. Hence there always exists at least one common zero ("nullstellen" in German) for all the polynomials contained in a proper ideal (over an algebraically closed field).

We next see that the affine algebraic sets define a topology on affine n-space. Recall that a *topological space* is any set X together with a collection of subsets $\mathcal{T}$ of X, called the *closed sets* in X, satisfying the following axioms:

(i) an arbitrary intersection of closed sets is closed: if $S_i \in \mathcal{T}$ for i in any index set, then $\cap S_i \in \mathcal{T}$,

(ii) a finite union of closed sets is closed: if $S_1, \ldots, S_q \in \mathcal{T}$ then $S_1 \cup \cdots \cup S_q \in \mathcal{T}$, and

(iii) the empty set and the whole space are closed: $\emptyset, X \in \mathcal{T}$.

A subset U of X is called *open* if its complement, $X - U$, is closed (i.e., $X - U \in \mathcal{T}$). The axioms for a topological space are often (equivalently) phrased in terms of the collection of open sets in X.

There are many examples of topological spaces, and a wealth of books on topology. A fixed set X may have a number of different topologies on it, and the collections of closed sets need not be related in these different structures. On any set X there are always at least two topologies: the so-called discrete topology in which every subset of X is closed (i.e., $\mathcal{T}$ is the collection of *all* subsets of X), and the so-called trivial topology in which the only closed sets are $\emptyset$ and X required by axiom (iii).

Suppose now that $X = \mathbb{A}^n$ is affine n-space over an arbitrary field k. Then the collection $\mathcal{T}$ consisting of all the affine algebraic sets in $\mathbb{A}^n$ satisfies the three axioms for a topological space — these are precisely properties (3), (4) and (5) of algebraic sets in the preceding section. It follows that these sets can be taken to be the closed sets in a topology on $\mathbb{A}^n$:

Definition. The *Zariski topology* on affine n-space over an arbitrary field k is the topology in which the closed sets are the affine algebraic sets in $\mathbb{A}^n$.

The Zariski topology is quite "coarse" in the sense that there are "relatively few" closed (or open) sets. For example, for the Zariski topology on $\mathbb{A}^1$ the only closed sets are $\emptyset$, k and the finite sets (cf. Exercise 14 in Section 1), and so the nonempty open sets are the complements of finite sets. If k is an infinite field it follows that in the Zariski topology any two nonempty open sets in $\mathbb{A}^1$ have nonempty intersection. In the language of point-set topology, the Zariski topology is always T_1 (points are closed sets), but for infinite fields the Zariski topology is never T_2 (Hausdorff), i.e., two distinct points never belong to two disjoint open sets (cf. the exercises). For example, when $k = \mathbb{R}$, a nonempty Zariski open set is just the real line $\mathbb{R}$ with some finite number of points removed, and any two such sets have (infinitely many) points in common. Note also that the Zariski open (respectively, closed) sets in $\mathbb{R}$ are also open (respectively, closed) sets with respect to the usual Euclidean topology. The converse is not true; for example the interval [0,1] is closed in the Euclidean topology but is not closed in the Zariski topology. In this sense the Euclidean topology on $\mathbb{R}$ is much "finer"; there are many more open sets in the Euclidean topology, in fact the collection of Euclidean open (respectively, closed) sets properly contains the collection of Zariski open (respectively, closed) sets.

The Zariski topology on $\mathbb{A}^n$ is defined so that the affine algebraic subsets of $\mathbb{A}^n$ are closed. In other words, the topology is defined by the zero sets of the ideals in the coordinate ring of $\mathbb{A}^n$. A similar definition can be used to define a Zariski topology on *any* algebraic set V in $\mathbb{A}^n$, as follows. If $k[V]$ is the coordinate ring of V, then the distinct elements of $k[V]$ define distinct k-valued functions on V and there is a natural way of defining

$$\mathcal{Z} : \{\text{ideals in } k[V]\} \longrightarrow \{\text{algebraic subsets of } V\}$$
$$\mathcal{I} : \{\text{subsets of } V\} \longrightarrow \{\text{ideals in } k[V]\}$$

just as for the case $V = \mathbb{A}^n$. For example, if $\overline{J}$ is an ideal in $k[V]$, then $\mathcal{Z}(\overline{J})$ is the set of elements in V that are common zeros of all the functions in the ideal $\overline{J}$. It is easy to verify that the resulting zero sets in V satisfy the three axioms for a topological space, defining a *Zariski topology on* V, where the closed sets are the algebraic subsets, $\mathcal{Z}(\overline{J})$, for any ideal $\overline{J}$ of $k[V]$. By the Lattice Isomorphism Theorem, the ideals of $k[V]$ are the ideals of $k[x_1, \ldots, x_n]$ that contain $\mathcal{I}(V)$ taken mod $\mathcal{I}(V)$. If J is the complete preimage in $k[x_1, \ldots, x_n]$ of $\overline{J}$, then the locus of J in $\mathbb{A}^n$ is the same as the locus of $\overline{J}$ in V. It follows that this definition of the Zariski topology on V is just the *subspace topology* for $V \subseteq \mathbb{A}^n$. (Recall that in a topological space X, the closed sets with respect to the subspace topology of a subspace Y are defined to be the sets $C \cap Y$, where C is a closed set in X.) The advantage to the definition of the Zariski topology on V above is that it is defined intrinsically in terms of the coordinate ring $k[V]$ of V, and since the isomorphism type of $k[V]$ does not depend on the affine space $\mathbb{A}^n$ containing V, the Zariski topology on V also depends only on V and not on the ambient affine space in which V may be embedded.

If V and W are two affine algebraic spaces, then since a morphism $\varphi : V \to W$ is defined by polynomial functions, it is easy to see that φ is *continuous* with respect to the Zariski topologies on V and W (cf. Exercise 27 in Section 1, which shows that the inverse image of a Zariski closed set under a morphism is Zariski closed). In fact the Zariski topology is the coarsest topology in which points are closed and for which polynomial maps are continuous. There exist maps that are continuous with respect to the Zariski topology that are not morphisms, however (cf. Exercise 17).

We have the usual topological notions of closure and density with respect to the Zariski topology.

Definition. For any subset A of $\mathbb{A}^n$, the *Zariski closure* of A is the smallest algebraic set containing A. If $A \subseteq V$ for an algebraic set V then A is *Zariski dense* in V if the Zariski closure of A is V.

For example, if $k = \mathbb{R}$, the algebraic sets in $\mathbb{A}^1$ are $\emptyset$, $\mathbb{R}$, and finite subsets of $\mathbb{R}$ by Exercise 14 in Section 1. The Zariski closure of any infinite set A of real numbers is then all of $\mathbb{A}^1$ and A is Zariski dense in $\mathbb{A}^1$.

Proposition 15. The Zariski closure of a subset A in $\mathbb{A}^n$ is $\mathcal{Z}(\mathcal{I}(A))$.

Proof: Certainly $A \subseteq \mathcal{Z}(\mathcal{I}(A))$. Suppose V is any algebraic set containing A: $A \subseteq V$. Then $\mathcal{I}(V) \subseteq \mathcal{I}(A)$ and $\mathcal{Z}(\mathcal{I}(A)) \subseteq \mathcal{Z}(\mathcal{I}(V)) = V$, so $\mathcal{Z}(\mathcal{I}(A))$ is the smallest algebraic set containing A.

If $\varphi : V \to W$ is a morphism of algebraic sets, the image $\varphi(V)$ of V need not be an algebraic subset of W, i.e., need not be Zariski closed in W. For example the projection of the hyperbola $V = \mathcal{Z}(xy - 1)$ in $\mathbb{R}^2$ onto the x-axis has image $\mathbb{R}^1 - \{0\}$, which as we have just seen is not an affine algebraic set.

The next result shows that the Zariski closure of the image of a morphism is determined by the kernel of the associated k-algebra homomorphism.

Proposition 16. Suppose $\varphi : V \to W$ is a morphism of algebraic sets and $\widetilde{\varphi} : k[W] \to k[V]$ is the associated k-algebra homomorphism of coordinate rings. Then
  (1) The kernel of $\widetilde{\varphi}$ is $\mathcal{I}(\varphi(V))$.
  (2) The Zariski closure of $\varphi(V)$ is the zero set in W of $\ker \widetilde{\varphi}$. In particular, the homomorphism $\widetilde{\varphi}$ is injective if and only if $\varphi(V)$ is Zariski dense in W.

Proof: Since $\widetilde{\varphi} = f \circ \varphi$, we have $\widetilde{\varphi}(f) = 0$ if and only if $(f \circ \varphi)(P) = 0$ for all $P \in V$, i.e., $f(Q) = 0$ for all $Q = \varphi(P) \in \varphi(V)$, which is the statement that $f \in \mathcal{I}(\varphi(V))$, proving the first statement. Since the Zariski closure of $\varphi(V)$ is the zero set of $\mathcal{I}(\varphi(V))$ by the previous proposition, the first statement in (2) follows.

If $\widetilde{\varphi}$ is injective then the Zariski closure of $\varphi(V)$ is $\mathcal{Z}(0) = W$ and so $\varphi(V)$ is Zariski dense. Conversely, suppose $\varphi(V)$ is Zariski dense in W, i.e., $\mathcal{Z}(\mathcal{I}(\varphi(V))) = W$. Then $\mathcal{I}(\varphi(V)) = \mathcal{I}(\mathcal{Z}(\mathcal{I}(\varphi(V)))) = \mathcal{I}(W) = 0$ and so $\ker \widetilde{\varphi} = 0$.

By Proposition 16 the ideal of polynomials defining the Zariski closure of the image of a morphism φ is the kernel of the corresponding k-algebra homomorphism $\widetilde{\varphi}$ in Theorem 6. Proposition 8(1) allows us to compute this kernel using Gröbner bases.

Example: (Implicitization)

A morphism $\varphi : \mathbb{A}^n \to \mathbb{A}^m$ is just a map

$$\varphi((a_1, a_2, \ldots, a_n)) = (\varphi_1(a_1, a_2, \ldots, a_n), \ldots, \varphi_m(a_1, a_2, \ldots, a_n))$$

where φ_i is a polynomial. If k is an infinite field, then $\mathcal{I}(\mathbb{A}^m)$ and $\mathcal{I}(\mathbb{A}^n)$ are both 0, so we may write $k[\mathbb{A}^m] = k[y_1, \ldots, y_m]$ and $k[\mathbb{A}^n] = k[x_1, \ldots, x_n]$. The k-algebra homomorphism $\widetilde{\varphi} : k[\mathbb{A}^m] \to k[\mathbb{A}^n]$ corresponding to φ is then defined by mapping y_i to $\varphi_i = \varphi_i(x_1, \ldots, x_n)$. The image $\varphi(\mathbb{A}^n)$ consists of the set of points $(b_1, \ldots, b_m)$ with

$$b_1 = \varphi_1(a_1, a_2, \ldots, a_n)$$
$$b_2 = \varphi_2(a_1, a_2, \ldots, a_n)$$
$$\vdots$$
$$b_m = \varphi_m(a_1, a_2, \ldots, a_n)$$

where $a_i \in k$. This is the collection of points in $\mathbb{A}^m$ *parametrized* by the functions $\varphi_1, \ldots, \varphi_m$ (with the a_i as parameters). In general such a parametrized collection of points is not an algebraic set. Finding the equations for the smallest algebraic set containing these points is referred to as *implicitization*, since it amounts to finding a ('smallest') collection of equations satisfied by the b_i (the 'implicit' algebraic relations).

By Proposition 16, this algebraic set is the Zariski closure of $\varphi(\mathbb{A}^n)$ and is the zero set of $\ker \widetilde{\varphi}$. By Proposition 8 this kernel is given by $\mathcal{A} \cap k[y_1, \ldots, y_m]$, where $\mathcal{A}$ is the ideal in $k[x_1, \ldots, x_n, y_1, \ldots, y_m]$ generated by the polynomials $y_1 - \varphi_1, \ldots, y_m - \varphi_m$. If we compute the reduced Gröbner basis G for $\mathcal{A}$ with respect to the lexicographic monomial ordering $x_1 > \cdots > x_n > y_1 > \cdots > y_m$, then the polynomials of G lying in $k[y_1, \ldots, y_m]$ generate $\ker \tilde{\pi}$. The zero set of these polynomials defines the Zariski closure of $\varphi(\mathbb{A}^n)$ and therefore give the implicitization.

For an explicit example, consider the points $A = \{(a^2, a^3) \mid a \in \mathbb{R}\}$ in $\mathbb{R}^2$. Using coordinates x, y for $\mathbb{R}^2$ and t for $\mathbb{R}^1$, the ideal $\mathcal{A}$ in $\mathbb{R}[x, y, z, t]$ is $(x - t^2, y - t^3)$. The only element of the reduced Gröbner basis for $\mathcal{A}$ for the ordering $t > x > y$ lying in $\mathbb{R}[x, y]$ is $x^3 - y^2$, so $\mathcal{Z}(x^3 - y^2)$ is the smallest algebraic set in $\mathbb{R}^2$ containing A.

Example: (Projections of Algebraic Sets)

Suppose $V \subseteq \mathbb{A}^n$ is an algebraic set and $m < n$. Let $\pi : V \to \mathbb{A}^m$ be the morphism projecting onto the first m coordinates:

$$\pi((a_1, a_2, \ldots, a_n)) = (a_1, a_2, \ldots, a_m).$$

If we use coordinates $x_1, \ldots, x_n$ in $k[V]$ and coordinates $y_1, \ldots, y_m$ in $k[\mathbb{A}^m]$, the k-algebra homomorphism corresponding to π is given by the map

$$\tilde{\pi} : k[y_1, \ldots, y_m] \longrightarrow k[x_1, \ldots, x_n]/\mathcal{I}(V)$$
$$y_i \longmapsto x_i.$$

Suppose $V = \mathcal{Z}(I)$ and $I = (f_1, \ldots, f_s)$. The Zariski closure of $\pi(V)$ is the zero set of $\ker \tilde{\pi} = \mathcal{A} \cap k[y_1, \ldots, y_m]$ where $\mathcal{A}$ is the ideal in $k[x_1, \ldots, x_n, y_1, \ldots, y_m]$ generated by the polynomials $y_1 - x_1, \ldots, y_m - x_m$ together with a set of generators for $\mathcal{I}(V)$. The polynomials involving only $y_1, \ldots, y_m$ in the reduced Gröbner basis G for $\mathcal{A}$ with respect to the lexicographic monomial ordering $x_1 > \cdots > x_n > y_1 > \cdots > y_m$ are generators for the Zariski closure of $\pi(V)$.

If k is algebraically closed we can actually do better with the help of the Nullstellensatz, which gives $\mathcal{I}(V) = \mathrm{rad}\, I$. Then it is straightforward to see that we obtain the same zero set if in the ideal $\mathcal{A}$ we replace the generators for $\mathcal{I}(V)$ by the generators $f_1, \ldots, f_s$ of I (cf. Exercise 46).

For an explicit example, consider projection onto the first two coordinates of $V = \mathcal{Z}(xy - z^2, xz - y, x^2 - z)$ in $\mathbb{C}^3$. Using u, v as coordinates in $\mathbb{C}^2$, we find the reduced Gröbner basis G for the ideal $(u - x, v - y, xy - z^2, xz - y, x^2 - z)$ for the ordering $x > y > z > u > v$ contains only the polynomial $u^3 - v$ in $\mathbb{C}[u, v]$. The smallest algebraic set containing $\pi(V)$ is then the cubic $v = u^3$.

Affine Varieties

We next consider the question of whether an algebraic set can be decomposed into smaller algebraic sets and the corresponding algebraic formulation in terms of its coordinate ring.

Definition. A nonempty affine algebraic set V is called *irreducible* if it cannot be written as $V = V_1 \cup V_2$, where V_1 and V_2 are proper algebraic sets in V. An irreducible affine algebraic set is called an affine *variety*.

Equivalently, an algebraic set (which is a closed set in the Zariski topology) is irreducible if it cannot be written as the union of two proper, closed subsets.

Proposition 17.

(1) The affine algebraic set V is irreducible if and only if $\mathcal{I}(V)$ is a prime ideal.
(2) Every nonempty affine algebraic set V may be written uniquely in the form
$$V = V_1 \cup V_2 \cup \cdots \cup V_q$$
where each V_i is irreducible, and $V_i \not\subseteq V_j$ for all $j \neq i$ (i.e., the decomposition is "minimal" or "irredundant").

Proof: Let $I = \mathcal{I}(V)$ and suppose first that $V = V_1 \cup V_2$ is reducible, where V_1 and V_2 are proper closed subsets. Since $V_1 \neq V$, there is some function f_1 that vanishes on V_1 but not on V, i.e., $f_1 \in \mathcal{I}(V_1) - I$. Similarly, there is a function $f_2 \in \mathcal{I}(V_2) - I$. Then $f_1 f_2$ vanishes on $V_1 \cup V_2 = V$, so $f_1 f_2 \in I$ which shows that I is not a prime ideal. Conversely, if I is not a prime ideal, there exists $f_1, f_2 \in k[\mathbb{A}^n]$ such that $f_1 f_2 \in I$ but neither f_1 nor f_2 belongs to I. Let $V_1 = \mathcal{Z}(f_1) \cap V$ and $V_2 = \mathcal{Z}(f_2) \cap V$. Since the intersection of closed sets is closed, V_1 and V_2 are algebraic sets. Since neither f_1 nor f_2 vanishes on V, both V_1 and V_2 are proper subsets of V. Because $f_1 f_2 \in I$, $V \subseteq \mathcal{Z}(f_1 f_2) = \mathcal{Z}(f_1) \cup \mathcal{Z}(f_2)$, and so V is reducible. This proves (1).

To prove (2), let $\mathcal{S}$ be the collection of nonempty algebraic sets that cannot be written as a finite union of irreducible algebraic sets, and suppose by way of contradiction that $\mathcal{S} \neq \emptyset$. Let I_0 be a maximal element of the corresponding set of ideals, $\{\mathcal{I}(V) \mid V \in \mathcal{S}\}$, which exists (by Theorem 2) since $k[\mathbb{A}^n]$ is Noetherian. Then $V_0 = \mathcal{Z}(I_0)$ is a *minimal* element of $\mathcal{S}$. Since $V_0 \in \mathcal{S}$, it cannot be irreducible by the definition of $\mathcal{S}$. On the other hand, if $V_0 = V_1 \cup V_2$ for some proper, closed subsets V_1, V_2 of V_0, then by the minimality of V_0 both V_1 and V_2 may be written as finite unions of irreducible algebraic sets. Then V_0 may be written as a finite union of irreducible algebraic sets, a contradiction. This proves $\mathcal{S} = \emptyset$, i.e., every affine algebraic set has a decomposition into affine varieties.

To prove uniqueness, suppose V has two decompositions into affine varieties (where redundant terms have been removed from each decomposition):
$$V = V_1 \cup V_2 \cup \cdots \cup V_r = U_1 \cup U_2 \cup \cdots \cup U_s.$$
Then V_1 is contained in the union of the U_i. Since $V_1 \cap U_i$ is an algebraic set for each i, we obtain a decomposition of V_1 into algebraic subsets:
$$V_1 = (V_1 \cap U_1) \cup (V_1 \cap U_2) \cup \cdots \cup (V_1 \cap U_s).$$
Since V_1 is irreducible, we must have $V_1 = V_1 \cap U_j$ for some j, i.e., $V_1 \subseteq U_j$. By the symmetric argument we have $U_j \subseteq V_{j'}$ for some j'. Thus $V_1 \subseteq V_{j'}$, so $j' = 1$ and $V_1 = U_j$. Applying a similar argument for each V_i it follows that $r = s$ and that $\{V_1, \ldots, V_r\} = \{U_1, \ldots, U_s\}$. This completes the proof.

Corollary 18. An affine algebraic set V is a variety if and only if its coordinate ring $k[V]$ is an integral domain.

Proof: This follows immediately since $\mathcal{I}(V)$ is a prime ideal if and only if the quotient $k[V] = k[\mathbb{A}^n]/\mathcal{I}(V)$ is an integral domain (Proposition 13 of Chapter 7).

Definition. If V is a variety, then the field of fractions of the integral domain $k[V]$ is called the field of *rational functions* on V and is denoted by $k(V)$. The *dimension* of a variety V, denoted dim V, is defined to be the transcendence degree of $k(V)$ over k.

Examples

(1) Single points in $\mathbb{A}^n$ are affine varieties since their corresponding ideals in $k[\mathbb{A}^n]$ are maximal ideals. The coordinate ring of a point is isomorphic to k, which is also the field of rational functions. The dimension of a single point is 0. Any finite set is the union of its single point subsets, and this is its unique decomposition into affine subvarieties.

(2) The x-axis in $\mathbb{R}^2$ is irreducible since it has coordinate ring $\mathbb{R}[x, y]/(y) \cong \mathbb{R}[x]$, which is an integral domain. Similarly, the y-axis and, more generally, lines in $\mathbb{R}^2$ are also irreducible (cf. Exercise 23 in Section 1). Linear sets in $\mathbb{R}^n$ are affine varieties. The field of rational functions on the x-axis is the quotient field $\mathbb{R}(x)$ of $\mathbb{R}[x]$, which is why $\mathbb{R}(x)$ is called a rational function field. The dimension of the x-axis (or, more generally, any line) is 1.

(3) The union of the x and y axes in $\mathbb{R}^2$, namely $\mathcal{Z}(xy)$, is not a variety: $\mathcal{Z}(xy) = \mathcal{Z}(x) \cup \mathcal{Z}(y)$ is its unique decomposition into subvarieties. The corresponding coordinate ring $\mathbb{R}[x, y]/(xy)$ contains zero divisors.

(4) The hyperbola $xy = 1$ in $\mathbb{R}^2$ is a variety since we saw in Section 1 that its coordinate ring is the integral domain $\mathbb{R}[x, 1/x]$. Note that the two disjoint branches of the hyperbola (defined by $x > 0$ and $x < 0$) are not subvarieties (cf. also Exercises 12–13).

(5) If $V = \mathcal{Z}(l_1, l_2, \ldots, l_m)$ is the zero set of *linear* polynomials $l_1, \ldots, l_m$ in $k[x_1, \ldots, x_m]$ and $V \neq \emptyset$, then V is an affine variety (called a *linear variety*). Note that determining whether $V \neq \emptyset$ is a linear algebra problem.

We end this section with some general ring-theoretic results that were originally motivated by their connection with decomposition questions in geometry.

Primary Decomposition of Ideals in Noetherian Rings

The second statement in Proposition 17 shows that any ideal of the form $\mathcal{I}(V)$ in $k[\mathbb{A}^n]$ may be written uniquely as a finite intersection of prime ideals, and by Hilbert's Nullstellensatz this applies in particular to all radical ideals when k is algebraically closed. In a large class of commutative rings (including all Noetherian rings) every ideal has a *primary decomposition*, which is a similar decomposition but allows ideals that are analogous to "prime powers" (but see the examples below). This decomposition can be considered as a generalization of the factorization of an integer $n \in \mathbb{Z}$ into the product of prime powers. We shall be primarily concerned with the case of Noetherian rings.

Definition. A proper ideal Q in the commutative ring R is called *primary* if whenever $ab \in Q$ and $a \notin Q$, then $b^n \in Q$ for some positive integer n. Equivalently, if $ab \in Q$ and $a \notin Q$, then $b \in \operatorname{rad} Q$.

Some of the basic properties of primary ideals are given in the following proposition.

Proposition 19. Let R be a commutative ring with 1.
   (1) Prime ideals are primary.
   (2) The ideal Q is primary if and only if every zero divisor in R/Q is nilpotent.
   (3) If Q is primary then rad Q is a prime ideal, and is the unique smallest prime ideal containing Q.
   (4) If Q is an ideal whose radical is a maximal ideal, then Q is a primary ideal.
   (5) Suppose M is a maximal ideal and Q is an ideal with $M^n \subseteq Q \subseteq M$ for some $n \geq 1$. Then Q is a primary ideal with rad $Q = M$.

Proof: The first two statements are immediate from the definition of a primary ideal. For (3), suppose $ab \in$ rad Q. Then $a^m b^m = (ab)^m \in Q$, and since Q is primary, either $a^m \in Q$, in which case $a \in$ rad Q, or $(b^m)^n \in Q$ for some positive integer n, in which case $b \in$ rad Q. This proves that rad Q is a prime ideal, and it follows that rad Q is the smallest prime ideal containing Q (Proposition 12).

To prove (4) we pass to the quotient ring R/Q; by (2), it suffices to show that every zero divisor in this quotient ring is nilpotent. We are reduced to the situation where $Q = (0)$ and $M =$ rad $Q =$ rad(0), which is the nilradical, is a maximal ideal. Since the nilradical is contained in every prime ideal (Proposition 12), it follows that M is the unique prime ideal, so also the unique maximal ideal. If d were a zero divisor, then the ideal (d) would be a proper ideal, hence contained in a maximal ideal. This implies that $d \in M$, hence every zero divisor is indeed nilpotent.

Finally, suppose $M^n \subseteq Q \subseteq M$ for some $n \geq 1$ where M is a maximal ideal. Then $Q \subseteq M$ so rad $Q \subseteq$ rad $M = M$. Conversely, $M^n \subseteq Q$ shows that $M \subseteq$ rad Q, so rad $Q = M$ is a maximal ideal, and Q is primary by (4).

Definition. If Q is a primary ideal, then the prime ideal $P =$ rad Q is called the *associated prime* to Q, and Q is said to *belong* to P (or to be *P-primary*).

It is easy to check that a finite intersection of P-primary ideals is again a P-primary ideal (cf. the exercises).

Examples
   (1) The primary ideals in $\mathbb{Z}$ are 0 and the ideals (p^m) for p a prime and $m \geq 1$.
   (2) For any field k, the ideal (x) in $k[x, y]$ is primary since it is a prime ideal. For any $n \geq 1$, the ideal $(x, y)^n$ is primary since it is a power of the maximal ideal (x, y).
   (3) The ideal $Q = (x^2, y)$ in the polynomial ring $k[x, y]$ is primary since we have $(x, y)^2 \subseteq (x^2, y) \subseteq (x, y)$. Similarly, $Q' = (4, x)$ in $\mathbb{Z}[x]$ is a $(2, x)$-primary ideal.
   (4) Primary ideals need not be powers of prime ideals. For example, the primary ideal Q in the previous example is not the power of a prime ideal, as follows. If $(x^2, y) = P^k$ for some prime ideal P and some $k \geq 1$, then $x^2, y \in P^k \subseteq P$ so $x, y \in P$. Then $P = (x, y)$, and since $y \notin (x, y)^2$, it would follow that $k = 1$ and $Q = (x, y)$. Since $x \notin (x^2, y)$, this is impossible.
   (5) If R is Noetherian, and Q is a primary ideal belonging to the prime ideal P, then
$$P^m \subseteq Q \subseteq P$$
for some $m \geq 1$ by Proposition 14. If P is a maximal ideal, then the last statement in Proposition 19 shows that the converse also holds. This is not necessarily true if P

is a prime ideal that is *not maximal*. For example, consider the ideal $I = (x^2, xy)$ in $k[x, y]$. Then $(x^2) \subset I \subset (x)$, and (x) is a prime ideal, but I is not primary: $xy \in I$ and $x \notin I$, but no positive power of y is an element of I. This example also shows that an ideal whose radical is prime (but not maximal as in (4) of the proposition) is not necessarily primary.

(6) Powers of prime ideals need not be primary. For example, consider the quotient ring $R = \mathbb{R}[x, y, z]/(xy - z^2)$, the coordinate ring of the cone $z^2 = xy$ in $\mathbb{R}^3$, and let $P = (\bar{x}, \bar{z})$ be the ideal generated by $\bar{x}$ and $\bar{z}$ in R. This is a prime ideal in R since the quotient is $R/(\bar{x}, \bar{z}) \cong \mathbb{R}[x, y, z]/(x, z) \cong \mathbb{R}[y]$ (because $(xy - z^2) \subset (x, z)$). The ideal

$$P^2 = (\bar{x}^2, \bar{x}\bar{z}, \bar{z}^2) = (\bar{x}^2, \bar{x}\bar{z}, \bar{x}\bar{y}) = \bar{x}(\bar{x}, \bar{y}, \bar{z}),$$

however, is not primary: $\bar{x}\bar{y} = \bar{z}^2 \in P^2$, but $\bar{x} \notin P^2$, and no power of $\bar{y}$ is in P^2. Note that P^2 is another example of an ideal that is not primary whose radical is prime.

(7) Suppose R is a U.F.D. If π is an irreducible element of R then it is easy to see that the powers (π^n) for $n = 1, 2, \ldots$ are (π)-primary ideals. Conversely, suppose Q is a (π)-primary ideal, and let n be the largest integer with $Q \subseteq (\pi^n)$ (such an integer exists since, for example, $\pi^k \in Q$ for some $k \geq 1$, so $n \leq k$). If q is an element of Q not contained in (π^{n+1}), then $q = r\pi^n$ for some $r \in R$ and $r \notin (\pi)$. Since $r \notin (\pi)$ and Q is (π)-primary, it follows that $\pi^n \in Q$. This shows that $Q = (\pi^n)$.

In the examples above, the ideal (x^2, xy) in $k[x, y]$ is not a primary ideal, but it can be written as the intersection of primary ideals: $(x^2, xy) = (x) \cap (x, y)^2$.

Definition.
  (1) An ideal I in R has a *primary decomposition* if it may be written as a finite intersection of primary ideals:

  $$I = \bigcap_{i=1}^{m} Q_i \qquad Q_i \text{ a primary ideal.}$$

  (2) The primary decomposition above is *minimal* and the Q_i are called the *primary components of I* if
      (a) no primary ideal contains the intersection of the remaining primary ideals, i.e., $Q_i \not\supseteq \cap_{j \neq i} Q_j$ for all i, and
      (b) the associated prime ideals are all distinct: $\text{rad } Q_i \neq \text{rad } Q_j$ for $i \neq j$.

We now prove that in a Noetherian ring every proper ideal has a minimal primary decomposition. This result is often called the Lasker–Noether Decomposition Theorem, since it was first proved for polynomial rings by the chess master Emanuel Lasker and the proof was later greatly simplified and generalized by Emmy Noether.

Definition. A proper ideal I in the commutative ring R is said to be *irreducible* if I cannot be written nontrivially as the intersection of two other ideals, i.e., if $I = J \cap K$ with ideals J, K implies that $I = J$ or $I = K$.

It is easy to see that a prime ideal is irreducible (see Exercise 11 in Section 7.4). The ideal $(x, y)^2$ in $k[x, y]$ in Example 2 earlier shows that primary ideals need not be irreducible since it is the intersection of the ideals $(x) + (x, y)^2 = (x, y^2)$ and $(y)+(x, y)^2 = (y, x^2)$. In a Noetherian ring, however, irreducible ideals are necessarily primary:

Proposition 20. Let R be a Noetherian ring. Then
(1) every irreducible ideal is primary, and
(2) every proper ideal in R is a finite intersection of irreducible ideals.

Proof: To prove (1) let Q be an irreducible ideal and suppose that $ab \in Q$ and $b \notin Q$. It is easy to check that for any fixed n the set of elements $x \in R$ with $a^n x \in Q$ is an ideal, A_n, in R. Clearly $A_1 \subseteq A_2 \subseteq \ldots$ and since R is Noetherian this ascending chain of ideals must stabilize, i.e., $A_n = A_{n+1} = \ldots$ for some $n > 0$. Consider the two ideals $I = (a^n) + Q$ and $J = (b) + Q$ of R, each containing Q. If $y \in I \cap J$ then $y = a^n z + q$ for some $z \in R$ and $q \in Q$. Since $ab \in Q$, it follows that $aJ \subseteq Q$, and in particular $ay \in Q$. Then $a^{n+1}z = ay - aq \in Q$, so $z \in A_{n+1} = A_n$. But $z \in A_n$ means that $a^n z \in Q$, so $y \in Q$. It follows that $I \cap J = Q$. Since Q is irreducible and $(b) + Q \neq Q$ (since $b \notin Q$), we must have $a^n \in Q$, which shows that Q is primary.

The proof of (2) is the same as the proof of the second statement in Proposition 17. Let $\mathcal{S}$ be the collection of ideals of R that cannot be written as a finite intersection of irreducible ideals. If $\mathcal{S}$ is not empty, then since R is Noetherian, there is a maximal element I in $\mathcal{S}$. Then I is not itself irreducible, so $I = J \cap K$ for some ideals J and K distinct from I. Then $I \subset J$ and $I \subset K$ and the maximality of I implies that neither J nor K is in $\mathcal{S}$. But this means that both J and K can be written as finite intersections of irreducible ideals, hence the same would be true for I. This is a contradiction, so $\mathcal{S} = \emptyset$, which completes the proof of the proposition.

It is immediate from the previous proposition that in a Noetherian ring every proper ideal has a primary decomposition. If any of the primary ideals in this decomposition contains the intersection of the remaining primary ideals, then we may simply remove this ideal since this will not change the intersection. Hence we may assume the decomposition satisfies (a) in the definition of a minimal decomposition. Since a finite intersection of P-primary ideals is again P-primary (Exercise 31), replacing the primary ideals in the decomposition with the intersections of all those primary ideals belonging to the same prime, we may also assume the decomposition satisfies (b) in the definition of a minimal decomposition. This proves the first statement of the following:

Theorem 21. *(Primary Decomposition Theorem)* Let R be a Noetherian ring. Then every proper ideal I in R has a minimal primary decomposition. If

$$I = \bigcap_{i=1}^{m} Q_i = \bigcap_{i=1}^{n} Q_i'$$

are two minimal primary decompositions for I then the sets of associated primes in the two decompositions are the same:

$$\{ \operatorname{rad} Q_1, \operatorname{rad} Q_2, \ldots, \operatorname{rad} Q_m \} = \{ \operatorname{rad} Q_1', \operatorname{rad} Q_2', \ldots, \operatorname{rad} Q_n' \}.$$

Moreover, the primary components Q_i belonging to the minimal elements in this set of associated primes are uniquely determined by I.

Proof: The proof of the uniqueness of the set of associated primes is outlined in the exercises, and the proof of the uniqueness of the primary components associated to the minimal primes will be given in Section 4.

Definition. If I is an ideal in the Noetherian ring R then the associated prime ideals in any primary decomposition of I are called the *associated prime ideals of I*. If an associated prime ideal P of I does not contain any other associated prime ideal of I then P is called an *isolated prime ideal*; the remaining associated prime ideals of I are called *embedded prime ideals*.

The prime ideals associated to an ideal I provide a great deal of information about the ideal I (cf. for example Exercises 41 and 43):

Corollary 22. Let I be a proper ideal in the Noetherian ring R.
(1) A prime ideal P contains the ideal I if and only if P contains one of the associated primes of I, hence if and only if P contains one of the isolated primes of I, i.e., the isolated primes of I are precisely the minimal elements in the set of all prime ideals containing I. In particular, there are only finitely many minimal elements among the prime ideals containing I.
(2) The radical of I is the intersection of the associated primes of I, hence also the intersection of the isolated primes of I.
(3) There are prime ideals $P_1, \ldots, P_n$ (not necessarily distinct) containing I such that $P_1 P_2 \cdots P_n \subseteq I$.

Proof: The first statement in (1) is an exercise (cf. Exercise 37), and the remainder of (1) follows. Then (2) follows from (1) and Proposition 12, and (3) follows from (2) and Proposition 14.

The last statement in Theorem 21 states that not only the isolated primes, but also the primary components belonging to the isolated primes, are uniquely determined by I. In general the primary decomposition of an ideal I is itself not unique.

Examples

(1) Let $I = (x^2, xy)$ in $\mathbb{R}[x, y]$. Then
$$(x^2, xy) = (x) \cap (x, y)^2 = (x) \cap (x^2, y)$$

are two minimal primary decompositions for I. The associated primes for I are (x) and $\mathrm{rad}((x, y)^2) = \mathrm{rad}((x^2, y)) = (x, y)$. The prime (x) is the only isolated prime since $(x) \subset (x, y)$, and (x, y) is an embedded prime. A prime ideal P contains I if and only if P contains (x). The (x)-primary component of I corresponding to this isolated prime is just (x) and occurs in both primary decompositions; the (x, y)-primary component of I corresponding to this embedded prime is not uniquely determined — it is $(x, y)^2$ in the first decomposition and is (x^2, y) in the second. The radical of I is the isolated prime (x).

This example illustrates the origin of the terminology: in general the irreducible components of the algebraic space $\mathcal{Z}(I)$ defined by I are the zero sets of the isolated primes for I, and the zero sets of the embedded primes are irreducible subspaces of these components (so are "embedded" in the irreducible components). In this example, $\mathcal{Z}(I)$ is the set of points with $x^2 = xy = 0$, which is just the y-axis in $\mathbb{R}^2$. There is only one irreducible component of this algebraic space (namely the y-axis), which is the locus for the isolated prime (x). The locus for the embedded prime (x, y) is the origin $(0, 0)$, which is an irreducible subspace embedded in the y-axis.

(2) Suppose R is a U.F.D. If $a = p_1^{e_1} \cdots p_m^{e_m}$ is the unique factorization into distinct prime powers of the element $a \in R$, then $(a) = (p_1)^{e_1} \cap \cdots \cap (p_m)^{e_m}$ is the minimal primary decomposition of the principal ideal (a). The associated primes to (a) are $(p_1), \ldots, (p_m)$ and are all isolated. The primary decomposition of ideals is a generalization of the factorization of elements into prime powers. See also Exercise 44 for a characterization of U.F.D.s in terms of minimal primary decompositions.

For any Noetherian ring, an ideal I is radical if and only if the primary components of a minimal primary decomposition of I are all *prime* ideals (in which case this primary decomposition is unique), cf. Exercise 43. This generalizes the observation made previously that Proposition 17 together with Hilbert's Nullstellensatz shows that any radical ideal in $k[\mathbb{A}^n]$ may be written uniquely as a finite intersection of prime ideals when the field k is algebraically closed — this is the algebraic statement that an algebraic set can be decomposed uniquely into the union of irreducible algebraic sets.

EXERCISES

1. Prove (3) of Corollary 22 directly by considering the collection $\mathcal{S}$ of ideals that do not contain a finite product of prime ideals. [If I is a maximal element in $\mathcal{S}$, show that since I is not prime there are ideals J, K properly containing I (hence not in $\mathcal{S}$) with $JK \subseteq I$.]

2. Let I and J be ideals in the ring R. Prove the following statements:
    (a) If $I^k \subseteq J$ for some $k \geq 1$ then rad $I \subseteq$ rad J.
    (b) If $I^k \subseteq J \subseteq I$ for some $k \geq 1$ then rad $I =$ rad J.
    (c) $\text{rad}(IJ) = \text{rad}(I \cap J) = $ rad $I \cap$ rad J.
    (d) $\text{rad}(\text{rad } I) = $ rad I.
    (e) rad $I +$ rad $J \subseteq \text{rad}(I + J)$ and $\text{rad}(I + J) = \text{rad}(\text{rad } I + \text{rad } J)$.

3. Prove that the intersection of two radical ideals is again a radical ideal.

4. Let $I = \mathfrak{m}_1 \mathfrak{m}_2$ be the product of the ideals $\mathfrak{m}_1 = (x, y)$ and $\mathfrak{m}_2 = (x - 1, y - 1)$ in $\mathbb{F}_2[x, y]$. Prove that I is a radical ideal. Prove that the ideal $(x^3 - y^2)$ is a radical ideal in $\mathbb{F}_2[x, y]$.

5. If $I = (xy, (x - y)z) \subset k[x, y, z]$ prove that rad $I = (xy, xz, yz)$. For this ideal prove directly that $\mathcal{Z}(I) = \mathcal{Z}(\text{rad } I)$, that $\mathcal{Z}(I)$ is not irreducible, and that rad I is not prime.

6. Give an example to show that over a field k that is not algebraically closed the containment $I \subseteq \mathcal{I}(\mathcal{Z}(I))$ can be proper even when I is a radical ideal.

7. Suppose R and S are rings and $\varphi : R \to S$ is a ring homomorphism. If I is an ideal of R show that $\varphi(\text{rad } I) \subseteq \text{rad}(\varphi(I))$. If in addition φ is surjective and I contains the kernel of φ show that $\varphi(\text{rad } I) = \text{rad}(\varphi(I))$.

8. Suppose the prime ideal P contains the ideal I. Prove that P contains the radical of I.

9. Prove that for any field k the map $\mathcal{Z}$ in the Nullstellensatz is always surjective and the map $\mathcal{I}$ in the Nullstellensatz is always injective. [Use property (10) of the maps $\mathcal{Z}$ and $\mathcal{I}$ in Section 1.] Give examples (over a field k that is not algebraically closed) where $\mathcal{Z}$ is not injective and $\mathcal{I}$ is not surjective.

10. Prove that for k a finite field the Zariski topology is the same as the discrete topology: every subset is closed (and open).

11. Let V be a variety in $\mathbb{A}^n$ and let U_1 and U_2 be two subsets of $\mathbb{A}^n$ that are open in the Zariski topology. Prove that if $V \cap U_1 \neq \emptyset$ and $V \cap U_2 \neq \emptyset$ then $V \cap U_1 \cap U_2 \neq \emptyset$. Conclude that *any* nonempty open subset of a variety is *everywhere dense* in the Zariski topology (i.e., its closure is all of V).

12. Use the fact that nonempty open sets of an affine variety are everywhere dense to prove that an affine variety is connected in the Zariski topology. (A topological space is *connected* if it is not the union of two disjoint, proper, open subsets.)

13. Prove that the affine algebraic set V is connected in the Zariski topology if and only if $k[V]$ is not a direct sum of two nonzero ideals. Deduce from this that a variety is connected in the Zariski topology.

14. Prove that if k is an infinite field, then the varieties in $\mathbb{A}^1$ are the empty set, the whole space, and the one point subsets. What are the varieties in $\mathbb{A}^1$ in the case of a finite field k?

15. Suppose V is a hypersurface in $\mathbb{A}^n$ and $\mathcal{I}(V) = (f)$ for some nonconstant polynomial $f \in k[x_1, x_2, \ldots, x_n]$. Prove that V is a variety if and only if f is irreducible.

16. Suppose $V \subseteq \mathbb{A}^n$ is an affine variety and $f \in k[V]$. Prove that the *graph* of f (cf. Exercise 25 in Section 1) is an affine variety.

17. Prove that any permutation of the elements of a field k is a continuous map from $\mathbb{A}^1$ to itself in the Zariski topology on $\mathbb{A}^1$. Deduce that if k is an infinite field, there are Zariski continuous maps from $\mathbb{A}^1$ to itself that are not polynomials.

18. Let V be an affine algebraic set in $\mathbb{A}^n$ over $k = \mathbb{C}$.
  (a) Prove that morphisms of algebraic sets over $\mathbb{C}$ are continuous in the Euclidean topology (the topology on $\mathbb{C}^n$ obtained by identifying $\mathbb{C}^n$ with $\mathbb{R}^{2n}$ with its usual Euclidean topology).
  (b) Prove that V is a closed set in the Euclidean topology on $\mathbb{C}^n$ (so the Zariski closed sets of $\mathbb{A}^n$ over $\mathbb{C}$ are also Euclidean closed).
  (c) Give an example of a set that is closed in the Euclidean topology but is not closed in the Zariski topology, i.e., is not an affine algebraic set (so the Euclidean topology is "finer" than the Zariski topology).

19. Give an example of an injective k-algebra homomorphism $\widetilde{\varphi} : k[W] \to k[V]$ whose associated morphism $\varphi : V \to W$ is not surjective.

20. Suppose $\varphi : V \to W$ is a surjective morphism of affine algebraic sets. Prove that if V is a variety then W is a variety.

21. Let V be an algebraic set in $\mathbb{A}^n$ and let $f \in k[V]$. Define $V_f = \{v \in V \mid f(v) \neq 0\}$.
  (a) Show that V_f is a Zariski open set in V (called a *principal open set* in V).
  (b) Let J be the ideal in $k[x_1, \ldots, x_n, x_{n+1}]$ generated by $\mathcal{I}(V)$ and $x_{n+1}f - 1$, and let $W = \mathcal{Z}(J) \subseteq \mathbb{A}^{n+1}$. Show that $J = \mathcal{I}(W)$ and that the map $\pi : \mathbb{A}^{n+1} \to \mathbb{A}^n$ by projection onto the first n coordinates is a Zariski continuous bijection from W onto V_f (so the principal *open* set V_f in V may be embedded as a *closed* set in some (larger) affine space).
  (c) If U is any open set in V show that $U = V_{f_1} \cup \cdots \cup V_{f_m}$ for some $f_1, \ldots, f_m \in k[V]$. (This shows that the principal open sets form a *base* for the Zariski topology.)

22. Prove that $GL_n(k)$ is an open affine algebraic set in $\mathbb{A}^{n^2}$ and can be embedded as a closed affine algebraic set in $\mathbb{A}^{n^2+1}$. In particular, deduce that the set $k^\times$ of nonzero elements in

$\mathbb{A}^1$ embeds into $\mathbb{A}^2$ as the hyperbola $xy = 1$. [Use the preceding exercise.]

23. Show that if k is infinite then $\{(a, a^2, a^3) \mid a \in k\} \subset \mathbb{A}^3$ is an affine algebraic variety. If k is finite show that this set is always reducible.

24. Let $V = \mathcal{Z}(xz - y^2, yz - x^3, z^2 - x^2y) \subset \mathbb{A}^3$. Show that if k is infinite then V is an affine variety. [Use Exercise 26 of Section 1 and Exercise 20.]

25. Suppose $f(x) = x^3 + ax^2 + bx + c$ is an irreducible cubic in $\mathbb{Q}[x]$ of discriminant D. Let $I = (x + y + z + a, xy + xz + yz - b, xyz + c)$ in $\mathbb{Q}[x, y, z]$.
   (a) Prove that I is a prime ideal if and only if D is not a square in $\mathbb{Q}$, in which case I is a maximal ideal and $\mathbb{Q}[x, y, z]/I$ is a splitting field for $f(x)$ over $\mathbb{Q}$.
   (b) If $D = r^2$, prove that the primary decomposition of I is $I = Q_+ \cap Q_-$ where $Q_{\pm} = (I, (x - y)(x - z)(y - z) \pm r)$. Prove Q_+ and Q_- are maximal ideals, and $\mathbb{Q}[x, y, z]$ modulo Q_+ or Q_- is a splitting field for $f(x)$ over $\mathbb{Q}$.

26. A topological space X is called *quasicompact* if whenever any collection of closed subsets V_i of X has empty intersection, then some finite number of these also has empty intersection, i.e.,

$$\text{whenever } \bigcap_i V_i = \emptyset \text{ there exists } V_{i_1}, V_{i_2}, \dots, V_{i_N} \text{ such that } \bigcap_{t=1}^{N} V_{i_t} = \emptyset.$$

Prove that every affine algebraic set is quasicompact. [Translate the definition into a property of ideals in $k[x_1, \dots, x_n]$.] (A quasicompact and Hausdorff space is called *compact*.)

27. When k is an infinite field prove that the Zariski topology on k^2 is not the same as taking the Zariski topology on k and then forming the product topology on $k \times k$. [By Exercise 14 of Section 1, in the product topology on $k \times k$ the Zariski closed sets in $k \times k$ are finite unions of sets of the form $\{a\} \times \{b\}$, $\{a\} \times k$ and $k \times \{b\}$, for any $a, b \in k$.]

28. Prove that each of the following rings have infinitely many minimal prime ideals, and that (0) is not the intersection of any finite number of these (so (0) does not have a primary decomposition in these rings):
   (a) the infinite direct product ring $\mathbb{Z}/2\mathbb{Z} \times \mathbb{Z}/2\mathbb{Z} \times \cdots$ (which is a Boolean ring, cf. Exercise 23 in Section 7.4).
   (b) $k[x_1, x_2, \dots]/(x_1 x_2, x_3 x_4, \dots, x_{2i-1} x_{2i}, \dots)$, where $x_1, x_2, \dots$ are independent variables over the field k.

29. Suppose that A and B are ideals with $AB \subseteq Q$ for a primary ideal Q. Prove that if $A \nsubseteq Q$ then $B \subset \operatorname{rad} Q$.

30. Let Q be a P-primary ideal and suppose A is an ideal not contained in Q. Define $A' = \{r \in R \mid rA \subseteq Q\}$ to be the elements of R that when multiplied by elements of A give elements of Q. Prove that A' is a P-primary ideal.

31. Prove that if Q_1 and Q_2 are primary ideals belonging to the same prime ideal P, then $Q_1 \cap Q_2$ is a primary ideal belonging to P. Conclude that a finite intersection of P-primary ideals is again P-primary.

32. Prove that if Q_1 and Q_2 are primary ideals belonging to the same *maximal* ideal M, then $Q_1 + Q_2$ and $Q_1 Q_2$ are primary ideals belonging to M. Conclude that finite sums and finite products of M-primary ideals are again M-primary.

33. Let $I = (x^2, xy, xz, yz)$ in $k[x, y, z]$. Prove that a primary decomposition of I is $I = (x, y) \cap (x, z) \cap (x, y, z)^2$, determine the isolated and embedded primes of I, and find $\operatorname{rad} I$.

34. Suppose $\varphi : R \to S$ is a surjective ring homomorphism. Prove that an ideal Q in R containing the kernel of φ is primary if and only if $\varphi(Q)$ is primary in S, and when this is

the case the prime associated to $\varphi(Q)$ is the image $\varphi(P)$ of the prime P associated to Q.

35. Suppose $\varphi : R \to S$ is a ring homomorphism.
  (a) Suppose I is an ideal of R containing $\ker \varphi$ with minimal primary decomposition $I = Q_1 \cap \cdots \cap Q_m$ with $\operatorname{rad} Q_i = P_i$. If φ is a surjective homomorphism prove that $\varphi(I) = \varphi(Q_1) \cap \cdots \cap \varphi(Q_m)$, where $\operatorname{rad} \varphi(Q_i)$ is given by $\varphi(P_i)$, is a minimal primary decomposition of $\varphi(I)$. [Use the previous exercise.]
  (b) Suppose I is an ideal of S with minimal primary decomposition $I = Q_1 \cap \cdots \cap Q_m$ with $\operatorname{rad} Q_i = P_i$. Prove that $\varphi^{-1}(I) = \varphi^{-1}(Q_1) \cap \cdots \cap \varphi^{-1}(Q_m)$, where $\operatorname{rad} \varphi^{-1}(Q_i)$ is given by $\varphi^{-1}(P_i)$, is a primary decomposition of $\varphi^{-1}(I)$, and is minimal if φ is surjective.

36. Let $I = (xy, x - yz)$ in $k[x, y, z]$. Prove that $(x, z) \cap (y^2, x - yz)$ is a minimal primary decomposition of I. [Consider the ring homomorphism $\varphi : k[x, y, z] \to k[y, z]$ given by mapping x to yz, y to y, and z to z and use the previous exercise.]

37. Prove that a prime ideal P contains the ideal I if and only if P contains one of the associated primes of a minimal primary decomposition of I. [Use Exercise 3 and Exercise 11 in Section 7.4.]

38. Show that every associated prime ideal for a radical ideal is isolated. [Suppose that $P_2 = \operatorname{rad} Q_2 \subseteq P_1 = \operatorname{rad} Q_1$ in the decomposition of Theorem 21 for the radical ideal I. Show that if $a \in Q_2 \cap \cdots \cap Q_m \subseteq P_2$ then $a^n \in I$ for some $n \geq 1$, conclude that $a \in Q_1$ and derive a contradiction to the minimality of the primary decomposition.]

39. Fix an element a in the ring R. For any ideal I in the ring R let $I_a = \{r \in R \mid ar \in I\}$.
  (a) Prove that I_a is an ideal and $I_a = R$ if and only if $a \in I$.
  (b) Prove that $(I \cap J)_a = I_a \cap J_a$ for ideals I and J.
  (c) Suppose that Q is a P-primary ideal and that $a \notin Q$. Prove that Q_a is a P-primary ideal and that $Q_a = Q$ if $a \notin P$.

40. With notation as in the previous exercise, suppose $I = Q_1 \cap \cdots \cap Q_m$ is a minimal primary decomposition of the ideal I and let P_i be the prime ideal associated to Q_i.
  (a) Prove that $I_a = (Q_1)_a \cap \cdots \cap (Q_m)_a$ and that $\operatorname{rad}(I_a) = \operatorname{rad}((Q_1)_a) \cap \cdots \cap \operatorname{rad}((Q_m)_a)$.
  (b) Prove that $\operatorname{rad}(I_a)$ is the intersection of the prime ideals P_i for which $a \notin Q_i$. [Use the previous exercise.]
  (c) Prove that if $\operatorname{rad}(I_a)$ is a prime ideal then $\operatorname{rad}(I_a) = P_j$ for some j. [Use the fact that prime ideals are irreducible.]
  (d) For each $i = 1, \ldots, m$, prove that $\operatorname{rad}(I_a) = P_i$ for some $a \in R$. [Show there exists an $a \in R$ with $a \notin Q_i$ but $a \in Q_j$ for all $j \neq i$.]
  (e) Show from (c) and (d) that the associated primes for a minimal primary decomposition are precisely the collection of prime ideals among the ideals $\operatorname{rad}(I_a)$ for $a \in R$, and conclude that they are uniquely determined by I independent of the minimal primary decomposition.

41. Let $P_1, \ldots, P_m$ be the associated prime ideals of the ideal (0) in the Noetherian ring R.
  (a) Show that $P_1 \cap \cdots \cap P_m$ is the collection of nilpotent elements in R. [Apply Corollary 22 to $I = (0)$.]
  (b) Show that $P_1 \cup \cdots \cup P_m$ is the collection of zero divisors in R. [Let $I = (0)$ in the previous exercise and show that the set of zero divisors is given by the set $\cup_{a \in R - \{0\}} (0)_a = \cup_{a \in R - \{0\}} \operatorname{rad}((0)_a)$.]

42. Suppose R is a Noetherian ring. Prove that R is either an integral domain, has nonzero nilpotent elements, or has at least two minimal prime ideals. [Use the previous exercise.]

43. Prove that the ideal I in the Noetherian ring R is radical if and only if the primary components of a minimal primary decomposition are all prime ideals, and conclude that in this case the minimal primary decomposition is unique. [If $I = Q_1 \cap \cdots \cap Q_m$ is radical with Q_i a P_i-primary component of a minimal decomposition, show that if $a \in P_1 \cap \cdots \cap P_m$ then some power of a is in I, hence $a \in I$ since I is radical. Deduce that $I = P_1 \cap \cdots \cap P_m$ and show that this is also a minimal primary decomposition, i.e., for any i there exists b with $b \notin P_i$, but $b \in P_j$ for $j \neq i$. If $a \in P_i$, show that $ab \in Q_i$, and that $a \in Q_i$. Conclude that $Q_i = P_i$.]

44. Prove that a Noetherian integral domain R is a U.F.D. if and only if for every $a \in R$ the isolated primes associated to the principal ideal (a) are principal ideals. [See Example 2 following Corollary 22. To prove R is a U.F.D., show that an irreducible $a \in R$ is prime and then follow the proof of Theorem 14 in Section 8.3.]

45. Let R be the ring of all real valued functions on the open interval $(-1, 1)$ that have derivatives of all orders (the ring of C^∞ functions). Let

$$F(x) = \begin{cases} e^{-1/x^4} & \text{if } x \neq 0 \\ 0 & \text{if } x = 0 \end{cases}$$

(you may assume $F \in R$ and $F^{(n)}(0) = 0$ for all $n \geq 0$). Let (F) be the principal ideal generated by F and let $A = \text{rad}((F))$. Let M be the (maximal) ideal of all functions in R that are zero at $x - 0$ and let $P - \cap_{n=1}^\infty M^n$.
  (a) Prove that $M = (x)$ is the ideal generated by the function x in R and that $M^n = (x^n)$ consists of the functions whose first $n - 1$ derivatives vanish at the origin.
  (b) Prove that R is not Noetherian (compare Exercise 33 in Section 7.4). [One approach is the following: Let $G(x)$ be the function that is 0 for $x < 0$ and is equal to $F(x)$ for $x \geq 0$. Let I_n be the ideal of functions in R vanishing for all $x \leq 1/n$. Use translates of $G(x)$ to show that $I_1 \subset I_2 \subset I_3 \subset \cdots$ is an infinite ascending chain.]
  (c) Prove that P consists of the functions all of whose derivatives are zero at $x = 0$ (i.e., the functions whose associated Taylor series at $x = 0$ is identically zero), and that P is a prime ideal.
  (d) Prove that $F \in P$ and deduce that $A \subseteq P$.
  (e) Prove that $A \neq P$. [Let $G(x) = e^{-1/x^2}$ when $x \neq 0$ and $G(0) = 0$. Show that $G \in P$ but $G \notin A$.]
  (f) Show that there is a prime ideal Q containing (F) with $Q \neq P, M$. Prove that $Q \subset P$ i.e., there are nonzero prime ideals properly contained in P.

46. Let $\mathcal{A}$ be any ideal in $R = k[x_1, \ldots, x_n, y_1, \ldots, y_m]$.
  (a) Show that $\text{rad}(\mathcal{A} \cap k[y_1, \ldots, y_m]) = \text{rad }\mathcal{A} \cap k[y_1, \ldots, y_m]$.
  (b) Suppose $(f_1, \ldots, f_s)$ is an ideal in $k[x_1, \ldots, x_n]$. Let $F_1, \ldots, F_t$ be generators for the radical of $(f_1, \ldots, f_s)$, computed in $k[x_1, \ldots, x_n]$. Suppose J is an ideal in R and let $\mathcal{A} = J + (f_1, \ldots, f_s)$, $\mathcal{B} = J + (F_1, \ldots, F_t)$ as ideals in R. Prove that $\text{rad }\mathcal{A} = \text{rad }\mathcal{B}$.
  (c) Conclude from (a) and (b) that $\mathcal{A} = (y_1 - x_1, \ldots, y_m - x_m, f_1, \ldots, f_s) \cap k[y_1, \ldots, y_m]$ and $\mathcal{B} = (y_1 - x_1, \ldots, y_m - x_m, F_1, \ldots, F_t) \cap k[y_1, \ldots, y_m]$ have the same zero sets over an algebraically closed field k. [Use Hilbert's Nullstellensatz.]

47. Determine the Zariski closure in $\mathbb{C}^3$ of the points on the curve $\{(a^2, a^3, a^4) \mid a \in \mathbb{C}\}$.

48. Show that $\mathcal{Z}(x^3 - xyz + z^2)$ is the smallest algebraic set in $\mathbb{R}^3$ containing the points $\{(st, s + t, s^2 t) \mid s, t \in \mathbb{R}\}$.

49. Show that $\mathcal{Z}(x^3 z^2 - 3xy^2 z^2 - y^6 - z^4)$ is the smallest algebraic set in $\mathbb{R}^3$ containing the points $\{(s^2 + t^2, st, s^3) \mid s, t \in \mathbb{R}\}$.

50. Find equations defining the Zariski closure of the set of points $\{(s^4, s^3t, st^3, t^4) \mid s, t \in \mathbb{R}\}$.

51. Show that $V = \mathcal{Z}(x^2 - y^2z)$ (the *Whitney umbrella surface*) is the smallest algebraic set in $\mathbb{R}^3$ containing the points $S = \{(st, s, t^2) \mid s, t \in \mathbb{R}\}$. Show that S is not Zariski closed in V (the missing points explain the name for the surface). Do the same over $\mathbb{C}$, but show that in this case $S = V$ is closed.

52. Let $V = \mathcal{Z}(xz^2 - w^3, xw^2 - y^4, y^4z^2 - w^5) \subset \mathbb{C}^4$. Determine the Zariski closure of the image of V under the projection $\pi((x, y, z, w)) = (x, y, z)$.

53. Let $V = \mathcal{Z}(xy - 1)$ in $\mathbb{A}^2$ and let S be the projection of V onto the x-axis in $\mathbb{A}^1$.
   (a) If $k = \mathbb{R}$, show that $\mathcal{I}(V) = (xy - 1) \subset \mathbb{R}[x, y]$ and that $(u - x, xy - 1) \cap \mathbb{R}[u] = 0$ in $\mathbb{R}[x, y, u]$. Use Propositions 8 and 16 to conclude that the Zariski closure of S is $\mathbb{A}^1$ and show that S is not itself closed.
   (b) If $k = \mathbb{F}_3$, show that $\mathcal{I}(V) = (xy - 1, x^3 - x, y^3 - y) \subset \mathbb{F}_3[x, y]$ and that $(u - x, xy - 1, x^3 - x, y^3 - y) \cap \mathbb{F}_3[u] = (u^2 - 1)$ in $\mathbb{F}_3[x, y, u]$. Use Propositions 8 and 16 to conclude that S is Zariski closed in $\mathbb{A}^1$.

54. Recall the *ideal quotient* $(I : J) = \{r \in R \mid rJ \in I\}$ of two ideals I, J in a ring R (cf. Exercise 34 *ff.* in Section 9.6). Clearly $I \subseteq (I : J)$.
   (a) Show that $\mathcal{Z}(I) - \mathcal{Z}(J)$, the set of elements of $\mathcal{Z}(I)$ not lying in $\mathcal{Z}(J)$, is contained in $\mathcal{Z}((I : J))$ and conclude that the Zariski closure of $\mathcal{Z}(I) - \mathcal{Z}(J)$ is contained in $\mathcal{Z}((I : J))$.
   (b) Show that if k is algebraically closed and I is a radical ideal then $\mathcal{Z}((I : J))$ is precisely the Zariski closure of $\mathcal{Z}(I) - \mathcal{Z}(J)$.
   (c) Show that if V and W are affine algebraic sets then $(\mathcal{I}(V) : \mathcal{I}(W)) = \mathcal{I}(V - W)$.

15.3 INTEGRAL EXTENSIONS AND HILBERT'S NULLSTELLENSATZ

In this section we consider the important concept of an integral extension of rings, which is a generalization to rings of algebraic extensions of fields. This leads to the definition of the "integers" in finite extensions of $\mathbb{Q}$ (the basic subject of the branch of mathematics called algebraic number theory) and is also related to the existence of tangent lines for algebraic curves.

Definition. Suppose R is a subring of the commutative ring S with $1 = 1_S \in R$.
   (1) An element $s \in S$ is *integral over* R if s is the root of a monic polynomial in $R[x]$.
   (2) The ring S is an *integral extension of* R or just *integral over* R if every $s \in S$ is integral over R.
   (3) The *integral closure* of R in S is the set of elements of S that are integral over R.
   (4) The ring R is said to be *integrally closed in* S if R is equal to its integral closure in S. The integral closure of an integral domain R in its field of fractions is called the *normalization of* R. An integral domain is called *integrally closed* or *normal* if it is integrally closed in its field of fractions.

Before giving some examples of integral extensions we prove some basic properties of integral elements analogous to those of algebraic elements over fields.

Proposition 23. Let R be a subring of the commutative ring S with $1 \in R$ and let $s \in S$. Then the following are equivalent:

    (1) s is integral over R,

    (2) $R[s]$ is a finitely generated R-module (where $R[s]$ is the ring of all R-linear combinations of powers of s), and

    (3) $s \in T$ for some subring T, $R \subseteq T \subseteq S$, that is a finitely generated R-module.

Proof: Suppose first that (1) holds and let s be a root of the monic polynomial $x^n + a_{n-1}x^{n-1} + \cdots + a_0 \in R[x]$. Then

$$s^n = -(a_{n-1}s^{n-1} + a_{n-2}s^{n-2} + \cdots + a_0)$$

and so s^n, and then all higher powers of s, can be expressed as R-linear combinations of $s^{n-1}, \ldots, s, 1$. Hence $R[s] = R1 + Rs + \cdots + Rs^{n-1}$ is finitely generated as an R-module, which gives (2).

If (2) holds, then (3) holds with $T = R[s]$.

Suppose that (3) holds and let $v_1, v_2, \ldots, v_n$ be a finite generating set for T. Then for $i = 1, 2, \ldots, n$ the element sv_i is an element of T since T is a ring, and so can be written as R-linear combinations of $v_1, \ldots, v_n$:

$$s v_i = \sum_{j=1}^{n} a_{ij} v_j,$$

i.e.,

$$0 = \sum_{j=1}^{n} (\delta_{ij} s - a_{ij}) v_j \qquad i = 1, 2, \ldots, n$$

where δ_{ij} is the Kronecker delta. If B is the $n \times n$ matrix whose i, j entry is $\delta_{ij} s - a_{ij}$, and v is the $n \times 1$ column vector whose entries are $v_1, \ldots, v_n$, then these equations are simply $Bv = 0$. It follows from Cramer's Rule that $(\det B)v_i = 0$ for all i (cf. Exercise 3, Section 11.4). Since $1 \in T$ is an R-linear combination of $v_1, \ldots, v_n$, it follows that $\det B = 0$. But $B = sI - A$, where A is the matrix (a_{ij}). Thus s is a root of the monic polynomial $\det(xI - A) \in R[x]$ (the characteristic polynomial of A), and so s is a root of a monic polynomial with coefficients in R, which gives (1), completing the proof.

Corollary 24. Let $R \subseteq S$ be as in Proposition 23 and let $s, t \in S$.

    (1) If s and t are integral over R then so are $s \pm t$ and st.

    (2) The integral closure of R in S is a subring of S containing R.

    (3) Integrality is transitive: let S be a subring of T; if T is integral over S and S is integral over R, then T is integral over R.

Proof: Let s and t be integral over R. By Proposition 23 both $R[s]$ and $R[t]$ are finitely generated R-modules, say

$$R[s] = Rs_1 + Rs_2 + \cdots + Rs_n$$
$$R[t] = Rt_1 + Rt_2 + \cdots + Rt_m.$$

Then
$$R[s, t] = Rs_1t_1 + \cdots + Rs_it_j + \cdots + Rs_nt_m$$
is a ring containing $s \pm t$ and st that is also a finitely generated R-module. Hence $s \pm t$ and st are also integral over R, which proves (1) and also (2).

To prove (3), let $t \in T$. Since t is integral over S, it is the root of some monic polynomial $p(x) = x^n + a_{n-1}x^{n-1} + \cdots + a_0 \in S[x]$. Since $a_i \in S$ is integral over R, each ring $R[a_i]$ is a finitely generated R-module and so the ring $R_1 = R[a_0, a_1, \ldots, a_{n-1}]$ is also a finitely generated R-module. Since the monic polynomial $p(x)$ has its coefficients in R_1, t is integral over R_1 and it follows that the ring $R_1[t] = R[a_0, a_1, \ldots, a_{n-1}, t]$ is a finitely generated R-module. By the proposition, this means that t is integral over R, which gives (3).

The second statement in Corollary 24 shows that taking the elements of S that are integral over R gives a (possibly larger) subring of S, and the last statement in the corollary shows that the process of taking the integral closure stops after one step:

Corollary 25. Let R be a subring of the commutative ring S with $1 \in R$. Then the integral closure of R in S is integrally closed in S.

Examples

(1) If R and S are fields then S is integral over R if and only if S is algebraic over R — if $s \in S$ is a root of the polynomial $p(x)$ with coefficients in R then it is a root of the monic polynomial obtained by dividing by the (nonzero) leading coefficient of $p(x)$.

(2) Suppose S is an integral extension of R and I is an ideal in S. Then S/I is an integral ring extension of $R/(R \cap I)$ (reducing the monic polynomial over R satisfied by $s \in S$ modulo I gives a monic polynomial satisfied by $\bar{s} \in S/I$ over $R/(R \cap I)$).

(3) If R is a U.F.D. then R is integrally closed, as follows. Suppose a/b is an element in the field of fractions of R (with $b \neq 0$ and a and b having no common factors) and satisfies $(a/b)^n + r_{n-1}(a/b)^{n-1} + \cdots + r_1(a/b) + r_0 = 0$ with $r_0, \ldots, r_{n-1} \in R$. Then
$$a^n = b(-r_{n-1}a^{n-1} - \cdots - r_1ab^{n-2} - r_0b^{n-1})$$
shows that any irreducible element dividing b divides a^n, hence divides a. Since a/b is in lowest terms, this shows that b must be a unit, i.e., $a/b \in R$.

(4) The polynomial ring $k[x, y]$ over the field k is integrally closed in its fraction field $k(x, y)$ by example (3) above. The ideal $(x^2 - y^3)$ is prime (cf. Exercise 14, Section 9.1), so the quotient ring $R = k[x, y]/(x^2 - y^3) = k[\bar{x}, \bar{y}]$ is an integral domain. This domain is not integrally closed, however, since $\bar{x}/\bar{y}$ is an element of the fraction field of R that is integral over R (since $(\bar{x}/\bar{y})^3 - \bar{x} = 0$), but is not an element of R. In particular, R is not a U.F.D. by the previous example.

We next consider the behavior of ideals in integral ring extensions.

Definition. Let $\varphi : R \to S$ be a homomorphism of commutative rings.

(a) If I is an ideal in R then the *extension* of I to S is the ideal $\varphi(I)S$ of S generated by the image of I.

(b) If J is an ideal of S, then the *contraction* in R of J is the ideal $\varphi^{-1}(J)$.

In the special case where R is a subring of S and φ is the natural injection, the extension of $I \subseteq R$ is the ideal IS in S and the contraction of $J \subseteq S$ is the ideal $J \cap R$ of R.

It is immediate from the definition that

    (1) $I \subseteq IS \cap R$, more generally, I is contained in the contraction of its extension to S, and

    (2) $(J \cap R)S \subseteq J$, more generally, J contains the extension of its contraction in R.

In general equality need not hold in either situation (cf. the exercises).

If Q is a prime ideal in S, then its contraction is prime in R (although the contraction of a maximal ideal need not be maximal). On the other hand, if P is a prime ideal in R, its extension need not be prime (or even proper) in S; moreover, it is not generally true that P is the contraction of a prime ideal of S (cf. the exercises). For integral ring extensions, however, the situation is more controlled:

Theorem 26. Let R be a subring of the commutative ring S with $1 \in R$ and suppose that S integral over R.
    (1) Assume that S is an integral domain. Then R is a field if and only if S is a field.
    (2) Let P be a prime ideal in R. Then there is a prime ideal Q in S with $P = Q \cap R$. Moreover, P is maximal if and only if Q is maximal.
    (3) *(The Going-up Theorem)* Let $P_1 \subseteq P_2 \subseteq \cdots \subseteq P_n$ be a chain of prime ideals in R and suppose there are prime ideals $Q_1 \subseteq Q_2 \subseteq \cdots \subseteq Q_m$ of S with $P_i = Q_i \cap R$, $1 \le i \le m$ and $m < n$. Then the ascending chain of ideals can be completed: there are prime ideals $Q_{m+1} \subseteq \cdots \subseteq Q_n$ in S such that $P_i = Q_i \cap R$ for all i.
    (4) *(The Going-down Theorem)* Assume that S is an integral domain and R is integrally closed in S. Let $P_1 \supseteq P_2 \supseteq \cdots \supseteq P_n$ be a chain of prime ideals in R and suppose there are prime ideals $Q_1 \supseteq Q_2 \supseteq \cdots \supseteq Q_m$ of S with $P_i = Q_i \cap R$, $1 \le i \le m$ and $m < n$. Then the descending chain of ideals can be completed: there are prime ideals $Q_{m+1} \supseteq \cdots \supseteq Q_n$ in S such that $P_i = Q_i \cap R$ for all i.

Proof: To prove (1) assume first that R is a field and let s be a nonzero element of S. Then s is integral over R, so

$$s^n + a_{n-1}s^{n-1} + \cdots + a_1 s + a_0 = 0$$

for some $a_0, a_1, \ldots, a_{n-1}$ in R. Since S is an integral domain, we may assume $a_0 \ne 0$ (otherwise cancel factors of s). Then

$$s(s^{n-1} + a_{n-1}s^{n-2} + \cdots + a_1) = -a_0$$

and since $(-1/a_0) \in R$, this shows that $(-1/a_0)(s^{n-1} + a_{n-1}s^{n-2} + \cdots + a_1)$ is an inverse for s in S, so S is a field. Conversely, suppose S is a field and r is a nonzero element of R. Since $r^{-1} \in S$ is integral over R we have

$$r^{-m} + a_{m-1}r^{-m+1} + \cdots + a_1 r^{-1} + a_0 = 0$$

for some $a_0, \ldots, a_{m-1} \in R$. Then $r^{-1} = -(a_{m-1} + \cdots + a_1 r^{m-2} + a_0 r^{m-1}) \in R$, so R is a field.

The proof of the first statement in (2) is given in Corollary 50. For the second statement, observe that the integral domain S/Q is an integral extension of R/P (Example 2 following Corollary 25). By (1), S/Q is a field if and only if R/P is a field, i.e., Q is maximal if and only if P is maximal.

To prove (3), it suffices by induction to prove that if $P_1 \subseteq P_2$ and Q_1 is a prime of S with $Q_1 \cap R = P_1$ then there is a prime Q_2 of S with $Q_1 \subseteq Q_2$ and $Q_2 \cap R = P_2$. Since $\overline{S} = S/Q_1$ is an integral extension of $\overline{R} = R/P_1$, the first part of (2) shows that there exists a prime $\overline{Q_2}$ of $\overline{S}$ with $\overline{Q_2} \cap \overline{R} = P_2/P_1$. Then the preimage Q_2 of $\overline{Q_2}$ in S is a prime ideal containing Q_1 with $Q_2 \cap R = P_2$.

The proof of (4) is outlined in Exercise 24 in Section 4.

Corollary 27. Suppose R is a subring of the ring S with $1 \in R$ and assume S is integral and finitely generated (as a ring) over R. If P is a maximal ideal in R then there is a nonzero and finite number of maximal ideals Q of S with $Q \cap R = P$.

Proof: There exists at least one maximal ideal Q lying over P by (2) of the theorem, so we must see why there are only finitely many such maximal ideals in S. If Q is a maximal ideal of S with $Q \cap R = P$ then S/Q is a field containing the field R/P. To prove that there are only finitely many possible Q it suffices to prove that there are only finitely many homomorphisms from S to a field containing R/P that extend the homomorphism from R to R/P. Let $S = R[s_1, \ldots, s_n]$, where the elements s_i are integral over R by assumption, and let $p_i(x)$ be a monic polynomial with coefficients in R satisfied by s_i. If Q is a maximal ideal of S then $S/Q = (R/P)[\bar{s}_1, \ldots, \bar{s}_n]$ is the field extension of the field R/P with generators $\bar{s}_1, \ldots, \bar{s}_n$. The element $\bar{s}_i$ is a root of the monic polynomial $\bar{p}_i(x)$ with coefficients in R/P obtained by reducing the coefficients of $p_i(x)$ mod P. There are only a finite number of possible roots of this monic polynomial (in a fixed algebraic closure of R/P), and so only finitely many possible field extensions of the form $(R/P)[\bar{s}_1, \ldots, \bar{s}_n]$, which proves the corollary.

Algebraic Integers

We can use the concept of an integral ring extension to define the "integers" in extension fields of the rational numbers $\mathbb{Q}$:

Definition. Let K be an extension field of $\mathbb{Q}$.
  (1) An element $\alpha \in K$ is called an *algebraic integer* if α is integral over $\mathbb{Z}$, i.e., if α is the root of some monic polynomial with coefficients in $\mathbb{Z}$.
  (2) The integral closure of $\mathbb{Z}$ in K is called the *ring of integers* of K, and is denoted by $\mathcal{O}_K$.

An algebraic integer is clearly algebraic over $\mathbb{Q}$, so the ring of all algebraic integers is the ring of integers in $\overline{\mathbb{Q}}$, an algebraic closure of $\mathbb{Q}$. Examples of algebraic integers include $\sqrt{2}, \sqrt{-1}, \sqrt[3]{5}$, etc. since these elements are certainly roots of monic polynomials with coefficients in $\mathbb{Z}$. The definition of an algebraic integer α is that α be a root of *some* monic polynomial in $\mathbb{Z}[x]$, a condition which seems difficult to check. The next proposition gives a simple criterion for α to be an algebraic integer in terms of the minimal polynomial for α.

Proposition 28. An element α in some field extension of $\mathbb{Q}$ is an algebraic integer if and only if α is algebraic over $\mathbb{Q}$ and its minimal polynomial $m_{\alpha,\mathbb{Q}}(x)$ has integer coefficients. In particular, the algebraic integers in $\mathbb{Q}$ are the integers $\mathbb{Z}$, i.e., $\mathcal{O}_{\mathbb{Q}} = \mathbb{Z}$.

Proof: If α is algebraic over $\mathbb{Q}$ with $m_{\alpha,\mathbb{Q}}(x) \in \mathbb{Z}[x]$, then by definition α is integral over $\mathbb{Z}$. Conversely, assume α is integral over $\mathbb{Z}$, and let $f(x)$ be a monic polynomial in $\mathbb{Z}[x]$ of minimum degree having α as a root. If f were reducible in $\mathbb{Q}[x]$, then by Gauss' Lemma $f(x) = g(x)h(x)$ for some monic polynomials $g(x), h(x)$ in $\mathbb{Z}[x]$ of degree smaller than the degree of f. But then α would be a root of either g or h, contradicting the minimality of f. Hence f is irreducible in $\mathbb{Q}[x]$, so $f(x) = m_{\alpha,\mathbb{Q}}(x)$ and so the minimal polynomial for α has coefficients in $\mathbb{Z}$. Finally, the minimal polynomial of $\alpha = a/b \in \mathbb{Q}$ (a/b reduced to lowest terms and $b > 0$) is $bx - a$, which is monic if and only if $b = 1$, so $\alpha \in \mathbb{Q}$ is an algebraic integer if and only if $\alpha \in \mathbb{Z}$.

Because the integers $\mathbb{Z}$ are the algebraic integers in $\mathbb{Q}$, for emphasis (and clarity) the elements of $\mathbb{Z}$ are sometimes referred to as the "rational integers" to distinguish them from the "integers" in extensions of finite degree over $\mathbb{Q}$ (called *number fields*). The next result gives some of the basic structure of the ring of integers in a general number field.

Theorem 29. Let K be a number field of degree n over $\mathbb{Q}$.
   (1) The ring $\mathcal{O}_K$ of integers in K is a Noetherian ring and is a free $\mathbb{Z}$-module of rank n.
   (2) For every $\beta \in K$ there is some nonzero $d \in \mathbb{Z}$ such that $d\beta$ is an algebraic integer. In particular, K is the field of fractions of $\mathcal{O}_K$.
   (3) If $\beta_1, \beta_2, \ldots, \beta_n$ is any $\mathbb{Q}$-basis of K, then there is an integer d such that $d\beta_1, d\beta_2, \ldots, d\beta_n$ is a basis for a free $\mathbb{Z}$-submodule of $\mathcal{O}_K$ of rank n. Any basis of the $\mathbb{Z}$-module $\mathcal{O}_K$ is also a basis for K as a vector space over $\mathbb{Q}$.

Proof: Note first that any $\mathbb{Z}$-linear dependence relation among elements in $\mathcal{O}_K$ is a $\mathbb{Q}$-linear dependence relation in K, and multiplying a $\mathbb{Q}$-linear dependence relation of elements of $\mathcal{O}_K$ in K by a common denominator for the coefficients yields a $\mathbb{Z}$-linear dependence relation in $\mathcal{O}_K$. Let β be any element of K and let $x^k + a_{k-1}x^{k-1} + \cdots + a_0$ be the minimal polynomial of β over $\mathbb{Q}$. If d is a common denominator for the coefficients, then multiplying through by d^k shows that

$$(d\beta)^k + da_{k-1}(d\beta)^{k-1} + \cdots + d^{k-1}a_1(d\beta) + d^k a_0 = 0,$$

and $d^k a_0, d^{k-1}a_1, \ldots, da_{k-1} \in \mathbb{Z}$. Hence $d\beta$ is an algebraic integer, which proves the first part of (2) and then the second statement in (2) follows immediately.

If $\beta_1, \ldots, \beta_n$ are a $\mathbb{Q}$-basis for K over $\mathbb{Q}$, then there is a nonzero integer d such that $d\beta_1, \ldots, d\beta_n$ all lie in $\mathcal{O}_K$. These elements are still linearly independent over $\mathbb{Q}$, so in particular are independent over $\mathbb{Z}$, hence generate a free submodule of $\mathcal{O}_K$ of rank n, which proves the first statement in (3).

Since $\mathcal{O}_K$ is a subring of the field K, it is a torsion free $\mathbb{Z}$-module. If $\mathcal{O}_K$ were contained in some finitely generated $\mathbb{Z}$-module it would follow that $\mathcal{O}_K$ is also finitely generated over $\mathbb{Z}$, hence is a free $\mathbb{Z}$-module. If L is the Galois closure of K, then $\mathcal{O}_K \subseteq \mathcal{O}_L$ and so it suffices to see that $\mathcal{O}_L$ is contained in a finitely generated $\mathbb{Z}$-module. Let $\alpha_1, \ldots, \alpha_m$ be a $\mathbb{Q}$-basis for L. Multiplying by an integer $d \in \mathbb{Z}$, if necessary, we may assume that each α_i is an algebraic integer, i.e., $\alpha_1, \ldots, \alpha_m \in \mathcal{O}_L$. For each fixed $\theta \neq 0$ in L, the map

$$T_\theta : L \to \mathbb{Q} \quad \text{defined by} \quad T_\theta(\alpha) = \text{Tr}_{L/\mathbb{Q}}(\theta\alpha)$$

(where $\text{Tr}_{L/\mathbb{Q}}$ denotes the trace map from L to $\mathbb{Q}$, cf. Exercise 18 in Section 14.2) is a $\mathbb{Q}$-linear transformation from L to $\mathbb{Q}$. This linear transformation is nonzero because $T_\theta(\theta^{-1}) = \text{Tr}_{L/\mathbb{Q}}(1) = m$. It follows that the map from L to $\text{Hom}_{\mathbb{Q}}(L, \mathbb{Q})$ mapping θ to T_θ is an injective homomorphism of vector spaces over $\mathbb{Q}$. Since both spaces have the same dimension over $\mathbb{Q}$, the map is an isomorphism. Put another way, every linear functional on L is of the form T_θ for some $\theta \in L$. In particular, there are elements $\alpha_1', \ldots, \alpha_m'$ in L whose corresponding linear transformations $T_{\alpha_i'}$ give the dual basis of $\alpha_1, \ldots, \alpha_m$, i.e.,

$$\text{Tr}_{L/\mathbb{Q}}(\alpha_i'\alpha_j) = \begin{cases} 1, & \text{if } i = j \\ 0, & \text{otherwise.} \end{cases}$$

Since $\alpha_1', \ldots, \alpha_m'$ are linearly independent, they give a basis for L over $\mathbb{Q}$. Hence every element $\beta \in \mathcal{O}_L$ can be written

$$\beta = a_1\alpha_1' + \cdots + a_i\alpha_i' + \cdots + a_m\alpha_m'$$

with $a_1, \ldots, a_m \in \mathbb{Q}$. Multiplying by α_j and taking the trace shows that

$$\text{Tr}_{L/\mathbb{Q}}(\beta\alpha_j) = a_1\text{Tr}_{L/\mathbb{Q}}(\alpha_1'\alpha_j) + \cdots + a_i\text{Tr}_{L/\mathbb{Q}}(\alpha_i'\alpha_j) + \cdots + a_m\text{Tr}_{L/\mathbb{Q}}(\alpha_m'\alpha_j) = a_j.$$

But β and α_j are both elements of $\mathcal{O}_L$, so also $\beta\alpha_j$ is an element of $\mathcal{O}_L$, and this implies that $a_j = \text{Tr}_{L/\mathbb{Q}}(\beta\alpha_j)$ is an element of $\mathbb{Z}$ (cf. Exercise 18(d) of Section 14.2). It follows that

$$\mathcal{O}_L \subseteq \mathbb{Z}\alpha_1' + \cdots + \mathbb{Z}\alpha_m'$$

so that $\mathcal{O}_L$ is contained in a finitely generated $\mathbb{Z}$-module, proving that $\mathcal{O}_K$ (and also $\mathcal{O}_L$) is a free $\mathbb{Z}$-module.

Since K has dimension n as a vector space over $\mathbb{Q}$, it follows that $\mathcal{O}_K$ is a free $\mathbb{Z}$-module of rank at most n (by Theorem 5 of Section 12.1). Because $\mathcal{O}_K$ also contains a free $\mathbb{Z}$-submodule of rank n, it follows that the $\mathbb{Z}$-rank of $\mathcal{O}_K$ is precisely n, proving (1), and then the second statement in (3) follows by the remarks on $\mathbb{Z}$-linear and $\mathbb{Q}$-linear dependence relations.

Finally, any ideal I in $\mathcal{O}_K$ is a $\mathbb{Z}$-submodule of a free $\mathbb{Z}$-module of rank n, so is a free $\mathbb{Z}$-module of rank at most n, and a set of $\mathbb{Z}$-module generators for I is also a set of $\mathcal{O}_K$-generators. Hence every ideal of $\mathcal{O}_K$ can be generated by at most n elements, which implies that $\mathcal{O}_K$ is a Noetherian ring and completes the proof.

Definition. An *integral basis* for the number field K is a basis of the ring of integers in K considered as a free $\mathbb{Z}$-module of rank $[K : \mathbb{Q}]$.

If P is a nonzero prime ideal in the ring of integers $\mathcal{O}_K$ of a number field K then $P \cap \mathbb{Z}$ is a prime ideal in $\mathbb{Z}$. If $\alpha \in P$, then the constant term of the minimal polynomial for α over $\mathbb{Q}$ is then an element in $P \cap \mathbb{Z}$, which shows that $P \cap \mathbb{Z} = p\mathbb{Z}$ is also a nonzero prime ideal in $\mathbb{Z}$. By Theorem 26, every prime ideal (p) in $\mathbb{Z}$ arises in this way. Since $p\mathbb{Z}$ is a maximal ideal, it also follows from (2) in Theorem 26 that *nonzero prime ideals in $\mathcal{O}_K$ are maximal*, and then by Corollary 27, there are finitely many prime ideals P in $\mathcal{O}_K$ with $P \cap \mathbb{Z} = p\mathbb{Z}$. We shall see later (Corollary 16 in Section 16.3) that *every nonzero ideal in the ring of integers of a number field can be written uniquely as the product of prime ideals*, and in the case of the ideal $p\mathcal{O}_K$ the distinct prime factors are precisely the finitely many ideals P in $\mathcal{O}_K$ with $P \cap \mathbb{Z} = p\mathbb{Z}$. This property replaces the unique factorization of *elements* in $\mathcal{O}_K$ into primes (which need not hold since $\mathcal{O}_K$ *need not be a U.F.D.*). We shall also see that primary ideals in $\mathcal{O}_K$ are powers of prime ideals (in fact this is equivalent to the unique factorization of ideals of $\mathcal{O}_K$ into products of prime ideals, cf. the exercises).

Example: (The Ring of Integers in Quadratic Extensions of $\mathbb{Q}$)

If K is a quadratic extension of $\mathbb{Q}$ then $K = \mathbb{Q}(\sqrt{D})$ for some squarefree integer D. Then

$$\mathcal{O}_{\mathbb{Q}(\sqrt{D})} = \mathbb{Z}[\omega] = \mathbb{Z} \cdot 1 + \mathbb{Z} \cdot \omega,$$

with integral basis $1, \omega$, where

$$\omega = \begin{cases} \sqrt{D}, & \text{if } D \equiv 2, 3 \bmod 4 \\ \dfrac{1 + \sqrt{D}}{2}, & \text{if } D \equiv 1 \bmod 4. \end{cases}$$

This is the quadratic integer ring introduced in Section 7.1. Since ω satisfies $\omega^2 - D = 0$ (respectively, $\omega^2 - \omega + (1 - D)/4$) for $D \equiv 2, 3 \bmod 4$ (respectively, $D \equiv 1 \bmod 4$), it follows that ω is an algebraic integer in K and so $\mathbb{Z}[\omega] \subseteq \mathcal{O}_K$. To prove that this is the full ring of integers in K, let $\alpha = a + b\sqrt{D}$ with $a, b \in \mathbb{Q}$, and suppose that α is an algebraic integer. If $b = 0$, then $\alpha \in \mathbb{Q}$ and so $a \in \mathbb{Z}$. If $b \neq 0$, the minimal polynomial of α is $x^2 - 2ax + (a^2 - b^2 D)$. Then Proposition 28 shows that $2a$ and $a^2 - b^2 D$ are elements of $\mathbb{Z}$. Then $4(a^2 - b^2 D) = (2a)^2 - (2b)^2 D \in \mathbb{Z}$, hence $4b^2 D \in \mathbb{Z}$. Since D is squarefree it follows that $2b$ is an integer. Write $a = x/2$ and $b = y/2$ for some integers x, y. Since $a^2 - b^2 D$ is an integer, $x^2 - y^2 D \equiv 0 \pmod 4$. Since 0 and 1 are the only squares mod 4 and D is not divisible by 4, it is easy to check that the only possibilities are the following:

    (i) $D \equiv 2$ or $3 \pmod 4$ and x, y are both even, or

    (ii) $D \equiv 1 \pmod 4$ and x, y are both even or both odd.

In case (i), $a, b \in \mathbb{Z}$ and $\alpha \in \mathbb{Z}[\omega]$. In case (ii), $a + b\sqrt{D} = r + s\omega$ where $r = (x - y)/2$ and $s = y$ are both integers, so again $\alpha \in \mathbb{Z}[\omega]$.

Example: (The Ring of Integers in Cyclotomic Fields)

The ring of integers in the cyclotomic field $\mathbb{Q}(\zeta_n)$ of nth roots of unity is $\mathbb{Z}[\zeta_n]$, where ζ_n is any primitive n^{th} root of 1. The elements $1, \zeta_n, \ldots, \zeta_n^{\varphi(n)-1}$ are an integral basis. It is clear that ζ_n is an algebraic integer since it is a root of $x^n - 1$, so the ring $\mathbb{Z}[\zeta_n]$ is contained in the ring of integers. The proof that this is the full ring of algebraic integers in $\mathbb{Q}(\zeta_n)$ involves techniques from algebraic number theory beyond the scope of the material here.

Noether's Normalization Lemma and Hilbert's Nullstellensatz

We now apply some of the techniques from the algebraic theory of integral ring extensions to affine geometry.

Definition. If k is a field the elements $y_1, y_2, \ldots, y_q$ in some k-algebra are called *algebraically independent* over k if there is no nonzero polynomial p in q variables over k such that $p(y_1, y_2, \ldots, y_q) = 0$.

Thus $y_1, y_2, \ldots, y_q$ are algebraically independent if and only if the k-algebra homomorphism from the polynomial ring $k[x_1, \ldots, x_q]$ to $k[y_1, \ldots, y_q]$ defined by $x_i \mapsto y_i$ is an isomorphism. Elements in a field extension of k are algebraically independent if and only if they are independent transcendentals over k.

Theorem 30. *(Noether's Normalization Lemma)* Let k be a field and suppose that $A = k[r_1, r_2, \ldots, r_m]$ is a finitely generated k-algebra. Then for some q, $0 \le q \le m$, there are algebraically independent elements $y_1, y_2, \ldots, y_q \in A$ such that A is integral over $k[y_1, y_2, \ldots, y_q]$.

Proof: Proceed by induction on m. If $r_1, \ldots, r_m$ are algebraically independent over k then take $y_i = r_i$, $i = 1, \ldots, m$. Otherwise, there exists $f(x_1, \ldots, x_m) \in k[x_1, \ldots, x_m]$ such that $f(r_1, \ldots, r_m) = 0$. The polynomial f is a sum of monomials of the form $a x_1^{e_1} x_2^{e_2} \cdots x_m^{e_m}$, where the degree of this monomial is $e_1 + \cdots + e_m$ and the degree, d, of f is the maximum of the degrees of its monomials. Renumbering the variables if necessary, we may assume that f is a nonconstant polynomial in x_m with coefficients in the ring $k[x_1, x_2, \ldots, x_{m-1}]$. We now perform a change of variables that transforms (or "normalizes") f into a *monic* polynomial in x_m with coefficients from a subring of A which is generated over k by $m - 1$ elements, at which point we shall be able to apply induction.

Define integers $\alpha_i = (1 + d)^i$ and new variables $X_i = x_i - x_m^{\alpha_i}$ for $1 \le i \le m - 1$. Let

$$g(X_1, X_2, \ldots, X_{m-1}, x_m) = f(X_1 + x_m^{\alpha_1}, X_2 + x_m^{\alpha_2}, \ldots, X_{m-1} + x_m^{\alpha_{m-1}}, x_m),$$

so $g \in k[X_1, \ldots, X_{m-1}, x_m]$. Each monomial term of f contributes a single term of the form a constant times x_m^e to g. It is also easy to check that the choice of α_i ensures that distinct monomials in f give different values of e (for example by viewing the degrees of the monomials in the new variables as integers expressed in base $b = d + 1$). If N is the highest power of x_m that occurs, then it follows that

$$g = c x_m^N + \sum_{i=0}^{N-1} h_i(X_1, \ldots, X_{m-1}) x_m^i$$

for some nonzero $c \in k$. If now $s_i = r_i - r_m^{\alpha_i}$ then

$$\frac{1}{c} g(s_1, s_2, \ldots, s_{m-1}, r_m) = \frac{1}{c} f(r_1, r_2, \ldots, r_{m-1}, r_m) = 0,$$

which shows that r_m is integral over $B = k[s_1, \ldots, s_{m-1}]$. Each r_i for $1 \le i \le m - 1$ is integral over $B[r_m]$ since r_i is a root of the monic polynomial $x - s_i - r_m^{\alpha_i}$, so A is integral over $B[r_m]$. By transitivity of integrality, A is therefore integral over B. Since B is a k-algebra generated by $m - 1$ elements, induction completes the proof.

A more "geometric" interpretation of Noether's Normalization Lemma is indicated in Exercise 15. We next use the Normalization Lemma to prove that if k is an algebraically closed field then the maximal ideals of the polynomial ring $k[x_1, x_2, \ldots, x_n]$ are of the form $(x_1 - a_1, \ldots, x_n - a_n)$ for some $a_1, \ldots, a_n \in k$. Viewing $k[x_1, x_2, \ldots, x_n]$ as the ring of polynomial functions on $\mathbb{A}^n$, this says that the maximal ideals correspond to the kernels of evaluation maps at points of $\mathbb{A}^n$ — similar to the corresponding result for rings of continuous functions on a compact set (cf. Exercises 33, 34 in Section 7.4).

Theorem 31. (*Hilbert's Nullstellensatz — Weak Form*) Let k be an algebraically closed field. Then M is a maximal ideal in the polynomial ring $k[x_1, x_2, \ldots, x_n]$ if and only if $M = (x_1 - a_1, \ldots, x_n - a_n)$ for some $a_1, \ldots, a_n \in k$. Equivalently, the maps $\mathcal{Z}$ and $\mathcal{I}$ give a bijective correspondence

$$\{\text{points in } \mathbb{A}^n\} \quad \underset{\mathcal{Z}}{\overset{\mathcal{I}}{\rightleftarrows}} \quad \{\text{maximal ideals in } k[\mathbb{A}^n]\}.$$

Moreover, if I is any proper ideal in $k[x_1, x_2, \ldots, x_n]$ then $\mathcal{Z}(I) \neq \emptyset$.

Proof: Certainly $(x_1 - a_1, \ldots, x_n - a_n)$ is a maximal ideal in $k[x_1, x_2, \ldots, x_n]$. Conversely, for any maximal ideal M in $k[x_1, x_2, \ldots, x_n]$, let $E = k[x_1, x_2, \ldots, x_n]/M$. Then E is a field containing k that is finitely generated over k (by $\bar{x}_i, \ldots, \bar{x}_n$). By Noether's Normalization Lemma, E is integral over a polynomial ring $k[y_1, \ldots, y_q]$. Then $k[y_1, \ldots, y_q]$ is a field by Theorem 26(1), and since a polynomial ring in one or more variables is never a field, it follows that $q = 0$. Hence E is integral over k, so E is algebraic over k. Because k is algebraically closed, $E = k$, i.e., $\bar{x}_i \in k$ for $1 \le i \le n$. Hence for $i = 1, \ldots, n$ there is some $a_i \in k$ such that $x_i - a_i \in M$. This means that the maximal ideal $(x_1 - a_1, \ldots, x_n - a_n)$ is contained in M, so $M = (x_1 - a_1, \ldots, x_n - a_n)$. Finally, if I is any nonzero ideal in $k[x_1, x_2, \ldots, x_n]$ then I is contained in a maximal ideal $M = (x_1 - a_1, \ldots, x_n - a_n)$, and so $(a_1, \ldots, a_n) \in \mathcal{Z}(I)$.

Theorem 32. (*Hilbert's Nullstellensatz*) Let k be an algebraically closed field. Then $\mathcal{I}(\mathcal{Z}(I)) = \text{rad } I$ for every ideal I of $k[x_1, x_2, \ldots, x_n]$. Moreover, the maps $\mathcal{Z}$ and $\mathcal{I}$ define inverse bijections

$$\{\text{affine algebraic sets}\} \quad \underset{\mathcal{Z}}{\overset{\mathcal{I}}{\rightleftarrows}} \quad \{\text{radical ideals}\}.$$

Proof: Since $\text{rad } I \subseteq \mathcal{I}(\mathcal{Z}(I))$ it remains to prove the reverse inclusion. By Hilbert's Basis Theorem, $I = (f_1, f_2, \ldots, f_m)$. Let $g \in \mathcal{I}(\mathcal{Z}(I))$. Introduce a new variable x_{n+1} and consider the ideal I' generated by $f_1, \ldots, f_m$ and $x_{n+1}g - 1$ in $k[x_1, \ldots, x_n, x_{n+1}]$. At any point of $\mathbb{A}^{n+1}$ where $f_1, \ldots, f_m$ vanish the polynomial g also vanishes since $g \in \mathcal{I}(\mathcal{Z}(I))$, so that $x_{n+1}g - 1$ is nonzero. Hence $\mathcal{Z}(I') = \emptyset$ in $\mathbb{A}^{n+1}$. By the Weak Form of the Nullstellensatz, I' cannot be a proper ideal, i.e., $1 \in I'$. Write

$$1 = a_1 f_1 + \cdots + a_m f_m + a_{m+1}(x_{n+1}g - 1) \qquad \text{for some } a_i \in k[x_1, \ldots, x_{n+1}].$$

Letting $y = 1/x_{n+1}$ and multiplying by a high power of y in this equation shows that

$$y^N = c_1 f_1 + \cdots + c_m f_m + c_{m+1}(g - y) \qquad \text{for some } c_i \in k[x_1, \ldots, x_n, y].$$

Substituting g for y in this polynomial equation shows that $g^N \in I$ (in $k[x_1, \ldots, x_n]$), i.e., $g \in \text{rad } I$. Hence $\mathcal{I}(\mathcal{Z}(I)) \subseteq \text{rad } I$ and so $\mathcal{I}(\mathcal{Z}(I)) = \text{rad } I$, completing the proof.

It follows directly from Proposition 12 and Theorem 26(2) that if S is an integral extension of R with $1 \in R$ and if I is an ideal of R, then

$$(\text{rad}_S\, IS) \cap R = \text{rad}_R\, I$$

where IS is the ideal generated by I in S, and the subscript indicates the ring in which the radicals are being computed. This has the following geometric interpretation.

Corollary 33. *(Variant of Hilbert's Nullstellensatz)* If k is any field with algebraic closure $\bar{k}$ and I is an ideal in $k[x_1, x_2, \ldots, x_n]$, then $\mathcal{I}_k(\mathcal{Z}_{\bar{k}}(I)) = \text{rad } I$, where $\mathcal{Z}_{\bar{k}}(I)$ is the zero set in $\bar{k}^n$ of the polynomials in I and $\mathcal{I}_k(\mathcal{Z}_{\bar{k}}(I))$ is the ideal of polynomials in $k[x_1, x_2, \ldots, x_n]$ vanishing at all the points in $\mathcal{Z}_{\bar{k}}(I)$. In particular, $I = (1)$ if and only if there are no common zeros in $\bar{k}^n$ of the polynomials in I.

Proof: Since $\bar{k}[x_1, x_2, \ldots, x_n]$ is an integral extension of $k[x_1, x_2, \ldots, x_n]$ (generated by the integral elements $\bar{k}$), the corollary follows immediately from Theorem 32 and the remarks on radicals above.

From the Nullstellensatz we now have a dictionary between geometric and ring-theoretic objects over the algebraically closed field k:

Geometry	Algebra
affine algebraic set V	coordinate ring $k[V]$
points of V	maximal ideals of $k[V]$
affine algebraic subsets in V	radical ideals of $k[V]$
subvarieties in V	prime ideals in $k[V]$
morphism $\varphi : V \to W$	k-algebra homomorphism $\widetilde{\varphi} : k[W] \to k[V]$

Computing Radicals

There are algorithms for computing radicals and primary decompositions in polynomial rings using Gröbner bases. While they are relatively elementary, they are somewhat technical and so we limit our discussion here to some preliminary results.

For hypersurfaces $V = \mathcal{Z}(f)$ defined by a single polynomial $f \in k[x_1, \ldots, x_n]$, determining $\mathcal{I}(V) = \text{rad}(f)$ is straightforward. Since $k[x_1, \ldots, x_n]$ is a U.F.D., f factors uniquely as the product of powers of nonassociate irreducibles: $f = p_1^{a_1} \cdots p_s^{a_s}$ and then $\text{rad}(f)$ is generated by $p_1 \cdots p_s$ (the 'squarefree part' of f).

Example

Suppose $W = \mathcal{Z}(J)$ with $J = (u^3 - uv^2 + v^3) \in \mathbb{Q}[u, v]$. The polynomial $x^3 - x + 1$ is irreducible over $\mathbb{Q}$, so $f = u^3 - uv^2 + v^3$ is irreducible in $\mathbb{Q}[u, v]$. Hence rad $J = J$ and $\mathcal{I}(W) = J$.

For nonprincipal ideals I, determining rad I is more complicated. The following proposition (based on Hilbert's Nullstellensatz) gives a criterion determining when an element is contained in rad I.

Proposition 34. Suppose k is any field. If $I = (f_1, \ldots, f_s)$ is a proper ideal in $k[x_1, \ldots, x_n]$, then $f \in$ rad I if and only if $(f_1, \ldots, f_s, 1 - yf) = k[x_1, \ldots, x_n, y]$.

Proof: By Corollary 33, $(f_1, \ldots, f_s, 1 - yf) = k[x_1, \ldots, x_n, y]$ if and only if the equations

$$1 - yf(x_1, \ldots, x_n) = 0, \quad f_1(x_1, \ldots, x_n) = 0, \quad \ldots, \quad f_s(x_1, \ldots, x_n) = 0$$

have no common zero over the algebraic closure $\bar{k}$ of k. For a given $(a_1, \ldots, a_n) \in \bar{k}^n$, the equation $1 - yf(a_1, \ldots, a_n) = 0$ has a solution y unless $f(a_1, \ldots, a_n) = 0$. Hence, the system of equations has no common zero if and only if for every $(a_1, \ldots, a_n) \in \bar{k}^n$ with $f_1(a_1, \ldots, a_n) = \cdots = f_s(a_1, \ldots, a_n) = 0$ we also have $f(a_1, \ldots, a_n) = 0$. Equivalently, if $(a_1, \ldots, a_n) \in \mathcal{Z}_{\bar{k}}(I)$, then also $f(a_1, \ldots, a_n) = 0$, i.e., we have $f \in \mathcal{I}_k(\mathcal{Z}_{\bar{k}}(I)) =$ rad I, by Corollary 33.

Since the reduced Gröbner basis (with respect to any fixed monomial ordering) for an ideal is unique, we immediately obtain the following algorithmic method for determining when a polynomial lies in the radical of an ideal.

Corollary 35. Suppose $I = (f_1, \ldots, f_s)$ in $k[x_1, \ldots, x_n]$. Then $f \in$ rad I if and only if $\{1\}$ is the reduced Gröbner basis for the ideal $(f_1, \ldots, f_s, 1 - yf)$ in $k[x_1, \ldots, x_n, y]$ with respect to any monomial ordering.

Example

Consider $I = (x^2 - y^2, xy)$ in $k[x, y]$. The reduced Gröbner basis for $(x^2 - y^2, xy, 1 - tx)$ in $k[x, y, t]$ with respect to the order $x > y > t$ is $\{1\}$, showing $x \in$ rad(I). To determine the smallest power of x lying in I, we find that the ideal $(x^2 - y^2, xy, x^3)$ in $k[x, y]$ has the same reduced Gröbner basis as I (namely $\{x^2 - y^2, xy, y^3\}$), but $(x^2 - y^2, x^2, xy)$ has basis $\{x^2, xy, y^2\}$. It follows that $x^3 \in I$ and $x^2 \notin I$ (alternatively, x^3 leaves a nonzero remainder after general polynomial division by $\{x^2 - y^2, xy, y^3\}$, but x^3 has a remainder of 0). By a similar computation (or by symmetry), $y \in$ rad I, with $y^3 \in I$ but $y^2 \notin I$. Since $(x, y) \subseteq$ rad I, it follows that rad $I = (x, y)$.

Some additional results for computing radicals are presented in the exercises.

EXERCISES

Let R be a subring of the commutative ring S with $1 \in R$.

1. Use the fact that a U.F.D. is integrally closed to prove that the Gaussian integers, $\mathbb{Z}[i]$, is the ring of integers in $\mathbb{Q}(i)$.

2. Suppose k is a field and let $t = \bar{x}/\bar{y}$ in the field of fractions of the integral domain $R = k[x, y]/(x^2 - y^3)$. Prove that $K = k(t)$ is the fraction field of R and $k[t]$ is the integral closure of R in K.

3. Suppose k is a field and i and j are relatively prime positive integers. Find the normalization of the integral domain $R = k[x, y]/(x^i - y^j)$ (cf. Exercise 14, Section 9.1).

4. Suppose k is a field and let P be the ideal $(y^2 - x^3 - x^2)$ in the polynomial ring $k[x, y]$. Prove that P is a prime ideal and find the normalization of the integral domain $R = k[x, y]/P$. [To prove P is prime, show that $y^2 - x^3 - x^2$ is irreducible in the U.F.D. $k[x, y]$. Then consider $t = \bar{y}/\bar{x} \in R$.]

5. If R is an integral domain with field of fractions F, show that F is a finitely generated R-module if and only if $R = F$.

6. For each of the following give specific rings $R \subseteq S$ and explicit ideals in these rings that exhibit the specified relation:
   (a) an ideal I of R such that $I \neq SI \cap R$ (so the contraction of the extension of an ideal I need not equal I)
   (b) a prime ideal P of R such that there is no prime ideal Q of S with $P = Q \cap R$
   (c) a maximal ideal M of S such that $M \cap R$ is not maximal in R
   (d) a prime ideal P of R whose extension PS to S is not a prime ideal in S
   (e) an ideal J of S such that $J \neq (J \cap R)S$ (so the extension of the contraction of an ideal J need not equal J).

7. Let $\mathcal{O}_K$ be the ring of integers in a number field K.
   (a) Suppose that every nonzero ideal I of $\mathcal{O}_K$ can be written as the product of powers of prime ideals. Prove that an ideal Q of $\mathcal{O}_K$ is P-primary if and only if $Q = P^m$ for some $m \geq 1$. [Show first that since nonzero primes in $\mathcal{O}_K$ are maximal that $P_1^{m_1} \subseteq P_2^{m_2}$ for distinct nonzero primes P_1, P_2 implies $P_1 = P_2$.]
   (b) Suppose that an ideal Q of $\mathcal{O}_K$ is P-primary if and only if $Q = P^m$ for some $m \geq 1$. Assuming all of Theorem 21, prove that every nonzero ideal I of $\mathcal{O}_K$ can be written uniquely as the product of powers of prime ideals. [Prove that $P_1^{m_1}$ and $P_2^{m_2}$ are comaximal ideals if P_1 and P_2 are distinct nonzero prime ideals and use the Chinese Remainder Theorem.]

8. Prove that if $s_1, \ldots, s_n \in S$ are integral over R, then the ring $R[s_1, \ldots, s_n]$ is a finitely generated R-module.

9. Suppose that S is integral over R and that P is a prime ideal in R. Prove that every element s in the ideal PS generated by P in S satisfies an equation $s^n + a_{n-1}s^{n-1} + \cdots + a_1 s + a_0 = 0$ where the coefficients $a_0, a_1, \ldots, a_{n-1}$ are elements of P. [If $s = p_1 s_1 + \cdots + p_m s_m \in PS$, show that $T = R[s_1, \ldots, s_m]$ satisfies the hypotheses in Proposition 23(3). Follow the proof in Proposition 23 that s is integral, noting that $s \in PT$ so that the a_{ij} are elements of P.]

10. Prove the following generalization of Proposition 28: Suppose R is an integrally closed integral domain with field of fractions k and α is an element of an extension field K of k. Show that α is integral over R if and only if α is algebraic over k and the minimal polynomial $m_{\alpha,k}(x)$ for α over k has coefficients in R. [If α is integral prove the conjugates

of α, i.e., the roots of $m_{\alpha,k}(x)$, are also integral, so the elementary symmetric functions of the conjugates are elements of k that are integral over R.]

11. Suppose R is an integrally closed integral domain with field of fractions k and $p(x) \in R[x]$ is a monic polynomial. Show that if $p(x) = a(x)b(x)$ with monic polynomials $a(x), b(x) \in k[x]$ then $a(x), b(x) \in R[x]$ (compare to Gauss' Lemma, Proposition 5, Section 9.3). [See the previous exercise.]

12. Suppose S is an integral domain that is integral over a ring R as in the previous exercise. If P is a prime ideal in R, let s be any element in the ideal PS generated by P in S. Prove that, with the exception of the leading term, the coefficients of the minimal polynomial $m_{s,k}(x)$ for s over k are elements of P. [By Exercise 10, $m_{s,k}(x) \in R[x]$. Exercise 9 shows that s is a root of a monic polynomial $p(x) = x^n + a_{n-1}x^{n-1} + \cdots + a_0$ with $a_0, \ldots, a_{n-1} \in P$. Use the previous exercise to show that $p(x) = m_{s,k}(x)b(x)$ with $b(x)$ in $R[x]$, and consider this equation in the integral domain $(R/P)[x]$.]

The next two exercises extend Exercise 6 in Section 7.5 by characterizing fields that are not fields of fractions of any of their proper subrings.

13. Let K be a field of characteristic 0 and let A be a subring of K maximal with respect to $1/2 \notin A$. (Such A exists by Zorn's Lemma.) Let F be the field of fractions of A in K.
  (a) Show that K is algebraic over F. [If t is transcendental over F, show that $1/2 \notin A[t]$.]
  (b) Show that A is integrally closed in K. [Show that $1/2$ is not in the integral closure of A in K.]
  (c) Deduce from (a) and (b) that $K = F$.

14. Show that a field K is the field of fractions of some proper subring of K if and only if K is not a subfield of the algebraic closure of a finite field. [If K contains t transcendental over $\mathbb{F}_p$ argue as in the preceding exercise with $1/t$ in place of $1/2$ to show that K is the quotient field of some proper subring.]

The next exercise gives a "geometric" interpretation of Noether's Normalization Lemma, showing that every affine algebraic set is a *finite covering* of some affine n-space.

15. Let V be an affine algebraic set over an algebraically closed field k. Prove that for some n there is a surjective morphism from V onto $\mathbb{A}^n$ with finite fibers, and that if V is a variety, then n can be taken to be the dimension of V. [By Noether's Normalization Lemma the finitely generated k-algebra $S = k[V]$ contains a polynomial subalgebra $R = k[x_1, x_2, \ldots, x_n]$ such that S is integral over R. Apply Theorem 6 to the inclusion of R in S to obtain a morphism φ from V to $\mathbb{A}^n$. To see that φ is surjective with finite fibers, apply Corollary 27 to the maximal ideal $(x_1 - a_1, \ldots, x_n - a_n)$ of R corresponding to a point $(a_1, \ldots, a_n)$ of $\mathbb{A}^n$.]

16. Let V be an affine algebraic set in $\mathbb{C}^n$. Prove that V is compact in the Euclidean topology (i.e., closed and bounded) if and only if it is finite. [Use Exercise 18 in Section 2, the previous exercise, and the behavior of compact sets with respect to continuous functions.]

17. Let R be a subring of the commutative ring S with $1_S \in R$ and suppose that S is integral over R. This exercise proves that R and S have the same *Krull dimension*, cf. Section 16.1.
  (a) If $P_1 \subset P_2 \subset \cdots \subset P_n$ is a chain of distinct prime ideals in R prove that there is a chain $Q_1 \subset Q_2 \subset \cdots \subset Q_n$ of distinct prime ideals in S with $Q_i \cap R = P_i$.
  (b) Prove conversely that if $Q_1 \subset Q_2 \subset \cdots \subset Q_n$ is a chain of distinct prime ideals in S and $P_i = Q_i \cap R$ then $P_1 \subset P_2 \subset \cdots \subset P_n$ is a chain of distinct prime ideals in R. [To prove the P_i are distinct, pass to a quotient and reduce the problem to showing that if Q is a nonzero prime ideal in the integral domain S then $Q \cap R$ is a nonzero prime

ideal in R. In this case, if $s \in Q$ is nonzero, show that the constant coefficient of a polynomial of minimal degree in $R[x]$ satisfied by s is a nonzero element in $Q \cap R$.]

18. Let $V = \mathcal{Z}(I)$ and $W = \mathcal{Z}(J)$ where I is the ideal $(uv + v) \subset \mathbb{C}[u, v]$ and J is the ideal $(-2y - y^2 + 2z + z^2, 2x - yz - z^2) \subset \mathbb{C}[x, y, z]$.
   (a) Show that I and J are prime ideals. Conclude that $I = \mathcal{I}(V)$ and $J = \mathcal{I}(W)$ and that V and W are varieties.
   (b) Show that the map $\varphi : V \to W$ defined by $\varphi((a_1, a_2)) = (a_1^2 + a_2, a_1 + a_2, a_1 - a_2)$ is an isomorphism.

19. Let $I = (x^3 + y^3 + z^3, x^2 + y^2 + z^2, (x + y + z)^3) \subset k[x, y, z]$. Use Gröbner bases to show that $x, y, z \in \mathrm{rad}\, I$ if $\mathrm{ch}(k) \neq 2, 3$.

20. Let $I = (x^3 + y^3 + z^3, xy + xz + yz, xyz) \subset k[x, y, z]$. Use Gröbner bases to show that $x, y, z \in \mathrm{rad}\, I$.

21. Let $I = (x^4 + y^4 + z^4, x + y + z) \subset k[x, y, z]$.
   (a) Use Gröbner bases to show that $xy + xz + yz \in \mathrm{rad}\, I$ if $\mathrm{ch}(k) \neq 2$ and determine the smallest power of $xy + xz + yz$ contained in I. Show that none of x, y or z is contained in $\mathrm{rad}\, I$.
   (b) If $J = (x^4 + y^4 + z^4, x + y + z, xy + xz + yz)$ show that the reduced Gröbner basis of J relative to the lexicographic ordering $x > y > z$ is $\{x + y + z, y^2 + yz + z^2\}$. Deduce that $k[x, y, z]/J \cong k[y, z]/(y^2 + yz + z^2)$ and that J is radical if $\mathrm{ch}(k) \neq 3$.
   (c) If $\mathrm{ch}(k) \neq 2, 3$, show that $\mathrm{rad}\, I = J$.
   (d) If $\mathrm{ch}(k) = 3$, show that $\mathrm{rad}\, I = (x - y, y - z)$.
   (e) If $\mathrm{ch}(k) = 2$, show that $I = (x + y + z)$ is a prime, hence radical, ideal.

22. Let $I = (x^2y + z^3, x + y^3 - z, 2y^4z - yz^2 - z^3) \subset k[x, y, z]$. Use Gröbner bases to show that $x, y, z \in \mathrm{rad}\, I$ and conclude that $\mathrm{rad}\, I = (x, y, z)$. Show that x^9, y^7, z^9 are the smallest powers of x, y, z, respectively, lying in I.

23. Let $V = \mathcal{Z}(x^3 - x^2z - y^2z)$ and $W = \mathcal{Z}(x^2 + y^2 - z^2)$ in $\mathbb{C}^3$. Show that $\mathcal{I}(V) = (x^3 - x^2z - y^2z)$ and $\mathcal{I}(W) = (x^2 + y^2 - z^2)$ in $\mathbb{C}[x, y, z]$.

24. Let $V = \mathcal{Z}(x^3 + y^3 + 7z^3) \subset \mathbb{C}^3$. Show that $\mathcal{I}(V) = (x^3 + y^3 + 7z^3)$ in $\mathbb{C}[x, y, z]$.

25. Let $I = (xz + y^2 + z^2, xy - xz + yz - 2z^2)$ and let $K = I + (x^2 - 3y^2 + yz) \subset \mathbb{C}[x, y, z]$.
   (a) By Exercise 46 in Section 1, there is an injective $\mathbb{C}$-algebra homomorphism from $\mathbb{C}[x, y, z]/K$ to $\mathbb{C}[u, v]/(u^3 - uv^2 + v^3)$. Use this together with the example preceding Proposition 34 to prove that K is a radical ideal and deduce that $\mathrm{rad}\, I \subseteq K$.
   (b) Show that $\mathrm{rad}\, I \subseteq (y, z)$.
   (c) Show that $K \cap (y, z) = I$ and deduce that I is radical, so that $\mathcal{I}(V) = I$ if $V = \mathcal{Z}(I)$.
   (d) Show that $y(x^2 - 3y^2 + yz)$ and $z(x^2 - 3y^2 + yz)$ are elements of I but none of y, z, or $x^2 - 3y^2 + yz$ is contained in I.

26. Let I be an ideal in $k[x_1, \ldots, x_n]$. Prove that the following are equivalent (an ideal satisfying any of these conditions is called a *zero-dimensional ideal* because of (d)):
   (a) The quotient $k[x_1, \ldots, x_n]/I$ has finite dimension as a vector space over k.
   (b) $I \cap k[x_i] \neq 0$ for each $i = 1, 2, \ldots, n$.
   (c) If G is any reduced Gröbner basis for I then for each $i = 1, \ldots, n$, there is a $g_i \in G$ with leading term $x_i^{n_i}$ for some $n_i \geq 1$.
   (d) The set of common zeros $\mathcal{Z}_{\bar{k}}(I)$ of the polynomials in I in an algebraic closure $\bar{k}$ of k is finite.

   [For (a) implies (b) use the injection $k[x_i]/(I \cap k[x_i]) \hookrightarrow k[x_1, \ldots, x_n]/I$. For (b) implies (c) note some $LT(g_i)$ divides the leading term of a generator for $I \cap k[x_i]$. For (c) implies (a) use Exercise 37 in Section 9.6. Show (b) implies (d). For (d) implies (b) show the product $m_{a_1,k}(x_i) \ldots m_{a_N,k}(x_i)$ of the minimal polynomials of the ith coordinates $a_1, \ldots, a_N$ of the points in $\mathcal{Z}_{\bar{k}}(I)$ is a nonzero polynomial in $\mathcal{I}(\mathcal{Z}_{\bar{k}}(I))$ and apply Corollary 33.]

27. Let I be a zero-dimensional ideal in $k[x_1, \ldots, x_n]$ and let I' be the ideal generated by I in $\bar{k}[x_1, \ldots, x_n]$ where $\bar{k}$ is the algebraic closure of k. Let $\mathcal{Z}(I)$ be the zero set of I in k^n and let $\mathcal{Z}_{\bar{k}}(I)$ be the zero set of I (equivalently, of I') in $\bar{k}^n$.
    (a) Prove that $|\mathcal{Z}_{\bar{k}}(I)| = \dim_{\bar{k}} \bar{k}[x_1, \ldots, x_n]/\operatorname{rad} I'$. [Show that rad I' is the product of the maximal ideals corresponding to the points in $V_{\bar{k}}$ and use the Chinese Remainder Theorem.]
    (b) Show $|\mathcal{Z}(I)| \le \dim_k k[x_1, \ldots, x_n]/I$. [One approach: use Exercise 43 in Section 1 and observe that $\dim_{\bar{k}} \bar{k}[x_1, \ldots, x_n]/\operatorname{rad} I' \le \dim_{\bar{k}} \bar{k}[x_1, \ldots, x_n]/I'$.]

28. Suppose I is a zero-dimensional ideal in $k[x_1, \ldots, x_n]$, and suppose $I \cap k[x_i]$ is generated by the nonzero polynomial h_i (cf. Exercise 26). Let r_i be the product of the irreducible factors of h_i (the 'squarefree part' of h_i).
    (a) Prove that $I + (r_1, \ldots, r_n) \subseteq \operatorname{rad} I$.
    (b) *(Radicals of zero-dimensional ideals for perfect fields)* If k is a perfect field, prove that rad $I = I + (r_1, \ldots, r_n)$. [Use induction on n. Write $r_1 = p_1 \ldots p_t$ with distinct irreducibles p_i in $k[x_1]$. If $J = I + (r_1, \ldots, r_n)$ show that $J = J_1 \cap \cdots \cap J_t$ where $J_t = J + (p_t)$. Show for each i that reduction modulo p_i induces an isomorphism $k[x_1, \ldots, x_n]/J_i \cong K[x_2, \ldots, x_n]/J_i'$ where K is the extension field $k[x]/(p_i)$ and $J_i' \subseteq K[x_2, \ldots, x_n]$ is the reduction of the ideal J_i modulo (p_i). Use Exercise 11 of Section 13.5 to show that the image of r_j in $J_i' \cap K[x_j]$ remains a nonzero squarefree polynomial for each $j = 2, \ldots, n$ since k is perfect. Conclude by induction that J_i' is a radical ideal. Deduce that J_i is a radical ideal, and finally that J is a radical ideal.]
    (c) Find the radicals of $(x^7 + x + y^3, x^4 + y^3 + y)$, $(x^3 - xy^2 + x, x^2y + y^3)$, and $(x^4 + y^3, x^3 - xy + y^2)$ in $\mathbb{Q}[x, y]$ and of $(x^2 + y^2z, x^2y^2 + z^3, y^2 + z^2)$ in $\mathbb{Q}[x, y, z]$.
    (d) Let $k = \mathbb{F}_p(t)$. Show that $I = (x^p + t, y^p - t)$ is a zero-dimensional ideal in $k[x, y]$ such that both $I \cap k[x]$ and $I \cap k[y]$ contain nonzero squarefree polynomials, but that I is not a radical ideal (so the result in (b) need not hold if k is not perfect). [Show that $x + y \in \operatorname{rad} I$ but $x + y \notin I$.]

15.4 LOCALIZATION

The idea of "localization at a prime" in a ring is an extremely powerful and pervasive tool in algebra for isolating the behavior of the ideals in a ring. It is an algebraic analogue of the familiar idea of localizing at a point when considering questions of, for example, the differentiability of a function $f(x)$ on the real line. In fact one of the important applications (and also one of the original motivations for the development) of this technique is to translate such "local" properties in the geometry of affine algebraic spaces to corresponding properties of their coordinate rings.

We first consider a very general construction of "rings of fractions." Let D be a multiplicatively closed subset of R containing 1 (i.e., $1 \in D$ and $ab \in D$ if $a, b \in D$). The next result constructs a new ring $D^{-1}R$ which is the "smallest" ring in which the elements of D become units. This generalizes the construction of rings of fractions in Section 7.5 by allowing D to contain zero or zero divisors, and so in this case R need not embed as a subring of $D^{-1}R$.

Theorem 36. Let R be a commutative ring with $1 \neq 0$ and let D be a multiplicatively closed subset of R containing 1. Then there is a commutative ring $D^{-1}R$ and a ring homomorphism $\pi : R \to D^{-1}R$ satisfying the following universal property: for any homomorphism $\psi : R \to S$ of commutative rings that sends 1 to 1 such that $\psi(d)$ is a unit in S for every $d \in D$, there is a unique homomorphism $\Psi : D^{-1}R \to S$ such that $\Psi \circ \pi = \psi$.

Proof: The proof is very similar to the proof of Theorem 15 in Section 7.5. In this case we define a relation on $R \times D$ by

$$(r, d) \sim (s, e) \quad \text{if and only if} \quad x(er - ds) = 0 \quad \text{for some } x \in D.$$

This relation is clearly reflexive and symmetric. If $(r, d) \sim (s, e)$ and $(s, e) \sim (t, f)$ then $x(er - ds) = 0$ and $y(fs - et) = 0$ for some $x, y \in D$. Multiplying the first equation by fy and the second by dx and adding gives $exy(fr - dt) = 0$. Since D is closed under multiplication, $(r, d) \sim (t, f)$ and so $\sim$ is transitive.

Let r/d denote the equivalence class of (r, d) under $\sim$ and let $D^{-1}R$ be the set of these equivalence classes. Define addition and multiplication in $D^{-1}R$ by

$$\frac{a}{b} + \frac{c}{d} = \frac{ad + bc}{bd} \qquad \text{and} \qquad \frac{a}{b} \times \frac{c}{d} = \frac{ac}{bd}.$$

It is an exercise to check that these operations are well defined and make $D^{-1}R$ into a commutative ring with $1 = 1/1$. For each $d \in D$, $d/1$ is a unit in $D^{-1}R$ (even in the degenerate case when $D^{-1}R$ is the zero ring).

Finally, define $\pi : R \to D^{-1}R$ by $\pi(r) = r/1$. It follows easily that π is a ring homomorphism. Suppose that $\psi : R \to S$ is a homomorphism of commutative rings that sends 1 to 1 such that $\psi(d)$ is a unit in S for every $d \in D$. Define

$$\Psi : D^{-1}R \to S \qquad \text{by} \qquad \Psi\left(\frac{r}{d}\right) = \psi(r)\psi(d)^{-1}.$$

This map is well defined because if $r/d = s/e$ then $x(er - ds) = 0$ for some $x \in D$. Then $\psi(x)(\psi(er) - \psi(ds)) = 0$ in S, so $\psi(er) - \psi(ds) = 0$ since $\psi(x)$ is a unit in S, and therefore $\psi(r)\psi(d)^{-1} = \psi(s)\psi(e)^{-1}$. It is immediate that Ψ is a ring homomorphism and $\Psi \circ \pi = \psi$.

Finally, Ψ is unique because every element of $D^{-1}R$ can be written as a product $(r/1)(d/1)^{-1}$. The value of Ψ on each element of the form $x/1$ is uniquely determined by ψ, namely $\Psi(x/1) = \Psi(\pi(x)) = \psi(x)$. Since Ψ is a ring homomorphism, its value on u^{-1} for any unit u is uniquely determined by $\Psi(u)$. Thus Ψ is uniquely determined on every element of $D^{-1}R$, completing the proof.

Corollary 37. In the notation of Theorem 36,
  (1) $\ker \pi = \{r \in R \mid xr = 0 \text{ for some } x \in D\}$; in particular, $\pi : R \to D^{-1}R$ is an injection if and only if D contains neither zero nor any zero divisors of R, and
  (2) $D^{-1}R = 0$ if and only if $0 \in D$, hence if and only if D contains nilpotent elements.

Proof: By definition, we have $\pi(r) = 0$ if and only if $(r, 1) \sim (0, 1)$, i.e., if and only if $xr = 0$ for some $x \in D$, which is (1). For (2), note that $D^{-1}R = 0$ if and only if the 1 of this ring is zero, i.e., $(1, 1) \sim (0, 1)$. This occurs if and only if $x1 = 0$ for some $x \in D$, i.e., if and only if $0 \in D$.

Definition. The ring $D^{-1}R$ is called the *ring of fractions of R with respect to D* or the *localization of R at D*.

Examples

(1) Let R be an integral domain and let $D = R - \{0\}$. Then $D^{-1}R$ is the field of fractions, Q, of R described in Section 7.5. More generally, if D is any multiplicatively closed subset of $R - \{0\}$, then $D^{-1}R$ is the subring of Q consisting of elements r/d with $r \in R$ and $d \in D$.

(2) Let R be any commutative ring with 1 and let f be any element of R. Let D be the multiplicative set $\{f^n \mid n \geq 0\}$ of nonnegative powers of f in R. Define $R_f = D^{-1}R$. Note that $R_f = 0$ if and only if f is nilpotent. If f is not nilpotent, then f becomes a unit in R_f. It is not difficult to see that

$$R_f \cong R[x]/(xf - 1),$$

where $R[x]$ is the polynomial ring in the variable x (cf. the exercises). Note also that R_f and R_{f^n} are naturally isomorphic for any $n \geq 1$ since both f and f^n are units in both rings. If f is a zero divisor then $\pi : R \to R_f$ does not embed R into R_f. For example, let $R = k[x, y]/(xy)$, and take $f = x$. Then x is a unit in R_x and y is mapped to 0 by the first part of the corollary (explicitly: $y = xy/x = 0$ in R_x). In this case $\pi(R) = k[x] \subset R_f = k[x, x^{-1}]$.

(3) (*Localizing at a Prime*) Let P be a prime ideal in any ring R and let $D = R - P$. By definition of a prime ideal D is multiplicatively closed. Passing to the ring $D^{-1}R$ in this case is called *localizing R at P* and the ring $D^{-1}R$ is denoted by R_P. Every element of R not in P becomes a unit in R_P. For example, if $R = \mathbb{Z}$ and $P = (p)$ is a prime ideal, then

$$\mathbb{Z}_{(p)} = \{\frac{a}{b} \in \mathbb{Q} \mid p \nmid b\} \subseteq \mathbb{Q}$$

and every integer b not divisible by p is a unit.

(4) If V is any nonempty set and k is a field, let R be any ring of k-valued functions on V containing the constant functions (for instance, the ring of all continuous real valued functions on the closed interval $[0, 1]$). For any $a \in V$ let M_a be the ideal of functions in R that vanish at a. Then M_a is the kernel of the ring homomorphism from R to the field k given by evaluating each function in R at a. Since R contains the constant functions, evaluation is surjective and so M_a is a maximal (hence also prime) ideal. The localization of R at this prime ideal is then

$$R_{M_a} = \left\{\frac{f}{g} \mid f, g \in R, \ g(a) \neq 0\right\}.$$

Each function in R_{M_a} can then be evaluated at a by $(f/g)(a) = f(a)/g(a)$, and this value does not depend on the choice of representative for the class f/g, so R_{M_a} becomes a ring of k-valued "rational functions" defined at a.

We next consider extensions and contractions of ideals with respect to the map $\pi : R \to D^{-1}R$ in Theorem 36. To ease some of the notation, if I is an ideal of R, let ^eI denote the extension of I to $D^{-1}R$ (instead of the more cumbersome $D^{-1}R \, \pi(I)$), and if J is an ideal of $D^{-1}R$, let ^cJ denote the contraction of J to R.

If I is an ideal of R then it is easy to see that every element of ^eI can be written in the form a/d for some $a \in I$ and $d \in D$, so the extension of I to $D^{-1}R$ is also frequently denoted by $D^{-1}I$.

Proposition 38. In the preceding notation we have
(1) For any ideal J of $D^{-1}R$ we have $J = {}^e({}^cJ)$. In particular, every ideal of $D^{-1}R$ is the extension of some ideal of R, and distinct ideals of $D^{-1}R$ have distinct contractions in R.
(2) For any ideal I of R we have

$$^c({}^eI) = \{r \in R \mid dr \in I \text{ for some } d \in D\}.$$

Also, $^eI = D^{-1}R$ if and only if $I \cap D \neq \emptyset$.
(3) Extension and contraction give a bijective correspondence

$$\left\{ \begin{matrix} \text{prime ideals } P \text{ of } R \\ \text{with } P \cap D = \emptyset \end{matrix} \right\} \quad \begin{matrix} e \\ \longrightarrow \\ \longleftarrow \\ c \end{matrix} \quad \left\{ \text{prime ideals of } D^{-1}R \right\}.$$

(4) If R is Noetherian (or Artinian) then $D^{-1}R$ is Noetherian (Artinian, respectively).

Proof: We always have $^e({}^cJ) \subseteq J$. For the reverse inclusion let $a/d \in J$. Then $a/1 = d(a/d) \in J$, and so $a \in \pi^{-1}(J) = {}^cJ$. Thus $a/1 \in {}^e({}^cJ)$, so we also have $(a/1)(1/d) = a/d \in {}^e({}^cJ)$, hence $J = {}^e({}^cJ)$. This proves the first statement in (1) and the second statement follows immediately.

Let $I' = \{r \in R \mid dr \in I \text{ for some } d \in D\}$. We first show $I' \subseteq {}^c({}^eI)$. If $r \in I'$ then there is some $d \in D$ such that $dr = a \in I$. Then $r/1 = a/d \in {}^eI$, so $r \in {}^c({}^eI)$. To show the reverse containment $^c({}^eI) \subseteq I'$, let $r \in {}^c({}^eI)$ so that $r/1 = a/d$ for some $a \in I$ and $d \in D$. Then $x(dr - a) = 0$ for some $x \in D$, so $xdr = xa \in I$, and because $xd \in D$ it follows that $r \in I'$. This proves the first assertion of (2). Now $^eI = D^{-1}R$ if and only if $1/1 \in {}^eI$, if and only if $1 \in {}^c({}^eI) = I'$. The second assertion of (2) then follows from the definition of I'.

To prove (3) observe first that if Q is a prime ideal in $D^{-1}R$, then its preimage under any homomorphism sending 1 to 1 is a prime ideal (cf. Exercise 13, Section 7.4), so c maps prime ideals of $D^{-1}R$ to prime ideals of R disjoint from D. In the reverse direction, let P be a prime ideal of R disjoint from D and let $Q = {}^eP$ and suppose $(a/d_1)(b/d_2) \in Q$. Then $(ab)/(d_1d_2) \in Q$, so $ab/(d_1d_2) = c/d$ for some $c \in P$ and $d \in D$. Then $x(dab - d_1d_2c) = 0$ for some $x \in D$. Since $c \in P$ we have $xdab \in P$, and since P is a prime ideal disjoint from D we have $ab \in P$. Since P is prime, either $a \in P$ or $b \in P$, hence a/d_1 or b/d_2 is in Q. This proves Q is a prime ideal and shows that e maps prime ideals of R disjoint from D to prime ideals of $D^{-1}R$. Finally, it follows immediately from (2) that $P = {}^c({}^eP)$ for every prime ideal of R disjoint from D. Thus c and e are inverse correspondences, hence are bijections between these sets of prime ideals. This establishes (3).

By (1) every ascending (respectively, descending) chain of distinct ideals in $D^{-1}R$ contracts to an ascending (respectively, descending) chain of distinct ideals in R, giving (4) and completing the proof.

Because $1 \in D$, first localizing the ideal I and then contracting that localization as in (2) results in an ideal in R containing I: $I \subseteq {}^c({}^eI)$.

Definition. Suppose R is a commutative ring with 1 and D is a multiplicatively closed subset containing 1. The *saturation* of the ideal I in R with respect to D is the ideal ${}^c({}^eI)$ in R, where contraction and extension are computed with respect to $\pi : R \mapsto D^{-1}R$. If $I = {}^c({}^eI)$ then I is said to be *saturated* with respect to D.

Loosely speaking, (2) of Proposition 38 shows that the saturation of I consists of elements of R that would lie in I if we allowed denominators from D. The ideal is saturated with respect to D if we don't obtain any additional elements even if we allow denominators from D.

We can apply our results on localization to give an algorithm for determining whether an ideal P in the polynomial ring $k[x_1, \ldots, x_n]$ with coefficients in the field k is prime. The basic idea is to use the fact that $k[x_1, \ldots, x_i] = k[x_1, \ldots, x_{i-1}][x_i]$ to consider inductively whether the ideals $P_i = P \cap k[x_1, \ldots, x_i]$ are prime.

In general, suppose R is a commutative ring. If P is a prime ideal in $R[x]$ then $P \cap R$ is a prime ideal in R and so $S = R/(P \cap R)$ is an integral domain. Let F denote its quotient field. We then have two natural ring homomorphisms:

$$R[x] \longrightarrow (R/P \cap R)[x] = S[x] \longrightarrow F[x]$$

where the first is the natural projection homomorphism and the second is the natural inclusion induced by $S \subseteq F$. Note that $F[x]$ is the localization of $S[x]$ with respect to the multiplicatively closed set $D = S - \{0\}$. The next proposition shows that the image of P under the first homomorphism is a prime ideal in $S[x]$ that is saturated with respect to D and extends to a prime ideal in $F[x]$, and that, conversely, we can determine whether an ideal is prime in $R[x]$ by these properties.

Proposition 39. Suppose R is a commutative ring with 1 and I is an ideal in $R[x]$. Then I is a prime ideal in $R[x]$ if and only if
   i. $J = I \cap R$ is a prime ideal in R, i.e., $S = R/J$ is an integral domain, and
   ii. if $\overline{I}$ is the image of I in $S[x]$ then $\overline{I} F[x]$ is a prime ideal in $F[x]$ satisfying $\overline{I} F[x] \cap S[x] = \overline{I}$.

Proof: Suppose I is a prime ideal in $R[x]$, so that $J = I \cap R$ is a prime ideal in R and $S = R/J$ is an integral domain. By Proposition 2 in Chapter 9, the kernel of the reduction homomorphism $R[x] \mapsto S[x] = (R/J)[x]$ is $J[x]$, which is contained in $I[x]$, so we have a ring isomorphism $R[x]/I \cong S[x]/\overline{I}$. Since $R[x]/I$ is an integral domain, it follows that $\overline{I}$ is a prime ideal in the integral domain $S[x]$. The elements of $\overline{I} \cap S$ are the images of the elements in $R \cap I$, so $\overline{I} \cap S = 0$. Since the ring $F[x]$ is the localization of $S[x]$ with respect to the multiplicatively closed set $S - \{0\}$, condition (ii) follows by Proposition 38(3).
   Conversely, if I is not prime, then either J is not prime in R or J is prime in R but $\overline{I}$ is not prime in $S[x]$. In the latter case either $\overline{I} F[x]$ is not prime in $F[x]$ or, again

by Proposition 38(3), $\overline{I}$ is not saturated. Thus, if I is not prime, either (i) or (ii) fails, completing the proof.

Since $F[x]$ is a Euclidean Domain, the ideal $\overline{I} F[x] = (h(x))$ in Proposition 39 is principal, and is prime if and only if $h(x)$ is either 0 or is irreducible in $F[x]$. Suppose $h(x)$ is an element in I whose image in $S[x]$ has leading coefficient $a \in S$. The next proposition shows that a gives a bound on the denominators necessary for the saturation $\overline{I} F[x] \cap S[x]$ and can be used to compute this saturation.

Proposition 40. Let S be an integral domain with fraction field F and let A be a nonzero ideal in $S[x]$. Suppose $AF[x] = (h(x))$ where $h(x)$ is a polynomial in $S[x]$ with leading coefficient $a \in S$. Let S_a be the localization of S with respect to the powers of a. Then

  (1) $AF[x] \cap S[x] = AS_a[x] \cap S[x]$, and
  (2) if $\mathcal{A}$ denotes the ideal generated by A and $1 - at$ in the polynomial ring $S[x, t]$, then $AS_a[x] \cap S[x] = \mathcal{A} \cap S[x]$.

Proof: We first show $AF[x] \cap S_a[x] = AS_a[x]$. Since $S_a \subseteq F$, the containment $AS_a[x] \subseteq AF[x] \cap S_a[x]$ is immediate. Suppose now that $f(x) \in AF[x] \cap S_a[x]$. If the leading term of $f(x)$ is sx^N and the leading term of $h(x)$ is ax^m, then since $AF[x] = (h(x))$ we have $N \geq m$. Then the polynomial $f(x) - (s/a)x^{N-m}h(x)$ is again in $AF[x] \cap S_a[x]$ and is of lower degree than $f(x)$. Iterating, we see that $f(x)$ can be written as a polynomial in $S_a[x]$ times $h(x)$, so $f(x) \in AS_a[x]$. Intersecting both sides of $AF[x] \cap S_a[x] = AS_a[x]$ with $S[x]$ gives the first statement in the proposition.

To prove the second statement, suppose first that $f(x) \in \mathcal{A} \cap S[x]$. Then we can write $f(x) = f_1(x, t)b(x) + f_2(x, t)(1 - at)$ for some polynomials $b(x) \in A$ and $f_1, f_2 \in S[x, t]$. Substituting $t = 1/a$ gives $f(x) = f_1(x, 1/a)b(x)$, and since $f_1(x, 1/a) \in S_a[x]$, we obtain $f(x) \in AS_a[x] \cap S[x]$. Conversely, suppose that $f(x) = b(x)g(x) \in S[x]$ where $g(x) \in S_a(x)$ and $b(x) \in A$. If a^N is the largest power of a appearing in the denominators of the coefficients of $g(x)$ then $a^N g(x) \in S[x]$. Writing $f(x) = (at)^N f(x) + (1 - (at)^N)f(x) = b(x)t^N(a^N g(x)) + (1 - (at)^N)f(x)$ we see that $f(x) \in \mathcal{A} \cap S[x]$, giving the reverse containment and completing the proof.

Suppose now that P is an ideal in $k[x_1, \ldots, x_n]$. Let P_i for $i = 1, \ldots, n$ be the intersection of P with $k[x_1, \ldots, x_i]$. We use Propositions 39 and 40 to determine inductively whether $P_1, P_2, \ldots, P_n = P$ are prime ideals in their respective polynomial rings.

The ideal P_1 will be prime in the Euclidean Domain $k[x_1]$ if and only if it is 0 or is generated by an irreducible polynomial. Suppose now that $i \geq 2$ and we have already proved that P_{i-1} is a prime ideal in $k[x_1, \ldots, x_{i-1}]$, so that the quotient ring $S = k[x_1, \ldots, x_{i-1}]/P_{i-1}$ is an integral domain. If F denotes the quotient field of S, then by Proposition 39, P_i is a prime ideal in $k[x_1, \ldots, x_i]$ if and only if its image in $(k[x_1, \ldots, x_{i-1}]/P_{i-1})[x_i] = S[x_i]$ is a saturated ideal whose extension to the Euclidean Domain $F[x_i]$ is a prime ideal. Suppose $h(x_i) \in S[x_i]$ is a generator for this ideal and a is the leading coefficient of $h(x_i)$. Then $(h(x_i))$ is a prime ideal in $F[x_i]$ if and only if

$h(x_i) = 0$ or $h(x_i)$ is an irreducible polynomial. By Proposition 40, the image of P_i in $S[x_i]$ will be saturated if and only if it equals $\mathcal{A} \cap S[x_i]$ where $\mathcal{A}$ is the ideal generated by P_i and $1 - at$ in $S[x_i, t]$. This latter condition can be checked in $k[x_1, \ldots, x_i, t]$: it is equivalent to checking that the intersection of the ideal generated by P_i and $1 - at$ in $k[x_1, \ldots, x_i, t]$ with $k[x_1, \ldots, x_i]$ is just P_i (cf. Exercise 3).

Combining these observations with our results on Gröbner bases from Chapter 9 we obtain the following algorithm for determining whether the ideal P in $k[x_1, \ldots, x_n]$ is prime (or, equivalently, whether the associated affine algebraic set is a variety).

Algorithm for Determining when an Ideal in $k[x_1, \ldots, x_n]$ is Prime

(1) Compute the reduced Gröbner basis $G = \{g_1, \ldots, g_m\}$ for P with respect to the lexicographic monomial ordering $x_n > \cdots > x_1$.

By Proposition 29 in Section 9.6 the elements of G lying in $k[x_1, \ldots, x_i]$ will be the reduced Gröbner basis $\{g_1, \ldots, g_{m_i}\}$ for $P_i = P \cap k[x_1, \ldots, x_i]$.

(2) Determine whether P_1 is a prime ideal in $k[x_1]$ by checking that $P_1 = 0$ or the nonzero generator of P_1 is irreducible in $k[x_1]$.

For each $i \geq 2$, suppose P_{i-1} has been determined to be a prime ideal in $k[x_1, \ldots, x_{i-1}]$ (otherwise, P is not a prime ideal in $k[x_1, \ldots, x_n]$). Let $S = k[x_1, \ldots, x_{i-1}]/P_{i-1}$ and let F be the fraction field of S. Apply steps (3) and (4) to determine whether P_i is a prime ideal in $k[x_1, \ldots, x_i]$.

(3) If $m_i = m_{i-1}$ then P_i maps to the zero ideal in $S[x_i]$, hence is prime. Otherwise the image of P_i in $S[x_i]$ and in $F[x_i]$ is a nonzero ideal, and is generated by the images of $g_{m_{i-1}+1}, \ldots, g_{m_i}$. Apply the Euclidean algorithm in $F[x_i]$ to these generators to find an element $h(x_i)$ in P_i whose image in $F[x_i]$ generates the image of P_i in $F[x_i]$. Determine whether $h(x_i)$ is irreducible in $F[x_i]$—if not then P_i and P are not prime ideals.

(Note that after applying the Euclidean algorithm to the generators of the image of P_i in $F[x_i]$ we can multiply by a single element of S to 'clear denominators' in each equation so that all remainders (and in particular the last nonzero remainder $h(x_i)$) will be elements in the image of P_i.)

(4) Let $a \in k[x_1, \ldots, x_{i-1}]$ be the leading coefficient of $h(x_i)$ (as a polynomial in x_i). Compute the reduced Gröbner basis in $k[x_1, \ldots, x_i, t]$ for the ideal generated by P_i and $1 - at$ with respect to the lexicographic monomial ordering $t > x_i > \cdots > x_1$. Determine whether the elements of this reduced basis that lie in $k[x_1, \ldots, x_i]$ are $\{g_1, \ldots, g_{m_i}\}$—if so, then P_i is a prime ideal in $k[x_1, \ldots, x_i]$ and if not then P_i and P are not prime ideals.

Finally, we note that similar ideas (together with some minor modifications to extend results on Gröbner bases to polynomial rings $R[x_1, \ldots, x_n]$ with coefficients in an integral domain R) can be used to provide algorithms for determining when an ideal in, for example, $\mathbb{Z}[x_1, \ldots, x_n]$ is prime.

Examples

(1) Consider the ideal $P = (xz - y^2, yz - x^3, z^2 - x^2y)$ in $k[x, y, z]$ for any infinite field k. It follows from Exercise 26 in Section 1 that P is a prime ideal since there is an injection of $k[x, y, z]/P$ into the integral domain $k[\mathbb{A}^1]$ (cf. Exercise 24 in Section 2). Here we prove $P \subset \mathbb{Q}[x, y, z]$ is prime using the ideas in this section. The reduced Gröbner basis for P with respect to the lexicographic monomial ordering $x > y > z$ is $\{x^3 - yz, x^2y - z^2, xy^3 - z^3, xz - y^2, y^5 - z^4\}$. Hence $P_1 = P \cap \mathbb{Q}[z] = (0)$, and $P_2 \cap \mathbb{Q}[y, z] = (y^5 - z^4)$. Since $P_1 = 0$, the ideal P_1 is prime in $\mathbb{Q}[z]$.

We next check P_2 is prime in $\mathbb{Q}[y, z]$, which can be done directly (cf. Exercise 4 or Exercise 14 in Section 9.1). In this case $S = \mathbb{Q}[z]$ and $F = \mathbb{Q}(z)$. The image of P_2 in $F[y]$ is generated by $h(y) = y^5 - z^4$, which is irreducible in $\mathbb{Q}(z)[y]$. The leading coefficient of $h(y)$ is 1, and the reduced Gröbner basis for $(y^5 - z^4, 1 - t)$ in $\mathbb{Q}[y, z, t]$ with respect to the lexicographic monomial ordering $t > y > z$ is $\{y^5 - z^4, 1 - t\}$. The element in the reduced Gröbner basis for P_2 is the only element of this basis lying in $\mathbb{Q}[y, z]$ so P_2 is a prime ideal in $\mathbb{Q}[y, z]$.

We now use the fact that P_2 is prime to prove that P is prime. In this case S is the integral domain $\mathbb{Q}[y, z]/P_2 = \mathbb{Q}[y, z]/(y^5 - z^4)$ with quotient field F given by

$$S = \mathbb{Q}[\bar{z}] + \mathbb{Q}[\bar{z}]\bar{y} + \mathbb{Q}[\bar{z}]\bar{y}^2 + \mathbb{Q}[\bar{z}]\bar{y}^3 + \mathbb{Q}[\bar{z}]\bar{y}^4$$

$$F = \mathbb{Q}(\bar{z}) + \mathbb{Q}(\bar{z})\bar{y} + \mathbb{Q}(\bar{z})\bar{y}^2 + \mathbb{Q}(\bar{z})\bar{y}^3 + \mathbb{Q}(\bar{z})\bar{y}^4$$

where $\bar{y}^5 = \bar{z}^4$. The image of P in $S[x]$ is the ideal $\overline{P}$ generated by the elements $g_1 = x^3 - \bar{y}\bar{z}$, $g_2 = \bar{y}x^2 - \bar{z}^2$, $g_3 = \bar{y}^3x - \bar{z}^3$, $g_4 = \bar{z}x - \bar{y}^2$, and $\bar{y}^5 - \bar{z}^4 = 0$.

The greatest common divisor in $F[x]$ of g_1, g_2, g_3, g_4 generating the image of P in $F[x]$ is the irreducible polynomial $x - \bar{y}^2/\bar{z}$. The polynomial $h(x) = zx - y^2$ in P has image generating the same ideal in $F[x]$, so we may take $a = z$ in (4) of the algorithm. The reduced Gröbner basis for $(xz - y^2, yz - x^3, z^2 - x^2y, 1 - zt)$ with respect to the lexicographic monomial ordering $t > x > y > z$ consists of the reduced Gröbner basis for P together with the elements $ty^2 - x$ and $tz - 1$ involving t, so P is a prime ideal in $\mathbb{Q}[x, y, z]$.

(2) Consider the ideal $P = (xz - y^3, xy - z^2)$ in $\mathbb{Q}[x, y, z]$, with reduced Gröbner basis for the lexicographic monomial ordering $x > y > z$ given by $\{xy - z^2, xz - y^3, y^4 - z^3\}$. Here $P_1 = 0$ and $P_2 = P \cap \mathbb{Q}[y, z] = (y^4 - z^3)$ are prime ideals as in Example 1. In this case $S = \mathbb{Q}[y, z]/P_2$ is given by

$$S = \mathbb{Q}[\bar{z}] + \mathbb{Q}[\bar{z}]\bar{y} + \mathbb{Q}[\bar{z}]\bar{y}^2 + \mathbb{Q}[\bar{z}]\bar{y}^3$$

with $\bar{y}^4 = \bar{z}^3$, with quotient field F similar to the previous example, and $\overline{P} = (g_1, g_2)$ in $S[x]$ where $g_1 = \bar{y}x - \bar{z}^2$ and $g_2 = \bar{z}x - \bar{y}^3$. The extension of $\overline{P}$ to $F[x]$ is generated by the irreducible polynomial $\bar{y}x - \bar{z}^2$, and $h(x) = yx - z^2$ is an element of P having the same image in $F[x]$, with leading coefficient $a = y$. The reduced Gröbner basis for the ideal $(xz - y^3, xy - z^2, 1 - yt)$ in $\mathbb{Q}[x, y, z, t]$ using the lexicographic ordering $t > x > y > z$ is $\{x^2 - y^2z, xy - z^2, xz - y^3, y^4 - z^3, ty - 1, tz^2 - x\}$, containing the element $x^2 - y^2z$ not in the reduced Gröbner basis for P, so P is *not* a prime ideal in $\mathbb{Q}[x, y, z]$. This computation not only shows P is not a prime ideal, it does so by explicitly showing the image of P in $S[x]$ is not saturated using the localization S_a. The computation of $a = y$ allows us to find an explicit pair of elements not in P whose product is in P: $f = x^2 - y^2z \notin P$ and $y \notin P$, but some power of y times f lies in P. In this case a quick computation verifies that $yf \in P$.

Localizations of Modules

Suppose now that M is an R-module and D is a multiplicatively closed subset of R containing 1 as above. Then the ideas used in the construction of $D^{-1}R$ can be used to construct a $D^{-1}R$-module $D^{-1}M$ from M in a similar fashion, as follows. Define the relation on $D \times M$ by

$$(d, m) \sim (e, n) \quad \text{if and only if} \quad x(dn - em) = 0 \quad \text{for some } x \in D,$$

which is easily checked to be an equivalence relation. Let m/d denote the equivalence class of (d, m) and let $D^{-1}M$ denote the set of equivalence classes. It is then straightforward to verify that the operations

$$\frac{m}{d} + \frac{n}{e} = \frac{em + dn}{de} \qquad \text{and} \qquad \left(\frac{r}{d}\right)\left(\frac{m}{e}\right) = \frac{rm}{de}$$

are well defined and give $D^{-1}M$ the structure of a $D^{-1}R$-module.

Definition. The $D^{-1}R$-module $D^{-1}M$ is called the *module of fractions of M with respect to D* or the *localization of M at D*.

Note that the localization $D^{-1}M$ is also an R-module (since each $r \in R$ acts by $r/1$ on $D^{-1}M$), and there is an R-module homomorphism

$$\pi : M \to D^{-1}M \quad \text{defined by} \quad \pi(m) = \frac{m}{1}.$$

It follows directly from the definition of the equivalence relation that

$$\ker \pi = \{m \in M \mid dm = 0 \text{ for some } d \in D\}.$$

The homomorphism π has a universal property analogous to that in Theorem 36. Suppose N is an R-module with the property that left multiplication on N by d is a bijection of N for every $d \in D$. If $\psi : M \to N$ is any R-module homomorphism then there is a unique R-module homomorphism $\Psi : D^{-1}M \to N$ such that $\Psi \circ \pi = \psi$.

If M and N are R-modules and $\varphi : M \to N$ is an R-module homomorphism, then for any multiplicative set D in R it is easy to check that there is an induced $D^{-1}R$-module homomorphism from $D^{-1}M$ to $D^{-1}N$ defined by mapping m/d to $\varphi(m)/d$.

The next result shows that the localization of M at D is related to the tensor product.

Proposition 41. Let D be a multiplicatively closed subset of R containing 1 and let M be an R-module. Then $D^{-1}M \cong D^{-1}R \otimes_R M$ as $D^{-1}R$-modules, i.e., $D^{-1}M$ is the $D^{-1}R$-module obtained by extension of scalars from the R-module M.

Proof: The map from $D^{-1}R \times M$ to $D^{-1}M$ defined by mapping $(r/d, m)$ to rm/d is well defined and R-balanced, so induces a homomorphism from $D^{-1}R \otimes_R M$ to $D^{-1}M$. The map sending m/d to $(1/d) \otimes m$ gives a well defined inverse homomorphism (if $m/d = m'/d'$ in $D^{-1}M$ then $x(d'm - dm') = 0$ for some $x \in D$, and then $(1/d) \otimes m$ can be written as $(1/xd'd) \otimes (xd'm) = (1/xd'd) \otimes (xdm') = (1/d') \otimes m'$). Hence $D^{-1}M$ is isomorphic to $D^{-1}R \otimes_R M$ as an R-module since these inverse isomorphisms are also $D^{-1}R$-module homomorphisms.

Localizing a ring R or an R-module M at D behaves very well with respect to algebraic operations on rings and modules, as the following proposition shows:

Proposition 42. Let R be a commutative ring with 1 and let $D^{-1}R$ be its localization with respect to the multiplicatively closed subset D of R containing 1.

(1) Localization commutes with finite sums and intersections of ideals: If I and J are ideals of R, then

$$D^{-1}(I + J) = D^{-1}(I) + D^{-1}(J) \quad \text{and} \quad D^{-1}(I \cap J) = D^{-1}(I) \cap D^{-1}(J).$$

Localization commutes with quotients:

$$D^{-1}R / D^{-1}I \cong D^{-1}(R/I),$$

(where the localization on the right is with respect to the image of D in the quotient R/I).

(2) Localization commutes with taking radicals: If N is the nilradical of R, then $D^{-1}N$ is the nilradical of $D^{-1}R$. If I is an ideal in R, then $\text{rad}(D^{-1}I)$ is $D^{-1}(\text{rad } I)$.

(3) Primary ideals correspond to primary ideals in the correspondence (3) of Proposition 38. More precisely, suppose Q is a P-primary ideal in R. If $D \cap P \neq \emptyset$ then $D^{-1}Q = D^{-1}R$. If $D \cap P = \emptyset$ then $D^{-1}P$ is a prime ideal, the extension $D^{-1}Q$ of Q is a $D^{-1}P$-primary ideal in $D^{-1}R$, and the contraction back to R of $D^{-1}Q$ is Q.

(4) Localization commutes with finite sums, intersections and quotients of modules: If L and N are submodules of the R-module M, then
   (a) $D^{-1}(L + N) = D^{-1}L + D^{-1}N$ and $D^{-1}(L \cap N) = D^{-1}L \cap D^{-1}N$,
   (b) $D^{-1}N$ is a submodule of $D^{-1}M$ and $D^{-1}M / D^{-1}N = D^{-1}(M/N)$.

(5) Localization commutes with finite direct sums of modules: If M and N are R-modules, then $D^{-1}(M \oplus N) \cong D^{-1}M \oplus D^{-1}N$.

(6) Localization is exact (i.e., $D^{-1}R$ is a flat R-module): If $0 \to L \xrightarrow{\psi} M \xrightarrow{\varphi} N \to 0$ is a short exact sequence of R-modules, then the induced sequence $0 \to D^{-1}L \xrightarrow{\psi'} D^{-1}M \xrightarrow{\varphi'} D^{-1}N \to 0$ of $D^{-1}R$-modules is also exact.

Proof: We first prove (6). Suppose that $0 \to L \xrightarrow{\psi} M \xrightarrow{\varphi} N \to 0$ is a short exact sequence of R-modules. Every element of $D^{-1}N$ is of the form n/d for some $n \in N$ and $d \in D$. Since φ is surjective, $n = \varphi(m)$ for some $m \in M$, so $\varphi'(m/d) = \varphi(m)/d = n/d$ and $\varphi' : D^{-1}M \to D^{-1}N$ is surjective. If m/d is in the kernel of φ' then $d_1\varphi(m) = 0$ for some $d_1 \in D$. Then $\varphi(d_1 m) = 0$ implies $d_1 m = \psi(l)$ for some $l \in L$ by the exactness of the original sequence at M, so $m/d = d_1 m/(d_1 d) = \psi(l)/(d_1 d) = \psi'(l/(d_1 d))$ and $\ker(\varphi') \subseteq \text{image}(\psi')$. If $\psi(l)/d \in \text{image}(\psi')$ then $\varphi'(\psi(l)/d) = \varphi(\psi(l))/d = 0$, which shows the reverse inclusion $\text{image}(\psi') \subseteq \ker(\varphi')$, and we have exactness of the induced sequence at $D^{-1}M$. Finally, suppose $\psi'(l/d) = 0$. Then $d_2\psi(l) = 0$ for some $d_2 \in D$, i.e., $\psi(d_2 l) = 0$, so $d_2 l = 0$ by the injectivity of ψ. Hence $l/d = d_2 l/(d_2 d) = 0$ and ψ' is injective. This proves that the sequence $0 \to D^{-1}L \xrightarrow{\psi'} D^{-1}M \xrightarrow{\varphi'} D^{-1}N \to 0$ is exact.

To prove the first statement in (1), note that $(i + j)/d = i/d + j/d$ for $i \in I$, $j \in J$ and $d \in D$ shows $D^{-1}(I+J) \subseteq D^{-1}(I)+D^{-1}(J)$; and $i/d_1+j/d_2 = (d_2 i+d_1 j)/(d_1 d_2)$ for $i \in I$, $j \in J$ and $d_1, d_2 \in D$ shows $D^{-1}(I) + D^{-1}(J) \subseteq D^{-1}(I + J)$. For the second statement, the inclusion $D^{-1}(I \cap J) \subseteq D^{-1}(I) \cap D^{-1}(J)$ is immediate. If

$a/d \in D^{-1}(I) \cap D^{-1}(J)$ then $d_1 a \in I$ and $d_2 a \in J$ for some $d_1, d_2 \in D$. Then $d_1 d_2 a \in I \cap J$ and $a/d = (d_1 d_2 a)/(d_1 d_2 d)$ gives the inclusion $D^{-1}(I) \cap D^{-1}(J) \subseteq D^{-1}(I \cap J)$. The last statement in (1) follows by applying (6) to the exact sequence $0 \to I \overset{\psi}{\to} R \overset{\varphi}{\to} R/I \to 0$.

To prove (2), suppose first that $a \in \text{rad } I$, so that $a^n \in I$ for some $n \geq 1$. Then $(a/d)^n = a^n/d^n \in D^{-1}I$ so $D^{-1}(\text{rad } I) \subseteq \text{rad}(D^{-1}I)$. Conversely, if $a/d \in \text{rad}(D^{-1}I)$ then $(a/d)^n \in D^{-1}I$ for some $n \geq 1$, i.e., $d_1 a^n \in I$ for some $d_1 \in D$. Hence $(d_1 a)^n = d_1^{n-1}(d_1 a^n) \in I$, so $d_1 a \in \text{rad } I$ and then $a/d = d_1 a/(d_1 d) \in D^{-1}(\text{rad } I)$ shows that $\text{rad}(D^{-1}I) \subseteq D^{-1}(\text{rad } I)$. This proves the second statement in (2), and the first statement follows by applying this to the ideal $I = (0)$.

For (3), note first that $D \cap P = \emptyset$ if and only if $D \cap Q = \emptyset$ (one inclusion is obvious and the other follows since $d \in D \cap P$ implies $d^n \in D \cap Q$ for some n). The statement for $D \cap P \neq \emptyset$ and the fact that $D^{-1}P$ is a prime ideal for $D \cap P = \emptyset$ were proved in Proposition 38. To see that $D^{-1}Q$ is a primary ideal in $D^{-1}R$, suppose that $(a/d_1)(b/d_2) \in D^{-1}Q$ and $a/d_1 \notin D^{-1}Q$. Then there is some element $d \in D$ so that $dab \in Q$, and since $a \notin Q$ and Q is primary, we have $(db)^n \in Q$ for some $n \geq 1$. Then $(b/d_2)^n = d^n b^n/(d^n d_2^n) \in D^{-1}Q$, so that $D^{-1}Q$ is primary. The radical of $D^{-1}Q$ is $D^{-1}P$ by (2). Finally, by (2) of Proposition 38, the contraction of $D^{-1}Q$ is an ideal of R containing Q and consists precisely of the elements $r \in R$ with $dr \in Q$ for some $d \in D$. Since Q is P-primary, the definition of primary implies that if $dr \in Q$ and $d \notin P$, then $r \in Q$, hence the contraction of $D^{-1}Q$ is Q.

The proof of (4) is essentially the same as the proof of (1) and is left as an exercise.

It is easy to see that if the exact sequence $0 \to L \overset{\psi}{\to} M \overset{\varphi}{\to} N \to 0$ of R-modules splits, then the exact sequence $0 \to D^{-1}L \overset{\psi'}{\to} D^{-1}M \overset{\varphi'}{\to} D^{-1}N \to 0$ of $D^{-1}R$-modules also splits, which gives (5).

Proposition 38 shows that localizing at the multiplicatively closed set D emphasizes the ideals of R not containing any elements of D since the other ideals of R become trivial when extended to $D^{-1}R$. The following proposition provides a more precise statement in terms of the effect of localization on primary decomposition of ideals.

Proposition 43. Let R be a Noetherian ring and let

$$I = Q_1 \cap \cdots \cap Q_m$$

be a minimal primary decomposition of the proper ideal I, where Q_i is a P_i-primary ideal. Suppose D is a multiplicatively closed set of R containing 1 and the primary ideals $Q_1, \ldots, Q_m$ are numbered so that $D \cap P_i = \emptyset$ for $1 \leq i \leq t$ and $D \cap P_i \neq \emptyset$ for $t + 1 \leq i \leq m$. Then

$$D^{-1}I = D^{-1}Q_1 \cap \cdots \cap D^{-1}Q_t$$

is a minimal primary decomposition of $D^{-1}I$ in $D^{-1}R$ and $D^{-1}Q_i$ is a $D^{-1}P_i$-primary ideal. Further, the contraction of $D^{-1}Q_i$ back to R is Q_i for $1 \leq i \leq t$ and

$$^c(D^{-1}I) = Q_1 \cap \cdots \cap Q_t$$

is a minimal primary decomposition of the contraction of $D^{-1}I$ back to R.

Proof: By (3) of Proposition 42, $D^{-1}Q_i = D^{-1}R$ for $t+1 \le i \le m$, and $D^{-1}Q_i$ is a $D^{-1}P_i$-primary ideal with pullback Q_i for $1 \le i \le t$. By (1) of the same proposition, $D^{-1}I = D^{-1}Q_1 \cap \cdots \cap D^{-1}Q_t$, and (3) shows that this is a primary decomposition. Contracting to R shows that $^c(D^{-1}I) = Q_1 \cap \cdots \cap Q_t$, which also implies that the decompositions are minimal.

In particular we can finish the proof of Theorem 21:

Corollary 44. The primary ideals belonging to the isolated primes in a minimal primary decomposition of I are uniquely defined by I.

Proof: Let P be a minimal element in the set $\{P_1, \dots, P_m\}$ of primes belonging to I, and take $D = R - P$ in Proposition 43. Then $D \cap P_i = \emptyset$ only for $P = P_i$, so the contraction of the localization of I at D is precisely the primary ideal Q belonging to the minimal prime P. Since the prime ideals $\{P_1, \dots, P_m\}$ of primes belonging to I are uniquely determined by I, it follows that the primary ideals Q belonging to the isolated primes of I are also uniquely determined by I.

The effect of isolating in on certain prime ideals by localization is particularly precise in the case of localizing at a prime P (considered in Example 3 following Corollary 37 above). We first recall the definition of an important type of ring (cf. Exercises 37–39 in Section 7.4).

Definition. A commutative ring with 1 that has a unique maximal ideal is called a *local ring*.

Proposition 45. Let R be a commutative ring with 1. Then the following are equivalent:
  (1) R is a local ring with unique maximal ideal M
  (2) if M is the set of elements of R that are not units, then M is an ideal
  (3) there is a maximal ideal M of R such that every element $1 + m$ with $m \in M$ is a unit in R.

Proof: If $a \in R$ then the ideal (a) is either R, in which case a is a unit, or is a proper ideal, in which case (a) is contained in a maximal ideal (Proposition 11 of Section 7.4). It follows that if R is a local ring and M is its unique maximal ideal then every $a \notin M$ is a unit, so M consists precisely of the set of nonunits in R, showing that (1) implies (2). It also follows that if the set M of nonunits in R is an ideal then this ideal must be the unique maximal ideal in R, so that (2) implies (1).

Suppose now that (3) is satisfied. If a is an element of R not contained in the maximal ideal M, then $(a) + M = R$, so that $ab + m = 1$ for some $b \in R$ and $m \in M$. Then $ab = 1 - m$ is a unit by assumption, so a is also a unit. This shows that M is the unique maximal ideal in R, so (3) implies (1). Conversely, if R is a local ring, then $1 + m \notin M$ for any $m \in M$, so $1 + m$ is a unit, so (1) implies (3).

Proposition 46. For any commutative ring R with 1, let R_P be the localization of R at the prime ideal P and let ^eP be the extension of P to R_P.

    (1) The ring R_P is a local ring with unique maximal ideal ^eP. The contraction of ^eP to R is P, i.e., $^c(^eP) = P$, and the map from R to R_P induces an injection of the integral domain R/P into $R_P/^eP$. The quotient $R_P/^eP$ is a field and is isomorphic to the fraction field of the integral domain R/P.

    (2) If R is an integral domain, then R_P is an integral domain. The ring R injects into the local ring R_P, and, identifying R with its image in R_P, the unique maximal ideal of R_P is PR_P.

    (3) The prime ideals in R_P are in bijective correspondence with the prime ideals of R contained in P.

    (4) If P is a minimal nonzero prime ideal of R then R_P has a unique nonzero prime ideal.

    (5) If $P = M$ is a maximal ideal and I is any M-primary ideal of R then $R_M/^eI \cong R/I$. In particular, $R_M/^eM \cong R/M$ and $(^eM)/(^eM)^n \cong M/M^n$ for all $n \geq 1$.

Proof: If P' is a prime ideal of R, then $P' \cap (R - P) = \emptyset$ if and only if $P' \subseteq P$, so (3) is immediate from (3) in Proposition 38, and (4) follows. Since $^eP \neq R_P$ by (2) of Proposition 38, it follows from (3) that R_P is a local ring with unique maximal ideal ^eP, which proves the first statement in (1).

By Proposition 38(2) the contraction $^c(^eP)$ is the set $\{r \in R \mid dr \in P \text{ for some } d \in R - P\}$, and since P is prime, $dr \in P$ with $d \notin P$ implies $r \in P$. This shows that $^c(^eP) = P$, which is the second statement in (1).

The kernel of the map from R to $R_P/^eP$ is $^c(^eP) = P$, so the induced map from R/P into $R_P/^eP$ is injective. The quotient $R_P/^eP$ is a field by the first part of (1), so there is an induced homomorphism from the fraction field of the integral domain R/P into $R_P/^eP$. The universal property of the localization R_P shows there is an inverse homomorphism from $R_P/^eP$ to the fraction field of R/P (since every element of R not in P maps to a unit in R/P). It follows that $R_P/^eP$ is isomorphic to the fraction field of R/P.

If R is an integral domain, then $R - P$ has no zero divisors, so R injects into R_P by Corollary 37; if R is identified with its image in R_P then $^eP = PR_P$, so (2) follows.

To prove (5), by Proposition 42(1) we may pass to the quotient R/I and so reduce to the case $I = 0$. In this case the maximal ideal $P = M$ in R is the nilradical of R, hence is the unique maximal ideal of R. By Proposition 45 every element of $R - M$ is a unit, so $R_P = R$, and each of the statements in (5) follows immediately, completing the proof of the proposition.

Example

The results of (5) of the proposition are not true in general if P is a prime ideal that is not maximal. For example, $P = (0)$ in $R = \mathbb{Z}$ has $R/P = \mathbb{Z}$ and $R_P/PR_P = \mathbb{Q}$; in this case $(PR_P)/(PR_P)^n \cong P/P^n = 0$ for all $n \geq 1$ (cf. the exercises).

Definition. Let M be an R-module, let P be a prime ideal of R and set $D = R - P$. The R_P-module $D^{-1}M$ is called the *localization of M at P*, and is denoted by M_P.

By Proposition 41, M_P can also be identified with the tensor product $R_P \otimes_R M$. When R is an integral domain and $P = (0)$, then $M_{(0)}$ is a module over the field of fractions F of R, i.e., is a vector space over F.

The element $m/1$ is zero in M_P if and only if $rm = 0$ for some $r \in R - P$, so localizing at P annihilates the P'-torsion elements of M for primes P' not contained in P. In particular, *localizing at (0) over an integral domain annihilates the torsion subgroup of M.*

Definition. If R is an integral domain, then the *rank* of the R-module M is the dimension of the localization $M_{(0)}$ as a vector space over the field of fractions of R.

It is easy to see that this definition of rank agrees with the notion of rank introduced in Chapter 12.

Example

Let $R = \mathbb{Z}$ and let $\mathbb{Z}_{(p)}$ be the localization of $\mathbb{Z}$ at the nonzero prime ideal (p). Any abelian group M is a $\mathbb{Z}$-module so we may localize M at (p) by forming $M_{(p)}$. This abelian group is the same as the quotient of M with respect to the subgroup of elements whose order is finite and not divisible by p. If M is a finite (or, more generally, torsion) abelian group, then $M_{(p)}$ is a p-group, and is the Sylow p-subgroup or p-primary component of M. The localization $M_{(0)}$ of M at (0) is the trivial group. For a specific example, let $M = \mathbb{Z}/6\mathbb{Z}$ be the cyclic group of order 6, considered as a $\mathbb{Z}$-module. Then the localization of M at $p = 2$ is $\mathbb{Z}/2\mathbb{Z}$, at $p = 3$ is $\mathbb{Z}/3\mathbb{Z}$, and reduces to 0 at all other prime ideals of $\mathbb{Z}$.

Localization of a module M at a prime P in general produces a simpler module M_P whose properties are easier to determine. It is then of interest to translate these "local" properties of M_P back into "global" information about the module M itself. For example, the most basic question of whether a module M is 0 can be answered locally:

Proposition 47. Let M be an R-module. Then the following are equivalent:
  (1) $M = 0$,
  (2) $M_P = 0$ for all prime ideals P of R, and
  (3) $M_{\mathfrak{m}} = 0$ for all maximal ideals $\mathfrak{m}$ of R.

Proof: The implications (1) implies (2) implies (3) are obvious, so it remains to prove that (3) implies (1). Suppose m is a nonzero element in M, and consider the annihilator I of m in R, i.e., the ideal of elements $r \in R$ with $rm = 0$. Since m is nonzero I is a proper ideal in R. Let $\mathfrak{m}$ be a maximal ideal of R containing I and consider the element $m/1$ in the corresponding localization $M_{\mathfrak{m}}$ of M. If this element were 0, then $rm = 0$ for some $r \in R - \mathfrak{m}$. But then r would be an element in I not contained in $\mathfrak{m}$, a contradiction. Hence $M_{\mathfrak{m}} \neq 0$, which proves that (3) implies (1).

It is not in general true that a property shared by all of the localizations of a module M is also shared by M. For example, all of the localizations of a ring R can be integral domains without R itself being an integral domain (for example, $\mathbb{Z}/6\mathbb{Z}$ above). Nevertheless, a great deal of information *can* be ascertained from studying the various possible localizations, and this is what makes this technique so useful. If R is an integral domain, for example, then each of the localizations R_P can be considered as a subring of the fraction field F of R that contains R; the next proposition shows that the elements of R are the only elements of F contained in every localization.

Proposition 48. Let R be an integral domain. Then R is the intersection of the localizations of R: $R = \cap_P R_P$. In fact, $R = \cap_{\mathfrak{m}} R_{\mathfrak{m}}$ is the intersection of the localizations of R at the maximal ideals $\mathfrak{m}$ of R.

Proof: As mentioned, $R \subseteq \cap_{\mathfrak{m}} R_{\mathfrak{m}}$. Suppose now that a is an element of the fraction field F of R that is contained in $R_{\mathfrak{m}}$ for every maximal ideal $\mathfrak{m}$ of R, and consider

$$I_a = \{d \in R \mid da \in R\}.$$

It is easy to check that I is an ideal of R, and that $a \in R$ if and only if $1 \in I_a$, i.e., $I_a = R$. Suppose that $I_a \neq R$. Then there is a maximal ideal $\mathfrak{m}$ containing I_a, and since $a \in R_{\mathfrak{m}}$ we have $a = r/d$ for some $r \in R$ and $d \in R - \mathfrak{m}$. But then $d \in I_a$ and $d \notin \mathfrak{m}$, a contradiction. Hence $a \in R$, so $\cap_{\mathfrak{m}} R_{\mathfrak{m}} \subseteq R$, and we have proved the second assertion in the proposition. The first is then immediate.

Another important property of a ring R that can be detected locally is normality:

Proposition 49. Let R be an integral domain. Then the following are equivalent:
  (1) R is normal, i.e., R is integrally closed (in its field of fractions)
  (2) R_P is normal for all prime ideals P of R
  (3) $R_{\mathfrak{m}}$ is normal for all maximal ideals $\mathfrak{m}$ of R.

Proof: Let F be the field of fractions of R, so all of the various localizations of R may be considered as subrings of F.

Assume first that R is integrally closed and suppose $y \in F$ is integral over R_P. Then y is a root of a monic polynomial of degree n with coefficients of the form a_i/d_i for some $d_i \notin P$. The element $y' = y(d_0 d_1 \cdots d_{n-1})^n$ is then a root of a monic polynomial of degree n with coefficients from R, i.e., y' is integral over R. Since R is assumed normal, this implies $y' \in R$, and so $y = y'/(d_0 \cdots d_{n-1}) \in R_P$, which proves that (1) implies (2). The implication (2) implies (3) is trivial. Suppose now that $R_{\mathfrak{m}}$ is normal for all maximal ideals $\mathfrak{m}$ of R and let y be an element of F that is integral over R. Since $R \subseteq R_{\mathfrak{m}}$, y is in particular also integral over $R_{\mathfrak{m}}$ and so $y \in R_{\mathfrak{m}}$ for every maximal ideal by assumption. Then $y \in R$ by the previous proposition, which proves that (3) implies (1).

We now may easily prove the first part of the Going–up Theorem (cf. Section 3) that was used in the proof of Corollary 27.

Corollary 50. Let R be a subring of the commutative ring S with $1 \in R$, and assume that S is integral over R. If P is a prime ideal in R, then there is a prime ideal Q of S with $P = Q \cap R$.

Proof: Let $D = R - P$ so that D is a multiplicatively closed subset of both R and S. Then the following diagram commutes:

$$
\begin{array}{ccc}
R & \xrightarrow{\ \pi\ } & D^{-1}R = R_P \\
\downarrow{\iota} & & \downarrow{\iota} \\
S & \xrightarrow{\ \pi\ } & D^{-1}S
\end{array}
$$

where the vertical maps are inclusions. It is easy to see that $D^{-1}S$ is integral over R_P (Exercise 20). Let $\mathfrak{m}$ be any maximal ideal of $D^{-1}S$. Then $\mathfrak{m} \cap R_P$ is a maximal ideal in R_P by the second statement in Theorem 26(2) (note that the first part of Theorem 26(2) was not used in the proof of the second statement). By Proposition 46(1), $\mathfrak{m} \cap R_P$ is the extension of P to the local ring R_P, and the contraction of this ideal to R is just P. Put another way, the preimage of $\mathfrak{m}$ by the maps along the top and right of the diagram above is P. If $Q \subset S$ denotes the preimage of $\mathfrak{m}$ by the map along the bottom of the diagram, then Q is a prime ideal by Proposition 38(3). Since $Q \cap R$ is the pullback of Q by the map along the left of the diagram above, the commutativity of the diagram shows that $Q \cap R = P$.

Local Rings of Affine Algebraic Varieties

For the remainder of this section, let k be an algebraically closed field and let V be an affine variety over k with coordinate ring $k[V]$. Then $k[V]$ is an integral domain, so we may form its field of fractions:

$$k(V) = \{f/g \mid f, g \in k[V], \ g \neq 0\}.$$

The elements of $k(V)$ are called *rational functions* on V and $k(V)$ is called the *field of rational functions* on V. When $k[V]$ is a Unique Factorization Domain there is an essentially unique representative for f/g that is in "lowest terms," but in general each fraction $f/g \in k(V)$ has many representations as a ratio of two elements of $k[V]$. Since $k[V]$ is an integral domain, $f/g = f_1/g_1$ if and only if $fg_1 = f_1g$.

The elements of $k[V]$ can be considered as k-valued functions on V, and if the denominator doesn't vanish the same is true for an element of $k(V)$ (which helps to explain the terminology for this field). Since the same element of $k(V)$ may be written in the form f/g in several ways, we make the following definition:

Definition. We say f/g is *regular at v* or *defined at the point $v \in V$* if there is some $f_1, g_1 \in k[V]$ with $f/g = f_1/g_1$ and $g_1(v) \neq 0$.

If f_2, g_2 is another such pair with $g_2(v) \neq 0$, then $f_1(v)/g_1(v) = f_2(v)/g_2(v)$ as elements of k, so whenever f/g is regular at v there is a well defined way of specifying its value in k at v.

Example

The variety $V = \mathcal{Z}(xz - yw)$ in $\mathbb{A}^4$ has coordinate ring $k[V] = k[x, y, z, w]/(xz - yw)$. Consider the element $f = \bar{x}/\bar{y}$ in the quotient field $k(V)$ of $k[V]$. Since $\bar{x}\bar{z} = \bar{y}\bar{w}$ in $k[V]$, the element f can also be written as $\bar{w}/\bar{z}$. From the first expression for f it follows that f

is regular at all points of V where $\bar{y} \neq 0$, and from the second expression it follows that f is regular at all points of V where $\bar{z} \neq 0$. It is not too difficult to show that these are all the points of V where f is regular. Furthermore, there is no single expression $f = a/b$ for f with $a, b \in k[V]$ such that $b(v) \neq 0$ for every v where f is regular (cf. Exercise 25).

If $f/g \in k(V)$ is regular at the point v, say $f/g = f_1/g_1$ with $g_1(v) \neq 0$, then f/g is also regular at all the points v in the Zariski open neighborhood V_{g_1} of v where $g_1 \neq 0$. As a k-valued function on V this means that if f/g is defined at v, then it is also defined in a (Zariski open) neighborhood of v. Since any nonempty open set of an affine variety is Zariski dense (cf. Exercise 11 in Section 2), we see that every rational function on V is defined at a dense set of points in V (so "almost everywhere" in a suitable sense). Also, each pair f_1/g_1 and f_2/g_2 representing f/g agree as functions on the open neighborhood $V_{g_1} \cap V_{g_2}$ of v, but the "size" of this neighborhood depends on g_1 and g_2 — there is in general not a common open neighborhood of v where *all* representatives of f/g with nonzero denominator at v are simultaneously defined.

If v is a fixed point in V, then a rational function f/g is regular at v if and only if $f/g = f_1/g_1$ for some $f_1, g_1 \in k[V]$ with $g_1 \notin \mathcal{I}(v)$, the ideal of functions on V that are zero at v. This means that the set of rational functions that are defined at v is the same as the localization of $k[V]$ at the maximal ideal $\mathcal{I}(v)$:

Definition. For each point $v \in V$ the collection of rational functions on V that are defined at v,
$$\mathcal{O}_{v,V} = \{f/g \in k(V) \mid f/g \text{ is regular at } v\},$$
is called the *local ring of V at v*. Equivalently, the local ring of V at v is the localization of $k[V]$ at the maximal ideal $\mathcal{I}(v)$.

In particular, $\mathcal{O}_{v,V}$ is a local ring with unique maximal ideal $\mathfrak{m}_{v,V}$, where
$$\mathfrak{m}_{v,V} = \{f/g \in \mathcal{O}_{v,V} \mid f/g = f_1/g_1 \text{ with } f_1(v) = 0, \ g_1(v) \neq 0\}$$
is the set of rational functions on V that are defined and equal to 0 at v. Since $\mathcal{O}_{v,V}$ is a localization of the Noetherian integral domain $k[V]$ at a prime ideal, $\mathcal{O}_{v,V}$ is also a Noetherian integral domain. Note also that $\mathcal{O}_{v,V}/\mathfrak{m}_{v,V} \cong k[V]/\mathcal{I}(v) \cong k$ by Proposition 46(5).

Recall that the polynomial maps from V to k are also referred to as the *regular* maps of V to k. This is because these are precisely the rational functions on V that are regular everywhere:

Proposition 51. If V is an affine variety over an algebraically closed field k then the rational functions on V that are regular at all points of V are precisely the polynomial functions $k[V]$.

Proof: This follows from Proposition 48, which shows that the intersection (in $k(V)$) of all of the localizations of $k[V]$ at the maximal ideals of $k[V]$ is precisely $k[V]$.

Since the maximal ideals of $k[V]$ are in bijective correspondence with the points of V, the fact that the local ring $\mathcal{O}_{v,V}$ is the same as the localization of $k[V]$ at the maximal ideal corresponding to v shows that $\mathcal{O}_{v,V}$ depends intrinsically on the ring $k[V]$ and is independent of the embedding of V in a particular affine space.

Suppose $\varphi : V \to W$ is a morphism of affine varieties with associated k-algebra homomorphism $\widetilde{\varphi} : k[W] \to k[V]$. If $v \in V$ is mapped to $w \in W$ by φ, then it is straightforward to show that $\widetilde{\varphi}$ induces a homomorphism (also denoted by $\widetilde{\varphi}$) between the corresponding local rings:

$$\widetilde{\varphi} : \mathcal{O}_{w,W} \to \mathcal{O}_{v,V} \quad \text{where} \quad \widetilde{\varphi}(h/k) = \widetilde{\varphi}(h)/\widetilde{\varphi}(k),$$

and that under this homomorphism, $\widetilde{\varphi}^{-1}(\mathfrak{m}_{v,V}) = \mathfrak{m}_{w,W}$ (a homomorphism of local rings having this property is called a *local homomorphism*). Note that $\widetilde{\varphi}$ does not in general extend to a field homomorphism from *all* of $k(W)$ into $k(V)$ since elements of $k[W]$ lying in the kernel of $\widetilde{\varphi}$ do not map to invertible elements in $k(V)$. It is also easy to check that if $\psi \circ \varphi$ is a composition of morphisms then on the local rings $\widetilde{\psi \circ \varphi} = \widetilde{\varphi} \circ \widetilde{\psi}$.

The local ring $\mathcal{O}_{v,V}$ can be used to provide an algebraic definition of the "smoothness" (in the sense of the existence of tangents) of V at v, as we now indicate. Suppose first that $V = \mathcal{Z}(f)$ is the hypersurface variety in $\mathbb{A}^n$ defined by the zeros of an irreducible polynomial f in $k[x_1, \ldots, x_n]$. For any point $v = (v_1, \ldots, v_n)$ on V let $D_v(f)(x_1, \ldots, x_n)$ be the linear polynomial:

$$D_v(f)(x_1, \ldots, x_n) = \sum_{i=1}^{n} \frac{\partial f}{\partial x_i}(v)\, x_i,$$

where the partial derivative of f with respect to x_i is given by the usual formal rule for the derivative of a polynomial in x_i (with all other variables considered constant). The polynomial $D_v(f)(x_1 - v_1, \ldots, x_n - v_n)$ is the first order Taylor polynomial of the function f at v, so gives the best linear approximation to $f(x_1, \ldots, x_n) \in k[x_1, \ldots, x_n]$ at v. It follows that if $\mathbf{T}$ is the linear variety $\mathcal{Z}(D_v(f)(x_1, \ldots, x_n))$ consisting of those points where $D_v(f)$ is zero, then the translate $v + \mathbf{T}$ is "tangent" to the hypersurface $\mathcal{Z}(f)$ at v.

Example

Suppose $f = x^2 - y \in k[x, y]$, so that $V = \mathcal{Z}(f)$ is just the parabola $y = x^2$. We have $\partial f/\partial x = 2x$ and $\partial f/\partial y = -1$, which at $v = (3, 9)$ are equal to 6 and -1, respectively. Then

$$D_{(3,9)}(f)(x, y) = 6x - y,$$

and the corresponding linear variety $\mathbf{T}$ is the line $y = 6x$ through the origin. The translate $(3, 9) + \mathbf{T}$ is the usual tangent line to the parabola at $(3, 9)$. The Taylor expansion of $x^2 - y$ at $(3, 9)$ is $x^2 - y = [\,6(x - 3) - (y - 9)\,] + (x - 3)^2$. The first order terms are $D_{(3,9)}(f)(x - 3, y - 9)$ and give the best linear approximation to $x^2 - y$ near $(3,9)$.

It is straightforward to extend these notions to any affine variety V in $\mathbb{A}^n$.

Definition. Define the *tangent space to V at v* to be the linear variety

$$\mathbb{T}_{v,V} = \mathcal{Z}(\{D_v(f)(x_1, \ldots, x_n) \mid f \in \mathcal{I}(V)\}).$$

The formal partial derivatives are k-linear and obey the usual product rule for derivatives, so the tangent space may be computed from the generators for $\mathcal{I}(V)$:

$$\text{if} \quad \mathcal{I}(V) = (f_1, f_2, \ldots, f_m) \quad \text{then} \quad \mathbb{T}_{v,V} = \bigcap_{i=1}^{m} \mathcal{Z}(D_v(f_i)).$$

Note that $\mathbb{T}_{v,V}$ is an intersection of vector spaces, so is a vector subspace of k^n.

This definition of the tangent space $\mathbb{T}_{v,V}$, while making apparent the connection with tangents to the variety V, seems to depend on the embedding of V in $\mathbb{A}^n$. In fact the tangent space can be defined entirely in terms of the local ring $\mathcal{O}_{v,V}$, as the next proposition proves.

Proposition 52. Let V be an affine variety over the algebraically closed field k and let v be a point on V with local ring $\mathcal{O}_{v,V}$ and corresponding maximal ideal $\mathfrak{m}_{v,V}$. Then there is a k-vector space isomorphism

$$(\mathbb{T}_{v,V})^* \cong \mathfrak{m}_{v,V}/\mathfrak{m}_{v,V}^2$$

where $(\mathbb{T}_{v,V})^*$ denotes the vector space dual (cf. Section 11.3) of the tangent space $\mathbb{T}_{v,V}$ to V at v.

Proof: Let $(k^n)^*$ denote the n-dimensional vector space dual to k^n. Since each $D_v(f)$ is a linear function, D_v is a linear transformation from $k[x_1, \ldots, x_n]$ to $(k^n)^*$. Let M_v be the maximal ideal in $k[x_1, \ldots, x_n]$ generated by the set $x_i - v_i$ for $1 \le i \le n$. The image $M_v/\mathcal{I}(V)$ of M_v in $k[V]$ is the ideal $\mathcal{I}(v)$ of functions on V that are zero at v and $\mathcal{I}(v)^2 = M_v^2 + \mathcal{I}(V)$. Then $\mathcal{O}_{v,V}$ is the localization of $k[V]$ at $\mathcal{I}(v)$; and identifying $\mathcal{I}(v)$ with its image in $\mathcal{O}_{v,V}$ we have $\mathfrak{m}_{v,V} = \mathcal{I}(v)\mathcal{O}_{v,V}$ (Proposition 46(2)). By definition of D_v we have $D_v(x_i - v_i) = x_i$, and since these linear functions form a basis of $(k^n)^*$, it follows that D_v maps M_v surjectively onto $(k^n)^*$. The kernel of D_v consists of the elements of $k[x_1, \ldots, x_n]$ whose Taylor expansion at v starts in degree at least 2 and these are just the elements in M_v^2. Hence D_v defines an isomorphism

$$D_v : M_v/M_v^2 \stackrel{\sim}{\to} (k^n)^*.$$

The tangent space $\mathbb{T}_{v,V}$ is a vector subspace of k^n, so every linear function on k^n restricts to a linear function on $\mathbb{T}_{v,V}$. Composing D_v with this restriction map gives a linear transformation

$$D : M_v \xrightarrow{D_v} (k^n)^* \xrightarrow{\text{res}} (\mathbb{T}_{v,V})^*$$

which is surjective since the individual maps are each surjective. We have already seen that $\mathcal{I}(v)^2 = M_v^2 + \mathcal{I}(V)$, so $\mathcal{I}(v)/\mathcal{I}(v)^2 \cong M_v/(M_v^2 + \mathcal{I}(V))$. It follows by Proposition 46(5) that $\mathfrak{m}_{v,V}/\mathfrak{m}_{v,V}^2 \cong \mathcal{I}(v)/\mathcal{I}(v)^2$. To prove the proposition it is therefore sufficient to show that $\ker D = M_v^2 + \mathcal{I}(V)$, since then

$$\mathfrak{m}_{v,V}/\mathfrak{m}_{v,V}^2 \cong M_v/(M_v^2 + \mathcal{I}(V)) = M_v/\ker D \cong (\mathbb{T}_{v,V})^*.$$

The polynomial f is in $\ker D$ if and only if $D_v(f)$ is zero on $\mathbb{T}_{v,V}$, i.e., if and only if the linear term of the Taylor polynomial of f expanded about v lies in $\mathcal{I}(\mathbb{T}_{v,V})$. Since the linear terms of the functions in $\mathcal{I}(V)$ generate the ideal $\mathcal{I}(\mathbb{T}_{v,V})$, it follows that f is in $\ker D$ if and only if $f - g$ has zero linear term for some g in $\mathcal{I}(V)$. But this is equivalent to $f \in \mathcal{I}(V) + M_v^2$, so $\ker D = \mathcal{I}(V) + M_v^2$, completing the proof of the proposition.

Recall that the *dimension* of a variety V is by definition the transcendence degree of the field $k(V)$ over k. Since each local ring $\mathcal{O}_{v,V}$ has $k(V)$ as its field of fractions, the dimension of V is determined by the transcendence degree over k of the field of fractions of any of its local rings.

Definition. We say V is *nonsingular* at the point $v \in V$ (or v is a *nonsingular point* of V) if the dimension of the k-vector space $\mathbb{T}_{v,V}$ is $\dim V$. Equivalently (by Proposition 52), v is a nonsingular point of V if $\dim_k(\mathfrak{m}_{v,V}/\mathfrak{m}_{v,V}^2) = \dim V$. Otherwise the point v is called a *singular point*. The variety V is *nonsingular* or *smooth* if it is nonsingular at every point.

The geometric picture is that at a nonsingular point v there are as many independent tangents as one would expect: a tangent line on a curve, a tangent plane on a surface, etc.

Whether a variety V is nonsingular at a point v can be determined from properties of the local ring $\mathcal{O}_{v,V}$, namely whether $\dim_k(\mathfrak{m}_{v,V}/\mathfrak{m}_{v,V}^2) = \dim \mathcal{O}_{v,V}$. A local ring having this property is said to be a *regular local ring*. In particular, the notion of singularity does not depend on the embedding of V in a specific affine space. This algebraic interpretation can be used to *define* smoothness for abstract algebraic varieties, where the geometric intuition of tangent planes to surfaces (for example) is not as obvious.

If $f_1, \ldots, f_m$ are generators for $\mathcal{I}(V)$ defining V in $\mathbb{A}^n$, then the dimension of V can be determined from a Gröbner basis for $\mathcal{I}(V)$ (cf. Exercise 29). Determining the dimension of the tangent space $\mathbb{T}_{v,V}$ as a vector space over k is a linear algebra problem: this vector space is the set of solutions of the m linear equations $D_v(f_i)(x_1, \ldots, x_n) = 0$. If r is the rank of the $m \times n$ matrix of coefficients $\partial f_i / \partial x_j(v)$ of this system of equations, then $\mathbb{T}_{v,V}$ is a vector space of dimension $n - r$. Using this it is not too difficult to establish the following:

1. We have $\dim V \le \dim_k(\mathbb{T}_{v,V}) \le n$ for every point v in $V \subseteq \mathbb{A}^n$.

2. The set of singular points of V is a proper Zariski closed subset of V. The set of nonsingular points of V is a nonempty open subset of V; in particular the nonsingular points of V are dense in V (so "most" points of V are nonsingular).

We also state without proof the following result which further relates the local geometry of V to the algebraic properties of the local rings of V:

3. If v is a nonsingular point, then the local ring $\mathcal{O}_{v,V}$ is a Unique Factorization Domain; in particular, $\mathcal{O}_{v,V}$ is integrally closed (cf. Example 3 following Corollary 25).

The variety V is said to be *factorial* if $\mathcal{O}_{v,V}$ is a U.F.D. for every point $v \in V$, and is said to be a *normal* variety if $\mathcal{O}_{v,V}$ is integrally closed for every $v \in V$ (which by Proposition 49 is equivalent to $k[V]$ being integrally closed). By (3) above we have

$$\text{smooth varieties} \quad \subseteq \quad \text{factorial varieties} \quad \subseteq \quad \text{normal varieties}.$$

In general each of the above containments is proper. In the case when V has dimension 1, i.e., V is an *affine curve*, however, these three properties are in fact equivalent: we shall prove later that an irreducible affine curve is smooth if and only if it is normal or factorial (cf. Corollary 13 in Section 16.2). It follows that over an algebraically closed field k,

an irreducible affine curve C is smooth if and only if $k[C]$ is integrally closed.

For any irreducible affine curve C the integral closure, S, of $k[V]$ in $k(V)$ is also the coordinate ring of an irreducible affine curve $\widetilde{C}$. Then S is integral over $k[V]$ and, by Theorem 30 and Corollary 27 it follows that there is a morphism from the smooth curve $\widetilde{C}$ onto C that has finite fibers. The curve $\widetilde{C}$ is called the *normalization* or the *nonsingular model* of C, and one can show that it is unique up to isomorphism. Note how the existence of a smooth curve mapping finitely to C (a problem in "geometry") is solved by the existence of integral closures in ring extensions (a problem in "algebra").

We shall give another characterization of smoothness for irreducible affine curves at the end of Section 16.2.

EXERCISES

As usual R is a commutative ring with 1 and D is a multiplicatively closed set in R.

1. Suppose M is a finitely generated R-module. Prove that $D^{-1}M = 0$ if and only if $dM = 0$ for some $d \in D$.

2. Let I be an ideal in R, let D be a multiplicatively closed subset of R with ring of fractions $D^{-1}R$, and let $^c(^eI) = R$ be the saturation of I with respect to D.
   (a) Prove that $^c(^eI) = R$ if and only if $^eI = D^{-1}R$ if and only if $I \cap D \neq \emptyset$.
   (b) Prove that $I = {}^c(^eI)$ is saturated if and only if for every $d \in D$, if $da \in I$ then $a \in I$.
   (c) Prove that extension and contraction define inverse bijections between the ideals of R saturated with respect to D and the ideals of $D^{-1}R$.
   (d) Let $I = (2x, 3y) \subset \mathbb{Z}[x, y]$. Show the saturation of I with respect to $\mathbb{Z} - \{0\}$ is (x, y).

3. If I is an ideal in the commutative ring R let $\varphi : R[x_1, \ldots, x_n] \cong (R/I)[x_1, \ldots, x_n]$ be the ring homomorphism with kernel $I[x_1, \ldots, x_n]$ given by reducing coefficients modulo I. If $\overline{A}$ is an ideal in $(R/I)[x_1, \ldots, x_n]$, let A denote the inverse image of $\overline{A}$ under φ.
   (a) For any $i \geq 1$ show that the inverse image under φ of the subring $(R/I)[x_1, \ldots, x_i]$ is $R[x_1, \ldots, x_i] + I[x_1, \ldots, x_n]$.
   (b) Prove that $\varphi(A \cap R[x_1, \ldots, x_i]) = \overline{A} \cap (R/I)[x_1, \ldots, x_i]$

4. Let $f = y^5 - z^4$, viewed as a polynomial in y with coefficients in $\mathbb{Q}[z]$.
   (a) Prove that f has no roots in $\mathbb{Q}[z]$.
   (b) Suppose $f = (y^2 + ay + b)(y^3 + cy^2 + dy + e)$. Show that a, b, c, d, e satisfy the system of equations

   $$a + c = 0, \quad ac + b + d = 0, \quad ad + bc + e = 0, \quad ae + bd = 0, \quad be - z^4 = 0.$$

   Deduce that $e^5 = z^{12}$ and conclude that f is irreducible in $\mathbb{Q}[y, z]$. [Use elimination.]

5. Suppose R is a U.F.D. with field of fractions F and $p \in R[x]$ is a monic polynomial.
   (a) Show that the ideal $pR[x]$ generated by p in $R[x]$ is prime if and only if the ideal $pF[x]$ generated by p in $F[x]$ is prime. [Use Gauss' Lemma.]
   (b) Show that $pR[x]$ is saturated, i.e., that $pF[x] \cap R[x] = pR[x]$.

6. Show that $I = (y^3 - xz, xy^2 - z^2)$ is not a prime ideal in $\mathbb{Q}[x, y, z]$ and find explicit elements $a, b \in \mathbb{Q}[x, y, z]$ with $ab \in I$ but $a \notin I$ and $b \notin I$.

7. Show that $P = (y^3 - xz, xy^2 - z^2, x^2 - yz)$ is a prime ideal in $\mathbb{Q}[x, y, z]$.

8. Show that $P = (x^2 - yz, w^2 - x^4z)$ is a prime ideal in $\mathbb{Q}[x, y, z, w]$.

9. Show that $P = (xz^2 - w^3, xw^2 - y^4, y^4z^2 - w^5)$ is a prime ideal in $\mathbb{Q}[x, y, z, w]$.

10. Show that $I = (xy - w^3, y^2 - zw)$ is not a prime ideal in $\mathbb{Q}[x, y, z, w]$ and find a, b with $ab \in I$ but $a, b \notin I$.

11. Let R_P be the localization of R at the prime P. Prove that if Q is a P-primary ideal of R then $Q = {}^c({}^eQ)$ with respect to the extension and contraction of Q to R_P. Show the same result holds if Q is P'-primary for some prime P' contained in P.

12. Let $R = \mathbb{R}[x, y, z]/(xy - z^2)$, let $P = (\bar{x}, \bar{z})$ be the prime ideal generated by the images of x and y in R, and let R_P be the localization of R at P. Prove that $P^2R_P \cap R = (\bar{x})$ and is strictly larger than P^2.

13. Prove that if N and N' are two R-submodules of an R-module M with $N_P = N'_P$ in the localization M_P for every prime ideal P of R (or just for every maximal ideal) then $N = N'$.

14. Suppose $\varphi : M \to N$ is an R-module homomorphism. Prove that φ is injective (respectively, surjective) if and only if the induced R_P-module homomorphism $\varphi : M_P \to N_P$ is injective (respectively, surjective) for every prime ideal P of R (or just for every maximal ideal of R).

15. Let $R = \mathbb{Z}[\sqrt{-5}]$ be the ring of integers in the quadratic field $\mathbb{Q}(\sqrt{-5})$ and let I be the prime ideal $(2, 1 + \sqrt{-5})$ of R generated by 2 and $1 + \sqrt{-5}$ (cf. Exercise 5, Section 8.2). Recall that every nonzero prime ideal P of R contains a prime $p \in \mathbb{Z}$.
   (a) If P is a prime ideal of R not containing 2 prove that $I_P = R_P$.
   (b) If P is a prime ideal of R containing 2 prove that $P = I$ and that $I_P = (1 + \sqrt{-5})R_P$.
   (c) Prove that $I_P \cong R_P$ as R_P-modules for every prime ideal P of R but that I and R are not isomorphic R-modules. (This example shows that it is important in Exercise 14 to be *given* the R-module homomorphism φ.) [Observe that $I \cong R$ as R-modules if and only if I is a *principal* ideal.]

16. Prove that localization commutes with tensor products: there is a unique isomorphism of $D^{-1}R$-modules $\varphi : (D^{-1}M) \otimes_{D^{-1}R} (D^{-1}N) \cong D^{-1}(M \otimes_R N)$ with $\varphi((m/d) \otimes (n/d'))$ given by $(m \otimes n)/dd'$ for any R-modules M, N, and multiplicatively closed set D in R.

17. Prove that the R-module A is a flat R-module if and only if A_P is a flat R_P-module for every prime ideal P of R (or just for every maximal ideal of R). [Use Proposition 41, Exercises 14 and 16, and the exactness properties of localization.]

18. In the notation of Example 2 following Corollary 37, prove that $R_f \cong R[x]/(fx - 1)$ iff f is not nilpotent in R. [Show that the map $\varphi : R[x] \to R_f$ defined by $\varphi(r) = r/1$ and $\varphi(x) = 1/f$ gives a surjective ring homomorphism and the universal property in Theorem 36 gives an inverse.]

19. Prove that if R is an integrally closed integral domain and D is any multiplicatively closed subset of R containing 1, then $D^{-1}R$ is integrally closed.

20. Suppose that R is a subring of the ring S with $1 \in R$ and that S is integral over R. If D is any multiplicatively closed subset of R, prove that $D^{-1}S$ is integral over $D^{-1}R$.

21. Suppose $\varphi : R \to S$ is a ring homomorphism with $\varphi(1_R) = 1_S$ and D' is a multiplicatively closed subset of S. Let $D = \varphi^{-1}(D')$. Prove D is a multiplicatively closed subset of R and the map $\varphi' : D^{-1}R \to D'^{-1}S$ given by $\varphi'(r/d) = \varphi(r)/\varphi(d)$ is a ring homomorphism.

22. Suppose $P \subseteq Q$ are prime ideals in R and let R_Q be the localization of R at Q. Prove that the localization R_P is isomorphic to the localization of R_Q at the prime ideal PR_Q (cf. the preceding exercise).

23. Let $\varphi : A \to B$ be a homomorphism of commutative rings with $\varphi(1_A) = 1_B$, and let P be a prime ideal of A. Let contraction and extension of ideals with respect to φ be denoted by superscripts c and e respectively. Prove that P is the contraction of a prime ideal in B if and only if $P = (P^e)^c$. [Localize B at $\varphi(A - P)$.]

24. *(The Going-down Theorem)* Let S be an integral domain, let R be an integrally closed subring of S containing 1_S, and let k be the field of fractions of R. Suppose that $P_2 \subseteq P_1$ are prime ideals in R and that Q_1 is a prime ideal in S with $Q_1 \cap R = P_1$. Let S_{Q_1} be the localization of S at Q_1.
   (a) Show that $P_2 \subseteq P_2 S_{Q_1} \cap R$.
   (b) Suppose that $a \in P_2 S_{Q_1} \cap R$ and write $a = s/d$ with $s \in P_2 S$ and $d \in S$, $d \notin Q_1$. If the minimal polynomial of s over k is $x^n + a_{n-1}x^{n-1} + \cdots + a_1 x + a_0$ with $a_0, \ldots, a_{n-1} \in P_2$ (cf. Exercise 12 in Section 3) show that the minimal polynomial of d over k is $x^n + b_{n-1}x^{n-1} + \cdots + b_1 x + b_0$ where $b_i = a_i/a^{n-i}$ and conclude that $b_i \in R$. [Use Exercise 10 in Section 3.]
   (c) Show that $a \in P_2$ and conclude that $P_2 S_{Q_1} \cap R = P_2$. [Show $a \notin P_2$ implies $b_i \in P_2$ for $i = 0, 1, \ldots, n-1$, which would imply $d^n \in P_2 S \subseteq P_1 S \subseteq Q_1$ and so $d \in Q_1$.]
   (d) Prove that $P_2 S_{Q_1}$ is contained in a prime ideal P of S_{Q_1} with $P \cap R = P_2$. [Use (c) and the previous exercise for $\varphi : R \to S_{Q_1}$.]
   (e) Let $Q_2 = P \cap S$. Prove that $Q_2 \subseteq Q_1$ and that $Q_2 \cap R = P_2$.
   (f) Use induction together with the previous result to prove the Going-down Theorem: Theorem 26(4).

25. Let k be an algebraically closed field and let $V = \mathcal{Z}(xz - yw) \subset \mathbb{A}^4$. Prove that the set of points v where $f = \bar{x}/\bar{y} \in k(V)$ is regular is precisely the set of points (x, y, z, w) where $y \neq 0$ or $z \neq 0$. [If $f = \bar{a}/\bar{b}$ show that $ay - bx \in (xz - yw)$ as polynomials in $k[x, y, z, w]$ and conclude that $b \in (y, z)$.] Prove that there is no function $a/b \in k(V)$ with $b(v) \neq 0$ for every v where f is regular.

26. *(Differentials of Morphisms)* Let $\varphi : V \to W$ be a morphism of affine varieties over the algebraically closed field k and suppose $\varphi(v) = w$.
   (a) Show that φ induces a linear map from the k-vector space M_w/M_w^2 to the k-vector space M_v/M_v^2, and use this to show that φ induces a linear map $d\varphi$ (called the *differential* of φ) from the k-vector space $\mathbb{T}_{v,V}$ to the k-vector space $\mathbb{T}_{w,W}$.
   (b) Prove that if $V \subseteq \mathbb{A}^n$, $W \subseteq \mathbb{A}^m$ and $\varphi = (F_1(x_1, \ldots, x_n), \ldots, F_m(x_1, \ldots, x_n))$ then $d\varphi : \mathbb{T}_{v,V} \to \mathbb{T}_{w,W}$ is given explicitly by

$$(d\varphi)(a_1, \ldots, a_n) = (D_v(F_1)(a_1, \ldots, a_n), \ldots, D_v(F_m)(a_1, \ldots, a_n)).$$

[If $g = g(y_1, \ldots, y_m)$ show that the chain rule implies

$$\frac{\partial(g \circ \varphi)}{\partial x_i}(v) = \sum_{j=1}^{m} \frac{\partial g}{\partial y_j}(w) \frac{\partial F_j}{\partial x_i}(v),$$

so that $D_v(g \circ \varphi)(a_1, \ldots, a_n) = D_w(g)(b_1, \ldots, b_m)$ where $b_j = D_v(F_j)(a_1, \ldots, a_n)$. Then use the fact that $g \circ \varphi \in \mathcal{I}(V)$ if $g \in \mathcal{I}(W)$.]

(c) If $\psi : U \to V$ is another morphism with $\psi(u) = v$, prove that the associated $d(\varphi \circ \psi) : \mathbb{T}_{u,U} \to \mathbb{T}_{w,W}$ is the same as $d\varphi \circ d\psi$.

(d) Prove that if φ is an isomorphism then $d\varphi$ is a vector space isomorphism from $\mathbb{T}_{v,V}$ to $\mathbb{T}_{w,W}$ for every $\varphi(v) = w$.

27. Let $V = \mathbb{A}^1$ and $W = \mathcal{Z}(xz - y^2, yz - x^3, z^2 - x^2 y) \subset \mathbb{A}^3$. Let $\varphi : V \to W$ be the surjective morphism $\varphi(t) = (t^3, t^4, t^5)$ (cf. Exercise 26 in Section 1). For each $t \in \mathbb{A}^1$ describe the differential $d\varphi : \mathbb{T}_{t,\mathbb{A}^1} \to \mathbb{T}_{(t^3,t^4,t^5),W}$ in the previous exercise explicitly; in particular prove that $d\varphi$ is an isomorphism of vector spaces for all $t \neq 0$ and is the zero map for $t = 0$. Use this to prove that V and W are not isomorphic.

28. If k is a field, the quotient $k[x]/(x^2)$ is called the *ring of dual numbers* over k. If V is an affine algebraic set over k, show that a k-algebra homomorphism from $k[V]$ to $k[x]/(x^2)$ is equivalent to specifying a point $v \in V$ with $\mathcal{O}_{v,V}/\mathfrak{m}_{v,V} = k$ (called a *k-rational point* of V) together with an element in the tangent space $\mathbb{T}_{v,V}$ of V at v.

29. (*Computing the dimension of a variety*) Let P be a prime ideal in $k[x_1, \ldots, x_n]$, set $P_0 = 0$ and let $P_i = P \cap k[x_1, \ldots, x_i]$. Define the varieties $V_i = \mathcal{Z}(P_i) \subseteq \mathbb{A}^i$ with V_0 the zero dimensional variety consisting of a single point and coordinate ring k.

(a) Show that $\dim V_{i-1} \leq \dim V_i \leq \dim V_{i-1} + 1$. [First exhibit an injection from $k[V_{i-1}]$ into $k[V_i]$; then show that $k[V_i]$ is a k-algebra generated by $k[V_{i-1}]$ and one additional generator.]

(b) If the ideal generated by P_{i-1} in $k[x_1, \ldots, x_i]$ equals P_i, show that $V_i \cong V_{i-1} \times \mathbb{A}^1$ and deduce that $\dim V_i = \dim V_{i-1} + 1$.

(c) If the ideal generated by P_{i-1} in $k[x_1, \ldots, x_i]$ is properly contained in P_i, show that $\dim V_i = \dim V_{i-1}$.

(d) Show that $\dim V$ equals the number of $i \in \{1, 2, \ldots, n\}$ such that the ideal generated by P_{i-1} in $k[x_1, \ldots, x_i]$ equals the ideal P_i. Deduce that if G is the reduced Gröbner basis for P with respect to the lexicographic monomial ordering $x_n > \cdots > x_1$ and $G_i = G \cap k[x_1, \ldots, x_i]$ where $G_0 = \emptyset$, and N is the number of i with $G_i \neq G_{i-1}$ for $1 \leq i \leq n$, then $\dim V = n - N$.

The following eleven exercises introduce the notion of the *support* of an R-module M and its relation to the associated primes of M. Cf. also Exercises 29 to 35 in Section 1 and Exercises 25 to 30 in Section 5.

Definition. If M is an R-module, then the set of prime ideals P of R for which the localization M_P is nonzero is called the *support* of M, denoted Supp(M).

30. Prove that $M = 0$ if and only if Supp(M) = $\emptyset$. [Use Proposition 47.]

31. If $0 \to L \to M \to N \to 0$ is an exact sequence of R-modules, prove that the localization M_P is nonzero if and only if one of the localizations N_P and L_P is nonzero and deduce that Supp(M) = Supp(L) $\cup$ Supp(N). In particular, if $M = M_1 \oplus \cdots \oplus M_n$ prove that Supp(M) = Supp(M_1) $\cup \cdots \cup$ Supp(M_n).

32. Suppose $P \subseteq Q$ are prime ideals in R and that M is an R-module. Prove that the localization of the R-module M_Q at P is the localization M_P, i.e., $(M_Q)_P = M_P$. [Argue directly, or use Proposition 41 and the associativity of the tensor product.]

33. Suppose $P \subseteq Q$ are prime ideals in R and that M is an R-module. Prove that if $P \in$ Supp(M) then $Q \in$ Supp(M). [Use the previous exercise.]

34. (a) Suppose $M = Rm$ is a cyclic R-module. Prove that $M_P = 0$ if and only if there is

an element $r \in R$, $r \notin P$ with $rm = 0$. Deduce that $P \in \text{Supp}(M)$ if and only if P contains the annihilator of m in R (cf. Exercise 10 in Section 10.1).

(b) If $M = Rm_1 + \cdots + Rm_n$ is a finitely generated R-module prove that $P \in \text{Supp}(M)$ if and only if P is contained in $\text{Supp}(Rm_i)$ for some $i = 1, \ldots, n$. [Use Proposition 42.] Deduce that $P \in \text{Supp}(M)$ if and only if P contains the annihilator $\text{Ann}(M)$ of M in R. [Note $\text{Ann}(M) = \cap_{i=1}^{n} \text{Ann}(Rm_i)$, then use (a) and Exercise 11 of Section 7.4.]

35. Suppose P is a prime ideal of R with $P \cap D = \emptyset$. Prove that if $P \in \text{Ass}_R(M)$ then $D^{-1}P \in \text{Ass}_{D^{-1}R}(D^{-1}M)$. [Use Proposition 38(3) and Proposition 42.]

36. Suppose $D^{-1}P \in \text{Ass}_{D^{-1}R}(D^{-1}M)$ where $P = (a_1, \ldots, a_n)$ is a finitely generated prime ideal in R with $P \cap D = \emptyset$.
   (a) Suppose $m/d \in D^{-1}M$ has annihilator $D^{-1}P$ in $D^{-1}R$. Show that $d_i a_i m = 0 \in R$ for some $d_1, \ldots, d_n \in D$.
   (b) Let $d' = d_1 d_2 \ldots d_n$. Show that $P = \text{Ann}(d'm)$ and conclude that $P \in \text{Ass}_R(M)$. [The inclusion $P \subseteq \text{Ann}(d'm)$ is immediate. For the reverse inclusion, show that $b \in \text{Ann}(d'm)$ implies that $b/1$ annihilates m/d in $D^{-1}M$, hence $b/1 \in D^{-1}P$, and conclude $b \in P$.]

37. Suppose M is a module over the Noetherian ring R. Use the previous two exercises to show that under the bijection of Proposition 38(3) the prime ideals P of $\text{Ass}_R(M)$ with $P \cap D = \emptyset$ correspond bijectively with the prime ideals of $\text{Ass}_{D^{-1}R}(D^{-1}M)$.

38. Suppose M is a module over the Noetherian ring R and D is a multiplicatively closed subset of R. Let $\mathcal{S}$ be the subset of prime ideals P in $\text{Ass}_R(M)$ with $P \cap D \neq \emptyset$. This exercise proves that the kernel N of the localization map $M \to D^{-1}M$ is the unique submodule N of M with $\text{Ass}_R(N) = \mathcal{S}$ and $\text{Ass}_R(M/N) = \text{Ass}_R(M) - \mathcal{S}$.
   (a) If N' is a submodule of M with $\text{Ass}_R(N') = \mathcal{S}$ and $\text{Ass}_R(M/N') = \text{Ass}_R(M) - \mathcal{S}$ as in Exercise 35 in Section 1, prove that the diagram

$$
\begin{array}{ccc}
M & \xrightarrow{\ \pi\ } & M/N' \\
\varphi \downarrow & & \downarrow \varphi' \\
D^{-1}M & \xrightarrow{\ \pi'\ } & D^{-1}(M/N')
\end{array}
$$

is commutative, where π and π' are the natural projections (cf. Proposition 42(6)) and φ, φ' are the localization homomorphisms.
   (b) Show that $\text{Ass}_{D^{-1}R}(D^{-1}N') = \emptyset$ and conclude that $D^{-1}N' = 0$ and that π' is injective. [Use the previous exercise, the definition of $\mathcal{S}$, and Exercise 34 in Section 1.]
   (c) If x is the kernel K of φ' show that $\text{Ann}(x) \cap D \neq \emptyset$ and that $\text{Ass}_R(K) \subseteq \mathcal{S}$. Show that $\text{Ass}_R(K) \subseteq \text{Ass}_R(M/N')$ implies that $\text{Ass}_R(K) = \emptyset$, and deduce that $K = 0$.
   (d) Prove φ and π have the same kernel, i.e., $N = N'$, and this submodule of M is unique.

The next two exercises establish a fundamental relation between the sets $\text{Ass}_R(M)$ and $\text{Supp}(M)$ of prime ideals related to the R-module M.

39. Prove that $\text{Ass}_R(M) \subseteq \text{Supp}(M)$. [If $Rm \cong R/P$ use Proposition 42(4) and Proposition 46(1) to show that $0 \neq (Rm)_P \subseteq M_P$.]

40. Suppose that R is Noetherian and M is an R-module.
   (a) If $P \in \text{Supp}(M)$ prove that P contains a prime ideal Q with $Q \in \text{Ass}_R(M)$.
   (b) If P is a minimal prime in $\text{Supp}(M)$, show that $P \in \text{Ass}_R(M)$. [Use Exercise 33 in Section 1 to show that $\text{Ass}_{R_P}(M_P) \neq \emptyset$ and then use Exercise 37.]
   (c) Conclude that $\text{Ass}_R(M) \subseteq \text{Supp}(M)$ and that these two sets have the same minimal elements.

15.5 THE PRIME SPECTRUM OF A RING

Throughout this section the term "ring" will mean commutative ring with 1 and all ring homomorphisms $\varphi : R \to S$ will be assumed to map 1_R to 1_S.

We have seen that most of the geometric properties of affine algebraic sets V over k can be translated into algebraic properties of the associated coordinate rings $k[V]$ of k-valued functions on V. For example, the morphisms from V to W correspond to k-algebra ring homomorphisms from $k[W]$ to $k[V]$. When the field k is an algebraically closed field this translation is particularly precise: Hilbert's Nullstellensatz establishes a bijection between the points v of V and the maximal ideals $M = \mathcal{I}(v)$ of $k[V]$, and if $\varphi : V \to W$ is a morphism then $\varphi(v) \in W$ corresponds to the maximal ideal $\widetilde{\varphi}^{-1}(M)$ in $k[W]$. In this development we have generally started with geometric properties of the affine algebraic sets and then seen that many of the algebraic properties common to the associated coordinate rings can be defined for arbitrary commutative rings. Suppose now we try to reverse this, namely start with a general commutative ring as the algebraic object and attempt to define a corresponding "geometric" object by analogy with $k[V]$ and V.

Given a commutative ring R, perhaps the most natural analogy with $k[V]$ and V would suggest defining the collection of maximal ideals M of R as the "points" of the associated geometric object. Under this definition, if $\widetilde{\varphi} : R' \to R$ is a ring homomorphism, then $\widetilde{\varphi}^{-1}(M)$ should correspond to the maximal ideal M. Unfortunately, the inverse image of a maximal ideal by a ring homomorphism in general need not be a maximal ideal. Since the inverse image of a *prime* ideal under a ring homomorphism (that maps 1 to 1) *is* prime, this suggests that a better definition might include the prime ideals of R. This leads to the following:

Definition. Let R be a commutative ring with 1. The *spectrum* or *prime spectrum* of R, denoted Spec R, is the set of all prime ideals of R. The set of all maximal ideals of R, denoted mSpec R, is called the *maximal spectrum* of R.

Examples

(1) If R is a field then Spec R = mSpec R = $\{(0)\}$.

(2) The points in Spec $\mathbb{Z}$ are the prime ideal (0) and the prime ideals (p) where $p > 0$ is a prime, and mSpec $\mathbb{Z}$ consists of all the prime ideals of Spec $\mathbb{Z}$ except (0).

(3) The elements of Spec $\mathbb{Z}[x]$ are the following:

    (a) (0)

    (b) (p) where p is a prime in $\mathbb{Z}$

    (c) (f) where $f \neq 1$ is a polynomial of content 1 (i.e., the g.c.d. of its coefficients is equal to 1) that is irreducible in $\mathbb{Q}[x]$

    (d) (p, g) where p is a prime in $\mathbb{Z}$ and g is a monic polynomial that is irreducible mod p.

    The elements of mSpec $\mathbb{Z}[x]$ are the primes in (d) above.

In the analogy with $k[V]$ and V when k is algebraically closed, the elements $f \in k[V]$ are functions on V with values in k, obtained by evaluating f at the point v in V. Note that "evaluation at v" defines a homomorphism from $k[V]$ to k with kernel $\mathcal{I}(v)$, and that the value of f at v is the element of k representing f in the quotient

$k[V]/\mathcal{I}(v) \cong k$. Put another way, the value of $f \in k[V]$ at $v \in V$ can be viewed as the element $\bar{f} \in k[V]/\mathcal{I}(v) \cong k$. A similar definition can be made in general:

Definition. If $f \in R$ then the *value* of f at the point $P \in \operatorname{Spec} R$ is the element $f(P) = \bar{f} \in R/P$.

Note that the values of f at different points P in general lie in *different* integral domains. Note also that in general $f \in R$ is not uniquely determined by its values, rather f is determined only up to an element in the nilradical of R (cf. Exercise 3).

There are analogues of the maps $\mathcal{Z}$ and $\mathcal{I}$ and also for the Zariski topology. For any subset A of R define

$$\mathcal{Z}(A) = \{P \in X \mid A \subseteq P\} \subseteq \operatorname{Spec} R,$$

the collection of prime ideals containing A. It is immediate that $\mathcal{Z}(A) = \mathcal{Z}(I)$, where $I = (A)$ is the ideal generated by A so there is no loss simply in considering $\mathcal{Z}(I)$ where I is an ideal of R. Note that, by definition, $P \in \mathcal{Z}(I)$ if and only if $I \subseteq P$, which occurs if and only if $f \in P$ for every $f \in I$. Viewing $f \in R$ as a function on $\operatorname{Spec} R$ as above, this says that $P \in \mathcal{Z}(I)$ if and only if $f(P) = f \bmod P = 0 \in R/P$ for all $f \in I$. In this sense, $\mathcal{Z}(I)$ consists of the points in $\operatorname{Spec} R$ at which all the functions in I have the value 0.

For any subset Y of $\operatorname{Spec} R$ define

$$\mathcal{I}(Y) = \bigcap_{P \in Y} P,$$

the intersection of the prime ideals in Y.

Proposition 53. Let R be a commutative ring with 1. The maps $\mathcal{Z}$ and $\mathcal{I}$ between R and $\operatorname{Spec} R$ defined above satisfy
  (1) for any ideal I of R, $\mathcal{Z}(I) = \mathcal{Z}(\operatorname{rad}(I)) = \mathcal{Z}(\mathcal{I}(\mathcal{Z}(I)))$, and $\mathcal{I}(\mathcal{Z}(I)) = \operatorname{rad} I$,
  (2) for any ideals I, J of R, $\mathcal{Z}(I \cap J) = \mathcal{Z}(IJ) = \mathcal{Z}(I) \cup \mathcal{Z}(J)$, and
  (3) if $\{I_j\}$ is an arbitrary collection of ideals of R, then $\mathcal{Z}(\cup I_j) = \cap \mathcal{Z}(I_j)$.

Proof: If P is a prime ideal containing the ideal I then P contains $\operatorname{rad} I$ (Exercise 8, Section 2), which implies $\mathcal{Z}(I) = \mathcal{Z}(\operatorname{rad}(I))$. Since $\operatorname{rad} I$ is the intersection of all the prime ideals containing I (Proposition 12), the definition of $\mathcal{I}(I)$ gives $\mathcal{Z}(\operatorname{rad}(I)) = \mathcal{Z}(\mathcal{I}(I))$. Similarly,

$$\mathcal{I}(\mathcal{Z}(I)) = \bigcap_{P \in \mathcal{Z}(I)} P = \bigcap_{I \subseteq P} P = \operatorname{rad} I,$$

which completes the proof of (1). It is immediate that $\mathcal{Z}(I \cap J) = \mathcal{Z}(I) \cup \mathcal{Z}(J)$. Suppose the prime ideal P contains IJ. If P does not contain I then there is some element $i \in I$ with $i \notin P$. Since $iJ \subseteq P$, it follows that $J \subseteq P$. This proves $\mathcal{Z}(IJ) = \mathcal{Z}(I) \cup \mathcal{Z}(J)$ and completes the proof of (2). The proof of (3) is immediate.

The first statement in the proposition shows that every set $\mathcal{Z}(I)$ in $\operatorname{Spec} R$ occurs for some *radical* ideal I, and since $\mathcal{I}(\mathcal{Z}(I)) = \operatorname{rad} I$, this radical ideal is unique.

The second two statements in the proposition show that the collection

$$\mathcal{T} = \{\mathcal{Z}(I) \mid I \text{ is an ideal of } R\}$$

satisfies the three axioms for the closed sets of a topology on Spec R as in Section 2.

Definition. The topology on Spec R defined by the closed sets $\mathcal{Z}(I)$ for the ideals I of R is called the *Zariski topology* on Spec R.

By definition, the closure in the Zariski topology of the singleton set $\{P\}$ in Spec R consists of all the prime ideals of R that contain P. In particular, a point P in Spec R is closed in the Zariski topology if and only if the prime ideal P is not contained in any other prime ideals of R, i.e., if and only if P is a maximal ideal (so the Zariski topology on Spec R is not generally Hausdorff). These points are given a name:

Definition. The maximal ideals of R are called the *closed points* in Spec R.

In terms of the terminology above, the points in Spec R that are closed in the Zariski topology are precisely the points in mSpec R.

A closed subset of a topological space is *irreducible* if it is not the union of two proper closed subsets, or, equivalently, if every nonempty open set is dense. Arguments similar to those used to prove Proposition 17 show that the closed subset $Y = \mathcal{Z}(I)$ in Spec R is irreducible if and only if $\mathcal{I}(Y) = \text{rad } I$ is prime (cf. Exercise 16).

The following proposition summarizes some of these results:

Proposition 54. The maps $\mathcal{Z}$ and $\mathcal{I}$ define inverse bijections

$$\{\text{Zariski closed subsets of Spec } R\} \quad \overset{\mathcal{I}}{\underset{\mathcal{Z}}{\rightleftarrows}} \quad \{\text{radical ideals of } R\}.$$

Under this correspondence the closed points in Spec R correspond to the maximal ideals in R, and the irreducible subsets of Spec R correspond to the prime ideals in R.

Examples

(1) If $X = \text{Spec } \mathbb{Z}$ then X is irreducible and the nonzero primes give closed points in X. The point (0) is not a closed point, in fact the closure of (0) is all of X, i.e., (0) is *dense* in Spec $\mathbb{Z}$. For this reason the element (0) is called a *generic point* in Spec $\mathbb{Z}$.

Since every ideal of $\mathbb{Z}$ is principal, the Zariski closed sets in Spec $\mathbb{Z}$ are $\emptyset$, Spec $\mathbb{Z}$ and any finite set of nonzero prime ideals in $\mathbb{Z}$.

(2) Suppose $X = \text{Spec } \mathbb{Z}[x]$ as in Example 3 previously. For each integer prime p the Zariski closure of the element $(p) \in X$ consists of the maximal ideals (p, g) of type (d). Likewise for each $\mathbb{Q}$-irreducible polynomial f of type (c), the Zariski closure of the element (f) is the collection of prime ideals of type (d) where g is some divisor of f in $\mathbb{Z}/p\mathbb{Z}[x]$.

Example: (Affine k-algebras)

Suppose $R = k[V]$ is the coordinate ring of some affine algebraic set $V \subseteq \mathbb{A}^n$ over an algebraically closed field k. Then $R = k[x_1, \ldots, x_n]/\mathcal{I}(V)$ where $\mathcal{I}(V)$ is a radical ideal in $k[x_1, \ldots, x_n]$. In particular R is a finitely generated k-algebra and since $\mathcal{I}(V)$ is radical, R contains no nonzero nilpotent elements.

Definition. A finitely generated algebra over an algebraically closed field k having no nonzero nilpotent elements is called an *affine k-algebra*.

If R is an affine k-algebra, then by Corollary 5 there is a surjective k-algebra homomorphism $\pi : k[x_1, \ldots, x_n] \to R$ whose kernel $I = \ker \pi$ must be a radical ideal since R has no nonzero nilpotent elements. Let $V = \mathcal{Z}(I) \subseteq \mathbb{A}^n$. Then $R \cong k[x_1, \ldots, x_n]/I = k[V]$ is the coordinate ring of an affine algebraic set over k. Hence *affine k-algebras are precisely the rings arising as the rings of functions on affine algebraic sets over algebraically closed fields*.

By the Nullstellensatz, the points of mSpec R are in bijective correspondence with V, and the points of Spec R are in bijective correspondence with the subvarieties of V. By Theorem 6, morphisms between two affine algebraic sets correspond bijectively with (k-algebra) homomorphisms of affine k-algebras. In the language of categories these results show that over an algebraically closed field k there is an equivalence of categories

$$\left\{ \begin{array}{c} \text{affine algebraic sets} \\ \text{morphisms of algebraic sets} \end{array} \right\} \longleftrightarrow \left\{ \begin{array}{c} \text{affine } k\text{-algebras} \\ k\text{-algebra homomorphisms} \end{array} \right\}.$$

The map from left to right sends the affine algebraic set V to its coordinate ring $k[V]$. The map from right to left sends the affine k-algebra R to mSpec R. The pair $(\text{mSpec } R, R)$ is sometimes called the *canonical model* of the affine k-algebra R.

Over an algebraically closed field k, a k-algebra homomorphism $\varphi : R \to S$ between two affine k-algebras as in the previous example has the property (by the Nullstellensatz) that the inverse image of a maximal ideal in S is a maximal ideal in R. As previously mentioned, one reason for considering Spec R rather than just mSpec R for more general rings is that inverse images of maximal ideals under ring homomorphisms are not in general maximal ideals. When R is an affine k-algebra corresponding to an affine algebraic set V, the space Spec R contains not only the "geometric points" of V (in the form of the closed points in Spec R), but also the non-closed points corresponding to all of the subvarieties of V (in the form of the non-closed points in Spec R, i.e., the prime ideals P of R that are not maximal).

In general, if $\varphi : R \to S$ is a ring homomorphism mapping 1_R to 1_S and P is a prime ideal in S then $\varphi^{-1}(P)$ is a prime ideal in R. This defines a map $\varphi^* : \text{Spec } S \to \text{Spec } R$ with $\varphi^*(P) = \varphi^{-1}(P)$. If $\mathcal{Z}(I) \subseteq \text{Spec } R$ is a Zariski closed subset of Spec R, then it is easy to show that $(\varphi^*)^{-1}(\mathcal{Z}(I))$ is the Zariski closed subset $\mathcal{Z}(\varphi(I)S)$ defined by the ideal generated by $\varphi(I)$ in S. Since the inverse image of a closed subset in Spec R is a closed subset in Spec S, the induced map φ^* is continuous in the Zariski topology. This proves the following proposition.

Proposition 55. Every ring homomorphism $\varphi : R \to S$ mapping 1_R to 1_S induces a map $\varphi^* : \text{Spec } S \to \text{Spec } R$ that is continuous with respect to the Zariski topologies on Spec R and Spec S.

While the generalization from affine algebraic sets to Spec R for general rings R has made matters slightly more complicated, there are (at least) two very important benefits gained by this more general setting. The first is that Spec R can be considered even for commutative rings R containing nilpotent elements; the second is that Spec R need not be a k-algebra for any field k, and even when it is, the field k need not be algebraically closed. The fact that many of the properties found in the situation of affine k-algebras hold in more general settings then allows the application of "geometric" ideas to these situations (for example, to Spec R when R is finite).

Examples

(1) The natural inclusion $\varphi : \mathbb{Z} \to \mathbb{Z}[i]$ induces a map $\varphi^* : \text{Spec } \mathbb{Z}[i] \to \text{Spec } \mathbb{Z}$. The fiber of φ^* over the nonzero prime P in $\mathbb{Z}$ consists of the prime ideals of $\mathbb{Z}[i]$ containing P. If $P = (p)$ where $p = 2$ or p is a prime congruent to 3 mod 4, then there is only one element in this fiber; if p is a prime congruent to 1 mod 4, then there are two elements in the fiber: the primes (π) and (π') where $p = \pi\pi'$ in $\mathbb{Z}[i]$, cf. Proposition 18 in Section 8.3. This can be represented pictorially in the following figure:

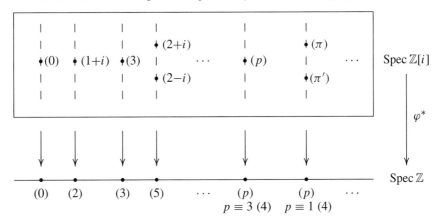

(2) If k is an algebraically closed field then Spec $k[x]$ consists of (0) and the ideals $(x - a)$ for $a \in k$; the natural inclusion $\varphi : k[x] \to k[x, y]$ induces the Zariski continuous map $\varphi^* : \text{Spec } k[x, y] \to \text{Spec } k[x]$. The elements of Spec $k[x, y]$ are

  (a) (0),
  (b) (f) where f is an irreducible polynomial in $k[x, y]$, and
  (c) $(x - a, y - b)$ with $a, b \in k$

(cf. Exercise 4). The prime (0) is Zariski dense in Spec $k[x, y]$; the Zariski closure of the primes in (b) consists of the primes $(x - a, y - b)$ in (c) with $f(a, b) = 0$; the closed points, i.e., the elements of mSpec $k[x, y]$, are the primes in (c).

By the Nullstellensatz, each prime ideal P in Spec $k[x, y]$ is uniquely determined by the corresponding zero set $\mathcal{Z}(P)$. The prime $(0) \in k[x, y]$ corresponds to $\mathbb{A}^2$. The prime (f) corresponds to the points where $f(x, y) = 0$, and $P = (f)$ is the intersection of all the maximal ideals containing P. The maximal ideal $(x - a, y - b)$ corresponds to the point $(a, b) \in \mathbb{A}^2$. Fibered over Spec $k[x]$ by the map φ^* these primes can be pictured geometrically as in the diagram on the following page.

In this diagram, the prime $(x - a)$ in Spec $k[x]$ is identified with the element $a \in k$. The prime $(x) \in \text{Spec } k[x, y]$ corresponds to the points in $\mathbb{A}^2$ with $x = 0$, i.e.,

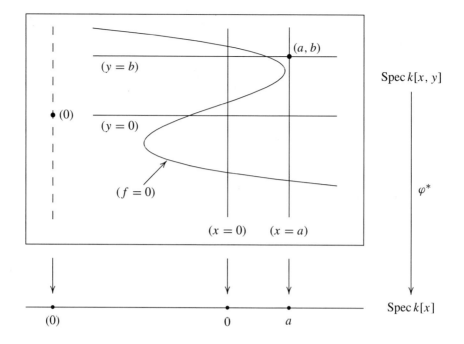

with the y-axis in $\mathbb{A}^2$; the prime $(y) \in \operatorname{Spec} k[x, y]$ similarly corresponds to the x-axis. The prime $(f) \in \operatorname{Spec} k[x, y]$ corresponds to the irreducible curve $f(x, y) = 0$ in $\mathbb{A}^2$; the points $(a, b) \in \mathbb{A}^2$ lying on this curve correspond to the maximal ideals $(x - a, y - b) \in \operatorname{Spec} k[x, y]$ containing (f). The closed point $(x - a, y - b) \in \operatorname{Spec} k[x, y]$ corresponds to the "geometric point" $(a, b) \in \mathbb{A}^2$.

Note that $\operatorname{Spec} k[x, y]$ captures all of the geometry of algebraic sets in $\mathbb{A}^2$: every algebraic set in $\mathbb{A}^2$ is the finite union of some subset of the irreducible algebraic sets corresponding to the elements of $\operatorname{Spec} k[x, y]$ pictured above. With the exception of the everywhere dense point (0), the "geometric" picture of $\operatorname{Spec} k[x, y]$ is precisely the usual geometry of the affine plane $\mathbb{A}^2$. When k is not algebraically closed the situation is slightly more complicated, but the picture is similar, cf. Exercise 4.

(3) The situation for $\operatorname{Spec} \mathbb{Z}[x]$, viewed as fibered over $\operatorname{Spec} \mathbb{Z}$ by the natural inclusion $\mathbb{Z} \to \mathbb{Z}[x]$ is very similar to the situation of $\operatorname{Spec} k[x, y]$ in the previous example. The elements of $\operatorname{Spec} \mathbb{Z}[x]$ were discussed in Example 2 following Proposition 54 and can be pictured as in the diagram on the following page.

The element (0) is Zariski dense in $\operatorname{Spec} \mathbb{Z}[x]$. The closure of (p) consists of (p) and all the closed points (p, g) where g is a monic polynomial in $\mathbb{Z}[x]$ that is irreducible mod p. The closure of (f) consists of (f) together with the maximal ideals (p, g) that contain (f), which is the same as saying that the image of f in the quotient $\mathbb{Z}[x]/(p, g)$ is 0, i.e., the irreducible polynomial g is a factor of f mod p. The closed points, $\operatorname{mSpec} \mathbb{Z}[x]$, are the maximal ideals (p, g).

Note that the maximal ideals (p, g) containing (f) are precisely the closed points in $\operatorname{mSpec} \mathbb{Z}[x]$ in the diagram above where the "function" f on $\operatorname{Spec} \mathbb{Z}[x]$ (taking the prime P to $f(P) = f \bmod P \in \mathbb{Z}[x]/P$) is zero. For example, the polynomial $f = x^3 - 4x^2 + x - 9 \in \mathbb{Z}[x]$ fits the diagram above: f is irreducible in $\mathbb{Z}[x]$, and

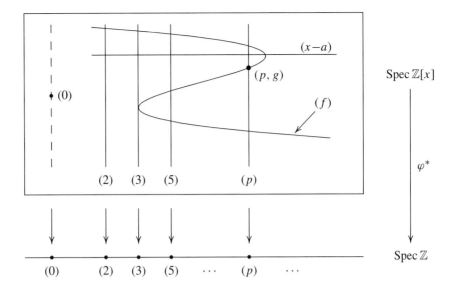

over $\mathbb{F}_p$ factors into irreducibles as follows:

$$f \equiv x^3 + x + 1 \bmod 2$$
$$f \equiv x(x+1)^2 \bmod 3$$
$$f \equiv (x+1)(x+2)(x+3) \bmod 5.$$

There is one point in the fiber over (2) intersecting (f), namely the closed point $(2, x^3 + x + 1)$. There are two closed points in the fiber over (3) given by $(3, x)$ and $(3, x + 1)$ (with some "multiplicity" at the latter point). Over (5) there are three closed points: $(5, x + 1)$, $(5, x + 2)$, and $(5, x + 3)$. For the diagram above, the prime p might be $p = 53$, since this is the first prime p greater than 5 for which this polynomial has three irreducible factors mod p. Note that while the prime (f) is drawn as a smooth curve in this diagram to emphasize the geometric similarity with the structure of $\operatorname{Spec} k[x, y]$ in the previous example, the fibers above the primes in $\operatorname{Spec} \mathbb{Z}$ are discrete, so some care should be exercised. For example, since f factors as $(x + 2)(x^2 + x + 6) \bmod 7$, the intersection of (f) with the fiber above (7) contains only the two points $(7, x + 2)$ and $(7, x^2 + x + 6)$, each with multiplicity one.

The possible number of closed points in (f) lying in a fiber over $(p) \in \operatorname{Spec} \mathbb{Z}$ is controlled by the Galois group of the polynomial f over $\mathbb{Q}$ (cf. Section 14.8). For example, $f = x^4 + 1$ has one closed point in the fiber above (2) and either two or four closed points in a fiber above (p) for p odd (cf. Exercise 8).

The space $\operatorname{Spec} R$ together with its Zariski topology gives a geometric generalization for arbitrary commutative rings of the points in a variety V. We now consider the question of generalizing the ring of rational functions on V.

When V is a variety over the algebraically closed field k the elements in the quotient field $k(V)$ of the coordinate ring $k[V]$ define the rational functions on V. Each element α in $k(V)$ can in general be written as a quotient a/f of elements $a, f \in k[V]$ in many different ways. The set of points U at which α is regular is an open subset of V; by definition, it consists of all the points $v \in V$ where α can be represented by some quotient a/f with $f(v) \neq 0$, and then the representative a/f defines an element in the local ring $\mathcal{O}_{v,V}$. Note also that the same representative a/f defines α not only at v, but also at all the other points where f is nonzero, namely on the open subset $V_f = \{w \in V \mid f(w) \neq 0\}$ of V. These open sets V_f (called principal open sets, cf. Exercise 21 in Section 2) for the various possible representatives a/f for α give an open cover of U. The example of the function $\alpha = \bar{x}/\bar{y}$ for $V = \mathcal{Z}(xz - yw) \subset \mathbb{A}^4$ preceding Proposition 51 shows that in general a single representative for α does not suffice to determine all of U — for this example, $U = V_{\bar{y}} \cup V_{\bar{z}}$, and U is not covered by any single V_f (cf. Exercise 25 of Section 4).

This interpretation of rational functions as functions that are regular on open subsets of V can be generalized to Spec R. We first define the analogues X_f in $X = \operatorname{Spec} R$ of the sets V_f and establish their basic properties.

Definition. For any $f \in R$ let X_f denote the collection of prime ideals in $X = \operatorname{Spec} R$ that do not contain f. Equivalently, X_f is the set of points of Spec R at which the value of $f \in R$ is nonzero. The set X_f is called a *principal* (or *basic*) *open set* in Spec R.

Since X_f is the complement of the Zariski closed set $\mathcal{Z}(f)$ it is indeed an open set in Spec R as the name implies. Some basic properties of the principal open sets are indicated in the next proposition. Recall that a map between topological spaces is a *homeomorphism* if it is continuous and bijective with continuous inverse.

Proposition 56. Let $f \in R$ and let X_f be the corresponding principal open set in $X = \operatorname{Spec} R$. Then
  (1) $X_f = X$ if and only if f is a unit, and $X_f = \emptyset$ if and only if f is nilpotent,
  (2) $X_f \cap X_g = X_{fg}$,
  (3) $X_f \subseteq X_{g_1} \cup \cdots \cup X_{g_n}$ if and only if $f \in \operatorname{rad}(g_1, \ldots, g_n)$; in particular $X_f = X_g$ if and only if $\operatorname{rad}(f) = \operatorname{rad}(g)$,
  (4) the principal open sets form a basis for the Zariski topology on Spec R, i.e., every Zariski open set in X is the union of some collection of principal open sets X_f,
  (5) the natural map from R to R_f induces a homeomorphism from Spec R_f to X_f, where R_f is the localization of R at f,
  (6) the spectrum of any ring is quasicompact (i.e., every open cover has a finite subcover); in particular, X_f is quasicompact, and
  (7) if $\varphi : R \to S$ is any homomorphism of rings (with $\varphi(1_R) = 1_S$) then under the induced map $\varphi^* : Y = \operatorname{Spec} S \to \operatorname{Spec} R$ the full preimage of the principal open set X_f in X is the principal open set $Y_{\varphi(f)}$ in Y.

Proof: Parts (1), (2) and (7) are left as easy exercises. For (3), observe that, by definition, $X_{g_1} \cup \cdots \cup X_{g_n}$ consists of the primes P not containing at least one of $g_1, \ldots, g_n$. Hence $X_{g_1} \cup \cdots \cup X_{g_n}$ is the complement of the closed set $\mathcal{Z}((g_1, \ldots, g_n))$ consisting of the primes P that contain the ideal generated by $g_1, \ldots, g_n$. If $(g_1, \ldots, g_n) = R$ then $X_{g_1} \cup \cdots \cup X_{g_n} = X$ and there is nothing to prove. Otherwise, $X_f \subseteq X_{g_1} \cup \cdots \cup X_{g_n}$ if and only if every prime P with $f \notin P$ also satisfies $P \notin \mathcal{Z}((g_1, \ldots, g_n))$. This latter condition is equivalent to the statement that if the prime P contains the ideal

$(g_1, \ldots, g_n)$ then P also contains f, i.e., f is contained in the intersection of all the prime ideals P containing $(g_1, \ldots, g_n)$. Since this intersection is $\mathrm{rad}(g_1, \ldots, g_n)$ by Proposition 12, this proves (3).

If $U = X - \mathcal{Z}(I)$ is a Zariski open subset of X, then U is the union of the sets X_f with $f \in I$, which proves (4).

The natural ring homomorphism from R to the localization R_f establishes a bijection between the prime ideals in R_f and the prime ideals in R not containing (f) (Proposition 38). The corresponding Zariski continuous map from $\mathrm{Spec}\, R_f$ to $\mathrm{Spec}\, R$ is therefore continuous and bijective. Since every ideal of R_f is the extension of some ideal of R (cf. Proposition 38(1)), it follows that the inverse map is also continuous, which proves (5).

In (6), every open set is the union of principal open sets by (4), so it suffices to prove that if X is covered by principal open sets X_{g_i} (for i in some index set $\mathcal{J}$) then X is a finite union of some of the X_{g_i}. If the ideal I generated by the g_i were a proper ideal in R, then I would be contained in some maximal ideal P. But in this case the element P in $X = \mathrm{Spec}\, R$ would not be contained in any principal open set X_{g_i}, contradicting the assumption that X is covered by the X_{g_i}. Hence $I = R$ and so $1 \in R$ can be written as a finite sum $1 = a_1 g_{i_1} + \cdots + a_n g_{i_n}$ with $i_1, \ldots, i_n \in \mathcal{J}$. Consider the finite union $X_{g_1} \cup \cdots \cup X_{g_n}$. Any point P in X not contained in this union would be a prime in R that contains $g_{i_1}, \ldots, g_{i_n}$, hence would contain 1, a contradiction. It follows that $X = X_{g_1} \cup \cdots \cup X_{g_n}$ as needed. The second part of (6) follows from (5).

We now define an analogue for $X = \mathrm{Spec}\, R$ of the rational functions on a variety V. As we observed, for the variety V a rational function $\alpha \in k(V)$ is a regular function on some open set U. At each point $v \in U$ there is a representative a/f for α with $f(v) \neq 0$, and this representative is an element in the localization $\mathcal{O}_{v,V} = k[V]_{\mathcal{I}(v)}$. In this way the regular function α on U can be considered as a function from U to the disjoint union of these localizations: the point $v \in U$ is mapped to the representative $a/f \in k[V]_{\mathcal{I}(v)}$. Furthermore the same representative can be used simultaneously not only at v but on the whole Zariski neighborhood V_f of v (so, "locally near v," α is given by a single quotient of elements from $k[V]$). Note that a/f is an element in the localization $k[V]_f$, which is contained in each of the localizations $k[V]_{\mathcal{I}(w)}$ for $w \in V_f$.

We now generalize this to $\mathrm{Spec}\, R$ by considering the collection of functions s from the Zariski open subset U of $\mathrm{Spec}\, R$ to the disjoint union of the localizations R_P for $P \in U$ such that $s(P) \in R_P$ and such that s is given locally by quotients of elements of R. More precisely:

Definition. Suppose U is a Zariski open subset of $\mathrm{Spec}\, R$. If $U = \emptyset$, define $\mathcal{O}(U) = 0$. Otherwise, define $\mathcal{O}(U)$ to be the set of functions $s : U \to \bigsqcup_{Q \in U} R_Q$ from U to the disjoint union of the localizations R_Q for $Q \in U$ with the following two properties:

    (1) $s(Q) \in R_Q$ for every $Q \in U$, and

    (2) for every $P \in U$ there is an open neighborhood $X_f \subseteq U$ of P in U and an element a/f^n in the localization R_f defining s on X_f, i.e., $s(Q) = a/f^n \in R_Q$ for every $Q \in X_f$.

If s, t are elements in $\mathcal{O}(U)$ then $s + t$ and st are also elements in $\mathcal{O}(U)$ (cf. Exercise 18), so each $\mathcal{O}(U)$ is a ring. Also, every $a \in R$ gives an element in $\mathcal{O}(U)$

defined by $s(Q) = a \in R_Q$, and in particular $1 \in R$ gives an identity for the ring $\mathcal{O}(U)$. If U' is an open subset of U, then there is a natural restriction map from $\mathcal{O}(U)$ to $\mathcal{O}(U')$ which is a homomorphism of rings (cf. Exercise 19).

Definition. Let R be a commutative ring with 1, and let $X = \mathrm{Spec}\, R$.
  (1) The collection of rings $\mathcal{O}(U)$ for the Zariski open sets of X together with the restriction maps $\mathcal{O}(U) \to \mathcal{O}(U')$ for $U' \subseteq U$ is called the *structure sheaf* on X, and is denoted simply by $\mathcal{O}$ (or $\mathcal{O}_X$).
  (2) The elements s of $\mathcal{O}(U)$ are called the *sections of $\mathcal{O}$* over U. The elements of $\mathcal{O}(X)$ are called the *global sections* of $\mathcal{O}$.

The next proposition generalizes the result of Proposition 51 that the only rational functions on a variety V that are regular everywhere are the elements of the coordinate ring $k[V]$.

Proposition 57. Let $X = \mathrm{Spec}\, R$ and let $\mathcal{O} = \mathcal{O}_X$ be its structure sheaf. The global sections of $\mathcal{O}$ are the elements of R, i.e., $\mathcal{O}(X) \cong R$. More generally, if X_f is a principal open set in X for some $f \in R$, then $\mathcal{O}(X_f)$ is isomorphic to the localization R_f.

Proof: Suppose that a/f^n is an element of the localization R_f. Then the map defined by $s(Q) = a/f^n \in R_Q$ for $Q \in X_f$ gives an element in $\mathcal{O}(X_f)$, and it is immediate that the resulting map ψ from R_f to $\mathcal{O}(X_f)$ is a ring homomorphism. Suppose that $a/f^n = b/f^m$ in R_Q for every $Q \in X_f$, i.e., $g(af^m - bf^n) = 0$ in R for some $g \notin Q$. If I is the ideal in R of elements $r \in R$ with $r(af^m - bf^n) = 0$, it follows from $g \in I$ that I is not contained in Q for any $Q \in X_f$. Put another way, every prime ideal of R containing I also contains f. Hence f is contained in the intersection of all the prime ideals of R containing I, which is to say that $f \in \mathrm{rad}\, I$. Then $f^N \in I$ for some integer $N \geq 0$, and so $f^N(af^m - bf^n) = 0$ in R. But this shows that $a/f^n = b/f^m$ in R_f and so the map ψ is injective. Suppose now that $s \in \mathcal{O}(X_f)$. Then by definition X_f can be covered by principal open sets X_{g_i} on which $s(Q) = a_i/g_i^{n_i} \in R_Q$ for every $Q \in X_{g_i}$. By (6) of Proposition 56, we may take a finite number of the g_i and then by taking different a_i we may assume all the n_i are equal (since $a_i/g_i^{n_i} = (a_i g_i^{n-n_i})/g_i^n$ if n is the maximum of the n_i). Since $s(Q) = a_i/g_i^n = a_j/g_j^n$ in R_Q for all $Q \in X_{g_i g_j} = X_{g_i} \cap X_{g_j}$, the injectivity of ψ (applied to $R_{g_i g_j}$) shows that $a_i/g_i^n = a_j/g_j^n$ in $R_{g_i g_j}$. This means that $g_i g_j^N(a_i g_j^n - a_j g_i^n) = 0$, i.e.,

$$a_i g_i^N g_j^{n+N} = a_j g_i^{n+N} g_j^N$$

in R for some $N \geq 0$, and we may assume N sufficiently large that this holds for every i and j. Since X_f is the union of the $X_{g_i} = X_{g_i^{n+N}}$, f is contained in the radical of the ideal generated by the g_i^n by (3) of Proposition 56, say

$$f^M = \sum_i b_i g_i^{n+N}$$

for some $M \geq 1$ and $b_i \in R$. Define $a = \sum b_i a_i g_i^N \in R$. Then

$$g_j^N a_j f^M = \sum_i b_i (a_j g_i^{n+N} g_j^N) = \sum_i b_i (a_i g_i^N g_j^{n+N}) = g_j^{n+N} a.$$

It follows that $a/f^M = a_j/g_j^n$ in R_{g_j}, and so the element in $\mathcal{O}(X_f)$ defined by a/f^M in R_f agrees with s on every X_{g_j}, and so on all of X_f since these open sets cover X_f. Hence the map ψ gives an isomorphism $R_f \cong \mathcal{O}(X_f)$. Taking $f = 1$ gives $R \cong \mathcal{O}(X)$, completing the proof.

In the case of affine varieties V the local ring $\mathcal{O}_{v,V}$ at the point $v \in V$ is the collection of all the rational functions in $k(V)$ that are defined at v. Put another way, $\mathcal{O}_{v,V}$ is the union of the rings of regular functions on U for the open sets U containing P, where this union takes place in the function field $k(V)$ of V. In the more general case of $X = \mathrm{Spec}\, R$, the rings $\mathcal{O}(U)$ for the open sets containing $P \in \mathrm{Spec}\, R$ are not contained in such an obvious common ring. In this case we proceed by considering the collection of pairs (s, U) with U an open set of X containing P and $s \in \mathcal{O}(U)$. We identify two pairs (s, U) and (s', U') if there is an open set $U'' \subseteq U \cap U'$ containing P on which s and s' restrict to the same element of $\mathcal{O}(U'')$. In the situation of affine varieties, this says that two functions defined in Zariski neighborhoods of the point v define the same regular function at v if they agree in some common neighborhood of v. The collection of equivalence classes of pairs (s, U) defines the *direct limit* of the rings $\mathcal{O}(U)$, and is denoted $\varinjlim \mathcal{O}(U)$ (cf. Exercise 8 in Section 7.6).

Definition. If $P \in X = \mathrm{Spec}\, R$, then the direct limit, $\varinjlim \mathcal{O}(U)$, of the rings $\mathcal{O}(U)$ for the open sets U of X containing P is called the *stalk* of the structure sheaf at P, and is denoted $\mathcal{O}_P$.

Proposition 58. Let $X = \mathrm{Spec}\, R$ and let $\mathcal{O} = \mathcal{O}_X$ be its structure sheaf. The stalk of $\mathcal{O}$ at the point $P \in X$ is isomorphic to the localization R_P of R at P: $\mathcal{O}_P \cong R_P$. In particular, the stalk $\mathcal{O}_P$ is a local ring.

Proof: If (s, U) represents an element in the stalk $\mathcal{O}_P$, then $s(P)$ is an element of the localization R_P. By the definition of the direct limit, this element does not depend on the choice of representative (s, U), and so gives a well defined ring homomorphism φ from $\mathcal{O}_P$ to R_P. If $a, f \in R$ with $f \notin P$, then the map $s(Q) = a/f \in R_Q$ defines an element in $\mathcal{O}(X_f)$. Then the class of (s, X_f) in the stalk $\mathcal{O}_P$ is mapped to a/f in R_P by φ, so φ is a surjective map. To see that φ is also injective, suppose that the classes of (s, U) and (s', U') in $\mathcal{O}_P$ satisfy $s(P) = s'(P)$ in R_P. By definition of $\mathcal{O}(U)$, $s = a/g^n$ on X_g for some $g \notin P$. Similarly, $s' = b/(g')^m$ on $X_{g'}$ for some $g' \notin P$. Since $a/g^n = b/(g')^m$ in R_P, there is some $h \notin P$ with $h(a(g')^m - bg^n) = 0$ in R. If $Q \in X_{gg'h} = X_g \cap X_{g'} \cap X_h$ this last equality shows that $a/g^n = b/(g')^m$ in R_Q, so that s and s' agree when restricted to $X_{gg'h}$. By definition of the direct limit, (s, U) and (s', U') define the same element in the stalk $\mathcal{O}_P$, which proves that φ is injective and establishes the proposition.

Proposition 58 shows that the algebraically defined localization R_P for $P \in \mathrm{Spec}\, R$ plays the role of the local ring $\mathcal{O}_{v,V}$ of regular functions at v for the affine variety V. If $\mathfrak{m}_P$ denotes the maximal ideal $P R_P$ in R_P and $k(P) = R_P/\mathfrak{m}_P$ denotes the corresponding quotient field (which by Proposition 46(1) is also the fraction field of R/P), then the *tangent space* at P is defined to be the $k(P)$-vector space dual of $\mathfrak{m}_P/\mathfrak{m}_P^2$.

This is an algebraic definition that generalizes the definition of the tangent space $\mathbb{T}_{v,V}$ to a variety V at a point v (by Proposition 52). This can now be used to define what it means for a point in Spec R to be nonsingular: the point $P \in$ Spec R is *nonsingular* or *smooth* if the local ring R_P is what is called a "regular local ring" (cf. Section 16.2).

Proposition 58 also suggests a nice geometric view of the structure sheaf on Spec R. If we view each point $P \in$ Spec R as having the local ring R_P above it, then above the open set U in $X =$ Spec R is a "sheaf" (in the sense of a "bundle") of these "stalks" (in the sense of a "stalk of wheat"), which helps explain some of the terminology. A section s in the structure sheaf $\mathcal{O}(U)$ is a map from U to this bundle of stalks. The image of U under such a section s is indicated by the shaded region in the following figure.

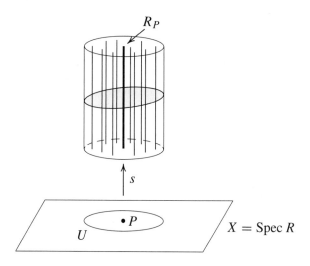

Definition. Let R be a commutative ring with 1. The pair (Spec R, $\mathcal{O}_{\text{Spec } R}$), consisting of the space Spec R with the Zariski topology together with the structure sheaf $\mathcal{O}_{\text{Spec } R}$, is called an *affine scheme*.

The notion of an affine scheme gives a completely algebraic generalization of the geometry of affine algebraic sets valid for arbitrary commutative rings, and is the starting point for modern algebraic geometry.

Examples

(1) If F is any field then $X =$ Spec $F = \{(0)\}$. In this case there are only two open sets X and $\emptyset$, both of which are principal open sets: $X = X_1$ and $\emptyset = X_0$. The global sections are $\mathcal{O}(X) = F$. There is only one stalk: $\mathcal{O}_{(0)} = F_0 = F$.

(2) If $R = \mathbb{Z}$ then because R is a P.I.D. every open set in $X =$ Spec $\mathbb{Z}$ is principal open:

$$X_n = \{(p) \mid p \nmid n\} \qquad \text{and}$$

$$\mathcal{O}(X_n) = \mathbb{Z}_n = \mathbb{Z}[1/n] = \{a/b \in \mathbb{Q} \mid \text{if the prime } p \mid b \text{ then } p \mid n\}.$$

For nonzero p the stalk at (p) is the local ring $\mathbb{Z}_{(p)}$, and the stalk at (0) is $\mathbb{Q}$. All the restriction maps as well as the maps from sections to stalks are the natural inclusions.

(3) For a general integral domain R with quotient field F the stalks and sections are

$$\mathcal{O}(U) = \{a/b \in F \mid b \notin P \text{ for all } P \in U\}$$

$$\mathcal{O}_P = R_P = \{a/b \in F \mid b \notin P\}$$

where the stalk at (0) is F, i.e., $\mathcal{O}_{(0)} = F$. Again, the restriction maps and the maps to the stalks are all inclusions.

(4) For the local ring $R = \mathbb{Z}_{(2)} = \{a/b \in \mathbb{Q} \mid b \text{ odd}\}$ we have $\operatorname{Spec} R = \{(0), (2)\}$ with (2) the only closed point and $\{(0)\} = X_2$ a principal open set. The sections $\mathcal{O}(\{(0)\})$ are $R_2 = \mathbb{Q}$, and the stalks are $\mathcal{O}_{(0)} = R_{(0)} = \mathbb{Q}$ and $\mathcal{O}_{(2)} = R_{(2)} = R$.

We next consider the relationship of the affine schemes corresponding to rings R and S with respect to a ring homomorphism from R to S.

Suppose that $\varphi : R \to S$ is a ring homomorphism. We have already seen in Proposition 56(7) that there is an induced continuous map φ^* from $Y = \operatorname{Spec} S$ to $X = \operatorname{Spec} R$ and that under this map the full preimage of the principal open set X_g for $g \in R$ is the principal open set $Y_{\varphi(g)}$. It follows that φ also induces a map on corresponding sections, as follows. Let $Q' \in Y$ be any element in $\operatorname{Spec} S$ and let $Q = \varphi^*(Q') = \varphi^{-1}(Q') \in X$ be the corresponding element in $\operatorname{Spec} R$. If U is a Zariski open set in X containing Q, then $U' = (\varphi^*)^{-1}(U)$ is a Zariski open set in Y containing Q'. Note that φ induces a natural ring homomorphism, φ_Q say, from the localization R_Q to the localization $S_{Q'}$ defined by $\varphi_Q(a/f) = \varphi(a)/\varphi(f) \in S_{Q'}$ for $f \notin Q$. Let $s \in \mathcal{O}_X(U)$ be a section of the structure sheaf of X given locally in the neighborhood X_g of $P \in X$ by a/g^n. It is easy to check that the composite

$$s' : U' \xrightarrow{\varphi^*} U \xrightarrow{s} \bigsqcup_{Q \in U} R_Q \xrightarrow{\varphi} \bigsqcup_{Q' \in U} S_{Q'}$$

defines a map given locally in the neighborhood $Y_{\varphi(g)}$ by the element $\varphi(a)/\varphi(g)^n$, so that $s' \in \mathcal{O}_Y(U')$ is a section of the structure sheaf of Y. It is then straightforward to check that the resulting map $\varphi^\# : \mathcal{O}_X(U) \to \mathcal{O}_Y(U')$ is a ring homomorphism (mapping $1 \in \mathcal{O}_X(U)$ to $1 \in \mathcal{O}_Y(U')$) that is compatible with the restriction maps on $\mathcal{O}_X$ and $\mathcal{O}_Y$ (cf. Exercise 20). It also follows that there is an induced ring homomorphism on the stalks: $\varphi^\# : \mathcal{O}_{X,P} \to \mathcal{O}_{Y,P'}$ for any point $P' \in \operatorname{Spec} S$ and corresponding point $P = \varphi^*(P') \in \operatorname{Spec} R$. Under the isomorphism in Proposition 58, the homomorphism $\varphi^\#$ from $R_P \cong \mathcal{O}_{X,P}$ to $S_{P'} \cong \mathcal{O}_{Y,P'}$ is just the natural ring homomorphism φ_P on the localizations induced by the homomorphism φ. In particular, the inverse image under $\varphi^\#$ of the maximal ideal in the local ring $\mathcal{O}_{Y,P'}$ is the maximal ideal in the local ring $\mathcal{O}_{X,P}$.

Definition. Suppose $(\operatorname{Spec} R, \mathcal{O}_{\operatorname{Spec} R})$ and $(\operatorname{Spec} S, \mathcal{O}_{\operatorname{Spec} S})$ are two affine schemes. A *morphism of affine schemes* from $(\operatorname{Spec} S, \mathcal{O}_{\operatorname{Spec} S})$ to $(\operatorname{Spec} R, \mathcal{O}_{\operatorname{Spec} R})$ is a pair $(\varphi^*, \varphi^\#)$ such that

(1) $\varphi^* : \operatorname{Spec} S \to \operatorname{Spec} R$ is Zariski continuous,
(2) there are ring homomorphisms $\varphi^\# : \mathcal{O}(U) \to \mathcal{O}(\varphi^{*-1}(U))$ for every Zariski open subset U in $\operatorname{Spec} R$ that commute with the restriction maps, and

(3) if $P' \in \operatorname{Spec} S$ with corresponding point $P = \varphi^*(P) \in \operatorname{Spec} R$, then under the induced homomorphism on stalks $\varphi^\# : \mathcal{O}_{\operatorname{Spec} R, P} \to \mathcal{O}_{\operatorname{Spec} S, P'}$ the preimage of the maximal ideal of $\mathcal{O}_{\operatorname{Spec} S, P'}$ is the maximal ideal of $\mathcal{O}_{\operatorname{Spec} R, P}$.

A homomorphism $\psi : A \to B$ from the local ring A to the local ring B with the property that the preimage of the maximal ideal of B is the maximal ideal of A is called a *local homomorphism* of local rings. The third condition in the definition is then the statement that the induced homomorphism on stalks is required to be a local homomorphism.

With this terminology, the discussion preceding the definition shows that a ring homomorphism $\varphi : R \to S$ induces a morphism of affine schemes from $(\operatorname{Spec} S, \mathcal{O}_{\operatorname{Spec} S})$ to $(\operatorname{Spec} R, \mathcal{O}_{\operatorname{Spec} R})$.

Conversely, suppose $(\varphi^*, \varphi^\#)$ is a morphism of affine schemes from $(\operatorname{Spec} S, \mathcal{O}_{\operatorname{Spec} S})$ to $(\operatorname{Spec} R, \mathcal{O}_{\operatorname{Spec} R})$. Then in particular, for $U = \operatorname{Spec} R$, $(\varphi^*)^{-1}(U) = \operatorname{Spec} S$, so by assumption there is a ring homomorphism $\varphi^\# : \mathcal{O}_{\operatorname{Spec} R}(\operatorname{Spec} R) \to \mathcal{O}_{\operatorname{Spec} S}(\operatorname{Spec} S)$ defined on the global sections. By Proposition 57, we have $\mathcal{O}_{\operatorname{Spec} R}(\operatorname{Spec} R) \cong R$ and $\mathcal{O}_{\operatorname{Spec} S}(\operatorname{Spec} S) \cong S$ as rings. Composing with these isomorphisms shows that $\varphi^\#$ gives a ring homomorphism $\varphi : R \to S$. By Proposition 58 we have a local homomorphism $\varphi^\# : R_P \to S_{P'}$, and by the compatibility with the restriction homomorphisms it follows that the diagram

$$
\begin{array}{ccc}
R & \xrightarrow{\varphi} & S \\
\downarrow & & \downarrow \\
R_P & \xrightarrow{\varphi^\#} & S_{P'}
\end{array}
$$

commutes, where the two vertical maps are the natural localization homomorphisms. Since $\varphi^\#$ is assumed to be a local homomorphism, $(\varphi^\#)^{-1}(P'S_{P'}) = PR_P$, from which it follows that $\varphi^{-1}(P') = P$. Hence the continuous map from $\operatorname{Spec} S$ to $\operatorname{Spec} R$ induced by φ is the same as φ^*, and it follows easily that φ also induces the homomorphism $\varphi^\#$. This shows that there is a ring homomorphism $\varphi : R \to S$ inducing both φ^* and $\varphi^\#$ as before.

We summarize this in the following proposition:

Theorem 59. Every ring homomorphism $\varphi : R \to S$ induces a morphism

$$(\varphi^*, \varphi^\#) : (\operatorname{Spec} S, \mathcal{O}_{\operatorname{Spec} S}) \to (\operatorname{Spec} R, \mathcal{O}_{\operatorname{Spec} R})$$

of affine schemes. Conversely, every morphism of affine schemes arises from such a ring homomorphism φ.

Theorem 59 is the analogue for $\operatorname{Spec} R$ of Theorem 6, which converted geometric questions relating to affine algebraic sets to algebraic questions for their coordinate rings.

The condition that the homomorphism on stalks be a local homomorphism in the definition of a morphism of affine schemes is necessary: a continuous map on the spectra together with a set of compatible ring homomorphisms on sections (hence also on stalks) is not sufficient to force these maps to come from a ring homomorphism.

Example

Let $R = \mathbb{Z}_{(2)}$ and $S = \mathbb{Q}$ as in the preceding set of examples. Define $\varphi^* : \operatorname{Spec} \mathbb{Q} \to \operatorname{Spec} \mathbb{Z}_{(2)}$ by $\varphi^*((0)) = (2)$ (which is Zariski continuous). Define $\varphi^\# : \mathcal{O}(\operatorname{Spec} R) \to \mathcal{O}(\operatorname{Spec} S)$ to be the inclusion map $\mathbb{Z}_{(2)} \hookrightarrow \mathbb{Q}$ and define $\varphi^\#$ for all other $U \subseteq \operatorname{Spec} R$ simply to be the zero map. It is straightforward to check that these homomorphisms commute with the restriction maps. This family of maps does *not* arise from a ring homomorphism, however, because on the stalks for $(0) \in \operatorname{Spec} S$ and $\varphi^*((0)) = (2) \in \operatorname{Spec} R$ the induced homomorphism

$$\varphi^\# : \mathcal{O}_{\operatorname{Spec} R, (2)} \hookrightarrow \mathcal{O}_{\operatorname{Spec} S, (0)}$$

is the injection $\mathbb{Z}_{(2)} \hookrightarrow \mathbb{Q}$, which is not a *local* homomorphism (the inverse image of (0) is (0) and not the maximal ideal $2\mathbb{Z}_{(2)}$).

The proof of Theorem 59 shows that a morphism $(\varphi^*, \varphi^\#)$ of affine schemes necessarily comes from the ring homomorphism defined by $\varphi^\#$ on global sections. In this example, the homomorphism on global sections is the inclusion map of R into S. The inclusion map from R to S defines a map from $\operatorname{Spec} S$ to $\operatorname{Spec} R$ that maps $(0) \in \operatorname{Spec} S$ to $(0) \in \operatorname{Spec} R$ and not to $(2) \in \operatorname{Spec} R$, so this map does not agree with the original map φ^*.

The previous example shows that the converse in Theorem 59 would not be true without the third (local homomorphism) condition in the definition of a morphism of affine schemes. As a result, Theorem 59 shows that the appropriate place to view affine schemes is in the category of *locally ringed spaces*. Roughly speaking, a locally ringed space is a topological space X together with a collection of rings $\mathcal{O}(U)$ for each open subset of X (with a compatible set of homomorphisms from $\mathcal{O}(U)$ to $\mathcal{O}(U')$ if $U' \subseteq U$ and with some local conditions on the sections) such that the stalks $\mathcal{O}_P = \varinjlim \mathcal{O}(U)$ for $P \in U$ are local rings. The morphisms in this category are continuous maps between the topological spaces together with ring homomorphisms between corresponding $\mathcal{O}(U)$ with precisely the same conditions as imposed in the definition of a morphism of affine schemes.

A *scheme* is a locally ringed space in which each point lies in a neighborhood isomorphic to an affine scheme (with some compatibility conditions between such neighborhoods), and is a fundamental object of study in modern algebraic geometry. The affine schemes considered here form the building blocks that are "glued together" to define general schemes in the same way that ordinary Euclidean spaces form the building blocks that are "glued together" to define manifolds in analysis.

EXERCISES

All rings are assumed commutative with identity, and all ring homomorphisms are assumed to map identities to identities.

1. If N is the nilradical of R, prove that $\operatorname{Spec} R$ and $\operatorname{Spec} R/N$ are homeomorphic. [Show that the natural homomorphism from R to R/N induces a Zariski continuous isomorphism from $\operatorname{Spec} R/N$ to $\operatorname{Spec} R$.]

2. Let I be an ideal in the ring R. Prove that the continuous map from $\operatorname{Spec} R/I$ to $\operatorname{Spec} R$ induced by the canonical projection homomorphism $R \to R/I$ maps $\operatorname{Spec} R/I$ homeomorphically onto the closed set $\mathcal{Z}(I)$ in $\operatorname{Spec} R$.

3. Prove that two elements $f, g \in R$ have the same values at all elements P in Spec R if and only if $f - g$ is contained in the nilradical of R. In particular, prove that an element in an affine k-algebra is uniquely determined by its values.

4. Let k be an arbitrary field, not necessarily algebraically closed. Prove that the prime ideals in $k[x, y]$ (i.e., the elements of Spec $k[x, y]$) are
   (i) (0),
   (ii) (f) where f is an irreducible polynomial in $k[x, y]$, and
   (iii) $(p(x), g(x, y))$ where $p(x)$ is an irreducible polynomial in $k[x]$ and $g(x, y)$ is an irreducible polynomial in $k[x, y]$ that is irreducible modulo $p(x)$, i.e., $g(x, y)$ remains irreducible in the quotient $k[x, y]/(p(x))$.

   Prove that mSpec $k[x, y]$ consists of the primes in (iii). [Use Exercise 20 in Section 1.]

5. Let $\mathfrak{m} = (p(x), g(x, y))$ be a maximal ideal in $k[x, y]$ as in the previous exercise. Show that $K = k[x, y]/\mathfrak{m}$ is an algebraic field extension of k, so that $k[x, y]$ can also be viewed as a subring of $K[x, y]$. If x, y are mapped to $\alpha, \beta \in K$, respectively, under the canonical homomorphism $k[x, y] \to k[x, y]/\mathfrak{m}$, prove that $\mathfrak{m} = k[x, y] \cap (x - \alpha, y - \beta) \subseteq K[x, y]$.

6. Describe the elements in Spec $\mathbb{R}[x]$ and Spec $\mathbb{C}[x]$. Describe the elements in Spec $\mathbb{Z}_{(2)}[x]$ where $\mathbb{Z}_{(2)} = \{a/b \in \mathbb{Q} \mid b \text{ is odd}\}$ is the localization of $\mathbb{Z}$ at the prime (2).

7. Let $(f) = (x^5 + x + 1)$ in Spec $\mathbb{Z}[x]$ viewed as fibered over Spec $\mathbb{Z}$ as in Example 3 following Proposition 55. Show that there are two closed points in the fiber over (2), three closed points in the fiber over (5), four closed points in the fiber over (19), and five closed points in the fiber over (211).

8. Let $(f) = (x^4 + 1)$ in Spec $\mathbb{Z}[x]$ viewed as fibered over Spec $\mathbb{Z}$ as in Example 3 following Proposition 55. Prove that there is one closed point in the fiber over (2), four closed points in the fiber over p for p odd, $p \equiv 1 \bmod 8$, and two closed points in the fiber over p for all other odd primes p (cf. Corollary 16 in Section 3 of Chapter 14).

9. Prove that the elements in the fiber over (p) of the Zariski continuous map from Spec $\mathbb{Z}[x]$ to Spec $\mathbb{Z}$ are homeomorphic with the elements in Spec$(\mathbb{Z}[x] \otimes_{\mathbb{Z}} \mathbb{F}_p)$.

10. Let $X = $ Spec R and let X_f be the principal open set corresponding to $f \in R$. Prove that $X_f \cap X_g = X_{fg}$. Prove that $X_f = X$ if and only if f is a unit in R, and that $X_f = \emptyset$ if and only if f is nilpotent.

11. If X_f and X_g are principal open sets in $X = $ Spec R, prove that the open set $X_f \cup X_g$ is the complement of the closed set $\mathcal{Z}(I)$ where $I = (f, g)$ is the ideal in R generated by f and g.

12. Prove that a Zariski open subset U of $X = $ Spec R is quasicompact if and only if U is a finite union of principal open subsets. Give an example of a ring R, a Zariski open subset U of Spec R, and a Zariski open covering of U that cannot be reduced to a finite subcovering.

13. Let $\varphi : R \to S$ be a homomorphism of rings. Prove that under the induced map φ^* from $Y = $ Spec S to $X = $ Spec R the full preimage of the principal open set X_f in X is the principal open set $Y_{\varphi(f)}$ in Y.

14. Suppose that $R = R_1 \times R_2$ is the direct product of the rings R_1 and R_2. Prove that $X = $ Spec R is the disjoint union of open subspaces X_1, X_2 (which are therefore also closed), where X_1 is homeomorphic to Spec R_1 and X_2 is homeomorphic to Spec R_2.

15. Prove that $X = $ Spec R is not connected if and only if R is the direct product of two nonzero rings if and only if R contains an idempotent e with $e \neq 0, 1$ (cf. the previous exercise).

16. Prove that $X = \operatorname{Spec} R$ is irreducible (i.e., any two nonempty open subsets have a nontrivial intersection) if and only if $X_f \cap X_g \neq \emptyset$ for any two nonempty principal open sets X_f and X_g. Deduce that $X = \operatorname{Spec} R$ is irreducible if and only if the nilradical of R is a prime ideal. [Use Exercise 10.]

17. Let $G = \langle \sigma \rangle$ be a group of order 2, let $R = \mathbb{Z}[G] = \{a + b\sigma \mid a, b \in \mathbb{Z}\}$ be the corresponding group ring, and let $X = \operatorname{Spec} R$.
    (a) Prove that the nilradical of R is (0) but is not a prime ideal. Prove that $X = X^+ \cup X^-$ where $X^+ = \mathcal{Z}(1 - \sigma)$ and $X^- = \mathcal{Z}(1 + \sigma)$. [Use $(1 + \sigma)(1 - \sigma) = 0$.]
    (b) Prove that the homomorphism $\mathbb{Z}[G] \to \mathbb{Z}$ defined by mapping σ to 1 induces a homeomorphism of X^+ with $\operatorname{Spec} \mathbb{Z}$, and the homomorphism mapping σ to -1 induces a homeomorphism of X^- with $\operatorname{Spec} \mathbb{Z}$.
    (c) Prove that $X^+ \cap X^-$ consists of the single element $\mathfrak{m} = (1 + \sigma, 1 - \sigma) = (2, 1 - \sigma)$ and that this is a closed point in X.
    (d) Show that $(1 - \sigma)$ and $(1 + \sigma)$ are the unique non-closed points in X, with closures X^+ and X^-, respectively. Describe the closed points, $\operatorname{mSpec} R$, in X and prove that $\operatorname{Spec} \mathbb{Z}[\langle \sigma \rangle]$ can be pictured as follows:

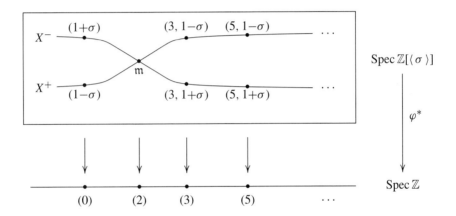

18. Let $\mathcal{O}$ be the structure sheaf on $X = \operatorname{Spec} R$, let U be an open set in X, and suppose $s, t \in \mathcal{O}(U)$. If $s = a/f_1^n$ on X_{f_1} and $t = b/f_2^m$ on X_{f_2}, show that

$$st = (abf_1^m f_2^n)/(f_1 f_2)^{n+m} \quad \text{and} \quad s + t = (af_1^m f_2^{m+n} + bf_1^{m+n} f_2^n)/(f_1 f_2)^{n+m}$$

on $X_{f_1 f_2}$. Deduce that $\mathcal{O}(U)$ is a commutative ring with identity.

19. Let $\mathcal{O}$ be the structure sheaf on $X = \operatorname{Spec} R$, let $V \subseteq U$ be open sets in X, and let $s \in \mathcal{O}(U)$. Suppose $P \in V$ and that $s = a/f^n$ on $X_f \subseteq U$.
    (a) Show that there is a principal open set $X_{f'} \subseteq V \cap X_f$ containing P.
    (b) Show that $(f')^m = bf$ for some $b \in R$.
    (c) Show that $s = (ab^n)/(f')^{mn}$ on $X_{f'}$ and conclude that restricting s to V gives a well defined ring homomorphism from $\mathcal{O}(U)$ to $\mathcal{O}(V)$.

20. Let $\varphi : R \to S$ be a homomorphism of rings, let $X = \operatorname{Spec} R$, $Y = \operatorname{Spec} S$, and let $V \subseteq U$ be Zariski open subsets of X. Set $V' = (\varphi^*)^{-1}(V)$ and $U' = (\varphi^*)^{-1}(U)$, the corresponding Zariski open subsets of Y with respect to the continuous map $\varphi^* : Y \to X$ induced by φ. Prove that the induced map $\varphi^\# : \mathcal{O}_X(U) \to \mathcal{O}_Y(U')$ on sections is a ring homomorphism. Prove that $V' \subseteq U'$ and that $\varphi^\#$ is compatible with restriction i.e., that

the diagram

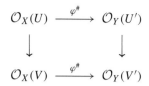

is commutative, where the vertical maps are the restriction homomorphisms.

21. Suppose D is a multiplicatively closed subset of R. Show that the localization homomorphism $R \to D^{-1}R$ induces a homeomorphism from $\mathrm{Spec}(D^{-1}R)$ to the collection of prime ideals P of R with $P \cap D = \emptyset$.

22. Show that $\mathrm{Spec}\, k[x, y]/(xy)$ is connected but is the union of two proper closed subsets each homeomorphic to $\mathrm{Spec}\, k[x]$, hence is not irreducible (cf. Exercise 16).

23. For each of the following rings R exhibit the elements of $\mathrm{Spec}\, R$, the open sets U in $\mathrm{Spec}\, R$, the sections $\mathcal{O}(U)$ of the structure sheaf for $\mathrm{Spec}\, R$ for each open U, and the stalks $\mathcal{O}_P$ at each point $P \in \mathrm{Spec}\, R$:
    (a) $\mathbb{Z}/4\mathbb{Z}$ **(b)** $\mathbb{Z}/6\mathbb{Z}$ **(c)** $\mathbb{Z}/2\mathbb{Z} \times \mathbb{Z}/3\mathbb{Z}$ **(d)** $\mathbb{Z}/2\mathbb{Z} \times \mathbb{Z}/2\mathbb{Z} \times \mathbb{Z}/2\mathbb{Z}$.

24. **(a)** If every ideal of R is principal, show every open set in $\mathrm{Spec}\, R$ is a principal open set.
    (b) Show that if $R = \mathbb{Z}[x]/(4, x^2)$ then R contains a nonprincipal ideal, but every open set in $\mathrm{Spec}\, R$ is a principal open set.

25. **(a)** If M is an R-module prove that $\mathrm{Supp}(M)$ is a Zariski closed subset of $\mathrm{Spec}\, R$. [Use Exercise 33 of Section 4.]
    (b) If M is a finitely generated R-module prove that $\mathrm{Supp}(M) = \mathcal{Z}(\mathrm{Ann}(M)) \subseteq \mathrm{Spec}\, R$. [Use Exercise 34 of Section 4.]

26. Suppose M is a finitely generated module over the Noetherian ring R.
    (a) Prove that there are finitely many minimal primes $*P_1, \ldots, P_n$ containing $\mathrm{Ann}(M)$. [Use Corollary 22.]
    (b) Prove that $\{P_1, \ldots, P_n\}$ is also the set of minimal primes in $\mathrm{Ass}_R(M)$ and that $\mathrm{Supp}(M)$ is the union of the Zariski closed sets $\mathcal{Z}(P_1), \ldots, \mathcal{Z}(P_n)$ in $\mathrm{Spec}\, R$. [Use the previous exercise and Exercise 40 in Section 4.]

The previous exercise gives a geometric view of a finitely generated module M over a Noetherian ring R: over each point P in $\mathrm{Spec}\, R$ is the localization M_P (the *stalk* over P). The stalk is nonzero precisely over the points in the Zariski closed subsets $\mathcal{Z}(P_1), \ldots, \mathcal{Z}(P_n)$ where the P_i are the minimal primes in $\mathrm{Ass}_R(M)$. These ideas lead to the notion of the *(coherent) module sheaf on* $\mathrm{Spec}\, R$ *associated to* M (with a picture similar to that of the structure sheaf following Proposition 58), which is a powerful tool in modern algebraic geometry.

27. Let $R = k[x, y]$ and let M be the ideal (x, y) in R. Prove that $\mathrm{Supp}(M) = \mathrm{Spec}\, R$ and $\mathrm{Ass}_R(M) = \{0\}$.

The next two exercises show that the associated primes for an ideal I in a Noetherian ring R in the sense of primary decomposition are the associated primes for I in the sense of $\mathrm{Ass}_R(R/I)$.

28. This exercise proves that the ideal Q in a Noetherian ring R is P-primary if and only if $\mathrm{Ass}_R(R/Q) = \{P\}$.
    (a) Suppose Q is a P-primary ideal and let M be the R-module R/Q. If $0 \neq m \in M$, show that $Q \subseteq \mathrm{Ann}(m) \subseteq P$ and that $\mathrm{rad}\,\mathrm{Ann}(m) = P$. Deduce that if $\mathrm{Ann}(m)$ is a prime ideal then it is equal to P and hence that $\mathrm{Ass}_R(R/Q) = \{P\}$. [Use Exercise 33 in Section 1.]

(b) For any ideal Q of R, let $0 \neq M \subseteq R/Q$. Prove that the radical of $\text{Ann}(M)$ is the intersection of the prime ideals in $\text{Supp}(M)$. [Use Proposition 12 and Exercise 25.]

(c) For M as in (b), prove that the radical of $\text{Ann}\,M$ is also the intersection of the prime ideals in $\text{Ass}_R(M)$. [Use Exercise 26(b).]

(d) If Q is an ideal of R with $\text{Ass}_R(R/Q) = \{P\}$ prove that $\text{rad}\,Q = P$. [Use the fact that $Q = \text{Ann}(R/Q)$ and (c).]

(e) If Q is an ideal of R with $\text{Ass}_R(R/Q) = \{P\}$ prove that Q is P-primary. [If $ab \in Q$ with $a \notin Q$ consider $0 \neq M = (Ra + Q)/Q \subseteq R/Q$ and show that b is contained in $\text{Ann}\,M \subseteq \text{rad}\,\text{Ann}(M)$. Use Exercises 33–34 in Section 1, to show that $\text{Ass}_R(M) = \{P\}$, then use (c) to show that $\text{rad}\,\text{Ann}(M) = P$, and conclude finally that $b \in P$.]

29. Suppose $I = Q_1 \cap \cdots \cap Q_n$ is a minimal primary decomposition of the ideal I in the Noetherian ring R with $P_i = \text{rad}\,Q_i$, $i = 1, \ldots, n$. This exercise proves that $\text{Ass}_R(R/I) = \{P_1, \ldots, P_n\}$.

(a) Prove that the natural projection homomorphisms induce an injection of R/I into $R/Q_1 \oplus \cdots \oplus R/Q_n$ and deduce that $\text{Ass}_R(R/I) \subseteq \{P_1, \ldots, P_n\}$. [Use Exercise 34 in Section 1 and the previous exercise.]

(b) Let $Q_i' = \cap_{j \neq i} Q_j$. Show that the minimality of the decomposition implies that $0 \neq Q_i'/I = (Q_i' + Q_i)/Q_i \subseteq R/Q_i$. Deduce that $\text{Ass}_R(Q_i'/I) = \{P_i\}$. [Use Exercises 33–34 in Section 1 and the previous exercise.] Deduce that $\{P_i\} \in \text{Ass}_R(R/I)$, so that $\text{Ass}_R(R/I) = \{P_1, \ldots, P_n\}$. [Use $Q_i'/I \subseteq R/I$ and Exercise 34 in Section 1.]

30. Let I be the ideal (x^2, xy, xz, yz) in $R = k[x, y, z]$. Prove that $\text{Ass}_R(R/I)$ consists of the primes $\{(x, y), (x, z), (x, y, z)\}$.

31. *(Spec for Quadratic Integer Rings)* Let R be the ring of integers in the quadratic field $K = \mathbb{Q}(\sqrt{D})$ where D is a squarefree integer and let P be a nonzero prime ideal in R. This exercise shows how the prime ideals in R are determined explicitly from the primes (p) in $\mathbb{Z}$, giving in particular a description of Spec R fibered over Spec $\mathbb{Z}$.

As in the discussion and example following Theorem 29, we have $R = \mathbb{Z}[\omega]$ where $\omega = \sqrt{D}$ if $D \equiv 2, 3 \bmod 4$ (respectively, $\omega = (1 + \sqrt{D})/2$ if $D \equiv 1 \bmod 4$), with minimal polynomial $m_\omega(x) = x^2 - D$ (respectively, $m_\omega(x) = x^2 - x + (1 - D)/4$), and $P \cap \mathbb{Z} = p\mathbb{Z}$ is a nonzero prime ideal of $\mathbb{Z}$.

(a) For any prime p in $\mathbb{Z}$ show that $R/pR \cong \mathbb{Z}[x]/(p, m_\omega(x)) \cong \mathbb{F}_p[x]/(\overline{m}_\omega(x))$ as rings, where $\overline{m}_\omega(x)$ is the reduction of $m_\omega(x)$ modulo p. Deduce that there is a prime ideal P in R with $P \cap \mathbb{Z} = (p)$ (this gives an alternate proof of Theorem 26(2) in this case).

(b) Use the isomorphism in (a) to prove that P is determined explicitly by the factorization of $m_\omega(x)$ modulo p:

(i) If $\overline{m}_\omega(x) \equiv (x - a)^2 \bmod p$ where $a \in \mathbb{Z}$ then $P = (p, \omega - a)$ and $pR = P^2$. Show that this case occurs only for the finitely many primes p dividing the discriminant of $m_\omega(x)$.

(ii) If $\overline{m}_\omega(x) \equiv (x - a)(x - b) \bmod p$ with integers $a, b \in \mathbb{Z}$ that are distinct modulo p then P is either $P_1 = (p, \omega - a)$ or $P_2 = (p, \omega - b)$ and P_1, P_2 are distinct prime ideals in R with $pR = P_1 P_2$.

(iii) If $\overline{m}_\omega(x)$ is irreducible modulo p then $P = pR$.

(c) Show that the picture for Spec R over Spec $\mathbb{Z}$ for any D is similar to that for the case $R = \mathbb{Z}[i]$ when $D = -1$: there is precisely one nonclosed point $(0) \in \text{Spec}\,R$ over $(0) \in \text{Spec}\,\mathbb{Z}$, precisely one closed point $P \in \text{Spec}\,R$ over each of the primes (p) in Spec $\mathbb{Z}$ in (i) (called *ramified* primes) and over the primes in (iii) (called *inert* primes), and precisely two closed points over the primes in (ii) (called *split* primes).

CHAPTER 16

Artinian Rings, Discrete Valuation Rings, and Dedekind Domains

Throughout this chapter R will denote a commutative ring with $1 \neq 0$.

16.1 ARTINIAN RINGS

In this section we shall study the basic theory of commutative rings that satisfy the descending chain condition (D.C.C.) on ideals, the Artinian rings (named after E. Artin). While one might at first expect that these rings have properties analogous to those for the commutative rings satisfying the ascending chain condition (the Noetherian rings), in fact this is not the case. The structure of Artinian rings is very restricted; for example an Artinian ring is necessarily also Noetherian (Theorem 3). Noncommutative Artinian rings play a central role in Representation Theory (cf. Chapters 18 and 19).

Definition. For any commutative ring R the *Krull dimension* (or simply the *dimension*) of R is the maximum possible length of a chain $P_0 \subset P_1 \subset P_2 \subset \cdots \subset P_n$ of distinct prime ideals in R. The dimension of R is said to be infinite if R has arbitrarily long chains of distinct prime ideals.

A ring with finite dimension must satisfy both the ascending and descending chain conditions on prime ideals (although not necessarily on all ideals). A field has dimension 0 and a Principal Ideal Domain that is not a field has dimension 1.

We shall see shortly that rings with D.C.C. on ideals always have dimension 0 (i.e., primes are maximal). If R is an integral domain that is also a finitely generated k–algebra over a field k, then the dimension of R is equal to the transcendence degree over k of the field of fractions of R (cf. Exercise 11). In particular, the Krull dimension agrees with the definition introduced earlier for the dimension of an affine variety. The advantage of the definition above is that it does not refer to any k–algebra structure and applies to arbitrary commutative rings R.

Definition. The *Jacobson radical* of R is the intersection of all maximal ideals of R and is denoted by Jac R.

The Jacobson radical is analogous to the Frattini subgroup of a group, and it enjoys some corresponding properties (cf. Exercise 24 in Section 6.1):

Proposition 1. Let $\mathcal{J}$ be the Jacobson radical of the commutative ring R.
   (1) If I is a proper ideal of R, then so is $(I, \mathcal{J})$, the ideal generated by I and $\mathcal{J}$.
   (2) The Jacobson radical contains the nilradical of R: $\operatorname{rad} 0 \subseteq \operatorname{Jac} R$.
   (3) An element x belongs to $\mathcal{J}$ if and only if $1 - rx$ is a unit for all $r \in R$.
   (4) *(Nakayama's Lemma)* If M is any finitely generated R-module and $\mathcal{J}M = M$, then $M = 0$.

    Proof: If I is a proper ideal in R, then $I \subseteq M$ for some maximal ideal M. Since $\mathcal{J} \subseteq M$, also $(I, \mathcal{J}) \subseteq M$, which proves (1).

    Part (2) follows from the definitions of the two radicals and Proposition 12 in Section 15.2 since maximal ideals are prime.

    Suppose $1 - rx$ is not a unit and let M be a maximal ideal containing $1 - rx$. Since $1 \notin M, rx \notin M$, so x cannot belong to $\mathcal{J}$ because $\mathcal{J} \subseteq M$. Conversely, suppose $x \notin \mathcal{J}$, i.e., there is a maximal ideal M with $x \notin M$. Then $R = (x, M)$, hence $1 = rx + y$ for some $y \in M$. Thus $1 - rx = y \in M$ and so $1 - rx$ is not a unit, which proves (3).

    To prove (4), assume $M \neq 0$ and let n be the smallest integer such that M is generated by n elements, say $m_1, \ldots, m_n$. Since $M = \mathcal{J}M$ we have

$$m_n = r_1 m_1 + r_2 m_2 + \cdots + r_n m_n \qquad \text{for some } r_1, r_2, \ldots, r_n \in \mathcal{J}.$$

Thus $(1 - r_n)m_n = r_1 m_1 + \cdots + r_{n-1} m_{n-1}$. By (3), $1 - r_n$ is a unit, so m_n lies in the module generated by $m_1, \ldots, m_{n-1}$, contradicting the minimality of n. Hence $M = 0$, completing the proof.

Definition. A commutative ring R is said to be *Artinian* or to satisfy the *descending chain condition on ideals* (or *D.C.C. on ideals*) if there is no infinite decreasing chain of ideals in R, i.e., whenever $I_1 \supseteq I_2 \supseteq I_3 \supseteq \cdots$ is a decreasing chain of ideals of R, then there is a positive integer m such that $I_k = I_m$ for all $k \geq m$. Similarly, an R-module M is said to be Artinian if it satisfies D.C.C. on submodules.

    It is immediate from the Lattice Isomorphism Theorem that every quotient R/I of an Artinian ring R by an ideal I is again an Artinian ring.

    The following result for Artinian rings is parallel to results in Theorem 15.2. The proof is completely analogous, and so is left as an exercise.

Proposition 2. The following are equivalent:
   (1) R is an Artinian ring.
   (2) Every nonempty set of ideals of R contains a minimal element under inclusion.

    The next result gives the main structure theorem for Artinian rings.

Theorem 3. Let R be an Artinian ring.

(1) There are only finitely many maximal ideals in R.

(2) The quotient $R/(\text{Jac } R)$ is a direct product of a finite number of fields. More precisely, if $M_1, \ldots, M_n$ are the finitely many maximal ideals in R then

$$R/(\text{Jac } R) \cong k_1 \times \cdots \times k_n,$$

where k_i is the field R/M_i for $1 \leq i \leq n$.

(3) Every prime ideal of R is maximal, i.e., R has Krull dimension 0. The Jacobson radical of R equals the nilradical of R and is a nilpotent ideal: $(\text{Jac } R)^m = 0$ for some $m \geq 1$.

(4) The ring R is isomorphic to the direct product of a finite number of Artinian local rings.

(5) Every Artinian ring is Noetherian.

Proof: To prove (1), let $\mathcal{S}$ be the set of all ideals of R that are the intersection of a finite number of maximal ideals. By Proposition 2, $\mathcal{S}$ has a minimal element, say $M_1 \cap M_2 \cap \cdots \cap M_n$. Then for any maximal ideal M we have

$$M \cap M_1 \cap M_2 \cap \cdots \cap M_n = M_1 \cap M_2 \cap \cdots \cap M_n,$$

so $M \supseteq M_1 \cap M_2 \cap \cdots \cap M_n$. By Exercise 11 in Section 7.4, $M \supseteq M_i$ for some i. Thus $M = M_i$ and so $M_1, \ldots, M_n$ are all the maximal ideals of R.

The proof of (2) is immediate from the Chinese Remainder Theorem (Section 7.6) applied to $M_1, \ldots, M_n$, since these maximal ideals are clearly pairwise comaximal and their intersection is $\text{Jac } R$.

For (3), we first prove $\mathcal{J} = \text{Jac } R$ is nilpotent. By D.C.C. there is some $m > 0$ such that $\mathcal{J}^m = \mathcal{J}^{m+i}$ for all positive i. By way of contradiction assume $\mathcal{J}^m \neq 0$. Let $\mathcal{S}$ be the set of proper ideals I such that $I\mathcal{J}^m \neq 0$, so $\mathcal{J} \in \mathcal{S}$. Let I_0 be a minimal element of $\mathcal{S}$. There is some $x \in I_0$ such that $x\mathcal{J}^m \neq 0$, so by minimality we must have $I_0 = (x)$. But now $((x)\mathcal{J})\mathcal{J}^m = x\mathcal{J}^{m+1} = x\mathcal{J}^m$, so it follows by minimality of (x) that $(x) = (x)\mathcal{J}$. By Nakayama's Lemma above, $(x) = 0$, a contradiction. This proves $\text{Jac } R$ is nilpotent.

Since $\text{Jac } R$ is nilpotent, in particular $\text{Jac } R \subseteq \text{rad } 0$, so these two ideals are equal by the second statement in Proposition 1.

Every prime ideal P in R contains the nilradical of R, hence contains $\text{Jac } R$ by what has already been proved,. The image of P is a prime ideal in the quotient ring $R/(\text{Jac } R) = k_1 \times \cdots \times k_n$. But in a direct product of rings $R_1 \times R_2$ (where each R_i has a 1) every ideal is of the form $I_1 \times I_2$, where I_j is an ideal of R_j for $j = 1, 2$ (cf. Exercise 3 in Section 7.6). It follows that a prime ideal in $k_1 \times \cdots \times k_n$ consists of the elements that are 0 in one of the components. In particular, such a prime ideal is also a maximal ideal in $k_1 \times \cdots \times k_n$ and it follows that P was a maximal ideal in R, which finishes the proof of (3).

Let $M_1, \ldots, M_n$ be all the distinct maximal ideals of R and let $(\text{Jac } R)^m = 0$ as in (3). Then

$$\prod_{i=1}^{n} M_i^m \subseteq \left(\prod_{i=1}^{n} M_i \right)^m \subseteq (\text{Jac } R)^m = 0.$$

By the Chinese Remainder Theorem it follows that

$$R \cong (R/M_1^m) \times (R/M_2^m) \times \cdots \times (R/M_n^m),$$

and each R/M_i^m is an Artinian ring with unique maximal ideal M_i/M_i^m, proving (4).

To prove (5), it suffices by (4) to prove that an Artinian local ring is Noetherian, so assume R is Artinian with unique maximal ideal M. In this case we have $M = \text{Jac } R$, so $M^m = (\text{Jac } R)^m = 0$ for some positive m. Then $R \cong R/M^m$, and in this case it is an exercise to see that R/M^m is Noetherian if and only if it is Artinian (cf. Exercise 8).

Corollary 4. The ring R is Artinian if and only if R is Noetherian and has Krull dimension 0.

Proof: The forward implication was proved in Theorem 3. Suppose now that R is Noetherian and that R has Krull dimension 0, i.e., that prime ideals of R are maximal. Since R is Noetherian, by Corollary 22(3) in Section 15.2, the ideal $(0) = P_1 \cdots P_n$ is the product of (not necessarily distinct) prime ideals, and these prime ideals are then maximal since R has dimension 0. By the Chinese Remainder Theorem, R is isomorphic to the direct product of a finite number of Noetherian rings of the form R/M^m where M is a maximal ideal in R. As in the proof of (5) of the theorem, R/M^m is Artinian, and it follows that R is Artinian.

Examples

(1) Let $n > 1$ be an integer. Since the ring $R = \mathbb{Z}/n\mathbb{Z}$ is finite, it is Artinian. If $n = p_1^{a_1} p_2^{a_2} \cdots p_s^{a_s}$ is the unique factorization of n into distinct prime powers, then

$$\mathbb{Z}/n\mathbb{Z} \cong (\mathbb{Z}/p_1^{a_1}\mathbb{Z}) \times (\mathbb{Z}/p_2^{a_2}\mathbb{Z}) \times \cdots \times (\mathbb{Z}/p_s^{a_s}\mathbb{Z}).$$

Each $\mathbb{Z}/p_i^{a_i}\mathbb{Z}$ is an Artinian local ring with unique maximal ideal $(p_i)/(p_i^{a_i})$, so this is the decomposition of $\mathbb{Z}/n\mathbb{Z}$ given by Theorem 3(4). The Jacobson radical of R is the ideal generated by $p_1 p_2 \cdots p_s$, the squarefree part of n and $R/(\text{Jac } R) \cong (\mathbb{Z}/p_1\mathbb{Z}) \times \cdots \times (\mathbb{Z}/p_s\mathbb{Z})$ is a direct product of fields. The ideals generated by p_i for $i = 1, \ldots, s$ are the maximal ideals of R.

(2) For any field k, a k-algebra R that is finite dimensional as a vector space over k is Artinian because ideals in R are in particular k-subspaces of R, hence the length of any chain of ideals in R is bounded by $\dim_k R$.

(3) Suppose f is a nonzero polynomial in $k[x]$ where k is a field. Then the quotient ring $R = k[x]/(f(x))$ is Artinian by the previous example. The decomposition of R as a direct product of Artinian local rings is given by

$$k[x]/(f(x)) \cong k[x]/(f_1(x)^{a_1}) \times \cdots \times k[x]/(f_s(x)^{a_s})$$

where $f(x) = f_1(x)^{a_1} \cdots f_s(x)^{a_s}$ is the factorization of $f(x)$ into powers of distinct irreducibles in $k[x]$ (cf. Proposition 16 in Section 9.5). The Jacobson radical of R is the ideal generated by the squarefree part of $f(x)$ and the maximal ideals of R are the ideals generated by the irreducible factors $f_i(x)$ for $i = 1, \ldots, s$ similar to Example 1.

EXERCISES

Let R be a commutative ring with 1 and let $\mathcal{J}$ be its Jacobson radical.

1. Suppose R is an Artinian ring and I is an ideal in R. Prove that R/I is also Artinian.

2. Show that every finite commutative ring with 1 is Artinian.

3. Prove that an integral domain of Krull dimension 0 is a field.

4. Prove that an Artinian integral domain is a field.

5. Suppose I is a nilpotent ideal in R and $M = IM$ for some R-module M. Prove that $M = 0$.

6. Suppose that $0 \longrightarrow M' \longrightarrow M \longrightarrow M'' \longrightarrow 0$ is an exact sequence of R-modules. Prove that M is an Artinian R-module if and only if M' and M'' are Artinian R-modules.

7. Suppose $R = F$ is a field. Prove that an R-module M is Artinian if and only if it is Noetherian if and only if M is a finite dimensional vector space over F.

8. Let M be a maximal ideal of the ring R and suppose that $M^n = 0$ for some $n \geq 1$. Prove that R is Noetherian if and only if R is Artinian. [Observe that each successive quotient $M^i/M^{i+1}, i = 0, \ldots, n-1$ in the filtration $R \supseteq M \supseteq \cdots \supseteq M^{n-1} \supseteq M^n = 0$ is a module over the field $F = R/M$. Then use the previous two exercises and Exercise 6 of Section 15.1.]

9. Let M be a finitely generated R-module. Prove that if $x_1, \ldots, x_n$ are elements of M whose images in $M/\mathcal{J}M$ generate $M/\mathcal{J}M$, then they generate M. Deduce that if R is Noetherian and the images of $a_1, \ldots, a_n$ in $\mathcal{J}/\mathcal{J}^2$ generate $\mathcal{J}/\mathcal{J}^2$, then $\mathcal{J} = (a_1, \ldots, a_n)$. [Let N be the submodule generated by $x_1, \ldots, x_n$ and apply Nakayama's Lemma to the module $A = M/N$.]

10. Let $R = \mathbb{Z}_{(2)}$ be the localization of $\mathbb{Z}$ at the prime ideal (2). Prove that Jac $R = (2)$ is the ideal generated by 2. If $M = \mathbb{Q}$, prove that $M/2M$ is a finitely generated R-module but that M is not finitely generated over R. Why doesn't this contradict the previous exercise? [Note the hypotheses in Nakayama's Lemma.]

11. Let V be an affine variety over a field k and let $R = k[V]$ be its coordinate ring. Let $d_t(R)$ denote the transcendence degree of the field of fractions $k(V)$ over k, and let $d_p(R)$ be the Krull dimension of R defined in terms of chains of prime ideals. This exercise shows $d_t(R) = d_p(R)$. By Noether's Normalization Lemma there is a polynomial subring $R_1 = k[y_1, \ldots, y_m]$ of R such that R is integral over R_1.
    (a) Show that $d_t(R_1) = d_t(R) = m$ and that $d_p(R_1) = d_p(R)$. Deduce that we may assume $R = R_1$. [Use the Going-up and Going-down Theorems (cf. Theorem 26, Section 15.3) to prove the second equality.]
    (b) When $R = R_1$ show that $d_p(R) \geq d_t(R)$ by exhibiting an explicit chain of prime ideals of length m.
    (c) When $R = R_1$ show that any nonzero prime ideal of R contains an element f such that $R(f)$ is transcendental over R of transcendence degree 1. Use induction to show that $d_p(R) \leq d_t(R)$, and deduce that $d_p(R) = d_t(R)$.

12. Let R be a Noetherian local ring with maximal ideal M.
    (a) The quotient M/M^2 is a module (i.e., vector space) over the field R/M. Prove that $d = \dim_{R/M}(M/M^2)$ is finite.
    (b) Prove that M can be generated as an ideal in R by d elements and by no fewer. [Use Exercise 9.]
    (c) Let $R = k[x_1, \ldots, x_n]_{(x_1, \ldots, x_n)}$ be the localization of the polynomial ring $k[x_1, \ldots, x_n]$ over the field k at the maximal ideal $(x_1, \ldots, x_n)$, and let M be the maximal ideal in

R. Prove that dim $_{R/M}(M/M^2) = n = $ dim R. [Cf. the previous exercise.]

It can be shown that dim $_{R/M}(M/M^2) \geq$ dim R for any Noetherian local ring R with maximal ideal M. A Noetherian local ring R is called a *regular local ring* if dim $_{R/M}(M/M^2) = $ dim R. It is a fact that a regular local ring is necessarily an integral domain and is also integrally closed.

13. If R is a Noetherian ring, prove that the Zariski topology on Spec R is discrete (i.e., every subset is Zariski open and also Zariski closed) if and only if R is Artinian.

14. Suppose I is the ideal $(x_1, x_2^2, x_3^3, \dots)$ in the polynomial ring $k[x_1, x_2, x_3, \dots]$ where k is a field and let R be the quotient ring $k[x_1, x_2, x_3, \dots]/I$. Prove that the image of the ideal $(x_1, x_2, x_3, \dots)$ in R is the unique prime ideal in R but is not finitely generated. Deduce that R is a local ring of Krull dimension 0 but is not Artinian.

16.2 DISCRETE VALUATION RINGS

In the previous section we showed that the Artinian rings are the Noetherian rings having Krull dimension 0. We now consider the easiest Noetherian rings of dimension 1, the Discrete Valuation Rings first introduced in Section 8.1:

Definition.
  (1) A *discrete valuation* on a field K is a function $v : K^\times \to \mathbb{Z}$ satisfying
    (i) v is surjective,
    (ii) $v(xy) = v(x) + v(y)$ for all $x, y \in K^\times$,
    (iii) $v(x + y) \geq \min\{v(x), v(y)\}$ for all $x, y \in K^\times$ with $x + y \neq 0$.
  The subring $\{x \in K \mid v(x) \geq 0\} \cup \{0\}$ is called the *valuation ring* of v. ·
  (2) An integral domain R is called a *Discrete Valuation Ring* (D.V.R.) if R is the valuation ring of a discrete valuation v on the field of fractions of R.

The valuation v is often extended to all of K by defining $v(0) = +\infty$, in which case (ii) and (iii) hold for all $a, b \in K$.

Examples
  (1) The localization $\mathbb{Z}_{(p)}$ of $\mathbb{Z}$ at any nonzero prime ideal (p) is a D.V.R. with respect to the discrete valuation v_p on $\mathbb{Q}$ defined as follows (cf. Exercise 27, Section 7.1). Every element $a/b \in \mathbb{Q}^\times$ can be written uniquely in the form $p^n(a_1/b_1)$ where $n \in \mathbb{Z}$, $a_1/b_1 \in \mathbb{Q}^\times$ and both a_1 and b_1 are relatively prime to p. Define

$$v_p\left(\frac{a}{b}\right) = v_p\left(p^n\frac{a_1}{b_1}\right) = n.$$

  One easily checks that the axioms for a D.V.R. are satisfied. We call v_p the *p-adic valuation* on $\mathbb{Q}$. The corresponding valuation ring is the set of rational numbers with $n \geq 0$ together with 0, i.e., the rational numbers a/b where b is not divisible by p, which is $\mathbb{Z}_{(p)}$.
  (2) For any field F, let f be an irreducible polynomial in $F[x]$. Every nonzero element in the field $F(x)$ can be written uniquely in the form $f^n(a/b)$ where $n \in \mathbb{Z}$, $a/b \in F[x]^\times$ and both a and b are relatively prime to f. Then

$$v_f\left(f^n\frac{a}{b}\right) = n$$

defines a valuation on $F(x)$ and the corresponding valuation ring is the localization $F[x]_f$ of $F[x]$ at f consisting of the rational functions in $F(x)$ whose denominator is not divisible by f. When $f = x - \alpha$ is a polynomial of degree 1 in $F[x]$, the valuation v_f gives the *order of the zero* (if $n \geq 0$) or *pole* (if $n < 0$) of the element in $F(x)$ at $x = \alpha$.

(3) The ring of formal Laurent series $F((x))$ with coefficients in the field F has a discrete valuation v defined by

$$v \left(\sum_{i \geq n}^{\infty} a_i x^i \right) = n$$

(cf. Exercise 5, Section 7.2). The corresponding D.V.R. is the ring $F[[x]]$ of power series in x with coefficients in F.

Note that $v(1) = v(1) + v(1)$ implies that $v(1) = 0$, so every Discrete Valuation Ring R is a ring with identity $1 \neq 0$. Since R is a subring of a field by definition, R is in particular an integral domain. It is easy to see that a D.V.R. is a Euclidean Domain (cf. Example 4 in Section 8.1), so in particular is also a P.I.D. and a U.F.D. In fact the factorization and ideal structure of a D.V.R. is very simple, as the next proposition shows.

Proposition 5. Suppose R is a Discrete Valuation Ring with respect to the valuation v, and let t be any element of R with $v(t) = 1$. Then

(1) A nonzero element $u \in R$ is a unit if and only if $v(u) = 0$.

(2) Every nonzero element $r \in R$ can be written in the form $r = ut^n$ for some unit $u \in R$ and some $n \geq 0$. Every nonzero element x in the field of fractions of R can be written in the form $x = ut^n$ for some unit $u \in R$ and some $n \in \mathbb{Z}$.

(3) Every nonzero ideal of R is a principal ideal of the form (t^n) for some $n \geq 0$. In particular, R is a Noetherian ring.

Proof: If u is a unit, then $uv = 1$ for some $v \in R$ and then $v(u) + v(v) = v(uv) = v(1) = 0$ with $v(u) \geq 0$ and $v(v) \geq 0$ shows that $v(u) = 0$. Conversely, if u is nonzero and $v(u) = 0$ then $u^{-1} \in K$ satisfies $v(u^{-1}) + v(u) = v(1) = 0$. Hence $v(u^{-1}) = 0$ and $u^{-1} \in R$, so u is a unit. This proves (1).

For (2), note that if $v(x) = n$ then $v(xt^{-n}) = 0$, so $xt^{-n} = u$ is a unit in R by (1). Hence $x = ut^n$, where $x \in R$ if and only if $n = v(x) \geq 0$.

If I is a nonzero ideal in R, let $r \in I$ be an element with $v(r)$ minimal. If $v(r) = n$, then r differs from t^n by a unit by (2), so $t^n \in I$ and $(t^n) \subseteq I$. If now a is any nonzero element of I, then $v(a) \geq n$ by choice of n. Then $v(at^{-n}) \geq 0$ and so $at^{-n} \in R$, which shows that $a \in (t^n)$. Hence $I = (t^n)$, proving the first statement in (3). It is then clear that ascending chains of ideals in R are finite, proving that R is Noetherian and completing the proof.

Definition. If R is a D.V.R. with valuation v, then an element t of R with $v(t) = 1$ is called a *uniformizing* (or *local*) *parameter* for R.

Corollary 6. Let R be a Discrete Valuation Ring.

(1) The ring R is an integrally closed local ring with unique maximal ideal given by the elements with strictly positive valuation: $M = \{r \in R \mid v(r) > 0\}$. Every nonzero ideal in R is of the form M^n for some integer $n \geq 0$.

(2) The only prime ideals of R are M and 0, i.e., Spec $R = \{0, M\}$. In particular, a D.V.R. has Krull dimension 1.

Proof: Any U.F.D. is integrally closed in its fraction field (Example 3 in Section 15.3), so R is integrally closed. The remainder of the statements follow immediately from the description of the ideals of R in Proposition 5.

The definition of a Discrete Valuation Ring is extremely explicit in terms of a valuation on the fraction field, and as a result it appears that it might be difficult to recognize whether a given ring R is a D.V.R. from purely "internal" algebraic properties of R. In fact, the ring-theoretic properties in Proposition 5 and Corollary 6 characterize Discrete Valuation Rings. The following theorem gives several alternate algebraic descriptions of Discrete Valuation Rings in which there is no explicit mention of the valuation.

Theorem 7. The following properties of a ring R are equivalent:

(1) R is a Discrete Valuation Ring,

(2) R is a P.I.D. with a unique maximal ideal $P \neq 0$,

(3) R is a U.F.D. with a unique (up to associates) irreducible element t,

(4) R is a Noetherian integral domain that is also a local ring whose unique maximal ideal is nonzero and principal,

(5) R is a Noetherian, integrally closed, integral domain that is also a local ring of Krull dimension 1 i.e., R has a unique nonzero prime ideal: Spec $R = \{0, M\}$.

Proof: That (1) implies each of the other properties was proved above.

If (2) holds then (3) is immediate since irreducible elements generate prime ideals in a U.F.D. (Proposition 12, Section 8.3).

If (3) holds, then every nonzero element in R can be written uniquely in the form ut^n for some unit u and some $n \geq 0$. Then every nonzero element in the fraction field of R can be written uniquely in the form ut^n for some unit u and some $n \in \mathbb{Z}$. It is now straightforward to check that the map $v(ut^n) = n$ is a discrete valuation on the field of fractions of R, and R is the valuation ring of v, and (1) holds.

Suppose (4) holds, let $M = (t)$ be the unique maximal ideal of R, and let $M_0 = \cap_{i=1}^{\infty} M^i$. Then $M_0 = MM_0$, and since R is Noetherian M_0 is finitely generated. By hypothesis $M = $ Jac R, so by Nakayama's Lemma $M_0 = 0$. If I is any proper, nonzero ideal of R then there is some $n \geq 0$ such that $I \subseteq M^n$ but $I \nsubseteq M^{n+1}$. Let $a \in I - M^{n+1}$ and write $a = t^n u$ for some $u \in R$. Then $u \notin M$, and so u is a unit in the local ring R. Thus $(a) = (t^n) = M^n$ for every $a \in I - M^{n+1}$. This shows that $I = (t^n)$, and so every ideal of R is principal, which shows that (2) holds.

We have shown that (1), (2), (3) and (4) are equivalent, and that each of these implies (5). To complete the proof we show that (5) implies (4), which amounts to showing that the ideal M in (5) is a principal ideal. Since $0 \neq M = $ Jac R and M is finitely generated because R is Noetherian, by Nakayama's Lemma (Proposition 1(4)), $M \neq M^2$. Let $t \in M - M^2$. We argue that $M = (t)$. By Proposition 12 in Section 15.2, the assumption that M is the unique nonzero prime ideal in R implies that $M = \mathrm{rad}\,(t)$, and then Proposition 14 in Section 15.2 implies that some power of M is contained in (t). Proceeding by way of contradiction, assume $(t) \neq M$, so that $M^n \subseteq (t)$ but $M^{n-1} \not\subseteq (t)$ for some $n \geq 2$. Then there is an element $x \in M^{n-1} - (t)$ such that $xM \subseteq (t)$. Note that $t \neq 0$ so $y = x/t$ belongs to the field of fractions of R. Also, $y \notin R$ because $x = ty \notin (t)$. However, by choice of x we have $yM \subseteq R$, and then one checks that yM is an ideal in R. If $yM = R$ then $1 = ym$ for some $m \in M$. This leads to a contradiction because we would then have $t = xm \in M^2$, contrary to the choice of t. Thus yM is a proper ideal, hence is contained in the unique maximal ideal of R, namely $yM \subseteq M$. Now M is a finitely generated R-module on which y acts by left multiplication as an R-module homomorphism. By the same (determinant) method as in the proof of Proposition 23 in Section 15.3 there is a monic polynomial p with coefficients in R such that $p(y)m = 0$ for all $m \in M$. Since $p(y)$ is an element of a field containing R and M, we must have $p(y) = 0$. Hence y is integral over R. Since R is integrally closed by assumption, it follows that $y \in R$, a contradiction. Hence $M = (t)$ is principal, so (5) implies (4), completing the proof of the theorem.

Corollary 8. If R is any Noetherian, integrally closed, integral domain and P is a minimal nonzero prime ideal of R, then the localization R_P of R at P is a Discrete Valuation Ring.

Proof: By results in Section 15.4, the localization R_P is a Noetherian (Proposition 38(4)), integrally closed (Proposition 49), integral domain (Proposition 46(2)), that is a local ring with unique nonzero prime ideal (Proposition 46(4)), so R_P satisfies (5) in the theorem.

Examples

(1) If R is any Principal Ideal Domain then every localization R_P of R at a nonzero prime ideal $P = (p)$ is a Discrete Valuation Ring. This follows immediately from Corollary 8 since R is integrally closed (being a U.F.D., cf. Example 3 in Section 15.3) and nonzero prime ideals in a P.I.D. are maximal (Proposition 8.7). Note that the quotient field K of R_P is the same as the quotient field of R, so each nonzero prime p in R produces a valuation v_p on K, given by the formula

$$v\left(p^n \frac{a}{b}\right) = n$$

where a and b are elements of R not divisible by p. This generalizes both Examples 1 and 2 above.

(2) The ring $\mathbb{Z}_p$ of p-adic integers is a Discrete Valuation Ring since it is a P.I.D. with unique maximal ideal $p\mathbb{Z}_p$ (cf. Exercise 11, Section 7.6). The fraction field of $\mathbb{Z}_p$ is called the *field of p-adic numbers* and is denoted $\mathbb{Q}_p$. The element p is a uniformizing parameter for $\mathbb{Z}_p$, so every nonzero element in $\mathbb{Q}_p$ can be written uniquely in the form $p^n u$ for some $n \in \mathbb{Z}$ and unit $u \in \mathbb{Z}_p^\times$, (where $u = a_0 + a_1 p + a_2 p^2 + \ldots$ with $0 < a_0 < p$ as in Exercise 11(c), Section 7.6). The corresponding *p-adic valuation* v_p on $\mathbb{Q}_p$ is then given by $v_p(p^n u) = n$.

A discrete valuation v on a field K defines an associated *metric* (or "distance function"), d_v, on K as follows: fix any real number $\beta > 1$ (the actual value of β does not matter for verifying the axioms of a metric), and for all $a, b \in K$ define

$$d_v(a, b) = \| a - b \|_v \qquad \text{where} \qquad \| a \|_v = \beta^{-v(a)}$$

and where we set $d_v(a, a) = 0$. It is easy to check that d_v satisfies the three axioms for a metric:

  (i) $d_v(a, b) \geq 0$, with equality holding if and only if $a = b$,
  (ii) $d_v(a, b) = d_v(b, a)$, i.e., d_v is symmetric,
  (iii) $d_v(a, b) \leq d_v(a, c) + d_v(c, b)$, for all $a, b, c \in K$, i.e., d_v satisfies the "triangle inequality."

The triangle inequality is a consequence of axiom (iii) of the discrete valuation. Indeed, a stronger version of the triangle inequality holds:

  (iii)′ $d_v(a, b) \leq \max\{d_v(a, c), d_v(c, b)\}$, for all $a, b, c \in K$.

For this reason d_v is sometimes called an *ultrametric*. One may now use Cauchy sequences to form the *completion* of K with respect to d_v, denoted by K_v, in the same way that the real numbers $\mathbb{R}$ are constructed from the rational numbers $\mathbb{Q}$. It is not difficult to show that K_v is also a field with a discrete valuation that agrees with v on the dense subset K of K_v.

Examples

  (1) Consider the p-adic valuation v_p on $\mathbb{Q}$ and take $\beta = p$. Write $\| a \|_p$ for $\| a \|_{v_p}$, so that for a, b relatively prime to p,

$$\| p^n \frac{a}{b} \|_p = p^{-n}.$$

Note that integers (or rational numbers) have small p-adic absolute value if they are divisible by a large power of p. For example, the sequence $1, p, p^2, p^3, \ldots$ converges to zero in the p-adic metric.

It is not too difficult to see that the completion of $\mathbb{Q}$ with respect to the p-adic metric is the field $\mathbb{Q}_p$ of p-adic numbers, and the completion of $\mathbb{Z}$ is the ring $\mathbb{Z}_p$ of p-adic integers. One way to see this is to check that each element a of the completion may be represented as a *p-adic Laurent series*:

$$a = \sum_{n=n_0}^{\infty} a_i p^i \qquad \text{where } n_0 \in \mathbb{Z} \text{ and } a_i \in \{0, 1, \ldots, p - 1\} \text{ for all } i,$$

and then use Example 2 previously. In terms of this expansion, the p-adic valuation is given by $v_p(a) = n_0$ (when $a_{n_0} \neq 0$).
  (2) In a similar way, the completion of $F(x)$ with respect to the valuation v_x in Example 2 at the beginning of this section gives the field $F((x))$ with corresponding valuation ring $F[[x]]$ in Example 3 in the same set of examples.

The completion of a field K with respect to a discrete valuation v is a field K_v in which the elements can be easily described in terms of a uniformizing parameter. In addition, K_v is a topological space where the topology is defined by the metric d_v. Furthermore, Cauchy sequences of elements in K_v converge to elements of K_v (i.e., K_v

is *complete* in the ν-adic topology). This is similar to the situation of the completion $\mathbb{R}$ of $\mathbb{Q}$ with respect to the usual Euclidean metric. This allows the application of ideas from analysis to the study of such rings, and is an important tool in the study of algebraic number fields and in algebraic geometry.

Fractional Ideals

We complete our discussion of Discrete Valuation Rings by giving another characterization of D.V.R.s in terms of "fractional ideals," which can be defined for any integral domain:

Definition. For any integral domain R with fraction field K, a *fractional ideal* of R is an R-submodule A of K such that $dA \subseteq R$ for some nonzero $d \in R$ (equivalently, a submodule of the form $d^{-1}I$ for some nonzero $d \in R$ and ideal I of R).

The equivalence of these two definitions follows from the observation that dA is an R-submodule (i.e., an ideal) of R.

The notion of a fractional ideal in K depends on the ring R. Loosely speaking, a fractional ideal is an ideal of R up to a fixed "denominator" d. The ideals of R are also fractional ideals of R (with denominator $d = 1$) and are the fractional ideals that are contained in R. For clarity these are occasionally called the *integral ideals* of R. When R is a Noetherian integral domain, a fractional ideal of R is the same as a finitely generated R-submodule of K (cf. Exercise 6).

For any $x \in K$ the (cyclic) R-module $Rx = \{rx \mid r \in R\}$ is called the *principal fractional ideal* generated by x.

If A and B are fractional ideals, their product, AB, is defined to be the set of all finite sums of elements of the form ab where $a \in A$ and $b \in B$. If $A = d^{-1}I$ and $B = (d')^{-1}J$ for ideals I, J in R and nonzero $d, d' \in R$, then $AB = (dd')^{-1}IJ$ where IJ is the usual product ideal. In particular, this shows that the product of two fractional ideals is a fractional ideal.

Definition. The fractional ideal A is said to be *invertible* if there exists a fractional ideal B with $AB = R$, in which case B is called the *inverse* of A and denoted A^{-1}.

If A is an invertible fractional ideal, the fractional ideal B with $AB = R$ is unique: $AB = AC = R$ implies $B = B(AC) = (BA)C = C$.

Proposition 9. Let R be an integral domain and let A be a fractional ideal of R.
  (1) If A is a nonzero principal fractional ideal then A is invertible.
  (2) If A is nonzero then the set $A' = \{x \in K \mid xA \subseteq R\}$ is a fractional ideal of R. In general we have $AA' \subseteq R$ and $AA' = R$ if and only if A is invertible, in which case $A^{-1} = A'$.
  (3) If A is an invertible fractional ideal of R then A is finitely generated.
  (4) The set of invertible fractional ideals is an abelian group under multiplication with identity R. The set of nonzero principal fractional ideals is a subgroup of the invertible fractional ideals.

Proof: If $A = xR$ is a nonzero principal fractional ideal, then taking $B = x^{-1}R$ shows that A is invertible, proving (1).

One easily sees that A' is an R-submodule of K. If A is a nonzero fractional ideal there is some nonzero element $d \in R$ such that $dA \subseteq R$, so A contains nonzero elements of R. Let a be any nonzero element of A contained in R. Then by definition of A' we have $aA' \subseteq R$, so A' is a fractional ideal. Also by definition, $AA' \subseteq R$. If $AA' = R$ then A is invertible with inverse $A^{-1} = A'$. Conversely, if $AB = R$, then $B \subseteq A'$ by definition of A'. Then $R = AB \subseteq AA' \subseteq R$, showing that $AA' = R$, proving (2).

If A is invertible, then $AA' = R$ by (2) and so $1 = a_1 a_1' + \cdots + a_n a_n'$ for some $a_1, \ldots, a_n \in A$ and $a_1', \ldots, a_n' \in A'$. If $a \in A$, then $a = (aa_1')a_1 + \cdots + (aa_n')a_n$, where each $aa_i' \in R$ by definition of A'. It follows that A is generated over R by $a_1, \ldots, a_n$ and so A is finitely generated, proving (3).

Finally, it is clear that the product of two invertible fractional ideals is again invertible. This product is commutative, associative, and $RA = A$ for any fractional ideal. The inverse of an invertible fractional ideal is an invertible fractional ideal by definition, proving the first statement in (4). The second statement in (4) is immediate since the product of xR and yR is $(xy)R$ and the inverse of xR is $x^{-1}R$.

Definition. If R is an integral domain, then the quotient of the group of invertible fractional ideals of R by the subgroup of nonzero principal fractional ideals of R is called the *class group* of R. The order of the class group of R is called the *class number* of R.

The class group of R is the trivial group and the class number of R is 1 if and only if R is a P.I.D. The class group of R measures how close the ideals of R are to being principal.

Whether a fractional ideal A of R is invertible is also related to whether A is *projective* as an R-module. Recall that an R-module M is projective over R if and only if M is a direct summand of a free module (Proposition 30, Section 10.5). Equivalently, M is projective if and only if there is a free R-module F and R-module homomorphisms $f : F \to M$ and $g : M \to F$ with $f \circ g = 1$ (Proposition 25, Section 10.5).

Proposition 10. Let R be an integral domain with fraction field K and let A be a nonzero fractional ideal of R. Then A is invertible if and only if A is a projective R-module.

Proof: Assume first that A is invertible, so $\sum_{i=1}^{n} a_i a_i' = 1$ for some $a_i \in A$ and $a_i' \in A'$ as in (2) of Proposition 9. Let F be the free R-module on $y_1, \ldots, y_n$. Define $f : F \to A$ by $f(\sum_{i=1}^{n} r_i y_i) = \sum_{i=1}^{n} r_i a_i$ and $g : A \to F$ by $f(c) = \sum_{i=1}^{n} (ca_i') y_i$. It is immediate that both f and g are R-module homomorphisms (note that $ca_i' \in R$ by definition of A'). Since

$$(f \circ g)(c) = f\left(\sum_{i=1}^{n} (ca_i') y_i\right) = \sum_{i=1}^{n} (ca_i') a_i = c\left(\sum_{i=1}^{n} a_i a_i'\right) = c,$$

so $f \circ g = 1$ and A is a direct summand of F, hence is projective.

Conversely, suppose that A is nonzero and projective, so there is a free R-module F and R-homomorphisms $f : F \to A$ and $g : A \to F$ with $f \circ g = 1$. Fix any $0 \neq a \in A$ and suppose $g(a) = \sum_{i=1}^{n} \tilde{a}_i y_i$ where $\tilde{a}_i \in R$ and $y_1, \ldots, y_n$ is part of a set of free generators for F. Define $a_i = f(y_i)$ and $a_i' = \tilde{a}_i/a \in K$ for $i = 1, \ldots, n$. For any $b \in A$ we have $bg(a) = ag(b) = g(ab)$ since g is an R-module homomorphism. Write $g(b) = \sum_{i=1}^{n} \tilde{b}_i y_i + \sum_{j \in \mathcal{J}} \tilde{b}_j y_j$ where $\{y_j\}$ for $j \in \mathcal{J}$ are the remaining elements in the set of free generators for F. Then

$$\sum_{i=1}^{n} (b\tilde{a}_i) y_i = \sum_{i=1}^{n} (a\tilde{b}_i) y_i + \sum_{j \in \mathcal{J}} (a\tilde{b}_j) y_j .$$

We may equate coefficients of the elements in the free R-module basis for F in this equation and it follows that $g(b) = \sum_{i=1}^{n} \tilde{b}_i y_i$ where $\tilde{b}_i \in R$ and that $b\tilde{a}_i = a\tilde{b}_i$ for $i = 1, \ldots, n$. In particular, it follows from the definition of a_i' that $ba_i' = b(\tilde{a}_i/a) = \tilde{b}_i$ is an element of R for every element b of A. This shows that $a_i' \in A'$ for $i = 1, \ldots, n$. Since $f \circ g = 1$, we have

$$a = f \circ g(a) = f\left(\sum_{i=1}^{n} \tilde{a}_i y_i\right) = \sum_{i=1}^{n} \tilde{a}_i a_i = \sum_{i=1}^{n} (aa_i') a_i = a\left(\sum_{i=1}^{n} a_i a_i'\right),$$

and so $\sum_{i=1}^{n} a_i a_i' = 1$. It follows that $AA' = R$ and so A is invertible by Proposition 9, completing the proof.

The next result shows that if the integral domain R is also a local ring, then whether fractional ideals are invertible determines whether R is a D.V.R.

Proposition 11. Suppose the integral domain R is a local ring that is not a field. Then R is a Discrete Valuation Ring if and only if every nonzero fractional ideal of R is invertible.

Proof: If R is a D.V.R. with uniformizing parameter t, then by Proposition 5 every nonzero ideal of R is of the form (t^n) for some $n \geq 0$ and every element d in R can be written in the form ut^m for some unit $u \in R$ and some $m \geq 0$. It follows that every nonzero fractional ideal of R is of the form $t^N R$ for some $N \in \mathbb{Z}$, so is a principal fractional ideal and hence invertible by the previous proposition.

Conversely, suppose that every nonzero fractional ideal of R is invertible. Then every nonzero ideal of R is finitely generated by (3) of Proposition 9, so R is Noetherian. Let M be the unique maximal ideal of R. If $M = M^2$ then $M = 0$ by Nakayama's Lemma, and then R would be a field, contrary to hypothesis. Hence there is an element t with $t \in M - M^2$. By assumption M is invertible, and since $t \in M$, the fractional ideal tM^{-1} is a nonzero ideal in R. If $tM^{-1} \subseteq M$, then $t \in M^2$, contrary to the choice of t. Hence $tM^{-1} = R$, so $(t) = M$, and M is a nonzero principal ideal. It follows by the equivalent condition 4 of Theorem 7 that R is a D.V.R., completing the proof.

We end this section with an application to algebraic geometry.

Nonsingularity and Local Rings of Affine Plane Curves

Let k be an algebraically closed field and let C be an irreducible affine *curve* over k. In other words, C is an affine algebraic set whose coordinate ring $k[C]$ is an integral domain and whose field of rational functions $k(C)$ has transcendence degree 1 over k (cf. Section 15.4).

Recall that, by definition, the point v on C is nonsingular if $\mathfrak{m}_{v,C}/\mathfrak{m}_{v,C}^2$ is a 1-dimensional vector space over k, where $\mathfrak{m}_{v,C}$ is the unique maximal ideal in the local ring $\mathcal{O}_{v,C}$ of rational functions on C defined at v.

Proposition 12. Let v be a point on the irreducible affine curve C over k. Then C is nonsingular at v if and only if the local ring $\mathcal{O}_{v,C}$ is a Discrete Valuation Ring.

Proof: Suppose first that v is nonsingular. Then $\dim_k(\mathfrak{m}_{v,C}/\mathfrak{m}_{v,C}^2) = 1$, and since $\mathcal{O}_{v,C}$ is Noetherian, it follows from Exercise 12 in Section 1 that $\mathfrak{m}_{v,C}$ is principal. Hence $\mathcal{O}_{v,C}$ is a D.V.R. by Theorem 7(4). Conversely, suppose $\mathcal{O}_{v,C}$ is a D.V.R. and t is a uniformizing element for $\mathcal{O}_{v,C}$. Then every element in $\mathfrak{m}_{v,C}$ can be written uniquely in the form at for some a in $\mathcal{O}_{v,C}$. The map from $\mathfrak{m}_{v,C}$ to $\mathcal{O}_{v,C}/\mathfrak{m}_{v,C}$ defined by mapping at to $a \bmod \mathfrak{m}_{v,C}$ is easily checked to be a surjective $\mathcal{O}_{v,C}$-module homomorphism with kernel $\mathfrak{m}_{v,C}^2$. Hence $\mathfrak{m}_{v,C}/\mathfrak{m}_{v,C}^2$ is isomorphic as an $\mathcal{O}_{v,C}/\mathfrak{m}_{v,C}$-module to $\mathcal{O}_{v,C}/\mathfrak{m}_{v,C}$. Since $\mathcal{O}_{v,C}/\mathfrak{m}_{v,C} \cong k$ (Proposition 46(5) in Section 15.4), it follows that $\dim_k(\mathfrak{m}_{v,C}/\mathfrak{m}_{v,C}^2) = 1$, and so v is a nonsingular point on C.

Definition. If v is a nonsingular point on C with corresponding discrete valuation ν_v defined on $k(C)$, then $\nu_v(f) = n$ for $f \in k(V)$ is the *order of zero of f at v* (if $n \geq 0$) or the *order of the pole of f at v* (if $n < 0$).

Using the criterion for nonsingularity for points on curves in Proposition 12 we can prove a result first mentioned in Section 15.4:

Corollary 13. An irreducible affine curve C over an algebraically closed field k is smooth if and only if its coordinate ring $k[C]$ is integrally closed.

Proof: The curve C is smooth if and only if every localization $\mathcal{O}_{v,C}$ is a D.V.R. Since $k[C]$ has Krull dimension 1 (Exercise 11 in Section 1), the same is true for each $\mathcal{O}_{v,C}$. It then follows by Theorem 7(5) that every localization $\mathcal{O}_{v,C}$ is a D.V.R. if and only if $\mathcal{O}_{v,C}$ is integrally closed. By Proposition 49 in Section 15.4, this in turn is equivalent to the statement that $k[C]$ is integrally closed, which proves the corollary.

EXERCISES

1. Suppose R is a Discrete Valuation Ring with respect to the valuation v on the fraction field K of R. If $x, y \in K$ with $v(x) < v(y)$ prove that $v(x + y) = \min(v(x), v(y))$. [Note that $x + y = x(1 + y/x)$.]

2. Suppose R is a Discrete Valuation Ring with unique maximal ideal M and quotient $F = R/M$. For any $n \geq 0$ show that M^n/M^{n+1} is a vector space over F and that $\dim_F(M^n/M^{n+1}) = 1$.

3. Suppose R is an integral domain that is also a local ring whose unique maximal ideal $M = (t)$ is nonzero and principal, and suppose that $\cap_{n \geq 1} (t^n) = 0$. Prove that R is a Discrete Valuation Ring. [Show that every nonzero element in R can be written in the form ut^n for some unit $u \in R$ and some $n \geq 0$.]

4. Suppose R is a Noetherian local ring whose unique maximal ideal $M = (t)$ is principal. Prove that either R is a Discrete Valuation Ring or $t^n = 0$ for some $n \geq 0$. In the latter case show that R is Artinian.

5. Suppose that R is a Noetherian integral domain that is also a local ring of Krull dimension 1. Let M be the unique maximal ideal of R and let $F = R/M$, so that M/M^2 is a vector space over F.
   (a) Prove that if $\dim_F (M/M^2) = 1$ then R is a Discrete Valuation Ring.
   (b) If every nonzero ideal of R is a power of M prove that R is a Discrete Valuation Ring.

6. Let R be an integral domain with fraction field K. Prove that every finitely generated R-submodule of K is a fractional ideal of R. If R is Noetherian, prove that A is a fractional ideal of R if and only if R is a finitely generated R-submodule of K.

7. If R is an integral domain and A is a fractional ideal of R, prove that if A is projective then A is finitely generated. Conclude that every integral domain that is not Noetherian contains an ideal that is not projective.

8. Suppose R is a Noetherian integral domain that is also a local ring with nonzero maximal ideal M. Prove that R is a D.V.R. if and only if the only M-primary ideals in R are the powers of M.

9. Let $C = \mathcal{Z}(xz - y^2, yz - x^3, z^2 - x^2 y) \subset \mathbb{A}^3$ over the algebraically closed field k. If $v = (0, 0, 0) \in C$, prove that $\dim_k (\mathfrak{m}_{v,C}/\mathfrak{m}_{v,C}^2) = 3$ so that v is singular on C. Conclude that $k[C]$ is not integrally closed in $k(C)$ and determine its integral closure. [cf. Exercise 27, Section 15.4.]

16.3 DEDEKIND DOMAINS

In the previous section we showed that Discrete Valuation Rings are the local rings that are integrally closed Noetherian integral domains of Krull dimension 1. In this section we consider the effect of relaxing the condition that the ring be a local ring:

Definition. A *Dedekind Domain* is a Noetherian, integrally closed, integral domain of Krull dimension 1.

Equivalently, R is a Dedekind Domain if R is a Noetherian, integrally closed, integral domain that is not a field in which every nonzero prime ideal is maximal.

The first result shows that Dedekind Domains are a generalization of the class of Principal Ideal Domains. We shall see later (Theorem 22) that there is a structure theorem for finitely generated modules over a Dedekind Domain extending the corresponding result for P.I.D.s proved in Section 12.1.

Proposition 14.
   (1) Every Principal Ideal Domain is a Dedekind Domain.
   (2) The ring of integers in an algebraic number field is a Dedekind Domain.

Proof: A P.I.D. is clearly Noetherian, is integrally closed since it is a U.F.D. (Example 3, Section 15.3), and nonzero prime ideals are maximal (Proposition 7 in Section 8.2), which proves (1). Let $\mathcal{O}_K$ be the ring of integers in the number field K, i.e., the integral closure of $\mathbb{Z}$ in K. Then Corollary 25 in Section 15.3 shows that $\mathcal{O}_K$ is integrally closed, $\mathcal{O}_K$ is Noetherian by Theorem 29 in Section 15.3, and the fact that nonzero prime ideals in $\mathcal{O}_K$ are maximal was proved in the discussion following the same theorem. This proves (2).

The following theorem gives a number of important equivalent characterizations of Dedekind Domains. Recall that the basic properties of fractional ideals were developed in the previous section.

Theorem 15. Suppose R is an integral domain with fraction field $K \neq R$. The following are equivalent conditions for R to be a Dedekind Domain:
  (1) The ring R is Noetherian, integrally closed, and every nonzero prime ideal is maximal.
  (2) The ring R is Noetherian and for each nonzero prime P of R the localization R_P is a Discrete Valuation Ring.
  (3) Every nonzero fractional ideal of R in K is invertible.
  (4) Every nonzero fractional ideal of R in K is a projective R-module.
  (5) Every nonzero proper ideal I of R can be written as a finite product of prime ideals: $I = P_1 P_2 \cdots P_n$ (not necessarily distinct).
      When the condition in (5) holds, the set of primes $\{P_1, \ldots, P_n\}$ is uniquely determined and so every nonzero proper ideal I of R can be written uniquely (up to order) as a product of powers of prime ideals.

Proof: If R satisfies (1), then R_P is a D.V.R. by Corollary 8, so (1) implies (2). Conversely, assume each R_P is a D.V.R. Then R is integrally closed by Proposition 49 in Section 15.4 and every nonzero prime ideal is maximal by Proposition 46(3) in Section 15.4, so (2) implies (1).

Suppose now that (1) is satisfied and that A is a nonzero fractional ideal of R. Let $A' = \{x \in K \mid xA \subseteq R\}$ as in Proposition 9. For any prime ideal P of R the behavior of R-modules under localization shows that $(AA')_P = A_P(A')_P = A_P(A_P)'$ (cf. Exercise 4). Since R_P is a D.V.R. by what has already been shown, $A_P(A_P)' = R_P$ by Proposition 11. Hence $(AA')_P = R_P$ for all nonzero primes P of R, so $AA' = R$ (Exercise 13 in Section 15.4), and A is invertible, showing (1) implies (3). Conversely, suppose every nonzero fractional ideal of R is invertible. Then every ideal in R is finitely generated by Proposition 9(3), so R is Noetherian. Every localization R_P of R at a nonzero prime P is a local ring in which the nonzero fractional ideals are invertible (cf. Exercise 4), hence is a D.V.R. by Proposition 11. Hence (3) implies (2) and so (1), (2) and (3) are equivalent. The equivalence of these with (4) is given by Proposition 10.

Suppose now that (1) is satisfied, and let I be any nonzero proper ideal in R. Since R is Noetherian, I has a minimal primary decomposition $I = Q_1 \cap \cdots \cap Q_n$ as in Theorem 21 of Section 15.2. The associated primes $P_i = \operatorname{rad} Q_i$ for $i = 1, \ldots, n$ are all distinct, and since primes are maximal in R by hypothesis, the associated primes are all pairwise comaximal, and it follows easily that the same is true for the Q_i (Exercise

5). It follows that $Q_1 \cap \cdots \cap Q_n = Q_1 \cdots Q_n$ (Theorem 17 in Section 7.6) so that I is the product of primary ideals. The P-primary ideals of R correspond bijectively with the $P R_P$-primary ideals in the localization R_P (Proposition 42(3) in Section 15.4), and since R_P is a D.V.R. (because (1) implies (2)), it follows from Corollary 6 that if Q is a P-primary ideal in R then $Q = P^m$ for some integer $m \geq 1$. Applying this to Q_i, $i = 1, \ldots, n$ shows that I is the product of powers of prime ideals, which gives the first implication in (5).

Conversely, suppose that all the nonzero proper ideals of R can be written as a product of prime ideals. We first show for any integral domain that a factorization of an ideal into *invertible* prime ideals is unique, i.e., if $P_1 \cdots P_n = \tilde{P}_1 \cdots \tilde{P}_m$ are two factorizations of I into invertible prime ideals then $n = m$ and the two sets of primes $\{P_1, \ldots, P_n\}$ and $\{\tilde{P}_1, \ldots, \tilde{P}_m\}$ are equal. Suppose $\tilde{P}_1$ is a minimal element in the set $\{\tilde{P}_1, \ldots, \tilde{P}_m\}$. Since $P_1 \cdots P_n \subseteq \tilde{P}_1$, the prime ideal $\tilde{P}_1$ contains one of the primes $P_1, \ldots, P_n$, say $P_1 \subseteq \tilde{P}_1$. Similarly P_1 contains $\tilde{P}_i$ for some $i = 1, \ldots, m$. Then $\tilde{P}_i \subseteq P_1 \subseteq \tilde{P}_1$ and by the minimality of $\tilde{P}_1$ it follows that $\tilde{P}_i = P_1 = \tilde{P}_1$, so the factorization becomes $P_1 P_2 \cdots P_n = P_1 \tilde{P}_2 \cdots \tilde{P}_m$. Since P_1 is invertible, multiplying by the inverse ideal shows that $P_2 \cdots P_n = \tilde{P}_2 \cdots \tilde{P}_m$ and an easy induction finishes the proof. In particular, the uniqueness statement in (5) now follows from the first statement in (5) since in a Dedekind domain every fractional ideal, in particular every prime ideal of R, is invertible.

We next show that *invertible* primes in R are maximal. Suppose then that P is an invertible prime ideal in R and take $a \in R$, $a \notin P$. We want to show that $P + aR = R$. By assumption, the two ideals $P + aR$ and $P + a^2 R$ can be written as a product of prime ideals, say $P + aR = P_1 \cdots P_n$ and $P + a^2 R = \tilde{P}_1 \cdots \tilde{P}_m$. Note that $P \subseteq P_i$ for $i = 1, \ldots, n$ and also $P \subseteq \tilde{P}_j$ for $j = 1, \ldots, m$. In the quotient R/P, which is an integral domain, we have the factorization $(\bar{a}) = (P_1/P) \cdots (P_n/P)$, and each P_i/P is a prime ideal in R/P. Since the product is a principal ideal, each P_i/P is also an invertible R/P-ideal (cf. Exercise 2). Similarly, $(\bar{a}^2) = (\tilde{P}_1/P) \cdots (\tilde{P}_m/P)$ is a factorization into a product of invertible prime ideals. Then $(\bar{a})^2 = (P_1/P)^2 \cdots (P_n/P)^2 = (\tilde{P}_1/P) \cdots (\tilde{P}_m/P)$ give two factorizations into a product of invertible prime ideals in the integral domain R/P, so by the uniqueness result in the previous paragraph, $m = 2n$ and $\{P_1/P, P_1/P, \ldots, P_n/P, P_n/P\} = \{\tilde{P}_1/P, \ldots, \tilde{P}_m/P\}$. It follows that the set of primes $\tilde{P}_1, \ldots, \tilde{P}_m$ in R consists of the primes $P_1, \ldots, P_n$, each repeated twice. This shows that $P + a^2 R = (P + aR)^2$. Since $P \subseteq P + a^2 R$ and $(P + aR)^2 \subseteq P^2 + aR$, we have $P \subseteq P^2 + aR$, so every element x in P can be written in the form $x = y + az$ where $y \in P^2$ and $z \in R$. Then $az = x - y \in P$ and since $a \notin P$, we have $z \in P$, which shows that $P \subseteq P^2 + aP$. Clearly $P^2 + aP \subseteq P$ and so $P = P^2 + aP = P(P + aR)$. Since P is assumed invertible, it follows that $R = P + aR$ for any $a \in R - P$, which proves that P is a maximal ideal.

We now show that every nonzero prime ideal is invertible. If P is a nonzero prime ideal, let a be any nonzero element in P. By assumption, $Ra = P_1 \cdots P_n$ can be written as a product of prime ideals, and $P_1, \ldots, P_n$ are invertible since their product is principal (by Exercise 2 again). Since $P_1 \cdots P_n = Ra \subseteq P$, the prime ideal P contains P_i for some $1 \leq i \leq n$. Since P_i is maximal by the previous paragraph, it follows that

$P = P_i$ is invertible.

Finally, since every nonzero proper ideal of R is a product of prime ideals, it follows that every nonzero ideal of R is invertible, and since every fractional ideal of R is of the form $(d^{-1})I$ for some ideal in R, also every fractional ideal of R is invertible. This proves that (5) implies (3), and complete the proof of the theorem.

The following corollary follows immediately from Proposition 14:

Corollary 16. If $\mathcal{O}_K$ is the ring of integers in an algebraic number field K then every nonzero ideal I in $\mathcal{O}_K$ can be written uniquely as the product of powers of distinct prime ideals:

$$I = P_1^{e_1} P_2^{e_2} \cdots P_n^{e_n},$$

where $P_1, \ldots, P_n$ are distinct prime ideals and $e_i \geq 1$ for $i = 1, \ldots, n$.

Remark: The development of Dedekind Domains given here reverses the historical development. As mentioned in Section 9.3, the unique factorization of nonzero *ideals* into a product of prime *ideals* replaces the failure of unique factorization of nonzero *elements* into products of prime *elements* in rings of integers of number fields. This property of rings of integers in Corollary 16 is what led originally to the definition of an ideal, and Dedekind originally defined what we now call Dedekind Domains by property 5 in Theorem 15. It was Noether who observed that they can also be characterized by property (1), which we have taken as the initial definition of a Dedekind Domain.

The unique factorization into prime ideals in Dedekind Domains can be used to explicitly define the valuations v_P on R with respect to which the valuation rings are the localizations R_P in Theorem 15(2) (cf. Exercise 6). We now indicate how unique factorization for ideals can be used to define a divisibility theory for ideals similar to the divisibility of integers in $\mathbb{Z}$.

Definition. If A and B are ideals in the integral domain R then B is said to *divide* A (and A is *divisible by* B) if there is an ideal C in R with $A = BC$.

If B divides A then certainly $A \subseteq B$. If R is a Dedekind Domain, the converse is true: $A \subseteq B$ implies $C = AB^{-1} \subseteq BB^{-1} = R$ so C is an ideal in R with $BC = A$.

We can also define the notion of the *greatest common divisor* (A, B) of two ideals A and B: (A, B) divides both A and B and any ideal dividing both A and B divides (A, B). The second statement in the next proposition shows that this greatest common divisor always exists for integral ideals in a Dedekind Domain and gives a formula for it similar to the formula for the greatest common divisor of two integers.

Proposition 17. Suppose R is a Dedekind Domain and A, B are two nonzero ideals in R, with prime ideal factorizations $A = P_1^{e_1} \cdots P_n^{e_n}$ and $B = P_1^{f_1} \cdots P_n^{f_n}$ (where $e_i, f_i \geq 0$ for $i = 1, \ldots, n$). Then
    (1) $A \subseteq B$ if and only if B divides A (i.e., "to contain is to divide") if and only if $f_i \leq e_i$ for $i = 1, \ldots, n$,

(2) $A + B = (A, B) = P_1^{\min(e_1, f_1)} \cdots P_n^{\min(e_n, f_n)}$, so in particular A and B are relatively prime, $A + B = R$, if and only if they have no prime ideal factors in common.

Proof: We proved the first statement in (1) above. If each $f_i \leq e_i$, then taking $C = P_1^{e_1 - f_1} \cdots P_n^{e_n - f_n} \subseteq R$ shows that B divides A. Conversely, if B divides A, then writing C as a product of prime ideals in $A = BC$ shows that $f_i \leq e_i$ for all i, which proves all of (1). Since $A + B$ is the smallest ideal containing both A and B, (2) now follows from (1).

Proposition 18. *(Chinese Remainder Theorem)* Suppose R is a Dedekind Domain, $P_1, P_2, \ldots, P_n$ are distinct prime ideals in R and $a_i \geq 0$ are integers, $i = 1, \ldots, n$. Then

$$R/P_1^{a_1} \cdots P_n^{a_n} \cong R/P_1^{a_1} \times R/P_2^{a_2} \times \cdots \times R/P_n^{a_n}.$$

Equivalently, for any elements $r_1, r_2, \ldots, r_n \in R$ there exists an element $r \in R$, unique up to an element in $P_1^{a_1} \cdots P_n^{a_n}$, with

$$r \equiv r_1 \bmod P_1^{a_1}, \quad r \equiv r_2 \bmod P_2^{a_2}, \quad \ldots, \quad r \equiv r_n \bmod P_n^{a_n}.$$

Proof: This is immediate from Theorem 17 in Section 7.6 since the previous proposition shows that the $P_i^{a_i}$ are pairwise comaximal ideals.

Corollary 19. Suppose I is an ideal in the Dedekind Domain R. Then
    (1) there is an ideal J of R relatively prime to I such that the product $IJ = (a)$ is a principal ideal,
    (2) if I is nonzero then every ideal in the quotient R/I is principal; equivalently, if I_1 is an ideal of R containing I then $I_1 = I + Rb$ for some $b \in R$, and
    (3) every ideal in R can be generated by two elements; in fact if I is nonzero and $0 \neq a \in I$ then $I = Ra + Rb$ for some $b \in I$.

Proof: Suppose $I = P_1^{e_1} \cdots P_n^{e_n}$ is the prime ideal factorization of I in R. For each $i = 1, \ldots, n$, let r_i be an element of $P_i^{e_i} - P_i^{e_i + 1}$. By the proposition, there is an element $a \in R$ with $a \equiv r_i \bmod P_i^{e_i + 1}$ for all i. Hence $a \in P_i^{e_i} - P_i^{e_i + 1}$ for all i, so the power of P_i in prime ideal factorization of (a) is precisely e_i by (1) of Proposition 17:

$$(a) = P_1^{e_1} \cdots P_n^{e_n} P_{n+1}^{e_{n+1}} \cdots P_m^{e_m}$$

for some prime ideals $P_{n+1}, \ldots, P_m$ distinct from $P_1, \ldots, P_n$. Letting $J = P_{n+1}^{e_{n+1}} \cdots P_m^{e_m}$ gives (1). For (2), by the Chinese Remainder Theorem it suffices to prove that every ideal in R/P^m is principal in the case of a power of a prime ideal P, and this is immediate since $R/P^m \cong R_P/P^m R_P$ and the localization R_P is a P.I.D. Finally, (3) follows from (2) by taking $I = Ra$.

The first statement in Corollary 19 shows that there is an integral ideal J relatively prime to I lying in the inverse class of I in the class group of R. One can even impose additional conditions on J, cf. Exercise 11.

Corollary 20. If R is a Dedekind Domain then R is a P.I.D. (i.e., R has class number 1) if and only if R is a U.F.D.

Proof: Every P.I.D. is a U.F.D., so suppose that R is a U.F.D. and let P be any prime ideal in R. Then $P = Ra + Rb$ for some $a \neq 0$ and b in R by Corollary 19. We have $(a') \subseteq P$ for one of the irreducible factors a' of a since their product is an element in the prime P, and then P divides (a') in R by Proposition 17(1). It follows that $P = (a')$ is principal since (a') is a prime ideal (Proposition 12 in Section 8.3). Since every ideal in R is a product of prime ideals, every ideal of R is principal, i.e., R is a P.I.D.

Corollary 20 shows that the class number of a Dedekind domain R gives a measure of the failure of unique factorization of elements. It is a fundamental result in algebraic number theory that the class number of the ring of integers of an algebraic number field is finite. For general Dedekind Domains, however, the class number need not be finite. In fact, for any abelian group A (finite or infinite) there is a Dedekind Domain whose class group is isomorphic to A.

Modules over Dedekind Domains and the Fundamental Theorem of Finitely Generated Modules

We turn next to the consideration of modules over Dedekind Domains R. Every fractional ideal of R is an R-module and the first statement in the following proposition shows that two fractional ideals of R are isomorphic as R-modules if and only if they represent the same element in the class group of R.

Proposition 21. Let R be a Dedekind Domain with fraction field K.
  (1) Suppose I and J are two fractional ideals of R. Then $I \cong J$ as R-modules if and only if I and J differ by a nonzero principal ideal: $I = (a)J$ for some $0 \neq a \in K$.
  (2) More generally, suppose $I_1, I_2, \ldots, I_n$ and $J_1, J_2, \ldots, J_m$ are nonzero fractional ideals in the fraction field K of the Dedekind Domain R. Then

$$I_1 \oplus I_2 \oplus \cdots \oplus I_n \cong J_1 \oplus J_2 \oplus \cdots \oplus J_m$$

  as R-modules if and only if $n = m$ and the product ideals $I_1 I_2 \cdots I_n$ and $J_1 J_2 \cdots J_n$ differ by a principal ideal:

$$I_1 I_2 \cdots I_n = (a) J_1 J_2 \cdots J_n$$

  for some $0 \neq a \in K$.
  (3) In particular,

$$I_1 \oplus I_2 \oplus \cdots \oplus I_n \cong \underbrace{R \oplus \cdots \oplus R}_{n-1 \text{ factors}} \oplus (I_1 I_2 \cdots I_n)$$

  and $R^n \oplus I \cong R^n \oplus J$ if and only if I and J differ by a principal ideal: $I = (a)J$, $a \in K$.

Proof: Multiplication by $0 \neq a \in K$ gives an R-module isomorphism from J to $(a)J$, so if $I = (a)J$ we have $I \cong J$ as R-modules. For the converse, observe that we may assume $J \neq 0$ and then $I \cong J$ implies $R \cong J^{-1}I$. But this says that $J^{-1}I = aR$ is principal (with generator a given by the image of $1 \in R$), i.e., $I = (a)J$, proving (1).

We next show that for any nonzero fractional ideals I and J that $I \oplus J \cong R \oplus IJ$. Replacing I and J by isomorphic R-modules aI and bJ, if necessary, we may assume that I and J are integral ideals that are relatively prime (cf. Exercise 12), so that $I + J = R$ and $I \cap J = IJ$. It is easy to see that the map from $I \oplus J$ to $I + J = R$ defined by mapping (x, y) to $x + y$ is a surjective R-module homomorphism with kernel $I \cap J = IJ$, so we have an exact sequence

$$0 \longrightarrow IJ \longrightarrow I \oplus J \longrightarrow R \longrightarrow 0$$

of R-modules. This sequence splits since R is free, so $I \oplus J \cong R \oplus IJ$, as claimed.

The first statement in (3) now follows by induction, and combining this statement with (1) shows that if $I_1 \cdots I_n = (a)J_1 \cdots J_n$ for some nonzero $a \in K$ then $I_1 \oplus \cdots \oplus I_n$ is isomorphic to $J_1 \oplus \cdots \oplus J_n$. This proves the "if" statement in (2). It remains to prove the "only if" statement in (2) since the corresponding statement in (3) is a special case. So suppose $I_1 \oplus I_2 \oplus \cdots \oplus I_n \cong J_1 \oplus J_2 \oplus \cdots \oplus J_m$ as R-modules.

Since $I \otimes_R K$ is the localization of the ideal I in K (cf. Proposition 41 in Section 15.4) it follows that $I \otimes_R K \cong K$ for any nonzero fractional ideal I of K. Since tensor products commute with direct sums, $(I_1 \oplus \cdots \oplus I_n) \otimes_R K \cong K^n$ is an n-dimensional vector space over K. Similarly, $J_1 \oplus \cdots \oplus J_m \otimes_R K \cong K^m$, from which it follows that $n = m$.

Note that replacing I_1 by the isomorphic fractional ideal $a_1^{-1}I_1$ for any nonzero element $a_1 \in I_1$ does not effect the validity of the statements in (2). Hence we may assume I_1 contains R, and similarly we may assume that each of the fractional ideals in (2) contains R. Let φ denote the R-module isomorphism from $I_1 \oplus \cdots \oplus I_n$ to $J_1 \oplus \cdots \oplus J_n$. For $i = 1, 2, \ldots, n$ define

$$\varphi((0, \ldots, 0, 1, 0, \ldots, 0)) = (a_{1,i}, a_{2,i}, \ldots, a_{n,i}) \in J_1 \oplus J_2 \oplus \cdots \oplus J_n$$

where $1 \in I_i$ on the left hand side occurs in position i. Since φ is an R-module homomorphism it follows that

$$J_j = a_{j,1}I_1 + a_{j,2}I_2 + \cdots + a_{j,i}I_i + \cdots + a_{j,n}I_n$$

for each $j = 1, 2, \ldots, n$. Taking the product of these ideals for $j = 1, 2, \ldots, n$ it follows that

$$(a_{j_1,1}a_{j_2,2} \cdots a_{j_n,n})I_1 I_2 \cdots I_n \subseteq J_1 J_2 \cdots J_n$$

for any permutation $\{j_1, j_2, \ldots, j_n\}$ of $\{1, 2, \ldots, n\}$. Hence

$$dI_1 I_2 \cdots I_n \subseteq J_1 J_2 \cdots J_n$$

where d is the determinant of the matrix $(a_{i,j})$, since the determinant is the sum of terms $\epsilon(\sigma)a_{1,\sigma(1)} \cdots a_{n,\sigma(n)}$ where $\epsilon(\sigma)$ is the sign of the permutation σ of $\{1, 2, \ldots, n\}$. Similarly, for $j = 1, \ldots, n$, define

$$\varphi^{-1}((0, \ldots, 0, 1, 0, \ldots, 0)) = (b_{1,j}, b_{2,j}, \ldots, b_{n,j}) \in I_1 \oplus I_2 \oplus \cdots \oplus I_n$$

where $1 \in J_j$ on the left hand side occurs in position j. The product of the two matrices $(a_{i,j})$ and $(b_{i,j})$ is just the identity matrix, so $d \neq 0$ and the determinant of the matrix $(b_{i,j})$ is d^{-1}. As above we have

$$d^{-1}J_1 J_2 \cdots J_n \subseteq I_1 I_2 \cdots I_n,$$

which shows that $I_1 I_2 \cdots I_n = (a) J_1 J_2 \cdots J_n$, where $0 \neq a = d^{-1} \in K$, completing the proof of the proposition.

We now consider finitely generated modules over Dedekind Domains and prove a structure theorem for such modules extending the results in Chapter 12 for finitely generated modules over P.I.D.s.

Recall that the *rank* of M is the maximal number of R-linearly independent elements in M, or, equivalently, the dimension of $M \otimes_R K$ as a K-vector space, where K is the fraction field of R (cf. Exercises 1–4, 20 in Section 12.1).

Theorem 22. Suppose M is a finitely generated module over the Dedekind Domain R. Let $n \geq 0$ denote the rank of M and let $\text{Tor}(M)$ be the torsion submodule of M. Then

$$M \cong \underbrace{R \oplus R \oplus \cdots \oplus R \oplus I}_{n \text{ factors}} \oplus \text{Tor}(M)$$

for some ideal I of R, and

$$\text{Tor}(M) \cong R/P_1^{e_1} \times R/P_2^{e_2} \times \cdots \times R/P_s^{e_s}$$

for some $s \geq 0$ and powers $P_i^{e_i}$, $e_1 \geq 1$, of (not necessarily distinct) prime ideals. The ideals $P_i^{e_i}$ for $i = 1, \ldots, s$ are unique and the ideal I is unique up to multiplication by a principal ideal.

Proof: Suppose first that M is a finitely generated torsion free module over R, i.e., $\text{Tor}(M) = 0$. Then the natural R-module homomorphism from M to $M \otimes_R K$ is injective, so we may view M as an R-submodule of the vector space $M \otimes_R K$. If M has rank n over R, then $M \otimes_R K$ is a vector space over K of dimension n. Let $x_1, \ldots, x_n$ be a basis for $M \otimes_R K$ over K and let $m_1, \ldots, m_s$ be R-module generators for M. Each m_i, $i = 1, \ldots, s$ can be written as a K-linear combination of $x_1, \ldots, x_n$. Let $0 \neq d \in R$ be a common denominator for all the coefficients in K of these linear combinations, and set $y_i = x_i/d$, $i = 1, \ldots, n$. Then

$$M \subseteq R y_1 + \cdots + R y_n \subset K x_1 + \cdots + K x_n$$

which shows that M is contained in a *free* R-submodule of rank n and every element m in M can be written uniquely in the form

$$m = a_1 y_1 + \cdots + a_n y_n$$

with $a_1, \ldots, a_n \in R$. The map $\varphi : M \to R$ defined by $\varphi(a_1 y_1 + \cdots + a_n y_n) = a_n$ is an R-module homomorphism, so we have an exact sequence

$$0 \longrightarrow \ker \varphi \longrightarrow M \overset{\varphi}{\longrightarrow} I_1 \longrightarrow 0$$

where I_1 is the image of φ in R, hence is an ideal in R. The submodule $\ker \varphi$ is also a torsion free R-module whose rank is at most $n - 1$ (since it is contained in $R y_1 + \cdots + R y_{n-1}$), and it follows by comparing ranks that I_1 is nonzero and that $\ker \varphi$ has rank precisely $n - 1$. By (4) of Theorem 15, I_1 is a projective R-module, so this sequence splits:

$$M \cong I_1 \oplus (\ker \varphi).$$

By induction on the rank, we see that a finitely generated torsion free R-module is isomorphic to the direct sum of n nonzero ideals of R:

$$M \cong I_1 \oplus I_2 \oplus \cdots \oplus I_n.$$

Since $I_1, \ldots, I_n$ are each projective R-modules, it follows that any finitely generated torsion free R-module is projective.

If now M is any finitely generated R-module, the quotient $M/\mathrm{Tor}(M)$ is finitely generated and torsion free, hence projective by what was just proved. The exact sequence

$$0 \longrightarrow \mathrm{Tor}(M) \longrightarrow M \longrightarrow M/\mathrm{Tor}(M) \longrightarrow 0$$

therefore splits, and so

$$M \cong \mathrm{Tor}(M) \oplus (M/\mathrm{Tor}(M)).$$

By the results in the previous paragraph $M/\mathrm{Tor}(M)$ is isomorphic to a direct sum of n nonzero ideals of R, and by Proposition 21 we obtain

$$M \cong \underbrace{R \oplus R \oplus \cdots \oplus R \oplus I}_{n \text{ factors}} \oplus \mathrm{Tor}(M)$$

for some ideal I of R. The uniqueness statement regarding the ideal I is also immediate from the uniqueness statement in Proposition 21(3).

It remains to prove the statements regarding the torsion submodule $\mathrm{Tor}(M)$. Suppose then that N is a finitely generated torsion R-module. Let $I = \mathrm{Ann}(N)$ be the annihilator of N in R and suppose $I = P_1^{e_1} \cdots P_t^{e_t}$ is the prime ideal factorization of I in R, where $P_1, \ldots, P_t$ are distinct prime ideals. Then N is a module over R/I, and

$$R/I \cong R/P_1^{e_1} \times R/P_2^{e_2} \times \cdots \times R/P_t^{e_t}.$$

It follows that

$$N \cong (N/P_1^{e_1} N) \times (N/P_2^{e_2} N) \times \cdots \times (N/P_t^{e_t} N)$$

as R-modules. Each $N/P^e N$ is a finitely generated module over $R/P^e \cong R_P/P^e R_P$ where R_P is the localization of R at the prime P, i.e., is a finitely generated module over R_P that is annihilated by $P^e R_P$. Since R is a Dedekind Domain, each R_P is a P.I.D. (even a D.V.R.), so we may apply the Fundamental Theorem for Finitely Generated Modules over a P.I.D. to see that each $N/P^e N$ is isomorphic as an R_P-module to a direct sum of finitely many modules of the form $R_P/P^f R_P$ where $f \le e$. It follows that each $N/P^e N$ is isomorphic as an R-module to a direct sum of finitely many modules of the form $R/P^f R$ where $f \le e$. This proves that N is isomorphic to the direct sum of finitely many modules of the form $R/P_i^{f_i}$ for various prime ideals P_i. Hence $\mathrm{Tor}(M)$ can be decomposed into a direct sum as in the statement in the theorem.

Finally, it remains to prove that the ideals $P_i^{e_i}$ for $i = 1, \ldots, s$ in the decomposition of $\mathrm{Tor}(M)$ are unique. This is similar to the uniqueness argument in the proof of Theorem 10 in Section 12.1 (cf. also Exercises 11–12 in Section 12.1): for any prime ideal P of R, the quotient $P^{i-1} M / P^i M$ is a vector space over the field R/P and the difference $\dim_{R/P} P^{i-1} M / P^i M - \dim_{R/P} P^i M / P^{i+1} M$ is the number of direct summands of M isomorphic to R/P^i, hence is uniquely determined by M. This concludes the proof of the theorem.

If M is a finitely generated module over the Dedekind Domain R as in Theorem 22, then the isomorphism type of M as an R-module is determined by the *rank n*, the prime powers $P_i^{e_i}$ for $i = 1, \ldots, s$ (called the *elementary divisors* of M, and the class of the ideal I in the class group of R (called the *Steinitz class* of M). Note that a P.I.D. is the same as a Dedekind Domain whose class number is 1, in which case every nonzero ideal I of R is isomorphic as an R-module simply to R. In this case, Theorem 22 reduces to the elementary divisor form of the structure theorem for finitely generated modules over P.I.D.s in Chapter 12. There is also an invariant factor version of the description of the torsion R-modules in Theorem 22 (cf. Exercise 14).

The next result extends the characterization of finitely generated projective modules over P.I.D.s (Exercise 21 in Section 12.1) to Dedekind Domains.

Corollary 23. A finitely generated module over a Dedekind Domain is projective if and only if it is torsion free.

Proof: We showed that a finitely generated torsion free R-module is projective in the proof of Theorem 22, so by the decomposition of M in Theorem 22, M is projective if and only if $\text{Tor}(M)$ is projective (cf. Exercise 3 in Section 10.5). To complete the proof it suffices to show that no nonzero torsion R-module is projective, which is left as an exercise (cf. Exercise 15).

EXERCISES

1. If R is an integral domain, show that every fractional ideal of R is invertible if and only if every integral ideal of R is invertible.

2. Suppose R is an integral domain with fraction field K and $A_1, A_2, \ldots, A_n$ are fractional ideals of R whose product is a nonzero principal fractional ideal: $A_1 A_2 \cdots A_n = Rx$ for some $0 \neq x \in K$. For each $i = 1, \ldots, n$ prove that A_i is an invertible fractional ideal with inverse $(x^{-1}) A_1 \cdots A_{i-1} A_{i+1} \cdots A_n$.

3. Suppose R is an integral domain with fraction field K and P is a nonzero prime ideal in R. Show that the fractional ideals of R_P in K are the R_P-modules of the form $A R_P$ where A is a fractional ideal of R.

4. Suppose R is an integral domain with fraction field K and A is a fractional ideal of R in K. Let $A' = \{x \in K \mid xA \subseteq R\}$ as in Proposition 9.
   (a) For any prime ideal P in R prove that the localization $(A')_P$ of A' at P is a fractional ideal of R_P in K.
   (b) If A is a finitely generated R-module, prove that $(A')_P = (A_P)'$ where $(A_P)'$ is the fractional R_P ideal $\{x \in K \mid xA_P \subseteq R_P\}$ corresponding to the localization A_P.

5. If Q_1 is a P_1-primary ideal and Q_2 is a P_2-primary ideal where P_1 and P_2 are comaximal ideals in a Noetherian ring R, prove that Q_1 and Q_2 are also comaximal. [Use Proposition 14 in Section 15.2.]

6. Suppose R is a Dedekind Domain with fraction field K.
   (a) Prove that every nonzero fractional ideal of R in K can be written uniquely as the product of distinct prime powers $P_1^{a_1} \cdots P_n^{a_n}$ where the a_i are nonzero integers, possibly negative.

(b) If $0 \neq x \in K$, let $P^{\nu_P(x)}$ be the power of the prime P in the factorization of the principal ideal (x) as in (a) (where $\nu_P(x) = 0$ if P is not one of the primes occurring). Prove ν_P is a valuation on K with valuation ring R_P, the localization of R at P.

7. Suppose R is a Noetherian integral domain that is not a field. Prove that R is a Dedekind Domain if and only if for every maximal ideal M of R there are no ideals I of R with $M^2 \subset I \subset M$. [Use Exercise 12 in Section 1 and Theorems 7 and 15.]

8. Suppose R is a Noetherian integral domain with Krull dimension 1. Prove that every nonzero ideal I in R can be written uniquely as a product of primary ideals whose radicals are all distinct. [Cf. the proof of Theorem 15. Use the uniqueness of the primary components belonging to the isolated primes in a minimal primary decomposition (Theorem 21 in Section 15.2).]

9. Suppose R is an integral domain. Prove that R_P is a D.V.R. for every nonzero prime ideal P if and only if R_M is a D.V.R. for every nonzero maximal ideal.

10. Suppose R is a Noetherian integral domain that is not a field. Prove that R is a Dedekind Domain if and only if nonzero primes M are maximal and every M-primary ideal is a power of M.

11. If I and J are nonzero ideals in the Dedekind Domain R show there exists an integral ideal I_1 in R that is relatively prime to both I and J such that $I_1 I$ is a principal ideal in R.

12. If I and J are nonzero fractional ideals for the Dedekind Domain R prove there are elements $\alpha, \beta \in K$ such that αI and βJ are nonzero integral ideals in R are relatively prime.

13. Suppose I and J are nonzero ideals in the Dedekind Domain R. Prove that there is an ideal $I_1 \cong I$ that is relatively prime to J. [Use Corollary 19 to find an ideal I_2 with $I_2 I = (a)$ and $(I_2, J) = R$. If $I_2 = P_1^{e_1} \cdots P_n^{e_n}$, choose $b \in R$ with $b \in P_i^{e_i} - P_i^{e_i+1}$ and $b \equiv 1 \bmod P$ for every prime P dividing J. Show that $(b) = I_2 I_1$ for some ideal I_1 and consider $(a)I_1$ to prove that $I_1 \cong I$.]

14. Prove that every finitely generated torsion module over a Dedekind Domain R is isomorphic to a direct sum $R/I_1 \oplus R/I_2 \oplus \cdots \oplus R/I_n$ with unique nonzero ideals $I_1, \ldots, I_n$ of R satisfying $I_1 \subseteq I_2 \subseteq \cdots \subseteq I_n$ (called the *invariant factors* of M). [cf. Section 12.1.]

15. If P is a nonzero prime ideal in the Dedekind Domain R prove that R/P^n is not a projective R-module for any $n \geq 1$. [Consider the exact sequence $0 \to P^n/P^{n+1} \to R/P^{n+1} \to R/P^n \to 0$.] Conclude that if $M \neq 0$ is a finitely generated torsion R-module then M is not projective. [cf. Exercise 3, Section 10.5.]

16. Prove that the class number of the Dedekind Domain R is 1 if and only if every finitely generated projective R-module is free.

17. Suppose R is a Dedekind Domain.
    (a) Show that $I \sim J$ if and only if $I \cong J$ as R-modules defines an equivalence relation on the set of nonzero fractional ideals of R. Let $C(R)$ be the corresponding set of R-module isomorphism classes and let $[I] \in C(R)$ denote the equivalence class containing the fractional ideal I of R.
    (b) Show that the multiplication $[I][J] = [I \oplus J]$ gives a well defined binary operation with respect to which $C(R)$ is an abelian group with identity $1 = [R]$.
    (c) Prove that the abelian group $C(R)$ in (b) is isomorphic to the class group of R.

18. If R is a Dedekind Domain and I is any nonzero ideal, prove that R/I contains only finitely many ideals. In particular, show that R/I is an Artinian ring.

19. Suppose I is a nonzero fractional ideal in the Dedekind Domain R. Explicitly exhibit I as a direct summand of a free R-module to show that I is projective. [Consider $I \oplus I^{-1}$

and use Proposition 21.]

20. Suppose I and J are two nonzero fractional ideals in the Dedekind Domain R and that $I^n = J^n$ for some $n \neq 0$. Prove that $I = J$.

21. Suppose K is an algebraic number field and $\mathcal{O}_K$ is the ring of integers in K. If P is a nonzero prime ideal in $\mathcal{O}_K$ prove that $P = (p, \pi)$ for some prime $p \in \mathbb{Z}$ and algebraic integer $\pi \in \mathcal{O}_K$.

22. Suppose $K = \mathbb{Q}(\sqrt{D})$ is a quadratic extension of $\mathbb{Q}$ where D is a squarefree integer and $\mathcal{O}_K$ is the ring of integers in K.
  (a) Prove that $|\mathcal{O}_K/(p)| = p^2$. [Observe that $\mathcal{O}_K \cong \mathbb{Z}^2$ as an abelian group.]
  (b) Use Corollary 16 to show that there are 3 possibilities for the prime ideal factorization of (p) in $\mathcal{O}_K$:
  (i) $(p) = P$ is a prime ideal with $|\mathcal{O}_K/P| = p^2$,
  (ii) $(p) = P_1 P_2$ with distinct prime ideals P_1, P_2 and $|\mathcal{O}_K/P_1| = |\mathcal{O}_K/P_2| = p$,
  (iii) $(p) = P^2$ for some prime ideal P with $|\mathcal{O}_K/P| = p$.

(In cases (i), (ii), and (iii) the prime p is said to be *inert*, *split*, or *ramified* in $\mathcal{O}_K$, respectively. The set of ramified primes is finite: the primes p dividing D if $D \equiv 1, 2 \bmod 4$; $p = 2$ and the primes p dividing D if $D \equiv 3 \bmod 4$. Cf. Exercise 31 in Section 15.5.)
  (c) Determine the prime ideal factorizations of the primes $p = 2, 3, 5, 7, 11$ in the ring of integers $\mathcal{O}_K = \mathbb{Z}[\sqrt{-5}]$ of $K = \mathbb{Q}(\sqrt{-5})$.

23. Let $\mathcal{O}$ be the ring of integers in the algebraic closure $\overline{\mathbb{Q}}$ of $\mathbb{Q}$.
  (a) Show that the infinite sequence of ideals in $\mathcal{O}$ $(2) \subseteq (\sqrt{2}) \subseteq (\sqrt[4]{2}) \subseteq (\sqrt[8]{2}) \subseteq \cdots$ is strictly increasing, and so $\mathcal{O}$ is not Noetherian.
  (b) Show that $\mathcal{O}$ has Krull dimension 1. [Use Theorem 26 in Section 15.3.]
  (c) Let K be a number field and let I be any ideal in $\mathcal{O}_K$. Show that there is some finite extension L of K such that I becomes principal when extended to $\mathcal{O}_L$, i.e., the ideal $I\mathcal{O}_L$ is principal (where L depends on I)—you may use the theorem that the class group of K is a finite group. [cf. Exercise 20.]
  (d) Prove that $\mathcal{O}$ is a Bezout Domain (cf. Section 8.1).

24. Suppose F and K are algebraic number fields with $\mathbb{Q} \subseteq F \subseteq K$, with rings of integers $\mathcal{O}_F$ and $\mathcal{O}_K$, respectively. Since $\mathcal{O}_F \subseteq \mathcal{O}_K$, the ring $\mathcal{O}_K$ is naturally a module over $\mathcal{O}_F$.
  (a) Prove $\mathcal{O}_K$ is a torsion free $\mathcal{O}_F$-module of rank $n = [K : F]$. [Compute ranks over $\mathbb{Z}$.] If $\mathcal{O}_K$ is *free* over $\mathcal{O}_F$ then $\mathcal{O}_K$ is said to have a *relative integral basis* over $\mathcal{O}_F$.
  (b) Prove that if F has class number 1 then $\mathcal{O}_K$ has a relative integral basis over $\mathcal{O}_F$.

If $K = \mathbb{Q}(\sqrt{-5}, \sqrt{2})$ then the ring of integers $\mathcal{O}_K$ is given by

$$\mathcal{O}_K = \mathbb{Z} + \mathbb{Z}\sqrt{-5} + \mathbb{Z}\sqrt{-10} + \mathbb{Z}\omega \qquad \text{where } \omega = (\sqrt{-10} + \sqrt{2})/2.$$

  (c) If $F_1 = \mathbb{Q}(\sqrt{2})$ prove that $\mathcal{O}_K$ has a relative integral basis over $\mathcal{O}_{F_1}$ and find an explicit basis $\{\alpha, \beta\}$: $\mathcal{O}_K = \mathcal{O}_{F_1} \cdot \alpha + \mathcal{O}_{F_1} \cdot \beta$.
  (d) If $F_2 = \mathbb{Q}(\sqrt{-5})$, show that $P_3 = (3, 1 - \sqrt{-5}) = (3, 5 + \sqrt{-5})$ is a prime ideal of $\mathcal{O}_{F_2}$ that is not principal and that $\mathcal{O}_K = \mathcal{O}_{F_2} \cdot 1 + (1/3)P_3 \cdot \omega$. [Check that $\sqrt{-10} = (5 + \sqrt{-5})\omega/3$.] Conclude that the Steinitz class of $\mathcal{O}_K$ as a module over $\mathcal{O}_{F_2}$ is the nontrivial class of P_3 in the class group of $\mathcal{O}_{F_2}$ and so there is no relative integral basis of $\mathcal{O}_K$ over $\mathcal{O}_{F_2}$.
  (e) Determine whether $\mathcal{O}_K$ has a relative integral basis over the ring of integers of the remaining quadratic subfield $F_3 = \mathbb{Q}(\sqrt{-10})$ of K.

25. Suppose C is a nonsingular irreducible affine curve over an algebraically closed field k. Prove that the coordinate ring $k[C]$ is a Dedekind Domain.

CHAPTER 17

Introduction to Homological Algebra and Group Cohomology

Let R be a ring with 1. In Section 10.5 we saw that a short exact sequence

$$0 \longrightarrow L \xrightarrow{\psi} M \xrightarrow{\varphi} N \longrightarrow 0 \tag{17.1}$$

of R-modules gives rise to an exact sequence of abelian groups

$$0 \longrightarrow \operatorname{Hom}_R(N, D) \xrightarrow{\varphi'} \operatorname{Hom}_R(M, D) \xrightarrow{\psi'} \operatorname{Hom}_R(L, D) \tag{17.2}$$

for any R-module D and that the homomorphism ψ' is in general not surjective so this sequence cannot always be extended to a short exact sequence. Equivalently, homomorphisms from L to D cannot in general be lifted to homomorphisms from M into D. In this chapter we introduce some of the techniques of "homological algebra," which provide a method of extending some exact sequences in a natural way. For the situation above one obtains an infinite exact sequence involving the "cohomology groups" $\operatorname{Ext}_R^n(\_\_, D)$ (cf. Theorem 8), and these groups provide a measure of the set of homomorphisms from L into D that cannot be extended to M. We then consider the analogous questions for the other two functors considered in Section 10.5, namely taking homomorphisms *from D* into the terms of the sequence (1) and tensoring the sequence (1) with D.

In the subsequent sections we concentrate on an important special case of this general type of homological construction—the "cohomology of finite groups." We make explicit the computations in this case and indicate some applications of these techniques to establish some new results in group theory. In this sense, Sections 2–4 may be considered as an explicit "example" illustrating some uses of the general theory in Section 1.

Cohomology and homology groups occur in many areas of mathematics. The formal notions of homology and cohomology groups and the general area of homological algebra arose from algebraic topology around the middle of the 20th century in the study of the relation between the higher homotopy groups and the fundamental group of a topological space, although the study of certain specific cohomology groups, such as Schur's work on group extensions (described in Section 4), predates this by half a century. As with much of algebra, the ideas common to a number of different areas were abstracted into general theories. Much of the language of homology and cohomology reflects its topological origins: homology groups, chains, cycles, boundaries, etc.

17.1 INTRODUCTION TO HOMOLOGICAL ALGEBRA—EXT AND TOR

In this section we describe some general terminology and results in homological algebra leading to the so called Long Exact Sequence in Cohomology. We then define certain (cohomology) groups associated to the sequence (2) and apply the general homological results to obtain a long exact sequence extending this sequence at the right end. We then indicate the corresponding development for sequences obtained by taking homomorphisms from D to the terms in (1) or by tensoring the terms with D.

We begin with a generalization of the notion of an exact sequence, namely a sequence of abelian group homomorphisms where successive maps compose to zero (i.e., the image of one map is contained in the kernel of the next):

Definition. Let $\mathcal{C}$ be a sequence of abelian group homomorphisms:

$$0 \longrightarrow C^0 \xrightarrow{d_1} C^1 \longrightarrow \cdots \longrightarrow C^{n-1} \xrightarrow{d_n} C^n \xrightarrow{d_{n+1}} \cdots. \tag{17.3}$$

(1) The sequence $\mathcal{C}$ is called a *cochain complex* if the composition of any two successive maps is zero: $d_{n+1} \circ d_n = 0$ for all n.
(2) If $\mathcal{C}$ is a cochain complex, its nth *cohomology group* is the quotient group $\ker d_{n+1} / \operatorname{image} d_n$, and is denoted by $H^n(\mathcal{C})$.

There is a completely analogous "dual" version in which the homomorphisms are between groups in *decreasing* order, in which case the sequence corresponding to (3) is written $\cdots \xrightarrow{d_{n+1}} C_n \xrightarrow{d_n} \cdots \xrightarrow{d_1} C_0 \to 0$. Then if the composition of any two successive homomorphisms is zero, the complex is called a *chain complex*, and its *homology groups* are defined as $H_n(\mathcal{C}) = \ker d_n / \operatorname{image} d_{n+1}$. For chain complexes the notation is often chosen so that the indices appear as subscripts and are decreasing, whereas for cochain complexes the indices are superscripts and are increasing. We shall instead use a uniform notation for the maps on both, since it will be clear from the context whether we are dealing with a chain or a cochain complex.

Chain complexes were the first to arise in topological settings, with cochain complexes soon following. With our applications in Section 2 in mind, we shall concentrate on cochains and cohomology, although all of the general results in this section have similar statements for chains and homology. We shall also be interested in the situation where each C^n is an R-module and the homomorphisms d_n are R-module homomorphisms (referred to simply as a *complex of R-modules*), in which case the groups $H^n(\mathcal{C})$ are also R-modules.

Note that if $\mathcal{C}$ is a cochain (respectively, chain) complex then $\mathcal{C}$ is an exact sequence if and only if all its cohomology (respectively, homology) groups are zero. Thus the nth cohomology (respectively, homology) group measures the failure of exactness of a complex at the nth stage.

Definition. Let $\mathcal{A} = \{A^n\}$ and $\mathcal{B} = \{B^n\}$ be cochain complexes. A *homomorphism of complexes* $\alpha : \mathcal{A} \to \mathcal{B}$ is a set of homomorphisms $\alpha_n : A^n \to B^n$ such that for every n the following diagram commutes:

$$\begin{array}{ccc}
\cdots \longrightarrow A^n \longrightarrow A^{n+1} \longrightarrow \cdots \\
\quad\quad\quad \downarrow \alpha_n \qquad\qquad \downarrow \alpha_{n+1} \\
\cdots \longrightarrow B^n \longrightarrow B^{n+1} \longrightarrow \cdots
\end{array} \qquad (17.4)$$

Proposition 1. A homomorphism $\alpha : \mathcal{A} \to \mathcal{B}$ of cochain complexes induces group homomorphisms from $H^n(\mathcal{A})$ to $H^n(\mathcal{B})$ for $n \geq 0$ on their respective cohomology groups.

Proof: It is an easy exercise to show that the commutativity of (4) implies that the images and kernels at each stage of the maps in the first row are mapped to the corresponding images and kernels for the maps in the second row, thus giving a well defined map on the respective quotient (cohomology) groups.

Definition. Let $\mathcal{A} = \{A^n\}$, $\mathcal{B} = \{B^n\}$ and $\mathcal{C} = \{C^n\}$ be cochain complexes. A *short exact sequence* of complexes $0 \to \mathcal{A} \overset{\alpha}{\to} \mathcal{B} \overset{\beta}{\to} \mathcal{C} \to 0$ is a sequence of homomorphisms of complexes such that $0 \to A^n \overset{\alpha_n}{\to} B^n \overset{\beta_n}{\to} C^n \to 0$ is short exact for every n.

One of the main features of cochain complexes is that they lead to long exact sequences in cohomology, which is our first main result:

Theorem 2. *(The Long Exact Sequence in Cohomology)* Let $0 \to \mathcal{A} \overset{\alpha}{\to} \mathcal{B} \overset{\beta}{\to} \mathcal{C} \to 0$ be a short exact sequence of cochain complexes. Then there is a long exact sequence of cohomology groups:

$$0 \to H^0(\mathcal{A}) \to H^0(\mathcal{B}) \to H^0(\mathcal{C}) \overset{\delta_0}{\to} H^1(\mathcal{A})$$
$$\to H^1(\mathcal{B}) \to H^1(\mathcal{C}) \overset{\delta_1}{\to} H^2(\mathcal{A}) \to \cdots \qquad (17.5)$$

where the maps between cohomology groups at each level are those in Proposition 1. The maps δ_n are called *connecting homomorphisms*.

Proof: The details of this proof are somewhat lengthy. For each n the verification that the sequence $H^n(\mathcal{A}) \to H^n(\mathcal{B}) \to H^n(\mathcal{C})$ is exact is a straightforward check of the definition of exactness of each map, similar to the proof of Theorem 33 in Section 10.5. The construction of a connecting homomorphism δ_n is outlined in Exercise 2. Some work is then needed to show that δ_n is a homomorphism, and that the sequence is exact at δ_n.

One immediate consequence of the existence of the long exact sequence in Theorem 2 is the fact that if any two of the cochain complexes $\mathcal{A}$, $\mathcal{B}$, $\mathcal{C}$ are exact, then so is the third (cf. Exercise 6).

Homomorphisms and the Groups $Ext_R^n(A, B)$

To apply Theorem 2 to analyze the sequence (2), we try to produce a cochain complex whose first few cohomology groups in the long exact sequence (5) agree with the terms in (2). To do this we introduce the notion of a "resolution" of an R-module:

Definition. Let A be any R-module. A *projective resolution* of A is an exact sequence

$$\cdots \longrightarrow P_n \xrightarrow{d_n} P_{n-1} \longrightarrow \cdots \xrightarrow{d_1} P_0 \xrightarrow{\epsilon} A \longrightarrow 0 \qquad (17.6)$$

such that each P_i is a projective R-module.

Every R-module has a projective resolution: Let P_0 be any free (hence projective) R-module on a set of generators of A and define an R-module homomorphism ϵ from P_0 onto A by Theorem 6 in Chapter 10. This begins the resolution $\epsilon : P_0 \to A \to 0$. The surjectivity of ϵ ensures that this sequence is exact. Next let $K_0 = \ker \epsilon$ and let P_1 be any free module mapping onto the submodule K_0 of P_0; this gives the second stage $P_1 \to P_0 \to A$ which, by construction, is also exact. We can continue this way, taking at the nth stage a free R-module P_{n+1} that maps surjectively onto the submodule $\ker d_n$ of P_n, obtaining in fact a *free* resolution of A.

One of the reasons that *projective* modules are used in the resolution of A is that this makes it possible to lift various maps (cf. the proof of Proposition 4 following, for instance).

In general a projective resolution is infinite in length, but if A is itself projective, then it has a very simple projective resolution of finite length, namely $0 \to A \xrightarrow{1} A \to 0$ given by the identity map from A to itself.

Given the projective resolution (6), we may form a related sequence by taking homomorphisms of each of the terms into D, keeping in mind that this reverses the direction of the homomorphisms. This yields the sequence

$$0 \longrightarrow \text{Hom}_R(A, D) \xrightarrow{\epsilon} \text{Hom}_R(P_0, D) \xrightarrow{d_1} \text{Hom}_R(P_1, D) \xrightarrow{d_2} \cdots$$

$$\cdots \xrightarrow{d_{n-1}} \text{Hom}_R(P_{n-1}, D) \xrightarrow{d_n} \text{Hom}_R(P_n, D) \xrightarrow{d_{n+1}} \cdots \qquad (17.7)$$

where to simplify notation we have denoted the induced maps from $\text{Hom}_R(P_{n-1}, D)$ to $\text{Hom}_R(P_n, D)$ for $n \geq 1$ again by d_n and similarly for the map induced by ϵ (cf. Section 10.5). This sequence is not necessarily exact, however it *is* a cochain complex (this is part of the proof of Theorem 33 in Section 10.5). The corresponding cohomology groups have a special name.

Definition. Let A and D be a R-modules. For any projective resolution of A as in (6) let $d_n : \text{Hom}_R(P_{n-1}, D) \to \text{Hom}_R(P_n, D)$ for all $n \geq 1$ as in (7). Define

$$\text{Ext}_R^n(A, D) = \ker d_{n+1} / \text{image } d_n$$

where $\text{Ext}_R^0(A, D) = \ker d_1$. The group $\text{Ext}_R^n(A, D)$ is called the nth *cohomology group derived from the functor* $\text{Hom}_R(\_, D)$. When $R = \mathbb{Z}$ the group $\text{Ext}_{\mathbb{Z}}^n(A, D)$ is also denoted simply $\text{Ext}^n(A, D)$.

Note that the groups $\mathrm{Ext}_R^n(A, D)$ are also the cohomology groups of the cochain complex obtained from (7) by replacing the term $\mathrm{Hom}_R(A, D)$ with zero (which does not effect the cochain property), i.e., they are the cohomology groups of the cochain complex $0 \to \mathrm{Hom}_R(P_0, D) \to \cdots$.

We shall show below that these cohomology groups do not depend on the choice of projective resolution of A. Before doing so we identify the 0th cohomology group and give some examples.

Proposition 3. For any R-module A we have $\mathrm{Ext}_R^0(A, D) \cong \mathrm{Hom}_R(A, D)$.

Proof: Since the sequence $P_1 \xrightarrow{d_1} P_0 \xrightarrow{\epsilon} A \to 0$ is exact, it follows that the corresponding sequence $0 \to \mathrm{Hom}_R(A, D) \xrightarrow{\epsilon} \mathrm{Hom}_R(P_0, D) \xrightarrow{d_1} \mathrm{Hom}_R(P_1, D)$ is also exact by Theorem 33 in Section 10.5 (noting the first comment in the proof). Hence $\mathrm{Ext}_R^0(A, D) = \ker d_1 = \mathrm{image}\, \epsilon \cong \mathrm{Hom}_R(A, D)$, as claimed.

Examples

(1) Let $R = \mathbb{Z}$ and let $A = \mathbb{Z}/m\mathbb{Z}$ for some $m \geq 2$. By the proposition we have $\mathrm{Ext}_{\mathbb{Z}}^0(\mathbb{Z}/m\mathbb{Z}, D) \cong \mathrm{Hom}_{\mathbb{Z}}(\mathbb{Z}/m\mathbb{Z}, D)$, and it follows that $\mathrm{Ext}_{\mathbb{Z}}^0(\mathbb{Z}/m\mathbb{Z}, D) \cong {}_mD$, where ${}_mD = \{d \in D \mid md = 0\}$ are the elements of D that have order dividing m. For the higher cohomology groups, we use the simple projective resolution

$$0 \longrightarrow \mathbb{Z} \xrightarrow{m} \mathbb{Z} \longrightarrow \mathbb{Z}/m\mathbb{Z} \longrightarrow 0$$

for A given by multiplication by m on $\mathbb{Z}$. Taking homomorphisms into a fixed $\mathbb{Z}$-module D gives the cochain complex

$$0 \longrightarrow \mathrm{Hom}_{\mathbb{Z}}(\mathbb{Z}/m\mathbb{Z}, D) \longrightarrow \mathrm{Hom}_{\mathbb{Z}}(\mathbb{Z}, D) \xrightarrow{m} \mathrm{Hom}_{\mathbb{Z}}(\mathbb{Z}, D) \longrightarrow 0 \longrightarrow \cdots.$$

We have $D \cong \mathrm{Hom}_{\mathbb{Z}}(\mathbb{Z}, D)$ (cf. Example 4 following Corollary 32 in Section 10.5) and under this isomorphism we have $\mathrm{Ext}_{\mathbb{Z}}^1(\mathbb{Z}/m\mathbb{Z}, D) \cong D/mD$ for any abelian group D. It follows immediately from the definition and the cochain complex above that $\mathrm{Ext}_{\mathbb{Z}}^n(\mathbb{Z}/m\mathbb{Z}, D) = 0$ for all $n \geq 2$ and any abelian group D, which we summarize as

$$\mathrm{Ext}_{\mathbb{Z}}^0(\mathbb{Z}/m\mathbb{Z}, D) \cong {}_mD$$
$$\mathrm{Ext}_{\mathbb{Z}}^1(\mathbb{Z}/m\mathbb{Z}, D) \cong D/mD$$
$$\mathrm{Ext}_{\mathbb{Z}}^n(\mathbb{Z}/m\mathbb{Z}, D) = 0, \quad \text{for all } n \geq 2.$$

(2) The same abelian groups may be modules over several different rings R and the Ext_R cohomology groups depend on R. For example, suppose $R = \mathbb{Z}/m\mathbb{Z}$ for some integer $m \geq 1$. An R-module D is the same as an abelian group D with exponent dividing m, i.e., $mD = 0$. In particular, for any divisor d of m, the group $\mathbb{Z}/d\mathbb{Z}$ is an R-module, and

$$\cdots \xrightarrow{m/d} \mathbb{Z}/m\mathbb{Z} \xrightarrow{d} \mathbb{Z}/m\mathbb{Z} \xrightarrow{m/d} \mathbb{Z}/m\mathbb{Z} \xrightarrow{d} \mathbb{Z}/m\mathbb{Z} \longrightarrow \mathbb{Z}/d\mathbb{Z} \longrightarrow 0$$

is a projective (in fact, free) resolution of $\mathbb{Z}/d\mathbb{Z}$ as a $\mathbb{Z}/m\mathbb{Z}$-module, where the final map is the natural projection mapping $x \bmod m$ to $x \bmod d$. Taking homomorphisms into the $\mathbb{Z}/m\mathbb{Z}$-module D, using the isomorphism $\mathrm{Hom}_{\mathbb{Z}/m\mathbb{Z}}(\mathbb{Z}/m\mathbb{Z}, D) \cong D$, and removing the first term gives the cochain complex

$$0 \longrightarrow D \xrightarrow{d} D \xrightarrow{m/d} D \xrightarrow{d} D \xrightarrow{m/d} \cdots.$$

Hence

$$\operatorname{Ext}^0_{\mathbb{Z}/m\mathbb{Z}}(\mathbb{Z}/d\mathbb{Z}, D) \cong {}_d D,$$

$$\operatorname{Ext}^n_{\mathbb{Z}/m\mathbb{Z}}(\mathbb{Z}/d\mathbb{Z}, D) \cong {}_{(m/d)} D/dD, \quad n \text{ odd}, n \geq 1,$$

$$\operatorname{Ext}^n_{\mathbb{Z}/m\mathbb{Z}}(\mathbb{Z}/d\mathbb{Z}, D) \cong {}_d D/(m/d)D, \quad n \text{ even}, n \geq 2,$$

where ${}_k D = \{d \in D \mid kd = 0\}$ denotes the set of elements of D killed by k. In particular, $\operatorname{Ext}^n_{\mathbb{Z}/p^2\mathbb{Z}}(\mathbb{Z}/p\mathbb{Z}, \mathbb{Z}/p\mathbb{Z}) \cong \mathbb{Z}/p\mathbb{Z}$ for all $n \geq 0$, whereas, for example, $\operatorname{Ext}^n_{\mathbb{Z}}(\mathbb{Z}/p\mathbb{Z}, \mathbb{Z}/p\mathbb{Z}) = 0$ for all $n \geq 2$.

In order to show that the cohomology groups $\operatorname{Ext}^n_R(A, D)$ are independent of the choice of projective resolution of A we shall need to be able to "compare" resolutions. The next proposition shows that an R-module homomorphism from A to B lifts to a homomorphism from a projective resolution of A to a projective resolution of B — this lifting property is one instance where the projectivity of the modules in the resolution is important.

Proposition 4. Let $f : A \to A'$ be any homomorphism of R-modules and take projective resolutions of A and A', respectively. Then for each $n \geq 0$ there is a lift f_n of f such that the following diagram commutes:

$$
\begin{array}{ccccccccc}
\cdots & \xrightarrow{d_2} & P_1 & \xrightarrow{d_1} & P_0 & \xrightarrow{\epsilon} & A & \longrightarrow & 0 \\
& & \downarrow{f_1} & & \downarrow{f_0} & & \downarrow{f} & & \\
\cdots & \xrightarrow{d'_2} & P'_1 & \xrightarrow{d'_1} & P'_0 & \xrightarrow{\epsilon'} & A' & \longrightarrow & 0
\end{array}
\tag{17.8}
$$

where the rows are the projective resolutions of A and A', respectively.

Proof: Given the two rows and map f in (8), then since P_0 is projective we may lift the map $f\epsilon : P_0 \to A'$ to a map $f_0 : P_0 \to P'_0$ in such a way that $\epsilon' f_0 = f\epsilon$ (Proposition 30(2) in Section 10.5). This gives the first lift of f. Proceeding inductively in this fashion, assume f_n has been defined to make the diagram commutative to that point. Thus image $f_n d_{n+1} \subseteq \ker d'_n$. The projectivity of P_{n+1} implies that we may lift the map $f_n d_{n+1} : P_{n+1} \to P'_n$ to a map $f_{n+1} : P_{n+1} \to P'_{n+1}$ to make the diagram commute at the next stage. This completes the proof.

The commutative diagram in Proposition 4 implies that the induced diagram

$$
\begin{array}{ccccccccc}
0 & \longrightarrow & \operatorname{Hom}_R(A, D) & \longrightarrow & \operatorname{Hom}_R(P_0, D) & \longrightarrow & \operatorname{Hom}_R(P_1, D) & \longrightarrow & \cdots \\
& & \uparrow{f} & & \uparrow{f_0} & & \uparrow{f_1} & & \\
0 & \longrightarrow & \operatorname{Hom}_R(A', D) & \longrightarrow & \operatorname{Hom}_R(P'_0, D) & \longrightarrow & \operatorname{Hom}_R(P'_1, D) & \longrightarrow & \cdots
\end{array}
\tag{17.9}
$$

is also commutative. The two rows of this diagram are cochain complexes, and this commutative diagram depicts a homomorphism of these cochain complexes. By Proposition 1 we have an induced map on their cohomology groups:

Proposition 5. Let $f : A \to A'$ be a homomorphism of R-modules and take projective resolutions of A and A' as in Proposition 4. Then for every n there is an induced group homomorphism $\varphi_n : \text{Ext}_R^n(A', D) \to \text{Ext}_R^n(A, D)$ on the cohomology groups obtained via these resolutions, and the maps φ_n depend only on f, not on the choice of lifts f_n in Proposition 4.

Proof: The existence of the map on the cohomology groups Ext_R^n follows from Proposition 1 applied to the homomorphism of cochain complexes (9). The more difficult part is showing these maps do not depend on the choice of lifts f_n in Proposition 4. This is easily seen to be equivalent to showing that if f is the zero map, then the induced maps on cohomology groups are also all zero. Assume then that $f = 0$. By the projectivity of the modules P_i one may inductively define R-module homomorphisms $s_n : P_n \to P'_{n+1}$ with the property that for all n,

$$f_n = d'_{n+1}s_n + s_{n-1}d_n \tag{17.10}$$

so the maps s_n give reverse downward diagonal arrows across the squares in (8). (The collection of maps $\{s_n\}$ is called a *chain homotopy* between the chain homomorphism given by the f_n and the zero chain homomorphism, cf. Exercise 4.) Taking homomorphisms into D gives diagram (9) with additional upward diagonal arrows from the homomorphisms induced by the s_n, and these induced homomorphisms satisfy the relations in (10) (i.e., they form a homotopy between cochain complex homomorphisms). It is now an easy exercise using the diagonal maps added to (9) to see that any element in $\text{Hom}_R(P'_n, D)$ representing a coset in $\text{Ext}_R^n(A', D)$ maps to the zero coset in $\text{Ext}_R^n(A, D)$ (cf. Exercise 4). This completes the argument.

One may also check that the homomorphism $\varphi_0 : \text{Ext}_R^0(A', D) \to \text{Ext}_R^0(A, D)$ in Proposition 5 is the same as the map $f : \text{Hom}_R(A', D) \to \text{Hom}_R(A, D)$ defined in Section 10.5 once the corresponding groups have been identified via the isomorphism in Proposition 3.

Theorem 6. The groups $\text{Ext}_R^n(A, D)$ depend only on A and D, i.e., they are independent of the choice of projective resolution of A.

Proof: In the notation of Proposition 4 let $A' = A$, let $f : A \to A'$ be the identity map and let the two rows of (8) be two projective resolutions of A. For any choice of lifts of the identity map, the resulting homomorphisms on cohomology groups $\varphi_n : \text{Ext}_R^n(A', D) \to \text{Ext}_R^n(A, D)$ are seen to be isomorphisms as follows. Add a third row to the diagram (8) by copying the projective resolution in the top row below the second row. Let g be the identity map from A' to A and lift g to maps $g_n : P'_n \to P_n$ by Proposition 4. Let $\psi_n : \text{Ext}_R^n(A, D) \to \text{Ext}_R^n(A', D)$ be the resulting map on cohomology groups. The maps $g_n \circ f_n : P_n \to P_n$ are now a lift of the identity map $g \circ f$, and they are seen to induce the homomorphisms $\varphi_n \circ \psi_n$ on the cohomology groups. However, since the first and third rows are identical, taking the identity map from P_n to itself for all n is a particular lift of $g \circ f$, and this choice clearly induces the identity map on cohomology groups. The last assertion of Proposition 5 then implies that $\varphi_n \circ \psi_n$ is also the identity on $\text{Ext}_R^n(A, D)$. By a symmetric argument $\psi_n \circ \varphi_n$ is the identity on $\operatorname{Ext}_R^n(A', D)$. This shows the maps φ_n and ψ_n are isomorphisms, as needed to complete the proof.

For a fixed R-module D and fixed integer $n \geq 0$, Proposition 5 and Theorem 6 show that $\operatorname{Ext}_R^n(\_, D)$ defines a (contravariant) functor from the category of R-modules to the category of abelian groups.

The next result shows that projective resolutions for a submodule and corresponding quotient module of an R-module M can be fit together to give a projective resolution of M.

Proposition 7. *(Simultaneous Resolution)* Let $0 \to L \to M \to N \to 0$ be a short exact sequence of R-modules, let $L = A$ have a projective resolution as in (6) above, and let N have a similar projective resolution where the projective modules are denoted by $\overline{P}_n$. Then there is a resolution of M by the projective modules $P_n \oplus \overline{P}_n$ such that the following diagram commutes:

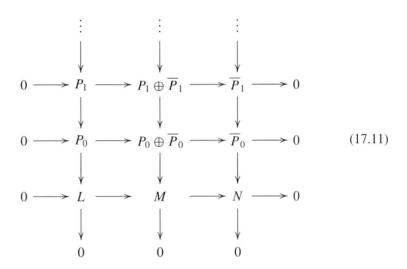

$$(17.11)$$

Moreover, the rows and columns of this diagram are exact and the rows are split.

Proof: The left and right nonzero columns of (11) are exact by hypothesis. The modules in the middle column are projective (cf. Exercise 3, Section 10.5) and the row maps are the obvious ones to make each row a split exact sequence. It remains then to define the vertical maps in the middle column in such a way as to make the diagram commute. This is accomplished in a straightforward manner, working inductively from the bottom upward — the first step in this process is outlined in Exercise 5.

Theorem 2 and Proposition 7 now yield the long exact sequence for Ext_R that extends the exact sequence (2).

Theorem 8. Let $0 \to L \to M \to N \to 0$ be a short exact sequence of R-modules. Then there is a long exact sequence of abelian groups

$$0 \to \mathrm{Hom}_R(N, D) \to \mathrm{Hom}_R(M, D) \to \mathrm{Hom}_R(L, D) \overset{\delta_0}{\to} \mathrm{Ext}^1_R(N, D)$$
$$\to \mathrm{Ext}^1_R(M, D) \to \mathrm{Ext}^1_R(L, D) \overset{\delta_1}{\to} \mathrm{Ext}^2_R(N, D) \to \cdots \quad (17.12)$$

where the maps between groups at the same level n are as in Proposition 5 and the connecting homomorphisms δ_n are given by Theorem 2.

Proof: Take a simultaneous projective resolution of the short exact sequence as in Proposition 7 and take homomorphisms into D. To obtain the cohomology groups Ext^n_R from the resulting diagram, as noted in the discussion preceding Proposition 3 we replace the lowest nonzero row in the transformed diagram with a row of zeros to get the following commutative diagram:

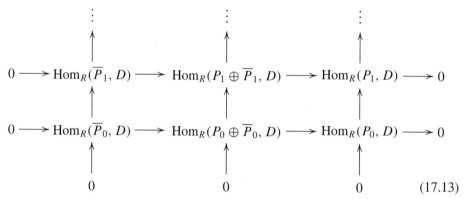

$$(17.13)$$

The columns of (13) are cochain complexes, and the rows are split by Proposition 29(2) of Section 10.5 and the discussion following it. Thus (13) is a short exact sequence of cochain complexes. Theorem 2 then gives a long exact sequence of cohomology groups whose terms are, by definition, the groups $\mathrm{Ext}^n_R(\_\_, D)$, for $n \geq 0$. The 0th order terms are identified by Proposition 3, completing the proof.

Theorem 8 shows how the exact sequence (2) can be extended in a natural way and shows that the group $\mathrm{Ext}^1_R(N, D)$ is the first measure of the failure of (2) to be exact on the right — in fact (2) can be extended to a short exact sequence on the right if and only if the connecting homomorphism δ_0 in (12) is the zero homomorphism. In particular, if $\mathrm{Ext}^1_R(N, D) = 0$ for all R-modules N, then (2) will be exact on the right for *every* exact sequence (1). We have already seen (Corollary 35 in Section 10.5) that this implies the R-module D is injective. Part of the next result shows that the converse is also true and characterizes injective modules in terms of Ext_R groups.

Proposition 9. For an R-module Q the following are equivalent:
  (1) Q is injective,
  (2) $\mathrm{Ext}^1_R(A, Q) = 0$ for all R-modules A, and
  (3) $\mathrm{Ext}^n_R(A, Q) = 0$ for all R-modules A and all $n \geq 1$.

Proof: We showed (2) implies (1) above, and (3) implies (2) is trivial, so it remains to show that if Q is injective then $\text{Ext}_R^n(A, Q) = 0$ for all R-modules A and all $n \geq 1$. Take a projective resolution

$$\cdots \longrightarrow P_n \longrightarrow P_{n-1} \longrightarrow \cdots \longrightarrow P_0 \longrightarrow A \longrightarrow 0$$

for A. Since Q is injective, the sequence

$$0 \to \text{Hom}_R(A, Q) \to \text{Hom}_R(P_0, Q) \to \cdots \to \text{Hom}_R(P_{n-1}, Q) \to \text{Hom}_R(P_n, Q) \to \cdots$$

is still exact (Corollary 35 in Section 10.5), so all of the cohomology groups for this cochain complex are 0. In particular, the groups $\text{Ext}_R^n(A, Q)$ for $n \geq 1$ are all trivial, which is (3).

For a fixed R-module D, the result in Theorem 8 can be viewed as explaining what happens to the short exact sequence $0 \to L \to M \to N \to 0$ on the right after applying the left exact functor $\text{Hom}_R(\_\_, D)$. This is why the (contravariant) functors $\text{Ext}_R^n(\_\_, D)$ are called the *right derived functors* for the functor $\text{Hom}_R(\_\_, D)$.

One can also consider the effect of applying the left exact functor $\text{Hom}_R(D, \_\_)$, i.e., by taking homomorphisms *from* D rather than *into* D. The next theorem shows that in fact the same Ext_R groups define the (covariant) right derived functors for $\text{Hom}_R(D, \_\_)$ as well.

Theorem 10. Let $0 \to L \to M \to N \to 0$ be a short exact sequence of R-modules. Then there is a long exact sequence of abelian groups

$$0 \to \text{Hom}_R(D, L) \to \text{Hom}_R(D, M) \to \text{Hom}_R(D, N) \overset{\gamma_0}{\to} \text{Ext}_R^1(D, L)$$
$$\to \text{Ext}_R^1(D, M) \to \text{Ext}_R^1(D, N) \overset{\gamma_1}{\to} \text{Ext}_R^2(D, L) \to \cdots.$$

(17.14)

Proof: Let $0 \to L \to M \to N \to 0$ be a short exact sequence of R-modules. By taking a projective resolution of D and then applying $\text{Hom}_R(\_\_, L)$, $\text{Hom}_R(\_\_, M)$ and $\text{Hom}_R(\_\_, N)$ to this resolution one obtains the columns in a commutative diagram similar to (13), but with L, M and N in the second positions rather than the first. Applying the Long Exact Sequence Theorem to this array gives (14).

Theorem 10 shows that the group $\text{Ext}_R^1(D, L)$ measures whether the exact sequence

$$0 \longrightarrow \text{Hom}_R(D, L) \longrightarrow \text{Hom}_R(D, M) \longrightarrow \text{Hom}_R(D, N)$$

can be extended to a short exact sequence — it can be extended if and only if γ_0 is the zero homomorphism. In particular, this will always be the case if the module D has the property that $\text{Ext}_R^1(D, B) = 0$ for all R-modules B; in this case it follows by Corollary 32 in Section 10.5 that D is a projective R-module. As in the situation of injective R-modules in Proposition 9, the vanishing of these cohomology groups in fact characterizes projective R-modules:

Proposition 11. For an R-module P the following are equivalent:

(1) P is projective,

(2) $\text{Ext}^1_R(P, B) = 0$ for all R-modules B, and

(3) $\text{Ext}^n_R(P, B) = 0$ for all R-modules B and all $n \geq 1$.

Proof: We proved (2) implies (1) above, and (3) implies (2) is trivial, so it remains to prove that (1) implies (3). If P is a projective R-module, then the simple exact sequence

$$0 \longrightarrow P \xrightarrow{\ 1\ } P \longrightarrow 0$$

given by the identity map on P is a projective resolution of P. Taking homomorphisms into B gives the simple cochain complex

$$0 \rightarrow \text{Hom}_R(P, B) \xrightarrow{\ 1\ } \text{Hom}_R(P, B) \rightarrow 0 \rightarrow \cdots \rightarrow 0 \rightarrow \cdots$$

from which it follows by definition that $\text{Ext}^n_R(P, B) = 0$ for all $n \geq 1$, which gives (3).

Examples

(1) Since $\mathbb{Z}^m$ is a free, hence projective, $\mathbb{Z}$-module, it follows from Proposition 11 that

$$\text{Ext}^n_{\mathbb{Z}}(\mathbb{Z}^m, B) = 0$$

for all abelian groups B, all $m \geq 1$, and all $n \geq 1$.

(2) It is not difficult to show that $\text{Ext}^n_R(A_1 \oplus A_2, B) \cong \text{Ext}^n_R(A_1, B) \oplus \text{Ext}^n_R(A_2, B)$ for all $n \geq 0$ (cf. Exercise 10), so the previous example together with the example following Proposition 3 determines $\text{Ext}^n_{\mathbb{Z}}(A, B)$ for all finitely generated abelian groups A. In particular, $\text{Ext}^n_{\mathbb{Z}}(A, B) = 0$ for all finitely generated groups A, all abelian groups B, and all $n \geq 2$.

We have chosen to define the cohomology group $\text{Ext}^n_R(A, B)$ using a projective resolution of A. There is a parallel development using an *injective resolution* of B:

$$0 \rightarrow B \rightarrow Q_0 \rightarrow Q_1 \rightarrow \cdots$$

where each Q_i is injective. In this situation one defines $\text{Ext}^n_R(A, B)$ as the nth cohomology group of the cochain sequence obtained by applying $\text{Hom}_R(A, \_)$ to the resolution for B. The theory proceeds in a manner analogous to the development of this section. Ultimately one shows that there is a natural isomorphism between the groups $\text{Ext}^n_R(A, B)$ constructed using both methods.

Examples

(1) Suppose $R = \mathbb{Z}$ and A and B are $\mathbb{Z}$-modules, i.e., are abelian groups. Recall that a $\mathbb{Z}$-module is injective if and only if it is divisible (Proposition 36 in Section 10.5). The group B can be embedded in an injective $\mathbb{Z}$-module Q_0 (Corollary 37 in Section 10.5) and the quotient, Q_1, of Q_0 by the image of B is again injective. Hence we have an injective resolution

$$0 \longrightarrow B \longrightarrow Q_0 \longrightarrow Q_1 \longrightarrow 0$$

of B. Applying $\text{Hom}_{\mathbb{Z}}(A, \underline{})$ to this sequence gives the cochain complex

$$0 \longrightarrow \text{Hom}_{\mathbb{Z}}(A, B) \longrightarrow \text{Hom}_{\mathbb{Z}}(A, Q_0) \longrightarrow \text{Hom}_{\mathbb{Z}}(A, Q_1) \longrightarrow 0 \longrightarrow \cdots$$

from which it follows immediately that

$$\text{Ext}_{\mathbb{Z}}^n(A, B) = 0$$

for all abelian groups A and B and all $n \geq 2$, showing that the result of the previous example holds also when A is not finitely generated.

(2) Suppose A is a torsion abelian group. Then we have $\text{Ext}^0(A, \mathbb{Z}) \cong \text{Hom}(A, \mathbb{Z}) = 0$ since $\mathbb{Z}$ is torsion free. The sequence $0 \to \mathbb{Z} \to \mathbb{Q} \to \mathbb{Q}/\mathbb{Z} \to 0$ gives an injective resolution of $\mathbb{Z}$. Applying $\text{Hom}(A, \underline{})$ gives the cochain complex

$$0 \longrightarrow \text{Hom}(A, \mathbb{Z}) \longrightarrow \text{Hom}(A, \mathbb{Q}) \longrightarrow \text{Hom}(A, \mathbb{Q}/\mathbb{Z}) \longrightarrow 0 \longrightarrow \cdots$$

and since $\mathbb{Q}$ is also torsion free, this shows that

$$\text{Ext}_{\mathbb{Z}}^1(A, \mathbb{Z}) \cong \text{Hom}_{\mathbb{Z}}(A, \mathbb{Q}/\mathbb{Z}).$$

The group $\text{Hom}(A, \mathbb{Q}/\mathbb{Z})$ is called the *Pontriagin dual group* to A. If A is a finite abelian group the Pontriagin dual of A is isomorphic to A (cf. Exercise 14, Section 5.2). In particular, $\text{Ext}^1(A, \mathbb{Z}) \cong A$ is nonzero for all nonzero finite abelian groups A. We have $\text{Ext}^n(A, \mathbb{Z}) = 0$ for all $n \geq 2$ by the previous example.

We record an important property of Ext_R^1, which helps to explain the name for these cohomology groups. Recall that equivalent extensions were defined at the beginning of Section 10.5.

Theorem 12. For any R-modules N and L there is a bijection between $\text{Ext}_R^1(N, L)$ and the set of equivalence classes of extensions of N by L.

Although we shall not prove this result, in Section 4 we establish a similar bijection between equivalence classes of group extensions of G by A and elements of a certain cohomology group, where G is any finite group and A is any $\mathbb{Z}G$-module.

Example

Suppose $R = \mathbb{Z}$ and $A = B = \mathbb{Z}/p\mathbb{Z}$. We showed above that $\text{Ext}_R^1(\mathbb{Z}/p\mathbb{Z}, \mathbb{Z}/p\mathbb{Z}) \cong \mathbb{Z}/p\mathbb{Z}$, so by Theorem 12 there are precisely p equivalence classes of extensions of $\mathbb{Z}/p\mathbb{Z}$ by $\mathbb{Z}/p\mathbb{Z}$. These are given by the direct sum $\mathbb{Z}/p\mathbb{Z} \oplus \mathbb{Z}/p\mathbb{Z}$ (which corresponds to the trivial class in $\text{Ext}_R^1(\mathbb{Z}/p\mathbb{Z}, \mathbb{Z}/p\mathbb{Z})$) and the $p-1$ extensions

$$0 \longrightarrow \mathbb{Z}/p\mathbb{Z} \longrightarrow \mathbb{Z}/p^2\mathbb{Z} \overset{i}{\longrightarrow} \mathbb{Z}/p\mathbb{Z} \longrightarrow 0$$

defined by the map $i(x) = ix \bmod p$ for $i = 1, 2, \ldots, p-1$. Note that while these are inequivalent as extensions, they all determine the same group $\mathbb{Z}/p^2\mathbb{Z}$.

Tensor Products and the Groups $\mathrm{Tor}_n^R(A, B)$

The cohomology groups $\mathrm{Ext}_R^n(A, B)$ determine what happens to short exact sequences on the right after applying the left exact functors $\mathrm{Hom}_R(D, \_\_)$ and $\mathrm{Hom}_R(\_\_, D)$. One may similarly ask for the behavior of short exact sequences on the left after applying the right exact functor $D \otimes_R \_\_$ or the right exact functor $\_\_ \otimes_R D$. This leads to the Tor (homology) groups (whose name derives from their relation to torsion submodules), and we now briefly outline the development of these left derived functors. In some respects this theory is "dual" to the theory for Ext_R. We concentrate on the situation for $D \otimes_R \_\_$ when D is a right R-module. When D is a left R-module there is a completely symmetric theory for $\_\_ \otimes_R D$; when R is commutative and all R-modules have the same left and right R action the homology groups resulting from both developments are isomorphic.

Suppose then that D is a right R-module. Then for every left R-module B the tensor product $D \otimes_R B$ is an abelian group and the functor $D \otimes \_\_$ is covariant and right exact, i.e., for any short exact sequence (1) of left R-modules,

$$D \otimes L \longrightarrow D \otimes M \longrightarrow D \otimes N \longrightarrow 0$$

is an exact sequence of abelian groups. This sequence may be extended at the left end to a long exact sequence as follows. Let

$$\cdots \longrightarrow P_n \xrightarrow{d_n} P_{n-1} \longrightarrow \cdots \xrightarrow{d_1} P_0 \xrightarrow{\epsilon} B \longrightarrow 0$$

be a projective resolution of B, and take tensor products with D to obtain

$$\cdots \longrightarrow D \otimes P_n \xrightarrow{1 \otimes d_n} D \otimes P_{n-1} \longrightarrow \cdots \xrightarrow{1 \otimes d_1} D \otimes P_0 \xrightarrow{1 \otimes \epsilon} D \otimes B \longrightarrow 0. \quad (17.15)$$

It follows from the argument in Theorem 39 of Section 10.5 that (15) is a chain complex — the composition of any two successive maps is zero — so we may form its homology groups.

Definition. Let D be a right R-module and let B be a left R-module. For any projective resolution of B by left R-modules as above let $1 \otimes d_n : D \otimes P_n \to D \otimes P_{n-1}$ for all $n \geq 1$ as in (15). Then

$$\mathrm{Tor}_n^R(D, B) = \ker(1 \otimes d_n) / \mathrm{image}(1 \otimes d_{n+1})$$

where $\mathrm{Tor}_0^R(D, B) = (D \otimes P_0) / \mathrm{image}(1 \otimes d_1)$. The group $\mathrm{Tor}_n^R(D, B)$ is called the nth *homology group derived from the functor* $D \otimes \_\_$. When $R = \mathbb{Z}$ the group $\mathrm{Tor}_n^{\mathbb{Z}}(D, B)$ is also denoted simply $\mathrm{Tor}_n(D, B)$.

Thus $\mathrm{Tor}_n^R(D, B)$ is the nth homology group of the chain complex obtained from (15) by removing the term $D \otimes B$.

A completely analogous proof to Proposition 3 (but relying on Theorem 39 in Section 10.5) implies the following:

Proposition 13. For any left R-module B we have $\mathrm{Tor}_0^R(D, B) \cong D \otimes B$.

Example

Let $R = \mathbb{Z}$ and let $B = \mathbb{Z}/m\mathbb{Z}$ for some $m \geq 2$. By the proposition, $\mathrm{Tor}_0^{\mathbb{Z}}(D, \mathbb{Z}/m\mathbb{Z})$ is isomorphic to $D \otimes \mathbb{Z}/m\mathbb{Z}$, so we have $\mathrm{Tor}_0^{\mathbb{Z}}(D, \mathbb{Z}/m\mathbb{Z}) \cong D/mD$ (Example 8 following Corollary 12 in Section 10.4). For the higher groups we apply $D \otimes \_\_$ to the projective resolution

$$0 \longrightarrow \mathbb{Z} \overset{m}{\longrightarrow} \mathbb{Z} \longrightarrow \mathbb{Z}/m\mathbb{Z} \longrightarrow 0$$

of B and use the isomorphisms $D \otimes \mathbb{Z} \cong D$ and $D \otimes \mathbb{Z}/m\mathbb{Z} \cong D/mD$. This gives the chain complex

$$\cdots \longrightarrow 0 \longrightarrow D \overset{m}{\longrightarrow} D \longrightarrow D/mD \longrightarrow 0.$$

It follows that $\mathrm{Tor}_1^{\mathbb{Z}}(D, \mathbb{Z}/m\mathbb{Z}) \cong {}_mD$ is the subgroup of D annihilated by m and that $\mathrm{Tor}_n^{\mathbb{Z}}(D, \mathbb{Z}/m\mathbb{Z}) = 0$ for all $n \geq 2$, which we summarize as

$$\mathrm{Tor}_0(D, \mathbb{Z}/m\mathbb{Z}) \cong D/mD,$$
$$\mathrm{Tor}_1(D, \mathbb{Z}/m\mathbb{Z}) \cong {}_mD,$$
$$\mathrm{Tor}_n(D, \mathbb{Z}/m\mathbb{Z}) = 0, \quad \text{for all } n \geq 2.$$

As for Ext, the Tor groups depend on the ring R (cf. Exercise 20).

Following a similar development to that for Ext_R, one shows:

Proposition 14.
   (1) The homology groups $\mathrm{Tor}_n^R(D, B)$ are independent of the choice of projective resolution of B, and
   (2) for every R-module homomorphism $f : B \to B'$ there are induced maps $\psi_n : \mathrm{Tor}_n^R(D, B) \to \mathrm{Tor}_n^R(D, B')$ on homology groups (depending only on f).

There is a Long Exact Sequence in Homology analogous to Theorem 2, except that all the arrows are reversed, whose proof follows mutatis mutandis from the argument for cohomology. This together with Simultaneous Resolution gives:

Theorem 15. Let $0 \to L \to M \to N \to 0$ be a short exact sequence of left R-modules. Then there is a long exact sequence of abelian groups

$$\cdots \to \mathrm{Tor}_2^R(D, N) \overset{\delta_1}{\to} \mathrm{Tor}_1^R(D, L) \to \mathrm{Tor}_1^R(D, M) \to$$

$$\mathrm{Tor}_1^R(D, N) \overset{\delta_0}{\to} D \otimes L \to D \otimes M \to D \otimes N \to 0$$

where the maps between groups at the same level n are as in Proposition 14 (and the maps δ_n are called connecting homomorphisms).

There is a characterization of flat modules corresponding to Propositions 9 and 11 whose proof is very similar and is left as an exercise.

Proposition 16. For a right R-module D the following are equivalent:
  (1) D is a flat R-module,
  (2) $\text{Tor}_1^R(D, B) = 0$ for all left R-modules B, and
  (3) $\text{Tor}_n^R(D, B) = 0$ for all left R-modules B and all $n \geq 1$.

We have defined $\text{Tor}_n^R(A, B)$ as the homology of the chain complex obtained by tensoring a projective resolution of B on the left with A. The same groups are obtained by taking the homology of the chain complex obtained by tensoring a projective resolution of A on the right by B. Put another way, the $\text{Tor}_n^R(A, B)$ groups define the (covariant) left derived functors for both of the right exact functors $A \otimes_R \underline{}$ and $\underline{} \otimes_R B$: if D is a left R-module, then the short exact sequence $0 \to L \to M \to N \to 0$ of right R-modules gives rise to the long exact sequence

$$\cdots \to \text{Tor}_2^R(N, D) \xrightarrow{\gamma_1} \text{Tor}_1^R(L, D) \to \text{Tor}_1^R(M, D) \to$$

$$\text{Tor}_1^R(N, D) \xrightarrow{\gamma_0} L \otimes_R D \to M \otimes_R D \to N \otimes_R D \to 0$$

of abelian groups. In particular, the left R-module D is flat if and only if $\text{Tor}_1^R(A, D) = 0$ for all right R-modules A.

When R is commutative, $A \otimes_R B \cong B \otimes_R A$ (Proposition 20 in Section 10.4) for any two R-modules A and B with the standard R-module structures, and it follows that $\text{Tor}_n^R(A, B) \cong \text{Tor}_n^R(B, A)$ as R-modules. When R is commutative the Tor long exact sequences are exact sequences of R-modules.

Examples

  (1) If $R = \mathbb{Z}$, then since $\mathbb{Z}^m$ is free, hence flat (Corollary 42, Section 10.5), we have $\text{Tor}_n(A, \mathbb{Z}^m) = 0$ for all $n \geq 1$ and all abelian groups A.
  (2) Since $\text{Tor}_n^R(A, B_1 \oplus B_2) \cong \text{Tor}_n^R(A, B_1) \oplus \text{Tor}_n^R(A, B_2)$ (cf. Exercise 10), the previous two examples together determine $\text{Tor}_n^R(A, B)$ for all abelian groups A and all finitely generated abelian groups B.
  (3) As a particular case of the previous example, $\text{Tor}_1(A, B)$ is a torsion group and $\text{Tor}_n(A, B) = 0$ for every abelian group A, every finitely generated abelian group B, and all $n \geq 2$. In fact these results hold without the condition that B be finitely generated.
  (4) The exact sequence $0 \to \mathbb{Z} \to \mathbb{Q} \to \mathbb{Q}/\mathbb{Z} \to 0$ gives the long exact sequence

$$\cdots \to \text{Tor}_1(D, \mathbb{Q}) \to \text{Tor}_1(D, \mathbb{Q}/\mathbb{Z}) \to D \otimes \mathbb{Z} \to D \otimes \mathbb{Q} \to D \otimes \mathbb{Q}/\mathbb{Z} \to 0.$$

Since $\mathbb{Q}$ is a flat $\mathbb{Z}$-module (Example 2 following Corollary 42 in Section 10.5), the proposition shows that we have an exact sequence

$$0 \longrightarrow \text{Tor}_1(D, \mathbb{Q}/\mathbb{Z}) \longrightarrow D \longrightarrow D \otimes \mathbb{Q}$$

and so $\text{Tor}_1(D, \mathbb{Q}/\mathbb{Z})$ is isomorphic to the kernel of the natural map from D into $D \otimes \mathbb{Q}$, which is the torsion subgroup of D (cf. Exercise 9 in Section 10.4).

The following results show that, for $R = \mathbb{Z}$, the Tor groups are closely related to torsion subgroups. The Tor groups first arose in applications of torsion abelian groups in topological settings, which helps explain the terminology.

Proposition 17. Let A and B be $\mathbb{Z}$-modules and let $t(A)$ and $t(B)$ denote their respective torsion submodules. Then $\text{Tor}_1(A, B) \cong \text{Tor}_1(t(A), t(B))$.

Proof: In the case where A and B are finitely generated abelian groups this follows by Examples 3 and 4 above. For the general case, cf. Exercise 16.

Corollary 18. If A is an abelian group then A is torsion free if and only if $\text{Tor}_1(A, B) = 0$ for every abelian group B (in which case A is flat as a $\mathbb{Z}$-module).

Proof: By the proposition, if A has no elements of finite order then we have $\text{Tor}_1(A, B) = \text{Tor}_1(t(A), B) = \text{Tor}_1(0, B) = 0$ for every abelian group B. Conversely, if $\text{Tor}_1(A, B) = 0$ for all B, then in particular $\text{Tor}_1(A, \mathbb{Q}/\mathbb{Z}) = 0$, and this group is isomorphic to the torsion subgroup of A by the example above.

The results of Proposition 17 and Corollary 18 hold for any P.I.D. R in place of $\mathbb{Z}$ (cf. Exercise 26 in Section 10.5 and Exercise 16).

Finally, we mention that the cohomology and homology theories we have described may be developed in a vastly more general setting by axiomatizing the essential properties of R-modules and the Hom_R and tensor product functors. This leads to the general notions of *abelian categories* and *additive functors*. In the case of the abelian category of R-modules, any additive functor $\mathcal{F}$ to the category of abelian groups gives rise to a set of *derived functors*, $\mathcal{F}_n$, also from R-modules to abelian groups, for all $n \geq 0$. Then for each short exact sequence $0 \to L \to M \to N \to 0$ of R-modules there is a long exact sequence of (cohomology or homology) groups whose terms are $\mathcal{F}_n(L)$, $\mathcal{F}_n(M)$ and $\mathcal{F}_n(N)$, and these long exact sequences reflect the exactness properties of the functor $\mathcal{F}$. If $\mathcal{F}$ is left or right exact then the 0th derived functor $\mathcal{F}_0$ is naturally equivalent to $\mathcal{F}$ (hence the 0th degree groups $\mathcal{F}_0(X)$ are isomorphic to $\mathcal{F}(X)$), and if $\mathcal{F}$ is an exact functor then $\mathcal{F}_n(X) = 0$ for all $n \geq 1$ and all R-modules X.

EXERCISES

1. Give the details of the proof of Proposition 1.
2. This exercise defines the connecting map δ_n in the Long Exact Sequence of Theorem 2 and proves it is a homomorphism. In the notation of Theorem 2 let $0 \to \mathcal{A} \xrightarrow{\alpha} \mathcal{B} \xrightarrow{\beta} \mathcal{C} \to 0$ be a short exact sequence of cochain complexes, where for simplicity the cochain maps for $\mathcal{A}$, $\mathcal{B}$ and $\mathcal{C}$ are all denoted by the same d.
   (a) If $c \in C^n$ represents the class $x \in H^n(\mathcal{C})$ show that there is some $b \in B^n$ with $\beta_n(b) = c$.
   (b) Show that $d_{n+1}(b) \in \ker \beta_{n+1}$ and conclude that there is a unique $a \in A^{n+1}$ such that $\alpha_{n+1}(a) = d_{n+1}(b)$. [Use $c \in \ker d_{n+1}$ and the commutativity of the diagram.]
   (c) Show that $d_{n+2}(a) = 0$ and conclude that a defines a class $\bar{a}$ in the quotient group $H^{n+1}(\mathcal{A})$. [Use the fact that α_{n+2} is injective.]
   (d) Prove that $\bar{a}$ is independent of the choice of b, i.e., if b' is another choice and a' is its unique preimage in A^{n+1} then $\bar{a} = \bar{a'}$, and that $\bar{a}$ is also independent of the choice of c representing the class x.
   (e) Define $\delta_n(x) = \bar{a}$ and prove that δ_n is a group homomorphism from $H^n(\mathcal{C})$ to $H^{n+1}(\mathcal{A})$. [Use the fact that $\delta_n(x)$ is independent of the choices of c and b to compute $\delta_n(x_1 + x_2)$.]

3. Suppose

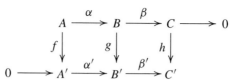

is a commutative diagram of R-modules with exact rows.

(a) If $c \in \ker h$ and $\beta(b) = c$ prove that $g(b) \in \ker \beta'$ and conclude that $g(b) = \alpha'(a')$ for some $a' \in A'$. [Use the commutativity of the diagram.]

(b) Show that $\delta(c) = a' \bmod \text{image } f$ is a well defined R-module homomorphism from $\ker h$ to the quotient $A'/\text{image } f$.

(c) *(The Snake Lemma)* Prove there is an exact sequence

$$\ker f \longrightarrow \ker g \longrightarrow \ker h \xrightarrow{\delta} \text{coker } f \longrightarrow \text{coker } g \longrightarrow \text{coker } h$$

where coker f (the *cokernel* of f) is $A'/(\text{image } f)$ and similarly for coker g and coker h.

(d) Show that if α is injective and β' is surjective (i.e., the two rows in the commutative diagram above can be extended to short exact sequences) then the exact sequence in (c) can be extended to the exact sequence

$$0 \longrightarrow \ker f \longrightarrow \ker g \longrightarrow \ker h \xrightarrow{\delta} \text{coker } f \longrightarrow \text{coker } g \longrightarrow \text{coker } h \longrightarrow 0$$

4. Let $\mathcal{A} = \{A^n\}$ and $\mathcal{B} = \{B^n\}$ be cochain complexes, where the maps $A^n \to A^{n+1}$ and $B^n \to B^{n+1}$ in both complexes are denoted by d_{n+1} for all n. Cochain complex homomorphisms α and β from $\mathcal{A}$ to $\mathcal{B}$ are said to be *homotopic* if for all n there are module homomorphisms $s_n : A^{n+1} \to B^n$ such that the maps $\alpha_n - \beta_n$ from A^n to B^n satisfy

$$\alpha_n - \beta_n = d_n s_{n-1} + s_n d_{n+1}.$$

The collection of maps $\{s_n\}$ is called a *cochain homotopy* from α to β. One may similarly define chain homotopies between chain complexes.

(a) Prove that homotopic maps of cochain complexes induce the same maps on cohomology, i.e., if α and β are homotopic homomorphisms of cochain complexes then the induced group homomorphisms from $H^n(\mathcal{A})$ to $H^n(\mathcal{B})$ are equal for every $n \geq 0$. (Thus "homotopy" gives a sufficient condition for two maps of complexes to induce the same maps on cohomology or homology; this condition is not in general necessary.) [Use the definition of homotopy to show $(\alpha_n - \beta_n)(z) \in \text{image } d_n$ for every $z \in \ker d_{n+1}$.]

(b) Prove that the relation $\alpha \sim \beta$ if α and β are homotopic is an equivalence relation on any set of cochain complex homomorphisms.

5. Establish the first step in the Simultaneous Resolution result of Proposition 7 as follows: assume the first two nonzero rows in diagram (11) are given, except for the map from $P_0 \oplus \overline{P}_0$ to M (where the maps along the row of projective modules are the obvious injection and projection for this split exact sequence). Let $\mu : \overline{P}_0 \to M$ be a lifting to $\overline{P}_0$ of the map $\overline{P}_0 \to N$ (which exists because $\overline{P}_0$ is projective). Let λ be the composition $P_0 \to L \to M$ in the diagram. Define

$$\pi : P_0 \oplus \overline{P}_0 \to M \qquad \text{by} \qquad \pi(x, y) = \lambda(x) + \mu(y).$$

Show that with this definition the first two nonzero rows of (11) form a commutative diagram.

6. Let $0 \to \mathcal{A} \xrightarrow{\alpha} \mathcal{B} \xrightarrow{\beta} \mathcal{C} \to 0$ be a short exact sequence of cochain complexes. Prove that if any two of $\mathcal{A}, \mathcal{B}, \mathcal{C}$ are exact, then so is the third. [Use Theorem 2.]

7. Prove that a finitely generated abelian group A is free if and only if $\text{Ext}^1(A, \mathbb{Z}) = 0$.

8. Prove that if $0 \to L \to M \to N \to 0$ is a split short exact sequence of R-modules, then for every $n \geq 0$ the sequence $0 \to \text{Ext}^n_R(N, D) \to \text{Ext}^n_R(M, D) \to \text{Ext}^n_R(L, D) \to 0$ is also short exact and split. [Use a splitting homomorphism and Proposition 5.]

9. Show that

$$ 0 \longrightarrow \mathbb{Z}/d\mathbb{Z} \longrightarrow \mathbb{Z}/m\mathbb{Z} \xrightarrow{d} \mathbb{Z}/m\mathbb{Z} \xrightarrow{m/d} \mathbb{Z}/m\mathbb{Z} \xrightarrow{d} \mathbb{Z}/m\mathbb{Z} \xrightarrow{m/d} \cdots $$

is an injective resolution of $\mathbb{Z}/d\mathbb{Z}$ as a $\mathbb{Z}/m\mathbb{Z}$-module. [Use Proposition 36 in Section 10.5.] Use this to compute the groups $\text{Ext}^n_{\mathbb{Z}/m\mathbb{Z}}(A, \mathbb{Z}/d\mathbb{Z})$ in terms of the dual group $\text{Hom}_{\mathbb{Z}/m\mathbb{Z}}(A, \mathbb{Z}/m\mathbb{Z})$. In particular, if $m = p^2$ and $d = p$, give another derivation of the result $\text{Ext}^n_{\mathbb{Z}/p^2\mathbb{Z}}(\mathbb{Z}/p\mathbb{Z}, \mathbb{Z}/p\mathbb{Z}) \cong \mathbb{Z}/p\mathbb{Z}$.

10. **(a)** Prove that an arbitrary direct sum $\oplus_{i \in I} P_i$ of projective modules P_i is projective and that an arbitrary direct product $\prod_{j \in J} Q_j$ of injective modules Q_j is injective.

   (b) Prove that an arbitrary direct sum of projective resolutions is again projective and use this to show $\text{Ext}^n_R(\oplus_{i \in I} A_i, B) \cong \prod_{i \in I} \text{Ext}^n_R(A_i, B)$ for any collection of R-modules A_i ($i \in I$). [cf. Exercise 12 in Section 10.5.]

   (c) Prove that an arbitrary direct product of injective resolutions is an injective resolution and use this to show $\text{Ext}^n_R(A, \prod_{j \in J} B_j) \cong \prod_{j \in J} \text{Ext}^n_R(A, B_j)$ for any collection of R-modules B_j ($j \in J$). [cf. Exercise 12 in Section 10.5.]

   (d) Prove that $\text{Tor}^R_n(A, \oplus_{j \in J} B_j) \cong \oplus_{j \in J} \text{Tor}^R_n(A, B_j)$ for any collection of R-modules B_j ($j \in J$).

11. *(Bass' Characterization of Noetherian Rings)* Suppose R is a commutative ring.

   (a) If R is Noetherian, and I is any nonzero ideal in R show that the image of any R-module homomorphism $f : I \to \oplus_{j \in \mathcal{J}} Q_j$ from I into a direct sum of injective R-modules Q_j ($j \in \mathcal{J}$) is contained in some finite direct sum of the Q_j.

   (b) If R is Noetherian, prove that an arbitrary direct sum $\oplus_{j \in \mathcal{J}} Q_j$ of injective R-modules is again injective. [Use Baer's Criterion (Proposition 36) and Exercise 4 in Section 10.5 together with (a).]

   (c) Let $I_1 \subseteq I_2 \subseteq \ldots$ be an ascending chain of ideals of R with union I and let $I/I_i \to Q_i$ for $i = 1, 2, \ldots$ be an injection of the quotient I/I_i into an injective R-module Q_i (by Theorem 38 in Section 10.5). Prove that the composition of these injections with the product of the canonical projection maps $I \to I/I_i$ gives an R-module homomorphism $f : I \to \oplus_{i = 1, 2, \ldots} Q_i$.

   (d) Prove the converse of (b): if an arbitrary direct sum $\oplus_{j \in \mathcal{J}} Q_j$ of injective R-modules is again injective then R is Noetherian. [If the direct sum in (c) is injective, use Baer's Criterion to lift f to a homomorphism $F : R \to \oplus_{i = 1, 2, \ldots} Q_i$. If the component of $F(1)$ in Q_i is 0 for $i \geq n$ prove that $I = I_n$ and the ascending chain of ideals is finite.]

12. Prove Proposition 13: $\text{Tor}^R_0(D, A) \cong D \otimes_R A$. [Follow the proof of Proposition 3.]

13. Prove Proposition 16 characterizing flat modules.

14. Suppose $0 \to A \to B \to C \to 0$ is a short exact sequence of R-modules. Prove that if C is a flat R-module, then A is flat if and only if B is also flat. [Use the Tor long exact sequence.] Give an example to show that if A and B are flat then C need not be flat.

15. (a) If I is an ideal in R and M is an R-module, prove that $\mathrm{Tor}_1^R(M, R/I)$ is isomorphic to the kernel of the map $M \otimes_R I \to M$ that maps $m \otimes i$ to mi for $i \in I$ and $m \in M$. [Use the Tor long exact sequence associated to $0 \to I \to R \to R/I \to 0$ noting that R is flat.]

(b) *(A Flatness Criterion using Tor)* Prove that the R-module M is flat if and only if $\mathrm{Tor}_1^R(M, R/I) = 0$ for every finitely generated ideal I of R. [Use Exercise 25 in Section 10.5.]

16. Suppose R is a P.I.D. and A and B are R-modules. If $t(B)$ denotes the torsion submodule of B show that $\mathrm{Tor}_1^R(A, t(B)) \cong \mathrm{Tor}_1^R(A, B)$ and deduce that $\mathrm{Tor}_1^R(A, B)$ is isomorphic to $\mathrm{Tor}_1^R(t(A), t(B))$. [Use Exercise 26 in Section 10.5 to show that $B/t(B)$ is flat over R, then use the Tor long exact sequence with $D = A$ applied to the short exact sequence $0 \to t(B) \to B \to B/t(B) \to 0$ and the remarks following Proposition 16.]

17. Let $A = \mathbb{Z}/2\mathbb{Z} \oplus \mathbb{Z}/3\mathbb{Z} \oplus \mathbb{Z}/4\mathbb{Z} \oplus \cdots$. Prove that $\mathrm{Ext}^1(A, B) \cong (B/2B) \times (B/3B) \times (B/4B) \times \cdots$ for any abelian group B. [Use Exercise 10.] Prove that $\mathrm{Ext}^1(A, B) = 0$ if and only if B is divisible.

18. Prove that $\mathbb{Z}/2\mathbb{Z}$ is a projective $\mathbb{Z}/6\mathbb{Z}$-module and deduce that $\mathrm{Tor}_1^{\mathbb{Z}/6\mathbb{Z}}(\mathbb{Z}/2\mathbb{Z}, \mathbb{Z}/2\mathbb{Z}) = 0$.

19. Suppose $r \neq 0$ is not a zero divisor in the commutative ring R.

(a) Prove that multiplication by r gives a free resolution $0 \to R \xrightarrow{r} R \to R/rR \to 0$ of the quotient R/rR.

(b) Prove that $\mathrm{Ext}_R^0(R/rR, B) = {}_r B$ is the set of elements $b \in B$ with $rb = 0$, that $\mathrm{Ext}_R^1(R/rR, B) \cong B/rB$, and that $\mathrm{Ext}_R^n(R/rR, B) = 0$ for $n \geq 2$ for every R-module B.

(c) Prove that $\mathrm{Tor}_0^R(A, R/rR) = A/rA$, that $\mathrm{Tor}_1^R(A, R/rR) = {}_r A$ is the set of elements $a \in A$ with $ra = 0$, and that $\mathrm{Tor}_n^R(A, R/rR) = 0$ for $n \geq 2$ for every R-module A.

20. Prove that $\mathrm{Tor}_0^{\mathbb{Z}/m\mathbb{Z}}(A, \mathbb{Z}/d\mathbb{Z}) \cong A/dA$, that $\mathrm{Tor}_n^{\mathbb{Z}/m\mathbb{Z}}(A, \mathbb{Z}/d\mathbb{Z}) \cong {}_d A/(m/d)A$ for n odd, $n \geq 1$, and that $\mathrm{Tor}_n^{\mathbb{Z}/m\mathbb{Z}}(A, \mathbb{Z}/d\mathbb{Z}) \cong {}_{(m/d)} A/dA$ for n even, $n \geq 2$. [Use the projective resolution in Example 2 following Proposition 3.]

21. Let $R = k[x, y]$ where k is a field, and let I be the ideal (x, y) in R.

(a) Let $\alpha : R \to R^2$ be the map $\alpha(r) = (yr, -xr)$ and let $\beta : R^2 \to R$ be the map $\beta((r_1, r_2)) = r_1 x + r_2 y$. Show that

$$0 \longrightarrow R \xrightarrow{\alpha} R^2 \xrightarrow{\beta} R \longrightarrow k \longrightarrow 0$$

where the map $R \to R/I = k$ is the canonical projection, gives a free resolution of k as an R-module.

(b) Use the resolution in (a) to show that $\mathrm{Tor}_2^R(k, k) \cong k$.

(c) Prove that $\mathrm{Tor}_1^R(k, I) \cong k$. [Use the long exact sequence corresponding to the short exact sequence $0 \to I \to R \to k \to 0$ and (b).]

(d) Conclude from (c) that the torsion free R-module I is not flat (compare to Exercise 26 in Section 10.5).

22. *(Flat Base Change for Tor)* Suppose R and S are commutative rings and $f : R \to S$ is a ring homomorphism making S into an R-module as in Example 6 following Corollary 12 in Section 10.4. Prove that if S is flat as an R-module, then $\mathrm{Tor}_n^R(A, B) \cong \mathrm{Tor}_n^S(S \otimes_R A, B)$ for all R-modules A and all S-modules B. [Show that since S is flat, tensoring an R-module projective resolution for A with S gives an S-module projective resolution of $S \otimes_R A$.]

23. *(Localization and Tor)* Let $D^{-1}R$ be the localization of the commutative ring R with respect to the multiplicative subset D of R. Prove that localization commutes with Tor, i.e., $D^{-1}\mathrm{Tor}_n^R(A, B) \cong \mathrm{Tor}_n^{D^{-1}R}(D^{-1}A, D^{-1}B)$ for all R-modules A and B and all $n \geq 0$. [Use the previous exercise and the fact that $D^{-1}R$ is flat over R, cf. Proposition 42(6) in Section 15.4.]

24. *(Flatness is local)* Suppose R is a commutative ring. Prove that an R-module M is flat if and only if every localization M_P is a flat R_P-module for every maximal (hence also for every prime) ideal in R. [Use the previous exercise together with the characterization of flatness in terms of Tor.]

25. If R is an integral domain with field of fractions F, prove that $\mathrm{Tor}_1^R(F/R, B) \cong t(B)$ for any R-module B, where $t(B)$ denotes the R-torsion submodule of B.

An R-module M is said to be *finitely presented* if there is an exact sequence

$$R^s \longrightarrow R^t \longrightarrow M \longrightarrow 0$$

of R-modules for some integers s and t. Equivalently, M is finitely generated by t elements and the kernel of the corresponding R-module homomorphism $R^t \to M$ can be generated by s elements.

26. (a) Prove that every finitely generated module over a Noetherian ring R is finitely presented. [Use Exercise 8 in Section 15.1.]
  (b) Prove that an R-module M is finitely presented and projective if and only if M is a direct summand of R^n for some integer $n \geq 1$.

27. Suppose that M is a finitely presented R-module and that $0 \to A \xrightarrow{\alpha} B \xrightarrow{\beta} M \to 0$ is an exact sequence of R-modules. This exercise proves that if B is a finitely generated R-module then A is also a finitely generated R-module.

  (a) Suppose $R^s \xrightarrow{\psi} R^t \xrightarrow{\varphi} M \to 0$ and $e_1, \ldots, e_t$ is an R-module basis for R^t. Show that there exist $b_1, \ldots, b_t \in B$ so that $\beta(b_i) = \varphi(e_i)$ for $i = 1, \ldots, t$.
  (b) If f is the R-module homomorphism from R^t to B defined by $f(e_i) = b_i$ for $i = 1, \ldots, t$, show that $f(\psi(R^s)) \subseteq \ker \beta$. [Use $\varphi \circ \psi = 0$.] Conclude that there is a commutative diagram

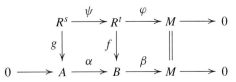

of R-modules with exact rows.
  (c) Prove that $A/\mathrm{image}\, g \cong B/\mathrm{image}\, f$ and use this to prove that A is finitely generated. [For the isomorphism, use the Snake Lemma in Exercise 3. Then show that image g and $A/\mathrm{image}\, g$ are both finitely generated and apply Exercise 7 of Section 10.3.]
  (d) If I is an ideal of R conclude that R/I is a finitely presented R-module if and only if I is a finitely generated ideal.

28. Suppose R is a local ring with unique maximal ideal $\mathfrak{m}$ and M is a finitely presented R-module. Suppose $m_1, \ldots, m_s$ are elements in M whose images in $M/\mathfrak{m}M$ form a basis for $M/\mathfrak{m}M$ as a vector space over the field $R/\mathfrak{m}$.
  (a) Prove that $m_1, \ldots, m_s$ generate M as an R-module. [Use Nakayama's Lemma.]
  (b) Conclude from (a) that there is an exact sequence $0 \to \ker \varphi \to R^s \xrightarrow{\varphi} M \to 0$ that maps a set of free generators of R^s to the elements $m_1, \ldots, m_s$. Deduce that there is

an exact sequence

$$\mathrm{Tor}_1^R(M, R/\mathfrak{m}) \longrightarrow (\ker \varphi)/\mathfrak{m}(\ker \varphi) \longrightarrow 0.$$

[Use the Tor long exact sequence with respect to tensoring with $R/\mathfrak{m}$, using the fact that $N \otimes R/\mathfrak{m} \cong N/\mathfrak{m}N$ for any R-module N (Example 8 following Corollary 12 in Section 10.4] and the fact that $\varphi : (R/\mathfrak{m})^s \cong M/\mathfrak{m}M$ is an isomorphism by the choice of $m_1, \dots, m_s$.]

(c) Prove that if $\mathrm{Tor}_1^R(M, R/\mathfrak{m}) = 0$ then $m_1, \dots, m_s$ are a set of *free* R-module generators for M. [Use the previous exercise and Nakayama's Lemma to show that $\ker \varphi = 0$.]

29. Suppose R is a local ring with unique maximal ideal $\mathfrak{m}$. This exercise proves that a finitely generated R-module is flat if and only if it is free.

   (a) Prove that $M = F/K$ is the quotient of a finitely generated free module F by a submodule K with $K \subseteq \mathfrak{m}F$. [Let F be a free module with $F/\mathfrak{m}F \cong M/\mathfrak{m}M$.]

   (b) Suppose $x \in K$ and write $x = a_1 e_1 + \cdots + a_n e_n$ where $e_1, \dots, e_n$ are an R-basis for F. Let $I = (a_1, \dots, a_n)$ be the ideal of R generated by $a_1, \dots, a_n$). Prove that if M is flat, then $I = \mathfrak{m}I$ and deduce that $K = 0$, so M is free. [Use Exercise 25(d) of Section 10.5 to see that $x \in IK \subseteq \mathfrak{m}IF$ and conclude that $I \subseteq \mathfrak{m}I$. Then apply Nakayama's Lemma to the finitely generated ideal I.]

30. Suppose R is a local ring with unique maximal ideal $\mathfrak{m}$, M is an R-module, and consider the following statements:

   (i) M is a free R-module,
   (ii) M is a projective R-module,
   (iii) M is a flat R-module, and
   (iv) $\mathrm{Tor}_1^R(M, R/\mathfrak{m}) = 0$.

   (a) Prove that (i) implies (ii) implies (iii) implies (iv).

   (b) Prove that (i), (ii), and (iii) are equivalent if M is finitely generated. (Exercise 34 below shows (iii) need not imply (i) or (ii) if M is finitely generated but R is not local.) [Use the previous exercise.]

   (c) Prove that (i), (ii), (iii), and (iv) are equivalent if M is finitely presented. (Exercise 35 below shows that (iv) need not imply (i), (ii) or (iii) if M is finitely generated but not finitely presented.) [Use Exercise 28.]

   Remark: It is a theorem of Kaplansky (cf. *Projective Modules*, Annals of Mathematics, 68(1958), pp. 372-377) that (i) and (ii) are equivalent without the condition that M be finitely generated.

31. (*Localization and* Hom *for Finitely Presented Modules*) Suppose $D^{-1}R$ is the localization of the commutative ring R with respect to the multiplicative subset D of R, and let M be a finitely presented R-module.

   (a) For any R-modules A and B prove there is a unique $D^{-1}R$-module homomorphism from $D^{-1}\mathrm{Hom}_R(A, B)$ to $\mathrm{Hom}_{D^{-1}R}(D^{-1}A, D^{-1}B)$ that maps $\varphi \in \mathrm{Hom}_R(A, B)$ to the homomorphism from $D^{-1}A$ to $D^{-1}B$ induced by φ.

   (b) For any R-module N and any $m \geq 1$ show that $\mathrm{Hom}_R(R^m, N) \cong N^m$ as R-modules and deduce that $D^{-1}\mathrm{Hom}_R(R^m, N) \cong (D^{-1}N)^m$ as $D^{-1}R$-modules.

   (c) Suppose $R^s \longrightarrow R^t \longrightarrow M \longrightarrow 0$ is exact. Prove there is a commutative diagram

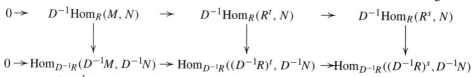

   of $D^{-1}R$-modules with exact rows. [For the first row first take R-module homomorphisms from the terms in the presentation for M into N using Theorem 33 of Section 10.5 (noting the first comment in the proof) and then tensor with the flat R-module $D^{-1}R$, cf. Propositions 41 and 42(6) in Section 15.4. For the second row first tensor the presentation with $D^{-1}R$ and then take $D^{-1}R$-module homomorphisms into $D^{-1}N$.]

   (d) Use (b) to prove that localization commutes with taking homomorphisms when M is finitely presented, i.e., $D^{-1}\text{Hom}_R(M, N) \cong \text{Hom}_{D^{-1}R}(D^{-1}M, D^{-1}N)$ as $D^{-1}R$-modules. [Show the second two vertical maps in the diagram above are isomorphisms and deduce that the left vertical map is also an isomorphism.] (This result is not true in general if M is not finitely presented.)

32. *(Localization and* Ext *for Finitely Presented Modules)* Suppose $D^{-1}R$ is the localization of the commutative ring R with respect to the multiplicative subset D of R. Prove that if M is a finitely presented R-module then $D^{-1}\text{Ext}^n_R(M, N) \cong \text{Ext}^n_{D^{-1}R}(D^{-1}M, D^{-1}N)$ as $D^{-1}R$-modules for every R-module N and every $n \geq 0$. [Use a projective resolution of N and the previous exercise, noting that tensoring the resolution with $D^{-1}R$ gives a projective resolution for the $D^{-1}R$-module $D^{-1}N$.]

33. Suppose R is a commutative ring and M is a finitely presented R-module (for example a finitely generated module over a Noetherian ring, or a quotient, R/I, of R by a finitely generated ideal I, cf. Exercises 26 and 27). Prove that the following are equivalent:
   (a) M is a projective R-module,
   (b) M is a flat R-module,
   (c) M is locally free, i.e., each localization M_P is a free R_P-module for every maximal (hence also for every prime) ideal P of R.

In particular show that finitely generated projective modules are the same as finitely presented flat modules. [Exercises 24 and 30 show that (b) is equivalent to (c). Use the Ext criterion for projectivity and Exercises 30 and 32 to see that (a) is equivalent to (c).]

34. **(a)** Prove that *every* R-module for the commutative ring R is flat if and only if every finitely generated ideal I of R is a direct summand of R, in which case every finitely generated ideal of R is principal and projective (such a ring is said to be *absolutely flat*). [Use Exercise 15, the previous exercise applied to the finitely presented R-module R/I, and the remarks following Proposition 16.]
   (b) Prove that every Boolean ring is absolutely flat. [Use Exercise 24 in Section 7.4, noting that if $I = Rx$ then x is an idempotent so $R = Rx \oplus R(1-x)$.]
   (c) Let R be the direct product and I the direct sum of countably many copies of $\mathbb{Z}/2\mathbb{Z}$. Prove that I is an ideal of the Boolean ring R that is not finitely generated and that the cyclic R-module $M = R/I$ is flat but not projective (so finitely generated flat modules need not be projective).

35. Let R be the local ring obtained by localizing the ring of C^∞ functions on the open interval $(-1, 1)$ at the maximal ideal of functions that are 0 at $x = 0$ (cf. Exercise 45 of Section 15.2), let $\mathfrak{m} = (x)$ be the unique maximal ideal of R and let P be the prime ideal $\cap_{n \geq 1}\mathfrak{m}^n$. Set $M = R/P$.
   (a) Prove that $\text{Tor}^R_1(M, R/\mathfrak{m}) = 0$. [Use Exercise 19 applied with $r = x$, noting that R/P is an integral domain.]
   (b) Prove that M is not flat (hence not projective). [Let F be as in Exercise 45 of Section 15.2. Show that the sequence $0 \to R \to R \to R/(F) \to 0$ induced by multiplication by F is exact, but is not exact after tensoring with M.]

17.2 THE COHOMOLOGY OF GROUPS

In this section we consider the application of the general techniques of the previous section in an important special case.

Let G be a group.

Definition. An abelian group A on which G acts (on the left) as automorphisms is called a *G-module*.

Note that a G-module is the same as an abelian group A and a homomorphism $\varphi : G \to \operatorname{Aut}(A)$ of G into the group of automorphisms of A. Since an abelian group is the same as a module over $\mathbb{Z}$, it is also easy to see that a G-module A is the same as a module over the integral group ring,$\mathbb{Z}G$, of G with coefficients in $\mathbb{Z}$. When G is an infinite group the ring $\mathbb{Z}G$ consists of all the finite formal sums of elements of G with coefficients in $\mathbb{Z}$.

As usual we shall often use multiplicative notation and write ga in place of $g \cdot a$ for the action of the element $g \in G$ on the element $a \in A$.

Definition. If A is a G-module, let $A^G = \{a \in A \mid ga = a \text{ for all } g \in G\}$ be the elements of A fixed by all the elements of G.

Examples

    (1) If $ga = a$ for all $a \in A$ and $g \in G$ then G is said to act *trivially* on A. In this case $A^G = A$. The abelian group $\mathbb{Z}$ will always be assumed to have trivial G-action for any group G unless otherwise stated.

    (2) For any G-module A the fixed points A^G of A under the action of G is clearly a $\mathbb{Z}G$-submodule of A on which G acts trivially.

    (3) If V is a vector space over the field F of dimension n and $G = GL_n(F)$ then V is naturally a G-module. In this case $V^G = \{0\}$ since any nonzero element in V can be taken to any other nonzero element in V by some linear transformation.

    (4) A semidirect product $E = A \rtimes G$ as in Section 5.5 in the case where A is an abelian normal subgroup gives a G-module A where the action of G is given by the homomorphism $\varphi : G \to \operatorname{Aut}(A)$. The subgroup A^G consists of the elements of A lying in the center of E. More generally, if A is any abelian normal subgroup of a group E, then E acts on A by conjugation and this makes A into a E-module and also an E/A-module. In this case $A^E = A^{E/A}$ also consists of the elements of A lying in the center of E.

    (5) If K/F is an extension of fields that is Galois with Galois group G then the additive group K is naturally a G-module, with $K^G = F$. Similarly, the multiplicative group $K^\times$ of nonzero elements in K is a G-module, with fixed points $(K^\times)^G = F^\times$.

The fixed point subgroups in this last example played a central role in Galois Theory in Chapter 14. In general, it is easy to see that a short exact sequence

$$0 \longrightarrow A \longrightarrow B \longrightarrow C \longrightarrow 0$$

of G-modules induces an exact sequence

$$0 \longrightarrow A^G \longrightarrow B^G \longrightarrow C^G \tag{17.15}$$

that in general cannot be extended to a short exact sequence (in general a coset in the quotient C that is fixed by G need not be represented by an *element* in B fixed by G). One way to see that (15) is exact is to observe that A^G can be related to a Hom group:

Lemma 19. Suppose A is a G-module and $\text{Hom}_{\mathbb{Z}G}(\mathbb{Z}, A)$ is the group of all $\mathbb{Z}G$-module homomorphisms from $\mathbb{Z}$ (with trivial G-action) to A. Then $A^G \cong \text{Hom}_{\mathbb{Z}G}(\mathbb{Z}, A)$.

Proof: Any G-module homomorphism α from $\mathbb{Z}$ to A is uniquely determined by its value on 1. Let α_a denote the G-module homomorphism with $\alpha(1) = a$. Since α_a is a G-module homomorphism, $a = \alpha_a(1) = \alpha_a(g \cdot 1) = g \cdot \alpha_a(1) = g \cdot a$ for all $g \in G$, so that a must lie in A^G. Likewise, for any $a \in A^G$ it is easy to check that the map $\alpha_a \mapsto a$ gives an isomorphism from $\text{Hom}_{\mathbb{Z}G}(\mathbb{Z}, A)$ to A^G.

Combined with the results of the previous section, the lemma not only shows that the sequence (15) is exact, it shows that any projective resolution of $\mathbb{Z}$ considered as a $\mathbb{Z}G$-module will give a long exact sequence extending (15). One such projective resolution is the *standard resolution* or *bar resolution* of $\mathbb{Z}$:

$$\cdots \longrightarrow F_n \xrightarrow{d_n} F_{n-1} \rightarrow \cdots \xrightarrow{d_1} F_0 \xrightarrow{\text{aug}} \mathbb{Z} \longrightarrow 0. \qquad (17.16)$$

Here $F_n = \mathbb{Z}G \otimes_{\mathbb{Z}} \mathbb{Z}G \otimes_{\mathbb{Z}} \cdots \otimes_{\mathbb{Z}} \mathbb{Z}G$ (where there are $n+1$ factors) for $n \geq 0$, which is a G-module under the action defined on simple tensors by $g \cdot (g_0 \otimes g_1 \otimes \cdots \otimes g_n) = (gg_0) \otimes g_1 \otimes \cdots \otimes g_n$. It is not difficult to see that F_n is a free $\mathbb{Z}G$-module of rank $|G|^n$ with $\mathbb{Z}G$ basis given by the elements $1 \otimes g_1 \otimes g_2 \otimes \cdots \otimes g_n$, where $g_i \in G$. The map $\text{aug} : F_0 \to \mathbb{Z}$ is the *augmentation map* $\text{aug}(\sum_{g \in G} \alpha_g g) = \sum_{g \in G} \alpha_g$, and the map d_1 is given by $d_1(1 \otimes g) = g - 1$. The maps d_n for $n \geq 2$ are more complicated and their definition, together with a proof that (16) is a projective (in fact, free) resolution can be found in Exercises 1–3.

Applying ($\mathbb{Z}G$-module) homomorphisms from the terms in (16) to the G-module A (replacing the first term by 0) as in the previous section, we obtain the cochain complex

$$0 \longrightarrow \text{Hom}_{\mathbb{Z}G}(F_0, A) \xrightarrow{d_1} \text{Hom}_{\mathbb{Z}G}(F_1, A) \xrightarrow{d_2} \text{Hom}_{\mathbb{Z}G}(F_2, A) \xrightarrow{d_3} \cdots, \quad (17.17)$$

the cohomology groups of which are, by definition, the groups $\text{Ext}^n_{\mathbb{Z}G}(\mathbb{Z}, A)$. Then, as in Theorem 8, the short exact sequence $0 \longrightarrow A \longrightarrow B \longrightarrow C \longrightarrow 0$ of G-modules gives rise to a long exact sequence whose first terms are given by (15) and whose higher terms are the cohomology groups $\text{Ext}^n_{\mathbb{Z}G}(\mathbb{Z}, A)$.

To make this more explicit, we can reinterpret the terms in this cochain complex without explicit reference to the standard resolution of $\mathbb{Z}$, as follows. The elements of $\text{Hom}_{\mathbb{Z}G}(F_n, A)$ are uniquely determined by their values on the $\mathbb{Z}G$ basis elements of F_n, which may be identified with the n-tuples $(g_1, g_2, \ldots, g_n)$ of elements g_i of G. It follows for $n \geq 1$ that the group $\text{Hom}_{\mathbb{Z}G}(F_n, A)$ may be identified with the set of functions from $G \times \cdots \times G$ (n copies) to A. For $n = 0$ we identify $\text{Hom}_{\mathbb{Z}G}(\mathbb{Z}G, A)$ with A.

Definition. If G is a finite group and A is a G-module, define $C^0(G, A) = A$ and for $n \geq 1$ define $C^n(G, A)$ to be the collection of all maps from $G^n = G \times \cdots \times G$ (n copies) to A. The elements of $C^n(G, A)$ are called *n-cochains (of G with values in A)*.

Each $C^n(G, A)$ is an additive abelian group: for $C^0(G, A) = A$ given by the group structure on A; for $n \geq 1$ given by the usual pointwise addition of functions: $(f_1 + f_2)(g_1, g_2, \ldots, g_n) = f_1(g_1, g_2, \ldots, g_n) + f_2(g_1, g_2, \ldots, g_n)$. Under the identification of $\mathrm{Hom}_{\mathbb{Z}G}(F_n, A)$ with $C^n(G, A)$ the cochain maps d_n in (17) can be given very explicitly (cf. also Exercise 3 and the following comment):

Definition. For $n \geq 0$, define the nth *coboundary* homomorphism from $C^n(G, A)$ to $C^{n+1}(G, A)$ by

$$d_n(f)(g_1, \ldots, g_{n+1}) = g_1 \cdot f(g_2, \ldots, g_{n+1})$$
$$+ \sum_{i=1}^{n} (-1)^i f(g_1, \ldots, g_{i-1}, g_i g_{i+1}, g_{i+2}, \ldots, g_{n+1})$$
$$+ (-1)^{n+1} f(g_1, \ldots, g_n) \qquad (17.18)$$

where the product $g_i g_{i+1}$ occupying the ith position of f is taken in the group G.

It is immediate from the definition that the maps d_n are group homomorphisms. It follows from the fact that (17) is a projective resolution that $d_n \circ d_{n-1} = 0$ for $n \geq 1$ (a self contained direct proof just from the definition of d_n above can also be given, but is tedious).

Definition.
(1) Let $Z^n(G, A) = \ker d_n$ for $n \geq 0$. The elements of $Z^n(G, A)$ are called n-*cocycles*.
(2) Let $B^n(G, A) = \mathrm{image}\, d_{n-1}$ for $n \geq 1$ and let $B^0(G, A) = 1$. The elements of $B^n(G, A)$ are called n-*coboundaries*.

Since $d_n \circ d_{n-1} = 0$ for $n \geq 1$ we have $\mathrm{image}\, d_{n-1} \subseteq \ker d_n$, so that $B^n(G, A)$ is always a subgroup of $Z^n(G, A)$.

Definition. For any G-module A the quotient group $Z^n(G, A)/B^n(G, A)$ is called the nth *cohomology group of G with coefficients in A* and is denoted by $H^n(G, A)$, $n \geq 0$.

The definition of the cohomology group $H^n(G, A)$ in terms of cochains will be particularly useful in the following two sections when we examine the low dimensional groups $H^1(G, A)$ and $H^2(G, A)$ and their application in a variety of settings. It should be remembered, however, that $H^n(G, A) \cong \mathrm{Ext}_{\mathbb{Z}G}^n(\mathbb{Z}, A)$ for all $n \geq 0$. In particular, these groups can be computed using *any* projective resolution of $\mathbb{Z}$.

Examples
(1) For $f = a \in C^0(G, A)$ we have $d_0(f)(g) = g \cdot a - a$ and so $\ker d_0$ is the set $\{a \in A \mid g \cdot a = a \text{ for all } g \in G\}$, i.e., $Z^0(G, A) = A^G$ and so

$$H^0(G, A) = A^G,$$

for any group G and G-module A.

(2) Suppose $G = 1$ is the trivial group. Then $G^n = \{(1, 1, \ldots, 1)\}$ is also the trivial group, so $f \in C^n(G, A)$ is completely determined by $f(1, 1, \ldots, 1) = a \in A$. Identifying $f = a$ we obtain $C^n(G, A) = A$ for all $n \geq 0$. Then, if $f = a \in A$,

$$d_n(f)(1, 1, \ldots, 1) = a + \sum_{i=1}^{n}(-1)^i a + (-1)^{n+1}a = \begin{cases} 0 & \text{if } n \text{ is even} \\ a & \text{if } n \text{ is odd} \end{cases},$$

so $d_n = 0$ if n is even and $d_n = 1$ is the identity if n is odd. Hence

$$H^0(1, A) = A^G = A$$

$$H^n(1, A) = 0 \text{ for all } n \geq 1.$$

Example: (Cohomology of a Finite Cyclic Group)

Suppose G is cyclic of order m with generator σ. Let $N = 1 + \sigma + \sigma^2 + \cdots + \sigma^{m-1} \in \mathbb{Z}G$. Then $N(\sigma - 1) = (\sigma - 1)N = \sigma^m - 1 = 0$, and so we have a particularly simple free resolution

$$\cdots \xrightarrow{\sigma-1} \mathbb{Z}G \xrightarrow{N} \mathbb{Z}G \xrightarrow{\sigma-1} \cdots \xrightarrow{N} \mathbb{Z}G \xrightarrow{\sigma-1} \mathbb{Z}G \xrightarrow{\text{aug}} \mathbb{Z} \longrightarrow 0$$

where aug denotes the augmentation map (cf. Exercise 8). Taking $\mathbb{Z}G$-module homomorphisms from the terms of this resolution to A (replacing the first term by 0) and using the identification $\text{Hom}_{\mathbb{Z}G}(\mathbb{Z}G, A) = A$ gives the chain complex

$$0 \longrightarrow A \xrightarrow{\sigma-1} A \xrightarrow{N} A \xrightarrow{\sigma-1} A \xrightarrow{N} \cdots$$

whose cohomology computes the groups $H^n(G, A)$:

$$H^0(G, A) = A^G, \text{ and } H^n(G, A) = \begin{cases} A^G/NA & \text{if } n \text{ is even, } n \geq 2 \\ {}_NA/(\sigma - 1)A & \text{if } n \text{ is odd, } n \geq 1 \end{cases}$$

where ${}_NA = \{a \in A \mid Na = 0\}$ is the subgroup of A annihilated by N, since the kernel of multiplication by $\sigma - 1$ is A^G.

If in particular $G = \langle \sigma \rangle$ acts trivially on A, then $N \cdot a = ma$, so that in this case $H^0(G, A) = A$, with $H^n(G, A) = A/mA$ for even $n \geq 2$, and $H^n(G, A) = {}_mA$, the elements of A of order dividing m, for odd $n \geq 1$. Specializing even further to $m = 1$ gives Example 2 previously.

Proposition 20. Suppose $mA = 0$ for some integer $m \geq 1$ (i.e., the G-module A has exponent dividing m as an abelian group). Then

$$mZ^n(G, A) = mB^n(G, A) = mH^n(G, A) = 0 \quad \text{for all } n \geq 0.$$

In particular, if A has exponent p for some prime p then the abelian groups $Z^n(G, A)$, $B^n(G, A)$ and $H^n(G, A)$ have exponent dividing p and so these groups are all vector spaces over the finite field $\mathbb{F}_p = \mathbb{Z}/p\mathbb{Z}$.

Proof: If $f \in C^n(G, A)$ is an n-cochain then $f \in A$ (if $n = 0$), in which case $mf = 0$, or f is a function from G^n to A (if $n \geq 1$), in which case mf is a function from G^n to $mA = 0$, so again $mf = 0$. Hence $mZ^n(G, A) = mB^n(G, A) = 0$ since these are subgroups of $C^n(G, A)$. Then $mH^n(G, A) = 0$ since $mZ^n(G, A) = 0$, and the remaining statements in the proposition are immediate.

By Example 1, the long exact sequence in Theorem 10 written in terms of the cohomology groups $H^n(G, A)$ becomes

Theorem 21. *(Long Exact Sequence in Group Cohomology)* Suppose

$$0 \longrightarrow A \longrightarrow B \longrightarrow C \longrightarrow 0$$

is a short exact sequence of G-modules. Then there is a long exact sequence:

$$0 \longrightarrow A^G \longrightarrow B^G \longrightarrow C^G \xrightarrow{\delta_0} H^1(G, A) \longrightarrow H^1(G, B) \longrightarrow H^1(G, C) \xrightarrow{\delta_1} \cdots$$

$$\cdots \xrightarrow{\delta_{n-1}} H^n(G, A) \longrightarrow H^n(G, B) \longrightarrow H^n(G, C) \xrightarrow{\delta_n} H^{n+1}(G, A) \longrightarrow \cdots$$

of abelian groups.

Among many other uses of the long exact sequence in Theorem 21 is a technique called *dimension shifting* which makes it possible to analyze the cohomology group $H^{n+1}(G, A)$ of dimension $n + 1$ for A by instead considering a cohomology group of dimension n for a different G-module. The technique is based on finding a G-module almost all of whose cohomology groups are zero. Such modules are given a name:

Definition. A G-module M is called *cohomologically trivial for G* if $H^n(G, M) = 0$ for all $n \geq 1$.

Corollary 22. *(Dimension Shifting)* Suppose $0 \to A \to M \to C \to 0$ is a short exact sequence of G-modules and that M is cohomologically trivial for G. Then there is an exact sequence

$$0 \longrightarrow A^G \longrightarrow M^G \longrightarrow C^G \longrightarrow H^1(G, A) \longrightarrow 0$$

and

$$H^{n+1}(G, A) \cong H^n(G, C) \text{ for all } n \geq 1.$$

Proof: Since M is cohomologically trivial for G, the portion

$$H^n(G, M) \longrightarrow H^n(G, C) \longrightarrow H^{n+1}(G, A) \longrightarrow H^{n+1}(G, M)$$

of the long exact sequence in Theorem 21 reduces to

$$0 \longrightarrow H^n(G, C) \longrightarrow H^{n+1}(G, A) \longrightarrow 0$$

which shows that $H^n(G, C) \cong H^{n+1}(G, A)$ for $n \geq 1$. Similarly, the first portion of the long exact sequence in Theorem 21 gives the first statement in the corollary.

We now indicate a natural construction that produces a G-module given a module over a subgroup H of G. When $H = 1$ is the trivial group this construction produces a cohomologically trivial module M and an exact sequence as in Corollary 22 for any G-module A.

Definition. If H is a subgroup of G and A is an H-module, define the *induced G-module* $M_H^G(A)$ to be $\text{Hom}_{\mathbb{Z}H}(\mathbb{Z}G, A)$. In other words, $M_H^G(A)$ is the set of maps f from G to A satisfying $f(hx) = hf(x)$ for every $x \in G$ and $h \in H$.

The action of an element $g \in G$ on $f \in M_H^G(A)$ is given by $(g \cdot f)(x) = f(xg)$ for $x \in G$ (cf. Exercise 10 in Section 10.5).

Recall that if H is a subgroup of G and A is an H-module, then the module $\mathbb{Z}G \otimes_{\mathbb{Z}H} A$ obtained by extension of scalars from $\mathbb{Z}H$ to $\mathbb{Z}G$ is a G-module. For a finite group G, or more generally if H has finite index in G, we have $M_H^G(A) \cong \mathbb{Z}G \otimes_{\mathbb{Z}H} A$ (cf. Exercise 10). When G is infinite this need no longer be the case (cf. Exercise 11). The module $\mathbb{Z}G \otimes_{\mathbb{Z}H} A$ is sometimes called the *induced G-module* and the module $M_H^G(A)$ is sometimes referred to as the *coinduced G-module*. For finite groups, associativity of the tensor product shows that $M_H^G(M_K^H(A)) = M_K^G(A)$ for subgroups $K \le H \le G$, and the same result holds in general (this follows from the definition using Exercise 7).

Examples

(1) If H is a subgroup of G and $0 \to A \to B \to C \to 0$ is a short exact sequence of H-modules then $0 \to M_H^G(A) \to M_H^G(B) \to M_H^G(C) \to 0$ is a short exact sequence of G-modules, since $M_H^G(A) \cong \mathbb{Z}G \otimes_{\mathbb{Z}H} A$ and $\mathbb{Z}G$ is free, hence flat, over $\mathbb{Z}H$.

(2) When G is finite and A is the trivial H-module $\mathbb{Z}$, the module $M_H^G(\mathbb{Z})$ is a free $\mathbb{Z}$-module of rank $m = |G : H|$. There is a basis $b_1, \dots, b_m$ such that G permutes these basis elements in the same way it permutes the left cosets of H in G by left multiplication, i.e., if we let $b_i \leftrightarrow g_i H$ then $gb_i = b_j$ if and only if $gg_i H = g_j H$. The module $M_H^G(\mathbb{Z})$ is the *permutation module* over $\mathbb{Z}$ for G with stabilizer H. A special case of interest is when $G = S_m$ and $H = S_{m-1}$ where S_m permutes $\{1, 2, \dots, m\}$ as usual. Permutation modules and induced modules over fields are studied in Part VI.

(3) Any abelian group A is an H-module when $H = 1$ is the trivial group. The corresponding induced G-module $M_1^G(A)$ is just the collection of all maps f from G into A. For $g \in G$ the map $g \cdot f \in M_1^G(A)$ satisfies $(g \cdot f)(x) = f(xg)$ for $x \in G$.

(4) Suppose A is a G-module. Then there is a natural map

$$\varphi : A \longrightarrow M_1^G(A)$$

from A into the induced G-module $M_1^G(A)$ in the previous example defined by mapping $a \in A$ to the function f_a with $f_a(x) = xa$ for all $x \in G$. It is clear that φ is a group homomorphism, and $f_{ga}(x) = x(ga) = (xg)a = f_a(xg) = (g \cdot f_a)(x)$ shows that φ is a G-module homomorphism as well. Since $f_a(1) = a$, it follows that f_a is the zero function on G if and only if $a = 0$ in A, so that φ is an injection. Hence we may identify A as a G-submodule of the induced module $M_1^G(A)$.

(5) More generally, if A is a G-module and H is any subgroup of G then the function $f_a(x)$ in the previous example is an element in the subgroup $M_H^G(A)$ since we have $f_a(hx) = (hx)(a) = h(xa) = hf_a(x)$ for all $h \in H$. The associated map from A to $M_H^G(A)$ is an injective G-module homomorphism.

(6) The fixed points $(M_H^G(A))^G$ are maps f from G to A with $gf = f$ for all $g \in G$, i.e., with $(gf)(x) = f(x)$ for all $g, x \in G$. By definition of the G-action on $M_H^G(A)$, this is the equation $f(xg) = f(x)$ for all $g, x \in G$. Taking $x = 1$ shows that f is constant on all of G: $f(g) = f(1) = a \in A$. The constant function $f = a$ is an element of $M_H^G(A)$ if and only if $a = f(hx) = hf(x) = ha$ for all $h \in H$, so $(M_H^G(A))^G \cong A^H$.

An element $f_a(x)$ in the previous example is contained in the subgroup $(M_H^G(A))^G$ if and only if xa is constant for $x \in G$, i.e., if and only if $a \in A^G$.

One of the important properties of the G-module $M_H^G(A)$ induced from the H-module A is that its cohomology with respect to G is the same as the cohomology of A with respect to H:

Proposition 23. *(Shapiro's Lemma)* For any subgroup H of G and any H-module A we have $H^n(G, M_H^G(A)) \cong H^n(H, A)$ for $n \geq 0$.

Proof: Let $\cdots \to P_n \to \cdots \to P_0 \to \mathbb{Z} \to 0$ be a resolution of $\mathbb{Z}$ by projective G-modules (for example, the standard resolution). The cohomology groups $H^n(G, M_H^G(A))$ are computed by taking homomorphisms from this resolution into $M_H^G(A) = \mathrm{Hom}_{\mathbb{Z}H}(\mathbb{Z}G, A)$. Since $\mathbb{Z}G$ is a free $\mathbb{Z}H$-module it follows that this G-module resolution is also a resolution of $\mathbb{Z}$ by projective H-modules, hence by taking homomorphisms into A the same resolution may be used to compute the cohomology groups $H^n(H, A)$. To see that these two collections of cohomology groups are isomorphic, we use the natural isomorphism of abelian groups

$$\Phi : \mathrm{Hom}_{\mathbb{Z}G}(P_n, \mathrm{Hom}_{\mathbb{Z}H}(\mathbb{Z}G, A)) \cong \mathrm{Hom}_{\mathbb{Z}H}(P_n, A)$$

given by $\Phi(f)(p) = f(p)(1)$, for all $f \in \mathrm{Hom}_{\mathbb{Z}G}(P_n, \mathrm{Hom}_{\mathbb{Z}H}(\mathbb{Z}G, A))$ and $p \in P_n$. The inverse isomorphism is defined by taking $\Psi(f')(p)$ to be the map from $\mathbb{Z}G$ to A that takes $g \in G$ to the element $f'(gp)$ in A for all $f' \in \mathrm{Hom}_{\mathbb{Z}H}(P_n, A)$ and $p \in P_n$, i.e., $(\Psi(f')(p))(g) = f'(gp)$. Note this is well defined because P_n is a G-module. (These maps are a special case of an Adjoint Associativity Theorem, cf. Exercise 7.) Since these isomorphisms commute with the cochain maps, they induce isomorphisms on the corresponding cohomology groups, i.e., $H^n(G, M_H^G(A)) \cong H^n(H, A)$, as required.

Corollary 24. For any G-module A the module $M_1^G(A)$ is cohomologically trivial for G, i.e., $H^n(G, M_1^G(A)) = 0$ for all $n \geq 1$.

Proof: This follows immediately from the proposition applied with $H = 1$ together with the computation of the cohomology of the trivial group in Example 2 preceding Proposition 20.

By the corollary, the fourth example above gives us a short exact sequence of G-modules

$$0 \longrightarrow A \xrightarrow{\varphi} M \longrightarrow C \longrightarrow 0$$

where $M = M_1^G(A)$ is cohomologically trivial for G and where C is the quotient of $M_1^G(A)$ by the image of A. The dimension shifting result in Corollary 22 then becomes:

Corollary 25. For any G-module A we have $H^{n+1}(G, A) \cong H^n(G, M_1^G(A)/A)$ for all $n \geq 1$.

We next consider several important maps relating various cohomology groups. Some applications of the use of these homomorphisms appear in the following two sections.

In general, suppose we have two groups G and G' and that A is a G-module and A' is a G'-module. If $\varphi : G' \to G$ is a group homomorphism then A becomes a G'-module by defining $g' \cdot a = \varphi(g')a$ for $g' \in G'$ and $a \in A$. If now $\psi : A \to A'$ is a homomorphism of abelian groups then we consider whether ψ is a G'-module homomorphism:

Definition. Suppose A is a G-module and A' is a G'-module. The group homomorphisms $\varphi : G' \to G$ and $\psi : A \to A'$ are said to be *compatible* if ψ is a G'-module homomorphism when A is made into a G'-module by means of φ, i.e., if $\psi(\varphi(g')a) = g'\psi(a)$ for all $g' \in G'$ and $a \in A$.

The point of compatible homomorphisms is that they induce group homomorphisms on associated cohomology groups, as follows.

If $\varphi : G' \to G$ and $\psi : A \to A'$ are homomorphisms, then φ induces a homomorphism $\varphi^n : (G')^n \to G^n$, and so a homomorphism from $C^n(G, A)$ to $C^n(G', A)$ that maps f to $f \circ \varphi^n$. The map ψ induces a homomorphism from $C^n(G', A)$ to $C^n(G', A')$ that maps f to $\psi \circ f$. Taken together we obtain an induced homomorphism

$$\lambda_n : C^n(G, A) \longrightarrow C^n(G', A')$$
$$f \longmapsto \psi \circ f \circ \varphi^n.$$

If in addition φ and ψ are *compatible* homomorphisms, then it is easy to check that the induced maps λ_n commute with the coboundary operator:

$$\lambda_{n+1} \circ d_n = d_n \circ \lambda_n$$

for all $n \geq 0$. It follows that λ_n maps cocycles to cocycles and coboundaries to coboundaries, hence induces a group homomorphism on cohomology:

$$\lambda_n : H^n(G, A) \longrightarrow H^n(G', A')$$

for $n \geq 0$.

We consider several instances of such maps:

Examples

(1) Suppose $G = G'$ and φ is the identity map. Then to say that the group homomorphism $\psi : A \to A'$ is compatible with φ is simply the statement that ψ is a G-module homomorphism. Hence any G-module homomorphism from A to A' induces a group homomorphism

$$H^n(G, A) \longrightarrow H^n(G, A') \quad \text{for } n \geq 0.$$

In particular, if $0 \to A \to B \to C \to 0$ is a short exact sequence of G-modules we obtain induced homomorphisms from $H^n(G, A)$ to $H^n(G, B)$ and from $H^n(G, B)$ to $H^n(G, C)$ for $n \geq 0$. These are simply the homomorphisms in the long exact sequence of Theorem 21.

(2) *(The Restriction Homomorphism)* If A is a G-module, then A is also an H-module for any subgroup H of G. The inclusion map $\varphi : H \to G$ of H into G and the identity

map $\psi : A \to A$ are compatible homomorphisms. The corresponding induced group homomorphism on cohomology is called the *restriction homomorphism*:

$$\mathrm{Res} : H^n(G, A) \longrightarrow H^n(H, A), \quad n \geq 0.$$

The terminology comes from the fact that the map on cochains from $C^n(G, A)$ to $C^n(H, A)$ is simply restricting a map f from G^n to A to the subgroup H^n of G^n.

(3) *(The Inflation Homomorphism)* Suppose H is a normal subgroup of G and A is a G-module. The elements A^H of A that are fixed by H are naturally a module for the quotient group G/H under the action defined by $(gH)\cdot a = g\cdot a$. It is then immediate that the projection $\varphi : G \to G/H$ and the inclusion $\psi : A^H \to A$ are compatible homomorphisms. The corresponding induced group homomorphism on cohomology is called the *inflation homomorphism*:

$$\mathrm{Inf} : H^n(G/H, A^H) \longrightarrow H^n(G, A), \quad n \geq 0.$$

(4) *(The Corestriction Homomorphism)* Suppose that H is a subgroup of G of index m and that A is a G-module. Let $g_1, \ldots, g_m$ be representatives for the left cosets of H in G. Define a map

$$\psi : M_H^G(A) \longrightarrow A \quad \text{by} \quad f \longmapsto \sum_{i=1}^{m} g_i \cdot f(g_i^{-1}).$$

Note that if we change any coset representative g_i by $g_i h$, then $(g_i h) f((g_i h)^{-1}) = g_i h f(h^{-1} g_i^{-1}) = g_i h h^{-1} f(g_i^{-1}) = g_i f(g_i^{-1})$ so the map ψ is independent of the choice of coset representatives. It is easy to see that ψ is a G-module homomorphism (and even that it is surjective), so we obtain a group homomorphism from $H^n(G, M_H^G(A))$ to $H^n(G, A)$, for all $n \geq 0$. Since A is also an H-module, by Shapiro's Lemma we have an isomorphism $H^n(G, M_H^G(A)) \cong H^n(H, A)$. The composition of these two homomorphisms is called the *corestriction homomorphism*:

$$\mathrm{Cor} : H^n(H, A) \longrightarrow H^n(G, A), \quad n \geq 0.$$

This homomorphism can be computed explicitly by composing the isomorphism Ψ in the proof of Shapiro's Lemma for any resolution of $\mathbb{Z}$ by projective G-modules P_n (note these are G-modules and not simply H-modules) with the map ψ, as follows. For a cocycle $f \in \mathrm{Hom}_{\mathbb{Z}H}(P_n, A)$ representing a cohomology class $c \in H^n(H, A)$, a cocycle $\mathrm{Cor}(f) \in \mathrm{Hom}_{\mathbb{Z}G}(P_n, A)$ representing $\mathrm{Cor}(c) \in H^n(G, A)$ is given by

$$\mathrm{Cor}(f)(p) = \sum_{i=1}^{m} g_i \cdot \Psi(f)(p)(g_i^{-1}) = \sum_{i=1}^{m} g_i f(g_i^{-1} p),$$

for $p \in P_n$. When $n = 0$ this is particularly simple since we can take $P_0 = \mathbb{Z}G$. In this case $f \in \mathrm{Hom}_{\mathbb{Z}H}(\mathbb{Z}G, A) = M_H^G(A)$ is a cocycle if $f = a$ is constant on G for some $a \in A^H$ and then $\mathrm{Cor}(f)$ is the constant function with value $\sum_{i=1}^{m} g_i \cdot a \in A^G$:

$$\mathrm{Cor} : H^0(H, A) = A^H \longrightarrow A^G = H^0(G, A)$$

$$a \longmapsto \sum_{i=1}^{m} g_i \cdot a.$$

The next result establishes a fundamental relation between the restriction and corestriction homomorphisms.

Proposition 26. Suppose H is a subgroup of G of index m. Then $\text{Cor} \circ \text{Res} = m$, i.e., if c is a cohomology class in $H^n(G, A)$ for some G-module A, then

$$\text{Cor}(\text{Res}(c)) = mc \in H^n(G, A) \quad \text{for all } n \geq 0.$$

Proof: This follows from the explicit formula for corestriction in Example 4 above, as follows. If $f \in \text{Hom}_{\mathbb{Z}H}(P_n, A)$ were in $\text{Hom}_{\mathbb{Z}G}(P_n, A)$, i.e., if f were also a G-module homomorphism, then $g_i f(g_i^{-1} p) = g_i g_i^{-1} f(p) = f(p)$, for $1 \leq i \leq m$. Since restriction is the induced map on cohomology of the natural inclusion of $\text{Hom}_{\mathbb{Z}G}(P_n, A)$ into $\text{Hom}_{\mathbb{Z}H}(P_n, A)$, for such an f we obtain

$$\text{Hom}_{\mathbb{Z}G}(P_n, A) \xrightarrow{\text{Res}} \text{Hom}_{\mathbb{Z}H}(P_n, A) \xrightarrow{\text{Cor}} \text{Hom}_{\mathbb{Z}G}(P_n, A)$$
$$f \longmapsto f \longmapsto mf.$$

It follows that $\text{Res} \circ \text{Cor}$ is multiplication by m on the cohomology groups as well.

Corollary 27. Suppose the finite group G has order m. Then $m H^n(G, A) = 0$ for all $n \geq 1$ and any G-module A.

Proof: Let $H = 1$, so that $[G : H] = m$, in Proposition 26. Then for any class $c \in H^n(G, A)$ we have $mc = \text{Cor}(\text{Res}(c))$. Since $\text{Res}(c) \in H^n(H, A) = H^n(1, A)$, we have $\text{Res}(c) = 0$ for all $n \geq 1$ by the second example preceding Proposition 20. Hence $mc = 0$ for all $n \geq 1$, which is the corollary.

Corollary 28. If G is a finite group then $H^n(G, A)$ is a torsion abelian group for all $n \geq 1$ and all G-modules A.

Proof: This is immediate from the previous corollary.

Corollary 29. Suppose G is a finite group whose order is relatively prime to the exponent of the G-module A. Then $H^n(G, A) = 0$ for all $n \geq 1$. In particular, if A is a finite abelian group with $(|G|, |A|) = 1$ then $H^n(G, A) = 0$ for all $n \geq 1$.

Proof: This follows since the abelian group $H^n(G, A)$ is annihilated by $|G|$ by the previous corollary and is annihilated by the exponent of A by Proposition 20.

Note that the statements in the preceding corollaries are not in general true for $n = 0$, since then $H^0(G, A) = A^G$, which need not even be torsion.

We mention without proof the following result. Suppose that H is a normal subgroup of G and A is a G-module. The cohomology groups $H^n(H, A)$ can be given the structure of G/H-modules (cf. Exercise 17). It can be shown that there is an exact sequence

$$0 \to H^1(G/H, A^H) \xrightarrow{\text{Inf}} H^1(G, A) \xrightarrow{\text{Res}} H^1(H, A)^{G/H} \xrightarrow{\text{Tra}} H^2(G/H, A^H) \xrightarrow{\text{Inf}} H^2(G, A)$$

where $H^1(H, A)^{G/H}$ denotes the fixed points of $H^1(H, A)$ under the action of G/H and Tra is the so-called *transgression homomorphism*. This exact sequence relates the cohomology groups for G to the cohomology groups for the normal subgroup H and for the quotient group G/H. Put another way, the cohomology for G is related to the cohomology for the factors in the filtration $1 \leq H \leq G$ for G. More generally, one could try to relate the cohomology for G to the cohomology for the factors in a longer filtration for G. This is the theory of *spectral sequences* and is an important tool in homological algebra.

Galois Cohomology and Profinite Groups

One important application of group cohomology occurs when the group G is the Galois group of a field extension K/F. In this case there are many groups of interest on which G acts, for example the additive group of K, the multiplicative group $K^\times$, etc. The Galois group $G = \mathrm{Gal}(K/F)$ is the inverse limit $\varprojlim \mathrm{Gal}(L/F)$ of the Galois groups of the finite extensions L of F contained in K and is a compact topological group with respect to its Krull topology (i.e., the group operations on G are continuous with respect to the topology defined by the subgroups $\mathrm{Gal}(K/L)$ of G of finite index), cf. Section 14.9. In this situation it is useful (and often essential) to take advantage of the additional topological structure of G. For example the subfields of K containing F correspond bijectively with the *closed* subgroups of $G = \mathrm{Gal}(K/F)$, and the example of the composite of the quadratic extensions of $\mathbb{Q}$ discussed in Section 14.9 shows that in general there are many subgroups of G that are not closed. Fortunately, the modifications necessary to define the cohomology groups in this context are relatively minor and apply to arbitrary inverse limits of finite groups (the *profinite* groups). If G is a profinite group then $G = \varprojlim G/N$ where the inverse limit is taken over the open normal subgroups N of G (cf. Exercise 23).

Definition. If G is a profinite group then a *discrete G-module A* is a G-module A with the discrete topology such that the action of G on A is continuous, i.e., the map $G \times A \to A$ mapping (g, a) to $g \cdot a$ is continuous.

Since A is given the discrete topology, every subset of A is open, and in particular every element $a \in A$ is open. The continuity of the action of G on A is then equivalent to the statement that the stabilizer G_a of a in G is an open subgroup of G, hence is of finite index since G is compact (cf. Exercise 22). This in turn is equivalent to the statement that $A = \cup A^H$ where the union is over the open subgroups H of G.

Some care must be taken in defining the cohomology groups $H^n(G, A)$ of a profinite group G acting on a discrete G-module A since there are not enough projectives in this category. For example, when G is infinite, the free G-module $\mathbb{Z}G$ is not a discrete G-module (G does not act continuously, cf. Exercise 25). Nevertheless, the explicit description of $H^n(G, A)$ given in this section (occasionally referred to as the *discrete* cohomology groups) can be easily modified — it is only necessary to require the cochains $C^n(G, A)$ to be *continuous* maps from G^n to A. The definition of the coboundary maps d_n in equation (18) is precisely the same, as is the definition of the groups of cocycles, coboundaries, and the corresponding cohomology groups. It is customary not to introduce a separate notation for these cohomology groups, but to specify which cohomology is meant in the terminology.

Definition. If G is a profinite group and A is a discrete G-module, the cohomology groups $H^n(G, A)$ computed using continuous cochains are called the *profinite* or *continuous* cohomology groups. When $G = \mathrm{Gal}(K/F)$ is the Galois group of a field extension K/F then the *Galois cohomology groups* $H^n(G, A)$ will always mean the cohomology groups computed using continuous cochains.

When G is a finite group, every G-module is a discrete G-module so the discrete and continuous cohomology groups of G are the same. When G is infinite, this need not be the case as shown by the example mentioned previously of the free G-module $\mathbb{Z}G$ when G is an infinite profinite group. All the major results in this section remain valid for the continuous cohomology‘ groups when "G-module" is replaced by "discrete G-module" and "subgroup" is replaced by "closed subgroup." For example, the Long Exact Sequence in Group Cohomology remains true as stated, the restriction homomorphism requires the subgroup H of G to be a closed subgroup (so that the restriction of a continuous map on G^n to H^n remains continuous), Proposition 26 requires H to be closed, etc.

We can write $G = \varprojlim(G/N)$ and $A = \cup A^N$ where N runs over the open normal subgroups of G (necessarily of finite index in G since G is compact). Then A^N is a discrete G/N-module and it is not difficult to show that

$$H^n(G, A) = \varinjlim_N H^n(G/N, A^N) \tag{17.19}$$

where the cohomology groups are continuous cohomology and the direct limit is taken over the collection of all open normal subgroups N of G (cf. Exercise 24). Since G/N is a finite group, the continuous cohomology groups $H^n(G/N, A^N)$ in this direct limit are just the (discrete) cohomology groups considered earlier in this section. The computation of the continuous cohomology for a profinite group G can therefore always be reduced to the consideration of finite group cohomology where there is no distinction between the continuous and discrete theories.

EXERCISES

1. Let $F_n = \mathbb{Z}G \otimes_{\mathbb{Z}} \mathbb{Z}G \otimes_{\mathbb{Z}} \cdots \otimes_{\mathbb{Z}} \mathbb{Z}G$ ($n+1$ factors) for $n \geq 0$ with G-action defined on simple tensors by $g \cdot (g_0 \otimes g_1 \otimes \cdots \otimes g_n) = (gg_0) \otimes g_1 \otimes \cdots \otimes g_n$.
   (a) Prove that F_n is a free $\mathbb{Z}G$-module of rank $|G|^n$ with $\mathbb{Z}G$ basis $1 \otimes g_1 \otimes g_2 \otimes \cdots \otimes g_n$ with $g_i \in G$.

   Denote the basis element $1 \otimes g_1 \otimes g_2 \otimes \cdots \otimes g_n$ in (a) by $(g_1, g_2, \ldots, g_n)$ and define the G-module homomorphisms d_n for $n \geq 1$ on these basis elements by $d_1(g_1) = g_1 - 1$ and

$$d_n(g_1, \ldots, g_n) = g_1 \cdot (g_2, \ldots, g_n) + \sum_{i=1}^{n-1} (-1)^i (g_1, \ldots, g_{i-1}, g_i g_{i+1}, g_{i+2}, \ldots, g_n)$$
$$+ (-1)^n (g_1, \ldots, g_{n-1}),$$

   for $n \geq 2$. Define the $\mathbb{Z}$-module *contracting homomorphisms*

$$\mathbb{Z} \xrightarrow{s_{-1}} F_0 \xrightarrow{s_0} F_1 \xrightarrow{s_1} F_2 \xrightarrow{s_2} \cdots$$

   on a $\mathbb{Z}$ basis by $s_{-1}(1) = 1$ and $s_n(g_0 \otimes \cdots \otimes g_n) = 1 \otimes g_0 \otimes \ldots \otimes g_n$.

(b) Prove that

$$\epsilon s_{-1} = 1, \qquad d_1 s_0 + s_{-1}\epsilon = 1, \qquad d_{n+1}s_n + s_{n-1}d_n = 1, \text{ for all } n \geq 1$$

where the map aug : $F_0 \to \mathbb{Z}$ is the augmentation map $\text{aug}(\sum_{g \in G} \alpha_g g) = \sum_{g \in G} \alpha_g$.

(c) Prove that the maps s_n are a chain homotopy (cf. Exercise 4 in Section 1) between the identity (chain) map and the zero (chain) map from the chain

$$\cdots \longrightarrow F_n \xrightarrow{d_n} F_{n-1} \xrightarrow{d_{n-1}} \cdots \xrightarrow{d_1} F_0 \xrightarrow{\text{aug}} \mathbb{Z} \longrightarrow 0 \qquad (*)$$

of $\mathbb{Z}$-modules to itself.

(d) Deduce from (c) that all $\mathbb{Z}$-module homology groups of $(*)$ are zero, i.e., $(*)$ is an exact sequence of $\mathbb{Z}$-modules. Conclude that $(*)$ is a projective G-module resolution of $\mathbb{Z}$.

2. Let P_n denote the free $\mathbb{Z}$-module with basis $(g_0, g_1, g_2, \ldots, g_n)$ with $g_i \in G$ and define an action of G on P_n by $g \cdot (g_0, g_1, \ldots, g_n) = (gg_0, gg_1, \ldots, gg_n)$. For $n \geq 1$ define

$$d_n(g_0, g_1, g_2, \ldots, g_n) = \sum_{i=0}^{n}(-1)^i (g_0, \ldots, \hat{g}_i, \ldots, g_n),$$

where $(g_0, \ldots, \hat{g}_i, \ldots, g_n)$ denotes the term $(g_0, g_1, g_2, \ldots, g_n)$ with g_i deleted.

(a) Prove that P_n is a free $\mathbb{Z}G$-module with basis $(1, g_1, g_2, \ldots, g_n)$ where $g_i \in G$.

(b) Prove that $d_{n-1} \circ d_n = 0$ for $n \geq 1$. [Show that the term $(g_0, \ldots, \hat{g}_j, \ldots, \hat{g}_k, \ldots, g_n)$ missing the entries g_j and g_k occurs twice in $d_{n-1} \circ d_n(g_0, g_1, g_2, \ldots, g_n)$, with opposite signs.]

(c) Prove that $\varphi : P_n \to F_n$ defined by

$$\varphi((g_0, g_1, g_2, \ldots, g_n)) = g_0 \otimes (g_0^{-1}g_1) \otimes (g_1^{-1}g_2) \ldots \otimes (g_{n-1}^{-1}g_n)$$

is a G-module isomorphism with inverse $\psi : P_n \to F_n$ given by

$$\psi(g_0 \otimes g_1 \otimes \ldots \otimes g_n) = (g_0, g_0g_1, g_0g_1g_2, \ldots, g_0g_1g_2 \cdots g_n).$$

(d) Prove that if $\epsilon(g_0) = 1$ for all $g_0 \in G$ then

$$\cdots \longrightarrow P_n \xrightarrow{d_n} P_{n-1} \xrightarrow{d_{n-1}} \cdots \xrightarrow{d_1} P_0 \xrightarrow{\epsilon} \mathbb{Z} \longrightarrow 0 \qquad (**)$$

is a free G-module resolution of $\mathbb{Z}$. [Show that the isomorphisms in (c) take the G-module resolutions $(**)$ and $(*)$ of the previous exercise into each other.]

3. Let F_n and P_n be as in the previous two exercises and let A be a G-module.

(a) Prove that $\text{Hom}_{\mathbb{Z}G}(F_n, A)$ can be identified with the collection $C^n(G, A)$ of maps from $G \times G \times \cdots \times G$ (n copies) to A and that under this identification the associated coboundary maps from $C^n(G, A)$ to $C^{n+1}(G, A)$ are given by equation (18).

(b) Prove that $\text{Hom}_{\mathbb{Z}G}(P_n, A)$ can be identified with the collection of maps f from $n+1$ copies $G \times G \times \cdots \times G$ to A that satisfy $f(gg_0, gg_1, \ldots, gg_n) = gf(g_0, g_1, \ldots, g_n)$.

The group $C^n(G, A)$ is sometimes called the group of *inhomogeneous n-cochains of G in A*, and the group in (b) of the previous exercise is called the group of *homogeneous n-cochains of G in A*. The inhomogeneous cochains are easier to describe since there is no restriction on the maps from G^n to A, but the coboundary map d_n on homogeneous cochains is less complicated (and more naturally suggested in topological contexts) than the coboundary map on inhomogeneous cochains. The results of the previous exercises show that the cohomology groups $H^n(G, A)$ defined using either homogeneous or inhomogeneous cochains are the same and indicate the origin of the coboundary maps d_n used in the text. Historically, $H^n(G, A)$ was originally defined using homogeneous cochains.

4. Suppose H is a normal subgroup of the group G and A is a G-module. For every $g \in G$ prove that the map $f(a) = ga$ for $a \in A^H$ defines an automorphism of the subgroup A^H.

5. Suppose the G-module A decomposes as a direct sum $A = A_1 \oplus A_2$ of G-submodules. Prove that for all $n \geq 0$, $H^n(G, A) \cong H^n(G, A_1) \oplus H^n(G, A_2)$.

6. Suppose $0 \rightarrow A \rightarrow M_1 \rightarrow M_2 \rightarrow \cdots \rightarrow M_k \rightarrow C \rightarrow 0$ is an exact sequence of G-modules where $M_1, M_2, \ldots, M_k$ are cohomologically trivial. Prove that $H^{n+k}(G, A) \cong H^n(G, C)$ for all $n \geq 1$. [Decompose the exact sequence into a succession of short exact sequences and use Corollary 22. For example, if $0 \rightarrow A \stackrel{\alpha}{\rightarrow} M_1 \stackrel{\beta}{\rightarrow} M_2 \stackrel{\gamma}{\rightarrow} C \rightarrow 0$ is exact, show that $0 \rightarrow A \rightarrow M_1 \rightarrow B \rightarrow 0$ and $0 \rightarrow B \rightarrow M_2 \rightarrow C \rightarrow 0$ are both exact, where $B = M_1/\operatorname{image} \alpha = M_1/\ker \beta \cong \operatorname{image} \beta = \ker \gamma$.]

7. (*Adjoint Associativity*) Let S and T be rings with 1, let P be a left S-module, let N be a (T, S)-bimodule, and let A be a left T-module. Prove that

$$\Phi : \operatorname{Hom}_S(P, \operatorname{Hom}_T(N, A)) \longrightarrow \operatorname{Hom}_T(N \otimes_S P, A)$$

defined by $\Phi(f)(n \otimes p) = f(p)(n)$ is an isomorphism of abelian groups. (See also Theorem 43 in Section 10.5).

8. Suppose G is cyclic of order m with generator σ and let $N = 1+\sigma+\sigma^2+\cdots+\sigma^{m-1} \in \mathbb{Z}G$.
   (a) Prove that the *augmentation* map $\operatorname{aug}(\sum_{i=0}^{m-1} a_i\sigma^i) = \sum_{i=0}^{m-1} a_i$ is a G-module homomorphism from $\mathbb{Z}G$ to $\mathbb{Z}$.
   (b) Prove that multiplication by N and by $\sigma - 1$ in $\mathbb{Z}G$ define a free G-module resolution of $\mathbb{Z}$: $\ldots \stackrel{\sigma-1}{\longrightarrow} \mathbb{Z}G \stackrel{N}{\longrightarrow} \mathbb{Z}G \stackrel{\sigma-1}{\longrightarrow} \ldots \stackrel{N}{\longrightarrow} \mathbb{Z}G \stackrel{\sigma-1}{\longrightarrow} \mathbb{Z}G \stackrel{\operatorname{aug}}{\longrightarrow} \mathbb{Z} \longrightarrow 0.$

9. Suppose G is an infinite cyclic group with generator σ.
   (a) Prove that multiplication by $\sigma - 1 \in \mathbb{Z}G$ defines a free G-module resolution of $\mathbb{Z}$: $0 \longrightarrow \mathbb{Z}G \stackrel{\sigma-1}{\longrightarrow} \mathbb{Z}G \longrightarrow \mathbb{Z} \longrightarrow 0.$
   (b) Show that $H^0(G, A) \cong A^G$, that $H^1(G, A) \cong A/(\sigma-1)A$, and that $H^n(G, A) = 0$ for all $n \geq 2$. Deduce that $H^1(G, \mathbb{Z}G) \cong \mathbb{Z}$ (so free modules need not be cohomologically trivial).

10. Suppose H is a subgroup of finite index m in the group G and A is an H-module. Let $x_1, \ldots, x_m$ be a set of left coset representatives for H in G: $G = x_1 H \cup \cdots \cup x_m H$.
    (a) Prove that $\mathbb{Z}G = \bigoplus_{i=1}^m x_i \mathbb{Z}H = \bigoplus_{i=1}^m \mathbb{Z}Hx_i^{-1}$ and $\mathbb{Z}G \otimes_{\mathbb{Z}H} A = \bigoplus_{i=1}^m (x_i \otimes A)$ as abelian groups.
    (b) Let $f_{i,a}$ be the function from $\mathbb{Z}G$ to A defined by

$$f_{i,a}(x) = \begin{cases} ha & \text{if } x = hx_i^{-1} \text{ with } h \in H \\ 0 & \text{otherwise.} \end{cases}$$

   Prove that $f_{i,a} \in M_H^G(A) = \operatorname{Hom}_{\mathbb{Z}H}(\mathbb{Z}G, A)$, i.e., $f_{i,a}(h'x) = h'f_{i,a}(x)$ for $h' \in H$.
    (c) Prove that the map $\varphi(f) = \sum_{i=1}^m x_i \otimes f(x_i^{-1})$ from $M_H^G(A)$ to $\mathbb{Z}G \otimes_{\mathbb{Z}H} A$ is a G-module homomorphism. [Write $x_i^{-1}g = h_i x_{i'}^{-1}$ for $i = 1, \ldots, m$ and observe that $x_i \otimes f(x_i^{-1}g) = x_i \otimes h_i f(x_{i'}^{-1}) = x_i h_i \otimes f(x_{i'}^{-1}) = gx_{i'} \otimes f(x_{i'}^{-1})$.]
    (d) Prove that φ gives a G-module isomorphism $\varphi : M_H^G(A) \cong \mathbb{Z}G \otimes_{\mathbb{Z}H} A$. [For the injectivity observe that an H-module homomorphism is 0 if and only if $f(x_i^{-1}) = 0$ for $i = 1, \ldots, m$. For the surjectivity prove that $\varphi(f_{i,a}) = x_i \otimes a$.]

11. Prove that the isomorphism $M_H^G(A) \cong \mathbb{Z}G \otimes_{\mathbb{Z}H} A$ in (d) of the previous exercise need not hold if H is not of finite index in G. [If G is an infinite cyclic group show that Shapiro's Lemma implies $H^1(G, M_1^G(\mathbb{Z})) = 0$ while $H^1(G, \mathbb{Z}G) \cong \mathbb{Z}$ by Exercise 9.]

12. If H is a subgroup of G and A is an abelian group let $M_{G/H}(A)$ denote the abelian group of all maps from the left cosets gH of H in G to A.

(a) Prove that $M_1^G(A) \cong M_1^H(M_{G/H}(A))$ as H-modules. [If $\{g_i\}_{i \in \mathcal{I}}$ is a choice of left coset representatives of H in G define the correspondence between $f \in M_1^G(A)$ and $F : H \to M_{G/H}(A)$ by $F(h)(g_i H) = f(g_i h)$, and check that this is an isomorphism of H-modules.]

(b) A G-module A such that $H^n(H, A) = 0$ for all $n \geq 1$ and all subgroups H of G is called *cohomologically trivial*. Prove that $M_1^G(A)$ is cohomologically trivial for any abelian group A.

(c) If G is finite, prove that $\mathbb{Z}G \otimes_{\mathbb{Z}} A$ is cohomologically trivial for all abelian groups A.

13. Suppose A is a G-module and H is a subgroup of G. Prove that the group homomorphism from $H^n(G, A)$ to $H^n(G, M_H^G(A))$ for all $n \geq 0$ induced from the G-module homomorphism from A to $M_H^G(A)$ in Example 3 following Corollary 22 composed with the isomorphism $H^n(G, M_H^G(A)) \cong H^n(H, A)$ of Shapiro's Lemma is the restriction homomorphism from $H^n(G, A)$ to $H^n(H, A)$.

14. Suppose $\varphi : H \to G$ is the inclusion map of the subgroup H of G into G. If A is an H-module and $M_H^G(A)$ the associated induced G-module, define the group homomorphism $\psi : M_H^G(A) \to A$ by mapping f to its value at 1: $\psi(f) = f(1)$.

(a) Prove that φ and ψ are compatible homomorphisms.

(b) Prove that the induced group homomorphism from $H^n(G, M_H^G(A))$ to $H^n(H, A)$ for $n \geq 0$ is the isomorphism in Shapiro's Lemma.

15. Suppose H is a normal subgroup of G and A is a G-module. For fixed $g \in G$, let $\psi(a) = ga$ and $\varphi(h) = g^{-1}hg$ for $h \in H$.

(a) Prove that φ and ψ are compatible homomorphisms.

(b) For each $n \geq 0$, prove that the homomorphism θ_g from $H^n(H, A)$ to $H^n(H, A)$ induced by the compatible homomorphisms φ and ψ is an automorphism of $H^n(H, A)$. [Observe that both φ and ψ have inverses.]

(c) Show that θ_g acting on $H^0(H, A)$ is the automorphism in Exercise 4.

16. Let A be a G-module and for $g \in G$ let θ_g denote the automorphism of $H^n(G, A)$ defined in the previous exercise.

(a) Prove that θ_g acting on $H^0(G, A) = A^G$ is the identity map.

(b) Prove that θ_g acting on $H^n(G, A)$ is the identity map for $n \geq 1$. [By induction on n and dimension shifting. For $n = 1$, use the exact sequence in Corollary 22, together with (a) applied to θ_g on C^G. For $n \geq 2$ use the isomorphism $H^{n+1}(G, A) \cong H^n(G, C)$ in Corollary 22.]

17. Suppose that H is a normal subgroup of G and A is a G-module. For $n \geq 0$ prove that $H^n(H, A)$ is a G/H-module where gH acts by the automorphism θ_g induced by conjugation by g on H and the natural action of g on A as in Exercise 15. [Use the previous exercise to show this action of a coset is well defined.]

18. Suppose that G is cyclic of order m, that H is a subgroup of G of index d, and that $\mathbb{Z}$ is a trivial G-module. Use the projective G-module resolution in Exercise 8 to prove

(a) that Cor $: H^n(H, \mathbb{Z}) \to H^n(G, \mathbb{Z})$ is multiplication by d from $\mathbb{Z}$ to $\mathbb{Z}$ for $n = 0$, from 0 to 0 if n is odd, and from $\mathbb{Z}/(m/d)\mathbb{Z}$ to $\mathbb{Z}/m\mathbb{Z}$ if n is even, $n \geq 2$, and

(b) that Res $: H^n(G, \mathbb{Z}) \to H^n(H, \mathbb{Z})$ is the identity map from $\mathbb{Z}$ to $\mathbb{Z}$ for $n = 0$, and is the natural projection map from $\mathbb{Z}/m\mathbb{Z}$ to $\mathbb{Z}/(m/d)\mathbb{Z}$ or from 0 to 0, depending on the parity of $n \geq 1$.

19. Let p be a prime and let P be a Sylow p-subgroup of the finite group G. Show that for

any G-module A and all $n \geq 0$ the map Res : $H^n(G, A) \to H^n(P, A)$ is injective on the p-primary component of $H^n(G, A)$. Deduce that if $|A| = p^a$ then the restriction map is injective on $H^n(G, A)$. [Use Proposition 26.]

20. Let p be a prime, let $G = \langle \sigma \rangle$ be cyclic of order p^m and let W be a vector space of dimension $d > 0$ over $\mathbb{F}_p$ on which σ acts as a linear transformation. Assume W has a basis such that the matrix of σ is a $d \times d$ elementary Jordan block with eigenvalue 1.
    (a) Prove that $d \leq p^m$. [Use facts about the minimal polynomial of an elementary Jordan block.]
    (b) Prove that $\dim_{\mathbb{F}_p} W^G = 1$.
    (c) Prove that $\dim_{\mathbb{F}_p} (\sigma - 1)W = d - 1$.
    (d) If $N = 1 + \sigma + \cdots + \sigma^{p^m - 1}$ is the usual norm element, prove that NW is of dimension 1 if $d = p^m$ (respectively, of dimension 0 if $d < p^m$) and that the dimension of $_NW$ is $d - 1$ (respectively, d). [Let R be the group ring $\mathbb{F}_p G$, and show that every nonzero R-submodule of R contains N. Note that W is a cyclic R-module and let $\varphi : R \to W$ be a surjective homomorphism. Conclude that if φ is not an isomorphism then $N \in \ker \varphi$.]
    (e) Deduce that if $d = p^m$ then $H^n(G, W) = 0$, and if $d < p^m$ then $H^n(G, W)$ has order p, for all $n \geq 1$ (i.e., these cohomology groups are zero if and only if W is a free $\mathbb{F}_p G$-module).

21. Let p be a prime, let $G = \langle \sigma \rangle$ be cyclic of order p^m and let V be a G-module of exponent p. Let $V = V_1 \oplus V_2 \oplus \cdots \oplus V_k$ be a decomposition of V giving the Jordan Canonical Form of σ, where each V_i is σ-invariant and a matrix of σ on V_i is an $d_i \times d_i$ elementary Jordan block with eigenvalue 1, $d_i \geq 1$ (cf. Section 12.3). Prove that $|V^G| = p^k$ and $|H^n(G, V)| = p^s$ where s is the number of V_i of dimension less than p^m over $\mathbb{F}_p$, for all $n \geq 1$. [Use the preceding exercise and Exercise 5.]

22. Suppose G is a topological group, i.e., there is a topology on G such that the maps $G \times G \to G$ defined by $(g_1, g_2) \mapsto g_1 g_2$ and $G \to G$ defined by $g \mapsto g^{-1}$ are continuous.
    (a) If H is an open subgroup of G and $g \in G$, prove that the cosets gH and Hg and the subgroup $g^{-1}Hg$ are also open.
    (b) Prove that any open subgroup is also closed. [The complement is the union of cosets as in (a).]
    (c) Prove that a closed subgroup of finite index is open.
    (d) If G is compact prove that every open subgroup H is of finite index.

23. Suppose G is a compact topological group. Prove the following are equivalent:
    (i) G is profinite, i.e., $G = \varprojlim G_i$ is the inverse limit of finite groups G_i.
    (ii) There exists a family $\{N_i\}$ ($i \in \mathcal{I}$) of open normal subgroups N_i in G such that $\cap_i N_i = 1$ and in this case $G \cong \varprojlim (G/N_i)$.
    (iii) There exists a family $\{H_j\}$ ($j \in \mathcal{J}$) of open subgroups H_j in G such that $\cap_j H_j = 1$.
    [To show (iii) implies (ii), let H be open in G and use (d) of the previous exercise to show that $N = \cap_{g \in G} g^{-1}Hg$ is a finite intersection and conclude that $N \subseteq H \subseteq G$ and N is open and normal in G.]

24. Suppose N and N' are open normal subgroups of the profinite group G and $N' \subseteq N$. Prove that the projection homomorphism $\varphi : G/N' \to G/N$ and the injection $\psi : A^N \to A^{N'}$ are compatible homomorphisms and deduce there is an induced homomorphism from $H^n(G/N, A^N)$ to $H^n(G/N', A^{N'})$.

25. If G is an infinite profinite group show that G does not act continuously on $A = \mathbb{Z}G$. [Show that the stabilizer of $a \in A$ is not always of finite index in G.]

17.3 CROSSED HOMOMORPHISMS AND $H^1(G,A)$

In this section we consider in greater detail the cohomology group $H^1(G, A)$ where G is a group and A is a G-module. From the definition of the coboundary map d_1 in equation (18), if $f \in C^1(G, A)$ then

$$d_1(f)(g_1, g_2) = g_1 \cdot f(g_2) - f(g_1 g_2) + f(g_1).$$

Thus any function $f : G \to A$ is a 1-cocycle if and only if it satisfies the identity

$$f(gh) = f(g) + gf(h) \qquad \text{for all } g, h \in G. \tag{17.20}$$

Equivalently, a 1-cocycle is determined by a collection $\{a_g\}_{g \in G}$ of elements in A satisfying $a_{gh} = a_g + ga_h$ for $g, h \in G$ (and then the 1-cocycle f is the function sending g to a_g). Note that if 1 denotes the identity of G, then $f(1) = f(1^2) = f(1) + 1 \cdot f(1) = 2f(1)$, so $f(1) = 0$ is the identity in A. Thus 1-cocycles are necessarily "normalized" at the identity. It then follows from the cocycle condition that $f(g^{-1}) = -g^{-1}f(g)$ for all $g \in G$.

If A is a G-module on which G acts trivially, then the cocycle condition (20) is simply $f(gh) = f(g) + f(h)$, i.e., f is simply a *homomorphism* from the multiplicative group G to the additive group A. Because of this the functions from G to A satisfying (20) are called *crossed homomorphisms*.

A 1-cochain f is a 1-coboundary if there is some $a \in A$ such that

$$f(g) = g \cdot a - a \qquad \text{for all } g \in G, \tag{17.21}$$

(equivalently, $a_g = ga - a$ in the notation above). Note that since $-a \in A$, the coboundary condition in (21) can also be phrased as $f(g) = a - g \cdot a$ for some fixed $a \in A$ and all $g \in G$. The 1-coboundaries are called *principal crossed homomorphisms*. With this terminology the cohomology group $H^1(G, A)$ is the group of crossed homomorphisms modulo the subgroup of principal crossed homomorphisms.

Example: (Hilbert's Theorem 90)

Suppose $G = \text{Gal}(K/F)$ is the Galois group of a finite Galois extension K/F of fields. Then the multiplicative group $K^\times$ is a G-module and $H^1(G, K^\times) = 0$. To see this, let $\{\alpha_\sigma\}$ be the values $f(\sigma)$ of a 1-cocycle f, so that $\alpha_\sigma \in K^\times$ and $\alpha_{\sigma\tau} = \alpha_\sigma \sigma(\alpha_\tau)$ (the cocycle condition written multiplicatively for the group $K^\times$). By the linear independence of automorphisms (Corollary 8 in Section 14.2), there is an element $\gamma \in K$ such that

$$\beta = \sum_{\tau \in G} \alpha_\tau \tau(\gamma)$$

is nonzero, i.e., $\beta \in K^\times$. Then for any $\sigma \in G$ we have

$$\sigma(\beta) = \sum_{\tau \in G} \sigma(\alpha_\tau)\, \sigma\tau(\gamma) = \alpha_\sigma^{-1} \sum_{\tau \in G} \alpha_{\sigma\tau}\, \sigma\tau(\gamma) = \alpha_\sigma^{-1}\beta$$

where the second equality comes from the cocycle condition. Hence $\alpha_\sigma = \beta/\sigma(\beta)$, which is the multiplicative form of the coboundary condition (21) (for the element $a = \beta^{-1}$). Since every 1-cocycle is a 1-coboundary, we have $H^1(G, K^\times) = 0$. The same result holds for infinite Galois extensions by equation (19) in the previous section since $H^1(G, K^\times)$ is the direct limit of trivial groups.

As a special case, suppose K/F is a Galois extension with cyclic Galois group G having generator σ. The cohomology groups for G were computed explicitly in the previous section, and in particular, $H^1(G, A) = {}_NA/(\sigma - 1)A$ for any G-module A (written additively). Since this group is trivial in the present context, we see that an element α in K is in the kernel of the norm map, i.e., $N_{K/F}(\alpha) = 1$ if and only if $\alpha = \sigma(\beta)/\beta$ for some $\beta \in K$. (For a direct proof of this result in the cyclic case, cf. Exercise 23 in Section 14.2.)

This famous result for cyclic extensions was first proved by Hilbert and appears as "Theorem 90" in his book (known as the *"Zahlbericht"*) on number theory in 1897. As a result, the more general result $H^1(G, K^\times) = 0$ is referred to in the literature as "Hilbert's Theorem 90." In general, the higher dimensional cohomology groups $H^n(G, K^\times)$ for $n \geq 2$ can be nontrivial (cf. Exercise 13).

Example

Suppose $G = \text{Gal}(K/F)$ is the Galois group of a finite Galois extension K/F of fields as in the previous example. Then the additive group K is also a G-module and $H^n(G, K) = 0$ for all $n \geq 2$. The proof of this in general uses the fact that there is a *normal basis* for K over F, i.e., there is an element $\alpha \in K$ whose Galois conjugates give a basis for K as a vector space over F, or, equivalently, $K \cong \mathbb{Z}G \otimes_\mathbb{Z} F$ as G-modules. The latter isomorphism shows that K is induced as a G-module, and then $H^n(G, K) = 0$ follows from Corollary 24 in Section 2. For a direct proof in the case where G is cyclic, cf. Exercise 26 in Section 14.2.

If G acts trivially on A, then $g \cdot a - a = 0$, so 0 is the only principal crossed homomorphism, i.e., $B^1(G, A) = 0$. This proves the following result:

Proposition 30. If A is a G-module on which G acts trivially then $H^1(G, A) = \text{Hom}(G, A)$, the group of all group homomorphisms from G to A.

If G is a profinite group, then the same result holds for the continuous cohomology group $H^1(G, A)$ provided one takes the group of continuous homomorphisms from G into A.

Examples

(1) If G acts trivially on A then $H^1(G, A) = H^1(G/[G, G], A)$ since any group homomorphism from G to the abelian group A factors through the commutator subgroup $[G, G]$ (cf. Proposition 7(5) in Section 5.4), so computing H^1 for trivial G-action reduces to computing H^1 for some abelian group.

(2) If G is a finite group acting trivially on $\mathbb{Z}$, then $H^1(G, \mathbb{Z}) = 0$ because $\mathbb{Z}$ has no nonzero elements of finite order so there is no nonzero group homomorphism from G to $\mathbb{Z}$.

(3) If A is cyclic of prime order p and G is a p-group then G must act trivially on A (since the automorphism group of A has order $p - 1$), so in this case one always has $H^1(G, A) = \text{Hom}(G, A)$.

(4) If G is a finite group that acts trivially on $\mathbb{Q}/\mathbb{Z}$ then $H^1(G, \mathbb{Q}/\mathbb{Z}) = \text{Hom}(G, \mathbb{Q}/\mathbb{Z}) = \hat{G}$ is the *dual group* of G (cf. Exercise 14 in Section 5.2.). Since $\mathbb{Q}/\mathbb{Z}$ is abelian, any homomorphism of G into $\mathbb{Q}/\mathbb{Z}$ factors through the commutator quotient $G^{\text{ab}} = G/[G, G]$ of G, so $\text{Hom}(G, \mathbb{Q}/\mathbb{Z}) = \text{Hom}(G^{\text{ab}}, \mathbb{Q}/\mathbb{Z})$. It follows that $\text{Hom}(G, \mathbb{Q}/\mathbb{Z}) \cong \hat{G}^{\text{ab}}$ (which by cf. Exercise 14 again is noncanonically isomorphic to G^{ab}).

If $0 \to A \to B \to C \to 0$ is a short exact sequence of G-modules then the long exact sequence in group cohomology in Theorem 21 of the previous section begins with terms

$$0 \longrightarrow A^G \longrightarrow B^G \longrightarrow C^G \xrightarrow{\delta_0} H^1(G, A) \longrightarrow \cdots.$$

The connecting homomorphism δ_0 is given explicitly as follows: if $c \in C^G$ then there is an element $b \in B$ mapping to c and then $\delta_0(c)$ is the class in $H^1(G, A)$ of the 1-cocycle given by

$$\delta_0(c) : G \longrightarrow A$$
$$g \longmapsto g \cdot b - b.$$

Note that $g \cdot b - b$ is (the image in B of) an element of A for all $g \in G$ since $c \in C^G$. To verify directly that $f = \delta_0(c)$ satisfies the cocycle condition in (20), we compute

$$f(gh) = gh \cdot b - b = (g \cdot b - b) + g \cdot (h \cdot b - b) = f(g) + gf(h).$$

From the explicit expression $f = g \cdot b - b$ it is also clear that $\delta_0(c) \in H^1(G, A)$ maps to 0 in the next term $H^1(G, B)$ of the long exact sequence above since f is the coboundary for the element $b \in B$.

Example: (Kummer Theory)

Suppose that F is a field of characteristic 0 containing the group μ_n of all n^{th} roots of unity for some $n \geq 1$. Let K be an algebraic closure of F and let $G = \text{Gal}(K/F)$. The group G acts trivially on μ_n since $\mu_n \subset F$ by assumption, i.e., $\mu_n \cong \mathbb{Z}/n\mathbb{Z}$ as G-modules. Hence the Galois cohomology group $H^1(G, \mu_n)$ is the group $\text{Hom}_c(G, \mathbb{Z}/n\mathbb{Z})$ of continuous homomorphisms of G into $\mathbb{Z}/n\mathbb{Z}$. If χ is such a continuous homomorphism, then $\ker \chi \subseteq G$ is a closed normal subgroup of G, hence corresponds by Galois theory to a Galois extension L_χ/F. Then $\text{Gal}(L_\chi/F) \cong \text{image } \chi$, so L_χ is a cyclic extension of F of degree d dividing n, and d is the order of χ. If i is relatively prime to d, then χ^i has the same kernel as χ, so defines the same cyclic extension of F. Conversely, every cyclic extension L of F of degree d dividing n defines such elements in $\text{Hom}_c(G, \mathbb{Z}/n\mathbb{Z})$–the elements trivial on $\text{Gal}(K/L)$ that map a generator of $\text{Gal}(L/F)$ to an element of order d in $\mathbb{Z}/n\mathbb{Z}$. This defines a bijection between the cyclic subgroups of order dividing n of the Galois cohomology group $H^1(G, \mu_n)$ and the cyclic extensions of F of degree dividing n.

The homomorphism of raising to the nth power is surjective on $K^\times$ (since we can always extract n^{th} roots in K) and has kernel μ_n. Hence the sequence

$$1 \longrightarrow \mu_n \longrightarrow K^\times \xrightarrow{n} K^\times \longrightarrow 1$$

is an exact sequence of discrete G-modules. The associated long exact sequence in Galois cohomology gives

$$1 \longrightarrow \mu_n^G \longrightarrow (K^\times)^G \xrightarrow{n} (K^\times)^G \longrightarrow H^1(G, \mu_n) \longrightarrow H^1(G, K^\times) \longrightarrow \cdots.$$

We have $\mu_n^G = \mu_n$ and $(K^\times)^G = F^\times$ by Galois theory, and $H^1(G, K^\times) = 0$ by Hilbert's Theorem 90, so this exact sequence becomes

$$1 \longrightarrow \mu_n \longrightarrow F^\times \xrightarrow{n} F^\times \longrightarrow H^1(G, \mu_n) \longrightarrow 0,$$

which in turn is equivalent to the isomorphism

$$H^1(G, \mu_n) \cong F^\times/F^{\times n}$$

where $F^{\times n}$ denotes the group of nth powers of elements of $F^\times$. This isomorphism is made explicit using the explicit form for the connecting homomorphism given above: for every $\alpha \in F^\times$ and $\sigma \in G$, the element $\sqrt[n]{\alpha}$ in $K^\times$ maps to α in the exact sequence and

$$\chi(\sigma) = \frac{\sigma(\sqrt[n]{\alpha})}{\sqrt[n]{\alpha}}$$

defines an element in $H^1(G, \mu_n)$ (cf. Exercise 11). The kernel of this homomorphism χ is the field $F(\sqrt[n]{\alpha})$. By the results of the previous paragraph, when F contains the nth roots of unity, an extension L/F is Galois with cyclic Galois group of order d dividing n if and only if $L = F(\sqrt[n]{\alpha})$ for some $\alpha \in F^\times$. Such an extension is called a *Kummer extension*, cf. Exercise 12 and Section 14.7. The subgroup generated by α in $F^\times/F^{\times n}$ is unique, i.e., $L = F(\sqrt[n]{\beta})$ if and only if for some i relatively prime to d, β differs from α^i by an nth power of an element in F (cf. Exercise 7 in Section 14.7).

If the characteristic of F is a prime p, the same argument applies when n is not divisible by p, replacing the algebraic closure of F with the separable closure of F (the largest separable algebraic extension of F).

Example: (The Transfer Homomorphism)

Suppose G is a finite group and H is a subgroup. The corestriction defines a homomorphism from $H^1(H, \mathbb{Q}/\mathbb{Z})$ to $H^1(G, \mathbb{Q}/\mathbb{Z})$, which by Example 4 above gives a homomorphism from $\hat{H}^{\text{ab}}$ to $\hat{G}^{\text{ab}}$. This gives a homomorphism

$$\text{Ver} : G^{\text{ab}} \longrightarrow H^{\text{ab}}$$

called the *transfer* (or *Verlagerungen*) homomorphism (cf. Exercise 14). To make this homomorphism explicit, consider the exact sequence

$$0 \longrightarrow \mathbb{Q}/\mathbb{Z} \longrightarrow M_1^G(\mathbb{Q}/\mathbb{Z}) \longrightarrow C \longrightarrow 0 \qquad (17.22)$$

defined by the homomorphism mapping $a \in \mathbb{Q}/\mathbb{Z}$ to $f_a \in M_1^G(\mathbb{Q}/\mathbb{Z})$ in Example 4 preceding Proposition 23 in the previous section (so $f_a(g) = g \cdot a$ for $g \in G$). This is a short exact sequence of G-modules and hence also of H-modules. The first portions of the associated long exact sequences for the cohomology with respect to H and then G give the rows in the commutative diagram

$$
\begin{array}{ccccccc}
\cdots & \longrightarrow & C^H & \xrightarrow{\delta_0} & H^1(H, \mathbb{Q}/\mathbb{Z}) & \longrightarrow & 0 \\
& & \downarrow{\scriptstyle\text{Cor}} & & \downarrow{\scriptstyle\text{Cor}} & & \\
\cdots & \longrightarrow & C^G & \xrightarrow{\delta_0} & H^1(G, \mathbb{Q}/\mathbb{Z}) & \longrightarrow & 0
\end{array}
$$

since $H^1(H, M_1^G(\mathbb{Q}/\mathbb{Z})) = H^1(G, M_1^G(\mathbb{Q}/\mathbb{Z})) = 0$ (cf. Exercise 12 in Section 2). Let $\chi \in H^1(H, \mathbb{Q}/\mathbb{Z})$ and suppose that $c \in C^H$ is an element mapping to χ by the surjective connecting homomorphism δ_0 in the first row of the diagram above. By the commutativity, $\chi' = \text{Cor}(\chi)$ is the image under the connecting homomorphism δ_0 of $c' = \text{Cor}(c) \in C^G$ in the second row of the diagram. By our explicit formula for the coboundary map δ_0, if $F \in M_1^G(\mathbb{Q}/\mathbb{Z})$ is any element mapping to c' in (22) then $g \cdot F - F = f_{a'}$ for a unique $a' \in \mathbb{Q}/\mathbb{Z}$, and we have $\chi'(g) = \delta_0(c')(g) = a'$ for $g \in G$. Since $f_{a'}(x) = x \cdot a' = a'$ for any $x \in G$ because G acts trivially on $\mathbb{Q}/\mathbb{Z}$, the function $g \cdot F - F$ in fact has the constant value a', and so can be evaluated at any $x \in G$ to determine the value of $\chi'(g)$.

Since $c' = \sum_{i=1}^{m} g_i \cdot c \in C^G$ where $g_1, \ldots, g_m$ are representatives of the left cosets of H in G (cf. Example 4 preceding Proposition 26), such an element F is given by

$$F = \sum_{i=1}^{m} g_i \cdot f,$$

where $f \in M_1^G(\mathbb{Q}/\mathbb{Z})$ is any element mapping to c in (22). This f can be used to compute the explicit coboundary of c as before: $h \cdot f - f = f_a$ for a unique $a \in \mathbb{Q}/\mathbb{Z}$ and $\chi(h) = a$ for $h \in H$. As before, the function $h \cdot f - f = f_a$ has the constant value a and so can be evaluated at any element x of G to determine the value of $\chi(h)$.

Computing $g \cdot F - F$ on the element $1 \in G$ it follows that

$$\chi'(g) = \sum_{i=1}^{m} f(gg_i) - \sum_{i=1}^{m} f(g_i).$$

For $i = 1, \ldots, m$, write

$$gg_i = g_j h(g, g_i) \qquad \text{with } h(g, g_i) \in H, \tag{17.23}$$

noting that the resulting set of g_j is some permutation of $\{g_1, \ldots, g_m\}$. Then

$$\sum_{i=1}^{m} f(gg_i) - \sum_{i=1}^{m} f(g_i) = \sum_{i=1}^{m}[f(g_j h(g, g_i)) - f(g_j)] = \sum_{i=1}^{m} \chi(h(g, g_i))$$

since as noted above, $\chi(h) = f(xh) - f(x)$ for any $x \in G$. Hence

$$\chi'(g) = \chi\left(\prod_{i=1}^{m} h(g, g_i)\right)$$

and so the transfer homomorphism is given by the formula

$$\mathrm{Ver}(g) = \prod_{i=1}^{m} h(g, g_i) \tag{17.24}$$

with the elements $h(g, g_i) \in H$ defined by equation (23). Note that this proves in particular that the map defined in (24) is a homomorphism from G^{ab} to H^{ab} that is independent of the choice of representatives g_i for H in G in (23). Proving that this map is a homomorphism directly is not completely trivial. The same formula also defines the transfer homomorphism when G is infinite and H is a subgroup of finite index in G.

As an example of the transfer, suppose $H = n\mathbb{Z}$ and $G = \mathbb{Z}$ and choose $0, 1, 2, \ldots, n-1$ as coset representatives for H in G. If $g = 1$, then all the elements $h(g, g_i)$ are 0 for $i = 1, 2, \ldots, n-1$ and $h(1, n-1) = n$. Hence the transfer map from $\mathbb{Z}$ to $n\mathbb{Z}$ maps 1 to n, so is simply multiplication by the index. Similarly, the transfer map from any cyclic group G to a subgroup H of index n is the nth power map. See also Exercise 8.

For the cyclic group $\mathbb{F}_p^\times$ for an odd prime p and subgroup $\{\pm 1\}$, it follows that the transfer map is the homomorphism $\mathrm{Ver} : \mathbb{F}_p^\times \to \{\pm 1\}$ given by

$$\mathrm{Ver}(a) = a^{(p-1)/2} = \left(\frac{a}{p}\right) = \begin{cases} +1 & \text{if } a \text{ is a square} \\ -1 & \text{if } a \text{ is not a square} \end{cases}$$

(the symbol $\left(\dfrac{a}{p}\right)$ is called the *Legendre symbol* or the *quadratic residue symbol*). If instead we take the elements $1, 2, \ldots, (p-1)/2$ as coset representatives for $\{\pm 1\}$ in $\mathbb{F}_p^\times$ we see that

$$\left(\frac{a}{p}\right) = (-1)^{m(a)}$$

where $m(a)$ is the number of elements among $a, 2a, \ldots, (p-1)a/2$ whose least positive remainder modulo p is greater than $(p-1)/2$ (in which case the element differs by -1 from one of our chosen coset representatives and contributes one factor of -1 to the product in (24)). This result is known as *Gauss' Lemma* in elementary number theory and can be used to prove Gauss' celebrated Quadratic Reciprocity Law (cf. also Exercise 15).

Next we give two important interpretations of $H^1(G, A)$ in terms of semidirect products. If A is a G-module, let E be the semidirect product $E = A \rtimes G$, where A is normal in E and the action of G (viewed as a subgroup of E) on A by conjugation is the same as its G-module action: $gag^{-1} = g \cdot a$. In the notation of Section 5.5, $E = A \rtimes_\varphi G$, where φ is the homomorphism of G into $\text{Aut}(A)$ given by the G-module action. In particular, E will be the direct product of A and G if and only if G acts trivially on A. As in Section 5.5, we shall write the elements of E as (a, g) where $a \in A$ and $g \in G$, with group operation

$$(a_1, g_1)(a_2, g_2) = (a_1 + g_1 \cdot a_2, g_1 g_2).$$

Note that A is written additively, while G and E are written multiplicatively.

Definition. Let X be any group and let Y be a normal subgroup of X. The *stability group* of the series $1 \trianglelefteq Y \trianglelefteq X$ is the group of all automorphisms of X that map Y to itself and act as the identity on both of the factors Y and X/Y, i.e.,

$$\text{Stab}(1 \trianglelefteq Y \trianglelefteq X) = \{\sigma \in \text{Aut}(X) \mid \sigma(y) = y \text{ for all } y \in Y,$$

$$\text{and } \sigma(x) \equiv x \bmod Y \text{ for all } x \in X\}.$$

In the special case where Y is an *abelian* normal subgroup of X, conjugation by elements of Y induce (inner) automorphisms of X that stabilize the series $1 \trianglelefteq Y \trianglelefteq X$, and in this case $Y/C_Y(X)$ is isomorphic to a subgroup of $\text{Stab}(1 \trianglelefteq Y \trianglelefteq X)$ (where $C_Y(X)$ is the elements of Y in the center of X).

Proposition 31. Let A be a G-module and let E be the semidirect product $A \rtimes G$. For each cocycle $f \in Z^1(G, A)$ define $\sigma_f : E \to E$ by

$$\sigma_f((a, g)) = (a + f(g), g).$$

Then the map $f \to \sigma_f$ is a group isomorphism from $Z^1(G, A)$ onto $\text{Stab}(1 \trianglelefteq A \trianglelefteq E)$. Under this isomorphism the subgroup $B^1(G, A)$ of coboundaries maps onto the subgroup $A/C_A(E)$ of the stability group.

Proof: It is an exercise to see that the cocycle condition implies σ_f is an automorphism of E that stabilizes the chain $1 \trianglelefteq A \trianglelefteq E$. Likewise one checks directly that $\sigma_{f_1+f_2} = \sigma_{f_1} \circ \sigma_{f_2}$, so the map $f \mapsto \sigma_f$ is a group homomorphism. By definition of σ_f this map is injective. Conversely, let $\sigma \in \text{Stab}(1 \trianglelefteq A \trianglelefteq E)$. Since σ acts trivially on E/A, each element $(0, g)$ in this semidirect product maps under σ to another element (a, g) in the same coset of A; define $f_\sigma : G \to A$ by letting $f_\sigma(g) = a$. If we identify A with the elements of the form $(a, 1)$ in E, then the group operation in E shows that

$$f_\sigma(g) = \sigma((0, g))(0, g)^{-1}.$$

Because σ is a stability automorphism of E, it is easy to check that f_σ satisfies the cocycle condition. It follows immediately from the definitions that $f_{\sigma_f} = f$, so the map $f \mapsto \sigma_f$ is an isomorphism.

Now f is a coboundary if and only if there is some $x \in A$ such that $f(g) = x - g \cdot x$ for all $g \in G$. Thus f is a coboundary if and only if $\sigma_f((a, g)) = (a + x - g \cdot x, g)$. But conjugation in E by the element $(x, 1)$ maps (a, g) to the same element $(a + x - g \cdot x, g)$, so the automorphism σ_f is conjugation by $(x, 1)$. This proves the remaining assertion of the proposition.

Corollary 32. In the notation of Proposition 31 let φ_a denote the automorphism of E given by conjugation by a for any $a \in A$. Then the cocycles f_1 and f_2 are in the same cohomology class in $H^1(G, A)$ if and only if $\sigma_{f_1} = \varphi_a \circ \sigma_{f_2}$, for some $a \in A$.

The proposition and corollary show that 1-cocycles may be computed by finding automorphisms of E that stabilize the series $1 \trianglelefteq A \trianglelefteq E$, and vice versa. The first cohomology group is then given by taking these automorphisms modulo inner automorphisms, i.e., is the group of "outer stability automorphisms" of this series.

Example

Let $G = Z_2$ act by inversion on $A = \mathbb{Z}/4\mathbb{Z}$. The corresponding semidirect product $E = A \rtimes G$ is the dihedral group of order 8, which has automorphism group isomorphic to D_8; viewing E as a normal (index 2) subgroup of D_{16}, conjugation in the latter group restricted to E exhibits 8 distinct automorphisms of E (cf. Proposition 17 in Section 4.4). The subgroup A of E is characteristic in E, hence every automorphism of E sends A to itself, and therefore also acts on E/A (necessarily trivially since $|E/A| = 2$). Half the automorphisms of E invert A and half centralize A; in fact, the cyclic subgroup of order 8 in D_{16} (which contains A) maps to a cyclic group of order 4 of automorphisms centralizing A. Thus $\mathrm{Stab}(1 \trianglelefteq A \trianglelefteq E) \cong Z_4 \cong Z^1(G, A)$. Since the center of E is a subgroup of A of order 2, $|A/Z(E)| = 2 = |B^1(G, A)|$. This proves $|H^1(G, A)| = 2$.

In the semidirect product E the subgroup G is a complement to A, i.e., $E = AG$ and $A \cap G = 1$; moreover, every E-conjugate of G is also a complement to A. But A may have complements in E that are not conjugate to G in E. Our second interpretation of $H^1(G, A)$ shows that this cohomology group characterizes the E-conjugacy classes of complements of A in E.

Proposition 33. Let A be a G-module and let E be the semidirect product $A \rtimes G$. For each 1-cocycle f let

$$G_f = \{(f(g), g) \mid g \in G\}.$$

Then G_f is a subgroup complement to A in E. The map $f \mapsto G_f$ is a bijection from $Z^1(G, A)$ to the set of complements to A in E. Two complements are conjugate in E if and only if their corresponding 1-cocycles are in the same cohomology class in $H^1(G, A)$, so there is a bijection between $H^1(G, A)$ and the set of E-conjugacy classes of complements to A.

Proof: By the cocycle condition,

$$(f(g), g)(f(h), h) = (f(g) + gf(h)g^{-1}, gh) = (f(g) + g \cdot f(h), gh) = (f(gh), gh),$$

and it follows that G_f is closed under the group operation in E. As observed earlier, each cocycle necessarily has $f(1) = 0$, so G_f contains the identity $(0, 1)$ of E. The inverse to $(f(g), g)$ in E is $(f(g^{-1}), g^{-1})$, so G_f is closed under inverses. This proves G_f is a subgroup of E. Since the distinct elements of G_f represent the distinct cosets of A in E, G_f is a complement to A in E. Distinct cocycles give different coset representatives, hence they determine different complements.

Conversely, if C is any complement to A in G, then C contains a unique coset representative $a_g g$ of Ag for each $g \in G$. Since C is closed under the group operation the element $(a_g g)(a_h h) = (a_g g a_h g^{-1})gh$ represents the coset Agh, and so a_{gh} is $a_g g a_h g^{-1} = a_g (g \cdot a_h)$ (written additively in A this becomes $a_{gh} = a_g + (g \cdot a_h)$). This shows that the map $f : G \to A$ given by $f(g) = a_g$ is a cocycle, and so $C = G_f$. Hence there is a bijection between 1-cocycles and complements to A in E.

Since $\mathrm{Stab}(1 \trianglelefteq A \trianglelefteq E)$ normalizes A it permutes the complements to A in E. In the notation of Proposition 31, for 1-cocycles f_1 and f_2 it follows immediately from the definition that $\sigma_{f_1}(G_{f_2}) = G_{f_1 + f_2}$. This shows that the permutation action of $\mathrm{Stab}(1 \trianglelefteq A \trianglelefteq E)$ on the set of complements to A in E is the (left) regular representation of this group. Furthermore, if $a \in A$ and φ_a is the stability automorphism conjugation by a, then

$$aG_f a^{-1} = \varphi_a(G_f) = G_{f + \beta_a} \tag{17.25}$$

where β_a is the 1-coboundary $\beta_a : g \mapsto a - g \cdot a$. Since G_f is a complement to A, any $e \in E$ may be written as ag for some $a \in A$ and $g \in G_f$. Then $eG_f e^{-1} = aG_f a^{-1}$, i.e., the E-conjugates of G_f are the just the A-conjugates of G_f. Now the complements G_{f_1} and G_{f_2} are conjugate in E if and only if $G_{f_2} = aG_{f_1} a^{-1} = G_{f_1 + \beta_a}$ for some $a \in A$ by (25). This shows two complements are conjugate in E if and only if their corresponding cocycles differ by a coboundary, i.e., represent the same cohomology class in $H^1(G, A)$, which completes the proof.

Corollary 34. Under the notation of Proposition 33, all complements to A are conjugate in E if and only if $H^1(G, A) = 0$.

Corollary 35. If A is a finite abelian group whose order is relatively prime to $|G|$ then all complements to A in any semidirect product $E = A \rtimes G$ are conjugate in E.

Examples

(1) Let $A = \langle a \rangle$ and $G = \langle g \rangle$ both be cyclic of order 2. The group G must act trivially on A, hence $A \rtimes G = A \times G$ is a Klein 4-group. Here $A \rtimes G$ is abelian, so every subgroup is conjugate only to itself, and since $H^1(G, A) = \mathrm{Hom}(Z_2, \mathbb{Z}/2\mathbb{Z})$ has order 2, there are precisely two complements to A in E, namely $\langle g \rangle$ and $\langle ag \rangle$.

(2) If $A = \langle a \rangle$ is cyclic of order 2 and $G = \langle x \rangle \times \langle y \rangle$ is a Klein 4-group, then as before G must act trivially on A, so $H^1(G, A) = \mathrm{Hom}(Z_2 \times Z_2, \mathbb{Z}/2\mathbb{Z})$ has order 4. The four complements to A in $A \times G$ are G, $\langle ax, y \rangle$, $\langle x, ay \rangle$ and $\langle ax, ay \rangle$.

(3) Proposition 33 can also be used to compute $H^1(G, A)$. Let $A = \langle r \rangle$ be cyclic of order 4 and let $G = \langle s \rangle$ be cyclic of order 2 acting on A by inversion: $srs^{-1} = r^{-1}$ as in the Example following Corollary 32. Then $A \rtimes G$ is the dihedral group D_8 of order 8. The subgroup A has four complements in D_8, namely the groups generated

by each of the four elements of order 2 not in A: $\langle s \rangle$, $\langle r^2 s \rangle$, $\langle rs \rangle$ and $\langle r^3 s \rangle$. The former pair and the latter pair are conjugate in D_8 (in both cases via r), but $\langle s \rangle$ is not conjugate to $\langle rs \rangle$. Thus A has 2 conjugacy classes of complements in $A \rtimes G$ and hence $H^1(Z_2, \mathbb{Z}/4\mathbb{Z})$ has order 2. This also follows from the computation of the cohomology of cyclic groups in Section 2.

EXERCISES

1. Let G be the cyclic group of order 2 and let A be a G-module. Compute the isomorphism types of $Z^1(G, A)$, $B^1(G, A)$ and $H^1(G, A)$ for each of the following:
   (a) $A = \mathbb{Z}/4\mathbb{Z}$ (trivial action),
   (b) $A = \mathbb{Z}/2\mathbb{Z} \times \mathbb{Z}/2\mathbb{Z}$ (trivial action),
   (c) $A = \mathbb{Z}/2\mathbb{Z} \times \mathbb{Z}/2\mathbb{Z}$ (any nontrivial action).

2. Let p be a prime and let P be a p-group.
   (a) Show that $H^1(P, \mathbb{F}_p) \cong P/\Phi(P)$, where $\Phi(P)$ is the Frattini subgroup of P (cf. the exercises in Section 6.1).
   (b) Deduce that the dimension of $H^1(P, \mathbb{F}_p)$ as a vector space over $\mathbb{F}_p$ equals the minimum number of generators of P. [Use Exercise 26(c), Section 6.1.]

3. If G is the cyclic group of order 2 acting by inversion on $\mathbb{Z}$ show that $|H^1(G, \mathbb{Z})| = 2$. [Show that in $E = \mathbb{Z} \rtimes G$ every element of $E - \mathbb{Z}$ has order 2, and there are two conjugacy classes in this coset.]

4. Let A be the Klein 4-group and let $G = \text{Aut}(A) \cong S_3$ act on A in the natural fashion. Prove that $H^1(G, A) = 0$. [Show that in the semidirect product $E = A \rtimes G$, G is the normalizer of a Sylow 3-subgroup of E. Apply Sylow's Theorem to show all complements to A in E are conjugate.]

5. Let G be the cyclic group of order 2 acting on an elementary abelian 2-group A of order 2^n. Show that $H^1(G, A) = 0$ if and only if $n = 2k$ and $|A^G| = 2^k$. [In $E = A \rtimes G$ show that (a, x) is an element of order 2 if and only if $a \in A^G$, where $G = \langle x \rangle$. Then compare the number of complements to A with the number of E-conjugates of x.]

6. *(Thompson Transfer Lemma)* Let G be a finite group of even order, let T be a Sylow 2-subgroup of G, let $M \leq T$ with $|T : M| = 2$, and let x be an element of order 2 in G. Show that if G has no subgroup of index 2 then M contains some G-conjugate of x as follows:
   (a) Let $\text{Ver} : G/[G, G] \to T/[T, T]$ be the transfer homomorphism. Show that
   $$\text{Ver}(x) = \prod_g g^{-1} x g \bmod [T, T]$$
   where the product is over representatives of the cosets gT that are fixed under left multiplication by x.
   (b) Show that under left multiplication x fixes an odd number of left cosets of T in G.
   (c) Show that if G has no subgroup of index 2 then $\text{Ver}(x) \in M/[T, T]$. Deduce that for some $g \in G$ we must have $g^{-1} x g \in M$. [Consider the product $\text{Ver}(x)$ in the group T/M of order 2.]

7. Let H be a subgroup of G and let $x \in G$. The transfer $\text{Ver} : G/[G, G] \to H/[H, H]$ may be computed as follows: let $\mathcal{O}_1, \mathcal{O}_2, \ldots, \mathcal{O}_k$ be the distinct orbits of x acting by left multiplication on the left cosets of H in G, let $\mathcal{O}_i$ have length n_i and let $g_i H$ be any representative of $\mathcal{O}_i$.

(a) Show that $\mathcal{O}_i = \{g_i H, xg_i H, x^2 g_i H, \ldots, x^{n_i-1} g_i H\}$ and that $g_i^{-1} x^{n_i} g_i \in H$.

(b) Show that $\text{Ver}(x) = \prod_{i=1}^{k} g_i^{-1} x^{n_i} g_i \bmod [H, H]$.

8. Assume the center, $Z(G)$, of G is of index m. Prove that $\text{Ver}(x) = x^m$, for all $x \in G$, where Ver is the transfer homomorphism from $G/[G, G]$ to $Z(G)$. [Use the preceding exercise.]

9. Let p be a prime, let $n \geq 3$, and let V be an n-dimensional vector space over $\mathbb{F}_p$ with basis $v_1, v_2, \ldots, v_n$. Let V be a module for the symmetric group S_n, where each $\pi \in S_n$ permutes the basis in the natural way: $\pi(v_i) = v_{\pi(i)}$.

(a) Show that $|H^1(S_n, V)| = \begin{cases} 0, & \text{if } p \neq 2 \\ 2, & \text{if } p = 2 \end{cases}$. [Use Shapiro's Lemma.]

(b) Show that $H^1(A_n, V) = 0$ for all primes p.

10. Let V be the natural permutation module for S_n over $\mathbb{F}_2$, $n \geq 3$, as described in the preceding exercise, and let $W = \{a_1 v_1 + \cdots + a_n v_n \mid a_1 + \cdots + a_n = 0\}$ (the "trace zero" submodule of V). Show that if n is even then $H^1(A_n, W) \neq 0$. [Show that in the semidirect product $V \rtimes A_n$ the element v_1 induces a nontrivial outer automorphism on $E = W \rtimes A_n$ that stabilizes the series $1 \trianglelefteq W \trianglelefteq E$.]

11. Let F be a field of characteristic not dividing n and let α be any nonzero element in F. Let K be a Galois extension of F containing the splitting field of $x^n - a$, and let $\sqrt[n]{\alpha}$ be a fixed n^{th} root of α in K.

(a) Prove that $\sigma(\sqrt[n]{\alpha})/\sqrt[n]{\alpha}$ is an nth root of unity.

(b) Prove that the function $f(\sigma) = \sigma(\sqrt[n]{\alpha})/\sqrt[n]{\alpha}$ is a 1-cocycle of G with values in the group μ_n of nth roots of unity in K (note μ_n is not assumed to be contained in F).

(c) Prove that the 1-cocycle obtained by a different choice of nth root of α in K differs from the 1-cocycle in (b) by a 1-coboundary.

12. Let F be a field of characteristic not dividing n that contains the nth roots of unity, and suppose L/F is a Galois extension with abelian Galois group of exponent dividing n. Prove that L is the composite of cyclic extensions of F whose degrees are divisors of n and use this to prove that there is a bijection between the subgroups of the multiplicative group $F^\times / F^{\times n}$ and such extensions L.

13. The Galois group of the extension $\mathbb{C}/\mathbb{R}$ is the cyclic group $G = \langle \tau \rangle$ of order 2 generated by complex conjugation τ. Prove that $H^2(G, \mathbb{C}^\times) \cong \mathbb{R}^\times/\mathbb{R}^+ \cong \mathbb{Z}/2\mathbb{Z}$ where $\mathbb{R}^+$ denotes the positive real numbers.

14. For any group G let $\hat{G} = \text{Hom}(G, \mathbb{Q}/\mathbb{Z})$ denote its dual group.

(a) If $\varphi : G_1 \to G_2$ is a group homomorphism prove that composition with φ induces a homomorphism $\hat{\varphi} : \hat{G}_2 \to \hat{G}_1$ on their dual groups.

(b) For any fixed g in G, show that evaluation at g gives a homomorphism φ_g from $\hat{G}$ to $\mathbb{Q}/\mathbb{Z}$.

(c) Prove that the map taking $g \in G$ to φ_g in (b) defines a homomorphism from G to its double dual $(\hat{\hat{G}})$.

(d) Prove that if G is a finite abelian group then the homomorphism in (c) is an isomorphism of G with its double dual. (By Exercise 14 in Section 5.2 the group G is (noncanonically) isomorphic to its dual $\hat{G}$. This shows that G is *canonically* isomorphic to its double dual — the isomorphism is independent of any choice of generators for G.)

(e) If $\psi : \hat{G}_2 \to \hat{G}_1$ is a homomorphism where G_1 and G_2 are finite abelian groups, then by (a) and (d) there is an induced homomorphism $\varphi : G_1 \to G_2$. Prove that

$$\varphi(g_1) = g_2 \text{ if } \chi(g_2) = \chi'(g_1) \text{ for } \chi' = \psi(\chi).$$

15. Use Gauss' Lemma in the computation of the transfer map for $\mathbb{F}_p^\times$ to $\{\pm 1\}$ to prove that 2 is a square modulo the odd prime p if and only if $p \equiv \pm 1 \bmod 8$. [Count how many elements in $2, 4, \ldots, p - 1$ are greater than $(p - 1)/2$.]

17.4 GROUP EXTENSIONS, FACTOR SETS AND $H^2(G,A)$

If A is a G-module then from the definition of the coboundary map d_2 in equation (18) a function f from $G \times G$ to A is a 2-cocycle if it satisfies the identity

$$f(g, h) + f(gh, k) = g \cdot f(h, k) + f(g, hk) \qquad \text{for all } g, h, k \in G. \qquad (17.26)$$

Equivalently, a 2-cocycle is determined by a collection of elements $\{a_{g,h}\}_{g,h \in G}$ of elements in A satisfying $a_{g,h} + a_{gh,k} = g \cdot a_{h,k} + a_{g,hk}$ for $g, h, k \in G$ (and then the 2-cocycle f is the function sending (g, h) to $a_{g,h}$).

A 2-cochain f is a coboundary if there is a function $f_1 : G \to A$ such that

$$f(g, h) = gf_1(h) - f_1(gh) + f_1(g), \qquad \text{for all } g, h \in G \qquad (17.27)$$

i.e., f is the image under d_1 of the 1-cochain f_1.

One of the main results of this section is to make a connection between the 2-cocycles $Z^2(G, A)$ and the *factor sets* associated to a group extension of G by A, which arise when considering the effect of choosing different coset representatives in defining the multiplication in the extension. In particular, we shall show that there is a bijection between equivalence classes of group extensions of G by A (with the action of G on A fixed) and the elements of $H^2(G, A)$.

We first observe some basic facts about extensions. Let E be any group extension of G by A,

$$1 \longrightarrow A \overset{\iota}{\longrightarrow} E \overset{\pi}{\longrightarrow} G \longrightarrow 1. \qquad (17.28)$$

The extension (28) determines an action of G on A, as follows. For each $g \in G$ let e_g be an element of E mapping onto g by π (the choice of such a set of representatives for G in E is called a set-theoretic *section* of π). The element e_g acts by conjugation on the normal subgroup $\iota(A)$ of E, mapping $\iota(a)$ to $e_g \iota(a) e_g^{-1}$. Any other element in E that maps to g is of the form $e_g \iota(a_1)$ for some $a_1 \in A$, and since $\iota(A)$ is abelian, conjugation by this element on $\iota(A)$ is the same as conjugation by e_g, so is independent of the choice of representative for g. Hence G acts on $\iota(A)$, and so also on A since ι is injective. Since conjugation is an automorphism, the extension (28) defines A as a G-module.

Recall from Section 10.5 that two extensions $1 \to A \overset{\iota_1}{\to} E_1 \overset{\pi_1}{\to} G \to 1$ and $1 \to A \overset{\iota_2}{\to} E_2 \overset{\pi_2}{\to} G \to 1$ are *equivalent* if there is a group isomorphism $\beta : E_1 \to E_2$ such that the following diagram commutes:

$$(17.29)$$

In this case we simply say β is the equivalence between the two extensions. As noted in Section 10.5, equivalence of extensions is reflexive, symmetric and transitive. We also observe that

equivalent extensions define the same G-module structure on A.

To see this assume (29) is an equivalence, let g be any element of G and let e_g be any element of E_1 mapping onto g by π_1. The action of g on A given by conjugation in E_1 maps each a to $\iota_1^{-1}(e_g \iota_1(a) e_g^{-1})$. Let $e_g' = \beta(e_g)$. Since the diagram commutes, $\pi_2(e_g') = g$, so the action of g on A in the second extension is given by conjugation by e_g'. This conjugation maps a to $\iota_2^{-1}(e_g' \iota_2(a) e_g'^{-1})$. Since ι_1, ι_2 and β are injective, the two actions of g on a are equal if and only if they result in the same image in E_2, i.e., $\beta \circ \iota_1(\iota_1^{-1}(e_g \iota_1(a) e_g^{-1})) = e_g' \iota_2(a) e_g'^{-1}$. This equality is now immediate from the definition of e_g' and the commutativity of the diagram.

We next see how an extension as in (28) defines a 2-cocycle in $Z^2(G, A)$. For simplicity we identify A as a subgroup of E via ι and we identify G as E/A via π.

Definition. A map $\mu : G \to E$ with $\pi \circ \mu(g) = g$ and $\mu(1) = 0$, i.e., so that for each $g \in G$, $\mu(g)$ is a representative of the coset Ag of E and the identity of E (which is the zero of A) represents the identity coset, is called a *normalized section* of π.

Fix a section μ of π in (28). Each element of E may be written uniquely in the form $a\mu(g)$, where $a \in A$ and $g \in G$. For $g, h \in G$ the product $\mu(g)\mu(h)$ in E lies in the coset Agh, so there is a unique element $f(g, h)$ in A such that

$$\mu(g)\mu(h) = f(g, h)\mu(gh) \qquad \text{for all } g, h \in G. \tag{17.30}$$

If in addition μ is normalized at the identity we also have

$$f(g, 1) = 0 = f(1, g) \qquad \text{for all } g \in G. \tag{17.31}$$

Definition. The function f defined by equation (30) is called the *factor set* for the extension E associated to the section μ. If f also satisfies (31) then f is called a *normalized* factor set.

We shall see in the examples following that it is possible for different sections μ to give the same factor set f.

We now verify that the factor set f is in fact a 2-cocycle. First note that the group operation in E may be written

$$\begin{aligned}
(a_1\mu(g))(a_2\mu(h)) &= (a_1 + \mu(g)a_2\mu(g)^{-1})\mu(g)\mu(h) \\
&= (a_1 + g \cdot a_2)(\mu(g)\mu(h)) \\
&= (a_1 + g \cdot a_2 + f(g, h))\mu(gh)
\end{aligned} \tag{17.32}$$

where $g \cdot a_2$ denotes the G-module action of g on a_2 given by conjugation in E. Now use (32) and the associative law in E to compute the product $\mu(g)\mu(h)\mu(k)$ in two different ways:

$$\begin{aligned}
(\mu(g)\mu(h))\mu(k) &= (f(g, h) + f(gh, k))\mu(ghk) \\
\mu(g)(\mu(h)\mu(k)) &= (gf(h, k) + f(g, hk))\mu(ghk).
\end{aligned} \tag{17.33}$$

It follows that the factors in A of the two right hand sides in (33) are equal for every $g, h, k \in G$, and this is precisely the 2-cocycle condition (26) for f. This shows that the factor set associated to the extension E and any choice of section μ is an element in $Z^2(G, A)$.

We next see how the factor set f depends on the choice of section μ. Suppose μ' is another section for the same extension E in (28), and let f' be its associated factor set. Then for all $g \in G$ both $\mu(g)$ and $\mu'(g)$ lie in the same coset Ag, so there is a function $f_1 : G \to A$ such that $\mu'(g) = f_1(g)\mu(g)$ for all g. Then

$$\mu'(g)\mu'(h) = f'(g, h)\mu'(gh) = (f'(g, h) + f_1(gh))\mu(gh).$$

We also have

$$\mu'(g)\mu'(h) = (f_1(g)\mu(g))(f_1(h)\mu(h)) = (f_1(g) + g \cdot f_1(h))(\mu(g)\mu(h))$$
$$= (f_1(g) + g \cdot f_1(h) + f(g, h))\mu(gh).$$

Equating the factors in A in these two expressions for $\mu'(g)\mu'(h)$ shows that

$$f'(g, h) = f(g, h) + (gf_1(h) - f_1(gh) + f_1(g)) \qquad \text{for all } g, h \in G,$$

in other words f and f' differ by the 2-coboundary of f_1 as in (27).

We have shown that the factor sets associated to the extension E corresponding to different choices of sections give 2-cocycles in $Z^2(G, A)$ that differ by a coboundary in $B^2(G, A)$. Hence associated to the extension E is a well defined cohomology class in $H^2(G, A)$ determined by the factor set in (30) for any choice of section μ.

If the extension E of G by A is a *split* extension (which is to say that $E = A \rtimes G$ is the semidirect product of G by A with the given conjugation action of G on A), then there is a section μ of G that is a *homomorphism* from G to E. In this case the factor set f in (30) is identically 0: $f(g, h) = 0$ for all $g, h \in G$. Hence the cohomology class in $H^2(G, A)$ defined by a split extension is the trivial class.

Suppose now that β is an equivalence between the extension in (28) and an extension E':

$$
\begin{array}{ccccccccc}
1 & \longrightarrow & A & \overset{\iota}{\longrightarrow} & E & \overset{\pi}{\longrightarrow} & G & \longrightarrow & 1 \\
& & \downarrow{\scriptstyle \text{id}} & & \downarrow{\scriptstyle \beta} & & \downarrow{\scriptstyle \text{id}} & & \\
1 & \longrightarrow & A & \overset{\iota'}{\longrightarrow} & E' & \overset{\pi'}{\longrightarrow} & G & \longrightarrow & 1.
\end{array}
$$

If μ is a section of π, then $\mu' = \beta \circ \mu$ is a section of π', so what we have just proved can be used to determine the cohomology class in $H^2(G, A)$ corresponding to E'. Applying the homomorphism β to equation (30) gives

$$\beta(\mu(g))\beta(\mu(h)) = \beta(f(g, h))\beta(\mu(gh)) \qquad \text{for all } g, h \in G.$$

Since β restricts to the identity map on A, this is

$$\mu'(g)\mu'(h) = f(g, h)\mu'(gh) \qquad \text{for all } g, h \in G,$$

which shows that the factor set for E' associated to μ' is the same as the factor set for E associated to μ. This proves that equivalent extensions define the same cohomology class in $H^2(G, A)$.

826

We next show how this procedure may be reversed: Given a class in $H^2(G, A)$ we construct an extension E_f whose corresponding factor set is in the given class in $H^2(G, A)$. The process generalizes the semidirect product construction of Section 5.5 (which is the special case when f is the zero cocycle representing the trivial class).

Note first that any 2-cocycle arising from the factor set of an extension as above where the section μ is normalized satisfies the condition in (31).

Definition. A 2-cocycle f such that $f(g, 1) = 0 = f(1, g)$ for all $g \in G$ is called a *normalized* 2-cocycle.

The construction of E_f is a little simpler when f is a normalized cocycle and for simplicity we indicate the construction in this case (the minor modifications necessary when f is not normalized are indicated in Exercise 4).

We first see that any 2-cocycle f lies in the same cohomology class as a normalized 2-cocycle. Let $d_1 f_1$ be the 2-coboundary of the constant function f_1 on G whose value is $f(1, 1)$. Then $f(1, 1) = d_1 f_1(1, 1)$, and one easily checks from the 2-cocycle condition that $f - d_1 f_1$ is normalized.

We may therefore assume that our cohomology class in $H^2(G, A)$ is represented by the normalized 2-cocycle f. Let E_f be the set $A \times G$, and define a binary operation on E_f by

$$(a_1, g)(a_2, h) = (a_1 + g \cdot a_2 + f(g, h), gh) \qquad (17.34)$$

where, as usual, $g \cdot a_2$ denotes the module action of G on A. It is straightforward to check that the group axioms hold: Since f is normalized, the identity element is $(0,1)$ and inverses are given by

$$(a, g)^{-1} = (-g^{-1} \cdot a - f(g^{-1}, g), g^{-1}). \qquad (17.35)$$

The cocycle condition implies the associative law by calculations similar to (32) and (33) earlier — the details are left as exercises.

Since f is a normalized 2-cocycle, $A^* = \{(a, 1) \mid a \in A\}$ is a subgroup of E_f, and the map $\iota^* : a \mapsto (a, 1)$ is an isomorphism from A to A^*. Moreover, from (34) and (35) it follows that

$$(0, g)(a, 1)(0, g)^{-1} = (g \cdot a, 1) \qquad \text{for all } g \in G \text{ and all } a \in A. \qquad (17.36)$$

Since E_f is generated by A^* together with the set of elements $(0, g)$ for $g \in G$, (36) implies that A^* is a normal subgroup of E_f. Furthermore, it is immediate from (34) that the map $\pi^* : (a, g) \mapsto g$ is a surjective homomorphism from E_f to G with kernel A^*, i.e., $E_f/A^* \cong G$. Thus

$$1 \longrightarrow A \overset{\iota^*}{\longrightarrow} E_f \overset{\pi^*}{\longrightarrow} G \longrightarrow 1 \qquad (17.37)$$

is a specific extension of G by A, where (36) ensures also that the action of G on A by conjugation in this extension is the module action specified in determining the 2-cocycle f in $H^2(G, A)$. The extension sequence (37) shows that this extension has the normalized section $\mu(g) = (0, g)$ whose corresponding normalized factor set is f. Note that this proves not only that every cohomology class in $H^2(G, A)$ arises from some extension E, but that every normalized 2-cocycle arises as the normalized factor set of some extension.

Finally, suppose f' is another normalized 2-cocycle in the same cohomology class in $H^2(G, A)$ as f and let $E_{f'}$ be the corresponding extension. If f and f' differ by the coboundary of $f_1 : G \to A$ then $f(g, h) - f'(g, h) = g f_1(h) - f_1(gh) + f_1(g)$ for all $g, h \in G$. Setting $g = h = 1$ shows that $f_1(1) = 0$. Define

$$\beta : E_f \longrightarrow E_{f'} \qquad \text{by} \qquad \beta((a, g)) = (a + f_1(g), g).$$

It is immediate that β is a bijection, and

$$\begin{aligned}
\beta((a_1, g)(a_2, h)) &= \beta((a_1 + g \cdot a_2 + f(g, h), gh)) \\
&= (a_1 + g \cdot a_2 + f(g, h) + f_1(gh), gh)) \\
&= (a_1 + f_1(g) + g \cdot (a_2 + f_1(h)) + f'(g, h), gh) \\
&= (a_1 + f_1(g), g)(a_2 + f_1(h), h) = \beta((a_1, g))\beta((a_2, h))
\end{aligned}$$

shows that β is an isomorphism from E_f to $E_{f'}$.

The restriction of β to A is given by $\beta((a, 1)) = (a + f_1(1), 1) = (a, 1)$, so β is the identity map on A. Similarly β is the identity map on the second component of (a, g), so β induces the identity map on the quotient G. It follows that β defines an equivalence between the extensions E_f and $E_{f'}$. This shows that the equivalence class of the extension E_f depends only on the cohomology class of f in $H^2(G, A)$.

We summarize this discussion in the following theorem.

Theorem 36. Let A be a G-module. Then

    (1) A function $f : G \times G \to A$ is a normalized factor set of some extension E of G by A (with conjugation given by the G-module action on A) if and only if f is a normalized 2-cocycle in $Z^2(G, A)$.

    (2) There is a bijection between the equivalence classes of extensions E as in (1) and the cohomology classes in $H^2(G, A)$. The bijection takes an extension E into the class of a normalized factor set f for E associated to any normalized section μ of G into E, and takes a cohomology class c in $H^2(G, A)$ to the extension E_f defined by the extension (37) for any normalized cocycle f in the class c.

    (3) Under the bijection in (2), split extensions correspond to the trivial cohomology class.

Corollary 37. Every extension of G by the abelian group A splits if and only if $H^2(G, A) = 0$.

Corollary 38. If A is a finite abelian group and $(|A|, |G|) = 1$ then every extension of G by A splits.

Proof: This follows immediately from Corollary 29 in Section 2.

We can use Corollary 38 to prove the same result without the restriction that A be an abelian group.

828

Theorem 39. (*Schur's Theorem*) If E is any finite group containing a normal subgroup N whose order and index are relatively prime, then N has a complement in E.

Remark: Recall that a subgroup whose order and index are relatively prime is called a *Hall subgroup*, so Schur's Theorem says that every normal Hall subgroup has a complement that splits the group as a semidirect product.

Proof: We use induction on the order of E. Since we may assume $N \neq 1$, let p be a prime dividing $|N|$ and let P be a Sylow p-subgroup of N. Let E_0 be the normalizer in E of P and let $N_0 = N \cap E_0$. By Frattini's Argument (Proposition 6 in Section 6.1) $E = E_0 N$. It follows from the Second Isomorphism Theorem that N_0 is a (normal) Hall subgroup of E_0 and $|E_0 : N_0| = |E : N|$ (cf. Exercise 10 of Section 3.3).

If $E_0 < E$, then by induction applied to N_0 in E_0 we obtain that E_0 contains a complement K to N_0. Since $|K| = |E_0 : N_0|$, K is also a complement to N in E, as needed. Thus we may assume $E_0 = E$, i.e., P is normal in E.

Since the center of P, $Z(P)$, is characteristic in P, it is normal in E (cf. Section 4.4). If $Z(P) = N$, then N is abelian and the theorem follows from Corollary 38. Thus we may assume $Z(P) \neq N$. Let bars denote passage to the quotient group $E/Z(P)$. Then $\overline{N}$ is a normal Hall subgroup of $\overline{E}$. By induction it has a complement $\overline{K}$ in $\overline{E}$. Let E_1 be the complete preimage of $\overline{K}$ in E. Then $|E_1| = |\overline{K}||Z(P)| = |E/N||Z(P)|$, so $Z(P)$ is a normal Hall subgroup of E_1. By induction $Z(P)$ has a complement in E_1 which is seen by order considerations to also be a complement to N in E. This completes the proof.

Examples

(1) If $G = Z_2$ and $A = \mathbb{Z}/2\mathbb{Z}$ then G acts trivially on A and so $H^2(G, A) = A^G/NA = \mathbb{Z}/2\mathbb{Z}$ by the computation of the cohomology of cyclic groups in Section 2, so by Theorem 36 there are precisely two inequivalent extensions of G by A. These are the cyclic group of order 4 and the Klein 4-group, the latter being split and hence corresponding to the trivial class in H^2.

(2) If $G = \langle g \rangle \cong Z_2$ and $A = \langle a \rangle \cong \mathbb{Z}/4\mathbb{Z}$ is a group of order 4 on which G acts trivially, then $H^2(G, A) = A/2A \cong \mathbb{Z}/2\mathbb{Z}$ by the computation of the cohomology of cyclic groups. As in the previous example there are two inequivalent extensions of G by A; evidently these are the groups Z_8 and $Z_4 \times Z_2$, the latter split extension corresponding to the trivial cohomology class.

If $E = \langle r \rangle \times \langle s \rangle$ denotes the split extension of G by A, where $|r| = 4$ and $|s| = 2$, then $\mu_i(g) = r^i s$ for $i = 0, \ldots, 3$ give the four normalized sections of G in E. The sections μ_0, μ_2 both give the zero factor set f. The sections μ_1, μ_3 both give the factor set f' with $f'(g, g) = a^2 \in A$. Both f and f' give normalized 2-cocycles lying in the trivial cohomology class of $H^2(G, A)$. The extension E_f corresponding to the zero 2-cocycle f is the group with the elements $(a, 1)$ and $(1, g)$ as the usual generators (of orders 4 and 2, respectively) for $Z_4 \times Z_2$. In $E_{f'}$, however, $(a, 1)$ has order 4 but so does $(1, g)$ since $(1, g)^2 = (f'(g, g), g^2) = (a^2, 1)$. The 2-cocycles f and f' differ by the coboundary f_1 with $f_1(1) = 1$ and $f_1(g) = r$. The isomorphism $\beta(a, g) = (a + f_1(g), g)$ from E_f to $E_{f'}$ maps the generators $(a, 1)$ and $(1, g)$ of E_f to the generators $(a, 1)$ and (a, g) of $E_{f'}$ and gives the explicit equivalence of these two extensions.

The situation where G acts on A by inversion is handled in Exercise 3.

(3) Suppose $G = Z_2$ and A is the Klein 4-group. If G acts nontrivially on A then G interchanges two of the nonidentity elements, say a and b, of A and fixes the third nonidentity element c. Then $A^G = NA = \{1, c\}$ and so $H^2(G, A) = 0$, and so every extension E of G by A splits. This can be seen directly, as follows. Since the action is nontrivial, such a group must be nonabelian, hence must be D_8. From the lattice of D_8 in Section 2.5 one sees that for each Klein 4-group there is a subgroup of order 2 in D_8 not contained in the 4-group and that subgroup splits the extension.

If G acts trivially on A then $H^2(G, A) = A/2A \cong A$, so there are 4 inequivalent extensions of G by A in this case. These are considered in Exercise 1.

Example: (Groups of Order 8 and $H^2(Z_2 \times Z_2, \mathbb{Z}/2\mathbb{Z})$)

Let $G = \{1, a, b, c\}$ be the Klein 4-group and let $A = \mathbb{Z}/2\mathbb{Z}$. The 2-group G must act trivially on A. The elements of $H^2(G, A)$ classify extensions E of order 8 which has a quotient group by some Z_2 subgroup that is isomorphic to the Klein 4-group. Although there are, up to group isomorphism, only four such groups, we shall see that there are *eight* inequivalent extensions.

Since $G \times G$ has 16 elements, we have $|C^2(G, A)| = 2^{16}$. The cocycle condition (26) here reduces to

$$f(g, h) + f(gh, k) = f(h, k) + f(g, hk) \qquad \text{for all } g, h, k \in G. \tag{17.38}$$

The following relations hold for the subgroup $Z^2(G, A)$ of cocycles:

(1) $f(g, 1) = f(1, g) = f(1, 1)$, for all $g \in G$
(2) $f(g, 1) + f(g, a) + f(g, b) + f(g, c) = 0$, for all $g \in G$
(3) $f(1, h) + f(a, h) + f(b, h) + f(c, h) = 0$, for all $h \in G$.

The first of these come from (38) by setting $h = k = 1$ and by setting $g = h = 1$. The other two relations come from (38) by setting $g = h$ and $h = k$, respectively, using relations (1) and (2). It follows that every 2-cocycle f can be represented by a vector $(\alpha, \beta, \gamma, \delta, \epsilon)$ in $\mathbb{F}_2$ where

$$\alpha = f(1, g) = f(g, 1), \quad \text{for all } g \in G,$$
$$\beta = f(a, a), \quad \gamma = f(a, b), \quad \delta = f(b, a), \quad \epsilon = f(b, b)$$

because the relations above then determine the remaining values of f:

$$f(a, c) = \alpha + \beta + \gamma \qquad f(b, c) = \alpha + \delta + \epsilon \qquad f(c, a) = \alpha + \beta + \delta$$
$$f(c, b) = \alpha + \gamma + \epsilon \qquad f(c, c) = \alpha + \beta + \gamma + \epsilon.$$

It follows that $|Z^2(G, A)| \leq 2^5$. Although one could eventually show that every function satisfying these relations is a 2-cocycle (hence the order is exactly 32), this will follow from other considerations below.

A cocycle f is a coboundary if there is a function $f_1 : G \to A$ such that

$$f(g, h) = f_1(h) - f_1(gh) + f_1(g), \qquad \text{for all } g, h \in G.$$

This coboundary condition is easily seen to be equivalent to the conditions:

(i) $f(g, 1) = f(1, g) = f(g, g)$ for all $g \in G$, and
(ii) $f(g, h) = f(g', h')$ whenever g, h are distinct nonidentity elements and so are g', h'.

These relations are equivalent to $\alpha = \beta = \epsilon$ and $\gamma = \delta$. Thus $B^2(G, A)$ consists of the vectors $(\alpha, \alpha, \gamma, \gamma, \alpha)$, and so $H^2(G, A)$ has dimension at most 3 (i.e., order at most $2^3 = 8$). It is easy to see that $\{(0, \beta, \gamma, 0, \epsilon)\}$ with β, γ, and ϵ in $\mathbb{F}_2$ gives a set of representatives for $Z^2(G, A)/B^2(G, A)$, and each of these representative cocycles is normalized. We now prove $|H^2(G, A)| = 8$ (and also that $|Z^2(G, A)| = 2^5$) by explicitly exhibiting eight inequivalent group extensions,

Suppose E is an extension of G by A, where for simplicity we assume $A \leq E$. If $\mu : G \to E$ is a section, the factor set for E associated to μ satisfies

$$\mu(g)\mu(h) = f(g, h)\mu(gh).$$

The group E is generated by $\mu(a)$, $\mu(b)$ and A, and A is contained in the center of E since G acts trivially on A. Hence E is abelian if and only if $\mu(a)\mu(b) = \mu(b)\mu(a)$, which by the relation above occurs if and only if $f(a, b) = f(b, a)$. If g is a nonidentity element in G, we also see from the relation above that $\mu(g)$ is an element of order 2 in E if and only if $f(g, g) = 0$. Because A is contained in the center of E, both elements in any nonidentity coset $A\mu(g)$ have the same order (either 2 or 4).

There are four groups of order 8 containing a normal subgroup of order 2 with quotient group isomorphic to the Klein 4-group: $Z_2 \times Z_2 \times Z_2$, $Z_4 \times Z_2$, D_8, and Q_8.

The group $E \cong Z_2 \times Z_2 \times Z_2$ is the split extension of G by A and has $f = 0$ as factor set.

When $E \cong Q_8$, in the usual notation for the quaternion group $A = \langle -1 \rangle$. In this (non-abelian) group every nonidentity coset consists of elements of order 4, and this property is unique to Q_8, so the resulting factor set f satisfies $f(g, g) \neq 0$ for all nonidentity elements in G.

When $E \cong Z_4 \times Z_2 = \langle x \rangle \times \langle y \rangle$ we must have $A = \langle x^2 \rangle$. The cosets Ax and Axy both consist of elements of order 4, and the coset Ay consists of elements of order 2, so exactly one of $\mu(a)$, $\mu(b)$ or $\mu(c)$ is an element of order 2 and the other two must be of order 4. This suggests three homomorphisms from E to G, defined on generators by

$$\begin{aligned} \pi_1(y) &= a & \pi_1(x) &= b \\ \pi_2(y) &= b & \pi_2(x) &= a \\ \pi_3(y) &= c & \pi_3(x) &= a \end{aligned}$$

Each of these homomorphisms maps surjectively onto G, has A as kernel, and has $\mu(a)$ (respectively, $\mu(b)$, $\mu(c)$) an element of order 2 in E. Any isomorphism of E with itself that is the identity on A must take the unique nonidentity coset Ay of A consisting of elements of order 2 to itself. Hence any extension equivalent to the extension E_1 defined by π_1 also maps y to a (since the equivalence is the identity on G). It follows that the three extensions defined by π_1, π_2 and π_3 are inequivalent.

The situation when $E \cong D_8 = \langle r, s \rangle$ is similar. In this case $A = \langle r^2 \rangle$, the cosets As and Asr consist of elements of order 2, and the coset Ar consists of elements of order 4. In this case exactly one of $\mu(a)$, $\mu(b)$ or $\mu(c)$ is an element of order 4 and the other two are of order 2, suggesting the three homomorphisms defined on generators by

$$\begin{aligned} \pi_1(r) &= a & \pi_1(s) &= b \\ \pi_2(r) &= b & \pi_2(s) &= a \\ \pi_3(r) &= c & \pi_3(s) &= a \end{aligned}$$

As before, the corresponding extensions are inequivalent.

The existence of 8 inequivalent extensions of G by A proves that $|H^2(G, A)| = 8$, and hence that these are a complete list of all the inequivalent extensions. In particular, the extension $E_1' \cong Z_4 \times Z_2$ defined by the homomorphism π_1' mapping y to a and x to c must be equivalent to the extension E_1 above (and similarly for the other two extensions isomorphic to $Z_4 \times Z_2$ and the three extensions for D_8). This proves the existence of certain outer automorphisms for these groups, cf. Exercise 9.

Remark: For any prime p the cohomology groups of the elementary abelian group E_{p^m} with coefficients in the finite field $\mathbb{F}_p$ may be determined by relating them to the cohomology groups of the factors in the direct product as mentioned at the end of Section 2. In general, $H^2(E_{p^m}, \mathbb{F}_p)$ is a vector space over $\mathbb{F}_p$ of dimension $\frac{1}{2}m(m+1)$. When $p = 2$ and $m = 2$ this is the result $H^2(Z_2 \times Z_2, \mathbb{Z}/2\mathbb{Z}) \cong (\mathbb{Z}/2\mathbb{Z})^3$ above.

Crossed Product Algebras and the Brauer Group

Suppose F is a field. Recall that an F-algebra B is a ring containing the field F in its center and the identity of B is the identity of F, cf. Section 10.1.

Definition. An F-algebra A is said to be *simple* if A contains no nontrivial proper (two sided) ideals. A *central simple F-algebra* A is a simple F-algebra whose center is F.

Among the easiest central simple F-algebras are the matrix algebras $M_n(F)$ of $n \times n$ matrices with coefficients in F.

If K/F is a finite Galois extension of fields with Galois group $G = \text{Gal}(K/F)$, then we can use the normalized 2-cocycles in $Z^2(G, K^\times)$ to construct certain central simple K-algebras. The construction of these algebras from 2-cocycles and their classification in terms of $H^2(G, K^\times)$ (cf. Theorem 42 below) are important applications of cohomological methods in number theory. Their construction in the case when G is cyclic was one of the precursors leading to the development of abstract cohomology.

Suppose $f = \{a_{\sigma,\tau}\}_{\sigma,\tau \in G}$ is a normalized 2-cocycle in $Z^2(G, K^\times)$. Let B_f be the vector space over K having basis u_σ for $\sigma \in G$:

$$B_f = \left\{ \sum_{\sigma \in G} \alpha_\sigma u_\sigma \mid \alpha_\sigma \in K \right\}. \tag{17.39}$$

Define a multiplication on B_f by

$$u_\sigma \alpha = \sigma(\alpha) u_\sigma \qquad u_\sigma u_\tau = a_{\sigma,\tau} u_{\sigma\tau} \tag{17.40}$$

for $\alpha \in K$ and $\sigma, \tau \in G$. The second equation shows that the $a_{\sigma,\tau}$ give a "factor set" for the elements u_σ in B_f and is one reason this terminology is used. Using this multiplication we find

$$(u_\sigma u_\tau) u_\rho = a_{\sigma,\tau} a_{\sigma\tau,\rho} u_{\sigma\tau\rho} \qquad \text{and} \qquad u_\sigma (u_\tau u_\rho) = \sigma(a_{\tau,\rho}) a_{\sigma,\tau\rho} u_{\sigma\tau\rho}.$$

Since $a_{\sigma,\tau} a_{\sigma\tau,\rho} = \sigma(a_{\tau,\rho}) a_{\sigma,\tau\rho}$ is the multiplicative form of the cocycle condition (26), it follows that the multiplication defined in (40) is associative.

Since the cocycle is normalized we have $a_{1,\sigma} = a_{\sigma,1} = 1$ for all $\sigma \in G$ and it follows from (40) that the element u_1 is an identity in B_f. Identifying K with the elements αu_1 in B_f, we see that B_f is an F-algebra containing the field K and having dimension n^2 over F if $n = [K : F] = |G|$.

Proposition 40. The F-algebra B_f with K-vector space basis u_σ in (39) and multiplication defined by (40) is a central simple F-algebra.

Proof: It remains to show that the center of B_f is F and that B_f contains no nonzero proper ideals. Suppose $x = \sum_{\sigma \in G} \alpha_\sigma u_\sigma$ is an element in the center of B_f. Then $x\beta = \beta x$ for $\beta \in K$ shows that $\sigma(\beta) = \beta$ if $\alpha_\sigma \neq 0$. Since there is an element $\beta \in K$ not fixed by σ for any $\sigma \neq 1$, this shows that $a_\sigma = 0$ for all $\sigma \neq 1$, so $x = \alpha_1 u_1$. Then $x u_\tau = u_\tau x$ if and only if $\tau(\alpha_1) = \alpha_1$, so if this is true for all τ then we must have $\alpha_1 = a \in K$. Hence $x = a u_1$ and the center of B_f is F.

To show that B_f is simple, suppose I is a nonzero ideal in B_f and let

$$x = \alpha_{\sigma_1} u_{\sigma_1} + \cdots + \alpha_{\sigma_m} u_{\sigma_m}$$

be a nonzero element of I with the minimal number m of nonzero terms. If $m > 1$ there is an element $\beta \in K^\times$ with $\sigma_m(\beta) \neq \sigma_{m-1}(\beta)$. Then the element $x - \sigma_m(\beta) x \beta^{-1}$ would be an element of the ideal I with the nonzero element $(1 - \sigma_m(\beta)\sigma_{m-1}(\beta)^{-1})\alpha_{\sigma_{m-1}}$ as coefficient of $u_{\sigma_{m-1}}$, and would have fewer nonzero terms than x since the coefficient of u_{σ_m} is 0. It follows that $m = 1$ and $x = \alpha u_\sigma$ for some $\alpha \in K$ and some σ. This element is a unit, with inverse $\sigma^{-1}(\alpha^{-1}) u_{\sigma^{-1}}$, so $I = B_f$, completing the proof.

Definition. The central simple F-algebra B_f defined by (39) and (40) is called the *crossed product algebra* for the factor set $\{a_{\sigma,\tau}\}$.

If $f' = a'_{\sigma,\tau}$ is a normalized cocycle in the same cohomology class in $H^2(G, K^\times)$ as $a_{\sigma,\tau}$ then there are elements $b_\sigma \in K^\times$ with

$$a'_{\sigma,\tau} = a_{\sigma,\tau}(\sigma(b_\tau)b_{\sigma\tau}^{-1}b_\sigma)$$

(the multiplicative form of the coboundary condition (27)). If $B_{f'}$ is the F-algebra with K-basis v_σ defined from this cocycle as in (39) and (40), then the K-vector space homomorphism φ defined by mapping u'_σ to $b_\sigma u_\sigma$ satisfies

$$\varphi(u'_\sigma u'_\tau) = \varphi(a'_{\sigma,\tau} u'_{\sigma\tau}) = a'_{\sigma,\tau} b_{\sigma\tau} u_{\sigma\tau} = b_\sigma \sigma(b_\tau) u_\sigma u_\tau$$
$$= (b_\sigma u_\sigma)(b_\tau u_\tau) = \varphi(u'_\sigma)\varphi(u'_\tau).$$

It follows that φ is an F-algebra isomorphism from $B_{f'}$ to B_f.

We have shown that every cohomology class c in $H^2(G, K^\times)$ defines an isomorphism class of central simple F-algebras, namely the isomorphism class of any crossed product algebra for a normalized cocycle $\{a_{\sigma,\tau}\}$ representing the class c. The next result shows that the trivial cohomology class corresponds to the isomorphism class containing $M_n(F)$.

Proposition 41. The crossed product algebra for the trivial cohomology class in $H^2(G, K^\times)$ is isomorphic to the matrix algebra $M_n(F)$ where $n = [K : F]$.

Proof: If $\alpha \in K$ then multiplication by α defines a linear transformation T_α of K viewed as an n-dimensional vector space over F. Similarly, every automorphism $\sigma \in G$ defines an F-linear transformation T_σ of K, and we may view both T_α and T_σ as elements of $M_n(F)$ by choosing a basis for K over F. If B_0 denotes the crossed product algebra for the trivial factor set ($a_{\sigma,\tau} = 1$ for all $\sigma, \tau \in G$), consider the additive map $\varphi : B_0 \to M_n(F)$ defined by $\varphi(\alpha u_\sigma) = T_\alpha T_\sigma$. Since $T_{a\alpha} = aT_\alpha$ for $a \in F$, the map φ is an F-vector space homomorphism. If $x \in K$, we have

$$T_\sigma T_\alpha(x) = T_\sigma(\alpha x) = \sigma(\alpha x) = \sigma(\alpha)\sigma(x) = T_{\sigma(\alpha)}T_\sigma,$$

so $T_\sigma T_\alpha = T_{\sigma(\alpha)}T_\sigma$ as linear transformations on K. It then follows from $u_\sigma u_\tau = u_{\sigma\tau}$ that

$$\varphi((\alpha u_\sigma)(\beta u_\tau)) = \varphi(\alpha\sigma(\beta) u_{\sigma\tau}) = T_{\alpha\sigma(\beta)}T_{\sigma\tau} = T_\alpha T_{\sigma(\beta)}T_\sigma T_\tau$$
$$= T_\alpha T_\sigma T_\beta T_\tau = \varphi(\alpha u_\sigma)\,\varphi(\beta u_\tau)$$

which shows that φ is an F-algebra homomorphism from B_0 to $M_n(F)$. Since $\ker \varphi$ is an ideal in B_0 and $\varphi \neq 0$, it follows from Proposition 40 that $\ker \varphi = 0$ and φ is an injection. Since both B_0 and $M_n(F)$ have dimension n^2 as vector spaces over F, it follows that φ is an F-algebra isomorphism, proving the proposition.

Example

If $K = \mathbb{C}$ and $F = \mathbb{R}$, then $G = \mathrm{Gal}(\mathbb{C}/\mathbb{R})$ is of order 2 and generated by complex conjugation τ. We have $|H^2(G, \mathbb{C}^\times)| = 2$. The central simple $\mathbb{R}$-algebra B_0 corresponding to the trivial class is $\mathbb{C}u_1 \oplus \mathbb{C}u_\tau$ with $u_\tau(a + bi) = (a - bi)u_\tau$ and $u_\tau^2 = u_1$. This is isomorphic to the matrix algebra $M_2(\mathbb{R})$ under the map

$$\varphi((a + bi)u_1 + (c + di)u_\tau) = aI + bT_i + cT_\tau + dT_iT_\tau = \begin{pmatrix} a + c & -b + d \\ b + d & a - c \end{pmatrix}.$$

A normalized cocycle f representing the nontrivial cohomology class is defined by the values $a_{1,1} = a_{1,\tau} = a_{\tau,1} = 1$ and $a_{\tau,\tau} = -1$. The corresponding central simple $\mathbb{R}$-algebra B_f is given by $\mathbb{C}v_1 \oplus \mathbb{C}v_\tau$. The element v_1 is the identity of B_f, and we have the relations $v_\tau(a + bi) = (a - bi)v_\tau$ and $v_\tau^2 = -v_1$. Letting $v_1 = 1$ and $v_\tau = j$ we see that B_f is isomorphic as an $\mathbb{R}$-algebra to the real Hamilton Quaternions $\mathbb{R} + \mathbb{R}i + \mathbb{R}j + \mathbb{R}k$.

There is a rich theory of simple algebras and we mention without proof the following results. Let A be a central simple F-algebra of finite dimension over F.

I. If $F \subseteq B \subseteq A$ where B is a simple F-algebra define the *centralizer* B^c of B in A to be the elements of A that commute with all the elements of B. Define the *opposite algebra* B^{opp} to be the set B with opposite multiplication, i.e., the product $b_1 b_2$ in B^{opp} is given by the product $b_2 b_1$ in B. Both B^c and B^{opp} are simple F-algebras and we have
   a. $(\dim_F B)(\dim_F B^c) = \dim_F A$
   b. $A \otimes_F B^{opp} \cong M_r(B^c)$ as F-algebras, where $r = \dim_F B$
   c. $B \otimes_F B^c \cong A$ if B is a central simple F-algebra.
II. If A' is an Artinian (satisfies D.C.C. on left ideals) simple F-algebra, then $A \otimes_F A'$ is an Artinian simple F-algebra with center $(A')^c$.
III. We have $A \cong M_r(\Delta)$ for some division ring Δ whose center is F and some integer $r \geq 1$. The division ring Δ and r are uniquely determined by A. The same statement holds for any Artinian simple F-algebra.

The last result is part of Wedderburn's Theorem described in greater detail in the following chapter.

Definition. If A is a central simple F-algebra then a field L containing F is said to *split* A if $A \otimes_F L \cong M_m(L)$ for some $m \geq 1$.

It follows from (II) that every maximal commutative subalgebra of Δ is a field E with $E = E^c = E^{opp}$; if $[E : F] = m$ we obtain $\dim_F \Delta = m^2$. Applying (II) to $A = \Delta$ and $B = E$ we also see that $\Delta \otimes_F E \cong M_m(E)$. It can also be shown that a maximal subfield E of the central simple F-algebra A also satisfies $E = E^c = E^{opp}$ and so again by (II) it follows that $A \otimes_F E \cong M_r(E)$ $(r^2 = \dim_F A)$.

If $A = M_r(\Delta)$ then the field L splits A if and only if L splits Δ, as follows. If $\Delta \otimes_F L \cong M_n(L)$ then

$$A \otimes_F L \cong M_r(\Delta) \otimes_F L \cong M_r(\Delta \otimes_F L) \cong M_r(M_n(L)) \cong M_{rn}(L).$$

Conversely if $A \otimes_F L \cong M_n(L)$ then

$$M_n(L) \cong M_r(\Delta) \otimes_F L \cong M_r(\Delta \otimes_F L).$$

By (II) and (III), $\Delta \otimes_F L \cong M_s(\Delta')$ for some division ring Δ'. Together with the previous isomorphism, the uniqueness statement in (III) shows that $\Delta' \cong L$ and then the isomorphism $\Delta \otimes_F L \cong M_s(L)$ shows that L splits Δ.

We see from the discussion above that a maximal commutative subfield of Δ splits both Δ and $A \cong M_r(\Delta)$ for any $r \geq 1$. It is not too difficult to show from this that every central simple F-algebra of finite dimension over F can be split by a finite Galois extension of F.

Applying (I) by taking A to be the crossed product algebra B_f and taking $B = K$ shows that $K = K^c = K^{opp}$ and $B_f \otimes_F K \cong M_n(K)$. In particular, the crossed product algebras B_f are always split by K.

Example

In the example of the Hamilton Quaternions above we have $B_f \otimes_{\mathbb{R}} \mathbb{C} \cong M_2(\mathbb{C})$. We have $B_f \otimes_{\mathbb{R}} \mathbb{C} = \mathbb{C} + \mathbb{C}i + \mathbb{C}j + \mathbb{C}k$ and an explicit isomorphism φ to $M_2(\mathbb{C})$ is given by

$$\varphi(i) = \begin{pmatrix} \sqrt{-1} & 0 \\ 0 & -\sqrt{-1} \end{pmatrix} \qquad \varphi(j) = \begin{pmatrix} 0 & -1 \\ 1 & 0 \end{pmatrix}$$

and extending $\mathbb{C}$-linearly.

By (III) every central simple F-algebra A is isomorphic as an F-algebra to $M_r(\Delta)$ for some division ring Δ uniquely determined up to F-isomorphism, called the *division ring part* of A.

Definition. Two central simple F-algebras A and B are *similar* if $A \cong M_r(\Delta)$ and $B \cong M_s(\Delta)$ for the same division ring Δ, i.e., if A and B have the same division ring parts.

Let $[A]$ denote the similarity class of A. By (II), if A and B are central simple F-algebras then $A \otimes_F B$ is again a central simple F-algebra, so we may define a multiplication on similarity classes by $[A][B] = [A \otimes_F B]$. The class $[F]$ is an identity for this multiplication and associativity of the tensor product shows that the multiplication is associative. By (Ib) applied with $B = A$ (so then $B^c = F$ since A is central) we have $[A][A^{opp}] = [F]$, so inverses exist with this multiplication.

Definition. The group of similarity classes of central simple F-algebras with multiplication $[A][B] = [A \otimes_F B]$ is called the *Brauer group* of F and is denoted $Br(F)$.

If L is any extension field of F then by (II) the algebra $A \otimes_F L$ is a central simple L-algebra. It is easy to check that the map $[A] \to [A \otimes_F L]$ is a well defined homomorphism from $Br(F)$ to $Br(L)$. The kernel of this homomorphism consists of the classes of the algebras A with $A \otimes_F L \cong M_m(L)$ for some $m \geq 1$, i.e., the algebras A that are split by L.

Definition. If L/F is a field extension then the *relative Brauer group* $Br(L/F)$ is the group of similarity classes of central simple F-algebras that are split by L. Equivalently, $Br(L/F)$ is the kernel of the homomorphism $[A] \to [A \otimes_F L]$ from $Br(F)$ to $Br(L)$.

The following theorem summarizes some major results in this area and shows the fundamental connection between Brauer groups and the crossed product algebras constructed above.

Theorem 42. Suppose K/F is a Galois extension of degree n with $G = \text{Gal}(K/F)$.

(1) The central simple F-algebra A with $\dim_F A = n^2$ is split by K if and only if $A \otimes_F K \cong M_n(K)$ if and only if A is isomorphic to a crossed product algebra B_f as in (39) and (40).

(2) There is a bijection between the F-isomorphism classes of central simple F-algebras A with $A \otimes_F K \cong M_n(K)$ and the elements of $H^2(G, K^\times)$. Under this bijection the class $c \in H^2(G, K^\times)$ containing the normalized cocycle f corresponds to the isomorphism class of the crossed product algebra B_f defined in (39) and (40), and the trivial cohomology class corresponds to $M_n(F)$.

(3) Every central simple F-algebra of finite dimension over F and split by K is similar to one of dimension n^2 split by K. The bijection in (2) also establishes a bijection between $Br(K/F)$ and $H^2(G, K^\times)$ which is also an isomorphism of groups.

(4) There is a bijection between the collection of F-isomorphism classes of central simple division algebras over F that are split by K and $H^2(G, K^\times)$.

As previously mentioned, every central simple F-algebra of finite dimension over F can be split by some finite Galois extension of F, and it follows that

$$Br(F) = \bigcup_K Br(K/F)$$

where the union is over all finite Galois extensions of F. It follows that there is a bijection between $Br(F)$ and $H^2(\text{Gal}(F^s/F), (F^s)^\times)$ where F^s denotes a separable algebraic closure of F. Here $\text{Gal}(F^s/F)$ is considered as a profinite group and the cohomology group refers to continuous Galois cohomology.

One consequence of this result and Theorem 42 is that a full set of representatives for the F-isomorphism classes of central simple division algebras Δ over F can be obtained from the division algebra parts of the crossed product algebras for finite Galois extensions of F. Those division algebras that are split over K occur for the crossed product algebras for K/F.

Example

Since $H^2(\text{Gal}(\mathbb{F}_{q^d}/\mathbb{F}_q), \mathbb{F}_{q^d}^\times) = 0$ (cf. Exercise 10), we have $Br(\mathbb{F}_{q^d}/\mathbb{F}_q) = 0$ and hence also $Br(\mathbb{F}_q) = 0$. As a consequence, every finite division algebra is a field (cf. Exercise 13 in Section 13.6 for a direct proof), and every finite central simple algebra $\mathbb{F}_q$-algebra is isomorphic to a full matrix ring $M_r(\mathbb{F}_q)$.

EXERCISES

1. Let $A = \{1, a, b, c\}$ be the Klein 4-group and let $G = \langle g \rangle$ be the cyclic group of order 2 acting trivially on A.
   (a) Prove that $|C^2(G, A)| = 2^8$.
   (b) Show that coboundaries are constant functions, and deduce that $|B^2(G, A)| = 4$.
   (c) Use the cocycle condition to show that $|Z^2(G, A)| \le 2^4$.
   (d) If $E = Z_4 \times Z_2 = \langle x \rangle \times \langle y \rangle$, prove that the extensions $1 \to A \xrightarrow{\iota_i} E \xrightarrow{\pi} G \to 1$ defined by $\pi(x) = g$, $\pi(y) = 1$ and $\iota_1(a) = x^2$, $\iota_1(b) = y$ (respectively, $\iota_2(b) = x^2$, $\iota_2(a) = y$, and $\iota_3(c) = x^2$, $\iota_3(a) = y$), together with the split extension $Z_2 \times Z_2 \times Z_2$ give 4 inequivalent extensions of Z_2 by the Klein 4-group. Deduce that $H^2(G, A)$ has order 4 by explicitly exhibiting the corresponding cocycles.

2. Let $A = \mathbb{Z}/4\mathbb{Z}$ and let G be the cyclic group of order 2 acting trivially on A.
   (a) Prove that $|C^2(G, A)| = 2^8$.
   (b) Use the coboundary condition to show that $|B^2(G, A)| = 2^3$.
   (c) Use the cocycle condition to show that $|Z^2(G, A)| \le 2^4$.
   (d) Show that $|H^2(G, A)| = 2$ by exhibiting two inequivalent extensions of G by A and their corresponding cocycles.

3. Let $A = \mathbb{Z}/4\mathbb{Z}$ and let G be the cyclic group of order 2 acting by inversion on A.
   (a) Show that there are four coboundaries and that only the zero coboundary is normalized.
   (b) Prove by a direct computation of cocycle and coboundary groups that $|H^2(G, A)| = 2$.
   (c) Exhibit two distinct cohomology classes and their corresponding extension groups.
   (d) Show that for a given extension of G by A with extension group isomorphic to D_8 there are four normalized sections, all of which have the zero 2-cocycle as their factor set.
   (e) Show that for a given extension of G by A with extension group isomorphic to Q_8 there are sixteen sections, four of which are normalized, and all of the latter have the same factor set.

4. For a *non*-normalized 2-cocycle f one defines the extension group E_f on the set $A \times G$ by the same binary operation in equation (34). Verify two of the group axioms in this case by showing that identity is now $(-f(1, 1), 1)$ and inverses are given by

$$(a, x)^{-1} = (-x^{-1} \cdot a - f(x^{-1}, x) - f(1, 1), x^{-1}).$$

(Verification of the associative law is essentially the same as for normalized 2-cocycles.) Prove also that the set $A^{**} = \{(a - f(1, 1), 1) \mid a \in A\}$ is a subgroup of E_f and the map $\iota^{**} : a \mapsto (a - f(1, 1), 1)$ is an isomorphism from A to A^{**}. Show that this extension E_f, with the injection ι^{**} and the usual projection map π^* onto G, is equivalent to an extension derived from a normalized cocycle in the same class as f.

5. Show that the set of equivalences of a given extension $1 \to A \xrightarrow{\iota} E \xrightarrow{\pi} G \to 1$ with itself form a group under composition, and that this group is isomorphic to the stability group

Stab$(1 \trianglelefteq \iota(A) \trianglelefteq E)$. (Thus Proposition 31 implies $Z^1(G, A)$ is the group of equivalences of the extension with itself).

6. *(Gaschütz's Theorem)* Let p be a prime, let A be an abelian normal p-subgroup of a finite group G, and let P be a Sylow p-subgroup of G. Prove that G is a split extension of G/A by A if and only if P is a split extension of P/A by A. (Note that $A \leq P$ by Exercise 37 in Section 4.5). [Use Sylow's Theorem to show if G splits over A then so too does P. Conversely, show that a normalized 2-cocycle associated to the extension of P/A by A via Theorem 36 is the image of a normalized 2-cocycle in $H^2(G/A, A)$ under the restriction homomorphism Res : $H^2(G/A, A) \to H^2(P/A, A)$. Then use Proposition 26 and the fact that multiplication by $|G : P|$ is an automorphism of A.]

7. (a) Prove that $H^2(A_4, \mathbb{Z}/2\mathbb{Z}) \neq 0$ by exhibiting a nonsplit extension of A_4 by a cyclic group of order 2. [See Exercise 11, Section 4.5.]
   (b) Prove that $H^2(A_5, \mathbb{Z}/2\mathbb{Z}) \neq 0$ by showing that $SL_2(\mathbb{F}_5)$ is a nonsplit extension of A_5 by a cyclic group of order 2. [Use Propositions 21 and 23 in Section 4.5.]

8. The *Schur multiplier* of a finite group G is defined as the group $H^2(G, \mathbb{C}^\times)$, where the multiplicative group $\mathbb{C}^\times$ of complex numbers is a trivial G-module. Prove that the Schur multiplier is a finite group. [Show that every cohomology class contains a cocycle whose values lie in the nth roots of unity, where $n = |G|$, as follows: If f is any cocycle then by Corollary 27, $f^n \in B^2(G, \mathbb{C}^\times)$. Define $k \in C^2(G, \mathbb{C}^\times)$ by $k(g_1, g_2) = f(g_1, g_2)^{1/n}$ (take any n^{th} roots). Show that $k \in B^2(G, \mathbb{C}^\times)$ and fk^{-1} takes values in the group of nth roots of 1.]

9. Use the classification of the extensions of the Klein 4-group by Z_2 in the example following Theorem 39 to prove the following (in the notation of that example):
   (a) There is an (outer) automorphism of $Z_4 \times Z_2$ which interchanges the cosets Ax and Axy and fixes the coset Ay.
   (b) There is an outer automorphism of D_8 which interchanges the cosets As and Asr and fixes the coset Ar.

10. Suppose $\mathbb{F}_q$ is a finite field with $G = \text{Gal}(\mathbb{F}_{q^d}/\mathbb{F}_q) = \langle \sigma_q \rangle$ where σ_q is the Frobenius automorphism, and let N be the usual norm element for the cyclic group G.
   (a) Use Hilbert's Theorem 90 to prove that $|N(\mathbb{F}_{q^d}^\times)| = (q^d - 1)/(q - 1)$, and deduce that the norm map from $\mathbb{F}_{q^d}$ to $\mathbb{F}_q$ is surjective.
   (b) Prove that $H^n(G, \mathbb{F}_{q^d}^\times) = 0$ for all $n \geq 1$.

Part VI

INTRODUCTION TO
THE REPRESENTATION THEORY
OF FINITE GROUPS

The final two chapters are an introduction to the representation theory of finite groups together with some applications. We have already seen in Part I how actions of groups on sets, namely permutation representations, are a fundamental tool for unravelling the structure of groups. Cayley's Theorem and Sylow's Theorem as well as many of the results and applications in Sections 6.1 and 6.2 are based on groups acting on sets. The chapter on Galois Theory developed one of the most beautiful correspondences in mathematics where the action of a group as automorphisms of a field gives rise to a correspondence between the lattice of subgroups of the Galois group and the lattice of subfields of a Galois extension of fields. In these final two chapters we study groups acting as linear transformations on vector spaces. We shall be primarily interested in utilizing these linear actions to provide information about the groups themselves.

In Part III we saw that modules are the "representation objects" for rings in the sense that the axioms for an R-module specify a "ring action" of R on some abelian group M which preserves the abelian group structure of M. In the case where M was an $F[x]$-module, x acted as a linear transformation from the vector space M to itself. In Chapter 12 the classification of finitely generated modules over Principal Ideal Domains gave us a great deal of information about these linear transformations of M (e.g., canonical forms). In Chapter 16 we used the ideal structure in Dedekind Domains to generalize the results of Chapter 12 to the classification of finitely generated modules over such domains. In this part we follow a process similar to the study of $F[x]$-modules, replacing the polynomial ring with the group ring FG of G and classifying all finitely generated FG-modules for certain fields F (Wedderburn's Theorem). We then use this classification to derive some results about finite groups such as Burnside's Theorem on the solvability of groups of order $p^a q^b$ in Chapter 19.

CHAPTER 18

Representation Theory and Character Theory

18.1 LINEAR ACTIONS AND MODULES OVER GROUP RINGS

For the remainder of the book the groups we consider will be finite groups, unless explicitly mentioned otherwise. Throughout this section F is a field and G is a finite group. We first introduce the basic terminology. Recall that if V is a vector space over F, then $GL(V)$ is the group of nonsingular linear transformations from V to itself (under composition), and if $n \in \mathbb{Z}^+$, then $GL_n(F)$ is the group of invertible $n \times n$ matrices with entries from F (under matrix multiplication).

Definition. Let G be a finite group, let F be a field and let V be a vector space over F.
   (1) A *linear representation* of G is any homomorphism from G into $GL(V)$. The *degree* of the representation is the dimension of V.
   (2) Let $n \in \mathbb{Z}^+$. A *matrix representation* of G is any homomorphism from G into $GL_n(F)$.
   (3) A linear or matrix representation is *faithful* if it is injective.
   (4) The *group ring* of G over F is the set of all formal sums of the form

$$\sum_{g \in G} \alpha_g g, \qquad \alpha_g \in F$$

with componentwise addition and multiplication $(\alpha g)(\beta h) = (\alpha\beta)(gh)$ (where α and β are multiplied in F and gh is the product in G) extended to sums via the distributive law (cf. Section 7.2).

Unless we are specifically discussing permutation representations the term "representation" will always mean "linear representation." When we wish to emphasize the field F we shall say F-representation, or representation of G on V over F.

Recall that if V is a finite dimensional vector space of dimension n, then by fixing a basis of V we obtain an isomorphism $GL(V) \cong GL_n(F)$. In this way any linear representation of G on a finite dimensional vector space gives a matrix representation and vice versa. For the most part our linear representations will be of finite degree and we shall pass freely between linear representations and matrix representations (specifying a basis when we wish to give an explicit correspondence between the two). Furthermore, given a linear representation $\varphi : G \to GL(V)$ of finite degree, a corresponding matrix representation provides numerical invariants (such as the determinant of $\varphi(g)$ for $g \in G$) which are independent of the choice of basis giving the isomorphism between $GL(V)$ and $GL_n(F)$. The exploitation of such invariants will be fundamental to our development.

Before giving examples of representations we recall the group ring FG in greater detail (group rings were introduced in Section 7.2, and some notation and examples were discussed in that section). Suppose the elements of G are $g_1, g_2, \ldots, g_n$. Each element of FG is of the form

$$\sum_{i=1}^{n} \alpha_i g_i, \qquad \alpha_i \in F.$$

Two formal sums[1] are equal if and only if all corresponding coefficients of group elements are equal. Addition and multiplication in FG are defined as follows:

$$\sum_{i=1}^{n} \alpha_i g_i \quad + \quad \sum_{i=1}^{n} \beta_i g_i = \sum_{i=1}^{n} (\alpha_i + \beta_i) g_i$$

$$\left(\sum_{i=1}^{n} \alpha_i g_i \right) \left(\sum_{i=1}^{n} \beta_i g_i \right) = \sum_{k=1}^{n} \left(\sum_{\substack{i,j \\ g_i g_j = g_k}} \alpha_i \beta_j \right) g_k$$

where addition and multiplication of the coefficients α_i and β_j is performed in F. Note that by definition of multiplication,

FG is a commutative ring if and only if G is an abelian group.

The group G appears in FG (identifying g_i with $1g_i$) and the field F appears in FG (identifying β with βg_1, where g_1 is the identity of G). Under these identifications

$$\beta \left(\sum_{i=1}^{n} \alpha_i g_i \right) = \sum_{i=1}^{n} (\beta \alpha_i) g_i, \qquad \text{for all } \beta \in F.$$

In this way

FG is a vector space over F with the elements of G as a basis.

In particular, FG is a vector space over F of dimension equal to $|G|$. The elements of F commute with all elements of FG, i.e., F is in the *center* of FG. When we wish to emphasize the latter two properties we shall say that FG is an *F-algebra* (in general, an F-algebra is a ring R which contains F in its center, so R is both a ring and an F-vector space).

Note that the operations in FG are similar to those in the F-algebra $F[x]$ (although $F[x]$ is infinite dimensional over F). In some works FG is denoted by $F[G]$, although the latter notation is currently less prevalent.

---

[1] The formal sum displayed above is a way of writing the function from G to F which takes the value α_i on the group element g_i. This same "formality" was used in the construction of free modules (see Theorem 6 in Section 10.3).

Examples

(1) If $G = \langle g \rangle$ is cyclic of order $n \in \mathbb{Z}^+$, then the elements of FG are of the form

$$\sum_{i=0}^{n-1} \alpha_i g^i.$$

The map $F[x] \to F\langle g \rangle$ which sends x^k to g^k for all $k \geq 0$ extends by F-linearity to a surjective ring homomorphism with kernel equal to the ideal generated by $x^n - 1$. Thus

$$F\langle g \rangle \cong F[x]/(x^n - 1).$$

This is an isomorphism of F-algebras, i.e., is a ring isomorphism which is F-linear.

(2) Under the notation of the preceding example let $r = 1 + g + g^2 + \cdots + g^{n-1}$, so r is a nonzero element of $F\langle g \rangle$. Note that $rg = g + g^2 + \cdots + g^{n-1} + 1 = r$, hence $r(1 - g) = 0$. Thus the ring $F\langle g \rangle$ contains zero divisors (provided $n > 1$). More generally, if G is any group of order > 1, then for any nonidentity element $g \in G$, $F\langle g \rangle$ is a subring of FG, so FG also contains zero divisors.

(3) Let $G = S_3$ and $F = \mathbb{Q}$. The elements $r = 5(1\,2) - 7(1\,2\,3)$ and $s = -4(1\,2\,3) + 12(1\,3\,2)$ are typical members of $\mathbb{Q}S_3$. Their sum and product are seen to be

$$r + s = 5(1\ 2) - 11(1\ 2\ 3) + 12(1\ 3\ 2)$$
$$rs = -20(2\ 3) + 28(1\ 3\ 2) + 60(1\ 3) - 84$$

(recall that products (compositions) of permutations are computed from right to left). An explicit example of a sum and product of two elements in the group ring $\mathbb{Q}D_8$ appears in Section 7.2.

Before giving specific examples of representations we discuss the correspondence between representations of G and FG-modules (after which we can simultaneously give examples of both). This discussion closely parallels the treatment of $F[x]$-modules in Section 10.1.

Suppose first that $\varphi : G \to GL(V)$ is a representation of G on the vector space V over F. As above, write $G = \{g_1, \ldots, g_n\}$, so for each $i \in \{1, \ldots, n\}$, $\varphi(g_i)$ is a linear transformation from V to itself. Make V into an FG-module by defining the action of a ring element on an element of V as follows:

$$\left(\sum_{i=1}^{n} \alpha_i g_i \right) \cdot v = \sum_{i=1}^{n} \alpha_i \varphi(g_i)(v), \qquad \text{for all } \sum_{i=1}^{n} \alpha_i g_i \in FG, \ v \in V.$$

We verify a special case of axiom 2(b) of a module (see Section 10.1) which shows precisely where the fact that φ is a group homomorphism is needed:

$$
\begin{aligned}
(g_i g_j) \cdot v &= \varphi(g_i g_j)(v) && \text{(by definition of the action)} \\
&= (\varphi(g_i) \circ \varphi(g_j))(v) && \text{(since φ is a group homomorphism)} \\
&= \varphi(g_i)(\varphi(g_j)(v)) && \text{(by definition of a composition of linear} \\
& && \text{transformations)} \\
&= g_i \cdot (g_j \cdot v) && \text{(by definition of the action).}
\end{aligned}
$$

This argument extends by linearity to arbitrary elements of FG to prove that axiom 2(b) of a module holds in general. It is an exercise to check that the remaining module axioms hold.

Note that F is a subring of FG and the action of the field element α on a vector is the same as the action of the ring element $\alpha 1$ on a vector i.e., the FG-module action extends the F action on V.

Suppose now that conversely we are given an FG-module V. We obtain an associated vector space over F and representation of G as follows. Since V is an FG-module, it is an F-module, i.e., it is a vector space over F. Also, for each $g \in G$ we obtain a map from V to V, denoted by $\varphi(g)$, defined by

$$\varphi(g)(v) = g \cdot v \qquad \text{for all } v \in V,$$

where $g \cdot v$ is the given action of the ring element g on the element v of V. Since the elements of F commute with each $g \in G$ it follows by the axioms for a module that for all $v, w \in V$ and all $\alpha, \beta \in F$ we have

$$\begin{aligned}
\varphi(g)(\alpha v + \beta w) &= g \cdot (\alpha v + \beta w) \\
&= g \cdot (\alpha v) + g \cdot (\beta w) \\
&= \alpha(g \cdot v) + \beta(g \cdot w) \\
&= \alpha \varphi(g)(v) + \beta \varphi(g)(w),
\end{aligned}$$

that is, for each $g \in G$, $\varphi(g)$ is a linear transformation. Furthermore, it follows by axiom 2(b) of a module that

$$\varphi(g_i g_j)(v) = (\varphi(g_i) \circ \varphi(g_j))(v)$$

(this is essentially the calculation above with the steps reversed). This proves that φ is a group homomorphism (in particular, $\varphi(g^{-1}) = \varphi(g)^{-1}$, so every element of G maps to a nonsingular linear transformation, i.e., $\varphi : G \to GL(V)$).

This discussion shows there is a bijection between FG-modules and pairs (V, φ):

$$\left\{ V \text{ an } FG\text{-module} \right\} \longleftrightarrow \left\{ \begin{array}{c} V \text{ a vector space over } F \\ \text{and} \\ \varphi : G \to GL(V) \text{ a representation} \end{array} \right\}.$$

Giving a representation $\varphi : G \to GL(V)$ on a vector space V over F is therefore equivalent to giving an FG-module V. Under this correspondence we shall say that the module V *affords* the representation φ of G.

Recall from Section 10.1 that if a vector space M is made into an $F[x]$-module via the linear transformation T, then the $F[x]$-submodules of M are precisely the T-stable subspaces of M. In the current situation if V is an FG-module affording the representation φ, then a subspace U of V is called *G-invariant* or *G-stable* if $g \cdot u \in U$ for all $g \in G$ and all $u \in U$ (i.e., if $\varphi(g)(u) \in U$ for all $g \in G$ and all $u \in U$). It follows easily that

the FG-submodules of V are precisely the G-stable subspaces of V.

Examples

(1) Let V be a 1-dimensional vector space over F and make V into an FG-module by letting $gv = v$ for all $g \in G$ and $v \in V$. This module affords the representation $\varphi : G \to GL(V)$ defined by $\varphi(g) = I$ = the identity linear transformation, for all $g \in G$. The corresponding matrix representation (with respect to any basis of V) is the homomorphism of G into $GL_1(F)$ which sends every group element to the 1×1 identity matrix. We shall henceforth refer to this as the *trivial representation* of G. The trivial representation has degree 1 and if $|G| > 1$, it is not faithful.

(2) Let $V = FG$ and consider this ring as a left module over itself. Then V affords a representation of G of degree equal to $|G|$. If we take the elements of G as a basis of V, then each $g \in G$ permutes these basis elements under the left regular permutation representation:

$$g \cdot g_i = gg_i.$$

With respect to this basis of V the matrix of the group element g has a 1 in row i and column j if $gg_j = g_i$, and has 0's in all other positions. This (linear or matrix) representation is called the *regular representation* of G. Note that each nonidentity element of G induces a nonidentity permutation on the basis of V so the regular representation is always faithful.

(3) Let $n \in \mathbb{Z}^+$, let $G = S_n$ and let V be an n-dimensional vector space over F with basis $e_1, e_2, \ldots, e_n$. Let S_n act on V by defining for each $\sigma \in S_n$

$$\sigma \cdot e_i = e_{\sigma(i)}, \qquad 1 \le i \le n$$

i.e., σ acts by permuting the subscripts of the basis elements. This provides an (injective) homomorphism of S_n into $GL(V)$ (i.e., a faithful representation of S_n of degree n), hence makes V into an FS_n-module. As in the preceding example, the matrix of σ with respect to the basis $e_1, \ldots, e_n$ has a 1 in row i and column j if $\sigma \cdot e_j = e_i$ (and has 0 in all other entries). Thus σ has a 1 in row i and column j if $\sigma(j) = i$.

For an example of the ring action, consider the action of FS_3 on the 3-dimensional vector space over F with basis e_1, e_2, e_3. Let σ be the transposition $(1\,2)$, let τ be the 3-cycle $(1\,2\,3)$ and let $r = 2\sigma - 3\tau \in FS_3$. Then

$$r \cdot (\alpha e_1 + \beta e_2 + \gamma e_3) = 2(\alpha e_{\sigma(1)} + \beta e_{\sigma(2)} + \gamma e_{\sigma(3)}) - 3(\alpha e_{\tau(1)} + \beta e_{\tau(2)} + \gamma e_{\tau(3)})$$
$$= 2(\alpha e_2 + \beta e_1 + \gamma e_3) - 3(\alpha e_2 + \beta e_3 + \gamma e_1)$$
$$= (2\beta - 3\gamma)e_1 - \alpha e_2 + (2\gamma - 3\beta)e_3.$$

(4) If $\psi : H \to GL(V)$ is any representation of H and $\varphi : G \to H$ is any group homomorphism, then the composition $\psi \circ \varphi$ is a representation of G. For example, let V be the FS_n-module of dimension n described in the preceding example. If $\pi : G \to S_n$ is any permutation representation of G, the composition of π with the representation above gives a linear representation of G. In other words, V becomes an FG-module under the action

$$g \cdot e_i = e_{\pi(g)(i)}, \qquad \text{for all } g \in G.$$

Note that the regular representation, (2), is just the special case of this where $n = |G|$ and π is the left regular permutation representation of G.

(5) Any homomorphism of G into the multiplicative group $F^\times = GL_1(F)$ is a degree 1 (matrix) representation. For example, suppose $G = \langle g \rangle \cong Z_n$ is the cyclic group of order n and ζ is a fixed n^{th} root of 1 in F. Let $g^i \mapsto \zeta^i$, for all $i \in \mathbb{Z}$. This representation of $\langle g \rangle$ is a faithful representation if and only if ζ is a primitive n^{th} root of 1.

(6) In many situations it is easier to specify an explicit matrix representation of a group G rather than to exhibit an FG-module. For example, recall that the dihedral group D_{2n} has the presentation

$$D_{2n} = \langle r, s \mid r^n = s^2 = 1, \; rs = sr^{-1} \rangle.$$

If R and S are any matrices satisfying the relations $R^n = S^2 = I$ and $RS = SR^{-1}$ then the map $r \mapsto R$ and $s \mapsto S$ extends uniquely to a homomorphism from D_{2n} to the matrix group generated by R and S, hence gives a representation of D_{2n}. An explicit example of matrices $R, S \in M_2(\mathbb{R})$ may be obtained as follows. If a regular n-gon is drawn on the x, y plane centered at the origin with the line $y = x$ as one of its lines of symmetry then the matrix R that rotates the plane through $2\pi/n$ radians and the matrix S that reflects the plane about the line $y = x$ both send this n-gon onto itself. It follows that these matrices act as symmetries of the n-gon and so satisfy the above relations. These matrices are readily computed (cf. Exercise 25, Section 1.6) and so the maps

$$r \mapsto R = \begin{pmatrix} \cos 2\pi/n & -\sin 2\pi/n \\ \sin 2\pi/n & \cos 2\pi/n \end{pmatrix} \quad \text{and} \quad s \mapsto S = \begin{pmatrix} 0 & 1 \\ 1 & 0 \end{pmatrix}$$

extend uniquely to a (degree 2) representation of D_{2n} into $GL_2(\mathbb{R})$. Since the matrices R and S have orders n and 2 respectively, it follows that they generate a subgroup of $GL_2(\mathbb{R})$ of order $2n$ and hence this representation is faithful.

(7) By using the usual generators and relations for the quaternion group

$$Q_8 = \langle i, j \mid i^4 = j^4 = 1, \; i^2 = j^2, \; i^{-1}ji = j^{-1} \rangle$$

one may similarly obtain (cf. Exercise 26, Section 1.6) a representation φ from Q_8 to $GL_2(\mathbb{C})$ defined by

$$\varphi(i) = \begin{pmatrix} \sqrt{-1} & 0 \\ 0 & -\sqrt{-1} \end{pmatrix} \quad \text{and} \quad \varphi(j) = \begin{pmatrix} 0 & -1 \\ 1 & 0 \end{pmatrix}.$$

This representation of Q_8 is faithful.

(8) A 4-dimensional representation of the quaternion group Q_8 may be obtained from the real Hamilton quaternions, $\mathbb{H}$ (cf. Section 7.1). The group Q_8 is a subgroup of the multiplicative group of units of $\mathbb{H}$ and each of the elements of Q_8 acts by left multiplication on the 4-dimensional real vector space $\mathbb{H}$. Since the real numbers are in the center of $\mathbb{H}$ (i.e., since $\mathbb{H}$ is an $\mathbb{R}$-algebra), left multiplication is $\mathbb{R}$-linear. This linear action thus gives a homomorphism from Q_8 into $GL_4(\mathbb{R})$. One can easily write out the explicit matrices of each of the elements of Q_8 with respect to the basis $1, i, j, k$ of $\mathbb{H}$. For example, left multiplication by i acts by $1 \mapsto i, i \mapsto -1, j \mapsto k$ and $k \mapsto -j$ and left multiplication by j acts by $1 \mapsto j, i \mapsto -k, j \mapsto -1$ and $k \mapsto i$ so

$$i \mapsto \begin{pmatrix} 0 & -1 & 0 & 0 \\ 1 & 0 & 0 & 0 \\ 0 & 0 & 0 & -1 \\ 0 & 0 & 1 & 0 \end{pmatrix} \quad \text{and} \quad j \mapsto \begin{pmatrix} 0 & 0 & -1 & 0 \\ 0 & 0 & 0 & 1 \\ 1 & 0 & 0 & 0 \\ 0 & -1 & 0 & 0 \end{pmatrix}.$$

This representation of Q_8 is also faithful.

(9) Suppose that H is a normal subgroup of the group G and suppose that H is an elementary abelian p-group for some prime p. Then $V = H$ is a vector space over $\mathbb{F}_p$, where the scalar a acts on the vector v by $av = v^a$ (see Section 10.1). The action of each element of G by conjugation on V is $\mathbb{F}_p$-linear because $gv^a g^{-1} = (gvg^{-1})^a$ and this action of G on V makes V into an $\mathbb{F}_p G$-module (the automorphisms of elementary abelian p-groups were discussed in Sections 4.4 and 10.1). The kernel of this representation is the set of elements of G that commute with every element of H, $C_G(H)$ (which always contains the abelian group H itself). Thus the action of a group on subsets of itself often affords linear representations over finite fields. Representations of groups over finite fields are called *modular representations* and these are fundamental to the study of the internal structure of groups.

(10) For an example of an FG-submodule, let $G = S_n$ and let V be the FS_n-module described in Example 3. Let N be the subspace of V consisting of vectors all of whose coordinates are equal, i.e.,

$$N = \{\alpha_1 e_1 + \alpha_2 e_2 + \cdots + \alpha_n e_n \mid \alpha_1 = \alpha_2 = \cdots = \alpha_n\}$$

(this is a 1-dimensional S_n-stable subspace). Each $\sigma \in S_n$ fixes each vector in N so the submodule N affords the trivial representation of S_n. As an exercise, one may show that if $n \geq 3$ then N is the *unique* 1-dimensional subspace of V which is S_n-stable, i.e., N is the unique 1-dimensional FS_n-submodule (N is called the *trace* submodule of FS_n).

Another FS_n-submodule of V is the subspace I of all vectors whose coordinates sum to zero:

$$I = \{\alpha_1 e_1 + \alpha_2 e_2 + \cdots + \alpha_n e_n \mid \alpha_1 + \alpha_2 + \cdots + \alpha_n = 0\}.$$

Again I is an S_n-stable subspace (since each $\sigma \in S_n$ permutes the coordinates of each vector in V, each σ leaves the sum of the coefficients unchanged). Since I is the kernel of the linear transformation from V onto F which sends a vector to the sum of its coefficients (called the augmentation map — cf. Section 7.3), I has dimension $n - 1$.

(11) If $V = FG$ is the regular representation of G described in Example 2 above, then V has FG-submodules of dimensions 1 and $|G| - 1$ as in the preceding example:

$$N = \{\alpha_1 g_1 + \alpha_2 g_2 + \cdots + \alpha_n g_n \mid \alpha_1 = \alpha_2 = \cdots = \alpha_n\}$$
$$I = \{\alpha_1 g_1 + \alpha_2 g_2 + \cdots + \alpha_n g_n \mid \alpha_1 + \alpha_2 + \cdots + \alpha_n = 0\}.$$

In fact N and I are 2-sided ideals of FG (not just left ideals — note that N is in the center of FG). The ideal I is called the *augmentation ideal* of FG and N is called the *trace ideal* of FG.

Recall that in the study of a linear transformation T of a vector space V to itself we made V into an $F[x]$-module (where x acted as T on V); our goal was to decompose V into a direct sum of cyclic submodules. In this way we were able to find a basis of V for which the matrix of T with respect to this basis was in some *canonical* form. Changing the basis of V did not change the module V but changed the matrix representation of T by similarity (i.e., changed the isomorphism between $GL(V)$ and $GL_n(F)$). We introduce the analogous terminology to describe when two FG-modules are the same up to a change of basis.

Definition. Two representations of G are *equivalent* (or *similar*) if the FG-modules affording them are isomorphic modules. Representations which are not equivalent are called *inequivalent*.

Suppose $\varphi : G \to GL(V)$ and $\psi : G \to GL(W)$ are equivalent representations (here V and W must be vector spaces over the same field F). Let $T : V \to W$ be an FG-module isomorphism between them. Since T is, in particular, an F-module isomorphism, T is a vector space isomorphism, so V and W must have the same dimension. Furthermore, for all $g \in G$, $v \in V$ we have $T(g \cdot v) = g \cdot (T(v))$, since T is an isomorphism of FG-modules. By definition of the action of ring elements this means $T(\varphi(g)v) = \psi(g)(T(v))$, that is

$$T \circ \varphi(g) = \psi(g) \circ T \qquad \text{for all } g \in G.$$

In particular, if we identify V and W as vector spaces, then two representations φ and ψ of G on a vector space V are equivalent if and only if there is some $T \in GL(V)$ such that $T \circ \varphi(g) \circ T^{-1} = \psi(g)$ for all $g \in G$. This T is a *simultaneous* change of basis for all $\varphi(g)$, $g \in G$.

In matrix terminology, two representations φ and ψ are equivalent if there is a fixed invertible matrix P such that

$$P\varphi(g)P^{-1} = \psi(g) \qquad \text{for all } g \in G.$$

The linear transformation T or the matrix P above is said to *intertwine* the representations φ and ψ (it gives the "rule" for changing φ into ψ).

In order to study the decomposition of an FG-module into (direct sums of) submodules we shall need some terminology. We state these definitions for arbitrary rings since we shall be discussing direct sum decompositions in greater generality in the next section.

Definition. Let R be a ring and let M be a nonzero R-module.
(1) The module M is said to be *irreducible* (or *simple*) if its only submodules are 0 and M; otherwise M is called *reducible*.
(2) The module M is said to be *indecomposable* if M cannot be written as $M_1 \oplus M_2$ for any nonzero submodules M_1 and M_2; otherwise M is called *decomposable*.
(3) The module M is said to be *completely reducible* if it is a direct sum of irreducible submodules.
(4) A representation is called *irreducible, reducible, indecomposable, decomposable* or *completely reducible* according to whether the FG-module affording it has the corresponding property.
(5) If M is a completely reducible R-module, any direct summand of M is called a *constituent* of M (i.e., N is a constituent of M if there is a submodule N' of M such that $M = N \oplus N'$).

An irreducible module is, by definition, both indecomposable and completely reducible. We shall shortly give examples of indecomposable modules that are not irreducible.

If $R = FG$, an irreducible FG-module V is a nonzero F-vector space with no nontrivial, proper G-invariant subspaces. For example, if $\dim_F V = 1$ then V is necessarily irreducible (its only subspaces are 0 and V).

Suppose V is a finite dimensional FG-module and V is reducible. Let U be a G-invariant subspace. Form a basis of V by taking a basis of U and enlarging it to a basis of V. Then for each $g \in G$ the matrix, $\varphi(g)$, of g acting on V with respect to this basis is of the form

$$\varphi(g) = \begin{pmatrix} \varphi_1(g) & \psi(g) \\ 0 & \varphi_2(g) \end{pmatrix}$$

where $\varphi_1 = \varphi|_U$ (with respect to the chosen basis of U) and φ_2 is the representation of G on V/U (and ψ is not necessarily a homomorphism — $\psi(g)$ need not be a square matrix). So reducible representations are those with a corresponding matrix representation whose matrices are in block upper triangular form.

Assume further that the FG-module V is decomposable, $V = U \oplus U'$. Take for a basis of V the union of a basis of U and a basis of U'. With this choice of basis the matrix for each $g \in G$ is of the form

$$\varphi(g) = \begin{pmatrix} \varphi_1(g) & 0 \\ 0 & \varphi_2(g) \end{pmatrix}$$

(i.e., $\psi(g) = 0$ for all $g \in G$). Thus decomposable representations are those with a corresponding matrix representation whose matrices are in block diagonal form.

Examples

(1) As noted above, all degree 1 representations are irreducible, indecomposable and completely reducible. In particular, this applies to the trivial representation and to the representations described in Example 5 above.

(2) If $|G| > 1$, the regular representation of G is reducible (the augmentation ideal and the trace ideal are proper nonzero submodules). We shall later determine the conditions under which this representation is completely reducible and how it decomposes into a direct sum.

(3) For $n > 1$ the FS_n-module described in Example 10 above is reducible since N and I are proper, nonzero submodules. The module N is irreducible (being 1-dimensional) and if the characteristic of the field F does not divide n, then I is also irreducible.

(4) The degree 2 representation of the dihedral group $D_{2n} = G$ described in Example 6 above is irreducible for $n \geq 3$. There are no G-invariant 1-dimensional subspaces since a rotation by $2\pi/n$ radians sends no line in $\mathbb{R}^2$ to itself. Similarly, the degree 2 complex representation of Q_8 described in Example 7 is irreducible since the given matrix $\varphi(i)$ has exactly two 1-dimensional eigenspaces (corresponding to its distinct eigenvalues $\pm\sqrt{-1}$) and these are not invariant under the matrix $\varphi(j)$. The degree 4 representation $\varphi : Q_8 \to GL_4(\mathbb{R})$ described in Example 8 can also be shown to be irreducible (see the exercises). We shall see, however, that if we view φ as a complex representation $\varphi : Q_8 \to GL_4(\mathbb{C})$ (just by considering the real entries of the matrices to be complex entries) then there is a *complex* matrix P such that $P^{-1}\varphi(g)P$ is a direct sum of 2×2 block matrices for all $g \in Q_8$. Thus an irreducible representation over a field F may become reducible when the field is extended.

(5) Let $G = \langle g \rangle$ be cyclic of order n and assume F contains all the nth roots of 1. As noted in Example 1 in the set of examples of group algebras, $F\langle g \rangle \cong F[x]/(x^n - 1)$. Thus the FG-modules are precisely the $F[x]$-modules annihilated by $x^n - 1$. The latter (finite dimensional) modules are described, up to equivalence, by the Jordan Canonical Form Theorem.

If the minimal polynomial of g acting on an $F\langle g \rangle$-module V has distinct roots in F, there is a basis of V such that g (hence all its powers) is represented by a diagonal matrix (cf. Corollary 25, Section 12.3). In this case, V is a completely reducible $F\langle g \rangle$-module (being a direct sum of 1-dimensional $\langle g \rangle$-invariant subspaces). In general, the minimal polynomial of g acting on V divides $x^n - 1$ so if $x^n - 1$ has distinct roots in F, then V is a completely reducible $F\langle g \rangle$-module. The polynomial $x^n - 1$ has distinct roots in F if and only if the characteristic of F does not divide n. This gives a sufficient condition for every $F\langle g \rangle$-module to be completely reducible.

If the minimal polynomial of g acting on V does *not* have distinct roots (so the characteristic of F does divide n), the Jordan canonical form of g must have an elementary Jordan block of size > 1. Since every linear transformation has a unique Jordan canonical form, g cannot be represented by a diagonal matrix, i.e., V is not completely reducible. It follows from results on cyclic modules in Section 12.3 that the (1-dimensional) eigenspace of g in any Jordan block of size > 1 admits no $\langle g \rangle$-invariant complement, i.e., V is reducible but not completely reducible.

Specifically, let p be a prime, let $F = \mathbb{F}_p$ and let g be of order p. Let V be the 2-dimensional space over $\mathbb{F}_p$ with basis v, w and define an action of g on V by

$$g \cdot v = v \quad \text{and} \quad g \cdot w = v + w.$$

This endomorphism of V does have order p (in $GL(V)$) and the matrix of g with respect to this basis is the elementary Jordan block

$$\varphi(g) = \begin{pmatrix} 1 & 1 \\ 0 & 1 \end{pmatrix}.$$

Now V is reducible (span$\{v\}$ is a $\langle g \rangle$-invariant subspace) but V is indecomposable (the above 2×2 elementary Jordan matrix is not similar to a diagonal matrix).

The first fundamental result in the representation theory of finite groups shows how Example 5 generalizes to noncyclic groups.

Theorem 1. *(Maschke's Theorem)* Let G be a finite group and let F be a field whose characteristic does not divide $|G|$. If V is any FG-module and U is any submodule of V, then V has a submodule W such that $V = U \oplus W$ (i.e., every submodule is a direct summand).

Remark: The hypothesis of Maschke's Theorem applies to any finite group when F has characteristic 0.

Proof: The idea of the proof of Maschke's Theorem is to produce an FG-module homomorphism

$$\pi : V \rightarrow U$$

which is a projection onto U, i.e., which satisfies the following two properties:
   (i) $\pi(u) = u$ for all $u \in U$
   (ii) $\pi(\pi(v)) = \pi(v)$ for all $v \in V$ (i.e., $\pi^2 = \pi$)
(in fact (ii) is implied by (i) and the fact that $\pi(V) \subseteq U$).

Suppose first that we can produce such an FG-module homomorphism and let $W = \ker \pi$. Since π is a module homomorphism, W is a submodule. We see that W is a direct sum complement to U as follows. If $v \in U \cap W$ then by (i), $v = \pi(v)$ whereas by definition of W, $\pi(v) = 0$. This shows $U \cap W = 0$. To show $V = U + W$ let v be an arbitrary element of V and write $v = \pi(v) + (v - \pi(v))$. By definition, $\pi(v) \in U$. By property (ii) of π,

$$\pi(v - \pi(v)) = \pi(v) - \pi(\pi(v)) = \pi(v) - \pi(v) = 0,$$

i.e., $v - \pi(v) \in W$. This shows $V = U + W$ and hence $V = U \oplus W$. To establish Maschke's Theorem it therefore suffices to find such an FG-module projection π.

Since U is a subspace it has a vector space direct sum complement W_0 in V (take a basis $\mathcal{B}_1$ of U, build it up to a basis $\mathcal{B}$ of V and let W_0 be the span of $\mathcal{B} - \mathcal{B}_1$). Thus $V = U \oplus W_0$ as vector spaces but W_0 need not be G-stable (i.e., need not be an FG-submodule). Let $\pi_0 : V \to U$ be the vector space projection of V onto U associated to this direct sum decomposition, i.e., π_0 is defined by

$$\pi_0(u + w) = u \qquad \text{for all } u \in U, \ w \in W_0.$$

The key idea of the proof is to "average" π_0 over G to form an FG-module projection π. For each $g \in G$ define

$$g\pi_0 g^{-1} : V \to U \qquad \text{by} \qquad g\pi_0 g^{-1}(v) = g \cdot \pi_0(g^{-1} \cdot v), \quad \text{for all } v \in V$$

(here $\cdot$ denotes the action of elements of the ring FG). Since π_0 maps V into U and U is stable under the action of g we have that $g\pi_0 g^{-1}$ maps V into U. Both g and g^{-1} act as F-linear transformations, so $g\pi_0 g^{-1}$ is a linear transformation. Furthermore, if u is in the G-stable space U then so is $g^{-1}u$, and by definition of π_0 we have $\pi_0(g^{-1}u) = g^{-1}u$. From this we obtain that for all $g \in G$,

$$g\pi_0 g^{-1}(u) = u \qquad \text{for all } u \in U$$

(i.e., $g\pi_0 g^{-1}$ is also a vector space projection of V onto U).

Let $n = |G|$ and view n as an element of F ($n = 1 + \cdots + 1$, n times). By hypothesis n is not zero in F and so has an inverse in F. Define

$$\pi = \frac{1}{n} \sum_{g \in G} g\pi_0 g^{-1}.$$

Since π is a scalar multiple of a sum of linear transformations from V to U, it is also a linear transformation from V to U. Furthermore, each term in the sum defining π restricts to the identity map on the subspace U and so $\pi|_U$ is $1/n$ times the sum of n copies of the identity. These observations prove the following:

$$\pi : V \to U \text{ is a linear transformation}$$
$$\pi(u) = u \qquad \text{for all } u \in U$$
$$\pi^2(v) = \pi(v) \qquad \text{for all } v \in V.$$

It remains to show that π is an FG-module homomorphism (i.e., is FG-linear). It suffices to prove that for all $h \in G$, $\pi(hv) = h\pi(v)$, for $v \in V$. In this case

$$\pi(hv) = \frac{1}{n} \sum_{g \in G} g\pi_0(g^{-1}hv)$$

$$= \frac{1}{n} \sum_{g \in G} h(h^{-1}g)\pi_0((g^{-1}h)v)$$

$$= \frac{1}{n} \sum_{\substack{k=h^{-1}g \\ g \in G}} h(k\pi_0(k^{-1}v) = h\pi(v)$$

(as g runs over all elements of G, so does $k = h^{-1}g$ and the module element h may be brought outside the summation by the distributive law in modules). This establishes the existence of the FG-module projection π and so completes the proof.

The applications of Maschke's Theorem will be to finitely generated FG-modules. Unlike the situation of $F[x]$-modules, however, finitely generated FG-modules are automatically finite dimensional vector spaces (the difference being that FG itself is finite dimensional, whereas $F[x]$ is not). Let V be an FG-module. If V is a finite dimensional vector space over F, then a fortiori V is finitely generated as an FG-module (any F basis gives a set of generators over FG). Conversely, if V is finitely generated as an FG-module, say by $v_1, \ldots, v_k$, then one easily sees that V is spanned as a vector space by the finite set $\{g \cdot v_i \mid g \in G, \ 1 \le i \le k\}$. Thus

an FG-module is finitely generated if and only if it is finite dimensional.

Corollary 2. If G is a finite group and F is a field whose characteristic does not divide $|G|$, then every finitely generated FG-module is completely reducible (equivalently, every F-representation of G of finite degree is completely reducible).

Proof: Let V be a finitely generated FG-module. As noted above, V is finite dimensional over F, so we may proceed by induction on its dimension. If V is irreducible, it is completely reducible and the result holds. Suppose therefore that V has a proper, nonzero FG-submodule U. By Maschke's Theorem U has an FG-submodule complement W, i.e., $V = U \oplus W$. By induction, each of U and W are direct sums of irreducible submodules, hence so is V. This completes the induction.

Corollary 3. Let G be a finite group, let F be a field whose characteristic does not divide $|G|$ and let $\varphi : G \to GL(V)$ be a representation of G of finite degree. Then there is a basis of V such that for each $g \in G$ the matrix of $\varphi(g)$ with respect to this basis is block diagonal:

$$\begin{pmatrix} \varphi_1(g) & & & \\ & \varphi_2(g) & & \\ & & \ddots & \\ & & & \varphi_m(g) \end{pmatrix}$$

where φ_i is an irreducible matrix representation of G, $1 \le i \le m$.

Proof: By Corollary 2 we may write $V = U_1 \oplus U_2 \oplus \cdots \oplus U_m$, where U_i is an irreducible FG-submodule of V. Let $\mathcal{B}_i$ be a basis of U_i and let $\mathcal{B}$ be the union of the $\mathcal{B}_i$'s. For each $g \in G$, the matrix of $\varphi(g)$ with respect to the basis $\mathcal{B}$ is of the form in the corollary, where $\varphi_i(g)$ is the matrix of $\varphi(g)|_{U_i}$ with respect to the basis $\mathcal{B}_i$.

The converse of Maschke's Theorem is also true. Namely, if the characteristic of F does divide $|G|$, then G possesses (finitely generated) FG-modules which are not completely reducible. Specifically, the regular representation (i.e., the module FG itself) is not completely reducible.

In Section 18.2 we shall discuss the question of uniqueness of the constituents in direct sum decompositions of FG-modules into irreducible submodules.

EXERCISES

Let F be a field, let G be a finite group and let $n \in \mathbb{Z}^+$.

1. Prove that if $\varphi : G \to GL(V)$ is any representation, then φ gives a faithful representation of $G / \ker \varphi$.

2. Let $\varphi : G \to GL_n(F)$ be a matrix representation. Prove that the map $g \mapsto \det(\varphi(g))$ is a degree 1 representation.

3. Prove that the degree 1 representations of G are in bijective correspondence with the degree 1 representations of the abelian group G / G' (where G' is the commutator subgroup of G).

4. Let V be a (possibly infinite dimensional) FG-module (G is a finite group). Prove that for each $v \in V$ there is an FG-submodule containing v of dimension $\leq |G|$.

5. Prove that if $|G| > 1$ then every irreducible FG-module has dimension $< |G|$.

6. Write out the matrices $\varphi(g)$ for every $g \in G$ for each of the following representations that were described in the second set of examples:
   (a) the representation of S_3 described in Example 3 (let $n = 3$ in that example)
   (b) the representation of D_8 described in Example 6 (i.e., let $n = 4$ in that example and write out the values of all the sines and cosines, for all group elements)
   (c) the representation of Q_8 described in Example 7
   (d) the representation of Q_8 described in Example 8.

7. Let V be the 4-dimensional permutation module for S_4 described in Example 3 of the second set of examples. Let $\pi : D_8 \to S_4$ be the permutation representation of D_8 obtained from the action of D_8 by left multiplication on the set of left cosets of its subgroup $\langle s \rangle$. Make V into an FD_8-module via π as described in Example 4 and write out the 4×4 matrices for r and s given by this representation with respect to the basis $e_1, \ldots, e_4$.

8. Let V be the FS_n-module described in Examples 3 and 10 in the second set of examples.
   (a) Prove that if v is any element of V such that $\sigma \cdot v = v$ for all $\sigma \in S_n$ then v is an F-multiple of $e_1 + e_2 + \cdots + e_n$.
   (b) Prove that if $n \geq 3$, then V has a unique 1-dimensional submodule, namely the submodule N consisting of all F-multiples of $e_1 + e_2 + \cdots + e_n$.

9. Prove that the 4-dimensional representation of Q_8 on $\mathbb{H}$ described in Example 8 in the second set of examples is irreducible. [Show that any Q_8-stable subspace is a left ideal.]

10. Prove that $GL_2(\mathbb{R})$ has no subgroup isomorphic to Q_8. [This may be done by direct computation using generators and relations for Q_8. Simplify these calculations by putting one generator in rational canonical form.]

11. Let $\varphi : S_n \to GL_n(F)$ be the matrix representation given by the permutation module described in Example 3 in the second set of examples, where the matrices are computed with respect to the basis $e_1, \ldots, e_n$. Prove that $\det \varphi(\sigma) = \epsilon(\sigma)$ for all $\sigma \in S_n$, where $\epsilon(\sigma)$ is the sign of the permutation σ. [Check this on transpositions.]

12. Assume the characteristic of F is not 2. Let H be the set of $T \in M_n(F)$ such that T has exactly one nonzero entry in each row and each column and zeros elsewhere, and the nonzero entries are ± 1. Prove that H is a subgroup of $GL_n(F)$ and that H is isomorphic to $E_{2^n} \rtimes S_n$ (semidirect product), where E_{2^n} is the elementary abelian group of order 2^n.

The next few exercises explore an important result known as Schur's Lemma and some of its consequences.

13. Let R be a ring and let M and N be simple (i.e., irreducible) R-modules.
    (a) Prove that every nonzero R-module homomorphism from M to N is an isomorphism. [Consider its kernel and image.]
    (b) Prove Schur's Lemma: if M is a simple R-module then $\operatorname{Hom}_R(M, M)$ is a division ring (recall that $\operatorname{Hom}_R(M, M)$ is the ring of all R-module homomorphisms from M to M, where multiplication in this ring is composition).

14. Let $\varphi : G \to GL(V)$ be a representation of G. The *centralizer* of φ is defined to be the set of all linear transformations, A, from V to itself such that $A\varphi(g) = \varphi(g)A$ for all $g \in G$ (i.e., the linear transformations of V which commute with all $\varphi(g)$'s).
    (a) Prove that a linear transformation A from V to V is in the centralizer of φ if and only if it is an FG-module homomorphism from V to itself (so the centralizer of φ is the same as the *ring* $\operatorname{Hom}_{FG}(V, V)$).
    (b) Show that if z is in the center of G then $\varphi(z)$ is in the centralizer of φ.
    (c) Assume φ is an irreducible representation (so V is a simple FG-module). Prove that if H is any finite *abelian* subgroup of $GL(V)$ such that $A\varphi(g) = \varphi(g)A$ for all $A \in H$ then H is cyclic (in other words, any finite abelian subgroup of the multiplicative group of units in the ring $\operatorname{Hom}_{FG}(V, V)$ is cyclic). [By the preceding exercise, $\operatorname{Hom}_{FG}(V, V)$ is a division ring, so this reduces to proving that a finite abelian subgroup of the multiplicative group of nonzero elements in a division ring is cyclic. Show that the division subring generated by an abelian subgroup of any division ring is a field and use Proposition 18, Section 9.5.]
    (d) Show that if φ is a faithful irreducible representation then the center of G is cyclic.
    (e) Deduce from (d) that if G is abelian and φ is any irreducible representation then $G/\ker \varphi$ is cyclic.

15. Exhibit all 1-dimensional complex representations of a finite cyclic group; make sure to decide which are inequivalent.

16. Exhibit all 1-dimensional complex representations of a finite abelian group. Deduce that the number of inequivalent degree 1 complex representations of a finite abelian group equals the order of the group. [First decompose the abelian group into a direct product of cyclic groups, then use the preceding exercise.]

17. Prove the following variant of Schur's Lemma for complex representations of abelian groups: if G is abelian, any irreducible complex representation, φ, of G is of degree 1 and $G/\ker \varphi$ is cyclic. [This can be done without recourse to Exercise 14 by using the observation that for any $g \in G$ the eigenspaces of $\varphi(g)$ are G-stable. Your proof that φ has degree 1 should also work for infinite abelian groups when φ has finite degree.]

18. Prove the following general form of Schur's Lemma for complex representations: if $\varphi : G \to GL_n(\mathbb{C})$ is an irreducible matrix representation and A is an $n \times n$ matrix com-

muting with $\varphi(g)$ for all $g \in G$, then A is a scalar matrix. Deduce that if φ is a faithful, irreducible, complex representation then the center of G is cyclic and $\varphi(z)$ is a scalar matrix for all elements z in the center of G. [As in the preceding exercise, the eigenspaces of A are G-stable.]

19. Prove that if G is an abelian group then any finite dimensional complex representation of G is equivalent to a representation into diagonal matrices (i.e., any finite group of commuting matrices over $\mathbb{C}$ can be simultaneously diagonalized). [This can be done without recourse to Maschke's Theorem by looking at eigenspaces.]

20. Prove that the number of degree 1 complex representations of any finite group G equals $|G : G'|$, where G' is the commutator subgroup of G. [Use Exercises 3 and 16.]

21. Let G be a noncyclic abelian group acting by conjugation on an elementary abelian p-group V, where p is a prime not dividing the order of G.
  (a) Prove that if W is an irreducible $\mathbb{F}_p G$-submodule of V then there is some nonidentity element $g \in G$ such that $W \leq C_V(g)$ (here $C_V(g)$ is the subgroup of elements of V that are fixed by g under conjugation).
  (b) Prove that V is generated by the subgroups $C_V(g)$ as g runs over all nonidentity elements of G.

22. Let p be a prime, let P be a p-group and let F be a field of characteristic p. Prove that the only irreducible representation of P over F is the trivial representation. [Do this for a group of order p first using the fact that F contains all p^{th} roots of 1 (namely 1 itself). If P is not of order p, let z be an element of order p in the center of P, prove that z is in the kernel of the irreducible representation and apply induction to $P/\langle z \rangle$.]

23. Let p be a prime, let P be a nontrivial p-group and let F be a field of characteristic p. Prove that the regular representation is not completely reducible. [Use the preceding exercise.]

24. Let p be a prime, let P be a nontrivial p-group and let F be a field of characteristic p. Prove that the regular representation is indecomposable.

18.2 WEDDERBURN'S THEOREM AND SOME CONSEQUENCES

In this section we give a famous classification theorem due to Wedderburn which describes, in particular, the structure of the group algebra FG when the characteristic of F does not divide the order of G. From this classification theorem we shall derive various consequences, including the fact that for each finite group G there are only a finite number of nonisomorphic irreducible FG-modules. This result, together with Maschke's Theorem, in some sense completes the Hölder Program for representation theory of finite groups over such fields. The remainder of the book is concerned with developing techniques for determining and working with the irreducible representations as well as applying this knowledge to obtain group-theoretic information.

Theorem 4. *(Wedderburn's Theorem)* Let R be a nonzero ring with 1 (not necessarily commutative). Then the following are equivalent:
  (1) every R-module is projective
  (2) every R-module is injective
  (3) every R-module is completely reducible
  (4) the ring R considered as a left R-module is a direct sum:

$$R = L_1 \oplus L_2 \oplus \cdots \oplus L_n,$$

where each L_i is a simple module (i.e., a simple left ideal) with $L_i = Re_i$, for some $e_i \in R$ with

      (i) $e_i e_j = 0$ if $i \neq j$

      (ii) $e_i^2 = e_i$ for all i

      (iii) $\sum_{i=1}^{n} e_i = 1$

(5) as rings, R is isomorphic to a direct product of matrix rings over division rings, i.e., $R = R_1 \times R_2 \times \cdots \times R_r$ where R_j is a two-sided ideal of R and R_j is isomorphic to the ring of all $n_j \times n_j$ matrices with entries in a division ring Δ_j, $j = 1, 2, \ldots, r$. The integer r, the integers n_j, and the division rings Δ_j (up to isomorphism) are uniquely determined by R.

Proof: A proof of Wedderburn's Theorem is outlined in Exercises 1 to 10

Definition. A ring R satisfying any of the (equivalent) properties in Theorem 4 is called *semisimple with minimum condition.*

Rings R satisfying any of the equivalent conditions of Theorem 4 also satisfy the *minimum condition* or *descending chain condition (D.C.C)* on left ideals:

$$\text{if } I_1 \supseteq I_2 \supseteq \cdots \quad \text{is a descending chain of left ideals of } R$$

$$\text{then there is an } N \in \mathbb{Z}^+ \text{ such that } I_k = I_N \text{ for all } k \geq N$$

(which explains the use of this term in the definition above). The rings we deal with will all have this minimum condition. For example, group algebras always have this property since in any strictly descending chain of ideals the vector space dimensions of the ideals (which are F-subspaces of FG) are strictly decreasing, hence the length of a strictly descending chain is at most the dimension of FG ($= |G|$). We shall therefore use the term "semisimple" to mean "semisimple with minimum condition." The rings R_i in conclusion (5) of Wedderburn's Theorem are called the *Wedderburn components* of R and the direct product decomposition of R is called its *Wedderburn decomposition.* Note that Wedderburn's Theorem for commutative rings is a consequence of the classification of Artinian rings in Section 16.1. A commutative semisimple ring with minimum condition is an Artinian ring with Jacobson radical equal to zero and so is a direct product of fields (which are its Wedderburn components).

One should note that condition (5) is a two-sided condition which describes the overall structure of R completely (the ring operations in this direct product of rings are componentwise addition and multiplication). In particular it implies that a semisimple ring also has the minimum condition on right ideals. A useful way of thinking of the elements of the direct product $R_1 \times \cdots \times R_r$ in conclusion (5) is as $n \times n$ (block diagonal) matrices of the form

$$\begin{pmatrix} A_1 & & & \\ & A_2 & & \\ & & \ddots & \\ & & & A_r \end{pmatrix}$$

where A_i is an arbitrary $n_i \times n_i$ matrix with entries from Δ_i (here $n = \sum_{i=1}^{r} n_i$).

Recall from Section 10.5 that an R-module Q is *injective* if whenever Q is a submodule of any R-module M, then M has a submodule N such that $M = Q \oplus N$. Maschke's Theorem therefore implies:

Corollary 5. If G is a finite group and F is a field whose characteristic does not divide $|G|$, then the group algebra FG is a semisimple ring.

Before obtaining more precise information about how the invariants n, r, Δ_j, etc., relate to invariants in group rings FG for certain fields F, we first study the structure of matrix rings (i.e., the rings described in conclusions (4) and (5) of Wedderburn's Theorem). We introduce some terminology which is used extensively in ring theory. Recall that the *center* of the ring R is the subring of elements commuting with all elements in R; it will be denoted by $Z(R)$ (the center will contain 1 if the ring has a 1).

Definition.
  (1) A nonzero element e in a ring R is called an *idempotent* if $e^2 = e$.
  (2) Idempotents e_1 and e_2 are said to be *orthogonal* if $e_1 e_2 = e_2 e_1 = 0$.
  (3) An idempotent e is said to be *primitive* if it cannot be written as a sum of two (commuting) orthogonal idempotents.
  (4) The idempotent e is called a *primitive central idempotent* if $e \in Z(R)$ and e cannot be written as a sum of two orthogonal idempotents in the ring $Z(R)$.

Proposition 6 describes the ideal structure of a matrix ring and Proposition 8 extends these results to direct products of matrix rings.

Proposition 6. Let Δ be a division ring, let $n \in \mathbb{Z}^+$, let R be the ring of all $n \times n$ matrices with entries from Δ and let I be the identity matrix (= the 1 of R).
  (1) The only two-sided ideals of R are 0 and R.
  (2) The center of R consists of the scalar matrices αI, where α is in the center of Δ: $Z(R) = \{\alpha I \mid \alpha \in Z(\Delta)\}$, and this is a field isomorphic to $Z(\Delta)$. In particular, if Δ is a field, the center of R is the subring of all scalar matrices. The only central idempotent in R is I (in particular, I is primitive).
  (3) Let e_i be the matrix with a 1 in position i, i and zeros elsewhere. Then $e_1, \ldots, e_n$ are orthogonal primitive idempotents and $\sum_{i=1}^n e_i = I$.
  (4) $L_i = Re_i$ is the left ideal consisting of arbitrary entries in column i and zeros in all other columns. L_i is a simple left R-module. Every simple left R-module is isomorphic to L_1 (in particular, all L_i are isomorphic R-modules) and as a left R-module we have $R = L_1 \oplus \cdots \oplus L_n$.

Before proving this proposition it will be useful to have the following result.

Lemma 7. Let R be an arbitrary nonzero ring.
  (1) If M and N are simple R-modules and $\varphi : M \to N$ is a nonzero R-module homomorphism, then φ is an isomorphism.
  (2) *(Schur's Lemma)* If M is a simple R-module, then $\mathrm{Hom}_R(M, M)$ is a division ring.

Proof of Lemma 7: To prove (1) note that since φ is nonzero, $\ker \varphi$ is a proper submodule of M. By simplicity of M we have $\ker \varphi = 0$. Similarly, the image of φ is a nonzero submodule of the simple module N, hence $\varphi(M) = N$. This proves φ is bijective, so (1) holds.

By part (1), every nonzero element of the ring $\text{Hom}_R(M, M)$ is an isomorphism, hence has an inverse. This gives (2).

Proof of Proposition 6 Let A be an arbitrary matrix in R whose i, j entry is a_{ij}. Let E_{ij} be the matrix with a 1 in position i, j and zeros elsewhere. The following straightforward computations are left as exercises:

(i) $E_{ij}A$ is the matrix whose i^{th} row equals the jth row of A and all other rows are zero.

(ii) AE_{ij} is the matrix whose j^{th} column equals the ith column of A and all other columns are zero.

(iii) $E_{pq}AE_{rs}$ is the matrix whose p, s entry is a_{qr} and all other entries are zero.

To prove (1) suppose J is any nonzero 2-sided ideal of R and let A be an element of J with a nonzero entry in position q, r. Given any $p, s \in \{1, \ldots, n\}$ we obtain from (iii) that

$$E_{ps} = \frac{1}{a_{qr}} E_{pq}AE_{rs} \in J.$$

Since the Δ-linear combinations of $\{E_{ps} \mid 1 \le p \le n, \ 1 \le s \le n\}$ give all of R, it follows that $J = R$. This proves (1).

To prove (2) assume $A \in Z(R)$. Thus for all i, j we have $E_{ij}A = AE_{ij}$. From (i) and (ii) above it follows immediately that all off-diagonal entries of A are zero and all diagonal entries of A are equal. Thus $A = \alpha I$ for some $\alpha \in \Delta$. Furthermore, A must also commute with the set of all scalar matrices βI, $\beta \in \Delta$, i.e., α must commute with all elements of Δ. Finally, since $Z(R)$ is a field, it is immediate that it contains a unique idempotent (namely I). This establishes all parts of (2).

In part (3) it is clear that $e_1, \ldots, e_n$ are orthogonal idempotents whose sum is I. We defer proving that they are primitive until we have established (4).

Next we prove (4). From (ii) above it follows that $Re_i = RE_{ii}$ is the set of matrices with arbitrary entries in the ith column and zeros in all other columns. Furthermore, if A is any nonzero element of Re_i, then certainly $RA \subseteq Re_i$. The reverse inclusion holds because if a_{pi} is a nonzero entry of A, then by (i) above

$$e_i = E_{ii} = \frac{1}{a_{pi}} E_{ip}A \in RA.$$

This proves $Re_i = RA$ for any nonzero element $A \in Re_i$, and so Re_i must be a simple R-module.

Let M be any simple R-module. Since $Im = m$ for all $m \in M$ and since $I = \sum_{i=1}^n e_i$, there exists some i and some $m \in M$ such that $e_i m \ne 0$. For this i and m the map $re_i \mapsto re_i m$ is a nonzero R-module homomorphism from the simple R-module Re_i to the simple module M. By Lemma 7(1) it is an isomorphism. By (ii), the map $r \mapsto rE_{i1}$ gives $Re_i \cong Re_1$. Finally, every matrix is the direct sum of its columns so $R = L_1 \oplus \cdots \oplus L_n$. This completes the proof of (4).

It remains to prove that the idempotents in part (3) are primitive. If $e_i = a + b$, for some orthogonal idempotents a and b, then we shall see that

$$L_i = Re_i = Ra \oplus Rb.$$

This will contradict the fact that L_i is a simple R-module. To establish the above direct sum note first that since $ab = ba = 0$, we have $ae_i = a \in Re_i$ and $be_i = b \in Re_i$. For all $r \in R$ we have $re_i = ra + rb$, hence $Re_i = Ra + Rb$. Moreover, $Ra \cap Rb = 0$ because if $ra = sb$ for some $r, s \in R$, then $ra = raa = sba = 0$ (recall $a = a^2$ and $ba = 0$). This completes all parts of the proof.

Proposition 8. Let $R = R_1 \times R_2 \times \cdots \times R_r$, where R_i is the ring of $n_i \times n_i$ matrices over the division ring Δ_i, for $i = 1, 2, \ldots, r$.
  (1) Identify R_i with the ith component of the direct product. Let z_i be the r-tuple with the identity of R_i in position i and zero in all other positions. Then $R_i = z_i R$ and for any $a \in R_i$, $z_i a = a$ and $z_j a = 0$ for all $j \neq i$. The elements $z_1, \ldots, z_r$ are all of the primitive central idempotents of R. They are pairwise orthogonal and $\sum_{i=1}^{r} z_i = 1$.
  (2) Let N be any left R-module and let $z_i N = \{z_i x \mid x \in N\}$, $1 \leq i \leq r$. Then $z_i N$ is a left R-submodule of N, each $z_i N$ is an R_i-module on which R_j acts trivially for all $j \neq i$, and

$$N = z_1 N \oplus z_2 N \oplus \cdots \oplus z_r N.$$

  (3) The simple R-modules are the simple R_i-modules on which R_j acts trivially for $j \neq i$ in the following sense. Let M_i be the unique simple R_i-module (cf. Proposition 6). We may consider M_i as an R-module by letting R_j act trivially for all $j \neq i$. Then $M_1, \ldots, M_r$ are pairwise nonisomorphic simple R-modules and any simple R-module is isomorphic to one of $M_1, \ldots, M_r$. Explicitly, the R-module M_i is isomorphic to the simple left ideal $(0, \ldots, 0, L^{(i)}, 0, \ldots, 0)$ of all elements of R whose i^{th} component, $L^{(i)}$, consists of matrices with arbitrary entries in the first column and zeros elsewhere.
  (4) For any R-module N the R-submodule $z_i N$ is a direct sum of simple R-modules, each of which is isomorphic to the module M_i in (3). In particular, if M is a simple R-module, then there is a unique i such that $z_i M = M$ and for this index i we have $M \cong M_i$; for all $j \neq i$, $z_j M = 0$.
  (5) If each Δ_i equals the field F, then R is a vector space over F of dimension $\sum_{i=1}^{r} n_i^2$ and $\dim_F Z(R) = r$.

  Proof: In part (1) since multiplication in the direct product of rings is componentwise it is clear that z_i times the element $(a_1, \ldots, a_r)$ of R is the r-tuple with a_i in position i and zeros elsewhere. Thus $R_i = z_i R$, z_i is the identity in R_i and $z_i a = 0$ if $a \in R_j$ for any $j \neq i$. It is also clear that $z_1, \ldots, z_r$ are pairwise orthogonal central idempotents whose sum is the identity of R. The central idempotents of R are, by definition, the idempotents in $Z(R) = F_1 \times F_2 \times \cdots \times F_r$, where F_i is the center of R_i. By Proposition 6, F_i is the field $Z(\Delta_i)$. If $w = (w_1, \ldots, w_r)$ is any central idempotent then $w_i \in F_i$ for all i, and since $w^2 = w$ we have $w_i^2 = w_i$ in the field F_i. Since 0 and 1 are the only solutions to $x^2 = x$ in a field, the only central idempotents in R are r-tuples whose entries are 0's and 1's. Thus $z_1, \ldots, z_r$ are primitive central idempotents and since every central idempotent is a sum of these, they are the complete set of primitive central idempotents of R. This proves (1).

To prove (2) let N be any left R-module. First note that for any $z \in Z(R)$ the set $\{zx \mid x \in N\}$ is an R-submodule of N. In particular, $z_i N$ is an R-submodule. Let $z_i x \in z_i N$ and let $a \in R_j$ for some $j \neq i$. By (1) we have that $a = az_j$ and so $az_i x = (az_j)(z_i x) = az_i z_j x = 0$ because $z_i z_j = 0$. Thus the R-submodule $z_i N$ is acted on trivially by R_j for all $j \neq i$. For each $x \in N$ we have by (1) that $x = 1x = z_1 x + \cdots + z_r x$, hence $N = z_1 N + \cdots + z_r N$. Finally, this sum is direct because if, for instance, $x \in z_1 N \cap (z_2 N + \cdots + z_r N)$, then $x = z_1 x$ whereas z_1 times any element of $z_2 N + \cdots + z_r N$ is zero. This proves (2).

In part (3) first note that an R_i-module M becomes an R-module when R_j is defined to act trivially on M for all $j \neq i$. For such a module M the R-submodules are the same as the R_i-submodules. Thus M_i is a simple R-module for each i since it is a simple R_i-module.

Next, let M be a simple R-module. By (2), $M = z_1 M \oplus \cdots \oplus z_r M$. Since M has no nontrivial proper R-submodules, there must be a unique i such that $M = z_i M$ and $z_j M = 0$ for all $j \neq i$. Thus the simple R-module M is annihilated by R_j for all $j \neq i$. This implies that the R-submodules of M are the same as the R_i-submodules of M, so M is therefore a simple R_i-module. By Proposition 6, M is isomorphic as an R_i-module to M_i. Since R_j acts trivially on both M and M_i for all $j \neq i$, it follows that the R_i-module isomorphism may be viewed as an R-module isomorphism as well.

Suppose $i \neq j$ and suppose $\varphi : M_i \to M_j$ is an R-module isomorphism. If $s_i \in M_i$ then $s_i = z_i s_i$ so

$$\varphi(s_i) = \varphi(z_i s_i) = z_i \varphi(s_i) = 0,$$

since $\varphi(s_i) \in M_j$ and z_i acts trivially on M_j. This contradicts the fact that φ is an isomorphism and proves that $M_1, \ldots, M_r$ are pairwise nonisomorphic simple R-modules.

Finally, the left ideal of R described in (3) is acted on trivially by R_j for all $j \neq i$ and, by Proposition 6, it is up to isomorphism the unique simple R_i-module. This left ideal is therefore a simple R-module which is isomorphic to M_i. This proves (3).

For part (4) we have already proved that if M is any simple R-module then there is a unique i such that $z_i M = M$ and $z_j M = 0$ for all $j \neq i$. Furthermore, we have shown that for this index i the simple R-module M is isomorphic to M_i. Now let N be any R-module. Then $z_i N$ is a module over R_i which is acted on trivially by R_j for all $j \neq i$. By Wedderburn's Theorem $z_i N$ is a direct sum of simple R-modules. Since each of these simple summands is acted on trivially by R_j for all $j \neq i$, each is isomorphic to M_i. This proves (4).

In part (5) if each Δ_i equals the field F, then as an F-vector space

$$R \cong M_{n_1}(F) \oplus M_{n_2}(F) \oplus \cdots \oplus M_{n_r}(F).$$

Each matrix ring $M_{n_i}(F)$ has dimension n_i^2 over F, hence R has dimension $\sum_{i=1}^{r} n_i^2$ over F. Furthermore, the center of each $M_{n_i}(F)$ is 1-dimensional (since by Proposition 6(2) it is isomorphic to F), hence $Z(R)$ has dimension r over F. This completes the proof of the proposition.

We now apply Wedderburn's Theorem (and the above ring-theoretic calculations) to the group algebra FG. First of all, in order to apply Wedderburn's Theorem we need the characteristic of F not to divide $|G|$. In fact, since we shall be dealing with numerical data in the sections to come it will be convenient to have the characteristic of F equal to 0. Secondly, it will simplify matters if we force all the division rings which will appear in the Wedderburn decomposition of FG to equal the field F — we shall prove that imposing the condition that F be algebraically closed is sufficient to ensure this. To simplify notation we shall therefore take $F = \mathbb{C}$ for most of the remainder of the text. The reader can easily check that any algebraically closed field of characteristic 0 (e.g., the field of all algebraic numbers) can be used throughout in place of $\mathbb{C}$.

By Corollary 5 the ring $\mathbb{C}G$ is semisimple so by Wedderburn's Theorem

$$\mathbb{C}G \cong R_1 \times R_2 \times \cdots \times R_r$$

where R_i is the ring of $n_i \times n_i$ matrices over some division ring Δ_i. Thinking of the elements of this direct product as $n \times n$ block matrices ($n = \sum_{i=1}^{r} n_i$) where the ith block has entries from Δ_i, the field $\mathbb{C}$ appears in this direct product as scalar matrices and is contained in the center of $\mathbb{C}G$. Note that each Δ_i is a vector space over $\mathbb{C}$ of dimension $\leq n$. The next result shows that this implies each $\Delta_i = \mathbb{C}$.

Proposition 9. If Δ is a division ring that is a finite dimensional vector space over an algebraically closed field F and $F \subseteq Z(\Delta)$, then $\Delta = F$.

Proof: Since $F \subseteq Z(\Delta)$, for each $\alpha \in \Delta$ the division ring generated by α and F is a field. Also, since Δ is finite dimensional over F the field $F(\alpha)$ is a finite extension of F. Because F is algebraically closed it has no nontrivial finite extensions, hence $F(\alpha) = F$ for all $\alpha \in \Delta$, i.e., $\Delta = F$.

This proposition proves that each R_i in the Wedderburn decomposition of $\mathbb{C}G$ is a matrix ring over $\mathbb{C}$:

$$R_i = M_{n_i}(\mathbb{C}).$$

Now Proposition 8(5) implies that

$$\sum_{i=1}^{r} n_i^2 = |G|.$$

The final application in this section is to prove that r (= the number of Wedderburn components in $\mathbb{C}G$) equals the number of conjugacy classes of G. To see this, first note that Proposition 8(5) asserts that $r = \dim_{\mathbb{C}} Z(\mathbb{C}G)$. We compute this dimension in another way.

Let $\mathcal{K}_1, \ldots, \mathcal{K}_s$ be the distinct conjugacy classes of G (recall that these partition G). For each conjugacy class $\mathcal{K}_i$ of G let

$$X_i = \sum_{g \in \mathcal{K}_i} g \quad \in \mathbb{C}G.$$

Note that X_i and X_j have no common terms for $i \neq j$, hence they are linearly independent elements of $\mathbb{C}G$. Furthermore, since conjugation by a group element permutes the elements of each class, $h^{-1}X_i h = X_i$, i.e., X_i commutes with all group elements. This proves that $X_i \in Z(\mathbb{C}G)$.

We show the X_i's form a basis of $Z(\mathbb{C}G)$, which will prove $s = \dim_{\mathbb{C}} Z(\mathbb{C}G) = r$. Since the X_i's are linearly independent it remains to show they span $Z(\mathbb{C}G)$. Let $X = \sum_{g \in G} \alpha_g g$ be an arbitrary element of $Z(\mathbb{C}G)$. Since $h^{-1}Xh = X$,

$$\sum_{g \in G} \alpha_g h^{-1} gh = \sum_{g \in G} \alpha_g g.$$

Since the elements of G form a basis of $\mathbb{C}G$ the coefficients of g in the above two sums are equal:

$$\alpha_{hgh^{-1}} = \alpha_g.$$

Since h was arbitrary, every element in the same conjugacy class of a fixed group element g has the same coefficient in X, hence X can be written as a linear combination of the X_i's.

We summarize these results in the following theorem.

Theorem 10. Let G be a finite group.
  (1) $\mathbb{C}G \cong M_{n_1}(\mathbb{C}) \times M_{n_2}(\mathbb{C}) \times \cdots \times M_{n_r}(\mathbb{C})$.
  (2) $\mathbb{C}G$ has exactly r distinct isomorphism types of irreducible modules and these have complex dimensions $n_1, n_2, \ldots, n_r$ (and so G has exactly r inequivalent irreducible complex representations of the corresponding degrees).
  (3) $\sum_{i=1}^{r} n_i^2 = |G|$.
  (4) r equals the number of conjugacy classes in G.

Corollary 11.
  (1) Let A be a finite abelian group. Every irreducible complex representation of A is 1-dimensional (i.e., is a homomorphism from A into $\mathbb{C}^\times$) and A has $|A|$ inequivalent irreducible complex representations. Furthermore, every finite dimensional complex matrix representation of A is equivalent to a representation into a group of diagonal matrices.
  (2) The number of inequivalent (irreducible) degree 1 complex representations of any finite group G equals $|G/G'|$.

Proof: If A is abelian, $\mathbb{C}A$ is a commutative ring. Since a $k \times k$ matrix ring is not commutative whenever $k > 1$ we must have each $n_i = 1$. Thus $r = |A|$ ($=$ the number of conjugacy classes of A). Since every $\mathbb{C}A$-module is a direct sum of irreducible submodules, there is a basis such that the matrices are diagonal with respect to this basis. This establishes the first part of the corollary.

For a general group G, every degree 1 representation, φ, is a homomorphism of G into $\mathbb{C}^\times$. Thus φ factors through G/G'. Conversely, every degree 1 representation of G/G' gives, by composition with the natural projection $G \to G/G'$, a degree 1 representation of G. The degree 1 representations of G are therefore precisely the irreducible representations of the abelian group G/G'. Part (2) is now immediate from (1).

Examples

(1) The irreducible complex representations of a finite abelian group A (i.e., the homomorphisms from A into $\mathbb{C}^\times$) can be explicitly described as follows: decompose A into a direct product of cyclic groups

$$A \cong C_1 \times \cdots \times C_n$$

where $|C_i| = |\langle x_i \rangle| = d_i$. Map each x_i to a (not necessarily primitive) d_i^{th} root of 1 and extend this to all powers of x_i to give a homomorphism. Since there are d_i choices for the image of each x_i, the number of distinct homomorphisms of A into $\mathbb{C}^\times = GL_1(\mathbb{C})$ defined by this process equals $|A|$. By Corollary 11, these are all the irreducible representations of A. Note that it is necessary that the field contain the appropriate roots of 1 in order to realize these representations. An exercise below explores the irreducible representations of cyclic groups over $\mathbb{Q}$.

(2) Let $G = S_3$. By Theorem 10 the number of irreducible complex representations of G is three (= the number of conjugacy classes of S_3). Since the sum of the squares of the degrees is 6, the degrees must be 1, 1 and 2. The two degree 1 representations are immediately evident: the trivial representation and the representation of S_3 into $\{\pm 1\}$ given by mapping a permutation to its sign (i.e., $\sigma \mapsto +1$ if σ is an even permutation and $\sigma \mapsto -1$ if σ is an odd permutation). The degree 2 representation can be found by decomposing the permutation representation on 3 basis vectors (described in Section 1) into irreducibles as follows: let S_3 act on the basis vectors e_1, e_2, e_3 of a vector space V by permuting their indices. The vector $t = e_1 + e_2 + e_3$ is a nonzero fixed vector, so t spans a 1-dimensional G-invariant subspace (which is a copy of the trivial representation). By Maschke's Theorem there is a 2-dimensional G-invariant complement, I. Note that the permutation representation is not a sum of degree 1 representations: otherwise it could be represented by diagonal matrices and the permutations would commute in their action — this is impossible since the representation is faithful and G is non-abelian. Thus I cannot be decomposed further, so I affords *the* irreducible 2-dimensional representation. Indeed, I is the "augmentation" submodule described in Section 1:

$$I = \{w \in V \mid w = \alpha_1 e_1 + \alpha_2 e_2 + \alpha_3 e_3 \quad \text{with} \quad \alpha_1 + \alpha_2 + \alpha_3 = 0\}.$$

Clearly $e_1 - e_2$ and $e_2 - e_3$ are independent vectors in I, hence they form a basis for this 2-dimensional space. With respect to this basis of I we obtain a matrix representation of S_3 and, for example, this matrix representation on two elements of S_3 is

$$(1\ 2) \mapsto \begin{pmatrix} -1 & 1 \\ 0 & 1 \end{pmatrix} \quad \text{and} \quad (1\ 2\ 3) \mapsto \begin{pmatrix} 0 & -1 \\ 1 & -1 \end{pmatrix}.$$

(3) We decompose the regular representation over $\mathbb{C}$ of an arbitrary finite group. Recall that this is the representation afforded by the left $\mathbb{C}G$-module $\mathbb{C}G$ itself. By Theorem 10, $\mathbb{C}G$ is first of all a direct product of two-sided ideals:

$$\mathbb{C}G \cong M_{n_1}(\mathbb{C}) \times M_{n_2}(\mathbb{C}) \times \cdots \times M_{n_r}(\mathbb{C}).$$

Now by Proposition 6(4) each $M_{n_i}(\mathbb{C})$ decomposes further as a direct sum of n_i isomorphic simple left ideals. These left ideals give a complete set of isomorphism classes of irreducible $\mathbb{C}G$-modules. Thus the regular representation (over $\mathbb{C}$) of G decomposes as *the direct sum of all irreducible representations of G, each appearing with multiplicity equal to the degree of that irreducible representation.*

We record one additional property of $\mathbb{C}G$ which we shall prove in Section 19.2.

Theorem 12. The degree of each complex irreducible representation of a finite group G divides the order of G, i.e., in the notation of Theorem 10, each n_i divides $|G|$ for $i = 1, 2, \ldots, r$.

In the next section we shall describe the primitive central idempotents of $\mathbb{C}G$ in terms of the group elements.

EXERCISES

Let G be a finite group and let R be a ring with 1.

1. Prove that conditions (1) and (2) of Wedderburn's Theorem are equivalent.

2. Prove that (3) implies (2) in Wedderburn's Theorem. [Let Q be a submodule of an R-module N. Use Zorn's Lemma to show there is a submodule M maximal with respect to $Q \cap M = 0$. If $Q + M = N$, then (2) holds; otherwise let M_1 be the complete preimage in N of some simple module in N/M not contained in $(Q + M)/M$, and argue that M_1 contradicts the maximality of M.]

3. Prove that (4) implies (3) in Wedderburn's Theorem. [Let N be a nonzero R-module. First show N contains simple submodules by considering a cyclic submodule. Then use Zorn's Lemma applied to the set of direct sums of simple submodules (appropriately ordered) to show that N contains a maximal completely reducible submodule M. If $M \neq N$ let M_1 be the complete preimage in N of a simple module in N/M and contradict the maximality of M.]

4. Prove that (5) implies (4) in Wedderburn's Theorem. [Use the methods in the proofs of Propositions 6 and 8 to decompose each R_i as a left R-module.]

The next six exercises establish some general results about rings and modules that imply the remaining implication of Wedderburn's Theorem: (2) implies (5). In these exercises assume R satisfies (2): every R-module is injective.

5. Show that R has the descending chain condition (D.C.C.) on left ideals. Deduce that R is a finite direct sum of left ideals. [If not, then show that as a left R-module R is a direct sum of an infinite number of nonzero submodules. Derive a contradiction by writing the element 1 in this direct sum.]

6. Show that $R = R_1 \times R_2 \times \cdots \times R_r$ where R_j is a 2-sided ideal and a simple ring (i.e., has no proper, nonzero 2-sided ideals). Show each R_j has an identity and satisfies D.C.C. on left ideals. [Use the preceding exercise to show R has a minimal 2-sided ideal R_1. As a left R-module $R = R_1 \oplus R'$ for some left ideal R'. Show R' is a right ideal and proceed inductively using D.C.C.]

7. Let S be a simple ring with 1 satisfying D.C.C. on left ideals and let L be a minimal left ideal in S. Show that $S \cong L^n$ as left S-modules, where $L^n = L \oplus \cdots \oplus L$ with n factors. [Argue by simplicity that $LS = S$ so $1 = l_1 s_1 + \cdots + l_n s_n$ for some $l_i \in L$ and $s_i \in S$ with n minimal. Show that the map $(x_1, \ldots, x_n) \mapsto x_1 s_1 + \cdots + x_n s_n$ is a surjective homomorphism of left S-modules; use the minimality of L and n to show it is an injection.]

8. Let A be any ring with 1, let L be any left A-module and let L^n be the direct sum of n copies of L with itself.
   (a) Prove the ring isomorphism $\text{Hom}_A(L^n, L^n) \cong M_n(D)$, where $D = \text{Hom}_A(L, L)$ (multiplication in the ring $\text{Hom}_A(X, X)$ is function composition, cf. Proposition 2(4) in Section 10.2).

(b) Deduce that if L is a simple A-module, then $\mathrm{Hom}_A(L^n, L^n)$ is isomorphic to a matrix ring over a division ring. [Use Schur's Lemma and (a).]

(c) Prove the ring isomorphism $\mathrm{Hom}_A(A, A) \cong A^{opp}$, where A^{opp} is the opposite ring to A (the elements and addition are the same as in A but the value of the product $x \cdot y$ in A^{opp} is yx, computed in A), cf. the end of Section 17.4. [Any homomorphism is determined by its value on 1.]

9. Prove that if S is a simple ring with 1 satisfying D.C.C. on left ideals then $S \cong M_n(\Delta)$ for some division ring Δ. (This result together with Exercise 6 completes the existence part of the proof that (2) implies (5) in Wedderburn's Theorem). [Use Exercises 7 and 8 to show $S^{opp} \cong \mathrm{Hom}_S(L^n, L^n) \cong M_n(D)$ for some division ring D. Then show $S \cong M_n(\Delta)$, where Δ is the division ring D^{opp}.]

10. Prove that Δ and n in the isomorphism $S \cong M_n(\Delta)$ of the previous exercise are uniquely determined by S (proving the uniqueness statement in Wedderburn's Theorem), as follows. Suppose $S = M_n(\Delta) \cong M_{n'}(\Delta')$ as rings, where Δ and Δ' are division rings.
    (a) Prove that $\Delta \cong \mathrm{Hom}_S(L, L)$ where L is a minimal left ideal in S. Deduce that $\Delta \cong \Delta'$. [Use Proposition 6(4).]
    (b) Prove that a finitely generated (left) module over a division ring Δ has a "basis" (a linearly independent generating set), and that any two bases have the same cardinality. Deduce that $n = n'$. [Mimic the proof of Corollary 4(2) of Section 11.1.]

11. Prove that if R is a ring with 1 such that every R-module is free then R is a division ring.

12. Let F be a field, let $f(x) \in F[x]$ and let $R = F[x]/(f(x))$. Find necessary and sufficient conditions on the factorization of $f(x)$ in $F[x]$ so that R is a semisimple ring. When R is semisimple, describe its Wedderburn decomposition. [See Proposition 16 in Section 9.5.]

13. Let G be the cyclic group of order n and let $R = \mathbb{Q}G$. Describe the Wedderburn decomposition of R and find the number and the degrees of the irreducible representations of G over $\mathbb{Q}$. In particular, show that if $n = p$ is a prime then G has exactly one nontrivial irreducible representation over $\mathbb{Q}$ and this representation has degree $p - 1$. [Recall from the first example in Section 1 that $\mathbb{Q}G = \mathbb{Q}[x]/(x^n - 1)$. Use Proposition 16 in Section 9.5 and results from Section 13.6.]

14. Let p be a prime and let $F = \mathbb{F}_p$ be the field of order p. Let G be the cyclic group of order 3 and let $R = FG$. For each of $p = 2$ and $p = 7$ describe the Wedderburn decomposition of R and find the number and the degrees of the irreducible representations of G over F.

15. Prove that if P is a p-group for some prime p, then P has a faithful irreducible complex representation if and only if $Z(P)$ is cyclic. [Use Exercise 18 in Section 1, Theorem 6.1(2) and Example 3.]

16. Prove that if V is an irreducible FG-module and F is an algebraically closed field then $\mathrm{Hom}_{FG}(V, V)$ is isomorphic to F (as a ring).

17. Let F be a field, let $R = M_n(F)$ and let M be the unique irreducible R-module. Prove that $\mathrm{Hom}_R(M, M)$ is isomorphic to F (as a ring).

18. Find all 2-sided ideals of $M_n(\mathbb{Z})$.

18.3 CHARACTER THEORY AND THE ORTHOGONALITY RELATIONS

In general, for groups of large order the representations are difficult to compute and unwieldy if not impossible to write down. For example, a matrix representation of degree 100 involves matrices with 10,000 entries, and a number of 100×100 matrices may be required to describe the representation, even on a set of generators for the group. There are, however, some striking examples where large degree representations have been computed and used effectively. One instance of this is a construction of the simple group J_1 by Z. Janko in 1965 (the existence problem for simple groups was discussed at the end of Section 6.2). Janko was investigating certain properties of simple groups and he found that if any simple group possessed these properties, then it would necessarily have order 175,560 and would be generated by two elements. Furthermore, he proved that a hypothetical simple group with these properties must have a 7-dimensional representation over the field $\mathbb{F}_{11}$ with two generators mapping to the two matrices

$$
\begin{pmatrix}
0 & 1 & 0 & 0 & 0 & 0 & 0 \\
0 & 0 & 1 & 0 & 0 & 0 & 0 \\
0 & 0 & 0 & 1 & 0 & 0 & 0 \\
0 & 0 & 0 & 0 & 1 & 0 & 0 \\
0 & 0 & 0 & 0 & 0 & 1 & 0 \\
0 & 0 & 0 & 0 & 0 & 0 & 1 \\
1 & 0 & 0 & 0 & 0 & 0 & 0
\end{pmatrix}
\quad \text{and} \quad
\begin{pmatrix}
-3 & 2 & -1 & -1 & -3 & -1 & -3 \\
-2 & 1 & 1 & 3 & 1 & 3 & 3 \\
-1 & -1 & -3 & -1 & -3 & -3 & 2 \\
-1 & -3 & -1 & -3 & -3 & 2 & -1 \\
-3 & -1 & -3 & -3 & 2 & -1 & -1 \\
1 & 3 & 3 & -2 & 1 & 1 & 3 \\
3 & 3 & -2 & 1 & 1 & 3 & 1
\end{pmatrix}
$$

(note that for any simple group S, every representation of S into $GL_n(F)$ which does not map all group elements to the identity matrix is a faithful representation, so S is isomorphic to its image in $GL_n(F)$). In particular, Janko's calculations showed that the simple group satisfying his properties was unique, if it existed. M. Ward was able to show that these two matrices do generate a subgroup of $GL_7(\mathbb{F}_{11})$ of order 175,560 and it follows that there does exist a simple group satisfying Janko's properties.

In a similar vein, S. Norton, R. Parker and J. Thackray constructed the simple group J_4 of order 86,775,571,046,077,562,880 using a 112-dimensional representation over $\mathbb{F}_2$. This group was shown to be generated by two elements, and explicit matrices in $GL_{112}(\mathbb{F}_2)$ for these two generators were computed in the course of their analysis.

In 1981, R. Griess constructed the largest of the sporadic groups, the so called *Monster*, of order

$$
2^{46} \cdot 3^{20} \cdot 5^9 \cdot 7^6 \cdot 11^2 \cdot 13^3 \cdot 17 \cdot 19 \cdot 23 \cdot 29 \cdot 31 \cdot 41 \cdot 47 \cdot 59 \cdot 71.
$$

His proof involves calculations of automorphisms of an algebra over $\mathbb{C}$ of dimension 196,884 and leads to a construction of the Monster by means of a representation of this degree.

By analogy, in general it is difficult to write out the explicit permutations associated to a permutation representation $\varphi : G \to S_n$ for large degrees n. There are, however, numerical invariants such as the signs and the cycle types of the permutations $\pi(g)$ and these numerical invariants might be easier to compute than the permutations themselves (i.e., it may be possible to determine the cycle types of elements without actually having to write out the permutations themselves, as in the computation of Galois groups over $\mathbb{Q}$ in Section 14.8). These invariants alone may provide enough information in a given situation to carry out some analysis, such as prove that a given group is not simple (as illustrated in Section 6.2). Furthermore, the invariants just mentioned do not depend on the labelling of the set $\{1, 2, \ldots, n\}$ (i.e., they are independent of a "change of basis" in S_n) and they are the same for elements that are conjugate in G.

In this section we show how to attach numerical invariants to linear representations. These invariants depend only on the equivalence class (isomorphism type) of the representation. In other words, for each representation $\varphi : G \to GL_n(F)$ we shall attach an element of F to each matrix $\varphi(g)$ and we shall see that this number can, in many instances, be computed without knowing the matrix $\varphi(g)$. Moreover, we shall see that these invariants are independent of the similarity class of φ (i.e., are the same for a fixed $g \in G$ if the representation φ is replaced by an equivalent representation) and that they, in some sense, characterize the similarity classes of representations of G.

Throughout this section G is a finite group and, for the moment, F is an arbitrary field. All representations considered are assumed to be finite dimensional.

Definition.
(1) A *class function* is any function from G into F which is constant on the conjugacy classes of G, i.e., $f : G \to F$ such that $f(g^{-1}xg) = f(x)$ for all $g, x \in G$.
(2) If φ is a representation of G afforded by the FG-module V, the *character* of φ is the function

$$\chi : G \to F \quad \text{defined by} \quad \chi(g) = \operatorname{tr} \varphi(g),$$

where $\operatorname{tr} \varphi(g)$ is the trace of the matrix of $\varphi(g)$ with respect to some basis of V (i.e., the sum of the diagonal entries of that matrix). The character is called *irreducible* or *reducible* according to whether the representation is irreducible or reducible, respectively. The *degree* of a character is the degree of any representation affording it.

In the notation of the second part of this definition we shall also refer to χ as the character afforded by the FG-module V. In general, a character is *not* a homomorphism from a group into either the additive or multiplicative group of the field.

Examples
(1) The character of the trivial representation is the function $\chi(g) = 1$ for all $g \in G$. This character is called the *principal* character of G.
(2) For degree 1 representations, the character and the representation are usually identified (by identifying a 1×1 matrix with its entry). Thus for abelian groups, irreducible complex representations and their characters are the same (cf. Corollary 11).
(3) Let $\Pi : G \to S_n$ be a permutation representation and let φ be the resulting linear representation on the basis $e_1, \ldots, e_n$ of the vector space V:

$$\varphi(g)(e_i) = e_{\Pi(g)(i)}$$

(cf. Example 4 of Section 1). With respect to this basis the matrix of $\varphi(g)$ has a 1 in the diagonal entry i, i if $\Pi(g)$ fixes i; otherwise, the matrix of $\varphi(g)$ has a zero in position i, i. Thus if π is the character of φ then

$$\pi(g) = \text{the number of fixed points of } g \text{ on } \{1, 2, \ldots, n\}.$$

In particular, if Π is the permutation representation obtained from left multiplication on the set of left cosets of some subgroup H of G then the resulting character is called the *permutation character* of G on H.

(4) The special case of Example 3 when Π is the regular permutation representation of G is worth recording: if φ is the regular representation of G (afforded by the module FG) and ρ is its character:

$$\rho(g) = \begin{cases} 0 & \text{if } g \neq 1 \\ |G| & \text{if } g = 1. \end{cases}$$

The character of the regular representation of G is called the *regular character* of G. Note that this provides specific examples where a character takes on the value 0 and is not a group homomorphism from G into either F or $F^\times$.

(5) Let $\varphi : D_{2n} \to GL_2(\mathbb{R})$ be the explicit matrix representation described in Example 6 in the second set of examples of Section 1. If χ is the character of φ then, by taking traces of the given 2×2 matrices one sees that $\chi(r) = 2\cos(2\pi/n)$ and $\chi(s) = 0$. Since φ takes the identity of D_{2n} to the 2×2 identity matrix, $\chi(1) = 2$.

(6) Let $\varphi : Q_8 \to GL_2(\mathbb{C})$ be the explicit matrix representation described in Example 7 in the second set of examples of Section 1. If χ is the character of φ then, by taking traces of the given 2×2 matrices, $\chi(i) = 0$ and $\chi(j) = 0$. Since the element $-1 \in Q_8$ maps to minus the 2×2 identity matrix, $\chi(-1) = -2$. Since φ takes the identity of Q_8 to the 2×2 identity matrix, $\chi(1) = 2$.

(7) Let $\varphi : Q_8 \to GL_4(\mathbb{R})$ be the matrix representation described in Example 8 in the second set of examples of Section 1. If χ is the character of φ then, by inspection of the matrices exhibited, $\chi(i) = \chi(j) = 0$. Since φ takes the identity of Q_8 to the 4×4 identity matrix, $\chi(1) = 4$.

For $n \times n$ matrices A and B, direct computation shows that $\operatorname{tr} AB = \operatorname{tr} BA$. If A is invertible, this implies that

$$\operatorname{tr} A^{-1}BA = \operatorname{tr} B.$$

Thus the character of a representation is independent of the choice of basis of the vector space affording it, i.e.,

$$\textit{equivalent representations have the same character.} \qquad (18.1)$$

Let φ be a representation of G of degree n with character χ. Since $\varphi(g^{-1}xg)$ is $\varphi(g)^{-1}\varphi(x)\varphi(g)$ for all $g, x \in G$, taking traces shows that

$$\textit{the character of a representation is a class function.} \qquad (18.2)$$

Since the trace of the $n \times n$ identity matrix is n and φ takes the identity of G to the identity linear transformation (or matrix),

$$\chi(1) \textit{ is the degree of } \varphi. \qquad (18.3)$$

If V is an FG-module whose corresponding representation has character χ, then each element of the group ring FG acts as a linear transformation from V to V. Thus each $\sum_{g \in G} \alpha_g g \in FG$ has a trace when it is considered as a linear transformation from V to V. The trace of $g \in G$ acting on V is, by definition, $\chi(g)$. Since the trace of any linear combination of matrices is the linear combination of the traces, the trace of $\sum_{g \in G} \alpha_g g$ acting on V is $\sum_{g \in G} \alpha_g \chi(g)$. Note that this trace function on FG is the unique extension of the character χ of G to an F-linear transformation from FG to F. In this way we shall consider characters of G as also being defined on the group ring FG.

Notice in Example 3 above that if the field F has characteristic $p > 0$, the values of the character mod p might be zero even though the number of fixed points is nonzero. In order to circumvent such anomalies and to use the consequences of Wedderburn's Theorem obtained when F is algebraically closed we again specialize the field to be the complex numbers (or any algebraically closed field of characteristic 0). By the results of the previous section

$$\mathbb{C}G \cong M_{n_1}(\mathbb{C}) \times M_{n_2}(\mathbb{C}) \times \cdots \times M_{n_r}(\mathbb{C}). \tag{18.4}$$

For the remainder of this section fix the following notation:

$$M_1, M_2, \ldots, M_r \text{ are the inequivalent irreducible } \mathbb{C}G\text{-modules,}$$
$$\chi_i \text{ is the character afforded by } M_i, \quad 1 \le i \le r. \tag{18.5}$$

Thus r is the number of conjugacy classes of G and we may relabel $M_1, \ldots, M_r$ if necessary so that the degree of χ_i is n_i for all i (which is also the dimension of M_i over $\mathbb{C}$).

Now every (finite dimensional) $\mathbb{C}G$-module M is isomorphic (equivalent) to a direct sum of irreducible modules:

$$M \cong a_1 M_1 \oplus a_2 M_2 \oplus \cdots \oplus a_r M_r, \tag{18.6}$$

where a_i is a nonnegative integer indicating the multiplicity of the irreducible module M_i in this direct sum decomposition, i.e.,

$$a_i M_i = \overbrace{M_i \oplus \cdots \oplus M_i}^{a_i \text{ times}}.$$

Note that if the representation φ is afforded by the module M and $M = M_1 \oplus M_2$, then we may choose a basis of M consisting of a basis of M_1 together with a basis of M_2. The matrix representation with respect to this basis is of the form

$$\varphi(g) = \begin{pmatrix} \varphi_1(g) & 0 \\ 0 & \varphi_2(g) \end{pmatrix}$$

where φ_i is the representation afforded by M_i, $i = 1, 2$. One sees immediately that if ψ is the character of φ and ψ_i is the character of φ_i, then $\psi(g) = \psi_1(g) + \psi_2(g)$, i.e., $\psi = \psi_1 + \psi_2$. By induction we obtain:

the character of a representation is the sum of the characters
of the constituents appearing in a direct sum decomposition. $\tag{18.7}$

If ψ is the character afforded by the module M in (6) above, this gives

$$\psi = a_1 \chi_1 + a_2 \chi_2 + \cdots + a_r \chi_r. \tag{18.8}$$

Thus every (complex) character is a nonnegative integral sum of irreducible (complex) characters. Conversely, by taking direct sums of modules one sees that every such sum of characters is the character of some complex representation of G.

We next prove that the correspondence between characters and equivalence classes of complex representations is *bijective*. Let $z_1, z_2, \ldots, z_r$ be the primitive central idempotents of $\mathbb{C}G$ described in the preceding section. Since these are orthogonal (or equivalently, since they are the r-tuples in the decomposition of $\mathbb{C}G$ into a direct product of r

subrings which have a 1 in one position and zeros elsewhere), $z_1, \ldots, z_r$ are $\mathbb{C}$-linearly independent elements of $\mathbb{C}G$. As above, each irreducible character χ_i is a function on $\mathbb{C}G$. By Proposition 8(3) we have

(a) if $j \neq i$ then $z_j M_i = 0$, i.e., z_j acts as the zero matrix on M_j, hence $\chi_j(z_i) = 0$, and

(b) z_i acts as the identity on M_i, hence $\chi_i(z_i) = n_i$.

Thus $\chi_1, \ldots, \chi_r$ are multiples of the dual basis to the independent set $z_1, \ldots, z_r$, hence are linearly independent functions. Now if the $\mathbb{C}G$-module M described in (6) above can be decomposed in a different fashion into irreducibles, say,

$$M \cong b_1 M_1 \oplus b_2 M_2 \oplus \cdots \oplus b_r M_r,$$

then we would obtain a relation

$$a_1 \chi_1 + a_2 \chi_2 + \cdots + a_r \chi_r = b_1 \chi_1 + b_2 \chi_2 + \cdots + b_r \chi_r.$$

By linear independence of the irreducible characters, $b_i = a_i$ for all $i \in \{1, \ldots, r\}$. Thus, in any decomposition of M into a direct sum of irreducibles, the multiplicity of the irreducible M_i is the same, $1 \leq i \leq r$. In particular,

two representations are equivalent if and only if they have the same character.

(18.9)

This uniqueness can be seen in an alternate way. First, use Proposition 8(2) to decompose an arbitrary finite dimensional $\mathbb{C}G$-module M uniquely as

$$M = z_1 M \oplus z_2 M \oplus \cdots \oplus z_r M.$$

By part (4) of the same proposition, $z_i M$ is a direct sum of simple modules, each of which is isomorphic to M_i. The multiplicity of M_i in a direct sum decomposition of $z_i M$ is, by counting dimensions, equal to $\dfrac{\dim z_i M}{\dim M_i}$. This proves that the multiplicity of M_i in any direct sum decomposition of M into simple submodules is uniquely determined.

Note that, as with decompositions of $F[x]$-modules into cyclic submodules, a $\mathbb{C}G$-module may have many direct sum decompositions into irreducibles — only the multiplicities are unique (see also the exercises). More precisely, comparing with the Jordan canonical form of a single linear transformation, the direct summand $a_i M_i = M_i \oplus \cdots \oplus M_i$ (a_i times) which equals the submodule $z_i M$ is the analogue of the generalized eigenspace corresponding to a single eigenvalue. This submodule of M is unique (as is a generalized eigenspace) and is called the χ_i^{th} *isotypic component* of M. Within the χ_i^{th} isotypic component, the summands M_i are analogous to the 1-dimensional eigenspaces and, just as with the eigenspace of an endomorphism there is no unique basis for the eigenspace. If $G = \langle g \rangle$ is a finite cyclic group, the isotypic components of M are the same as the generalized eigenspaces of g.

Observe that the vector space of all (complex valued) class functions on G has a basis consisting of the functions which are 1 on a given class and zero on all other classes. There are r of these, where r is the number of conjugacy classes of G, so the dimension of the complex vector space of class functions is r. Since the number of (complex) irreducible characters of G equals the number of conjugacy classes and these are linearly independent class functions, we see that

the irreducible characters are a basis for the space of all complex class functions.
$$(18.10)$$

The next step in the theory of characters is to put an Hermitian inner product structure on the space of class functions and prove that the irreducible characters form an orthonormal basis with respect to this inner product. For class functions θ and ψ define

$$(\theta, \psi) = \frac{1}{|G|} \sum_{g \in G} \theta(g) \overline{\psi(g)}$$

(where the bar denotes complex conjugation). One easily checks that $(\ ,\ )$ is Hermitian: for $\alpha, \beta \in \mathbb{C}$

(a) $(\alpha \theta_1 + \beta \theta_2, \psi) = \alpha(\theta_1, \psi) + \beta(\theta_2, \psi)$,
(b) $(\theta, \alpha \psi_1 + \beta \psi_2) = \overline{\alpha}(\theta, \psi_1) + \overline{\beta}(\theta, \psi_2)$,   and
(c) $(\theta, \psi) = \overline{(\psi, \theta)}$.

Our principal aim is to show that the irreducible characters form an orthonormal basis for the space of complex class functions with respect to this Hermitian form (we already know that they are a basis). This fact will follow from the orthogonality of the primitive central idempotents, once we have explicitly determined these in the next proposition.

Proposition 13. Let $z_1, \ldots, z_r$ be the orthogonal primitive central idempotents in $\mathbb{C}G$ labelled in such a way that z_i acts as the identity on the irreducible $\mathbb{C}G$-module M_i, and let χ_i be the character afforded by M_i. Then

$$z_i = \frac{\chi_i(1)}{|G|} \sum_{g \in G} \chi_i(g^{-1}) g.$$

Proof: Let $z = z_i$ and write

$$z = \sum_{g \in G} \alpha_g g.$$

Recall from Example 4 in this section that if ρ is the regular character of G then

$$\rho(g) = \begin{cases} 0 & \text{if } g \neq 1 \\ |G| & \text{if } g = 1 \end{cases} \qquad (18.11)$$

and recall from the last example in Section 2 that

$$\rho = \sum_{j=1}^{r} \chi_j(1) \chi_j. \qquad (18.12)$$

To find the coefficient α_g, apply ρ to zg^{-1} and use linearity of ρ together with equation (11) to obtain

$$\rho(zg^{-1}) = \alpha_g |G|.$$

Computing $\rho(zg^{-1})$ using (12) then gives

$$\sum_{j=1}^{r} \chi_j(1)\chi_j(zg^{-1}) = \alpha_g |G|. \tag{18.13}$$

Let φ_j be the irreducible representation afforded by M_j, $1 \leq j \leq r$. Since we may consider φ_j as an algebra homomorphism from $\mathbb{C}G$ into $\mathrm{End}(M_j)$, we obtain $\varphi_j(zg^{-1}) = \varphi_j(z)\varphi_j(g^{-1})$. Also, we have already observed that $\varphi_j(z)$ is 0 if $j \neq i$ and $\varphi_i(z)$ is the identity endomorphism on M_i. Thus

$$\varphi_j(zg^{-1}) = \begin{cases} 0 & \text{if } j \neq i \\ \varphi_i(g^{-1}) & \text{if } j = i. \end{cases}$$

This proves $\chi_j(zg^{-1}) = \chi_i(g^{-1})\delta_{ij}$, where δ_{ij} is zero if $i \neq j$ and is 1 if $i = j$ (called the Kronecker delta). Substituting this into equation (13) gives $\alpha_g = \dfrac{1}{|G|}\chi_i(1)\chi_i(g^{-1})$. This is the coefficient of g in the statement of the proposition, completing the proof.

The orthonormality of the irreducible characters will follow directly from the orthogonality of the central primitive idempotents via the following calculation:

$$z_i\delta_{ij} = z_i z_j$$

$$= \frac{\chi_i(1)}{|G|}\frac{\chi_j(1)}{|G|}\sum_{g,h\in G}\chi_i(g^{-1})\chi_j(h^{-1})gh$$

$$= \frac{\chi_i(1)}{|G|}\frac{\chi_j(1)}{|G|}\sum_{y\in G}\left[\sum_{x\in G}\chi_i(xy^{-1})\chi_j(x^{-1})\right]y$$

(to get the latter sum from the former substitute y for gh and x for h). Since the elements of G are a basis of $\mathbb{C}G$ we may equate coefficients with those of z_i found in Proposition 13 to get (the coefficient of g)

$$\delta_{ij}\frac{\chi_i(1)}{|G|}\chi_i(g^{-1}) = \frac{\chi_i(1)\chi_j(1)}{|G|^2}\sum_{x\in G}\chi_i(xg^{-1})\chi_j(x^{-1}).$$

Simplifying (and replacing g by g^{-1}) gives

$$\delta_{ij}\frac{\chi_i(g)}{\chi_j(1)} = \frac{1}{|G|}\sum_{x\in G}\chi_i(xg)\chi_j(x^{-1}) \quad \text{for all } g \in G. \tag{18.14}$$

Taking $g = 1$ in (14) gives

$$\delta_{ij} = \frac{1}{|G|}\sum_{x\in G}\chi_i(x)\chi_j(x^{-1}). \tag{18.15}$$

The sum on the right side would be precisely the inner product (χ_i, χ_j) if $\chi_j(x^{-1})$ were equal to $\overline{\chi_j(x)}$; this is the content of the next proposition.

Proposition 14. If ψ is any character of G then $\psi(x)$ is a sum of roots of 1 in $\mathbb{C}$ and $\psi(x^{-1}) = \overline{\psi(x)}$ for all $x \in G$.

Proof: Let φ be a representation whose character is ψ, fix an element $x \in G$ and let $|x| = k$. Since the minimal polynomial of $\varphi(x)$ divides $X^k - 1$ (hence has distinct roots), there is a basis of the underlying vector space such that the matrix of $\varphi(x)$ with respect to this basis is a diagonal matrix with k^{th} roots of 1 on the diagonal. Since $\psi(x)$ is the sum of the diagonal entries (and does not depend on the choice of basis), $\psi(x)$ is a sum of roots of 1. Moreover, if ϵ is a root of 1, $\epsilon^{-1} = \bar{\epsilon}$. Thus the inverse of a diagonal matrix with roots of 1 on the diagonal is the diagonal matrix with the complex conjugates of those roots of 1 on the diagonal. Since the complex conjugate of a sum is the sum of the complex conjugates, $\psi(x^{-1}) = \operatorname{tr} \varphi(x^{-1}) = \overline{\operatorname{tr} \varphi(x)} = \overline{\psi(x)}$.

Keep in mind that in the proof of Proposition 14 we first fixed a group element x and then chose a basis of the representation space so that $\varphi(x)$ was a diagonal matrix. It is always possible to diagonalize a single element but it is possible to *simultaneously* diagonalize all $\varphi(x)$'s if and only if φ is similar to a sum of degree 1 representations.

Combining the above proposition with equation (15) proves:

Theorem 15. *(The First Orthogonality Relation for Group Characters)* Let G be a finite group and let $\chi_1, \ldots, \chi_r$ be the irreducible characters of G over $\mathbb{C}$. Then with respect to the inner product $(,)$ above we have

$$(\chi_i, \chi_j) = \delta_{ij}$$

and the irreducible characters are an orthonormal basis for the space of class functions. In particular, if θ is any class function then

$$\theta = \sum_{i=1}^{r} (\theta, \chi_i) \chi_i.$$

Proof: We have just established that the irreducible characters form an orthonormal basis for the space of class functions. If θ is any class function, write $\theta = \sum_{i=1}^{r} a_i \chi_i$, for some $a_i \in \mathbb{C}$. It follows from linearity of the Hermitian product that $a_i = (\theta, \chi_i)$, as stated.

We list without proof the Second Orthogonality Relation; we shall not require it for the applications in this book.

Theorem 16. *(The Second Orthogonality Relation for Group Characters)* Under the notation above, for any $x, y \in G$

$$\sum_{i=1}^{r} \chi_i(x) \overline{\chi_i(y)} = \begin{cases} |C_G(x)| & \text{if } x \text{ and } y \text{ are conjugate in } G \\ 0 & \text{otherwise.} \end{cases}$$

Definition. For θ any class function on G the *norm* of θ is $(\theta, \theta)^{1/2}$ and will be denoted by $\|\theta\|$.

When a class function is written in terms of the irreducible characters, $\theta = \sum \alpha_i \chi_i$, its norm is easily calculated as $|| \theta || = (\sum \alpha_i^2)^{1/2}$. It follows that

a character has norm 1 if and only if it is irreducible.

Finally, observe that computations of the inner product of characters θ and ψ may be simplified as follows. If $\mathcal{K}_1, \ldots, \mathcal{K}_r$ are the conjugacy classes of G with sizes $d_1, \ldots, d_r$ and representatives $g_1, \ldots, g_r$ respectively, then the value $\theta(g_i)\overline{\psi(g_i)}$ appears d_i times in the sum for (θ, ψ), once for each element of $\mathcal{K}_i$. Collecting these terms gives

$$(\theta, \psi) = \frac{1}{|G|} \sum_{i=1}^{r} d_i \theta(g_i)\overline{\psi(g_i)},$$

a sum only over representatives of the conjugacy classes of G. In particular, the norm of θ is given by

$$|| \theta ||^2 = (\theta, \theta) = \frac{1}{|G|} \sum_{i=1}^{r} d_i |\theta(g_i)|^2.$$

Examples

(1) Let $G = S_3$ and let π be the permutation character of degree 3 described in the examples at the beginning of this section. Recall that $\pi(\sigma)$ equals the number of elements in $\{1, 2, 3\}$ fixed by σ. The conjugacy classes of S_3 are represented by 1, (1 2) and (1 2 3) of sizes 1, 3 and 2 respectively, and $\pi(1) = 3$, $\pi((1\ 2)) = 1$, $\pi((1\ 2\ 3)) = 0$. Hence

$$|| \pi ||^2 = \frac{1}{6} \left[1\,\pi(1)^2 + 3\,\pi((1\ 2))^2 + 2\,\pi((1\ 2\ 3))^2 \right]$$

$$= \frac{1}{6}(9 + 3 + 0) = 2$$

This implies that π is a sum of two distinct irreducible characters, each appearing with multiplicity 1. Let χ_1 be the principal character of S_3, so that $\chi_1(\sigma) = \overline{\chi_1(\sigma)} = 1$ for all $\sigma \in S_3$. Then

$$(\pi, \chi_1) = \frac{1}{6} \left[1\,\pi(1)\overline{\chi_1(1)} + 3\,\pi((1\ 2))\overline{\chi_1((1\ 2))} + 2\,\pi((1\ 2\ 3))\overline{\chi_1((1\ 2\ 3))} \right]$$

$$= \frac{1}{6}(3 + 3 + 0) = 1$$

so the principal character appears as a constituent of π with multiplicity 1. This proves $\pi = \chi_1 + \chi_2$ for some irreducible character χ_2 of S_3 of degree 2 (and agrees with our earlier decomposition of this representation). This also shows that the value of χ_2 on $\sigma \in S_3$ is the number of fixed points of σ minus 1.

(2) Let $G = S_4$ and let π be the natural permutation character of degree 4 (so again $\pi(\sigma)$ is the number of fixed points of σ). The conjugacy classes of S_4 are represented by 1, (1 2), (1 2 3), (1 2 3 4) and (1 2)(3 4) of sizes 1, 6, 8, 6 and 3 respectively. Again we compute:

$$|| \pi ||^2 = \frac{1}{24} \left[1\,\pi(1)^2 + 6\,\pi((1\ 2))^2 + 8\,\pi((1\ 2\ 3))^2 + 6\,\pi((1\ 2\ 3\ 4))^2 \right.$$

$$\left. + 3\,\pi((1\ 2)(3\ 4))^2 \right]$$

$$= \frac{1}{24}(16 + 24 + 8 + 0 + 0) = 2$$

so π has two distinct irreducible constituents. If χ_1 is the principal character of S_4, then

$$(\pi, \chi_1) = \frac{1}{24} \left[1\,\pi(1) + 6\,\pi((1\ 2)) + 8\,\pi((1\ 2\ 3)) \right.$$
$$\left. + 6\,\pi((1\ 2\ 3\ 4)) + 3\,\pi((1\ 2)(3\ 4)) \right]$$
$$= \frac{1}{24}(4 + 12 + 8 + 0 + 0) = 1.$$

This proves that the degree 4 permutation character is the sum of the principal character and an irreducible character of degree 3.

(3) Let $G = D_8$, where

$$D_8 = \langle r, s \mid s^2 = r^4 = 1,\ rs = sr^{-1} \rangle.$$

The conjugacy classes of D_8 are represented by $1, s, r, r^2$ and sr and have sizes 1, 2, 2, 1 and 2, respectively. Let φ be the degree 2 matrix representation of D_8 obtained as in Example 6 in Section 1 from embedding a square in $\mathbb{R}^2$:

$$\varphi(s) = \begin{pmatrix} 0 & 1 \\ 1 & 0 \end{pmatrix},\ \varphi(r) = \begin{pmatrix} 0 & -1 \\ 1 & 0 \end{pmatrix},\ \varphi(r^2) = \begin{pmatrix} -1 & 0 \\ 0 & -1 \end{pmatrix},\ \varphi(sr) = \begin{pmatrix} 1 & 0 \\ 0 & -1 \end{pmatrix}.$$

Let ψ be the character of this representation (where we consider the real matrices as a subset of the complex matrices). Again, since ψ is real valued one computes

$$\| \psi \|^2 = \frac{1}{8} \left[1\psi(1)^2 + 2\psi(s)^2 + 2\psi(r)^2 + 1\psi(r^2)^2 + 2\psi(sr)^2 \right]$$
$$= \frac{1}{8}(4 + 0 + 0 + 4 + 0) = 1.$$

This proves the representation φ is irreducible (even if we allow similarity transformations by complex matrices).

We have seen that the sum of two characters is again a character. Specifically, if ψ_1 and ψ_2 are characters of representations φ_1 and φ_2, then $\psi_1 + \psi_2$ is the character of $\varphi_1 + \varphi_2$.

Proposition 17. If ψ_1 and ψ_2 are characters, then so is their product $\psi_1\psi_2$.

Proof: Let V_1 and V_2 be $\mathbb{C}G$-modules affording characters ψ_1 and ψ_2 and define $W = V_1 \otimes_{\mathbb{C}} V_2$. Since each $g \in G$ acts as a linear transformation on V_1 and V_2, the action of g on simple tensors by $g(v_1 \otimes v_2) = (gv_1) \otimes (gv_2)$ extends by linearity to a well defined linear transformation on W by Proposition 17 in Section 11.2. One easily checks that this action also makes W into a $\mathbb{C}G$-module. By Exercise 38 in Section 11.2 the character afforded by W is $\psi_1\psi_2$.

The next chapter will contain further explicit character computations as well as some applications of group characters to proving theorems about certain classes of groups.

Some Remarks on Fourier Analysis and Group Characters

This brief discussion is intended to indicate some connections of the results above with other areas of mathematics.

The theory of group representations described to this point is a special branch of an area of mathematics called Harmonic Analysis. Readers may already be familiar with the basic theory of Fourier series which also falls into this realm. We make some observations which show how representation theory for finite groups corresponds to "Fourier series" for some infinite groups (in particular, to Fourier series on the circle). To be mathematically precise one needs the Lebesgue integral to ensure completeness of certain (Hilbert) spaces but readers may get the flavor of things by replacing "Lebesgue" by "Riemann."

Let G be the multiplicative group of points on the unit circle in $\mathbb{C}$:

$$G = \{z \in \mathbb{C} \mid |z| = 1\}.$$

We shall usually view G as the interval $[0, 2\pi]$ in $\mathbb{R}$ with the two end points identified, i.e., as the additive group $\mathbb{R}/2\pi\mathbb{Z}$ (the isomorphism is: the real number x corresponds to the complex number e^{ix}). Note that G has a translation invariant measure, namely the Lebesgue measure, and the measure of the circle is 2π. For finite groups, the counting measure is the translation invariant measure (so the measure of a subset H is the number of elements in that subset, $|H|$) and integrals on a finite group with respect to this counting measure are just finite sums.

The space

$$L^2(G) = \{f : G \to \mathbb{C} \mid f \text{ is measurable and } |f|^2 \text{ is integrable over } G \}$$

plays the role of the group algebra of the infinite group G. This space becomes a commutative ring with 1 under the convolution of functions: for $f, g \in L^2(G)$ the product $f * g : G \to \mathbb{C}$ is defined by

$$(f * g)(x) = \frac{1}{2\pi} \int_0^{2\pi} f(x - y)g(y)\, dy \qquad \text{for all } x \in G.$$

(Recall that for a finite group H, the group algebra is also formally the ring of $\mathbb{C}$-valued functions on H under a convolution multiplication and that these functions are written as formal sums – the element $\sum \alpha_g g \in \mathbb{C}G$ denotes the function which sends g to $\alpha_g \in \mathbb{C}$ for all $g \in G$.)

The complete set of continuous homomorphisms of G into $GL_1(\mathbb{C})$ is given by

$$e_n(x) = e^{inx}, \quad x \in [0, 2\pi], \quad n \in \mathbb{Z}.$$

(Recall that for a finite abelian group, all irreducible representations are 1-dimensional and for 1-dimensional representations, characters and representations may be identified.)

The ring $L^2(G)$ admits an Hermitian inner product: for $f, g \in L^2(G)$

$$(f, g) = \frac{1}{2\pi} \int_0^{2\pi} f(t)\overline{g(t)}\, dt.$$

Under this inner product, $\{e_n \mid n \in \mathbb{Z}\}$ is an orthonormal basis (where the term "basis" is used in the analytic sense that these are independent and 0 is the only function orthogonal to all of them). Moreover,

$$L^2(G) = \widehat{\bigoplus_{n \in \mathbb{Z}} E_n}$$

where E_n is the 1-dimensional subspace spanned by e_n, the hat over the direct sum denotes taking the closure of the direct sum in the L^2-topology, and equality indicates equality in the L^2 sense. (Recall that the group algebra of a finite abelian group is the direct sum of the irreducible 1-dimensional submodules, each occurring with multiplicity one.) These facts imply the well known result from Fourier analysis that every square integrable function $f(x)$ on $[0, 2\pi]$ has a Fourier series

$$\sum_{n=-\infty}^{\infty} c_n e^{inx}$$

where the Fourier coefficients, c_n, are given by

$$c_n = (f, e_n) = \frac{1}{2\pi} \int_0^{2\pi} f(t)e^{-int} \, dt.$$

This brief description indicates how the representation theory of finite groups extends to certain infinite groups and the results we have proved may already be familiar in the latter context. In fact, there is a completely analogous theory for arbitrary (not necessarily abelian) compact Lie groups — here the irreducible (complex) representations need not be 1-dimensional but they are all finite dimensional and $L^2(G)$ decomposes as a direct sum of them, each appearing with multiplicity equal to its degree. The emphasis (at least at the introductory level) in this theory is often on the importance of being able to represent functions as (Fourier) series and then using these series to solve other problems (e.g., solve differential equations). The underlying group provides the "symmetry" on which to build this "harmonic analysis," rather than being itself the principal object of study.

EXERCISES

Let G be a finite group. Unless stated otherwise all representations and characters are over $\mathbb{C}$.

1. Prove that tr $AB = $ tr BA for $n \times n$ matrices A and B with entries from any commutative ring.

2. In each of (a) to (c) let ψ be the character afforded by the specified representation φ.
   (a) Let φ be the degree 2 representation of D_{10} described in Example 6 in the second set of examples in Section 1 (here $n = 5$) and show that $\| \psi \|^2 = 1$ (hence φ is irreducible).
   (b) Let φ be the degree 2 representation of Q_8 described in Example 7 in the second set of examples in Section 1 and show that $\| \psi \|^2 = 1$ (hence φ is irreducible).
   (c) Let φ be the degree 4 representation of Q_8 described in Example 8 in the second set of examples in Section 1 and show that $\| \psi \|^2 = 4$ (hence even though φ is irreducible over $\mathbb{R}$, φ decomposes over $\mathbb{C}$ as twice an irreducible representation of degree 2).

3. If χ is an irreducible character of G, prove that the χ-isotypic subspace of a $\mathbb{C}G$-module is unique.

4. Prove that if N is any irreducible $\mathbb{C}G$-module and $M = N \oplus N$, then M has infinitely many direct sum decompositions into two copies of N.

5. Prove that a class function is a character if and only if it is a positive integral linear combination of irreducible characters.

6. Let $\varphi : G \to GL(V)$ be a representation with character ψ. Let W be the subspace $\{v \in V \mid \varphi(g)(v) = v \text{ for all } g \in G\}$ of V fixed pointwise by all elements of G. Prove that $\dim W = (\psi, \chi_1)$, where χ_1 is the principal character of G.

7. Assume V is a $\mathbb{C}G$-module on which G acts by permuting the basis $\mathcal{B} = \{e_1, \ldots, e_n\}$. Write $\mathcal{B}$ as a disjoint union of the orbits $\mathcal{B}_1, \ldots, \mathcal{B}_t$ of G on $\mathcal{B}$.
   (a) Prove that V decomposes as a $\mathbb{C}G$-module as $V_1 \oplus \cdots \oplus V_t$, where V_i is the span of $\mathcal{B}_i$.
   (b) Prove that if v_i is the sum of the vectors in $\mathcal{B}_i$ then the 1-dimensional subspace of V_i spanned by v_i is the unique $\mathbb{C}G$-submodule of V_i affording the trivial representation (in other words, any vector in V_i that is fixed under the action of G is a multiple of v_i). [Use the fact that G is transitive on $\mathcal{B}_i$. See also Exercise 8 in Section 1.]
   (c) Let $W = \{v \in V \mid \varphi(g)(v) = v \text{ for all } g \in G\}$ be the subspace of V fixed pointwise by all elements of G. Deduce that $\dim W = t = $ the number of orbits of G on $\mathcal{B}$.

8. Prove the following result (sometimes called Burnside's Lemma although its origin is with Frobenius): let G be a subgroup of S_n and for each $\sigma \in G$ let $\text{Fix}(\sigma)$ denote the number of fixed points of σ on $\{1, \ldots, n\}$. Let t be the number of orbits of G on $\{1, \ldots, n\}$. Then

$$t|G| = \sum_{g \in G} \text{Fix}(g).$$

[Use the preceding two exercises.]

9. Let G be a nontrivial, transitive group of permutations on the finite set Ω and let ψ be the character afforded by the linear representation over $\mathbb{C}$ obtained from Ω (cf. Example 4 in Section 1) so $\psi(\sigma)$ is the number of fixed points of σ on Ω. Now let G act on the set $\Omega \times \Omega$ by $g \cdot (\omega_1, \omega_2) = (g \cdot \omega_1, g \cdot \omega_2)$ and let π be the character afforded by the linear representation obtained from this action.
   (a) Prove that $\pi = \psi^2$.
   (b) Prove that the number of orbits of G on $\Omega \times \Omega$ is given by the inner product (ψ, ψ). [By the preceding exercises, the number of orbits on $\Omega \times \Omega$ is equal to (π, χ_1), where χ_1 is the principal character.]
   (c) Recall that G is said to be *doubly transitive* on Ω if it has precisely 2 orbits in its action on $\Omega \times \Omega$ (it always has at least 2 orbits since the diagonal, $\{(\omega, \omega) \mid \omega \in \Omega\}$, is one orbit). Prove that if G is doubly transitive on Ω then $\psi = \chi_1 + \chi_2$, where χ_1 is the principal character and χ_2 is a nonprincipal irreducible character of G.
   (d) Let $\Omega = \{1, 2, \ldots, n\}$ and let $G = S_n$ act on Ω in the natural fashion. Show that the character of the associated linear representation decomposes as the principal character plus an irreducible character of degree $n - 1$.

10. Let ψ be the character of any 2-dimensional representation of a group G and let x be an element of order 2 in G. Prove that $\psi(x) = 2, 0$ or -2. Generalize this to n-dimensional representations.

11. Let χ be an irreducible character of G. Prove that for every element z in the center of G we have $\chi(z) = \epsilon \chi(1)$, where ϵ is some root of 1 in $\mathbb{C}$. [Use Schur's Lemma.]

12. Let ψ be the character of some representation φ of G. Prove that for $g \in G$ the following hold:
   (a) if $\psi(g) = \psi(1)$ then $g \in \ker \varphi$, and

(b) if $|\psi(g)| = \psi(1)$ and φ is faithful then $g \in Z(G)$ (where $|\psi(g)|$ is the complex absolute value of $\psi(g)$). [Use the method of proof of Proposition 14.]

13. Let $\varphi : G \to GL(V)$ be a representation and let $\chi : G \to \mathbb{C}^\times$ be a degree 1 representation. Prove that $\chi\varphi : G \to GL(V)$ defined by $\chi\varphi(g) = \chi(g)\varphi(g)$ is a representation (note that multiplication of the linear transformation $\varphi(g)$ by the complex number $\chi(g)$ is well defined). Show that $\chi\varphi$ is irreducible if and only if φ is irreducible. Show that if ψ is the character afforded by φ then $\chi\psi$ is the character afforded by $\chi\varphi$. Deduce that the product of any irreducible character with a character of degree 1 is also an irreducible character.

The next few exercises study the notion of *algebraically conjugate* characters. These exercises may be considered as extensions of Proposition 14 and some consequences of these extensions. In particular we obtain a group-theoretic characterization of the conditions under which all irreducible characters of a group take values in $\mathbb{Q}$.

Let F be the subfield of $\mathbb{C}$ of all elements that are algebraic over $\mathbb{Q}$ (the field of algebraic numbers). Thus F is the algebraic closure of $\mathbb{Q}$ contained in $\mathbb{C}$ and all the results established over $\mathbb{C}$ hold without change over F.

14. Note that since $F \subseteq \mathbb{C}$, every representation $\varphi : G \to GL_m(F)$ may also be considered as a complex representation. Prove that if φ is a representation over F that is irreducible over F, then φ is also irreducible when considered over the larger field $\mathbb{C}$ (note that this is not true if F is not algebraically closed — cf. Exercise 2(c) above). Show that the set of irreducible characters of G over F is the same as the set of irreducible characters over $\mathbb{C}$ (i.e., these are exactly the same set of class functions on G). Deduce that every complex representation is equivalent to a representation over F. [Since F is algebraically closed of characteristic 0, the irreducible characters over either F or $\mathbb{C}$ are characterized by the first orthogonality relation.]

Let $\varphi : G \to GL_m(F)$ be any representation with character ψ. Let $\mathbb{Q}(\varphi)$ denote the subfield of F generated by all the entries of the matrices $\varphi(g)$ for all $g \in G$.

15. Prove that $\mathbb{Q}(\varphi)$ is a finite extension of $\mathbb{Q}$.

Now let K be any Galois extension of $\mathbb{Q}$ containing $\mathbb{Q}(\varphi)$ and let $\sigma \in \mathrm{Gal}(K/\mathbb{Q})$. In fact, since every automorphism of K extends to an automorphism of F, we may assume σ is any automorphism of F. The map $\varphi^\sigma : G \to GL_n(F)$ is defined by letting $\varphi^\sigma(g)$ be the $n \times n$ matrix whose entries are obtained by applying the field automorphism σ to the entries of the matrix $\varphi(g)$.

16. Prove that φ^σ is a representation. Prove also that the character of φ^σ is ψ^σ, where
$$\psi^\sigma(g) = \sigma(\psi(g)).$$

17. Prove that φ is irreducible if and only if φ^σ is irreducible.

The representation φ^σ (or character ψ^σ) is called the *algebraic conjugate* of φ by σ (or of ψ, respectively); two representations φ_1 and φ_2 (or characters ψ_1 and ψ_2) are said to be *algebraically conjugate* if there is some automorphism σ of F such that $\varphi_1^\sigma = \varphi_2$ (or $\psi_1^\sigma = \psi_2$, respectively). Some care needs to be taken with this (standard) notation since the exponential notation usually denotes a right action whereas automorphisms of F act on the left on representations: $\varphi^{(\sigma\tau)} = (\varphi^\tau)^\sigma$.

Let $\mathbb{Q}(\psi)$ be the subfield of F generated by the numbers $\psi(g)$ for all $g \in G$. Let $|G| = n$ and let ϵ be a primitive n^{th} root of 1 in F.

18. Prove that $\mathbb{Q}(\psi) \subseteq \mathbb{Q}(\epsilon)$. Deduce that $\mathbb{Q}(\psi)$ is a Galois extension of $\mathbb{Q}$ with abelian Galois group. [See Proposition 14.]

Recall from Section 14.5 that $\text{Gal}(\mathbb{Q}(\epsilon)/\mathbb{Q}) \cong (\mathbb{Z}/n\mathbb{Z})^\times$, where the Galois automorphisms are given on the generator ϵ by $\sigma_a : \epsilon \mapsto \epsilon^a$, where a is an integer relatively prime to n.

19. Prove that if $\sigma_a \in \text{Gal}(\mathbb{Q}(\epsilon)/\mathbb{Q})$ is the field automorphism defined above, then for all $g \in G$ we have $\psi^{\sigma_a}(g) = \psi(g^a)$. [Use the method of Proposition 14.]

20. Prove that if g is an element of G which is conjugate to g^a for all integers a relatively prime to n, then $\psi(g) \in \mathbb{Q}$, for every character ψ of G. [Use the preceding exercise and the fact that $\mathbb{Q}$ is the field fixed by all σ_a's.]

21. Prove that an element $g \in G$ is conjugate to g^a for all integers a relatively prime to $|G|$ if and only if g is conjugate to $g^{a'}$ for all integers a' relatively prime to $|g|$.

22. Show for any positive integer n that every character of the symmetric group S_n is rational valued (i.e., $\psi(g) \in \mathbb{Q}$ for all $g \in S_n$ and all characters ψ of S_n).

The next two exercises establish the converse to Exercise 20.

23. Prove that elements x and y are conjugate in a group G if and only if $\chi(x) = \chi(y)$ for all irreducible characters χ of G.

24. Let $g \in G$ and assume that every irreducible character of G is rational valued on g. Prove that g is conjugate to g^a for every integer a relatively prime to $|G|$. [If g is not conjugate to g^a for some a relatively prime to $|G|$ then by the preceding exercise there is an irreducible character χ such that $\chi(g) \neq \chi(g^a)$. Derive a contradiction from the hypothesis that $\chi(g) \in \mathbb{Q}$.]

25. Describe which irreducible characters of the cyclic group of order n are algebraically conjugate.

26. Prove that every irreducible character of both Q_8 and D_8 is rational valued. Prove that D_{10} has an irreducible character that is not rational valued.

27. Let $G = H \times K$ and let $\varphi : H \to GL(V)$ be an irreducible representation of H with character χ. Then $G \xrightarrow{\pi_H} H \xrightarrow{\varphi} GL(V)$ gives an irreducible representation of G, where π_H is the natural projection; the character, $\widetilde{\chi}$, of this representation is $\widetilde{\chi}((h, k)) = \chi(h)$. Likewise any irreducible character ψ of K gives an irreducible character $\widetilde{\psi}$ of G with $\widetilde{\psi}((h, k)) = \psi(k)$.
(a) Prove that the product $\widetilde{\chi}\widetilde{\psi}$ is an irreducible character of G. [Show it has norm 1.]
(b) Prove that every irreducible character of G is obtained from such products of irreducible characters of the direct factors. [Use Theorem 10, either (3) or (4).]

28. *(Finite subgroups of $GL_2(\mathbb{Q})$)* Let G be a finite subgroup of $GL_2(\mathbb{Q})$.
(a) Show that $GL_2(\mathbb{Q})$ does not contain an element of order n for $n = 5, 7$, or $n \geq 9$. Deduce that $|G| = 2^a 3^b$. [Use rational canonical forms.]
(b) Show that the Klein 4-group is the only noncyclic abelian subgroup of $GL_2(\mathbb{Q})$. Deduce from this and (a) that $|G| \mid 24$.
(c) Show that the only finite subgroups of $GL_2(\mathbb{Q})$ are the cyclic groups of order 1, 2, 3, 4, and 6, the Klein 4-group, and the dihedral groups of order 6, 8, and 12. [Use the classifications of groups of small order in Section 4.5 and Exercise 10 of Section 1 to restrict G to this list. Show conversely that each group listed has a 2-dimensional faithful rational representation.]

CHAPTER 19

Examples and Applications of Character Theory

19.1 CHARACTERS OF GROUPS OF SMALL ORDER

The *character table* of a finite group is the table of character values formatted as follows: list representatives of the r conjugacy classes along the top row and list the irreducible characters down the first column. The entry in the table in row χ_i and column g_j is $\chi_i(g_j)$. The character table of a finite group is unique up to a permutation of its rows and columns. It is customary to make the principal character the first row and the identity the first column and to list the characters in increasing order by degrees. In our examples we shall list the size of the conjugacy classes under each class so the entire table will have $r + 2$ rows and $r + 1$ columns (although strictly speaking, the character table is the $r \times r$ matrix of character values). This will enable one to easily check the "orthogonality of rows" using the first orthogonality relation: if the classes are represented by $g_1, \ldots, g_r$ of sizes $d_1, \ldots, d_r$ then

$$(\chi_i, \chi_j) = \frac{1}{|G|} \sum_{k=1}^{r} d_k \chi_i(g_k) \overline{\chi_j(g_k)}.$$

The second orthogonality relation says that the Hermitian product of any two distinct columns of a character table is zero (i.e., it gives an "orthogonality of columns").

A number of character tables are given in the *Atlas of Finite Groups* by Conway, Curtis, Norton, Parker and Wilson, Clarendon Press, 1985. These include the character table of the Monster simple group, M. The group M has 194 irreducible characters. The smallest degree of a nonprincipal irreducible character of M is 196883 and the largest degree is on the order of 2×10^{26}. Nonetheless, it is possible to compute the values of all these characters on all conjugacy classes of M.

For the first example of a character table let $G = \langle x \rangle$ be the cyclic group of order 2. Then G has 2 conjugacy classes and two irreducible characters:

classes:	1	x
sizes:	1	1
χ_1	1	1
χ_2	1	-1

Character Table of Z_2

The characters and representations of this abelian group are the same, and the irreducible representations of any abelian group are described in Example 1 at the end of Section 18.2.

Similarly, if $G = \langle x \rangle$ is cyclic of order 3, and ζ is a fixed primitive cube root of 1 (so $\zeta^2 = \bar{\zeta}$), then the character table of G is the following:

classes:	1	x	x^2
sizes:	1	1	1
χ_1	1	1	1
χ_2	1	ζ	ζ^2
χ_3	1	ζ^2	ζ

Character Table of Z_3

Next we construct the character table of S_3. Recall from Example 2 in Section 18.2 that S_3 has 3 irreducible characters whose values are described in that example and in Example 1 at the end of Section 18.3.

classes:	1	$(1\,2)$	$(1\,2\,3)$
sizes:	1	3	2
χ_1	1	1	1
χ_2	1	-1	1
χ_3	2	0	-1

Character Table of S_3

Next we consider D_8, adopting the notation of Example 3 of Section 18.3. By Corollary 11, D_8 has four characters of degree 1. Also, in Example 3 we constructed an irreducible degree 2 representation. Since the sum of the squares of the degrees of these representations is 8, this accounts for all irreducible representations (or, since there are 5 conjugacy classes, there are 5 irreducible representations). If we let bars denote passage to the commutator quotient group (which is the Klein 4-group), then $\bar{1} = \bar{r^2}$. The degree 1 representations (= their characters) are computed by sending generators $\bar{s}$ and $\bar{r}$ to ± 1 (and the product class is mapped to the product of the values). Matrices for the degree 2 irreducible representation were computed in Example 3 of Section 18.3 and the character of this representation can be read directly from these matrices. The character table of D_8 is therefore the following:

classes:	1	r^2	s	r	sr
sizes:	1	1	2	2	2
χ_1	1	1	1	1	1
χ_2	1	1	-1	1	-1
χ_3	1	1	1	-1	-1
χ_4	1	1	-1	-1	1
χ_5	2	-2	0	0	0

Character Table of D_8

Now we compute the character table of the quaternion group of order 8. We use the usual presentation

$$Q_8 = \langle i, j \mid i^4 = 1, \ i^2 = j^2, \ i^{-1}ji = j^{-1} \rangle$$

and let $k = ij$ and $i^2 = -1$. The conjugacy classes of Q_8 are represented by $1, -1, i$, j and k of sizes 1, 1, 2, 2 and 2, respectively. Since the commutator quotient of Q_8 is the Klein 4-group, there are four characters of degree 1. The one remaining irreducible character must have degree 2 in order that the sum of the squares of the degrees be 8. Let χ_5 be the degree 2 irreducible character of Q_8. One may check that the representation $\varphi : Q_8 \to GL_2(\mathbb{C})$ described explicitly in Example 7 in the second set of examples of Section 18.1 affords χ_5, but we show how the orthogonality relations give the values of χ_5 without knowing these explicit matrices. If φ is an irreducible representation of degree 2, by Schur's Lemma (cf. Exercise 18 in Section 18.1) $\varphi(-1)$ is a 2×2 scalar matrix and so is $\pm$ the identity matrix since -1 has order 2 in Q_8. Hence $\chi_5(-1) = \pm 2$. Let $\chi_5(i) = a$, $\chi_5(j) = b$ and $\chi_5(k) = c$. The orthogonality relations give

$$1 = (\chi_5, \chi_5) = \frac{1}{8}(2^2 + (\pm 2)^2 + 2a\bar{a} + 2b\bar{b} + 2c\bar{c}).$$

Since $a\bar{a}$, $b\bar{b}$ and $c\bar{c}$ are nonnegative real numbers, they must all be zero. Also, since χ_5 is orthogonal to the principal character we get

$$0 = (\chi_1, \chi_5) = \frac{1}{8}(2 + (\pm 2) + 0 + 0 + 0),$$

hence $\chi_5(-1) = -2$. The complete character table of Q_8 is the following:

classes:	1	−1	i	j	k
sizes:	1	1	2	2	2
χ_1	1	1	1	1	1
χ_2	1	1	−1	1	−1
χ_3	1	1	1	−1	−1
χ_4	1	1	−1	−1	1
χ_5	2	−2	0	0	0

Character Table of Q_8

Observe that D_8 and Q_8 have the same character table, hence

nonisomorphic groups may have the same character table.

Note that the values of the degree 2 representation of Q_8 could also have been easily calculated by applying the second orthogonality relation to each column of the character table. We leave this check as an exercise. Also note that although the degree 2 irreducible characters of D_8 and Q_8 have the same (real number) values the degree 2 representation of D_8 may be realized by real matrices whereas it may be shown that Q_8 has no faithful 2-dimensional representation over $\mathbb{R}$ (cf. Exercise 10 in Section 18.1).

For the next example we construct the character table of S_4. The conjugacy classes of S_4 are represented by 1, (1 2), (1 2 3), (1 2 3 4) and (1 2)(3 4) with sizes 1, 6, 8, 6, and 3 respectively. Since $S_4' = A_4$, there are two characters of degree 1: the principal character and the character whose values are the sign of the permutation.

To obtain a degree 2 irreducible character let V be the normal subgroup of order 4 generated by $(1\,2)(3\,4)$ and $(1\,3)(2\,4)$. Any representation φ of $S_4/V \cong S_3$ gives, by composition with the natural projection $S_4 \to S_4/V$, a representation of S_4; if the former is irreducible, so is the latter. Let φ be the composition of the projection with the irreducible 2-dimensional representation of S_3, and let χ_3 be its character. The classes of 1 and $(1\,2)(3\,4)$ map to the identity in the S_3 quotient, $(1\,2)$ and $(1\,2\,3\,4)$ map to transpositions and $(1\,2\,3)$ maps to a 3-cycle. The values of χ_3 can thus be read directly from the values of the character of degree 2 in the table for S_3.

Since S_4 has 5 irreducible characters and the sum of the squares of the degrees is 24, there must be two remaining irreducible characters, each of degree 3. In Example 2 of Section 18.3 one of these was calculated, call it χ_4. Recall that

$$\chi_4(\sigma) = (\text{the number of fixed points of } \sigma) - 1.$$

The remaining irreducible character, χ_5, is $\chi_4\chi_2$. One can either use Proposition 17 in Section 18.3 or Exercise 13 in Section 18.3 to see that this product is indeed a character. The first orthogonality relation verifies that it is irreducible.

classes:	1	$(1\,2)$	$(1\,2\,3)$	$(1\,2\,3\,4)$	$(1\,2)(3\,4)$
sizes:	1	6	8	6	3
χ_1	1	1	1	1	1
χ_2	1	-1	1	-1	1
χ_3	2	0	-1	0	2
χ_4	3	1	0	-1	-1
χ_5	3	-1	0	1	-1

Character Table of S_4

From the character table of S_4 one can easily compute the character table of A_4. Note that A_4 has 4 conjugacy classes. Also $|A_4 : A_4'| = 3$, so A_4 has three characters of degree 1 with $V = A_4'$ in the kernel of each degree 1 representation. The remaining irreducible character must have degree 3. One checks directly from the orthogonality relation applied in A_4 that the character χ_4 of S_4 restricted to $A_4 (= \chi_5|_{A_4})$ is irreducible. This irreducibility check is really necessary since an irreducible representation of a group need not restrict to an irreducible representation of a subgroup (for instance, the irreducible degree 2 representation of S_3 must become reducible when restricted to any proper subgroup, since these are all abelian). The character table of A_4 is the following

classes:	1	$(1\,2)(3\,4)$	$(1\,2\,3)$	$(1\,3\,2)$
sizes:	1	3	4	4
χ_1	1	1	1	1
χ_2	1	1	ζ	ζ^2
χ_3	1	1	ζ^2	ζ
χ_4	3	-1	0	0

Character Table of A_4

where ζ is a primitive cube root of 1 in $\mathbb{C}$.

As a final example we construct the following character table of S_5:

classes:	1	(1 2)	(1 2 3)	(1 2 3 4)	(1 2 3 4 5)	(1 2)(3 4)	(1 2)(3 4 5)
sizes:	1	10	20	30	24	15	20
χ_1	1	1	1	1	1	1	1
χ_2	1	−1	1	−1	1	1	−1
χ_3	4	2	1	0	−1	0	−1
χ_4	4	−2	1	0	−1	0	1
χ_5	5	−1	−1	1	0	1	−1
χ_6	5	1	−1	−1	0	1	1
χ_7	6	0	0	0	1	−2	0

Character Table of S_5

The conjugacy classes and their sizes were computed in Section 4.3. Since $|S_5 : S_5'| = 2$, there are two degree 1 characters: the principal character and the "sign" character.

The natural permutation of S_5 on 5 points gives rise to a permutation character of degree 5. As with S_4 and S_3 the orthogonality relations show that the square of its norm is 2 and it contains the principal character. Thus χ_3 is the permutation character minus the principal character (and, as with the smaller symmetric groups, $\chi_3(\sigma)$ is the number of fixed points of σ minus 1). As argued with S_4, it follows that $\chi_4 = \chi_3\chi_2$ is also an irreducible character.

To obtain χ_5 recall that S_5 has six Sylow 5-subgroups. Its action by conjugation on these gives a faithful permutation representation of degree 6. If ψ is the character of the associated linear representation, then since $\sigma \in S_5$ fixes a Sylow 5-subgroup if and only if it normalizes that subgroup, we have

$$\psi(\sigma) = \text{the number of Sylow 5-subgroups normalized by } \sigma.$$

The normalizer in S_5 of the Sylow 5-subgroup $\langle (1 2 3 4 5) \rangle$ is $\langle (1 2 3 4 5), (2 3 5 4) \rangle$ and all normalizers of Sylow 5-subgroups are conjugate in S_5 to this group. This normalizer contains only the identity, 5-cycles, 4-cycles and products of two disjoint transpositions. No other cycle type normalizes any Sylow 5-subgroup so on any other class, ψ is zero. To compute ψ on the remaining three nonidentity classes note (by inspection in S_6) that in any faithful action on 6 points the following hold: an element of order 5 must be a 5-cycle (hence fixes 1 point); any element of order 4 which fixes one point must be a 4-cycle (hence fixes 2 points); an element of order 2 which is the square of an element of order 4 fixes exactly 2 points also. This gives all the values of ψ. Now direct computation shows that

$$\| \psi \|^2 = 2 \qquad \text{and} \qquad (\chi_1, \psi) = 1.$$

Thus $\chi_5 = \psi - \chi_1$ is irreducible of degree 5. By the same theory as for χ_4 one gets that $\chi_6 = \chi_5\chi_2$ is another irreducible character.

Since there are 7 conjugacy classes, there is one remaining irreducible character and its degree is 6. Its values can be obtained immediately from the decomposition of the regular character, ρ (cf. Example 3 in Section 18.2 and Example 4 in Section 18.3):

$$\chi_7 = \frac{\rho - \chi_1 - \chi_2 - 4\chi_3 - 4\chi_4 - 5\chi_5 - 5\chi_6}{6}.$$

A direct calculation by the orthogonality relations checks that χ_7 is irreducible. Note that the values of the character χ_7 were computed without explicitly exhibiting a representation with this character.

EXERCISES

1. Calculate the character tables of $Z_2 \times Z_2$, $Z_2 \times Z_3$ and $Z_2 \times Z_2 \times Z_2$. Explain why the table of $Z_2 \times Z_3$ contains primitive 6th roots of 1.

2. Compute the degrees of the irreducible characters of D_{16}.

3. Compute the degrees of the irreducible characters of A_5. Deduce that the degree 6 irreducible character of S_5 is not irreducible when restricted to A_5. [The conjugacy classes of A_5 are worked out in Section 4.3.]

4. Using the character tables in this section, for each of parts (a) to (d) use the first orthogonality relation to write the specified permutation character (cf. Example 3, Section 18.3) as a sum of irreducible characters:
   (a) the permutation character of the subgroup A_3 of S_3
   (b) the permutation character of the subgroup $\langle (1\ 2\ 3\ 4) \rangle$ of S_4
   (c) the permutation character of the subgroup V_4 of S_4
   (d) the permutation character of the subgroup $\langle (1\ 2\ 3), (1\ 2), (4\ 5) \rangle$ of S_5 (this subgroup is the normalizer of a Sylow 3-subgroup of S_5).

5. Assume that for any character ψ of a group, ψ^2 is also a character (where $\psi^2(g) = (\psi(g))^2$) — this is a special case of Proposition 17 in Section 18.3. Using the character tables in this section, for each of parts (a) to (e) write out the values of the square, χ^2, of the specified character χ and use the first orthogonality relation to write χ^2 as a sum of irreducible characters:
   (a) $\chi = \chi_3$, the degree 2 character in the table of S_3
   (b) $\chi = \chi_5$, the degree 2 character in the table of Q_8
   (c) $\chi = \chi_5$, the last character in the table of S_4
   (d) $\chi = \chi_4$, the second degree 4 character in the table of S_5
   (e) $\chi = \chi_7$, the last character in the table of S_5.

6. Calculate the character table of A_5.

7. Show that S_6 has an irreducible character of degree 5.

8. Calculate the character table of D_{10}. (This table contains irrational entries.)

9. Calculate the character table of D_{12}.

10. Calculate the character table of $S_3 \times S_3$.

11. Calculate the character table of $Z_3 \times S_3$.

12. Calculate the character table of $Z_2 \times S_4$.

13. Calculate the character table of $S_3 \times S_4$.

14. Let n be an integer with $n \geq 3$. Show that every irreducible character of D_{2n} has degree 1 or 2 and find the number of irreducible characters of each degree. [The conjugacy classes of D_{2n} were found in Exercises 31 and 32 of Section 4.3 and its commutator subgroup was computed in Section 5.4.]

15. Prove that the character table is an invertible matrix. [Use the orthogonality relations.]

16. For each of A_5 and D_{10} describe which irreducible characters are algebraically conjugate (cf. the exercises in Section 18.3).

17. Let p be any prime and let P be a non-abelian group of order p^3 (up to isomorphism there are two choices for P; for odd p these were constructed when the groups of order p^3 were classified in Section 5.5). This exercise determines the character table of P and shows that both isomorphism types have the same character table (the argument includes the $p = 2$ case worked out in this section).

(a) Prove that P has p^2 characters of degree 1.

(b) Prove that P has $p - 1$ irreducible characters of degree p and that these together with the p^2 degree 1 characters are all the irreducible characters of P. [Use Theorem 10(3) and Theorem 12 in Section 18.2.]

(c) Deduce that (regardless of the isomorphism type) the group P has $p^2 + p - 1$ conjugacy classes, p of which are of size 1 (i.e., are central classes) and $p^2 - 1$ of which each have size p. Deduce also that the classes of size p are precisely the nonidentity cosets of the center of P (i.e., if $x \in P - Z(P)$ then the conjugacy class of x is the set of p elements in the coset $xZ(P)$).

(d) Prove that if χ is an irreducible character of degree p then the representation affording χ is faithful.

(e) Fix a generator, z, of the center of P and let ϵ be a fixed primitive p^{th} root of 1 in $\mathbb{C}$. Prove that if χ is an irreducible character of degree p then $\chi(z) = p\epsilon^i$ for some $i \in \{1, 2, \ldots, p - 1\}$. Prove further that $\chi(x) = 0$ for all $x \in P - Z(P)$. (Note then that the degree p characters are all algebraically conjugate.) [Use the same reasoning as in the construction of the character table of Q_8.]

(f) Prove that for each $i \in \{1, 2, \ldots, p - 1\}$ there is a unique irreducible character χ_i of degree p such that $\chi_i(z) = p\epsilon^i$. Deduce that the character table of P is uniquely determined, and describe it. [Recall from Section 6.1 that regardless of the isomorphism type, $P' = Z(P)$ and $P/P' \cong Z_p \times Z_p$. From this one can write out the degree 1 characters. Part (e) describes the degree p characters.]

19.2 THEOREMS OF BURNSIDE AND HALL

In this section we give a "theoretical" application of character theory: Burnside's $p^a q^b$ Theorem. We also prove Philip Hall's characterization of finite solvable groups, which is a group-theoretic proof relying on Burnside's Theorem as the first step in its induction.

Burnside's Theorem

The following result was proved by Burnside in 1904. Although purely group-theoretic proofs of it were discovered recently (see Theorem 2.8 in *Finite Groups III* by B. Huppert and N. Blackburn, Springer-Verlag, 1982) the original proof by Burnside presented here is very accessible, elegant, and quite brief (given our present knowledge of representation theory).

Theorem 1. (Burnside) For p and q primes, every group of order $p^a q^b$ is solvable.

Before undertaking the proof of Burnside's Theorem itself we establish some results of a general nature. An easy consequence of these preliminary propositions is that the degrees of the irreducible characters of any finite group divide its order. The particular results that lead directly to the proof of Burnside's Theorem appear in Lemmas 6 and 7.

It follows quite easily that a counterexample to Burnside's Theorem of minimal order is a non-abelian simple group, and it is these two character-theoretic lemmas that give the contradiction by proving the existence of a normal subgroup.

We first recall from Section 15.3 the definition of algebraic integers.

Definition. An element $\alpha \in \mathbb{C}$ is called an *algebraic integer* if it is a root of a monic polynomial with coefficients from $\mathbb{Z}$.

The basic results needed for the proof of Burnside's Theorem are:

Proposition 2. Let $\alpha \in \mathbb{C}$.
   (1) The following are equivalent:
       (i) α is an algebraic integer,
       (ii) α is algebraic over $\mathbb{Q}$ and the minimal polynomial of α over $\mathbb{Q}$ has integer coefficients, and
       (iii) $\mathbb{Z}[\alpha]$ is a finitely generated $\mathbb{Z}$-module (where $\mathbb{Z}[\alpha]$ is the subring of $\mathbb{C}$ generated by $\mathbb{Z}$ and α, i.e., is the ring of all $\mathbb{Z}$-linear combinations of nonnegative powers of α).
   (2) The algebraic integers in $\mathbb{C}$ form a ring and the algebraic integers in $\mathbb{Q}$ are the elements of $\mathbb{Z}$.

Proof: These are established in Section 15.3. (The portion of Section 15.3 consisting of integral extensions and properties of algebraic integers may be read independently from the rest of Chapter 15.)

Corollary 3. For every character ψ of the finite group G, $\psi(x)$ is an algebraic integer for all $x \in G$.

Proof: By Proposition 14 in Section 18.3, $\psi(x)$ is a sum of roots of 1. Each root of 1 is an algebraic integer, so the result follows immediately from Proposition 2(2).

We shall also need some preliminary character-theoretic lemmas before beginning the main proof. Adopt the following notation for the arbitrary finite group G: $\chi_1, \dots, \chi_r$ are the distinct irreducible (complex) characters of G, $\mathcal{K}_1, \dots, \mathcal{K}_r$ are the conjugacy classes of G and φ_i is an irreducible matrix representation whose character is χ_i for each i.

Proposition 4. Define the complex valued function ω_i on $\{\mathcal{K}_1, \dots, \mathcal{K}_r\}$ for each i by

$$\omega_i(\mathcal{K}_j) = \frac{|\mathcal{K}_j|\chi_i(g)}{\chi_i(1)}$$

where g is any element of $\mathcal{K}_j$. Then $\omega_i(\mathcal{K}_j)$ is an algebraic integer for all i and j.

Proof: We first prove that if I is the identity matrix, then

$$\sum_{g \in \mathcal{K}_j} \varphi_i(g) = \omega_i(\mathcal{K}_j)I. \tag{19.1}$$

To see this let X be the left hand side of (1). As we saw in Section 18.2, each $x \in G$ acting by conjugation permutes the elements of $\mathcal{K}_j$ and so X commutes with $\varphi_i(g)$ for all g. By Schur's Lemma (Exercise 18 in Section 18.1) X is a scalar matrix:

$$X = \alpha I \qquad \text{for some } \alpha \in \mathbb{C}.$$

It remains to show that $\alpha = \omega_i(\mathcal{K}_j)$. But

$$\operatorname{tr} X = \sum_{g \in \mathcal{K}_j} \operatorname{tr} \varphi_i(g) = \sum_{g \in \mathcal{K}_j} \chi_i(g) = |\mathcal{K}_j| \chi_i(g).$$

Thus $\alpha \chi_i(1) = \operatorname{tr} X = |\mathcal{K}_j| \chi_i(g)$, as needed to establish (1).

Now let g be a fixed element of $\mathcal{K}_s$ and define a_{ijs} to be the number of ordered pairs g_i, g_j with $g_i \in \mathcal{K}_i$, $g_j \in \mathcal{K}_j$ and $g_i g_j = g$. Notice that a_{ijs} is an integer. It is independent of the choice of g in $\mathcal{K}_s$ because if $x^{-1} g x$ is a conjugate of g, every ordered pair g_i, g_j whose product is g gives rise to an ordered pair $x^{-1} g_i x$, $x^{-1} g_j x$ whose product is $x^{-1} g x$ (and vice versa).

Next we prove that for all $i, j, t \in \{1, \dots, r\}$

$$\omega_t(\mathcal{K}_i) \omega_t(\mathcal{K}_j) = \sum_{s=1}^{r} a_{ijs} \omega_t(\mathcal{K}_s). \tag{19.2}$$

To see this note that by (1), the left hand side of (2) is the diagonal entry of the scalar matrix on the left of the following equation:

$$\left(\sum_{g \in \mathcal{K}_i} \varphi_t(g) \right) \left(\sum_{g \in \mathcal{K}_j} \varphi_t(g) \right) = \sum_{g_i \in \mathcal{K}_i} \sum_{g_j \in \mathcal{K}_j} \varphi_t(g_i g_j)$$

$$= \sum_{s=1}^{r} \sum_{g \in \mathcal{K}_s} a_{ijs} \varphi_t(g)$$

$$= \sum_{s=1}^{r} a_{ijs} \sum_{g \in \mathcal{K}_s} \varphi_t(g) \qquad \begin{array}{l} \text{(since } a_{ijs} \text{ is independent} \\ \text{of } g \in \mathcal{K}_s) \end{array}$$

$$= \sum_{s=1}^{r} a_{ijs} \omega_t(\mathcal{K}_s) I \qquad (\text{by (1)}).$$

Comparing entries of these scalar matrices gives (2).

Now (2) implies that the subring of $\mathbb{C}$ generated by $\mathbb{Z}$ and $\omega_t(\mathcal{K}_1), \dots, \omega_t(\mathcal{K}_r)$ is a finitely generated $\mathbb{Z}$-module for each $t \in \{1, \dots, r\}$ (it is generated as a $\mathbb{Z}$-module by $1, \omega_t(\mathcal{K}_1), \dots, \omega_t(\mathcal{K}_r)$). Since $\mathbb{Z}$ is a Principal Ideal Domain the submodule $\mathbb{Z}[\omega_t(\mathcal{K}_i)]$ is also a finitely generated $\mathbb{Z}$-module, hence $\omega_t(\mathcal{K}_i)$ is an algebraic integer by Proposition 2. This completes the proof.

Corollary 5. The degree of each complex irreducible representation of a finite group G divides the order of G, i.e., $\chi_i(1) \mid |G|$ for $i = 1, 2, \dots, r$.

Proof: Under the notation of Proposition 4 and with $g_j \in \mathcal{K}_j$ we have

$$\frac{|G|}{\chi_i(1)} = \frac{|G|}{\chi_i(1)}(\chi_i, \chi_i)$$

$$= \sum_{j=1}^{r} \frac{|\mathcal{K}_j|\chi_i(g_j)\overline{\chi_i(g_j)}}{\chi_i(1)}$$

$$= \sum_{j=1}^{r} \omega_i(\mathcal{K}_j)\overline{\chi_i(g_j)}.$$

The right hand side is an algebraic integer and the left hand side is rational, hence is an integer. This proves the corollary.

The next two lemmas lead directly to Burnside's Theorem.

Lemma 6. If G is any group that has a conjugacy class $\mathcal{K}$ and an irreducible matrix representation φ with character χ such that $(|\mathcal{K}|, \chi(1)) = 1$, then for $g \in \mathcal{K}$ either $\chi(g) = 0$ or $\varphi(g)$ is a scalar matrix.

Proof: By hypothesis there exist $s, t \in \mathbb{Z}$ such that $s|\mathcal{K}| + t\chi(1) = 1$. Thus

$$s|\mathcal{K}|\chi(g) + t\chi(1)\chi(g) = \chi(g).$$

Divide both sides of this by $\chi(1)$ and note that by Corollary 3 and Proposition 4 both $\chi(g)$ and $\dfrac{|\mathcal{K}|\chi(g)}{\chi(1)}$ are algebraic integers, hence so is $\dfrac{\chi(g)}{\chi(1)}$. Let $a_1 = \dfrac{\chi(g)}{\chi(1)}$ and let $a_1, a_2, \ldots, a_n$ be all its algebraic conjugates over $\mathbb{Q}$ (i.e., the roots of the minimal polynomial of a_1 over $\mathbb{Q}$). Since a_1 is a sum of $\chi(1)$ roots of 1 divided by the integer $\chi(1)$, each a_i is also a sum of $\chi(1)$ roots of 1 divided by $\chi(1)$. Thus a_i has complex absolute value ≤ 1 for all i. Now $b = \prod_{i=1}^{n} a_i \in \mathbb{Q}$ and b is an algebraic integer ($\pm b$ is the constant term of the irreducible polynomial of a_1), hence $b \in \mathbb{Z}$. But

$$|b| = \prod_{i=1}^{n} |a_i| \leq 1,$$

so $b = 0, \pm 1$. Since all a_i's are conjugate, $b = 0 \Leftrightarrow a_1 = 0 \Leftrightarrow \chi(g) = 0$. Also, $b = \pm 1 \Leftrightarrow |a_i| = 1$ for all i. Thus either $\chi(g) = 0$ or $|\chi(g)| = \chi(1)$. In the former situation the lemma is established, so assume $|\chi(g)| = \chi(1)$.

Let φ_1 be a matrix representation equivalent to φ in which $\varphi_1(g)$ is a diagonal matrix:

$$\varphi_1(g) = \begin{pmatrix} \epsilon_1 & & & \\ & \epsilon_2 & & \\ & & \ddots & \\ & & & \epsilon_n \end{pmatrix}.$$

Thus $\chi(g) = \epsilon_1 + \cdots + \epsilon_n$. By the triangle inequality if $\epsilon_i \neq \epsilon_j$ for any i, j, then $|\epsilon_1 + \cdots + \epsilon_n| < n = \chi(1)$. Since this is not the case we must have $\varphi_1(g) = \epsilon I$ (where $\epsilon = \epsilon_i$ for all i). Since scalar matrices are similar only to themselves, $\varphi(g) = \epsilon I$ as well. This completes the proof.

Lemma 7. If $|\mathcal{K}|$ is a power of a prime for some nonidentity conjugacy class $\mathcal{K}$ of G, then G is not a non-abelian simple group.

Proof: Suppose to the contrary that G is a non-abelian simple group and let $|\mathcal{K}| = p^c$. Let $g \in \mathcal{K}$. If $c = 0$ then $g \in Z(G)$, contrary to a non-abelian simple group having a trivial center. As above, let $\chi_1, \ldots, \chi_r$ be all the irreducible characters of G with χ_1 the principal character and let ρ be the regular character of G. By decomposing ρ into irreducibles we obtain

$$0 = \rho(g) = 1 + \sum_{i=2}^{r} \chi_i(1)\chi_i(g). \tag{19.3}$$

If $p \mid \chi_j(1)$ for every $j > 1$ with $\chi_j(g) \neq 0$, then write $\chi_j(1) = pd_j$. In this case (3) becomes

$$0 = 1 + p \sum_{j} d_j \chi_j(g).$$

Thus $\sum_j d_j \chi_j(g) = -1/p$ is an algebraic integer, a contradiction. This proves there is some j such that p does not divide $\chi_j(1)$ and $\chi_j(g) \neq 0$. If φ is a representation whose character is χ_j, then φ is faithful (because G is assumed to be simple) and, by Lemma 6, $\varphi(g)$ is a scalar matrix. Since $\varphi(g)$ commutes with all matrices, $\varphi(g) \in Z(\varphi(G))$. This forces $g \in Z(G)$, contrary to G being a non-abelian simple group. The proof of the lemma is complete.

We now prove Burnside's Theorem. Let G be a group of order $p^a q^b$ for some primes p and q. If $p = q$ or if either exponent is 0 then G is nilpotent hence solvable. Thus we may assume this is not the case. Proceeding by induction let G be a counterexample of minimal order. If G has a proper, nontrivial normal subgroup N, then by induction both N and G/N are solvable, hence so is G (cf. Section 3.4 or Proposition 6.10). Thus we may assume G is a non-abelian simple group. Let $P \in Syl_p(G)$. By Theorem 8 of Chapter 4 there exists $g \in Z(P)$ with $g \neq 1$. Since $P \leq C_G(g)$, the order of the conjugacy class of g (which equals $|G : C_G(g)|$) is prime to p, i.e., is a power of q. This violates Lemma 7 and so completes the proof of Burnside's Theorem.

Philip Hall's Theorem

Recall that a subgroup of a finite group is called a *Hall subgroup* if its order and index are relatively prime. For any subgroup H of a group G a subgroup K such that $G = HK$ and $H \cap K = 1$ is called a *complement* to H in G.

Theorem 8. (P. Hall) Let G be a group of order $p_1^{\alpha_1} p_2^{\alpha_2} \cdots p_t^{\alpha_t}$ where $p_1, \ldots, p_t$ are distinct primes. If for each $i \in \{1, \ldots, t\}$ there exists a subgroup H_i of G with $|G : H_i| = p_i^{\alpha_i}$, then G is solvable.

Hall's Theorem can also be phrased: if for each $i \in \{1, \ldots, t\}$ a Sylow p_i-subgroup of G has a complement, then G is solvable. The converse to Hall's Theorem is also true — this was Exercise 33 in Section 6.1.

We shall first need some elementary lemmas.

Lemma 9. If G is solvable of order > 1, then there exists $P \trianglelefteq G$ with P a nontrivial p-group for some prime p.

Proof: This is a special case of the exercise on minimal normal subgroups of solvable groups at the end of Section 6.1. One can see this easily by letting P be a nontrivial Sylow subgroup of the last nontrivial term, $G^{(n-1)}$, in the derived series of G (where G has solvable length n). In this case $G^{(n-1)}$ is abelian so P is a characteristic subgroup of $G^{(n-1)}$, hence is normal in G.

Lemma 10. Let G be a group of order $p_1^{\alpha_1} p_2^{\alpha_2} \cdots p_t^{\alpha_t}$ where $p_1, \ldots, p_t$ are distinct primes. Suppose there are subgroups H and K of G such that for each $i \in \{1, \ldots, t\}$, either $p_i^{\alpha_i}$ divides $|H|$ or $p_i^{\alpha_i}$ divides $|K|$. Then $G = HK$ and $|H \cap K| = (|H|, |K|)$.

Proof: Fix some $i \in \{1, \ldots, t\}$ and suppose first that $p_i^{\alpha_i}$ divides the order of H. Since HK is a disjoint union of right cosets of H and each of these right cosets has order equal to $|H|$, it follows that $p_i^{\alpha_i}$ divides $|HK|$. Similarly, if $p_i^{\alpha_i}$ divides $|K|$, since HK is a disjoint union of left cosets of K, again $p_i^{\alpha_i}$ divides $|HK|$. Thus $|G| \mid |HK|$ and so $G = HK$. Since

$$|HK| = \frac{|H||K|}{|H \cap K|},$$

it follows that $|H \cap K| = (|H|, |K|)$.

We now begin the proof of Hall's Theorem, proceeding by induction on $|G|$. Note that if $t = 1$ the hypotheses are trivially satisfied for any group ($H_1 = 1$) and if $t = 2$ the hypotheses are again satisfied for any group by Sylow's Theorem (H_1 is a Sylow p_2-subgroup of G and H_2 is a Sylow p_1-subgroup of G). If $t = 1$, G is nilpotent, hence solvable and if $t = 2$, G is solvable by Burnside's Theorem. Assume therefore that $t \geq 3$.

Fix i and note that by the preceding lemma, for all $j \in \{1, \ldots, t\} - \{i\}$,

$$|H_i : H_i \cap H_j| = p_j^{\alpha_j}.$$

Thus every Sylow p_j-subgroup of H_i has a complement in H_i: $H_j \cap H_i$. By induction H_i is solvable.

By Lemma 9 we may choose $P \trianglelefteq H_1$ with $|P| = p_i^a > 1$ for some $i > 1$. Since $t \geq 3$ there exists an index $j \in \{1, \ldots, t\} - \{1, i\}$. By Lemma 10

$$|H_1 \cap H_j| = p_2^{\alpha_2} \cdots p_{j-1}^{\alpha_{j-1}} p_{j+1}^{\alpha_{j+1}} \cdots p_t^{\alpha_t}.$$

Thus $H_1 \cap H_j$ contains a Sylow p_i-subgroup of H_1. Since P is a normal p_i-subgroup of H_1, P is contained in every Sylow p_i-subgroup of H_1 and so $P \leq H_1 \cap H_j$. By Lemma 10, $G = H_1 H_j$ so each $g \in G$ may be written $g = h_1 h_j$ for some $h_1 \in H_1$ and $h_j \in H_j$. Then

$$g H_j g^{-1} = (h_1 h_j) H_j (h_1 h_j)^{-1} = h_1 H_j h_1^{-1}$$

and so

$$\bigcap_{g \in G} g H_j g^{-1} = \bigcap_{h_1 \in H_1} h_1 H_j h_1^{-1}.$$

Now $P \leq H_j$ and $h_1 P h_1^{-1} = P$ for all $h_1 \in H_1$. Thus

$$1 \neq P \leq \bigcap_{h_1 \in H_1} h_1 H_j h_1^{-1}.$$

Thus $N = \cap_{g \in G} g H_j g^{-1}$ is a nontrivial, proper normal subgroup of G. It follows that both N and G/N satisfy the hypotheses of the theorem (cf. the exercises in Section 3.3). Both N and G/N are solvable by induction, so G is solvable. This completes the proof of Hall's Theorem.

EXERCISES

1. Show that every character of the symmetric group S_n is integer valued, for all n (i.e., $\psi(g) \in \mathbb{Z}$ for all $g \in S_n$ and all characters ψ of S_n). [See Exercise 22 in Section 18.3.]

2. Let G be a finite group with the property that every maximal subgroup has either prime or prime squared index. Prove that G is solvable. (The simple group $GL_3(\mathbb{F}_2)$ has the property that every maximal subgroup has index either 7 or 8, i.e., either prime or prime cubed index — cf. Section 6.2.). [Let p be the largest prime dividing $|G|$ and let P be a Sylow p-subgroup of G. If $P \trianglelefteq G$, apply induction to G/P. Otherwise let M be a maximal subgroup containing $N_G(P)$. Use Exercise 51 in Section 4.5 to show that $p = 3$ and deduce that $|G| = 2^a 3^b$.]

3. Assume G is a finite group that possesses an abelian subgroup H whose index is a power of a prime. Prove that G is solvable.

4. Repeat the preceding exercise with the word "abelian" replaced by "nilpotent."

5. Use the ideas in the proof of Philip Hall's Theorem to prove Burnside's $p^a q^b$ Theorem in the special case when all Sylow subgroups are abelian (without use of character theory.)

19.3 INTRODUCTION TO THE THEORY OF INDUCED CHARACTERS

Let G be a finite group, let H be a subgroup of G and let φ be a representation of the subgroup H over an arbitrary field F. In this section we show how to obtain a representation of G, called the induced representation, from the representation φ of its subgroup. We also determine a formula for the character of this induced representation, the induced character, in terms of the character of φ and we illustrate this formula by computing some induced characters in specific groups. Finally, we apply the theory of induced characters to prove that there are no simple groups of order $3^3 \cdot 7 \cdot 13 \cdot 409$, a group order which was discussed at the end of Section 6.2 in the context of the existence problem for simple groups. The theory of induced representations and induced characters marks the beginning of more advanced representation theory. This section is intended as an introduction rather than as a comprehensive treatment, and the results we have included were chosen to serve this purpose.

First observe that it may not be possible to extend a representation φ of the subgroup H to a representation Φ of G in such a way that $\Phi|_H = \varphi$. For example, $A_3 \leq S_3$ and A_3 has a faithful representation of degree 1 (cf. Section 1). Since every degree 1 representation of S_3 contains $A_3 = S_3'$ in its kernel, this representation of A_3 cannot be extended to a representation of S_3. For another example of a representation of a subgroup which cannot be extended to the whole group take G to be any simple group and let φ be any representation of H with the property that $\ker \varphi$ is a proper, nontrivial normal subgroup of H. If φ extended to a representation Φ of G then the kernel of Φ would be a proper, nontrivial normal subgroup of G, contrary to G being a simple group. We shall see that the method of induced characters produces a representation Φ of G from a given representation φ of its subgroup H but that $\Phi|_H \neq \varphi$ in general (indeed, unless $H = G$ the degree of Φ will be greater than the degree of φ).

We saw in Example 5 following Corollary 9 in Section 10.4 that because FH is a subring of FG, the ring FG is an (FG, FH)-bimodule; and so for any left FH-module V, the abelian group $FG \otimes_{FH} V$ is a left FG-module (called the extension of scalars from FH to FG for V). In the representation theory of finite groups this extension is given a special name.

Definition. Let H be a subgroup of the finite group G and let V be an FH-module affording the representation φ of H. The FG-module $FG \otimes_{FH} V$ is called the *induced module* of V and the representation of G it affords is called the *induced representation* of φ. If ψ is the character of φ then the character of the induced representation is called the *induced character* and is denoted by $\mathrm{Ind}_H^G(\psi)$.

Theorem 11. Let H be a subgroup of the finite group G and let $g_1, \ldots, g_m$ be representatives for the distinct left cosets of H in G. Let V be an FH-module affording the matrix representation φ of H of degree n. The FG-module $W = FG \otimes_{FH} V$ has dimension nm over F and there is a basis of W such that W affords the matrix representation Φ defined for each $g \in G$ by

$$\Phi(g) = \begin{pmatrix} \varphi(g_1^{-1}gg_1) & \cdots & \varphi(g_1^{-1}gg_m) \\ \vdots & \vdots & \vdots \\ \varphi(g_m^{-1}gg_1) & \cdots & \varphi(g_m^{-1}gg_m) \end{pmatrix}$$

where each $\varphi(g_i^{-1}gg_j)$ is an $n \times n$ block appearing in the i, j block position of $\Phi(g)$, and where $\varphi(g_i^{-1}gg_j)$ is defined to be the zero block whenever $g_i^{-1}gg_j \notin H$.

Proof: First note that FG is a free right FH-module:

$$FG = g_1 FH \oplus g_2 FH \oplus \cdots \oplus g_m FH.$$

Since tensor products commute with direct sums (Theorem 17, Section 10.4), as abelian groups we have

$$W = FG \otimes_{FH} V \cong (g_1 \otimes V) \oplus (g_2 \otimes V) \oplus \cdots \oplus (g_m \otimes V).$$

Since F is in the center of FG it follows that this is an F-vector space isomorphism as well. Thus if $v_1, v_2, \ldots, v_n$ is a basis of V affording the matrix representation φ, then $\{g_i \otimes v_j \mid 1 \leq i \leq m, \ 1 \leq j \leq n\}$ is a basis of W. This shows the dimension of W is mn. Order the basis into m sets, each of size n as

$$g_1 \otimes v_1, g_1 \otimes v_2, \ldots, g_1 \otimes v_n, g_2 \otimes v_1, \ldots, g_2 \otimes v_n, \ldots \ldots, g_m \otimes v_n.$$

We compute the matrix representation $\Phi(g)$ of each g acting on W with respect to this basis. Fix j and g, and let $gg_j = g_ih$ for some index i and some $h \in H$. Then for every k

$$g(g_j \otimes v_k) = (gg_j) \otimes v_k = g_i \otimes hv_k$$

$$= \sum_{t=1}^{n} a_{tk}(h)(g_i \otimes v_t)$$

where a_{tk} is the t, k coefficient of the matrix of h acting on V with respect to the basis $\{v_1, \ldots, v_n\}$. In other words, the action of g on W maps the jth block of n basis vectors of W to the ith block of basis vectors, and then has the matrix $\varphi(h)$ on that block. Since $h = g_i^{-1}gg_j$, this describes the block matrix $\Phi(g)$ of the theorem, as needed.

Corollary 12. In the notation of Theorem 11
    (1) if ψ is the character afforded by V then the induced character is given by

$$\text{Ind}_H^G(\psi)(g) = \sum_{i=1}^{m} \psi(g_i^{-1}gg_i)$$

    where $\psi(g_i^{-1}gg_i)$ is defined to be 0 if $g_i^{-1}gg_i \notin H$, and
    (2) $\text{Ind}_H^G(\psi)(g) = 0$ if g is not conjugate in G to some element of H. In particular, if H is a normal subgroup of G then $\text{Ind}_H^G(\psi)$ is zero on all elements of $G - H$.

Remark: Since the character ψ of H is constant on the conjugacy classes of H we have $\psi(g) = \psi(h^{-1}gh)$ for all $h \in H$. As h runs over all elements of H, xh runs over all elements of the coset xH. Thus the formula for the induced character may also be written

$$\text{Ind}_H^G(\psi)(g) = \frac{1}{|H|} \sum_{x \in G} \psi(x^{-1}gx)$$

where the elements x in each fixed coset give the same character value $|H|$ times (which accounts for the factor of $1/|H|$), and again $\psi(x^{-1}gx) = 0$ if $x^{-1}gx \notin H$.

    Proof: From the matrix of g computed above, the blocks $\varphi(g_i^{-1}gg_i)$ down the diagonal of $\Phi(g)$ are zero except when $g_i^{-1}gg_i \in H$. Thus the trace of the block matrix $\Phi(g)$ is the sum of the traces of the matrices $\varphi(g_i^{-1}gg_i)$ for which $g_i^{-1}gg_i \in H$. Since the trace of $\varphi(g_i^{-1}gg_i)$ is $\psi(g_i^{-1}gg_i)$, part (1) holds.

    If $g_i^{-1}gg_i \notin H$ for all coset representatives g_i then each term in the sum for $\text{Ind}_H^G(\psi)(g)$ is zero. In particular, if g is not in the normal subgroup H then neither is any conjugate of g, so $\text{Ind}_H^G(\psi)$ is zero on g.

Examples

    (1) Let $G = D_{12} = \langle r, s \mid r^6 = s^2 = 1, \ rs = sr^{-1} \rangle$ be the dihedral group of order 12 and let $H = \{1, s, r^3, sr^3\}$, so that H is isomorphic to the Klein 4-group and $|G : H| = 3$. Following the notation of Theorem 11 we exhibit the matrices for r and s of the induced representation of a specific representation φ of H. Let the representation of H on a 2-dimensional vector space over $\mathbb{Q}$ with respect to some basis v_1, v_2 be given by

$$\varphi(s) = \begin{pmatrix} -1 & 0 \\ 0 & 1 \end{pmatrix} = A, \quad \varphi(r^3) = \begin{pmatrix} 1 & 0 \\ 0 & -1 \end{pmatrix} = B, \quad \varphi(sr^3) = \begin{pmatrix} -1 & 0 \\ 0 & -1 \end{pmatrix} = C,$$

so $n = 2$, $m = 3$ and the induced representation Φ has degree $nm = 6$. Fix representatives $g_1 = 1$, $g_2 = r$, and $g_3 = r^2$ for the left cosets of H in G, so that $g_k = r^{k-1}$. Then

$$g_i^{-1} r g_j = r^{-(i-1)+1+(j-1)} = r^{j-i+1}, \text{ and}$$
$$g_i^{-1} s g_j = s r^{(i-1)+(j-1)} = s r^{i+j-2}.$$

Thus the 6×6 matrices for the induced representation are seen to be

$$\Phi(r) = \begin{pmatrix} 0 & 0 & B \\ I & 0 & 0 \\ 0 & I & 0 \end{pmatrix} \qquad \Phi(s) = \begin{pmatrix} A & 0 & 0 \\ 0 & 0 & C \\ 0 & C & 0 \end{pmatrix}$$

where the 2×2 matrices A, B and C are given above, I is the 2×2 identity matrix and 0 denotes the 2×2 zero matrix.

(2) If H is any subgroup of G and ψ_1 is the principal character of H, then $\mathrm{Ind}_H^G(\psi_1)(g)$ counts 1 for each coset representative g_i such that $g_i^{-1} g g_i \in H$. Since $g_i^{-1} g g_i \in H$ if and only if g fixes the left coset $g_i H$ under left multiplication, $\mathrm{Ind}_H^G(\psi_1)(g)$ is the number of points fixed by g in the permutation representation of g on the left cosets of H. Thus by Example 3 of Section 18.3 we see that: *if ψ_1 is the principal character of H then $\mathrm{Ind}_H^G(\psi_1)$ is the permutation character on the left cosets of H in G*. In the special case when $H = 1$, this implies *if χ_1 is the principal character of the trivial subgroup $H = 1$ then $\mathrm{Ind}_1^G(\chi_1)$ is the regular character of G*. This also shows that an induced character is not, in general, irreducible even if the character from which it is induced is irreducible.

(3) Let $G = S_3$ and let ψ be a nonprincipal linear character of $A_3 = \langle x \rangle$, so that $\psi(x) = \zeta$, for some primitive cube root of unity ζ (the character tables of $A_3 = Z_3$ and S_3 appear in Section 1). Let $\Psi = \mathrm{Ind}_{A_3}^{S_3}(\psi)$. Thus Ψ has degree $1 \cdot |S_3 : A_3| = 2$ and, by the corollary, Ψ is zero on all transpositions. If y is any transposition then 1, y is a set of left coset representatives of A_3 in S_3 and $y^{-1} x y = x^2$. Thus $\Psi(x) = \psi(x) + \psi(x^2)$ equals $\zeta + \zeta^2 = -1$. This shows that if ψ is either of the two nonprincipal irreducible characters of A_3 then the induced character of ψ is the (unique) irreducible character of S_3 of degree 2. In particular, different characters of a subgroup may induce the same character of the whole group.

(4) Let $G = D_8$ have its usual generators and relations and let $H = \langle s \rangle$. Let ψ be the nonprincipal irreducible character of H and let $\Psi = \mathrm{Ind}_H^G(\psi)$. Pick left coset representatives $1, r, r^2, r^3$ for H. By Theorem 11, $\Psi(1) = 4$. Since $\psi(s) = -1$, one computes directly that $\Psi(s) = -2$. By Corollary 12(2) we obtain $\Psi(r) = \Psi(r^2) = \Psi(sr) = 0$. In the notation of the character table of D_8 in Section 1, by the orthogonality relations we obtain $\Psi = \chi_2 + \chi_4 + \chi_5$ (which may be checked by inspection).

For the remainder of this section the field F is taken to be the complex numbers: $F = \mathbb{C}$.

Before concluding with an application of induced characters to simple groups we compute the characters of an important class of groups.

Definition. A finite group G is called a *Frobenius group* with *Frobenius kernel* Q if Q is a proper, nontrivial normal subgroup of G and $C_G(x) \le Q$ for all nonidentity elements x of Q.

In view of the application to simple groups mentioned at the beginning of this section we shall restrict attention to Frobenius groups G of order $q^a p$, where p and q are distinct primes, such that the Frobenius kernel Q is an elementary abelian q-group of order q^a and the cyclic group G/Q acts irreducibly by conjugation on Q. In other words, we shall assume Q is a direct product of cyclic groups of order q and the only normal subgroups of G that are contained in Q are 1 and Q, i.e., Q is a minimal normal subgroup of G. For example, A_4 is a Frobenius group of this type with Frobenius kernel V_4, its Sylow 2-subgroup. Also, if p and q are distinct primes with $p < q$ and G is a non-abelian group of order pq (one always exists if $p \mid q - 1$) then G is a Frobenius group whose Frobenius kernel is its Sylow q-subgroup (which is normal by Sylow's Theorem). We essentially determine the character table of these Frobenius groups. Analogous results on more general Frobenius groups appear in the exercises.

Proposition 13. Let G be a Frobenius group of order $q^a p$, where p and q are distinct primes, such that the Frobenius kernel Q is an elementary abelian q-group of order q^a and the cyclic group G/Q acts irreducibly by conjugation on Q. Then the following hold:

   (1) $G = QP$ where P is a Sylow p-subgroup of G. Every nonidentity element of G has order p or q. Every element of order p is conjugate to an element of P and every element of order q belongs to Q. The nonidentity elements of P represent the $p - 1$ distinct conjugacy classes of elements of order p and each of these classes has size q^a. There are $(q^a - 1)/p$ distinct conjugacy classes of elements of order q and each of these classes has size p.

   (2) $G' = Q$ so the number of degree 1 characters of G is p and every degree 1 character contains Q in its kernel.

   (3) If ψ is any nonprincipal irreducible character of Q, then $\mathrm{Ind}_Q^G(\psi)$ is an irreducible character of G. Moreover, every irreducible character of G of degree > 1 is equal to $\mathrm{Ind}_Q^G(\psi)$ for some nonprincipal irreducible character ψ of Q. Every irreducible character of G has degree either 1 or p and the number of irreducible characters of degree p is $(q^a - 1)/p$.

Proof: Note that QP equals G by order consideration. By definition of a Frobenius group and because Q is abelian, $C_G(h) = Q$ for every nonidentity element h of Q. If x were an element of order pq, then x^p would be an element of order q, hence would lie in the unique Sylow q-subgroup Q of G. But then x would commute with x^p and so x would belong to $C_G(x^p) = Q$, a contradiction. Thus G has no elements of order pq. By Sylow's Theorem every element of order p is conjugate to an element of P and every element of order q lies in Q. No two distinct elements of P are conjugate in G because if $g^{-1}xg = y$ for some $x, y \in P$ then $\overline{g^{-1}xg} = \overline{y}$ in the abelian group $\overline{G} = G/Q$ and so $\overline{x} = \overline{y}$. Then $x = y$ because $\overline{P} \cong P$. Thus there are exactly $p - 1$ conjugacy classes of elements of order p and these are represented by the nonidentity elements of P. If x is a nonidentity element of P, then $C_G(x) = P$ and so the conjugacy class of

x consists of $|G : P| = q^a$ elements. Finally, if h is a nonidentity element of Q, then $C_G(h) = Q$ and the conjugacy class of h is $\{h, h^x, \dots, h^{x^{p-1}}\}$, where $P = \langle x \rangle$. This proves all parts of (1).

Since G/Q is abelian, $G' \le Q$. Since G is non-abelian and Q is, by hypothesis, a minimal normal subgroup of G we must have $G' = Q$. Part (2) now follows from Corollary 11 in Section 18.2.

Let ψ be a nonprincipal irreducible character of Q and let $\Psi = \mathrm{Ind}_Q^G(\psi)$. We use the orthogonality relations to show that Ψ is irreducible. Let $1, x, \dots, x^{p-1}$ be coset representatives for Q in G. By Corollary 12, Ψ is zero on $G - Q$ so

$$|| \Psi ||^2 = \frac{1}{|G|} \sum_{h \in Q} \Psi(h)\overline{\Psi(h)}$$

$$= \frac{1}{|G|} \sum_{h \in Q} \sum_{i=0}^{p-1} \psi(x^i h x^{-i})\overline{\psi(x^i h x^{-i})}$$

$$= \frac{p}{|G|} \sum_{h \in Q} \psi(h)\overline{\psi(h)}$$

$$= \frac{p|Q|}{|G|} = 1,$$

where the second line follows from the definition of the induced character Ψ, the third line follows because each element of Q appears exactly p times in the sum in the second line, and the last line follows from the first orthogonality relation in Q because ψ is an irreducible character of Q. This proves Ψ is an irreducible character of G.

We prove that every irreducible character of G of degree > 1 is the induced character of some nonprincipal degree 1 character of Q by counting the number of distinct irreducible characters of G obtained this way. By parts (1) and (2) the number of irreducible characters of G (= the number of conjugacy classes) is $p + (q^a - 1)/p$ and the number of degree 1 characters is p. Thus the number of irreducible characters of G of degree > 1 is $(q^a - 1)/p$. The group P acts on the set $\mathcal{C}$ of nonprincipal irreducible characters of Q as follows: for each $\psi \in \mathcal{C}$ and each $x \in P$ let ψ^x be defined by

$$\psi^x(h) = \psi(xhx^{-1}) \qquad \text{for all } h \in Q.$$

Since ψ is a nontrivial homomorphism from Q into $\mathbb{C}^\times$ (recall that all irreducible characters of the abelian group Q have degree 1) it follows easily that ψ^x is also a homomorphism. Thus $\psi^x \in \mathcal{C}$ and so P permutes the elements of $\mathcal{C}$. Now let x be a generator for the cyclic group P. Then $1, x, \dots, x^{p-1}$ are representatives for the left cosets of Q in G. By Corollary 12 applied with this set of coset representatives we see that if $\psi \in \mathcal{C}$ then the value of $\mathrm{Ind}_Q^G(\psi)$ on any element h of Q is given by the sum $\psi(h) + \psi^x(h) + \cdots + \psi^{x^{p-1}}(h)$. Thus when the induced character $\mathrm{Ind}_Q^G(\psi)$ is restricted to Q it decomposes into irreducible characters of Q as

$$\mathrm{Ind}_Q^G(\psi)|_Q = \psi + \psi^x + \cdots + \psi^{x^{p-1}}.$$

If ψ_1 and ψ_2 are in different orbits of the action of P on $\mathcal{C}$ then the induced characters $\mathrm{Ind}_Q^G(\psi_1)$ and $\mathrm{Ind}_Q^G(\psi_2)$ restrict to distinct characters of Q (they have no irreducible constituents in common). Thus characters induced from elements of distinct orbits of P on C are distinct irreducible characters of G. The abelian group Q has $q^a - 1$ nonprincipal irreducible characters (i.e., $|C| = q^a - 1$) and $|P| = p$ so there are at least $(q^a - 1)/p$ orbits of P on C and hence at least this number of distinct irreducible characters of G of degree p. Since G has exactly $(q^a - 1)/p$ irreducible characters of degree > 1, every irreducible character of G of degree > 1 must have degree p and must be an induced character from some element of C. The proof is complete.

For the final example we shall require two properties of induced characters. These properties are listed in the next proposition and the proofs are straightforward exercises which follow easily from the formula for induced characters or from the definition of induced modules together with properties of tensor products.

Proposition 14. Let G be a group, let H be a subgroup of G and let ψ and ψ' be characters of H.
  (1) *(Induction of characters is additive)* $\operatorname{Ind}_H^G(\psi + \psi') = \operatorname{Ind}_H^G(\psi) + \operatorname{Ind}_H^G(\psi')$.
  (2) *(Induction of characters is transitive)* If $H \leq K \leq G$ then
$$\operatorname{Ind}_K^G(\operatorname{Ind}_H^K(\psi)) = \operatorname{Ind}_H^G(\psi).$$

It follows from part (1) of Proposition 14 that if $\sum_{i=1}^s n_i \psi_i$ is any integral linear combination of characters of H with $n_i \geq 0$ for all i then
$$\operatorname{Ind}_H^G\left(\sum_{i=1}^s n_i \psi_i\right) = \sum_{i=1}^s n_i \operatorname{Ind}_H^G(\psi_i). \tag{$*$}$$

A class function of H of the form $\sum_{i=1}^s n_i \psi_i$, where the coefficients are any integers (not necessarily nonnegative) is called a *generalized character* or *virtual character* of H. For a generalized character of H we define its induced generalized character of G by equation $(*)$, allowing now negative coefficients n_i as well. In this way the function Ind_H^G becomes a group homomorphism from the additive group of generalized characters of H to the additive group of generalized characters of G (which maps characters to characters). This implies that the formula for induced characters in Corollary 12 holds also if ψ is a generalized character of H.

Application to Groups of Order $3^3 \cdot 7 \cdot 13 \cdot 409$

We now conclude with a proof of the following result:

there are no simple groups of order $3^3 \cdot 7 \cdot 13 \cdot 409$.

As mentioned at the beginning of this section, simple groups of this order were discussed at the end of Section 6.2 in the context of the existence problem for simple groups. It is possible to prove that there are no simple groups of this order by arguments involving a permutation representation of degree 819 (cf. the exercises in Section 6.2). We include a character-theoretic proof of this since the methods illustrate some important ideas in the theory of finite groups. The approach is based on M. Suzuki's seminal paper *The nonexistence of a certain type of simple group of odd order*, Proc. Amer. Math. Soc.,

8(1957), pp. 686–695, which treats much more general groups. Because we are dealing with a specific group order, our arguments are simpler and numerically more explicit, yet they retain some of the key ideas of Suzuki's work. Moreover, Suzuki's paper and its successor, *Finite groups in which the centralizer of any non-identity element is nilpotent*, by W. Feit, M. Hall and J. Thompson, Math. Zeit., 74(1960), pp. 1–17, are prototypes for the lengthy and difficult Feit–Thompson Theorem (cf. Section 3.4). Our discussion also conveys some of the flavor of these fundamental papers. In particular, each of these papers follows the basic development in which the structure and embedding of the Sylow subgroups is first determined and then character theory (with heavy reliance on induced characters) is applied.

For the remainder of this section we assume G is a simple group of order $3^3 \cdot 7 \cdot 13 \cdot 409$. We list some properties of G which may be verified using the methods stemming from Sylow's Theorem discussed in Section 6.2. The details are left as exercises.

(1) Let $q_1 = 3$, let Q_1 be a Sylow 3-subgroup of G and let $N_1 = N_G(Q_1)$. Then Q_1 is an elementary abelian 3-group of order 3^3 and N_1 is a Frobenius group of order $3^3 \cdot 13$ with Frobenius kernel Q_1 and with N_1/Q_1 acting irreducibly by conjugation on Q_1.

(2) Let $q_2 = 7$, let Q_2 be a Sylow 7-subgroup of G and let $N_2 = N_G(Q_2)$. Then Q_2 is cyclic of order 7 and N_2 is the non-abelian group of order $7 \cdot 3$ (so N_2 is a Frobenius group with Frobenius kernel Q_2).

(3) Let $q_3 = 13$, let Q_3 be a Sylow 13-subgroup of G and let $N_3 = N_G(Q_3)$. Then Q_3 is cyclic of order 13 and N_3 is the non-abelian group of order $13 \cdot 3$ (so N_3 is a Frobenius group with Frobenius kernel Q_3).

(4) Let $q_4 = 409$, let Q_4 be a Sylow 409-subgroup of G and let $N_4 = N_G(Q_4)$. Then Q_4 is cyclic of order 409 and N_4 is the non-abelian group of order $409 \cdot 3$ (so N_4 is a Frobenius group with Frobenius kernel Q_4).

(5) Every nonidentity element of G has prime order and $Q_i \cap Q_i^g = 1$ for every $g \in G - N_i$, for each $i = 1, 2, 3, 4$. The nonidentity conjugacy classes of G are:
   (a) 2 classes of elements of order 3 (each of these classes has size $7 \cdot 13 \cdot 409$)
   (b) 2 classes of elements of order 7 (each of these classes has size $3^3 \cdot 13 \cdot 409$)
   (c) 4 classes of elements of order 13 (each of these classes has size $3^3 \cdot 7 \cdot 409$)
   (d) 136 classes of elements of order 409 (each of these classes has size $3^3 \cdot 7 \cdot 13$), and so there are 145 conjugacy classes in G.

Since each of the groups N_i is a Frobenius group satisfying the hypothesis of Proposition 13, the number of characters of N_i of degree > 1 may be read off from that proposition:
   (i) N_1 has 2 irreducible characters of degree 13
   (ii) N_2 has 2 irreducible characters of degree 3
   (iii) N_3 has 4 irreducible characters of degree 3
   (iv) N_4 has 136 irreducible characters of degree 3.

From now on, to simplify notation, for any subgroup H of G and any generalized character μ of H let

$$\mu^* = \mathrm{Ind}_H^G(\mu)$$

so a star will always denote induction from a subgroup to the whole group G and the subgroup will be clear from the context.

The following lemma is a key point in the proof. It shows how the vanishing of induced characters described in Corollary 12 (together with the *trivial intersection* property of the Sylow subgroups Q_i, namely the fact that $Q_i \cap Q_i^g = 1$ for all $g \in G - N_G(Q_i)$) may be used to relate inner products of certain generalized characters to the inner products of their induced generalized characters. For these computations it is important that the generalized characters are zero on the identity (which explains why we are considering *differences* of characters of the same degree).

Lemma 15. For any $i \in \{1, 2, 3, 4\}$ let $q = q_i$, let $Q = Q_i$, let $N = N_i$ and let $p = |N : Q|$. Let $\psi_1, \ldots, \psi_4$ be any irreducible characters of N of degree p (not necessarily distinct) and let $\alpha = \psi_1 - \psi_2$ and $\beta = \psi_3 - \psi_4$. Then α and β are generalized characters of N which are zero on every element of N of order not equal to q. Furthermore, α^* and β^* are generalized characters of G which are zero on every element of G of order not equal to q and

$$(\alpha^*, \beta^*)_G = (\alpha, \beta)_N$$

(where $(\ ,\ )_H$ denotes the usual Hermitian product of class functions computed in the group H). In other words, induction from N to G is an inner product preserving map on such generalized characters α, β of N.

Proof: By Proposition 13, there are nonprincipal characters $\lambda_1, \ldots, \lambda_4$ of Q of degree 1 such that $\psi_j = \text{Ind}_Q^N(\lambda_j)$ for $j = 1, \ldots, 4$. By Corollary 12 therefore, each ψ_j vanishes on $N - Q$, hence so do α and β. Note that since $\psi_j(1) = p$ for all j we have $\alpha(1) = \beta(1) = 0$. By the transitivity of induction, $\psi_j^* = \text{Ind}_N^G(\psi_j) = \text{Ind}_Q^G(\lambda_j)$ for all j. Again by Corollary 12 applied to the latter induced character we see that ψ_j^* vanishes on all elements not conjugate in G to some element of Q, hence so do both α^* and β^*. Since the induced characters ψ_j^* all have degree $|G : Q|$, the generalized characters α^* and β^* are zero on the identity. Thus α^* and β^* vanish on all elements of G which are not of order q. Finally, if $g_1, \ldots, g_m$ are representatives for the left cosets of N in G with $g_1 = 1$, then because $Q \cap Q^{g_k} = 1$ for all $k > 1$ (by (5) above), it follows immediately from the formula for induced (generalized) characters that $\alpha^*(x) = \alpha(x)$ and $\beta^*(x) = \beta(x)$ for all nonidentity elements $x \in Q$ (i.e., for all elements $x \in N$ of order q). Furthermore, by Sylow's Theorem every element of G of order q lies in a conjugate of Q, hence the collection of G-conjugates of the set $Q - \{1\}$ partition the elements of order q in G into $|G : N|$ disjoint subsets. Since α^* and β^* are class functions on G, the sum of $\alpha^*(x)\overline{\beta^*(x)}$ as x runs over any of these subsets is the same. These facts imply

$$(\alpha^*, \beta^*)_G = \frac{1}{|G|} \sum_{x \in G} \alpha^*(x)\overline{\beta^*(x)}$$

$$= \frac{1}{|G|} \sum_{\substack{x \in G \\ |x|=q}} \alpha^*(x)\overline{\beta^*(x)}$$

$$= \frac{1}{|G|} \sum_{\substack{x \in N \\ |x|=q}} |G : N|\alpha^*(x)\overline{\beta^*(x)}$$

$$= \frac{1}{|N|} \sum_{x \in N} \alpha(x) \overline{\beta(x)} = (\alpha, \beta)_N.$$

This completes the proof.

The next lemma sets up a correspondence between the irreducible characters of N_i of degree > 1 and some nonprincipal irreducible characters of G.

Lemma 16. For any $i \in \{1, 2, 3, 4\}$ let $q = q_i$, let $Q = Q_i$, let $N = N_i$ and let $p = |N : Q|$. Let $\psi_1, \ldots, \psi_k$ be the distinct irreducible characters of N of degree p. Then there are distinct irreducible characters $\chi_1, \ldots, \chi_k$ of G, all of which have the same degree, and a fixed sign $\epsilon = \pm 1$ such that $\psi_1^* - \psi_j^* = \epsilon(\chi_1 - \chi_j)$ for all $j = 2, 3, \ldots, k$.

Proof: Let $\alpha_j = \psi_1 - \psi_j$ for $j = 2, 3, \ldots, k$ so α_j satisfies the hypothesis of Lemma 15. Since $\psi_1 \neq \psi_j$, by Lemma 15

$$2 = || \alpha_j ||^2 = (\alpha_j, \alpha_j)_N = (\alpha_j^*, \alpha_j^*)_G = || \alpha_j^* ||^2$$

for all j. Thus α_j^* must have two distinct irreducible characters of G as its irreducible constituents. Since $\alpha_j^*(1) = 0$ it must be a difference of two distinct irreducible characters, both of which have the same degree. In particular, the lemma holds if $k = 2$ (which is the case for $q = 3$ and $q = 7$). Assume therefore that $k > 2$ and write

$$\alpha_2^* = \psi_1^* - \psi_2^* = \epsilon(\chi - \chi')$$
$$\alpha_3^* = \psi_1^* - \psi_3^* = \epsilon'(\theta - \theta')$$

for some irreducible characters $\chi, \chi', \theta, \theta'$ of G and some signs ϵ, ϵ'. As proved above, $\chi \neq \chi'$ and $\theta \neq \theta'$. Interchanging θ and θ' if necessary, we may assume $\epsilon = \epsilon'$. Thus

$$\alpha_3^* - \alpha_2^* = \psi_2^* - \psi_3^* = \epsilon(\theta - \theta' - \chi + \chi').$$

By Lemma 15, $\psi_2^* - \psi_3^* = (\psi_2 - \psi_3)^*$ also has exactly two distinct irreducible constituents, hence either $\theta = \chi$ or $\theta' = \chi'$. Replacing ϵ by $-\epsilon$ if necessary we may assume that $\theta = \chi$ so that now we have

$$\alpha_2^* = \psi_1^* - \psi_2^* = \epsilon(\chi - \chi')$$
$$\alpha_3^* = \psi_1^* - \psi_3^* = \epsilon(\chi - \theta')$$

where χ, χ' and θ are distinct irreducible characters of G and the sign ϵ is determined. Label $\chi = \chi_1, \chi' = \chi_2$ and $\theta = \chi_3$. Now one similarly checks that for each $j \geq 3$ there is an irreducible character χ_j of G such that

$$\alpha_j^* = \psi_1^* - \psi_j^* = \epsilon(\chi_1 - \chi_j)$$

and $\chi_1, \ldots, \chi_k$ are distinct. Since all χ_j's have the same degree as χ_1, the proof is complete.

We remark that it need not be the case that $\chi_j = \psi_j^*$ for any j, but only that the differences of irreducible characters of N induce to differences of irreducible characters of G.

The irreducible characters χ_j of G obtained via Lemma 16 are called *exceptional characters* associated to Q.

Lemma 17. The exceptional characters associated to Q_i are all distinct from the exceptional characters associated to Q_j for i and j distinct elements of $\{1, 2, 3, 4\}$.

Proof: Let χ be an exceptional character associated to Q_i and let θ be an exceptional character associated to Q_j. By construction, there are distinct irreducible characters ψ and ψ' of Q_i such that $\psi^* - \psi'^* = \chi - \chi'$ and there are distinct irreducible characters λ and λ' of Q_j such that $\lambda^* - \lambda'^* = \theta - \theta'$. Let $\alpha = \psi - \psi'$ and let $\beta = \lambda - \lambda'$. By Lemma 15, α^* is zero on all elements of G whose order is not equal to q_i (including the identity) and β^* is zero on all elements of G whose order is not equal to q_j. Thus clearly $(\alpha^*, \beta^*) = 0$. It follows easily that the two irreducible constituents of α^* are pairwise orthogonal to those of β^* as well. This establishes the lemma.

It is now easy to show that such a simple group G does not exist. By Lemma 16 and properties (i) to (iv) of G we can count the number of exceptional characters:

   (i) there are 2 exceptional characters associated to Q_1

   (ii) there are 2 exceptional characters associated to Q_2

   (iii) there are 4 exceptional characters associated to Q_3

   (iv) there are 136 exceptional characters associated to Q_4 .

Denote the common degree of the exceptional characters associated to Q_i by d_i for $i = 1, \ldots, 4$. By Lemma 17, the exceptional characters account for 144 nonprincipal irreducible characters of G hence these, together with the principal character, are all the irreducible characters of G (the number of conjugacy classes of G is 145). The sum of the squares of the degrees of the irreducible characters is the order of G:

$$1 + 2d_1^2 + 2d_2^2 + 4d_3^2 + 136d_4^2 = 1004913.$$

Simplifying this, we obtain

$$d_1^2 + d_2^2 + 2d_3^2 + 68d_4^2 = 502456. \qquad (19.4)$$

Finally, since each nonprincipal irreducible representation of the simple group G is faithful and since the smallest degree of a faithful representation of N_1 is 13, each $d_i \geq 13$. Since $d_4 < \sqrt{502456/68} < 86$ and d_4 divides $|G|$, it follows that

$$d_4 \in \{13, 21, 27, 39, 63\}.$$

Furthermore, each $d_i \mid |G|$ by Corollary 5 and so there are a small number of possibilities for each d_i. One now checks that equation (4) has no solution (this is particularly easy to do by computer). This contradiction completes the proof.

EXERCISES

Throughout the exercises all representations are over the complex numbers.

1. Let $G = S_3$, let $H = A_3$ and let V be the 3-dimensional $\mathbb{C}H$-module which affords the natural permutation representation of A_3. More explicitly, let V have basis e_1, e_2, e_3 and let $\sigma \in A_3$ act on V by $\sigma e_i = e_{\sigma(i)}$. Let 1 and (1 2) be coset representatives for the left cosets of A_3 in S_3 and write out the explicit matrices described in Theorem 11 for the action of S_3 on the induced module W, for each of the elements of S_3.

2. In each of parts (a) to (f) a character ψ of a subgroup H of a particular group G is specified. Compute the values of the induced character $\mathrm{Ind}_H^G(\psi)$ on all the conjugacy classes of G and use the character tables in Section 1 to write $\mathrm{Ind}_H^G(\psi)$ as a sum of irreducible characters:

(a) ψ is the unique nonprincipal degree 1 character of the subgroup $\langle\,(1\ 2)\,\rangle$ of S_3

(b) ψ is the degree 1 character of the subgroup $\langle\,r\,\rangle$ of D_8 defined by $\psi(r) = i$, where $i \in \mathbb{C}$ is a square root of -1

(c) ψ is the degree 1 character of the subgroup $\langle\,r\,\rangle$ of D_8 defined by $\psi(r) = -1$

(d) ψ is any of the nonprincipal degree 1 characters of the subgroup $V_4 = \langle\,(1\ 2),\ (3\ 4)\,\rangle$ of S_4

(e) $\psi = \chi_4$ is the first of the two characters of degree 3 in the character table of $H = S_4$ in Section 1 and H is a subgroup of $G = S_5$

(f) ψ is any of the nonprincipal degree 1 characters of the subgroup $V_4 = \langle\,(1\ 2),\ (3\ 4)\,\rangle$ of S_5.

3. Use Proposition 13 to explicitly write out the character table of each of the following groups:
   (a) the dihedral group of order 10
   (b) the non-abelian group of order 57
   (c) the non-abelian group of order 56 which has a normal, elementary abelian Sylow 2-subgroup.

4. Let H be a subgroup of G, let φ be a representation of H and suppose that N is a normal subgroup of G with $N \le H$ and N contained in the kernel of φ. Prove that N is also contained in the kernel of the induced representation of φ.

5. Let N be a normal subgroup of G and let ψ_1 be the principal character of N. Let Ψ be the induced character $\mathrm{Ind}_N^G(\psi_1)$ so that by the preceding exercise we may consider Ψ as the character of a representation of G/N. Prove that Ψ is the character of the regular representation of G/N.

6. Let Z be any subgroup of the center of G, let $|G : Z| = m$ and let ψ be a character of Z. Prove that
$$\mathrm{Ind}_Z^G(\psi)(g) = \begin{cases} m\psi(g) & \text{if } g \in Z \\ 0 & \text{if } g \notin Z. \end{cases}$$

7. Let φ be a matrix representation of the subgroup H of G and define matrices $\Phi(g)$ for every $g \in G$ by the displayed formula in the statement of Theorem 11. Prove directly that Φ is a representation by showing that $\Phi(xy) = \Phi(x)\Phi(y)$ for all $x, y \in G$.

8. Let G be a Frobenius group with Frobenius kernel Q. Assume that both Q and G/Q are abelian but G is not abelian (i.e., $G \neq Q$). Let $|Q| = n$ and $|G : Q| = m$.
   (a) Prove that G/Q is cyclic and show that $G = QC$ for some cyclic subgroup C of G with $C \cap Q = 1$ (i.e., G is a semidirect product of Q and C and $|C| = m$). [Let q be a prime divisor of n and let G/Q act by conjugation on the elementary abelian q-group $\{h \in Q \mid h^q = 1\}$. Apply Exercise 14(e) of Section 18.1 and the definition of a Frobenius group to an irreducible constituent of this $\mathbb{F}_q G/Q$-module.]
   (b) Prove that n and m are relatively prime. [If a prime p divides both the order and index of Q, let P be a Sylow p-subgroup of G. Then $P \cap Q \trianglelefteq P$ and $P \cap Q$ is a Sylow p-subgroup of Q. Consider the centralizer in G of the subgroup $Z(P) \cap Q$ (this intersection is nontrivial by Theorem 1 of Section 6.1).]
   (c) Show that G has no elements of order qp, where q is any nontrivial divisor of n and p is any nontrivial divisor of m. [Argue as in Proposition 13.]
   (d) Prove that the number of nonidentity conjugacy classes of G contained in Q is $(n-1)/m$ and that each of these classes has size m. [Argue as in Proposition 13.]
   (e) Prove that no two distinct elements of C are conjugate in G. Deduce that the non-identity elements of C are representatives for $m - 1$ distinct conjugacy classes of G and that each of these classes has size n. Deduce then that every element of $G - Q$

is conjugate to some element of C and that G has $m + (n-1)/m$ conjugacy classes.

(f) Prove that $G' = Q$ and deduce that G has m distinct characters of degree 1. [To show $Q \leq G'$ let $C = \langle x \rangle$ and argue that the map $h \mapsto [h, x] = x^{-1}h^{-1}xh$ is a homomorphism from Q to Q whose kernel is trivial, hence this map is surjective.]

(g) Show that if ψ is any nonprincipal irreducible character of Q, then $\mathrm{Ind}_Q^G(\psi)$ is an irreducible character of G. Show that every irreducible character of G of degree > 1 is equal to $\mathrm{Ind}_Q^G(\psi)$ for some nonprincipal irreducible character ψ of Q. Deduce that every irreducible character of G has degree either 1 or m and the number of irreducible characters of degree m is $(n-1)/m$. [Check that the proof of Proposition 13(3) establishes this more general result with the appropriate changes to the numbers involved.]

9. Use the preceding exercise to explicitly write out the character table of $\langle (1\ 2\ 3\ 4\ 5), (2\ 3\ 5\ 4) \rangle$, which is the normalizer in S_5 of a Sylow 5-subgroup (this group is a Frobenius group of order 20).

10. Let N be a normal subgroup of G, let ψ be a character of N and let $g \in G$. Define ψ^g by $\psi^g(h) = \psi(ghg^{-1})$ for all $h \in N$.

(a) Prove that ψ^g is a character of N (ψ and ψ^g are called G-*conjugate* characters of N). Prove that ψ^g is irreducible if and only if ψ is irreducible.

(b) Prove that the map $\psi \mapsto \psi^g$ is a right group action of G on the set of characters of N and N is in the kernel of this action.

(c) Prove that if ψ_1 and ψ_2 are G-conjugate characters of N, then $\mathrm{Ind}_N^G(\psi_1) = \mathrm{Ind}_N^G(\psi_2)$. Prove also that if ψ_1 and ψ_2 are characters of N that are not G-conjugate then $\mathrm{Ind}_N^G(\psi_1) \neq \mathrm{Ind}_N^G(\psi_2)$. [Use the argument in the proof of Proposition 13(3).]

11. Show that if $G = A_4$ and $N = V_4$ is its Sylow 2-subgroup then any two nonprincipal irreducible characters of N are G-conjugate (cf. the preceding exercise).

12. Let $G = D_{2n}$ be presented by its usual generators and relations. Prove that if ψ is any degree 1 character of $H = \langle r \rangle$ such that $\psi \neq \psi^s$, then $\mathrm{Ind}_H^G(\psi)$ is an irreducible character of D_{2n}. Show that every irreducible character of D_{2n} is the induced character of some degree 1 character of $\langle r \rangle$.

13. Prove both parts of Proposition 14.

14. Prove the following result known as *Frobenius Reciprocity*: let $H \leq G$, let ψ be any character of H and let χ be any character of G. Then

$$(\psi, \chi|_H)_H = (\mathrm{Ind}_H^G(\psi), \chi)_G.$$

[Expand the right hand side using the formula for the induced character $\mathrm{Ind}_H^G(\psi)$ or follow the proof of Shapiro's Lemma in Section 17.2.]

15. Assume G were a simple group of order $3^3 \cdot 7 \cdot 13 \cdot 409$ whose Sylow subgroups and their normalizers are described by properties (1) to (5) in this section. Prove that the permutation character of degree 819 obtained from the action of G on the left cosets of the subgroup N_4 decomposes as $\chi_0 + \gamma + \gamma'$, where χ_0 is the principal character of G and γ and γ' are distinct irreducible characters of G of degree 409. [Use Exercise 9 in Section 18.3 to show that this permutation character π has $||\pi||^2 = 3$.]

Cartesian Products and Zorn's Lemma

Section 1 of this appendix contains the definition of the Cartesian product of an arbitrary collection of sets. In the text we shall primarily be interested in products of finitely many (or occasionally countably many) sets. We indicate how the general definition agrees with the familiar "ordered n-tuple" notion of a Cartesian product in these cases. Section 2 contains a discussion of Zorn's Lemma and related topics.

1. CARTESIAN PRODUCTS

A set I is called an *indexing set* or *index set* if the elements of I are used to index some collection of sets. In particular, if A and I are sets, we can form the collection $\{A_i \mid i \in I\}$ by specifying that $A_i = A$ for all $i \in I$. Thus *any* set can be an indexing set; we use this term to emphasize that the elements are used as indices.

Definition.

    (1) Let I be an indexing set and let $\{A_i \mid i \in I\}$ be a collection of sets. A *choice function* is any function

$$f : I \to \bigcup_{i \in I} A_i$$

    such that $f(i) \in A_i$ for all $i \in I$.

    (2) Let I be an indexing set and for all $i \in I$ let A_i be a set. The *Cartesian product* of $\{A_i \mid i \in I\}$ is the set of all choice functions from I to $\cup_{i \in I} A_i$ and is denoted by $\prod_{i \in I} A_i$ (where if either I or any of the sets A_i are empty the Cartesian product is the empty set). The elements of this Cartesian product are written as $\prod_{i \in I} a_i$, where this denotes the choice function f such that $f(i) = a_i$ for each $i \in I$.

    (3) For each $j \in I$ the set A_j is called the jth *component* of the Cartesian product $\prod_{i \in I} A_i$ and a_j is the jth *coordinate* of the element $\prod_{i \in I} a_i$.

    (4) For $j \in I$ the *projection map* of $\prod_{i \in I} A_i$ onto the jth coordinate, A_j, is defined by $\prod_{i \in I} a_i \mapsto a_j$.

Each choice function f in the Cartesian product $\prod_{i \in I} A_i$ may be thought of as a way of "choosing" an element $f(i)$ from each set A_i.

If $I = \{1, 2, \dots, n\}$ for some $n \in \mathbb{Z}^+$ and if f is a choice function from I to $A_1 \cup \cdots \cup A_n$, where each A_i is nonempty, we can associate to f a unique (ordered) n-tuple:

$$f \to (f(1), f(2), \dots, f(n)).$$

Note that by definition of a choice function, $f(i) \in A_i$ for all i, so the n-tuple above has an element of A_i in the ith position for each i .

Conversely, given an n-tuple $(a_1, a_1, \ldots, a_n)$, where $a_i \in A_i$ for all $i \in I$, there is a unique choice function, f, from I to $\cup_{i \in I} A_i$ associated to it, namely

$$f(i) = a_i, \qquad \text{for all } i \in I.$$

It is clear that this map from n-tuples to choice functions is the inverse to the map described in the preceding paragraph. Thus *there is a bijection between ordered n-tuples and elements of* $\prod_{i \in I} A_i$. Henceforth when $I = \{1, 2, \ldots, n\}$ we shall write

$$\prod_{i=1}^{n} A_i \quad \text{or} \quad A_1 \times A_2 \times \cdots \times A_n$$

for the Cartesian product and we shall describe the elements as ordered n-tuples.

If $I = \mathbb{Z}^+$, we shall similarly write: $\prod_{i=1}^{\infty} A_i$ or $A_1 \times A_2 \times \cdots$ for the Cartesian product of the A_i's. We shall write the elements as ordered tuples: $(a_1, a_2, \ldots)$, i.e., as infinite sequences whose i^{th} terms are in A_i.

Note that when $I = \{1, 2, \ldots, n\}$ or $I = \mathbb{Z}^+$ we have used the natural ordering on I to arrange the elements of our Cartesian products into n-tuples. Any other ordering of I (or any ordering on a finite or countable index set) gives a different representation of the elements of the same Cartesian product.

Examples

(1) $A \times B = \{(a, b) \mid a \in A, b \in B\}$.

(2) $\mathbb{R}^n = \mathbb{R} \times \mathbb{R} \times \cdots \times \mathbb{R}$ (n factors) is the usual set of n-tuples with real number entries, Euclidean n-space.

(3) Suppose $I = \mathbb{Z}^+$ and A_i is the same set A, for all $i \in I$. The Cartesian product $\prod_{i \in \mathbb{Z}^+} A$ is the set of all (infinite) sequences $a_1, a_2, a_3 \ldots$ of elements of A. In particular, if $A = \mathbb{R}$, then the Cartesian product $\prod_{i \in \mathbb{Z}^+} \mathbb{R}$ is the set of all real sequences.

(4) Suppose I is any indexing set and A_i is the same set A, for all $i \in I$. The Cartesian product $\prod_{i \in I} A$ is just the set of all functions from I to A, where the function $f : I \to A$ corresponds to the element $\prod_{i \in I} f(i)$ in the Cartesian product. This Cartesian product is often (particularly in topology books) denoted by A^I. Note that for each fixed $j \in I$ the projection map onto the jth coordinate sends the function f to $f(j)$, i.e., is evaluation at j.

(5) Let R be a ring and let x be an indeterminate over R. The definition of the ring $R[x]$ of polynomials in x with coefficients from R may be given in terms of Cartesian products rather than in the more intuitive and familiar terms of "formal sums" (in Chapters 7 and 9 we introduced them in the latter form since this is the way we envision and work with them). Let I be the indexing set $\mathbb{Z}^+ \cup \{0\}$ and let $R[x]$ be the subset of the Cartesian product $\prod_{i=0}^{\infty} R$ consisting of elements $(a_0, a_1, a_2, \ldots)$ such that only finitely many of the a_i's are nonzero. If $(a_0, a_1, a_2, \ldots, a_n, 0, 0, \ldots)$ is such a sequence we represent it by the more familiar "formal sum" $\sum_{i=0}^{n} a_i x^i$. Addition and multiplication of these sequences is defined so that the usual rules for addition and multiplication of polynomials hold.

Proposition 1. Let I be a nonempty countable set and for each $i \in I$ let A_i be a set. The cardinality of the Cartesian product is the product of the cardinalities of the sets A_i, i.e.,

$$\left| \prod_{i \in I} A_i \right| = \prod_{i \in I} |A_i|,$$

(where if some A_i is an infinite set or if I is infinite and an infinite number of A_i's have cardinality ≥ 2, both sides of this equality are infinity). In particular,

$$|A_1 \times A_2 \times \cdots \times A_n| = |A_1| \times |A_2| \times \cdots \times |A_n|.$$

Proof: In order to count the number of choice functions note that each $i \in I$ may be mapped to any of the $|A_i|$ elements of A_i and for $i \neq j$ the values of choice functions at i and j may be chosen completely independently. Thus the number of choice functions is the product of the cardinalities of the A_i's, as claimed.

For Cartesian products of finitely many sets, $A_1 \times A_2 \times \cdots \times A_n$, one can see this easily from the n-tuple representation: the elements of $A_1 \times A_2 \times \cdots \times A_n$ are n-tuples $(a_1, a_2, \ldots, a_n)$ and each a_i may be chosen as any of the $|A_i|$ elements of A_i. Since these choices are made independently for $i \neq j$, there are $|A_1| \cdot |A_2| \cdots |A_n|$ elements in the Cartesian product.

EXERCISE

1. Let I and J be any two indexing sets and let A be an arbitrary set. For any function $\varphi : J \to I$ define

$$\varphi^* : \prod_{i \in I} A \to \prod_{j \in J} A \qquad \text{by} \qquad \varphi^*(f) = f \circ \varphi \quad \text{for all choice functions } f \in \prod_{i \in I} A.$$

   (a) Let $I = \{1, 2\}$, let $J = \{1, 2, 3\}$ and let $\varphi : J \to I$ be defined by $\varphi(1) = 2$, $\varphi(2) = 2$ and $\varphi(3) = 1$. Describe explicitly how an ordered pair in $A \times A$ maps to a 3-tuple in $A \times A \times A$ under this φ^*.

   (b) Let $I = J = \{1, 2, \ldots, n\}$ and assume φ is a permutation of I. Describe in terms of n-tuples in $A \times A \times \cdots \times A$ the function φ^*.

2. PARTIALLY ORDERED SETS AND ZORN'S LEMMA

We shall have occasion to use Zorn's Lemma as a form of "infinite induction" in a few places in the text where it is desirable to know the existence of some set which is *maximal* with respect to certain specified properties. For example, Zorn's Lemma is used to show that every vector space has a basis. In this situation a basis of a vector space V is a subset of V which is maximal as a set consisting of linearly independent vectors (the maximality ensures that these vectors span V). For finite dimensional spaces this can be proved by induction; however, for spaces of arbitrary dimension Zorn's Lemma is needed to establish this. By having results which hold in full generality the theory often becomes a little neater in places, although the main results of the text do not require its use.

A specific instance in the text where a maximal object which helps to simplify matters is constructed by Zorn's Lemma is the algebraic closure of a field. An algebraic closure of a field F is an extension of F which is maximal among any collection of algebraic extensions. Such a field contains (up to isomorphism) all elements which are algebraic over F, hence all manipulations involving such algebraic elements can be effected in this one larger field. In any particular situation the use of an algebraic closure can be avoided by adjoining the algebraic elements involved to the base field F, however this becomes tedious (and often obscures matters) in complicated proofs. For the specific fields appearing as examples in this text the use of Zorn's Lemma to construct an algebraic closure can be avoided (for example, the construction of an algebraic closure of any subfield of the complex numbers or of any finite field does not require it).

The first example of the use of Zorn's Lemma appears in the proof of Proposition 11 in Section 7.4.

In order to state Zorn's Lemma we need some terminology.

Definition. A *partial order* on a nonempty set A is a relation $\leq$ on A satisfying
   (1) $x \leq x$ for all $x \in A$ (reflexive),
   (2) if $x \leq y$ and $y \leq x$ then $x = y$ for all $x, y \in A$ (antisymmetric),
   (3) if $x \leq y$ and $y \leq z$ then $x \leq z$ for all $x, y, z \in A$ (transitive).

We shall usually say that A is a partially ordered set under the ordering $\leq$ or that A is partially ordered by $\leq$.

Definition. Let the nonempty set A be partially ordered by $\leq$.
   (1) A subset B of A is called a *chain* if for all $x, y \in B$, either $x \leq y$ or $y \leq x$.
   (2) An *upper bound* for a subset B of A is an element $u \in A$ such that $b \leq u$, for all $b \in B$.
   (3) A *maximal element* of A is an element $m \in A$ such that if $m \leq x$ for any $x \in A$, then $m = x$.

In the literature a chain is also called a *tower* or called a *totally ordered* or *linearly ordered* or *simply ordered* subset.

Some examples below highlight the distinction between upper bounds and maximal elements. Also note that if m is a *maximal* element of A, it is not necessarily the case that $x \leq m$ for all $x \in A$ (i.e., m is not necessarily a *maximum* element).

Examples

   (1) Let A be the power set (i.e., set of all subsets) of some set X and $\leq$ be set containment: $\subseteq$. Notice that this is only a *partial* ordering since some subsets of X may not be comparable, e.g. singletons: if $x \neq y$ then $\{x\} \not\subseteq \{y\}$ and $\{y\} \not\subseteq \{x\}$. In this situation an example of a chain is a collection of subsets of X such as

$$X_1 \subseteq X_2 \subseteq X_3 \subseteq \cdots.$$

Any subset B of A has an upper bound, b, namely,

$$b = \bigcup_{x \in B} x.$$

This partially ordered set A has a (unique) maximal element, X.

In many instances the set A consists of some (but not necessarily all) subsets of a set X (i.e., A is a subset of the power set of X) and with the ordering on A again being inclusion. The existence of upper bounds and maximal elements depends on the nature of A.

(2) Let A be the collection of all *proper* subsets of $\mathbb{Z}^+$ ordered under $\subseteq$. In this situation, chains need not have maximal elements, e.g. the chain

$$\{1\} \subseteq \{1, 2\} \subseteq \{1, 2, 3\} \subseteq \cdots$$

does not have an upper bound. The set A does have maximal elements: for example $\mathbb{Z}^+ - \{n\}$ is a maximal element of A for any $n \in \mathbb{Z}^+$.

(3) Let $A = \mathbb{R}$ under the usual $\leq$ relation. In this example every subset of A is a chain (including A itself). The notion of a subset of A having an upper bound is the same as the usual notion of a subset of $\mathbb{R}$ being bounded above by some real number (so some sets, such as intervals of finite length, have upper bounds and others, such as the set of positive reals, do not). The set A does not have a maximal element.

Zorn's Lemma If A is a nonempty partially ordered set in which every chain has an upper bound then A has a maximal element.

It is a nontrivial result that *Zorn's Lemma is independent of the usual (Zermelo–Fraenkel) axioms of set theory*[1] in the sense that if the axioms of set theory are consistent,[2] then so are these axioms together with Zorn's Lemma; and if the axioms of set theory are consistent, then so are these axioms together with the *negation* of Zorn's Lemma. The use of the term "lemma" in Zorn's Lemma is historical.

For the sake of completeness (and to relate Zorn's Lemma to formulations found in other courses) we include two other equivalent formulations of Zorn's Lemma.

The Axiom of Choice The Cartesian product of any nonempty collection of nonempty sets is nonempty. In other words, if I is any nonempty (indexing) set and A_i is a nonempty set for all $i \in I$, then there exists a choice function from I to $\cup_{i \in I} A_i$.

Definition. Let A be a nonempty set. A *well ordering* on A is a total ordering on A such that every nonempty subset of A has a minimum (or smallest) element, i.e., for each nonempty $B \subseteq A$ there is some $s \in B$ such that $s \leq b$, for all $b \in B$.

The Well Ordering Principle Every nonempty set A has a well ordering.

Theorem 2. Assuming the usual (Zermelo–Fraenkel) axioms of set theory, the following are equivalent:

(1) Zorn's Lemma

(2) the Axiom of Choice

(3) the Well Ordering Principle.

Proof: This follows from elementary set theory. We refer the reader to *Real and Abstract Analysis* by Hewitt and Stromberg, Springer-Verlag, 1965, Section 3 for these equivalences and some others.

---

[1] See P.J. Cohen's papers in: Proc. Nat. Acad. Sci., 50(1963), and 51(1964).

[2] This is not known to be the case!

EXERCISES

1. Let A be the collection of all finite subsets of $\mathbb{R}$ ordered by inclusion. Discuss the existence (or nonexistence) of upper bounds, minimal and maximal elements (where minimal elements are defined analogously to maximal elements). Explain why this is not a well ordering.

2. Let A be the collection of all infinite subsets of $\mathbb{R}$ ordered by inclusion. Discuss the existence (or nonexistence) of upper bounds, minimal and maximal elements. Explain why this is not a well ordering.

3. Show that the following partial orderings on the given sets are not well orderings:
   (a) $\mathbb{R}$ under the usual relation $\leq$.
   (b) $\mathbb{R}^+$ under the usual relation $\leq$.
   (c) $\mathbb{R}^+ \cup \{0\}$ under the usual relation $\leq$.
   (d) $\mathbb{Z}$ under the usual relation $\leq$.

4. Show that $\mathbb{Z}^+$ is well ordered under the usual relation $\leq$.

Category Theory

Category theory provides the language and the mathematical foundations for discussing properties of large classes of mathematical objects such as the class of "all sets" or "all groups" while circumventing problems such as Russell's Paradox. In this framework one may explore the commonality across classes of concepts and methods used in the study of each class: homomorphisms, isomorphisms, etc., and one may introduce tools for studying relations between classes: functors, equivalence of categories, etc. One may then formulate precise notions of a "natural" transformation and "natural" isomorphism, both within a given class or between two classes. (In the text we described "natural" as being "coordinate free.") A prototypical example of natural isomorphisms within a class is the isomorphism of an arbitrary finite dimensional vector space with its double dual in Section 11.3. In fact one of the primary motivations for the introduction of categories and functors by S. Eilenberg and S. MacLane in 1945 was to give a precise meaning to the notions of "natural" in cases such as this. Category theory has also played a foundational role for formalizing new concepts such as schemes (cf. Section 15.5) that are fundamental to major areas of contemporary research (e.g., algebraic geometry). Pioneering work of this nature was done by A. Grothendieck, K. Morita and others.

Our treatment of category theory should be viewed more as an introduction to some of the basic language. Since we have not discussed the Zermelo–Fraenkel axioms of set theory or the Gödel–Bernays axioms of classes we make no mention of the foundations of category theory. To remain consistent with the set theory axioms, however, we implicitly assume that there is a *universe* set **U** which contains all the sets, groups, rings, etc. that one would encounter in "ordinary" mathematics (so that the category of "all sets" implicitly means "all sets in **U**"). The reader is referred to books on set theory, logic, or category theory such as *Categories for the Working Mathematician* by S. MacLane, Springer–Verlag, 1971 for further study.

We have organized this appendix so that wherever possible the examples of each new concept use terminology and structures in the order that these appear in the body of the text. For instance, the first example of a functor involves sets and groups, the second example uses rings, etc. In this way the appendix may be read early on in one's study, and a greater appreciation may be gained through rereading the examples as one becomes conversant with a wider variety of mathematical structures.

1. CATEGORIES AND FUNCTORS

We begin with the basic concept of this appendix.

Definition. A *category* **C** consists of a class of *objects* and sets of *morphisms* between those objects. For every ordered pair A, B of objects there is a set $\text{Hom}_{\mathbf{C}}(A, B)$ of morphisms from A to B, and for every ordered triple A, B, C of objects there is a *law of composition* of morphisms, i.e., a map

$$\text{Hom}_{\mathbf{C}}(A, B) \times \text{Hom}_{\mathbf{C}}(B, C) \longrightarrow \text{Hom}_{\mathbf{C}}(A, C)$$

where $(f, g) \mapsto gf$, and gf is called the composition of g with f. The objects and morphism satisfy the following axioms: for objects A, B, C and D

(i) if $A \neq C$ or $B \neq D$, then $\text{Hom}_{\mathbf{C}}(A, B)$ and $\text{Hom}_{\mathbf{C}}(C, D)$ are disjoint sets,

(ii) composition of morphisms is associative, i.e., $h(gf) = (hg)f$ for every f in $\text{Hom}_{\mathbf{C}}(A, B)$, g in $\text{Hom}_{\mathbf{C}}(B, C)$ and h in $\text{Hom}_{\mathbf{C}}(C, D)$,

(iii) each object has an identity morphism, i.e., for every object A there is a morphism $1_A \in \text{Hom}_{\mathbf{C}}(A, A)$ such that $f 1_A = f$ for every $f \in \text{Hom}_{\mathbf{C}}(A, B)$ and $1_A g = g$ for every $g \in \text{Hom}_{\mathbf{C}}(B, A)$.

Morphisms are also called *arrows*. It is an exercise to see that the identity morphism for each object is unique (by the same argument that the identity of a group is unique). We shall write $\text{Hom}(A, B)$ for $\text{Hom}_{\mathbf{C}}(A, B)$ when the category is clear from the context.

The terminology we use throughout the text is common to all categories: a morphism from A to B will be denoted by $f : A \to B$ or $A \xrightarrow{f} B$. The object A is the *domain* of f and B is the *codomain* of f. A morphism from A to A is an endomorphism of A. A morphism $f : A \to B$ is an isomorphism if there is a morphism $g : B \to A$ such that $gf = 1_A$ and $fg = 1_B$.

There is a natural notion of a *subcategory* category $\mathbf{C}$ of $\mathbf{D}$, i.e., when every object of $\mathbf{C}$ is also an object in $\mathbf{D}$, and for objects A, B in $\mathbf{C}$ we have the containment $\text{Hom}_{\mathbf{C}}(A, B) \subseteq \text{Hom}_{\mathbf{D}}(A, B)$.

Examples

In each of the following examples we leave the details of the verification of the axioms for a category as exercises.

(1) **Set** is the category of all sets. For any two sets A and B, $\text{Hom}(A, B)$ is the set of all functions from A to B. Composition of morphisms is the familiar composition of functions: $gf = g \circ f$. The identity in $\text{Hom}(A, A)$ is the map $1_A(a) = a$, for all $a \in A$. This category contains the category of all finite sets as a subcategory.

(2) **Grp** is the category of all groups, where morphisms are group homomorphisms. Note that the composition of group homomorphisms is again a group homomorphism. A subcategory of **Grp** is **Ab**, the category of all abelian groups. Similarly, **Ring** is the category of all nonzero rings with 1, where morphisms are ring homomorphisms that send 1 to 1. The category **CRing** of all commutative rings with 1 is a subcategory of **Ring**.

(3) For a fixed ring R, the category R–**mod** consists of all left R-modules with morphisms being R-module homomorphisms.

(4) **Top** is the category whose objects are topological spaces and morphisms are continuous maps between topological spaces (cf. Section 15.2). Note that the identity (set) map from a space to itself is continuous in every topology, so $\text{Hom}(A, A)$ always has an identity.

(5) Let **0** be the empty category, with no objects and no morphisms. Let **1** denote the category with one object, A, and one morphism: $\text{Hom}(A, A) = \{1_A\}$. Let **2** be the category with two objects, A_1 and A_2, and only one nonidentity morphism:

$\mathrm{Hom}(A_1, A_2) = \{f\}$ and $\mathrm{Hom}(A_2, A_1) = \emptyset$. Note that the objects A_1 and A_2 and the morphism f are "primitives" in the sense that A_1 and A_2 are not defined to be sets and f is simply an arrow (literally) from A_1 to A_2; it is not defined as a set map on the elements of some set. One can continue this way and define **N** to be the category with N objects $A_1, A_2, \ldots, A_N$ with the only nonidentity morphisms being a unique arrow from A_i to A_j for every $j > i$ (so that composition of arrows is uniquely determined).

(6) If G is a group, form the category **G** as follows. The only object is G and $\mathrm{Hom}(G, G) = G$; the composition of two functions f and g is the product gf in the group G. Note that $\mathrm{Hom}(G, G)$ has an identity morphism: the identity of the group G.

Definition. Let **C** and **D** be categories.
  (1) We say $\mathcal{F}$ is a *covariant functor* from **C** to **D** if
    (a) for every object A in **C**, $\mathcal{F}A$ is an object in **D**, and
    (b) for every $f \in \mathrm{Hom}_{\mathbf{C}}(A, B)$ we have $\mathcal{F}(f) \in \mathrm{Hom}_{\mathbf{D}}(\mathcal{F}A, \mathcal{F}B)$,
  such that the following axioms are satisfied:
    (i) if gf is a composition of morphisms in **C**, then $\mathcal{F}(gf) = \mathcal{F}(g)\mathcal{F}(f)$ in **D**, and
    (ii) $\mathcal{F}(1_A) = 1_{\mathcal{F}A}$.
  (2) We say $\mathcal{F}$ is a *contravariant functor* from **C** to **D** if the conditions in (1) hold but property (b) and axiom (i) are replaced by:
    (b') for every $f \in \mathrm{Hom}_{\mathbf{C}}(A, B)$, $\mathcal{F}(f) \in \mathrm{Hom}_{\mathbf{D}}(\mathcal{F}B, \mathcal{F}A)$,
    (i') if gf is a composition of morphisms in **C**, then $\mathcal{F}(gf) = \mathcal{F}(f)\mathcal{F}(g)$ in **D**
  (i.e., contravariant functors reverse the arrows).

Examples

In each of these examples the verification of the axioms for a functor are left as exercises. Additional examples of functors appear in the exercises at the end of this section.
  (1) The identity functor $\mathcal{I}_C$ maps any category **C** to itself by sending objects and morphisms to themselves. More generally, if **C** is a subcategory of **D**, the *inclusion functor* maps **C** into **D** by sending objects and morphisms to themselves.
  (2) Let $\mathcal{F}$ be the functor from **Grp** to **Set** that maps any group G to the same set G and any group homomorphism φ to the same set map φ. This functor is called the *forgetful functor* since it "removes" or "forgets" the structure of the groups and the homomorphisms between them. Likewise there are forgetful functors from the categories **Ab**, R–**mod**, **Top**, etc., to **Set**.
  (3) The *abelianizing* functor maps **Grp** to **Ab** by sending each group G to the abelian group $G^{\mathrm{ab}} = G/G'$, where G' is the commutator subgroup of G (cf. Section 5.4). Each group homomorphism $\varphi : G \to H$ is mapped to the induced homomorphism on quotient groups:

$$\overline{\varphi} : G^{\mathrm{ab}} \to H^{\mathrm{ab}} \qquad \text{by} \qquad \overline{\varphi}(xG') = \varphi(x)H'.$$

  The definition of the commutator subgroup ensures that $\overline{\varphi}$ is well defined and the axioms for a functor are satisfied.
  (4) Let R be a ring and let D be a left R-module. For each left R-module N the set $\mathrm{Hom}_R(D, N)$ is an abelian group, and is an R-module if R is commutative (cf. Proposition 2 in Section 10.2). If $\varphi : N_1 \to N_2$ is an R-module homomorphism, then for every $f \in \mathrm{Hom}_R(D, N_1)$ we have $\varphi \circ f \in \mathrm{Hom}_R(D, N_2)$. Thus

$\varphi' : \mathrm{Hom}_R(D, N_1) \to \mathrm{Hom}_R(D, N_2)$ by $\varphi'(f) = \varphi \circ f$. This shows the map

$$\mathcal{H}om(D, \_) : N \longrightarrow \mathrm{Hom}_R(D, N)$$

$$\mathcal{H}om(D, \_) : \varphi \longrightarrow \varphi'$$

is a covariant functor from R–**Mod** to **Grp**. If R is commutative, it maps R–**Mod** to itself.

(5) In the notation of the preceding example, we observe that if $\varphi : N_1 \to N_2$, then for every $g \in \mathrm{Hom}_R(N_2, D)$ we have $g \circ \varphi \in \mathrm{Hom}_R(N_1, D)$. Thus $\varphi' : \mathrm{Hom}_R(N_2, D) \to \mathrm{Hom}_R(N_1, D)$ by $\varphi'(g) = g \circ \varphi$. In this case the map

$$\mathcal{H}om(\_, D) : N \longrightarrow \mathrm{Hom}_R(N, D)$$

$$\mathcal{H}om(\_, D) : \varphi \longrightarrow \varphi'$$

defines a *contravariant* functor.

(6) When D is a right R-module the map $D \otimes_R \_ : N \to D \otimes_R N$ defines a covariant functor from R–**Mod** to **Ab** (or to R–**Mod** when R is commutative). Here the morphism $\varphi : N_1 \to N_2$ maps to the morphism $1 \otimes \varphi$.

Likewise when D is a left R-module $\_ \otimes_R D : N \to N \otimes_R D$ defines a covariant functor from the category of right R-modules to **Ab** (or to R–**Mod** when R is commutative), where the morphism φ maps to the morphism $\varphi \otimes 1$.

(7) Let K be a field and let K–**fdVec** be the category of all finite dimensional vector spaces over K, where morphisms in this category are K-linear transformations. We define the *double dual* functor $\mathcal{D}^2$ from K–**fdVec** to itself. Recall from Section 11.3 that the dual space, V^*, of V is defined as $V^* = \mathrm{Hom}_K(V, K)$; the double dual of V is $V^{**} = \mathrm{Hom}_K(V^*, K)$. Then $\mathcal{D}^2$ is defined on objects by mapping a vector space V to its double dual V^{**}. If $\varphi : V \to W$ is a linear transformation of finite dimensional spaces, then

$$\mathcal{D}^2(\varphi) : V^{**} \to W^{**} \qquad \text{by} \qquad \mathcal{D}^2(\varphi)(E_v) = E_{\varphi(v)},$$

where E_v denotes "evaluation at v" for each $v \in V$. By Theorem 19 in Section 11.3, $E_v \in V^{**}$, and each element of V^{**} is of the form E_v for a unique $v \in V$. Since $\varphi(v) \in W$ we have $E_{\varphi(v)} \in W^{**}$, so $\mathcal{D}^2(\varphi)$ is well defined.

The functor $\mathcal{F}$ from **C** to **D** is called *faithful* (or is called *full*) if for every pair of objects A and B in **C** the map $\mathcal{F} : \mathrm{Hom}(A, B) \to \mathrm{Hom}(\mathcal{F}A, \mathcal{F}B)$ is injective (or surjective, respectively). Thus, for example, the forgetful functor is faithful but not full.

EXERCISES

1. Let N be a group and let **Nor**–N be the collection of all groups that contain N as a normal subgroup. A morphism between objects A and B is any group homomorphism that maps N into N.
   (a) Prove that **Nor**–N is a category.
   (b) Show how the projection homomorphism $G \mapsto G/N$ may be used to define a functor from **Nor**–N to **Grp**.

2. Let H be a group. Define a map $\mathcal{H}\times$ from **Grp** to itself on objects and morphisms as follows:

$$\mathcal{H}\times : G \to H \times G, \text{ and}$$

$$\text{if } \varphi : G_1 \to G_2 \quad \text{then } \mathcal{H}\times(\varphi) : H \times G_1 \to H \times G_2 \quad \text{by} \quad (h, g) \mapsto (h, \varphi(g)).$$

Prove that $\mathcal{H}\times$ is a functor.

3. Show that the map **Ring** to **Grp** by mapping a ring to its group of units (i.e., $R \mapsto R^\times$) defines a functor. Show by explicit examples that this functor is neither faithful nor full.

4. Show that for each $n \geq 1$ the map $\mathcal{GL}_n : R \to GL_n(R)$ defines a functor from **CRing** to **Grp**. [Define $\mathcal{GL}_n$ on morphisms by applying each ring homomorphism to the entries of a matrix.]

5. Supply the details that show the double dual map described in Example 7 satisfies the axioms of a functor.

2. NATURAL TRANSFORMATIONS AND UNIVERSALS

As mentioned in the introduction to this appendix, one of the motivations for the inception of category theory was to give a precise definition of the notion of "natural" isomorphism. We now do so, and see how some natural maps mentioned in the text are instances of the categorical concept. We likewise give the categorical definition of "universal arrows" and view some occurrences of universal properties in the text in this light.

Definition. Let **C** and **D** be categories and let $\mathcal{F}$, $\mathcal{G}$ be covariant functors from **C** to **D**. A *natural transformation* or *morphism of functors* from $\mathcal{F}$ to $\mathcal{G}$ is a map η that assigns to each object A in **C** a morphism η_A in $\mathrm{Hom}_\mathbf{D}(\mathcal{F}A, \mathcal{G}A)$ with the following property: for every pair of objects A and B in **C** and every $f \in \mathrm{Hom}_\mathbf{C}(A, B)$ we have $\mathcal{G}(f)\eta_A = \eta_B \mathcal{F}(f)$, i.e., the following diagram commutes:

$$
\begin{array}{ccc}
\mathcal{F}A & \xrightarrow{\ \eta_A\ } & \mathcal{G}A \\
{\scriptstyle \mathcal{F}(f)}\downarrow & & \downarrow{\scriptstyle \mathcal{G}(f)} \\
\mathcal{F}B & \xrightarrow{\ \eta_B\ } & \mathcal{G}B
\end{array}
$$

If each η_A is an isomorphism, η is called a *natural isomorphism* of functors.

Consider the special case where $\mathbf{C} = \mathbf{D}$ and **C** is a subcategory of **Set**, and where $\mathcal{F}$ is the identity functor. There is a natural transformation η from the identity functor to $\mathcal{G}$ if whenever $\mathcal{G}$ maps the object A to the object $\mathcal{G}A$ there is a morphism η_A from A to $\mathcal{G}A$, and whenever there is a morphism f from A to B the morphism $\mathcal{G}(f)$ is compatible with f as a map from $\mathcal{G}A$ to $\mathcal{G}B$. In fact $\mathcal{G}(f)$ is uniquely determined by f as a map from the subset $\eta_A(A)$ in $\mathcal{G}A$ to the subset $\eta_B(B)$ of $\mathcal{G}B$. If η is a natural isomorphism, then the value of $\mathcal{G}$ on every morphism is completely determined by η, namely $\mathcal{G}(f) = \eta_B f \eta_A^{-1}$. In this case the functor $\mathcal{G}$ is entirely specified by η. We shall see that some of the examples of functors in the preceding section arise this way.

Examples

(1) For any categories **C** and **D** and any functor $\mathcal{F}$ from **C** to **D** the identity is a natural isomorphism from $\mathcal{F}$ to itself: $\eta_A = 1_{\mathcal{F}A}$ for every object A in **C**.

(2) Let R be a ring and let $\mathcal{F}$ be any functor from R–**Mod** to itself. The zero map is a natural transformation from $\mathcal{F}$ to itself: $\eta_A = 0_A$ for every R-module A, where 0_A is the zero map from A to itself. This is not a natural isomorphism.

(3) Let $\mathcal{F}$ be the identity functor from **Grp** to itself, and let $\mathcal{G}$ be the abelianizing functor (Example 3) considered here as a map from **Grp** to itself. For each group G let $\eta_G : G \to G/G'$ be the usual projection map onto the quotient group. Then η is a natural transformation (but not an isomorphism) with respect to these two functors. (We call the maps η_G the *natural projection* maps.)

(4) Let $\mathcal{G} = \mathcal{D}^2$ be the double dual functor from the category of finite dimensional vector spaces over a field K to itself (Example 7). Then there is a natural isomorphism η from the identity functor to $\mathcal{G}$ given by

$$\eta_V : V \to V^{**} \qquad \text{by} \qquad \eta_V(v) = E_v$$

where E_v is "evaluation at v" for every $v \in V$.

(5) Let $\mathcal{GL}_n$ be the functor from **CRing** to **Grp** defined as follows. Each object (commutative ring) R is mapped by $\mathcal{GL}_n$ to the group $GL_n(R)$ of $n \times n$ invertible matrices with entries from R. For each ring homomorphism $f : R \to S$ let $\mathcal{GL}_n(f)$ be the map of matrices that applies f to each matrix entry. Since f sends 1 to 1 it follows that $\mathcal{GL}_n(f)$ sends invertible matrices to invertible matrices (cf. Exercise 4 in Section 1). Let $\mathcal{G}$ be the functor from **CRing** to **Grp** that maps each ring R to its group of units $R^\times$, and each ring homomorphism f to its restriction to the groups of units (also denoted by f). The *determinant* is a natural transformation from $\mathcal{GL}_n$ to $\mathcal{G}$ because the determinant is defined by the same polynomial for all rings so that the following diagram commutes:

$$
\begin{array}{ccc}
GL_n(R) & \xrightarrow{\ \det\ } & R^\times \\
{\scriptstyle \mathcal{GL}_n(f)}\downarrow & & \downarrow{\scriptstyle f} \\
GL_n(S) & \xrightarrow{\ \det\ } & S^\times
\end{array}
$$

Let **C**, **D** and **E** be categories, let $\mathcal{F}$ be a functor from **C** to **D**, and let $\mathcal{G}$ be a functor from **D** to **E**. There is an obvious notion of the composition of functors $\mathcal{GF}$ from **C** to **E**. When **E** = **C** the composition $\mathcal{GF}$ maps **C** to itself and $\mathcal{FG}$ maps **D** to itself. We say **C** and **D** are *isomorphic* if for some $\mathcal{F}$ and $\mathcal{G}$ we have $\mathcal{GF}$ is the identity functor $\mathcal{I}_C$, and $\mathcal{FG} = \mathcal{I}_D$. By the discussion in Section 10.1 the categories $\mathbb{Z}$–**Mod** and **Ab** are isomorphic. It also follows from observations in Section 10.1 that the categories of elementary abelian p-groups and vector spaces over $\mathbb{F}_p$ are isomorphic. In practice we tend to identify such isomorphic categories. The following generalization of isomorphism between categories gives a broader and more useful notion of when two categories are "similar."

Definition. Categories **C** and **D** are said to be *equivalent* if there are functors $\mathcal{F}$ from **C** to **D** and $\mathcal{G}$ from **D** to **C** such that the functor $\mathcal{GF}$ is naturally isomorphic to $\mathcal{I}_C$ (the identity functor of **C**) and $\mathcal{FG}$ is naturally isomorphic to the identity functor $\mathcal{I}_D$.

It is an exercise that equivalence of categories is reflexive, symmetric and transitive. The example of Affine k-algebras in Section 15.5 is an equivalence of categories (where one needs to modify the direction of the arrows in the definition of a natural transformation to accommodate the contravariant functors in this example). Another example (which requires some proving) is that for R a commutative ring with 1 the categories of left modules R–**Mod** and $M_{n \times n}(R)$–**Mod** are equivalent.

Finally, we introduce the concepts of universal arrows and universal objects.

Definition.

(1) Let **C** and **D** be categories, let $\mathcal{F}$ be a functor from **C** to **D**, and let X be an object in **D**. A *universal arrow* from X to $\mathcal{F}$ is a pair $(U(X), \iota)$, where $U(X)$ is an object in **C** and $\iota : X \to \mathcal{F}U(X)$ is a morphism in **D** satisfying the following property: for any object A in **C** if φ is any morphism from X to $\mathcal{F}A$ in **D**, then there exists a unique morphism $\Phi : U(X) \to A$ in **C** such that $\mathcal{F}(\Phi)\iota = \varphi$, i.e., the following diagram commutes:

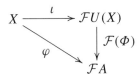

(2) Let **C** be a category and let $\mathcal{F}$ be a functor from **C** to the category **Set** of all sets. A *universal element* of the functor $\mathcal{F}$ is a pair (U, ι), where U is an object in **C** and ι is an element of the set $\mathcal{F}U$ satisfying the following property: for any object A in **C** and any element g in the set $\mathcal{F}A$ there is a unique morphism $\varphi : U \to A$ in **C** such that $\mathcal{F}(\varphi)(\iota) = g$.

Examples

(1) (*Universal Arrow: Free Objects*) Let R be a ring with 1. We translate into the language of universal arrows the statement that if $U(X)$ is the free R-module on a set X then any set map from X to an R-module A extends uniquely by R-linearity to an R-module homomorphism from $U(X)$ to A (cf. Theorem 6, Section 10.3): Let $\mathcal{F}$ be the forgetful functor from R–**Mod** to **Set**, so that $\mathcal{F}$ maps an R-module A to the set A, i.e., $A = \mathcal{F}A$ as sets. Let X be any set (i.e., an object in **Set**), let $U(X)$ be the free R-module with basis X, and let $\iota : X \to \mathcal{F}U(X)$ be the set map which sends each $b \in X$ to the basis element b in $U(X)$. Then the universal property of free R-modules is precisely the result that $(U(X), \iota)$ is a universal arrow from X to the forgetful functor $\mathcal{F}$.

Similarly, free groups, vector spaces (which are free modules over a field), polynomial algebras (which are free R-algebras) and the like are all instances of universal arrows.

(2) (*Universal Arrow: Fields of Fractions*) Let $\mathcal{F}$ be the forgetful functor from the category of fields to the category of integral domains, where the morphisms in both categories are *injective* ring homomorphisms. For any integral domain X let $U(X)$ be its field of fractions and let ι be the inclusion of X into $U(X)$. Then $(U(X), \iota)$ is a universal arrow from X to the functor $\mathcal{F}$ (cf. Theorem 15(2) in Section 7.5).

(3) (*Universal Object: Tensor Products*) This example refers to the construction of the tensor product of two modules in Section 10.4. Let $\mathbf{C} = R$–**Mod** be the category of R-modules over the commutative ring R, and let M and N be R-modules. For each R-module A let $\mathrm{Bilin}(M, N; A)$ denote the set of all R-bilinear functions from $M \times N$ to A. Define a functor from R–**Mod** to **Set** on objects by

$$\mathcal{F} : A \longrightarrow \mathrm{Bilin}(M, N; A),$$

and if $\varphi : A \to B$ is an R-module homomorphism then

$$\mathcal{F}(\varphi)(h) = \varphi \circ h \qquad \text{for every } h \in \text{Bilin}(M, N; A).$$

Let $U = M \otimes_R N$ and let ι be the bilinear function

$$\iota : M \times N \to M \otimes_R M \qquad \text{by} \qquad \iota(m, n) = m \otimes n,$$

so ι is an element of the set $\text{Bilin}(M, N; M \otimes_R N) = \mathcal{F}U$. Then $(M \otimes_R N, \iota)$ is a universal element of $\mathcal{F}$ because for any R-module A and for any bilinear map $g : M \times N \to A$ (i.e., any element of $\mathcal{F}A$) there is a unique R-module homomorphism $\varphi : M \otimes_R N \to A$ such that $g = \varphi \circ \iota = \mathcal{F}(\varphi)(\iota)$.

EXERCISES

1. Let **Nor**–N be the category described in Exercise 1 of Section 1, and let $\mathcal{F}$ be the inclusion functor from **Nor**–N into **Grp**. Describe a functor $\mathcal{G}$ from **Nor**–N into **Grp** such that the transformation η defined by $\eta_G : G \to G/N$ is a natural transformation from $\mathcal{F}$ to $\mathcal{G}$.

2. Let H and K be groups and let $\mathcal{H}\times$ and $\mathcal{K}\times$ be functors from **Grp** to itself described in Exercise 2 of Section 1. Let $\varphi : H \to K$ be a group homomorphism.
   (a) Show that the maps $\eta_A : H \times A \to K \times A$ by $\eta_A(h, a) = (\varphi(h), a)$ determine a natural transformation η from $\mathcal{H}\times$ to $\mathcal{K}\times$.
   (b) Show that the transformation η is a natural isomorphism if and only if φ is a group isomorphism.

3. Express the universal property of the commutator quotient group — described in Proposition 7(5) of Section 5.4 — as a universal arrow for some functor $\mathcal{F}$.

Index

free, 218, 354
  of a field extension, 513
  of a vector space, 408
Bass' Characterization of Noetherian Rings, 793
belongs to an ideal, 682
Berlekamp's Factorization Algorithm, 311, 589*ff.*
Betti number, 159, 464
Bezout Domain, 274, 283, 294, 302, 307, 775
bijection, 2
bilinear, 368*ff.*, 372, 436
bimodule, 366, 404
binary, operation, 16
  relation, 3
Binomial Theorem, 60, 249, 548
biquadratic, extension, 530, 582, 589
  polynomial, 617
block, 117
  diagonal, 423, 475
  upper triangular, 423
Boolean ring, 231, 232, 249, 250, 258, 267
Brauer group, 836
Buchberger's Algorithm, 324*ff.*
Buchberger's Criterion, 324*ff.*, 332
building, 212
Building–Up Lemma, 411
Burnside's Basis Theorem, 199
Burnside's Lemma, 877
Burnside's N/C-Theorem, 213
Burnside's $p^a q^b$ Theorem, 196, 886*ff.*

C

$C^n(G; A)$ — see cochains
cancellation laws, 20
canonical forms, 457, 472
canonical model, 734
Cardano's Formulas, 630*ff.*, 638*ff.*
cardinality, 1
Cartesian product, 1, 905*ff.*
Castelnuovo's Theorem, 646
Casus irreducibilis, 633, 637
category, 391, 911*ff.*
Cauchy's Theorem, 93, 96, 102, 146
Cayley–Hamilton Theorem, 478
Cayley's Theorem, 118*ff.*
center, of a group, 50, 84, 89, 124, 134, 198
  of a group ring, 239
  of a matrix ring, 239, 834, 856
  of a p-group, 125, 188
  of a ring, 231, 231, 344, 832*ff.*, 856
central idempotent, 357, 856
central product, 157, 169
central simple algebra, 832*ff.*
centralize, 94

centralizer, 49*ff.*, 123*ff.*, 133*ff.*
  of a cycle, 173
  of a representation, 853
chain complex, 777
  homotopy, 782
change of basis, 40, 419
changing the base — see extension of scalars
character, of a group, 568, 866
  of a representation, 866
character table, 880*ff.*
  of A_4, 883
  of D_8, 881
  of Q_8, 882
  of S_3, 881
  of S_4, 883
  of S_5, 884
  of $\mathbb{Z}/2\mathbb{Z}$, 880
  of $\mathbb{Z}/3\mathbb{Z}$, 881
characteristic, of a field, 510
  of a ring, 250
characteristic function, 249
characteristic p fields, 510
characteristic polynomial, 473
characteristic subgroup, 135*ff.*, 174
Chinese Remainder Theorem, 246, 265*ff.*, 313, 357, 768
choice function, 905
class equation, 122*ff.*, 556
class field theory, 600
class function, 866, 870
class group, 761, 774
class number, 761
Classical Greek Problems, 531*ff.*
classification theorems, 38, 142*ff.*, 181*ff.*
closed, topologically, 676
  under an operation, 16, 242, 528
closed points, 733
coboundaries, 800
cochain, 777, 799, 808
cochain complex, 777
cochain homotopy, 792
cocycle, 800
codomain, 1
coefficient matrix, 424
cofactor, 439
  Expansion Formula, 439
  Formula for the Inverse of a Matrix, 440
coherent module sheaf, 748
cohomologically trivial, 802, 804, 812
cohomology group, 777, 798*ff.*
coinduced module, 803, 811, 812
cokernel, 792
coloring graphs, 335
column rank, 418, 427, 434

comaximal ideals, 265
commutative, 16, 223
    diagram, 100
commutator, 89, 169
commutator series — see derived series
commutator subgroup, 89, 169, 195*ff*.
commute, diagram, 100
compact, 688
    support, 225
companion matrix, 475
compatible homomorphisms, 805
complement, 180, 453, 454, 820, 829, 890
complete, 759*ff*.
complete preimage, 83
completely reducible, 847
completion, 759*ff*.
complex conjugation, 345, 567, 603, 618, 654, 872
complex numbers, 1, 512, 515, 654
component of a direct product, 155, 338
composite extensions, 529, 591*ff*.
    of fields, 528
composition factors, 103
composition series, 103*ff*.
computing *k*-algebra homomorphisms, 664*ff*.
computing Galois groups, 640*ff*.
congruence class, 8*ff*.
congruent, 8
conjugacy class, 123*ff*., 489, 860
conjugate, algebraic, 573
    field, 573
    of a field element, 573
    of a group element, 82, 123*ff*.
    of a set, 123*ff*.
    of a subgroup, 134, 139*ff*.
conjugation, 45, 52, 122*ff*., 133
    in A_n, 127, 131
    in S_n, 125*ff*.
connected, 687
connecting homomorphisms, 778, 791
constituent of a module, 847
constructible, 532*ff*.
constructibility of a regular *n*-gon, 534*ff*., 601*ff*.
construction of cube roots, 535
construction of the regular 17-gon, 602*ff*.
continuous cohomology groups, 809
continuous group action, 808*ff*.
contracting homomorphisms, 809
contraction of ideals, 693, 708*ff*.
contravariant, 659
converge, 503
coordinate ring, 661
coprime — see relatively prime
corestriction homomorphism, 806, 807
corresponding group actions, 129

coset, 77*ff*., 89*ff*.
    representatives, 77
Cramer's Rule, 438
Criterion for the Solvability of a Quintic, 639
crossed homomorphisms, 814*ff*.
crossed product algebra, 833*ff*.
cubic equations, formulas for roots, 630*ff*.
curve, 726
cycle, 29, 30, 33, 106*ff*., 173
cycle decomposition, 29, 30, 115*ff*., 641
    algorithm, 30*ff*.
cycle type, 126*ff*.
    of automorphisms, 640
cyclic extensions, 625, 636
cyclic group, 22, 54*ff*., 90, 149, 192, 198, 539
    characters of, 880, 881
    cohomology of, 801, 811
cyclic module, 351, 462
cyclotomic extensions, 552*ff*., 596*ff*.
cyclotomic field, 540*ff*., 698
cyclotomic polynomial, 310, 489, 552*ff*.
cyclotomy, 598

D

D.C.C. — see descending chain condition
decomposable module, 847
Dedekind Domain, 764*ff*.
    modules over, 769*ff*.
Dedekind–Hasse Criterion, 281
Dedekind–Hasse norm, 281, 289, 294
degree, of a character, 866
    of a field element, 520
    of a field extension, 512
    of a monomial, 621
    of a polynomial, 234, 295, 297
    of a representation, 840
    of a symmetric group, 29
degree ordering, 331
dense, 677, 687
density of primes, 642
derivative, of a polynomial, 312, 546
    of a power series, 505
derived functors, 785
derived series, 195*ff*.
descending chain condition (D.C.C.), 331, 657, 751, 855
determinant, 248, 435*ff*., 450, 488
    computing, 441
determinant ideal, 671
diagonal subgroup, 49, 89
diagonalizable matrices criterion, 493, 494
Dickson's Lemma, 334
differential, 723

of a morphism, 728
dihedral group, 23*ff.*
    as Galois group, 617*ff.*
    characters of, 881, 885
    commutator subgroup of, 171
    conjugacy classes in, 132
dimension, of a ring, 750, 754*ff.*
    of a tensor product, 421
    of a variety, 681, 729
    of a vector space, 408, 411
    of $\mathcal{S}^k(V)$, 446
    of $\mathcal{T}^k(V)$, 443
    of $\bigwedge^k(V)$, 449
dimension shifting, 802
Diophantine Equations, 14, 245, 276, 278
direct factor, 455
direct limit, 268, 358, 741
direct product, characters of, 879
    infinite, 157, 357, 414
    of free modules, 358
    of groups, 18, 152*ff.*, 385, 593
    of injective modules, 793
    of injective resolutions, 793
    of modules, 353, 357, 358, 385
    of rings, 231, 233, 265*ff.*
direct sum, infinite, 158, 357, 414
    of injective modules, 403
    of modules, 351*ff.*, 357, 385
    of projective modules, 392, 403, 793
    of projective resolutions, 793
    of rings, 232
direct summand, 373, 385, 451
directed set, 268
Dirichlet's Theorem on Primes in Arithmetic
    Progressions, 557
discrete G-module, 808
discrete cohomology groups, 808*ff.*
discrete valuation, 232, 238, 272, 755
Discrete Valuation Ring, 232, 272, 755*ff.*, 762
discriminant, 610
    as resultant, 621
    of a cubic, 612
    of a polynomial, 610
    of a quadratic, 611
    of a quartic, 614
    of pth cyclotomic polynomial, 621
distributive laws, 34, 223
divides, 4, 252, 274
divisibility of ideals, 767
divisible, group, 66, 86, 167
    module, 397
Division Algorithm, 4, 270, 299
division ring, 224, 225, 834
divisor, 274

domain, 1
double coset, 117
double dual, 432, 823, 914
Doubling the Cube impossibility of, 531*ff.*
doubly transitive, 117, 877
dual basis, 432
dual group, 167, 815, 823
dual module, 404, 404
dual numbers, 729
dual vector space, 431

E

echelon, 425
eigenspace, 473
eigenvalue, 414, 423, 472
eigenvector, 414, 423, 472
Eisenstein's Criterion, 309*ff.*, 312
elementary abelian group, 136, 155, 339, 654
elementary divisor, 161*ff.*, 465*ff.*
    decomposition, 161*ff.*, 464
    decomposition algorithm, 495
elementary Jordan matrix, 492
elementary row and column operations, 424, 470*ff.*,
    479*ff.*
elementary symmetric functions, 607
elimination ideal, 328*ff.*
elimination theory, 327*ff.*
elliptic, curve, 14
    function, 600
    function field, 653
    integral, 14
embedded prime ideal, 685
embedding, 83, 359, 569
endomorphism, 347
    ring, 347
equivalence class, 3, 45, 114
equivalence of categories, 734, 916
equivalence of short exact sequences, 381
equivalence relation, 3, 45, 114
equivalent extensions, 381, 787, 824
equivalent representations, 846, 869
Euclidean Algorithm, 5, 271
Euclidean Domain, 270*ff.*, 299
    modules over, 470, 490
Euler φ-function, 7, 8, 11, 267, 315, 539*ff.*, 589
Euler's Theorem, 13, 96
evaluation homomorphism, 244, 255, 432*ff.*
exact, functor, 391, 396
    sequence, 378
exactness, of Hom, 389*ff.*, 393*ff.*
    of tensor products, 399
exceptional characters, 901
exponent of a group, 165*ff.*, 626

exponential map, 86
exponential notation, 20, 22
exponential of a matrix, 503*ff.*
Ext$_R^n(A, B)$, 779*ff.*
extension, of a map, 3, 386, 393
    of ideals, 693, 708*ff.*
    of modules, 378
    of scalars, 359*ff.*, 363*ff.*, 369, 373
extension field, 511*ff.*
extension problem, 104, 378, 776
Extension Theorem, for Isomorphisms of Fields,
    519, 541
exterior algebra, 446
exterior power, 446
exterior product — see wedge product
external, direct product, 172
    direct sum, 353

F

F-algebra — see algebra
factor group — see quotient group
factor set, 824*ff.*
factor through, homomorphism, 100, 365
factorial variety, 726
factorization, 283*ff.*
faithful, action, 43, 112*ff.*
    functor, 914
    representation, 840
Fano Plane, 210
Feit–Thompson Theorem, 104, 106, 149, 196, 212,
    899
Fermat primes, 601
Fermat's Little Theorem, 96
Fermat's Theorem on sums of squares, 291
fiber, 2, 73*ff.*, 240*ff.*
fiber product of homomorphisms, 407
fiber sum of homomorphisms, 407
field, 34, 224, 226, 510*ff.*
    of fractions, 260*ff.*
    of p-adic numbers, 759
    of rational functions, 264, 516, 530, 567, 585,
        647*ff.*, 681, 721
field extension, 511*ff.*
field generated by, 511, 516
field norm, 229
finite covering, 704
finite dimensional, 408, 411
finite extensions, 512*ff.*, 521, 526
finite fields, 34, 301, 529
    algebraic closure of, 588
    existence and uniqueness of, 549*ff.*
    Galois groups of, 566, 586
    of four elements, 516, 653

subfields of, 588
finite group, 17
finitely generated, field extension, 524*ff.*, 646
    group, 65, 158, 218*ff.*
    ideal, 251, 317
    k-algebra, 657
    module, 351*ff.*, 458
finitely presented, group, 218*ff.*
    module, 795*ff.*
First Order Diophantine Equation, 276, 278
First Orthogonality Relation, 872
Fitting ideal, 671
Fitting's Lemma, 668
fixed, element, 558
    field, 560
    set, 131, 798
fixed point free, 41, 132
flat module, 400*ff.*, 405*ff.*, 790, 795
form, 297
formal Laurent series, 238, 265, 756, 759
formal power series, 238, 258, 265, 668
formally real fields, 530
Fourier Analysis, 875*ff.*
fractional ideal, 760*ff.*
fractional linear transformations, 567, 647
Frattini subgroup, 198*ff.*
Frattini's Argument, 193
free, abelian group, 158, 355
    group, 215*ff.*
    module, 338, 352, 354*ff.*, 358, 400
    nilpotent group, 221
free generators, 218
    of a module, 354
free rank, 159, 218, 355, 460, 464
Frobenius automorphism, 549, 556, 566, 586, 589,
    604
Frobenius group, 168, 638, 643*ff.*, 896
    as Galois group, 638
    characters of, 896
Frobenius kernel, 896
Frobenius Reciprocity, 904
full functor, 914
function, 1
function field, 646, 653
functor, 391, 396, 398, 913
    contravariant, 395, 913
    covariant, 391, 398, 913
fundamental matrix, 506
Fundamental Theorem, of Algebra, 545, 615*ff.*
    of Arithmetic, 6, 289
    of Finitely Generated Abelian Groups, 158*ff.*,
        196, 468
    of Finitely Generated Modules over a
        Dedekind Domain, 769*ff.*

H

$H^n(G; A)$ — see cohomology group
Hall subgroup, 101, 200, 829, 890
Hall's Theorem, 105, 196, 890
Hamilton Quaternions, 224ff., 231, 237, 249, 299
Harmonic Analysis, 875
Heisenberg group, 35, 53, 174, 179, 187
Hilbert's Basis Theorem, 316, 334, 657
Hilbert's Nullstellensatz, 675, 700ff.
Hilbert's Specialization Theorem, 648
Hilbert's Theorem 90, 583, 814
    additive form, 584, 815
Hilbert's Zahlbericht, 815
Hölder Program, 103ff.
holomorph, 179, 186
Hom, of direct products, 404
    of direct sums, 388, 388, 404
$Hom_F(V, W)$, 416
$Hom_R(M, N)$, 345ff., 385ff.
homeomorphism, 738
homogeneous cochains, 810
homogeneous component, of a polynomial, 297
    of a graded ring, 443
homogeneous ideal, 299
homogeneous of degree m, 621
homogeneous polynomial, 297
homological algebra, 391, 655, 776ff.
homology groups, 777
homomorphism, of algebras, 343, 657
    of complexes, 777
    of fields, 253, 512
    of graded rings, 443
    of groups, 36, 73ff., 215
    of modules, 345ff.
    of rings, 239ff.
    of short exact sequences, 381ff.
    of tensor algebras, 450
homotopic, 792
hypernilpotent group, 191
hypersurface, 659

I

icosahedron — see Platonic solids
ideal quotient, 333, 691
ideal, 242ff.
    generated by set, 251
idempotent, 267, 856
idempotent linear transformation, 423
identity, of a group, 17
    matrix, 236
    of a ring, 223
image, of a map, 2

of a k-algebra homomorphism, computing, 665ff.
    of a linear transformation, computing, 429
implicitization, 678
incidence relation, 210
indecomposable module, 847
independence of characters, 569, 872
independent transcendentals, 645
index, of a subgroup, 90ff.
    of a field extension, 512
induced, character, 892ff., 898
    module, 363, 803, 811, 812, 893
    representation, 893
inductive limit — see direct limit
inequivalent extensions, 379ff.
inert prime, 749, 775
infinite cyclic group, 57, 811
infinite Galois groups, 651ff.
inflation homomorphism, 806
inhomogeneous cochains, 810
injective envelope — see injective hull
injective hull, 398, 405, 405
injective map, 2
injective module, 395ff., 403ff., 784
injective resolution, 786
injectively equivalent, 407
inner automorphism, 134
inner product of characters, 870ff.
inseparable degree, of a polynomial, 550
    of a field extension, 650
inseparable extension, 551, 566
inseparable polynomial, 546
insolvability of the quintic, 625, 629
integer, 1, 695ff.
integers mod n — see $\mathbb{Z}/n\mathbb{Z}$
integral basis, 698, 775
integral closure, 229, 691ff.
integral domain, 228, 235
integral element, 691
integral extension, 691ff.
integral group ring ($\mathbb{Z}G$), 237, 798
integral ideal, 760
integral Quaternions, 229
integrally closed, 691ff.
internal, direct product, 172
    direct sum, 354
intersection of ideals, computing, 330ff.
intertwine, 847
invariant factor, 159ff., 464, 774
    decomposition, 159ff., 462ff.
    of a matrix, 475, 477
Invariant Factor Decomposition Algorithm, 480
invariant subspace, 341, 843
inverse, of a map, 2
    of an element in a group, 17

inverse image, 2
inverse limit, 268, 358, 652*ff.*
inverse of a fractional ideal, 760
inverse of matrices, 427, 440
invertible fractional ideal, 760
irreducibility, criteria, 307*ff.*
　of a cyclotomic polynomial, 310
irreducible algebraic set, 679
irreducible character, 866, 870, 873
irreducible element, 284
　in $\mathbb{Z}[i]$, 289*ff.*
irreducible ideal, 683
irreducible module, 356, 847
irreducible polynomial, 287, 512*ff.*, 572
　of degree n over $\mathbb{F}_p$, 301, 586
irreducible topological space, 733
isolated prime ideal, 685
isomorphism, classes, 37
　of algebras, 343
　of cyclic groups, 56
　of groups, 37
　of modules, 345
　of rings, 239
　of short exact sequences, 381
　of vector spaces, 408
Isomorphism Theorems, for groups, 97*ff.*
　for modules, 349
　for rings, 243, 246
isomorphism type, 37
isotypic component, 869

J

Jacobson radical, 259, 750
join, 67, 88
Jordan block, 492
Jordan canonical form, 457, 472, 492*ff.*
Jordan–Hölder Theorem, 103*ff.*

K

k-stage Euclidean Domains, 294
k-tensors, 442
kernel, of a group action, 43, 51, 112*ff.*
　of a homomorphism, 40, 75, 239, 345
　of a k-algebra homomorphism, computing, 665*ff.*
　of a k-algebra homomorphism, 678
　of a linear transformation, computing, 429
Klein 4-group (Viergruppe), 68, 136, 155
Kronecker product, 421*ff.*, 431
Kronecker–Weber Theorem, 600
Krull dimension, 704, 750*ff.*, 754
Krull topology, 652

Krull's Theorem, 652
Kummer extensions, 627, 817
Kummer generators for cyclic extensions, 636
Kummer theory, 626, 816, 823

L

Lagrange resolvent, 626
Lagrange's Theorem, 13, 45, 89*ff.*, 460
lattice of subfields, 574
　of $\mathbb{Q}(\sqrt[3]{2}, \rho)$, 568
　of $\mathbb{Q}(\zeta_{13})$, 598
　of $\mathbb{Q}(2^{1/8}, i)$, 581
lattice of subgroups, 66*ff.*
　of A_4, 111
　of D_8, 69, 99
　of D_{16}, 70
　of Q_8, 69, 99
　of QD_{16}, 72, 580
　of S_3, 69
　of $\mathbb{Z}/2\mathbb{Z}$, 67
　of $\mathbb{Z}/4\mathbb{Z}$, 67
　of $\mathbb{Z}/6\mathbb{Z}$, 68
　of $\mathbb{Z}/8\mathbb{Z}$, 67
　of $\mathbb{Z}/12\mathbb{Z}$, 68
　of $\mathbb{Z}/n\mathbb{Z}$, 67
　of $\mathbb{Z}/p^n\mathbb{Z}$, 68
　of $\mathbb{Z}/2\mathbb{Z} \times \mathbb{Z}/2\mathbb{Z}$ (Klein 4-group), 68
　of $\mathbb{Z}/2\mathbb{Z} \times \mathbb{Z}/4\mathbb{Z}$, 71*ff.*
　of $\mathbb{Z}/2\mathbb{Z} \times \mathbb{Z}/8\mathbb{Z}$, 72
　of the modular group of order 16, 72
lattice of subgroups for quotient group, 98*ff.*
Laurent series — see formal Laurent series
leading coefficient, 234, 295
leading term, 234, 295, 318
　ideal of, 318*ff.*
least common multiple (l.c.m.), 4, 279, 293
least residue, 9
left derived functor, 788
left exact, 391, 395, 402
left group action, 43
left ideal, 242, 251, 256
left inverse, in a ring, 233
　of a map, 2
left module, 337
left multiplication, 44, 118*ff.*, 531
left Principal Ideal Domain, 302
left regular representation, 44, 120
left translation, 44
left zero divisor, 233
Legendre symbol, 818
length of a cycle, 30
lexicographic monomial ordering, 317*ff.*, 622
Lie groups, 505, 876

nonprincipal ideal, 252, 273, 298
nonsimple field extension, 595
nonsingular, point, 725, 742, 763
    variety, 725
nonsingular, linear transformation, 413
    matrix, 417
nonsingular curve, 775
nonsingular model, 726
norm, 232, 270, 299
    of a character, 872
    of an element in a field, 582, 585
normal basis, 815
normal complement, 385
normal extension, 537, 650
normal ring, 691
normal subgroup, 82*ff.*
normal variety, 726
normalization, 691, 726
normalize, 82, 94
normalized, cocycle, 827
    factor set, 825
    section, 825
normalizer, 50*ff.*, 123*ff.*, 134, 147, 206*ff.*
null space, 413
nullity, 413
number fields, 696

O

object, 911
opposite algebra, 834
orbit, 45, 115*ff.*, 877
order, of a permutation, 32
    of a set, 1
    of an element in a group, 20, 55, 57, 90
order of conductor f, 232
order of zero or pole, 756, 763
ordered basis, 409
orthogonal characters, 872
orthogonal idempotents, 377, 856, 870
orthogonality relations, 872
outer automorphism group, 137

P

p-adic integers, 269, 652, 758*ff.*
p-adic Laurent series, 759
p-adic valuation, 759
p-extensions, 596, 638
p-group, 139, 188
    characters of, 886
    representations of, 854, 864
p-primary component, 142, 358, 465
p^{th}-power map, 166, 174

P.I.D. — see Principal Ideal Domain
parabolic subgroup, 212
partition, of a set, 3
    of n, 126, 162
Pell's equation, 230
perfect field, 549
perfect group, 174
periods in cyclotomic fields, 598, 602, 604
permutation, 3, 29, 42
    even, 108*ff.*
    odd, 108*ff.*
    sign of, 108*ff.*, 436*ff.*
permutation character, 866, 877, 895
permutation group, 116, 120
permutation matrix, 157
permutation module, 803
permutation representation, 43, 112*ff.*, 203*ff.*, 840,
        844, 852, 877
pivotal element, 425
Platonic solids, symmetries of, 28, 45, 92, 111, 148
pole, 756
polynomial, 234
    map, 299, 662
    ring, 234*ff.*, 295*ff.*
polynomials with S_n as Galois group, 642*ff.*
Pontriagin dual group, 787
positive norm, 270
Postage Stamp Problem, 278
power of an ideal, 247
power series of matrices, 502*ff.*
power set, 232
preimage, 2
presentation, 26*ff.*, 39, 218*ff.*, 380
primary component — see p-primary component
Primary Decomposition Theorem, for abelian
        groups, 161
    for ideals, 681*ff.*, 716*ff.*
    for modules, 357, 465, 772
primary ideal, 260, 298, 748
prime, 6
prime element in a ring, 284
prime factorization, 6
    for ideals, 765*ff.*
prime ideal, 255*ff.*, 280, 674
    algorithm for determining, 710*ff.*
prime spectrum, 731*ff.*
prime subfield, 264, 511, 558
primes associated, to a module, 670
    to an ideal, 670
primitive central idempotent, 856, 870
primitive element, 517, 594
Primitive Element Theorem, 595
primitive idempotent, 856
primitive permutation group, 117

primitive roots of unity, 539*ff.*
principal character, 866
principal crossed homomorphisms, 814
principal fractional ideal, 760
principal ideal, 251
Principal Ideal Domain (P.I.D.), 279*ff.*, 284, 459
    characterization of, 281, 289, 294
    that is not Euclidean, 282
principal open set, 687, 738
product, of ideals, 247, 250
    of subgroups, 93*ff.*
profinite, 809, 813
projection, 83, 423, 453
    homomorphism, 153*ff.*
projections of algebraic sets, 679
projective limit — see inverse limit
projective module, 390*ff.*, 400, 403*ff.*, 761, 773, 786
projective plane, 210
projective resolution, 779
projectively equivalent, 407
Public Key Code, 279
pullback of a homomorphism, 407
purely inseparable, 649
purely transcendental, 646
pushout of a homomorphism, 407
Pythagoras' equation rational solutions, 584

Q

$\mathbb{Q}$, subgroups of, 65, 198
$\mathbb{Q}/\mathbb{Z}$, 86
quadratic, equation, 522, 533
    extensions, 522, 533
    field, 227, 698
    subfield of cyclic quartic fields, criterion, 638
    subfield of $\mathbb{Q}(\zeta_p)$, 621, 637
quadratic integer rings, 229*ff.*, 248, 271, 278, 286,
      293*ff.*, 698, 749
    that are Euclidean, 278
    that are P.I.D.s, 278
Quadratic Reciprocity Law, 819
quadratic residue symbol, 818
quartic equations, formulas for roots, 634*ff.*
quasicompact, 688, 738, 746
quasidihedral group, 71*ff.*, 186
    as Galois group, 579
quaternion group, 36
    as Galois group, 584
    characters of, 882
    generalized, 178
    representations of, 845, 852
Quaternion ring, 224, 229, 258
      (see also Hamilton Quaternions)
quintic, insolvability, 625, 629

quotient, computations in k-algebras, 672
    group, 15, 73*ff.*, 76, 574
    module, 348
    ring, 241*ff.*
    vector space, 408, 412
quotient field, 260*ff.*

R

radical extension, 625*ff.*
radical ideal, 258, 673, 689
radical of an ideal, 258, 673*ff.*, 701
    computing, 701
radical of a zero-dimensional ideal, 706*ff.*
radicals, 625
ramified prime, 749, 775
range, 2
rank, of a free module, 338, 354, 356, 358, 459
    of a group, 165, 218, 355
    of a linear transformation, 413
    of a module, 460, 468, 469, 471, 719, 773
rational canonical form, 457, 472*ff.*
    computing, 481*ff.*
rational functions — see field of rational functions
rational group ring, 237
rational numbers, 1, 260
rational valued characters, 879
real numbers, 1
    modulo 1, 21, 86
reciprocity, 229, 621
recognition theorem, 171, 180
reduced Gröbner basis, 326*ff.*
reduced row echelon form, 425
reduced word, 216*ff.*
reducible character, 866
reducible element, 284
reducible module, 847
reduction homomorphism, 245, 296, 300, 586
reduction mod n, 10, 243, 296, 640
reduction of polynomials mod p, 586, 589
reflexive, 3
regular at a point, 721
regular local ring, 725, 755
regular map, 662, 722
regular representation, 844, 862*ff.*
relations, 25*ff.*, 218*ff.*, 380
relations matrix, 470
relative Brauer group, 836
relative degree of a field extension, 512
relative integral basis, 775
relatively prime, 4, 282
remainder, 5, 270, 320*ff.*
Replacement Theorem, 410, 645
representation, 840*ff.*

extension, 645*ff.*
transfer homomorphism, 817, 822
transgression homomorphism, 807
transition matrix, 419
transitive, action, 115, 606, 640
    subgroups of S_5, 643
    subgroups of S_n, 640
transitive relation, 3
transpose, 434, 501
transposition, 107*ff.*
trilinear, 372, 436
Trisecting an Angle impossibility of, 531*ff.*
trivial, action, 43
    homomorphism, 79
    ideal, 243
    representation, 844
    ring, 224
    subgroup, 47
    submodule, 338
twisted polynomial ring, 302
two-sided ideal, 242, 251
two-sided inverse, 2

U

U.F.D. — see Unique Factorization Domain
ultrametric, 759
uniformizing parameter, 756
unipotent radical, 212
Unique Factorization Domain (U.F.D.), 283*ff.*, 303*ff.*, 690, 698, 769
unique factorization of ideals, 767
uniqueness of splitting fields, 542
unital module, 337
units, 226
    in $\mathbb{Z}/n\mathbb{Z}$, 10, 17, 61, 135, 267, 314, 596
universal property, of direct limits, 268
    of free groups, 215*ff.*
    of free modules, 354
    of inverse limits, 269
    of multilinear maps, 372, 442, 445, 447
    of tensor products, 361, 365
universal side divisor, 277
universe, 911
upper central series, 190
upper triangular matrices, 49, 174, 187, 236, 502

V

valuation ring, 232, 755*ff.*
value of f in Spec R, 732
Vandermonde determinant, 619
variety, 679*ff.*
vector space, 338, 408*ff.*, 512
Verlagerungen — see transfer homomorphism
virtual character, 898

W

Wedderburn components, 855
Wedderburn decomposition, 855
Wedderburn's Theorem on Finite Division Rings, 556*ff.*
Wedderburn's Theorem on Semisimple rings, 854*ff.*
wedge product, 447
    of ideals, 449, 455
    of a monomial, 621
well defined, 1, 77, 100
Well Ordering of $\mathbb{Z}$, 4, 8, 273, 909
Wilson's Theorem, 551
word, 215
wreath product, 187

Z

$Z^n(G; A)$ — see cocycles
$\mathbb{Z}[i]$ — see Gaussian integers
$\mathbb{Z}[\sqrt{2}\,]$, 278, 311
$\mathbb{Z}[\sqrt{-5}\,]$, 273, 279, 283*ff.*
$\mathbb{Z}[(1 + \sqrt{-19})/2]$, 277, 280, 282
$\mathbb{Z}/n\mathbb{Z}$, 8*ff.*, 17, 56, 75*ff.*, 226, 267
$(\mathbb{Z}/n\mathbb{Z})^{\times}$, 10, 18, 61, 135, 267, 314, 596
Zariski closed set, 676
Zariski closure, 677*ff.*, 691
Zariski dense, 677, 687
Zariski topology, 676*ff.*, 733
zero divisor, 226, 689
zero ring, 224
zero set, 659
zero-dimensional ideal, 705*ff.*
Zorn's Lemma, 65, 254, 414, 645, 907*ff.*